International Congress of Automotive and Transport Engineering (CAR 2017)

Mobility Engineering and Environment

IOP Conference Series: Materials Science and Engineering
Volume 252

Pitesti, Romania
8 – 10 November 2017

ISBN: 978-1-5108-5030-9
ISSN: 1757-8981

Printed from e-media with permission by:

Curran Associates, Inc.
57 Morehouse Lane
Red Hook, NY 12571

Printed by Curran Associates, Inc. (2017)

For permission requests, please contact the Institute of Physics
at the address below.

Institute of Physics
Dirac House, Temple Back
Bristol BS1 6BE UK

Phone: 44 1 17 929 7481
Fax: 44 1 17 920 0979

techtracking@iop.org

Additional copies of this publication are available from:

Curran Associates, Inc.
57 Morehouse Lane
Red Hook, NY 12571 USA
Phone: 845-758-0400
Fax: 845-758-2633
Email: curran@proceedings.com
Web: www.proceedings.com

Table of contents

Volume 252

CAR2017 International Congress of Automotive and Transport Engineering - Mobility Engineering and Environment

8–10 November 2017, Pitesti, Romania

Accepted papers received: 4 October 2017
Published online: 23 October 2017

Preface

Accident research and analysis. Active and passive safety. Vehicle dynamics

Advanced engineering methods

Advanced materials, manufacturing technologies and logistics

Advanced transport systems and road traffic

Hybrid and electric vehicles

Internal combustion engines

Vehicle electronics and software

IOP Conf. Series: Materials Science and Engineering **252** (2017) 011001 doi:10.1088/1757-899X/252/1/011001

FOREWORD

The Congress of Automotive and Road Transport Engineering (in short, CAR) is a traditional scientific event initiated by University of Pitești, Romania in 1978. CAR is reaching its 11th edition in 2017 and is part of the series of similar annual scientific events organized by Society of Automotive Engineers of Romania (SIAR) and various Romanian universities with interests in Automotive and Transport Engineering. Since its 7th edition from 1997, CAR gained the patronage of FISITA (Fédération Internationale des Sociétés d'Ingénieurs des Techniques de l'Automobile) and EAEC (European Automotive Engineers Cooperation).

Considering that now, more than ever, there is an obvious need to strike a balance between natural increase of road mobility and environmental protection, CAR2017 was focused on ***Mobility Engineering and Environment.*** Moreover, the motto of the Congress (***Academia, Industry and Government: together for automotive and transport engineering development***) was an indication of the organizers belief that there is a need for meetings and discussions about the current and future challenges of the automotive and transport world in order to find the best solutions. In other words, we are convinced that the challenges of the future can only be overcome if these tripartite discussions occur permanently.

The papers included in this volume are selected by the Scientific Committee amongst the almost 160 technical papers proposed to be presented at the Congress. The authors of these articles are experts from research, industry and universities coming from 12 countries. Their papers are covering the latest issues such as: fuel economy and environment, materials and technologies, automotive safety, traffic and road transport systems, advanced engineering methods, advanced powertrains and hybrid and electric drives.

Therefore, we hope these papers will generate fruitful discussions about an exciting topic: the automotive and road transport engineering in the light of the future challenges.

Adrian CLENCI *Ştefan TABACU*

car 2017

Patronage of FISITA

International Congress of Automotive and Road Transport Engineering,
Mobility Engineering and Environment,
www.car2017.ro
https://www.facebook.com/SIAR.CAR/
https://www.fisita.com/events/diary
car2017@upit.ro

ACKNOWLEDGEMENT

HONORARY COMMITTEE

Dan NICHOLSON – FISITA President
Chris MASON –FISITA' CEO
Ludek HYNCIK – FISITA VP Education
Karl SIEBERTZ – FISITA VP Strategic Planning and FORD Research & Advanced Engineering Germany
Rafael BORONAT – EAEC President
Gunter HOHL – Former EAEC President
Adrian CLENCI – SIAR President and Chairman of CAR2017
Eugen Mihail NEGRUȘ –SIAR Former President
Cedomir DUBOKA – JUMV President, Serbia
Daniel PFRIMMER – Managing director of SIA, France
Yves CARACTZANIS – General Manager, Renault Romania
Pascal CANDAU – General Manager, Renault Technologie Roumanie
Gerolf STROHMEIER – Managing Director AVL List Austria
Constantin STROE – Former ACAROM President
Gabriel SICOE – ACAROM President
George-Adrian DINCĂ – General manager of RAR (Romanian Auto Registry)
Radu DINESCU – General Secretary of UNTRR (National Union of Road Transporters from Romania)
Benone COSTEA – Director, Magic Engineering
Mihaela DIACONU – President of the Senate of University Pitești
Dumitru CHIRLEȘAN – Rector of the University of Pitești

SCIENTIFIC AND TECHNICAL COMMITTEE

President of the Scientific and Technical Committee:

Professor emeritus **Ion TABACU,** University of Pitești, Romania

Members:

Cristian ANDREESCU	University "Politehnica" of Bucharest - Romania Professor
Alexandru BOROIU	University of Pitesti - Romania Professor
Nicolae BURNETE	Technical University of Cluj-Napoca - Romania Professor
Anghel CHIRU	Transilvania University of Brasov - Romania Professor
Amélie DANLOS	Conservatoire National des Arts et Métiers de Paris - France Professor
Georges DESCOMBES	Conservatoire National des Arts et Métiers de Paris - France Professor
Ilie DUMITRU	University of Craiova - Romania Professor
Michel FEIDT	Université de Lorraine - France Professor
John FIELDHOUSE	University of Huddersfield - UK Professor

Edward RAKOȘI	Technical University "Gheorghe Asachi" of Iasi - Romania Professor
Adrian SACHELARIE	Technical University "Gheorghe Asachi" of Iasi - Romania Assoc. Professor
Corina SANDU	Virginia Tech, U.S.A. Professor
Tom SAVU	University "Politehnica" of Bucharest - Romania Professor
Nicolae Doru STANESCU	University of Pitesti - Romania Professor
Ștefan TABACU	University of Pitesti - Romania Professor
Reinhard TATSCHL	AVL Graz, Advanced Simulation Technology - Austria Dr. eng.
Adrian TODORUȚ	Technical University of Cluj-Napoca - Romania Professor
Bogdan VARGA	Technical University of Cluj-Napoca - Romania Professor

ORGANIZING COMMITTEE

Congress Chairman

Adrian CLENCI	Director of Department Automobiles and Transport, University of Pitești, Romania and SIAR President

Members

Minu MITREA	General Secretary of SIAR
Dinu COVACIU	Website Coordinator, Transilvania University of Brașov
Florian IVAN	Professor, University of Pitești, Department Automobiles and Transport
Alexandru BOROIU	Professor, University of Pitești, Department Automobiles and Transport
Viorel NICOLAE	Professor, University of Pitești, Department Automobiles and Transport
Dinel POPA	Professor, University of Pitești, Department Automobiles and Transport
Sebastian PÂRLAC	Professor, University of Pitești, Department Automobiles and Transport
Marinica STAN	Professor, University of Pitești, Department Automobiles and Transport
Ștefan TABACU	Professor, University of Pitești, Department Automobiles and Transport
Dănuț MARINESCU	Senior lecturer, University of Pitești, Department Automobiles and Transport
Elena NEAGU	Senior lecturer, University of Pitești, Department Automobiles and Transport
Ionel VIERU	Senior lecturer, University of Pitești, Department Automobiles and Transport
Sorin ILIE	Senior lecturer, University of Pitești, Department Automobiles and Transport
Rodica NICULESCU	Lecturer, University of Pitești, Department Automobiles and Transport
Cătălin ZAHARIA	Lecturer, University of Pitești, Department Automobiles and Transport
Mihaela ISTRATE	Lecturer, University of Pitești, Department Automobiles and Transport
Helene ȘUSTER	Lecturer, University of Pitești, Department Automobiles and Transport
Gabriela MITRAN	Lecturer, University of Pitești, Department Automobiles and Transport
Petre STAN	Lecturer, University of Pitești, Department Automobiles and Transport
Victor IORGA-SIMĂN	Researcher, University of Pitești, Department Automobiles and Transport
Eduard NIȚU	Professor, University of Pitești, Department Manufacturing and Industrial Management

Alin RIZEA	Senior lecturer, University of Pitești, Department Manufacturing and Industrial Management
Monica IORDACHE	Senior lecturer, University of Pitești, Department Manufacturing and Industrial Management
Gabriela PLAIAȘU	Senior lecturer, University of Pitești, Department Manufacturing and Industrial Management
Daniel ANGHEL	Lecturer, University of Pitești, Department Manufacturing and Industrial Management
Monica BÂLDEA	Lecturer, University of Pitești, Department Manufacturing and Industrial Management
Nadia BELU	Lecturer, University of Pitești, Department Manufacturing and Industrial Management
Ana GAVRILUȚĂ	Lecturer, University of Pitești, Department Manufacturing and Industrial Management

SPONSORS

The organizing committee would like to thank the following companies for their support.

GROUPE RENAULT

AVL

HORIBA Automotive Test Systems

FEV

ACAROM Asociația Constructorilor de Automobile din România

REGISTRUL AUTO ROMÂN

UNTRR Uniunea Națională a Transportatorilor Rutieri din România

MAGIC ENGINEERING

Compact Industrial

Nobel Automotive

Haulotte GROUP

CADWORKS

IOP Conf. Series: Materials Science and Engineering **252** (2017) 011003 doi:10.1088/1757-899X/252/1/011003

Peer review statement

All papers published in this volume of *IOP Conference Series: Materials Science and Engineering* have been peer reviewed through processes administered by the proceedings Editors. Reviews were conducted by expert referees to the professional and scientific standards expected of a proceedings journal published by IOP Publishing.

IOP Conf. Series: Materials Science and Engineering **252** (2017) 012001 doi:10.1088/1757-899X/252/1/012001

Vehicle-pedestrian collisions – Aspects regarding pedestrian kinematics, dynamics and biomechanics

L Petrescu[1] and Al Petrescu[1,2]

[1]Asig Tehnic Expert SRL, bvd. N. Grigorescu no.55, Bucharest, 030446, Romania
[1]Politehnica University of Bucharest, Splaiul Independentei no. 313, Bucharest, 060042, Romania

liviu1petrescu@gmail.com

Abstract. Vehicle-pedestrian collisions result in a substantial number of pedestrian fatalities and injuries worldwide. Concern continues to limit and reduce the tragic consequences suffered by pedestrians involved in road accidents, caused the vehicle-pedestrian accident reconstruction become an important area and distinctly outlined in the reconstruction of road incidents involving vehicle. This paper analyzes the dynamics of vehicle-pedestrian impact influence over pedestrian biomechanics, which is directly connected with the severity of injury after contact with the vehicle profile and with the place where the pedestrian is projected. The main goal of this paper is to highlight some features of reconstruction of road accidents involving pedestrian, looking at the kinematics and dynamics of pedestrian impact for a better understanding of the phenomena that occur. The study on the dynamics and biomechanics of the pedestrian hit by the vehicle is useful in order to understand how the injuries, including the lethal ones, are generated in the collision, what is essential in road accidents reconstruction.

1. Introduction

Pedestrians are the most vulnerable road users in traffic crashes. On road accidents the impact between pedestrian and motor vehicle generates, in most situations, severe injuries and even deaths among pedestrians.

During the last twenty years pedestrian safety has been widely studied [1; 2; 5; 7] and pedestrian protection becomes of increasing concern of the world, especially in the EU. Currently the pedestrian protection at accidents considers one of the most important challenges in automotive industry everywhere in the world due to the large number records of death, and severe injuries occurred for the pedestrians in road accidents.

This paper refers to significant aspects regarding the frontal collision between motor vehicles and pedestrians, being used real-world collisions (car-to-pedestrian collision filmed in real time) and analytical methods with numerical solving that allow computer-based reconstitution of the vehicle-pedestrian impact. In order to report the results of multibody numerical simulation of a vehicle crash with the pedestrian has been used specialized software PC-Crash 10.1. The results are analysed by calculating the values of the head injury criterion and the combined thoracic index. The pedestrian injuries are known from the medico-legal report and the theoretical results are assessed according to the real pedestrian injuries, emphasizing certain phases of the pedestrian projection movement on which the severity of injury in pedestrian depends.

2. Reconstruction of the specific parameters of the vehicle-pedestrian traffic accident

2.1. Circumstances of the traffic accident

The accident occurred on str. Mihai Bravu, in Bucharest, and was filmed by a video surveillance camera that records at a 150 fps frequency, with a frame size of 1920 x 1080 pixels. The camera was located on the side of the street with the sidewalk where the pedestrian had started to cross the road. The accident involved a VW Golf vehicle and a pedestrian, male, aged 50, with a height of 1.72 m and a weight of 78 kg.

Figure 1 shows the area where the accident occurred, the direction of the vehicle and that of the pedestrian, respectively.

Figure 1. View of the area where the accident occurred

The pedestrian crossed the street on the red color of the pedestrian traffic light and he entered the lane of the VW vehicle from the right side, as shown in figure 2.

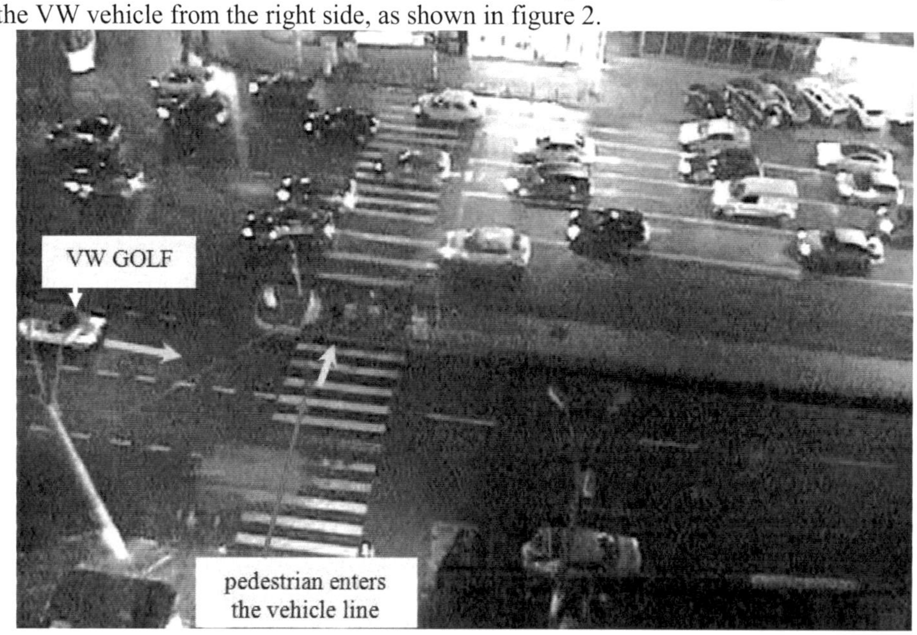

Figure 2. The pedestrian enters the lane of the VW vehicle from the right side

The pedestrian was hit by the mid frontal area of the vehicle as illustrated in figure 3 and he is struck by the hood and windshield, and ends up falling down on the road in front of the VW vehicle as it can be seen in figure 4.

Figure 3. The pedestrian was hit by the mid frontal area of the VW vehicle

Figure 4. The pedestrian ends up falling down on the road in front of the VW vehicle

The damage caused to the VW vehicle are shown in figure 5, which reads: A - deformed license plate; B - frontal grids between headlights is broken; C - traces of deformation of the engine bonnet; D - damage to the left side of the windscreen.

Figure 5. The damage caused to the VW vehicle after the impact with pedestrian

The configuration of the windshield damages presented in figure 6 show the existence of two damage areas, one of which is more pronounced, which can be corroborated with the video recording of the accident, and which indicates the fact that there wasn't only one, but two impacts of the pedestrian with this car part, as also resulted during the computer reconstruction of the accident.

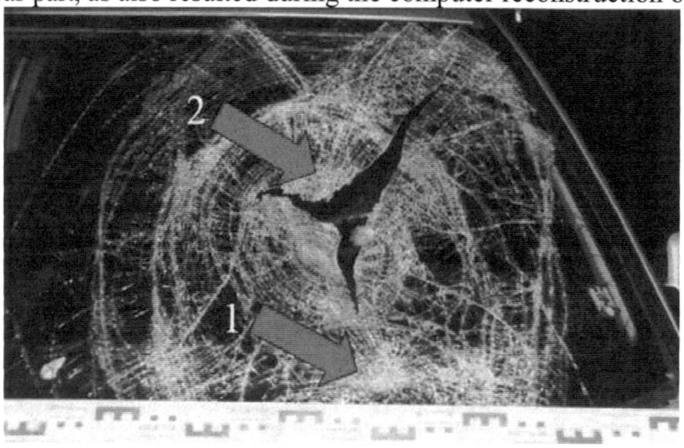

Figure 6. The configuration of the windshield damages

The consequences of the accident suffered by pedestrian are highlighted by the forensic findings, as follows:

- severe head trauma with multiple neurocranium and viscerocranium fractures on the right side, and cerebral hemorrhage;
- fracture of both bones in the left lower leg;
- chest trauma with left pulmonary contusion.

The real-time recording of the traffic accident provided the opportunity to establish the way it occurred, determining the time and place of the impact, as well as the momentums and the final positions when both vehicle and pedestrian stopped on the roadway, also allowing to ascertain the post-impact trajectory of the pedestrian and the relative position between the pedestrian and the vehicle on initial impact. The total duration of the accident was accurately determined at 3.5 s from the initial moment of the impact until the pedestrian and the VW vehicle stopped on the roadway.

2.2. Reconstruction of the specific parameters of the vehicle-pedestrian traffic accident using analytical methods with numerical solutions

In order to reconstruct the kinematics parameters specific to this traffic accident, it was used the compute-based collision reconstruction method, with the help of the PC-Crash software, version 10.1. With PC-Crash can be reconstruct complex accidents, using multibody dynamics formulations [4].

The dimensional parameters, as a whole, of the road section where the accident occurred and the positions where both the vehicle and the pedestrian stopped after the impact were identified by the measurements taken by the police at the accident site.

A sketch of the accident site is shown in figure 7. It contains the measurements taken by the police at the accident site, as well as the impact position and the final positions, on the roadway, of the vehicle and of the pedestrian.

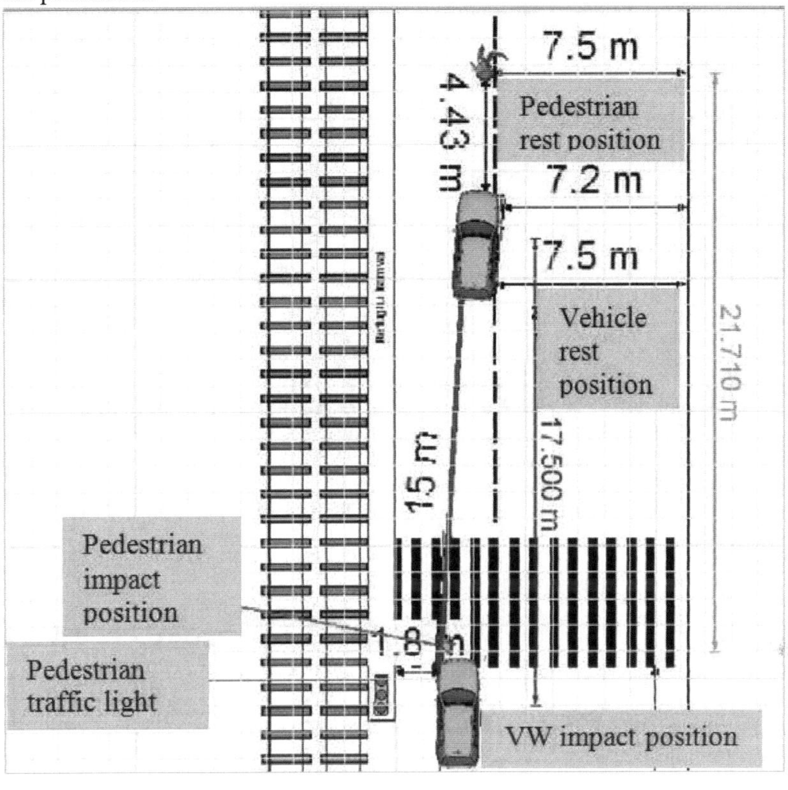

Figure 7. The sketch of the accident site

The distance between the impact position and the final position of the vehicle (relative to the centre of gravity of the vehicle) is of 17.50 m, and the distance between the impact position and the final position of the pedestrian (relative to the centre of gravity of the pedestrian) is of 21.71 m.

In the simulation generated using the computer-based reconstruction method, the impact speed of the vehicle was of 44 km/h and the pedestrian speed was of 6 km/h, given de fact that the vehicle and the pedestrian stopped, in the simulation, in the positions recorded by the police in their on-site investigation; the initial positions and trajectories of the vehicle and of the pedestrian are similar to those shown in the video recording.

The dynamics of the vehicle-pedestrian impact, achieved using the computer-generated reconstruction of the collision, provided by the PC-Crash software, are shown in figure 8.

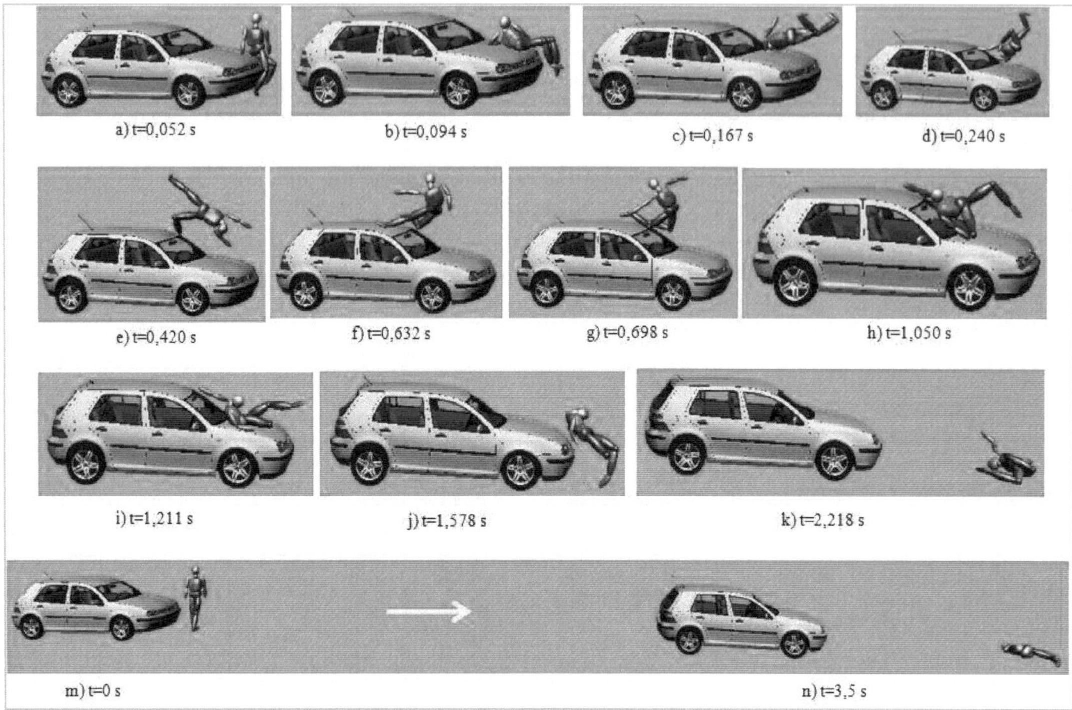

Figure 8. The dynamics of the vehicle-pedestrian impact

2.3. Pedestrian kinematics and biomechanics

Concerning the pedestrian biomechanics, below are shown the results obtained through the computer reconstruction of the collision.

Figure 9 shows the speed of the pedestrian, and figures 10 and 11 show the distances relative to the axes, travelled by the pedestrian (by his centre of gravity).

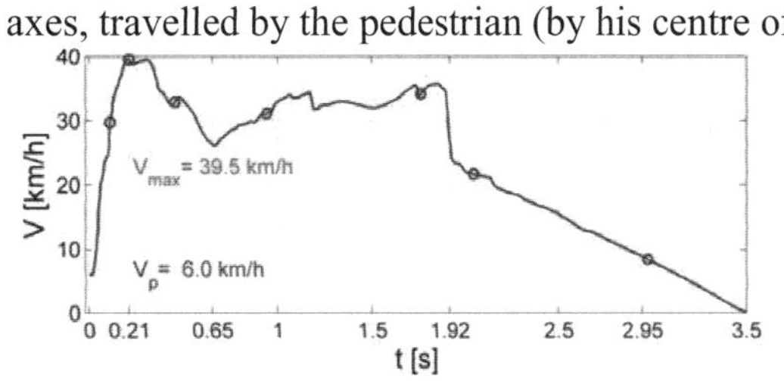

Figure 9. The speed of the pedestrian

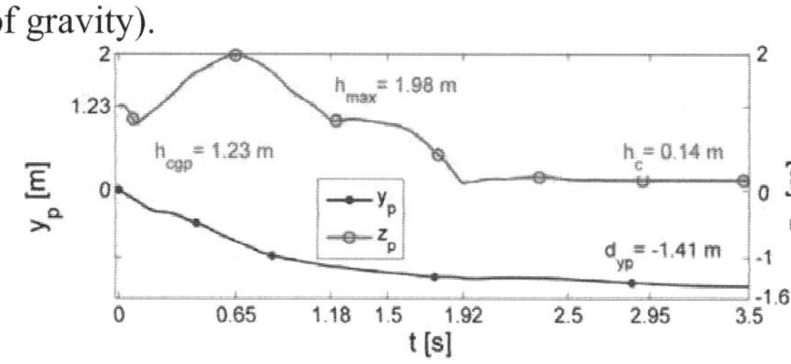

Figure 10. The distances on the Oy and Oz axes, travelled by the pedestrian (by his centre of gravity)

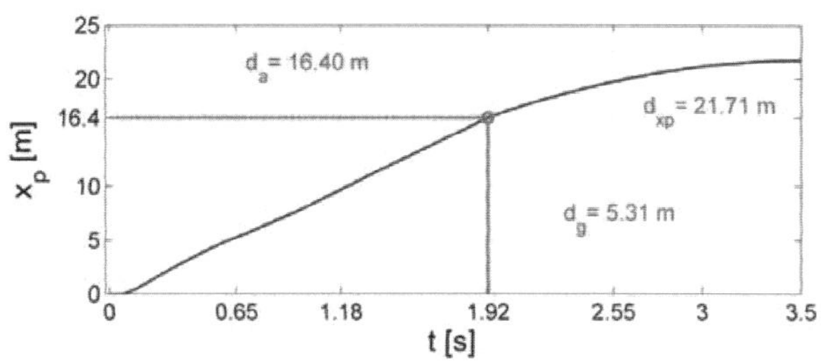

Figure 11. The distance on the Ox axis, travelled by the pedestrian (by his centre of gravity)

IOP Conf. Series: Materials Science and Engineering **252** (2017) 012001 doi:10.1088/1757-899X/252/1/012001

From the graphic presented in figure 9 can be ascertained that from the impact speed of the pedestrian, 6 km/h, the speed of the pedestrian after impact grew up to the value of 39.5 km/h, at time t=0.21 s.

From figure 10 it can be also ascertained that at time t=0.65 s, the pedestrian reached the maximum throwing height h_{max}=1.98 m. Because starting from this moment on the speed of the pedestrian increases (figure 9), results that after the pedestrian was thrown into air, he crashed again into the automobile, from which it absorbed another quantity of energy. From figure 10 can be ascertained that at time t=1.92 s the pedestrian fell down on the road, because the height z_p is minimum. Therefore, from this moment on and until stopping (at t=3.5 s), the pedestrian falls on the road. As results from figure 10 the falling of the pedestrian on the road lasts for 1.58 s and the distance travelled during this period of time is d_g =5.31 m. From the moment of the initial vehicle-pedestrian impact, until the pedestrian reached the road, has passed a period of time of 1.92 s, and the distance travelled in this period of time is d_a=16.4 m.

From figure 11 results that the distance travelled by the pedestrian on the Ox axle direction is d_{xp}=21.71 m, value that was also shown in figure 7.

In figure 12 are shown the values on the three coordinate axes of the accelerations/decelerations of the head (a_x, a_y and a_z), and in figure 13 is shown the value of the resultant and its gradient. In figure 12 are also shown the maximum levels of the accelerations values on the three axes through absolute values and through values related to the gravitational acceleration g. Also, in figure 13 are shown the maximum value of the resultant acceleration a_{max} and the limit values of its gradient, g_{max} and g_{min}.

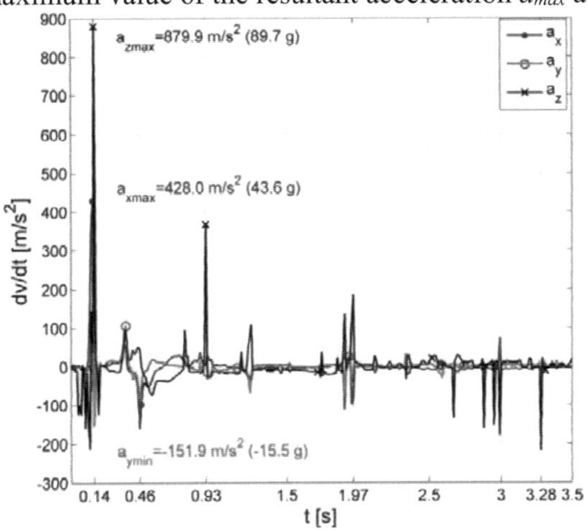

Figure 12. The derivatives speeds of the head (accelerations/decelerations), values on the three coordinate axes

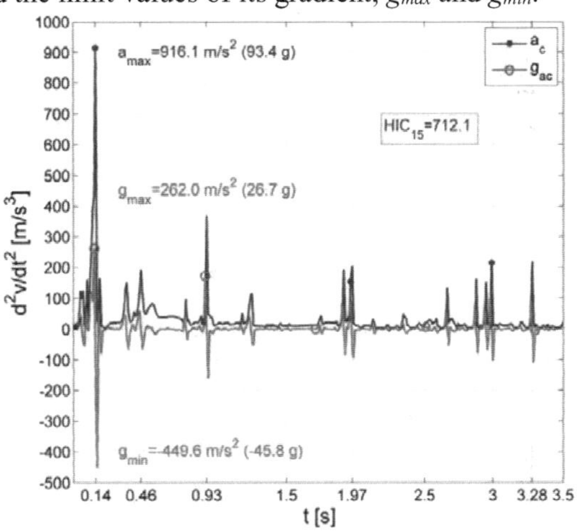

Figure 13. The resultant and gradient of pedestrian head acceleration

As it can be seen in figure 12 the longitudinal acceleration of the head a_x exceeds the maximum tolerated acceleration (43.6>30 g [3]), and the lateral acceleration a_y is lower than the acceleration tolerated by a human being (15.5>9 g [3]).

Therefore, the head of the pedestrian is under accelerations higher than the ones tolerated. Moreover, in figure 13 can be observed that the value of the HIC_{15} criterion is higher than the maximum tolerated one (712.1>700). For the persons aged six years and older is accepted the fact that the maximum limit of tolerance is HIC_{15}=700 [3].

Exceeding the tolerance criteria justifies the trauma suffered by the head of the pedestrian as they were highlighted in the consequences of the traffic accident (paragraph 2.1).

As known in the specialty literature [6], in order to appreciate the severity of the head injuries the most used criterion is the one named HIC (Head Injury Criteria), which is expressed through the relation:

$$\mathrm{HIC} = \left[\left(\frac{1}{t_2 - t_1} \int_{t_1}^{t_2} a_{rez} \mathrm{d}t \right)^{2,5} \cdot (t_2 - t_1) \right]_{max} \quad (1)$$

in which t_1 and t_2 are the limits of the period of time considered, and a_{rez} represents the resultant of the instant acceleration/deceleration related to the gravitational acceleration g.

The risk of fracturing the skull bones refers to the period of time t_2-t_1=15 ms and it's marked with HIC_{15}.

In figure 14 are shown the contact forces of the thorax (torso) of the pedestrian. From figure 15 results that in the thorax area there are rib fractures, because the contact force (the resultant force at the thorax level) is higher than the tolerated one (7186.1>6600 N [6]).

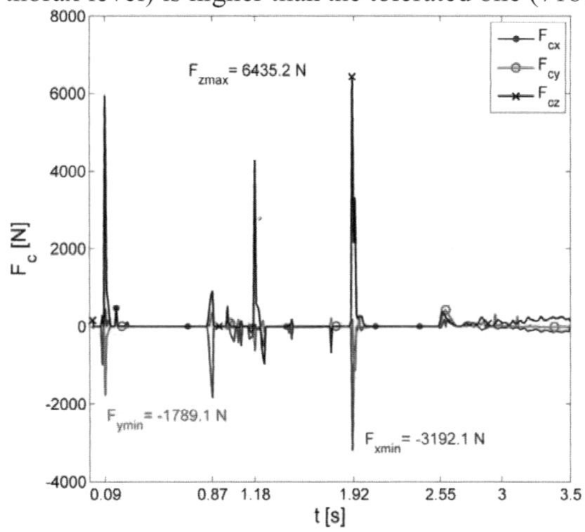

Figure 14. The contact forces of the thorax (torso) of the pedestrian, values on the three coordinate axes

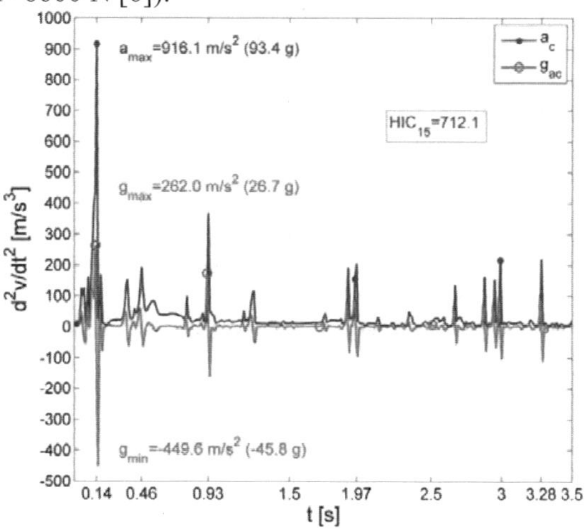

Figure 15. The resultant force at the thorax level of pedestrian

The contact force at the thorax level which is higher than the tolerated contact force justifies the existence of the thorax trauma with a left pulmonary contusion mentioned in the forensic findings and highlighted in the consequences of the traffic accident involving the pedestrian (paragraph 2.1).

Also, from figure 15 results that this trauma may have appeared at time t=1.92 s, namely when the pedestrian, after the separation from the vehicle, hit the road surface and started slipping on it (figure 10). The presence of the trauma at the moment of impact with the vehicle (at t=0.09 s) is less probable, because the contact force value is lower than the tolerated value (6226.9<6600 N [7]).

3. Conclusions

The current research methods can assess the level of human body damage, considered in its entirety. The assessment of kinematic and dynamic parameters for a **certain component or organ** inclined to be affected during the manifestation of certain reactions specific to the type of dynamic load and the comparative analysis related to the human tolerance at the force applied to the human body during traffic accidents, generate a **higher level of accuracy for the car accident reconstruction, including concerning how pedestrian injuries are occurring.**

The results obtained after analyzing the parameters related to the pedestrian kinematic and dynamic show that during the impact significant values of the contact forces and the accelerations were displayed in the thorax and head area, these values being superior to the ones tolerated by the human body. The size of the accelerations and the contact forces, as well as the duration of the impact represent the main factors that generate the severity of the lesions suffered by the victims. **The secondary impact with the ground proved to be another important factor that generated severe**

trauma.

The assessment methods of the kinematic and dynamic parameters on components seem to represent the best solutions for evaluating the potential for pedestrian injury resulted in the car accident, during the impact between the vehicle and the victim, as well as during the impact of the pedestrian with the ground.

The paper is the result of the research conducted during the preparation of the PhD thesis within the "Automotive Engineering" doctoral studies section of the Faculty of Engineering and Technology - University of Pitești.

References

[1] Burg H and Moser A 2013 *Handbook of accident reconstruction* Viewg&Teubner (Kippenheim Germany)
[2] Huston L R 2009 *Principles of biomechanics* CRC Press Taylor & Francis Group (Boca Raton USA)
[3] Gaiginschi R and Filip I 2009 *Reconstrucția și expertiza accidentelor rutiere* Ed. Tehnică București
[4] Moser A, Steffan H, Hoschopf H and Kasanicky G 2000 Validation of the PC-Crash Pedestrian Model *SAE Tehnical Paper* (2000-01-0847)
[5] Soica A 2008 *Caroserii si sisteme pentru siguranta pasiva II* Ed. Univ. Transilvania Brasov
[6] Stevenson T J 2006 Simulation Of Vehicle-Pedestrian Interaction *thesis submitted in partial fulfilment of the requirements for the Degree of Doctor of Philosophy in Engineering in the University of Canterbury* U.K
[7] Wood D, Simms C and Walsh D G 2005 Vehicle–pedestrian collisions: validated models for pedestrian impact and projection, *Proc. IMechE Journal of Automobile Engineering* **D** 219

IOP Conf. Series: Materials Science and Engineering **252** (2017) 012002 doi:10.1088/1757-899X/252/1/012002

Continuous monitoring the vehicle dynamics and driver behavior using navigation systems

George Ene

Claim Expert Coordinator, Bucharest, RO

ene_george@yahoo.com

Abstract: In all fields cost is very important and the increasing amount of data that are needed for active safety systems, like ADAS, lead to implementation of some complex and powerful unit for processing raw data. In this manner is necessary a cost-effective method to estimate the maximum available tire road friction during acceleration and braking by continuous monitoring the vehicle dynamics and driver behavior. The method is based on the hypothesis that short acceleration and braking periods can indicate vehicle dynamics, and thus the available tire road friction coefficient, and also human behavior and his limits. Support for this hypothesis is found in the literature and is supported by the result of experiments conducted under different conditions and seasons.

1. Objective

Developing a system that continuous monitoring the vehicle dynamics and driver behavior, and thus the quality and tire grip level of the road surface, without stereo cameras, adaptive suspension or other expensive sensors. Using verified algorithms and raw data taken from navigation systems it can be determined the level of road roughness and road friction, which means that we would predict how easily a vehicle will skid on a road. For a given wheel, the normalized traction force, μ, is:

$$\mu = \frac{\sqrt{F_x^2 + F_y^2}}{N_z} \quad (1)$$

When a vehicle of mass m has the same μ_{max} at all four of its tires, and we consider only longitudinal motion, so the side force F_y can be neglected, the largest longitudinal acceleration $|a|_{max}$ it can achieve, neglecting all resistances, is:

$$|a|_{max} = \max\left|\frac{F_{x11}+F_{x12}+F_{x21}+F_{x22}}{m}\right| \leq \mu_{max} \cdot g \quad (2)$$

Where F_{xij} is the longitudinal force at the xij wheel and g is the gravitational constant. Thus, an estimate of $|a|_{max}$= gives us an upper bound on the acceleration, in g's, that a vehicle can achieve. Note that vertical load will contribute to different contact patch area and tire deformation. In all studies vertical wheel load input are kept constant, which in real driving conditions it varies[8].

2. Applications

Anyone who uses a computer, tablet or even a smartphone will no doubt have faced informations from GPS system provided in vehicle by the infotainment systems. Results from collected informations can be used in over-the-air services, with greath functionality for autonomous vehicle in the future and no matter how many levels of security a system might have, can inform drivers, machines or people of dangerous conditions, so that they can change their driving speed or style to prevent dangerous emergency situations [1] [2]. Blending the navigation systems with active safety features have serious advantages because there is a lot of high-speed processing going on the vehicle computers and active-safety domain which includes the vehicle's radar, LiDAR, vision systems, driver-state monitoring and various related sensors and actuators.

3. Methodology. Acquiring and processing data.

The vehicle used is a front wheel drive with standard ABS, whitout traction or stability control unit, and it was equipped with Velocity BOX , VBOX, and PerformanceBox from Racelogic Ltd. The Racelogic VBOX, and PerformanceBox, is a family of data acquisition devices which is used by the majority of Car Manufacturers, Tyre Manufacturers and car magazines around the world to assess performance. Because it is very easy to edit the test ranges, is a very powerful tool for use in many different kinds of vehicle testing.

Figure 1. Instruments used for acquiring data

Velocity • Accuracy: 0.1 Km/h • Units: Km/h or Mph • Update rate: 20 Hz • Minimum velocity: 0.1 Km/h • Maximum velocity: 1800 km/h • Resolution: 0.01 Km/h	**Distance** • Accuracy: 0.05%(<50cm per Km) • Units: Km/h or Mph • Update rate: 20 Hz • Resolution: 0.05 cm • 2D Position: ±5m 95% CEP * • Height: 5 Metres 95% CEP * ***95% CEP (Circle of Error Probable)**
Acceleration • Accuracy: 0.5 % • Maximum: 4 G • Resolution: 0.01 G	**Heading** • Resolution: 0.01° s • Accuracy: ±0.2° s
Memory • Type SD Card • Recording time Dependent on card capacity*	**Environmental and Physical** • Weight 130 grams • Size 104.5 mm x 72.8 mm x 25.1 mm • Operating temperature -20 °C to +50°C • Storage temperature -30°C to +80°C

Tabel 1. VBOX Sport Specification

The data acquisition system is based on GPS 20Hz sensor, that offers very precise position and velocity information. For a GPS receiver, the position as latitude and longitude coordinates are required to plot the vehicle's path, and the heading is necessary to calculate lateral acceleration. The calculated acceleration values are affected by GPS positioning errors and low sampling rate. In case of longitudinal acceleration, the positioning information is not used, only speed and time, which are very accurate.

One of the test track is presented in figure 2, as it is in a Google Earth environment, with highlighted road and red line for the sample track.

Figure 2. Test track shown in Google Earth

The raw data for velocity and longitudinal with lateral accelerations, obtained from VBOX and GPS receiver is presented in figure 3, for the same recorded track. The values used for these diagrams are not filtered.

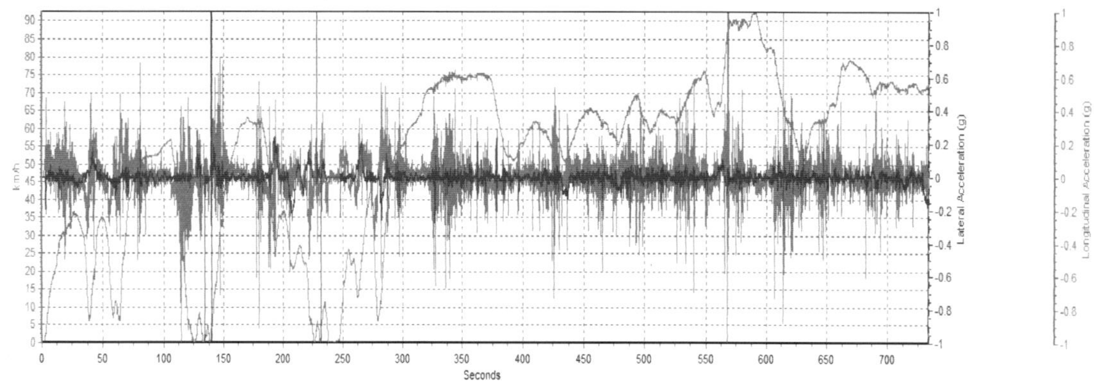

Figure 3. Velocity, longitudinal and lateral acceleration presented as raw data

The PerformanceTools software allows to view the driving data recorded by VBOX file. The software also allows you to display available channels, such as velocity, lateral acceleration, longitudinal acceleration, heading, height, satellites number, yaw rate, latitude, longitude, brake trigger, UTC Time, distance, time, radius of turn results, and to carry out detailed analysis of driver and vehicle performance. The value of velocity, presented in figure 3, is smooth enough but it is obvious that the values used to generate longitudinal and lateral acceleration, presented in the same graph need to be filtered [9] [10]. The software developer of VBOX indicates that smoothing level can be applied to each channel individually and is useful when displaying acceleration channels. Note that the Smoothing Level should therefore be carefully applied, too little and the data can be too 'noisy' for easy analysis and too much and valuable information can be lost.

4. Quantification of skill, rule and knowledge-based behavior in road traffic

Human and technical features of the road traffic system combined with their interactive compatibility influence primary safety and accident prevention potential. The enormous kinetic energy involved in the dynamic process of driving is coupled with danger of loss control. Such modeling approaches for human driver behavior in road traffic represent challenges for driver assistance systems [4]. The gaps between human performance limitations and physico-technical constraints of vehicle, road and traffic environment characteristics open up the most promising realms for the development of driver assistance systems. The quantificantion of skill, rule, and knowledge-based behavior in road traffic was intensified by initiating empirical acquisition of driving behavior data. The g-g diagram, longitudinal versus lateral accelerations, serve as an example to explain this approach. This diagram shows the range of longitudinal versus lateral accelerations at some drivers with average experience, applied during test runs on winding country roads and highways in everyday traffic [3]. The circumferential line present an 85 percentile line, all drivers remain below this outline in 85% of the test driving time, the rule-based range reaches up to the 95th percentile, and all driving states beyond this 95-percentile outline can be looked at as rare events and belong to the knowledge-based behavioral level.

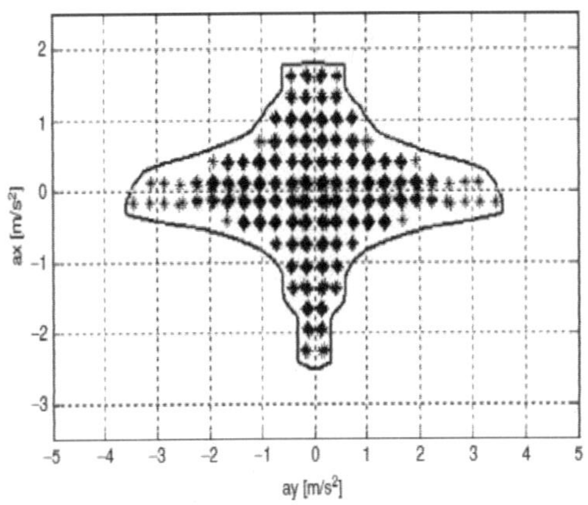

Figure 4. The g-g diagram of normaly experienced drivers

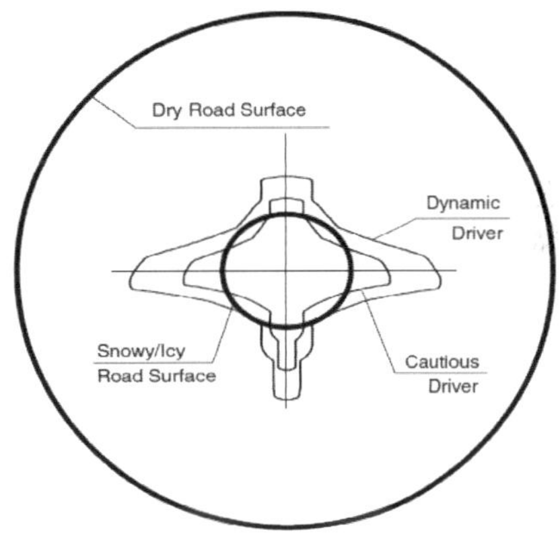

Figure 5. Driving behavior collectives compared to adhesion limit circles (differently skilled drivers, road surfaces with high and low friction potentials)

The empirical results at hand show consistently that drivers behavior collectives in traffic on public roads usually remain well below tire adhesion limits as long as the pavement is dry. Thus it may happen that even a cautious driver will exceed the limit of adhesion and hence the risk of accidents will grow considerably. Selecting the G-Circle produces a plot of the Longitudinal vs Latitudinal Acceleration data. Plotting out your own G-circle lets you know if you are exploiting your tyre's true potential. Ideally is a symmetrical plot, but the vehicle almost always generate more G under braking than acceleration, so the accelerating area will be shallower than the breaking part of the circle. All the experiments were performed under different conditions, in all four seasons with different tire types, in different road and atmospheric conditions.

The highly complex nature of many real-world conditions, though, often means that inventing specialized algorithms that will solve them perfectly every time is impractical, if not impossible. But a specially developed algorithm, like Machine Learning, detects the typical features for many different road conditions or driver behavior and this will always be a case with real-world data. Using that kind of algorithms to reach an accurate conclusion, involves many random samples as training data.

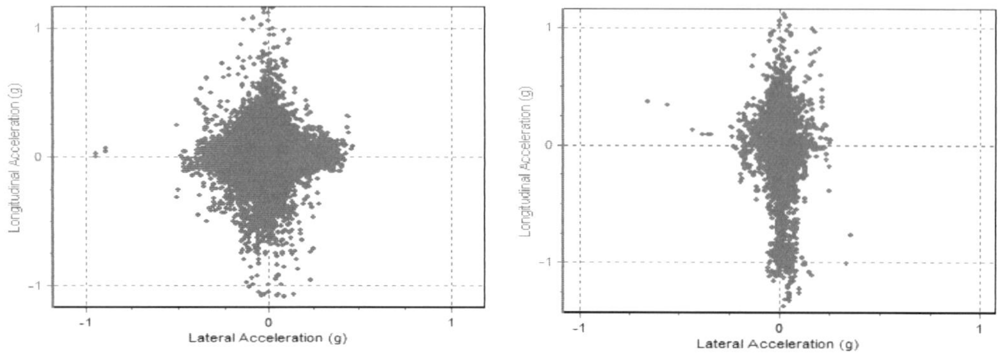

Figure 6. Presenting samples during warm wheather with dry pavement and summer tires

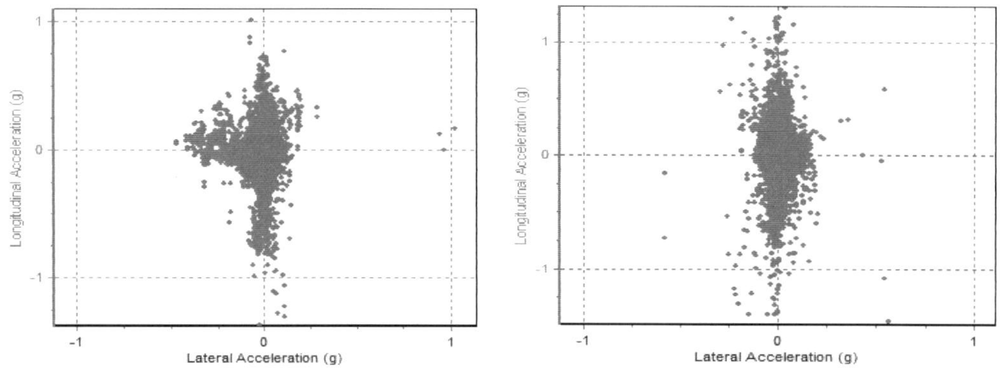

Figure 7. Presenting samples during cold wheather with wet pavement and summer tires

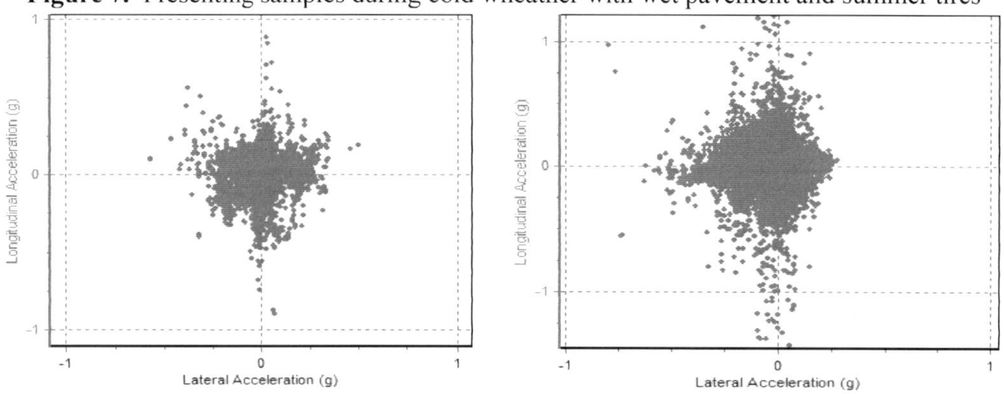

Figure 8. Presenting samples during cold wheather with wet and right pavement and winter tires

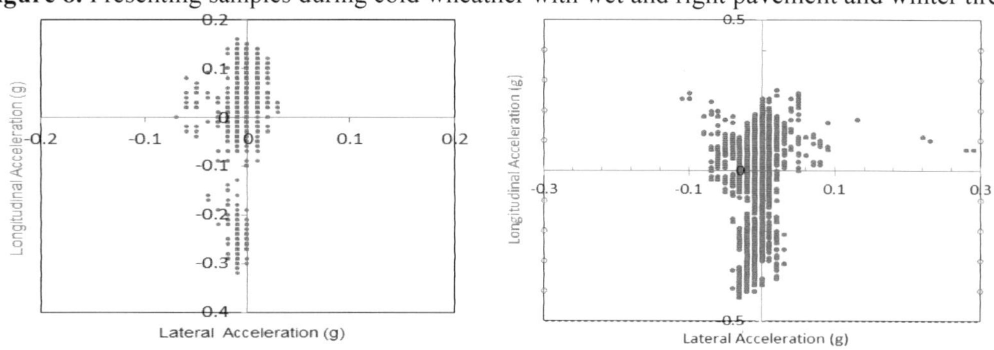

Figure 9. Presenting samples during cold wheather with snow and ice pavement and winter tires

5. Prediction capability and different approaches to estimate the μ_{max}

This paper evaluate the dynamic capability of the vehicle during acceleration or braking. In the normal driving conditions there is more acceleration then deceleration process, in this case the system can evaluate the vehicle dynamic capabilities and driving behavior, not just informations from tire or road conditions. Numerous vehicle parameters cause that an estimator for μ_{max} to be a hard operation, and finally μ_{max} to be a uncertain value. The parameters that need to take in account classifies as vehicle parameters like velocity, camber angle and wheel load; tire parameters like material, tire type, tread deapth and inflation pressure; road conditions like water, snow, ice films depth, temperatures; road parameters like road type, micro and macro geometry, drainage capacity. All this multiple parameters need to take in account for a system that estimate the μ_{max} .

5.1. Correlation method

The friction coefficient, μ, at a tire is related to the acceleration or deceleration obtained by the vehicle, implicitly to vehicle capacity for traction and braking maneuvers. Extensive testing on different road surfaces with different types of tires shows that this method can develop estimated values for g_{max} at braking using values of g_{max} at acceleration, and also allows for classification of roads as either dry, wet, snowy or icy. By using the raw data, acceleration and deceleration, g, from diferent traction and braking manouvres taked from on a variety of road conditions and surfaces, i demonstrated that it is a direct correlation between g_{max} at acceleration and g_{max} at deceleration obtained with same type of road and tyre condition.

Surfaces with lower friction coefficient generate low g_{max} at traction maneuvres and also low g_{max} at braking maneuvres, for example on icy road if the vehicle generate 0.15g on traction period can also generate 0,3g on brak ing period, on summer dry road if the vehicle generate 0.45g on traction period can also generate 1.1g on braking period and so on examples can continue on other road and tyre types. In figure 10 and 11 it is presented the model and the samples that sustain that observation.

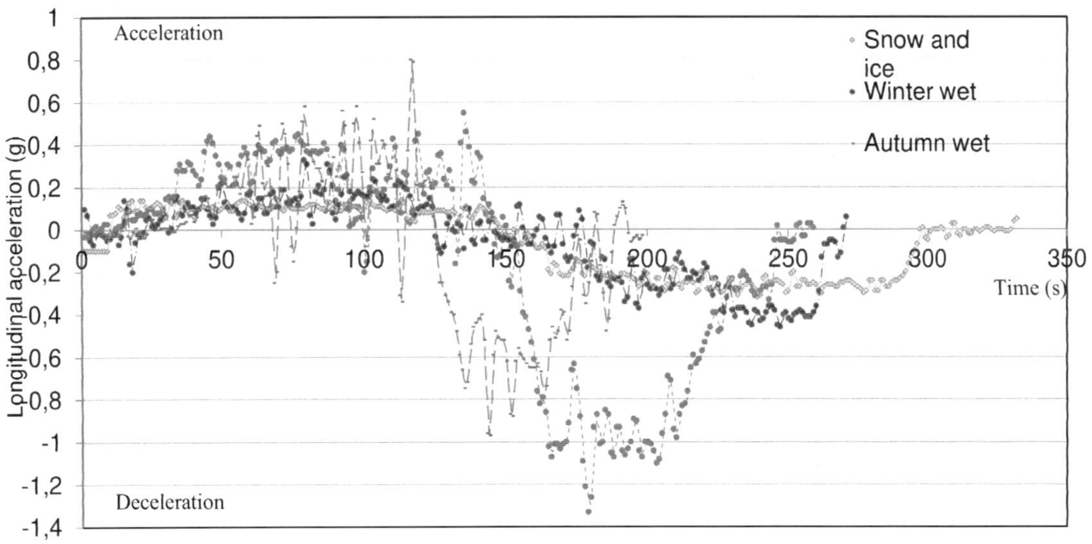

Figure 10. The reproduced model for this direct correlation, using the raw data from diferent traction and braking manouvres taked from on a variety of road conditions and surfaces.

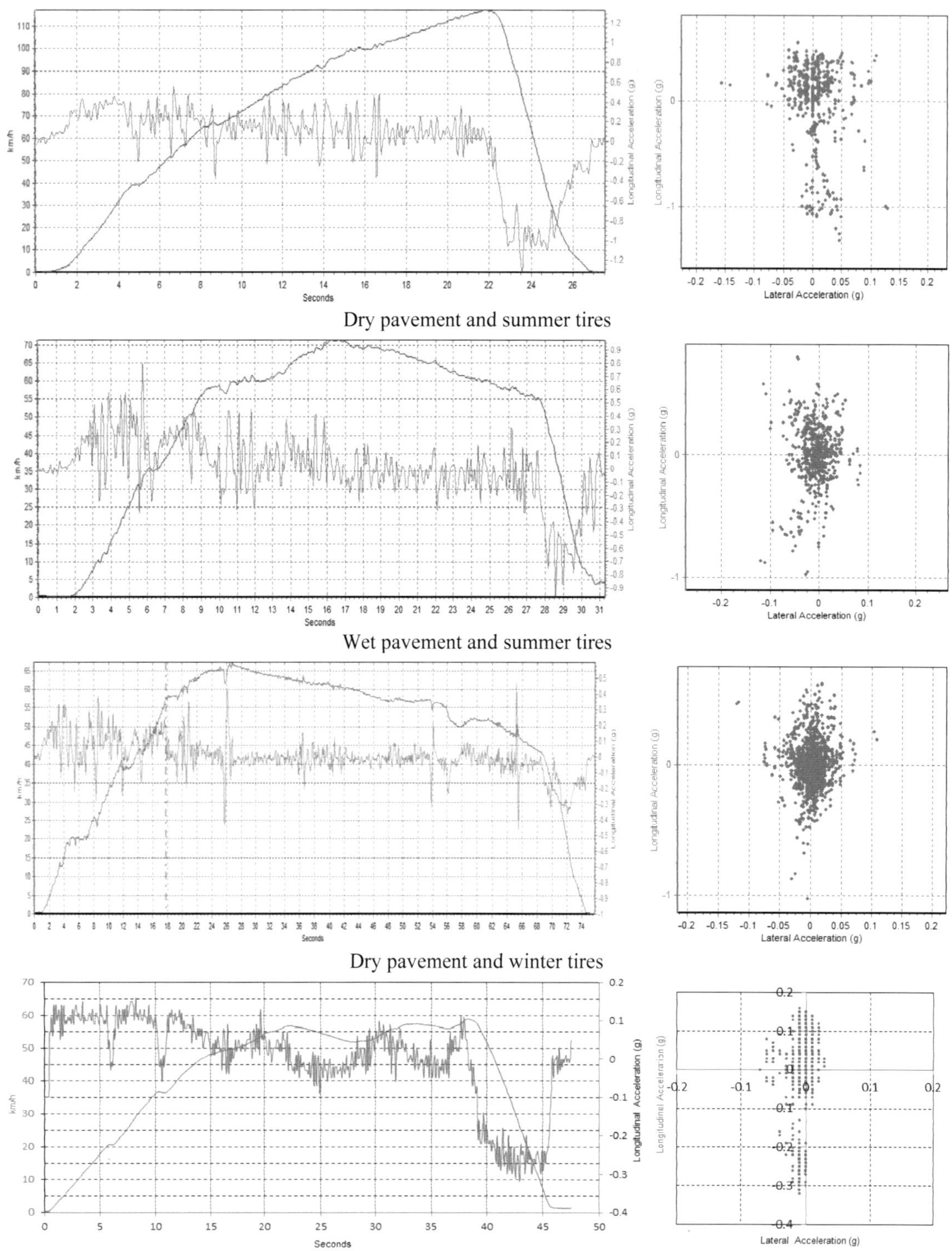

Figure 11. Samples from diferent traction and braking manouvres taked from on a variety of road conditions and surfaces.

5.2. Identification method

With Mathlab application, toolbox System Identification, we can approximate the acceleration and deceleration curve using the regression model. Another approach to estimate the vehicle dynamics is to collect raw data and then to use this data to estimate the entire nonlinear curve [5]. Determining μ_{max} is then facile because it is the maximum or minimum for braking or acceleration of this estimated curve. Experimental research with vehicles also has the role of obtaining the necessary data to establish mathematical models of their dynamics. Establishing mathematical models based on experimental data is the object of system identification. In present, a mathematical model is often theoretically prefigured and finalized based on experimental data.

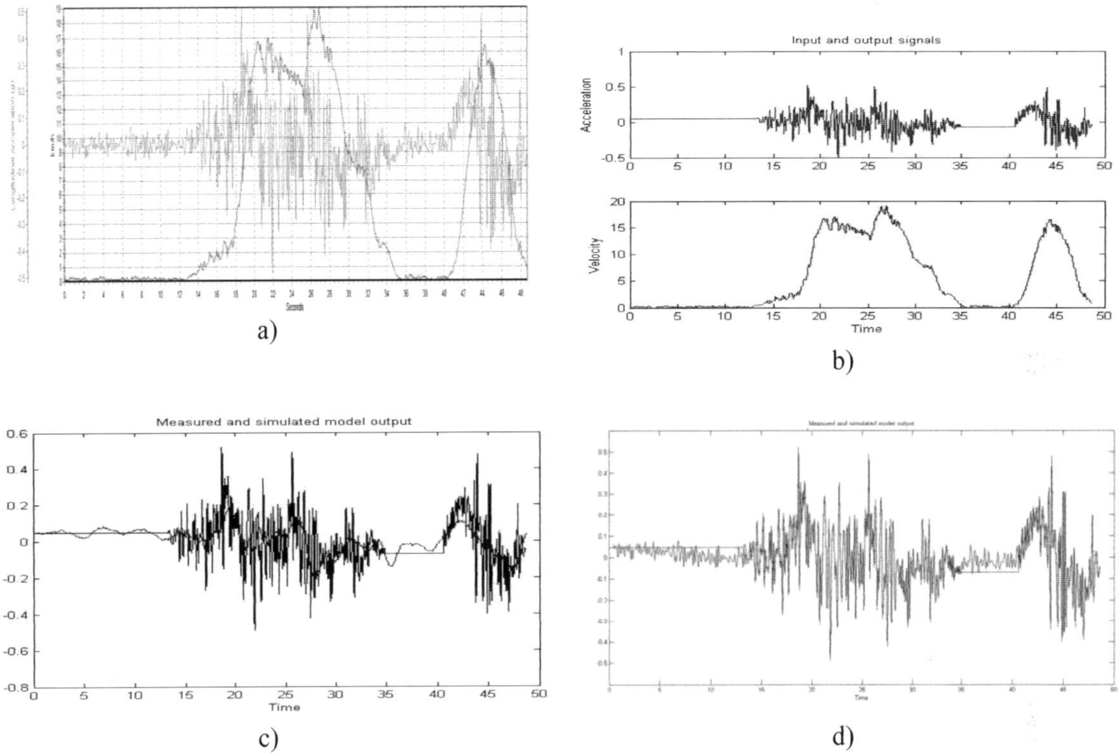

Figure 12. Samples data with inputs and outputs signals and different identification method, a) samples data, b) input and output signals, c) continuous-time identification, d) nonlinear identification.

Because experimental data is a discrete series, it is advisable to establish a discrete model and then a continuous model. To obtain a very low error value, which ensures high precision of modeling, is required to use nonlinear identification with multiple regressors. As can be seen from the discrete model expressions, for example, the relation (5) it is a autoregression, in this case with current regression k to k-10. Due to this fact, which makes the discrete model memory (taking into account regressors before the current one), a very high accuracy is obtained, even if the model is linear.

$$A(q)y(t) = \frac{B(q)}{F(q)}u(t-nk) + \frac{C(q)}{D(q)}e(t) \quad (3)$$

The limitation encountered to identify the mathematical models of vehicle dynamics, especially dynamic capabilities, is the high number of regressors which involves a powerful and complex program, therefore a delayed estimation about vehicle dynamic, then is needed a different approach.

5.3. Slope method

Another approach to monitor the vehicle dynamics is to find the slope of a linear curves during traction or braking manouvres and find the difference between the slopes of acceleration or deceleration curves on different road surface. Using the linear identification for traction or braking curves, it can be estimated the slope curves for different road surfaces [8].

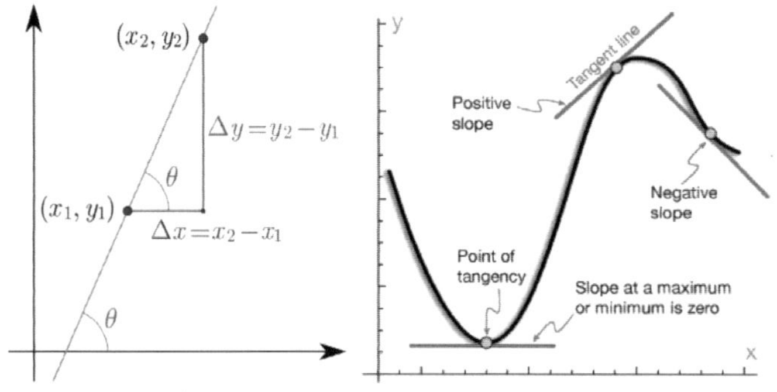

a) b)

The slope angle (θ) is the vertical angle formed by the inclined distance (Δ_y) with its horizontal projection (Δ_x). If the angle θ is positive, vehicle is on the accelerating process , and if it is negative, vehicle is on the braking process. The slope, m, and the slope angle is define by: $m = \frac{\Delta_y}{\Delta_x} = tg(\theta)$

$$\theta = arctg(\theta) \qquad (4)$$

Figure 13. Representation of slope angle (a) and orientation (b)

The slop angle is direct correlated with the parameters of polynomial equation. Using Mathlab, toolbox Curve Fitting, one-degree polynomial equations can be obtained, $f(x) = p_1.x + p_2$, with parameters that have 95% confidence bounds.

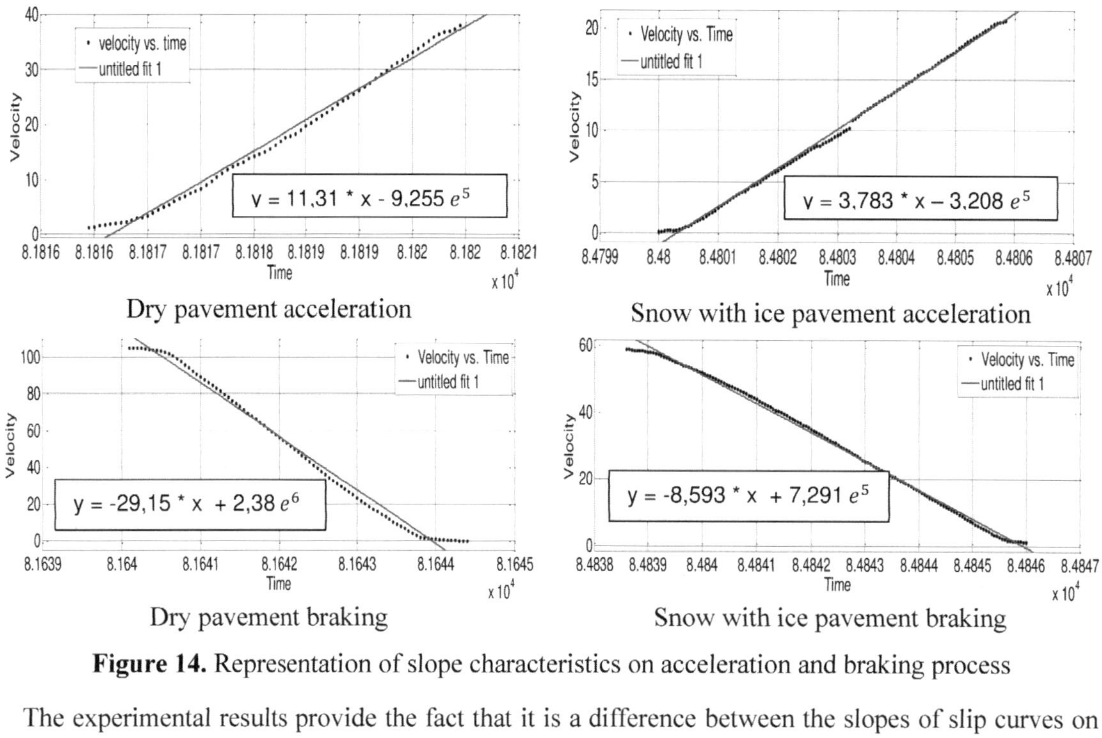

Figure 14. Representation of slope characteristics on acceleration and braking process

The experimental results provide the fact that it is a difference between the slopes of slip curves on wet and dry surface. The results obtained following the processing of experimental data show that the polynomial parameter represented the slope curve is higher on dry road and falls on wet or icy road. The slope method using the linear curve is viable indicator if the data points are grouped along on an axis.

6.Conclusions

Relatively low amount of data to process may can be an asset for many other researching methods to detect the driver and vehicle dynamics with available friction coefficient as an interaction between tires and the road surface in a proactive way and to use this information to make their vehicles safer. For example, the file presented in figure 3, have 1,19mb with more than 720 seconds of record, involving less than 2,5kb/s to process. By connecting vehicle to a cloud service, road data can be distributed to other vehicles, enabling drivers to adapt their driving style or change routes as they receive information about upcoming hazards or dangerous situations. This cloud data will then be possible to share with other drivers, road authorities and systems for Advanced Driver Assistance or for real time weather navigation data [6] [7].

Analyzing g-g diagrams for multiple subjects or teams form by human-machine, vehicle and software developers can create maps with human performance limitations and physico-technical constraints of vehicle.

Using various cost-effective methods, particularly correlation and slope method, to estimate adhesion limits can enhance prediction capabilities of autonomous vehicles.

Major problems for navigations systems represent reception interference like gaps in reception and multipath reception, and one of the solutions to counter this problems is to work in parallel with other data from CAN vehicle.

References

[1] Corolla D, Foster D, Kobayashi T and Vaughan N V 2004 *Encyclopedia of Automotive Engineering* John Wiley & Sons, Ltd pp 303-336
[2] Mastinu G and Poelchi M 2014 *Road and Off-Road Vehicle System Dynamics Handbook* Taylor & Francis Group pp 1564-1594
[3] Winner H, Hakuli S, Lotz F and Singer C 2016 *Handbook of Driver Assistance Systems* Springer International Publishing pp 23-45
[4] Meywerk M 2015 *Vehicle Dynamics* Willy pp 73-107
[5] Copae I, Cazacu C and Lespezeanu I 2006 *Vehicle Dynamics* Ericom pp 177-210
[6] Millikin M 2016 NIRA Dynamics, InfoCar expand availability of Road Surface Information software with OBD plug-in
[7] Millikin M 2017 Continental developing Road Condition Observer for active safety
[8] Uchanski M 2001 *Road Friction Estimation for Automobiles Using Digital Signal, PhD Thesis* University of California pp 11-14
[9] Racelogic Ltd 2014 *PERFORMANCE BOX Manual* Racelogic Ltd pp 65-67
[10] Racelogic Ltd 2016 *VBOXTools SOFTWARE Manual* Racelogic Ltd pp 55-76

Mathematical model for studying cyclist kinematics in vehicle-bicycle frontal collisions

OA Condrea, A Chiru, RL Chiriac, S Vlase
Transilvania University of Braşov, Romania
ovidiu.condrea@unitbv.ro

Abstract. For the development of effective vehicle related safety solutions to improve cyclist protection, kinematic predictions are essential. The objective of the paper was the elaboration of a simple mathematical model for predicting cyclist kinematics, with the advantage of yielding simple results for relatively complicated impact situations. Thus, the use of elaborated math software is not required and the calculation time is shortened. The paper presents a modelling framework to determine cyclist kinematic behaviour for the situations in which a M1 category vehicle frontally hits the rear part of a bicycle. After the primary impact between the vehicle front bumper and the bicycle, the cyclist hits the vehicle's bonnet, the windscreen or both the vehicle's bonnet and the windscreen in short succession. The head-windshield impact is often the most severe impact, causing serious and potentially lethal injuries. The cyclist is represented by a rigid segment and the equations of motion for the cyclist after the primary impact are obtained by applying Newton's second law of motion. The impact time for the contact between the vehicle and the cyclist is yielded afterwards by formulating and intersecting the trajectories for two points positioned on the cyclist's head/body and the vehicle's windscreen/bonnet while assuming that the cyclist's equations of motion after the primary impact remain the same. Postimpact kinematics for the secondary impact are yielded by applying linear and angular momentum conservation laws.

1. Introduction

Bicycling is an economical, healthy and environmentally friendly mode of transportation. In 2010, 7,4% of EU cyclists used the bicycle as a main mode of transport [1]. In Europe, as the number of bicycles and their usage in urban traffic is predicted to grow based on the increased policy interest in promoting cycling [2], new concerns arise regarding cyclist safety. Cyclists are non-motorized road users which present a high degree of vulnerability in traffic and a low degree of observability during night time. Accident statistics show that although driving safety significantly improved in the last few decades, cycling safety did not improve at the same rate [3]. Thus, additional efforts in accident analysis have to be made in order to understand cyclist kinematics (velocity, projection distance) and dynamics (interacting forces, motion and equilibrium) following a motor vehicle based collision.

Cyclist kinematics are characterized by a high degree of variability. One challenge here is the lack of current vehicle-cyclist impact formulations, given by the general assumption that cyclist and pedestrian kinematics, dynamics and injury causation mechanisms present a high degree of similarity. Therefore, safety standards and measures regarding the design of vehicles frontal profiles for pedestrian impacts are considered to cover a part of the cyclist impact typologies [4-10].

While the primary impact between the bumper and the bicycle/pedestrian leg differs in terms of inertia and geometry, a relatively high degree of similarity can be indeed established for the impact between the human head and the vehicle's bonnet, windshield or roof. However, the Wrap Around Distance (WAD) measured along the vehicle profile from the ground to the cyclist/pedestrian head impact location is higher for cyclists than for pedestrians [11], which overall increases the vertical coordinate of the point of impact, modifies the angle of impact and consequently leads to more severe head impacts.

A second challenge is the increased modelling complexity of vehicle-bicycle collisions in comparison to vehicle-pedestrian collisions, caused by the addition of the bicycle and by the interactions which appear between the bicycle and the cyclist.

The study presented here shows a simple two-dimensional approach for front-rear vehicle-bicycle collisions, divided into three stages: *1) Mathematical modelling of the primary vehicle-bicycle impact*
2) Determining post-primary impact cyclist kinematics and the time of the secondary impact
3) Mathematical modelling of the secondary cyclist-vehicle impact

2. Methods

In order to simplify the model and reduce the number of unknown variables, the following hypotheses were formulated:

1) The primary vehicle-bicycle impact:
a. during the primary impact, the cyclist weight force is applied only to the bicycle's saddle
b. handlebar and pedal forces are neglected
c. vehicle velocity losses are neglected
2) Post-primary vehicle-bicycle impact:
a. between the primary and secondary impact, no other contacts occur
b. after the primary impact, gravity is the only force acting on the cyclist
c. bicycle and vehicle post impact vertical velocities are neglected
d. air resistance has a negligible effect upon the cyclist
3) The secondary cyclist-vehicle impact:
a. the contact occurs for a very short time
b. the impact does not produce a moment around the vehicle's center of mass

The cyclist is considered a rigid body. Following the primary impact between the vehicle and the bicycle (fig 1.a.), the cyclist will be subjected to an air projection towards the vehicle's frontal profile. Depending on the height of the bicycle's saddle and on the vehicle's impact velocity and frontal profile geometry, the cyclist hits the vehicle's bonnet (fig 1.b.), the windscreen (fig 1.c.) or both the vehicle's bonnet and the windscreen in short succession. The objective of this paper is to study cyclist kinematic parameters for the first two cases.

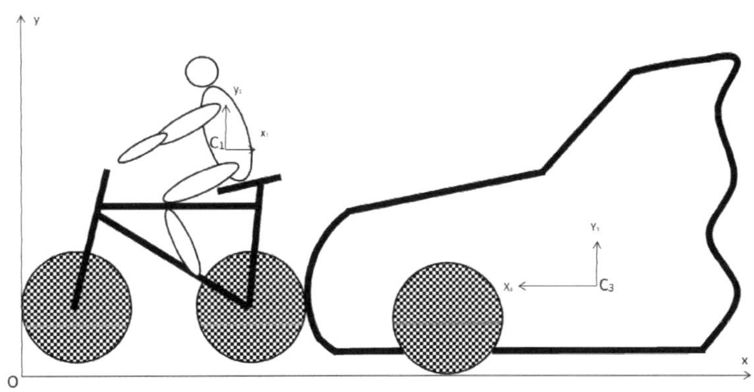

a. Primary impact schematised

IOP Conf. Series: Materials Science and Engineering **252** (2017) 012003 doi:10.1088/1757-899X/252/1/012003

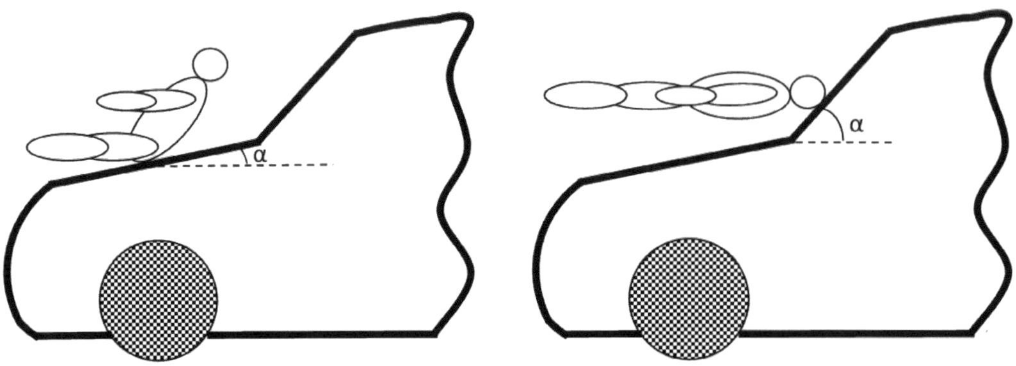

b. Cyclist-bonnet impact c. Cyclist-windscreen impact

Fig. 1 (a, b, c) Primary impact and possible secondary impact configurations

The global origin is positioned with the Ox axis at ground level and Oy axis vertically upwards, tangential to the bicycle's front wheel. The cyclist coordinate system origin is the centre of mass of the cyclist (C_1) during the instant of the primary impact between the bicycle and the vehicle (t_0=0). The origin of the vehicle's coordinate system is at its centre of mass (C_3) with the positive Ox axis direction aligned with the vehicle's travel direction and the positive Oy axis direction vertically upwards.

3. Mathematical modelling of the primary vehicle-bicycle impact

During the primary impact, the resultant impact force F_e is applied on the bicycle's rear wheel with its corresponding force pair acting upon the vehicle's bumper, and a pair of resultant contact forces F_b appear and interact with the bicycle's saddle and the cyclist's prop point (fig. 2).

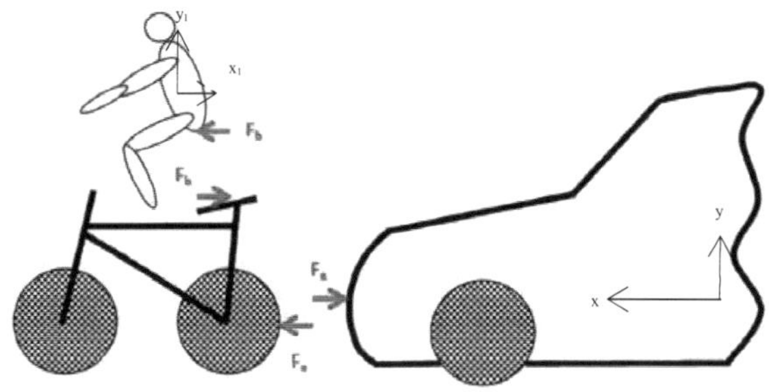

Fig. 2. Pairs of forces which appear during the primary impact

Upon the cyclist's center of mass (C1) acts its own weight force G_1 while in its prop point (A) acts the bicycle's saddle afferent normal force N_A (fig 3).

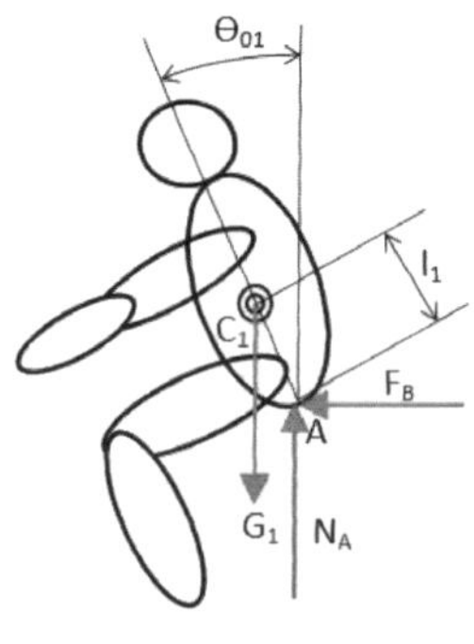

Fig. 3. Forces which act upon the cyclist during the primary impact

Considering the cyclist, the bicycle and the vehicle as rigid segments and applying the second principle of mechanics during the primary impact, the following system of equations results, where the 1, 2 and 3 indices stand for the cyclist,the bicycle and the vehicle respectively:

- For the cyclist

$$m_1\ddot{x}_1 = -F_b \ (1)$$
$$m_1\ddot{y}_1 = -G_1 + N_A \ (2)$$

- For the bicycle

$$m_2\ddot{x}_2 = F_b - F_e \ (3)$$

- For the vehicle

$$m_3\ddot{x}_3 = F_e \ (4)$$

Where $\ddot{x}_1$, $\ddot{x}_2$ and $\ddot{x}_3$ are the center of mass acceleration components on the Ox axis, $\ddot{y}_1$ is the cyclist's center of mass acceleration component on the Oy axis and m_1, m_2, and m_3 are physical body masses.

From the condition of moments equilibrium taken about the cyclist's center of mass C_1 point, the following equation is yielded:

$$J_1\ddot{\theta}_1 = -F_b l_1 cos\theta_{01} + N_A l_1 sin\theta_{01} \ (5)$$

All the equations can be written in the form:

$$\begin{bmatrix} m_1 & & & & \\ & m_1 & & 0 & \\ & & J_1 & & \\ & 0 & & m_2 & \\ & & & & m_3 \end{bmatrix} \begin{bmatrix} \ddot{x}_1 \\ \ddot{y}_1 \\ \ddot{\theta}_1 \\ \ddot{x}_2 \\ \ddot{x}_3 \end{bmatrix} = \begin{bmatrix} -F_b \\ -G_1 + N_A \\ -F_b l_1 cos\theta_1 + N_A l_1 sin\theta_1 \\ F_b - F_e \\ m_3\ddot{x}_3 = F_e \end{bmatrix} (6)$$

Where θ_{01} is the initial angular displacement of the cyclist, $\ddot{\theta}_{01}$ is the cyclist's angular acceleration, l_1 is the distance from the cyclist's center of mass to its prop point A on the saddle of the bicycle and J_1 is the cyclist's moment of inertia.

The cyclist's velocity is equal to the bicycle's velocity during the instant of the primary impact between the bicycle and the vehicle. Considering that the bicycle's and vehicle's movement in the aftermath of the primary impact is solely on the Ox axis and neglecting vertical velocities which may appear depending on the difference between the bicycle's and vehicle's centers of mass vertical coordinates ($\dot{z_2} = 0; \dot{z_3} = 0$), the following equation is yielded:

$$\dot{x_2} = \dot{x_A} \ (7)$$

A good approximation that can avoid solving the equation (1)-(4) is to consider that the vehicle transfers entirely its velocity to the point A of the cyclist:

$$\dot{x_2} = \dot{x_A} = \dot{x_3} \ (8)$$

In this case it can be considered that the vehicle's velocity at impact is the initial velocity (if there is no braking).

During the plastic impact of the bycicle, the kinematic conditions are:

$$x_1 = x_2 + l_1 sin\theta_1 \ (9)$$
$$y_1 = y_2 + l_1 cos\theta_1 \ (10)$$

and after differentiation:

$$\dot{x}_1 = \dot{x_2} + l_1\dot{\theta}_1 cos\theta_1 \ (11)$$
$$\dot{y}_1 = -l_1\dot{\theta}_1 sin\theta_1 \ (12)$$

and

$$\ddot{x_1} = \ddot{x_2} + l_1\ddot{\theta}_1 cos\theta_1 - l_1\dot{\theta}_1^2 sin\theta_1 \ (13)$$
$$\ddot{y}_1 = -l_1\ddot{\theta}_1 sin\theta_1 - l_1\dot{\theta}_1^2 cos\theta_1 \ (14)$$

$$\begin{bmatrix}\ddot{x}_1\\ \ddot{y}_1\\ \ddot{\theta}_1\\ \ddot{x}_2\\ \ddot{x}_3\end{bmatrix} = \begin{bmatrix} l_1 cos\theta_1 & 1 & 0\\ -l_1 sin\theta_1 & 0 & 0\\ 1 & 0 & 0\\ 0 & 1 & 0\\ 0 & 0 & 1\end{bmatrix}\begin{bmatrix}\ddot{\theta}_1\\ \ddot{x}_2\\ \ddot{x}_3\end{bmatrix} + \begin{bmatrix}-l_1 sin\theta_1\\ -l_1 cos\theta_1\\ 0\\ 0\\ 0\end{bmatrix}\dot{\theta}_1^2 \quad (15)$$

By substituting (15) in (7) and premultiplying the equations with

$$\begin{bmatrix} l_1 cos\theta_1 & 1 & 0\\ -l_1 sin\theta_1 & 0 & 0\\ 1 & 0 & 0\\ 0 & 1 & 0\\ 0 & 0 & 1\end{bmatrix}^T$$

The following equations are yielded:

$$\begin{bmatrix} m_1 l_1^2 cos^2\theta_1 + m_2 l_1^2 sin^2\theta_1 + J_1 & m_1 l_1 cos\theta_1 & 0\\ m_1 l_1 cos\theta_1 & m_1 + m_2 & 0\\ 0 & 0 & m_3\end{bmatrix}\begin{bmatrix}\ddot{\theta}_1\\ \ddot{x}_2\\ \ddot{x}_3\end{bmatrix} + \begin{bmatrix}-(m_1 + m_2) l_1^2 sin\theta_1 cos\theta_1\\ -m_1 l_1 sin\theta_1\\ 0\end{bmatrix}\dot{\theta}_1^2$$

$$= \begin{Bmatrix} G_1 l_1 sin\theta_1 - 2F_b l_1 cos\theta_1\\ -F_e\\ F_e\end{Bmatrix} (16)$$

It is possible to obtain the differential motion equations of the system. There are 3 differential equations with 3 unknowns x_2, θ_1 and x_3 .

4. Determining post-primary impact cyclist kinematics and the time of the secondary impact

Following the primary impact, the cyclist is launched towards the vehicle's frontal profile on a parabolic trajectory. This yields the following equations of motion for the cyclist's center of mass C_1:

$$x_1 = x_{01} + \dot{x}_1 t \quad (17)$$

$$y_1 = y_{01} - \frac{gt^2}{2} \quad (18)$$

Where t_l is the time of the secondary impact between the cyclist's head/body with the vehicle's windscreen/bonnet and g is the acceleration of gravity.

The secondary impact takes place when a point P positioned on the cyclist's head/body reaches a point Q positioned on the vehicle's windscreen/bonnet (fig 4).

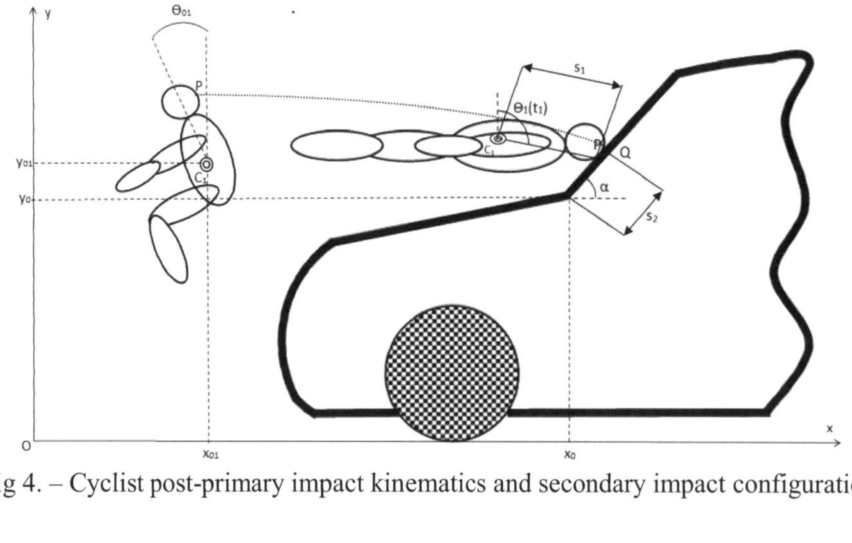

Fig 4. – Cyclist post-primary impact kinematics and secondary impact configuration

The secondary impact time can be determined by intersecting the trajectories of points P and Q. For point P the following equations of motion result:

$$x_P = x_1 + s_1 cos\theta \ (19)$$

$$y_P = y_1 + s_1 sin\theta \ (20)$$

$$\theta = \theta_{01} - \omega_0 t \ (21)$$

Where s_1 is the distance from the cyclist's center of mass C1 to the point P.

Considering the vehicle's bonnet or windscreen as a line of slope tan(α), the following equations of motion for point Q result:

$$x_Q = x_0 + s_2 cos\alpha \ (22)$$

$$y_Q = y_0 + s_2 sin\alpha \ (23)$$

Where s_2 is the distance from the vehicle's center of mass C_3 to the point Q and a is the horizontal distance from center of mass C_3 to the starting point of the windscreen/bonnet slope.

From (22) and (23) the following equation is yielded:

$$\frac{x_Q - x_0}{y_Q - y_0} = \frac{cos\alpha}{sin\alpha} = ctg\ \alpha \ (24)$$

The secondary impact time can be determined by applying the following conditions:

$$x_P = x_Q \ (25)$$

$$y_P = y_Q \ (26)$$

From (24), (25) and (26), the following equation results:

$$(x_P - x_0) = (y_P - y_0)\ ctg\ \alpha \ (27)$$

Equation (27) can be rewritten as a transcendent equation:

$$(x_{01} + \dot{x_1} t + s_1 cos\theta_1 - x_0) = \left(y_{01} - \frac{gt^2}{2} + s_1 sin\theta_1 - y_0\right) ctg\ \alpha \ (28)$$

Where θ_1 is the initial angular displacement of the cyclist at the time of the secondary impact t_1.

Equation (28) doesn't have close-form solutions, it is solved approximately using the bisection method, yielding the time of the secondary impact t_1.

5. Mathematical modelling of the secondary cyclist-vehicle impact

Considering the secondary impact is produced at t_1 between the cyclist's head and the windscreen as it is the case for collisions with moderate or high vehicle impact velocities, cyclist postimpact kinematic parameters can be determined by applying the laws of conservation of linear and angular momentum (29), (30), (31) and by formulating the equations for the relative motion of the cyclist in relation to the vehicle (32), (33):

$$m_1 \dot{x}_1' = m_1 \dot{x_1} + m_3 \dot{x_3} \ (29)$$

$$m_1 \dot{y}_1' + m_3 \dot{y}_3' = m_1 \dot{y_1} \ (30)$$

$$J_1 \dot{\theta}_1' + m_3 s_1 \dot{x}_1' cos\theta_1' - m_3 s_1 \dot{y}_1' sin\theta_1' = J_1 \dot{\theta}_1 + m_3 s_1 \dot{x}_1 cos\theta_1 - m_3 s_1 \dot{y}_1 sin\theta_1 \ (31)$$

$$\dot{x}_1' + \dot{\theta}_1' s_1 cos\theta_1' - \dot{x}_3' = 0 \ (32)$$

$$\dot{y}_1' - \dot{\theta}_1' s_1 sin\theta_1' - \dot{y}_3' = 0 \ (33)$$

Where $\dot{x}_1, \dot{y}_1$ are cyclist preimpact velocity components at t_1, $\dot{x}_3, \dot{y}_3$ are vehicle preimpact velocity components at t_1, $\dot{x}_1', \dot{y}_1'$ are cyclist postimpact velocity components, $\dot{x}_3'$ $\dot{y}_3'$ are vehicle postimpact velocity components, $\dot{\theta}_1$ is the cyclist's preimpact angular velocity at t_1, $\dot{\theta}_1'$ is the cyclist postimpact angular velocity, θ_1 is the cyclist's preimpact angular displacement at t_1 and θ_1' is the cyclist postimpact angular displacement.

The system of equations (29)-(33) is solved using numerical integration, yielding cyclist and vehicle postimpact kinematic parameters.

The modelling for the secondary impact described can also be applied to the contact between the cyclist's body and the vehicle's bonnet when this is the case, however an additional modelling phase would be required to yield the parameters of the tertiary impact between the cyclist's head and the vehicle's windscreen or roof.

6. Input data for the mathematical model

Input data for the model consists of the following constants and initial parameters: physical body masses m_1, m_2, m_3 , resulting force acting upon the primary impact F_e, cyclist moment of inertia J_1, windshield/ bonnet angle α, initial cyclist angular displacement θ_{01}, cyclist position at primary impact x_{01}, y_{01}, s_1, vehicle position at primary impact x_0, y_0, s_2 and vehicle, bicycle and cyclist initial velocities $\dot{x}_1, \dot{x}_2, \dot{x}_3$.

Example: Considering a vehicle having a velocity of 72 km/h. The cyclist has a velocity of 36 km/h. In the short time of the crash the vehicle's velocity practically doesn't change. The cyclist receives a momentum that changes the velocity of his center of mass, to 54 km/h. At the same time the cyclist receives a moment of momentum which causes a spin, and the velocity for the contact point between the cyclist and the bicycle becomes 72 km/h. The angular velocity of the cyclist, which receives a rotation motion around his mass center, becomes 50 rad/s. The time between the moment in which the bicycle is collided by the vehicle and the moment of the contact between cyclist and the vehicle (windscreen) is, for example, 0,04 s and the rotation angle of the cyclist in this time is approx. 2 rad. With this rotation (approximately 114°), the cyclist's head reaches and collides the windscreen at its base (as it is the case in the most situations for this accident typology).

7. Conclusions

A simple mathematical model for predicting the impact and cyclist kinematics after impact that provides insights into the crash dynamics was elaborated. The rigid body approach yields simple results for relatively complicated impact situations, does not require the use of elaborated math software and the calculation time is significantly shortened. The main advantage of this approach is the necessary time and cost to obtain some conclusions concerning an impact. Cyclist kinematic parameters are a result of the primary impact between the vehicle and the bicycle and are directly influenced by inertia, geometry of the vehicle and cyclist's riding position at impact.

The mathematical model developed has the following limits:

1) is a simple two-dimensional model with concentrated masses;

2) the cyclist is represented by a rigid body and therefore deformations cannot be calculated

3) the model takes into account only one collision between the cyclist and the vehicle's frontal profile.

IOP Conf. Series: Materials Science and Engineering **252** (2017) 012003 doi:10.1088/1757-899X/252/1/012003

References

[1] *** 2010 Flash Eurobarometer - Future of transport. Analytical report. Flash EB Series #312 Available: http://ec.europa.eu/commfrontoffice/publicopinion/flash/fl_312_en.pdf
[2] de Hartog J, Boogard H, Nijland H & Hoek G 2010 Do the health benefits of cycling outweigh the risks? *Environmental Health Perspectives,* **118,** pp 1109–1116
[3] *** 2011 Traffic safety basis facts 2011-cyclists DaCoTa Project, European Road Safety Observatory Available: https://ec.europa.eu/transport/road_safety/sites/roadsafety/files/pdf/statistics/dacota/dacota-3.5-asr-2011.pdf
[4] *** 2007 Proposed draft global technical regulation (GTR) on pedestrian protection. Available: http://www.unece.org/fileadmin/DAM/trans/doc/2004/wp29grsp/ps-116.pdf
[5] *** 2002 EEVC working group 17 report Improved test methods to evaluate pedestrian protection afforded by passenger cars
[6] *** 2003 ECDirective 2003/102/EC of the European Parliament and of the Council of the 17th November 2003. Protection of pedestrians and other vulnerable road users before and in the event of a collision with a motor vehicle and amending directive 70/156/EEC
[7] Simms C, Wood D 2009 *Pedestrian and cyclist impact – A biomechanical perspective*, ed G M L Gladwell (Dublin:Springer) chapter 7 pp 115-136
[8] Wismans J, Janssen EG 1994 Improved safety of pedestrians and cyclists, *XXV FISITA Congress*, (Beijing:International Acad. Publ.) pp 143-149
[9] Janssen EG, Wismans J 1985 Experimental and mathematical simulation of pedestrian-vehicle and cyclist vehicle accidents, *SAE Technical Paper 856113* pp 977-988
[10] Janssen EG, Wismans J 1987 Evaluation of vehicle-cyclist impacts through dummy and human cadaver tests *The 11th Technical Conf. on Experimental Safety Vehicles*
[11] Haight W, Eubanks J 1990 Trajectory analysis for collisions involving bicycles and automobiles *SAE Technical Paper 900368*, pp 143-160.

Study regarding seat's rigidity during rear end collisions using a MADYMO occupant model

RADU Alexandru Ionut[1], COFARU Corneliu[1], TOLEA Bogan[1], POPESCU Mihaela[1]

[1]University "TRANSILVANIA" of Brasov

Abstract – The aim of this paper is to study the effects of different front occupant backseat's rigidities in the case of a rear end collision using a multibody virtual model of an occupant. Simulation will be conducted in PC Crash, the most common accident reconstruction software using a MADYMO multibody occupant to simulate kinematics and dynamic of the passenger. Different backseat torques will be used to see how this will influence the acceleration in the head and torso of the occupant. Also, a real crash test is made to analyze the kinematics of the occupant. We believe that the softer seat's rigidity will reduce not only the head's acceleration but also reduces the effect of „whiplash" upon the neck due to the fact that the backseat will rotate backwards increasing its displacement and absorb some of the energy generated by the collision. Although a softer seat could reduce the head's acceleration, a broken seat will increase it due to the fact that the impact of the backseat with the vehicle's rear seats will generate a second collision. So, in order to achieve a lower acceleration, a controlled torque is recommended and a controlled angular displacement of the backseat is to be used.

1. Introduction

Rear-end collisions are the most frequent types of car-to-car accidents where some form of injury occurs upon the occupant [6]. Usually, in this type of collision the neck and the head suffer the most severe damage due to the phenomenon called "whiplash" that consists of the neck and head's moving back against the headrest, generating low injuries to the neck, but with long- term consequences [2], [11], [13],. One of the most negative effects of "whiplash" is the long-term injury that can be permanent in some cases. Only in Europe, over 10 million euros where paid by insurance companies to people that reported long-term injuries of "whiplash" [1], [5]. It was observed that, if the backseat moves in accordance with the occupant, a part of the energy is absorbed. Jakobsson designed and built a system to accomplish this, it is called WHIPS and it is mounted on some modern Volvo vehicles [7]. This is done by a mechanism that limits the movement of the backrest slightly, but also moves the backrest upwards to align the head with the headrest [10]. On the other hand, if the backrest's hinge breaks and moves too much, it will hit the backseat of the vehicle, generating a higher acceleration upon the head. Because of this, the study was realized in order to analyze different torque values of the backseat's hinge where the acceleration on the occupant's head varies.

2. Method used

In the simulations, different torque values of the backseat hinge have been chosen to observe the occupant's kinematics and the head's acceleration variation as well. This study was conducted in the virtual environment in the PC-CRASH 10.2 program using the MADYMO occupant's module integrated into a PC-CRASH [14]. The occupant's height is fixed to the 50th percentile, and the dashboard geometry is fixed. Its dimensions and seat position are shown in Figure 1.

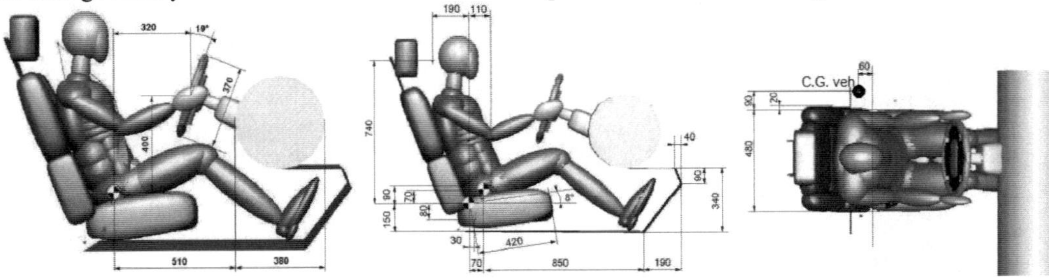

Figure 1. MADYMO occupant's position and dimensions [4]

The module allows modification of certain parameters regarding the seat-back angle, its rigidity, the head's restraint, occupant's mass data, and the airbag deployment times and seat belt pre-tensioning. The MADYMO interface is shown in Figure 2.

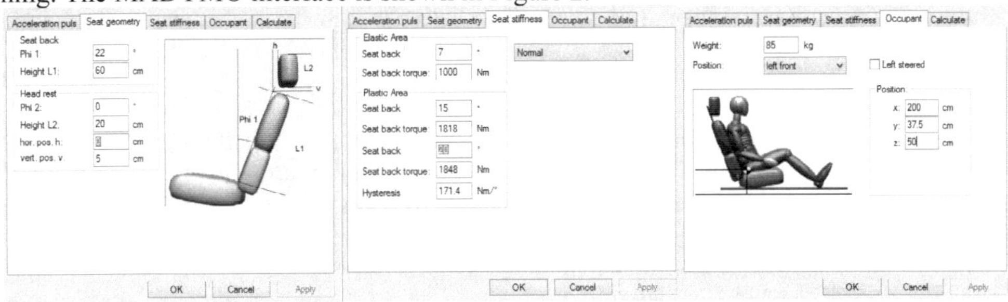

Figure 2. Interface used to modify MADYMO model parameters

A crash test was made in order to analyze the occupant's kinematics and to validate the simulation of the impact by using the MADYMO model. The test consisted of a vehicle-to-vehicle rear-end collision with an overlap of 100%. The crash configuration is presented in Figure 3.

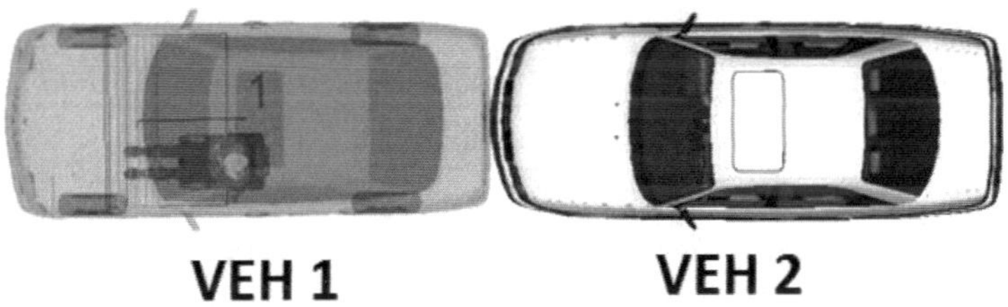

Figure 3. Crash test configuration

The crash test was filmed with a high-speed video camera, capable of filming at 1000 fps, in order to analyze each phase of the impact and to study the movement of the occupant. After this test had been conducted, a simulation in PC-Crash was also was done using the same configuration and compared to the crash test. In Figure 4 the comparison of the occupant's kinematics is presented between the crash test and the simulation.

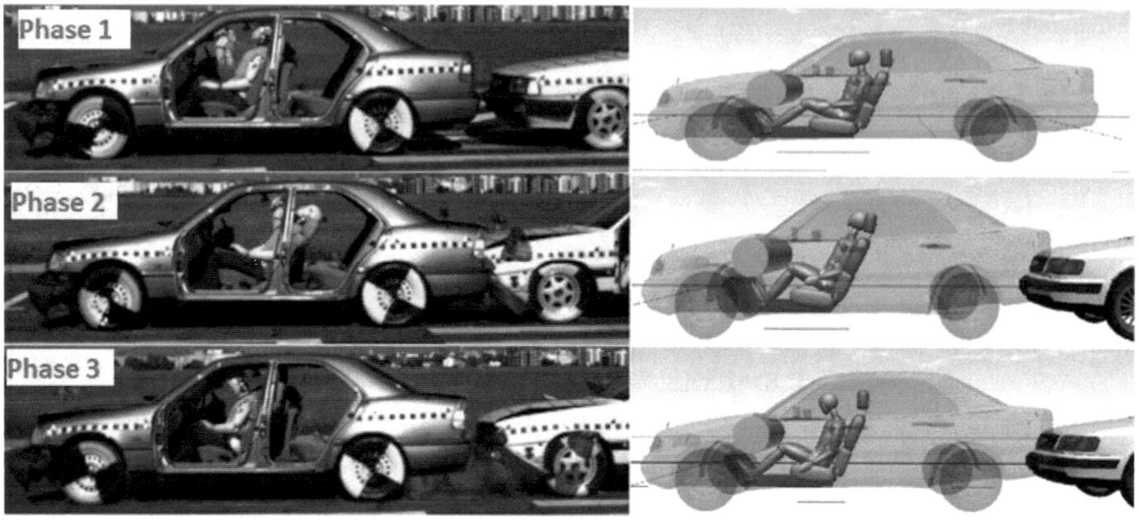

Figure 4. Occupant's kinematics comparison between the crash test and simulation

In phase 1, the posture of the occupants is seen before the collision, followed by phase 2, when the occupant begins to move towards the inside of the seat, and the head hits the headrest. Also, in this phase, the extension of neck occurs. In phase 3, the chest tends to rebound forward, but it is limited by the seatbelt. Also, the head flexes forward as well. This movement can be seen in both occupants. The parameter of interest in this study is the acceleration of the occupant's head. The comparative chart between the occupant acceleration in the experimental test and the MADYMO model is shown in Figure 5.

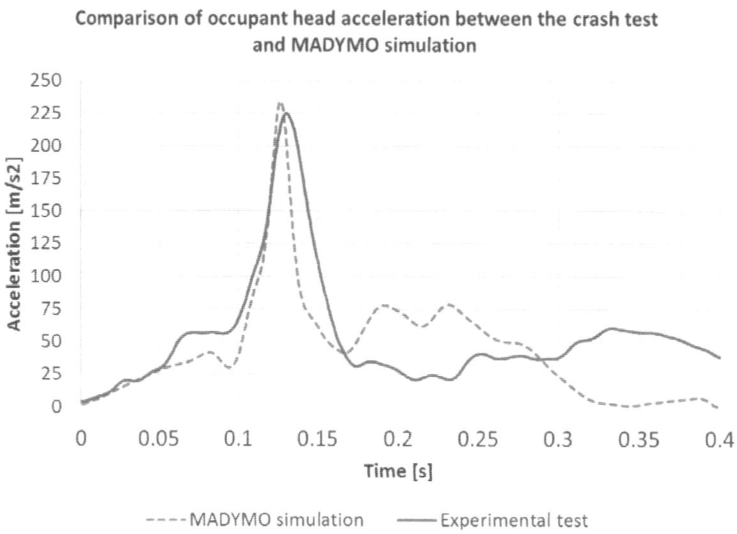

Figure 5. Comparison of occupant's head's acceleration between the crash test and simulation

The start of the collision takes place after 100 ms, point where the occupant's chest travel to the seat and the head reaches contact with the headrest at 130 ms, generating an acceleration peak of 230 m/s^2. It can be seen that, in both cases, the acceleration has a similar pattern, validating the simulation with an error of 12%.

The acceleration value, according to the impulse duration, shows a maximum limit of 40 g for the head. In the case of pedestrians the situation is more dangerous, at the impact with the ground, much higher head accelerations (120 - 200 g) can cause severe injuries. The head injury risk is assessed using HIC (Head Injury Criteria) criteria on a time interval of 36 ms for the occupant, respectively 15 ms for the pedestrian [8], [9].

The velocity of the vehicles was analyzed and presented in Figure 6. The collision velocity was 34 km/h for the striking vehicle (VEH 2) and 0 km/h (stationary) for the struck vehicle (VEH 1).

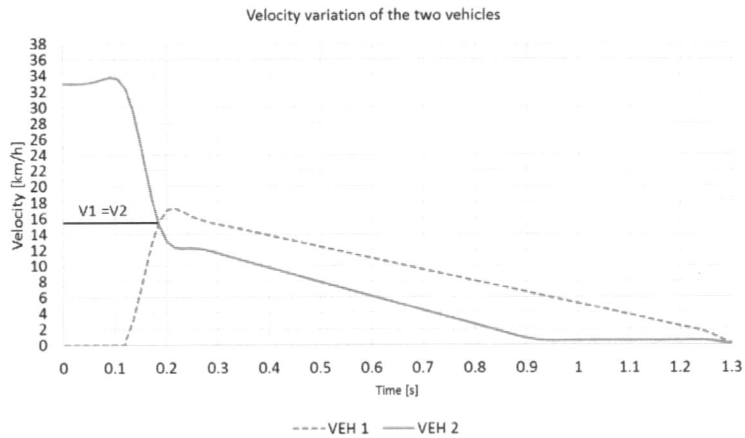

Figure 6. Velocity variation of the two vehicles

In can be observed that the collision takes place after 100 ms, when the velocity of the striking vehicle (VEH 2) drops from 34 km/h to 13 km/h at 180 ms, then, a low decrease rate to 0 km/h at 1000 ms occurs. Also, after 100 ms, the stroked vehicle (VEH 1) has an increase in velocity, from 0 km/h, up to 17 km/h at 200 ms, following a lower rate decrease. At 170 ms, it can be seen that both vehicles have the same velocity, of 15 km/h.

In the next phase, three simulations were conducted by using the MADYMO module, in which the torque between the seatback and the seat was set for the three situations and the stiffness was set to normal (1000 Nm), lower (550 Nm) and very low (250 Nm). The kinematics of the occupant and the variation of the back angle for the three situations are presented in Figure 7.

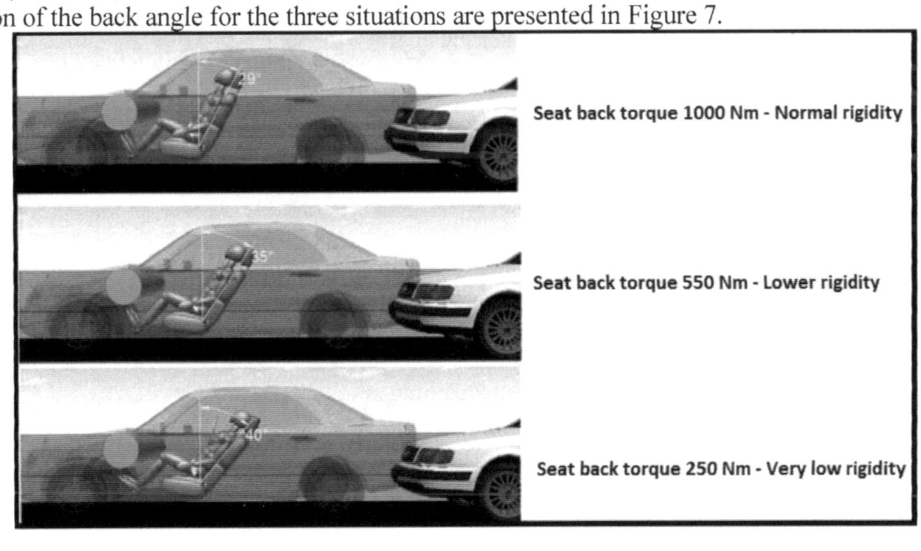

Figure 7. Variation of the inclination angle of the backrest according to its rigidity

In the case of normal rigidity, the torque is 1000 Nm [4] and the maximum inclination of the backrest is 29 degrees. To simulate a reduced rigidity, the torque was reduced to 550 Nm, resulting in a maximum angle during the collision of 35 degrees, and to simulate a very low stiffness, the torque was reduced to 250 Nm, resulting in a maximum angular displacement of 40 degrees. The comparative analysis of the occupant's head's acceleration for the three cases is presented in Figure 8.

Whiplash injuries can be predicted using the correlation between the head and torso rotation angle. NHTSA developed a method to predict the probability of whiplash in accordance with the head-torso rotation based on sled tests conducted in a controlled environment using crash test dummies and different seats and headrests. In Figure 8 the probability of whiplash is presented [12].

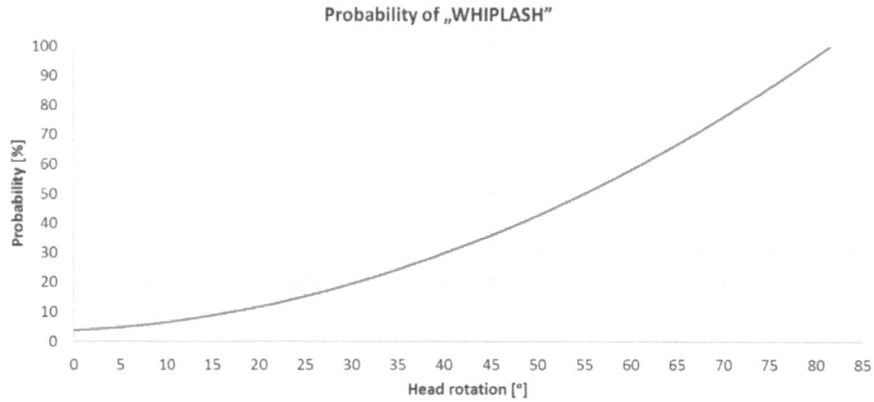

Figure 8. Probability of "whiplash" in accordance with the head-torso rotation angle [12]

In can observed that for a 100% whiplash probability the head-torso angle is around 84 degrees. At this angle, there is very large change that injuries will occur such as muscle tear, vertebra dislocation and fractures. These types of injuries are not life threating but they do generate long term symptoms and limitations such as pain, dizziness, anxiety, muscle stiffness and head rotation turn limit [3].

In the simulation, the headrest position was not altered (normal position) and as such, the whiplash effect was greatly reduced by the headrest. During the collision the head and torso were aligned and thus, the torso-head angles was almost 0. This can be observed in Figure 9.

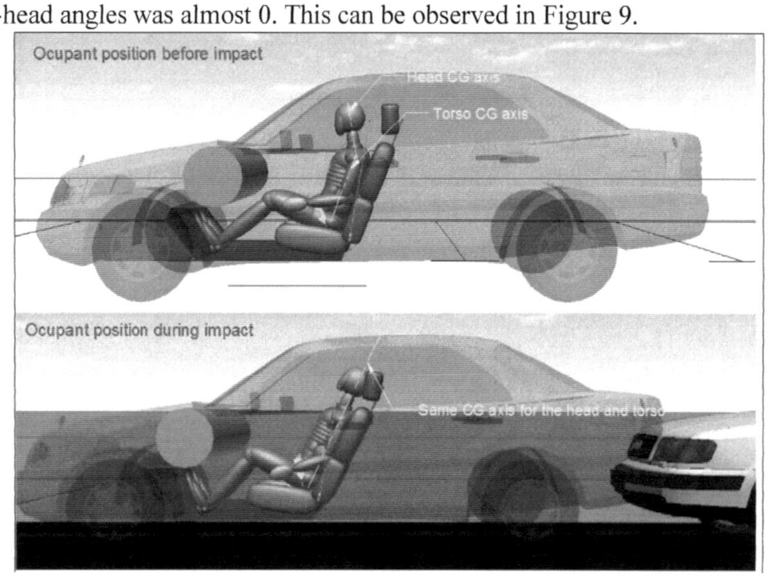

Figure 9. Center of gravity axis change for the head and torso of the occupant during the collision

Before the collision, the head and torso CG axis can be observed, they are at an angle of 23 degrees (Figure 10). When the collision occurs, the head moves back and it is stopped by the headrest, thus the two axis align and reduce the whiplash effect. In Figure 10 the movement of the head and torso is presented.

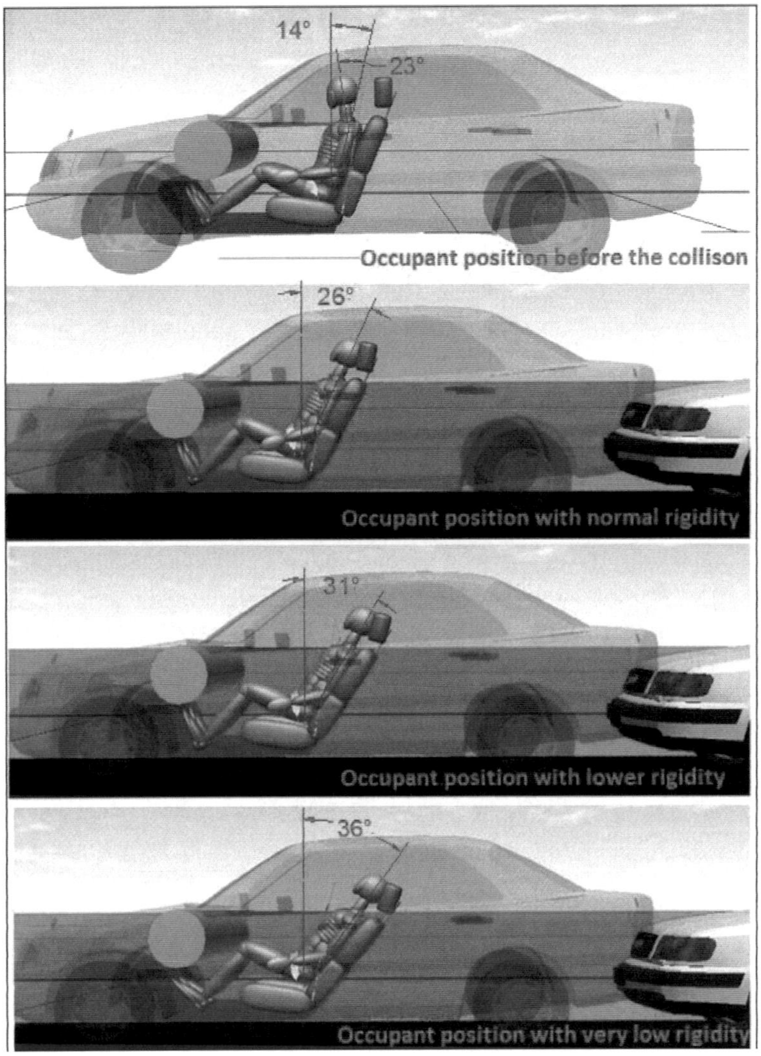

Figure 10. Head and torso angle variation during the collision

Because the torso and head have the same axis, the angle is counted for both of them in regards to a vertical axis. In the initial condition, the torso is at a 14 degree angle in the normal position and rotates at different angles for the different seat rigidities analysed (26 degrees – normal, 31 degrees – low rigidity and 36 degrees – very low rigidity).

From this point of view, the whiplash effect is minimal and can be neglected due to the fact that the occupant model had the headrest adjusted correctly and only the seat rigidity was modified. This is why the head acceleration was taken into account and analysed as the main parameter for the occupant.

3. Results

The results consist of the head's acceleration comparison between the three cases. In the graph presented in Figure 11, the head's acceleration was obtained, compared and analysed.

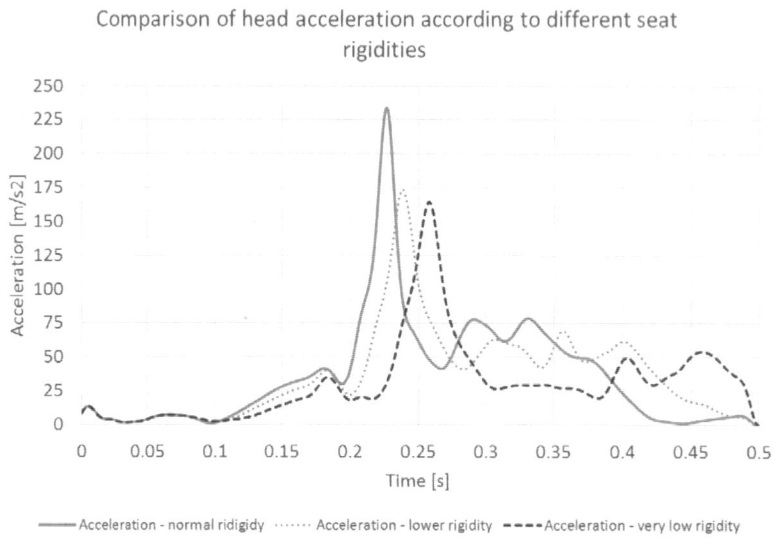

Figure 11. Comparative analysis of head's acceleration according to the torque of the backrest

In the case of normal rigidity, the head's acceleration is 230 m/s^2, at 130 ms. In the case of low rigidity, maximum head's acceleration is 170 m/s2 at 140 ms, and in the case of very low rigidity, the acceleration peak is 160 m/s2 at 160 ms. It can be seen that acceleration peaks are attained at different times due to the increased angular displacement of the occupant. The reduction in headline acceleration is shown in Figure 12.

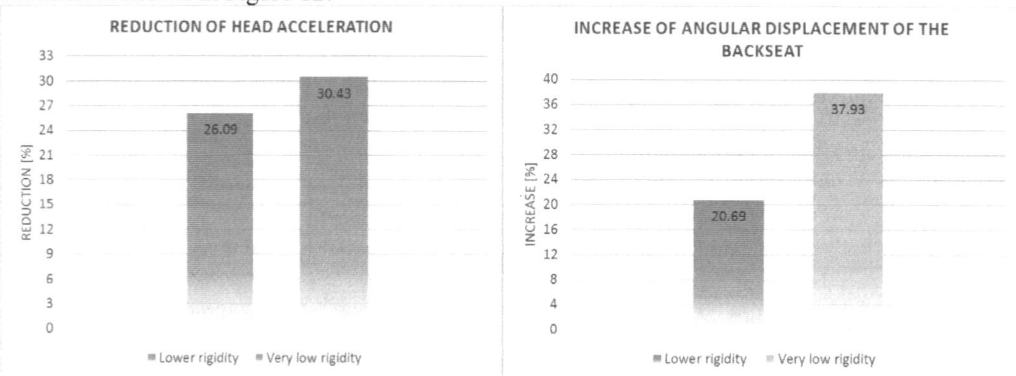

Figure 12. Percentage reduction of head's acceleration and percentage increase of occupant's angular displacement according to the backrest stiffness

It can be observed that, if the backrest stiffness is reduced, the acceleration is also reduced by 26%, while the angular displacement increases by 20%, and in the case of very low backrest stiffness, the acceleration decreases by up to 30% and the angular displacement increases by 38%.

4. Conclusions

After conducting the experimental test, the simulation was successfully validated, using the same impact configuration, with an error of just 12%. By modifying the backseat torque, a reduction in head's acceleration was obtained. In the case of a lower rigidity seat, the torque was of 550 Nm, the acceleration reduction was of 26% and in the case of a very low rigidity, the torque was of 250 Nm, the acceleration reduction was even higher, up to 30%. The reduction takes place due to the increase of the angular displacement of the backrest, resulting in the absorption of collision energy. Angular displacement of the backrest shows an increase of approximately 21% in the case of lower rigidity, and 38% in the case of very low rigidity. It can be concluded that, by reducing the torque of the backrest, the head's acceleration of the occupant is also reduced, but with an increase of an angular backrest's displacement.

References

[1] Cassidy JD, Carroll LJ, Côté P, Lemstra M, Berglund A, Nygren Å. Effect of eliminating compensation for pain and suffering on the outcome of insurance claims for whiplash injury. New England Journal of Medicine. 2000 Apr 20; 342(16):1179-86.
[2] Castro WH, Schilgen M, Meyer S, Weber M, Peuker C, Wörtler K. European Spine Society—The AcroMed prize for spinal research 1997. European Spine Journal. 1997 Dec 1;6(6):366-75.
[3] Chen HB, King HY, Wang ZG. Biomechanics of whiplash injury. Chinese Journal of Traumatology (English Edition). 2009 Oct 1; 12 (5):305-14.
[4] Datentechnik, Dr. Steffan. PC CRASH Operating and Technical Manual. DSD, 2015.
[5] Foreman SM, Croft AC. Whiplash injuries: the cervical acceleration/deceleration syndrome. Lippincott Williams & Wilkins; 2002.
[6] IIHS Statistics. 2015. http://www.iihs.org/iihs/topics/t/general-statistics (accessed 2017).
[7] Jakobsson L, Lundell B, Norin H, Isaksson-Hellman I. WHIPS–Volvo's whiplash protection study. Accident Analysis & Prevention. 2000 Mar 31; 32(2):307-19.
[8] Kang B, Moon SW, Lim J, Choi BH. Observation and simulations of an impact on the automotive windshield using the time-dependent dynamic failure criterion. Proceedings of the Institution of Mechanical Engineers, Part D: Journal of Automobile Engineering. 2015 Jun; 229(7):866-74.
[9] Kramer M. Sicherheitseigenschaften von verbundglasscheiben bei wechsel nden scheibentemperaturen. AUTOMOBILTECHN. Z. 1972 Oct; 74(10).
[10] Lundell B, Jakobsson L, Alfredsson B, Lindström M, Simonsson L. The WHIPS seat-a car seat for improved protection against neck injuries in rear-end impacts. InProc. 16th ESV Conference, Paper 1998 May 31 (No. 98-S7, pp. 0-08).
[11] Ono K, Kaneoka K. Motion analysis of human cervical vertebrae during low-speed rear impacts by the simulated sled. Traffic Injury Prevention. 1999 Sep 1;1(2):87-99.
[12] NHTSA. Correlation of Dynamic Test Procedure to Field Performance. 2nd Head Restraint Informal Working Group Meeting. 2005
[13] Svensson MY, Aldman B, Hansson HA, Lövsund P, Seeman T, Suneson A, Örtengren T. Pressure effects in the spinal canal during whiplash extension motion-a possible cause of injury to the cervical spinal ganglia. In Proceedings of the International Research Committee on the Biomechanics of Impacts (IRCOBI) conference, Eindhoven, Netherlands 1993 (pp. 189-200).
[14] TASS. „MADYMO Reference Manual." 2010. https://www1.ethz.ch/biomed.ee/intranet/support/madymo/Reference.pdf (accessed 2017).

IOP Conf. Series: Materials Science and Engineering **252** (2017) 012005 doi:10.1088/1757-899X/252/1/012005

Passenger head in impact with frontal airbag in OOP postures

Adrian Ovidiu SOICA[1], George-Radu TOGANEL[1]

[1]Automotive and Transport Engineering Department, Transilvania University of Brasov, Brasov, Romania

a.soica@unitbv.ro, g.toganel@unitbv.ro

Corresponding author: a.soica@unitbv.ro

Abstract. Road accidents represent an aspect of road traffic that may lead negative consequences. In order to solve the problems associated with such events, interdisciplinary knowledge is called for, complex teams of engineers, doctors, lawyers, experts working together in order to reduce the severity of such events. Road safety is a continuous concern for both experts and various government organizations with the aim of protecting the lives of the participants in traffic. It has been estimated that the costs of traffic accidents account for 1-3% of a country GDP, depending on the level of country development [26]. In this paper we analyze a particular class of cases of injuries caused to passengers caused by the inflation of the frontal airbag when they are with the passenger out of position. Head kinematics, accelerations, as well as the severity of injuries expressed by HIC, as related to the AIS scale have been analysed.

1. Introduction

Most of the participants in road traffic, occupants of a vehicle, use the seat belt and have a correct sitting posture in the chair. Nonetheless, there are situations when occupants do not wear a seatbelt or when their attention is diverted for various reasons and they are not correctly sat positioned in the chair. An imminent accident occurred at such a time usually has serious consequences upon the state of injury of vehicle occupants.

Occupants' posture called Out Of Position (OOP) increased risk of severe and lethal injury. In vehicle crash testing this term indicates a passenger position which is not the normal upright and forward-facing position, [9]. For example, a common case observed in crashes is the position of an occupant when reaching for the car radio, [16], or panic braking in unbelted passengers [25].

The concept is of interest because small changes in a passenger's position can have profound effects on the actual kinematic response, especially in rear impacts, as shown both by practical testing and theoretical models, [9].

According to Huelke, women and shorter drivers are more likely to receive facial injuries from an inflating airbag. Nearly 42 % of the female drivers in the sample suffered facial injuries from a deploying airbag, as compared with about 24 % of the male drivers. Likewise, half of the drivers under the height of 5'5" received facial injuries, while only about 18 percent of drivers under 5'11" or taller suffered facial injuries. [10].

There have been multiple reports of head and neck injuries related to airbags. The injuries in the head-neck area include facial trauma, [15], cervical spine fractures [12], [2], mandibular joint injury, [13] and even decapitation [11]. In addition, soft tissue injuries are also seen, including damage to the blood vessels, [7], [8], [17]. The eye seems to be particularly vulnerable to injury, especially if spectacles are worn [5]. Injuries include orbital fractures, [3], retinal detachment, [18] and lens rupture, [21]. The chemicals involved in inflating the bag have been implicated in eye injury [19], as have the cover components [6].

Similarly, there have been numerous reports of injuries directly attributable to airbags, with some calls for them to be removed or deactivated. While accepting that these injuries occur, most (up to 96%) are comparatively minor. Most occur with the larger system used in the US, although UK airbags have been associated with a wide range of injuries [22]. In this paper we analyzed the effects of airbag inflation through the injuries caused to the occupant's head, considering two seating postures of the passenger in a vehicle.

2. Method

About 85 % of vehicles involving in accidents are passenger cars, [20]. The first case refers to overweight people who tend to position the car seat as back as possible in order to have a relaxed posture during traveling. The second case shows the car seat positioned forwards toward the dashboard, providing a more cramped seating posture. In both cases we considered fixed seating positions, with the passenger not leaning back in his/her seat back but bent towards the dashboard at various distances from it, according to the scenarios reported in [16] and [25]. In both cases analyzed the occupant sizes were the same. Seven experimental tests have been carried out.

The seating posture of the passenger dummy was selected to represent a position which is not default but also not less realistic. Such a posture for example can occur when the passenger is looking for something in the glove box. Other, in some cases more extreme seating positions were proposed by the automotive industry for evaluating out-of-position vehicle occupant's interactions with inflating air bags.

The risk of injuries induced by the air bag increases in cases of extreme out-of-position situations [1]. Considering the vehicle stationary, the upper torso of the passenger is bent forwards at various distances from the dashboard, the airbag is triggered, inflating and hitting the face of the passenger who is in a out-of-position posture.

The dimensions and kinematics of the test device are meant to replicate unbelted adult passenger who is to an OOP posture.

In all cases studied it is noticed that in the first stages of airbag deployment, the airbag is projected like a lash towards the occupant's face. Although high levels of accelerations are not recorded (at this moment), this hit may cause injuries to the eyes and the soft tissues of the face, like in [5], [3], [18], [21].

The sudden changes in acceleration, as noticed on all three axes of measurement, ranging from negative to positive values, indicate a strong effort which can cause serious damage to the brain by shaking and extruding it out of the skull. Similarly, these accelerations are transmitted towards the cervical vertebrae that they can damage along with the spine.

For the calculation of HIC, in accordance with regulations from [4] we may use as a domain the entire duration of impact, lasting for 36 milliseconds, that is a HIC36 and an impact duration of 15 milliseconds, that is HIC15.

$$HIC = \max_{t_1, t_2} \left\{ (t_2 - t_1) \cdot \left[\frac{1}{t_2 - t_1} \int_{t_1}^{t_2} a(t)dt \right]^{2.5} \right\} \quad (1)$$

where

$$a = \sqrt{a_x^2 + a_y^2 + a_z^2} \quad (2)$$

is the resultant acceleration of the center of gravity of the head, measured in (g) units and t1, t2 is the time limits of crash, in second, for which the HIC is maximum.

According to European Regulation ECE/96/79 HIC36 $\leq$ 1000 and the resultant acceleration $\leq$ 80 (g) for an impact duration of 3 milliseconds.

3. Results

3.1. Cramped sitting posture

The acceleration chart reveals that the time zero was the trigger marking the hit given by the deploying airbag as felt by the head (lash). Thus, a correlation with the value of the maximum impact is noticed, depending on the distance between the head and the dashboard (passenger airbag). Therefore, for a distance = 240 mm the maximum impact occurs in the range 0-20 ms, for a distance = 280 mm the impact has maximum intensity and it occurs in the range 20-40 ms. For a distance = 310 mm maximum impact is felt in the range 60-80 ms. For a distance = 340 mm it is noticed that the hit was of minimum intensity.

For the cramped sitting posture the maximum head acceleration is obtained at a distance of 280 mm from the dashboard (airbag). The value of the HIC criterion was calculated and summarized in Table 1.

The maximum duration of impact between the deploying airbag and the occupant's head is of about 20-25 ms for this posture, depending on the distance between the head and dashboard, figure 1.

Table 1 Cramped sitting posture

Pos.	F=240 mm	F=280 mm	F=310 mm	F=340 mm
Val uess	axmax(g) =49.8 aymax(g) =18 azmax(g) =33.4 axmin(g)=-117.2 aymin(g) =-74.9 azmin(g) =-61.4	axmax(g) =111.4 aymax(g) =228.3 azmax(g) =86.5 axmin(g) =-133 aymin(g) =-96.3 azmin(g)=-293.4	axmax(g) =86 aymax(g)=108.8 azmax(g) =40 axmin(g) =-75.4 aymin(g) =-66 azmin(g) =-33.4	axmax(g) =19.4 aymax(g) =4 azmax(g) =5 axmin(g) =-16.8 aymin(g) =-12 azmin(g) =-25.4
	amax(g)=132	amax(g)=348	amax(g)=132.6	amax(g)=28
	HIC =198 40 ms HIC36 =200 HIC15 =248	HIC =1360 40 ms HIC36 =1520 HIC15 =3240	HIC =290 40 ms HIC36 =315 HIC15 =470	HIC =12 40 ms HIC36 =14 HIC15 =8.5

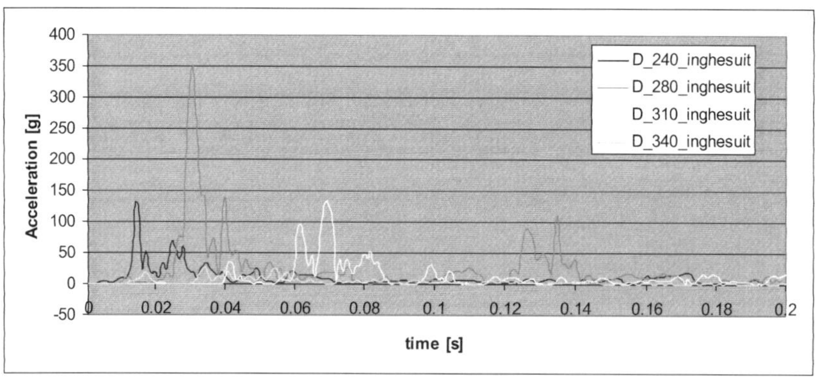

Figure 1. Resultant head acceleration for cramped sitting posture

At a distance of 340 mm the level of accelerations is minimized, in this case the occupant's head being only easily touched by the airbag, as there is not a strong flexion/extension of the neck like in other cases.

3.2. Relaxed sitting posture

In this case the inclination angle of the mobile arm of the test bench is greater, as related to the normal sitting position (with the back leaning against the seat back), figure 2.

Figure 2. Inclination of mobile arm in the two cases

Similar to the previous case, the time zero was the trigger marking the hit given by the deploying airbag as felt by the head (lash). Thus, a correlation with the value of the maximum impact is noticed, depending on the distance between the head and the dashboard (passenger airbag). Therefore, for a distance = 240 mm the maximum impact occurs in the range 20-55 ms, for a distance= 280 mm the impact has maximum intensity and it occurs in the range 25-70 ms. For a distance = 340 mm it is noticed that the hit was of minimum intensity.

For the relaxed sitting posture the maximum head acceleration was also obtained at a distance of 280 mm from the dashboard (airbag). The value of the HIC criterion and of both maximum and minimum acceleration was summarized in Table 2.

Table 2 Relaxed sitting posture

Pos.	F=240 mm	F=280 mm	F=310 mm	F=340 mm
Valuess	axmax(g) =38.3	axmax(g) =50.7	axmax(g) =n/a	axmax(g) =30.7
	aymax(g) =90	aymax(g) =49.3	aymax(g) = n/a	aymax(g) =17.9
	azmax(g) =92.3	azmax(g) =98	azmax(g) = n/a	azmax(g) =50.7
	axmin(g) =-95.5	axmin(g) =-62.4	axmin(g) = n/a	axmin(g) =-15.8
	aymin(g) =-29.8	aymin(g) =-70.8	aymin(g) =n/a	aymin(g) =-30.4
	azmin(g) =-58.4	azmin(g)=-122.3	azmin(g) =n/a	azmin(g) =-42
	amax(g)=121	amax(g)=133.6		amax(g)=51.8
	HIC =605	HIC =450		HIC =84
	50 ms	50 ms		50 ms
	HIC36 =550	HIC36 =545		HIC36 =70
	HIC15 =390	HIC15 =338		HIC15 =47

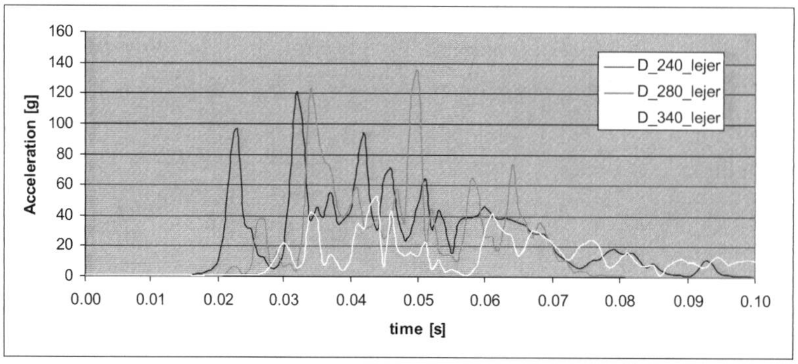

Figure 3. Resultant head acceleration for relaxed sitting posture

The maximum impact duration between the deploying airbag and the occupant's head is of about 40-45 ms for the relaxed sitting posture of the occupant in the car seat, figure 3.

4. Conclusions

The analysis of the two OOP sitting postures shows that the cramped posture causes greater injury to the occupant when the distance between the occupant's head and the dashboard is 280 mm.

For extreme postures at 240 mm and 340 mm from the dashboard, the cramped posture causes HIC values that are slightly lower than in case of the relaxed posture at same distances to the dashboard.

Impact duration between airbag deployment and dummy's head is shorter in the cramped sitting posture, which is reflected by the higher level of head acceleration.

In order to reduce the accidents injuries to vehicle occupants, found in OOP, an important role is held by the road traffic education, not only for drivers but also for passengers. Thus, education campaigns should be organized in this respect from an early age, educating young people in schools and not only.

Future advances in airbag technology will help to reduce injuries caused by these devices, but will need to be implemented in conjunction with advances in other restraint and sensing systems to be most effective, (Wallis and Greaves, 2002) [22].

References

[1] Alexander Berg F., Schmitt B., Epple J., Mattern R., Kallieris D. (1998). Results Of Full-Scale Crash Tests, Stationary Tests And Sled Tests To Analyse The Effects Of Air Bags On Passengers With Or Without Seat Belts In The Standard Sitting Position And In Out-Of-Position Situations, Institute of Forensic Medicine, University of Heidelberg, Germany, Paper Number 98-S5-O-10.

[2] Bailey H., Perez N., Blank-Reid C., et al. (2000). Atlanto-occipital dislocation: an unusual lethal airbag injury. J Emerg Med 2000;18:215–19.

[3] Cacciatori M., Bell R.W., Habib N.E., (1997). Blow-out fracture of the orbit associated with inflation of an airbag: a case report. Br J Oral Maxillofac Surg 1997; 35:241–2.

[4] Chicos D., de Vogel D., & others, (2005) Crash Analysis Criteria Description, Version 1.6.2, 2005.

[5] Duma S.M., Kress T.A., Porta D.J., et al. (1996). Airbag induced eye injuries: a report of 25 cases. J Trauma 1996;41:114–19.

[6] Duma S.M., Crandall J.R. (2000). Eye injuries from airbags with seamless module covers. J Trauma 2000; 48:786–9.

[7] Duncan M.A., Dowd N., Rawluk D., et al. (2000). Traumatic bilateral internal carotid artery dissection following airbag deployment in a patient with fibromuscular dysplasia. Br J

Anaesth 2000;85:476–8.
[8] Epperly N.A., Still J.T., Law E., et al. (1997). Supraglottic and subglottic airway injury due to deployment and rupture of an automobile airbag. Am Surg 1997;63:979–81.
[9] Foreman S. M.; Croft A. C. (2012). Whiplash Injuries: The Cervical Acceleration/Deceleration Syndrome. Lippincott Williams & Wilkins. pp. 104. ISBN 978-0-7817-2681-8. Retrieved 8 December 2012.
[10] De Groat B. (1997). News and Information Services. http://ur.umich.edu/9697/Apr01_97/artcl05.htm.
[11] Huff G.F., Bagwell S.P., Bachmann D. (1998). Airbag injuries in infants and children: a case report and review of the literature. Pediatrics 1998;102: electronic.
[12] Hart R.A., Mayberry J.C., Herzberg A.M. (2000). A cute cervical spinal cord injury secondary to air bag deployment without proper use of lap or shoulder harnesses. J Spinal Disord 2000;13:36–8.
[13] Levy Y., Hasson O., Zeltser R., et al. (1998). Temporo-mandibular joint derangement after air bag deployment: report of two cases. J Oral Maxillofac Surg 1998;56:1000–3.
[14] Mariotti G. V. and Bellavia G. (2007). Development Of An Anthropomorphic Model For Vehicle-Pedestrian Crash Test, XXI Science and Motor Vehicles 2007, JUMV international Conference with Exbition, 23-24 April 2007, Belgrade, Serbia, ISBN 978-86-80941-31-8.
[15] Murphy R.X., Birmingham K.L., Okunski W.J., et al. (2000). The influence of airbag and restraining devices on the patterns of facial trauma in motor vehicle collisions. Plast Reconstr Surg 2000;105:516–20.
[16] Nordhoff L. S. (2005). Motor Vehicle Collision Injuries: Biomechanics, Diagnosis, And Management. Jones & Bartlett Learning. pp. 513–. ISBN 978-0-7637-3335-3. Retrieved 8 December 2012.
[17] Perdikis G., Schmitt T., Chait D., et al. (2000). Blunt laryngeal fracture: another airbag injury. J Trauma 2000;48:544–6.
[18] Ruiz-Moreno JM. (1998) Air bag-associated retinal tear. Eur J Ophthalmol 1998;8:52–3.
[19] Smally A.J., Binzer A., Dolin S., et al. (1992). Alkaline chemical keratitis: eye injury from airbags. Ann Emerg Med 1992;21:1400–2.
[20] Shoojati, M., (2003). Correlation between injury risk and impact severity index ASI, 3rd Swiss Transport Research Conference, March 19-21, 2003.
[21] Zabriskie N.A., Hwang I.P., Ramsey J.F., et al. (1997). Anterior lens capsule rupture caused by air bag trauma. Am J Ophthalmol 1997;123:832–3.
[22] Wallis L. A. , Greaves I. (2002). Injuries associated with airbag deployment, Emerg Med J 2002;19:490-493 doi:10.1136/emj.19.6.490.
[23] CNN (2014) http://edition.cnn.com/2014/10/23/tech/ innovation/obese-crash-test-dummies-mci/.
[24] Integrated Conference of Better Air Quality (BAQ) 2014 & Intergovernmental Eighth Regional Environmentally Sustainable Transport (EST) Forum in Asia (2014). http://www.uncrd.or.jp/content/documents/19638EST-Concept-Note.pdf.
[25] Society of Automotive Engineers. (2004). Air bags and belt restraints. Society of Automotive Engineers. ISBN 978-0-7680-1412-9. Retrieved 8 December 2012.
[26] Wijen, W., Stipdonk, H. (2016), Social costs of road crashes: An international analysis, Accident Analysis and Prevention 94 (2016) 97–106.

Experimental research on pedestrian lower leg impact

B A Constantin[1], D M Iozsa[2] and C Stan[3]

POLITEHNICA University Bucharest, Automotive Department, Bucharest, Romania

E-mail: constantin.bogdan.alexandru@gmail.com[1], daniel_iozsa@yahoo.com[2], corraista@yahoo.com[3]

Abstract. The present paper is centred on the research of deceleration measured at the level of the lower leg during a pedestrian impact in multiple load cases. Basically, the used methodology for physical test setup is similar to EuroNCAP and European Union regulatory requirements. Due cost reduction reasons, it was not used a pneumatic system in order to launch the lower leg impactor in the direction of the vehicle front-end. During the test it was used an opposite solution, namely the vehicle being in motion, aiming the standstill lower leg impactor. The impactor has similar specifications to those at EU level, i.e. dimensions, materials, and principle of measurement of the deceleration magnitude. Therefore, all the results obtained during the study comply with the requirements of both EU regulation and EuroNCAP. As a limitation, due to unavailability of proper sensors in the equipment of the lower leg impactor, that could provide precise results, the bending angle, the shearing and the detailed data at the level of knee ligaments were not evaluated. The knee joint should be improved for future studies as some bending angles observed during the post processing of several impact video files were too high comparing to other studies. The paper highlights the first pedestrian impact physical test conducted by the author, following an extensive research in the field. Deceleration at the level of pedestrian knee can be substantially improved by providing enough volume between the bumper fascia and the front-end structure and by using pedestrian friendly materials for shock absorbers, such as foams.

1. Introduction

The European Union has had impressive success in achieving the highest pedestrian protection level on the globe. In 2013, 5.712 pedestrians were killed in road accidents in the EU, which is 22% of all fatalities. In the last decade, in the European Union, pedestrian fatalities were reduced by 37%, while the total number of fatalities were also reduced by almost 45%. The rate of pedestrian deaths in European countries varies from 3 pedestrian fatalities per million population in the Netherlands to more than 35 pedestrian fatalities per million population in Romania, a rate about 12 times higher. [2]

The present paper is focused on the research and development of an easier, cost-effective and more user-friendly tibia impact solution, for an overview analysis of decelerations at pedestrian knee joint level. The physical test results will be used in an advanced study for developing a method that can be used in the upstream stage of the vehicle project in order to predict the pedestrian impact performance and optimize the front bumper area.

2. Pedestrian regulations

A pedestrian crash can usually be divided into 4 stages: the vehicle initiates the contact with the pedestrian by touching his leg (tibia) with the front bumper, the front edge of the bonnet or headlight hits the upper leg (pelvis), the head of the pedestrian hits the bonnet or the windshield and finally the pedestrian is projected in the air and hits the ground. The first three types of load cases are described in the European Commission Regulation, each using different sub-systems impactors to represent the main phases of a car-to-pedestrian impact. The three types of impactors are: a leg form impactor representing the adult lower limb to indicate lateral knee-joint shear displacement, bending angle and tibia acceleration, caused by the contact with the bumper; an upper leg form impactor representing the adult pelvis, in order to record bending moments and forces caused by the contact of the bonnet leading edge (BLE); child and adult head form impactors, recording head accelerations caused by the contact with the bonnet.

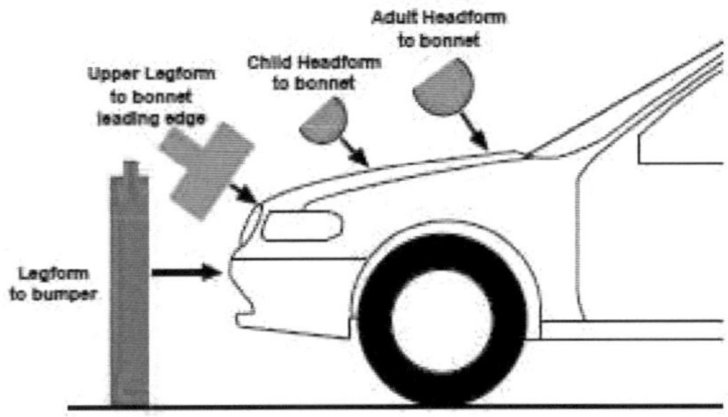

Figure 1. The sub-system tests used in EC Directive [1]

Additionally, vehicles are tested by specialized programs, like EuroNCAP (for Europe), JNCAP (for Japan), KNCAP (for Korea), etc., that publish the results, being a main tool for each customer who wants to buy a new vehicle to guide himself over the official reports in order to make a decision over the purchase. The exigencies for each of the programs mentioned above can be observed in table 1, and therefore it can be easily concluded that the most demanding in the lower leg impact case is EuroNCAP.

Table 1. Pedestrian lower leg impact test requirements analysed by geographical criteria [4]

Test method	Parameter	Europe Directives 78/09 & 631/09 - Ph2	EuroNCAP		JNCAP		KNCAP		GTR No.09
			max score	zero score	max score	zero score	max score	zero score	
EEVC lower legform impactor (TRL)	Velocity	40km/h	40km/h		-		40km/h		40km/h
	Impact angle	0°	0°				0°		0°
	Acceleration	170g	150g	200g			150g	170g	170g
	Bending	19°	15°	20°			15°	19°	19°
	Shearing	6mm	6mm	7mm			5mm	6mm	6mm
FlexPLI impactor	Velocity	-	40km/h		40km/h		-		40km/h
	Impact angle		0°		0°				0°
	Tibia bending		282Nm	340Nm	224Nm	380Nm			340Nm
	MCL elongation		19mm	22mm	16.4mm	22mm			22mm
	ACL/PCL elong.		10mm	10mm	0mm	13mm			13mm

The trial for lower leg impact, the case analyzed in this study, may be conducted with two types of impactors: the rigid one (TRL) and the optimized flexible one (FlexPLI).

The difference between the two impactors is that the first one does not allow the two leg segments to deform, thus to simulate the flexibility of the human bone. Also, it cannot predict neither the fracture of the bone, nor the breaking of the knee ligaments. In this situation, the new developed FlexPLI impactor was created, with four flexible sections for each segment of the leg, that can provide results for torques read at the level of the tibia and three displacements read at the level of the knee that can give an overview of the risk associated with ligament breakage.

The rigid specification of the impactor was used in the pedestrian impact research, the analysed parameter, the deceleration at knee joint level being predicted more accurate in this configuration. Also, the development of this specification is less expensive and faster to build and calibrate.

3. Experimental research on pedestrian lower leg impact

A Renault Clio 2 vehicle, manufactured in 2005, with modifications conducted to the front bumper and front cross member in order to accommodate an EPP absorber specification, was used as test stand.

The impactor used was developed in the same way as described by the specifications in Regulation EC631/2009 [1], consisting of two circular cross sections (bars) of Ø70mm in diameter and 500mm in length, a simplified rotational knee joint at the intersection of the two segments, three 15mm thick foam layers made of 80% NBR (rubber butadiene rubber) and 20% PVC (polyvinyl chloride) and an exterior layer 1.5mm thick made of self-adhesive neoprene. The impactor was equipped with an X200-4 accelerometer (figure 2), produced by Gulf Coast Data Concepts, with 3-axis measurement capabilities and ± 200g range, which can be used with a sampling rate of up to 3200Hz. In the study, sampling rates of 400Hz, 800Hz, 1600Hz and 3200Hz were used for initial tests and trials. In these cases, it can be assumed that at a value of 400Hz, an acceleration value will be read at about 2.5ms; for a 800Hz sampling value, an acceleration value will be read at about 1.2ms; for a value of 1600Hz will be read an acceleration value at about 0.6ms and for a value of 3200Hz, an acceleration value will be read at each 0.3ms. Each line of the output file will contain 4 columns: in the first column the time will be found, in the second column the acceleration in the direction of the X axis, in the third column the acceleration along the Y axis, and in the last column the acceleration along the Z axis.

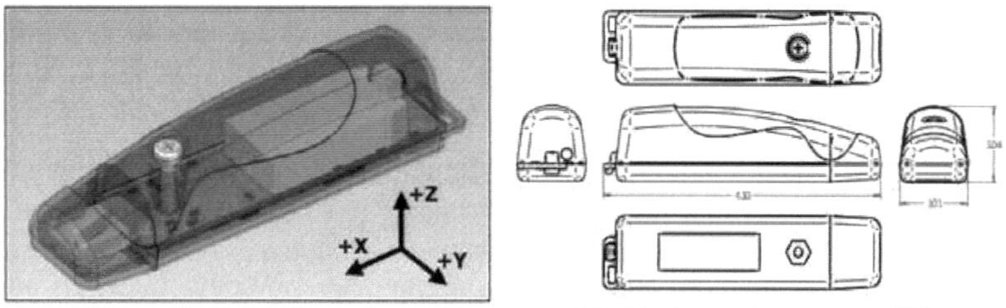

Figure 2. Gulf Coast Data Concepts X200-4 accelerometer [8]

The boundary conditions proposed for the physical impact test, were different from those specified in actual regulations. In the absence of a pneumatic system that launches the impactor in vehicle direction, it was chosen the reverted condition, the impactor being aimed by the vehicle. This configuration is closer to real accident conditions, but also has a number of limitations that need to be closely monitored during the tests in order to validate the results. The surface on which the car is running during the tests, shall be approximately flat, without unevenness, and the vehicle speed shall be measured and stabilized at least 3 seconds before the impact is produced. It has been chosen for instantaneous speed readings, vehicle OBD data, being one of the most accurate and simple methods. Additionally, speed measurement was carried out using a professional measuring system based on GPS satellite vehicle positioning, with an accuracy of 0.1km/h. The ambient temperature should be within the range of 23°C, ±3°C. At the time of the trials, on April 13th 2017, in the location where the tests took place, Bucharest, the outside temperature fluctuated between 20°C and 22°C. The zone where the impact has been achieved, should be evaluated and measured as soon as it's completed. The impactor must be positioned appropriately (the accelerometer positioning in the direction of impact can be tracked as a reference) and must not be affected by any external environmental factors.

Impact kinematics were captured on digital support, similar to the one used in the EuroNCAP impact tests. This allows to record a large amount of information, which are captured in a very short time, with camcorders capable of recording high-speed images. Using this process, the images were captured at a rate of 240 frames per second, with the help of a GoPro Hero 4 Black Edition video camera, later being played back by slow motion. This process slows down the rate of photogram playback at a perceptible speed by the human eye, or even by "freezing" a single instant frame, which allows extracting as much information as possible.

The analysed data from physical tests, are the deceleration curves read at the knee joint of the impactor.

The configurations used in the simulations are described in the list below:

1. Standard vehicle configuration – vehicle speed 42.6km/h
2. Modified vehicle configuration, with EPP pedestrian absorber – vehicle speed 38.4km/h
3. Modified vehicle configuration, with EPP pedestrian absorber – vehicle speed 39.1km/h
4. Configuration 2/3, without front bumper fascia – vehicle speed 38.7km/h
5. Configuration 2/3, without front bumper fascia – vehicle speed 39.3km/h

Below in figure 3, the impact test setup can be observed, in 4th and 5th configuration.

Figure 3. Pedestrian testing configuration

The first test was carried out using the vehicle's standard configuration, in order to assess the test conditions and initial vehicle impact performance. Two other impact tests were carried out in a modified EPP absorber configuration, Renault reference 62090JZ00B, modified accordingly to the volume available in the interior of the front bumper. The last two attempts were derived from the previous configuration, but without a front bumper fascia, in order to better observe the deformations and kinematics of the impact. After each test, the shock absorber used was replaced with a new one.

Following the pedestrian impact tests, there were observed two areas where the impactor slightly deformed the bonnet outer panel. In the first area, marked with the number 1 in figure 4, the impactor deformed the hood near the 225mm Y coordinate. The bonnet leading edge (BLE) was not affected and the effects on the pedestrian will not have a considerable magnitude, the local rigidity of the bonnet in that area being reduced. In the second area, marked with number 2 in figure 4, the impactor hasn't deformed the hood, but left visible traces of contact friction at the 135mm Y coordinate. The bonnet leading edge has not been deformed in this case either. During the impact tests, a single absorber was broken, in test configuration number 5, in which it was in direct contact with the lower leg impactor. The rupture was identified in the perpendicular direction of impact and propagated along the longitudinal axis of the vehicle. (figure 5)

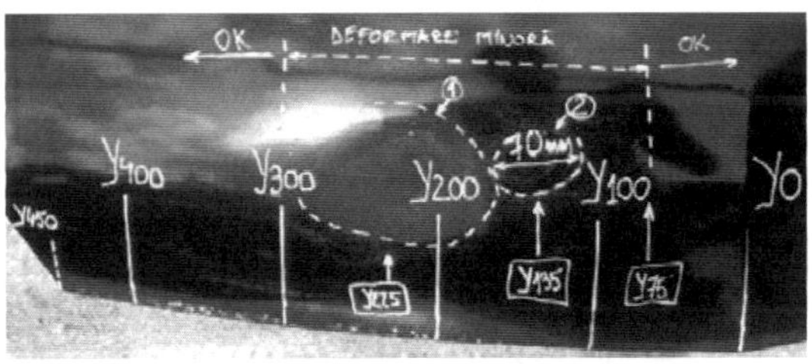

Figure 4. Bonnet deformation after pedestrian test

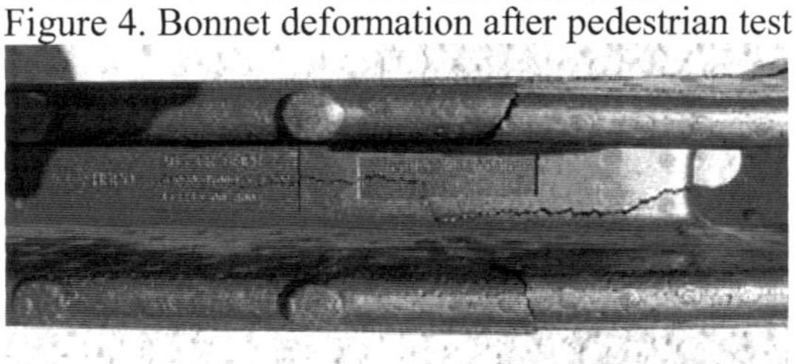

Figure 5. Absorber damage observed after test no. 5

The kinematics of the impact can be observed in figure 6. The chronology is similar to other physical tests conducted in the automotive field.

Figure 6. Pedestrian impact chronology

Decelerations obtained from the physical tests can be observed in figures 7-11, that contain the curve data. In the first load case, corresponding to the standard equipment of the vehicle, it can be noticed the ramp that the curve follows immediately after the initiation of the impact. This is most likely determined by the very rigid construction of the front bumper assembly, with a plastic honeycomb absorber and the presence of the three decorative elements, disposed integrally across the width of the front crossmember.

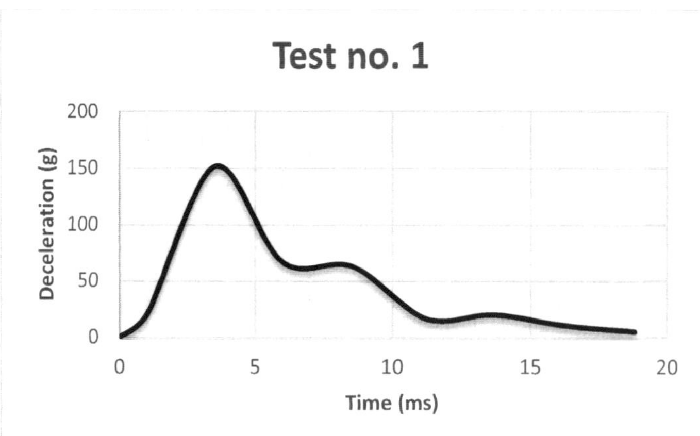

Figure 7. Deceleration curve – physical pedestrian impact test - configuration no.1

The following two test cases were carried out starting from the standard configuration of the test vehicle, modified by replacing the absorber with a specification of EPP material, adapted to be mounted in the volume between the bumper fascia and the cross member. In these two cases, the maximum deceleration value recorded during impact may be reduced by approximately 25g. The difference between the shapes of the curves can be given in this case by the differences between vehicle speed values at impact initiation and impact section positioning along the lateral axis of the vehicle.

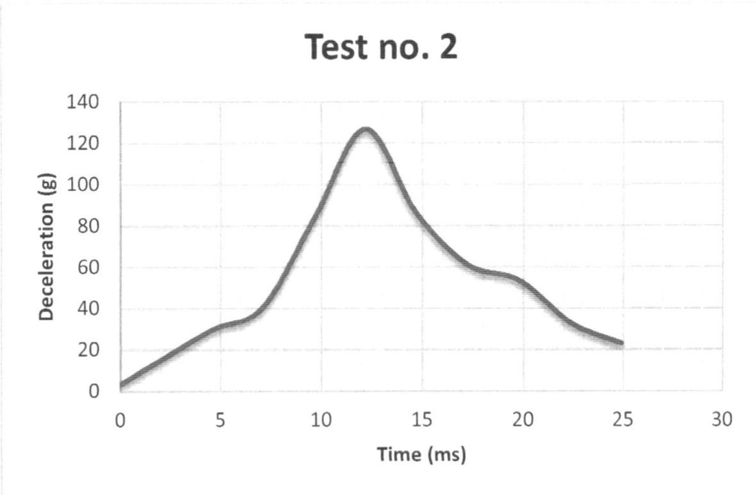

Figure 8. Deceleration curve – physical pedestrian impact test - configuration no.2

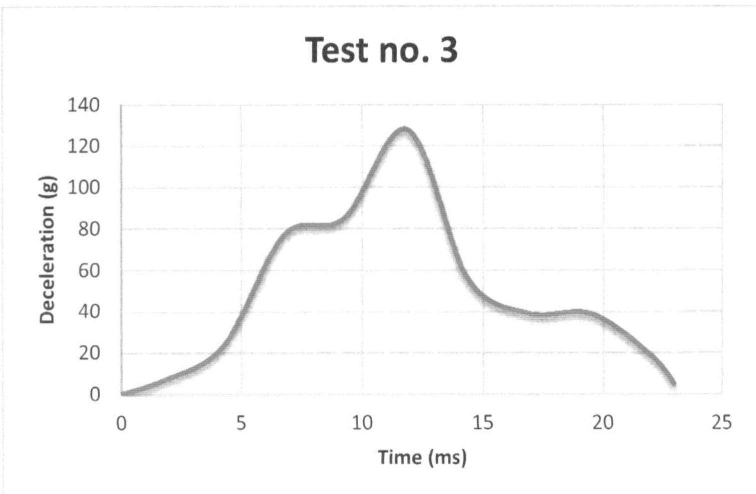

Figure 9. Deceleration curve – physical pedestrian impact test - configuration no.3

In order to better observe the deformations during the impact, and also to evaluate the influence in rigidity of the bumper fascia, the last two tests were launched and show a reduction of about 5-10 g in terms of the maximum values of the resulting decelerations, due to the more available volume for the absorber along the longitudinal axis of the vehicle.

IOP Conf. Series: Materials Science and Engineering **252** (2017) 012006 doi:10.1088/1757-899X/252/1/012006

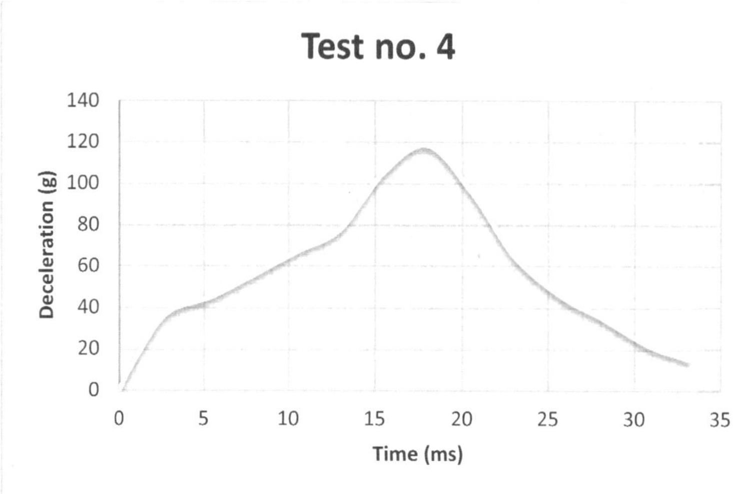

Figure 10. Deceleration curve – physical pedestrian impact test - configuration no.4

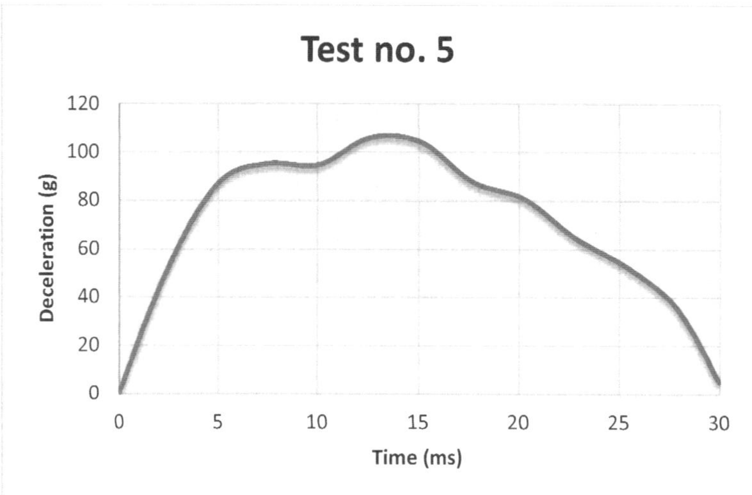

Figure 11. Deceleration curve – physical pedestrian impact test - configuration no.5

The maximal deceleration values obtained from the physical pedestrian impact test can be analyzed in table no.2 below.

Table 2. Pedestrian lower leg impact test deceleration test results

Test configuration	Maximal deceleration (g)
TEST 1	151.5
TEST 2	126.5
TEST 3	127.8
TEST 4	118.5
TEST 5	106.1
EuroNCAP limit	**150**

4. Conclusions

The field of passive safety, although set in the spotlight in recent decades, alongside other development directions in automotive industry and the number of road accident victims in the European Union, felling by about 25%, despite the significant increase in car fleets, some 40,000 people still lose their lives in traffic events, most of them pedestrians. Due to the continuous and very rapid evolution of motor vehicles, the variation in the intervals of the worldwide population anthropological parameters, the numerous scenarios of load cases that may be encountered in real life, the difficult manufacturing of devices that can precisely reproduce the biomechanical behavior of the human body and the limits to which human body can survive an impact, pedestrian protection will still be one of the most challenging research fields.

As the present study shows, the decelerations comply with the requirements of EuroNCAP. Using pedestrian friendly materials such as EPP foams, the decelerations measured during the impact are reduced by approximately 45g, compared to vehicle's standard absorber specification, calculating a gain up to 30%. The EPP material offers great advantages for fast and cost-effective modification of a vehicle. Most likely scenario might be the adaptation of existing vehicles according to new regulations or new market requirements. In this case, an opportunity can be identified for vehicles currently sold in developing countries without exigent regulatory requirements, to implement the proper solutions already existing and applied in Europe or North America, therefore to reduce the highest number of pedestrian fatalities worldwide, registered in low and medium income countries.

References

[1] Directive 2003/102/EC of the European Parliament and of the council of 17th November 2003
[2] Traffic safety basic facts 2015 – Pedestrians. European Road Safety Observatory Report, European Union, 2015
[3] Difference of TRL legform impactor/injury criteria and Flex pedestrian legform impactor/injury critieria, Informal document GRSP-49-24, 49th GRSP, 2011
[4] Safety companion 2014, carhs training gmbh, Alzenau, Germany, 2013
[5] Schuster P., Current trends in bumper design for pedestrian impact, SAE, 2006
[6] Lawrence G.J., Study on feasibility of measures relating to the protection of pedestrians and other vulnerable road users, TRL Ltd, 2004
[7] Burnaz E., Giubalca A., Pedestrian crash – Leg/Front bumper analysis, Renault Technologie Roumanie, The annals of Dunarea de jos University of Galati, Fascicle V, Technologies in machine building, 2009
[8] X200-4 USB Accelerometer Data Logger – User Manual, Revision B, Firmware version 1110, Gulf Coast Data Concepts, 20 May 2016
[9] Technical evaluation of the flexible pedestrian leg impactor, report no. NHTSA-2008-0145 – US Department of Transportation, National Highway Traffic Safety Administration, May 2014
[10] Directive 2005/66/EC of the European Parliament
[11] EuroNCAP Pedestrian testing protocol – EuroNCAP, 2015
[12] Material physical properties for foam (density range from 20g/l to 90g/l) molded from ARPRO Expanded Polypropylene EPP beads – publication JSPI/GEN-PHYS/20-90-2005/09, ARPRO, JSP, www.jsp.com
[13] Zander O., Wisch M., Gerhring D.U. – Improvement of the protection of lower extremities of vulnerable road users in the event of a collision with motor vehicles, paper number 15-0366, Germany, 2015
[14] Matsui Y., Takagi S, Yoshitomo T., Hosokawa N., Itoh F., Nakasato H., Watanabe N., Yonezawa H. – Characteristics of the TRL Pedestrian Legform and the Flexible Pedestrian Legform Impactors in Car-front Impact Tests, paper number 09-0206, National Traffic Safety and Enviroment Laboratory, Japan
[15] Murata S., Shioya S., Suffis B. – Expanded PolyPropylene (EPP) – A global solution for pedestrian safety bumper systems, JSP Automotive materials development group

Solutions for acceleration measurement in vehicle crash tests

D S Dima and D Covaciu

Department of Automotive and Transport Engineering, Transilvania University of Brasov, Romania

d.dima@unitbv.ro, dinu.covaciu@unitbv.ro

Abstract. Crash tests are useful for validating computer simulations of road traffic accidents. One of the most important parameters measured is the acceleration. The evolution of acceleration versus time, during a crash test, form a crash pulse. The correctness of the crash pulse determination depends on the data acquisition system used. Recommendations regarding the instrumentation for impact tests are given in standards, which are focused on the use of accelerometers as impact sensors. The goal of this paper is to present the device and software developed by authors for data acquisition and processing. The system includes two accelerometers with different input ranges, a processing unit based on a 32-bit microcontroller and a data logging unit with SD card. Data collected on card, as text files, is processed with a dedicated software running on personal computers. The processing is based on diagrams and includes the digital filters recommended in standards.

1. Introduction

At Transilvania University of Brasov are carried out, every year, crash tests between vehicles, vehicles with pedestrian or cyclist, vehicles with stationary obstacles (pole). The crashes are simulated also using specialized computer programs. The simulation should be as closed as possible to the real crash. By combining the results of computer simulation and crash tests, the young experts learn how to fine tune their software in order to reproduce a crash event. This will help them in reconstruction of real road accidents.

When a vehicle is involved in a collision, there is a crushing of metal, which reduces the speed of the vehicle until the vehicle is stopped or disengages from the vehicle/object. The soft crush of sheet metal produces low deceleration pulses in the vehicle. A hard metal (like structural members of the vehicle frame) produces more resistance, more deceleration, and more energy transfer. When the engine is engaged, it produces a high amplitude spike of deceleration transferred to the vehicle. The series of decelerations, with a duration of over a hundred or more milliseconds of a crash, may be called a crash pulse or deceleration curve [1]. The objective of the vehicle manufacturers is to keep decelerations low in order to minimize occupant injuries. The essence of crashworthiness is the energy management and the control of crash deceleration pulses.

The simulation of a road accident may be based on the evolution of acceleration versus time, during the crash. The crash pulse is characterized by shape, amplitude and duration [2]. The crash pulses used in simulation are obtained from measured acceleration time history, using analytical techniques. The stages of the crash, mentioned above, may depend by the type of vehicles and tests.

2. Data acquisition

A crash test includes instrumentation to measure the time variation of acceleration, force and other parameters.

Vehicles instrumentation and data processing are defined in standards like SAE J211 [3] and ISO 6487 [4]. The standards provide a meaningful comparison of test results from different sources. SAE J211 and ISO 6487 are similar documents published by the two standards organizations and are the main specification documents governing the performance requirements of crash test data acquisition systems [5].

2.1. Data acquisition devices

The standards include the main technical characteristics for accelerometers and parameters of digital filters used for preliminary data processing. The filtering procedure recommended by SAE J211 is to use the Channel Frequency Classes (CFC) of low-pass filters. Filtering is the most critical phase in the processing of impact signals, having the roles of eliminating the high-frequency noise and reducing the peaks in the signal. SAE J211 designate four CFC filters for crash data processing: CFC 60, 180, 600, and 1000 [3]. The CFC filters are derived from analog Butterworth filters that have the corner frequency (the frequency at which the signal loses one half of its power) equal to the CFC designation divided by 0.6 [7]. Selecting the frequency response class appropriate for each application is an important task in preparation of the crash test. The frequency class will determine also the sampling frequency.

The SAE requirements for data acquisition hardware include: a sampling rate of 10 kHz (10000 samples/seconds/channel), 12 bit resolution, and a determination of impact time within ±0.4 msec. The time base should have at least 1/100 seconds resolution - this don't affect the requirement for sampling rate; instead, it means it is possible to use the oversampling technique. Despite the increasing capabilities of sensors, recording media and software, it is not easy to meet all the standard requirements for crash data acquisition.

One of the challenges of the crash-test team from Transilvania University is to develop reliable and cost-affordable data acquisition devices to be used in tests, together with the professional recording devices (a reference device is PicDAQ [8]).

The measurement systems consists in one or more data channels. According to above mentioned standards, a data channel represents "all the instrumentation from, and including, a single transducer (or multiple transducers, the outputs of which are combined in some specified way) to, and including, any analysis procedures that may alter the frequency content or the amplitude content of data" [4]. In other words, a data channel is a combination of a sensor, data acquisition hardware and post-processing software. The combined performance of these components determines the accuracy of the acquired data.

2.2. Acceleration measurement

The first device in a data channel, used to convert a physical quantity to be measured into a second quantity (such as an electrical voltage), is the transducer. The transducers used in crash tests convert a mechanical movement of a sensing mass inside, in electrical voltage. The electronic sensor consists in a transducer to generate the crash signal, a circuitry to process the signal and a means to send out a triggering signal [9]. Although the transducer signal is not restricted to acceleration pulses only, the use of accelerometer dominates in vehicular applications. The most used transducers used in acceleration measurements are MEMS (Micro-Electro-Mechanical Systems) [10]. A MEMS accelerometer is integrating an accelerometer and the electronics into a single silicon chip. The accelerometer is essentially a capacitive or piezoresistive device consisting of a suspended pendulum proof mass/plate assembly. As acceleration acts on the proof mass, micro-machined capacitive or piezoresistive plates sense a change in acceleration from deflection of the plates [11].

The first phase of data processing is inside the transducer. Signals from crashes may contain high amplitude, high frequency components, containing very little useful information. These components may be reduced by a filter included by the manufacturer at the output of the accelerometer, like a 2-pole Bessel [12]. However, the accelerometer output does not look like a filtered version of the acceleration signal, because of signal saturation before the filter.

Depending by the type of output signal, the accelerometers can be digital or analogue. An important parameter to be considered when choosing the appropriate accelerometer for a certain test is the input range [7]. This can vary from ±2 g to ±200 g or even more. The accelerometer must be chosen depending by the expected values to be measured. The impact duration is very short, so the bandwidth is also important. The frequency response of the sensor is an indication of the sensor's ability to respond to changes in the measured displacement. Higher bandwidth means that it can be measured higher frequency motion, and will have an influence on the sampling frequency of the acquisition device. The MEMS accelerometers usually can have a bandwidth between 0.5 Hz and 1600 Hz (which can be set using external capacitors). In some tests this can be a limitation. Another parameter is the sensitivity, and depends by the input range and maximum output voltage. For example, if the input range of the accelerometer is ±5 g, and the output voltage is between 0 and 3.3V, this means the output resolution is 0.33 V for 1 g. The sensitivity is influenced also by the temperature.

2.3. Crash Data Logger

Based on MEMS accelerometers, a prototype of a crash data logger (named *CDL-01*) was built in laboratory. A simple block diagram of the data acquisition system is shown in Figure 1. There are two sensors used, one for low accelerations and another one for high accelerations. Both are analog 3-axis accelerometers, with signal conditioned voltage outputs. The first one is ADXL 337 (manufactured by Analog Devices [13]), with an input range of ± 3 g. This is to be used especially in accelerating and braking tests, but can also be useful in vehicle-pedestrian crash tests, mounted on the vehicle. The second sensor is ADXL 377 (from the same manufacturer), with an input range of ± 200 g. These are low cost accelerometers, and both have similar limitations in bandwidth and resolution. Bandwidths can be selected to suit the application, with a range (for ADXL 377) of 0.5 Hz to 1300 Hz for the x-axis and y-axis and a range of 0.5 Hz to 1000 Hz for the z-axis.

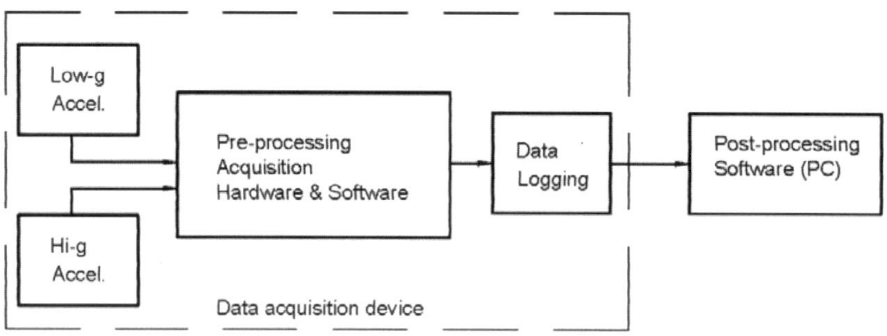

Figure 1. Basic schematic of the data acquisition system

The acquisition hardware uses the microcontroller AT91SAM3X8E (32 bit, ARM core), that operates at a frequency of 84 MHz and features up to 512 Kbytes of Flash and up to 100 Kbytes of SRAM. The microcontroller is installed on an Arduino Due board [14], which allows the use of all features of the microcontroller, including 12 analog inputs (12 bits).

The data logging module is a logging shield made for Arduino boards: Adafruit Data Logger Shield [15], with SD card interface, RTC (real time clock) and a prototyping area. An important limitation of the system is the writing speed of the SD card. This limits the sampling rate of the system to maximum 500 Hz, but the sampling rate may be increased by using the available memory of the microcontroller. In the available memory is possible to store about 3 seconds of data from 6 channels at a rate of 1 kHz. More information could be stored if using additional RAM modules.

The source code of the software stored in the microcontroller is written in C++. The tasks of the program are to pre-process raw data taken from the accelerometers, to store data on SD card, to offer a simple user interface (based on buttons and LEDs) and to communicate, through USB interface, with a personal computer (an advanced configuration interface). The main tasks are data pre-processing and storing. When a high acceleration value is measured, the system starts to record next data in the memory, at 1 kHz, as detail information. When the acquisition process is finished, the data from memory are stored also on the SD card. In order to avoid the start of detail acquisition when a false acceleration spike is measured, raw data measured are filtered using a low-pass filter. This filter is used only in real time measurement and is implemented so that a minimum processing time is required. It uses the function:

$$y(t) = x(t) \cdot \alpha + y(t-1) \cdot (1-\alpha) \tag{1}$$

where y(t) is the current calculated (estimated) value, y(t-1) is the previous calculated value, x(t) is the current measured value and α is an attenuation factor between 0 and 1.

The effect of this filter is not saved in the data file; only raw data are saved. The "raw data" represents accelerations measured, expressed as integer numbers that are part of the entire range of the analog input of the acquisition board. As example, a 10 bits input allow values from 0 to 1023. This means the zero level of an accelerometer will be around the value 512. Some sample lines are given in Table 1. In the example it can be seen the small variation of values when vehicle is stationary. The columns X1, Y1, Z1 are for the low-g channels, and X2, Y2, Z2 are for hi-g channels. Obviously, the difference between Z values and the other two axes is equal with g (gravity acceleration).

Table 1. Raw data samples

Time	X1	Y1	Z1	X2	Y2	Z2
0	506	499	612	515	516	519
3	508	499	612	514	516	519
5	510	498	612	514	515	519
7	508	498	613	514	516	519
9	507	497	614	515	515	520
11	508	495	613	515	515	518
13	507	498	614	514	515	519

In order to transform raw data to acceleration values, expressed in m/s^2 or g, it is necessary to identify the zero level as exactly as possible, and also a transformation scale (gain), which depends by the sensor resolution. An accurate and simple calibration method is to use two points per axis of interest [6]. When an axis is placed into a +1 g and −1 g field, the measured outputs are as follows:

$$A_{1g} = A_0 + (1g \cdot Gain)$$
$$A_{-1g} = A_0 + (-1g \cdot Gain) \tag{2}$$

where A_0 is the zero level of acceleration, or acceleration offset, and *Gain* is the transformation scale, which can be calculated from the measured levels of +1g and -1g. An example diagram of a 3-axis accelerometer calibration is shown in Figure 2, and the raw values for 1g/-1g can be identified and extracted from the raw data file (using the time stamp as reference). The gain and offset (zero level) values are then used in post-processing phase.

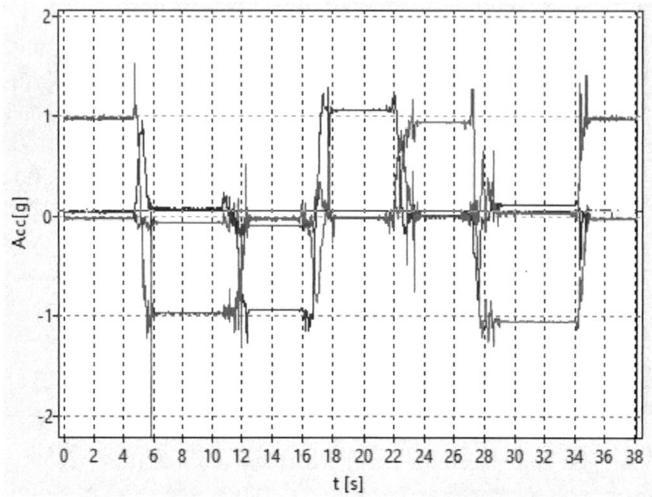

Figure 2. Example of a 3-axis accelerometer calibration

3. Data processing and results

The tool used for data processing is a dedicated program, developed for the *CDL* data acquisition system. From the program features, can be mentioned: possibility of data filtering using CFC or simple low-pass filters, display the processed data as diagrams (as shown in Figure 2), selection of the acceleration curves (channels) to be displayed, zoom and pan functions for diagrams, possibility to update the zero level for each channel, export data from the table (calculated) or as raw data as text files.

The user interface of the software is presented in Figure 3. The left side of the interface is covered mainly by the chart area. The size of the chart area is changing when the program window is resized, since the other areas remain constant. So, the chart area will be much bigger when the window is maximized, and is also interactive (a click on the chart will give information about the selected point on diagram).

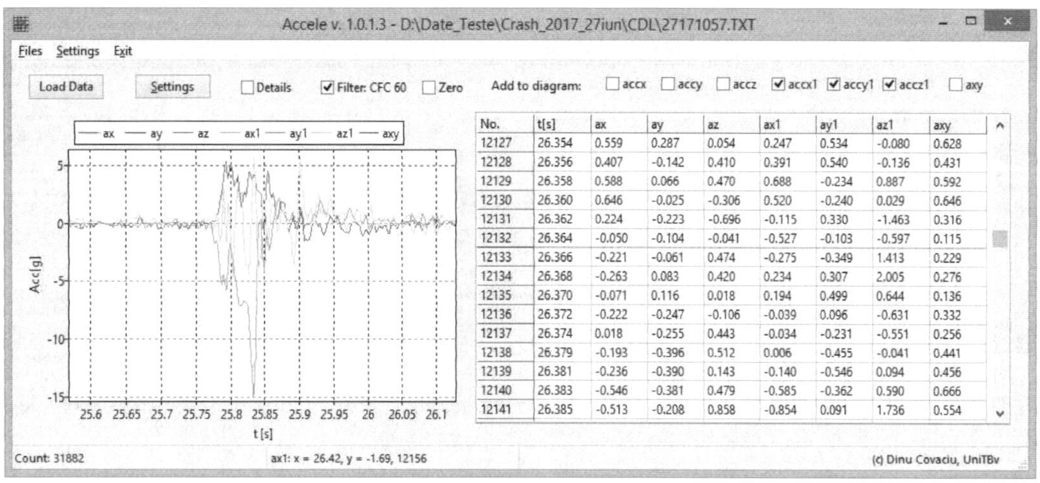

Figure 3. The interface of the data processing software

The digital filter used for crash tests data, as per SAE J211, is CFC (Channel Frequency Classes) filter (which is also a low-pass filter). The basic relation for a CFC filter is:

$$y(t) = a_0 \cdot x(t) + a_1 \cdot x(t-1) + a_2 \cdot x(t-2) + b_1 \cdot y(t-1) + b_2 \cdot y(t-2) \tag{3}$$

where the coefficients a_0, a_1, a_2, b_1, b_2 are calculated using the relations available in the standard [3] and many other sources, as [7], and they depend by the filter class and sampling frequency. The x values represents measured data (current, and two previous values) and y represents estimated data (two previous values).

3.1. *Vehicle-vehicle collision*

The prototype of the acquisition device was used in various crash tests, at different speeds. The device was installed with bolts on the vehicle floor, with one axis oriented in the direction of travel. It is not important which axis is aligned with the travel direction, as long as the position of the device is known (the accelerometer axis can also define an angle with the direction of travel, but this will make calculation more complicated).

The crash pulse measured on a stationary vehicle that was impacted by another vehicle is given in Figure 4, using raw and filtered data. The filter used is CFC60. Raw data used are from the Hi-g accelerometer (input range ±200 g).

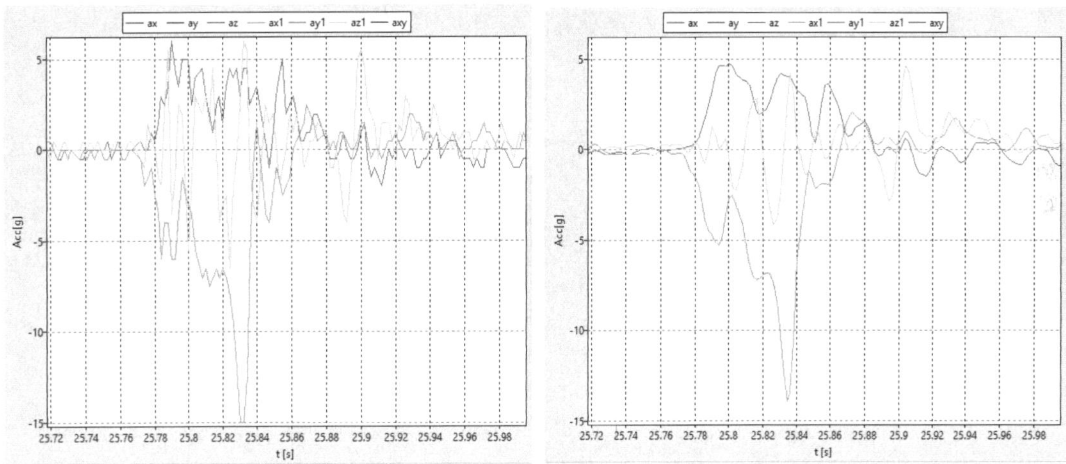

Figure 4. Example crash pulse from a vehicle-vehicle collision: left-unfiltered; right-filtered

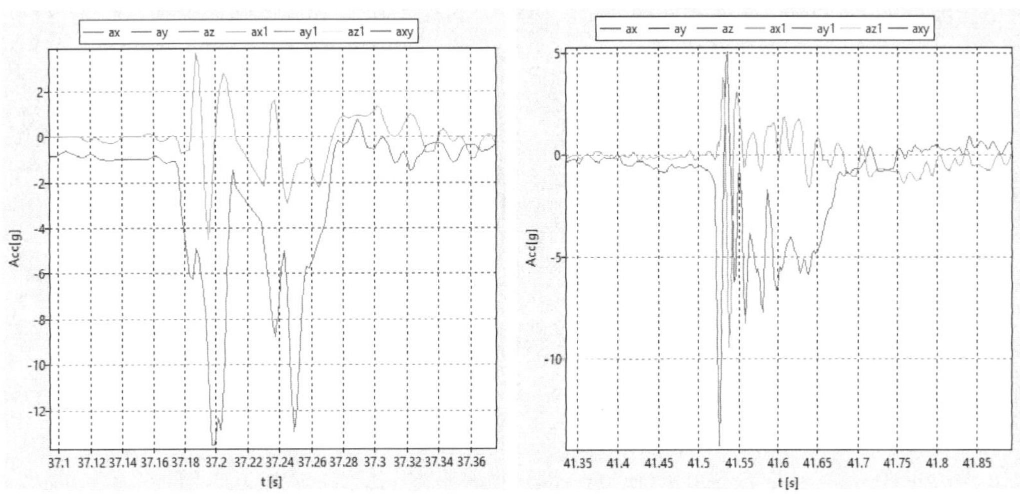

Figure 5. Example crash pulses from vehicle-vehicle collisions (the moving vehicle)

Two other crash tests given the diagrams shown in Figure 5. There are two different tests, but in both cases the crash data logger was installed on the moving vehicle. The impact speeds were between 35 and 40 km/h.

3.2. Vehicle-pedestrian or vehicle-cyclist collision

Other collision types for which the acquisition device was used are vehicle-pedestrian and vehicle-cyclist. These two types of crash tests are quite similar regarding the accelerations measured on the vehicle. The CDL acquisition device is installed on the vehicle. The raw data used are provided by the low-g accelerometer (input range ±3 g).

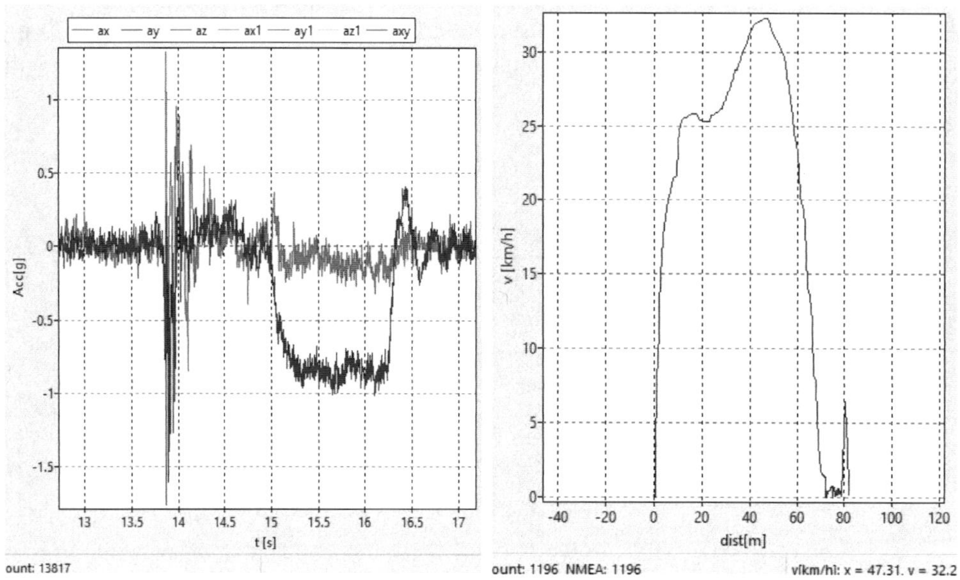

Figure 6. Vehicle-pedestrian collision: vehicle acceleration-left; vehicle speed-right

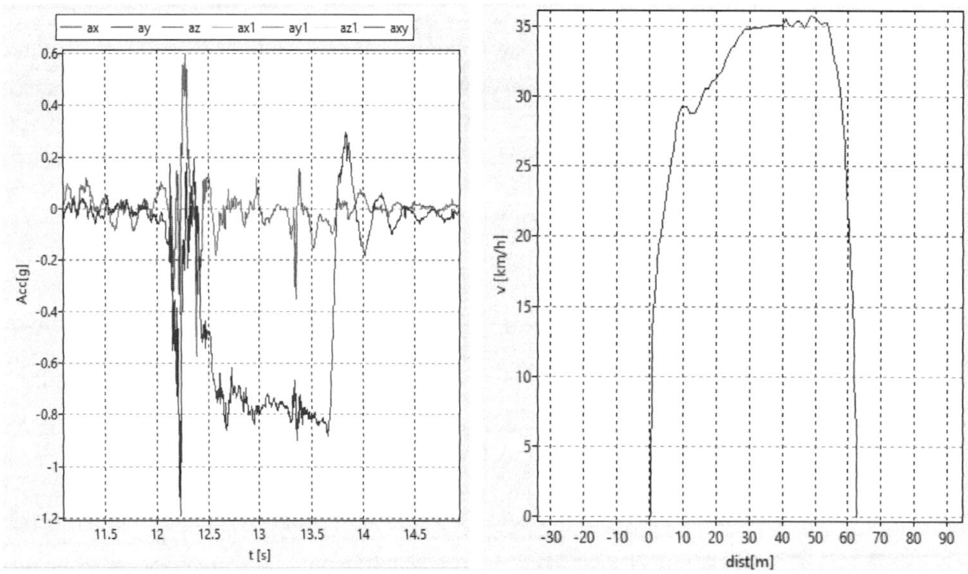

Figure 7. Vehicle-cyclist collision: vehicle acceleration-left; vehicle speed-right

Resulted diagram, for longitudinal and lateral accelerations (blue and red curves, respectively), is shown in Figure 6. The velocity of the vehicle, measured with a GPS based acquisition device, was about 32 km/h at the moment of impact. The velocity diagram is shown also in Figure 6 (as function of distance).

The diagrams of vehicle-cyclist collision are presented in Figure 7. The vehicle speed at the impact moment was about 35 km/h.

Both acceleration diagrams show the same profile: a short crash pulse, at the time of collision, then the braking sequence, when the driver presses the brake pedal. The maximum value of acceleration during the crash pulse is less than 2 g, which means that the data from the low-g accelerometer should be used.

4. Conclusion

The accelerations of vehicles involved in road accidents differ from an event to another. The crash tests can help the experts to identify the ranges of accelerations that may appear in various collision types. The correct sensors and data acquisition devices used in crash tests depend by the type of collisions. There are also different requirements for the sensors installed on vehicles or dummies.

The sensor characteristics specified in standards are difficult to obtain with cheap accelerometers. However, the shape of a crash pulse can be determined also using low cost devices, with limited bandwidth and resolution.

The original data acquisition device presented in this paper is affordable, easy to use and highly configurable. The dedicated data processing software allows user to generate the crash pulses quick and easy. For additional operation, if needed, the data can be exported to other software, like Excel, as text files.

References

[1] Peters G A and Peters B J 2002 *Automotive Vehicle Safety* (London: Taylor & Francis)
[2] Varat M S and Husher S E 2003 Crash pulse modelling for vehicle safety research *Proc: International Technical Conference on the Enhanced Safety of Vehicles* (National Highway Traffic Safety Administration)
[3] SAE 1995 J211- Instrumentation for impact tests, Rev. 03
[4] ISO 6487 2015 Road vehicles - Measurement techniques in impact tests - Instrumentation, Ed. 6
[5] DTS 2008 Vehicle Crash Test Data Acquisition - A Review of Standards *DTS Newsletter* (Online: http://dtsweb.com/news/newsletters2.php, last accessed: June 2017)
[6] Preda I 1994 *Dynamic Strains Studies of Off-Road Vehicle Transmissions for Optimization Purpose* (Brasov: Transilvania University of Brasov)
[7] Covaciu D and Dima D S 2016 Crash Tests Data Acquisition and Processing *Int. Congress of Automotive and Transport Engineering* (Springer International Publishing)
[8] DSD PicDAQ Technical Data (Online: http://www.dsd.at,[accessed May 2017)
[9] Amin M S et al. 2014 Preference and placement of vehicle crash sensors *Tehnicki vjesnik* **21.4** pp 889-896
[10] Dima D S and Covaciu D 2016 Use of MEMS sensors for data acquisition in crash tests, *EVU Int. Congress* (Bratislava: EVU)
[11] PRIME Faraday Partnership 2002 *An Introduction to MEMS* (Loughborough University)
[12] Analog Devices 2005 ADXL193 - Single-Axis, High-g, MEMS Accelerometers, Technical Datasheet
[13] Analog Devices website *MEMS Accelerometers* (Online: http://www.analog.com/en/products/mems/mems-accelerometers.htm, accessed June 2017)
[14] Arduino website *Arduino Due Technical Specifications* (Online: https://store.arduino.cc/arduino-due, accessed June 2017)
[15] Adafruit Industries website *Data Logger Shield* (Online: https://learn.adafruit.com/adafruit-data-logger-shield/overview, accessed June 2017)

IOP Conf. Series: Materials Science and Engineering **252** (2017) 012008 doi:10.1088/1757-899X/252/1/012008

Simulation of car collision with an impact block

R Kostek[1] **and P Aleksandrowicz**[1]
[1]Faculty of Mechanical Engineering, University of Science and Technology in Bydgoszcz, Al. prof. S. Kaliskiego 7, Poland

E-mail: robertkostek@op.pl, almot@op.pl

Abstract. This article presents the experimental results of crash test of Fiat Cinquecento performed by Allgemeiner Deutscher Automobil-Club (ADAC) and the simulation results obtained with program called V-SIM for default settings. At the next stage a wheel was blocked and the parameters of contact between the vehicle and the barrier were changed for better results matching. The following contact parameters were identified: stiffness at compression phase, stiffness at restitution phase, the coefficients of restitution and friction. The changes lead to various post-impact positions, which shows sensitivity of the results to contact parameters. V-SIM is commonly used by expert witnesses who tend to use default settings, therefore the companies offering simulation programs should identify those parameters with due diligence.

1. Introduction

Expert witnesses make computer simulations of car accidents using programs, which simulate crashes in the MBS (Multi Body System) convention [1,2,3]. Such simulations are not as accurate as simulations performed with the FEM (Finite Element Method), however, they are quickly performed using a PC [4,5]. Examples of computer simulations made with PC Crash and Virtual Crash were presented in the publications [6,7,8]. Authors compared real accidents with the results of simulations and they received a good agreement. The simplifications assumed in the MBS programs include a linear dependence between the vehicle deformation and contact force F_c, as well as constant mass parameters of the vehicle during the crash, whereas moments of inertia are usually reduced due to large deformations, and contact force F_c is a non-linear function of deformation [9]. Moreover, post-impact movement is sensitive to contact parameters, which is typical for non-linear systems [10]. It leads to differences between the experimental and simulation results. The sensitivity of post impact movement to contact parameters was not studied with V-SIM for frontal collision in the papers known to the authors, except [11,12], hence the coverage of the problem in this article. The key factor are identification contact parameters, which provides an opportunity to obtain a good agreement between experimental results and the results of the simulation. V-SIM is also employed to verify impact damages and explain circumstances of accidents which have practical implications [13,14].

2. Simulation of collision

The V-SIM program models the time history of the crash using the equations of motion of the rigid body. The four-wheel vehicle model has 10 degrees of freedom; 6 degrees of freedom correspond to the motion of the car body and the other 4 – to the circular motion of the wheels. The program also uses the TM-Easy tire models. The contact force is calculated based on the car body deformation and the force shows a linear dependence.

The subject of this research is sensitivity of post impact movement to contact parameters and the identification values of parameters. Thus results obtained for default (blue car) and identified data (red car) are depicted in Fig.1. The object of the study is a Fiat Cinquecento 700 ED. The vehicle curb weight was 675 kg, whereas drivers and passenger weight was 68 kg. The following contact parameters between the tire and the road were adopted: adhesive friction coefficient 0.98, slip friction coefficient 0.8, and rolling resistance coefficient 0.015. The initial speed was v_0= 50 km/h, and overlap: 0.595 m (40 %). The default parameters are presented in Table 1.

Table 1. Values of the default parameters.

Adopted parameters	Default data
Stiffness of the car body for compression phase	331 kN/m^3
Stiffness of the car body for restitution phase	331 kN/m^3
Restitution coefficient	0.24
Friction coefficient between car and impact block	3.00
Lowering of plane of contact force Δz	0.0 mm
Blocked left front wheel	-

Time t=0 was adopted for contact initiation between the vehicle and the impact block. The research started from a collision simulation for default parameters [15,16], which is depicted in Fig.1. for blue car. The images were compared in a way that at first the scale of the vehicles was adjusted, and then based on the edges of the barrier the images were fitted. The first car slowed down (t=0-60 ms) without significant rotation. Contact force F_c acts near the edge of the impact block, despite the fact that the front chassis leg is deformed, which is a reason of small rotation. Next, large contact force is observed for t=80 ms, and the vehicle rotates (100-120 ms). Then contact force drops (t=120 ms). The largest difference between the experimental results and the result of simulation is observed for 120 ms, as the real car rotates faster.

The side view provides an opportunity to study the linear vehicle displacement based on the recorded pictures. Black and white scale is helpful as its length is known, and thus the pixels position can be calculated (Fig.2). The vehicle position was calculated based on the position of the points located shortly behind the doors and the angle of rotation. This part behind the doors does not undergo large deformations and, as such, it can be used. Expert witnesses use this method to reconstruct collisions.

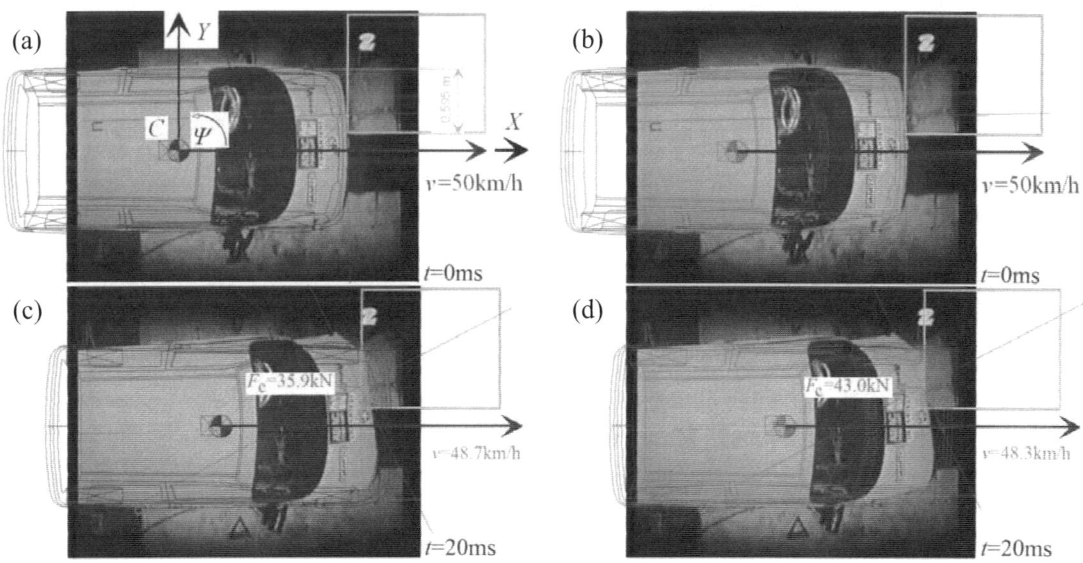

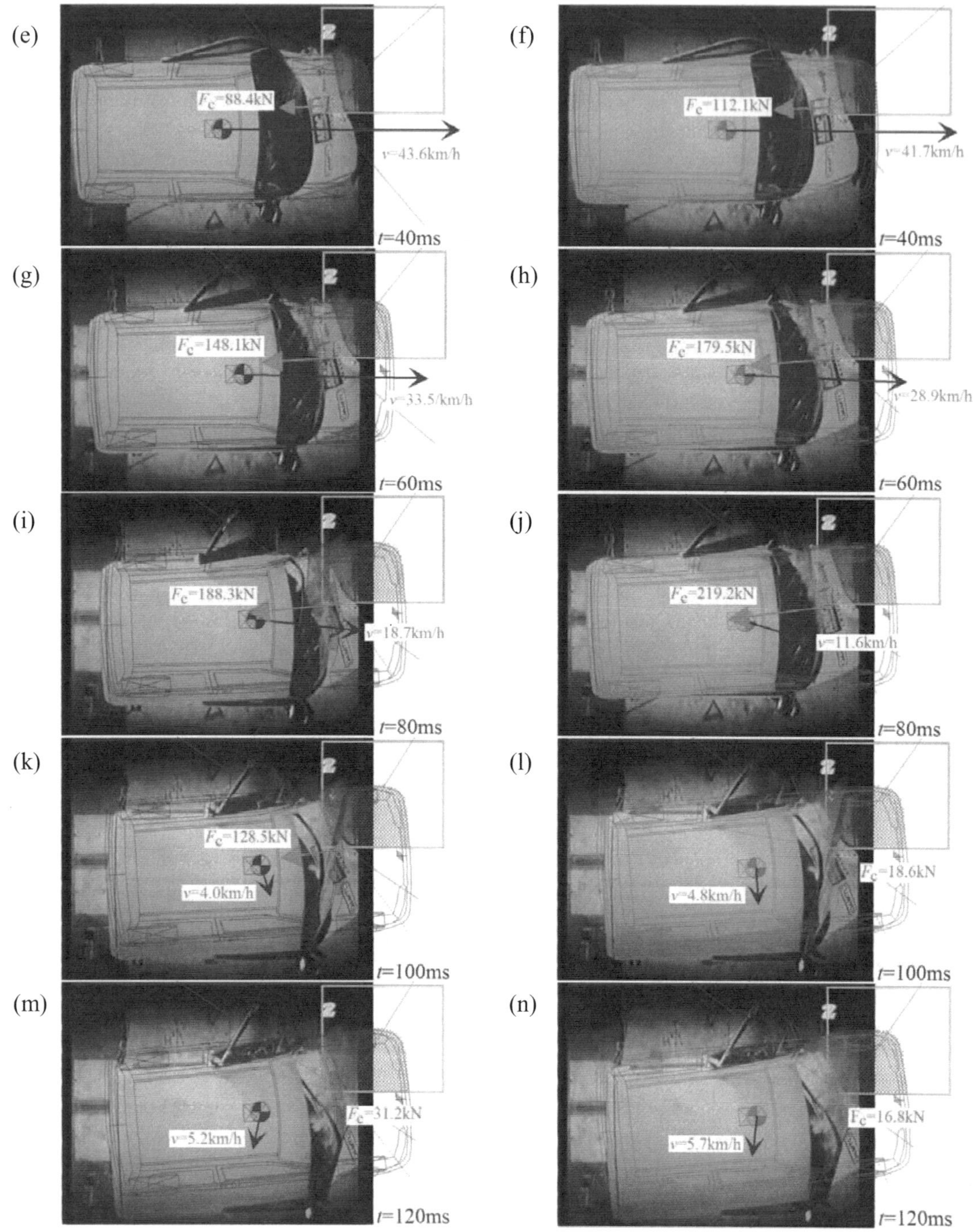

Fig. 1. Comparison of the crash test results published by ADAC with the results of computer simulation performed in V-SIM4 for default data (blue car - a,c,e,g,i,k,m) and identified data (red car - b,d,f,h,j,l,n).

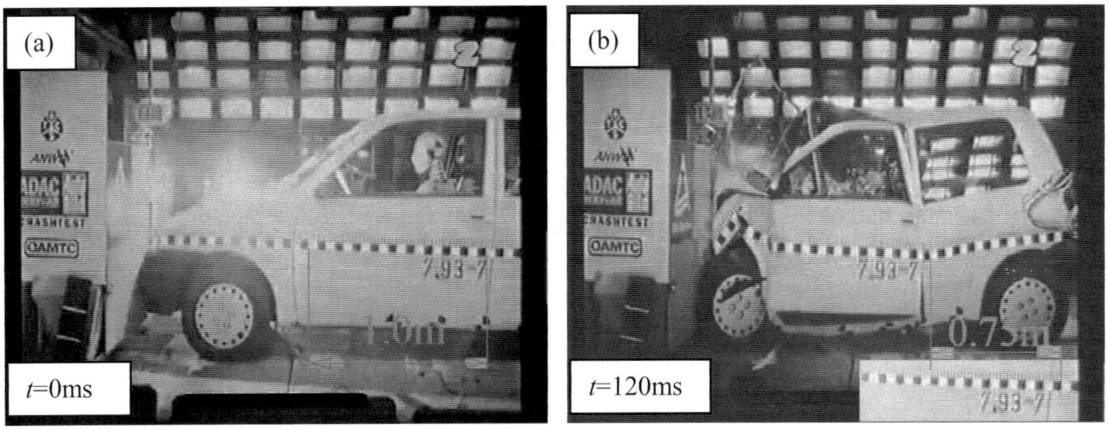

Fig. 2. Side view recorded for t=0 ms (a) and t=120 ms (b).

The linear displacement obtained for default data does not show a good agreement with experimental results and thus the values of the parameters should be identified – changed (Fig. 3c). First, the contact parameters were identified by a comparison of the simulated displacement of vehicle's mass centre x_C with the results measured based on the movie. Additionally, angle of rotation Ψ was compared for t=120 ms (Fig. 1n). Then wheel blocking was introduced to the left front wheel for time being longer than t>34 ms, as it was observed that the wheel did not rotate after this time [15,16]. Moreover the wheel is in the deformation zone (Fig. 1f). Finally, the vehicle's rear wheels detach (120 – 155 ms) and fall off (313 ms), which was observed during the crash test. To obtain this effect, the level the contact force acts at was lowered by 50 mm. For the parameters identified the vehicle's rear wheels detach for the time of 100-160 ms and fall off for t=320 ms, which was reconstructed. The identification was an iterative process because the parameters are coupled. The parameters of contact between the tires and the road were not changed, however, it should be mentioned that a lower friction coefficient value introduces an increase in rotation angle. The identified parameters are presented in Table 2.

Table 2. Values of the identified parameters.

Adopted parameters	Identified data
Stiffness of the car body for compression phase	426.45 kN/m^3
Stiffness of the car body for restitution phase	426.45 kN/m^3
Restitution coefficient	0.12
Friction coefficient between car and impact block	2.00
Lowering of plane of contact force Δz	50 mm
Blocked left front wheel	t>34 ms

At the first stage of impact (t=0-40 ms) displacements of vehicles x_C y_C and vehicle rotation angles Ψ obtained for default and identified data are similar (Figs. 1,3). Nevertheless, the components of the resultant force acting on the vehicle F_c and vertical component of the resultant moment acting on the vehicle M_{Cz} are various for both simulations, which results in various post-impact movements. For the identified data a smaller deformation and larger contact forces are observed as the red car is "stiffer", which results in shorter time of contact between the car and the impact block. The largest forces act at the compression phase, which takes about 100 ms (Fig. 3a). Next the red car bounced off at a lower speed as a result of a lower value of restitution coefficient. This leads to better agreement with the experimental results (Fig. 3c). Velocity vectors have various directions, which is obvious at this stage (100-120 ms). Finally, for the identified parameters the vehicle stopped faster since the left front wheel is blocked and the restitution coefficient is lower.

The time history of the resultant moment acting on the vehicle shows that for the compression phase the moment is above zero, which causes a vehicle rotation. Then for the restitution phase the moment is below zero, which reduces the speed of rotation. This phenomenon is caused by F_{ry} force. Various time histories of resultant moment acting on the vehicle result in various vehicle rotation angles Ψ in these two cases. For the identified parameters this angle is larger, which shows better agreement with experimental results (Fig. 1n). After the vehicle's rear wheels detach, the red vehicle rotates with almost constant speed, which makes the rotation angle larger. For a better presentation of the results, movies are presented [17].

Summarizing, the default values of the coefficients of stiffness and restitution are not accurate, which results in differences between measured and simulated displacement. An agreement between the experimental results and the simulation results was achieved after an increase in the stiffness coefficient of the body at the compression and restitution phases; introducing the left front wheel blocking task and a decrease in the value of the coefficient of restitution; lowering of the level in which the contact force acts. The deformation for default parameters was too large and real vehicles after the crash do not bounce off the barrier considerably and do not drive far away, which the Authors know from experience. Thanks to those changes a good agreement between the results was obtained. It shows that contact phenomena are important and mechanical systems are sensitive to contact parameters. Authors suppose that for various initial conditions, various parameters of contact are identified, which is a limitation of MBS programs.

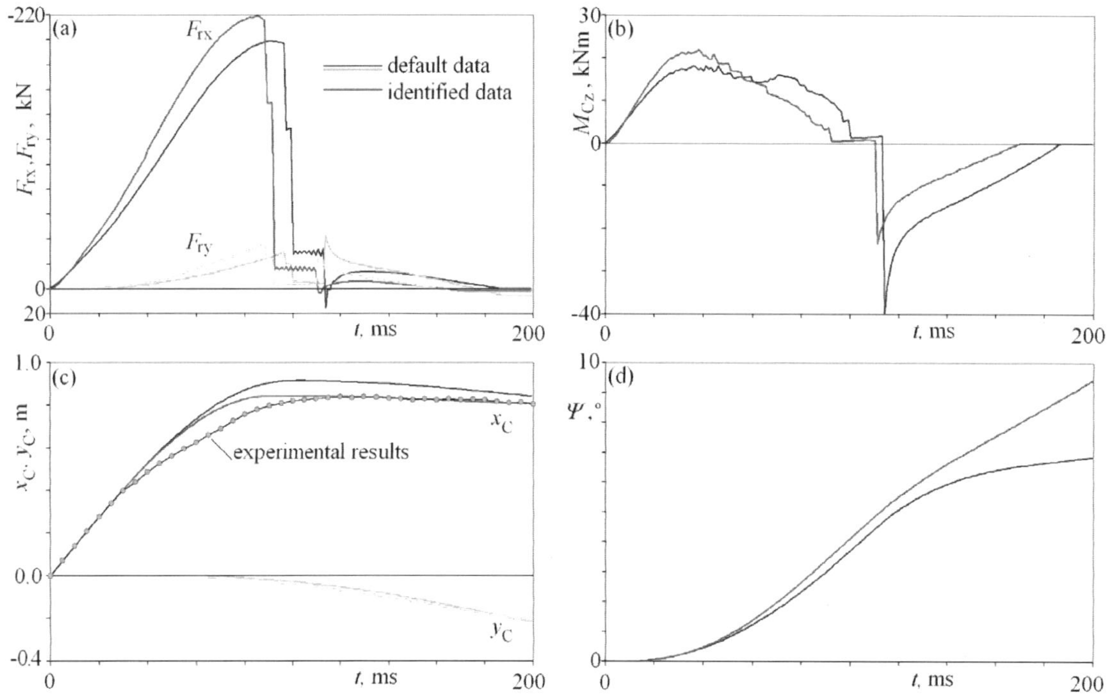

Fig. 3. Time histories of the components of the resultant force acting on the vehicle (a), displacements of vehicle mass centre (c), vertical component of the resultant moment acting on the vehicle (b), and vehicle rotation angle (d).

3. Conclusions

The obtained results led to a conclusion that for default settings there were achieved approximate results which differed from the experimental results. Moreover the pattern of the accident is sensitive to initial conditions and parameters, which are uncertain. The knowledge is not common among witness experts who consider the simulation results to be reliable. Identification is a difficult problem as parameters are coupled, and many sets of parameters provide similar results. Finally, companies offering simulation programs should identify the values of parameters and models with due diligence. Expert witnesses have no chance for an accurate reconstruction of the course of accident if the parameters are not correct.

References

[1] http://www.cyborgidea.com.pl (access 08.06.2017)
[2] http://www.pc-crash.com (access 08.06.2017)
[3] http://www.vcrashusa.com (access 08.06.2017)
[4] Dias de Meira A *et all* 2016 Numerical Analysis of an Intercity Bus Structure: A Simple Unifilar Model Proposal to Assess Frontal and Semifrontal Crash Scenarios *Latin American Journal of Solids and Structures* vol 13 no 9 pp 1616-1640
[5] http://www.lstc.com (access 08.06.2017)
[6] Andrews S *et al* 2009 A Comparision of Computer Modeling to Actual Data and Video of a Staged Rollover Collision NHTSA Paper no 09-0346
[7] Coufal T and Semeia M 2014 Research of Impact Parameters for Traffic Accident Analysis in Case of Small Overlap Crash Test *2nd Annual International Conference on Forensic Science - Criminalistics Research (Singapore)* pp 1-9
[8] Dima D S, Covaciu D and Chiru A 2014 Validation of Simulation and Optimization Reconstruction in PC Crash with Video Recorded Sample *3rd International Congress Science and Management of Automotive and Transportation Engineering (Craiova Romania)* SMAT Paper no 2014-MTSRT51 pp 207-212
[9] Kostek R The modelling of loading, unloading and reloading of the elastic-plastic contact of rough surfaces 2012 *Journal of Theoretical and Applied Mechanics* vol 50 issue 2 pp 509-530
[10] Kostek R 2014 Simulation and analysis of vibration of rolling bearing, *Key Engineering Materials* vol 588 pp 257-265
[11] Kostek R and Aleksandrowicz P 2017 Effect of Contact Parameters on the Pattern of Vehicle Collisions with a Round Pillar *23rd International Conference Engineering Mechanics (Svratka Czech Republic)* vol 1 pp 490-493
[12] Kostek R and Aleksandrowicz P 2017 Identification of the Parameters of Vehicle Contact with a Rigid Barrier from a Crash Test *23rd International Conference Engineering Mechanics (Svratka Czech Republic)* vol 1 pp 494-497
[13] Aleksandrowicz P 2017 Analysis of Vehicle Collisions with the SDC Method *23rd International Conference Engineering Mechanics (Svratka, Czech Republic)* vol 1 pp 78-81
[14] Aleksandrowicz P 2017 Verifying the Application of the Models of Crash and Collision Detection of Incompatible Vehicles *23rd International Conference Engineering Mechanics (Svratka Czech Republic)* vol 1 pp 82-85
[15] https://www.adac.de (access 08.06.2017)
[16] https://www.youtube.com/watch?v=UoaI-NuGBN4 (access 08.06.2017)
[17] https://www.youtube.com/watch?v=Z7fjoVRVZIg&list=PLHRuIvXAxu9nXAOyOJfsrL5D2oab7msev&index=3 (access 08.06.2017)

IOP Conf. Series: Materials Science and Engineering **252** (2017) 012009 doi:10.1088/1757-899X/252/1/012009

Experimental Validation of Various Temperature Modells for Semi-Physical Tyre Model Approaches

Andreas Hackl, Christoph Scherndl, Wolfgang Hirschberg & Cornelia Lex

Institute of Automotive Engineering, Graz University of Technology, Inffeldgasse 11/2, 8010 Graz, Austria

E-mail: andreas.hackl@tugraz.at

Abstract. With increasing level of complexity and automation in the area of automotive engineering, the simulation of safety relevant Advanced Driver Assistance Systems (ADAS) leads to increasing accuracy demands in the description of tyre contact forces. In recent years, with improvement in tyre simulation, the needs for coping with tyre temperatures and the resulting changes in tyre characteristics are rising significantly. Therefore, experimental validation of three different temperature model approaches is carried out, discussed and compared in the scope of this article. To investigate or rather evaluate the range of application of the presented approaches in combination with respect of further implementation in semi-physical tyre models, the main focus lies on the a physical parameterisation. Aside from good modelling accuracy, focus is held on computational time and complexity of the parameterisation process. To evaluate this process and discuss the results, measurements from a Hoosier racing tyre 6.0 / 18.0 10 LCO C2000 from an industrial flat test bench are used. Finally the simulation results are compared with the measurement data.

1. Introduction

With enhanced integration of simulation and modelling into vehicle development and assistance systems, the requirements for simulation models regarding precision and computational time are increasing rapidly. In this case, not only demands affecting vehicle modelling but also tyre modelling become more important. In recent years, with improvement in tyre simulation, the needs for coping with tyre temperatures and the resulting changes in tyre characteristics are rising significantly. These changes include, amongst other things, the temperature dependent behaviour of friction coefficients and the loss of cornering stiffness with increasing temperature. To account those effects in modern assistance systems, it will become mandatory to reliably estimate tyre temperatures with information gathered from sensors or state estimators the vehicle can provide. This needs to be done in real time and therefore a reasonable compromise between accuracy and computational time has to be achieved. To generate a comparative basis three different models are tested in regard of the fulfilment of the requirements mentioned. To keep the effort needed for parameterisation as low as possible, the models were designed to handle the simulation with little to no information about the car itself and instead focusing on properties derived from tyre measurements.

2. Basic investigation - mechanics of heat generation and distribution

Before describing the three different types of temperature model mentioned above, a basic overview of the mechanics of heat generation and distribution of a tyre is given in this chapter, see also Figure 1.

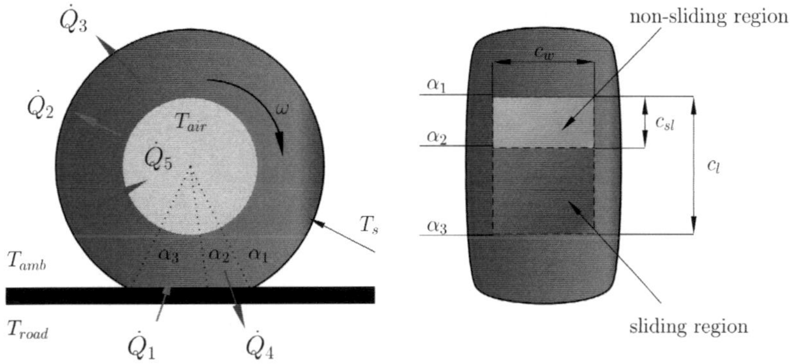

Figure 1: Different heat phenomena occurring at a tyre during operation and divided in heat generation and heat exchange, based on [1]

Heat generation is described by the different effects which increase the inner and/or outer temperature of a tyre. Two different types of heat generation are common used in the literature, see e.g. [2] or [3]. First, heat is generated by different sliding velocities between road and tyre, see $\dot{Q}_1$ in Figure 1, and second heat is generated inside due the hysteresis effect, see $\dot{Q}_2$. The heat in the sliding region of the contact patch $\dot{Q}_1$ is generated by an differential velocity between tyre and road and is mathematically described with

$$\dot{Q}_{1,Friction(x/y)} \cong F_{x/y} \cdot v_{s,x/y} \cdot \left(c_s \cdot \frac{T_{nom}}{T_{sur}} \cdot \frac{1}{2} \right), \tag{1}$$

where $F_{x/y}$ describes the longitudinal and lateral forces and $v_{s,x/y}$ the sliding velocities. Additionally a parameter c_s is needed which scales the distribution between sliding and non-sliding region of the contact patch depending on the current longitudinal slip and slip angle, see Figure 1. The term $\frac{T_{nom}}{T_{sur}}$ describes the power distribution between tyre and road where T_{nom} stands for a temperature equilibrium where the generated heat is distributed equally among the tyre and the roadway.

As mentioned above, the second type of heat is generated inside the tyre due to hysteresis effect $\dot{Q}_2$. In the course of the parameterisation process it could be observed that the heat generation of the tested tyre is governed by the impact of the frictional powers, therefore heat generation due to hysteresis is neglected in this article. Through multiple tests it could also be concluded that the temperature at the center area of the tyre is not influenced by inclination angle, therefore inclination angle is used in this article.

Heat exchange owing to conduction between the adhesive part of the contact patch and the roadway $\dot{Q}_4$ and outside the contact patch due to convection with the surrounding air $\dot{Q}_3$. Furthermore convection is happening constantly between the inner layer and the filling medium $\dot{Q}_5$. Equation 2 shows the mathematical approach to model this heat exchange

$$\dot{Q}_{3/4/5,convection/conduction} = \underbrace{f(v^{Exp_i}, \lambda_i, \rho_i, c_{p,(i)})}_{h_i} \cdot A_{(i)} \cdot \Delta T_{(i)}. \tag{2}$$

The parameter A describes the area and ΔT the temperature difference between the affected regions, where i = {area in contact with ambient air, track, filling medium}. The convection coefficient h_i is strongly dependent on the velocity v, the adjacent material properties (λ, ρ, c_p) and the way the tyre is mounted in the car (Exp). In open wheel racing the airflow is surrounding the tyre at all times whereas passenger car tyres are mostly covered by the wheel case. Because the examined tyre is used to compare measurements done on an industrial flat track test bench with constant tyre speed, this function is treated as a constant value h_i in this article. Investigations especially on the convection function will be done in the future.
The same is true for the conduction coefficient which is highly dependent on the chemical composition and the surface design of the tyre. Although knowing that the slick tyre tested is characterised by a very soft compound, almost no information about a reasonable value range could be obtained through literature research.

2.1. Single-layer temperature model

The single-layer model is characterised by consisting of the simplest structure in this article. The whole tyre consists of one layer with combined thermal properties from rubber and belt materials, [1]. For this single-layer, the first law of thermodynamics is applied using

$$\mathbf{\Delta}T = \frac{\Delta t}{m \cdot c_p} \cdot \left(\sum_i \dot{Q}_i \right), \tag{3}$$

where $\dot{Q}_i$ describes the different phenomena of heat generation and exchange, see Figure 1, m the tyre mass, c_p the combined thermal heat capacity and Δt the time difference between two measurement points. This energy balance contains the heat generation due to friction and also the heat emission to the surrounding environment and the filling medium. With these inputs, temperature alternation from an initial starting temperature and hence the temperature of the layer can be calculated.

2.2. Multi-layer temperature model

The multi-layer model is an improvement of the single-layer model, based on [2], where in this article the tyre is split in three layers with individual thermal properties and input sizes. These layers describe the main constituents of a tyre and for each of these layers the energy balance considers different heat mechanisms. The outer layer for example, is in direct contact with the road and the ambient air whereas the inner layer is exchanging heat with the filling medium. Additionally heat flows among the three layers are taken into account, these are modelled using an adapted form of the stationary heat conduction as shown in Figure 2 and given by

$$\dot{\mathbf{Q}}_{\mathbf{conduction}} = \frac{1}{\frac{\delta_i}{\lambda_i}} \cdot A \cdot \Delta T, \tag{4}$$

where A describes the area and ΔT the temperature difference between adjacent point masses. Combined with thickness δ_i and thermal conduction coefficients λ_i the thermal resistors are modelled properly. Thus the temperature for each zone can be calculated depending on the behaviour of the other zones and therefore a temperature distribution over the tyre depth can be shown. As example, Equation 5 shows the energy balance for the surface layer, where $\dot{Q}_{surface}$ describes the conduction term as described through Equation 4, this can be seen in Figure 3 and applied for all layers.

$$\mathbf{\Delta}T_{sur} = \frac{\Delta t}{m_{sur} \cdot c_{p,sur}} \cdot (\dot{Q}_{1,friction} + \dot{Q}_{3,convection} + \dot{Q}_{4,conduction} + \dot{Q}_{surface}) \tag{5}$$

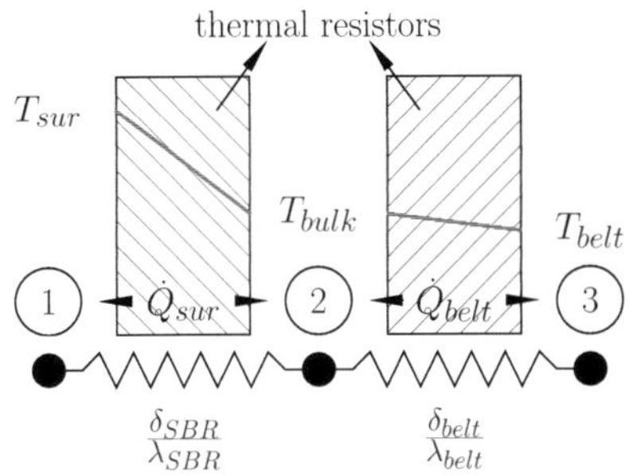

Figure 2: Multi-layer model described by lumped masses with intermediate thermal resistors

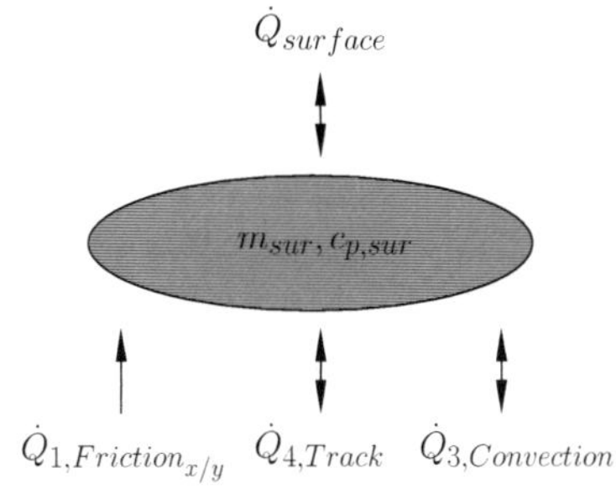

Figure 3: Energy balance for surface layer, based on [2]

2.3. Partial temperature model

The partial temperature model describes the problem of an unsteady heat conduction in the most accurate way compared to the other two models. It is described by an one dimensional partial differential Fourier heat equation, as described in [3] by

$$\lambda_i \frac{\partial^2 T}{\partial^2 \delta} = \rho_i c_{p,i} \frac{\partial T}{\partial t}. \tag{6}$$

The term $\frac{\partial^2 T}{\partial \delta^2}$ describes the directional derivative of the temperature over the tyre depth δ and ρ_i the corresponding density. An infinitesimal element as shown in Figure 4 revolves around the wheel axis and traverses all zones relevant for heat exchange $\dot{Q}_{1-5}$. For that reason the position regarding tyre depth and circumference are additional state variables. Heat flows are described as boundary conditions on the outer and inner surface, therefore, $\dot{Q}_{1-5}$ (Eq. 7/8) have to be adapted slightly compared to Equations 1 and 2. Because of the integration over the circumference of the tyre, areas regarding convection A_j do not occur in the respective equations whereas the heat generation has to be divided by the sliding area $A_{cp,sl}$ as seen in Equation 8.

$$\dot{\mathbf{Q}}_{\mathbf{3/4/5}} = -\lambda_i \frac{\partial T}{\partial \delta}\bigg|_{\delta=0} = h_i \cdot [T_\infty - T(0,t)] \tag{7}$$

$$\dot{\mathbf{Q}}_{\mathbf{1,x/y}} = -\lambda_i \frac{\partial T}{\partial \delta}|_{\delta=0} \cong \frac{1}{A_{cp,sl}} \cdot F_{x/y} \cdot v_{s,x/y} \cdot \left(c_s \cdot \frac{T_{nom}}{T_{sur}} \cdot \frac{1}{2} \right) \tag{8}$$

The material properties regarding the heat flux are changing along the tyre depth, thus modeling tyre geometry correctly, see Fig. 4. Because of the partial access, temperatures can be calculated steplessly throughout the tyre depth and circumference. This model is characterised through a distinctive temperature behaviour as shown in Figure 5. At each revolution the element enters the sliding zone where the temperature rises nearly step like. Afterwards, in the convection and conduction zone, the temperature decreases continuously throughout the remaining rotation until the sliding zone is reached again. The angular position to determine the needed boundary condition is calculated by means of the wheel rotational speed and compared to angles α_{1-3} as shown in Figure 1.

3. Parameterisation Process

The tyre tested is a Hoosier slick 6.0 / 18.0 10 LCO C2000 used by the Formula Student racing team of TU Graz. The tyre is characterised by a slim construction and a very soft compound. The testing procedure was carried out on a flat track test bench and consists of transient manoeuvres with changes in slip angle, slip ratio, contact force and inclination angle.

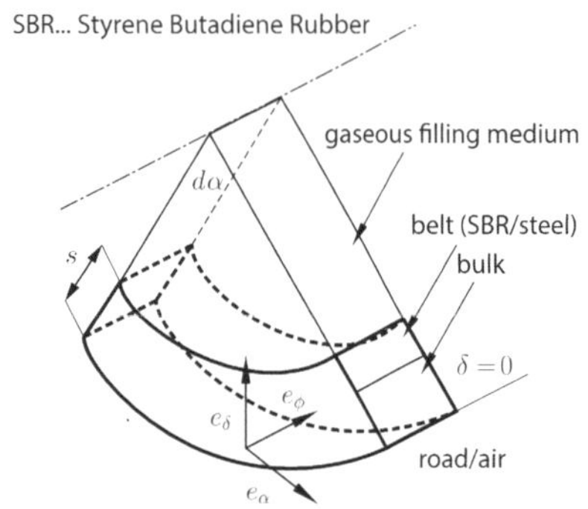

Figure 4: Infinitesimal element spinning around the tyre axis, based on [3]

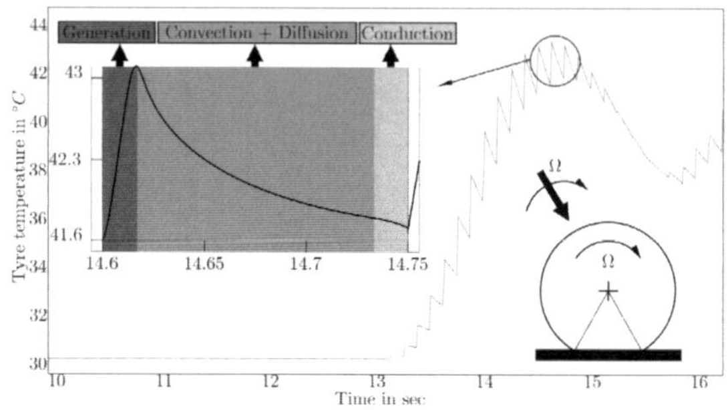

Figure 5: Temperature versus time trend of the outermost element, based on [3]

By measuring the geometric properties of the tyre and combining this information with heat transfer coefficients found in literature, [5], the internal heat flux was described. Because of the wide range of values for convection coefficients found in literature, e.g. [5], these parameters are optimised within a physically justifiable range. The same was done for the conduction coefficient in the adhesive part of the contact patch, with the difference that no explicit upper limit for this kind of material combination could be determined in literature. Also a reference temperature T_{nom} is optimised where the distribution of heat generation due to friction between tyre and road is in a thermal state of equilibrium. Table 1 summarises the parameters used for the optimisation process. Since IR-sensors, used to measure tyre temperature, scan only the outermost elements of the measured area, it is necessary to have a very responsive layer in the surface region. Therefore, the surface layer of the multi-layer model represents the first millimetre of rubber of the tyre. For the same reason a very fine resolution for the partial access of the PDE-tool (Partial Differential Equation Toolbox, Matlab [6]), used to solve the Fourier equation, was chosen. Because of that thin surface layer, for the multi-layer model an additional parameter p_4 was implemented, this coefficient scales the power generated due to friction between the surface and the bulk layer, where a value of one represents frictional power only occurring in the surface layer and vice versa. Figure 6 shows the temperature penetration of a tyre experiencing flash heating due to frictional powers, it can be seen that heat generated due to friction is not only penetrating the outermost elements of the tyre, but rather affects internal regions too. Therefore this parameter was necessary to control the temperature balance between the three layers of the multi-layer model. Overall, the parameter solving was performed using the least squares method where the error between measured and simulated temperature was minimised. In terms of the three layer model, this temperature was represented by the surface layer, whereas for the partial-model the temperature of the outermost element was used.

4. Results

Table 2 summarises the parameters solved in the course of the fitting process. Overall the parameters identified are approximately the same size and satisfactorily match the expected values. The convection coefficient is situated at the bottom end of the allowed value range. This can be attributed to the fact that the boundaries found in [5] refer to a passenger car tyre where the turbulent boundary air layer is more pronounced compared to a slick tyre, therefore this coefficient is estimated to be higher at those type of tyres. With respect to the vague information regarding the conduction coefficient between road and tyre the solved values seem reasonable. The single-layer model reached the upper limit set for this parameter which was chosen because of the values identified through the other models. This limit was necessary

Table 1: Parameters used in the fitting process

Description	h_{road}	h_{amb}	T_{nom}	F_{fric}
Partial model	p_1	p_2	p_3	
Single-layer model	p_1	p_2	p_3	
Multi-layer model	p_1	p_2	p_3	p_4

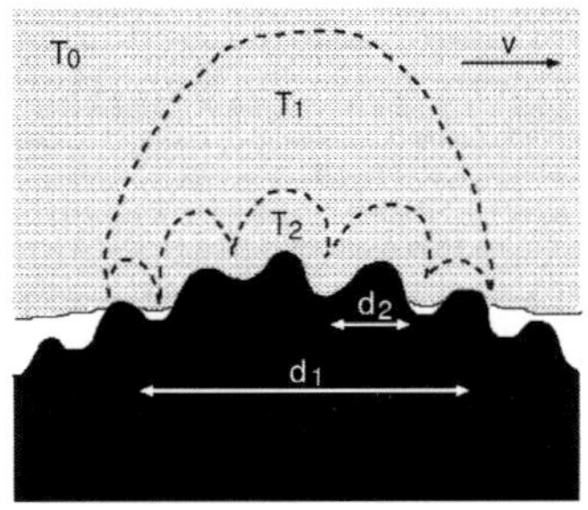

Figure 6: Temperature penetration due to flash heating from frictional power, [4]

to keep the convection coefficient in an physically meaningful value range. The reference temperature found by the optimisation tool is a few degrees above the road temperature and can be interpreted as an averaged contact temperature between road and tyre, this comes in handy because to calculate this temperature much information would be needed about the track properties, which is unavailable in most cases. Figure 9 shows the results of the temperature simulation in lateral-, longitudinal and combined direction. Although all three models have a satisfactory average error, only the multi-layer and the partial model can reproduce the behaviour of the surface region, in terms of flash temperature, realistically. Additionally the partial model is able to picture the cooling condition of the tyre the most accurate way and can be modified to deal with full braking lock scenarios, where only a specific region of the surface is affected. When comparing the inner regions of the partial and the multi-layer model a difference can be observed. This comes from the different approaches on how the power is introduced into the tyre, where at the partial model the frictional power enters the tyre through a boundary condition whereas at the multi-layer model parts of this power are directly entering the bulk region. Another reason is the different heat distribution characteristic of these approaches as seen in Figure 7. Since there is no measurement data of the inner parts of the tyre, no clear statement could be made which model is more accurate. Another interesting feature of the partial model are three dimensional plots over tyre depth, time and temperature, which can be really illuminating, as evidenced by Figure 8. Concerning computational time, all three models are capable of running online, although it should be noted that the chosen resolution of the partial model had to be adjusted carefully to ensure that premise without loosing accuracy. Table 2 summarise the used material specific values used for the simulation.

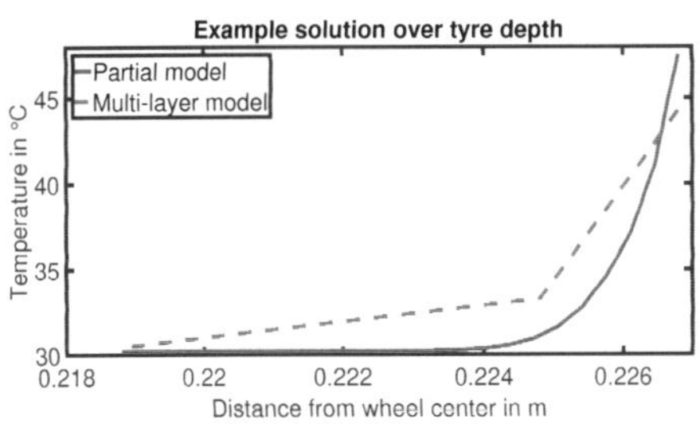

Figure 7: Temperature distribution over tyre depth comparison between the partial and the multi-layer model

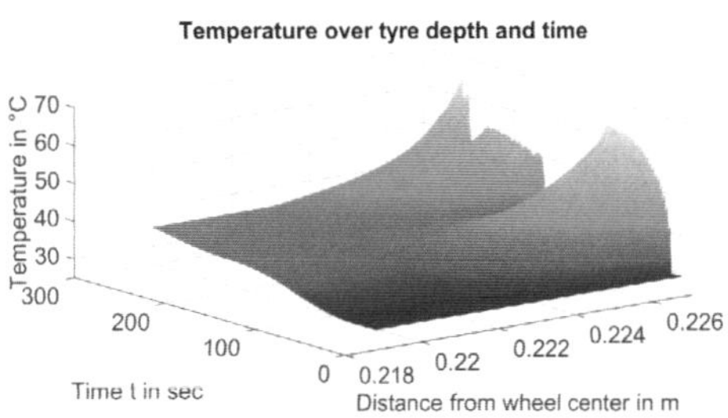

Figure 8: Temperature over tyre depth and time modelled by the partial model

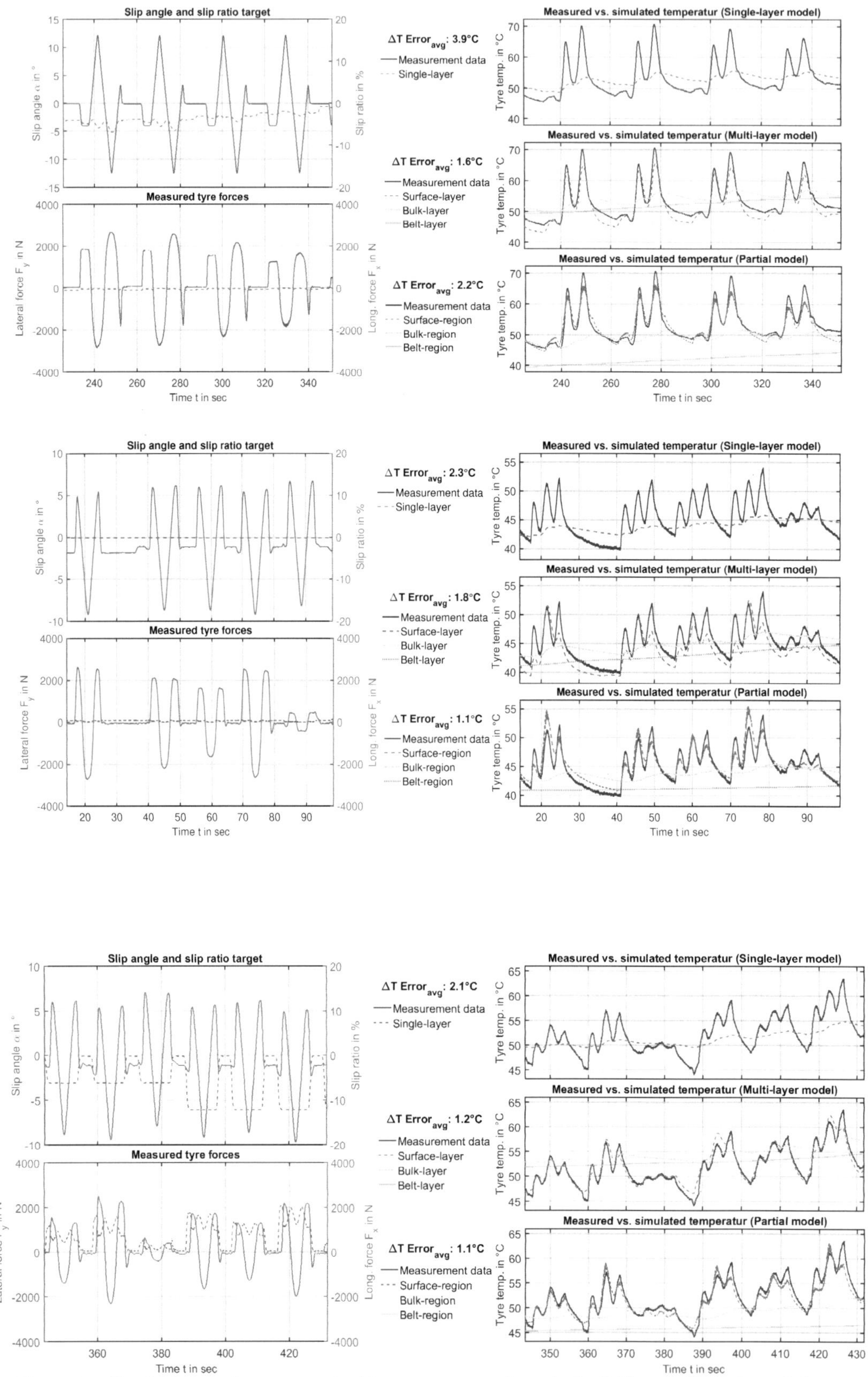

Figure 9: Simulation results compared with measurement data. Left: Different slip ratio and slip angle targets. Right: Temperature comparison by using different model approaches

Table 2: Parameters identified in the fitting process

Description	h_{road} in W/m^2 K	h_{amb} in W/m^2 K	T_{nom} in °C	F_{fric} in -
Partial model	1282.1	36.7	30.2	
Single-layer model	2000.0	18.8	34.7	
Multi-layer model	1496.3	17.9	31.9	0.55

Table 3: Material specific values used to simulate the presented temperature models

Symbol	Description	Unit	Value
λ_{sbr}	Heat conduction coefficient SBR	W/m K	0.25
λ_{steel}	Heat conduction coefficient steel	W/m K	10
ρ_{sbr}	Density SBR	kg/m^3	1550
ρ_{steel}	Density steel	kg/m^3	7850
$c_{p,sbr}$	specific heat capacity SBR	J/kg K	1880
$c_{p,steel}$	specific heat capacity steel	J/kg K	450
h_{fill}	convection coefficient inner liner/filling medium	W/m^2 K	1

5. Conclusion

The influence of tyre and road temperature on the tyre‘s contact force characteristics is a well-known phenomenon. Different model approaches, with respect of possible implementation in semi-physical tyre models, were compared in model accuracy, computational time and complexity of the parameterisation process in this article. To evaluate this process and discuss the results, measurements from a Hoosier racing tyre 6.0 / 18.0 10 LCO C2000 from an industrial flat test bench are used. It is presented that a single-layer model is easy to parametrise but has lack in accuracy especially on dynamic manoeuvres and investigations depending on the tyre depth. A multi-layer model with three layers shows quite good evaluation results but brings the disadvantage of additional model parameters. A partial-model, based on the Fourier heat equations, brings long calculation time but shows the most accurate results and is furthermore able to model the temperature depending on the circumference of the wheel. Summarised, depending on the application, the presented model approaches are able to model the tyre temperature in an accurate way. To evaluate the model including the parameterisation process investigations will be done on different types of passenger cars tyres.

References

[1] Tremlett A. J. & Limebeer D. J. N. (2016), Optimal tyre usage for a Formula One car, *Vehicle System Dynamics*, 54:10, 1448-1473, DOI: 10.1080/00423114.2016.1213861

[2] Buettner F., Unterreiner M. & Bortolussi P. (2015), An effective method to identify thermodynamic tire characteristics through driving maneuvers. *15. Internationales Stuttgarter Symposium. Proceedings. Springer Vieweg*, Wiesbaden, DOI: 10.1007/978-3-658-08844-6_62

[3] De Rosa R., Di Stazio F., Giordano D., Russo M. & Terzo M. (2008), ThermoTyre: tyre temperature distribution during handling manoeuvres, *Vehicle System Dynamics*, 46:9, 831-844, DOI: 10.1080/00423110701684579

[4] Lorenz B., Persson B. N. J., Fortunato G., Giustiniano M. & Baldoni F. (2013), Rubber friction for tyre tread compound on road surfaces, *Journal of Physics: Condensed Matter*, DOI: 0953-8984-25-9-095007

[5] Yokota K., Higuchi E. & Kitagawa M. (2012), Estimation of Tire Temperature Distribution and Rolling Resistance under Running Conditions Including Environmental Factors, *SAE Technical Paper* 2012-01-0796, DOI:10.4271/2012-01-0796.

[6] The MathWorks & Partial Differential Equation Toolbox (2015b), *The MathWorks Inc.* R2015b (8.6.0.267246)

Pedal force determination respect to ride comfort

Slavica Mačužić[1] , Jovanka Lukić[1] , Jasna Glišović[1] and Danijela Miloradović[1]

[1]University of Kragujevac, Faculty of Engineering, Sestre Janjić 6, 34000 Kragujevac, Serbia

E-mail: macuzicslavica@gmail.com

Abstract. Automotive ergonomics is a set of knowledge which has a task to design a vehicle to make the passengers feel comfortable. Interior packaging represents an important stage in the vehicle design process, in order to enable the driver to every important aspect of movement. During the process of driving, the driver performs various movements of arms and legs, leading to a certain fatigue. Each seating position in the vehicle, contain certain boundary conditions, and for that reason it was necessary to examine how the seating position affects the driver possibilities. In this paper, the pedal forces were determined by application of Ramsis human model. Different human populations were taken into account. Correlation between subjects' anthropometrics measures and the foot pedal force pedal was observed. Obtained results were significant input data for vehicle packaging.

1. Introduction

Car manufacturers design the interior of the vehicle in order to better meet the needs of the driver. Ergonomics is a science that helps manufacturers makes cars as comfortable as possible. Every designer must use certain norms and regulations when designing a car interior. First, it is necessary to increase efficiency and productivity of production, then the improvement of health, and finally the safety and comfort man in his work environment [1]. Using knowledge of this science, there is a possibility to know in advance the possibility of movement during driving. During the driving process, due to certain movements, driver fatigue occurs. This is particularly the result of a long distance driving and often pedal activation. In this study, the force required to press the pedal in a driving position was studied.

This study investigated the ergonomic position of the mannequin in the virtual environment of the interior of the vehicle VW UP! (figure 1). The anthropometric data of male and female of nineteen populations were analysed. The software packages Catia v5 R18 and Ramsis (*Rechnergestutztes Anthropometerisches Mathematisches System zur Insassen-Simulation*) [2] were used. It is a leading program in the field of 3D CAD ergonomics, and more than 70% of all car manufacturers worldwide now use this software, which have ability to set a mannequin at a specific place and to manipulate his movements [3].

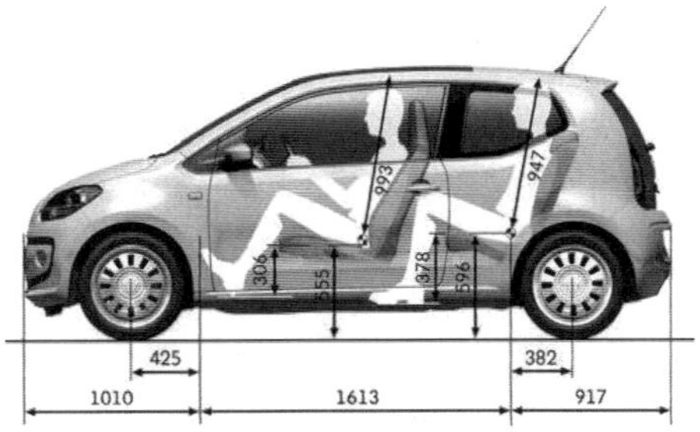

Figure 1. Volkswagen Up! dimensions [4].

2. Literature review

In the automotive industry, a very competitive market forces car manufacturers to develop their products better and faster. Application of computer-aided design (CAD) and numerical simulations is possible to apply ergonomics in the early stage of vehicle creation, using digital human models [5]. The assessment of discomfort was done through a discomfort assessment method such as RULA [6], REBA [7], etc. In the paper published in 2012 in the journal Applied Ergonomics [8], it has been shown that near to 105.000 papers were published using the term discomfort in their title, while only a few studies dealt with differences in comfort and discomfort. In the mid-nineties of the last century, three authors [9] suggested that comfort and discomfort are not two opposing concepts on a continuous scale. Discomfort is primarily associated with physiological and biomechanical factors such as fatigue or pain, while comfort is associated with physical ease and well-being. A few years later, in [10] model of comfort and discomfort in which he connects the physical environment to perceived discomfort is proposed.

Pedals of brake and clutch are one of the most used controls used in vehicles. The force exertion capacity depends on the influence of many factors [11], [12]:

- Subject factors (gender, age, anthropometry, etc.),
- Product factors (size, weight, material of contact area, etc.),
- Environmental factors (vibration, temperature, etc.),
- Posture factors (body parts in contact, body parts used for the force exertion, etc.) and
- Task factors (required force level, speed of motion, direction of force, frequency, etc.).

The first research of necessary activation force of a pedal began in the seventies of the last century. Author [13] showed that maximum static pedal force depended on human model position, seat and pedal type. The required force value can be achieved if the backrest is well adjusted and when the pedal is positioned so that there are small flexies of the knee (between 20° and 40°). Also, the authors [14] claim that changes in knee angles significantly affect the value of the maximum force on the pedals. Another group of authors [15] showed that the maximum pedal force can be achieved at the stretched leg position of the observed subject (for the angles of the knee between 35° and 45°). Psychological parameters and characteristics of the passengers in a vehicle can affect the force of effort. A large sample of female and male drivers took part in the measurement of pedal force [16]. It was concluded that women drivers are weaker than men drivers and achieve less force on pedals. The aim of this paper is to determine the force on the clutch pedal using digital models by observing certain boundary conditions, in the case of male and female digital models.

3. Methods

Digital human models have been used for many years in the automotive industry to test how the car's design and the position of the various components meet drivers` needs. In this work the first task was to create a working environment of the mannequin in the vehicle. The Catia v5 R18 software package and the Part Design module were used for seat and interior modeling. The interior of a small city car was used (Volkswagen UP!). A small car was chosen due to the growing crowds in the cities and city driving regimes required frequent pedal activation. The next task was to place the human model in the seat and set certain boundary conditions (figure 2) using Ramsis software. This software has the option to work with two types of models - kinematic and geometric. The kinematic model takes into account the human skeleton and is wire frame model, while the second represents surfaces. In the paper, the geometric model of the mannequin with five fingers, and foot legs with shoes is used.

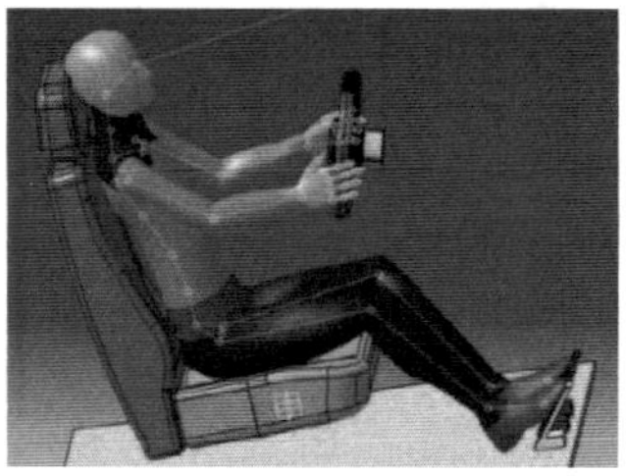

Figure 2. Mannequin in the passenger compartment after setting boundary conditions.

This study included the analysis of activation force on the pedal of nineteen different populations - male and female which anthropometric data are presented in tables 1 and 2 [17]. These populations belong to the age group of 18-70 years old, for the 1% and 99% male and female population.

Table 1. The anthropometric data of male and female 1% population [17].

Population	Height (mm)		Sitting height (mm)		Foot length (mm)	
	male	female	male	female	male	female
Japan	1585	1476	813	760	224	204
France	1607	1481	853	797	237	209
North Africa	1527	1454	807	770	237	217
South America	1608	1478	846	783	227	205
North America	1627	1494	860	803	239	217
West Africa	1507	1402	734	720	234	206
Spain and Portugal	1533	1465	804	780	228	198
North India	1535	1412	800	750	222	199
Eastern Europe	1615	1502	840	814	237	217
North Europe	1668	1541	880	823	232	217
Australia	1607	1521	860	810	244	212
South East Europe	1595	1485	830	790	237	212
Central Europe	1575	1518	870	803	232	212
South East Africa	1545	1442	790	750	232	202
Middle East	1582	1496	813	780	232	214
South India	1485	1351	743	723	217	194
North India	1506	1469	814	780	218	199
South China	1590	1406	770	720	224	204
South East Asia	1495	1402	763	730	214	201

Table 2. The anthropometric data of male and female 99% population [17].

Population	Height (mm)		Sitting height (mm)		Foot length (mm)	
	male	female	male	female	male	female
Japan	1855	1704	1027	960	266	246
France	1933	1779	1007	923	293	261
North Africa	1853	1766	933	910	293	273
South America	1892	1762	1014	937	293	275
North America	1953	1806	1000	957	291	273
West Africa	1833	1658	906	860	286	244
Spain and Portugal	1710	1735	890	920	270	292
North India	1805	1668	940	890	278	241
Eastern Europe	1885	1758	980	926	293	273
North Europe	1952	1839	1020	977	288	283
Australia	1933	1819	1000	950	286	268
South East Europe	1865	1755	970	930	293	268
Central Europe	1845	1802	1010	957	298	268
South East Africa	1815	1698	930	890	288	258
Middle East	1838	1724	967	920	288	266
South India	1755	1649	897	877	273	236

North India	1874	1711	986	920	285	251
South China	1730	1634	910	860	266	246
South East Asia	1765	1658	917	870	256	239

For the better development of an efficient braking system, as well as the clutch mechanism, a good foot pedal design is needed. The activation force on the pedal can be active and passive. The active forces is realized only by muscle activity, while the passive forces is generated by muscle activity and additionally support forces which are generated at the start of the joint chain to the environment [2]. Foot ankle joint load represents torque loads of the individual joints in the chain; in this case joint of the foot ankle. The recommended values for activating pedal brakes are in the range of 25 to 400 N [19]. These are the values recommended in 2006 issue. A change in the anthropometric dimensions of the population affects pedal activation forces. The goal of the car maker is to make vehicles accessible to wider a population. The aim of this work was to investigate proper range values of pedal activation force and foot ankle joint load for different populations in order to design pedal mechanisim.

4. Results

The RAMSIS software is able to predict the posture of a digital human model. This is based on statistical analysis of results obtained from the conducted experiments, [18]. By collecting all anthropometric data of the virtual driver population, digital human models were generated with appropriate boundary conditions. By using this software and ECE/TRANS/180/Add.3 regulations [19], the analysis of the interaction driver foot-foot pedal was conducted, in order to determine the activation pedal force. Partial results of performed analysis are given in figure 3, a) analyses conducted for the male driver model of the West Africa 1% population, and b) analyses conducted on the Spain and Portuguese 99% population.

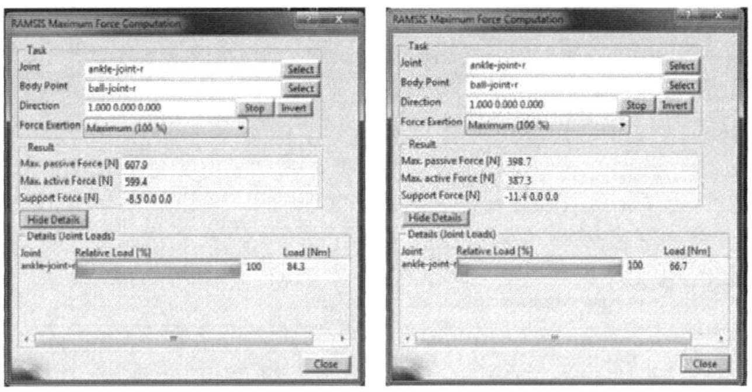

a) b)

Figure 3. Maximum force of the computation for male driver model; a) West Africa 1% population, b) Spain and Portuguese 99% population.

Table 3 Descriptive statistics of results obtained for male population.

	Height [mm]		Sitting height (mm)		Foot length (mm)		Max. passive force [N]		Max. active force [N]		Torque load [Nm]	
	1%	**99%**	**1%**	**99%**	**1%**	**99%**	**1%**	**99%**	**1%**	**99%**	**1%**	**99%**
Max	1668	1953	880	1027	224	298	607.9	398.7	599.4	387.3	84.3	66.7
Min	1490	1710	734	890	214	256	322.3	147.7	297.4	136.3	38.6	24.4
Mean	1568.2	1846.8	815.2	963.3	229.8	283.3	473.3	289.1	461.7	274.6	62.1	47.1
St. deviation	50.3	71.6	42.21	45.9	8.24	11.7	91.28	75.3	94.82	78.3	14.72	12.1

Table 4 Descriptive statistics of results obtained for female population.

	Height [mm]		Sitting height (mm)		Foot length (mm)		Max. passive force [N]		Max. active force [N]		Torque load [Nm]	
	1%	**99%**	**1%**	**99%**	**1%**	**99%**	**1%**	**99%**	**1%**	**99%**	**1%**	**99%**
Max	1541	1839	823	977	217	292	244.1	262	231.5	259.9	28.2	41.7
Min	1351	1634	720	860	194	236	150.8	119.6	141.4	114	19.4	18.7
Mean	1402.8	1732.8	772.9	917.5	207.3	261.1	190.6	191.7	182.1	183.6	23.4	28.8
St. deviation	49.2	61.8	33.1	35.1	7.4	15.9	19.3	42.2	18.7	42	2.1	6.4

The result of performed analysis showed that maximum passive force for nineteen male 1% populations (figure 3a) is 607.9 N (West Africa population), while the maximum active force is 599.4 N for maximum effort of 100 %. Foot ankle joint load of nineteen 1% male populations is 84.3 Nm. The same analysis of 99% male population is shown in figure 3b (Spain and Portuguese population). The maximum passive force is 398.7 N, while the maximum active force is 387.3 N for maximum effort of 100 % (table 3). Differences are primarily derived from different anthropometric dimensions.

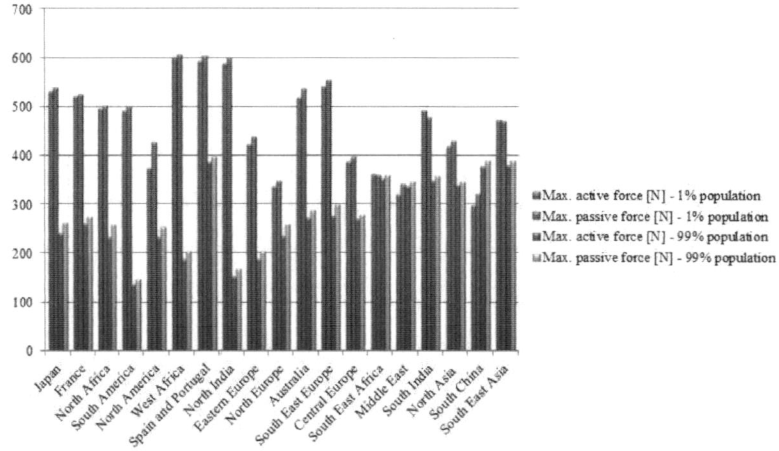

Figure 4. Comparison of the results obtained for male 1% and 99% populations.

The figure 4 showed that the highest values of active and passive forces on the pedal were recorded by the 1% male population from West Africa, Spain and Portugal and North India. The results for 99% male population showed that the highest values of active and passive forces on the pedal were recorded by the 1% male population from Spain and Portugal, South China and South East Asia.

Analysis of these nineteen populations shows that with an increase in the height of the male driver decreases maximum active force that he can achieve on the pedals. Dependency between maximum active pedal force of the male drivers and height was showed of figure 5.

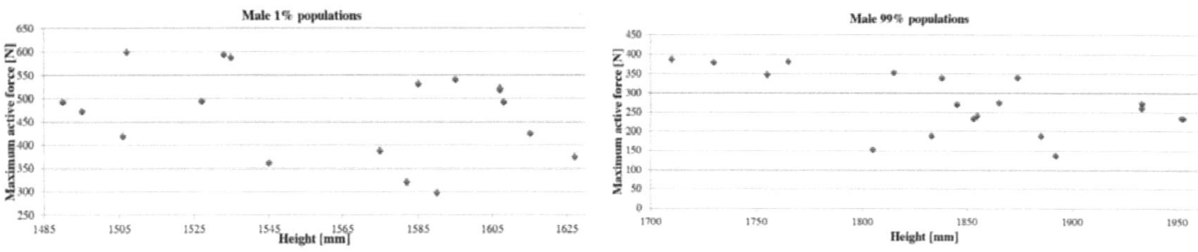

a) b)

Figure 5. Maximum active pedal force of the male drivers vs. height.

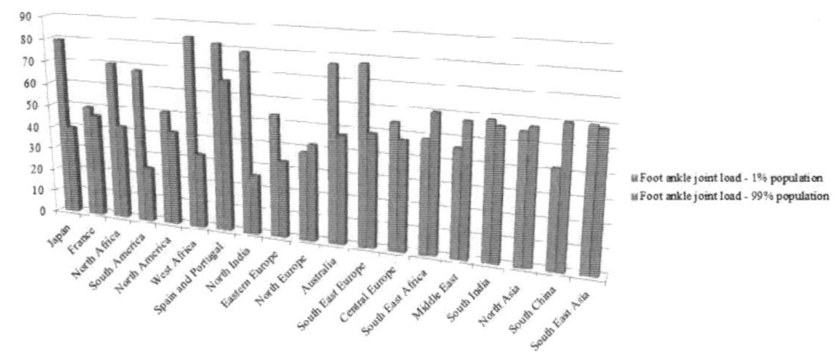

Figure 6. The foot ankle joint load analysis of nineteen male populations.

Figure 7 showed the results of analysis of nineteen female populations, for both cases: 1% and 99% populations. The seat is moved forward so that those feel more comfortable. The result of performed analysis showed that maximum passive force for nineteen 1% female populations is 244.1 N, while the maximum active force is 231.5 N for maximum effort of 100%. Foot ankle joint load of nineteen 1% female populations is 28.2 Nm. These values refer to the female population of South India. An analysis of all 99% female populations is conducted in the same way. The maximum passive force was obtained for the Japanese population and is 262 N, while the maximum active force is 259.9 N for maximum effort of 100%. Foot ankle joint load of nineteen 99% female populations is 41.7 Nm.

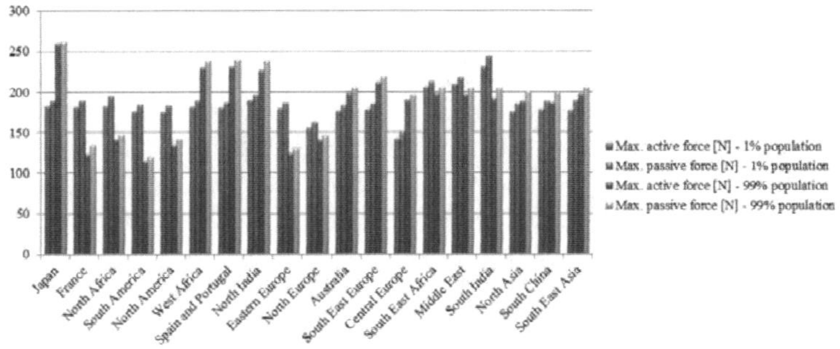

Figure 7. Comparison of the results obtained for female 1% and 99% populations.

This analysis showed that for a certain number of populations, higher force values are obtained in the case of 99% female population, which was not the case with male models. It can be concluded that there are considerably less differences in the results obtained between 1% and 99% of female populations, in regard to the same results of male drivers.

Dependence between height and active pedal force for female driver is showed on figure 8.

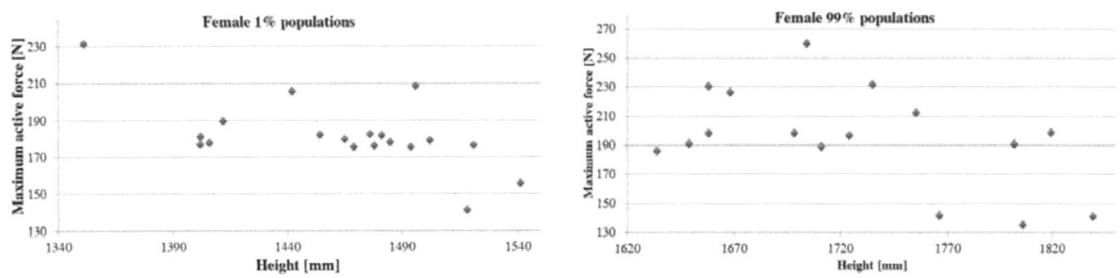

a) b)

Figure 8. Maximum active pedal force of the female drivers that can be achieved on the pedal vs. height.

Figure 9 showed the results of foot ankle joint load analysis of nineteen female populations. The highest values of foot ankle joint load are obtained for 1% female population of South East Africa and South India and are 38.1 Nm and 28.2 Nm, respectively. In the case of 99% female population, the highest recorded values of this ankle joint load is obtained for female populations of Japan (41.7 Nm), Spain and Portugal (39.5 Nm) and South East Europe (33.7 Nm).

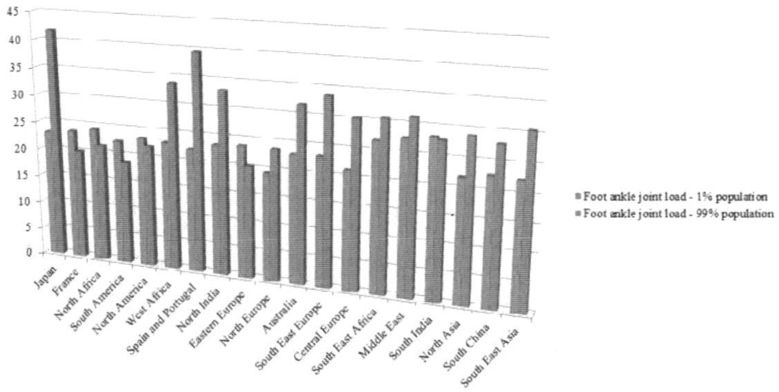

Figure 9. The foot ankle joint load analysis of nineteen female populations.

Poor negative correlation between anthropometric data of analyzed human population between body height and foot pedal forces and ankle joint load is obtained. In the case of the 1% male population negative correlation coefficient between body height and foot pedal forces is -0.384, and correlation between height and ankle joint load is -0.371. Foot length did not correlate with foot pedal forces. Female 1% population has negative correlation coefficient between body height and foot pedal forces -0.607 and correlation coefficient between body height and ankle joint load is -0.440. Analysis of 99% male population shows negative correlation between height and foot pedal forces -0.574 and between height and ankle joint load -0.507. Results obtained for 99% female population, compared to the previous 1% female population, have smaller negative correlation coefficient between body height and foot pedal forces and ankle joint load (-0.522 ÷ -0.387). Foot length did not correlate with ankle joint load, for both populations.

5. Conclusion

The aim of this study was to determine the force required to activate foot pedal in a driving position. The anthropometric dimensions of nineteen 1% and 99% populations - male and female, were used. Testing was performed by application of the software Ramsis. Validation of vehicle interior packaging can be performed, before realization of the first prototype by application of Ramsis. This paper has shown that digital human model, both male and female, in a driving position, can be used to determine the pedal activation force. The increase of the height of the driver male and female decrease a maximum value of an active pedal force as well as passive pedal force. In the case of 1% male population, the maximal active pedal force varied from 297.4 N to 599.4 N, while the maximum active force varied from 141.4 N to 231.5 N for 1% female population. The maximum active pedal force varied from 136.3 N to 387.3 N when male are drivers, while in the case of women drivers, the maximum active force varied from 114 N to 259.9 N. Based on this, it can be concluded that there are considerably less differences in the results obtained between 1% and 99% of female populations, in regard to the same results of male drivers.

Comparison of the results showed that there are less difference between 1% and 99% female population than results obtained for 1% and 99% male population. The significance of this research during the development and production of vehicles can save a money and time, and also lead to the final product driver friendly designed. The influence of position and indication angle of pedal activation force will be analysed in further research.

Acknowledgment
The research is supported by the Ministry of Education, Science and Technological Development of Republic of Serbia through Grant TR 35041.

References

[1] Bridger R S 2009 *Introduction to Ergonomics* Boca Raton: Taylor & Francis Group, LLC
[2] http://www.human-solutions.com/mobility/front_content.php?idcat=252, Human solutions 2010 RAMSIS model applications accessed 07.04.2017
[3] http://www.intrinsys.com/software/ramsis?src=applied accessed 02.07.2017
[4] https://www.netcarshow.com/volkswagen/2017-up/ accessed 02.07.2017
[5] Duffy V G 2008 *Handbook of Digital Human Modeling: Research for Applied Ergonomics and Human Factors Engineering* CRC Press Inc. Boca Raton FL USA
[6] McAtamney L and Corlett E N 1993 Rula: a survey method for the investigation of workrelated upper limb disorders Applied Ergonomics 24 (2) pp 91–99
[7] Hignett S and McAtamney L 2000 Rapid entire body assessment (REBA) Applied Ergonomics 31 (2) pp 201–205
[8] Vink P and Hallbeck S 2012 Editorial: Comfort and discomfort studies demonstrate the need for a new model Applied Ergonomics 43 (2) pp 271–276
[9] Zhang L Helander M and Drury C 1996 Identifying factors of comfort and discomfort Human Factors 38 pp 377–389
[10] De Looze M P Kuijt-Evers L F M and Van Dieën J H 2003 Sitting comfort and discomfort and the relationships with objective measures Ergonomics 46 pp 985–997
[11] Haslegrave C M 2004 Force exertion. In: Dellman N J Haslegrave C M and Chaffin, D B Working postures and movements: tools for evaluation and engineering CRC Press London pp 367-402
[12] Kumar S 2004 Muscle strength Boca Raton CRC Press
[13] Kroemer K H E 1971 Foot operation of controls Ergonomics 14(3) pp 333–361
[14] Lara-Lopez A Aguilera-Cortes L and Barbosa-Castillo F 1999 Measurement of forces applied to handgrips and pedals for a sample population of Mexican males Applied Ergonomics 30 pp 173-176
[15] Mehta C R Tiwari P S Rokade S Pandey M M Pharade S C Gite L G and Yadav S B 2007 Leg strength of Indian operators in the operation of tractor pedals International journal of industrial Ergonomics 37 pp 283-289
[16] Mortimer R G Segel L Dugoff H Campbell J D Jorgeson C M and Murphy R W 1974 Brake force requirement study: driver-vehicle braking performance as a function of brake system design variables National Highway Safety Bureau U.S Department of transportation
[17] http://dined.io.tudelft.nl/en accessed 01.05.2017
[18] Heiner Bubb H Bengler K Grünen R and Vollrath M 2015 Automobilergonomie, Springer Vieweg
[19] https://www.unece.org/trans/main/wp29/wp29wgs/wp29gen/wp29glob_registry.html, ECE/TRANS/180/Add.3: Agreement concerning the establishing of global technical regulations for wheeled vehicles, equipment and parts which can be fitted and/or be used on wheeled vehicles, 21 December 2006, International Organization for Standardization, Geneva, Switzerland, accessed 07.06.2017

IOP Conf. Series: Materials Science and Engineering **252** (2017) 012011 doi:10.1088/1757-899X/252/1/012011

Evaluation of the tire pressure influence on the lateral forces that occur between tire and road

N Cordoş[1], A Todoruţ[1], I Barabás[1]

[1] Department of Automotive Engineering and Transports, Technical University of Cluj-Napoca, Faculty of Mechanical Engineering, Cluj-Napoca, Romania

E-mail: Nicolae.Cordos@auto.utcluj.ro

Abstract. The main objective of the paper is to capture the behavior of a vehicle on a race circuit, depending on the different inflation pressures of the tires (underpressure and overpressure). Taking into account that in the cornering the forces and the moments of inertia rise due the vehicle mass, and an inertial force decomposes in two components, one in the longitudinal plane and one in the transverse plane, the work aims the assessing the lateral forces that appear, to the contact between the tires and the tread, depending on the inflation pressure. The results have a graphic interpretation, enabling a comparative study of them. Results have been obtained regarding the lateral tire forces that occur between the tire and the road. The differences between these forces were particularly noticeable in cornering, and the differences between these forces were interpreted according to the tire inflation pressure.

1.Introduction

The wheels and the tires are the connecting elements of the vehicle with the tread. Wheel components allow consideration of several variables that influence its effect on the vehicle running state [1]. The role of the tire is to take up the vertical loads and to generate, along with the track, the tangential forces to accelerate or decelerate the vehicle and the side forces to guide it. Overpressure or underpressure in the tire reduces its ability to have adequate grip. Not only will it contribute to excessive and asymmetrical wear on the tire, but it will also have a substantial influence on the tire, which can create hazards especially at high-speed corners.

The values of the tire pressure and of the loads on the wheel are parameters which have an impact on the contact area between tire and the road [2,3]. Depending on the rim size, the tire size and on the load on the wheel, the tire manufacturer offers recommended pressures for the vehicle for the front or rear wheels [4]. The lower is the tire pressure, the greater is the contact surface between the tire and the road, and thus the rolling resistance is greater. This has a negative effect on the tire and may occur the excessive tire heating and the premature wear [5]. The high pressure in the tire causes greater wear of the central section of the tire and under the extreme conditions may lead to the delamination of the tire [6].

The lateral tire forces that occur between the tires and the road play an important role in the vehicles dynamics, determining the slip angle of the wheel and therefore the vehicle's skidding angle and the vehicle yaw rate.

The lateral tire force is the force required to keep the vehicle on the cornering trajectory. This force is generated by the deformation of the tire which is in contact with the road surface. The lateral tire force is generally shown according to the sideslip angle [7].

Tire pressure can affect the tire's characteristics in a variety of ways. Being that the stiffness coefficients are essentially a measure of the elasticity of the tire, an increase in tire pressure will increase the stiffness of the tire [8]. This will in turn increase the lateral force. If an inflation pressure 70% above the design value is used, a 20% gain in tire stiffness can be obtained. Increasing pressure above this will then decrease stiffness [9]. Due to inflation pressure carcass stiffness is increases but simultaneously reduces the contact length of the tire it influences on cornering stiffness of tire [10].

Various vehicle dynamic simulation software programs have been developed. The paper [11] underlines a number of advantages for simulation software programs: reduce costs to build a model able to change the structure of the model at any stage and adapt the characteristics of the model requirements imposed in a short time. In [12], the estimated tire forces were close to the simulation values during all the process. The simulation and results analysis showed that the proposed tire model can estimate both the longitudinal and lateral tire forces under different driving conditions, which demonstrated its validity in vehicle dynamics analysis. The vehicle dynamics analyzing model was established by using Matlab/Simulink and CarSim platform.

The paper aims to evaluate the lateral tire forces that occur at the contact between the tire and the road in the case of underpressure, overpressure and the optimal inflation pressure of the tires for a vehicle running on a race circuit, by computer simulation using the software *IPG CarMaker.*

2. Material and methods

To analyze the effect of the inflation tire on the vehicle transversal stability, the simulations are performed using the numerical simulation tool *IPG CarMaker*. To run the simulation in *IPG CarMaker* software, the road must be loaded, with the desired manoeuvres [13]. The main graphical interface of the software, it can be seen in Figure 1.

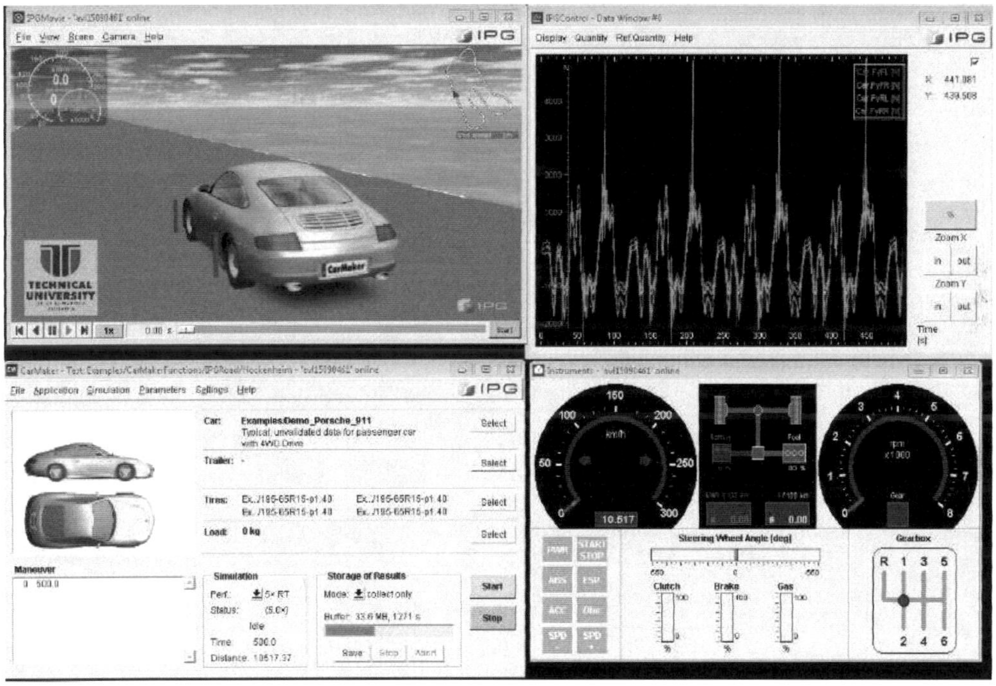

Figure 1. The graphical interface of the *IPG CarMaker* software

Within the framework of the presented paper, the simulations are carried out with the Porsche 911 Turbo vehicle. The main parameters of the vehicle are given in Table 1.

Table 1. The main parameters of the vehicle

Vehicle type	Porsche 911 Turbo
Unladen weight (EC)	1470 kg
Axle load front/rear	743/727 kg
Engine	3800 cm^3; (397 kW; 660 Nm at 1950 – 5000 rpm)
Tyre type	Continental
Tire dimension	195/45 R16 V
Length	4507 mm
Width	1880 mm
Height	1297 mm
Wheelbase	2450 mm

In *IPG CarMaker* software, three main axis systems are used [14] - see Figure 2:
- Frame Fr0: CarMaker inertial axis system – fixed origin, (0.X.Y) is the horizontal driving plane, and (Z) directed upwards.
- Frame Fr1: axis system of the vehicle where which respect the ISO sign convention where (X) points the forward driving direction, (Y) the left, and (Z) directed upwards.
- Frame Fr2: carrier axis system where (O) is the centre of the wheels, (X) the forward driving direction, (Y) is along the wheel spin axis and (Z) directed upwards. For every wheel there is a mountpoint (Mnt) defined within the Fr1 system

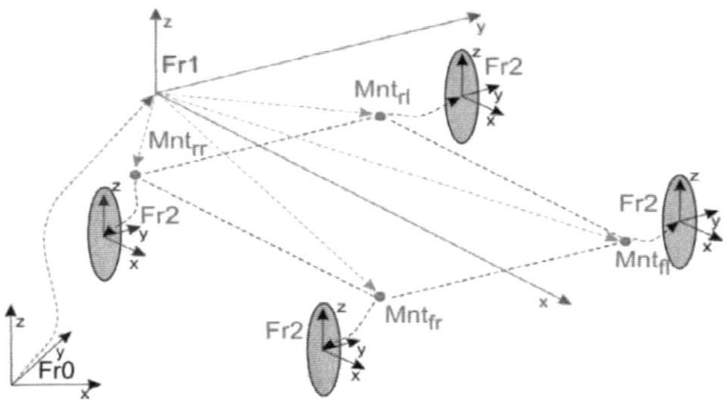

Figure 2. The main axis systems in *IPG CarMaker* software

Lateral force was analyzed on a short race circuit Hockenheimring (Germany) - 2.63 km (see Figure 3) for three different tire pressures: 1.4 bar, 2.2 bar (nominal pressure) and 2.9 bar.

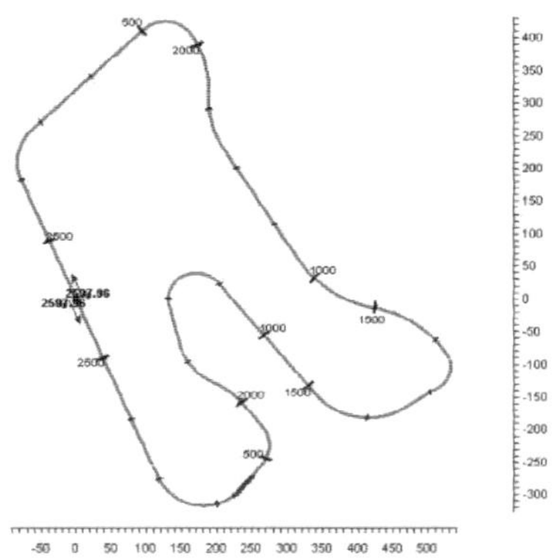

Figure 3. The short race circuit Hockenheimring

The example of choosing the tire inflation pressure in the *IPG Camaker* software it can be seen in the Figure 4. The condition and the nature of the road was dry asphalt with the friction coefficient of 0.8.

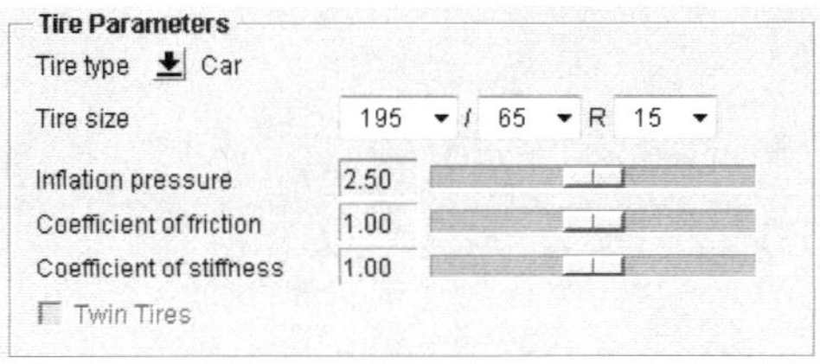

Figure 4. The graphical interface for selecting the tire pressure

3.Results

The dependence between the lateral tire force and the lateral deflection angle that occurr between the tire and the road is of fundamental importance for the directional control and for the vehicles stability. Inflation pressure affects the tire's flexibility and the rolling resistance in various ways. Generally, the rolling resistance decreases as the inflation tire pressure increases. This is because, with increasing the tire pressures, the deformation of the tire decreases, respectively decreases the contact patch between the tire and the road. When the tire is in cornering at a certain angle, the contact patch deforms resulting a lateral force. The slip angle increases by increasing the lateral force. Figure 5 shows the evolution of the lateral tire force for all four wheels of the vehicle and for the three inflation pressures with which the simulations were made (FL-front left wheel, FR- front right wheel, RL- rear left wheel, RR rear right wheel). The lateral tire forces were captured for one-lap race. We can see a tendency to increase the lateral tire forces that appear on all the wheels for 1.4 bar inflation pressures. The lateral tire forces at 2.2 bar inflation pressure and respectively for 2.9 bar have values close to each other.

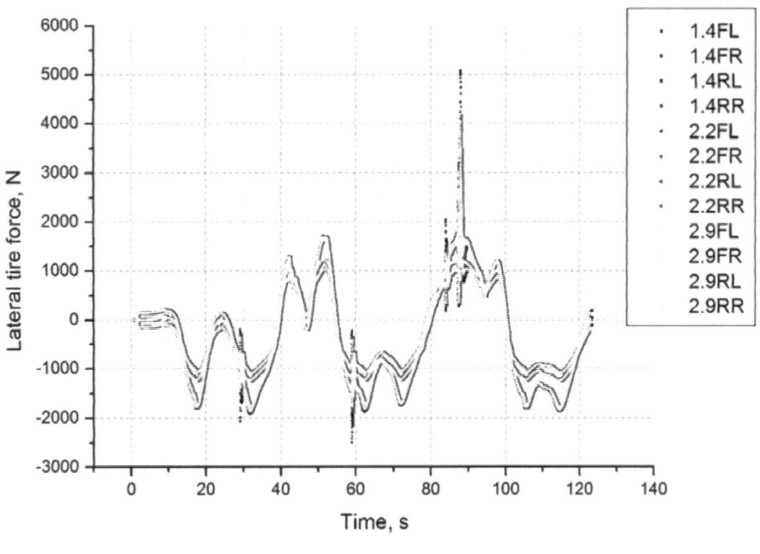

Figure 5. The evolution of the lateral tire force for all wheels at different tire inflation pressure

Figure 6 shows the lateral forces on the right front wheel of the vehicle for the three inflation pressures of the tires. The lateral tire force is greater for the inflation pressures of 1.4 bar up to 1000 N, reaching in tight corners to a maximum value of approximately 5000 N. This is due to the low tire inflation pressure, the contact patch between the tire and the road is greater than the tire with the nominal inflation pressure, or in the case of overpressure.

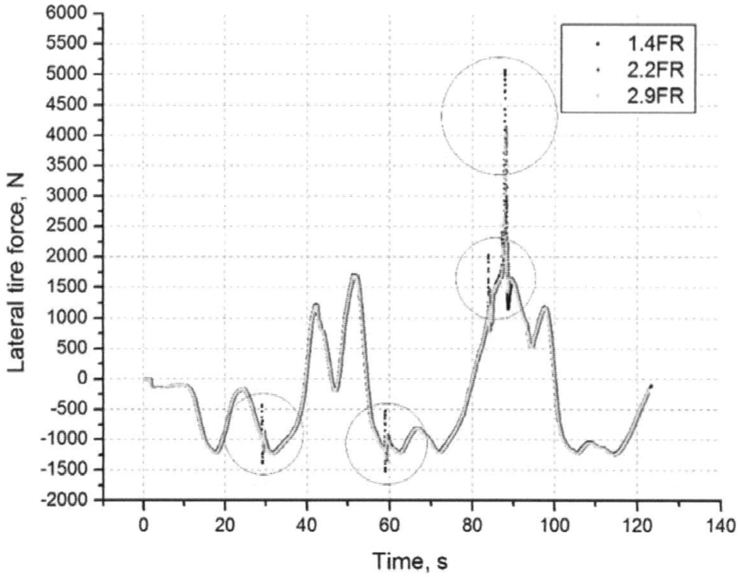

Figure 6. The lateral tire force for the right front wheel

The slip angle increases by increasing the lateral tire force. The maximum value of the lateral tire forces in the case of 2.2 bar respectively 2.9 bar have very close values reaching the maximum value of approximately 4000 N.

4.Conclusions

As a result of the computerized simulations, it can be noted the influence of the tire inflation pressure on the lateral tire forces that appear at the wheels of a car running on a race circuit. A vehicle with low inflation pressure of the tire will tend to lose the grip and will appear less receptive in cornering. The tire develops lateral forces due to the side slip angle between the tire and the road. The side slip angle of the tire increases at low tire pressures. The decreasing of the tire inflation pressure will increase the load on the tire, thus increasing the deformation of the tire and the tire solicitations, which will lead to the increasing of the losses energy, to the increasing of the rolling resistance, respectively to the increasing of lateral forces at the contact between the tire and the road.
The results of the simulations can be developed by following the effects of the tire inflation pressure on the forces at the wheels that occur on different natures and states of the road and the effects of the air temperature on the tires on these forces.

Acknowledgment

The simulations presented in the paper were done using the software AVL InMotion and IPG CarMaker supported by AVL List GmbH, Austria.

References

[1] Varga B O, Mariaşiu F, Moldovanu D and Iclodean C 2015 *Electric and Plug-In Hybrid Vehicles – Advanced Simulation Methodologies* © Springer International Publishing Switzerland ISBN: 978-3-319-18638-2 524
[2] Guan D, Shang Jin, Yam L H 2002 *Study on Tire Dynamic Cornering Properties using Experimental Modal Parameters*, Vehicle SystemDynamics, 37 129-144.

[3] Waluś K J, Wolniewicz R M, Olszewski Z, Drożyński M 2009 *Analysis of pressure and load influence on geometric parameters of tire understatic conditions*, 17th French-Polish Seminar of Mechanics, Lille, France, pp. 31-36
[4] Polasik J, Waluś K J, Warguła Ł 2017 *Experimental Studies of the Size Contact Area of a Summer Tire as a Function of Pressure and the Load*, Procedia Engineering, Volume 177, Pages 347-351, ISSN 1877-7058, https://doi.org/10.1016/j.proeng.2017.02.203
[5] Yam L H, Guan D H, Shang J, Zhang A Q 2000 *Study on tire rolling resistance using experimental modal analysis*, International Journal ofVehicle Design, 30, 251 - 262
[6] Adetan D A, Oledajo K A, Fasogbon S K, 2008 *Redesigning the manual automobile tyre bead breaker*, Technology in Society, 30, 184-193
[7] Doumiati M, Victorino A, Charara A and Lechner D 2010 *A method to estimate the lateral tire force and the sideslip angle of a vehicle: Experimental validation*, Proceedings of the 2010 American Control Conference, Baltimore, pp. 6936-6942.doi: 10.1109/ACC.2010.5531319
[8] Smith N D 2004 *Understanding parameters influencing tire modeling*. Department of Mechanical Engineering, Colorado State University.
[9] Dixon J C 1996 Tires, Suspension and Handling 2nd ed. Warrendale: Society of Automotive Engineers
[10] Virkar D S, Thombare D G 2013 *Parametric Study And Experimental Evaluation Of Vehicle Tire Performance.*
[11] Varga B O, Burnete N, Iclodean C 2016 *Validation Procedure for Worldwide Harmonized Light Vehicles Test Cycle via Hardware in the Loop - Real Time Testing*, CONAT 2016 International Congress of Automotive and Transport Engineering, pp. 128-135, Springer International Publishing, ISBN: 978-3-319-45447-4; http://link.springer.com/book/10.1007%2F978-3-319-45447-4
[12] Bian M, Chen L, Luo Y, and Li K 2014 *A Dynamic Model for Tire/Road Friction Estimation under Combined Longitudinal/Lateral Slip Situation*," SAE Technical Paper 2014-01-0123, doi:10.4271/2014-01-0123
[13] Varga B O and Iclodean C 2015 *Electric buses for urban transportation: assessments on cost, infrastructure and exploitation* Fascicle of Management and Technological Engineering 1 pp 253-258
[14] IPG Automotive GmbH, "CarMaker Programmer's Guide", Version 2.1.11.

Analysis of car's frontal collision against pole

N Ispas[1*] and M Nastasoiu[2]

[1,2]Automotive and Transportation Department, Transilvania University of Brasov, Romania

*inicu@unitbv.ro

Abstract. Reducing the effects of traffic accidents over the occupants is a major objective of collision attempts. Impacts between the car and the pole are very dangerous for the physical integrity of the car's occupants. To minimalize the effects of such events on the passengers of a vehicle, a whole series of efforts by both designers and experienced engineers led to increasingly the vehicles safety. The main aim of these paper is to quantify the influences over the car passengers of loads involved by car against pole collisions using the same car model at different speeds. Also, this kind on occupant influences were study using a small car model. Other goal of the paper was the study of the cars stiffness in frontal collision against the pole. The paper's experimental results were obtained by support of DSD, Dr. Steffan Datentechnik GmbH - Linz, Austria. The described tests were performed in full test facility of DSD Linz, in "Easter 2016 PC-Crash Seminar". Cars accelerations, velocities, rotations angles after pole impact were registered and discussed. The novelty of the paper consists in studies referred for the same car model involved in car against pole collisions at different impact speeds. Paper's conclusions can be future used for car safety improvement.

1. Introduction

Collisions between cars and poles are common both in the areas inside the localities and in the areas of roads outside the localities. Frequently, collisions of this type, from inhabited areas, take place between the car and the poles serving various public utilities (lighting, support for electric power network, telephony or cable TV) or on which traffic signs or traffic lights are fixed. Outside the localities, collisions of this type occur frequently with the trees on the roadside or with the poles of the electric or phone networks. For example, in the USA, utility pole collisions are the second most frequent type of fatal fixed-object crashes after trees. [1]. Moreover, almost 40 percent of all crashes involving utility poles involve some type of non-fatal injury. [1] Each year more than 1,000 deaths occur as a result of collisions with utility poles in the United States alone. [1]. The safety of the occupants of a car involved in such frontal collisions depends on the speed of impact, the configuration/dimensions of the pole/tree and on the safety solutions involved in the construction of the motor vehicle. RD (Research Development) in the field of occupant safety often involves crash tests that are strong tools for improve both occupants and road vehicle safety. For energy absorption developed in frontal collisions (including car to pole collisions) vehicle structure has been improved thanks to materials, new improved restraint systems and new structural solution for crash energy management, that provide a better understanding of the front energy absorption. Front design is at the cross road of numerous contradictory constraints: self-protection of occupants, protection of vulnerable users such as pedestrians, reparability, styling, aerodynamics, engine cooling and so on. [2. 3, 4]. For the goals of the paper experimental results were obtained by support of DSD, Dr. Steffan Datentechnik GmbH - Linz, Austria. The described tests were performed in full test facility of DSD Linz, in "Easter 2016 PC-Crash Seminar". The main technical data of the cars used in the crashes experiments together with the cars shapes are presented in the Table 1. In figure 1 on can see characteristics of the experimental pole used in all front collisions test. [5]

Table 1. Car to pole tests details.

Cars involved					
	Collision 1	Collision 2	Collision 3	Collision 4	Collision 5
Label	P01	P02	P03	P04	P05
Make	Fiat	Fiat	Fiat	Fiat	Simpa
Model	Punto 55	Punto 55	Punto 55	Punto 55	JDM Type 97
Year	1995	1997	1998	1997	1998
Velocity [km/h]	39.4	60	85.9	102.8	39.3
Weight [kg]	850	865	900	885	340
Length [m]	3.76	3.76	3.76	3.76	2.57
Width [m]	1.63	1.63	1.63	1.63	1.36
Height [m]	1.46	1.46	1.46	1.46	1.38
Wheelbase [m]	2.45	2.45	2.45	2.45	1.74
Trackwidth [m]	1.35	1.35	1.35	1.35	1.19
Front overhang [m]	0.78	0.78	0.78	0.78	0.44
Axle load front [kg]	530	535	555	545	
Axle load rear [kg]	320	330	345	340	
Dist. COG to front axle [m]	0.92	0.93	0.94	0.94	

Figure 1. Experimental pole construction used in all collisions tests.

2. Experimental research

4 crash tests (39.4, 60, 85.9 and 102.3 km/h) with practically the same car type (FIAT Punto 55) against a pole will be performed. Another extra collision test was conducted using a minicar (SIMPA JDM Type 97), which hit the same pole at a speed of 39.3 km/h. For the pole tests the DSD full scale facility accelerates the driving vehicles to the appropriate test velocity. Approximately 10 m in front of the collision point, the vehicle was disconnected from the guiding sled and the vehicle crashes against the pole. The ignition of the vehicles was turned on. Velocities, rotations and accelerations of the cars were

also measured. Figures 1-5 show the details of each collision and the amplitude of damage caused by the front collision with the pole. In 4 of the 5 cases, the main damage picture was seen from top and bottom view. The upper and lower views of the damage allow the maximum depth of penetration to be determined in each collision test. The DAQ5 PIC device was developed by DSD Austria, and has the role of recording the parameters of acceleration, speed, angular speed and position of the car. The device supports an external triaxial sensor, being able to record external acceleration data. This sensor was mounted in the centre of gravity (COG) of the car

A great advantage offered by this device is that it is not based on an external source of power, but it works with batteries, making it a mobile device that is very easy to use. Data is stored on an SD card, and can be easily transferred to a computer. The data recording mode can be set by the user so that the data can be recorded in a "user loop" pre-set time, or until user intervention (infinite time). The sampling rate is at the maximum of 10 kHz and is very useful for field-type collisions where a high frequency of data recording is required.

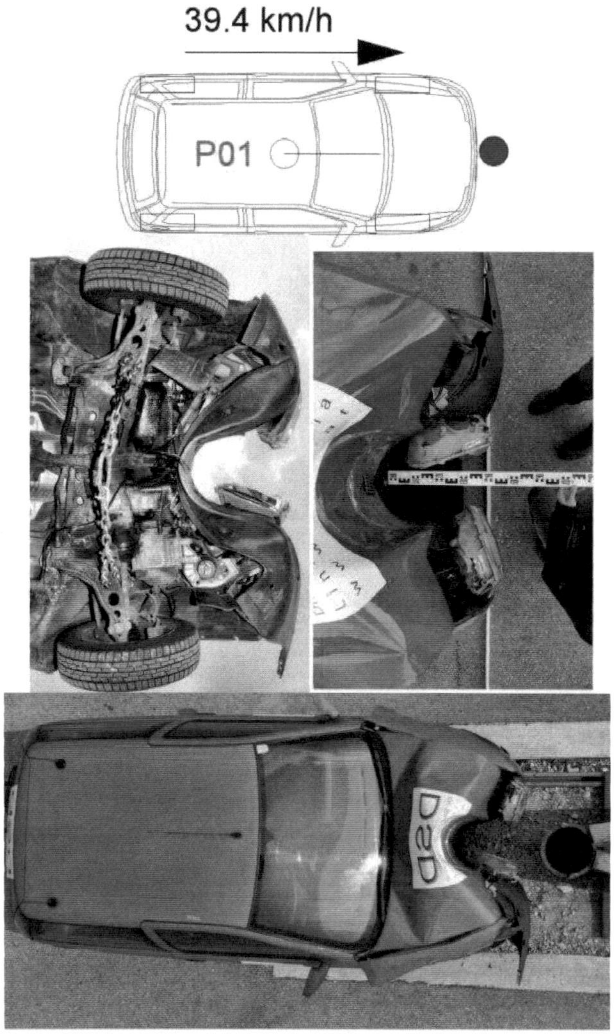

Figure 2. First car to pole collision and main damage details (top and bottom view).

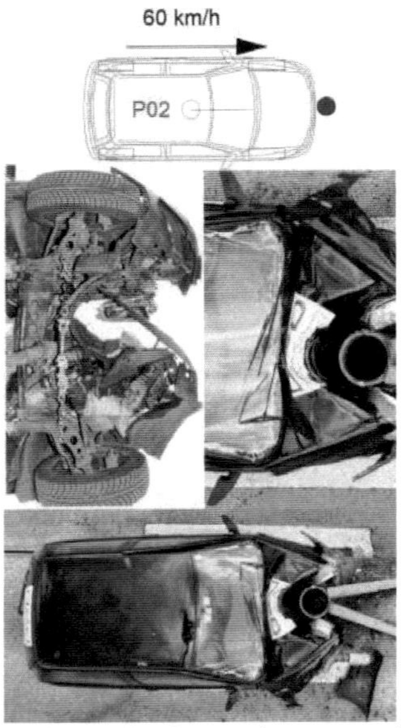

Figure 3. Second car to pole collision and main damage details (top and bottom view).

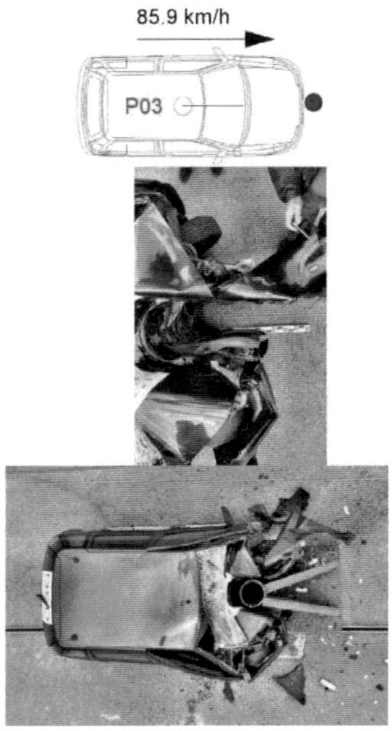

Figure 4. Third collision and main damage details (top view).

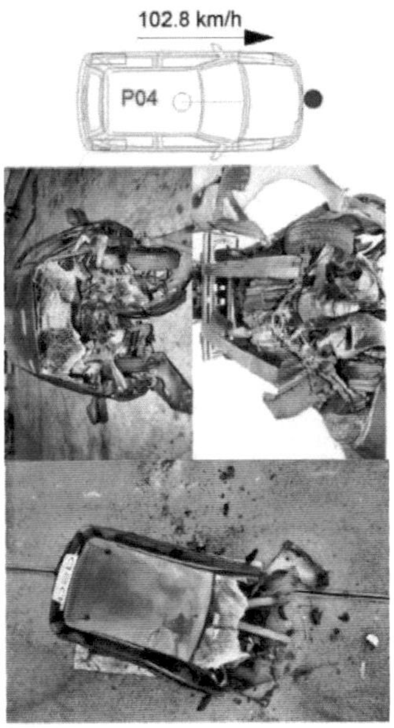

Figure 5. Fourth car to pole collision and main damage details (top and bottom view).

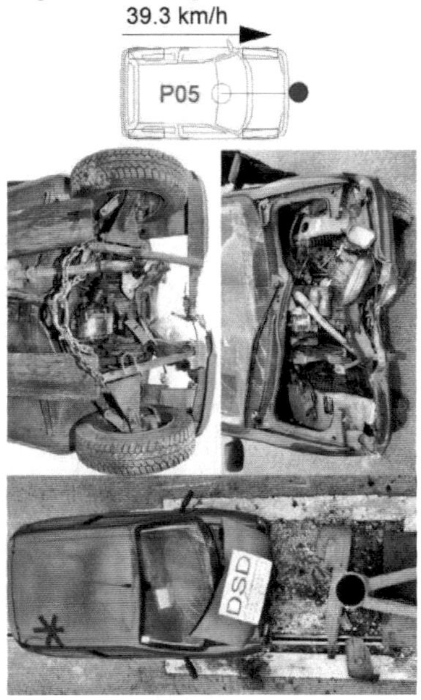

Figure 6. Fifth minicar to pole collision and main damage details (top and bottom view).

3. Results and Discussion

For each collision case velocities (figures 7, 10, 13, 16 and 19) rotations angles (figures 8, 11, 14, 17 and 20) and acceleration (figures 9, 12, 15, 18 and 21) are represented in the next pages.

IOP Conf. Series: Materials Science and Engineering **252** (2017) 012012 doi:10.1088/1757-899X/252/1/012012

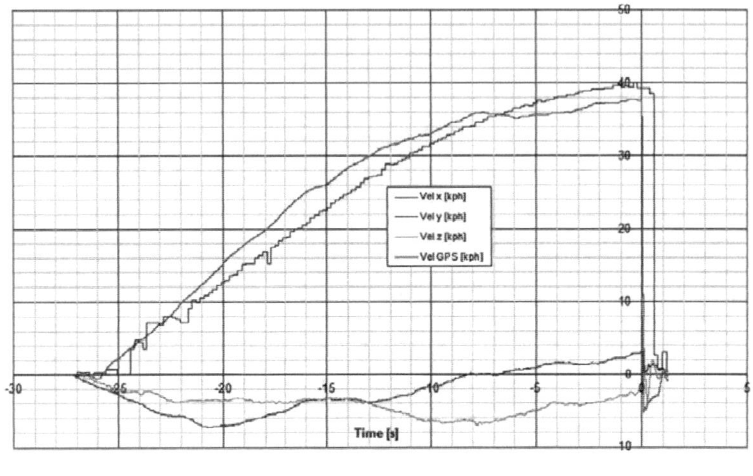

Figure 7. Car velocities as time functions in the first car to pole collision.

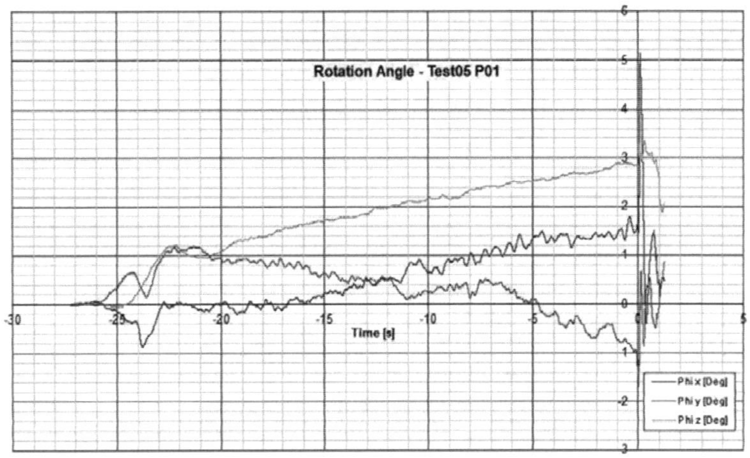

Figure 8. Car rotation angles as time functions in the first car to pole collision.

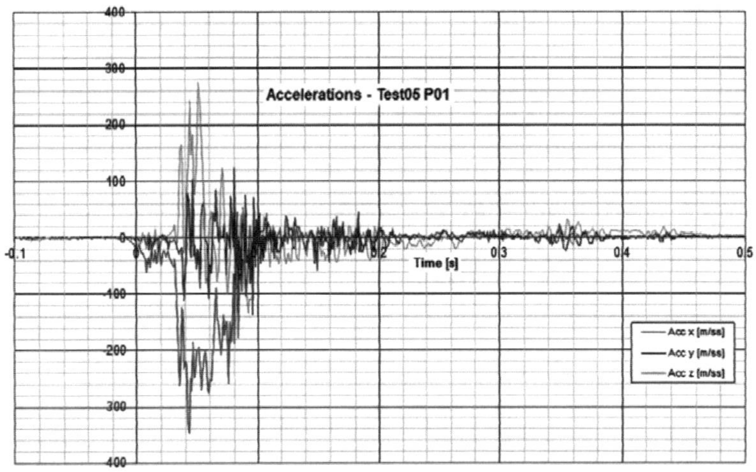

Figure 9. Car accelerations as time functions in the first car to pole collision.

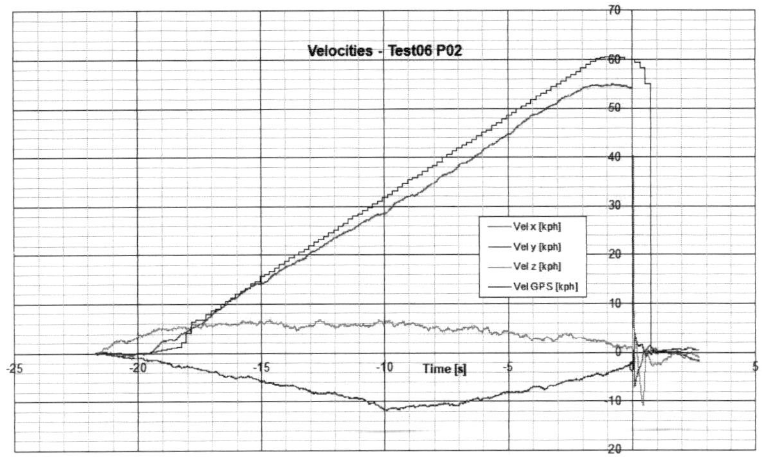

Figure 10. Car velocities as time functions in the second car to pole collision.

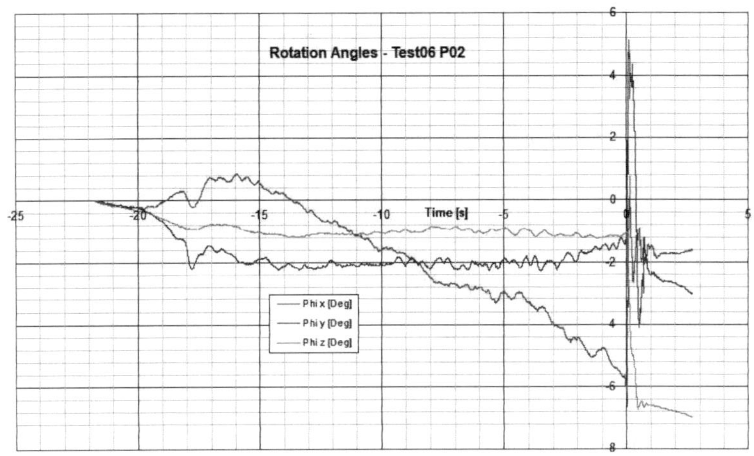

Figure 11. Car rotation angles as time functions in the second car to pole collision.

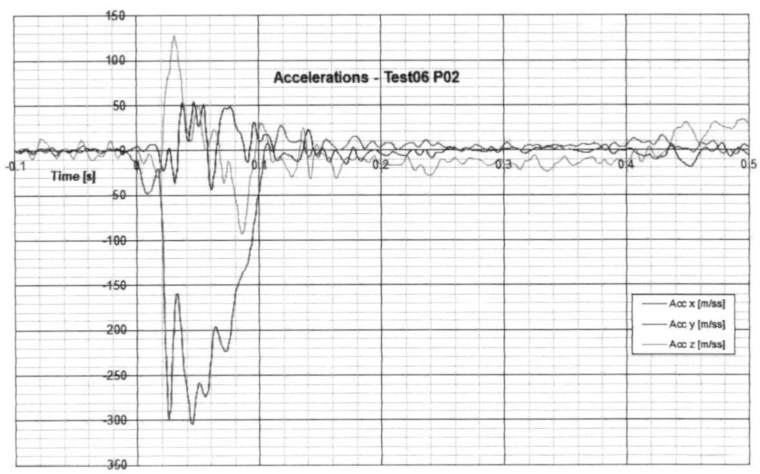

Figure 12. Car accelerations as time functions in the second car to pole collision.

IOP Conf. Series: Materials Science and Engineering **252** (2017) 012012 doi:10.1088/1757-899X/252/1/012012

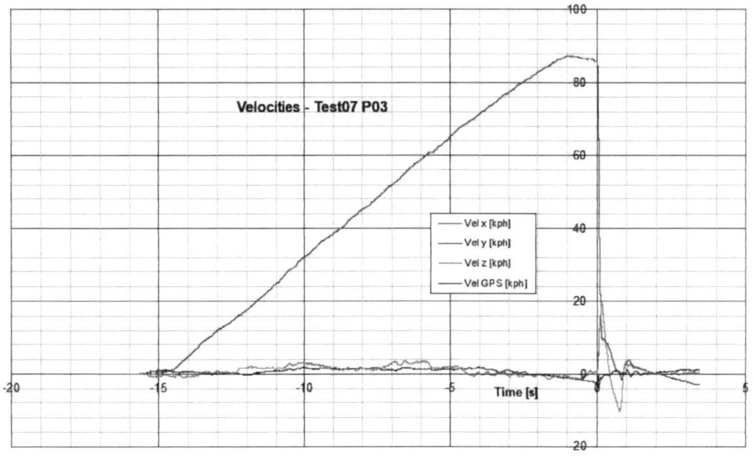

Figure 13. Car velocities as time functions in the third car to pole collision.

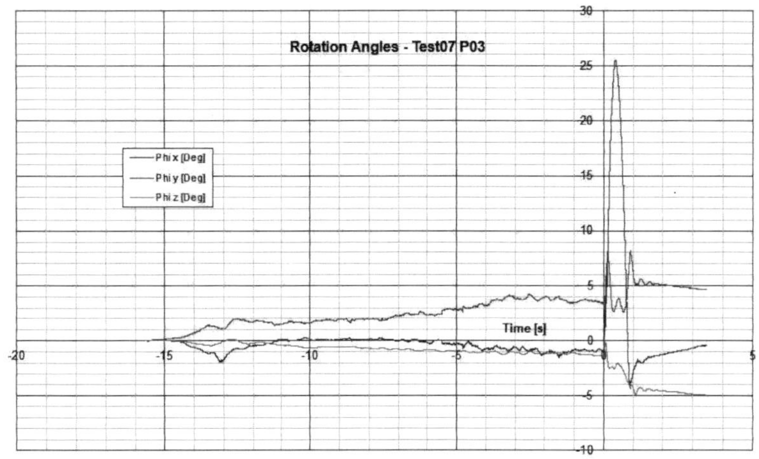

Figure 14. Car rotation angles as time functions in the third car to pole collision.

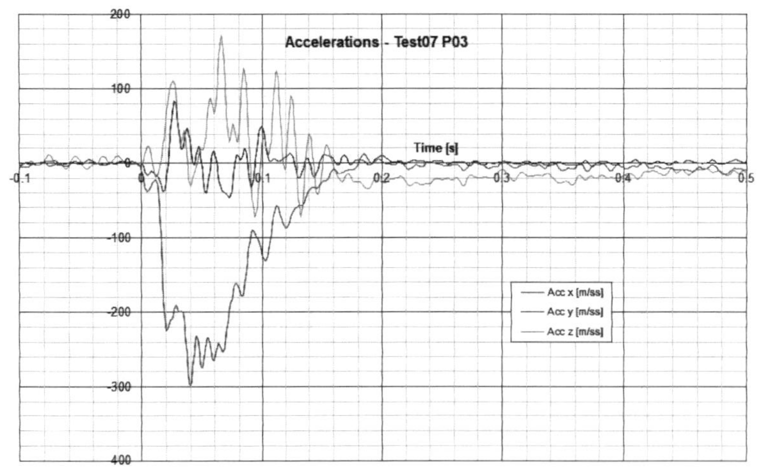

Figure 15. Car accelerations as time functions in the third car to pole collision.

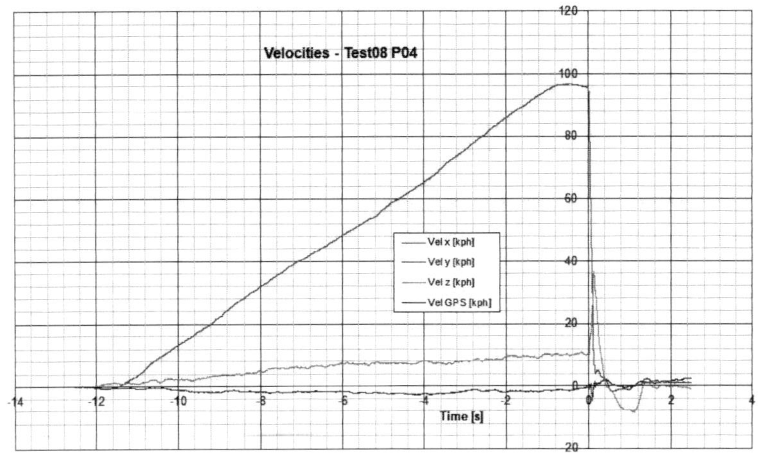

Figure 16. Car velocities as time functions in the fourth car to pole collision.

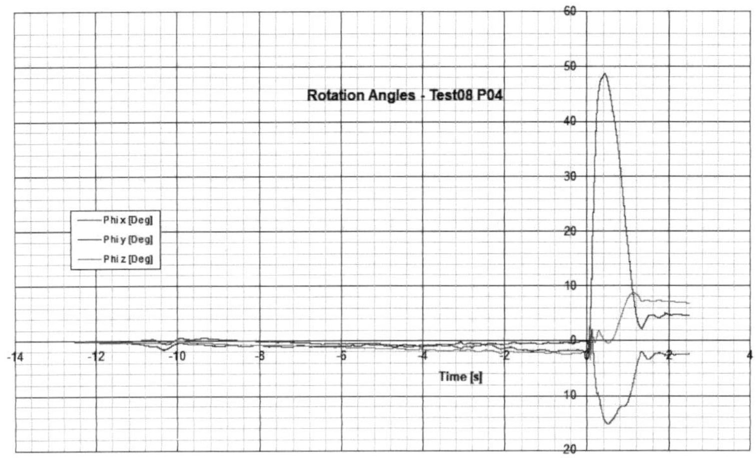

Figure 17. Car rotation angles as time functions in the fourth car to pole collision.

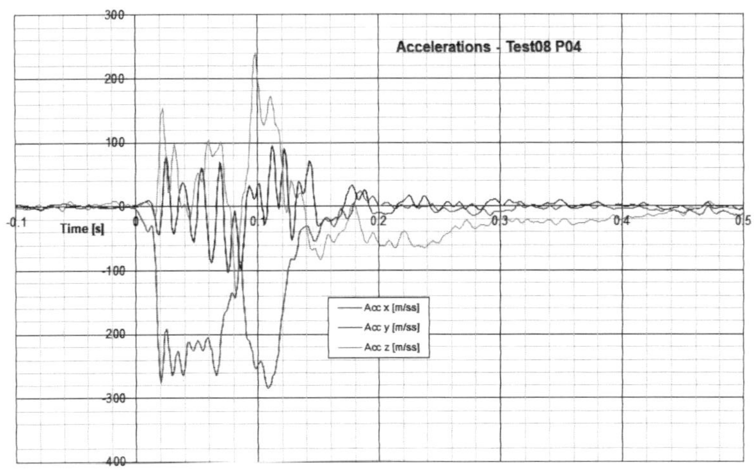

Figure 18. Car accelerations as time functions in the fourth car to pole collision.

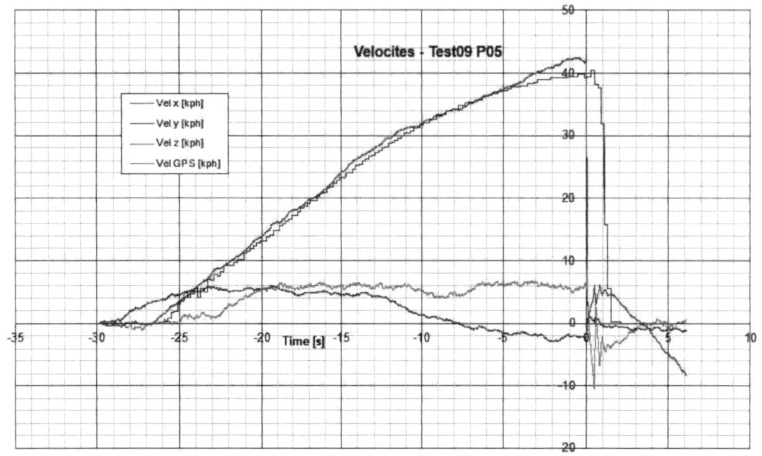

Figure 19. Car velocities as time functions in the fifth car to pole collision.

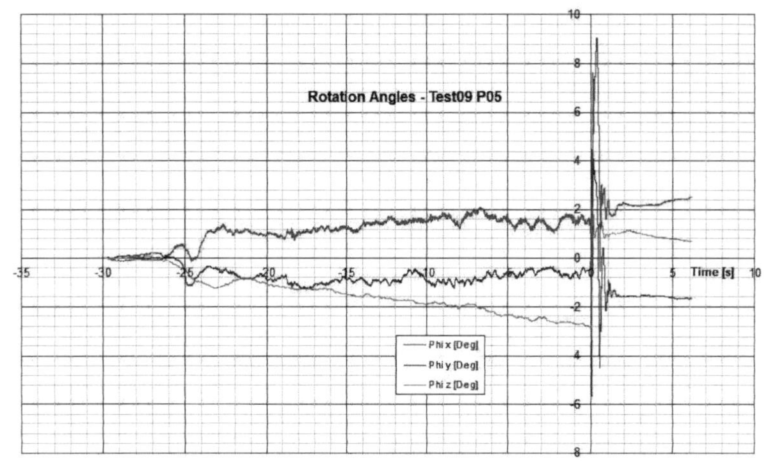

Figure 20. Car rotation angles as time functions in the fifth car to pole collision.

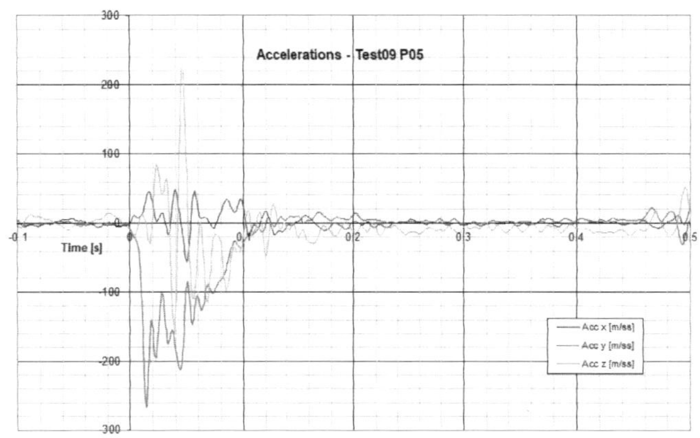

Figure 21. Car accelerations as time functions in the fifth car to pole collision.

In the first four collision tests with the pole, the same car model, with approximately the same age, was basically used. This has enabled the behaviour of the front areas of the vehicle at various collision speeds. On the other hand, the pole used in all collision tests very well simulates the characteristics of a column used in utilities in localities, made by metal or reinforced concrete. It is a very rigid construction, quite different from the characteristics of a tree, with the same diameter of the trunk, and the collision tests with this object were very similar to the plastic collisions. At a trunk diameter of 0.3 m, depending on the wood essence, in real-life crashes, it is possible to break the tree at high collision speeds. When the tree breaks up and a barrier/pole travel occurs during collision, the features of an elastic collision may also be more prominent. For the stiffness of the construction and the simulation of the utility pole fixing in the ground or in the concrete foundation, the pole used in the five collision tests has two reinforcements. Their presence influences on some extent the deformation of the front part of the car at higher speeds of collision. The presence of these reinforcements, at high collision speeds, results in an elongation of the form of penetration of the pillar inside the front of the car and perhaps an increase in the ability to hitch the structures of the car's bumper into the reinforcement profile.

4. Conclusions

Using the same car model in the first four car-to-pole collision tests has made it possible to get some important details about the front end behaviour of an old car involved in this type of collision. Although the solutions used for impact energy absorption were not so elaborate, neither the materials nor the technologies as advanced as they are today, a relatively good behaviour of the front structure of the cars is observed. It is very important to note that the level of mechanical stress transmitted to the occupants of the vehicles is relatively constant, regardless of the collision speed. However, at high collision speeds, due to the high penetration depth of the column into the passenger compartment, the engine and / or the gearbox may be pushed into the passenger compartment causing serious injury to the occupants. Geometric, all the Fiat Punto 55 cars involved in the first four tests were the same. They also had the same trackwidth (1.35 m) and the same wheelbase (2.45 m).
In particular, for each of the five collision cases, the following conclusions can be drawn:

A. First crash test:

1. The Fiat Punto 55 involved was the oldest (1995) of the four and had the smallest weight (850 kg);
2. The car had the smallest (out of 4) front axle load - 530 kg;
3. The distance from the centre of gravity (COG) to the front axe of the car involved was 0.92 m;
4. The impact velocity in this first car-to-pole collision was relatively low - 39.4 km/h;
5. From the analysis of speeds, rotation angles and accelerations, it can be seen that the time required for maximum penetration of the pole at the front of the car was about 0.2 s;
6. After the moment $t = 0$, at which the collision begins, COG of the car has a rotational movement, of small amplitude, around the Oy axis, of about 5.18 degrees;
7. The maximum deceleration value at the COG level of the vehicle is reached after approximately 0.042 s from the initial collision, in the "X" direction, with a value of 350 m/s^2;
8. In the same direction, after about 0.002 s from the moment of maximum, three deceleration peaks are recorded with values of approximately 270 m/s^2 (the first two) and 260 m/s^2 (third);
9. In the "Z" direction, after about 0.05 s from $t = 0$, an acceleration peak of 280 m/s^2 is recorded:
10. In the "Y" direction, after about 0.08 s from $t = 0$, an acceleration peak of 100 m/s^2 is recorded:
11. The high COG deceleration of the car in the "X" direction corresponds to the successive passage of the pole through the different stiffness of the front of the car.

B. Second crash test:

1. The Fiat Punto 55 involved was made in 1997 and had a weight of 850 kg;
2. The car had 535 kg front axle load;
3. The distance from the centre of gravity (COG) to the front axe of the car involved was 0.93 m;
4. The impact velocity in this second car-to-pole collision was medium - 60 km/h;

5. From the analysis of speeds, rotation angles and accelerations, it can be seen that the time required for maximum penetration of the pole at the front of the car was about 0.16 s;
6. After the moment t = 0, at which the collision begins, COG of the car has a rotational movement, of small amplitude, around the Oy axis, of about 5 degrees;
7. The maximum deceleration value at the COG level of the vehicle is reached after approximately 0.041 s from the initial collision, in the "X" direction, with a value of 305 m/s^2;
8. In the same direction, after about 0.022 s from t=0 was recorded a peaks of deceleration of 300 m/s^2;
9. In the "Z" direction, after about 0.025 s from t = 0, an acceleration peak of 130 m/s^2 is recorded:
10. In the "Y" direction, after about 0.042 s from t = 0, an acceleration peak of 55 m/s^2 is recorded:
11. The high COG deceleration of the car in the "X" direction corresponds to the successive passage of the pole through the different stiffness of the front of the car.

C. Third crash test:

1. The Fiat Punto 55 involved was made in 1998 and had a weight of 900 kg;
2. The car had 555 kg front axle load;
3. The distance from the centre of gravity (COG) to the front axe of the car involved was 0.94 m;
4. The impact velocity in this third car-to-pole collision was 85.9 km/h;
5. From the analysis of speeds, rotation angles and accelerations, it can be seen that the time required for maximum penetration of the pole at the front of the car was about 0.16 s;
6. After the moment t = 0, at which the collision begins, COG of the car has a rotational movement, of small amplitude, around the Oy axis, of about 25.5 degrees;
7. The maximum deceleration value at the COG level of the vehicle is reached after approximately 0.04 s from the initial collision, in the "X" direction, with a value of 300 m/s^2;
8. In the same direction, after about 0.05 and 0.6 s from t=0 was recorded two peaks of deceleration of 275 m/s^2 and 265 m/s^2;
9. In the "Z" direction, after about 0.062 s from t = 0, an acceleration peak of 170 m/s^2 is recorded:
10. In the "Y" direction, after about 0.023 s from t = 0, an acceleration peak of 82 m/s^2 is recorded:
11. The high COG deceleration of the car in the "X" direction corresponds to the successive passage of the pole through the different stiffness of the front of the car.

D. Fourth crash test:

1. The Fiat Punto 55 involved was made in 1997 and had a weight of 885 kg;
2. The car had 545 kg front axle load;
3. The distance from the centre of gravity (COG) to the front axe of the car involved was 0.94 m;
4. The impact velocity in this car-to-pole collision was high – 102.8 km/h;
5. From the analysis of speeds, rotation angles and accelerations, it can be seen that the time required for maximum penetration of the pillar at the front of the car was about 0.2 s;
6. After the moment t = 0, at which the collision begins, COG of the car has a rotational movement, of small amplitude, around the Oy axis, of about 49 degrees;
7. The maximum deceleration value at the COG level of the vehicle is reached after approximately 0.13 s from the initial collision, in the "X" direction, with a value of 285 m/s^2;
8. In the same direction, after about 0.02, 0.03, 0.038 and 0.043 s s from t=0 was recorded some peaks of deceleration of 275 m/s^2, 263 m/s^2, 263 m/s^2 and 265 m/s^2
9. In the "Z" direction, after about 0.095 s from t = 0, an acceleration peak of 240 m/s^2 is recorded:
10. In the "Y" direction, after about 0.068 s from t = 0, a deceleration peak of 102 m/s^2 is recorded:
11. The high COG deceleration of the car in the "X" direction corresponds to the successive passage of the pole through the different stiffness of the front of the car.

E. Fifth crash test:

1. The SIMPA JDM Type 97 is a minicar different from Fiat Punto 55, made in 1997 and had a weight of 340 kg;
2. The impact velocity in this first car-to-pole collision was low – 39.3 km/h;
3. From the analysis of speeds, rotation angles and accelerations, time required for maximum penetration of the pillar at the front of the car was about 0.16 s;
4. After the moment t = 0, at which the collision begins, COG of the car has a rotational movement, of small amplitude, around the Ox axis, of about 8.6 degrees;
5. The maximum deceleration value at the COG level of the vehicle is reached after approximately 0.018 s from the initial collision, in the "X" direction, with a value of 268 m/s^2;
6. In the same direction, after about 0.046 s from t=0 was recorded a peak of deceleration of 214 m/s^2;
7. In the "Z" direction, after about 0.042 s from t = 0, an acceleration peak of 223 m/s^2 is recorded:
8. In the "Y" direction, after about 0.05 s from t = 0, a deceleration peak of 48 m/s^2 is recorded:
9. Given the small size of the front of this minicar, the maximum pole penetration to the passenger compartment is low, the presence of the engine and other systems in the short area in front of the passenger compartment preventing the entry to a greater depth..

References

[1] Pfefer R. and Slack K. NCHRP Report 500 Volume 8: A Guide for Reducing Collisions Involving Utility Poles. Washington D.C. TRB, National Research Board, 2004.
[2] Du Bois P, and all Vehicle Crashworthiness and Occupant protection, American Iron and Steel Institute, 2000 Town Center, Southfield, Michigan 4875, 2004.
[3] Ispas N and Năstăsoiu M 2016 Crash Tests and the Loads over Driver Head in Different Side Impact Cases Current Solutions in Mechanical Engineering pp 181-186.
[4] Ispas N and Năstăsoiu M 2015 A Study Concerning the Loads Over Drivers Heads in the Cases of Cars Crashes with Involved Cars of the Same or Different Generation Proceed-ings of the European Automotive Congress EAEC-ESFA 2015 Springer pp 325-336.
[5] DVD-ROM: DSD Crash Tests March 2016, "Easter 2016 PC-Crash Seminar", © DSD, Linz 2016.

Simulation of the right-angle car collision based on identified parameters

R Kostek[1] **and P Aleksandrowicz**[1]
[1]Faculty of Mechanical Engineering, University of Science and Technology in Bydgoszcz, Al. prof. S. Kaliskiego 7, Poland

E-mail: robertkostek@op.pl, almot@op.pl

Abstract. This article presents an influence of contact parameters on the collision pattern of vehicles. In this case a crash of two Fiat Cinquecentos with perpendicular median planes was simulated. The first vehicle was driven with a speed 50 km/h and crashed into the other one, standing still. It is a typical collision at junctions. For the first simulation, the default parameters of the V-SIM simulation program were assumed and then the parameters identified from the crash test of a Fiat Cinquecento, published by ADAC (Allgemeiner Deutscher Automobil-Club) were used. Various post-impact movements were observed for both simulations, which demonstrates a sensitivity of the simulation results to the assumed parameters. Applying the default parameters offered by the program can lead to inadequate evaluation of the collision part due to its only approximate reconstruction, which in consequence, influences the court decision. It was demonstrated how complex it is to reconstruct the pattern of the vehicles' crash and what problems are faced by expert witnesses who tend to use default parameters.

1. Introduction

For crash simulations, research institutes use Finite Element Method (FEM), which provides an opportunity to conduct accurate computations [1]. However, in court practice, simpler IT tools which do not require a high computational capacity and introducing e.g. material constants are used, as in the FEM programs [2,3]. The programs perform computations in MBS (Multi Body System) convention, and use obvious simplifications; for example, they assume linear dependencies between vehicle deformation and force, as well as constant mass parameters during the crash [4,5,6]. The MBS programs are applied by expert witnesses to the analysis of collisions, as well as in the verification of the vehicles damage circumstances [7]. The results influence the judgment and, as such, they should be accurate. Examples of simulations made with Virtual Crash and PC Crash were presented in the publications [8,9,10], showing a good agreement with experimental results even for rollover. In Poland MBS program called V-SIM is widely employed by experts, and it is used in this article.

Due to a lack of additional data verified experimentally, expert witnesses usually assume default parameters and do not change them. With that in mind, a practical problem was investigated; sensitivity of the results to the assumed input data [11,12,13]. In literature known to the authors, side collision was not studied with V-SIM apart from [14]; which is studied in this article.

2. Results of simulation

The V-SIM employs a few co-ordinate systems described below (Fig. 1a). A global co-ordinate system x, y, z is used to describe the displacement of vehicles and the position of objects around cars. Another system is related to the vehicle x', y', z', and its origin is in mass centre C. Axis x' is directed forwards and parallel to the ground, whereas axis *z'* is directed vertically upwards. The next co-ordinate system is related to the vehicle and its origin is in the mass centre too, it is central principal co-ordinate system *x", y", z"*. This system is not presented in V-SIM. MBS programs model the car with rigid bodies, thus equations describing a rigid body motion are applied. The equations are presented below:

$$a_{x''} = F_{Rx''} / m \qquad (1)$$

$$a_{y''} = F_{Ry''} / m \qquad (2)$$

$$a_{z''} = F_{Rz''} / m \qquad (3)$$

$$I_{x''}\, \varepsilon_{x''} + (I_{z''} - I_{y''})\, \omega_{y''}\, \omega_{z''} = M_{Cx''} \qquad (4)$$

$$I_{y''}\, \varepsilon_{y''} + (I_{x''} - I_{z''})\, \omega_{x''}\, \omega_{z''} = M_{Cy''} \qquad (5)$$

$$I_{z''}\, \varepsilon_{z''} + (I_{y''} - I_{x''})\, \omega_{x''}\, \omega_{y''} = M_{Cz''} \qquad (6)$$

where: $a_{x''}$, $a_{y''}$, $a_{z''}$ - denote the acceleration of the mass centre in central principal co-ordinate system x'',y'',z'', $F_{Rx''}$, $F_{Ry''}$, $F_{Rz''}$ - projections of resultant force onto axes x'',y'',z'', m - is the body mass, $I_{x''},I_{y''},I_{z''}$ - denote the central principal moments of inertia, $\omega_{x''},\omega_{y''},\omega_{z''}$ - the projections of the vector of angular velocity onto axes x'',y'',z'', $\varepsilon_{x''},\varepsilon_{y''},\varepsilon_{z''}$ - the projections of angular acceleration vector onto axes x'',y'',z''; $M_{Cx''},M_{Cy''},M_{Cz''}$ - the projections of resultant moment onto axes x'',y'',z''.

The car model employed in V-SIM has ten degrees of freedom. Six degrees of freedom describe the car body motion, whereas four degrees of freedom describe the rotation of wheels. In consequence, the vehicle motion is described with ten degrees of freedom.

(a) (b)

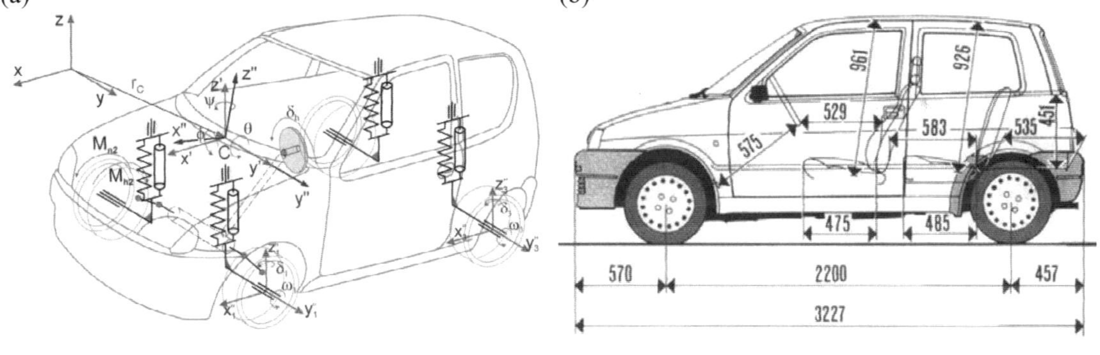

Fig. 1. Co-ordinate systems used in V-SIM (a) and side view of Fiat Cinquecento (b) [15,16].

The object of the study is a Fiat Cinquecento (Fig. 1b); its dimensions are: length 3.227 m, width 1.487 m, height 1.435 m, wheel space 2.2 m, whereas the total mass of the car was 811 kg. The tires model employed was TM-Easy. The first vehicle was driven with a speed 50 km/h and crashed into the other one, standing still. The vehicles were positioned in a way as to receive a 40 % overlap (0.595 m). Time *t*=0 ms was adopted for the initiation of the contact between vehicles (Fig. 2a). Then, the contact forces between the cars were calculated for the 2D collision detection model [18]. The first simulation was conducted for default data, whereas the second one – for identified data. The data is presented in Table 1. The default parameters of the crash were assumed following the data offered by the program producer, while the identified parameters were received from the crash test [17]. The contact parameters identification involved an introduction of such changes in parameters which lead to a similar pattern of the crash and similar vehicles positions. The overlap was evaluated from the analysis of the images recorded; due to its complexity, it is not presented here. The second simulation (for identified data) was performed for a lowered level with contact forces acting between vehicles due to the detachment of the rear-axle wheels. The front wheels of vehicles being in deformation zones were blocked for time longer than $t \geq 70$ ms as their movement is suppressed (Fig. 2b). Unfortunately the program does not suggest such operations. Then a simulation was made.

Table 1. Values of parameters adopted to simulations.

Adopted parameters	Default data	Identified data
Adhesive friction coefficient	0.90	0.90
Slip friction coefficient	0.80	0.80
Rolling resistance coefficient	0.015	0.015
Stiffness of the car body for compression phase	331 kN/m^3	426.45 kN/m^3
Stiffness of the car body for restitution phase	331 kN/m^3	426.45 kN/m^3
Restitution coefficient	0.26	0.12
Friction coefficient between cars	3.63	2.00
Lowering of plane of contact force Δz	0.0 mm	50 mm
Blocking front wheel	-	t>70 ms

First, vehicle positions for default and identified data are similar, however velocities are slightly varied as acting forces are various (Fig. 2b-d). At this stage contact forces play a significant role. For t=200 ms the difference between the speeds of the crashed vehicles is 1.6 km/h. Gradually, various forces lead to various vectors of velocities and, finally, to various displacements – different post-impact movements (Fig. 2e-j). Various directions of movement are a result of various car rotations. At this stage blocked wheels cause larger car decelerations for identified parameters, which is obvious for the crashed vehicle (Fig. 2e-j). Large differences in linear velocity were observed for crashing vehicles as well (t=2100 ms, 14.2 km/h). At the same time, differences in vehicle displacements are a few meters and get larger. The results for the experimentally identified data seem to be more reliable than the results for default data, as the vehicles do not drive away that far. To illustrate these phenomena, a film showing this collision for a longer period was presented [19].

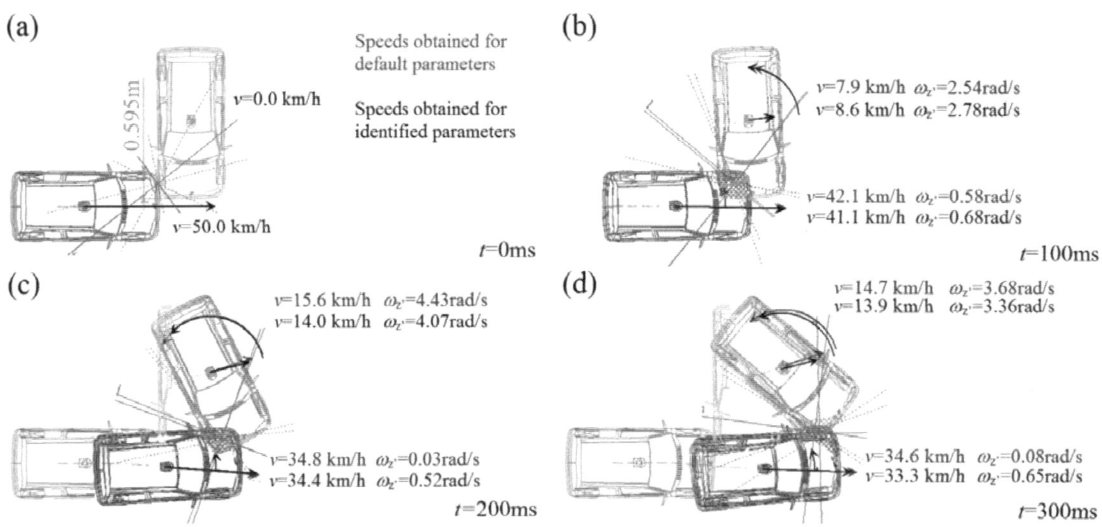

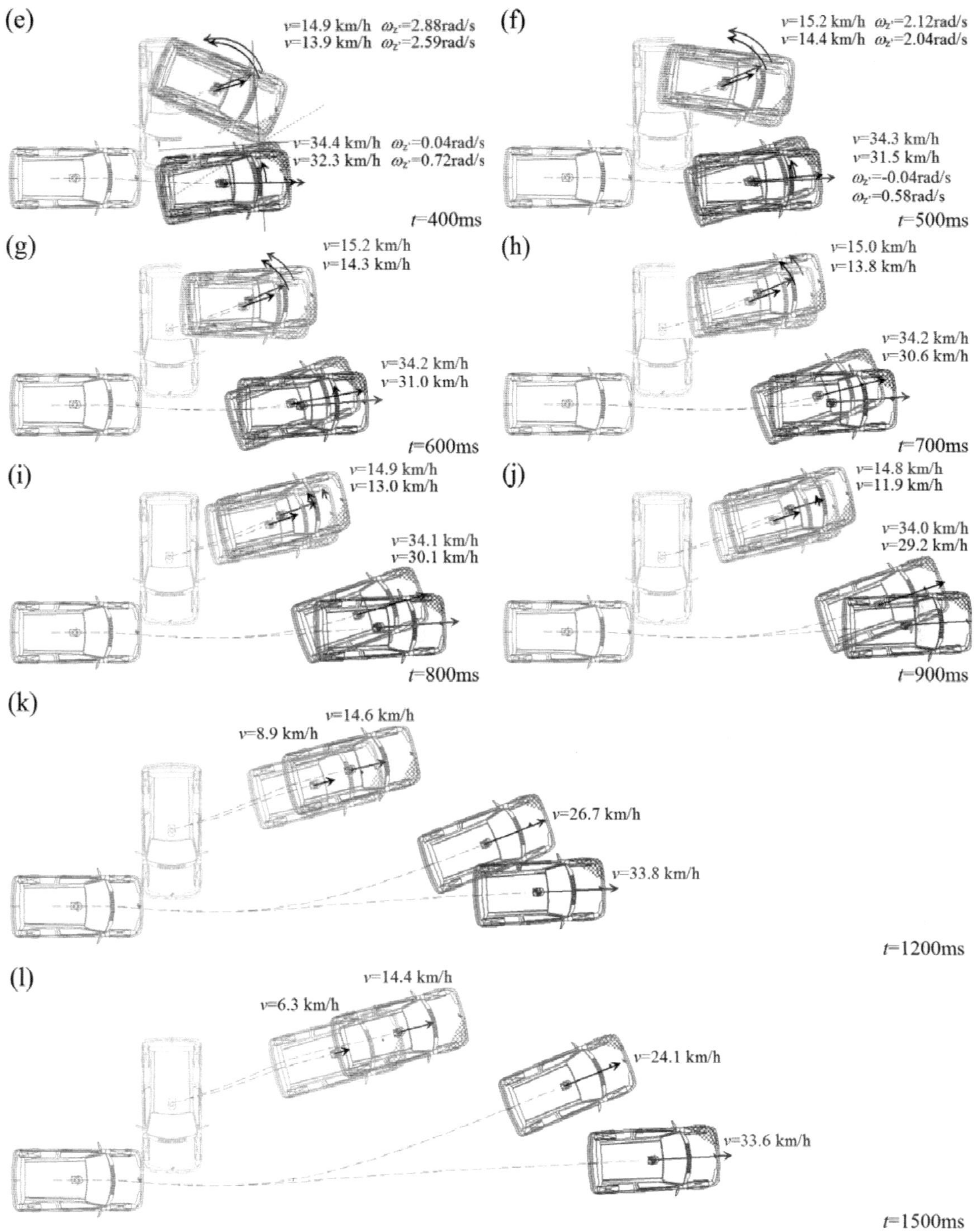
(e)
v=14.9 km/h ωz=2.88rad/s
v=13.9 km/h ωz=2.59rad/s
v=34.4 km/h ωz=0.04rad/s
v=32.3 km/h ωz=0.72rad/s
t=400ms
(f)
v=15.2 km/h ωz=2.12rad/s
v=14.4 km/h ωz=2.04rad/s
v=34.3 km/h
v=31.5 km/h
ωz=-0.04rad/s
ωz=0.58rad/s
t=500ms
(g)
v=15.2 km/h
v=14.3 km/h
v=34.2 km/h
v=31.0 km/h
t=600ms
(h)
v=15.0 km/h
v=13.8 km/h
v=34.2 km/h
v=30.6 km/h
t=700ms
(i)
v=14.9 km/h
v=13.0 km/h
v=34.1 km/h
v=30.1 km/h
t=800ms
(j)
v=14.8 km/h
v=11.9 km/h
v=34.0 km/h
v=29.2 km/h
t=900ms
(k)
v=14.6 km/h
v=8.9 km/h
v=26.7 km/h
v=33.8 km/h
t=1200ms
(l)
v=14.4 km/h
v=6.3 km/h
v=24.1 km/h
v=33.6 km/h
t=1500ms

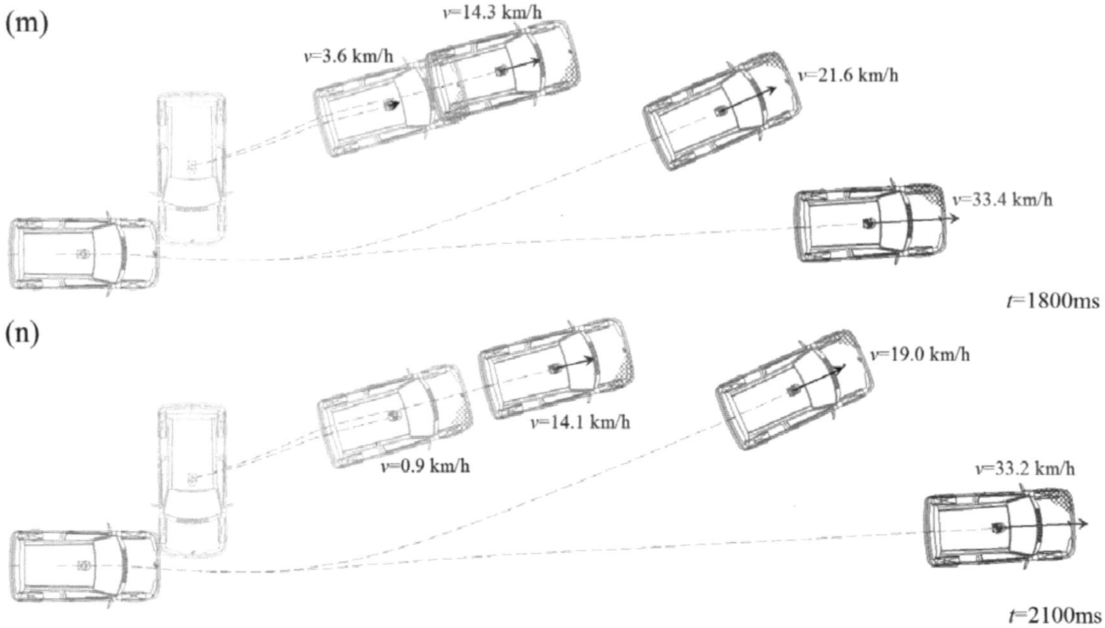

Fig. 2. Consecutive stages of collision obtained for default (blue) and identified parameters (red).

Acting forces and resultant moment cause various post-impact movements; their time histories are depicted in Fig. 3. As one can see, an intense contact between vehicles lasts around 150 ms, as a consequence, that time determines the post-impact movement. Time of contact is shorter and forces are larger for identified parameters as the red cars are "stiffer". Time histories of forces are similar, whereas differences between resultant moments are obvious. This resulted in various angles of rotation (Fig. 2). These differences lead to various post-impact positions, which shows sensitivity to input data.

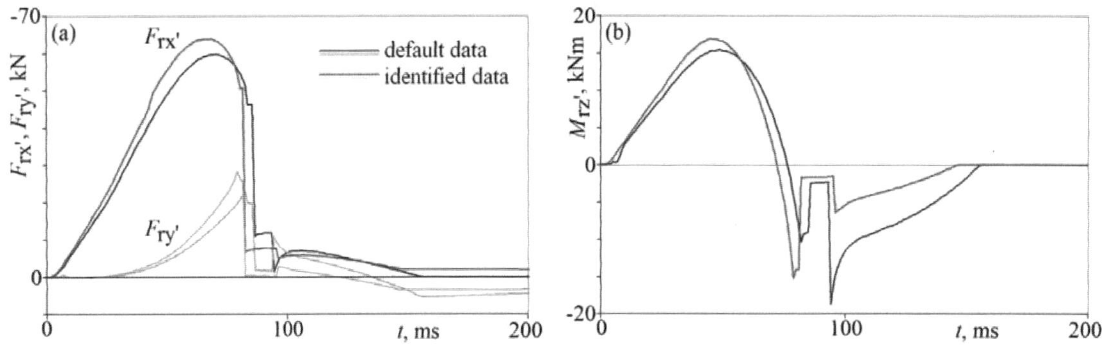

Fig. 3. Time histories of the components of the resultant force F_r acting on the vehicle (a), and vertical component of resultant moment M_r acting on the vehicle (b).

3. Conclusions

Differences were observed between the simulation results obtained for default and identified data, and the differences were getting larger with time. For the default parameters, the rotation of the crashing vehicle around axis *z'* is smaller and the speed is greater, thus the vehicles drive further in another direction. One should note that for t=4148 ms, the crashing vehicle for identified parameters stops, whereas for default parameters, it drives away, still with the velocity of v=31.9 km/h; which confirms the sensitivity of the simulation results to input data. This leads to the conclusion that applying default parameters offered by programs can lead to wrong results. A careful reconstruction requires many computations, and data derived from crash tests and literature. Moreover, the program should suggest some actions like blocking wheels. Finally, expert witnesses are not able to reconstruct accidents adequately without credible default values of parameters.

References

[1] http://www.lstc.com (access 08.06.2017)
[2] Qi W *et al* 2006 Improvement of Energy-absorbing Structures of a Commercial Vehicle for Crashworthiness Using Finite Element Method *The International Journal of Advanced Manufacturing Technology* vol 30 issue 11 pp 1001–1009
[3] Zhang X Y *et al* 2008 Vehicle Crash Accident Reconstruction Based on the Analysis 3D Deformation of the Auto-body *Advances in Engineering Software* vol 39 issue 6 pp 459-465
[4] http://www.cyborgidea.com.pl (access 08.06.2017)
[5] http://www.pc-crash.com (access 08.06.2017)
[6] http://www.vcrashusa.com (access 08.06.2017)
[7] Aleksandrowicz P 2017 Analysis of Vehicle Collisions with the SDC Method *23rd International Conference Engineering Mechanics (Svratka, Czech Republic)* vol 1 pp 78-81
[8] Andrews S *et al* 2009 A Comparision of Computer Modeling to Actual Data and Video of a Staged Rollover Collision NHTSA Paper no **09**-0346
[9] Coufal T and Semeia M 2014 Research of Impact Parameters for Traffic Accident Analysis in Case of Small Overlap Crash Test *2nd Annual International Conference on Forensic Science - Criminalistics Research (Singapore)* pp 1-9
[10] Dima D S, Covaciu D and Chiru A 2014 Validation of Simulation and Optimization Reconstruction in PC Crash with Video Recorded Sample *3rd International Congress Science and Management of Automotive and Transportation Engineering (Craiova, Romania)* SMAT Paper no 2014-MTSRT51 pp 207-212
[11] http://www.autoview.at (access 08.06.2017)
[12] Kostek R and Aleksandrowicz P 2017 Effect of Contact Parameters on the Pattern of Vehicle Collisions with a Round Pillar *23rd International Conference Engineering Mechanics (Svratka, Czech Republic)* vol 1 pp 490-493
[13] Kostek R and Aleksandrowicz P 2017 Identification of the Parameters of Vehicle Contact with a Rigid Barrier from a Crash Test *23rd International Conference Engineering Mechanics (Svratka, Czech Republic)* vol 1 pp 494-497
[14] Dąbrowski F and Gidlewski M 2015 Right-Angle Collision of Motor Vehicles Participating in Road Traffic– Comparative Analysis of Results of Experimental and Model Tests *Proceedings of The Institute of Vehicles* vol 4 pp 49-62
[15] https://www.autocentrum.pl (access 08.06.2017)
[16] V-SIM4.0 – User manual 2016 CIBID *(Cracow, Poland)*
[17] https://www.youtube.com/watch?v=UoaI-NuGBN4 (access 08.06.2017)
[18] Aleksandrowicz P 2017 Verifying the Application of the Models of Crash and Collision Detection of Incompatible Vehicles *23rd International Conference Engineering Mechanics (Svratka, Czech Republic)* vol 1 pp 82-85
[19] https://youtu.be/APsCfoS6180 (access 08.06.2017)

IOP Conf. Series: Materials Science and Engineering **252** (2017) 012014 doi:10.1088/1757-899X/252/1/012014

The measurement of dynamic radii for passenger car tyre

G Anghelache, R Moisescu

University POLITEHNICA of Bucharest, Splaiul Independentei 313, 060042, Bucharest, Romania

gabriel.anghelache@upb.ro

Abstract. The tyre dynamic rolling radius is an extremely important parameter for vehicle dynamics, for operation of safety systems as ESP, ABS, TCS, etc., for road vehicle research and development, as well as for validation or as an input parameter of automotive simulations and models. The paper investigates the dynamic rolling radii of passenger car tyre and the influence of rolling speed and inflation pressure on their magnitude. The measurement of dynamic rolling radii has been performed on a chassis dynamometer test rig. The dynamic rolling radii have been measured indirectly, using longitudinal rolling speed and angular velocity of wheel. Due to the subtle effects that the parameters have on rolling radius magnitude, very accurate equipment has to be used. Two different methods have been chosen for measuring the wheel angular velocity: the stroboscopic lamp and the incremental rotary encoder. The paper shows that the stroboscopic lamp has an insufficient resolution, therefore it was no longer used for experimental investigation. The tyre dynamic rolling radii increase with rolling speed and with tyre inflation pressure, but the effect of pressure is more significant. The paper also makes considerations on the viability of simplified formulae from literature for calculating the tyre dynamic rolling radius.

1. Introduction

The dynamic radius of rolling tyre is an extremely important parameter for vehicle dynamics, for operation of safety systems as ESP, ABS, TCS, etc., for road vehicle research and development, as well as for validation or as an input parameter of automotive simulations and models. The dynamic rolling radius is influenced by many factors, determined not only by the tyre, but also by the working conditions, such as inflation pressure, load, speed etc. [1]. The study of influencing parameters is essential also for controlling the operation of tyre pressure monitoring systems [2].

The study of the rolling speed influence on tyre dynamic rolling radii is rarely mentioned in literature; the few existing investigations [3], [4] are mainly concerning large tractor tyres with limited speed range.

The authors have extended experience in tyre-road phenomena research and modelling, and they have frequently needed very accurate information about the tyre dynamic rolling radii. The paper investigates through experiments the dynamic rolling radius defined according to [5], and the influence of rolling speed and inflation pressure on the dynamic rolling radii of passenger car tyre.

2. Experimental setup

The measurement of dynamic rolling radii has been performed for a passenger car tyre, in laboratory conditions, using a chassis dynamometer in the Automotive Engineering Department of the University POLITEHNICA of Bucharest. A Dacia Logan, equipped with tyres Bridgestone Turanza ER300,

185/65R15 88H, has been installed with the wheels of the front axle rolling on the dynamometer rollers, as it can be seen in Figure 1.

The tyre rolling on the drum has a slightly curved contact patch, which is different compared to the real rolling conditions, thus influencing the tyre dynamic rolling radii to some extent. Despite this fact, the mentioned influencing parameters (inflation pressure and rolling speed) affect in the same way the tyre dynamic rolling radii.

The dynamic rolling radii r_d have been measured indirectly, using two input parameters: longitudinal rolling speed V and angular velocity of wheel ω, as expressed in equation (1). Due to slight effects that the influencing parameters have on the dynamic rolling radius magnitude, very accurate equipment has to be used.

$$r_d = \frac{V}{3.6 \cdot \omega} \ [m] \quad (1)$$

The longitudinal rolling speed has been precisely controlled and measured using the MAHA AIP ECDM chassis dynamometer in the "constant velocity" mode, as shown in Figure 2. The passenger car front wheels have been driven by the chassis dynamometer.

Figure 1. Passenger car used for experimental research on the chassis dynamometer

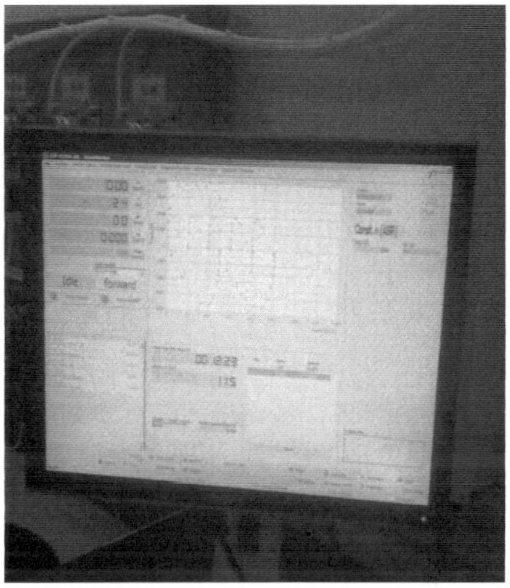

Figure 2. Measurement of rolling speed using the chassis dynamometer

Two different methods have been chosen for measuring the wheel angular velocity: the stroboscopic lamp and the incremental rotary encoder. The equipment used for measuring the wheel angular velocity through the two different methods is shown in Figure 3. The stroboscopic lamp method allows measuring the wheel angular velocity without contact, but requires a visible mark traced on the passenger car tyre. The incremental rotary encoder needs to be mounted on the rim nuts using a special flange with hexagonal adapters, and also needs to be attached to the car body using a mounting system with suction cups. The pulse signal has been recorded using a counter on the data acquisition card National Instruments USB-6211 and a LabVIEW application, shown in Figure 4.

Tyre inflation pressure has been measured using a pressure gauge with digital display, having a resolution of 1 kPa, shown in Figure 5. The tyre pressure has been adjusted after warm-up before each measurement and has been checked after performing the tests.

Measurements have been performed at different values of tyre inflation pressure, between 140 kPa and 260 kPa, and at different values of rolling speed, between 20 km/h and 120 km/h.

Figure 3. Measurement of wheel angular velocity using the stroboscopic lamp and the rotary encoder

Figure 4. Data acquisition system for measuring wheel angular velocity

Figure 5. Measurement of tyre inflation pressure using a pressure gauge with digital display

Tyre temperature has an important effect on inflation pressure, and consequently all the measurements have been performed in quasi-stationary conditions, after a time of rolling for warm-up.

3. Measurement results

3.1. The resolution depending on the measurement method

The resolution of the stroboscope method is 1 rev/min. The resolution of the encoder method, recorded through measuring system, is theoretically of 1 Hz. Due to rolling tyre and wheel vibrations, this better resolution can practically be read only at lower speeds. At higher speeds, the resolution of encoder method is 5 Hz or even 10 Hz.

The resolution of dynamic rolling radius is obtained from the previous mentioned values, but taking into consideration the speed values obtained from rollers of chassis dynamometer. The resolution depends both on the speed and on the tyre inflation pressure, as shown in Table 1.

Analysing the data in Table 1, some conclusions could be drawn referring to the measuring resolution of stroboscope method:

- considering that the variation in dynamic rolling radius is small depending on speed and tyre inflation pressure, the resolution values obtained are insufficient;
- the resolution of dynamic rolling radius is strongly influenced by the speed, providing better values at higher speed;
- the tyre inflation pressure has a very slight influence on the resolution of dynamic rolling radius, better resolution values being obtained at lower pressure.

Table 1. Resolution of the dynamic rolling radius measurement

Inflation pressure (kPa)	Speed (km/h)	Dynamic rolling radius resolution (mm)	
		stroboscope method	encoder method
140	20	1.67	0.10
140	100	0.34	0.10
260	20	1.70	0.10
260	100	0.34	0.10

Analysing the measuring resolution of encoder method, some conclusions could be drawn:

- the resolution of encoder method remains relatively constant, regardless of speed and tyre inflation pressure;
- a resolution of 0.1 mm allows investigating the influence of speed and tyre inflation pressure on the tyre dynamic rolling radius.

Comparing the resolution values obtained through the two different methods, it can be observed that the encoder method provides a resolution at least three times better than the stroboscope method.

Figure 6 gives an example of the influence of resolution on the dynamic rolling radius measurement. The compared results have been obtained by simultaneously measuring through the two different methods.

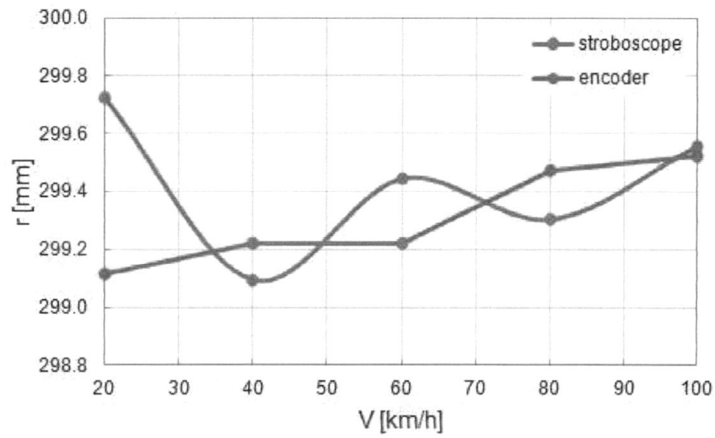

Figure 6. Comparative example of dynamic rolling radius measurement through the two methods

The dynamic rolling radii are similar at 100 km/h, since the stroboscope method provides better resolution values at higher speed. When speed decreases, the differences between the results of the two methods tend to increase, so the most important gap, of 0.6 mm, occurs at 20 km/h. This gap is about one third of the resolution of stroboscope method at the mentioned speed (about 1.7 mm according to Table 1), and consequently the measured value is uncertain. The sign of the difference between results of the two methods varies, thus at some speeds the difference is positive, at other speed values it is negative.

The resolution of the stroboscope method does not allow highlighting the influence of speed and tyre inflation pressure on the dynamic rolling radius, because the errors generated by insufficient resolution are higher than the effect of changing the rolling speed. However, it can be noticed that the resolution of the encoder method allows emphasizing the increase in dynamic rolling radius due to growing rolling speed and tyre inflation pressure. Therefore, from this point, the paper analyses only the results obtained using the encoder method.

3.2. The influence of rolling speed and inflation pressure

The measured values of dynamic rolling radius are presented in Figure 7 as a function of rolling speed. The speed has been adjusted between 20 km/h and 120 km/h, in steps of 20 km/h. Measurements have been performed at several values of tyre inflation pressure, between 140 kPa and 260 kPa.

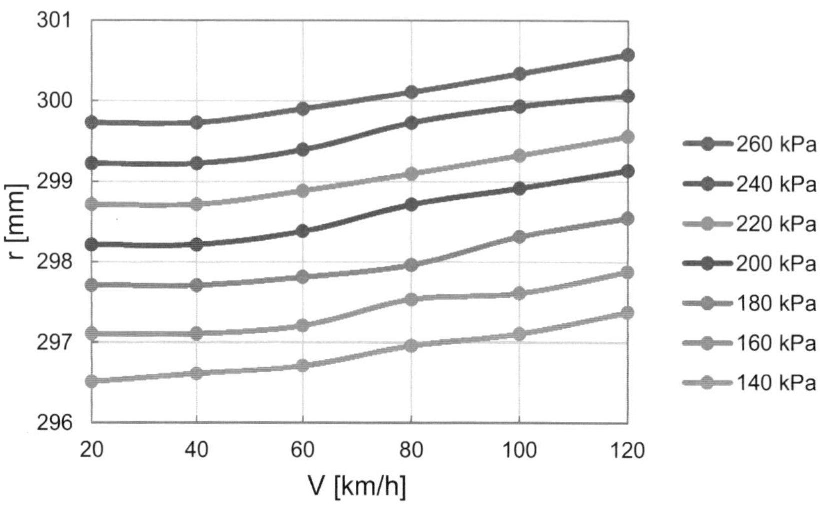

Figure 7. The influence of rolling speed on the tyre dynamic rolling radius measured at different inflation pressures

The values of dynamic rolling radius at 20 km/h and 40 km/h tend to remain constant or change insignificantly, but at rolling speeds above 40 km/h they tend to increase with speed. In some cases, such as those measured at 220 kPa and at 260 kPa, a linear behaviour is noticed, but the rest of the curves are slightly nonlinear, which could be caused by the encoder method resolution.

The absolute and relative increases of dynamic rolling radius corresponding to a growth of rolling speed from 20 km/h to 120 km/h are presented in Table 2. The increase of dynamic rolling radius has absolute values between 0.77 mm and 0.92 mm, therefore the stroboscope method would not allow accurately detecting the variation of dynamic rolling radius.

Table 2. Increase of dynamic rolling radius due to changing rolling speed from 20 km/h to 120 km/h

Inflation pressure (kPa)	Increase of dynamic rolling radius	
	absolute (mm)	relative (%)
140	0.86	0.29
160	0.77	0.26
180	0.84	0.28
200	0.92	0.31
220	0.84	0.28
240	0.85	0.28
260	0.85	0.28

The increase of dynamic rolling radius has relative values between 0.26 % and 0.31 %, although the speed increases six times, from 20 km/h to 120 km/h. Therefore, for all the considered values of inflation pressure, the rolling speed has a very small influence on the tyre dynamic rolling radius.

The measured values of dynamic rolling radius are presented in Figure 8 as a function of inflation pressure. The inflation pressure has been adjusted between 140 kPa and 260 kPa, in steps of 20 kPa. Measurements have been performed at several values of speed, between 20 km/h and 120 km/h.

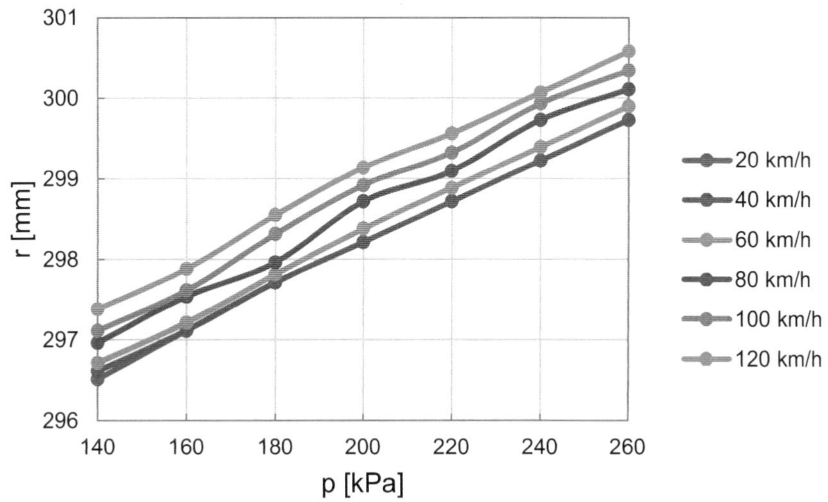

Figure 8. The influence of inflation pressure on the tyre dynamic rolling radius measured at different rolling speeds

The values of dynamic rolling radius clearly increase when inflation pressure is increased. Although several of the measured curves are linear, there are also some cases in which a slightly nonlinear behaviour is noticed, such as those measured at 80 km/h and at 100 km/h, possibly because of the encoder method resolution. It can be seen that the curves measured at 20 km/h and 40 km/h are overlapping and that they are almost linear.

The absolute and relative increases of dynamic rolling radius corresponding to a growth of inflation pressure from 140 kPa to 260 kPa are presented in Table 3, for all the considered values of speed.

Table 3. Increase of dynamic rolling radius due to changing inflation pressure from 140 kPa to 260 kPa

Speed (km/h)	Increase of dynamic rolling radius	
	absolute (mm)	relative (%)
20	2.71	0.91
40	2.61	0.88
60	2.68	0.90
80	2.77	0.93
100	2.82	0.95
120	2.69	0.91

The increase of dynamic rolling radius due to changing inflation pressure from 140 kPa to 260 kPa has absolute values between 2.61 mm and 2.82 mm, and these results are almost independent of speed.

When the inflation pressure is changed by ± 30 % with respect to the nominal value, the increase of dynamic rolling radius has relative values ranging between 0.88 % and 0.95 %, therefore close to 1 %. It can be asserted that, for all the considered values of rolling speed, the inflation pressure has an important influence on the tyre dynamic rolling radius.

The measured values of dynamic rolling radius are presented in Figure 9 as a function of both rolling speed and inflation pressure.

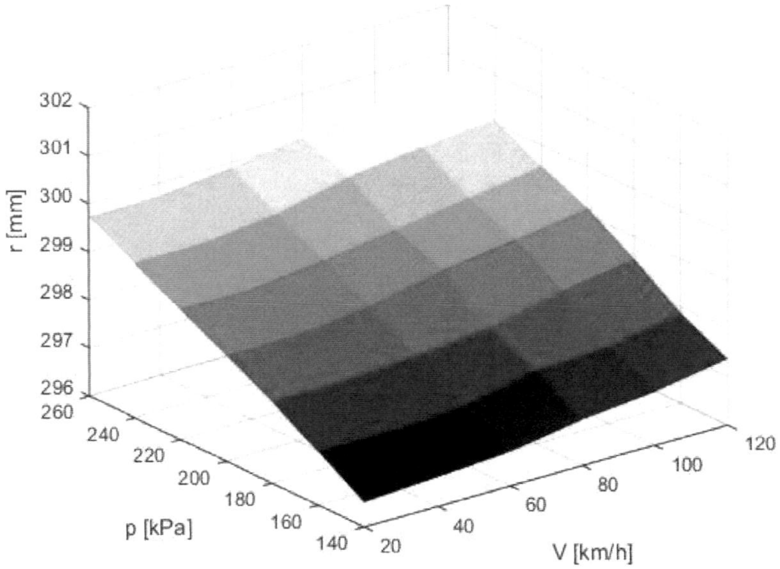

Figure 9. The influence of rolling speed and inflation pressure on the tyre dynamic rolling radius

It can be noticed that the influence of inflation pressure on the measured tyre dynamic rolling radius is stronger than the influence of rolling speed, with absolute values almost three times larger, although the speed changes up to six times while the inflation pressure is changed by only ± 30 % with respect to the nominal value.

3.3. Aspects regarding validation of technical knowledge

The tyre catalogue corresponding to the investigated tyre [6] includes specifications based on which the dynamic rolling radius can be calculated, obtaining the value of 301.4 mm. Comparing with the results shown in Figure 9, it can be observed that all the results for tyre dynamic rolling radius are smaller than that mentioned above. The main reason refers to the cylindrical shape of the tyre contact patch achieved on chassis dynamometer rollers, which leads to dynamic rolling radii smaller than those obtained when rolling on the real roads.

Another aspect concerns checking the validity of the equations (2.38) from [7] and (2.28) from [8] (but also from other references) which provide an approximate calculus for the tyre rolling radius:

$$r_r = \lambda \cdot r_0 \tag{2}$$

where λ is the tyre deformation coefficient, ranging from 0.930 to 0.935 for low pressure tyres, and from 0.945 to 0.950 for high pressure tyres, and r_0 is the tyre unloaded radius [7], [8].

For the tyre tested in this paper, the catalogue indicates an unloaded radius of 310.5 mm. When equation (2) is applied for the tested tyre, dynamic rolling radii of 288.8 mm to 295.0 mm are calculated. Irrespective of the terminology referring to the tyre radius definitions, the calculated values are considerably smaller than those obtained experimentally and shown in chapters 3.2 and 3.3. Taking into consideration that in real rolling conditions the tyre dynamic rolling radii are greater than those on the chassis dynamometer, the tyre radii calculated with equation (2) are significantly lower than in reality. This aspect shows the inadequacy of results obtained using equation (2) as input parameter for analytical and numerical tyre / automotive models.

In the book [9], the value of 0.98 is indicated for the coefficient λ that leads to 304.3 mm dynamic rolling radius, which is closer to the experimental values.

4. Conclusions

Since almost every researcher uses different definitions for the same radius, it is useful to unify the terminology in terms of tyre radii.

The tyre dynamic rolling radii measured on the drum are smaller than those measured in the real rolling conditions because of the slightly curved contact patch generated by the drum. Despite this fact, the tyre dynamic rolling radii are influenced by the inflation pressure and rolling speed in the same manner as they would be on the road.

Indirect measurements allow measuring the tyre dynamic rolling radii. The resolution of the stroboscope method does not allow highlighting the influence of speed and tyre inflation pressure on the dynamic rolling radius, as the resolution of measuring method should be smaller than 0.1 mm. The pressure gauge should have a resolution of at least 1 kPa to allow accurately adjusting the inflation pressure.

The tyre internal pressure for rolling tyre changes in time and with rolling speed. For obtaining valid results, quasi-stationary conditions are required for the measurements.

Many references give simplified formulae for calculating the tyre dynamic rolling radius, but calculated results do not match the real values.

The values of dynamic rolling radius slightly increase with speed. The values of dynamic rolling radius clearly increase with inflation pressure. The influence of inflation pressure on the measured tyre dynamic rolling radius is almost three times larger than the influence of rolling speed.

Acknowledgement
The research activities presented in this paper were performed within the scientific research contract “Computerized system for testing automotive steering boxes in view of increasing the safety of traffic participants - SITECH”, National Research Development and Innovation Plan PNCDI III - Program 2 Subprogram 2.1 – “Bridge Grant”, financed by the Romanian Ministry of Research and Innovation.

References
[1] Genta G 2006 *Motor vehicle dynamics: modeling and simulation* World Scientific Publishing
[2] Rau M 2007 *Koordination aktiver Fahrwerk-Regelsysteme zur Beeinflussung der Querdynamik mittels Verspannungslenkung* PhD Thesis - Universität Stuttgart
[3] Kising A and Göhlich H 1988 Ackerschlepper-Reifendynamik. Teil 1: Fahrbahn-und Prüfstandsergebnisse *Grundlagen der Landtechnik* 38.3 (1988)
[4] Böhler H and Schlotter V 1999 Dynamikparameter von Grünlandreifen *Landtechnik–Agricultural Engineering* 54.3 (1999): 164-165
[5] DIN 70020-5:1986-12 Kraftfahrzeugbau; Reifen und Räder; Begriffe und Meßbedingungen *Automotive engineering; tyres and wheels; concepts and measuring conditions* 1986-12
[6] Bridgestone Europe 2008 *Bridgestone Pattern Digest 2008 Version* Zaventem, Belgium
[7] Stoicescu AP 1980 *Dinamica Autovehiculelor, Vol.1* Litografia IPB, Bucharest
[8] Andreescu Cr 2010 *Dinamica autovehiculelor pe roți* Editura Politehnica Press, Bucharest
[9] Jazar RN 2008 *Vehicle dynamics. Theory and Applications* Riverdale, NY: Springer Science Business Media

IOP Conf. Series: Materials Science and Engineering **252** (2017) 012015 doi:10.1088/1757-899X/252/1/012015

Investigation of the influence of vertical force on the contact between truck tyre and road using finite element analyses

Alexandra-Raluca Moisescu, Gabriel Anghelache

University POLITEHNICA of Bucharest, Automotive Engineering Department, Splaiul Independentei 313, sector 6, 060042, Bucharest, Romania

raluca.moisescu@upb.ro

Abstract. In the modern context of automobile integration with the emerging technologies of the interconnected society, the interaction between tyre and road is an element of major importance for automobile safety systems such as the intelligent tyres, as well as for passenger comfort, fuel economy, environmental protection, infrastructure and vehicle durability. The tyre-road contact generates the distribution of forces exerted on each unit area in the contact patch, therefore the distribution of contact stresses on three orthogonal directions. The numerical investigation of stresses distribution in the contact patch requires the development of finite element models capable of accurately describing the interaction between tyre and rolling surface. The complex finite element model developed for the 11R22.5 truck tyre has been used for investigating the influence of vertical force on the distributions of contact stresses. In addition to these contributions, the paper presents aspects related to the simulation of truck tyre radial stiffness. The influence of tyre rolling has not been taken into consideration, as the purpose of the current research is the investigation of tyre-road contact in stationary conditions.

1. Introduction

The integration of the modern automobile in the evolution of the increasingly interconnected human society depends on the development of emerging smart technologies such as the intelligent tyre systems [1]-[8]. In this progressively complex context, the interaction between tyre and road is an element of crucial importance not only for automobile safety and fuel economy, but also for passenger comfort, environmental protection, durability of vehicle and infrastructure [9]-[11]. The tyre-road contact is concretized through the distribution of forces exerted on each unit area in the contact patch, therefore the distribution of contact stresses, which can be decomposed on three orthogonal directions.

The experimental research of tyre-road contact stresses distributions involves complex measuring equipment, embedded in the rolling surface [12]-[15] or mounted inside the tyre [16]. The numerical investigation of the distribution of contact stresses throughout the contact patch surface requires the development of finite element models capable of accurately describing the tyre behaviour and the interaction between tyre and rolling surface [17]-[21]. Several complex measuring systems, as well as numerous finite element models have been developed in the University POLITEHNICA of Bucharest for the investigation of tyre-road interaction [1], [12], [21]-[24].

Taking into consideration the non-uniformity and anisotropy that generate the highly non-linear behaviour of tyres, the finite element model of a radial truck tyre has been developed with particular attention to the complexity of the tyre structure and multitude of different components. The complex finite element model developed for the 11R22.5 truck tyre has been used for investigating the influence of vertical force, applied in stationary conditions, on the distributions of contact stresses.

In addition to these contributions, the current paper presents aspects related to the numerical simulation of tyre radial stiffness. The influences of rolling speed and rolling conditions have not been taken into consideration, as the purpose of the current research is the investigation of tyre-road contact in stationary conditions. Furthermore, the results presented in this paper are obtained taking into account constant inflation pressure and zero degrees camber angle for the radial tyre model.

2. Finite element model of radial truck tyre

The finite element model for investigation of the contact between truck tyre and road has been developed using Abaqus software. The profile coordinates have been measured on the exterior contour of an 11R22.5 truck tyre as well as on the exterior and interior contours of a corresponding truck tyre section. The 3D model shown in Figure 1 has been created accordingly [22], taking into consideration the components of real tyre structure, including the tread profile with circumferential grooves, the shape and position of beads, the real position and orientation of cord layers, and the different types of rubber components.

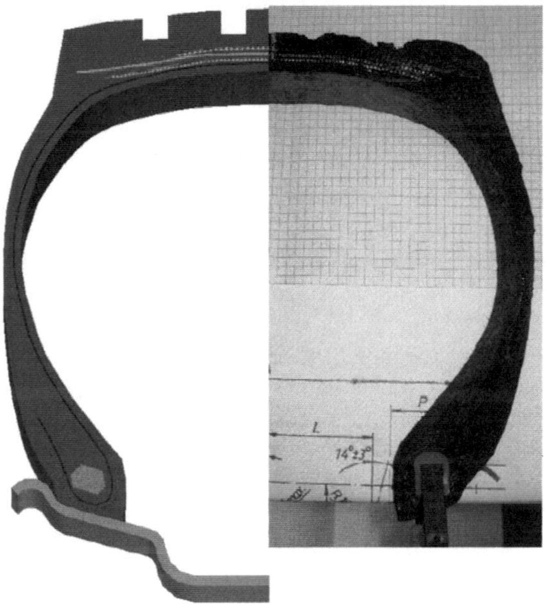

Figure 1. Perspective view of the 3D model compared to the image of the real tyre section

The finite element model includes the surface type elements for modelling the rebar layers, which have been defined with angle, spacing, section area and material properties, as well as the volume type elements for the different types of rubber components in the tyre structure [25]. Due to the orientation of belt plies, the model of tyre section is not symmetrical, and therefore the entire section has been modelled instead of using a symmetrical model for only half of the tyre section, which would have allowed reducing the number of elements and nodes [23]. The circumferential meshing is non-uniform, as it can be seen in Figure 2, in view of increasing the resolution in the contact patch [24], while maintaining a relatively small model size: 43214 nodes and 50441 elements.

Linear elastic properties have been attributed to the cord materials and bead materials [26], while hyperelastic behaviour has been defined using neo-Hookean model for the different types of rubber components [27]-[30]. Boundary conditions have been defined using a rigid body for modelling the rim, with all degrees of freedom suppressed excepting rotation around tyre axis, and a rigid planar surface representing the road, with one translational degree of freedom allowed on vertical direction.

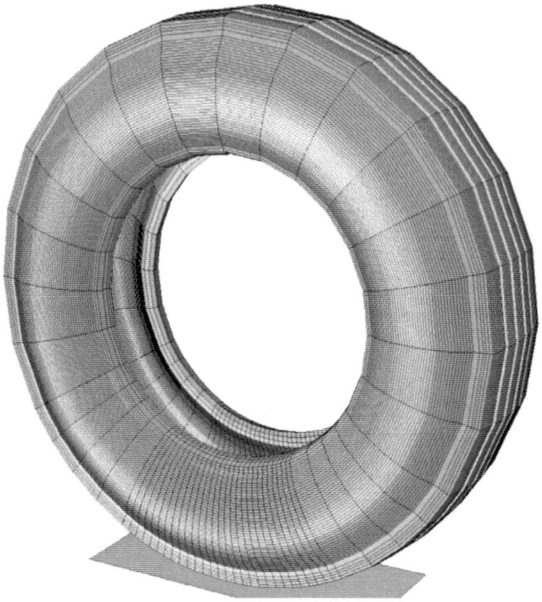

Figure 2. Three-dimensional tyre model with non-uniform circumferential meshing [31]

Tyre inflation pressure has been defined on the interior surface of the model, and subsequently road contact has been applied, with friction coefficient corresponding to real conditions [32],[33].

3. Steady-state analyses with vertical force applied between tyre and road surface

The steady-state analyses performed on the finite element model of the 11R22.5 truck tyre, with inflation pressure and with vertical force, allowed simulating the contact between tyre and road and obtaining the contact patch shape.

The contact patch shape resulting from the finite element analyses performed at 780 kPa inflation pressure with 20234 N vertical force has been compared, as shown in Figure 3, with the contact patch print determined experimentally in corresponding conditions. The tyre-road contact model is validated by the good similarity of these results, although there are some small differences of shape which can be explained by the mounting position with small camber angle of the real truck wheel. It can also be noticed that some ribs in the tread profile of the real tyre have very thin transversal sipes, which have not been included in the finite element model, because of limited meshing density.

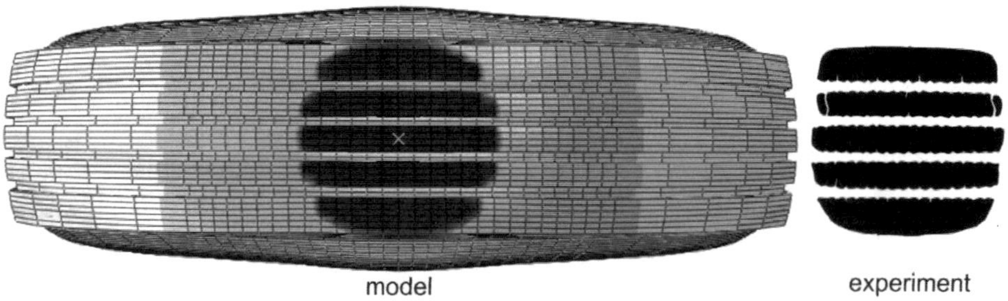

Figure 3. Contact patch shape of the truck tyre model at 780 kPa inflation pressure, compared to the contact print obtained experimentally

The radial stiffness curve of the truck tyre model shown in Figure 4 has been determined from the finite element analyses performed at 800 kPa inflation pressure, for a wide range of vertical loads. The highlighted point represents the radial stiffness obtained experimentally at 800 kPa inflation pressure, for a single value of vertical load, corresponding to the truck load in the road test conditions. It can be seen that the measured stiffness is relatively similar to the simulated value, but the real tyre stiffness is slightly higher than that of the finite element model. The radial stiffness of the tyre model tends to increase when the vertical load is higher [34], as it can be seen from the shape of the curve.

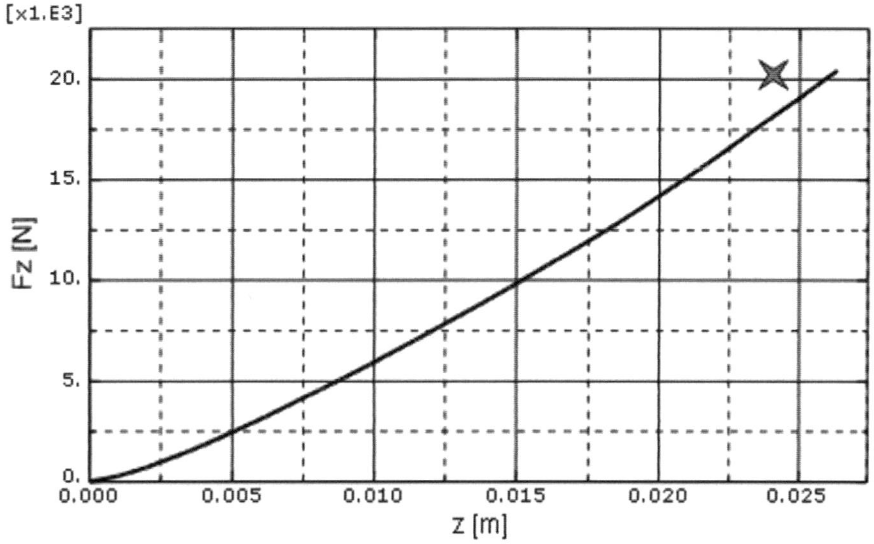

Figure 4. Radial stiffness curve of the truck tyre model at 800 kPa inflation pressure, compared to the value of radial stiffness obtained experimentally

The influence of vertical force on the shape of contact patch and on the distributions of contact stresses on three orthogonal directions obtained from simulations is presented in Figure 5; the comparison has been made between the results obtained at two different values of vertical force, 20445 N and 10223 N, applied in stationary conditions. The simulations have been performed on the finite element model of the 11R22.5 truck tyre with 780 kPa inflation pressure and 1° camber angle.

The dimensions of the contact patch are strongly influenced by the vertical force applied on the tyre model. The shape of the contact patch under smaller vertical load is closer to a circular shape.

The distributions of normal and lateral stresses are symmetrical with respect to the lateral plane containing the contact patch centre, while the longitudinal stresses are antisymmetrical with respect to this plane [35]. The symmetry of the distributions of normal and longitudinal stresses with respect to the longitudinal plane containing the contact patch centre is slightly affected by the camber angle. In stationary conditions, the maximum values of lateral stresses are located on the edges of tread ribs, and the global distribution of lateral stresses tends to be antisymmetrical with respect to the longitudinal plane containing the contact patch centre.

The values of longitudinal stresses are significantly smaller in the case of the lower vertical force, and the values of lateral stresses are also reduced in this case. The maximum values of normal stresses are approximately the same for the two different values of vertical force, and in both cases they are located at the contact patch centre. However the high values of normal stress are applied on a much smaller area of the contact patch in the case of the lower vertical force.

IOP Conf. Series: Materials Science and Engineering **252** (2017) 012015 doi:10.1088/1757-899X/252/1/012015

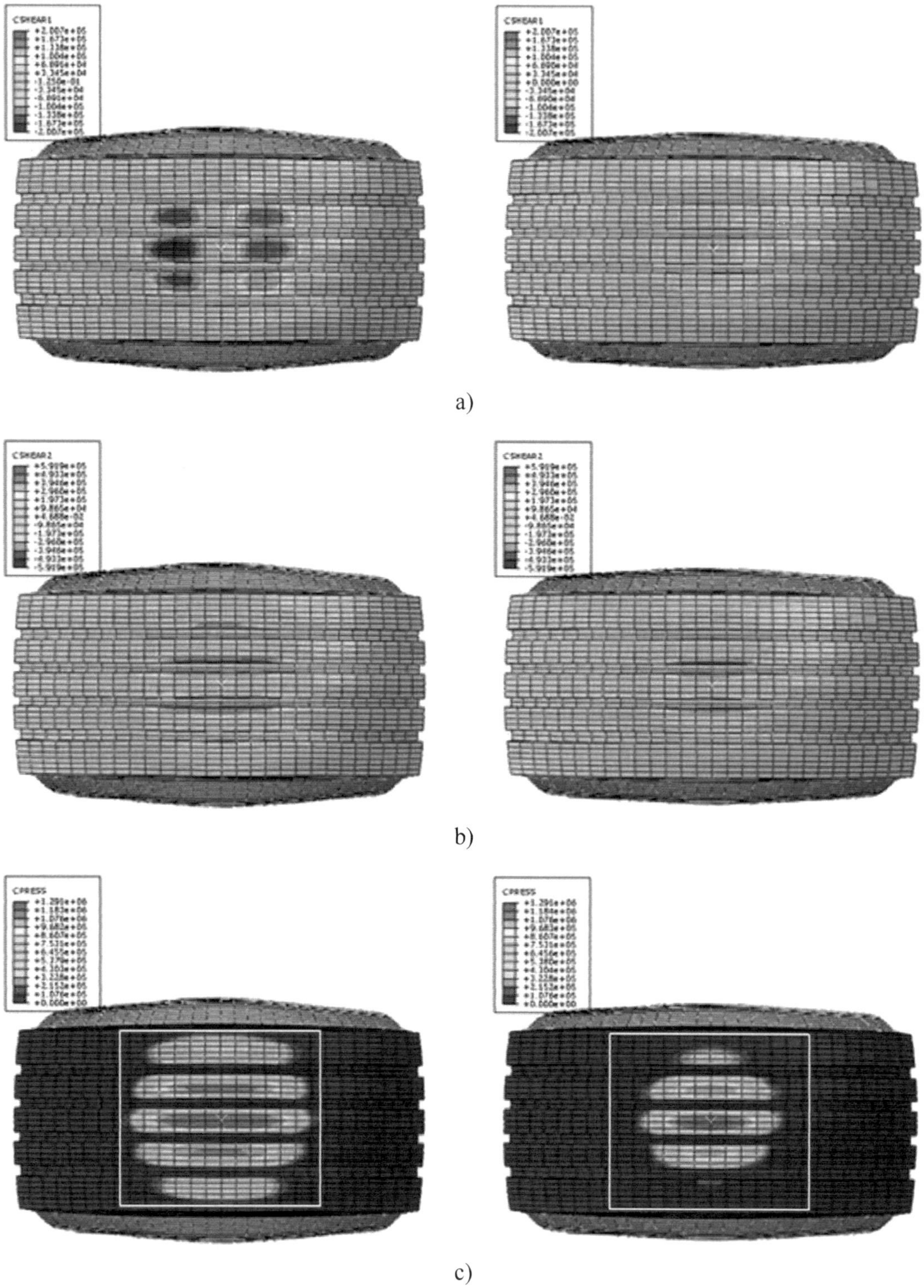

Figure 5. The distributions of contact stresses in [N/m^2] for finite element tyre model with 20445 N vertical force (left) and 10223 N vertical force (right) at 780 kPa inflation pressure: a) longitudinal stresses, b) lateral stresses, c) normal stresses

4. Conclusions

The finite element model developed for the 11R22.5 truck tyre has allowed investigating the contact patch shape and dimensions, as well as the tyre radial stiffness. These parameters have been used for finite element model validation against experimental results.

The influence of vertical force applied in stationary conditions on the distributions of contact stresses has also been emphasized by the results of finite element analyses. The dimensions of the contact patch are strongly influenced by the vertical force applied on the tyre model. The vertical force has negligible influence of on the peak values of normal stresses, but changes significantly the area of the contact patch on which the high values of normal stress are applied. The distribution of lateral stresses is also related to the vertical force applied on the tyre model, while the effect on the values of longitudinal stresses is even more significant, with major consequences on vehicle stability and safety.

Acknowledgement

The research activities presented in this paper were performed within the scientific research contract 'Experimental and Numerical Research on Tyre–Road Interaction in View of Increasing Automotive Safety and Road Transport Efficiency', National Research Development and Innovation Plan – IDEAS Program, financed by National Authority for Scientific Research of the Romanian Ministry of Education, Research and Innovation.

References

[1] Anghelache G 1999 Tyre-Road Interaction at High Passenger Car Rolling Speed *PhD Thesis - University POLITEHNICA of Bucharest*
[2] Yang X et al. 2013 Experimental Investigation of Tire Dynamic Strain Characteristics for Developing Strain-Based Intelligent Tire System *SAE International Journal of Passenger Cars-Mechanical Systems* 8;6(2013-01-0633):97-108
[3] Yang X et al. 2015 FE-Based Tire Loading Estimation for Developing Strain-Based Intelligent Tire System *SAE Technical Paper* 2015-01-0627 2015 doi:10.4271/2015-01-0627
[4] Garcia-Pozuelo D et al. 2017 A novel strain-based method to estimate tire conditions using fuzzy logic for intelligent tires *Sensors* 17.2 (2017): 350. doi: 10.3390/s17020350
[5] Garcia-Pozuelo D et al. 2017 A Strain-Based Method to Estimate Slip Angle and Tire Working Conditions for Intelligent Tires Using Fuzzy Logic *Sensors* 17.4 (2017): 874
[6] Buretea DL, Cormos CA 2009 Hybrid Car Battery Management *International Congress ESFA2009 Bucharest* Paper E09H102
[7] Buretea DL et al. 2006 Reducerea poluării prin optimizarea traficului rutier urban *International Conference CLEANPROD2007 Suceava*
[8] Nemtanu F et al. 2016 Mobility as a service in smart cities *Responsible Entrepreneurship Vision Development and Ethics* (2016): 425-435
[9] Li T, Burdisso R and Sandu C 2017 An Artificial Neural Network Model to Predict Tread Pattern-Related Tire Noise *SAE Technical Paper* 2017-01-1904 2017
[10] Huang H et al. 2015 Three-dimensional global pattern prediction for tyre tread wear *Proceedings of the Institution of Mechanical Engineers Part D: Journal of automobile engineering* 229.2 (2015): 197-213 http://dx.doi.org/10.1177/0954407014537640
[11] Wang C et al. 2016 The influence of the contact features on the tyre wear in steady-state conditions *Proceedings of the Institution of Mechanical Engineers Part D: Journal of Automobile Engineering* (2016): 0954407016671462. doi: 10.1177/0954407016671462
[12] Anghelache G et al. 2011 Measuring system for investigation of tri-axial stress distribution across the tyre-road contact patch *Measurement* 2011 Mar 31;44(3):559-68
[13] De Beer M, Fisher C 2013 Stress-In-Motion (SIM) system for capturing tri-axial tyre-road interaction in the contact patch. *Measurement* 46.7 (2013): 2155-2173
[14] Hernandez JA et al. 2013 Impact of tire loading and tire pressure on measured 3D contact stresses *Airfield and highway pavement 2013: Sustainable and efficient pavements* (551-560)

[15] Lundberg OE et al. 2017 A compact internal drum test rig for measurements of rolling contact forces between a single tread block and a substrate *Measurement* 103 (2017): 370-378
[16] Matilainen M, Tuononen A 2015 Tyre contact length on dry and wet road surfaces measured by three-axial accelerometer *Mechanical Systems and Signal Processing* 2015 28;52:548-58
[17] Hernandez JA et al. 2015 Numerical prediction of three-dimensional tire-pavement contact stresses *Illinois Center for Transportation Research Report No. ICT-17-004* ISSN0197-9191
[18] Olatunbosun OA, Bolarinwa EO 2002 Analysis of rolling tire dynamic response using a 3D finite element model *Twenty-first Annual Meeting and Conference on Tire Science and Technology Akron Ohio USA*
[19] Chang YP, El-Gindy M 2002 FEA Rotating Tire Modeling for Transient Response Simulations *ASME 2002 International Mechanical Engineering Congress and Exposition. American Society of Mechanical Engineers* 2002. doi: 10.1115/IMECE2002-33200
[20] Shokouhfar S et al. 2016 Modal analysis of a rolling truck tyre subjected to inflation pressure and vertical deflection *International Journal of Vehicle Systems Modelling and Testing* 2016;11(2):116-41
[21] Negrus EM et al. 1997 Finite Element Analysis and Experimental Analysis of Natural Frequencies and Mode Shapes for a Non-Rotating Tyre *Vehicle System Dynamics Supplement vol. 27* pp. 221-224. http://dx.doi.org/10.1080/00423119708969656
[22] Moisescu AR 2010 Research on Automotive Active Safety (The Tyre as an Element of Active Safety) *PhD Thesis - University POLITEHNICA of Bucharest*
[23] Moisescu AR, Anghelache G 2005 A Numerical Approach to the Contact between Tyre and Road *CAR2005 - The 9th International Congress on Automotive Pitești 2005* ISBN 973-690-450-4 Paper CAR20051099
[24] Moisescu AR, Fratila G 2011 Finite Element Model of Radial Truck Tyre for Analysis of Tyre-Road Contact Stress *University POLITEHNICA of Bucharest Scientific Bulletin Series D Mechanical Engineering* Nr. 3 ISSN 1454-2358 Bucharest
[25] Wei C et al. 2017 A finite-element-based approach to characterising FTire model for extended range of operation conditions *Vehicle System Dynamics* 55.3 (2017): 295-312
[26] Celigüeta-Azurmendi MC 2014 Modelización estructural de un neumático de competición por el método de los elementos finitos *MSc Thesis Universidad de Navarra*
[27] Yang X et al. 2010 Materials Testing for Finite Element Tire Model *SAE Int. J. Mater. Manuf.* 3(1):211-220 2010 doi: 10.4271/2010-01-0418.
[28] Anghelache G Moisescu AR 2008 Analysis of Rubber Elastic Behaviour and Influence on Modal Properties of Truck Tyre *Materiale Plastice* 45.2 (2008): 143-148
[29] Lopez Arteaga I 2011 Green's functions for a loaded rolling tyre *International Journal of Solids and Structures* 2011 Dec 15;48(25):3462-70
[30] Behroozi M et al. 2012 Finite element analysis of aircraft tyre – Effect of model complexity on tyre performance characteristics *Materials & Design* 35 (2012): 810-819
[31] Anghelache G, Moisescu AR 2015 Truck Tyre-Road Contact Stress Measurement and Modelling *Proceedings of the 4th International Tyre Colloquium University of Surrey Guildford UK 2015* http://epubs.surrey.ac.uk/id/eprint/807823
[32] Ghoreishy MH 2008 A state of the art review of the finite element modelling of rolling tyres *Iranian Polymer Journal* 2008 Aug 1;17(8):571-97
[33] Svendenius J 2007 Tire modeling and friction estimation *PhD Thesis - Department of Automatic Control - Lund Institute of Technology - Lund University*
[34] Li B, Yang X and Yang J 2014 Tire model application and parameter identification-A literature review *SAE International Journal of Passenger Cars-Mechanical System* 2014 Apr 1;7(2014-01-0872):231-43
[35] Anghelache G Moisescu AR 2006 Numerical and Experimental Research on Truck Tire-Road Contact Stress *FISITA Congress Yokohama Japan 2006* Paper No. F2006V162 Session 3

IOP Conf. Series: Materials Science and Engineering **252** (2017) 012016 doi:10.1088/1757-899X/252/1/012016

Upon the reconstruction of accidents triggered by tire explosion. Analytical model and case study

L.Gaiginschi, I.Agape, S.Talif

Faculty of Mechanical Engineering, Technical University "Gheorghe Asachi" Iasi, Romania

lidiagaiginschi@tuiasi.ro, iulian.agape@tuiasi.ro , stalif@yahoo.com

Abstract. Accident Reconstruction is important in the general context of increasing road traffic safety. In the casuistry of traffic accidents, those caused by tire explosions are critical under the severity of consequences, because they are usually happening at high speeds. Consequently, the knowledge of the running speed of the vehicle involved at the time of the tire explosion is essential to elucidate the circumstances of the accident. The paper presents an analytical model for the kinematics of a vehicle which, after the explosion of one of its tires, begins to skid, overturns and rolls. The model consists of two concurent approaches built as applications of the momentum conservation and energy conservation principles, and allows determination of the initial speed of the vehicle involved, by running backwards the sequences of the road event. The authors also aimed to both validate the two distinct analytical approaches by calibrating the calculation algorithms on a case study

1. Introduction

Overturning and rolling a vehicle involves a high enough energy input; it is taken from the kinetic energy, which means that such accidents occur at relatively high speeds. Overturning accidents have the specificity that the vehicle can no longer be controlled, it's movement being made by sliding and rolling the body. For these reasons, the consequences of these accidents are generally serious, both for the occupants of the overturned car and for the other road users.

Generally, the overturning of the vehicle has a unique cause, namely the action of a mechanical momentum generated by lateral forces whose arm is related to the road contact points of the wheels that remain in contact with the tread. The lateral forces that generate the overturning moment are the result of either the driver - road interaction or the defects that occurred during the vehicle's travel. In the second case, the most common accidents are generated by the explosion of a tire, as a consequence of dynamic imbalances generated in vehicle's movement. Sometimes, after the explosion, the evolution of the vehicle can also be marked by the driver's behaviour if he performs braking maneuvers and/or tight turns. The cause of such a roll is relatively easy to set up after the typical trace on the road as well as damage analysis after the accident.

In the study and analysis of road accidents is particularly important to establish kinematics parameters during events (vehicle speed in the main stages of movement of motor vehicles and victims, distances traveled by the vehicle during these stages) and kinematic parameters that would allow to avoid the event (the speed of the vehicle for the event could have been avoided).

Over the years there have been developed various models for calculating the vehicles speeds at different stages of an accident, patterns that differ depending on the type of event to which it is addressed (energetic brakes, collisions, pedestrian hits, etc.), the theoretical principles applied in establishing relationships calculation, the possibility of using traces aquired at the scene. Thus, an accident reconstruction will require a decision on the appropriate calculation model, and adapt it to the specific case of the respective accident.

The paper presents an analytical model for determining the initial velocity of a vehicle involved in an accident triggered by the explosion of one of its tires. Further movement of the vehicle was in the first phase of skidding, then overturning and rolling. The elaborated model was calibrated in the paper, being applied on a case study.

2. The determination of the speed of the vehicle at the moment of the accident

Between the moments of the tire explosion and the overturning, the vehicle is skidding. The determination of the vehicle's speed before skidding is based on an energy method, constituting in fact an adaptation of the energy conservation law applied between the initial and final moments of the skid phase.

The speed of the vehicle before skidding, w_0 [m/s], is determined with the equation:

$$w_0 = \sqrt{w_i^2 + 2 \cdot \varphi_t \cdot g \cdot S_d} \tag{1}$$

where: w_i [m/s] – vehicle speed at the end of skidding phase; φ_t [-] - transverse adhesion coefficient; $g=9.81$ [m/s^2] - gravitational acceleration; S_d [m] - skidding distance.

S_d is determined by measurements at the site of the accident taking into consideration the specific methodology for investigating and assessing skidding traces.

The approach of the skidding direction by considering the anisotropy of the adhesion coefficient in the contact surface between the tire and the track is original and is justified by the need to increase the precision of the speed determination. Alternatively, an experimentally determined resistance coefficient can be used, but it does not reflect the influence of skidding direction.

Adherence on the transverse direction is less than on the longitudinal direction, due to the fact that the forms and surfaces of contact spots are different when sliding friction occurs on the longitudinal or transverse direction, due to the tires construction. As a result, the transverse adhesion coefficient [1] will be determined with the equation:

$$\varphi_t = 0.8 \cdot \varphi \tag{2}$$

with φ [-] – longitudinal adhesion coefficient, which is expressed according to the road structure, road condition and tire type.

2.1. Determining the speed of the vehicle at the start of rolling phase

A first approach for determining the speed of the vehicle at the start of rollover is based on the law of impulse conservation. Special attention should be given to identifying the axle of the vehicle after which the vehicle is rolled.

In principle, the lateral movement component, produced by skidding combined with the initial longitudinal translation motion of the vehicle, determines its complex, parallel planar evolution in the road surface. Frequently, due to this complex motion, the rollover is initiated around an axis containing the contact points of the wheels of the same axle with the rolling path.

The accurate setting of the axis after which the overturning is initiated will be based on the acquisition and interpretation of the vehicle's traces and from the scene of the accident. Our model considers that rollover occurs around the transverse axis that contains the contact points of the front wheels of the vehicle with the road.

Generally, in the vehicle's lifting movement in neutral stability position (NSP), the mechanical rotating work of the vehicle around the above-mentioned axis is entirely transformed into the potential energy of the vehicle at the NSP.

Therefore, the vehicle energy when the rollover starts is given by:

$$\frac{J_{ty} \cdot \omega^2}{2} = M \cdot g \cdot \left[\sqrt{\left(\frac{A}{2}\right)^2 + h^2} - h\right] \tag{3}$$

where J_{ty} [kg·m^2] represents the mass moment of inertia of the vehicle relative to a transverse axis passing through the contact points between the road and the front wheels; ω[rad/s] - the angular velocity of the vehicle in relation to this axis; h[m] - the vehicle's center of gravity height; M[kg] - total weight of the vehicle (including the curb weight and mass of passengers); A[m] – the vehicle wheelbase.

For passenger vehicles the height h is determined, according to [1], from:

$$\frac{h}{A} = 0.195 \ldots 0.23 \tag{4}$$

higher values characterizing loaded vehicle.

The mass moment of inertia J_{ty} is determined using Steiner's theorem:

$$J_{ty} = J_{yy} + M \cdot \left[\left(\frac{A}{2}\right)^2 + h^2\right] \tag{5}$$

where J_{yy} [kg·m^2] represents the mass moment of inertia of the vehicle relative to the transverse axis passing through its center of mass.

The mass moment of inertia J_{yy} is determined, according to [2], with the relation:

$$J_{yy} = \frac{H+h}{k_v} \cdot A \cdot M \tag{6}$$

H [m] – the height of the vehicle; k_v [-] – a coefficient that takes into account vehicle class and axis after which is calculated the mass moment of inertia [2].

From equation (3) is determined the angular velocity ω [rad/s] of the vehicle relative to this axis:

$$\omega = \sqrt{2 \cdot M \cdot g \cdot \left[\sqrt{\left(\frac{A}{2}\right)^2 + h^2} - h\right] / J_{ty}} \tag{7}$$

Without taking into account the vehicle's deformations, speed w_i^* [m/s] at the beginning of rollover results from the equation of momentum conservation:

$$J_{ty} \cdot \omega = M \cdot w_i^* \cdot h \tag{8}$$

$$w_i^* = J_{ty} \cdot \omega / (M \cdot h) \tag{9}$$

The speed w_i [m/s] at the end of slippage phase is determined by the known relation:

$$w_i = \sqrt{w_i^{*2} + \frac{2 \cdot E_d}{M}} \tag{10}$$

where E_d [J] - the deformation energy of the vehicle body elements.

For front or rear collision it has been defined a deformation coefficient C [2], [3], established experimentally, with the same dimensions [kN/m] as the elasticity coefficient.

If is denoted by F(ζ)the function of force variation that produces deformation then we can write:

$$E_d = \int F(\zeta)\, d\zeta = \int C \cdot \zeta\, d\zeta \tag{11}$$

After integration, the deformation energy E_d of the vehicle is obtained by:

$$E_d = \frac{C_f}{2} \cdot \zeta_f^2 + \frac{C_s}{2} \cdot \zeta_s^2 \tag{12}$$

C_f [kN/m] – front stiffness coefficient; C_s [kN/m] – rear stiffness coefficient; ζ_f and ζ_S – medium deformations for the body vehicle at the front and rear part.

The structure of the vehicle and the deformations (depending on speed impact) have the greatest influences on the stiffness coefficients. Higher values for the C coefficients are characteristic for domains of low deformations or rear collisions. It has therefore been considered that the rollover described above produces deformation of the front and rear parts of the vehicle, similar to collinear collisions, both front and rear.

A second approach to the problem is purely energetic, considering the total dissipation of the kinetic energy of the vehicle from the initial moment of rolling, due to the rolling resistance of the vehicle. After t_{r0} [s], the vehicle will stop permanently from rolling after covering the distance S_r [m].

Thus, the w_i speed can be established also from the energy balance:

$$w_i = \sqrt{2 \cdot g \cdot f \cdot S_r} \tag{13}$$

Finally, one can determine the speed of the vehicle before slippage, w_0 from equation (1).
The time t_{r0} [s] in which was covered the distance S_r that the vehicle was rolling is determined based on the coefficient of running resistance, f:

$$t_{r0} = \frac{w_i}{f \cdot g} \tag{14}$$

Between the total angle of rotation θ (the angle of rotation of the vehicle during S_r) and angular velocity from the beginning of rotation exists the dependence:

$$\theta = \frac{2 \cdot \omega_r \cdot t_{r0}}{3} \tag{15}$$

where ω_r represents real angular velocity that must be entered in the momentum conservation equation:

$$\omega_r = \frac{M \cdot w_i \cdot h}{J_{ty}} \tag{16}$$

Since one rotation is the rotation of the car 360 degrees, finally is determined the total number of revolutions N, which is compared with data available from research at the accident site, checking the good agreement between reality and calculation methods :

$$N = \theta / 360° \tag{17}$$

We specify that the conditions of the road separation of the rear axle wheels and reaching the neutral stability position are assured implicitly due to the equation (3).

The algorithm neglects the energy consumed by the car to rotate 90 degrees during the skid, around the vertical axis (z - z) through its center of gravity, and the energy consumption of the suspension elements during the rollover phase. At the usual speeds of rollover accidents, these energy consumptions are really small and therefore negligible, and will slightly affect the precision of the calculation [4].

3. Model calibration. Case study

It was considered a real event, produced in Iasi County, an accident triggered by the explosion of the right rear tire of the vehicle. Further movement of the vehicle was in the first phase of skidding, then rolling.

The involved vehicle, Opel Vectra brand, has the following technical characteristics: wheelbase A = 2.64 [m]; Front track E = 1.435 [m]; Length L = 4.477 [m]; Width B = 1.707 [m]; Height H = 1.428 [m]; Own mass Mp = 1317 [kg].

At the time of the accident in the car there were 5 people, with a medium mass of 75 [kg], so the total mass of the car was: M = 1317 + 5 · 75 = 1692 [kg].

According to the investigation at the site of the accident, and the traces collected (figure 1), the car rolled around the transverse axis passing through its center of mass, the rolling being produced due to the tipping around the front wheels when they entered the rainwater drainage duct and they impacted in a tilted position the channel edge with the front side.

During the rolling, only the front and rear parts of the vehicle were in contact with the ground outside the road, blows that generated their intense deformation (the rear of the car was deformed inwards by a length ζ_s= 0.3 [m] and the front part was pushed inwards for a length ζ_f = 0.5 [m]) - figure 2.

Figure 1 – traces collected at the site of the accident

Between the traces printed on the road a distance of 1.6 [m] was measured. Considering the thickness of at least 0.2 [m] of a trace (at least equal to the width of the tread of a tire), it was determined the real distance between the middle of the traces: 1.6 + 0.2 = 1.8 [m], which is much larger than the front track E = 1.4 [m] of the car. It results with certainty that those traces are typical of skidding and not braking.

Figure 2 – the vehicle's front and rear part deformations

The slippage took place on a dry asphalt road with medium wear; in this case the longitudinal adhesion coefficient has a value φ = 0.65...0.7, given that this car was not equipped with ABS, and the transverse adhesion coefficient $\varphi_t = 0.8 \cdot \varphi = 0.8 \cdot (0.65...0.7) = 0.52...0.56$ [2]. From the moment when it began to print traces of skidding and until leaving with the front wheels the rainwater collecting channel, the car has traveled, according to the data acquired at the scene, a distance S_d = 38. 5 [m].

The rollover of a vehicle is carried out with a uniform deceleration. In such situations, the literature [5] recommends a coefficient of forward resistance f = 0.43 for rolling on the ground. According to the data acquired at the scene, and the accompanying sketch, the car rolled over a distance S_r = 36 [m]. For the analyzed car, the front stiffness coefficient C_f has a value of approx. 628 [kN/m] and the stiffness coefficient of the rear part is C_s = 467 [kN/m].

Applying the calculation algorithm described in the first part of the paper, it was determined the speed of the car before skidding, between 92.83 and 97 [km/h]. The total number of rotations N calculated was 1.45. In order for the car to stop on the roof (figure 3), 1.5 turns were needed and the calculated number is close to 1.5, with an error of only 3.3 [%], which reflects the good concordance between reality and calculation methodology.

IOP Conf. Series: Materials Science and Engineering **252** (2017) 012016 doi:10.1088/1757-899X/252/1/012016

Figure 3 – the final position of the car, after the rollover

4. Conclusions

The paper presents a method of precise determination for the kinematic parameters of a vehicle which, as a result of the explosion of a tire, begins to skid, overturns and rolls, as well as the determination of rolls, made by the car till stop.

A limitation of this method is classic in the reconstruction of accidents and refers to the dependence of the method's precision by the abundance and the quality of traces acquired at the site of the accident. The method has limited applicability in the case of a vehicle that is overturned around a longitudinal axis. In this case, the deformations are placed on the lateral sides, the experimental data regarding the lateral stiffness coefficients of the vehicle being very poor. Last but not least, the overall precision of the model lies in the precision of determining the average deformations of the vehicle.

The originality of the paper is the simple but unitary approach of the analytical determination of the vehicle kinematics in accidents caused by tire explosions. The paper is representative from the aspect of considering the skidding direction and, implicitly, of the adhesion anisotropy, and also the correct identification of the axle around which the vehicle initiates the supposed rotation.

By applying the analytical model proposed in the paper it is possible to determine with sufficient precision the initial speed of a vehicle involved in a road accident triggered by the explosion of a tire, the vehicle subsequently evolving through skidding and overturning / rolling.

References

[1] Danner, M., Halm, J. Technische - Analyse von Verkehrsunfallen. Eurotax (International) AG CH - 8808 Pfaffikon, 1994;

[2] Gaiginschi R., - Reconstructia si expertiza accidentelor rutiere, Editura Tehnica, Bucuresti 2009;

[3] Navin, F., MacNabb, M., and Miyasaki, G., "Elastic Properties of Selected Vehicles," SAE Technical Paper 880223, 1988

[4] Funk J., LuepKe P. – Trajectory Model of Occupants Ejected in Rollover Crashes. Accident Reconstruction, Warrendele, USA, 2007

[5] Brach R., Brach. M. – „Vehicle Accident Analysis and Reconstruction Methods”, SAE International, Warrendale, U.S.A. 2005

IOP Conf. Series: Materials Science and Engineering **252** (2017) 012017 doi:10.1088/1757-899X/252/1/012017

Dynamic models to analyse the influence of the seat belt in a frontal collision

Oţăt Oana, Dumitru Nicolae, Dumitru Ilie

University of Craiova, Faculty of Mechanical Engineering

Calea Bucuresti str, no. 107, 200512, Craiova, Dolj, Romania

Corresponding author's e-mail: otatoana@yahoo.com

Abstract. Traffic accidents are influenced by various factors, yet, the highest impacting ones are related to vehicle impact speed and collision type. Also, passive vehicle safety systems play a significant role upon the injuries suffered by vehicle occupants. Under the circumstances, a particularly important aspect to consider when using such systems is the position of the vehicle's driver and its occupants. In what follows we embark upon an in-depth analysis in order to investigate the contact effects between the seat belt and the driver, under a dynamic regime. We set out to identify the variation of the kinematic and dynamic parameters for both the driver and the seat belt via comparative analyses between the normal position of the driver and some other out of position instances, considered as critical.

1. Introduction

The identification and the analysis of the factors that concur to the occurrence of traffic accidents remains still a highly complex issue, given the large number of disturbing factors, the various traffic situations, and, also, the different behaviour and reactions manifested by each driver. An increased concern with regard to the development and the optimisation of virtual models to analyse the driver's behaviour at the moment of impact is highlighted within a series of research studies such as in [2,4,6,7]. The dynamic seatbelt simulations as well as vehicle overturning tests are outlined by Taheo et al. in [7], by means of the MADYMO software.

Further research aims regarding vehicle occupant behaviour analysis envisage the proper positioning of the dummy on the driver's seat have been addressed by a group of researchers from DYNAmore GmbH in [6]. Within this research study, two positioning methods of the dummy were highlighted. The cinematic and dynamic study of the driver's behaviour in the event of a frontal collision is put forward in [8]. Also, within this research approach, special attention was paid to the identification of an effective method to assess the injuries suffered by the occupant in the head, thorax and pelvis. In order to determine the differences regarding the head injury risk in [8], some vehicle - pedestrian and vehicle - bicycle collision scenarios were investigated.

The most frequently investigated part of the human body following traffic accidents is the thorax area. The research study carried out in [3] highlights the initial steps undertaken to develop a finite element model based on the LS-Dyna code in the event of a collision within the thorax area. In order to validate the results obtained further comparisons to PMHS (post mortem human subjects) tests provided by mainstream literature have been developed. Another reference study in the development of a thorax-area model is set forth in [1]. Based on the MRI images obtained via scanning, an anthropomorphically correct model of a 5th percentile female-type dummy was developed. Then, the model developed was tested in both lateral collisions (generating results with an increased bio-fidelity) and frontal collisions. The effect of a failed automobile front seat in the case of a rear-end collision and its effects on the occupants was studied in [5]. The design and the developing of two collision tests applied to the same conditions enabled a reliable comparison between a normal seat and a failed one.

The analysis of the influence triggered by the safety belt upon the driver has been and continues to be a focal topic dwelled upon within a series of previously published papers. The abdomen and the thorax are structural elements of the human body that come in contact with the safety belt, thus, most of the studies have been carried out to analyse normal driving positions. However, the novelty put forward in the present paper underpins a close analysis of the dynamic contact for the various positions of the driver in relation to the steering wheel, in order to identify some critical situations with reference to the influence exercised by the safety belt. Moreover, we aim at designing and developing a database encompassing virtual contact dynamic models to carry out a complex analysis upon the influence of the safety belt by reconsidering some material models as well as a series of kinematic and dynamic parameters that define the contact.

2. Methodology

The influence of the safety belt upon the driver's thorax and pelvis will be investigated by applying two interdependent methods, i.e. virtual prototyping and experimental tests.

The monitoring of the collision, as a kinematic and a dynamic structure, is carried out under the same conditions for both virtual prototyping and experimental tests.

2.1. Virtual prototyping

To design and run the virtual prototyping, we develop a frontal collision model between a medium-sized vehicle and a rigid obstacle, the vehicle speed is set at 50 km/h. The collision simulation is completed via Virtual Crash, dedicated software for traffic accidents modeling and reconstruction. Taking our investigation a step further, we have developed an interface between Virtual Crash and LS-Dyna that enables us to input virtual prototyping data in the LS-Dyna software. The virtual model includes the driver, the seat, the dashboard, and the passive safety systems (airbag and seat belt).

Aiming to analyse the influence of the safety belt on the driver, we had first to define the contact between the component parts. Thus, Figure 1 below illustrates the parameter window that we set to define a contact of an Automatic_nodes_to_surface -type. The two contacts were defined between the thorax area of the dummy and the seat belt and between the pelvis and the safety belt. Our selection with regard to this type of contact was based on some of the features (considerations) we had to take into account such as thickness, orientation of segments not needed, the contact from both sides, initial penetrations are detected, possible to change or scale contact thickness, friction and damping available [10].

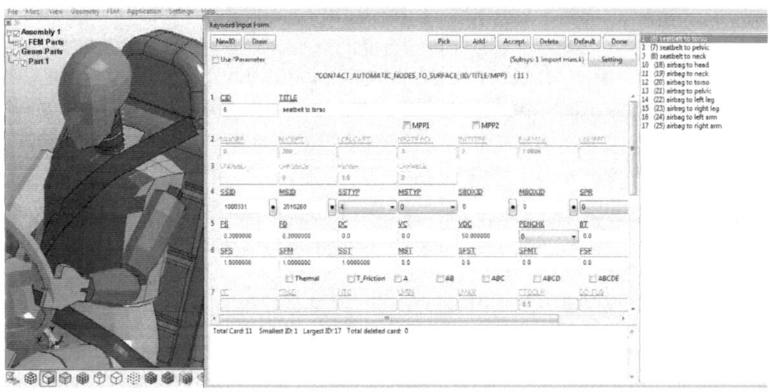

Figure.1 Defining the seat belt – thorax contact

To establishing the contact model, we took into consideration the type of the component materials as well. Hence, the material used in the thorax area has elastic and visco-elastic properties, of a plastic-kinematic and viscoelastic type. In the pelvis area, the material is of low density foam type. To model the seatbelt we used a seat belt type material, i.e. piecewise linear plasticity, the properties of which are illustrated in Figure 2.

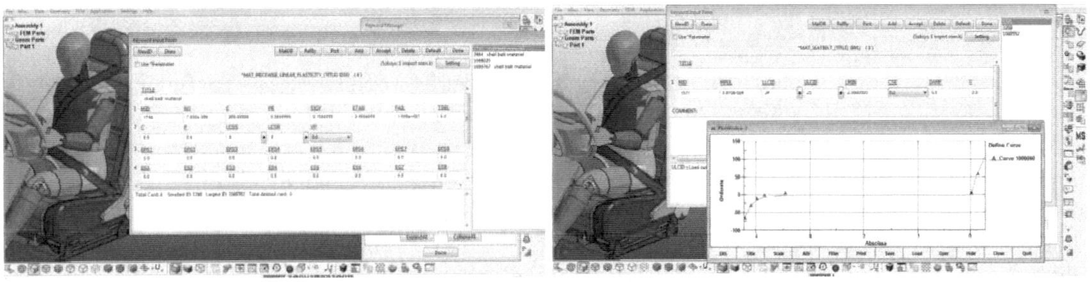

Figure 2. Material properties for the seat belt element

2.2. Experimental tests

The experimental tests, carried out to investigate the frontal collision, have been performed with two vehicles, under polygon conditions, using a dummy, data acquisition systems, sensors and other equipment meant to register acceleration values, force and torque rates. During the experimental test, we positioned the dummy on the driver's seat observing the procedures applied to reach the objectives set. The vehicle was subject to a frontal collision with another vehicle at a speed of 50 km/h. The restrain systems tested are not correlated with the head, thorax and lower limb transducers, if the biomechanical-based data benchmarks of the injury criteria are exceeded.

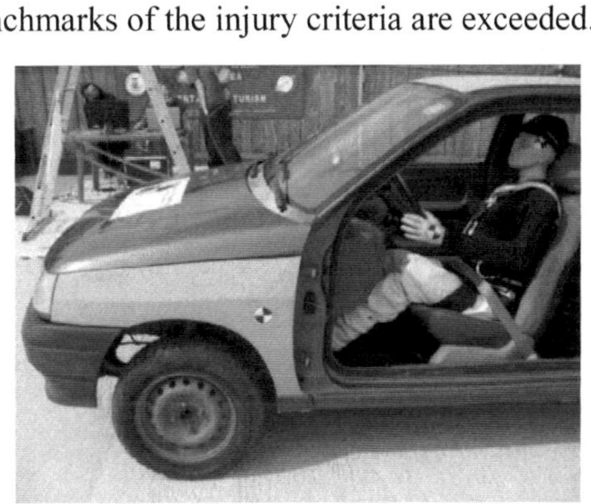

Figure 3. Positioning the dummy in the driving position

In the preliminary test phase, the sensor network, the measuring transducers and related wiring are mounted, observing the required safety and protection measures, hence reducing the disturbing possibilities that might occur during normal vehicle driving. The vehicle is equipped with redundant sensors and the information system selected to measure the variables of interest both on the dummy and on the vehicles involved in the collision. The characteristic points of the collision surface were appropriately marked with reflective cones. The equipment used for the dynamic measurements is of Xsens-type and DSD-PIC DAQ. This equipment is mounted on the vehicle before each test, see Figure 4 and Figure 5. To carry out the tests and to mount the measuring and the recording equipment we rested back the rear seat.

IOP Conf. Series: Materials Science and Engineering **252** (2017) 012017 doi:10.1088/1757-899X/252/1/012017

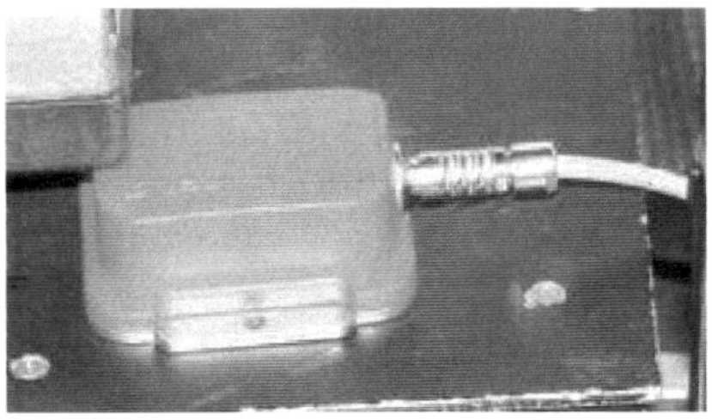

Figure 4. INS system of Xsens-type

Figure 5. DSD system with flexible modules

Based on sensitive measurement procedures we set up the experimental polygon and the measuring chains for each controlled variable of interest. The polygon where the collision tests have been carried out was foreseen with a fixed coordinate system, based on which we could measure the dynamic elements. In order to measure the force in the seat belt we used strain gauges, via which, following the calibration procedure, we determined the strains, and, subsequently, by means of equivalent relations, the corresponding forces.

2.3. Results validation

To validate our virtual model, we developed a comparative analysis between the modelling-based results and the results determined experimentally. Figures 6 and 7 indicate the seat belt force variations in the pelvis area and in the thorax area, determined both experimentally and virtually, via the LS-Dyna software.

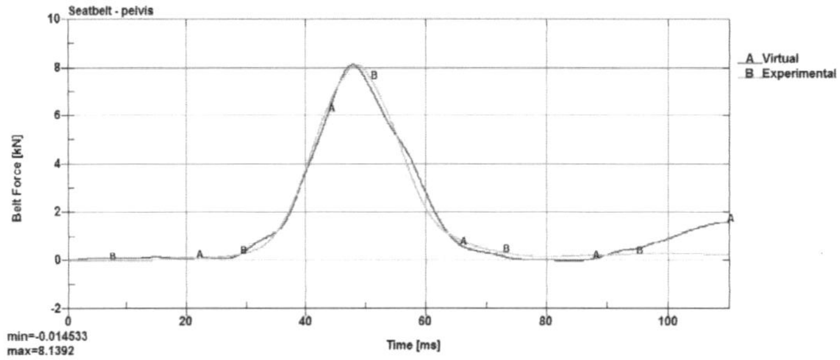

Figure 6. The force in the seat belt - pelvis

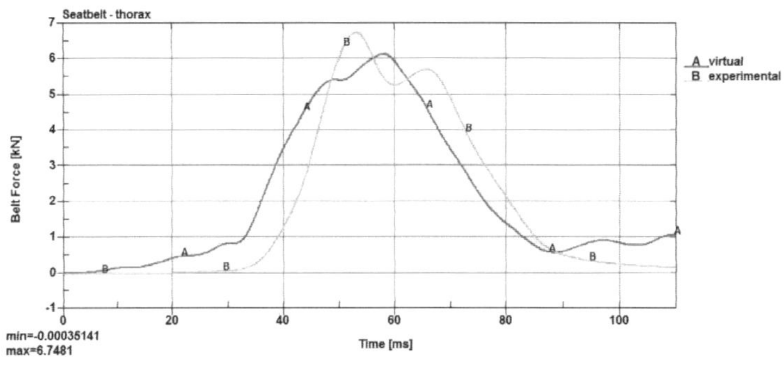

Figure 7. The force in the seat belt - thorax

As indicated in the figures above, the results obtained via a dynamic environment prototyping of the dummy-seat-safety systems assembly are very close to those established following experimental tests, hence validating our model.

Then, in order to analyse the influence of the safety belt on the driver, in a normal position of the dummy, we investigated the graphic representation of the contact forces generated between the seat belt and the pelvis, and respectively the thorax – see Figure 8.

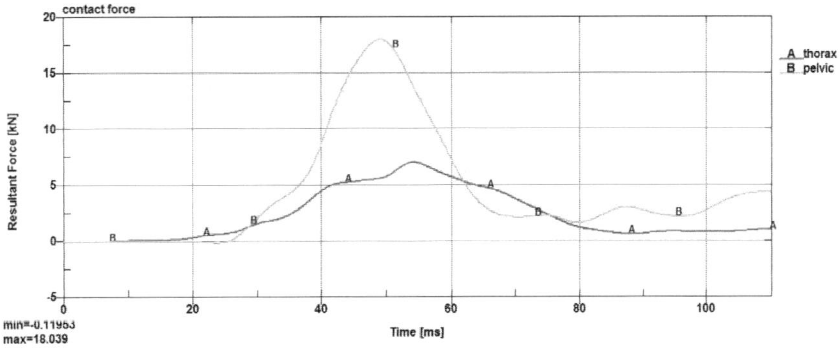

Figure 8. The contact force between seat belt – thorax – pelvis

In the pelvis area, the maximum contact force is of 18 kN, whereas in the thorax area, the maximum contact force is of 7 kN. Figures 9-12 illustrate the strains and Von Mises equivalent stresses in the pelvis and thorax areas:

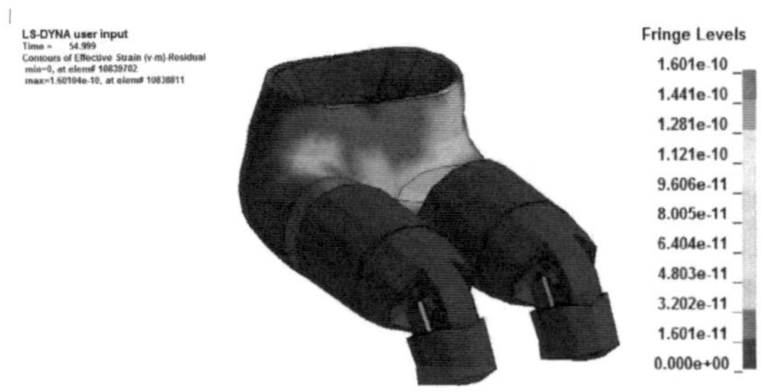

Figure 9. Effective strain – pelvis

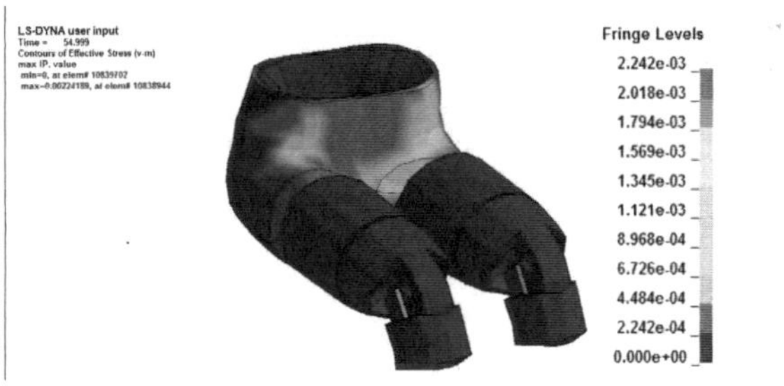

Figure 10. Effective stress Von Mises – pelvis [kN/mm^2]

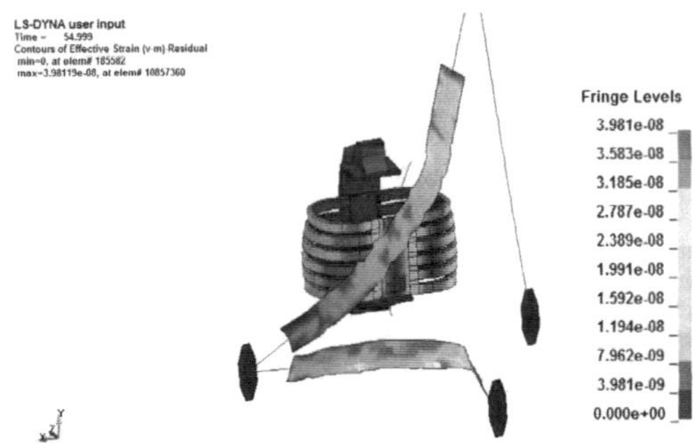

Figure 11. Effective strain – thorax

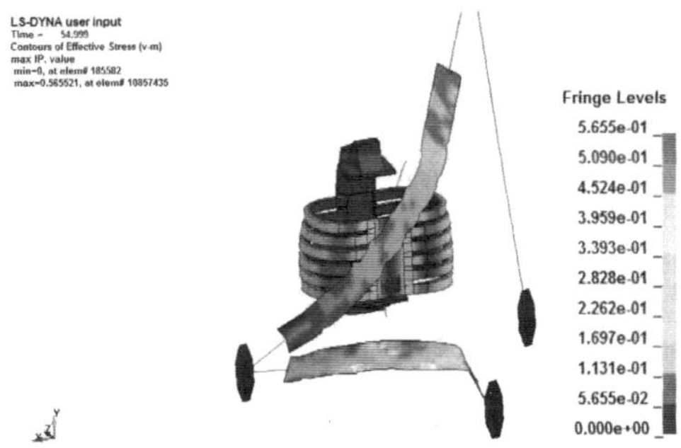

Figure 12. Effective stress Von Mises – thorax [kN/mm^2]

A close analysis of Figures 9, 11 and respectively 10, 12, indicates that the strains and the stresses generated by the contact with the seat belt are higher in the thorax area, although the contact force is higher in the pelvis area. Our results are reinforced by the geometric and material characteristics of the component elements, and, also, by the fact that during the impact, the airbag system also impacts upon the thorax area. Although the motions in the thorax are larger (see Figure 13), the analysis of the diagrams in Figure 14 indicates that the pelvis motion bears a more pronounced dynamic character, hence validating the triggering of the submarine phenomenon during the collision.

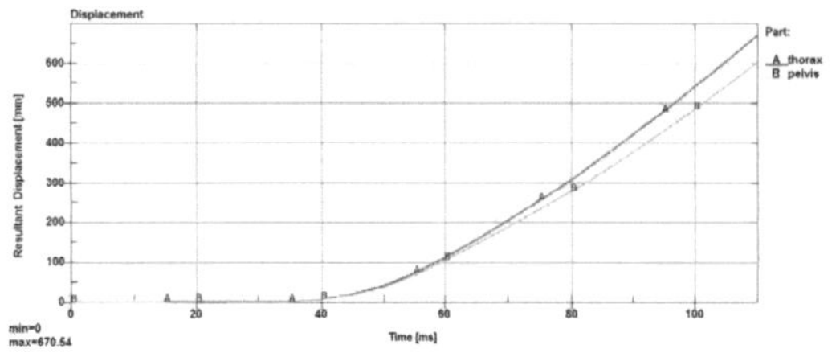

Fig.13. Resultant displacement – normal position

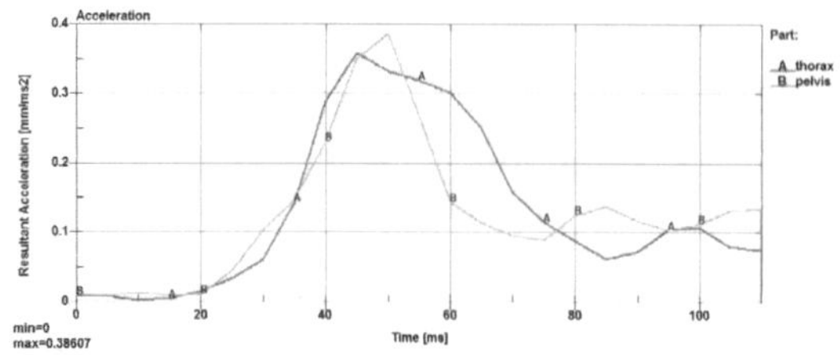

Fig.14. Resultant acceleration – normal position

3. Driver's out of position

For the first situation analysed we considered a normal position of the dummy, thus, the Hybrid III - 5th female percentile dummy was positioned with the thorax centered on the steering wheel at a distance of about 350 mm from the steering wheel. To position the dummy, the knee joint was set at an angle of -30 °, and the hands were placed on the steering wheel by modelling the elbow joint at an angle of -60 °. During subsequent tests we focused on the out of position (OOP) by changing the position of the dummy, both in terms of the distance from the steering wheel and the dummy's angular position in relation to it. In all the 11 analysed situations, the dummy was considered to be fixed by means of a three-point seat belt restraint system. Starting from the normal position of the dummy, in the first test scenario, for the first four OOPs, we reduced the distance between the thorax and the steering wheel to a minimum of 150 mm. Starting with the OOP5 position, we rotated the dummy from the H point (dummy's mass center) in relation to the steering wheel by 5, 10 and 15 degrees, first to the left and then to the right.

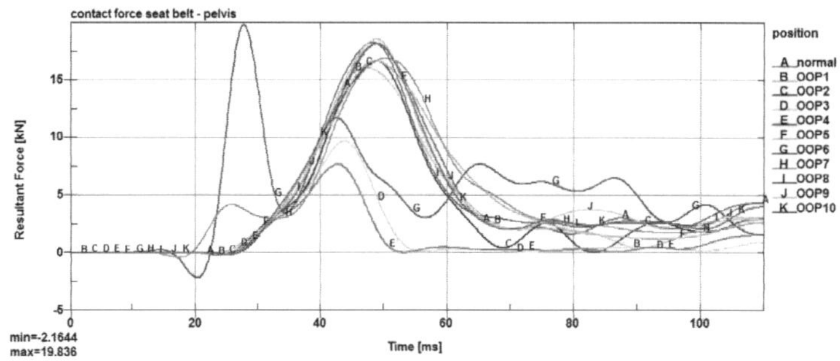

Figure 15. The contact force between seat belt and pelvis for the out of position

As illustrated in Figure 15, we can notice that the maximum value of the contact force recorded in the pelvis area was obtained for the OOP6 case, i.e. the rotating of the dummy's mass center to the left, hence causing stains, and implicitly leading to lesions on the right side of the pelvic area.

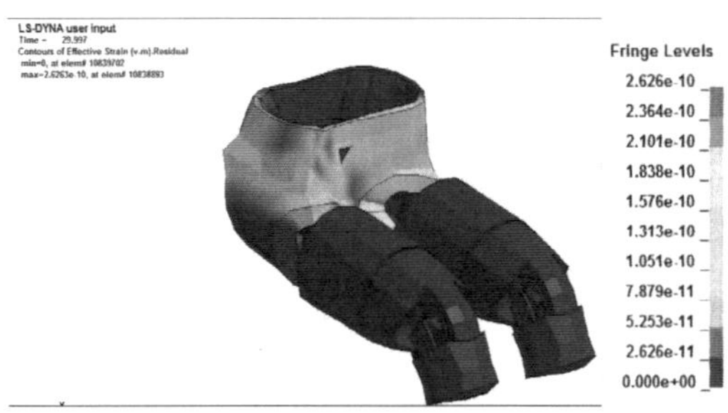

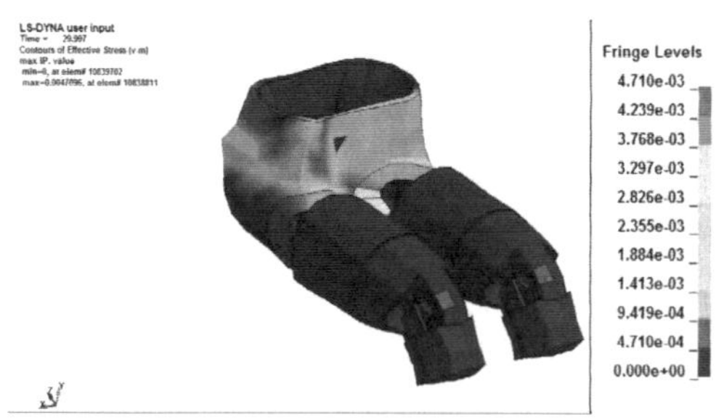

Figure 16. Effective strain – pelvis – OOP6 Figure 17. Effective stress Von Mises – pelvis – OOP6 [kN/mm^2]

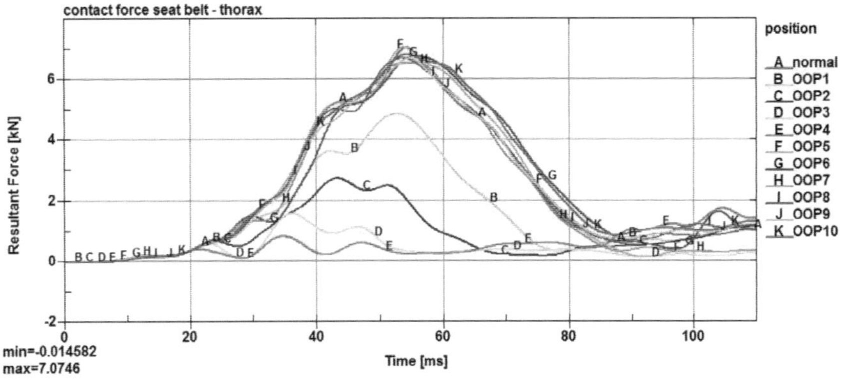

Figure 18. The contact force between seat belt and thorax for the out of position

Also, the maximum value of the contact force recorded in the thorax area was obtained in a dummy-out of position case, by rotating its mass center towards the left. Thus, we could identify the maximum stresses and strains values, recorded in the central-right area of the thorax, i.e. in the seat belt contact area.

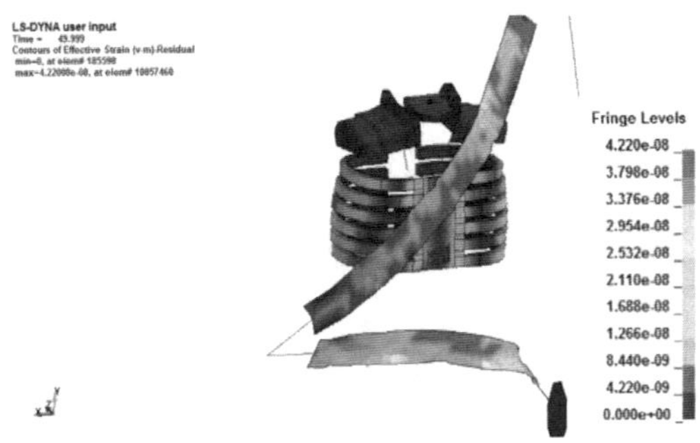

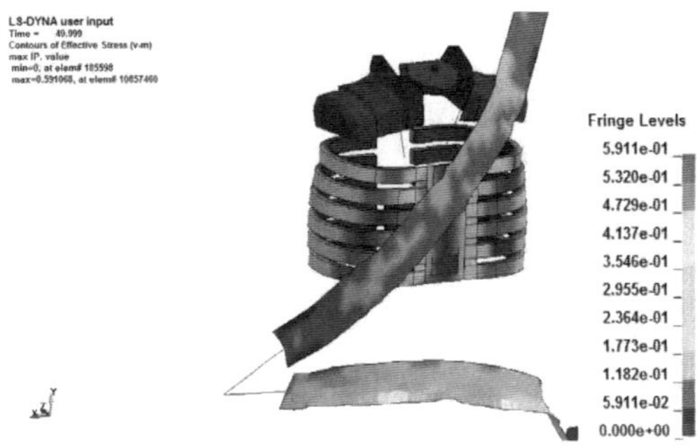

Figure 19. Effective strain – thorax – OOP5

Figure 20. Effective stress Von Mises – thorax – OOP5 [kN/mm^2]

4. Results

Following our research study, we could determine the time-related variation laws for the kinematic and dynamic parameters that allow for the monitoring of the driver's motion during the collision. Also, we established the motions, the strains and the stresses in all areas affected by the driver's seat belt, hence enabling us to identify the critical positions induced by the dynamic flexibility of the seat belt.

The variation of the dynamic parameters (contact forces, reactions) were analysed by means of comparison, applying the two methods previously mentioned, i.e. the numerical modelling and simulation and the experimental analysis.

The results obtained subsequent to our virtual prototyping were validated by the analysis and experimental tests, carried out mainly to investigate the driver's normal position. Thus, we were able to widen our modelling and numerical simulation of the frontal collision phenomenon as applied to different front positions of the driver in relation to steering wheel.

Although, the experimental testing of the collision phenomenon between two vehicles in the polygon is essential for the further development of the analysis, the quality of such trails in terms of results accuracy implies rather higher costs. Therefore, the virtual prototyping on a complete model, designed as a structure, enabled us to develop a complex approach to different types of collision.

5. Conclusion

Our analysis on the influence of the safety belt upon the driver was carried out via two methods, virtual prototyping and experimental tests.

The experimental tests, carried out under polygon conditions, implied the monitoring of a frontal collision between two medium-class vehicles, aimed at identifying the kinematics and dynamics of the dummy's behaviour during the impact.

The virtual prototype entailed the design of the virtual model for the assembly: dummy - dashboard - seat - passive safety systems, while defining the dynamic contact between the components of the dummy (thorax, pelvis) and the seat belt, as well as the airbag system.

Several dynamic models have been developed for the contact between the dummy and the passive safety systems, mainly the safety belt. The dynamic analysis and the results comparison between the virtual and the experimental models were carried out to investigate the normal position of the dummy in relation to the steering wheel, hence validating our virtual model. Next, our research was extended to monitor the dynamic response of the dummy's motion during a frontal collision for several out of position instances (OOP), identified as critical.

References

[1] Ghosh, P., Chitteti, R., Shah, P., Mayer, C., Kumar H.G., Influence of Ribcage Shape on Thoracic Response of Anthropometrically Correct 5th Percentile Female Finite Element Model, 13th LS-DYNA Forum, Bamberg, 2014

[2] Ming-Pei Lin, Chih-Min Chang, Cho-Hsuan, Tsai, Chia-Hui Tai, Chun-Te Lee, Usage of the LSTC_NCAC Hybrid III 50th Dummy in Frontal Occupant Simulation, 13th International LS-DYNA Users Conference, Session: Occupant Safety, 2014

[3] Nsiampa, N.N., Robbe, C., Papy, A., Development of a Thorax Finite Element Model for Thoracic Injury Assessment, 8th European LS-DYNA Users Conference, Strasbourg, mai 2011

[4] Pradeep Mohan, Chung-Kyu Park, Dhafer Marzougui, Cing-Dao Kan, LSTC/NCAC Dummy Model Development, 11th International LS-DYNA Users Conference, Session: Occupant Safety, 2010

[5] Radu, A.I., Cofaru, C., Tolea, B., Trusca, D.,D., Toganel, G., Research Regarding the Influence of Vehicle's Safety Restraint Systems on Its Occupants in Case of Rear-End Collision, CONAT 2016 International Congress of Automotive and Transport Engineering, DOI 10.1007/978-3-319-45447-4_89, ISBN 978-3-319-45446-7, pp 812-819, Springer International Publishing Switzerland, 2017

[6] Stahlschmidt, S., Gromer, A., D'Souza, R., Franz, U., Update in Dummy Model Enhancements and Effective Pre-processing, 13th International LS-DYNA Users Conference, Session: Occupant Safety, 2014

[7] Takeo, U., Yasuhito, T., Toshikazu, N., Applications of Occupant Safety Simulation using MADYMO, vol. 58 nr.165, Komatsu Tehnical Report, 2012

[8] Tso-Liang Tenga, Fwu-An Changb, Yung-Sheng Liuc, Cheng-Ping Pengd, Analysis of dynamic response of vehicle occupant in frontal crash using multibody dynamics method, Mathematical and Computer Modelling 48, pp. 1724–1736, 2008, Elsevier

[9] Tolea, B., Tarulescu, S., Trusca, D., D., Toganel, G., Radu, A., I., The Assessment of the Head Injury of a Pedestrian in Comparison with a Cyclist, CONAT 2016 International Congress of Automotive and Transport Engineering, DOI 10.1007/978-3-319-45447-4_89, ISBN 978-3-319-45446-7, pp 812-819, Springer International Publishing Switzerland, 2017

[10] http://www.solidmechanics.iei.liu.se/Examiners/Courses/Master_Level/TMHL19/intro_lsdyna_lsprepost.pdf , Short Introduction to LS-DYNA and LS-PrePost, Jimmy Forsberg, 2013

IOP Conf. Series: Materials Science and Engineering **252** (2017) 012018 doi:10.1088/1757-899X/252/1/012018

Indirect Tire Monitoring System - Machine Learning Approach

O Svensson[1], S Thelin[2], S Byttner and Y Fan

Department of Intelligent Systems and Digital Design, Halmstad University, Box 823, S301 18 Halmstad, Sweden

E-mail: [1]oskarsvensson.94@gmail.com, [2]simon.thelin90@gmail.com

{stefan.byttner, yuantao.fan}@hh.se

Abstract. The heavy vehicle industry has today no requirement to provide a tire pressure monitoring system by law. This has created issues surrounding unknown tire pressure and thread depth during active service. There is also no standardization for these kind of systems which means that different manufacturers and third party solutions work after their own principles and it can be hard to know what works for a given vehicle type. The objective is to create an indirect tire monitoring system that can generalize a method that detect both incorrect tire pressure and thread depth for different type of vehicles within a fleet without the need for additional physical sensors or vehicle specific parameters. The existing sensors that are connected communicate through CAN and are interpreted by the Drivec Bridge hardware that exist in the fleet. By using supervised machine learning a classifier was created for each axle where the main focus was the front axle which had the most issues. The classifier will classify the vehicles tires condition and will be implemented in Drivecs cloud service where it will receive its data. The resulting classifier is a random forest implemented in Python. The result from the front axle with a data set consisting of 9767 samples of buses with correct tire condition and 1909 samples of buses with incorrect tire condition it has an accuracy of 90.54% (0.96%). The data sets are created from 34 unique measurements from buses between January and May 2017. This classifier has been exported and is used inside a Node.js module created for Drivecs cloud service which is the result of the whole implementation. The developed solution is called Indirect Tire Monitoring System (ITMS) and is seen as a process. This process will predict bad classes in the cloud which will lead to warnings. The warnings are defined as incidents. They contain only the information needed and the bandwidth of the incidents are also controlled so incidents are created within an acceptable range over a period of time. These incidents will be notified through the cloud for the operator to analyze for upcoming maintenance decisions.

1. Introduction

Knowledge in component failures before they occur is important for the heavy duty vehicle industry. The industry demand products that work dynamically towards the customer to reduce costs and remain efficiency in the fleet. European-Union has decided upon a law[1] that is based on environmental grounds that all new passenger cars from November 1st 2014 are equipped with a Tire Pressure Monitoring System (TPMS). The system is based upon physical sensors installed inside the tires. Explicitly heavy duty vehicles are not included in the law.

NHTSA(National Highway Traffic Safety Administration) is an agency of the executive branch of the U.S. government. With a mission to save lives and prevent injuries combined with reducing vehicle related crashes. They have published a report[2] that conclude that 12.4%

of all passenger vehicles in the U.S. of model years 2004-2011 have at least one tire that is severely underinflated as defined by FMVSS No.138[3](25% or more below what the vehicle manufacturer recommend). The report also found that 23.1 percent vehicles in the study without TPMS had at least one severely underinflated tire.

Tire monitoring is time consuming and operators have problems with scheduling proper maintenance. The global market for TPMS is projected to reach 4 billion US dollars by 2020[4]. The primary goal of TPMS is to reduce underinflation in order to make vehicles safer to operate together with improved fuel economy. The resulting classifier and cloud module shown in this article is researched and developed for Drivec AB which is a company located in Helsingborg, Sweden. It show that it is possible to create a generalized indirect Tire Pressure Monitoring System by using supervised machine learning. The system created is targeting large fleet operators with different vehicle models. It enables them to view the vehicles tire condition on each axle from a cloud service, unlike the usual tire pressure monitoring systems.

2. Related Work

Remote diagnosis, maintenance and prognosis for advanced driver assistance systems is currently a hot topic. Mostafa Anwar Taie among others have created a framework[5] which predict the remaining useful life for the prognosis of the advanced driver assistance system's safety critical components by using machine learning algorithms. This framework can be used during development phase but mainly after production.

Niclas Persson, Fredrik Gustafsson and Markus Drevö wrote a thesis about Indirect Tire Pressure Monitoring Using Sensor Fusion[6]. Markus Drevö is working at the Swedish company Nira Dynamics which is one of the leading companies in indirect tire pressure monitoring systems. This thesis explains a system based on wheel radius and vibration analysis. With both approaches combined it can detect a pressure loss of 15% in one, two, three or four tires and independently detect a pressure loss of 25% within one minute. This solution does not use any supervised machine learning methods but instead tests values against a predetermined threshold.

Stefan Byttner, Thorsteinn Rögnvaldsson and Magnus Svensson created Consensus self-organized models for fault detection (COSMO)[7]. COSMO is an unsupervised approach which builds up knowledge over time. It uses an onboard self-organized search for models to find relations between the vehicles data values together with an offline server application which compares the parameters of the models. A significant benefit is that the vehicles normal behaviour is found under real operating conditions and not observed in a number of tests in laboratory environment. This has also formed the basis for several licentiate theses.

2.1. Industrial relation

Each vehicle in this fleet is equipped with Drivec Bridge that fetch CAN-bus data in real time, it is a part of the existing fleet management system. When the whole implementation is done for the cloud service, the application will analyze the data in real time from the vehicles and signal when faults occur. Projects mentioned above does not have a connection to an existing fleet management system. COSMO use the Volvo Analysis and Communication Tool (VACT), that is capable of recording data using telematics to communicate remotely. The existing solutions on direct TPMS and indirect TPMS does not deliver the data to a cloud service. The information is locked within the vehicle. The application of this project will analyze the result and make it viewable for the operators regardless of the vehicles location and without having to include the driver in the decision process.

3. Methodology

Drivec AB provide expert knowledge in CAN-bus interpretation thanks to their hardware Drivec Bridge. It has enabled the research and development of the result shown in this article.

3.1. Data

Data management is handled with data obtained from the vehicle itself and from o board measurements. The data from the vehicles contains vehicle specific information where every sample output includes an average sum and standard deviation for each parameter during a period of 10 seconds. The reason for this is that an interval of 10 seconds is the minimum amount of time required for receiving trustworthy GPS data. O board measurements include tire depth in millimeters and tire pressure in bar of all wheels of the vehicle. Together with a time stamp for the current date and time. From this data sets has been created to be used within the supervised machine learning process. The data sets need to contain data that is valid. To obtain validness in the data sets filtering has been done. The filtering makes sure that the received data only represent when the vehicle is driving with a constant speed without any heavy turns or heavy accelerations. It is done in both Python, for the training, and in Node.js with JavaScript for the cloud service. The parameters that the filtering is based upon are the vehicle speed, vehicle speed standard deviation, GPS speed, GPS course change standard deviation, individual wheel speed per wheel and the brake signal.

The targets are set per day and vehicle according to the o board measurements of the vehicles tire pressure and thread depth. From the beginning the targets were decided by looking if the difference in pressure of left and right tire was greater than 0.5 bar. This would set the target as bad for the vehicle. It was discovered that 0.5 bar was too low and resulted in too many vehicles with rather small differences being classed as bad. Together with Drivec it was decided that a vehicle need to have a major difference in either tire pressure or thread depth to be classed as bad. This to avoid causing false alarms or alarms for minor issues which do not yet need to be resolved. The resulting value for deciding whether the vehicle shall be classed as bad is if it has a tire pressure difference greater than 1.5 bar or if the difference in thread depth is greater than 6mm. When the targets have been determined the training will be executed within Python where the Scikit-learn module provides the classifier. The input from the data set and targets will be used to call the t method to generate the classifier that predicts the class to which the unseen samples belong. Random forest is used as classifier.

When solving a problem with a supervised machine learning method features are created. These features main purpose is to build a concise model from a distribution of class labels. The resulting classifier is then used to assign class labels to the testing instances where the values of the predictor features are known, but the value of the class label is unknown. The features are chosen from available parameters from the data set and need to have a physical buoyancy. The features used are related to different speed sources.

3.2. Random Forest Classifier

The random forest is an ensemble approach[8]. It is a divide-and-conquer approach used to improve performance. But with the regular tree in mind the random forest enhance this by combining trees. The tree may be referred to as weak learners but when assembling all these trees together they can possibly create a strong learner. If there exist a sample of N cases at random with a given replacement it will create a subset of the given data input. How large this subset should be may vary. At each node there should be a number X where X is the selected predictor variables at random from all the predictor variables. The predictor variable that provides the most accurate split is then later on used to perform a binary split on that specific node. When running random forest, the given input will go trough all the trees and the result may be different. It is crucial to know that if there is a large number of predictors the predictor will be more different between the nodes. The big benefits with random forest is that it is rapid and good at handling imbalanced data.

3.3. Optimization and Evaluation

To optimize the classifier it is crucial to use different tools ensure that the outcome is as good as possible and that it behaves in the way that is best for the given situation. Confusion matrix, seen in figure 1, is used to evaluate the quality of the output of a classifier on the data set. The diagonal of the matrix represent the number of points for which the predicted label is equal to the true label. While the o -diagonal elements are those that are mislabeled by the classifier. This will result in a false negative or a false positive. A system like this requires more false positive mislabels than false negative since a false negative would trigger a false alarm.

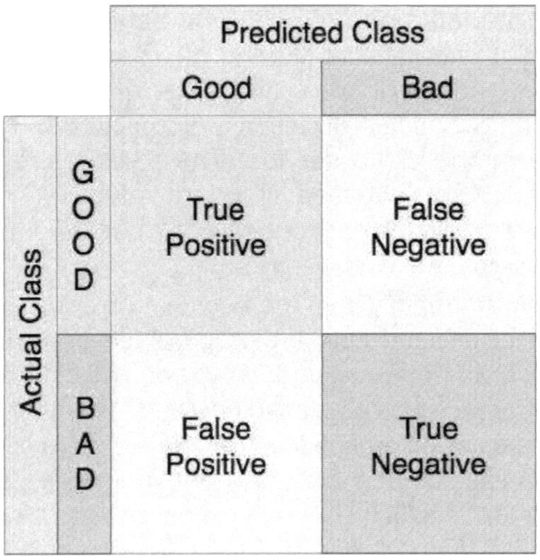

Figure 1: *Confusion Matrix*

Furthermore there is a need to evaluate how well the classifier perform on unknown data sets. This is solved by using stratified K-fold cross validation. It is a model variation technique for assessing how the result of a statistical analysis will generalize to an independent data set. K-fold cross validation is one way used to avoid over-fitting, where 90% from the unknown data set have good targets and 10% have bad targets in this case. The out-of-bag concept is used to estimate the generalization error which is the error rate of the out-of-bag classifier on the training set. An estimation of the error rate can be obtained based on the training data. This ensures that the classifier is built on about 63% of the available data and therefore pulling away 37% for testing.

3.4. Cloud

Drivec's products are all represented on a cloud service. The cloud is 100% scalable thus ensuring almost unlimited capacity in quantity of connected vehicles. The cloud provides better automatic software updates combined with a better disaster recovery. The resulting classifier from the supervised machine learning method which in this case has been the random forest. This classifier has been extracted in a format that been interpreted in the developed back-end module which is based on Node.js which becomes a process within the cloud service.

4. Results

It was concluded that the tires with the highest difference in both tire pressure and thread depth was the front axle. Therefore the focus was put to classify the tire condition of the vehicles front axle. The value of the vehicle speed that is used for the filtering was decided by analysis of the accuracy of the speed. Speeds under 30 km/h give a less reliable value hence the filtering is set to only accept data when the speed is above 30 km/h. Linearity between individual wheel speed and vehicle speed was found and resulted in a change of features in the classifier.

With a data set consisting of 9767 samples of buses with correct tire condition and 1909 samples of buses with incorrect tire condition from a total of 34 buses, the random forest classifier has an accuracy of 90.54% (0.96%) after its 10 folded cross validation. After creating and analyzing a given confusion matrix. It provided shows that 10.93% of the mislabels are false negatives and 89.89% are false positives. This ensures that more bad tires will be classed as good than the other way around which will avoid false alarms on the cloud service. Incidents in the module contain the information that is displayed in the reports in the cloud service. Incidents can be created and updated depending on their state. Incidents and updates of an incident occur asynchronously on different vehicles and axles at the same time. The whole cloud process is described in figure 4. The module also has a threshold value that has been set to 50%. In figure 2 the score predictions the classifier makes is shown over time and that the classifier is proven to catch the event of going from a bad condition to a good condition. When the scoring has been equal or over 50% more than 7 times during a set period of time, an incident will be created. In figure 3 it is shown how many incidents that are created with updates to the the drastic change where no more incidents are created. To verify this the workshop was contacted and it was proven that they received a report late in the evening of March 16th from a bus driver that stated that the tires of the front axle were in critically bad condition. The tires were changed the following morning of March 17th 07:45. The project group had no information about this when it happened but it has been researched together with the data analysis from the given vehicle.

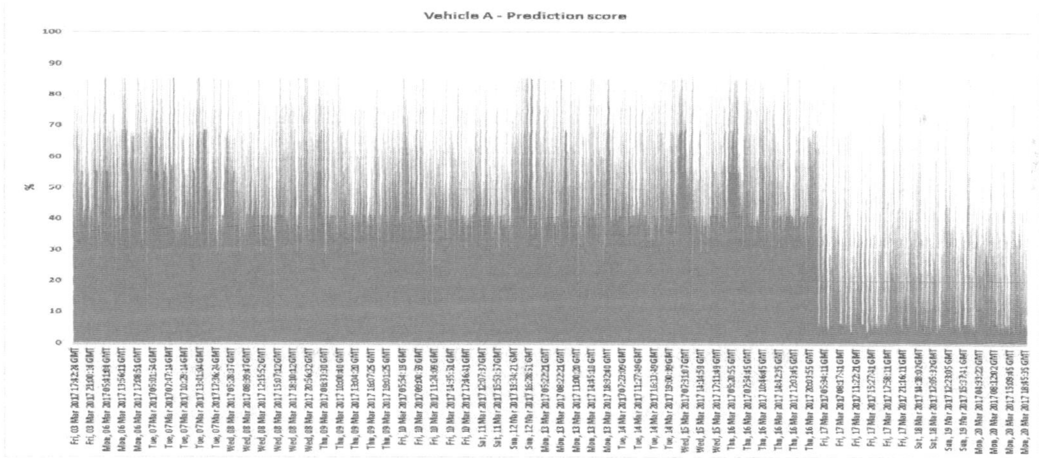

Figure 2: *The output created by the classifier from March 3rd to March 17th. The collection of outputs represent scores in percentage on the given dates. The scoring is fetched from bad to good with a distinct change on March 17th 2017. At this date the workshop changed tires without the classifier being aware of this fact.*

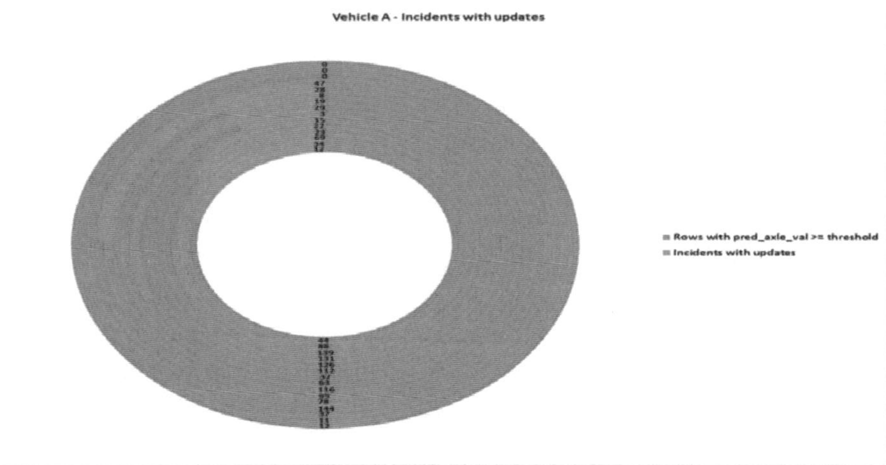

Figure 3: *Each date has a given collection of output scores above the threshold value of 50% which are represented as the blue area. During the reading of outputs, the ITMS module will perform a delay and check if the output score is above the threshold value. If the prediction meets the criteria, it will be added to the queue. If the number of bad predictions in the queue meet the given criteria an incident will be created seen as the orange area. If the number then get lower and then meet the requirement again, an update will occur for that specific incident. For example on March 3rd 44 output values were above the threshold. During the reading of the data 1 incident was created with 11 updates during that day. On March 17th no more incidents are created even though there may be values that meet the threshold value, however they do not occur close enough to each other to create an incident, which can be related to figure 2 and 4.*

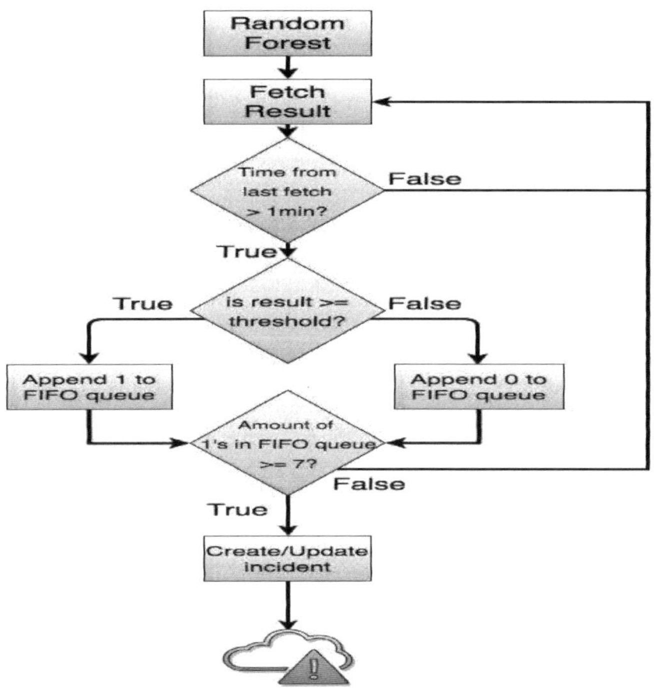

Figure 4: *Flowchart of ITMS process.*

5. Discussion

When doing the o board measurements we measure tire pressure and thread depth. The tire pressure was measured by putting a tire pressure gauge over the valve of the tire. Some tires of the boogie and drive axle did not have valves to put the gauge on. This cause some measurements of the tire pressure to be unknown. However on the boogie and drive axle there are always two tires per side. The measuring of thread depth was focused on the center of the wheel, therefore there are no measurements for tires that are worn out at the sides, only the depth from the center is being measured.

From the o board measurements we look at the difference in pressure between left and right wheel rather than the individual pressure of the wheel. This was considered as the better choice to start with and it also prevents users from disbelieving the first version of the system when warnings are created, rather receiving a more accurate warning for the whole axle than possibly incorrect warnings for the individual wheels. With the results from the classifier the warnings can be analyzed to create greater insight to solve problematic areas for further development when the time comes to try to pinpoint out specific wheel behaviour.

The system is currently an alpha version and is being tested internally at Drivec. For a more detailed description of the system see the Indirect Tire Monitoring System thesis [9]

6. Conclusions

This approach is using random forest as classifier and will receive real time data within a 10 second interval. The training of the classifier is done in Python and the classifier is exported to be used in a Node.js module in the cloud.

The practical study in this project is based on 34 MAN city buses with the fuel type of compressed natural gas. The buses have different lengths, 2 to 3 axles and are year models 2005 to 2011.

The European-Union law does not apply to heavy duty vehicles hence very few of them use a TPMS. It is difficult to say how well other solutions work for heavy duty vehicles but for passenger cars they work well since it has become a legal requirement. There is currently no product on the market that works in the same way as the developed tire monitoring system. The most similar product on the market is an indirect Tire Pressure Monitoring System (iTPMS) where the software use the vehicles ABS-system to receive data. Furthermore the data about the tires is kept in the vehicle and is only displayed for the driver. The most significant difference with this indirect tire monitoring system is that it will not display any information to the driver but will be a part of the fleet management system in the vehicle and show the data back-end for the operator. The solution is not oriented towards the driver of the vehicle but the vehicle itself. The driver is never exposed in the ITMS system and neither a part of the incident created. Therefore the integrity issue is put on the operator.

The most significant issue with random forest is overfitting. Random forest generally give good results which is why it is needed to be analyzed properly to avoid overfitting. Too large number of trees or too large depth of the trees is also an issue which may make the algorithm slow for real-time prediction.

Acknowledgments

We would like to thank Johan Göthe and the rest of the team at Drivec for the opportunity and the generous support throughout the whole project.

Oskar Svensson & Simon Thelin
Helsingborg 2017

References

[1] Jansen S, Schmeitz A, Maas S, Rodarius C, 2016 TNO 2014 R11423-v2 - Final report, Study on some safety-related aspect of tyre, TNO innovation for life 15

[2] National Highway Traffic Safety Administration, 2012 Evaluation of the Effectiveness of TPMS in Proper Tire Pressure, DOT HS 811 681 6

[3] National Highway Traffic Safety Administration, 2001 Tire Pressure Monitoring System, FMVSS No. 138

[4] Global Industry Analysts, Inc, 2015 Tire Pressure Monitoring Systems - A Global Strategic Business Report

[5] Taie M, Moawad E, Diab M, and ElHelw M, 2016 Remote Diagnosis, Maintenance and Prognosis for Advanced Driver Assistance Systems Using Machine Learning Algorithms 114{122

[6] Persson N, Gustafsson F and Drevö M 2002 Indirect tire pressure monitoring using sensor fusion

[7] Byttner S, Rögnvaldsson T and Svensson M 2011 Consensus self-organized models for fault detection (COSMO) 833{839

[8] Breiman L, 2001 random forests Machine Learning 45 5{32

[9] Svensson O, Thelin S. Indirect Tire Monitoring System - Machine Learning Approach *http://urn.kb.se/resolve?urn=urn:nbn:se:hh:diva-34290*

NUMERICAL SOLUTION OF THE PROCES OF HYDROPLANING AT WHEELS WITH ENVELOPE

M Stan[1], P Stan[2]

[1]University Of Pitesti, Arges, Romania, email: marinica.stan@upit.ro
[2]University Of Pitesti, Arges, Romania, email: petre.stan@upit.ro

Abstract. In this paper is presented a study for the envelope's properties settlement regarding the process of hydroplaning. The phenomenon of hydroplaning is calculable with the aid of the equations of motions for the viscid fluid in streamline regime for thin layers of fluid supposed to crush. Conclusively this paper offers the base of determinations prequisites to settle the envelope's properties concerning to the phenomenon of hydroplaning.

1. Characteristic quantities of the contact area between the envelope and the rolling track

If the envelope has inside air with pressure equal with atmospheric pressure, the experiment shows that in usual limits of quaquaversal distortion practically the envelope doesn't take over quaquaversal forces, that is to say that in this case it's acting like a perfectly elastic membrane at encasing. The experiment shows also that at loading of the envelope with charge G and at the appearance of the quaquaversal distortion between envelope R's ray and the radius of running r_r, the width B of the contact zone remains the same (figure. 1). For a tyre filling pressure given as p read at the pressure gauge it is seen that once with the charge G's growing it also grows the length of the contact zone having an almost rectangular shape of $A = BL$ area. The contact between the envelope [1] and the path is made through the bars of profile from area $A = BL$, but the effective contact area A_e between the bars of profiles and the path is smaller than area A and we denote

$$\lambda = \frac{A_e}{A} = \frac{A_e}{BL} < 1 \qquad (1)$$

the using of the contact zone's surface coefficient.

If the effective pressure between the bars of profiles and the path is denoted with p_e then the envelope's loading force G may be exprimed with relation

$$G = pA = p_e A_e \qquad (2)$$

From relations (1) and (2) may be exprimed the effective pressure

$$p_e = \frac{G}{A_e} = \frac{G}{\lambda A} = \frac{G}{\lambda BL}. \qquad (3)$$

The tyre studied contains profiles δ_1 and rectangulars δ_2 with dimensions mk and nk, where k is the pace of the meshwork for the numerical integration of the pressure [2] in the water course expelled from underthe profiles and m and n are natural numbers.

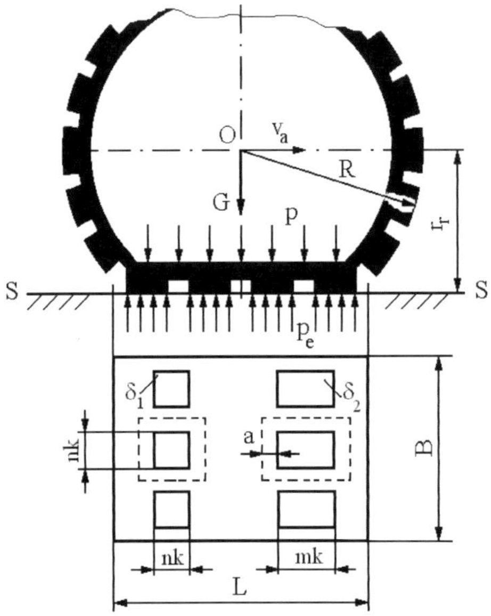

Figure 1. The model with the force G.

For each profile [3] recomes from surface A an enclosure of width a, so that the first output factor in the contact area is:

$$\lambda_1 = \frac{n^2k^2}{(nk+2a)^2}, \tag{4}$$

The second output factor in the contact area has the following expression:

$$\lambda_2 = \frac{mnk^2}{(mk+2a)(nk+2a)} \tag{5}$$

2. The liquid's expelling time from under the bar of circular profile

If on the bed bearer is found out a liquid δ with the dynamic viscosity η forming a layer of thickness H, then the circular basal plane $f-f$ of the bar of profile [4], before entering in contact with surface $S-S$ of the path of run, is due to expel through crushing this layer of liquid (figure. 2). In figure 3 are represented two particular situations of the profile block δ 's surface from figure 2 which are a square δ_1 and a rectangle δ_2, in which was layed a meshwork with pace k having $n=4$ for the square and $n=4$; $m=6$ for the rectangle.

The forces [5] that make the expell posible are:

$$F_1 = p_{e_1}n^2k^2;\; F_1 = p_{e_2}mnk^2 \tag{6}$$

The surface $S-S$ of the path of run is considered to be the flat surface that passes through the bed bearer's rugosity peaks. It is noticed that at the moment when the bar of profiles' circular basal plane $f-f$ is getting closer of surface $S-S$ in a parallel motion until the distance between them is deleted and the bed bearer's rugosity peaks are starting to get into the bar of profile's material, the liquid from under the surface f-f can still be expelled trough the bed bearer's random assigned micro channels.

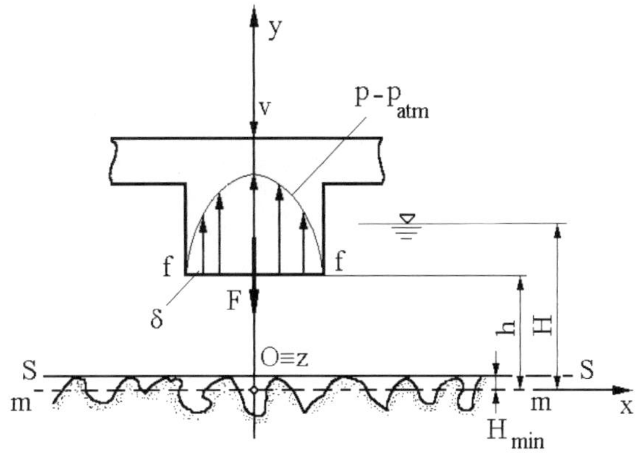

Figure 2. The rolling path with liquid.

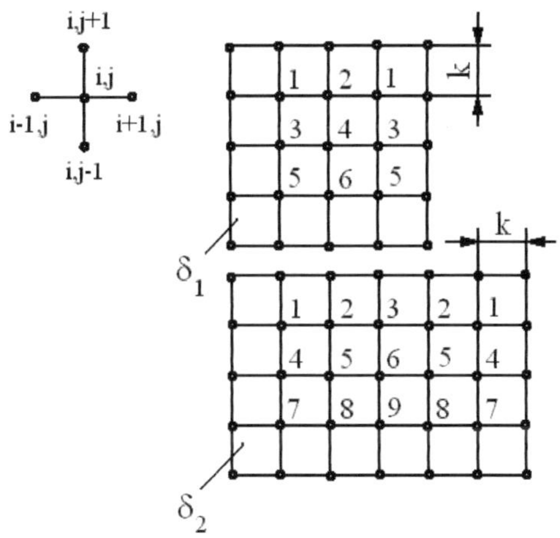

Figure 3. Particular situations of the surface of the profile block.

To express the expelling flow liquid from under the bar of profiles in this limit situation is entered a mathematical plan $m-m$, parallel with $S-S$ plan and situated at a distance H_{min} below this, so that the flow of expulsion expressed by computation among the parallel surfaces $m-m$ and respective $f-f$ to be equal with the real flow expulsion through the micro channels from the bed bearer's rugosity [6]. If in the axis system $Oxyz$ is defined a solid surface $\delta=\delta(x,z)$ which displaces after axis Ox with constant speed $\vec{u}=u\vec{i}$ and after axis Oy with constant speed $\vec{v}=v\vec{j}$, the space among δ and a plan also solid xOz being full with viscid fluid of η dynamic viscosity, with the very little value δ in report with the common zone of $\delta=\delta(x,z)$ and xOz, therefore is known the pressure equation [6] between δ and xOz is:

$$\frac{\partial}{\partial x}\left(\delta^3\frac{\partial p}{\partial x}\right)+\frac{\partial}{\partial z}\left(\delta^3\frac{\partial p}{\partial z}\right)=12\eta v-6\eta u\frac{\partial\delta}{\partial x}. \tag{7}$$

In the case from figure 2 we have $u=0$, $\delta=h=constant$ and $\vec{v}=-v\vec{j}$ and equation (6) becomes:

$$\frac{\partial^2 p}{\partial x^2}+\frac{\partial^2 p}{\partial z^2}=-\frac{12\eta v}{h^3}. \tag{8}$$

The expresion in limited differences of this equation on the meshwork from figure 3 for the arbitrary point i, j is:

$$\frac{p_{i+1,j} + p_{i-1,j} + p_{i,j+1} + p_{i,j-1} - 4p_{i,j}}{k^2} = -\frac{12\eta v}{h^3} \tag{9}$$

or with

$$v = -\frac{dh}{dt}, \tag{10}$$

$$p_{i+1,j} + p_{i-1,j} + p_{i,j+1} + p_{i,j-1} - 4p_{i,j} = \frac{12\eta k^2 dh}{h^3 dt} = T. \tag{11}$$

If is applied equation (11) in the δ_1 surface's meshwork points is obtained the equations system:

$$\begin{cases} p_2 + p_3 - 4p_1 = T \\ 2p_1 + p_4 - 4p_2 = T \\ p_1 + p_4 + p_5 - 4p_3 = T \\ p_2 + 2p_3 + p_6 - 4p_4 = T \\ p_3 + p_6 - 4p_5 = T \\ p_4 + 2p_5 - 4p_6 = T \end{cases} \tag{12}$$

It is working with relative pressures so on the δ_1 surface's contour the relative pressure is zero. It is kept the notation p also for the relative pressure for making it easier to write.

The previous system [7] solutions are:

$$p_1 = -0,6875T,\ p_2 = -0,875T, \tag{13}$$

$$p_3 = -0,875T,\ p_4 = -1,125T, \tag{14}$$

$$p_5 = -0,6875T,\ p_6 = -0,875T. \tag{15}$$

The medium pressure [8] under surface δ_1 is

$$p_{m_1} = \frac{2p_1 + p_2 + 2p_3 + p_4 + 2p_5 + p_6}{25} = 0,295T \tag{16}$$

If equation (10) is applied in the δ_2's surface meshwrk points is obtained the equation system:

$$\begin{cases} p_2 + p_4 - 4p_1 = T \\ p_1 + p_3 + p_5 - 4p_2 = T \\ 2p_2 + p_6 - 4p_3 = T \\ p_1 + p_5 + p_7 - 4p_4 = T \\ p_2 + p_4 + p_6 + p_8 - 4p_5 = T \\ p_3 + 2p_5 + p_9 - 4p_6 = T \\ p_4 + p_8 - 4p_7 = T \\ p_5 + p_7 + p_9 - 4p_8 = T \\ p_6 + 2p_8 - 4p_9 = T \end{cases} \tag{17}$$

We have:

$$p_1 = -0,775T;\ p_2 = -1,11T \tag{18}$$

$$p_3 = -1{,}19T, p_4 = -0{,}99T \tag{19}$$

$$p_5 = -1{,}475T;\ p_6 = -1{,}54T \tag{20}$$

$$p_7 = -0{,}775T; p_8 = -1{,}11T;\ p_9 = -1{,}19T. \tag{21}$$

The medium pressure under surface δ_2 is

$$p_{m_2} = \frac{2p_1 + 2p_2 + p_3 + 2p_4 + 2p_5}{35} + \frac{p_6 + 2p_7 + 2p_8 + p_9}{35} = -0{,}4682T \tag{22}$$

3. The equation of the evolution of the liquid feather from under the envelope and the hydroplaning phenomenon's definition

In figure 3 is represented the contact zone between the envelope and the path at rolling on wet path with ones for possible situation of the bar of profile temporally in which this is found out in the contact zone of length L. These, in position 1 the bar of profile entered into the contact zone of length L and below this is found out a liquid layer of thickness H which must expel it through crush on the periphery of profile δ and in this it can enter into contact with the bed bearer.

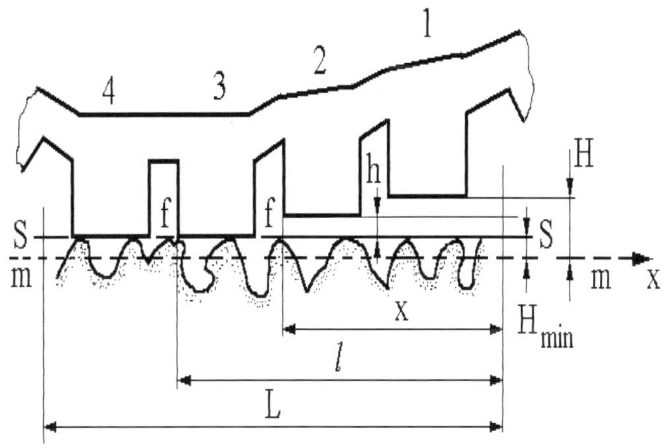

Figure 4. The contact zone between the tyre and the wet path.

From relations (6), (16) and (22) is obtained

$$p_{m_1} = p_{e_1} = -0{,}295T;$$
$$\tag{23}$$
$$p_{m_2} = p_{e_2} = -0{,}4682T,$$

and the with relation (11) we obtain the equations

$$p_{e_1} = -0{,}295\frac{12\eta k^2 dh}{h^3 dt}, \tag{24}$$

$$p_{e_1} = -0{,}4682\frac{12\eta k^2 dh}{h^3 dt}. \tag{25}$$

which are explined this way:

$$dt_1 = -\frac{0{,}295 \cdot 12\eta k^2}{p_{e_1}}\frac{dh}{h^3}, \tag{26}$$

respectively

$$dt_2 = -\frac{0,4682 \cdot 12\eta k^2}{p_{e_2}} \frac{dh}{h^3} \tag{27}$$

The time in which the layer of liquid's consistency decreases from value H to value $H_{\min}$ is in those two cases [8]

$$t_1 = -\frac{0,295 \cdot 12\eta k^2}{p_{e_1}} \int_H^{H_{\min}} \frac{dh}{h^3} = \frac{1,77\eta k^2}{p_{e_1}} \left(\frac{1}{H_{\min}^2} - \frac{1}{H^2} \right), \tag{28}$$

$$t_2 = -\frac{0,4682 \cdot 12\eta k^2}{p_{e_2}} \int_H^{H_{\min}} \frac{dh}{h^3} = \frac{2,8\eta k^2}{p_{e_2}} \left(\frac{1}{H_{\min}^2} - \frac{1}{H^2} \right) \tag{29}$$

The wheel's center *O*, represented in figure 1 is displacing with speed v_a and therefore the bar of profile's position is changing in the contact zone of length *L*, the bar abiding in the same time non-moving after direction $S - S$ belonging to the bed bearer surface's plan, so in position 2 from figure 3 is represented a certain situation in which the bar of profiles has position x in the contact zone and still has to expel a liquid layer of thickness *h*. After a time *t* from the entrance in the contact zone given by formula (28) and (29) position 3 of the bar of profile [9] in the contact zone is *l* given by

$$l_1 = v_a t_1;\ l_2 = v_a t_2, \tag{30}$$

situation in which the bar of profile's plan $f - f$ touched the plan $S - S$ of the bed bearer rugosity peaks, and from the viewpoint of the liquid from under plane $f - f$'s expelling phenomenon, there was an equivalent thickness of liquid H_{min} which can be expelled through the bed bearer rugosity's micro channels. For the reason that opening with position 3 and continuing then with position 4 the bar of profiles has expelled completely the liquid among planes $f - f$ and $S - S$, it is considerated that in these positions can be developed dry friction forces between surfaces $f - f$ and $S - S$.

On the contrary, on contact zone of length *l*, cannot be developed dry friction forces because the bars of profiles are suspended on a layer of liquid called feather of water, the phenomenon of this feather of water's appearance carrying the name of hydroplaning or hydroplaning.

If we replace the time from relation (30) into relation (28) and (29) it is obtained the feather of water evolution's equation depending on the wheel's run v_a:

$$l_1 = \frac{1,77\eta k^2}{p_{e_1}} \left(\frac{1}{H_{\min}^2} - \frac{1}{H^2} \right) v_a;\ l_2 = \frac{2,8\eta k^2}{p_{e_2}} \left(\frac{1}{H_{\min}^2} - \frac{1}{H^2} \right) v_a. \tag{31}$$

The envelope is behaving as a perfect elastic to flexure membrane also in the contact zone of length *L*, looker at the different positions of the bars of profiles on vertical direction concerning the variation of position *h* of these between *H* and H_{min}, fortiori how much values *h* are very little in report with length *L*.

4. The critical velocity of hydroplaning

From relation (28) and (29) it is noticed that the length *l* of the watery feather formed below the envelope grows in line with wheel' speed v_a. The critical velocity v_{crit} is defined for the situation when $l = L$, what means that the support of the envelope on bed bearer is made exclusively by dint of a liquid layer which separates completely the bars of profiles of the surface $S - S$ of the bed bearer and in this case, watching relation (26) the frictional forces F_f that can develop between the tyre and the bed bearer are null. The critical velocity of hydroplaning [10] is calculated from relation

$$L = \frac{1,77\eta k^2}{p_{e_1}}\left(\frac{1}{H_{\min}^2} - \frac{1}{H^2}\right)v_{crit_1}, \tag{32}$$

$$L = \frac{2,8\eta k^2}{p_{e_2}}\left(\frac{1}{H_{\min}^2} - \frac{1}{H^2}\right)v_{crit_2}. \tag{33}$$

The term $\frac{1}{H^2}$ can be neglected in report with term $\frac{1}{H_{min}^2}$ for all the situations that appear at hydroplaning and so the critical velocity has the expression:

$$v_{crit_1} = \frac{Lp_{e_1}H_{\min}^2}{1,77\eta k^2}. \tag{34}$$

The other critical velocity has the expression:

$$v_{crit_2} = \frac{Lp_{e_2}H_{\min}^2}{2,8\eta k^2}. \tag{35}$$

If we hold the expense of relation (3) we obtain for critical velocity the formula:

$$v_{crit_1} = \frac{GH_{\min}^2}{1,77\eta k^2\lambda_1 B}, \tag{36}$$

respectively

$$v_{crit_2} = \frac{BH_{\min}^2}{2,8\eta k^2\lambda_2 B}. \tag{37}$$

5. Numerical results

We efectuate a calculus [4] example for a tyre with width $B = 150mm$, filled with $G = 2500N$. We take water viscosity $\eta = 1,31 \cdot 10^{-3}\, Ns/m^2$ at $10\ ^0C$.

If we take pace $k = 6$mm then from figures and 3 results a square profile with leg $nk = 4 \cdot 6 = 24$mm and a rectangular profile with bigger leg $mk = 6 \cdot 6 = 36$mm and smaller leg $nk = 24$mm. With a value $a = 3$mm from formules (4) results $\lambda_1 = 0,64$ and $\lambda_2 = 0,68$.

We calculate the critical velocities of hydroplaning for the two profiles with formula (36), (37) and obtain:

$$v_{crit_1} = \frac{GH_{\min}^2}{1,77\eta k^2\lambda_1 B} = 44,92\frac{m}{s} = 161,7\frac{km}{h}, \tag{38}$$

$$v_{crit_2} = \frac{BH_{\min}^2}{2,8\eta k^2\lambda_2 B} = 26,72\frac{m}{s} = 96,22\frac{km}{h}. \tag{39}$$

6. Conclusions

The phenomenon of hydroplaning is calculable with the aid of the motional equations of the viscid fluid in streamline regime for thin layers of fluid supposed to crush. As physical characteristic size of this phenomenon is entered the minimal conventional thickness of the fluid H_{min}, who is expelled among the bar of profile and the bed bearer in the moment when the bar of profile touches the bed bearer rugosity's peaks. In this moment the expulsion is produced through the micro channels from the bed bearer and possible through the micro channels from the emplacement surface of the bar of profile. These expulsion possibilities are exprimed through thickness H_{min}, as through the planes that are approaching relatively, the emplacement plane of the bar of profile and the bed bearer's plane would not have rugosities and between them would be liquid of thickness H_{min}. Through the introduction of the evolution function of the watery feather from under the envelope at running on wet bed bearer and through the critical velocity's calculus this offered the possibility to express the diminution of the adherence to bed bearer of the envelope depending on the wheel's running speed as the complete detachment of the wheel from the path.

Conclusively this paper offers the scientific base of the experimentations prequisited to settle the envelope's properties concerning to the phenomenon of hydroplaning. The phenomenon of hydroplaning is calculable with the aid of the equations of motions for the viscid fluid in streamline regime for thin layers of fluid supposed to crush.

References

[1] Hara V and Stan M 2007 *The equations of the process of hydroplaning at wheels with envelope having the roller track from bars of detached profiles* (Pitesti: Pitesti University Press) p 180
[2] Stan M 2004 *Mechanics of Fluids* (Bucuresti: EDP Bucuresti) p134
[3] Stan M and Popa D 2005 *Fluid mechanics problems* (Bucuresti: EDP Bucuresti) p 240
[4] Persson B 2001 *Theory of rubber friction and contact mechanics* (Journal of Chemical Physics, vol. 115, no. 8) p 3840–3861
[5] Wies B and Roeger B 2009 *Influence of pattern void on hydroplaning and related target conflicts* (Tire Science & Technology, vol. 37, no. 3) p 187–206
[6] Kersys A 2013 *Research of the influence of tire hydroplaning on directional stability of vehicle* (American Review of Respiratory Disease, vol. 28) p. 374–380
[7] Wu Z and Zong Z 2014 *Mie-Gruneisen mixture Eulerian model for underwater explosion* (Engineering Computations, vol. 31, no. 3) p. 425–452
[8] Martin C S 1966 *Hydrodynamics of Tire Hydroplaning* (Georgia Institute of Technology, Final Report) p 166-187
[9] Bramono D P and Racz L M 1998 *Numerical Flow-Visulization of Slurry in a Chemical Mechanical Planarization Process* (CMP-MIC Conf) p 185-192
[10] Clark S K 1971 *Mechanics of Pneumatic Tires* (Monograph Number 122. National Bureau of Standards, Washinngton, D.C.) p 231-240

IOP Conf. Series: Materials Science and Engineering **252** (2017) 012020 doi:10.1088/1757-899X/252/1/012020

A late and failure of airbag deployment case study for drivers of passenger cars in rear-end collisions

George-Radu TOGANEL[1], Adrian Ovidiu SOICA[1]

[1]Automotive and Transport Engineering Department, Transilvania University of Brasov, Brasov, Romania

g.toganel@unitbv.ro, a.soica@unitbv.ro

Corresponding author: g.toganel@unitbv.ro

Abstract. The presented study was directed at two types of airbag miss-deployments: late deployment and non-deployment. Late deployment can be a product of override or underride road traffic accidents. Non-deployment can be a product of technical failure or trigger algorithm's inability to correctly assume the state of the accident to happen. In order to analyse the phenomena through physical tests, a specialized test device was used for a series of 8 non-deployment tests and a series of 4 airbag firing tests, totalling 12 tests. Acceleration based data was recorded and analysed for the movement of the device part simulating the driver head. High speed video recording was used to analyse the mechanics of airbag deployment and correlate with the acceleration based data. It has been determined, in the limitations of the laboratory testing environment, a significant variation of the time frame for the airbag deployments, despite using similar testing conditions and identical tested products. Also, the initial time frame for airbag deployment delay was overshadowed by other factors such as time to impact.

1. Introduction

Since its invention by John W. Hetrick and subsequent modification by David S. Breed in 1952 and 1967, respectively, the airbag revolutionized car safety in the 1970s [2]. During the 1980s, airbags have been introduced to Europe and by the mid-1990s most automotive manufacturers improved car safety by using an air cushion-based impact reduction system.[2]

It has been established that vehicle occupants form 64% of all road casualties. Airbags have been demonstrated to reduce car accident mortality by 19%, and reduce morbidity from facial fractures and chest injuries. In purely frontal crashes, airbags reduce the fatality risk by 34%. Over 3.3 million air bag deployments have occurred, with more than 6377 saved lives and countless injuries prevented [1], [2], [5].

However airbags produce their own patterns of injuries. The vast majority of those are minor (96.1%) and usually affect the face (42%). Still, there are several cases of airbag deployment causing more severe injuries and even death. [5]

As presented by Ahmad Z (2011, [2]), a search in PubMed brought up 602 citations with combined 'airbag + injury' terms. "Injuries ranging from burns and ocular injuries to major neurosurgical, vascular, thoracic and abdominal catastrophes have been described. In order to appreciate the damaging potential airbags have, it is important to understand how airbag deployment occurs, and which mechanisms are responsible for these injuries." [2]

It has been determined by a Braver et.al. study (2010, [4]), for U.S. data - NASS/CDS correlated with FARS for 1998-2006, a value of 9% (weighted estimate) for non-deployment airbags in cases of deceased front occupants in car accidents, a value from which about 89% represent actual unwarranted non-deployments based on crash characteristics, while the rest represent potential system failures where deployments would have been expected.

As determined through another study by Matthes et.al. (2006, [16]) there is in fact an increase of 7.2% in chest injuries for drivers with airbag deployment vs. non-deployment or non-existent. Also the mean sustained injury level, correlated with the abbreviated injury scale (AIS) was 2.3 for airbag deployment vs 1.6 for no deployment reflected to a difference of 21.1 vs 15.8 for mean injury severity scale (ISS).

U.S. regulations stipulate that, as of 1998, airbags – with depowering capability to inflate using 20% to 30% less energy - have been required as mandatory in all motor road vehicles and, as of vehicle model year 2007, certified advanced airbags – with dual-stage inflators with variable inflation by belt status and crash severity - have become the new norm for all passenger vehicles. [23], [4].

2. Background

The trigger algorithm becomes more and more complex as rules change and expand and different factors have to be taken into account for an airbag to fire. In example, an airbag fires up for a collision within a 60° frontal arc, with a force equal to running into a brick wall at 16—24 km/h, or a similar sized vehicle head-on at 32—48 km/h. [1], [2], [23]

Sodium azide (NaN_3), in combination with potassium nitrate (KNO_3), generates an explosive reaction during the inflation process, by creating nitrogen gas, in order to inflate the bag in a time threshold of approximately 10 to 50 ms. The volume within the vehicle taken up by the airbag as it inflates is known as the 'airbag deployment zone'.

The bigger the airbag, the quicker the deployment has to take place. European driver airbags hold 35 l of gas propellant and fully deploy within 25 ms, which means that they have to expand at anything up to 58 m/s, whilst American air bags typically holding 60 l of gas have to deploy even quicker.

Airbag deployment related injuries, as described by the literature, can be of the following – but not limited to - head & face - dislocation and fracture of mandibular condyle, teeth fracture [25], alkaline burns applied to cornea, facial and other tissue [26], blunt ocular trauma, corneal flaps and ectasia [15], [17], [23], DMD [19], retinal detachment [13]; neck - burns, aortic injuries, thyroid artery rupture [3]; thorax injuries: clavicle rupture, rib fractures [3], [24], sternal fracture [24], cardiac injuries - lacerations and/or ruptures of the intrapericardial inferior vena cava, the right ventricle, and the right pleuropericardium [24], [3]; a case of oesophageal rupture at the gastro-oesophageal junction [5].

The combustion of sodium azide, considered a potentially deadly chemical, is responsible to the generation of an alkali aerosolized sodium hydroxide and sodium carbonate compound that is partly released into the occupant area. [1], [23], while it is also known to react with cytochrome oxidase thus preventing cells from using oxygen, and has been noted to develop asthmatic reactions, reactive airway dysfunction syndrome in the presence or absence of associated chest trauma [1], chemical, thermal and mechanical friction burns [3], [26], melting of clothing from direct contact with the expelled gas or with the airbag itself [26].

Airbag deployment process must follow predetermined and controlled firing algorithm and most of the times it develops in timely manner. However there are exceptions when this process might suffer from several identified issues such as: early, late or aggressive deployment, wiring failure, non-deployment. On this occasions it has a great chance to add to the occupant injury degree instead of lowering it. The state of the airbag deployment process during an accident can be determined by analyzing post-crash data, recovered from black box devices, such as crash pulse and seat belt use. [29]

The presented study was directed at two of the miss-deployments: late deployment and non-deployment. Late deployment can be a product of override or underride road traffic accidents. [29]. Non-deployment can be a product of technical failure or trigger algorithm's inability to correctly assume the state of the accident to happen. As Ahmad Z. notes (2011, [2]), rear end shunts should not fire the airbag.

3. Method

In order to study the late –and failure of – airbag deployment case, a specialized test device was used in laboratory conditions.

The test device consists of a test dummy head attached to a pivoted arm, Figure 1. The arm has the role of human thorax and lower torso. The impact arm consists of a dummy head mounted to an arm that is pivoted at the opposite end.

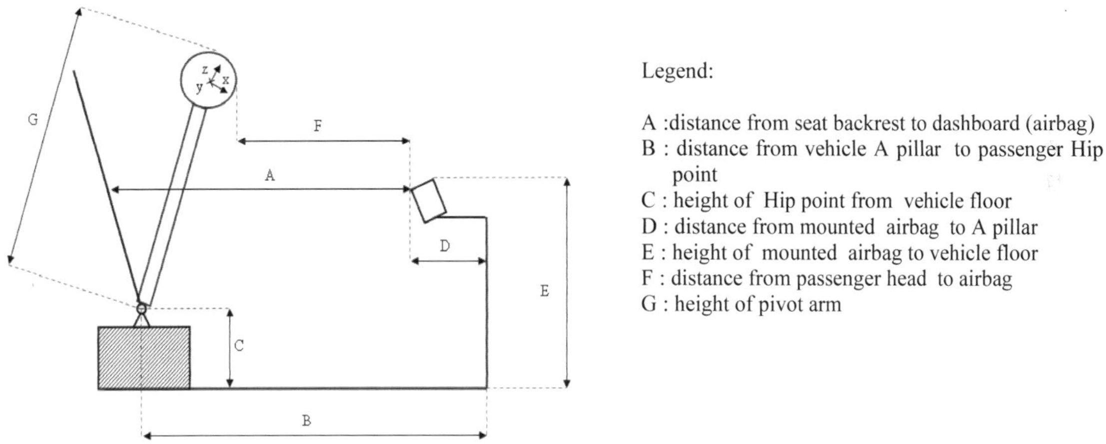

Figure 1. Test device diagram

The short event distance was able to accommodate a wired acquisition system. In order to measure the mechanical behavior of the driver head, two tri-axial accelerometers were mounted on the test device, one in the C.O.G. of the head impact device and one on the structure representing the spine, close to the neck mounting support. Voltage values were collected, parsed through dedicated amplifiers, digitized using a data acquisition board and then parsed through a Lab View testing and acquisition scenario. In total, 6 data channels were used: 1 for each accelerometer axis.

The events were recorded with a high fps camera using controlled light environment. The airbag deployment behavior was analyzed using post-processing imaging technique and was correlated for time with the head impact device behavior.

In order to study the impact effect on the occupant head, several specific head injury evaluation criterions were proposed, such as HIC_{15}, HIC_{36}, a_{3ms}, HPC. After reviewing the output data it has been determined that the criterion that filtered best our data was the 3 millisecond clip acceleration (a_{3ms}). This criterion is used in standardized automotive impact tests that simulate head impacts against interior structure and components. Standards containing procedures using the 3 millisecond clip acceleration criterions are ECE R21, ECE R25 for E.U. and (FMVSS) 201, (FMVSS) 208 for U.S. respectively. The 3 millisecond clip acceleration is a mean value and is measured against an interval of three milliseconds for which we have maximum consecutive values. In this respect, it is similar to measuring HIC.

The closest performance criterions, g-max and HIC_{15}, were also evaluated. Same model, same manufacturer airbags were used for results consistency.

Eight tests – 1, 2, 4 to 6, 8, 9 and 11– were done simulating airbag non-deployment, while four tests – 3, 7, 10 and 12 were done by deploying the airbag. Out of the last four, the first three simulated late deployment and the last one was chosen to simulate in-time deployment.

The moment the rear impulse, generated by the test device actuator, was transmitted through the pivoted arm was chosen as the time origin for our tests. The impulse was varied slightly and was recorded as acceleration data for the arm mounted sensor. These values can be found in Table 1, Column 2.

The late airbag deployment resulted through the variation of the moment of plastic cover rupture for the driver airbag unit, between 7 ms and 50 ms.

For each test, the raw data has been filtered, result acceleration curves have been calculated for the X, Y, Z axis values, the interest time frame has been chosen and values for a_{3ms}, HIC_{15} have thus been determined and compared against the threshold values.

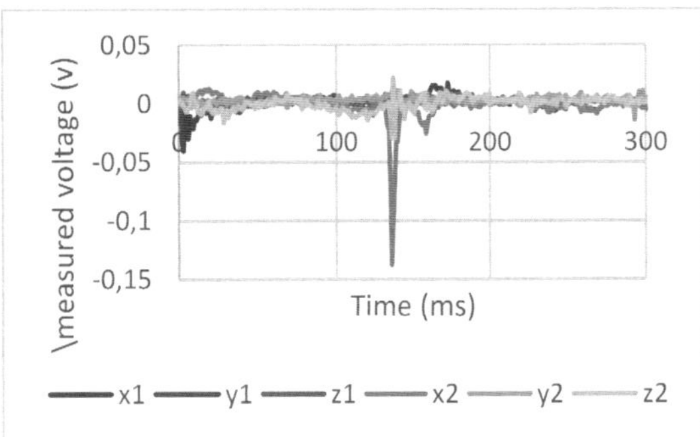

a) Unfiltered data for the event. Test 1

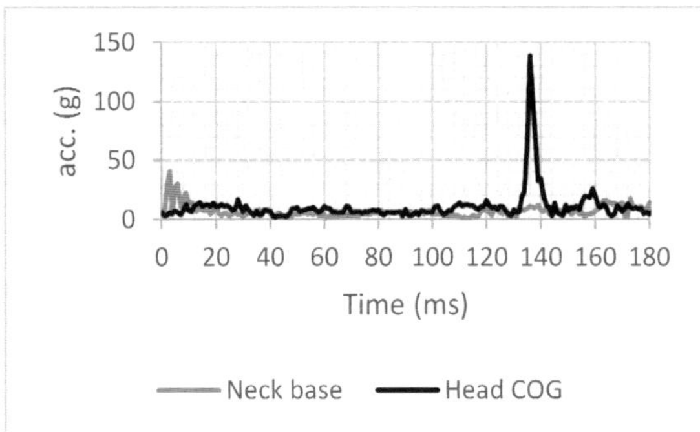

b) Result acceleration curves for head and neck base. Test 1

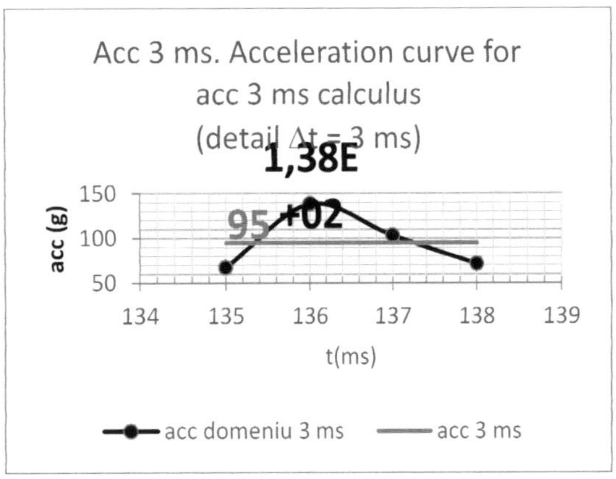

c) Maximum head acceleration (g-max) and a_{3ms} value. Test 1

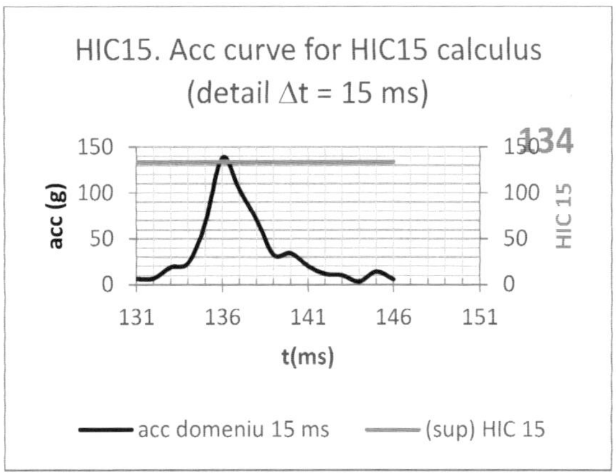

d) Acceleration curve determined for HIC_{15}. Test 1

Figure 2. Determining the values for two of the chosen performance criterions, a_{3ms} and HIC_{15}, respectively

4. Results

The time to impact was considered as the time from origin up to the maximum head acceleration value, with the exception of test number 10, where the impact was generated with the neck and thus the maximum neck acceleration was used. The mean time to impact has been determined to be 161.54 ms, ranging from 83 ms up to 272 ms, with a difference of between minimum and maximum values of 191 ms. Values for the time to impact can be found in Table 1, Column 5.

High speed recording data was analyzed against the accelerometers data and found to concur, with the exception of test number 7, for which sensor data registered only partially. In order to determine airbag deployment mechanics, sets of frames have been extracted from the relevant video footage for tests 3, 7, 10 and 12. Airbag related parameters were further determined.

Airbag inflation time, measured in milliseconds, – the time from the moment of rupture of the plastic cover up to the moment before gas expulsion, was found to average 31.25 ms, considering that the lowest inflation time was 17 ms and the highest was 43 ms. For both tests where these extreme conditions were registered, our main performance criterion registered extreme poor values. The variation in inflation time up to the point of gas release has been determined at 26 milliseconds – a time frame large enough to influence the crash mechanics and level of injury possibly to be sustained by the vehicle occupant.

Table 1. Airbag late and non-deployment measured data*

1	2	3	4	5	6	7	8	9	10
	IMPULSE Acc.	**AIRBAG deployment**	**AIRBAG inflation time ****	**TIME to impact**	**AIRBAG gas released to impact time**	**HEAD g_{-max}**	**HEAD a_{3ms}**	**HIC_{15}**	**High speed rec.**
	[g]	**[Type/-]**	**[ms]**	**[ms]**	**[ms]**	**[g]**	**[g]**		**[Yes/-]**
Threshold value	-	-	-	-	-	**< 200**	**< 80**	**< 700**	-
Test no.									
1	41	-	-	133	-	138	95	134	Yes
2	50	-	-	136	-	132	93	134	Yes
3	49	Late	30	110	28	421	124	257	Yes
4	52	-	-	271	-	125	108	431	-
5	45	-		272		133	118	462	Yes
6	55	-	-	270	-	74	58	84	Yes
7	-	Late	17	-	-	-	-	-	Yes
8	48	-	-	132	-	95	44	32	-
9	45	-	-	131	-	136	69	66	-
10	-	Late	43	83	16	522	-	-	Yes
11	47	-	-	129	-	133	88	138	Yes
12	38	In time	35	110	63	18	17	12	Yes

*Rounded to nearest integer, **prior to gas release

With respect to non-deployment tests, our main performance criterion, the a_{3ms}, has averaged at 84 g, with four out of eight tests showing a clear failure to respect the threshold of 80 g, one test coming close at 88 g and for three tests, the value was well within the accepted limits. The other two observed criterions, HIC_{15} and g-max have not been found to exceed their respective tolerated limits.

IOP Conf. Series: Materials Science and Engineering **252** (2017) 012020 doi:10.1088/1757-899X/252/1/012020

Test no.	3	7	10	12
Airbag inflated after... since event started	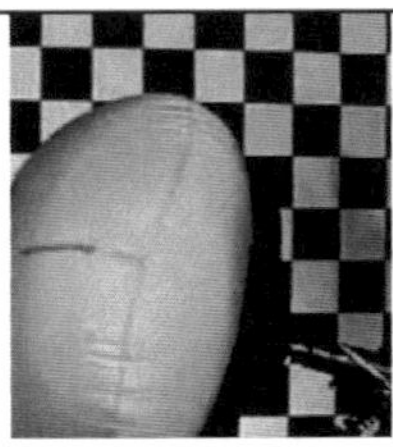	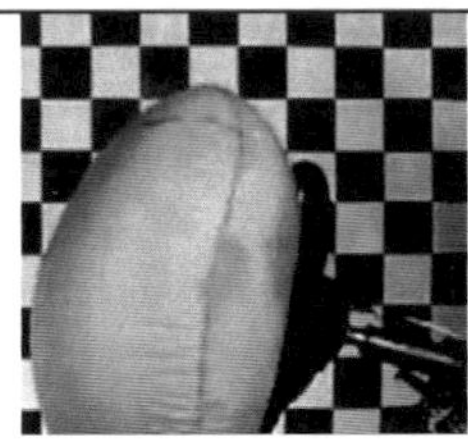	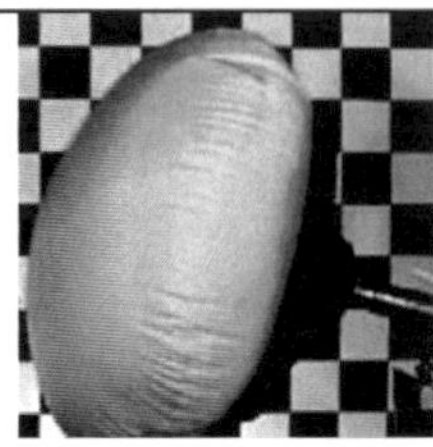	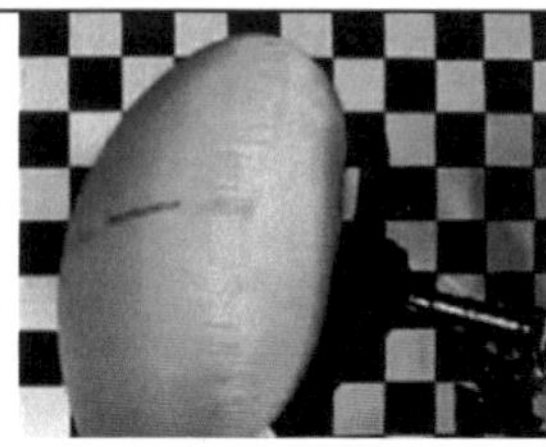
	80 ms	28 ms	65 ms	45 ms
Impact moment (maximum head acc.)	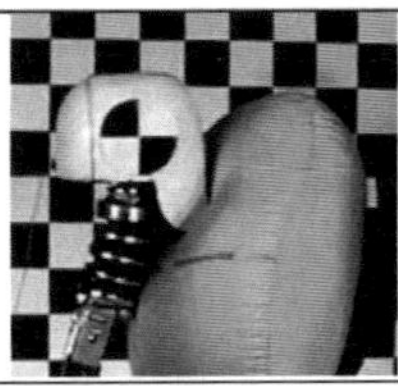	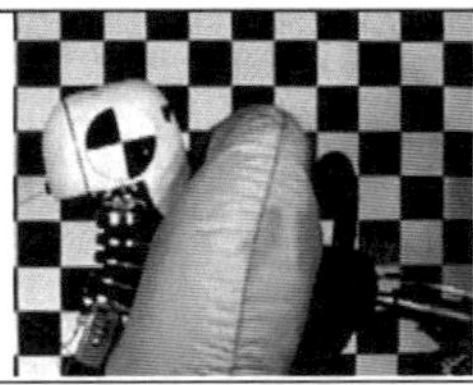	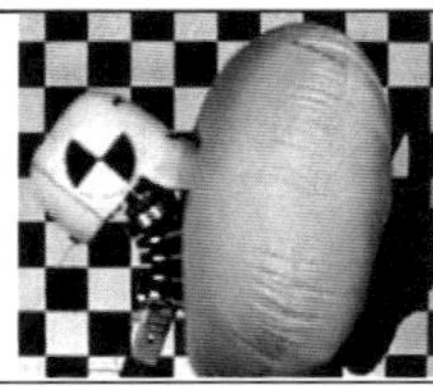	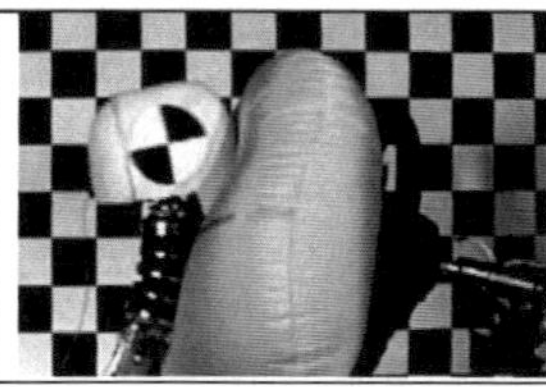
	110 ms	128 ms	83 ms	110 ms
Gas already releasing for... at the impact moment	30 ms	-	18 ms	65 ms

Figure 3. Airbag inflation stages for the moment prior to gas release (upper row) and the moment of maximum acceleration value – considered in our study the moment of impact.

With respect to airbag deployment tests, the results have shown that the in-time deployed airbag had been releasing gas for 65 ms before the moment of maximum head acceleration, twice to more than three times the period determined for our late-deployment tests.

For the three late-deployment tests, the a_{3ms} and g-max performance criterions took values far above their respective limits and above the values determined for non-deployment tests. For test 3 and 10, g-max values are 421 g and 522 g respectively, more than twice the limit of 200 g.

5. Conclusions

The time to impact variation of 191 ms greatly overshadowed the variation for the start of airbag deployment of 43 ms.

The a_{3ms} and g-max values for the late-deployment tests far exceeded that of no airbag deployment, at similar impulse acceleration values, proving that, - within the limits of our testing conditions -, a faulty airbag is more dangerous than no airbag.

We did not expect to find out such a large deviation from the mean airbag inflation time up to the point of gas release, for similar testing conditions and identical airbags -same manufacturer, model, production lot, and provider. The variation for the inflation time of 26 ms was more than half the variation for the start of airbag deployment.

While other external factors prior to out testing - such as storage, or testing factors such as contact period during airbag impact with occupant head, might have contributed to the difference noted, an increase of up to 152 percent in airbag inflation time can influence by itself the crash mechanics and the level of injury sustained by occupants of passenger cars.

The study here presented had the scope of analysing the deployment mechanics and the collision mechanism for late and non-airbag deployment in rear-end collisions for passenger cars. Four identical driver airbags were fired in similar conditions and with different results. Also eight non-deployment test were carried out. The somewhat unexpected results, while shedding light on the studied subject, also present themselves as driving factor for further analysis on the related phenomena.

IOP Conf. Series: Materials Science and Engineering **252** (2017) 012020 doi:10.1088/1757-899X/252/1/012020

References

[1] Aecio Almeida F., Vahid B., Pechet T.T.V., Costantini P.J., Farber J.L. (2006). Case Report. Lung injury after airbag deployment: Airbag lung. Injury Extra (2006) 37, 181—183, doi:10.1016/j.injury.2005.11.001.

[2] Ahmad Z. (2011). Case report. The unassuming danger of car airbags: Injuries secondary to airbag deployment, Injury Extra 42 (2011) 69–70, doi:10.1016/j.injury.2011.03.015.

[3] Boldin C., Peicha G., Passler J.M., Hauser H., Riccabona M. (2002). Case report. Inferior thyroid artery injury due to airbag deployment, Injury, Int. J. Care Injured 33 (2002) 283–284, PII: S0020-1383(01)00080-8.

[4] Braver E.R., McCartt A.T., Sherwood C.P., Zuby D.S., Blanar L., Scerbo M. (2010). Front air bag nondeployments in frontal crashes fatal to drivers or right-front passengers, Traffic Injury Prevention, 11:178–187, 2010, doi: 10.1080/15389580903473973.

[5] Cullinan M., Merriman T. (2001). Oesophageal rupture resulting from airbag deployment during a motor vehicle accident, ANZ J. Surg. (2001) 71, 554–555.

[6] Evans L. (1991). Airbag effectiveness in preventing fatalities predicted according to type of crash, driver age, and blood alcohol concentration*, Accident Analysis and Prevention, 1991, vol.23 no.6, pp.531-541.

[7] Fildes B.N., Cameron M.H., Vulcan A.P., Digges K.H., Taylor D. (1994). Airbag and facebag benefits and costs*, Accident Analysis and Prevention, 1994, vol.26 no.3, pp.339-541.

[8] Frieder R., Kumar S. (2007). Biomechanical analysis of late airbag deployment in motor vehicle crashes using computer simulation, Proceedings of the ASME 2007 Summer Bioengineering Conference (SBC2007), SBC2007-176654.

[9] Gao D., Wampler C.W. (2009). Head Injury Criterion. Assessing the Danger of Robot Impact, IEEE Robotics & Automation Magazine vol.16, dec.2009, 10.1109/MRA.2009.934824.

[10] Hazarika S., Willcox N., Porter K. (2007). Patterns of injury sustained by car occupants with relation to the direction of impact with motor vehicle trauma – evidence based review, Trauma 2007; 9: 145–150, 10.1177/1460408607084179.

[11] Hutchinson J., Kaiser M.J., Lankarani H.M. (1998). The Head Injury Criterion (HIC) functional, Applied Mathematics and Computation 96 (1998) 1-16, PII: S0096-3003(97) 10106-0.

[12] Kendall J., Solomon K.A. (2014). Air bag deployment criteria, The Forensic Examiner, 2014.

[13] Kenney K.S., Fanciullo L.M. (2005). Automobile air bags: friend or foe? A case of air bag-associated ocular trauma and a related literature review. Optometry, Volume 76 / Number 7 / July 2005, Optometry 2005;76:382-86.

[14] Keshavaraj R., Tock R.W., Nusholtz G.S. (1998). Simple numerical model (PRAM) for simulation of the passenger–bag interactions during deployment of an airbag, Journal of Applied Polymer Science, Vol. 67, 933–948 (1998).

[15] Liyanage S.E., Mearza A.A. (2009). Case report. Airbag-induced corneal flap, Contact Lens & Anterior Eye 32 (2009) 33–34.

[16] Matthes G., Schmucker U., Lignitz E., Huth M., Ekkernkamp A., Seifert J. (2006). Does the frontal airbag avoid thoracic injury?, Arch Orthop Trauma Surg (2006) 126:541–544, doi 10.1007/s00402-006-0181-y.

[17] Mearza A.A., Koufaki F.N., Aslanides I.M. (2008). Case report. Airbag induced corneal ectasia, Contact Lens & Anterior Eye 31 (2008) 38–40.

[18] Mercer G.N., Sidhu H.S. (2005). Modeling thermal burns due to airbag deployment. Burns 31 (2005) 977–980, DOI:10.1016/j.burns.2005.06.012.

[19] Moehnke T., Wagner H. (2009). Case report. Descemet's membrane detachment attributed to the mechanical forces of airbag deployment, Contact Lens & Anterior Eye 32 (2009) 27–30.

[20] Mon Y.J. (2007). Airbag controller designed by adaptive-network-based fuzzy inference system (ANFIS), Fuzzy Sets and Systems 158 (2007) 2706–2714.

[21] Segui-Gomez M. (2000). Driver air bag effectiveness by severity of the crash, American Journal of Public Health, 90:1575–1581, 2000.

[22] Stübig T., Brand S., Zeckey C., Beltran M.J., Otte D., et al. (2013) Thoracic injuries sustained by severely injured front-seat passengers and drivers: injury patterns and their relationship to crash characteristics, International Journal of Injury Control and Safety Promotion, 20:4, 313-320, doi: 10.1080/17457300.2012.724692
[23] Sugg J.R. Jr., Enzenauer R.W. (2002). Airbag-related Eye Trauma: A Review of Current Literature.
[24] Tanno K., Kohno M., Ono K., Ohashi N., Doy M., and all (2000). Fatal cardiovascular injuries to the unbelted occupant associated with airbag deployment: two case-reports. Legal Medicine (Legal Med) 2000 ; 4 : 227-31.
[25] Van Der Linden W.J. (2002). Dislocated fracture of the mandibular condylar process after airbag deployment: Report of a Case, J Oral Maxillofac Surg 60:113-115, 2002, doi:10.1053/joms.2002.29089.
[26] Whitaker I.S., Shokrollahi K., Mishra A., Potokar T.S., Dickson W.A. (2008). Paediatric alkaline and thermal burn injury due to airbag deployment: a public health risk, Injury Extra (2008) 39, 311—312, doi:10.1016/j.injury.2008.04.003.
[27] Yang S.K., Liu T.S. (1997). Failure analysis for an airbag inflator by petri nets, Quality And Reliability Engineering International, Vol. 13, 139–151 (1997).
[28] * Report, Unsigned document/N.H.T.S.A. (2007), Counts of frontal air bag related fatalities and seriously injured persons, U.S. Department of Transportation, National Highway Traffic Safety Administration, National Center for Statistics and Analysis, Crash Investigation Division, WDC, Jul.2007.
[29] *Web site. ARCCA. http://arcca.com/airbags-late-to-deploy/. Accessed on: August 10, 2017.

Most common road safety engineering deficiencies in South Eastern Europe as a part of safe system approach

D Jovanov[1], H J Vollpracht[2], H Beles[3], V Popa[4], B A Tolea[4]
[1]International Road Safety Centre (IRSC), Serbia
[2]World Road Association (PIARC), Germany
[3]University of Oradea, Romania
[4]University Transilvania of Brasov, Romania

Abstract. Most common road safety engineering deficiencies identified by the authors in South Eastern Europe, including Romania, have been collected together and presented in this paper as a part of road safety unbreakably connected to the safe system approach (driver-vehicle-road). In different South Eastern Europe countries Road Safety Audit (RSA), Road Safety Inspection (RSI), as well as Black Spot Management (BSM) was introduced and practical implementation experience enabled the authors to analyze the road safety problems. Typical road safety engineering deficiencies have been presented in 8 different subsections, based on PIARC (World Road Association) RSA approach. This paper presents collected common road safety problems with relevant illustrations (real pictures) with associated accident risks.

1. Introduction

After almost 15 years of experience with RSA and RSI and even more with BSM in South Eastern Europe, these procedures are now recognized as one of the most efficient road safety engineering tools. RSA/RSI are highly efficient and cost effective engineering tools for improvement of safety on roads, because it is much cheaper to identify road safety deficiencies in the process of design than later after construction is completed and accidents start to happen RSA/RSI are among the most cost-effective investments a Road Authority can undertake.

With its EU Directive No. 2008/96 on road infrastructure safety management, published in October 2008, the European Union (EU) made a clear decision that all three mentioned procedures will be mandatory for the Trans-European Road Network (TERN) in the coming years [11].

Unfortunately, in reality systematic application of RSA/RSI in South Eastern Europe is only partial. Where RSA/RSI are implemented they are mostly pushed by IFIs (WB, EIB, EBRD, etc.) and implemented by foreign consulting companies. The situation with BSM is better and the level of implementation is higher. However, even when procedures are undertaken, the proposed recommendations (measures) are not always implemented by the road authorities. EU funded Projects have tried to develop capacity for RSA/RSI implementation in each of the South Eastern Europe countries. Therefore, in South Eastern Europe important steps towards RSA/RSI implementation have been taken. Each country now has several trained auditors, a different Road Safety Audit manual has been produced and certain Pilot road sections have been audited. In almost all South Eastern Europe countries, RSA has been introduced into the legislation as a mandatory procedure. A similar or even better situation exists with BSM.

Within this paper, the authors try to make summary of their findings based on their implementation experience of RSA, RSI and BSM in South Eastern Europe countries. The paper follows the PIARC RSA/RSI approach and classifies identified road safety deficiencies into 8 broad groups or categories (subsections 3.1 to 3.8) [19].

2. Statistical data regarding the severe roads accidents in Romania

Despite some positive trends, road safety represents still a big challenge for Romania. Together with Leetonia, Bulgaria and Lithuania, Romania is one of the four European Union countries with the biggest road safety accidents rate where victims are involved. With a rate of 95 fatalities for one million inhabitants, the rate of road safety accidents in Romania is three times bigger than other performant countries in European Union, like Sweden, United Kingdom or Nederland [17].

In addition, the motorization degree of ~200 passenger cars for 1.000 inhabitants, Romania is much below the European Union average, but based on trends and estimations, they will drastically increase in the following years. This fact might contribute or increase the number of road accidents in the future, if preventive measures are not considered and implemented.

Considering the development of the specific dynamic indicators between 1990 – 2014 (figure 1), we could observe a strong growth trend for the period 2003 – 2008, which was then reversed into a decreasing trend. Roads accidents have dropped with 35%, from 2.796 in 2009 to 1.818 in 2014, but injuries have dropped with only 9% for the same period, from 15.295 to 13.237. In the same time, Romania registered a constant increase in the number of vehicles with a long term growth trend of 5% per year [17].

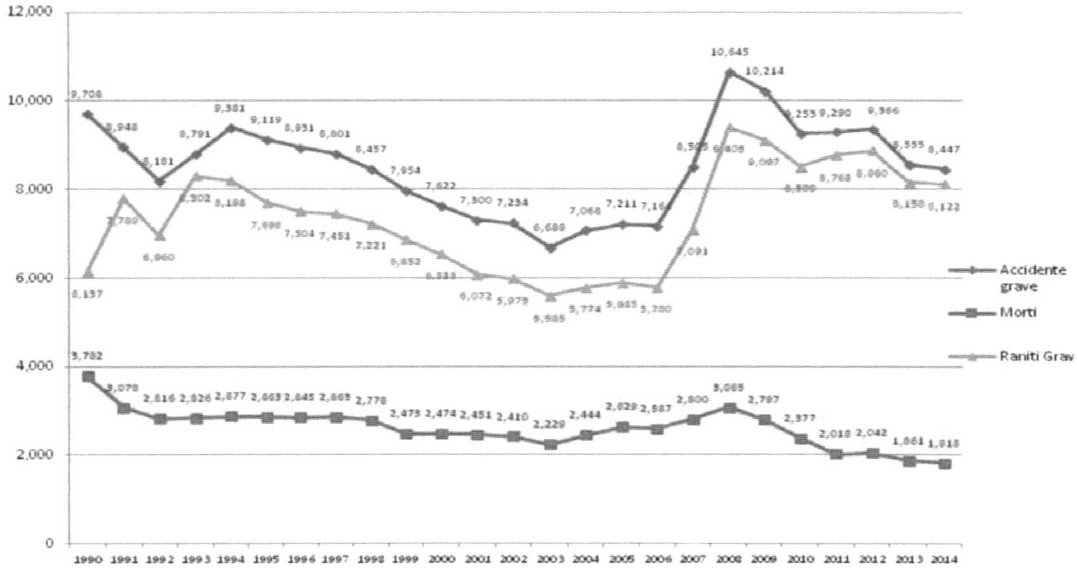

Figure 1. Specific dynamic indicators regarding the road accidents situation for Romania between 1990 – 2014 [17].

Based on the information published by the European Transport Safety Council (ETSC) in 2015 (figure 2), from 2001 (reference year when was set the objective for road accidents fatalities number reduction) to date, Romania has the worst performance, together with Malta, Bulgaria and Poland [17].

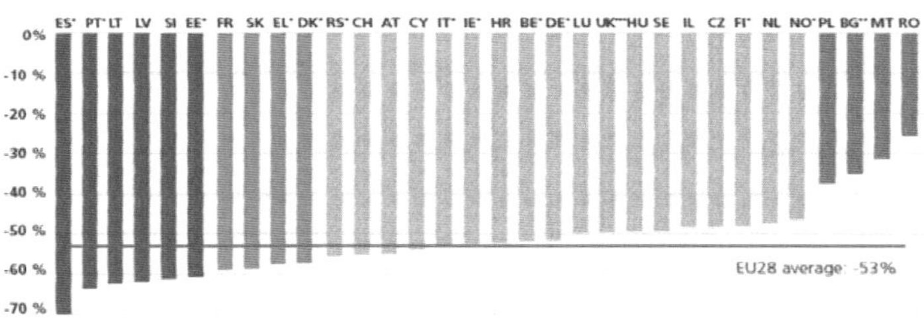

Figure 2. Fatalities number development as a consequence of road accidents in European Union between 2001 – 2014 [17].

There was an improvement of this indicator during 2010 – 2014, when the number of victims from road accidents decreased in Romania with 23,5%, above the European Union average of 18,2% (figure 3) [17].

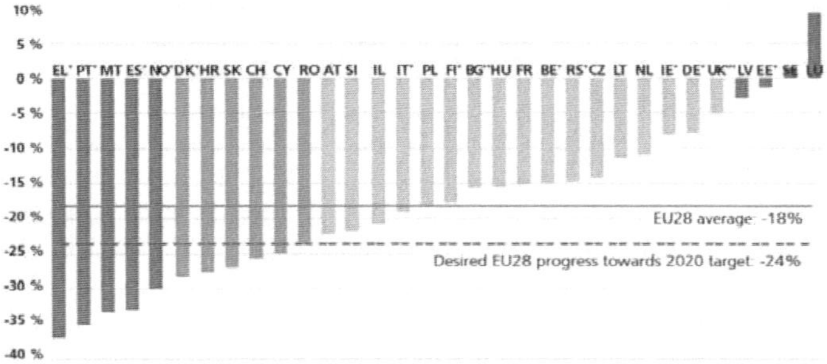

Figure 3. Fatalities number development recorded as a consequence of road accidents in European Union between 2010 – 2014 [17].

The reduction of fatalities number was a consequence of multiple factors: measures adopted for reaching the imposed legal requirements, improvement of the roads infrastructure, improvements in drivers education and requirements, improvements in the accidents intervention system.

3. Typical road safety engineering deficiencies

The most important road safety deficiencies identified within South Eastern Europe are briefly presented along with typical problems, illustrations of unsafe design and estimates of expected accidents that may occur in practice. Pictures from South Eastern Europe countries are used for illustrations of problems.

3.1. Road Function. Roads with mixed function (Linear settlements)

Typical problems: mixed road function (usage of the road as fast distributors for fast longer distance motorized traffic and as a route for slow local traffic) causes one of the major road safety problems in South Eastern Europe. This is one of the common problems where the rate of expansion of isolated communities along a road can rapidly reduce the effectiveness of a nationally or regionally important route as a result of the local traffic activities overwhelming the through route function of the road.

Examples of unsafe designs are presented in figures 4 and 5, for 2+2 road without median.

Figure 4. Example no. 1 of unsafe road designs.

Figure 5. Example no. 2 of unsafe road designs.

In such cases, the role of the road in the road hierarchy becomes confused. While the road is passing through settlements (where no by-pass exists) can it keep its geometry unchanged? Can it even be called International/Regional/National road, or does it become a street? This, simple planning (designing) confusion of local administrations, can cause tremendous problems in road safety. Once intense development has been allowed, it is very difficult to achieve improvements without major reconstruction on a new alignment. Often even when a bypass has been built, the village often, over time, extends out across to the new road. This is mainly an issue of poor access control and development control by relevant authorities [4], [5], [18].

Expected accidents: pedestrians crossing street outside a junction; pedestrians in the road; at least two vehicles - same direction - rear end collisions; at least two vehicles - head on collision in general; at least two vehicles - same road - opposite direction - turning left (right) in front of other vehicle, etc.

3.2. Cross Sections. Types of cross profiles (Width of the road)

Typical problems: a cross section will normally consist of the carriageway, shoulders or kerbs, drainage features and earthwork profiles. It may also include facilities for pedestrians, cyclists or other special user groups. There is some evidence to suggest that widening lane or carriageway width or widening shoulders up to a certain extent is beneficial in reducing certain types of accidents. However, beyond a certain point it can have negative effects on road safety (users will start using extended width as a regular lane). Dangerous cross sections of express roads and highways are frequently being used in South Eastern Europe participants, including of Romania. For example: four lane road without a crash barrier or two lane road with wide hard shoulders. A road with a wide hard shoulder can sometimes be misused by drivers as a very narrow four lane road, with disastrous results and very serious crashes [1], [2], [3].

Examples of unsafe designs regarding the cross sections of the road are presented in figures 6 and 7.

Figure 6. Too narrow acceleration lane.

Figure 7. 1+1 Width wide hard shoulders.

Cross sections, particularly on roads through built up areas, are often not uniform or consistent. Local developments may encroach onto the carriageway because of the lack of effective planning control. In rural conditions cross sections may be reduced at drainage structures causing sudden changes in width.

Steep side slopes, introduced for drainage purposes, do not allow a driver to recover in case he leaves the carriageway, and thereby add to the likelihood of an accident. Open channel drains can also increase the probability that driver error will result in an accident.

3.3. Alignment. Vertical and horizontal curves (Consistency)

Typical problems: unexpectedly tight horizontal curves can lead to accidents as drivers try to drive through them at too high a speed. A similar situation may occur on horizontal curves in other similar hazardous situations, such as steep gradient or after a long straight section where driver is encouraged or misled (by the approach geometry) to think that he can drive at higher speed than is safe for that location. The sight distances associated with larger curve radii may also encourage driver to overtake in unsafe conditions [6], [7], [8].

Poor co-ordination of the horizontal and vertical alignments can result in visual effects which contribute to the accidents and are detrimental to the road appearance. Unsafe combinations of horizontal and vertical curvature are likely to be misinterpreted by a driver and can result when horizontal and vertical curves of different length occur at the same location. These situations are particularly dangerous and are unfortunately frequently present in South Eastern Europe.

Two examples of unsafe designs, with respect to the alignment, are presented in figures 8 and 9 (for sharp curve).

Figure 8. Example no. 1 of unsafe designs.

Figure 9. Example no. 2 of unsafe designs.

Expected accidents: single vehicle accidents in a bend - going either side of the road; at least two vehicles - head on collisions in general; at least two vehicles - same direction - rear end collisions; etc.

3.4. Intersections. Channelization of traffic flows

Typical problems: channelization is a useful tool in traffic management. It should be applied to all junctions on high speed roads. This may require local widening but the small additional cost of this at the design stage will be offset by future safety benefits in almost every case. Consideration of the access needs of emergency and other priority vehicles is required, especially in the event of an accident or breakdown. If provision is not made for this, damage to kerbs will quickly occur.

Channelization guides the driver through the conflict points, provides safe areas for him to stop while making a maneuver and reduces conflicts between different flows of traffic.

Channelization by means of road markings, raised kerbs, traffic islands and bollards, can be used to guide vehicles along a specific path on the approach to and/or exit from a junction and to position them at the safest location to make their maneuver. The benefits of this are that movements are simplified, less confusion arises and the number of conflict points is minimized [9], [10].

Traffic islands have the added benefit of providing a refuge for pedestrians crossing the road. They also provide a convenient location for street furniture such as signs, street lighting and drainage covers.

Examples of unsafe designs, with respect to intersections, are presented in figures 10 and 11.

Figure 10. Confusion for drivers leads to mistakes.

Figure 11. Unsafe “Y” type of intersection.

Expected accidents: at least two vehicles - same road - opposite direction - turning left (right) in front of other vehicle; at least two vehicles - crossing (no turning) – different; at least two vehicles - head on collision in general; at least two vehicles - same direction - entering traffic; at least two vehicles - opposite direction no turning – others; etc.

3.5. Public and private services. Services along roadside

Typical problems: roadside facilities (rest places and petrol stations) are necessary to serve the long distance traffic between regions and towns (villages). Drivers need to rest at least once every 2 or 3 hours in order to maintain their concentration when driving. It is useful to combine rest areas with petrol and/or service stations at 30 – 50 km distances. Entrances and exits to and from Service and Rest areas can cause a disruption to traffic on the main carriageway if they are not separated well, and special attention should be given to design and maintenance of deceleration and acceleration lanes. It is important that sufficient rest areas are provided at around 10 km intervals but not too many to avoid constant disruption of the main flow of traffic by constantly exiting and merging traffic. Such rest areas may be used for selling goods by local farmers to minimize such activity occurring along the road itself. Farmers should be able to reach such areas from minor roads behind the service area.

In South Eastern Europe participants there some examples where roads are encroached upon by unacceptable commercial services or where there are unsuitable rest areas. This is dangerous for all road users, because of huge speed difference and mixture of different categories of road users. This can result in sudden vehicle stopping of vehicles and sudden entry of vehicles into the traffic as well as presence of unprotected pedestrians on high speed roads [12], [13].

Master plans, land usage, urban development and restrictions in access to the public road network are key elements that can be used for preventing these types of accidents.

Some examples of unsafe designs are presented in figures 12 and 13.

Figure 12. Unsafe Bus stop.

Figure 13. Rest area, lack of separation, no regular exit.

3.6. Vulnerable road users need. Pedestrian crossing

Typical problems: pedestrians should not have to walk at all along interurban roads. Hard shoulders are not intended for vulnerable road users but for emergency use by vehicles only. With the exception of roundabouts, pedestrian crossings should ideally be grade separated on major roads if large numbers of vulnerable road users are expected. At-grade pedestrian crossing on dual carriageways or multi-lane roads should be forbidden unless traffic signals are provided. To enable pedestrians to cross safely, crossings should be provided as underpasses or over bridges with ramps, not stairs. Any other solution significantly increases risks of pedestrian accidents. Even though it is not in accordance with any road standards/norms in the world, including modifications (upgrade) of old German (DIN) and ex-soviet SNiP and GOST standards used in South Eastern Europe, there are many such examples where pedestrian crossings have been placed at the same level on an international road.

In order to provide additional traffic capacity at junctions, local widening is sometimes carried out but this can often increase the crossing distances - again creating increased risk for pedestrians.

Heavy crossing demands may sometimes occur away from junctions where vehicle speeds are very high and this is often the case in South Eastern Europe. The provision of underpasses or over bridges however may be too expensive and may not be well used. Designers and the road authority need to provide crossings which the pedestrians will be willing to use [14], [15].

Examples of unsafe pedestrian crossing designs are presented in figures 14 and 15.

IOP Conf. Series: Materials Science and Engineering **252** (2017) 012021 doi:10.1088/1757-899X/252/1/012021

Figure 14. Example no. 1 of unsafe pedestrian crossing designs.

Figure 15. Example no. 2 of unsafe pedestrian crossing designs.

3.7. Traffic signing, marking and light. Signing

Typical problems: warning signs and warning markings are used to give advance notice of a potential hazard ahead or of any unexpected feature of the road geometry. The signs are used in specific situations when there is a change in the road, such as on a bend, on a high speed road or on the approach to junction. The location of signs is very important because they should provide adequate warning or information at sufficient distance, however they should not obscure important road features. Of great importance for the visibility of the signs is that they be located in positions where overgrown vegetation cannot obscure the visibility of the sign. Signs must be visible at all times, so reflective materials should be used for night-time visibility and urban signs may require to be lit internally or externally. In some South Eastern Europe countries, it is common practice for the signs to be missing (even at dangerous locations), not properly positioned, without reflectivity, non-standardized or even not consistent with International UN Conventions [16].

Examples of unsafe designs are presented in figures 16 and 17.

Figure 16. Hidden sign.

Figure 17. Hidden traffic light.

A recurring problem with signs is of them being obscured, either by permanent features such as street furniture and vegetation or by parked vehicles and, on dual carriageways, by moving vehicles in the nearside lane (if there is no repeated sign on the other side of the road). Too many signs can detract from their objective by overloading the driver with too much information too quickly, which leads to confusion or to a situation where the driver ignores certain signs. Signs may not be visible at night time because of poor illumination, lack of routine maintenance, reliability/continuity of power supply or inappropriate positioning (too high, set back out of road or turned away from driver). If reflective signs are not regularly cleaned, they may not retain their design properties.

Expected accidents: pedestrians crossing street outside a junction; pedestrians crossing street at a junction; at least two vehicles - crossing (no turning) - different; at least two vehicles - same direction - rear end collisions; at least two vehicles - same road - opposite direction - turning left (right); etc.

3.8. Roadside features and passive safety installations. Roadside obstacles (plants, trees, light poles, advertisements, etc.)

Typical problems: the presence of roadside obstacles, street furniture (for example, road signs and lighting columns) advertising signs and trees has two safety implications. The first is the potential danger of collision, and the second is their obstruction of visibility. Visibility is important not only for the driver, but also to other road users. Obstructions caused by trees, for example, may result in pedestrians making unsafe decisions.

Great care should be taken concerning the positioning of roadside features which may either obstruct visibility, lead to accidents or increase accident severity. Where obstructions cannot be practically removed, and could contribute to hazardous situations, consideration should be given to their replacement with equipment designed to collapse on impact, re-alignment of the road, or the introduction of barriers. Once a road is completed, care must be taken to ensure that new obstacles are not introduced by other institutions subsequently, such as telephone or electricity authorities.

Vegetation should be trimmed regularly and planning controls should be enforced to prevent stalls and structures being too close to the road edge. In some South Eastern Europe countries, trees are often planted adjacent to roads in order to provide shade for pedestrians, animals and parked vehicles and in other countries to prevent the wind from bringing snow onto the road [19].

Some examples of unsafe designs are presented in figures 18 and 19.

Figure 18. No guardrails at sharp curve.

Figure 19. Dangerous start point of guardrail.

Expected accidents: single vehicle accidents with obstacles on or above the road; single vehicle accidents with obstacles - others; single vehicle accident - leaving straight road - either side of the road; at least two vehicles - same direction - rear end collisions; pedestrians walking along the road, etc.

3.9. Temporary signing and marking at Work Zones

Typical problems: a work zone is an area of road or roadside where construction, maintenance or other works are performed and which may affect the safety and limit the free movement of road users through and in the vicinity of the Work Zone. Work zones are zones on the road with higher risk of accidents for both road users (vehicle occupants and vulnerable categories) and workers. A Traffic Management Plan (TMP) of good quality should be made and followed so that all participants in traffic are protected against risk of a traffic accident. Such TMP should contain all elements starting from design, placement, and maintenance to the removal of all elements regulating the road traffic [19].

To minimize the problems and increase safety, work zone layout (marking and signing) requires special consideration for the following reasons:

- Work zone is a section of road where, most often, geometrical characteristics of the road and the traffic conditions are changed to poorer conditions (less safe). The types of executed works are often road construction, rehabilitation and maintenance, but there are other types of work on the road that need the same treatment, for instance work with cables, pipes etc. located in the road area.

- Employees in work zones spend most of their working hours directly exposed to traffic. In accidents, happening in work zones, these employees are often the victims, and can often be at as much at risk as the road users.

The growing international transit traffic flow in South Eastern Europe implies the need for main traffic corridors to be constructed according to international standards and requires European standards and a widely recognized and consistent system for road works signing and work zone safety.

An example of unsafe designs is presented in the following figure 20.

Figure 20. Dangerous working zone

Expected accidents: single vehicle accidents with roadwork materials; hitting parked vehicles right (left) side of the road; at least two vehicles - head on collision in general; at least two vehicles - same direction - rear end collisions; pedestrians walking along the road, etc.

4. Conclusion

Education and training of the auditors/inspectors is probably one of the weakest points in the entire safe system approach in South Eastern Europe countries. The reasons for this are relatively short history of RSA in the region, lack of understanding of RSA methodology and procedures and lack of RSA literature in local languages. This is why the team of safety engineering specialists, who are very well acquainted with conditions in South Eastern Europe, prepared an extended version of this paper called: "Practical Guide for Road Safety Auditors and Inspectors" with the specific intention to help present and future auditors/inspectors in their work across the region.

The Practical Guide is based on actual traffic situations identified as road safety deficiencies and draws upon best international practice and proposals for improvement (treatment). As South Eastern Europe countries contain important transport links (corridors), harmonization of road standards and elimination of potential risks for the road users are of utmost importance.

The main reason for presenting this paper was to help new auditors/inspectors and clients to recognize typical road safety deficiencies in road engineering that have been occurring in past designs and to prevent them in the design/construction and maintenance processes. Also, in the long term, such collated knowledge about typical road safety deficiencies can help lead to the improvement of road design standards.

The objective of the national strategy for road safety in Romania is to reduce with 50% the number of fatalities from cars accidents until 2020 compared with 2010. The specific objective is to reach a maximum of 1188 fatalities in 2020, compared with 2377 in 2010. In the same time, the general objective of the Strategy is to transform Romania in a safe country for its citizens, investors and tourists with regard to road traffic, by progressive reduction of the injuries from road accidents in 2016 – 2020. The vision of this strategy is to reduce the number of fatalities and serious injuries from road transport system as close to zero in order to sustain a good level of social and economic life conditions [17].

Acknowledgments
This Paper presents a digest from Practical Guide for road safety auditors/inspectors which builds to a large extent on international best practice, direct experience of the authors and draws upon detailed guidance and concepts in the 5 key publications indicated below:

1. "Towards safer roads in developing countries", a guide for planners and engineers, developed by TRL, Ross Silcock partnership and ODA in 1991,
2. "The handbook of road safety measures", written by Rune Elvik and Truls Vaa, in 2004
3. "Catalogue of design safety problems and practical countermeasures", developed by World Road Association (PIARC) in 2009
4. "Romanian Manual for Road Safety Audits on Existing Roads (RSI) and on Road Design (RSA)", developed by World Road Association (PIARC) in 2016
5. "National Strategy for Road Safety for 2016 – 2022", developed by Transports Ministry of Romania in 2016.

The above 5 documents provide much more detailed guidance on all key aspects of safety engineering and authors recommend that road engineers should use these in planning and operation of roads to ensure safer road networks. We are grateful to the authors of the original documents for sharing their experience via these documents. All photographs used as illustration of the problems, or as best practices are provided by the authors.

REFERENCES

[1]. Belcher M, Proctor S and Cook R 2008 *Practical Road Safety Auditing* TMS 2nd Edition
[2]. Elvik R 2008 *State-Of-The-Art Approaches to Road Accident Black Spot Management and Safety Analysis of Road Networks*, RIPCORD ISEREST Report
[3]. Elvik R and Vaa T 2004 *The Handbook of Road Safety Measures* Elsevier Amsterdam,
[4]. *** 1991 *Towards Safer Roads in Developing Countries. A guide for Planners and Engineers* TRL Ross Silcock Partnership and Oda UK
[5]. *** 1996 *Low-Cost Road and Traffic Engineering Measures for Casualty Reduction* European Transport Safety Council (ETSC) Brussels
[6]. *** 2002 *Guidelines for Road Safety Audits* German Road and Transportation Research Association (FGSV)
[7]. *** 2002 *Good-Practice Guidelines to Infrastructural Road Safety* European Union Road Federation
[8]. *** 2005 *Sustainable Safe Road Design: A Practical Manual* World Bank
[9]. *** 2007 *Road Safety Audit Guideline* World Road Association (PIARC) Paris
[10]. *** 2007 *Road Safety Inspection Guideline* World Road Association (PIARC) Paris
[11]. *** 2008 *Directive on Road Infrastructure Safety Management No 96/2008*, European Parliament and of the Council, Brussels
[12]. *** 2008 *Road Safety Audit – Best Practice Guidelines, Qualification for Auditors and "Programming"*, RIPCORD-ISEREST Project – WP4 EU Project
[13]. *** 2008 *Road Safety Inspection – Best Practice and Implementation Plan*, RIPCORD-ISEREST Project – WP5 EU Project
[14]. *** 2009 *Catalogue of Design Safety Problems and Practical Countermeasures* World Road Association (PIARC) Paris
[15]. *** 2009 *Road Safety Audit Manual* South East Europe Transport Observatory (SEETO) EC/SEETO
[16]. *** 2009 *Road Safety Inspection Manual* South East Europe Transport Observatory (SEETO) EC/SEETO
[17]. *** 2016 *National Strategy for Road Safety for 2016 – 2022* Transport Ministry Romania
[18]. *** 2016 *Romanian Manual for Road Safety Audits on Existing Roads (RSI) and on Road Design (RSA)* World Road Association (PIARC) Paris
[19]. *** 2016 *Road Safety Manual* World Road Association (PIARC) Paris

A comparison of optimal semi-active suspension systems regarding vehicle ride comfort

Dimitrios Koulocheris, [1]Georgios Papaioannou, Emmanouil Chrysos

Vehicles Laboratory, School of Mechanical Engineering, National Technical University of Athens, 15780 Athens, Greece

[1]gpapaioan@central.ntua.g

Abstract. The aim of this work is to present a comparison of the main semi active suspension systems used in a passenger car, after having optimized the suspension systems of the vehicle model in respect with ride comfort and road holding. Thus, a half car model, equipped with controllable dampers, along with a seat and a driver was implemented. Semi-active suspensions have received a lot of attention since they seem to provide the best compromise between cost (energy consumption, actuators/sensors hardware) and performance in comparison with active and passive suspensions. In this work, the semi active suspension systems studied are comfort oriented and consist of (a) the two version of Skyhook control (two states skyhook and skyhook linear approximation damper), (b) the acceleration driven damper (ADD), (c) the power driven damper (PDD), (d) the combination of Skyhook and ADD (Mixed Skyhook-ADD) and (e) the combination of the two with the use of a sensor. The half car model equipped with the above suspension systems was excited by a road bump, and was optimized using genetic algorithms (GA) in respect with ride comfort and road holding. This study aims to highlight how the optimization of the vehicle model could lead to the best compromise between ride comfort and road holding, overcoming their well-known trade-off. The optimum results were compared with important performance metrics regarding the vehicle's dynamic behaviour in general.

1. Introduction

The primary purpose of a vehicle's suspension system is to provide stability, support, and comfort during the drive. Ride comfort and isolation from road disturbances are greatly influenced by the suspension system. Therefore, suspension design is a trade-off between the three conflicting criteria of road holding, load carrying and passenger comfort. In order to study the suspension system in vehicles, various mathematical models have been developed such as the quarter car, the half car and the full car model. The quarter model is dominant in the research studies due to its simplicity [1], however the half car model demonstrates more accurate results including the effect of either the pitch or the roll phenomena [2, 3]. Although, for greater accuracy the full car model is used including both the roll and the pitch phenomena, but it is more demanding computationally than the others and hence very few studies have been performed based on it. In addition to the vehicle model in recent studies, a seat [4] and a driver [5] have been added, so as to study the ride comfort of the passengers extensively and investigate in depth the effects of the vibrations induced by the road to the passengers.

The main focus of these models is the suspension system modeled with a spring and a damper. Vehicle suspensions are mainly classified into three types depending on their operation: passive, semi-active and active. Passive suspensions are older conventional systems having non-controlled springs and shock absorbing dampers with fixed parameters. On the contrary, active suspensions can adapt to the system variations based on online changes of the actuating force. Therefore, active suspensions have been extensively studied since 1960s and various approaches have been proposed in order to exploit their advantages However, active suspensions normally require important amounts of power supply, which is their main drawback, preventing them from being used extensively in practice. From 1970s, semi-active suspensions have received much more attention, since they can achieve greater performance than passive ones and consume less power than active ones. Thus, the dominant line of research lies in the semi-active suspensions since they provide the best trade-off between cost (energy-consumption, actuators/sensors hardware) and performance.

The research in this field follows two mainstreams: the study of new technologies of semi-active actuation of damping (like electro-hydraulic, electro-rheological and magnetorheological damper), and the design of semi-active control strategies. Bourmistrova et al. (2005) [6] applied evolutionary algorithms to the optimization of the control system parameters of a quarter car model. The multi objective fitness function which was a weighted sum of car body rate-of-change of acceleration and suspension travel was minimized. Poussot-Vassal et al. (2010) [9] aimed at providing a full analysis of the optimal performances both for the comfort and for the handling objectives. These optimal performances were approximated through a numerical optimization approach, using a Model Predictive Control model. Lajqi et al. (2012) [1] introduced a design and optimization procedure for active and semi-active non-linear suspension systems regarding terrain vehicles using a quarter car model. The excitation of the model was represented by double cosine road bumps. Savaresi et al. (2007) [10] presented a study of the performance limits of a semi-active suspension system mainly focused on the objective of comfort. The complementary characteristics of skyhook and acceleration driven damping (ADD) were combined and the mixed control of the two control laws was investigated. Poussot-Vassal et al. (2011) [11] evaluated the most classic semi-active control laws in time and frequency domain using a single corner model. Colina et al. (2014) [12] also investigated the basic semi-active control laws compared to the passive ones. The model was a quarter car and the general Sine Sweep test (ISO7401) was used as excitation for the models.

The objective of this paper is to optimize the parameters of passive and semi-active suspension systems using the Pareto Front optimization method in respect with ride comfort and road holding. The results are compared using performance metrics. The organization of this paper is as follows: in Chapter 2 a brief description of the models used for both the passive and semi-active cases is provided, in Chapter 3 the road excitation applied is illustrated, while in Chapter 4 the optimization procedure is outlined, its results are highlighted in Chapter 5 and, finally, in Chapter 6 conclusions are presented.

2. Mathematical models

2.1. *Half car model*

A half car model, equipped with a seat and a 7-DOF human-body model is considered for analysis. The model includes the front and rear axle allowing the observation of the pitch phenomena. As shown in Figure 1, the vehicle model consists of three basic subsystems: the tires, the suspension systems and the body of the vehicle. The latter is considered as a rigid body of mass m, equal to the half of the total mass of the vehicle. The distance of the front and rear unsprung masses (m_F and m_R) from the center of mass is equal to a_F and a_R, respectively. The front and rear tires are modelled as linear springs, which receive as input the irregularities of the road profile. The governing equations of the model have been described in detail in previous work [2] and the parameters with their values are show in Tables 1 and 2.

Additionally, the seat and the driver of a vehicle are modelled, as shown in Figure 2, in order to investigate in depth the ride comfort of the passenger. Different parts of the human body such as the pelvis, the diaphragm, the thorax etc. are described via several m – c – k subsystems. The equations of the model as well as the parameters were retrieved from the literature [5], [2].

One of the most important parameters of a vehicle model is the suspension travel, which is described as $z_{ST,F}=z_s-z_F-a_F\theta$ and $z_{ST,R}=z_s - z_R + a_R\theta$ for the front and rear suspension respectively, and it is used

IOP Conf. Series: Materials Science and Engineering **252** (2017) 012022 doi:10.1088/1757-899X/252/1/012022

extensively in the control laws of semi active suspension systems as part of their condition values. Moreover, another parameter crucial for the dynamic behavior of the vehicle is the tire deflection, described as$z_{T,F}=z_F-z_{Road_F}$ and $z_{T,R}=z_R-z_{Road_R}$ for the front and rear tire, respectively.

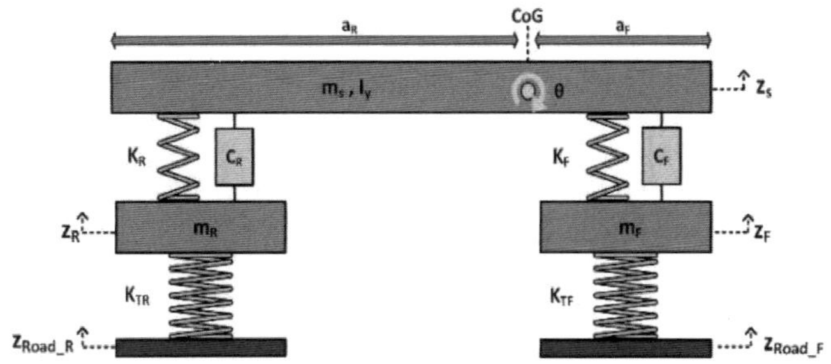

Figure 1. Half Car Model.

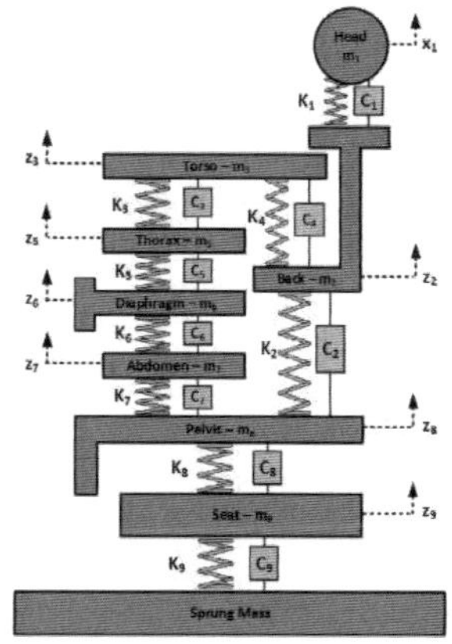

Figure 2. Seat – Driver Model.

Table 1. Nomenclature of vehicle parameters.

Parameters		**Subscripts**	
z	Vertical motion coordinate	**S**	Body
θ	Pitch motion coordinate	**F**	Front
z_{road}	Road excitation	**R**	Rear
m	Mass	**T**	Tire
C	Damper's coefficient	**ST**	Suspension deflection
K	Spring's stiffness coefficient		
a	Distance of the centre of the mass of the vehicle		

Table 2. Vehicle design parameters.

Parameter	**Value**	**Parameter**	**Value**
a_F [m]	0.91	**a_R [m]**	1.55
m_F [kg]	25	**m_R [kg]**	25
m_S [kg]	520	**K_{TF}, K_{TR} [N/m]**	$2.0\cdot10^5$
I [kg m^2]	473	**e [m]**	0.4

2.2. Semi active control laws

The different semi active controls applied to the suspension systems of the vehicle model are illustrated in this section. All of them are comfort oriented, mostly improving the comfort of the passengers by adjusting the damping coefficient of the suspension system according to the dynamic behavior of the vehicle and its response to the applied excitation. The same control law will be applied in each case in both the front and rear suspension system. Taking into consideration the results of previous studies, most of the control laws applied in this work were expanded from the ones proposed to the literature so as to minimize the changes in the damping factor and thus, avoid the energy cost because of unnecessary changes of the damper's state. In order to achieve this, a small positive value of the condition used in each control law, for both the front and rear suspension system (A_F and A_R), was set as the limit for the different states, leading the damper's coefficient to change not only regarding the sign of its condition but also based on this value.

2.2.1. Skyhook two state damper control (SH-2). The 2-states Skyhook control is an on/off strategy that switches between soft and stiff damping coefficient. This control law in literature consists of two states in which the damping factor C_i of the damper (i.e. its fluid viscosity, air resistance, etc.) changes according to the sign of the product of $\dot{z}_{ST,i}$ and $\dot{z}_S$ (equation 1).

$$C_i=\begin{cases} C_{min,i}\ ,\ \text{if}\ \|\dot{z}_{ST,i}\,\dot{z}_S\|\le A_i \\ C_{min,i}\ ,\ \text{if}\ \dot{z}_{ST,i}\,\dot{z}_S < -A_i \\ C_{max,i}\ ,\ \text{if}\ \dot{z}_{ST,i}\,\dot{z}_S > A_i \end{cases} \qquad i=F,\ R \qquad (1)$$

where $\dot{z}_{ST,F}$,$\dot{z}_{ST,R}$ are the stroke velocities of the front and rear, respectively, and $C_{min,i}$and $C_{max,i}$ are the minimal and maximal damping factors achievable by the considered controlled damper.

2.2.2. Skyhook linear approximation damper control (SH-L). An improved version of Skyhook control has been used to achieve variable damping for additional energy saving. The linear approximation of the Skyhook control includes the change of the damping factor C_i according to the sign of the product of $\dot{z}_{ST,i}$and $\dot{z}_S$:

$$C_i=\begin{cases} C_{min,i} & \text{if}\ \|\dot{z}_{ST,i}\,\dot{z}_S\|\le A_i \\ C_{min,i} & \text{if}\ \dot{z}_{ST,i}\,\dot{z}_S < -A_i \\ \frac{aC_{max,i}\dot{z}_{ST,i}+(1-a)C_{max}\dot{z}_S}{\dot{z}_{ST,i}} & \text{if}\ \dot{z}_{ST,i}\,\dot{z}_S > A_i \end{cases} \qquad i=F,\ R \qquad (2)$$

where a ∈ [0, 1] is a tuning parameter that modifies the closed-loop performances and enables the controller to adjust its stiff value ($C_{max,i}$) according to the needs of the application, saving more energy. More specifically, when a = 1, this control law is equivalent to the two-state Skyhook control. The difference relies on the fact that, according to the third expression (when $\dot{z}_{ST,i}\,\dot{z}_S > A_i$), such a control provides an infinite number of damping coefficients, providing the ability of tuning it to be continuous.

2.2.3. Power driven damper control (PDD). The power driven damper (PDD) is a semi-active control law which consists in changing the damping coefficient C_i as:

$$C_i=\begin{cases} C_{min,i}, & \|K_i\dot{z}_{ST,i}\,z_{ST,i}+C_{min,i}\,\dot{z}_{ST,i}\|\le A_i \\ C_{min,i}, & K_i\dot{z}_{def,i}\,z_{def,i}+C_{min,i}\,\dot{z}_{ST,i} > A_i \\ C_{max,i}, & K_i\dot{z}_{def,i}z_{def,i}+C_{max,i}\,\dot{z}_{ST,i} < -A_i \\ \frac{C_{min,i}+C_{max,i}}{2}, & (z_{ST,i}\neq 0\ \text{and}\ \dot{z}_{ST,i}=0) \\ \frac{-K_i z_{ST,i}}{\dot{z}_{ST,i}}, & \text{otherwise} \end{cases} \qquad i=F,\ R \qquad (3)$$

This control law depends on the value of the suspension travel and the stroke velocity. The damping coefficient may be the minimum, maximum, their mean value or the product of the spring stiffness with suspension travel divided by the stroke velocity.

2.2.4. Acceleration driven damper control (ADD). The ADD control approach is described by the equation:

$$C_i=\begin{cases} C_{min,i} & \text{if}\ \|\dot{z}_{ST,i}\,\ddot{z}_S\|\le A_i \\ C_{min,i} & \text{if}\ \dot{z}_{ST,i}\,\ddot{z}_S\le -A_i \\ C_{max,i} & \text{if}\ \dot{z}_{ST,i}\,\ddot{z}_S > A_i \end{cases} \qquad i=F,\ R \qquad (4)$$

Thus, this strategy is shown to be optimal in the sense that it minimizes the vertical body acceleration ($\ddot{z}_S$). In literature, it is stated that the "switching dynamic" may influence the closed-loop performances. The swift changing of values of the sprung acceleration leads to quick changes of the damping factor and this dynamic change creates a noise.

2.2.5. Mixed Skyhook-Acceleration driven damper control (SH-ADD). SH and ADD present complementary behaviors in terms of performance, with SH providing the best performance at low frequency (around the body resonance), and ADD ensuring optimality at mid and high frequency (beyond the body resonance). Therefore, a mixed control law has been proposed in the past using a very simple but effective frequency range selector ($\ddot{z}_S^2-a^2\dot{z}_S^2$). According to the sign of the main condition, either the ADD strategy is selected (mid and high frequency dynamics), or the SH strategy is used. The equation describing this control law is:

$$C_i=\begin{cases} C_{max,\,i} & \text{if } \left[\left(\ddot{z}_S^2\text{-}a^2\dot{z}_S^2\right)\leq 0 \text{ and } \dot{z}_{ST,\,i}\,\dot{z}_S>0\right] \text{ or } \left[\left(\ddot{z}_S^2\text{-}a^2\dot{z}_S^2\right)>0 \text{ and } \dot{z}_{ST,\,i}\,\dot{z}_S>0\right] \\ C_{min,\,i} & \text{otherwise} \end{cases} \quad (5)$$

The parameter a represents the frequency limit between the low and the high frequency ranges. Specifically, the value of a is set at the crossover frequency (in rad/s) between SH and ADD. This law is almost optimal as it provides a mix of the best performance of the SH and ADD.

2.2.6. Mixed Skyhook-Acceleration driven damper control with 1 sensor (SH-ADD-1s). The current law is similar to the mixed Skyhook-ADD, being based upon the complementary performances of soft and hard passive suspensions. Hard suspension is able to damp optimally the body resonance, but without a desiring filtering at high frequency. On the contrary, a soft suspension ensures the best filtering but with the drawback of a poorly damped body resonance. This complementarity resembles the complementarity between the SH and the ADD, described before, however the main difference is that at mid frequency the passive performance is far from the one achievable by the best possible semi-active suspension. The law is described by the following equation:

$$C_i=\begin{cases} C_{min,\,i}, & \text{if } \left\|\ddot{z}_S^2\text{-}a^2\dot{z}_S^2\right\|\leq A_i \\ C_{min,\,i}, & \text{if } \ddot{z}_S^2\text{-}a^2\dot{z}_S^2\geq A_i \\ C_{max,\,i}, & \text{if } \ddot{z}_S^2\text{-}a^2\dot{z}_S^2<\text{-}A_i \end{cases} \qquad i=F,\,R \quad (6)$$

According to the current value of $\ddot{z}_S^2$-$a^2\dot{z}_S^2$, the soft damping condition is selected when it is positive, otherwise the hard-damping condition is used. Similarly to the mixed SH-ADD, the amount $\ddot{z}_S^2$-$a^2\dot{z}_S^2$ can be considered as a simple "frequency-range selector", where the parameter a represents the limit between the ranges of low and high frequency.

3. Road excitation

In order to study the dynamic behavior of the vehicle an excitation for the model is required. In this study, a sinusoidal function was used as excitation representing a road bump, as it is shown in Figure 3. The geometrical aspects of the bump were chosen as follows: its height was set to 0.05m and its length was set to 5m. The velocity of the vehicle was considered constant, at 8 m/s. Finally, a time lag between the front and rear wheel of the vehicle was applied. Specifically, the two wheels follow the same trajectory with a time delay due to the distance a_F+a_R between them ($t_{distance}=\frac{a_F+a_R}{V}$) [3].

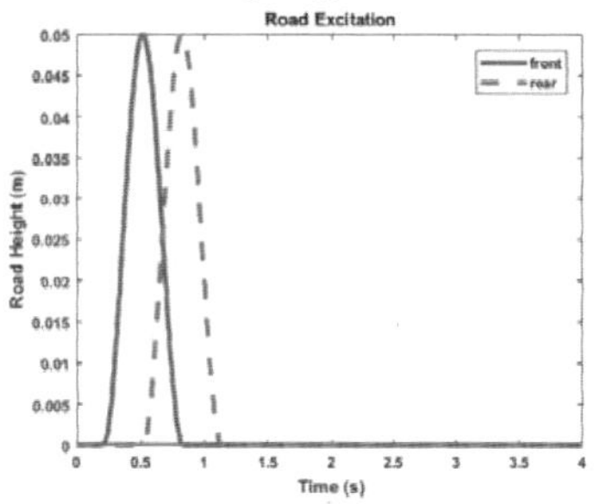

Figure 3. Road Bump.

4. Optimization Procedure

A multi objective optimization was applied to the model in respect to ride comfort and road holding using the Pareto front for several constraints. As design variables of the optimization, the controllable dampers and the spring stiffness were selected, as shown in equation 7:

$$\text{design variables}=[C_{min,\,F};\, C_{max,\,F};\, C_{min,\,R};\, C_{max,\,R};\, K_F;\, K_R] \quad (7)$$

where $C_{min,\,F}$, $C_{max,\,F}$, $C_{min,\,R}$ and $C_{max,\,R}$ were the minimum and maximum achievable damping coefficient of the front and rear damper, respectively, according to the semi active control law that is in use (Equations 1-6). Additionally, in semi active control laws SH-L, SH-ADD and SH-ADD-1s the tuning coefficient a was added as a design variable. The bounds of the design variables are illustrated in Table 3.

Moreover, the optimization of the passive model was carried out for two different bounds for the damping coefficient of the front and rear suspension. At first, the bounds of the minimum damping coefficient of the semi active laws were used (PS-LP) and then the ones of the maximum damping coefficient were applied to the design variables (PS-UP). The bounds of the springs' stiffness were the same. The objective functions of the optimization were selected so as to improve both the ride comfort and the road holding. Therefore, two targets were chosen: the root mean square of the seat's acceleration (f_1), representing the ride comfort, and the average of the variances of both the front and rear tire deflections (f_2), representing the road holding (Table 3). The constraints used in the optimization problem were selected so as to incorporate the practical considerations into the design process, such as the working space of the suspension system, as well as to enhance the optimization targets (Table 4).

Table 3. Design variables, bounds, objectives and constraints.

Design Variables	Lower Bounds	Upper Bounds	Objectives	Constraints
$C_{min,F}$, $C_{min,R}$(Ns/m)	500	2500	f_1=rms($\ddot{z}_9$)	$z_{ST,i} \leq 0.127$ m
$C_{max,F}$, $C_{max,F}$(Ns/m)	2500	5000		$z_{T,i} \leq 0.051$ m
K_F, K_R(N/m)	15000	70000	$f_2=\frac{1}{2}\left(\text{var}(z_{T,F})+\text{var}(z_{T,R})\right)$	$z_i \leq 0.07$ m
a (SH-ADD, SH-ADD-1s) (rad/s)	10	60		max($\ddot{z}_9$) ≤ 4.5 m/s^2

5. Results

The Pareto fronts of all the aforementioned cases are illustrated, in Figure 4. It is obvious that the Pareto fronts are located in three different areas. In the first area (A1) lay the Pareto fronts of four cases: SH-2, SH-L, SH-ADD and PS-UB, in the second one (A2) lays the Pareto front of SH-ADD-1s, while the Pareto fronts of ADD, PDD and PS-LB lay in the third area (A3) which is dominated by the previous ones, providing the "less optimal" solutions. The results belonging in A1 reach the lowest values of f1 (objective for ride comfort), whilst, in all the studied cases, regardless area, f_2 (objective for road holding) is in same range, apart from the PS-UB which has achieved the highest road holding of all cases (minimum value of f_2).

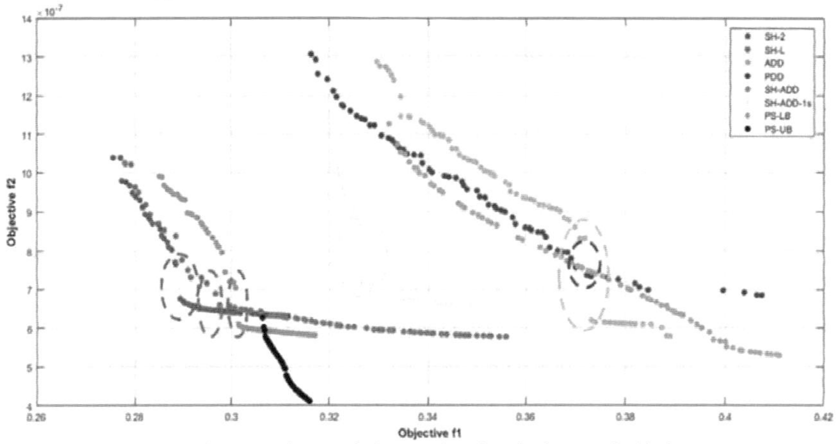

Figure 4. Comparison of the optimal solutions of all the cases

As it can be observed, in Figure 4, all the Pareto fronts of the semi active suspension laws are within the area created by the Pareto Front of PS-UB (black) and PS-LB (cyan), proving the main idea of semi active suspensions, which is the combination of a stiff and a soft damper. Another issue that should be mentioned is the difference between the Pareto of the passive case studies (PS) compared to the one of the semi-active case studies (SA). All the SA Pareto fronts reveal a different behavior during their attempt to increase the road holding of the vehicle. Instead of following a common trajectory like the Pareto of PS, which reveals the well-known conflict between ride comfort and road holding (increase of the one leads to the decrease of the other), they are demonstrating a vertical "jump off" for f_2 values (circled areas in Figure 4) and then, the Pareto front becomes almost parallel to the axis of f_1. This behavior illustrates the inability of the SA cases to further improve the road holding after a certain point, which is acceptable considering the fact that the applied control laws are comfort oriented.

The optimal design variables delivered from the Pareto front are illustrated in Figure 5. For each control law (Figures 5a-5f), eight curves of C-K are plotted. The four of them illustrate the optimal solutions of cases PS-LB and PS-UB (C_F-K_F and C_R-K_R for each case), while the other four demonstrate the optimal solutions of the SA cases (C_{Fmin}-K_F, C_{Fmax}-K_F, C_{Rmin}-K_R and C_{Rmax}-K_R). Each plot is divided in two areas with a red line and in the left one C_{Fmin}-K_F and C_{Rmin}-K_R are compared with the results of PS-LB, whilst in the right one C_{Fmax}-K_F and C_{Rmax}-K_R with the results of PS-UB. A "jump off" similar to the one of the Pareto fronts is again obvious. All the curves concerning the front suspension system of SA cases, shape an "L" regarding the values of C and K. This L-shape means that the optimization has concluded in various values of K for a specific value of C (vertical part of the L-shape), as well as in various values of C for a specific value of K (the horizontal part of the L-shape). The fact that the L-shape appears in the front suspension system, validates the necessity of using different suspension systems to the front and rear axles. On the other hand, in all the SA cases, the optimal solutions of the rear suspensions system have an almost fixed value for the spring stiffness and many alternatives for C_{Rmin} and C_{Rmax}, indicating a convergence to the optimization procedure.

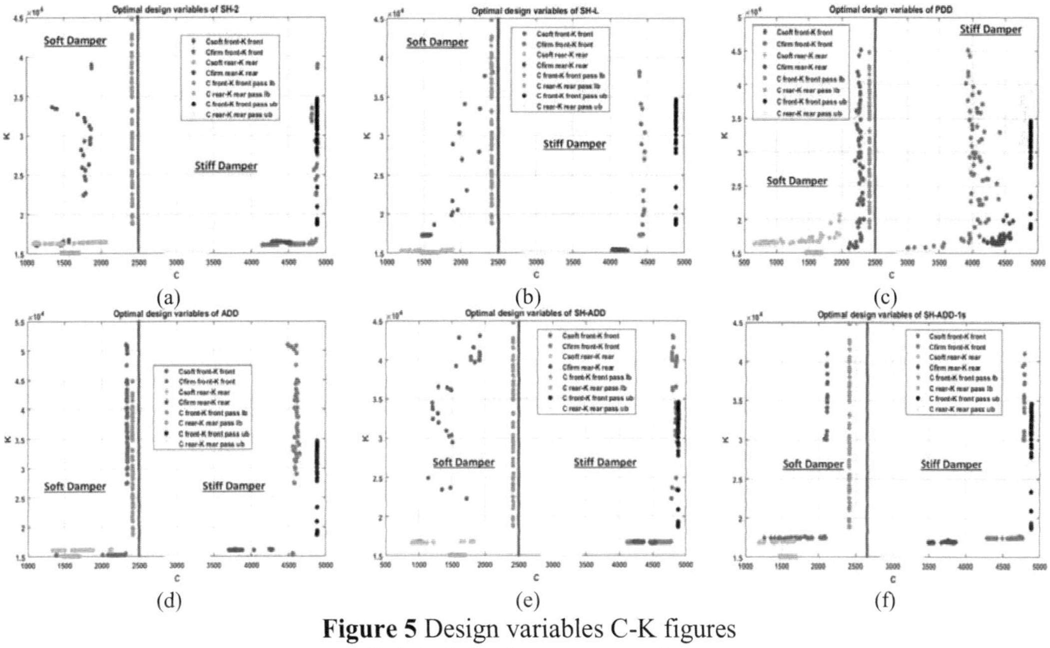

Figure 5 Design variables C-K figures

Table 4. Performance metrics

#	Metric	SH-2	SH-L	ADD	PDD	SH-ADD	SH-ADD-1s	Pass lb	Pass ub
1	**RMS($\ddot{z}_9$)[m/s^2]**	0.293	0.318	0.359	0.352	0.301	0.334	0.373	0.311
2	**$z_{ST,F}$[m]**	0.019	0.019	0.023	0.026	0.019	0.017	0.024	0.015
3	**$z_{ST,R}$[m]**	0.019	0.020	0.017	0.023	0.022	0.011	0.021	0.013
4	**VDV_{head}**	5.727	6.001	6.342	6.487	5.637	6.077	6.545	6.085
5	**CF_{head}**	3.691	3.715	3.544	3.580	3.607	3.540	3.456	3.340
6	**$z_{T,F}$ [m]**	0.004	0.004	0.005	0.005	0.004	0.004	0.004	0.003
7	**$z_{T,R}$ [m]**	0.002	0.002	0.003	0.003	0.003	0.002	0.003	0.002

Finally, some important metrics for the vehicle dynamics are illustrated in Table 4. The metrics calculated here are the root mean square value of the seat's acceleration, the front and rear suspension travel, the vibration dose value (VDV) and the crest factor of the head [2] and the front and rear tire deflection. The values included in Table 4 are the mean values of the optimal solutions in order to present the mean dynamic characteristics of all the solutions delivered from the Pareto front. The current comparison validates the conclusions stated previously. ADD and PDD do not deliver "good optimal" results, being unable to improve both the passengers' ride comfort (line 1 of Table 4) or the road holding of the vehicle (lines 6 and 7 of Table 4). Whereas, SH-2, SH-ADD and SH-ADD-1s have successfully converged to solutions either equal or better than the ones of PS-UB, which is the main reference case.

6. Conclusions

To sum up, the main aim of the current work was to present optimized semi-active suspensions along with optimized passive ones in order to compare them both in terms of the optimization results and the dynamic characteristics of the solutions. Moreover, in this work a different approach was proposed regarding the conditions used in the formulas of the control laws, so as to minimize the unnecessary changes in the damper's state and save energy for the system. Additionally, during the multi-objective optimization applied to the problem, remarks were illustrated regarding the well-known conflict of the suspension system and the attempt of the semi-active suspension systems to overcome it. Future work is in progress not only to expand this work for semi active control laws which are road holding oriented but also to implement decision algorithms to the Pareto solutions, so as to figure out the importance of the "jump" mentioned in the previous sections.

References

[1] Lajqi S and Pehan S 2012 Designs and optimizations of active and semi-active non linear suspension systems for a terrain vehicle Strojniški vestnik-Journal of Mechanical Engineering 58.12 732-43

[2] Koulocheris D, Papaioannou G and Christodoulou D, 2017 "Optimal design solution among pareto alternatives for vehicle nonlinear suspension system", *Proceedings of 26th International Automotive Conference "Science and Motor Vehicles" (26th JUMV),* Beograd, Serbia, 2017.

[3] Koulocheris D V, Papaioannou G D and Christodoulou D 2016 Assessment of the optimization procedure for the nonlinear suspension system of a heavy vehicle *Mobility and Vehicle Mechanics* **42** 17-35

[4] Shirahatt A, Prasad P S S, Panzade P and Kulkarni M M 2008 Optimal design of passenger car suspension for ride and road holding Journal of the Brazilian Society of Mechanical Sciences and Engineering 30.1 66-76

[5] Abbas W, Emam A, Badran S, Shebl M and Abouelatta O 2013 Optimal seat and suspension design for a half-car with driver model using genetic algorithm Intelligent Control and Automation 4.02 199

[6] Bourmistrova A, Storey I and Subic A 2005 December *Multiobjective optimisation of active and semi-active suspension systems with application of evolutionary algorithm* In The Proceedings of the 2005 International Conference on Modelling and Simulation Melbourne Australia

[7] Poussot-Vassal C, Savaresi S M, Spelta C, Sename O and Dugard L 2010 A methodology for optimal semi-active suspension systems performance evaluation In Decision and Control (CDC) 2010 49th IEEE Conference on pp 2892-97 IEEE

[8] Savaresi S M and Spelta C 2007 Mixed sky-hook and ADD: Approaching the filtering limits of a semi-active suspension Journal of dynamic systems, measurement, and control 129.4 382-92

[9] Poussot-Vassal C, Spelta C, Sename O, Savaresi S M and Dugard L 2011 Survey on some automotive semi-active suspension control methods: A comparative study on a single-corner model IFAC Proceedings Volumes 44.1 1802-07

[10] Colina A, Lerma G, Cabanes I and Iglesias I 2014 Modelling and control of a semi-active suspension system *In New Advances in Mechanisms, Transmissions and Applications* pp 25-32 Springer Netherlands

Nonlinear differential system applied of a mechanical plan model of the automotives used for the nonlinear stability analysis

Loreta Simniceanu, Bogdan Mihaela, Victor Otat and Mario Trotea

University of Craiova, Faculty of Mechanical Engineering, Romania

lsimniceanu@yahoo.com

Abstract. This paper proposes a plan mechanical model for the vehicles with two axles, taking into account the lateral deflection of the tire. For this mechanical model are determined two mathematical models under the nonlinear differential equations systems form without taking into account the action of the driver and taking into account. The analysis of driver-vehicle system consists in the mathematical description of vehicle dynamics, coupled with the possibilities and limits of the human factor. Description seeks to emphasize the significant influence of the driver in handling and stability analyzes of vehicles and vehicle-driver system stability until the advent of skidding. These mathematical models are seen as very useful tools to analyzing the vehicles stability. The paper analyzes the influence of some parameters of the vehicle on its behavior in terms of stability of dynamic systems.

1. Introduction

In this paper we are proposing a mathematical model for automotives moving, a model who includes some of the position parameters and moving parameters.

This mathematical model can be used into a preliminary stability analyzes by dynamic systems theory or by appling the classical mechanic theory ([4], [7], [8]).

Dynamic Systems Theory aims to analyze the stability of a system both in the field of stability and in the field of unstable movements or even chaotic movements. Chaotic movements do not occur for any parameter values, but only for certain values thereof. Hence, it is possible to choose parameter values other than those for which it is possible chaotic motion. In this way, artificial dynamic systems, designed and built, will have stable and predictable movements.

For this reasons is necessary to know the conditions that produce chaotic movements to avoid them.

2. The mathematical model of automotives moving

The paper takes into consideration the plan of a two-axle vehicle, for which the rolling and pitch oscillations are neglected.

The mechanical model is shown in Figure 1, and the notations are detailed below:

- C1, C2, C3, C4 - the center of contact area between tire and road;
- F_{x1}, F_{x2}, F_{x3}, F_{x4} - the longitudinal forces of contact area between tires and road;
- F_{y1}, F_{y2}, F_{y3}, F_{y4} - the lateral forces of contact area between tire and road;

- δ_1, δ_2 - the cornering angle for front axle;

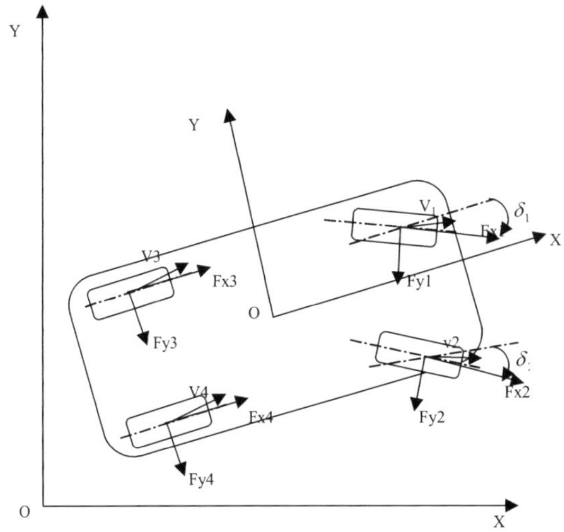

Fig. 1. The plan model of vehicle

The next references systems are taking into account:

a reference system is attaching to the suspended mass of the vehicle; this system has the origin in center mass point of this, and Ox, Oy axle making the plan of principal direction of displacement; the Oz axle is perpendicular to Ox, Oy axles;

a fixed $O_1x_1y_1z_1$ coordinate system presented in Fig. 1.

The next degree of freedom is taking into account:

x, define the displacement along longitudinal axle;

y, define the displacement along transversal axle;

u, define the displacement along vertical axle;

θ_1, define the circular displacement of suspended mass around Ox axle;

θ_2, define the circular displacement of suspended mass around Oy axle;

φ, define the circular displacement of suspended mass around Oz axle.

If we neglect the roll and pitch oscillations of the vehicle, forming a system that models the movement plane of the vehicle, and which follow the variables y, v_y, φ, ω. It is also considered a constant speed v_x = const. A system that modeling the vehicle plan moving are obtained, and this system has y, v_y, φ, ω variables and the velocity v_x is considerate constant ([11], [12]). We also intend to study the stability of the vehicle under the influence of disturbances that will be considered as a change in the steering angle of the wheels.

This change may be a periodic perturbation $\delta_p(t)$ applied to the steering axle tires, having the form:

$$\delta_p(t) = Q\cos(\omega_p t) \quad (1)$$

with Q, ω_d – amplitude and frequency of the disturbance. The origin of such disturbances can be steering or periodic deformations of the street.

In addition to this disruption may occur steering angle changes by the action of the driver, if he perceives a lateral deviation of the vehicle from the desired trajectory. The driver tries to control vehicle steering angle $\delta_s(t)$, as yet to reduce the deviation from the theoretical line of travel. This action of the driver can be modeled mathematically, inserting a feedback loop k, a response time Tr and the driver's front vision L, the driver's command can be described by a differential equation taking into account the state component y(t) as follows ([5], [6]):

$$\dot{\delta}_s = -\left[\frac{k}{T_r}\left(y + \frac{L}{v_x}\dot{y}\right) + \frac{\delta_s}{T_r}\right]. \tag{2}$$

The above will be supplemented by consideration of external perturbations and by taking into account the action of the driver [6] on the steering angle of the wheels.

The analysis of driver-vehicle system consists in the mathematical description of vehicle dynamics, coupled with the possibilities and limits of the human factor. Description seeks to emphasize the significant influence of the driver in handling and stability analyzes of motor vehicles and vehicle-driver system stability until the advent of skidding.

Given the assumptions stated above but considering the action of the driver, it is obtained the next system:

$$\dot{v}_y = \frac{1}{m}\cdot\left\{(c_{11}\cdot\left(\delta_s + Q\cos(\omega_d t) - arctg\frac{v_y + a\cdot\omega}{v_x - \frac{e}{2}\cdot\omega}\right)\right.$$

$$-c_{13}\cdot\left(\delta_s + Q\cos(\omega_d t) - arctg\frac{v_y + a\cdot\omega}{v_x - \frac{e}{2}\cdot\omega}\right)^3)\cdot\cos(\delta_s + Q\cos(\omega_d t))$$

$$+(c_{21}\cdot\left(\delta_s + Q\cos(\omega_d t) - arctg\frac{v_y + a\cdot\omega}{v_x + \frac{e}{2}\cdot\omega}\right) - c_{23}\cdot\left(\delta_s + Q\cos(\omega_d t) - arctg\frac{v_y + a\cdot\omega}{v_x + \frac{e}{2}\cdot\omega}\right)^3)$$

$$\cdot\cos(\delta_s + Q\cos(\omega_d t) + \left(F_{x1} + F_{x2}\right)\cdot\sin\left(\delta_s + Q\cos(\omega_d t)\right)$$

$$+\left(c_{31}\cdot\left(arctg\frac{v_y-b\cdot\omega}{v_x-\frac{e}{2}\cdot\omega}\right)-c_{33}\cdot\left(arctg\frac{v_y-b\cdot\omega}{v_x-\frac{e}{2}\cdot\omega}\right)^3\right)$$

$$+\left(c_{41}\cdot\left(arctg\frac{v_y-b\cdot\omega}{v_x+\frac{e}{2}\cdot\omega}\right)-c_{43}\cdot\left(arctg\frac{v_y-b\cdot\omega}{v_x+\frac{e}{2}\cdot\omega}\right)^3\right)\Bigg\}-\omega\cdot v_x$$

$$\dot{\omega}_z=\frac{1}{J_{zz}}\cdot\{\ -\frac{e}{2}\cdot(F_{x1}-F_{x2})\cdot\cos(\delta_s+Q\cos(\omega_p t))-a\cdot(F_{x1}+F_{x2})\cdot\sin(\delta_s+Q\cos(\omega_p t))$$

$$+\frac{e}{2}\cdot(-F_{x3}+F_{x4})-(c_{11}\cdot\left(\delta_s+Q\cos(\omega_p t)-arctg\frac{v_y+a\cdot\omega}{v_x-\frac{e}{2}\cdot\omega}\right)$$

$$-c_{13}\cdot\left(\delta_s+Q\cos(\omega_p t)-arctg\frac{v_y+a\cdot\omega}{v_x-\frac{e}{2}\cdot\omega}\right)^3)\cdot\left(-\frac{e}{2}\cdot\sin(\delta_s+Q\cos(\omega_p t))+a\cdot\cos(\delta_s+Q\cos(\omega_p t))\right)$$

$$+b\cdot\left(c_{31}\cdot\left(arctg\frac{v_y-b\cdot\omega}{v_x-\frac{e}{2}\cdot\omega}\right)-c_{33}\cdot\left(arctg\frac{v_y-b\cdot\omega}{v_x-\frac{e}{2}\cdot\omega}\right)^3\right)+$$

$$b\cdot\left(c_{41}\cdot\left(arctg\frac{v_y-b\cdot\omega}{v_x+\frac{e}{2}\cdot\omega}\right)-c_{43}\cdot\left(arctg\frac{v_y-b\cdot\omega}{v_x+\frac{e}{2}\cdot\omega}\right)^3\right)\}$$

$$\dot{y}=v_y\cos\varphi+v_x\sin\varphi$$

$$\dot{\varphi}=\omega \qquad (3)$$

$$\dot{\delta}_s=-\left[\frac{k}{T_r}\left(y+\frac{L}{v_x}\cdot(v_y\cos\varphi+v_x\sin\varphi)\right)+\frac{\delta_s}{T_r}\right]$$

Previous equations forming a system of linear differential equations of the general form

$\dot{X}=f(X,t,\gamma)$, $X\in R^5$ where: $X=(v_y,\omega,y,\varphi,\delta_s)^T$.

The parameters m, Iz, ., were grouped under the γ symbol.

3. Analysis of the influence of physical parameters on the stability of the system

The numerical parameters used in this mathematical analysis are:

For the case of a car with a load of 1 people: v=25 km/h, a= 1113 mm, b = 1457 mm, m=1358kg, Iz= 2611Kgm2;

For the case of a car with a load of 5 people: v=25 km/h, a = 1234 mm, b = 1336 mm, m=1667kg, Iz=2802 Kgm2;

Further, the numerical solutions of the proposed mathematical model are presented, in the case of a movement after a sinusoidal trajectory and the influence of the variation of the Iz parameter on the dynamic behavior of the vehicle-driver system is analyzed.

System solutions represent: S<0>=t , S<1>=vy, S<2>= ω , S<3>=y, S<4>= φ , S<5>= δ_s ;

The case of single-passenger vehicle loading:

System solutions are:

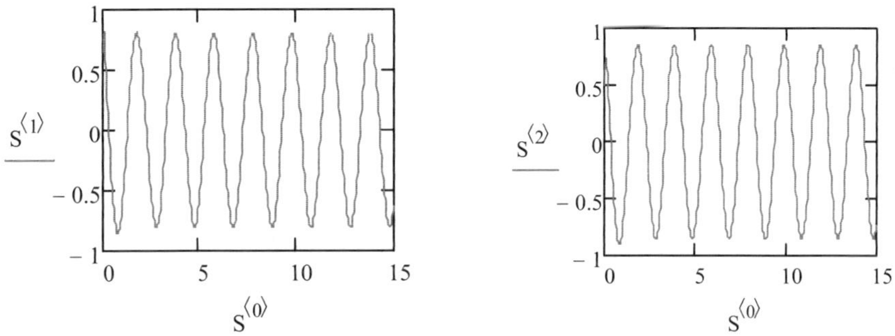

Diagram in the phase space:

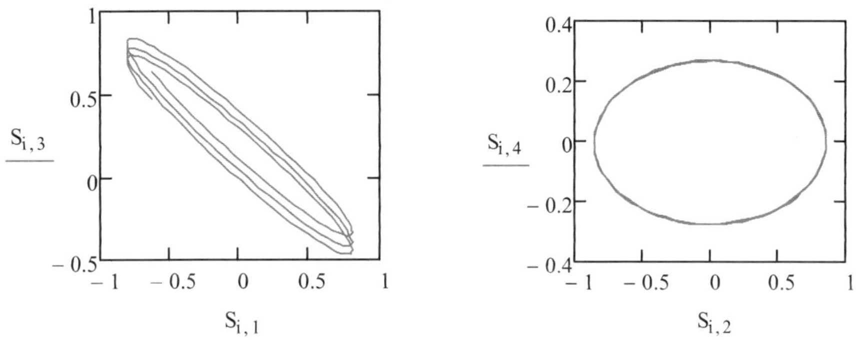

Wheel lateral forces:

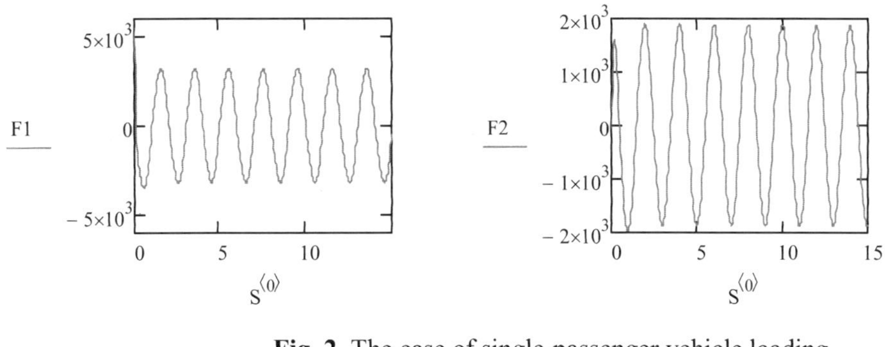

Fig. 2. The case of single-passenger vehicle loading

The movement is periodically stable, the amplitude of the lateral speed is 0.8, the amplitude of the angular velocity is 0.75 rad / sec and the wheel force amplitude is approximately 3500 (in case of F1) and 2000 (in case of F2).

The case of loading the vehicle with five passengers:

System solutions are:

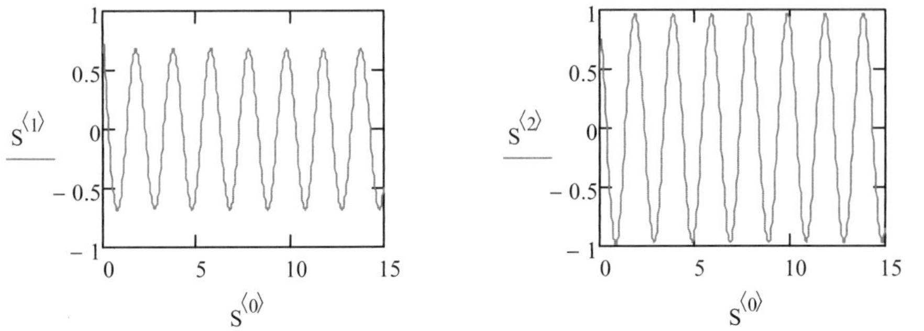

Diagram in the phase space:

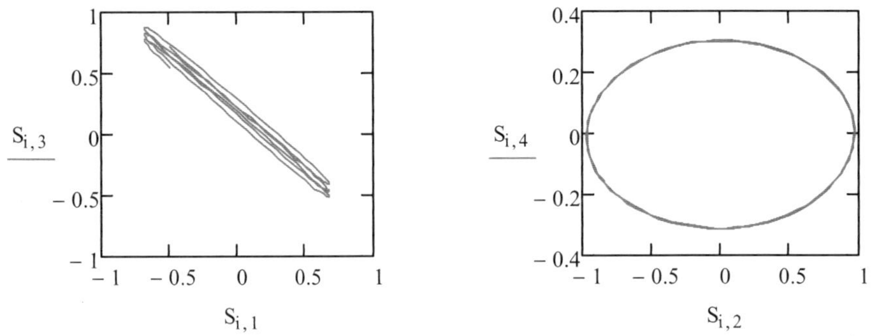

Wheel lateral forces:

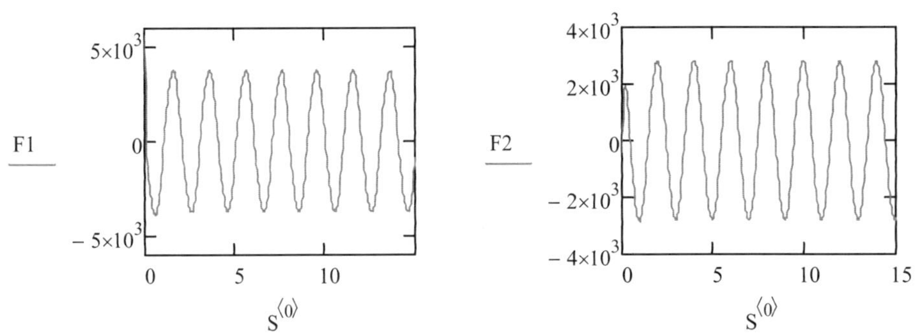

Fig. 3. The case of five-passengers vehicle loading

The movement is periodically stable, the amplitude of the lateral speed is 0.7, the amplitude of the angular velocity is 0.98 rad / sec and the wheel force amplitude is approximately 3800 (in case of F1) and 2800 (in case of F2).

IOP Conf. Series: Materials Science and Engineering **252** (2017) 012023 doi:10.1088/1757-899X/252/1/012023

4. Conclusions

For the proposed mathematical model and for the types of motion analyzed, variation of the moment of inertia (made by changing the load of the vehicle) changes both the movement character and the motion parameters, both by the amplitude of the angular velocity and the lateral velocity, and more particularly by the amplitude of the lateral force on the wheel.

Analyzing the previous results, we find the following: a variation of the inertia moment of approximately 20% leads to a variation of the lateral speed amplitude by 12%, a variation of the angular velocity of 26% and a variation of lateral forces up to 28%, reaching values comparable to the force of adhesion, which causes the vehicle to skid.

The variation of momentum of inertia produces a change in system behavior, from a periodic movement with 3 subarmons to a quasi-periodic movement with 5 subarmons.

These conclusions refer to the simplified plan layout where the center of the table is placed in the plane of the tread and on the symmetry axis of the car. If computational possibilities allow the analysis of some car models, including other physical and geometric parameters (unsuspended masses and suspended mass, rigidity of tires and suspension), the moment of inertia can influence the character of the movement in a considerable way.

References

[1] Baker L G and Gollub P J, *Chaotic dynamics an introduction*, Cambridge University Press, 1990.

[2] Ceaikovski I P, *Issledovanie protessa stabilizatii upravleaemîh koles primeneniem analogovoi E.V.M.*, Avtomobilnaia promîslennost, XXXIII, nr.6, 1967.

[3] Ciudakov E A, *Teoria automobilului*, Traducere din limba rusă. Institutul de documentare Bucuresti 1958.

[4] Guckenheimer J and Holmes Ph, *Nonlinear oscillations, dynamical systems and bifurcations of vector fields*, Springer-Verlag, New York Berlin Heidelberg Tokyo, 1983.

[5] Legouis D T, Modelisation et analyse de la dynamique du systeme vehicule pilot, *Ingenieurs de l'Automobile* oct. 1987, p.101-106.

[6] Legouis D T, Laneville A, Bourassa P and Payre G, Characterization of dynamic vehicle stability using two models of the human pilot behavior, *Vehicle System Dynamics 15*, 1986, 1–18.

[7] Voinea P R and Stroe I, *Sisteme dinamice*, Institutul Politehnic Bucuresti,1994.

[8] Simniceanu L, *Aplicatii ale teoriei sistemelor dinamice în dinamica automobilelor*, Institutul Politehnic Bucuresti, 2005.

[9] Oţăt V, Bolcu D-tru, Thierheimer W and Simniceanu L, *Dinamica autovehiculelor*, Universitaria Publishing, Craiova, 2005;

[10] Neagoe D-tru and Bolcu D-tru, *Studiul stabilitatii si maniabilitatii autovehiculelor. Cercetari experimentale si teoretice*, Universitaria Publishing, Craiova, 2008.

[11] Simniceanu L, Neagoe D-tru, Trotea M, Bogdan M and Constantinescu A, Mathematic Model for Analysing the Dynamic Behavior of a Vehicle in the Field of Chaotic Movements, *Applied Mechanics and Materials*, Vol. 822, pp. 54-59, 2016.

[12] Simniceanu L, Otat V and Trotea M, Determination to a plan model of the automotives moving used for the stability analysis, The 2015 International Conference on Mechatronics and Robotics, MEROSTA 2015, Grecia, ISBN 978-1-61804-325-2, *Recent Advances in Mechanics, Mechatronics and Civil, Chemical and Industrial Engineering*, 2015, pp 68 – 71.

[13] Oţăt O V, Dumitru N, Oţăt V and Dumitru I, Determination of Kinematic and Dynamic Behavior in the Driver's Skull upon the Impact with the Steering Wheel, CONAT 2016 International Congress of Automotive and Transport Engineering, , *Springer International Publishing Switzerland*, 2017, pp 812-819.

Development of a measuring system for the analysis of vibrations sent from engine to passenger seat

D. Voicu[1], M. Truta[1], L. Barothi[1], R. Stoica[1] and G. I. Dorobantu[1]
[1]Military Technical Academy, 39-49 George Cosbuc Blvd., Bucharest, Romania
Corresponding author's email: voi_dana_alina@yahoo.com

Abstract: The present paper aims at developing a measuring system for vibrations analysis of engine cylinder cover and passenger seat fixing device. Such a system is useful for analyzing the correlation, if any, between the vibrations generated by internal combustion engine and those felt by the passenger (which create a certain degree of discomfort).

1. Introduction

Considering the current existence of multiple and accessible sensors and acquisition boards, it is now possible to easily study the vibrations produced by an internal combustion engine. Moreover, due to the electronic control of engine parameters, we now have new possibilities of studying the engine's dynamical behavior. The present paper aims at developing a measuring system for vibrations analysis of engine cylinder cover and passenger seat fixing device. Such a system is useful for revealing the correlation, if any, between the vibrations generated by the internal combustion engine and those felt by the passenger (which create a certain degree of discomfort).

2. Experimental research

The main purpose of this paper is to develop a low price measuring system, which can allow the analysis of vibrations transmitted from internal combustion engine to passenger seat.

In order to accomplish this goal, there have been used the following constituents: 2 MEMS accelerometers of type ADXL345 and 2 Arduino Mega 2560 data acquisition boards. The first measuring system was placed on the engine, with the accelerometer being attached to the engine cylinder cover (figure 1a) and the second measuring system was placed inside the vehicle, with the accelerometer being attached to the passenger seat attachment rail (figure 1b).

To validate the accuracy of the measuring system, there have been considered multiple functional situations of engine such as: engine start, constant engine speed of 2000 rpm, constant engine speed of 3000 rpm. Also we have to mention that the tests were conducted considering a parked vehicle thus neglecting vibrations determined by road irregularities [3]. If vibrations would have been measured in case of a moving vehicle, it could have been used as a mathematical model the one developed in paper [7].

The measured data was processed using a software which allows a comparative analysis of vibrations obtained at engine level and at passenger seat level, considering the same functioning conditions [2], [4].

The obtained results revealed a correlation between the vibrations recorded by the two accelerometers, but, as expected, there were also some major differences between vibrations average values, which represents a validation of the proposed measuring system.

Figure 1 depicts the measuring system attached to engine (figure 1a) and passenger seat (figure 1b).

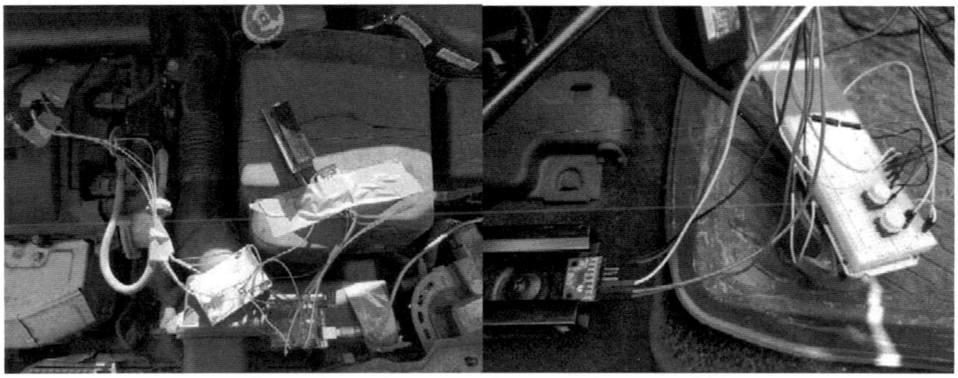

a) b)

Figure 1.

A series of limitations of the measuring system were introduced by the maximum data acquisition frequency of the microcontrollers (in this case of 1 Hz), on one hand, and by the accelerometers measuring precision, on the other hand. Also, considering the fact that our data acquisition board was not provided with internal storage capabilities, this led to the impossibility of real-time data storage and required a permanent connection between the board and a laptop, which represents a drawback of the proposed measuring system. However, this limitation could be eliminated by attaching a microSD card or a Bluetooth device to our board.

3. Data statistical processing

Engine vibrations are caused mainly by the following aspects [1]:

- The cylinder pressure due to combustion is not constant through the combustion cycle;
- The slider-crank mechanism does not output a smooth torque even if the pressure is constant (e.g., at top dead centre there is no torque generated);
- The motion of the piston mass and connecting rod mass generate alternating torques often referred to as "inertia" torques.

In terms of vibrations that affect passengers, specialty literature mentions multiple parameters specific to vibration phenomenon that should be taken into account when analyzing effects on human body [5]:

- Frequency;
- Amplitude, acceleration or vibration energy.

Due to the fact that the paper focuses mainly on the comparison of values of acceleration obtained from experimental tests, the formula of vibration acceleration is [5]:

$$a = A \cdot \omega^2 = (2 \cdot \pi \cdot f)^2 \cdot A \quad [m/s^2] \tag{1}$$

where A – vibration amplitude [*m*], ω – vibration pulsation [s^{-1}], f – vibration frequency [*Hz*].

Also, excitation frequency is given by formula [6]:

$$f = \frac{2 \times N \times I}{60 \times C} \quad [Hz] \tag{2}$$

where N – engine idle RPM, I – number of cylinders, C – engine cycle.

In figure 2 [7] are presented factors which induce vibrations (such as road irregularities, driveline excitation shaft, engine or transmission) and the vehicle's parts and elements which can damp vibrations (such as rubber mounts, vehicle suspension etc.). The graph highlights the track of different vibrations sent from point of origin to passenger or driver [7], [8].

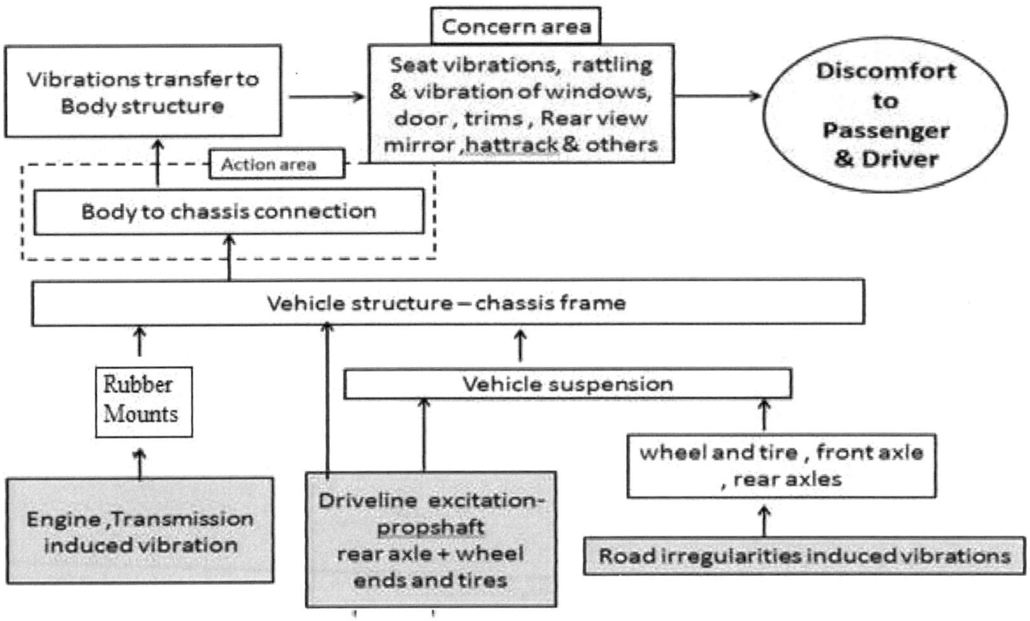

Figure 2.

Following, there are presented some graphical examples of vibrations produced by an internal combustion engine, considering different functional situations.

For example, in Figure 3 there are shown the acceleration values on the *OX* axis (vehicle's transverse direction), while in Figure 4 there are shown the acceleration values on the *OY* axis (vehicle's longitudinal direction) and on the *OZ* axis (vehicle's vertical direction), obtained in case of engine start.

If we analyze the two graphs, we can notice that the values corresponding to the accelerations on the vertical direction (*OZ* axis) are close to the value of the gravitational acceleration (approximately 9.81 m/s^2), while the accelerations measured on the other two directions are much higher at the engine level than those measured at the passenger seat level. For the analysis of the vibrations transmissivity from the engine to the passenger seat, we computed the mean value between the values recorded at the two placement locations of the accelerometers.

IOP Conf. Series: Materials Science and Engineering **252** (2017) 012024 doi:10.1088/1757-899X/252/1/012024

The average value of attenuation degree on *OX* axis, from engine to passenger seat, during tests, was approximately 5, denoting an attenuation of the vibrations sent from the engine to the passenger seat of 5 times. If the maximum value of acceleration recorded at the engine level, on the *OX* axis, was 4.5 m/s^2, the maximum value of acceleration recorded on the same direction but at the passenger seat level was 0.42 m/s^2. Comparing the values obtained on the *OX* and *OY* axes, it can be noticed that the values in the last case were higher, the maximum value of accelerations on the *OY* axis being 10.2 m/s^2, while at the passenger seat level the maximum value was 0.68 m/s^2. The time period when we recorded the highest acceleration values lasted for approximately 1 second.

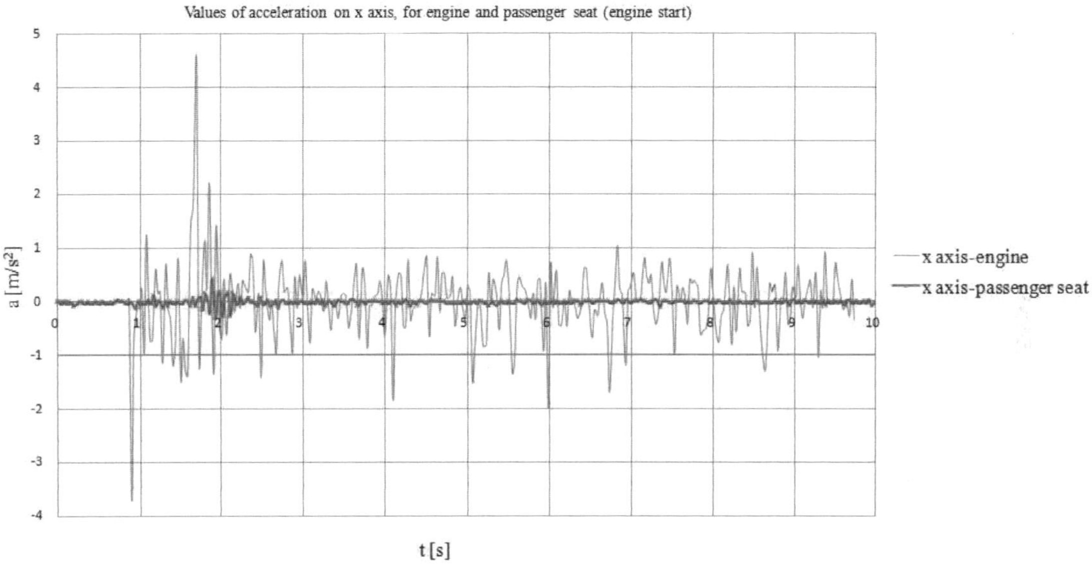

Figure 3.

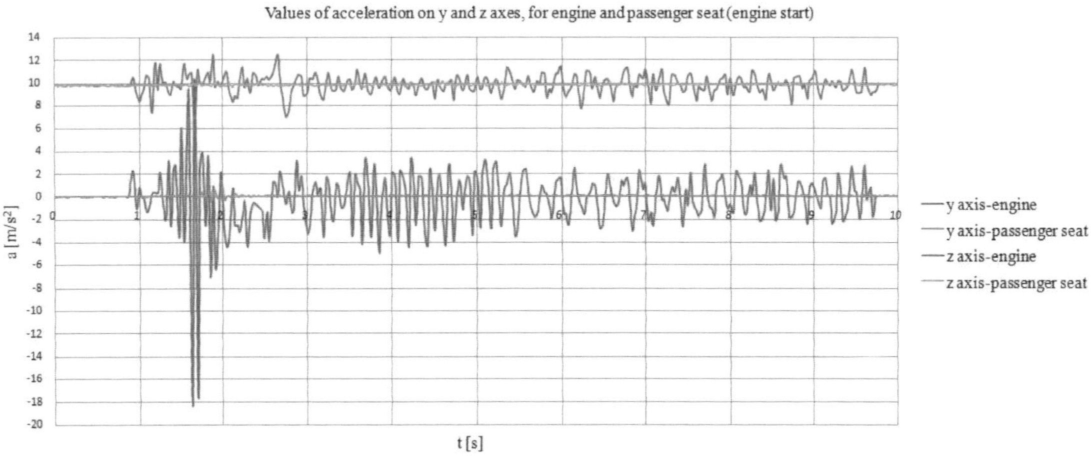

Figure 4.

In Figure 5 are presented the acceleration values on the *OX*, *OY* and *OZ* axes, obtained in case of an engine functioning at a constant engine speed of 2000 rev/min. From the analysis of the graph, it can be observed that the values recorded on the *OZ* axis (measured at the engine and passenger seat levels) were close to the value of the gravitational acceleration. Likewise, it can be noticed that the vibrations amplitude at the engine level was higher than the one measured at the passenger seat level. The maximum acceleration on the *OX* axis, at the engine level, was 2.567 m/s^2, while at the passenger seat level it was -0.192 m/s^2; on the *OY* axis, the maximum value of engine vibration was 6.781 m/s^2, and the maximum value of passenger
seat vibration was 0.076 m/s^2.

By analyzing the average acceleration values recorded on the two axes, it was observed an attenuation of the vibrations sent from engine to the passenger seat, on the *OX* direction, of approximately 32.91 times, and on the *OY* direction of approximately 40 times.

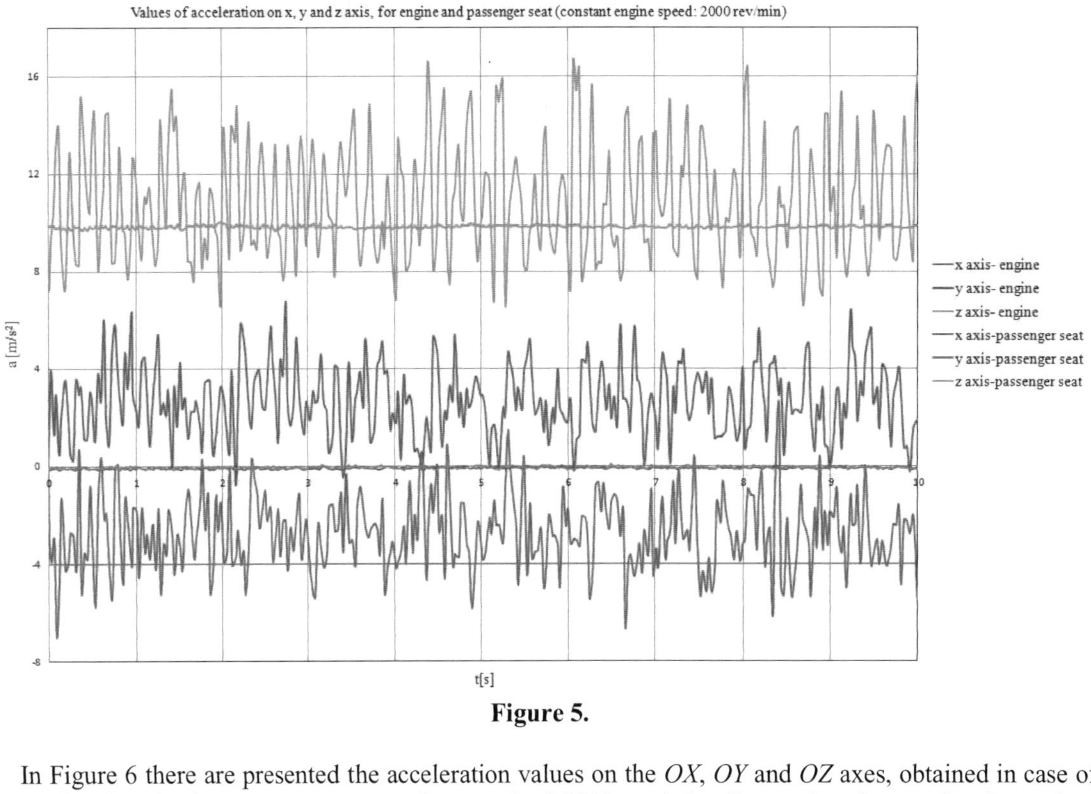

Figure 5.

In Figure 6 there are presented the acceleration values on the *OX*, *OY* and *OZ* axes, obtained in case of an engine functioning at a constant engine speed of 3000 rev/min. Comparing the acceleration values obtained at the two constant engine speeds (2000 rev/min and 3000 rev/min, respectively), it can be noticed that the values recorded in case of a constant engine speed of 3000 rev/min are higher; for example, the maximum acceleration value at the engine level, on the *OX* axis, was 7.086 m/s^2, while the maximum acceleration value at the passenger seat level was 0.115 m/s^2. On the *OZ* axis, we can observe that the acceleration values were close to the gravitational acceleration value, while on the *OY* direction the maximum acceleration value was 6.55 m/s^2 (at the engine level), and 0.076 m/s^2, respectively (at the passenger seat level).

By analyzing the average acceleration values recorded on the two axes (*OX*, and *OY*, respectively), we observed an attenuation of the vibrations sent from engine to the passenger seat, on the *OX* direction, of approximately 8.575 times, and of 9.05 times, on the *OY* direction.

This paper's novelty consists of the development of a measuring system with a decent accuracy related to its price. At the same time, the measuring system can be further used for performing a comparative analysis of the vibrations generated by an internal combustion engine running with conventional fuel (petrol) and alternative fuel (liquefied petroleum gas). Also, by making use of the included accelerometers, we can also measure the rolling and pitching motions of vehicles running on different road categories.

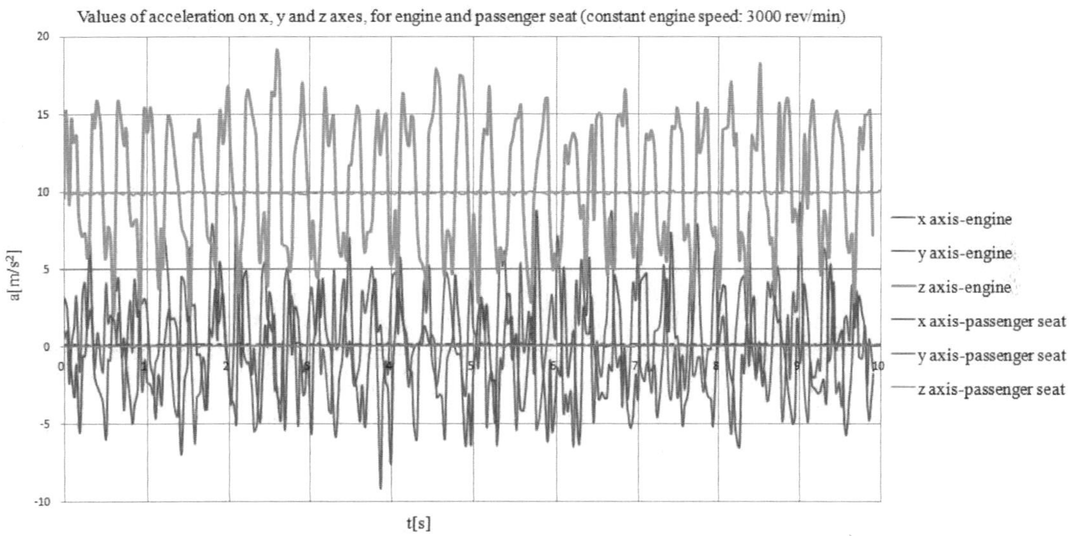

Figure 6.

4. Conclusions

In order to validate the adequate functioning of proposed measuring system there have been analyzed vibrations sent from engine to passenger seat, in different functional situations. Conclusions issued out of this study are the following:

- these boards (such as Arduino) have multiple advantages, starting with the fact that they are easy to program (using a programming language similar to C++);
- the price is accessible;
- there is a large number of sensors and actuators compatible with such platforms in the market, which allows measuring a wide variety of vehicle parameters;
- due to small sizes of data acquisition boards and sensors, it is easy to dispose them in the measuring area;
- besides all these advantages, these boards also have some limitations, especially regarding data acquisition frequency (which in this case was 1Hz);
- cables used to connect sensors to acquisition board have a drawback which is more common to appear in case of high vibrations, namely the cables may determine loss of connection.

The measuring system can be improved by using an external supply for the acquisition board (during these tests the acquisition board was powered directly from the laptop) and by attaching a Bluetooth device which can assure real time data transmission to a laptop or a device able to register data on a microSD card.

Data accuracy is considered acceptable, mainly in case of use for educational purpose. On the other hand, for research purpose, the measuring system is not recommended because of the poor measuring accuracy, needless to say that in the market there are more advanced equipments but at higher costs.

5. References

[1] Balachandran B and Magrab E B, Vibrations- Second edition, Library of Congress Control Number: 2008925925, ISBN-13: 978-0-534-55206-0

[2] Marinescu M, Vilău R and Becheru Gh - A signal filtering problem - "The XXXVIIth International Symposium of the National Agency for Research in Military Technologies", 25-26 May 2006, Bucureşti, CD edition, 6 pg., ISBN 10: 973-0-04429-5, ISBN 13: 978-973-0-04429-4

[3] Lespezeanu I, Vilau R, Radulescu V J, Trocan C, Neagu D and Copae I, Ion, 2016, Experimental Research of the Car Acceleration, Equipped with On-Board Computer, International Journal of Mechanical Engineering and Automation

[4] Marinescu M and Ilie C O, 2014, Filtering the signal of a measured mechanical parameter, 10th International Conference on COMMUNICATIONS (COMM), AGIR Publishing House, pp. 85-88, ISBN 978-1-4799-2385-4, IEEE catalog no. CFP1441J-AR

[5] Pavel C and Ene Gh, Introducere în tehnica izolării vibraţiilor si a zgomotului, Editura MATRIX ROM

[6] Torgal S and Saxena S, 2015, Methodology for Analysis and Design for Powertrain Vibration Reduction in Passenger Bus with Focused Engine and Transmission Mounts, International Journal of Innovative Research in Science, Engineering and Technology, Vol. 4, Issue 7, ISSN(Online) : 2319-8753

[7] Torgal S, 2016, Vibrations in passenger bus and its analysis, Engineering & Technology (IMPACT: IJRET) ISSN (E): 2321-8843; ISSN (P): 2347-4599 Vol. 4, Issue 2

[8] Zhou H and Qiu Y, 2015, A simple mathematical model of a vehicle with seat and occupant for studying the effect of vehicle dynamic parameters on ride comfort, Presented at the 50th United Kingdom Conference on Human Responses to Vibration, held at ISVR, University of Southampton, Southampton, England

IOP Conf. Series: Materials Science and Engineering **252** (2017) 012025 doi:10.1088/1757-899X/252/1/012025

Downforce variation dependence of angle of incidence modification for the rear wing of high speed vehicles

R Tarulescu[1], S Tarulescu[1] and C Leahu[1]

[1]Automotive and Transport Engineering Department, Transilvania University of Brasov, Brasov, Romania

E-mail: radu.tarulescu@unitbv.ro

Abstract. The conventional downforce devices (with fixed geometry) of high speed vehicles have parameters such as area, angle of incidence and head resistance coefficients, all with constant values. The downforce is proportional with the square of movement speed and the power consumed for the neutralization of aerodynamic road resistance is proportional with the cube of speed. The authors carried out an analytical study of downforce, adjustable/monitored by optimum incidence (modification of incidence angle of rear wing for performance improvement).

Introduction

The racing vehicle from figure 1 is provided with a rear wing inclined with the incidence angle i against the horizontal, having the downforce area S [m^2], lifting coefficient C_z and the head resistance coefficient C_x. The lifting/downforce F_z [N] and the head resistance force (drag) F_x [N] are given by the relations according to [1]:

$$F_z = C_z \cdot S \cdot \frac{\rho \cdot v^2}{2} \text{ [N]} \quad (1)$$

$$F_x = C_x \cdot S \cdot \frac{\rho \cdot v^2}{2} \text{ [N]} \quad (2)$$

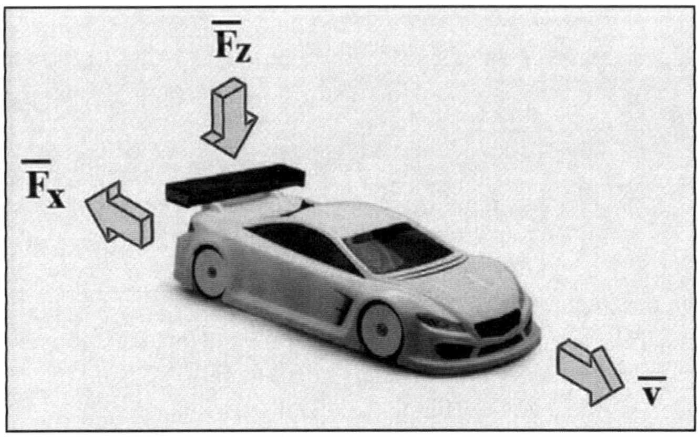

Figure 1. The acting forces on the racing vehicle

The power consumed to defeat the aerodynamic head resistance depending on the speed v [m/s] is:

$$N_x = v \cdot F_x = C_x \cdot S \cdot \frac{\rho \cdot v^3}{2} \text{ [W]} \tag{3}$$

The aerodynamic coefficients are to be adopted according to the polar diagram $C_z(C_x)_i$ associated with the adopted lifting profile [2]. In most cases, a rigid wing with the incidence angle i fixed, non-adjustable is used [3].

A model of rear wing with the variable incidence angle i is articulated around a horizontal axis, going through the leading edge [4].

The characteristics of the profile are given in table 1 and visually represented in figure 2. The paper examines two characteristic intervals: $i \in [0^o, 9^o]$ and $i \in [9^o, 18^o]$.

Table 1. The characteristics of the adopted profile of rear downforce wing

Incidence angle	Lifting coefficient	Head resistance coefficient
i [°]	$100 \cdot C_z$ [-]	$100 \cdot C_x$ [-]
-6	-14.2	5.25
-3	8.7	4.31
0	32.7	4.32
3	57.3	5.38
6	82.5	7.20
9	101.5	9.66
12	118.5	13.42
15	127.5	18.10
18	115.0	25.20

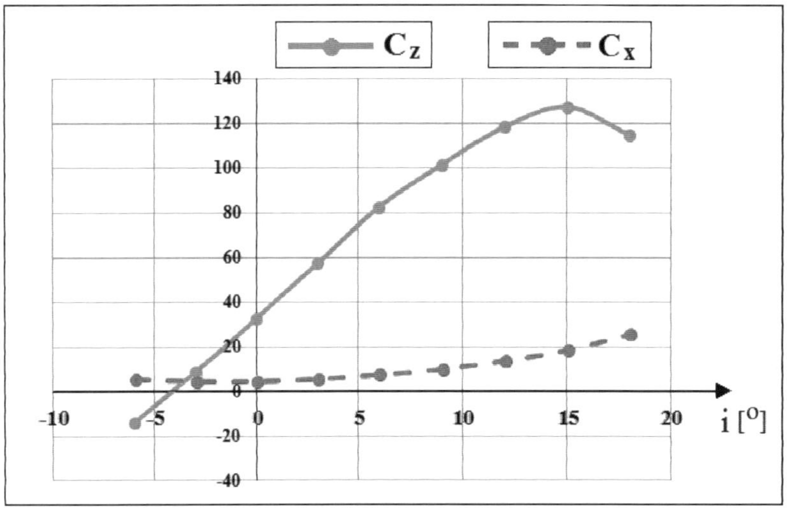

Figure 2. The lifting coefficient C_z and the head resistance coefficient C_x values at different incidence angles

Analytical study of the interval $i \in [0^o, 9^o]$

For this analysis, the authors suggest the following linear approximations for the aerodynamic coefficients depending on the incidence angle (independent variable):

$$100 \cdot C_x = 4.32 + 0.594 \cdot i, \tag{4}$$

$$100 \cdot C_z = 32.7 + 7.65 \cdot i, \tag{5}$$

$$100 \cdot C_z = 28.38 + 100 \cdot C_x + 7.056 \cdot i \tag{6}$$

The dependence on the speed v [m/s] and on the incidence angle i of the aerodynamic values is highlighted:

$$F_x = 10^{-2} \cdot \frac{\rho \cdot S}{2} \cdot (4.32 + 0.594 \cdot i) \cdot v^2 \text{ [N]}, \tag{7}$$

$$N_x = \frac{1}{200} \cdot \rho \cdot S \cdot (4.32 + 0.594 \cdot i) \cdot v^3 \text{ [W]} \tag{8}$$

and

$$F_z = \frac{1}{200} \cdot \rho \cdot S \cdot (32.7 + 7.65 \cdot i) \cdot v^2 \text{ [N]} \tag{9}$$

Relation (6) is written as:

$$C_x = C_z - 0.2838 - 7.056 \cdot 10^{-2} \cdot i = \frac{F_z}{S \cdot \frac{\rho \cdot v^2}{2}} - 0.2838 - 7.056 \cdot 10^{-2} \cdot i, \tag{10}$$

The aerodynamic values, head resistance force F_x [N] and power consumed in order to defeat aerodynamic head force N_x [W] are rewritten:

$$F_x = F_z - (0.2838 + 7.056 \cdot 10^{-2} \cdot i) \cdot S \cdot \frac{\rho \cdot v^2}{2} \text{ [N]} \tag{11}$$

and

$$N_x = v \cdot F_x = v \cdot F_z - K \cdot v^3 \text{ [W]}, \tag{12}$$

where K [N·s²/m²] is:

$$K = (0.2838 + 7.056 \cdot 10^{-2} \cdot i) \frac{\rho \cdot S}{2} \text{ [N·s}^2\text{/m}^2\text{]}, \tag{13}$$

Making an economic optimization (minimum value of the power consumed in order to defeat aerodynamic head resistance force N_{xmin} [W] and the optimum value of speed v_{opt} [m/s]), namely $\partial N_x / \partial v = 0$, we obtain v_{opt} (with variable i) [5]:

$$v_{opt} = \sqrt{\frac{F_z}{3 \cdot K}} = \sqrt{\frac{F_z}{(0.433 + 0.106 \cdot i) \cdot \rho \cdot S}} \text{ [m/s]}, \tag{14}$$

with the minimum value $\partial^2 N_x / \partial v^2 > 0$.

At relatively low speeds, taking into consideration a wing with $S = 1.5$ [m²], at a constant value of air density, $\rho_{air} = 1.2$ [kg/m³] and for a given lifting force $F_z = 2000$ [N], we calculate:

$$\frac{F_z}{\rho \cdot S} = 11.1 \cdot 10^2 \text{ [N]} \quad \text{and} \quad v_{opt} = \frac{33.3}{\sqrt{0.433 + 0.106 \cdot i}} \text{ [m/s]} \tag{15}$$

Adjusting the incidence angle $i = 0^\circ$, we obtain the value of the optimum speed $v_{opt} = 50.8$ [m/s] = 183 [km/h], adjusting the incidence angle $i = 2.45^\circ$ we obtain the value of the optimum speed $v_{opt} = 40$ [m/s] = 144 [km/h], adjusting the incidence angle $i = 4.5^\circ$ we obtain the value of the optimum speed $v_{opt} = 35$ [m/s] = 126 [km/h] and adjusting the incidence angle $i = 7.5^\circ$ we obtain the value of the optimum speed $v_{opt} = 30$[m/s] = 108 [km/h].

Analytical study of the interval $i \in [9^\circ, 18^\circ]$

For the interval $i \in [9^\circ, 18^\circ]$ of the incidence angle, the authors suggest for the head resistance coefficient $C_x(i)$ the increasing linear function:

$$100 \cdot C_x = 9.66 + 1.73 \cdot i, \tag{16}$$

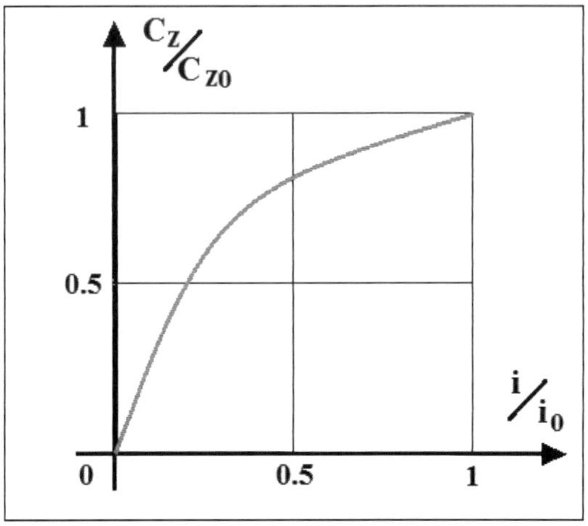

Figure 3. The transition function for the incidence angle $i \in [9^o, 15^o]$

For the smaller interval of the incidence angle $i \in [9^o, 15^o]$, the authors suggest the exponential approximation, the asymptotic increase, the transition function, represented in figure 3:

$$C_z = C_{z0} \cdot \frac{e^{\frac{i}{i_0} \cdot \ln 49} - 1}{e^{\frac{i}{i_0} \cdot \ln 49} + 1}, \tag{17}$$

with $i_0 = 15^o$, $C_{z0} = C_{z,\max} = 127.5 \cdot 10^{-2}$. By replacement in relation (17), results:

$$C_z = 127.5 \cdot 10^{-2} \cdot \frac{e^{3.89 \cdot \frac{i}{15}} - 1}{e^{3.89 \cdot \frac{i}{15}} + 1} = 127.5 \cdot 10^{-2} \cdot k_i, \tag{18}$$

Therefore, the dependence between the lifting coefficient and the drag coefficient $C_x(C_z)_i$, is given by the relation [6]:

$$C_x = C_z + 0.0966 + [0.0173 \cdot i - 1.275 \cdot k_i(i)], \tag{19}$$

with dimensionless coefficient

$$k_i(i) = \frac{e^{3.89 \cdot \frac{i}{15}} - 1}{e^{3.89 \cdot \frac{i}{15}} + 1} \tag{20}$$

In this case, the head resistance force F_x [N] and power consumed in order to defeat aerodynamic head force N_x [W] can be written [7]:

$$F_x = F_z + K \cdot v^2 \text{ [N]} \tag{21}$$

and

$$N_x = v \cdot F_x = v \cdot F_z + K \cdot v^3 \text{ [W]}, \tag{22}$$

with dimensional coefficient

$$K = [0.0966 + (0.0173 \cdot i - 1.275 \cdot k_i)] \cdot \frac{\rho \cdot S}{2} \text{ [N}\cdot\text{s}^2/\text{m}^2\text{]}, \tag{23}$$

Making for this case also an economic optimization $\partial N_x / \partial v = 0$,, we obtain v_{opt} (with variable i):

$$v_{opt} = \sqrt{\frac{F_z}{3 \cdot K}} = \sqrt{\frac{F_z}{[0.145 + (0.026 \cdot i - 1.92 \cdot k_i)] \cdot \rho \cdot S}} \text{ [m/s]}, \tag{24}$$

with the minimum value $\partial^2 N_x / \partial v^2 > 0$.

The interval $i \in [15^o, 18^o]$ of the incidence angle is not economical, having high values for the head resistance coefficient C_x and reduced values for the lifting coefficient C_z, which leads to the decrease of the head resistance force F_x [N] and increase of the power consumed in order to defeat aerodynamic the head resistance N_x [W].

At relatively high speeds, taking into consideration the same wing with $S = 1.5$ [m^2], at a constant value of air density, $\rho_{air} = 1.2$ [kg/m^3] and for a given lifting force $F_z = 3000$ [N], we calculate:

$$\frac{F_z}{\rho \cdot S} = 16.6 \cdot 10^2 \text{ [N]} \qquad \text{and} \qquad v_{opt} = \frac{40.7}{\sqrt{0.145 + (0.026 \cdot i - 1.92 \cdot k_i)}} \text{ [m/s]} \qquad (25)$$

Adjusting the incidence angle $i = 15^o$, $C_x = 0.181$, $C_{z,max} = 1.275$, we obtain the value of the optimum speed v_{opt}=68.5[m/s] = 247 [km/h], adjusting the incidence angle $i = 12^o$, $C_x = 0.1342$, $C_z = 1.185$, we obtain the value of the optimum speed v_{opt}=71[m/s] = 256 [km/h], adjusting the incidence angle $i = 9^o$, $C_x = 0.0966$, $C_z = 1.0155$, we obtain the value of the optimum speed v_{opt}= 77[m/s] = 277 [km/h]. More detailed values are presented in table 2 and table 3.

Table 2. The computed values for the increasing angle of incidence *i*

Speed	Incidence angle	Lifting coefficient	Head resistance coefficient	Lifting/ Downforce	Head resistance force	Power
v [m/s]	i [°]	$100 \cdot C_z$ [-]	$100 \cdot C_x$ [-]	F_z [kN]	F_x [kN]	N_x [kW]
30	0	32.7	4.32	0.264	0.035	1.05
40	3	57.3	5.38	0.825	0.078	3.12
50	6	82.5	7.20	1.850	0.152	7.60
60	9	101.5	9.66	3.280	0.314	18.84
70	12	118.5	13.42	5.250	0.590	41.30
80	15	127.5	18.10	7.350	1.040	83.20

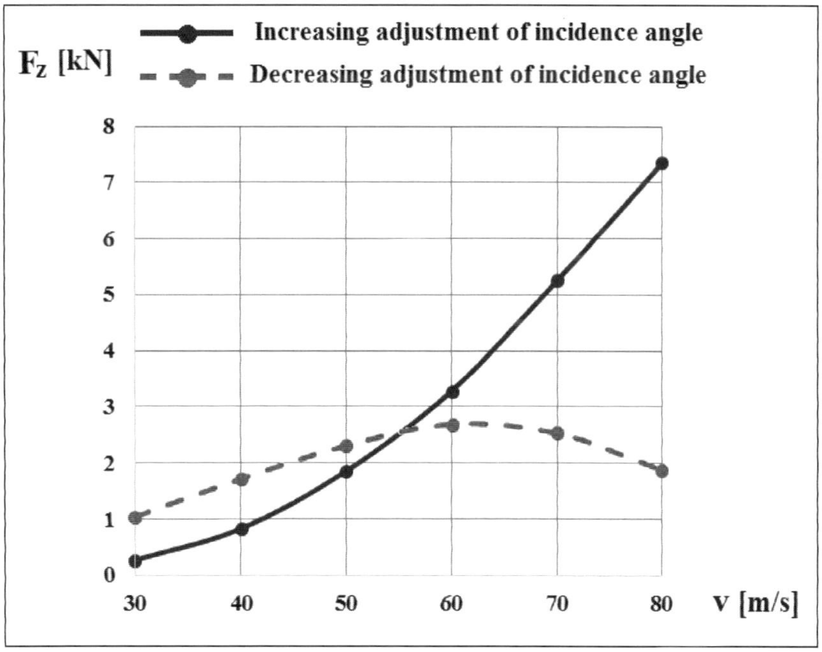

Figure 4. Obtained values of lifting/downforce F_z for increasing and decreasing angle of incidence

Table 3. The computed values for the decreasing angle of incidence *i*

Speed	Incidence angle	Lifting coefficient	Head resistance coefficient	Lifting/ Downforce	Head resistance force	Power
v [m/s]	i [°]	100 · C_z [-]	100 · C_x [-]	F_z [kN]	F_x [kN]	N_x [kW]
30	15	127.5	18.10	1.03	0.147	4.31
40	12	118.5	13.42	1.71	0.193	7.72
50	9	101.5	9.66	2.30	0.218	10.90
60	6	82.5	7.20	2.68	0.232	13.92
70	3	57.3	5.38	2.53	0.237	16.59
80	0	32.7	4.32	1.88	0.250	20.00

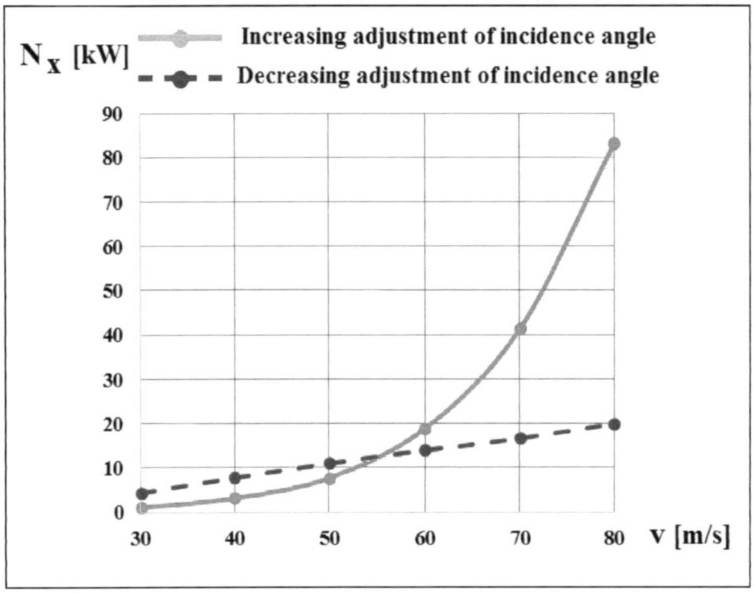

Figure 5. Obtained values of power consumed in order to defeat aerodynamic head force N_x for increasing and decreasing angle of incidence

Conclusions

Two manufacturers of high-speed vehicles have already made models with the adjustable rear wing. Porsche 911 Turbo model 2014 has an angle adjustable rear wing that elevates at speed and also a front chin spoiler that drops down in two phases. The other one is Bugatti Veyron, speed vehicle with changeable rear wing at high speed. The whole car exerts downforce and lower itself for two height settings with a checklist to adjust height and retract vents to achieve top speed.

References

[1] Anderson J. D, *Fundamentals of Aerodynamics*. 7th ed. McGraw-Hill. New York, 2016
[2] Katz J, *Race Car Aerodynamics: Designing for Speed.* 1st ed. Bentley Publishers, 1995
[3] Bertin J, Cumming R, Aerodynamics for Engineers (5th Edition), 2008
[4] Katz J, *Aerodynamics of race cars.* Annual Review of Fluid Mechanics, 2006
[5] Benche V, Ivănoiu M, *Mecanica fluidelor şi maşini hidropneumatice. Culegere de probleme*, Editura Universităţii Transilvania Brasov, 1989
[6] Barnard R H, *Road Vehicle Aerodynamic Design*, 2010
[7] Benche V, Tarulescu R, Tarulescu S, *Critic considerations about the topical aerodynamic similitude and simulation of the vehicles*, Bulletin of the Polytechnic Institute of Iasi, 2006

IOP Conf. Series: Materials Science and Engineering **252** (2017) 012026 doi:10.1088/1757-899X/252/1/012026

INFLUENCE OF HOUSING WALL COMPLIANCE ON SHOCK ABSORBERS IN THE CONTEXT OF VEHICLE DYNAMICS

G Pulvirenti and C Faria

Siemens Industry Software, Interleuvenlaan 68, Leuven, BE

E-mail: giorgio.pulvirenti@siemens.com

Abstract. Shock absorbers play a key role in vehicle dynamics. Researchers have spent significant effort in order to understand phenomena associated with this component, but there are still several issues to address, in part because new technology development and design trends continually lead to new challenges, among which weight reduction is crucial. For shock absorbers, weight reduction is related to the use of new materials (e.g. composite) or new design paradigms (e.g. more complex geometry, wall thickness, etc.). All of them are directly linked to wall compliance values higher than the actual ones. The present article proposes a first analysis of the phenomena introduced by a high wall compliance, through a modelling approach and various simulations in order to understand the vehicle behaviour changes. It is shown that high values of wall compliance lead to increased hysteresis in the force-velocity curve. However, comfort, handling and ride performances are not significantly affected by this designing parameter.

1. Introduction

Shock absorbers reduce the amplitude of vibrations of the sprung and unsprung masses by dissipating the energy arising from suspension motion. Furthermore, they control the rate of weight transfer during transient vehicle behavior (braking, corner entry and acceleration). These duties make them fundamental sub-systems of most vehicles [1].

In order to accomplish their role, they exploit viscous friction between the hydraulic fluid and solid components. The phenomena associated with this component are very complex, making their study challenging from a scientific point of view. Several of these phenomena have already been studied and characterized (for instance [2–5]), nevertheless a significant effort is needed to address the ones related to the latest technology development and design trends, such as weight reduction. Less mass leads to lower fuel consumption and energy saving, both fundamental topics for current research [6,7]. In the case of shock absorbers, weight reduction could be obtained through the use of new materials (e.g. composite materials, polymers) and also through new design paradigms such as high housing walls compliance (which, depending also on mechanical properties, is indirectly related to new materials too).

Several kinds of shock absorbers are available for automotive applications with twin-tube shock absorbers being the most common [1], and their detailed modeling is quite complex and does not attend the needs of the presented work. Therefore, the authors focus on a more simple through-rod shock absorber (which is also found in automotive application, e.g. [8]). Various simulations are carried out to compare their behaviors and to understand how the walls compliance affect vehicle dynamics.

The next section describes the methodology used for modeling shock absorbers (section 2), followed by simulations and their results (section 3), closing with conclusions and indications for future works (section 4).

2. Methodology

The walls compliance of a generic hydraulic chamber whose volume is forced to change due to external phenomena can be defined from the equation [13]:

$$V(p,T) = (V_0 + \sum_i V_i)[1 + w_c(p - p_{sat})] \tag{1}$$

representing the volume of the chamber in function of the induced volume variations $\sum_i V_i$ and in function of internal pressure p and external pressure p_{ext}. The term V_0 is the initial volume and w_c is the chamber wall compliance, the last one taken constant. With respect to an infinitely stiff shock absorber, the one with $w_c \neq 0$ has an increased volume (the increase is equal to $w_c(p - p_{ext})$ which lead to pressure changes in each chambers, due to the relation between pressure and density expressed by [13]:

$$\rho(p) = \rho(0)e^{\overline{B_{fluid}(p_{sat})}^{\,p}} \tag{2}$$

with ρ density and $B_{fluid}(p_{sat})$ fluid isothermal bulk modulus at saturation pressure.

The wall compliance is calculated assuming the volume variation of the hydraulic chamber equal to the one of an infinitely long tube with thin wall, thus neglecting any increase of volume in the axial direction and any border effects at the interface between lateral and top/bottom walls. w_c is the same for both top and bottom hydraulic chambers.

Shock absorbers have been modelled in the LMS Imagine.Lab AMESim environment, according to the scheme in figure 1. AMESim has been choose due to its ability in reproducing high nonlinear system with accuracy and allowing, at the same time, a significant save of time due to its computation efficiency [9]. Furthermore, several studies on shock absorber made use of this tool (for instance [8–12]), which can be considered validated by the scientific community. Models have linear valves with cracking pressure p_{crack} (which is the pressure necessary to lift the disc valve), leakage with viscous friction and wall compliance. For details about the equations refer to the AMESim help manual [13].

Two geometrically identical models of shock absorber with different valves parameters have been used for the rear and front suspension systems. The different valves parameters lead to different damping behaviour. Each simulation is carried out twice, varying only the wall compliance value of hydraulic chambers in shock absorber models ("low" and "high" wall compliance). The "low wall compliance" is obtained with steel mechanical properties, while the "high wall compliance" is roughly ten times higher (for all the models w_c is equal in both bottom and top hydraulic chambers).

3. Simulations

Several simulations were carried out in order to have a broad view of the shock absorber behaviour and its influence in the full vehicle dynamics when w_c significantly change.

3.1. Force-velocity curve analysis

One of the most common way to characterize shock absorbers is through the force-velocity curve. To obtain it, the piston rod is forced to move with an imposed velocity (in the direction of its axis). The force produced by the shock absorber is then plotted in function of the velocity itself [1].

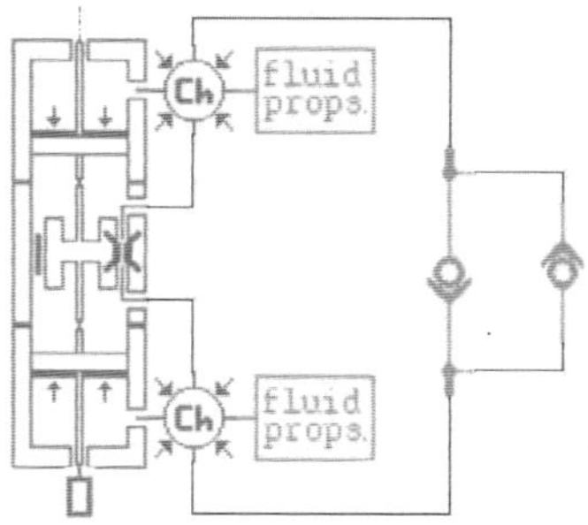

Figure 1: AMESim model.

Figure 2 shows the central area of the force-velocity curve of the two models (front suspension couple) obtained with a sinusoidal input (amplitude 1m/s, frequency 1Hz). The increased hysteresis for the high wall compliance model is clearly visible in figure 2(b).

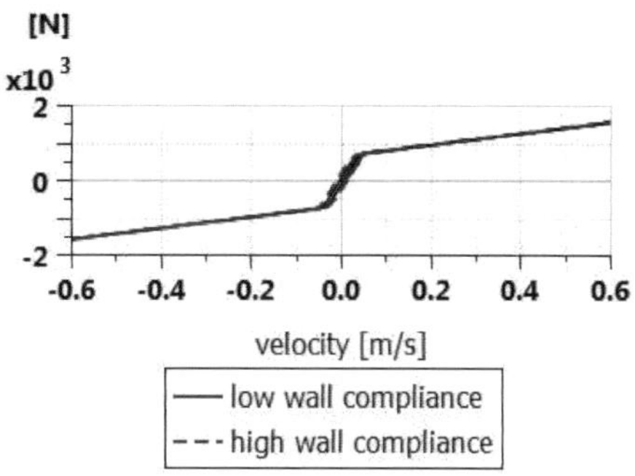

(a) Force velocity curve.

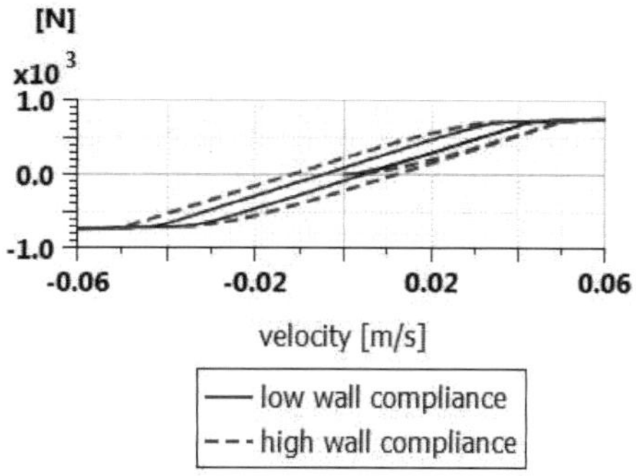

(b) Zoom in the low velocity area.

Figure 2: Force-velocity curve obtained at 0.6m/s, 1Hz.

The hysteresis is mainly caused by a delay in time-domain of some physical quantities responses to the input [14]. Among these quantities, pressures in both chambers play a significant role, because their difference is the main responsible of the reaction force developed by the shock absorber itself. Figure 3 shows pressure variation in both chambers for different amplitude of the input. The nonlinear behaviour of the shock absorber and its amplitude dependence are clear in the figure. The crucial differences between the two models (high and low wall compliances) are the delay in time domain (already clear from the increased hysteresis) and the maximum and minimum values of pressure reached. In a real component, the lower pressures in the high wall compliance model would be reason of lower mechanical stress in real applications, therefore might slightly affect the design.

Pressure difference is also related to the reaction force *F*, which is linked to the rod velocity v and to the average damping coefficient cd by the following relation:

$$F = c_d v. \tag{3}$$

In theory c_d should be different in the two models (due to diverse pressure difference). However, the maximum (minimum) forces reached difference are negligible, thus cd can be considered constant.

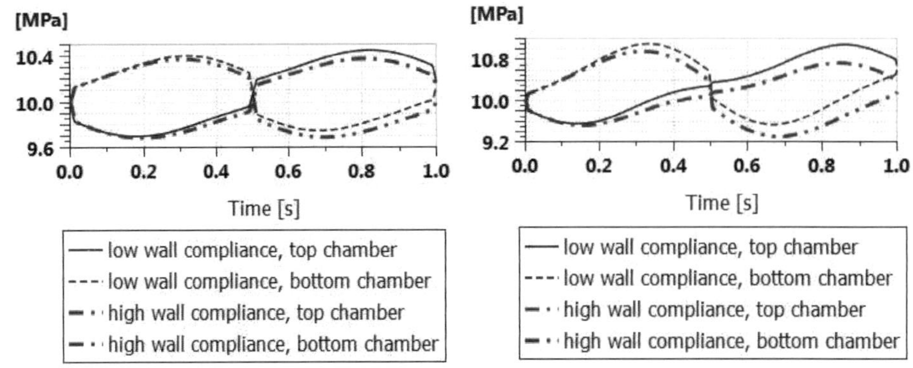

(a) Input amplitude: 0.6 m/s. (b) Input amplitude: 1.5 m/s.

Figure 3: Pressure time-history in both chambers for different levels of input amplitude.

Lastly, several parameters characterizing the fluid are pressure-dependent, thus measurable changes between the two models are present. Despite the difference relative to the variation is significant, most of the variables characterizing the fluid can be considered constant during the entire simulation. For example, the bulk modulus is shown in figure 4.

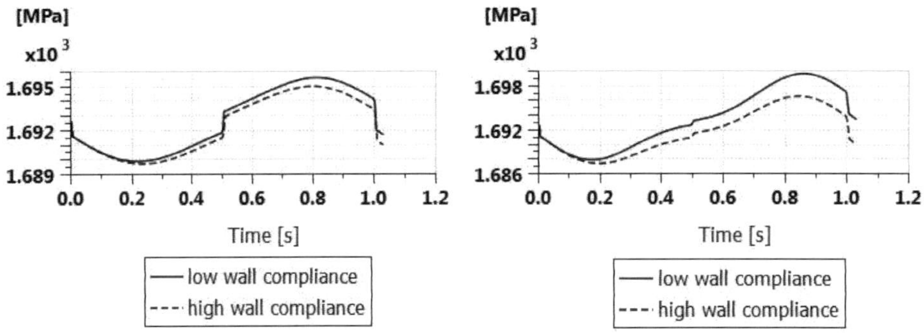

(a) Input amplitude: 0.6 m/s. (b) Input amplitude: 1.5 m/s.

Figure 4: Isothermal bulk modulus time-history for different levels of input amplitude.

3.2. K&C test bench simulation

In order to simulate the effect of an increased wall compliance in operative conditions, a K&C test has been simulated. A multi-body model of a McPherson suspension system has been modelled in AMESim, including:

- shock absorber;
- spring;
- spindle;
- lower wishbone;
- steering tie rod;
- steering rack;
- bushing (between strut and body, lower wishbone and body).

All the bodies are rigid. Input used to simulate the K&C test are:

- wheel vertical motion;

- rack motion;
- force (x and y direction) and moment (x, y and z direction) at wheel centre.

At first, lateral rack displacement was the predominant (motion) input, then it was wheel vertical motion. Figure 5(a) shows lateral (Y-axis) and vertical (Z-axis) wheel centre location in function of time, making clear when the vertical displacement assumes a predominant role; lateral displacement is always present due to interactions between rigid bodies. Forces produced by the two shock absorber models have been collected and confronted. Figure 5(b) shows a difference lower than 1% in the two cases.

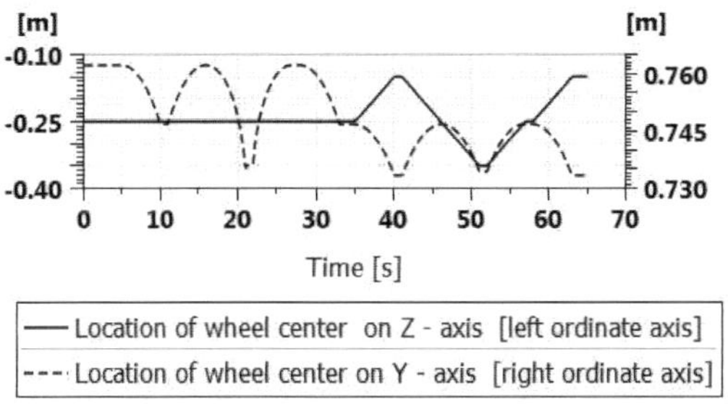

(a) Wheel centre absolute position in function of time.

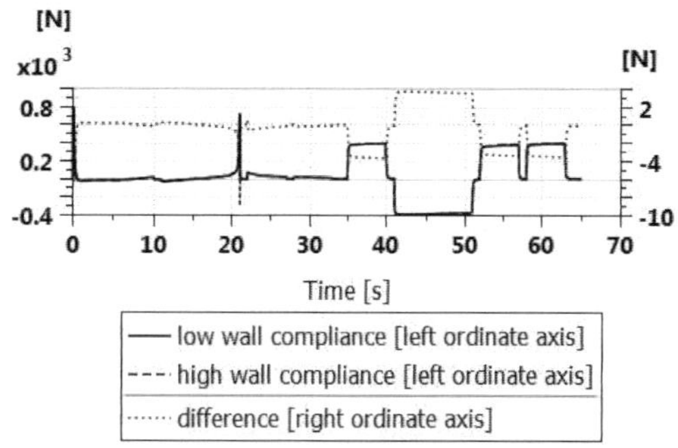

(b) Comparison of vertical forces at piston.

Figure 5: Selected inputs (a) and results (b) of K&C simulation.

3.3. *Comfort characterization*

Results of section 3.1 and 3.2 indicate that an increased wall compliance does not lead to significant changes into the operative behaviour of the shock absorber and of the suspension system. Nevertheless, this aspect is also investigated to evaluate the impact of high wall compliance into the full vehicle dynamic behaviour. At first, comfort has been studied, making use of the 7DOFs car model of AMESim. Performance in this context is characterized through the frequency response of pitch velocity and of (centre of gravity) vertical acceleration to sine wave inputs. Results for high and low wall compliance models are very similar and do not present relevant differences (figure 6 shows pitch angle response).

3.4. *Ride and handling manoeuvres*

Lastly, several manoeuvres done by car manufacturers have been simulated to evaluate the car behaviour for ride and handling purpose, focusing on several characteristics, among which:

- the car body yaw velocity rate in function of longitudinal velocity, evaluated during a manoeuvre with constant steering angle and increasing speed;
- over-steering at wheel, evaluated during a quasi-static steering manoeuvre in which the velocity is constant;
- the roll rate, which is also analysed during the quasi-static steering manoeuvre;
- steering torque during all the manoeuvres.

The 18DOFs AMESim full vehicle model has been used. Even in this case, all the more relevant metrics did not show any significant difference.

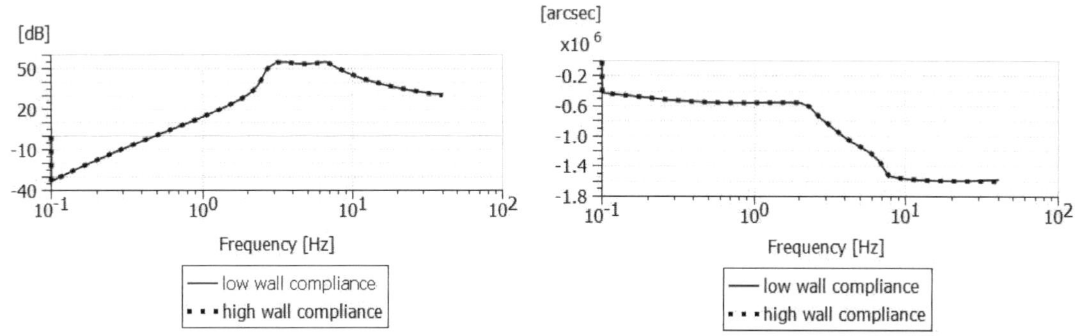

(a) Pitch angle amplitude. (b) Pitch angle phase.

Figure 6: Frequency response of pitch angle to road input.

4. Conclusions

Nowadays, thanks to new materials and advanced topology optimization techniques, it is theoretically possible to obtain values of walls compliance higher than reachable in the past, which is the main reason behind this study. In shock absorbers, high values of w_c would allow to reduce their weight, thus leading to a series of advantages such as less consumption and pollution, better performance in terms of sport driving. The aim of this study is to analyse the effect of w_c on the behaviour of shock absorbers and on vehicle dynamics through a simulation approach, in order to investigate eventual side effects that might compromise vehicle performance. Simulations have been chosen as a tool for this study because they allow to reduce both costs and time by avoiding the use of prototype and test campaign. The authors used the LMS Imagine.Lab AMESim environment due to its performance when studying nonlinear multiphysical system composed by several sub-components, such as shock absorbers. Furthermore, its tools, models and techniques have already been validated by the scientific community. This tool allow to study complex nonlinear

The force-velocity curve has been calculated for two shock absorber models different only for w_c, showing that an increased housing walls compliance leads to a higher hysteresis in the force-velocity curve. Furthermore, some physical quantities characterizing the shock absorber behaviour have been analysed; although many of them present differences, the only significant one was the pressure level, slightly lower in the shock absorber with high wall compliance. This is significant: lower operative pressure are related to lower mechanical stress into the component, and this should be considered during the design phase.

Despite the increased hysteresis, results of full vehicle simulations indicate that the entire system dynamic is not affected significantly, in the case of handling, primary ride and comfort. This is an important result, because it opens the doors to further investigation about this topic and to eventual practical applications.

References

[1] J.C. Dixon, 2007, *The shock absorber handbook*, 2nd edition

[2] Lang H.H.: 1977 A Study of the Characteristics of Automotive Hydraulic Dampers at High Stroking Frequencies PhD Dissertation University of Michigan

[3] Van-Kasteel R, Cheng-Guo W, Lixin Q, Jin-Zhao L and Guo-Hong Y 2005 A new shock absorber model for use in vehicle dynamics studies *Vehicle System Dynamics* **43:9** pp 613–631

[4] Lion A and Loose S 2002 A Thermomechanically Coupled Model for Automotive Shock Absorbers: Theory, Experiments and Vehicle Simulations on Test Tracks *Vehicle System Dynamics* **37:4** pp 241–261

[5] Ramosa J C, Rivasa A, Bierab J, Sacramento G and Salab J.A 2005 Development of a thermal model for automotive twin-tube shock absorbers *Applied Thermal Engineering* **25**, pp 1836–1853
[6] Joost W W 2012 Reducing Vehicle Weight and Improving U.S. Energy Efficiency Using Integrated Computational Materials Engineering *JOM* **64:9**, pp 1032–1038
[7] Miller W S, Zhuang L, Bottema J, WitteBrood A J, De Semet P Haszler A and Vieregge A 2000 Recent development in aluminum alloys for the automotive industry *Materials Science and Engineering* **A280**, pp 37–49
[8] Cui Y, Kurfess T R and Messman M 2013 Physical Modeling and Simulation Analysis of an Advanced Automotive Racing Shock Absorber using the 1D Simulation Tool AMESim *SAE International Journal of Passenger Cars - Mechanical Systems* **6** pp 7–17
[9] Dobre A, Hadaˇr A, Vasiliu D and Andreescu C 2016 Modelling and Simulation of the Dynamic Behaviour Automotives Suspension By AMESim *Proceedings of the European Automotive Congress EAEC-ESFA 2015* pp 317–323
[10] Vasiliu N, Hadaˇr A, Dobre A and Andreescu C 2016 Experimental Identification of the Automotive Magnetorheological Shock Absorbers *Proceedings of the European Automotive Congress EAEC-ESFA 2015* pp 61–67
[11] De Bruyne S, Van der Auweraer H and Anthonis J 2011 Improving active suspension performance by means of advanced vehicle state and parameter estimation, *IEEE International Conference on Mechatronics*
[12] De Bruyne S, Van der Auweraer H and Anthonis J 2011 Advanced State Estimator Design for an Active Suspension, *SAE Technical Paper*
[13] LMS Imagine.LabAMESim help manual.
[14] Hassani V, Tjahjowidodo T and Thanh Nho Do P 2014 A survey on hysteresis modeling, identification and control *Mechanical Systems and Signal Processing* **49:12** pp 209-233

Thermal and fluid simulation of the environment under the dashboard, compared with measurement data

C S Popescu[1], G M Sirbu[2] and I C Nita[2]

[1]Segula Technologies Romania, Romania
[2]Renault Technologie Roumanie, Romania

cristian-sorin.popescu-renexter@renault.com

Abstract. The development of vehicles during the last decade is related to the evolution of electronic systems added in order to increase the safety and the number of services available on board, such as advanced driver-assistance systems (ADAS). Cars already have a complex computer network, with electronic control units (ECUs) connected to each other and receiving information from many sensors. The ECUs transfer an important heat power to the environment, while proper operating conditions need to be provided to ensure their reliability at high and low temperature, vibration and humidity. In a car cabin, electronic devices are usually placed in the compartment under the dashboard, an enclosed space designed for functional purposes. In the early stages of the vehicle design it has become necessary to analyse the environment under dashboard, by the use of Computational Fluid Dynamics (CFD) simulations and measurements. This paper presents the cooling of heat sinks by natural convection, a thermal and fluid simulation of the environment under the dashboard compared with test data.

1. Introduction

The thermal management of the vehicle car cabin is mainly focused on the thermal comfort of passengers [1]. The air flow and temperature distribution were studied in the passenger compartment under thermal load, such as the solar heat transfer to the cabin of a parked car [2] and cooling of the cabin under solar radiation [3]. Another concern of thermal management of the car cabin is the environment of the electronic devices that provide safety functions and services available on board, such as the navigation and multimedia systems with electronic visual display.

The current study is focused on the under-dashboard compartment, where most of the electronic devices are located, a zone where there are different factors that can influence the thermal environment of the ECUs. The environment under the dashboard is a confined space, inside the vehicle car cabin, delimited by the panels placed in the lower region of the passengers' feet, the engine bay and the dashboard. The allocated volume of this region is designed mainly for functional purpose, considering other constraints than the ones usually taken into account for the car cabin comfort and ergonomics. Temperature and flow fields were studied for aircraft confined compartments that contain thermally dissipating equipment, such as the natural convection inside the crown compartment of a commercial aircraft [4].

The parts positioned in the under-dashboard compartment are from the following assemblies: structural, comfort, steering, electrical and electronic. The structural assembly includes the crossbar and supports for other parts; the comfort assembly includes the heating, ventilation and air conditioning unit (HVAC), the ventilation and defrosting ducts; the electrical and electronic include the wiring, the multimedia system (ECU and display), the cluster and the air conditioning/heater control panel.

The potential heat sources in the under-dashboard compartment are the ECUs and the heater, which is inside the HVAC unit, and its pipes. Depending on the external conditions, the region above the exhaust pipe, called the tunnel, can also generate some heat. Usually, the multimedia system is placed in the central upper region of the dashboard, adjacent with the other heat sources like the HVAC unit and the tunnel. The electronic systems are the sources of heat that have thermal constraints for proper operating conditions, therefore their environment requires a controlled temperature.

The cooling of electronic components is performed by the use of heat sinks can be passive or active and it has been widely studied [5]. The passive heat sink relies on an increased surface area in order to transfer the heat, while the active heat sink, such as a fan, requires a power supply. Depending on the type of the heat sink the cooling with air can be achieved by a free convection, in the case of passive heat sinks or by forced convection, in the case of active heat sinks. The current study is focused on an electronic device with a passive heat sink.

2. Measurement Test

A measurement test was performed on a vehicle, in the wind tunnel at an ambient temperature for a warm climate and the engine load at a cruising speed, considering common operating conditions for the multimedia system. The current stage of the test was performed for a duration that would allow stabilized conditions, with data acquisition of 23 thermocouples placed in the under-dashboard compartment at designated locations.

3. Numerical simulation

3.1. Case setup

The geometry of the case includes most of the parts under the dashboard: multimedia system, fuse box, wiring; HVAC unit, heater pipes, AC control panel; crossbar, supports. A general view of the model is presented in Figure 1.

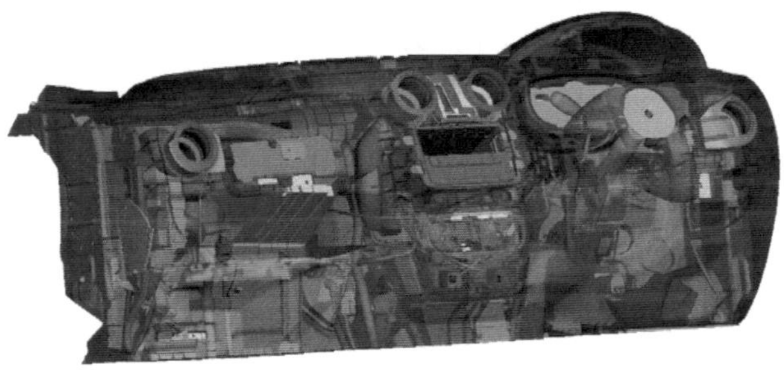

Figure 1. General view of the under-dashboard compartment.

The surface meshing of these parts is quite complex, considering the tight volume and the surface shapes. The geometrical details were simplified for most of the parts and fine mesh was applied on specific areas of the considered sources, such as the multimedia system. It is different than the common vehicle car cabin simulations where there are mainly large surfaces bounding a bigger volume. The under-dashboard computational volume was delimited by the cabin on the lateral sides and toward the engine bay, and by the lower region of the passengers' feet and the dashboard on the other sides. The simulation was performed with ANSYS Fluent-3D Release 15.0.

The current configuration has only passive heat sinks for the multimedia ECU, the natural convection being generated by the temperature difference between the heat sources and the ambient air. The heat sources considered in this simulation are: the multimedia system (casing and heat sink), the lower part of the HVAC including the heater and its pipes. All the other surfaces are considered as adiabatic walls. Pressure outlet boundary condition was used for the flow of air through the open zones.

3.2. Simulation results

A general view of the flow under the dashboard, represented by temperature pathlines, is presented in Figure 2.

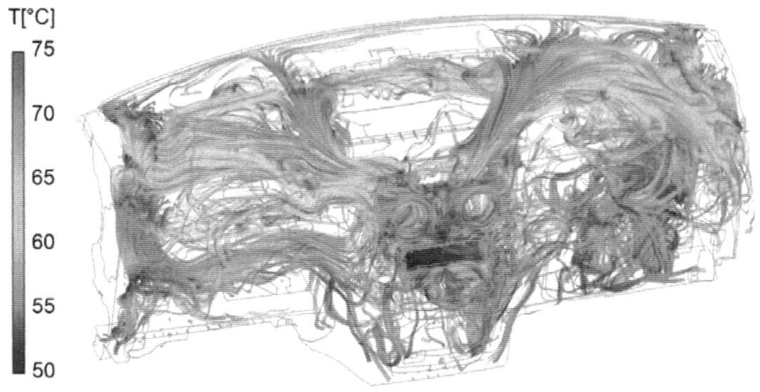

Figure 2. General view of the flow - temperature pathlines.

Samples of temperature results in the multimedia region are presented in Figure 3: downstream pathlines from casing and heat sink are represented in Figure 3a, volume with temperature above 65°C in Figure 3b and temperature contours in Figure 3c.

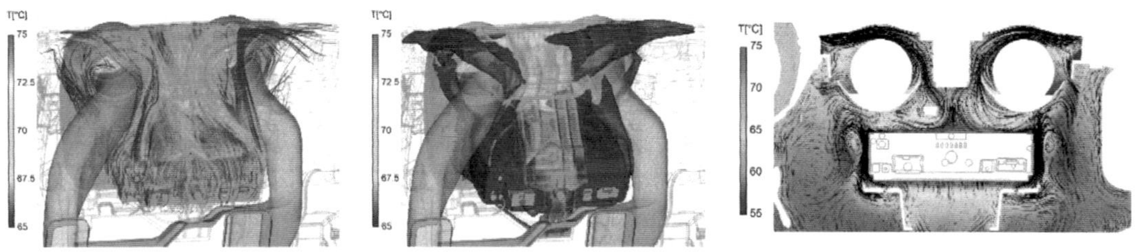

a b c

Figure 3. Multimedia region - temperature results.

4. Comparison between simulation results and test data

The comparison between the simulation results and test data is represented in Figure 4. Considering the 23 probes, the absolute error is approximately 5%, with an average temperature difference of 1.6 °C. Two probes have a temperature difference above the considered threshold of 3°C, with a maximum error of 15.5%.

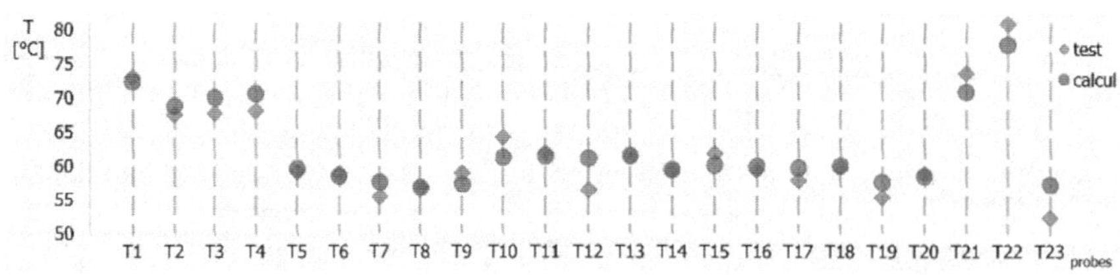

Figure 4. Comparison between simulation results and test data.

For a thorough analysis, the under-dashboard compartment was segregated into 6 regions of interest (see Table 1).

Table 1. Average temperature difference on regions.

Region	**Number of probes**	$\bar{\Delta}$ **[°C]**	$\bar{\varepsilon}$ **[%]**
Multimedia	6	1.3	4
Lower central - below multimedia	3	1.3	4
Upper central - under dashboard	4	2	6.4
Cross member level	5	0.9	2.8
Fuse box	2	1.3	4.1
Recycling region and heater pipes	3	3.6	11.3

5. Conclusions

The simulation results are compared with the test data, showing a good assessment in the zones of interest. The temperature values were analyzed on 23 probes and an absolute error of 5% was obtained.

The automobile manufacturer has to guarantee that the electronic systems are able to work in a wide temperature range, therefore this type of simulation was performed to ensure the performance and reliability of the electronic systems placed under the dashboard. This type of simulation provides a flexible method to identify the factors that influence the cooling of the electronic parts, by analyzing how redirecting the flow of air can improve the cooling performance.

References

[1] Alahmer A, Mayyas A, Mayyas A A, Omar M and Shan D 2011 Vehicular thermal comfort models; a comprehensive review, *Appl. Therm. Eng.* **31**(6) 995–1002.
[2] Al-Kayiem H H, Sidik M F B M and Munusammy Y R 2010 Study on the Thermal Accumulation and Distribution Inside a Parked Car Cabin, *Am. J. Appl. Sci.* **7**(6) 784-789
[3] Sevilgen G ans Kiliç M 2013 Investigation of transient cooling of an automobile cabin with a virtual manikin under solar radiation, *Therm. Sci.* **17**(2) 397-406
[4] Butler C and Newport D 2014 Experimental and numerical analysis of thermally dissipating equipment in an aircraft confined compartment, *Appl. Therm. Eng.* **73**(1) 869-878
[5] Shende D M and Mahalle A 2013 Cooling Of Electronic Equipment with Heat Sink: A Review of Literature, *IOSR J. Mech. and Civ. Eng.* **5**(2) 56-61

Research on some auxiliary mechanisms used in passenger cars

Daniela Antonescu[1], Daniel Iozsa[2], Ovidiu Antonescu[3] and Gh. Fratila[4]

[1] "Iuliu Maniu" High School, Bucharest, Romania, e-mail: daniela.mihalache@yahoo.com
[2] Dept. of Road Vehicles, Politehnica University of Bucharest, Romania
[3] Dept. of Mechanism and Robot Theory, Politehnica University of Bucharest, Romania
[4] Dept. of Road Vehicles, Politehnica University of Bucharest, Romania

Abstract. The paper presents the results of researches on the topological structure and geometrical analysis of the planar mechanisms with articulated bars, which are used for actuating the doors of cars. The main five types of car doors with rotate movement (folding) are presented, being described both as constructive structure and mode of operation, through suitable kinematic schemes. Some innovative solutions for vehicle door actuating mechanisms aim to use as little space as possible, which is beneficial for car parking. There are three types of car door movements: rotation, sliding and planar rotational-sliding. Most of the cars are equipped with folding doors, where the rotate movement is limited and operates horizontally. Almost all sliding doors are placed on the rear of the car (only for passengers, not for driver). Unlike rotate doors, the sliding doors require a minimum lateral space, which is an advantage, especially in parking places. In the end of the paper, a kinematic analysis of the canopy 4-bar mechanism has been performed, in order to increase the passenger comfort on the access into the vehicle.

1. Introduction

The automotive market is dominated by conventional vehicle doors (fig. 3a) with only one degree of mobility [1, 8], which represents a major discomfort for the passengers [2, 7], especially to parking on a tight surface or on a steep terrain. For moderated slopes, this issue specific for the vehicle doors has been overcome by means of using a variable movement of the door [2, 13]. Through an actuating device, for instance an electro-pneumatic actuator, the comfort and safety in operating the door can be increased. Thus, a collision prevention system can be developed using obstacle detection sensors. The use of multi-DOF (Degree Of Freedom) vehicle doors has been proposed, allowing an improved comfort on the access into the vehicle.
In figure 1 three types of unconventional vehicle doors are presented.

a *b* *c*

Figure 1. Unconventional car doors (*a* – suicide, *b* – scissors, *c* – gull wings) [11, 12]

IOP Conf. Series: Materials Science and Engineering **252** (2017) 012028 doi:10.1088/1757-899X/252/1/012028

2. The mechanism of the vehicle rotate door

The vehicle rotate door is mounted to the car body by means of some hinges (figure 2*a*) that are rotate kinematic joints, whose axis is usually vertical.

2.1. Vehicle rotate doors with vertical axis

The vehicle hinge is a rotate joint R single or double in car doors or, at least, double in fright vehicle doors.

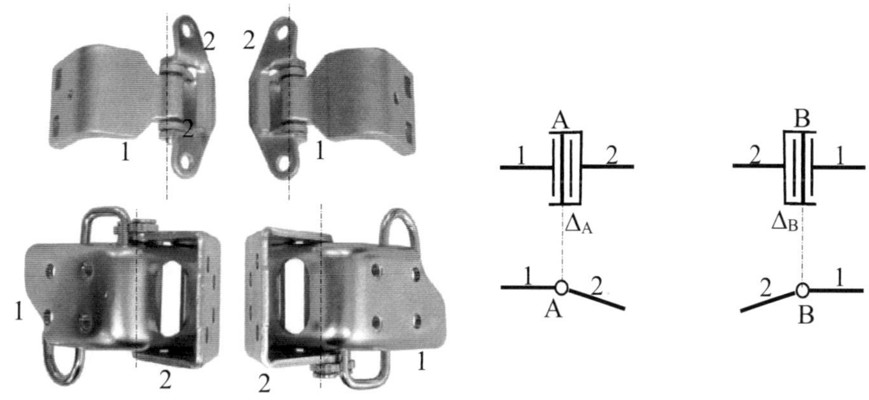

a *b*

Figure 2. Photos [12] of two pairs of hinges (*a*) and the kinematic schemes of two rotate joints (*b*)

The technical design solutions for hinges (figure 2*a*) vary a lot among manufacturers, but the schematic representation of them is that of a rotate kinematic joint R (figure 2*b*). In reality, the hinge axis (Δ_A, Δ_B) is not strictly vertical (figure 2*b*), the angle formed with the vertical axis being small and measured in a transversal vertical plane (figure 3*a*). Anyway, they are considered with vertical axis – figure 3*a*.

a *b*

Figure 3. Conventional vehicle rotate doors [11, 12] with vertical axis hinges (*a*) and front vehicle rotate door with inclined axis hinges (*b*)

2.2. Vehicle rotate doors with longitudinal-horizontal axis

In this case the vehicle door rotates by 90° in a transversal-vertical plane, around a horizontal axis of longitudinal direction (figure 4). Usually, this design solution is used to some coupe and sport vehicles (figure 4*a*) or only to the rear doors of some passenger cars (figure 4*b*).

a *b*

Figure 4. Doors [12] with horizontal axis, type *gull wings*: "coupe" (*a*) and at rear (*b*)

In order to maintain open the vehicle door, one or two devices type gas dampers are used (figure 4, 5).

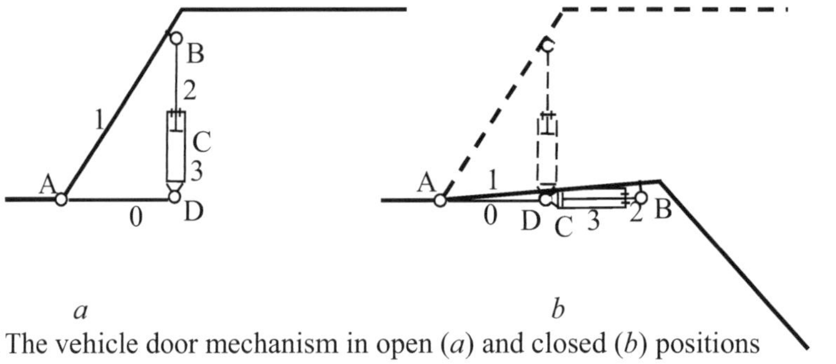

a *b*

Figure 5. The vehicle door mechanism in open (*a*) and closed (*b*) positions

The rotate joint A is the symbolic representation of the vehicle door hinge, whose rotate axis is horizontal, in longitudinal position related to the vehicle body (figure 5). In the open position of the door (figure 5*a*) the damper (piston 2 and cylinder 3) is in vertical position, and in the closed position of the door (figure 5*b*) the damper end up horizontally along a transversal direction.

Note that the damper has the actuator role due to the fluid pressure in cylinder 3, and its structure is a kinematic chain type dyad (0, 1, 2 and 3) with a structure RTR (figure 5*a*).

2.3. Vehicle rotate doors with inclined axis

Some of these vehicle doors have only a single hinge, whose axis is in a longitudinal-vertical plane (figure 6*b*), the inclining angle being about 45°.

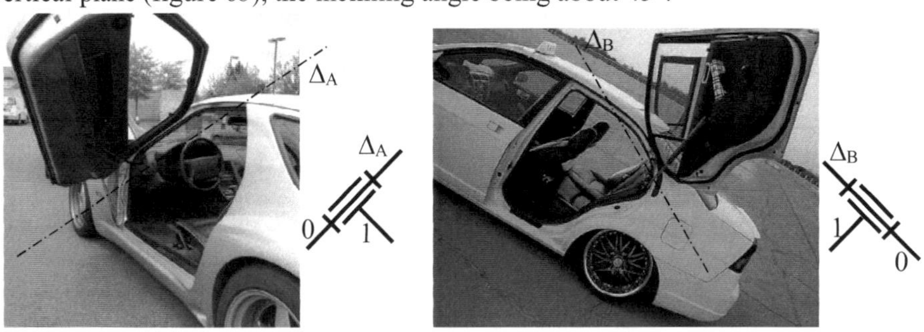

a *b*

Figure 6. Photos [11] of the vehicle rotate doors with inclined axis, type *butterfly*: "coupe" (*a*) and "sedan", at rear (*b*)

The vehicle door hinge makes the connection between the body 0 and the door 1, being graphically represented as a kinematic rotate joint R(0,1), whose axis is denoted with Δ_A (figure 6*a*) or Δ_B (figure 6*b*).
The inclined position of the hinge axis allows the rotation of the door in an inclined plane, this advantage permitting the passenger to get in/out the vehicle more easily in a narrow lateral space between the two parked vehicles (figure 6*a*).

2.4. Vehicle rotate doors in a longitudinal-vertical plane
These vehicle doors rotate by maximum 60° (figure 7) in a longitudinal-vertical plane, having a horizontal rotate axis on a transversal direction with respect to the body.
The rotate angle of the door appears in a real dimension in a longitudinal plane (figure 7*b*), in respect of which the rotate axis is perpendicular.

a *b*

Figure 7. Photos [12] with rotate doors in a longitudinal-vertical plane, type *scissors*: front/rear (*a*) and only in front (*b*)

In the case of actuating the vehicle door for opening, first the door performs a sliding motion toward outside (figure 8*a*), along a transversal direction, then the rotate motion in a longitudinal plane completes the task.

a *b*

Figure 8. The scheme of the open kinematic chain LC(0+1+2) [1]: T+R for opening (*a*) and R+T for closing (*b*)

The kinematic element 1 slides on the cylinder surface of the vehicle body 0, moving along with the door 2 by s_{10} in respect of a transversal direction Δ_{21} (figure 8).
When the vehicle door 2 is closing, first the door rotates by φ_{21} in a counter clockwise direction (figure 8*b*), and then it slides by s_{10} along the axis Δ_{21}.
Note that in order to open both the vehicle doors on the same side (figure 7*a*), the front door (AA_0) must slide more than the rear door (BB_0), as it can be observed in figure 9.

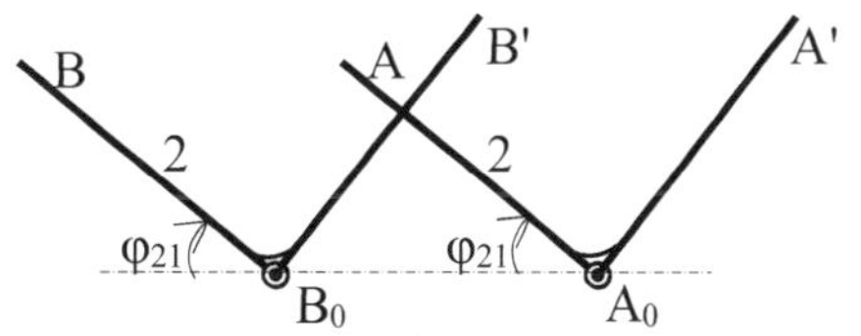

Figure 9. The kinematic scheme of the two same-side vehicle doors in open position

IOP Conf. Series: Materials Science and Engineering **252** (2017) 012028 doi:10.1088/1757-899X/252/1/012028

2.5. Vehicle rotate doors in a transversal plane

This type of vehicle door has a curved shape, having a cylindrical outer surface (figure 10), which it allows to roll into a circular guide under the body floor.
Note that these doors have no hinges, the opening/closing movement being performed through rotation in a transversal-vertical plane.

a *b*

Figure 10. Photos [11] of vehicle rotate door in two positions of closing phase

Note that the outer of the vehicle door is a cylindrical surface as the car side body is. Before the beginning of the door opening, its window lowers in the bottom (open) position. The rotation of the vehicle door is performed by means of an actuation electric system. The disadvantage of this kind of door is that it occupies too much room from vehicle inner space, but the advantage is obviously in parking places.

3. The vehicle sliding doors

These vehicle sliding doors are placed in front (figure 11 - coupe) and at rear (station wagon), being guided by means of longitudinal-horizontal slots.
Usually, the guiding slots are positioned at top, bottom and middle parts of the vehicle body.

Figure 11. Photo [11] of a vehicle sliding door

If the actuating of the vehicle sliding door is often and the driver is not willing to get out every time, the using of an electric adjusting mechanism is recommended [3]. The door opening is frequent when the vehicle, for instance a minibus, is used for urban passenger transportation, or a city van which delivers goods has the sliding door on the opposite side of the vehicle regarding the driver's seat, and the driver has to remain at steering wheel all the time. Since the vehicle rotate door can be actuated pneumatically, the sliding door is more suitable to be driven electrically. The opening/closing mechanism of the vehicle sliding door is using a cable as the driving component (figure 12).

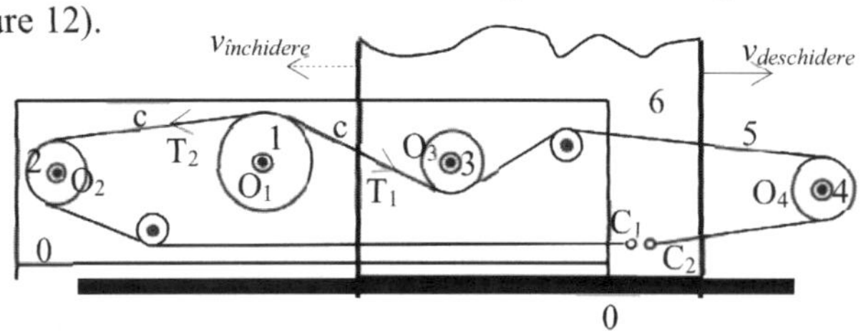

Figure 12. Kinematic scheme of the cable mechanism for actuating the sliding door

IOP Conf. Series: Materials Science and Engineering **252** (2017) 012028 doi:10.1088/1757-899X/252/1/012028

The mechanism actuation is performed by an electric motor which drives the reel 1 as driving element, on which the cable 5 is reeled having two branches tensioned with the forces T_1 and T_2 (figure 12). The pulleys 2, 3, 4 and the others are assembled on the body 0, and the cable ends are fixed on the sliding door in the points C_1 and C_2. When the reel 1 rotates in counter clockwise, the branch tensioned with the force T_1 (by the point C_2) moves the door 6 to the right for opening, or when the reel 1 rotates in clockwise, the branch T_2 linked to point C_1 drives the door 6 to the left for closing.

4. Articulated planar mechanisms used as hinges

4.1. The four-bar mechanism

The four-bar mechanism is used to lift/lower the vehicle roof which forms a single piece along with the doors and the windshield, named "canopy roof" (figure 13).

a *b*

Figure 13. Photos [11] of the *canopy* sport cars with a double 4-bar mechanism

Early mechanism was an articulated parallelogram (figure 13*a*), the roof performing a circular sliding motion. Late-on the linkage became a general four-bar mechanism (figure 13*b*) which allows a planar rotational-sliding movement of the canopy roof.
The four-bar mechanism is type planar linkage A_0ABB_0 (figure 14), on which the coupler 2 (AB) forms a rigid part along with the canopy roof.

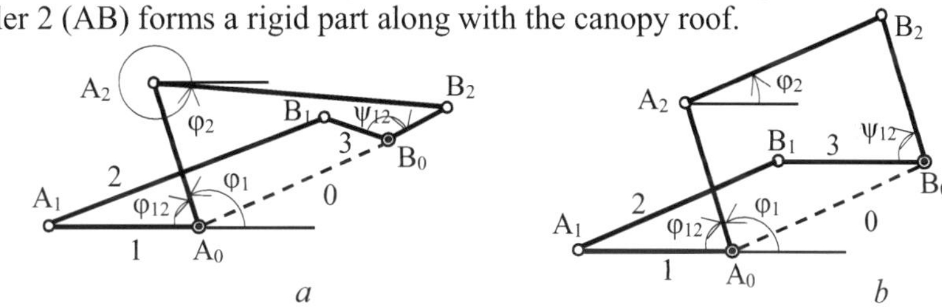

Figure 14. Four-bar mechanism [1, 5] - general case (*a*) and type parallelogram (*b*)

The fixed articulations A_0 and B_0 belong to the vehicle body, and about which the rockers 1 and 3 rotate by the angles φ_{12} and ψ_{12} in a vertical plane (figure 14*a*). In the particular case of the parallelogram mechanism (figure 14*b*) the coupler 2 performs a circular sliding motion alongside of the circle arcs A_1A_2 and B_1B_2. Actually, the four-bar mechanism (figure 15) is double, having the rockers 1 and 3 dephased in a transversal plane.

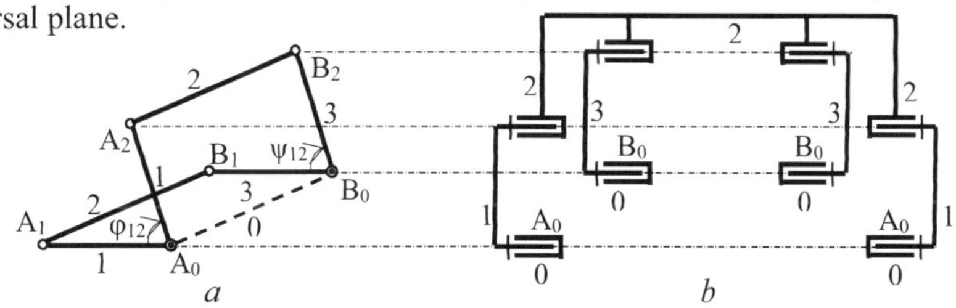

Figure 15. Parallelogram mechanism scheme - side view (*a*) and transversal view (*b*)

4.2. The bi-contour mechanism (Watt)

The Watt mechanism is an articulated planar linkage [1, 5] with two closed contours (figure 16*a*) and it is obtained by amplification of the 4-bar mechanism A_0ABB_0 with a kinematic chain type dyad LD(4,5) (figure 16*b*).

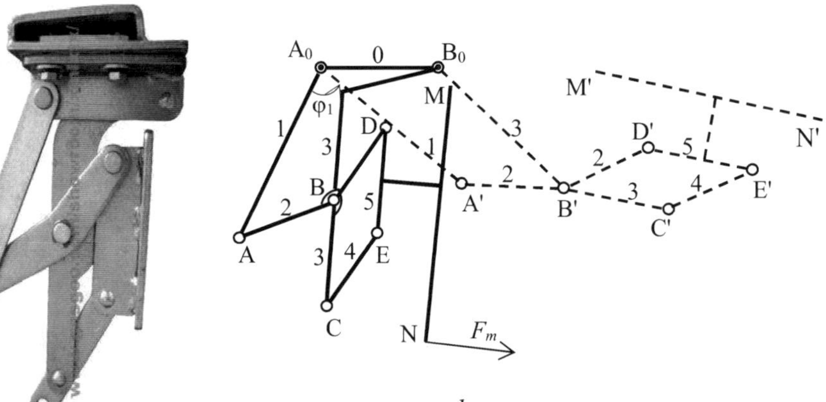

a *b*

Figure 16. Photo of the Watt mechanism (*a*) and its kinematic scheme (*b*)

5. Kinematics of the Canopy 4-bar mechanism

In figure 17 a computer program for calculation of the kinematics of canopy 4-bar mechanism is displayed. The TurboPascal 6.0 software for developing the programming source file has been used.

```
CANOPY - Notepad
File Edit Format View Help
program mec_RRR; {Kinematics of mechanism type canopy for n1=10 rpm !}
uses crt;
const l1=0.2;l2=0.6;l3=0.35;l0=0.6;l0p=0.2;xB0=0.6;yB0=-0.2;
var A2,B2,C2,t21,t22,fi21,fi22,A3,B3,C3,t31,t32,fi31,fi32:real;
        fi1,om1,fi2,fi3,om2,om3,ep2,ep3,u:real;
        xB,yB,xA,yA:real;
        i:integer; f:text;
function rg(u:real):real;
        begin rg:=u*180/pi; end;
function gr(u:real):real;
        begin gr:=u*pi/180; end;
begin
    assign(f,'D(2_3).lst'); rewrite(f);
        om1:=(10*pi)/30;
        writeln(f);
writeln(f,'                    TABLE OF RESULTS - D(2_3)                    ');
writeln(f,'  *************************************************************');
writeln(f,'   fi1    fi2      om2       ep2       fi3       om3       ep3  ');
writeln(f,'  *************************************************************');
 for i:=0 to 36 do
 begin
     fi1:=(i*pi)/18;
         A2:=2*l2*(l1*cos(fi1)-l0);
         B2:=2*l2*(l1*sin(fi1)+l0p);
         C2:=2*l1*(l0p*sin(fi1)-l0*cos(fi1))+sqr(l1)+sqr(l2)-sqr(l3)+sqr(l0)+sqr(l0p);
         t21:=(-B2+sqrt(sqr(A2)+sqr(B2)-sqr(C2)))/(C2-A2);
         t22:=(-B2-sqrt(sqr(A2)+sqr(B2)-sqr(C2)))/(C2-A2);
         fi21:=2*ArcTan(t21);
         fi22:=2*ArcTan(t22);
         fi2:=fi22+2*pi;
         A3:=2*l3*(l0-l1*cos(fi1));
         B3:=-2*l3*(l0p+l1*sin(fi1));
         C3:=2*l1*(l0p*sin(fi1)-l0*cos(fi1))+sqr(l1)-sqr(l2)+sqr(l3)+sqr(l0)+sqr(l0p);
         t31:=(-B3+sqrt(sqr(A3)+sqr(B3)-sqr(C3)))/(C3-A3);
```

Figure 17. Canopy 4-bar mechanism software for kinematic calculation

As it can be observed in figure above, a programming loop for the calculation of displacement, velocity and acceleration of the vehicle roof has been compiled and ran. Due to the fact that the velocity of actuating the canopy roof is low (10 rpm), it is shown only the angular displacement of the car roof (φ_2 – link 2, fig. 14a) in figure 18. As it can be observed in figure 18, the range of canopy roof angular displacement is about 40º from the closed position to the maximum open position. This means that the

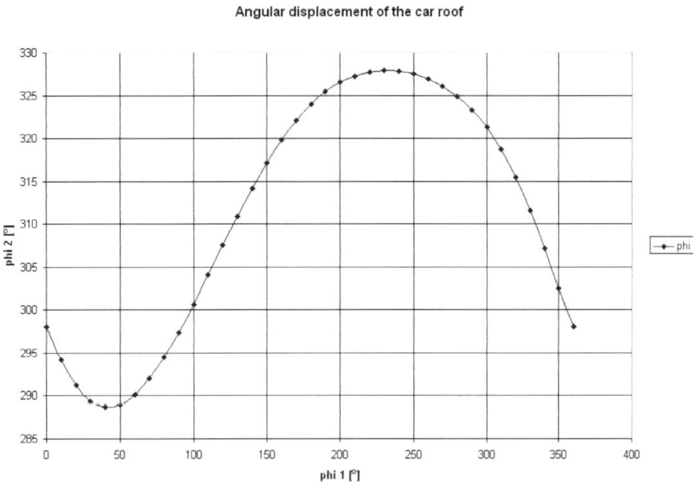

Figure 18. Variation of the angular displacement of the canopy roof

mechanism offers a pretty good comfort for the passenger to enter into the car. Yet, there is still some discomfort comparing to the classical rotate door with vertical axis. The greater is this angle, the better comfort it offers.

6. Conclusions

The most used types of vehicle doors are with rotate motion (hinges). Yet, the sliding doors are more and more present in passenger cars, not only in minibuses and vans. Following this idea, the paper presents the main five types of vehicle doors with rotate motion, the sliding door and, in the end, two linkage types used as door hinges. Some of the solutions are designed to save as much space as possible in narrow parking places. The sliding doors are the best from this point of view, but the disadvantages of them are the low safety, poor maneuvering and reliability, and the high-cost.
The paper ends with the developing of software for kinematic analysis of the 4-bar mechanism used in Canopy sport cars. It can be used in optimizing the angle of canopy roof for increasing the comfort when the passengers get in/out the car.

7. References

[1] Antonescu O Antonescu P 2016 *Mechanism and Machine Science*, Politehnica Press
[2] Fratila G 1977 *Calculation and construction of the automobiles* (RO), E.D.P., Bucharest
[3] Horwitz Alexander Netanya 2005 *Sliding door mechanism for vehicles*, EP1548221 A2
[4] Lichtenheldt W 1970 *Konstructionslehre der Getriebe*, Akad. Verlag, Berlin
[5] Manolescu N I et al. 1972 *Mechanism and Machine Theory* (RO), E.D.P., Bucharest
[6] Mass J Kern S 2007 *Mechatronic vehicle door assistant*, IEEE/ASME Internatioal Conference on Advanced Inteligent Mechatronics, pp. 1-5
[7] Mondiru C 1998 *Dacia vehicles - diagnosis, maintenance, repairing* (RO), E.T., Bucharest
[8] Oprisan C Popovici G A 2001 *Mechanisms* (in Romanian - RO), Ed. Tehnica, Chisinau
[9] Strolz M Morti A et al. *Development, Control and Evaluation of an Actuated Car Door*, Journal IEEE, pp. 1-11
[10] Tomohiko N Hiroyuki S 1993 *Opening and closing device of door for vehicle*, JP05112128A
[11] *https://en.wikipedia.org/wiki/List_of_cars_with_non-standard_door_designs*
[12] *https://en.wikipedia.org/wiki/Vehicle_door*
[13] *https://www.carid.com/lambo-doors.html*

Fatigue Analysis of Rotating Parts. A Case Study for a Belt Driven Pulley.

[1]**SANDU Ionela**, [1]**TABACU Stefan**, [1]**DUCU Catalin**,

[1]University of Pitesti, Romania

ionela.alina_sandu@yahoo.com; stefan.tabacu@upit.ro; catalin.ducu@upit.ro

Abstract. The present study is focused on the life estimation of a rotating part as a component of an engine assembly namely the pulley of the coolant pump. The goal of the paper is to develop a model, supported by numerical analysis, capable to predict the lifetime of the part. Starting from functional drawing, CAD Model and technical specifications of the part a numerical model was developed. MATLAB code was used to develop a tool to apply the load over the selected area. The numerical analysis was performed in two steps. The first simulation concerned the inertia relief due to rotational motion about the shaft (of the pump). Results from this simulation were saved and the stress – strain state used as initial conditions for the analysis with the load applied. The lifetime of a good part was estimated. A defect was created in order to investigate the influence over the working requirements. It was found that there is little influence with respect to the prescribed lifetime.

1. Introduction

In the present study, we assessed lifetime of pulley by numerical methods. The pulley is a component part of a belt drive. The drawing process was used to manufacture the pulley. The pulley works under both the fatigue compressive loading due to belt tension and the torque due to rotation. Therefore, to assure the safety of the belt drive system is very important to investigate the lifetime and optimization of the pulley [1]. According to [2], optimization of this part has been widely carried out through the stress analysis based approach. Common applications in the automotive industry include crankshafts, water pumps, air-conditioner compressors and power steering pumps. Although the shape and the usage of pulleys are very simple, evaluating the design of a pulley is difficult because of the loading conditions and installation environment. According to [1] Euler, first investigated the belt-drive mechanics and established the relationship between belt tension and coefficient of friction and Reynolds, developed creep theory to explain speed loss in power transmission belts. Furthermore, several researchers have studied the mechanics of V-belts, for example, [3] who described the relationship between ribs shear deformation and transmitted force. The applied stress distribution of the pulley under high-tension and torque was obtained by using FEA (Finite Element Analysis). The main features of fatigue analysis are listed below:

- Develop the three-dimensional model of the pulley, the FE (Finite Element) model;
- Perform critical stress/strain analysis under multiple loading;
- Process the maximum principal stress and strain history obtained by analysis with rotational angle;
- Input the probabilistic S–N (P–S–N) curve into an analytical tool;
- Compute the lifetime of the pulley at a specification level.

IOP Conf. Series: Materials Science and Engineering **252** (2017) 012029 doi:10.1088/1757-899X/252/1/012029

Pulleys are one of the main components used in a belt conveyor, supporting the loads generated from the belt tension according to its wrapping angle [7]. Few available reports [7,8] have shown that the majority of failure of pulleys are due to fatigue or overload failures, whereas few other investigators [9] have attributed a typical failure of a pulley to material and manufacturing process fault analyzing after detailed microstructural and finite element analysis. According to SS-EN 13306 [10], failure is the termination of the ability of an item (i.e., any part, component, device, subsystem, functional unit, component, or system that can be individually considered) to perform a required function. In other words, failure can be referred to as an event or process of component deterioration.

2. Model description and technical specifications

Starting from the CAD model of the part (Figure 1) a numerical model was developed. As specified in the functional drawing the bolts used to assembly the pulley on the pump's rotor were represented in the model. Figure 2 presents the numerical model that will be further used.

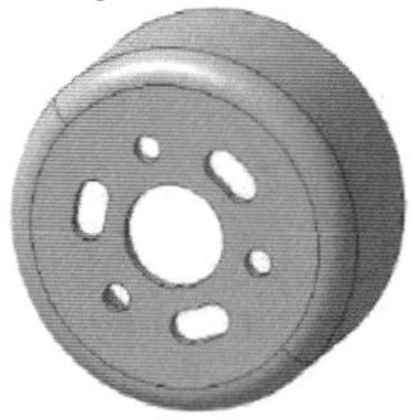

Figure 1. Pulley model (as generated)

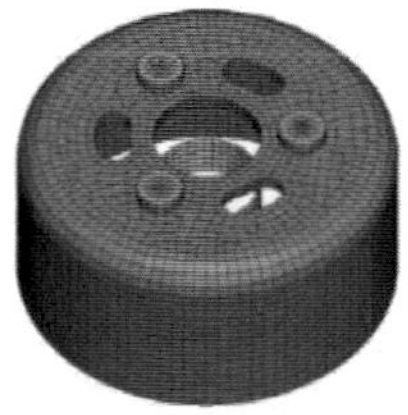

Figure 2. Numerical model

The part must resist 2.10^7 alternated stresses under a static load of 1016 N applied on an angle sector of 97° and 28mm width (Figure 3).
A custom MATLAB code was developed as a tool to apply the load over the selected area. The load is applied for each circumferential sector using a SIN function. The tool can also transform to load and apply it on the required area of the part (Figure 4).

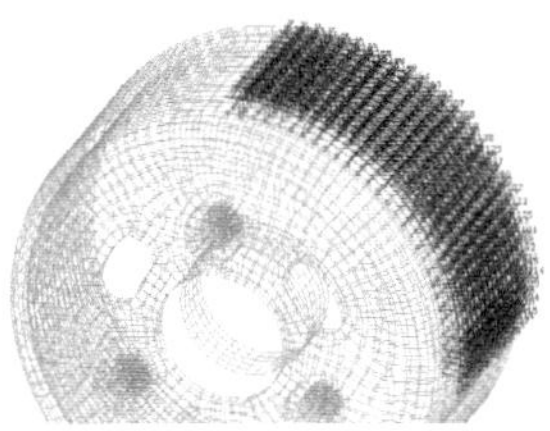

Figure 3. Area selected for the applied load

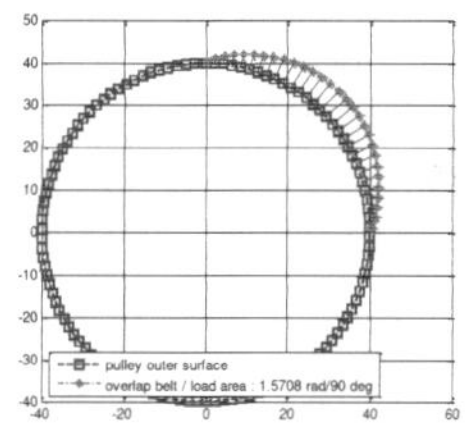

Figure 4. Load definition

For each node on the selected area a user defined coordinate system was created normal to the surface. The load defined by the *SIN* function and scaled accordingly was applied on the node along local Ox axis.

3. Fatigue Analysis

In the design of machine elements [4], three fundamental equations are used:

$$\sigma_{c,t} = \frac{P}{A}$$
$$\sigma_b = \frac{M_b \cdot y}{I} \quad (1)$$
$$\tau = \frac{M_t \cdot R}{J}$$

They are referring to the stress for compression/tension cases, bending or torsion. The notations used in set of equations (1) are: P - axial load; A - cross sectional area; M_b - bending moment; y - coordinates of a point on the cross section at which the stress is to be determined; I- area moment of inertia of the section; M_t - applied torque; R - radius of the section; J- torsion constant of the section.

3.1. S-N curve
The fatigue or endurance limit of a material is defined as the maximum amplitude of completely reversed stress that the standard specimen can sustain for an unlimited number of cycles without fatigue failure. The results of these tests are plotted by the means of the S-N curve. The S-N curve is the graphical representation of the stress (S_f) versus the number of stress cycles (N_f) before the fatigue failure on a log-log chart.

3.2. Endurance limit
When the laboratory data regarding the endurance limit of the materials are not available, the following procedure is adopted.

Two separate notations are used regarding the endurance limit (S_e') and (S_e) where:

- S_e' is the stress of a rotating beam specimen subjected to reversed bending stress;
- S_e is the stress of a particular mechanical component subjected to reversed bending stress.

There is an approximate relationship between the endurance limit and the ultimate tensile strength (S_{ut}) of the material (e.g. for steels):

$$S_e' = 0.5 \cdot S_{ut} \tag{2}$$

The relationship between (S_e') and (S_e) is as follows:

$$S_e = k_a \cdot k_b \cdot k_c \cdot k_d \cdot S_e' \tag{3}$$

where:

k_a is surface finish factor that for machined of cold – drawn parts is given by the equation $k_a = a \cdot (S_{ut})^b$ where $a = 4.51$ and $b = -0.265$.

k_b is the size factor which for larger the machine part, greater is the probability that a flaw exist somewhere in the volume and in this case of a value of 0.85 was assigned.

k_c is the reliability factor that the greater the likelihood that a part will survive, the more is the reliability thus a value of 0.659 was assigned.

k_d is the factor to account for stress concentration

4. Results. Structural analysis

4.1. Inertia Relief
The numerical simulation was performed in two steps with the first step focused on the inertia relief due to rotational motion about the shaft (of the pump). Results are presented in Figure 5.

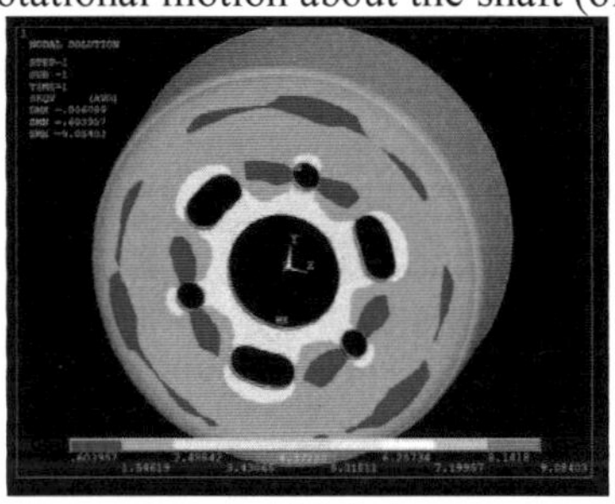

part with holes

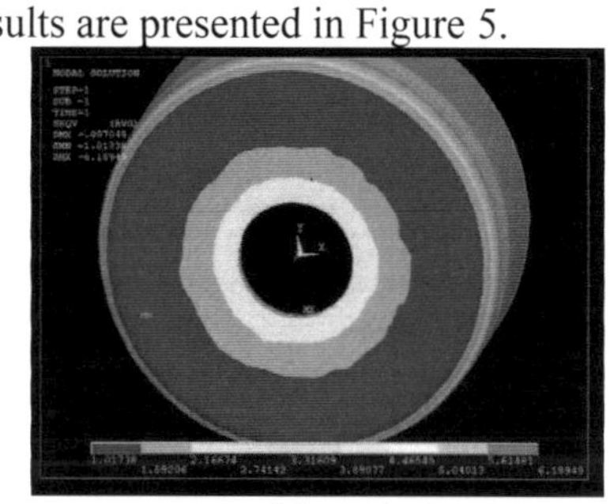

part without holes

Figure 5. Inertia relief. Stress analysis

IOP Conf. Series: Materials Science and Engineering **252** (2017) 012029 doi:10.1088/1757-899X/252/1/012029

The stress recorded on the bolt location has a value of about 7 MPa while the stress on the flange has a value of about 3 MPa. Results from this simulation were saved and the stress – strain state of the assembly will be further used as initial conditions for the analysis with the load applied.
Stress concentration is defined as the localization of high stresses due to the irregularities present in the component and the abrupt changes of the cross section. The modifying factor k_d to account for the effect of stress concentration is defined as:

$$k_d = \frac{1}{k_f} \tag{4}$$

where k_f is the stress concentration factor [4].

Based on results of the inertia relief analysis, the stress concentration factor can be evaluated as follows [4]:

$$k_{f,b} \approx \frac{7.0}{4.5} = 1.55$$
$$k_{f,f} = \frac{5.0}{3.5} = 1.45 \tag{5}$$

where:

- - $k_{f,b}$ is the stress concentration factor for the bolt holes;
- - $k_{f,f}$ is the stress concentration factor for the flange;

4.2. Analysis of an undamaged part

An repeated load cycle with an amplitude of 1 was defined for the load. The resulting (computed) values of the applied load are presented in Figure 6.

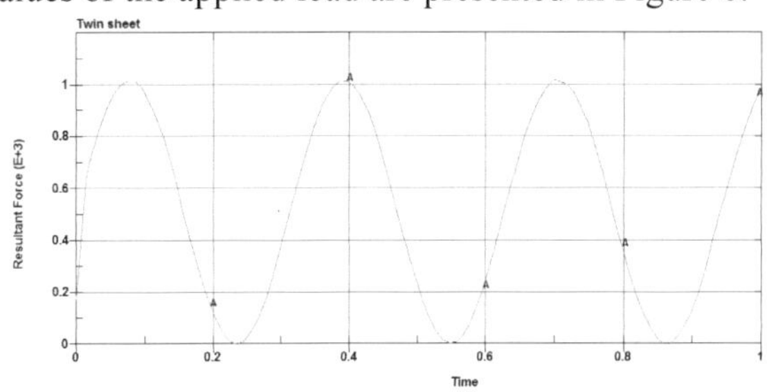

Figure 6. Resultant applied load (as computed)

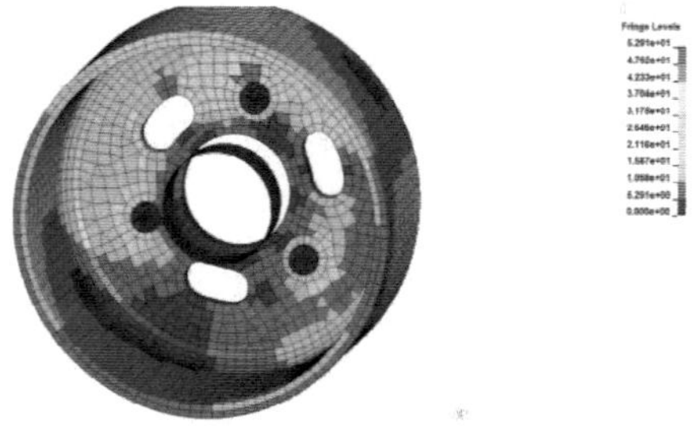

Figure 7. Stress state

A detailed view of the recorded stress of the area near the bolts is presented in Figure 8

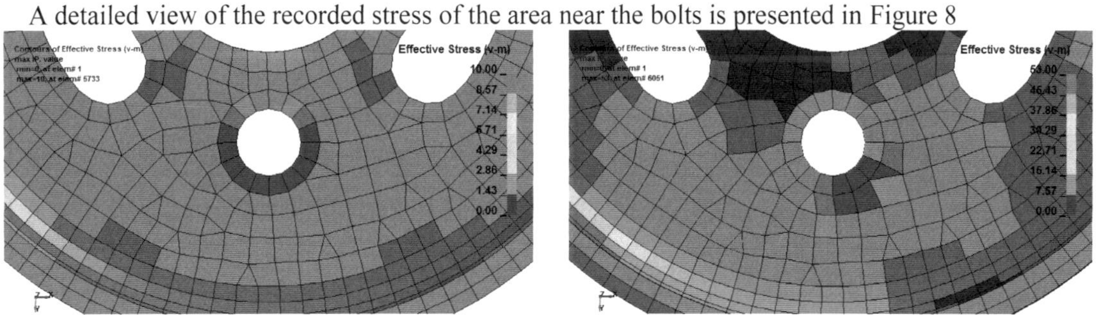

initial stress *peak stress*

Figure 8. Stress history at the bolt location.

A detailed view of the recorded stress on the flange of the pulley is presented in Figure 9.

IOP Conf. Series: Materials Science and Engineering **252** (2017) 012029 doi:10.1088/1757-899X/252/1/012029

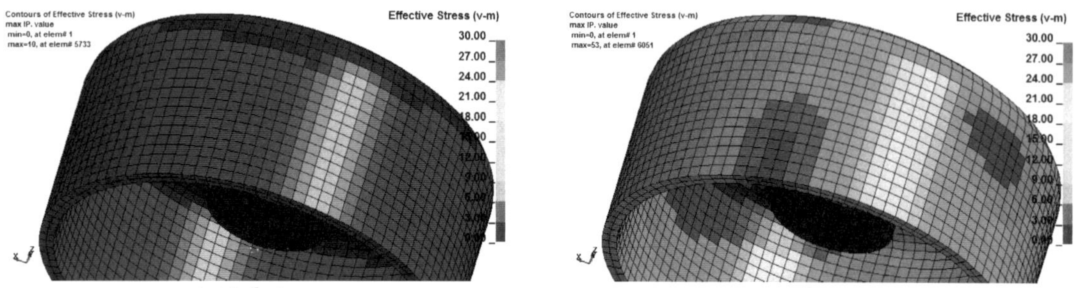

initial stress *peak stress*

Figure 9. Stress history on the flange.

Using the data plotted in Figure 7 and Figure 8 the following input parameters can be evaluated:

- the maximum stress: σ_{max} (**53 MPa** recorded **near the bolt hole** during loading / **27 MPa** recorded **on the flange** during loading).
- the minimum stress: σ_{min} (**7 MPa** recorded **near the bolt hole** during inertia relief / **3.5 MPa** recorded **on the flange** during inertia relief).

Therefore:

- the mean stress [4] (table 3):

$$\sigma_m = \frac{\sigma_{max} + \sigma_{min}}{2} \tag{6}$$

- the stress amplitude [4] (table 3):

$$\sigma_a = \frac{\sigma_{max} - \sigma_{min}}{2} \tag{7}$$

- yield stress [5]: $S_{yt} = 210\ MPa$ (material **DC 04** DIN EN 10130:2006)
- ultimate tensile stress [5]: $S_{ut} = 350\ MPa$ (material: **DC 04** DIN EN 10130:2006)
- surface finish factor [6, 4]: $k_a = 4.51 \cdot (350)^{-0.265} = 0.995$;
- size factor [6, 4]: $k_b = 1.0$;
- reliability factor [4] : $k_c = 0.659$;
- stress concentration factor [4]: k_d;
- endurance limit stress [4]

$$S_e' = 0.5 \cdot 350 = 175\ MPa \tag{8}$$

- endurance limit stress [4]

$$S_e = k_a \cdot k_b \cdot k_c \cdot k_d \cdot S_e' \tag{9}$$

The S-N curve (Figure 10) will be used to determine the number of working cycles before failure.

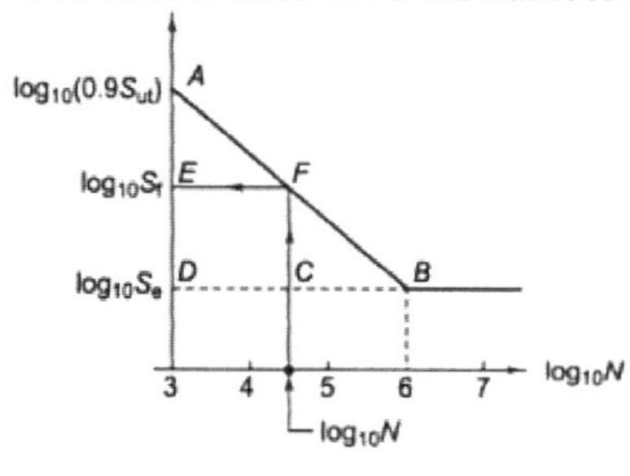

Figure 10. S-N curve [4]

The coordinates of failure point F can be determined using the equation of the $\overline{AB}$ line and the stress amplitude σ_a and the mean stress σ_m.

IOP Conf. Series: Materials Science and Engineering **252** (2017) 012029 doi:10.1088/1757-899X/252/1/012029

$$S_f = \frac{\sigma_a \cdot (0.9 \cdot S_{ut})}{(0.9 \cdot S_{ut}) - \sigma_m} \tag{10}$$

The abscissa points can be determined:

- $p_1 = log_{10}(0.9 \cdot S_{ut})$;
- $p_2 = log_{10}(S_e)$;
- $p_3 = log_{10}(S_f)$;

Table 1 presents the results of this study.

Table 1. Results

Parameter	Bolt location	Flange location
S_{yt}	210.000	210.000
S_{ut}	350.000	350.000
σ_{max}	**53**	**27**
σ_{min}	**7**	**3.5**
σ_m	30	15.25
σ_a	23	11.75
k_f	**1.550**	**1.450**
k_d	0.645	0.690
S'_e	175.000	175.000
S_e	60.397	64.562
S_f	25.421	12.348
$p_1 = log_{10}(0.9 \cdot S_{ut})$	2.4983	2.4983
$p_2 = log_{10}(S_e)$	1.7810	1.8100
$p_3 = log_{10}(S_f)$	1s.4051	1.0916
n_f	7.5717	9.1310
$N_f = 10^{n_f}$	**3.73E+07**	**1.35E+09**

Fatigue diagrams are presented in Figure 11 and Figure 12.

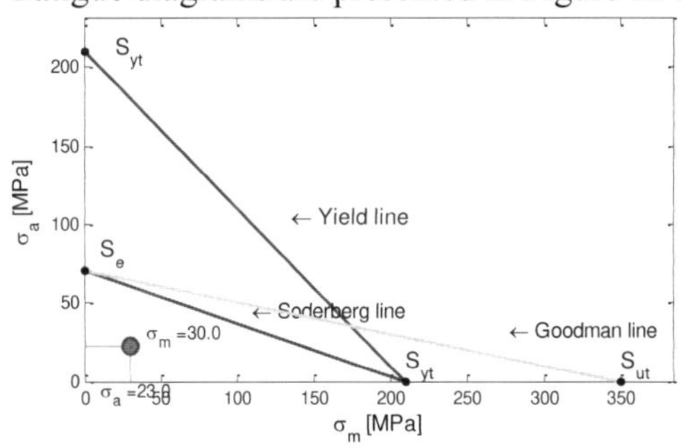

Figure 11. Fatigue diagram for bolt location

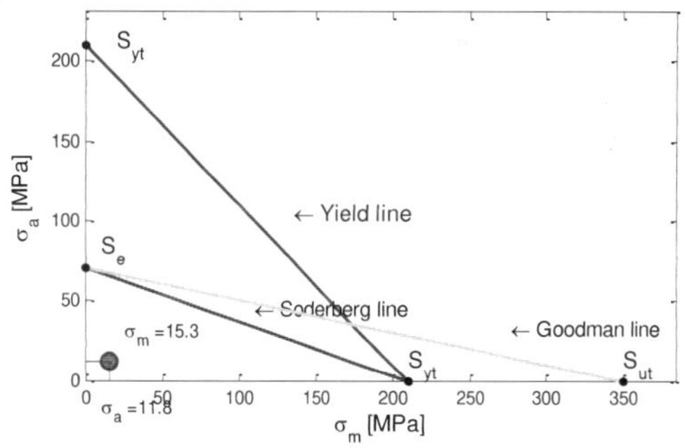

Figure 12. Fatigue diagram for the flange

4.3. Analysis of a damaged part

A simple damage was defined on the flange of the part. The damage simulates a crack with a length of 20 mm. In order to produce the damage the elements were disconnected. There were not used any specialized card to define the crack.
Figure 12 presents both the defect and the stress state at peak load on the flange of the part.

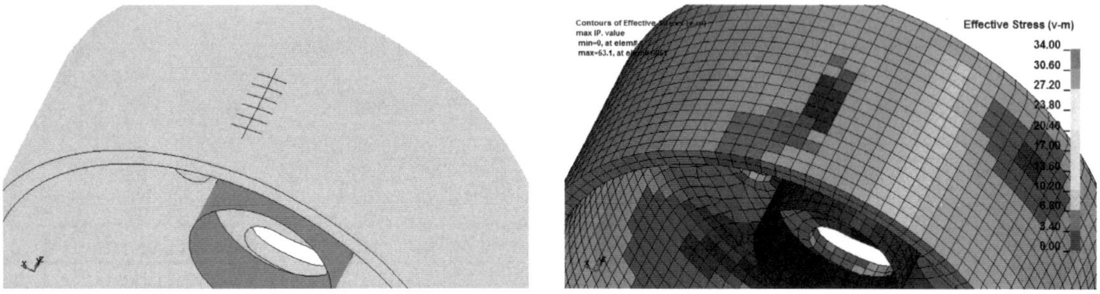

Defect *stress at peak*

Figure 13. Part with discontinuities

It can be noticed that there is a discontinuity in the stress on the area with the defect. Considering the above mentioned elements the life time can be estimated by the values listed in Table 2.

Table 2. Results

Parameter	Flange location
S_{yt}	210.000
S_{ut}	350.000
σ_{max}	**35**
σ_{min}	**3.5**
σ_m	19.25
σ_a	15.75
k_f	**1.450**
k_d	0.690
S_e'	175.000
S_e	64.562
S_f	16.775
$p_1 = log_{10}(0.9 \cdot S_{ut})$	2.4983
$p_2 = log_{10}(S_e)$	1.8100
$p_3 = log_{10}(S_f)$	1.2247
n_f	8.551
$N_f = 10^{n_f}$	**3.56E+08**

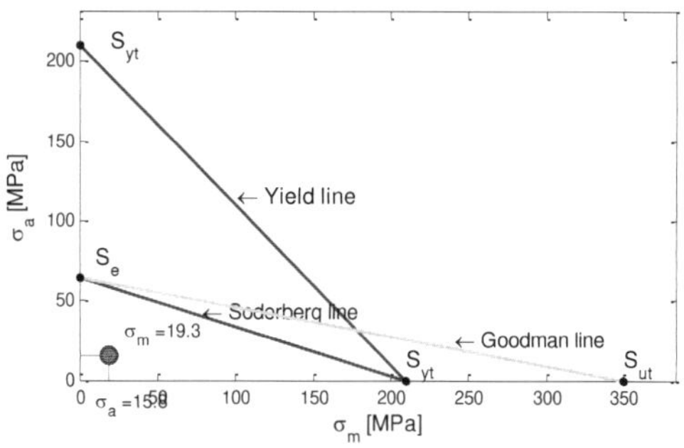

Figure 14. Fatigue diagram for the damaged flange

5. Conclusions

The paper presents a simple methodology useful to investigate the lifetime. Numerical analysis was used to obtain input data for the mathematical model. The analysis show that considering the requirements the part complies with the prescribed working conditions.

For the selected material **DC 04** the estimated number of working cycles is 3.73E+07 above the required limit of 2.1e7. There were analyzed points located on the top of the part (bolt fixation) and points on the flange (circumference).

A second model uses a simple method to define a defect of the part. The defect is of the type of a crack and is located on the flange. The numerical analysis showed, as expected, that there is discontinuity in the stress. Considering the previously used model the life estimation analysis was also performed.

Results show that this type of defect does not affect the functionality of the part. This was some additional rejection criteria can be defined that can account for the existence of defects on the part.

References

[1] Hee-Jin S and Jung-Kyu K 2008 *Cause of failure and optimization of a V-belt pulley considering fatigue life uncertainty in automotive applications,* Engineering Failure Analysis **16**: 1955–1963;
[2] Hee-Jin S and Jung-Kyu K 2006 *Consideration of fatigue life optimization of pulley in power-steering system,* Materials Science and Engineering A 483–484, 452–455;
[3] Amijima S, Fujii T, Kiji T, Tani K and Inukai M 1986 *Study on the V-ribbed belt.* Bull JSME; **29**:2317–22;
[4] V B Bhandari 2010 *Design of Machine Elements,* Tata McGraw-Hill Education, Third Edition;
[5] DIN EN 10130:2006 (E), *Cold rolled low carbon steel flat produts for cold forming – Technical delivery conditions;*
[6] Steven R S, Bernard J H and Bo O J 2014 *Fundamentals of Machine Elements, Third Edition: SI Version;*
[7] J A Martins, I Kövesdy and I Ferreira 2009 *Fracture analysis of collapsed heavy-duty pulley in a long-distance continuous conveyors application,* Engineering Failure Analysis **16**:2274-2280;
[8] Hee-Jin S and Jung-Kyu K 2009 *Cause of failure and optimization of a V-belt pulley considering fatigue life uncertainty in automotive applications,* Engineering Failure Analysis; **16**:1955–63;
[9] Yilmaz D, Çelik HK and Akinci I. 2009 *Finite element analysis of a failure in rear-mounted mower pulley,* J Food Agric Environ **7**:865–8;
[10] SS-EN 13306:2011 *Maintenance Terminology*, Swedish Standards Institute.

Vibration study of a vehicle suspension assembly with the finite element method

[1]Marinescu, Gabriel Cătălin*, [2]Castravete, Ştefan-Cristian, [1]Dumitru, Nicolae

[1]University of Craiova, Faculty of Mechanics
[2]Caelynx Europe

gabrielcatalinmarinescu@gmail.com

Abstract. The main steps of the present work represent a methodology of analysing various vibration effects over suspension mechanical parts of a vehicle. A McPherson type suspension from an existing vehicle was created using CAD software. Using the CAD model as input, a finite element model of the suspension assembly was developed. Abaqus finite element analysis software was used to pre-process, solve, and post-process the results. Geometric nonlinearities are included in the model. Severe sources of nonlinearities such us friction and contact are also included in the model. The McPherson spring is modelled as linear spring. The analysis include several steps: preload, modal analysis, the reduction of the model to 200 generalized coordinates, a deterministic external excitation, a random excitation that comes from different types of roads. The vibration data used as an input for the simulation were previously obtained by experimental means. Mathematical expressions used for the simulation were also presented in the paper.

1. Introduction

The suspension system of a vehicle is highly important for the comfort of the vehicle passengers. The way the suspension reacts to the road shape is reflected into the vibrations that can be felt inside a car.

The present paper is structured in eight major parts and approaches vibration aspects regarding the suspension from both theoretical and simulation point of view. There were considered two types of vibration sources that act upon the spring and damper assembly, first inducing vertical vibrations from the road through the wheel, and then vibration caused by the engine. Several other researches were done on vibrations acting on the suspension approaching various aspects. For example, in paper [1] there was analyzed and then performed an optimization for a suspension system generated with ADAMS CAR software. Paper [3] presents a mathematical model for the noise and vibration coming from suspension systems. The paper introduces the basic concepts of vibration theory and importance to vehicle design. It considers the role of the designer in vibration control and demonstrates methods for the control of vibration to help reduction of noise and harshness. Another vibration approach is paper [4], that presents the noise and vibration transmitted to the interior of the car. Paper [5] is a technical report that documents the results of extensive vehicle-based measurements on Sports Utility Vehicle, for the purpose of vibration source characterization. A sophisticated sensor and data acquisition system was fitted to a vehicle, supported by simultaneous video and CAN data recording. The paper presents a more experimental approach of road-vehicle vibration study.

Article [6] provides a discussion on the studies comprising active experiments conducted on selected structural elements of vehicles. The vibration excitation was achieved by applying an engine working on idle gear. The studies were conducted on vibration propagation for different constant rotational velocity. The changes of the signals were observed in time, frequency and simultaneously in time frequency domains. In [7] there are presented outcomes of a research project concerned with the development of a method for synthesizing, under controlled conditions in the laboratory.

The present paper is based on aspects that can be found in the papers mentioned before and studies the behavior of a city vehicle suspension assembly under two different external excitations: engine and road. The analysis includes several steps:

a) Preload. A static analysis where all the external forces are applied to the model which simulate the vehicle at its initial position. About that position, the model will oscillate.
b) Modal analysis. The model was linearized from the preloaded position and 200 frequencies were extracted. The model was reduced to 200 generalized coordinates.
c) A deterministic external excitation that is transmitted by the engine was applied. The model was tested for different engine rotational speeds.
d) A random excitation that comes from the wheel interaction with the road was applied. The model was tested for different types of roads.

Note that step c) and d) are perturbation steps and are independent on each other. Each are restarts from step b) modal analysis. The software used in this analysis is Dassault Simulia Abaqus 2017.

2. Theoretical background

A general finite element model is build which include geometric and boundary non-linearities (contact and friction). This model is used to run first step which include bolt preload and gravity. After this general step, a mode base dynamic system has been modelled. The model use mode shapes from a preceding modal analysis and is used in our case to perform sinusoidal and random forced vibration analyses. The analysis is linear and the response is studied in frequency domain. Few theoretical aspects about modal dynamics are showed below.

The finite element dynamic linear model has the following equilibrium equation [9, 11]:

$$\delta u^N (M^{NM}\ddot{u}^M + C^{NM}\dot{u}^M + K^{NM}u^M) = \delta u^N F^N \tag{1}$$

where M^{NM} is the mass matrix, C^{NM} is the damping matrix, K^{NM} is the stiffness matrix, F^N are the external loads, u^N is the value of degree of freedom N of the finite element model (displacement in our case), and δu^N is an arbitrary virtual variation.

For eigenvalue extraction, the following undamped system is solved:

$$\left(M^{NM}\omega_{\overline{\alpha}}^2 - K^{NM}\right)\phi_{\overline{\alpha}}^N = 0 \tag{2}$$

Note that the repeated subscripts and superscripts are assumed to be summed over the appropriate range. The modes are orthogonal across mass and stiffness matrices. The eigenmodes are normalized so the largest value of $\phi_{\overline{\alpha}}^N$ is 1.

Using eigenmodes from (2), the equation (1) projects into a set of uncoupled modal response equations,

$$m_{\overline{\alpha}}\ddot{q}_{\overline{\alpha}} + c_{\overline{\alpha}}\dot{q}_{\overline{\alpha}} + k_{\overline{\alpha}}q_{\overline{\alpha}} = f_{\overline{\alpha}} \tag{3}$$

where $f_{\overline{\alpha}} = \phi_{\overline{\alpha}}^N F^N$ is the generalized load for mode α and $q_{\overline{\alpha}}$ is the "generalized coordinate" (the modal amplitude) for mode α.

Steady-state excitation is of the form

$$f_{\overline{\alpha}} = Ae^{i2\pi ft} \tag{4}$$

and creates response of similar form that we write as

$$q_{\overline{\alpha}} = HAe^{i2\pi ft} \tag{5}$$

where $H(f)$ is the *complex frequency response function.*

In the case of sinusoidal forced vibration, excitation is given by equation (5). For random vibration, the excitation is characterized by cross-spectral density matrix $S^F_{NM}(f)$ which links all degree of freedom N and M. Projecting the matrix onto the modes, the cross-spectral density matrix for generalized coordinates is obtained:

$$S^f_{\overline{\alpha}\overline{\beta}}(f) = \phi^N_{\overline{\alpha}} S^F_{NM}(f) \phi^M_{\overline{\beta}} \tag{6}$$

The complex frequency response function then defines the response of the generalized coordinates as

$$S^q_{\overline{\alpha}\overline{\beta}}(f) = H_{\overline{\alpha}} S^f_{\overline{\alpha}\overline{\beta}}(f) H^*_{\overline{\beta}} \tag{7}$$

where $H^*_{\overline{\alpha}}$ is the complex conjugate of $H_{\overline{\alpha}}$.

Finally, the response of the physical variables is recovered from the modal responses as

$$S^u_{NM}(f) = \phi^N_{\overline{\alpha}} S^q_{\overline{\alpha}\overline{\beta}}(f) \phi^M_{\overline{\beta}} \tag{8}$$

so that the power spectral density of degree of freedom u^N is

$$S^u_{NN}(f) = \phi^N_{\overline{\alpha}} S^q_{\overline{\alpha}\overline{\beta}}(f) \phi^N_{\overline{\beta}} \tag{9}$$

3. Road profile

Road can be modelled as random excitation. Since we are going to run the road excitation analysis in spectral (frequency) domain, we need Power Spectral Density (PSD) of the road profile function of frequency. The road roughness PSD distribution can be approximated by [12]:

$$\Phi(\omega) = \frac{2\alpha V\sigma^2}{\omega^2+\alpha^2 V^2} \tag{10}$$

Where σ^2 is the road roughness variance, V the vehicle speed, α depends on the type of road surface, and ω is frequency in rad/s. Variance and α are adopted from Table 1 in function of road class.

Table 1. Road roughness standard deviation

Road class	$\sigma(10^{-3}m)$	$\Phi(\Omega_0)\ (10^{-6}m^3), \Omega_0 = 1$	α(rad/m)
A (very good)	2	1	0.127
B (good)	4	4	0.127
C (average)	8	16	0.127
D (poor)	16	64	0.127
E (very poor)	32	256	0.127

For a vehicle speed of 17.7 km/h the excitation PSD displacement is obtained and shown in Figure 1*a)*. Since the input in Abaqus require acceleration PSD for a random process, we obtained that by double derivation of displacement. The obtained curves are shown in Figure 1*b)* and are used in the road excitation analysis from paragraph 5.4.

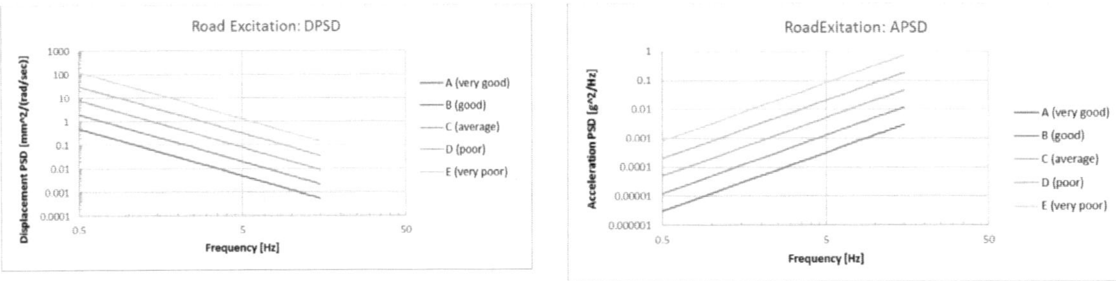

Figure 1. Road excitation
a) displacement PSD; *b)* aceleration PSD

4. Finite element model

The strut suspension main assembly was 3D modelled in order to use its geometry for the finite element analysis. Its most important elements can be seen in Figure 2. The representation in the left is the left side assembly of the front suspension seen from behind. The other is the side view of the same assembly.

The two components of the model are the suspension damper and the knuckle. The suspension damper was modelled by spring – dashpot connector with a stiffness of 25 N/mm and damping coefficient 0.0959. On its top mount it acts a mass of 328 kg that represents the static vertical load on one side of the front suspension in the case of a car with two passengers, each of them weighing 75 kg.

Both the damper support and the knuckle were modelled for the finite element analysis using tetrahedral elements [10]. The link between these two parts is done by two M12 bolt – nut assemblies. These were modelled by using hexahedral elements. The bolt axial preload is 40kN.

The tire that links the vehicle to the road surface was modelled by spring – dashpot connector elements with stiffness of 200 N/mm and damping coefficient 0.001562.

Table 2. Material models.

Component	Material	Density	Elasticity Modulus	Yield
Suspension damper support	Hard Steel	7.8 g/cmc	210000 MPa	1300 MPa
Knuckle	Forged Steel	7.8 g/cmc	210000 MPa	420 MPa

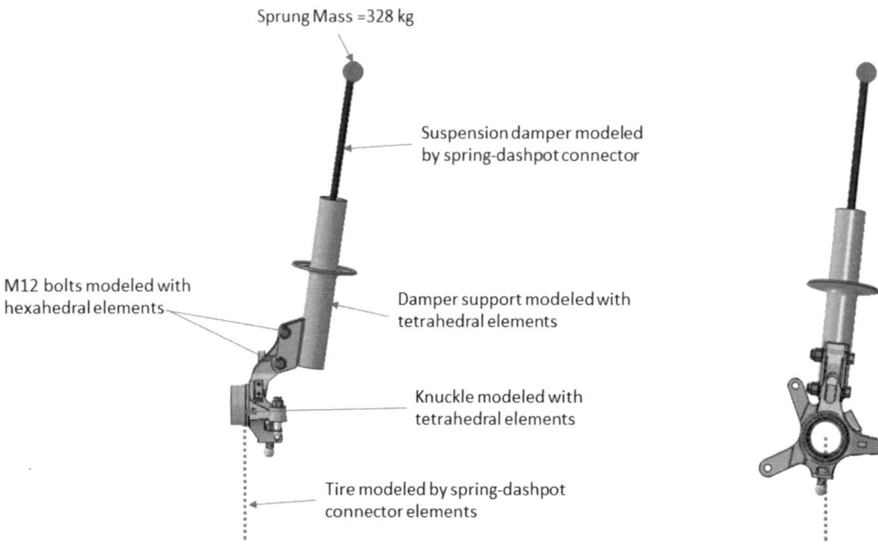

Figure 2. Quarter car suspension Finite Element Model.

5. Analysis

5.1. Preload

The preload step is a general nonlinear analysis. This step includes preload of the M12 bolts and gravity load. The model includes geometric and boundary nonlinearities. Geometric nonlinearities are modeled using Large Displacement Finite Element theory and boundary nonlinearities are due to contact and friction.

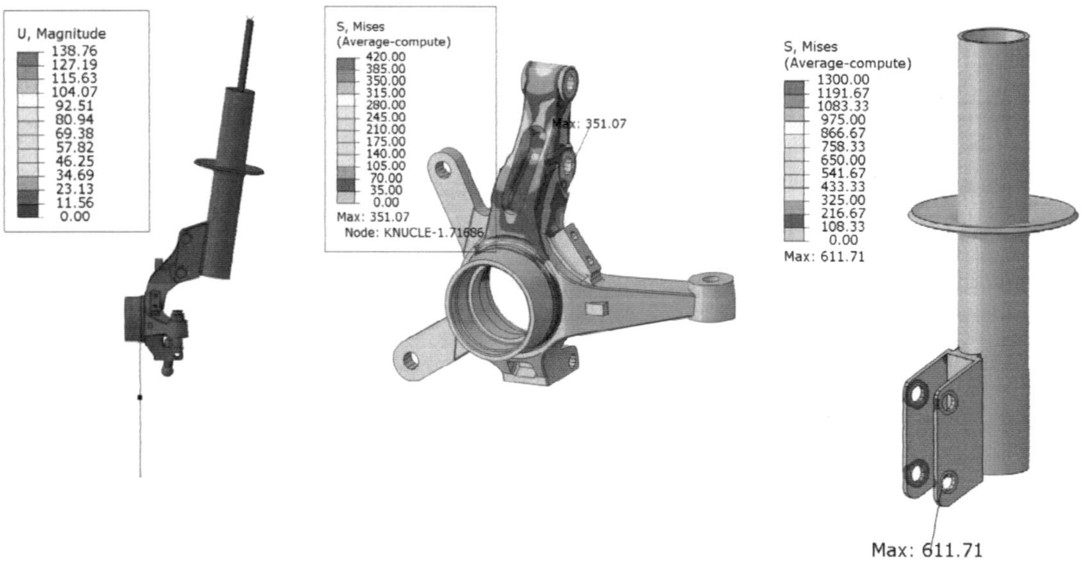

Figure 3. Preload.
a) displacement contour plot; *b)* VonMises contour plot: knuckle;
c) VonMises contour plot: spring-damper support

Figure 2 shows the displacement and stress contour plots after the preload step. Bolts are tightened at 40kN axial load. The stress obtained in this step is mainly compressive and is due to bolt preload.

The next step in the analysis is modal analysis. Modal analysis is a linear perturbation step. When considering nonlinear geometry, Abaqus starts the perturbation step from the previous general step. In our case the modal analysis starts from the displacement state in Figure 3*a)* taking into consideration the preload stress (Figures 3 *b)* and 3*c)*).

5.2. Modal analysis

A modal analysis is run first up to 200 modes. Table 3 shows first 7 modes.

Table 3. Mode Shapes.

Mode No.	Frequency [Hz]	Frequency [rpm]	*Observation*
1	1.3	78	Spring-damper axial motion
2	3.25	195	Tire lateral motion
3	3.3	198	Tire fore-aft motion
4	15	900	Tire vertical motion
5	27	1620	Twist around fore-aft axis
6	28	1680	Twist around lateral axis
7	521	31260	High frequency mode

For this analysis, first 6 modes are of interest since both road and engine excitation are within this frequency range. Above 7th mode, the frequency is too high for our analysis purpose. However, we extracted 200 modes to reduce the model in the subsequent modal dynamic analysis steps to 200 generalized coordinates.

The following analyses are:

- Engine excitation having the frequency range: 600-6000 rpm. Modes 4, 5 and 6 are in the excitation range of the engine
- Road excitation having the frequency range: 0.5-15 Hz. Modes 1-4 are in the road excitation range

5.3. Engine excitation analysis

This step is a modal dynamic steady state analysis and uses the modes from the frequency step (modal analysis) to build the reduced model. The analysis is done in frequency domain.

A constant acceleration excitation of 4g is considered in all three directions, spanned over a range of frequencies (angular velocity of the engine). We simulate an engine malfunction which generates sinusoidal excitations transmitted through vehicle body to the suspension.

Figure 4*a)* shows the direction of the excitation and boundary condition of the quarter-car suspension. Figure 4*b)* shows the excitation spanned over the engine angular velocity rage: 600-6000 rpm.

Figures 5*a)* and 5*b)* show the acceleration response at wheel center and spring base. The main resonant frequencies are 1620 and 1680 rpm which correspond to modes 5 and 6 of the suspension system. The resonant frequencies are caused by lateral and fore-aft excitation. It is observed a maximum acceleration of 15 g acceleration at wheel center and 10 g at spring base. There is also observed a smaller peek at 900 rpm (4th mode) in the case of vertical excitation. Note that mode 5 (1620 rpm) is excited by lateral excitation and mode 6 (1680 rpm) by fore-aft excitation.

IOP Conf. Series: Materials Science and Engineering **252** (2017) 012030 doi:10.1088/1757-899X/252/1/012030

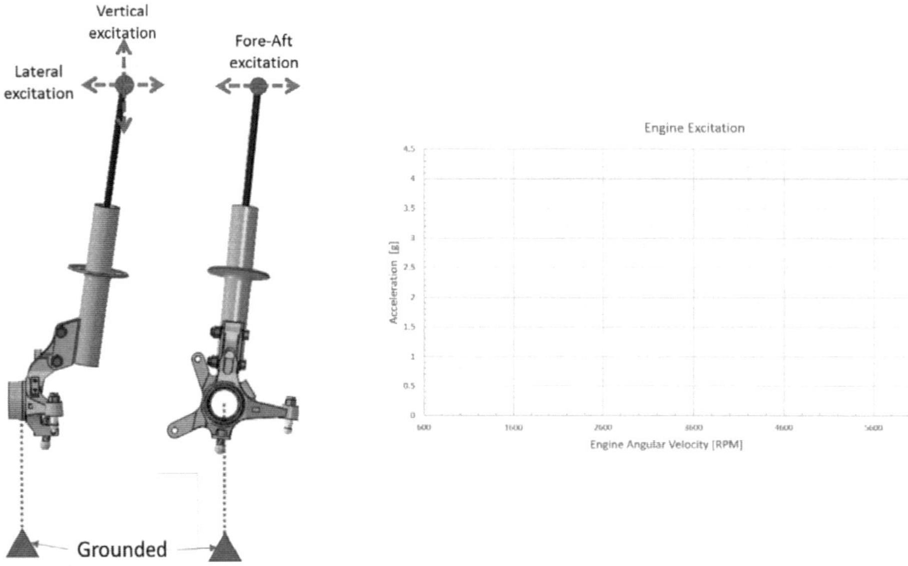

a) *b)*

Figure 4. Excitation.
a) Excitation and Boundary Conditions; *b)* Engine Excitation vs. angular velocity.

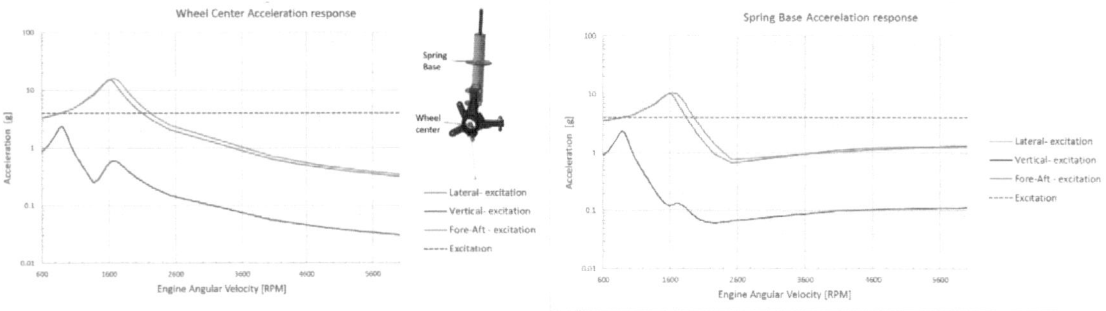

a) *b)*

Figure 5. Acceleration response.
a) Wheel center; *b)* Spring base

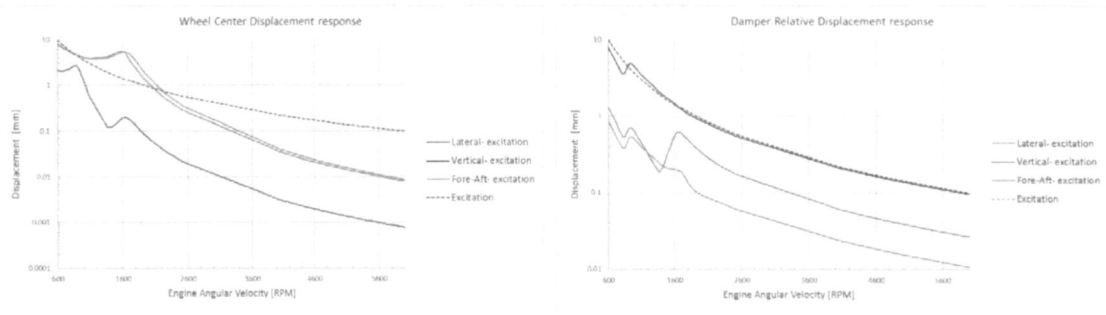

a) *b)*

Figure 6. Displacement Response.
a) Wheel center; *b)* Spring base

Figures 6*a)* and 6*b)* show the displacement response at wheel the center and at spring base. Still resonant frequencies at modes 5 and 6 for lateral and fore-aft excitation and mode 4 for vertical excitation.

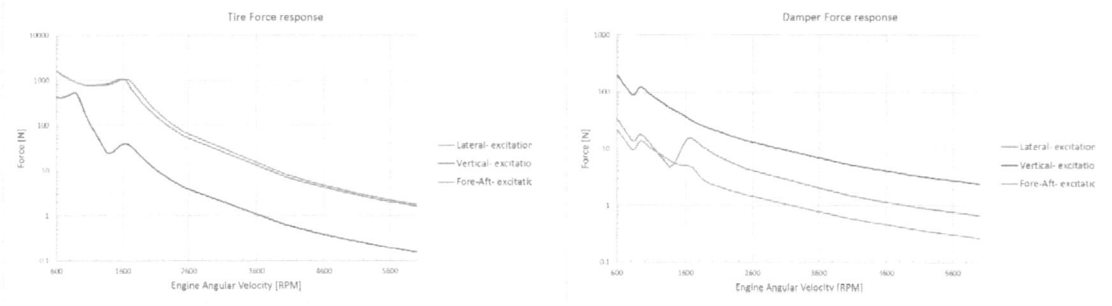

a) *b)*

Figure 7. Force response.
a) Tire spring-damper; *b)* Suspension Spring-Damper

Figures 7*a)* and 7*b)* show the forces at tire and suspension spring-damper. A maximum force of 1000 N is obtained at tire.

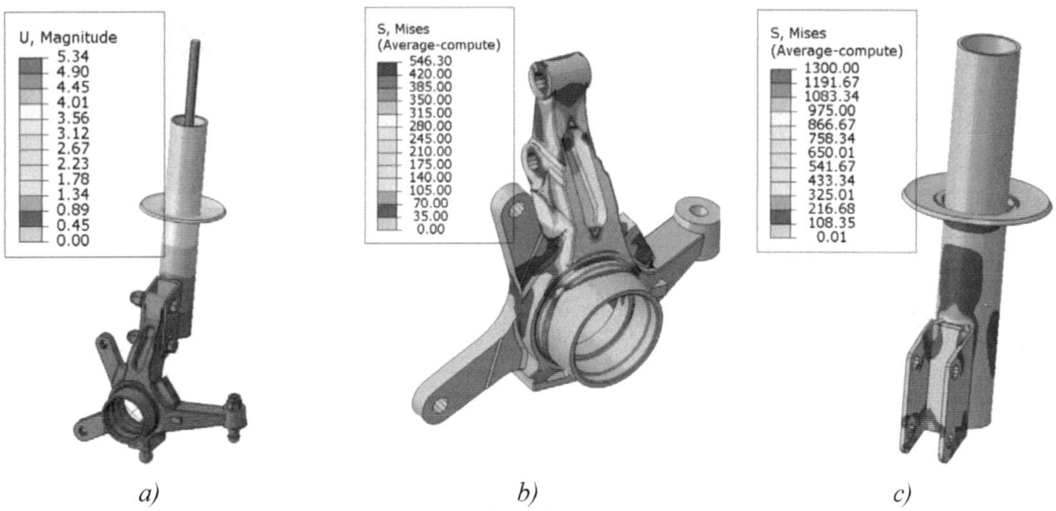

a) *b)* *c)*

Figure 8. Lateral Excitation at 1620 rpm
a) displacement contour plot; *b)* VonMises contour plot: knuckle;
c) VonMises contour plot: spring-damper support

Figures 8 show the displacement and Von Mises stress contour for Lateral Excitation at 1620 rpm. There can be observed the values of stress over yield in the knuckle. However, the values are localized in few spots and may not create major damage if the engine malfunction is sporadic. If the malfunction is persistent and the engine angular velocity is kept constant at 1620 rpm, permanent deformation or even crack might appear in the knuckle. Similar stress values are obtained in the fore-aft excitation.

5.4. Road excitation analysis

This step is a modal dynamic steady state analysis and uses the modes from the frequency step (modal analysis) to build the reduced model. The analysis is done in frequency domain. In this step, we consider a vehicle which travels at constant speed of 17.7 km/h on flat road.

IOP Conf. Series: Materials Science and Engineering **252** (2017) 012030 doi:10.1088/1757-899X/252/1/012030

Five types of roads are considered: very good, good, average, poor, very poor. The road is modeled as random excitation and it is described in paragraph 3.

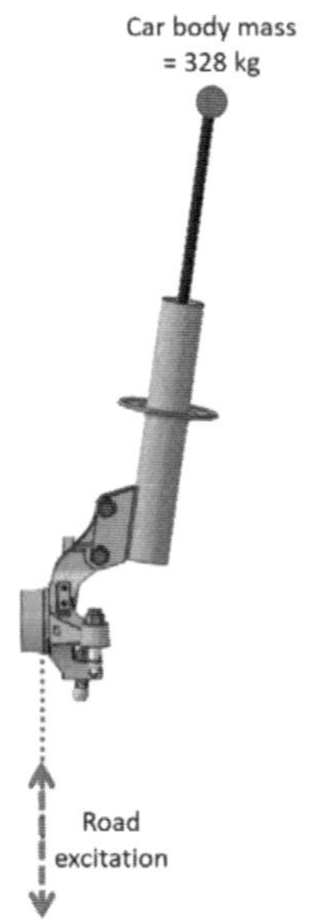

Figure 9. Road excitation and Boundary Conditions

Model showing the road excitation and boundary conditions is in Figure 9.

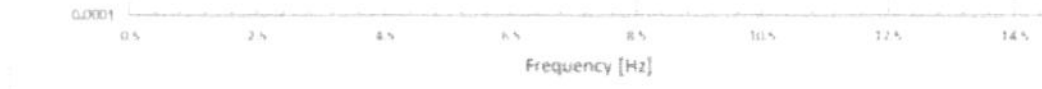

Figure 10. Acceleration response: Car body (sprung) mass

Figure 11. Displacement response: Car body (sprung) mass

We monitor the mass point that represents the car body mass called sprung mass. Figures 10 and 11 show the acceleration and displacement response at sprung mass. The suspension resonates at two frequencies 1.3 and 3.25 Hz, modes 1 and 2 respectively. Mode 1 is the spring-damper first mode and mode 2 is tire mode. Maximum acceleration is obtained at 3.25 Hz and its control depends only on the tire damping, any modification in the spring-damper system will not affect the acceleration response at this frequency. Maximum acceleration is 0.9g for the very poor road and 0.05g for the very good road. Regarding the displacement, it's maximum is found at 1.3 Hz and that can be controlled by varying the spring-damper parameters. Maximum displacement is 51mm for the very poor road and 3mm for the very good road.

IOP Conf. Series: Materials Science and Engineering **252** (2017) 012030 doi:10.1088/1757-899X/252/1/012030

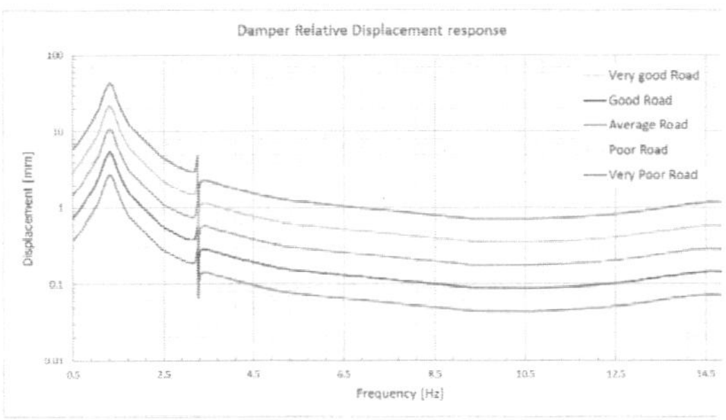

Figure 12. Relative displacement response: Damper

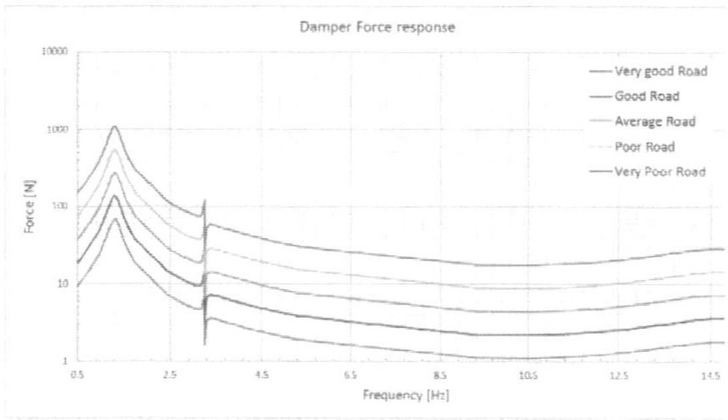

Figure 13. Force response: Damper

Figures 12 and 13 show the relative displacement and force in the damper. As expected the highest values are at damper first frequency of 1.3 Hz. The maximum relative displacement is 43 mm for the very poor road and 3 mm for the very good road. The maximum force is 1090N for the very poor road and 68N for the very good road. There can be observed large differences in values between very good roads and very poor roads. As expected, the comfort increases significantly with increase of road quality.

Next, we are going to show the effect of the road quality on the knuckle stress. Since we have a random excitation, the stress contour plot is in Root Mean Square (RMP) values and it is presented in the stress components: 11 (xx), 22, (yy) and 33 (zz). There is shown only normal stresses for the very poor road, average road and very good road for comparison purpose.

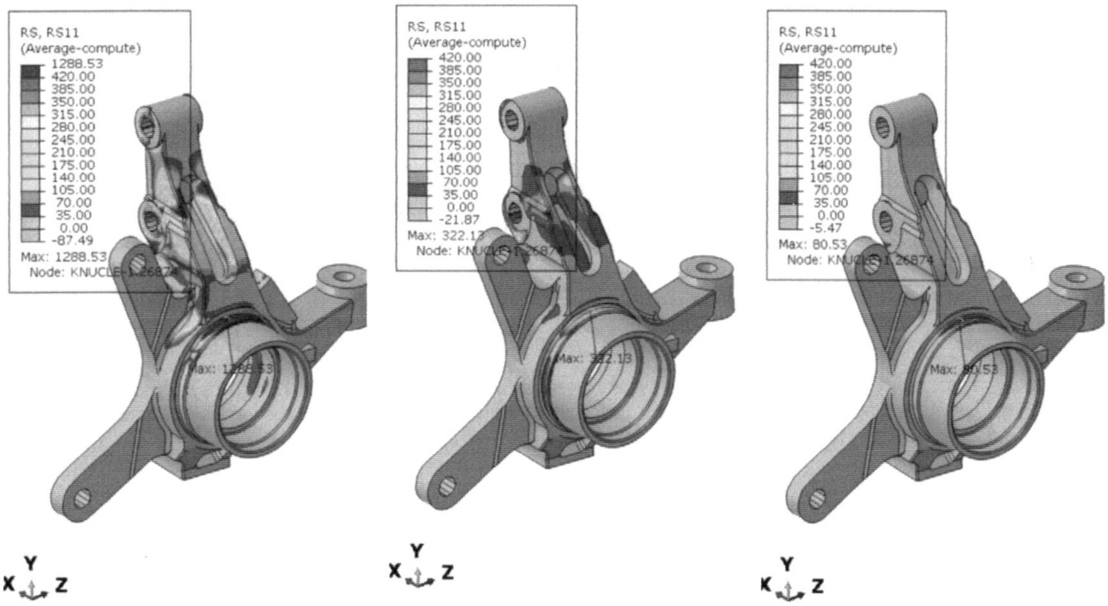

a) *b)* *c)*

Figure 14. RMS Stress Sxx.
a) very poor road; *b)* average road; very good road

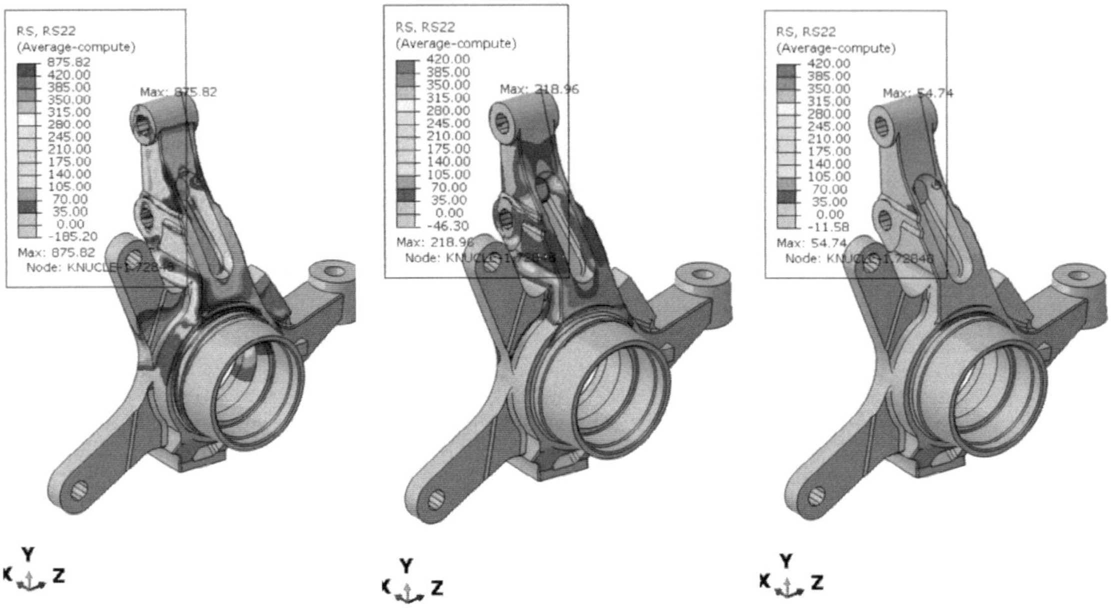

a) *b)* *c)*

Figure 15. RMS Stress Syy.
a) very poor road; *b)* average road; very good road

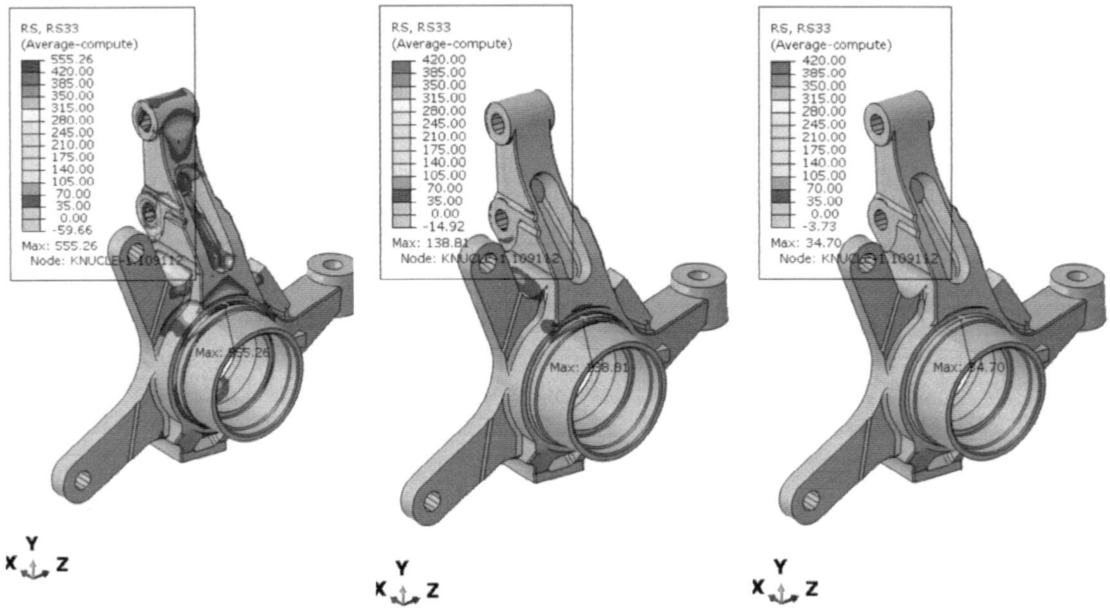

a) *b)* *c)*

Figure 16. RMS Stress Szz.
a) very poor road; *b)* average road; very good road

Figures 14-16 show knuckle RMS normal stresses on x, y and z directions for a very poor, average, and very good roads. As the vehicle travels with constant 17.7 km/h, we observe that on average and very good roads, the stress values are below yield. For very poor road the stress is above yield on

IOP Conf. Series: Materials Science and Engineering **252** (2017) 012030 doi:10.1088/1757-899X/252/1/012030

several spots and the knuckle can encounter permanent deformation or even crack. So even on low vehicle speed, the low-quality roads can cause serious damage on the suspension knuckle.

6. Conclusions

The results are part of a study regarding the effect of the random vibration generated by the road shape, combined with sinusoidal vibration generated by the engine. A goal was to compare the effect of engine vibration with the road vibration and to study the displacement, stress and strain for suspension system parts for various working conditions.

The authors performed several physical tests on the vehicle which the suspension is analyzed. The physical tests were used as inputs on the actual analysis. A realistic model which includes several types of nonlinearities was developed. The model was correlated with the actual tests performed on the vehicle. A nonlinear finite element model was developed and used it at the initial state for the following linear perturbations analyses. Both engine and road excitations were used in the analysis.

The paper shows the methodology of analyzing a suspension mechanism using Abaqus software. It starts from generating the virtual model and ends with interpreting results and effects of both random and sinusoidal vibration over stress, strain and displacements for various mechanical parts. Simulations were done with different input data for the two sources of excitation and a comparison between their effects over the passenger comfort was performed.

References

[1] Fernandes, E.S., Sheth S.S., Nawpute, K.K, Kadam V.V. and Mitra A.C., Optimization of Vehicle Suspension System to Improve Comfort, IOSR Journal of Mechanical and Civil Engineering (IOSR-JMCE) e-ISSN: 2278-1684, p-ISSN: 2320-334X ,www.iosrjournals.org 6th National Conference RDME 2017, March 2017.

[2] Longoria, R.G., Vehicle Vibration and Ride – 1 and 2, Vehicle System Dynamics and Control ME 360/390, Department of Mechanical Engineering, The University of Texas at Austin, Spring, 2016.

[3] http://web.iitd.ac.in/~achawla/public_html/736/14-NVH-V2.pdf

[4] Genuit, K., Vehicle Interior Noise – Combination of Sound, Vibration and Interactivity, HEAD acoustics GmbH, Herzogenrath, Germany, Sound & Vibration/December 2009, www.SandV.com

[5] Gordon T.J., Bareket Z., Vibration Transmission from Road Surface Features – Vehicle Measurement and Detection, Technical Report, The University of Michigan Transportation Research Institute (UMTRI), January 2007

[6] Burdzik, R., Doleček, R., Research of vibration distribution in vehicle constructive, Number 4,Volume VII, December 2012

[7] Rouillard, V., Statistical Models for Nonstationary and Non-Gaussian Road Vehicle Vibrations, Engineering Letters, 17:4, EL_17_4_03, Advance online publication, November 2009.

[8] Winberg M., Noise and Vibration Control of Combustion Engine Vehicles, Department of Signal Processing, Blekinge Institute of Technology, Ronneby, Sweden, April 2005

[9] Castravete, S.C., Bisu, C.F., Random and deterministic vibration of main spindle by finite element method, MVT 2012 4010.

[10] Marinescu G. C., Dumitru N., Geonea I., Dynamic Modelling and Simulation of an Auto Vehicle Steering Mechanism Considering its Elements as Flexible, Applied Mechanics and Materials Vol. 245, Trans Tech Publications, Switzerland, 2013.

[11] Abaqus 2017, Abaqus Theory Manual, Online Documentation, © Dassault Systèmes, 2017.

[12] Tyan, F., Hong, Yu-Fen, Tu, Shun-Hsu, Jeng, W. S., Generation of Random Road Profiles, ITRI Project: 5353C46000

The evaluation of the overall thermal comfort inside a vehicle

[1]Neacsu, Catalin*, [2]Tabacu, Ion, [2]Ivanescu, Mariana [2]Vieru, Ionel

1 S.C. Automobile Dacia SA
2 University of Pitesti, Romania

catalin.neacsu@renault.com

Abstract. The thermal comfort is one of the most important aspects of the modern vehicles that can influence the safety, the fuel consumption and the pollutions regulation. The objective of this paper is to compare the global and absolute thermal comfort indexes for two vehicles with different distribution air systems inside the car cockpit, one using only front air vents, and the other using both front and rear air vents. The methodology of calculus consists in using the 3D model of the interior vehicle, generally in a CAD format. Then, using a meshing software to create the finite element model of the interior surfaces inside the cockpit and the volume of internal air. Using the obtained finite element geometry, there will be conducted a Theseus FE calculus using the given boundary conditions. The results of the numerical simulation are presented in terms of graphs and figures and also PMV, PPD and DTS thermal comfort indexes. With the obtained results, we will then create the graphs that allows us to evaluate the global and absolute thermal comfort indexes. The results of the evaluation show us that the use of the method allow us to evaluate with a greater accuracy the thermal comfort for the whole vehicle, not only for each passenger, like the standard methods. This shows us that in terms of general and absolute thermal comfort, the vehicle that use front and rear systems is better than the version that use only a front system. The thermal comfort is an important aspect to be taken into account from the beginning of the design stage of a vehicle, by choosing the right air conditioning system. In addition, by using the numerical simulation, we are able to reduce the time needed for preliminary tests and be able to provide the vehicle to the market earlier, at a lower development cost.

1. Introduction

Due to increase demand of new vehicles [1], and taking into account the fact that thermal comfort is a must in the modern cars, all the manufacturers have to offer to their clients' cars equipped with efficient HVAC systems. This need for efficient HVAC systems has determined the R&D departments to develop new tools to predict the passenger's thermal comfort.

Tools that few years ago were used only for state of the art applications today represent the standard in the automotive industry. Therefore, the use of numerical simulation has reduced the time needed for the development of a new product or improving an existing one. The main advantages of these tools are represented by the speed, the reduced cost and one of the main aspects is the reliability of the results.

It is well known that one of the requirements to be fulfilled is that a person to be in thermal neutrality according to the comfort equation[9]. This is described and evaluated by the following indices: DTS (Dynamic Thermal Sensation), TS (Thermal Sensation), PMV (Predicted Mean Vote) and PPD (Predicted Percentage of Dissatisfied). The two first indices depend of the hypothalamus temperature and the mean skin temperature and PMV - PPD indices take into account the following six parameters: activity, clothing, air temperature, mean radiant temperature, air velocity and humidity [7].

In this paper, the authors will evaluate the overall thermal comfort inside a vehicle General Thermal Comfort Index (GTCI) and General Absolute Thermal Comfort Index (GATCI), presented in [2], [3]

2. Thermal comfort basic considerations

Human body's thermoregulatory system allows adjustment of physiological heat load and thermal comfort of the body in different conditions, in figure 1 being presented the processes that characterize the heat transfer between the human body and the environment. We can state that the three existent types of heat transfer affect the thermal comfort: conduction, convection and radiation.

There are six parameters that influence the thermal comfort and they can be split into environmental and personal factors [5],[7].

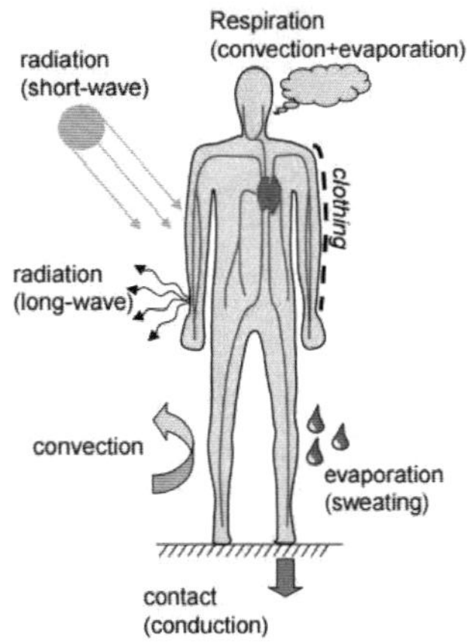

Fig. 1. Heat transfer between human body and the environment[4]

There is a variety of thermal comfort indexes, but the most used in the industry are PMV (Predicted Mean Vote), PPD (Predicted Percentage of Dissatisfied), DTS (Dynamic Thermal Sensation). These indexes are calculated by taking into account the parameters that influence the thermal comfort.

In [2] and [3], the authors present the development of the GTCI and GATCI indexes, that are used to evaluate the thermal comfort for the entire vehicle. So far, the thermal comfort was evaluated only for each occupant using in most of the cases the PMV, PPD, DTS and Equivalent temperature [8], [9], [10], [11]. The newly developed indexes are directly linked with the existing indexes PMV, PPD and DTS.

3. The use of the newly developed overall thermal comfort indexes

To use the newly developed thermal comfort indexes we have decided to realize a numerical simulation scenario, using a vehicle fully loaded with five occupants (four adults and one child), and this being something new in the field of the numerical simulation. In generally, the numerical simulation is conducted using only two occupants, generally placed on a vehicle diagonal (driver in front and right rear passenger).

We have chosen to use five manikins because, in the real life, the sun is not placed directly over the vehicle, but at a given angle.

The numerical simulation is done in various stages starting from the car CAD geometry we will create a shell finite element model that comprises the car interior components. The obtained shell mesh will be used, as it is to realize the Theseus-FE calculus. We will use the material proprieties for the interior vehicle parts.

The aim is to first calculate the PMV, PPD and DTS indexes for each occupant, and then, using the values obtained for the indexes to evaluate the overall thermal comfort using GTCI and GATCI thermal comfort indexes.

3.1. Model discretization

The first step of the numerical simulation is to mesh the CAD model of the interior of the vehicle, using Beta CAE ANSA software. With this program, we will add to each part of the interior an ID, used after

IOP Conf. Series: Materials Science and Engineering **252** (2017) 012031 doi:10.1088/1757-899X/252/1/012031

to assign the material properties in the Theseus FE software. In figure 2, can be seen the finite element model of the interior of the vehicle.

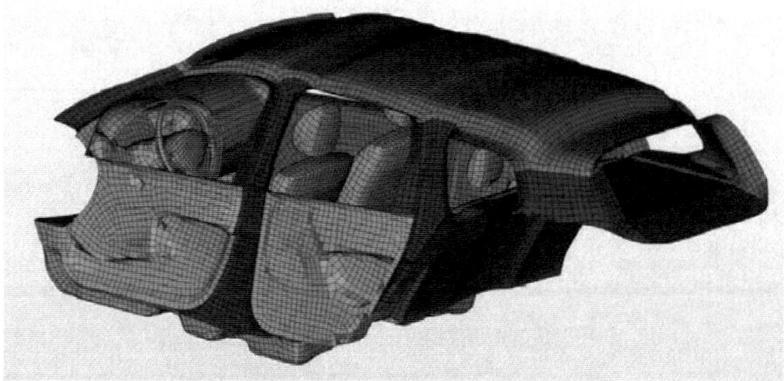

Fig. 2. Finite element model of the vehicle

3.2 Boundary conditions

With the discretized geometry, we will conduct the Theseus-FE simulation. Because we want to observe the overall thermal comfort indexes, we will realize two numerical simulations, using different boundary conditions.

The Theseus FE simulation, consist in two stages:

- Soak: the vehicle is placed under the environmental conditions without the AC on and without the manikins. Generally, this stage is about 3600-7200s.
- AC simulation: after the soak simulation, the manikins are placed inside the vehicle and the air conditioning system is turned on.

As exterior boundary conditions, we will use data from table 1. In Theseus FE, the azimuth angle represents the angle between the sun direction and the Ox axis of vehicle, as we can see in figure 3.

Table 1. Boundary conditions

Parameter	*Case 1*	*Case 2*
Solar radiation intensity [W/m^2]	850	1000
Diffuse solar intensity [W/m^2]	140	100
Sun azimuth angle [°]	135	315
Sun altitude angle [°]	63.5	78
Ground reflectance	0.2	0.25
Environmental temperature [°C]	30	40
Environment humidity [%]	40	10
Sky mean radiant temperature [°C]	10	12

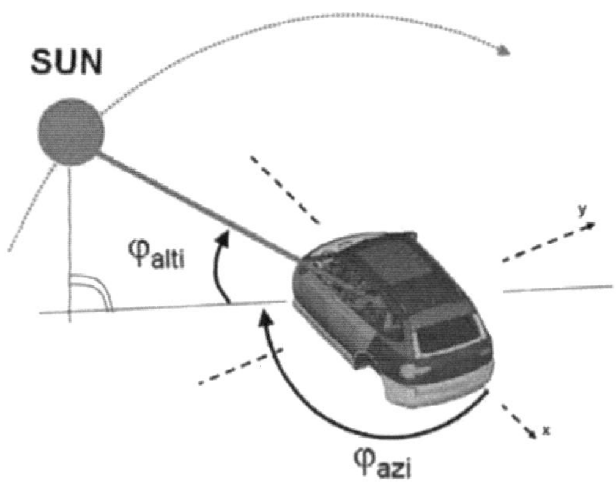

Fig. 3. Sun position in Theseus FE software

After the soak simulation, we will realize a simulation where we will start the HVAC system with five occupants inside the vehicle, as it can be seen in figure 4. The HVAC boundary conditions are presented in table 2.

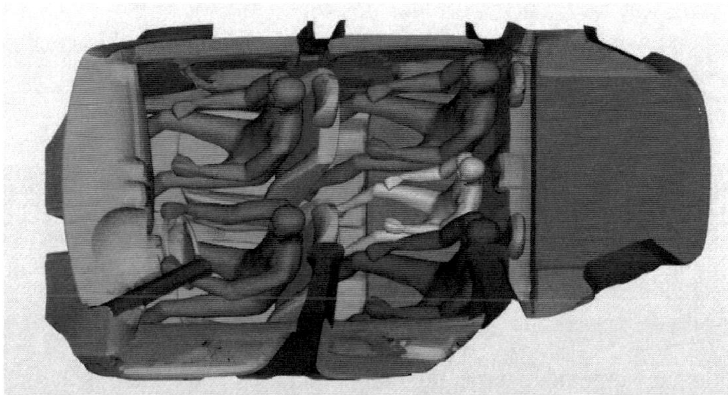

Fig. 4. The manikins placed inside the vehicle

Table 2. HVAC characteristics

Time [s]	Temperatue[°C]	Flow [m3/s]
3600	36	0,125
3750	20	0,125
3900	15	0,125
4150	7	0,125

3.3 Results

Using the boundary conditions presented earlier, the obtained results for the simulations are presented in table 3 and in figure 5.

Table 3. Average temperature inside the vehicle for the two simulations

Time (s)	0	600	1200	1800	2400	3000	3600	4200	4800	5400
Case 1 (°C)	30.0	37.1	41.8	44.9	47.3	49.0	50.4	20.1	18.0	17.3
Case 2 (°C)	40.0	47.5	52.6	56.0	58.6	60.5	62.0	23.4	20.9	20.0

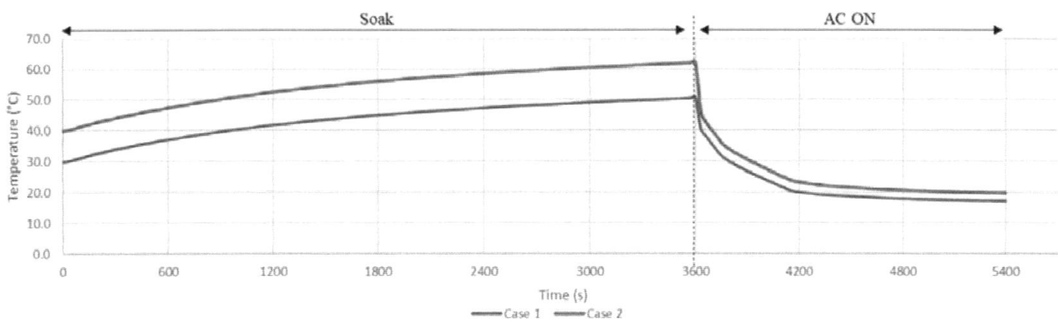

Fig. 5. The temperature inside the vehicle

As we can see from table 3 and figure 5, after the soak period, the temperature inside the car is 50,4°C for Case 1 and 62°C for Case 2. After starting the AC systems, in 10 minutes, we will have a drop in temperature inside the vehicle over 30°C for both cases. This proves that the AC system, on the maximum speed helps reduce the temperature inside with a great amount. The downside of this, is represented by the noise produced from the fan on the highest setting.

In figure 6 are presented two images with the temperature on the vehicle components at the end of the simulation.

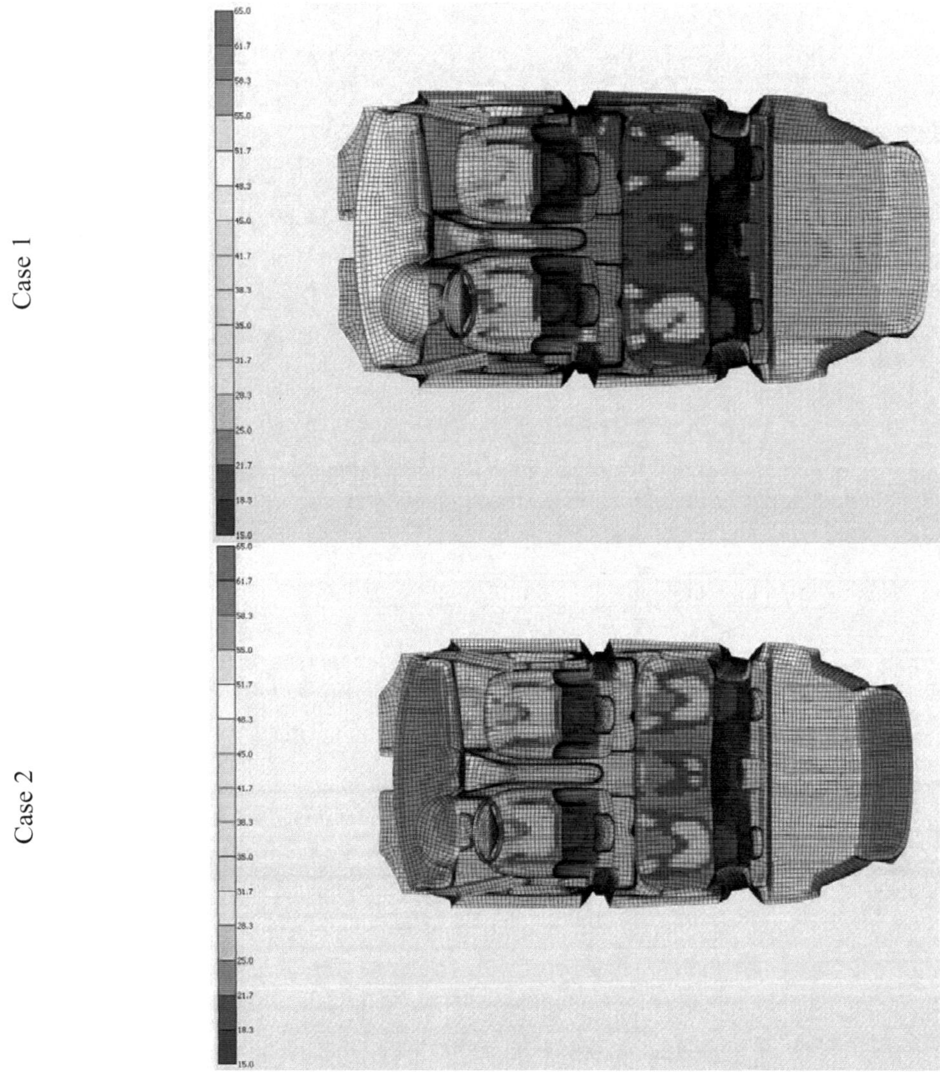

Figure 6. Temperature distribution on the vehicle parts

Althought, the exterior temperature and solar radiation intensity are bigger in the Case 2, compared to Case 1, the position of the sun at an altitude angle of 78° means that a bigger part of the radiation are received by the roof of the car, so that translates in a lower temperature on the interior at the end of the simulation

The first evaluation of the thermal comfort is done by analyzing the local comfort indexes according to ISO 7730[7]. As we can observe from the figure 7, the majority of parts are placed in the thermal comfort neutrality, the only parts that presents some kind of discomfort are the ones placed in the direct sun via the glass surfaces.

IOP Conf. Series: Materials Science and Engineering **252** (2017) 012031 doi:10.1088/1757-899X/252/1/012031

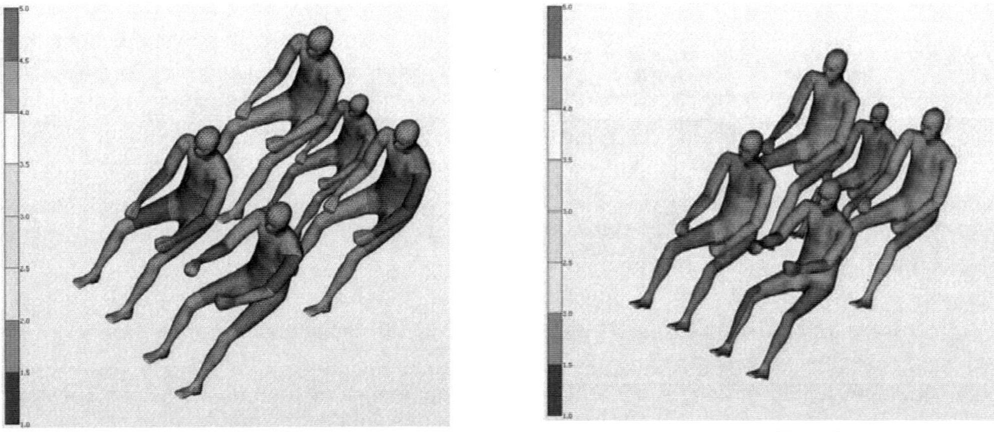

Case 1 Case 2

Figure 7.Local comfort indexes at the end of the simulations

The parameters that are needed to calculate the GTCI and GATCI are presented in table 4 and in figure 8.

Table 4. Thermal comfort indexes at the end of the simulation

	Driver			Front passenger			Rear left passenger		
	PMV	PPD(%)	DTS	PMV	PPD(%)	DTS	PMV	PPD(%)	DTS
Case 1	-0.1	5.4	-0.8	0.0	5.0	-0.7	-0.4	7.9	-1.2
Case 2	0.7	14.0	-0.1	0.3	7.5	-0.4	-0.3	6.4	-1.1
	Rear center passenger			Rear right passenger					
Case 1	-1.0	25.0	-1.7	-1.1	29.6	-1.8			
Case 2	-0.2	6.0	-1.0	0.2	5.7	-0.6			

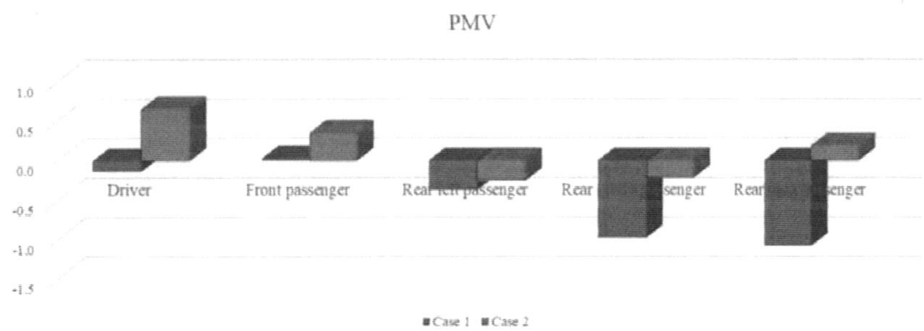

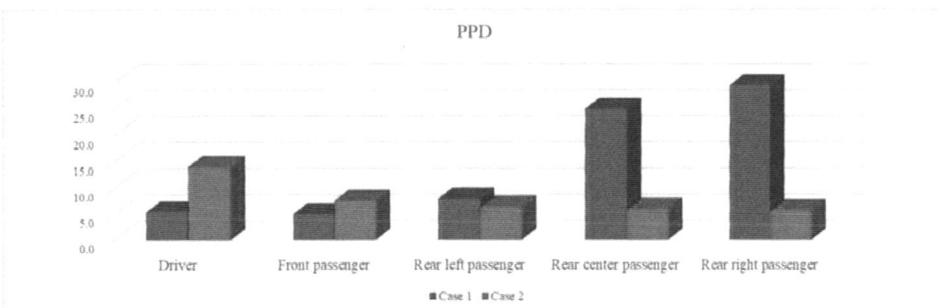

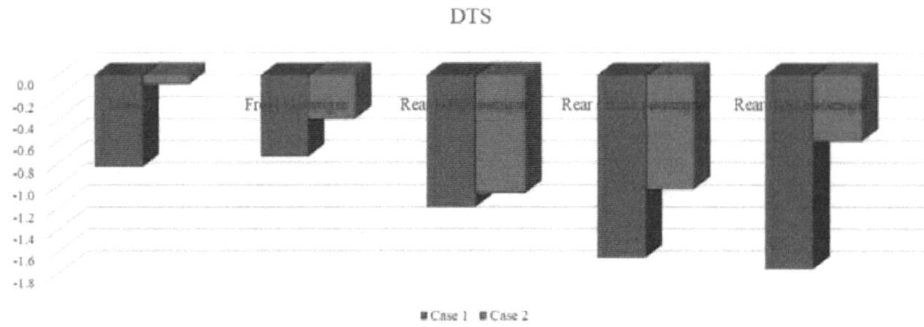

Figure 8. Thermal comfort indexes at the end of the simulation

Using the values presented in table 4 we will construct the diagrams for GTCI and GATCI index comfort. The diagrams are presented in figures 9 for GTCI and 10 for GATCI.

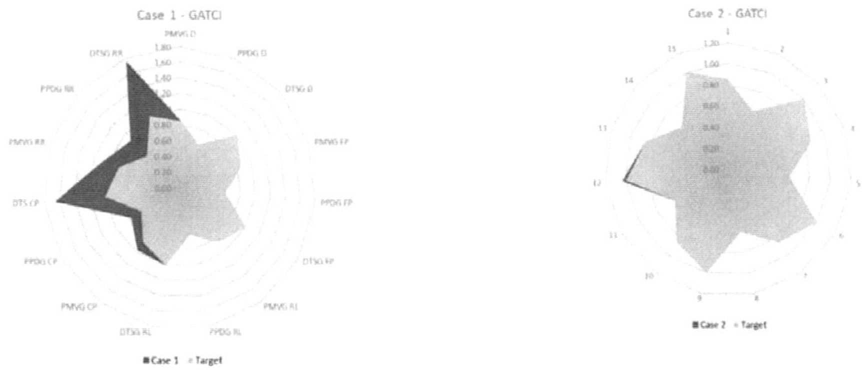

Figure 9. GTCI evaluation for both cases

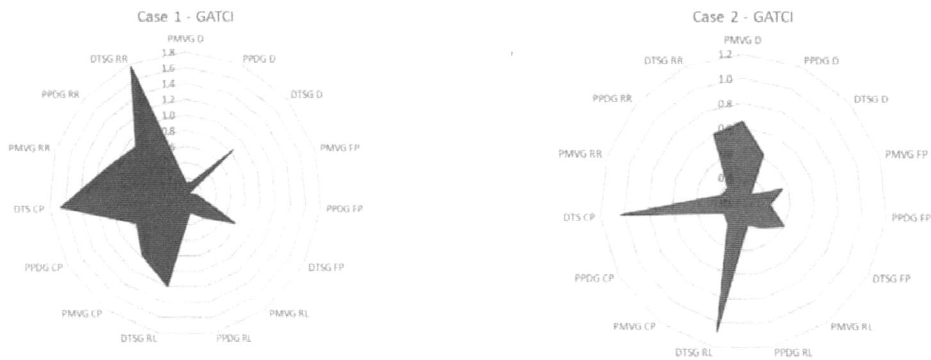

Figure 10. GATCI evaluation for both cases

Calculating the surfaces under the figures, we will obtain the results presented in table 5.

Table 5. GTCI and GATCI index values

Case	GTCI	GATCI
Case 1	2.801	1.756
Case 2	2.003	0.414
Target	1.993	

The results in terms of GTCI and GATCI are similar to those from the local comfort indexes. We can observe that for the Case 2, the GATCI value proves that all the manikins are near optimal thermal comfort.

In addition, another benefit of the newly developed index is represented by the GTCI index, that shows where, compared to the ideal condition, the vehicle is placed. In our case, we can observe that for the Case 2, all the passengers in the vehicle are situated near the optimal thermal comfort.

4. Conclusions

The thermal comfort is characterized by different indexes and we must define a link between those indexes and the general thermal comfort. The graphs are constructed taking into account the value of the thermal comfort indexes for all the vehicle occupants, and the GATCI and GTCI are defined by the area of the surfaces.

The use of those new thermal comfort indexes helps us to understand if all the occupants have reached optimal thermal comfort and also via GTCI and also compare two technical solutions for the HVAC systems by evaluating the GATCI.

Acknowledgement

This work was supported by a grant of the Romanian National Authority for Scientific Research, UEFISCDI, project number PN-II-PT-PCCA-2013-4-0569

References

[1] http://ec.europa.eu/eurostat/statistics-explained/index.php/Passenger_cars_in_the_EU
[2] Neacsu CA *et al* 2015 The Development of a New Thermal Comfort Indexes Proceedings of the European Automotive Congress EAEC-ESFA 2015 pp 703-714 - Springerlink
[3] Neacsu CA – 2011 – Contribuţii privind optimizarea confortului termic din habitaclu folosind simlarea numerică - Teza de doctorat
[4] Paulke S (2008) Thermal comfort design and assessment of vehicle cabins with THESEUS-FE, HdT seminar. München (14.10.08)
[5] ISO/TS 14505-1:2007 (2007) Ergonomics of the thermal environment—evaluation of thermal environments in vehicles—part 1: principles and methods for assessment of thermal stres
[6] Neacsu CA et al 2009 The influence of the solar radiation on the interior temperature of the car - Proceedings of the European Automotive Congress EAEC-ESFA 2009
[7] ISO 7730 (1984) Moderate thermal environments—determination of the PMV and PPD indices and specification for thermal comfort. International Standards Organization, Geneva
[8] Baker, P, Jenkins, M., Wagner, S., Ellinger, M. - An optimised thermal design and development process for passenger compartments, European Automotive Simulation Conference, Munich, Germany , 2009
[9] Nilsson, H, Holmér, M., Bohm, O. - Definition and theoretical background of the equivalent temperature. Assessment of thermal climate in operator's cabs - seminar in Florence 18 –19 November 1999
[10] Nillson, H. - Comfort Climate Evaluation with thermal manikin method and computer simulation models, ISBN 91-7283-693-8, 2004
[11] Nilsson, H. - Thermal comfort evaluation with virtual manikin method, International Journal of Building and Environment, 2007

Model for the study of active suspensions

Sorin MARCU[1], Dinel POPA[1], Nicolae–Doru STĂNESCU[1], Nicolae PANDREA[1]
[1]University of Pitesti, Romania

dinel_popa@yahoo.com

Abstract. In the beginning of the paper are briefly presented various types of semi-active and active suspensions. The advantages and disadvantages of the solutions analyzed are highlighted. There is a mathematical model for the study of an active suspension. In the next chapter we present a calculation model of active suspension controls based on elements of linear dynamic systems theory. The optimal synthesis of the dynamic system is based on a set performance criterion. At the end of the paper are presented the numerical results obtained for a two-degree system model used in several configurations.

1. introduction

Since 1970, the idea of suspension control has emerged. In the automotive industry, various concepts have emerged with regard to suspension control, so in the early 1980s, starting with the development of electronics, the concept of controlling road adaptability for the semi-active and active suspension of vehicles began to develop.

In this regard, following thorough studies, this topic becomes of great interest. Semi-active and active suspensions have been a subject of research for a long time because of their potential to improve vehicle performance. Many analytical and experimental studies have concluded that semi-active and active suspensions can provide substantial improvements over passive ones [10],[1].

Suspension passive -conventional suspension systems are passive suspensions consisting of a spring and shock absorber, making the link between the wheel and the body. The primary function of the arc is to store the energy of the vibrations, and that of the shock absorber is to dissipate the energy stored in the arc. The suspension system passively does not have flexibility [15].

Suspension semi-active - suspension systems have the ability to make changes in the damping coefficient and spring characteristics so that dampers can change the level of energy dissipation but do not provide energy to the system [7]. The damping coefficient can be controlled by a switch to increase comfort [15].

Suspension active – suspension systems are driven by an electric motor or a hydraulic cylinder, generating independent forces [7] having the ability to change the damping coefficient continuously.

2. Solutions used by car manufacturers in various models

2.1. Hydraulic or pneumatic suspension

The active or pneumatic suspension is made of a spring and a device that acts hydraulically or pneumatically by making damping forces between the suspended and the unsupported mass BMW has developed an active suspension with the hydraulic stabilizer bar. The system is made of: hydraulic pump, lateral acceleration sensors, command emitter, hydraulic suspension block and two active stabilizing bars with rotary hinged devices.

The advantages of this solution are:

Mercedes has developed the hydraulic active suspension system, consisting of a hydraulic arc, hydraulic structure, high pressure accumulator, hydraulic pump, shock absorbers and control unit. The control unit analyzes the signals received from the sensors, acting on the hydraulic system, while the vehicle is running. The hydraulic system independently controls the oil flow for each hydraulic spring [14].

AUDI has developed an active pneumatic suspension system, consisting of a pneumatic spring, reservoir, control unit, sensors, compressor, pressure sensor electromagnetic valve block. The active pneumatic suspension uses compressed air from a compressor or electric motor to perform damping [14]. The sensors measure certain parameters and the control unit determines the control mode in which the amount of air required for each shock absorber.

Ground clearance is done automatically, depending on the vehicle's driving conditions. The sensors constantly monitor the height of the car, so the amount of air in each shock absorber is adjusted to maintain the vehicle's height from the ground [4].

2.2. Active electromagnetic suspension

The electromagnetic active suspension system consists of an electromagnetic device and a mechanical spring. The device can function to recover energy from the suspension, thus reducing the energy consumption of the vehicle.

The following figure shows a BOSE suspension that uses a linear electromagnetic motor. A power amplifier provides power to the engine for vibration control. It also stores the energy generated at each compression of the suspension.

An electromagnetic damper compresses or relaxes much more, eliminating the discomfort of the passengers.

2.3. Advantages and disadvantages

Table 2. Advantages and disadvantages

Parameters	Passive suspension	Semi-active suspension	Active suspension
Construction	simple	complex	very complex
Cost	low	medium	high
Comfort	bad	good	very good
Dynamically perfected	passive	passive	good

3. Calculation methods used in the study of active suspensions

In the paper [3] there is a calculation model for the active suspensions.

This quarter car model allows research to increase or decrease the displacement of the suspended and unscrewed mass depending on the irregularities of the running track (fig. 1).

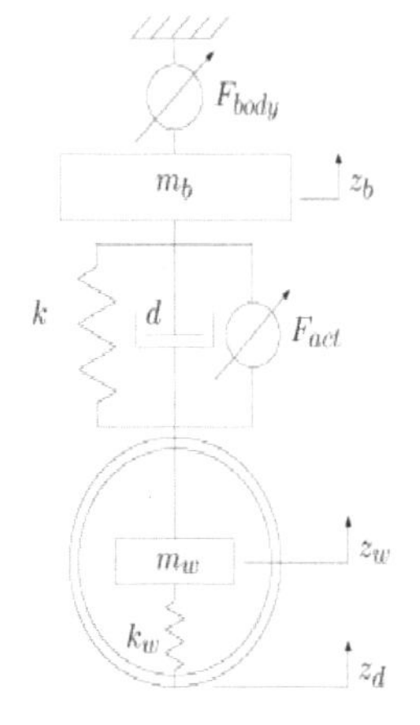

Figure 1. Suspension active.

In the paper [14] there are two calculation models for active suspensions. One refers to an active suspension in parallel and the second to a suspension (fig. 2).

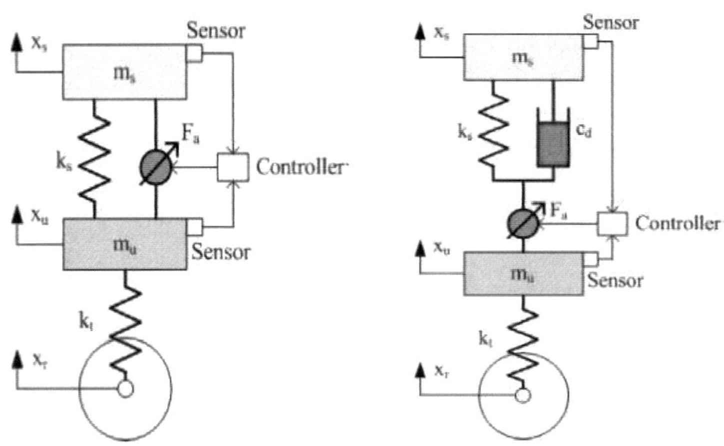

Figure 2. Suspension active.

4. Dynamic systems

It is called a dynamic system a mechanic system that evolves in time. The time evolution is described by a system of first order reductive differential equations,

$$\begin{cases} \dot{x}_i = f_i(x_1, x_2, ..., x_n, u_1, u_2, ..., u_m, \mathrm{t}) \\ i = 1, 2, ..., \mathrm{n} \end{cases} \tag{1}$$

where $(x_1, x_2, ..., x_n)$ they are called state vectors and $(u_1, u_2, ..., u_m)$ are called command vector.

If the functions f_i, depend directly on time, the system is unsparing and if it does not depend directly on time, the system is called autonomous.

A linear autonomous system with constant coefficients is written in the form,

$$\begin{aligned} \{\dot{x}\} &= [A]\{x\} + [B]\{u\} + [D]\{w\} \\ \{y\} &= [C]\{x\} \end{aligned} \tag{2}$$

where:

$$\{x\} = (x_1 \ x_2 \ .... \ x_n)^T \text{ represents the state vector;}$$
$$\{u\} = (u_1 \ u_2 \ .... \ u_m) \text{ is the command vector;}$$
$$\{y\} = (y_1 \ y_2 \ .... \ y_p) \text{ is the observable vector;}$$
$$\{w\} = (w_1 \ w_2 \ .... \ w_k) \text{ is the excitation vector;}$$
$$[A] = (a_{ij})_{\substack{1 \le i \le n \\ 1 \le j \le n}} \text{ is the matrix of coefficients;} \tag{3}$$
$$[B] = (b_{ij})_{\substack{1 \le i \le n \\ 1 \le j \le m}} \text{ is the command matrix;}$$
$$[C] = (c_{ij})_{\substack{1 \le i \le p \\ 1 \le j \le n}} \text{ is the observables matrix;}$$
$$[D] = (d_{ij})_{\substack{1 \le i \le n \\ 1 \le j \le k}} \text{ is the excitation matrix.}$$

The system

$$\begin{aligned} \{\dot{x}\} &= [A]\{x\} + [B]\{u\} \\ \{y\} &= [C]\{x\} \end{aligned} \tag{4}$$

is called stabilizable if there is a $\{u\}$ so as to achieve equality:

$$\lim_{t \to \infty} \{x\} = \{0\}. \tag{5}$$

Obviously, a system in which the matrix $[A]$ has its own negative real values or its own complex values with the negative real is stabilizable, because such a system can be taken $\{u\} = \{0\}$.

Either the system

$$\{\dot{x}\} = [A]\{x\} + [B]\{u\} \tag{6}$$

with the initial state $\{x_0\}$.

It is required to determine {u} the optimal control that drives the system $\{x_0\}$ from $\{0\}$ the state towards the time interval *T* minimizing the functional performance:

$$J = \int_0^T L(\{x\}, \{u\}, t)dt. \tag{7}$$

The most common performance criterion is:

$$J = \frac{1}{2}\int_0^T \left(\{x\}^T[Q]\{x\} + \{u\}^T[R]\{u\}\right)dt \tag{8}$$

where $[Q]$, $[R]$ represents weight matrix, $[Q]$ being positively semi defined, $\left(\{u\}^T[R]\{u\} \geq 0,\ \forall\{x\} \in R^n\right)$ and $[R]$ positively definite $\left(\{u\}^T[R]\{u\} > 0,\ \forall\{u\} \in R^p\right)$.

It is shown by varied calculation that the solution is

$$\{u\} = [K]\{x\} \tag{9}$$

the reaction matrix being

$$[K] = -[R]^{-1}[B]^T[P] \tag{10}$$

where the matrix $[P]$ is the solution of the Riccati algebra equation:

$$[P][A] + [A]^T[P] - [P][B][R]^{-1}[B]^T[P] + [Q] = [0]. \tag{11}$$

5. Numerical results

In the figure (figure 3) the following is an active suspension system with the following notations:

- m_1 - 1/4 car body mass,
- m_2 - suspension mass,
- k_1 - spring constant of suspension system,
- k_2 - spring constant of wheel and tire,
- c_1 - damping constant of suspension system,
- c_V - control force,
- z_1, z_2 - the displacements of the two masses,

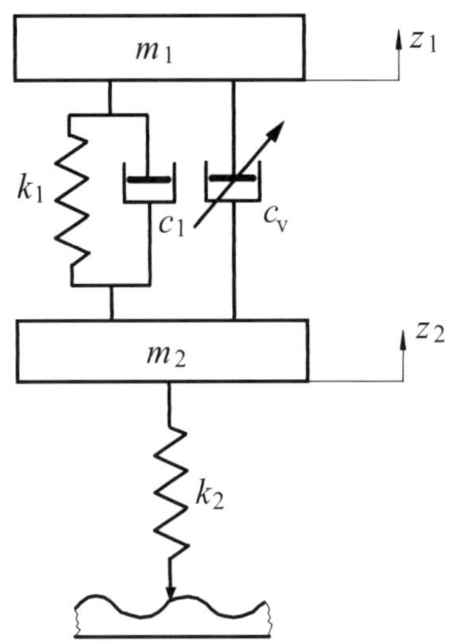

Figure 3. Suspension active.

In the initial step, the passive suspension response time will be analyzed by comparing the results of the analysis with the response time of an active suspension.

The following numerical data were used:

- $m_1 = 400$ kg,
- $m_2 = 200$ kg,

IOP Conf. Series: Materials Science and Engineering **252** (2017) 012032 doi:10.1088/1757-899X/252/1/012032

- $k_1 = 8000$ Nm,
- $k_2 = 80000$ Nm,
- $c_1 = 600$ Ns/m.

The active suspension response time is shown in Figure 5.

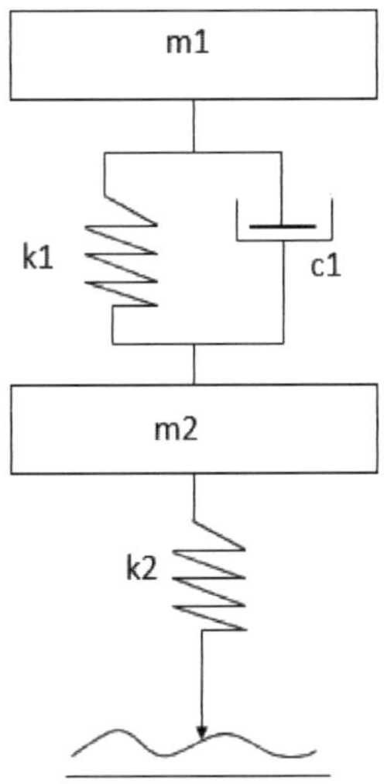

Figure 4. Suspension passive.

The passive suspension response time is shown in figure 5.

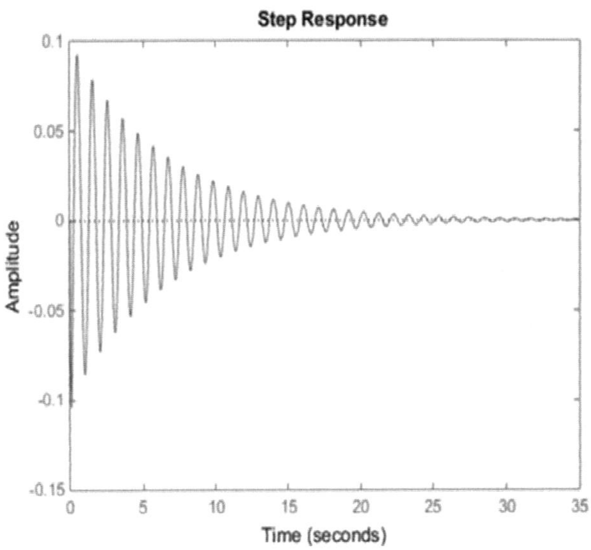

Figure 5. Response for suspension passive

In order to obtain numerical data for the active suspension, a calculation program was developed in Matlab where on the basis of the relations 1 ... 10 the graphs of Figures 6 - 8 were obtained.

The following numerical data were used:

- $m_1 = 200$[kg]
- $m_2 = 200$[kg],
- $k_1 = 8000$[N/m],
- $k_2 = 80000$[N/m].

For constants c_1 were used 3 values:

IOP Conf. Series: Materials Science and Engineering **252** (2017) 012032 doi:10.1088/1757-899X/252/1/012032

- $c_1 = 400$ [N*s/m], $c_1 = 600$ [N*s/m] and $c_1 = 1200$ [N*s/m].

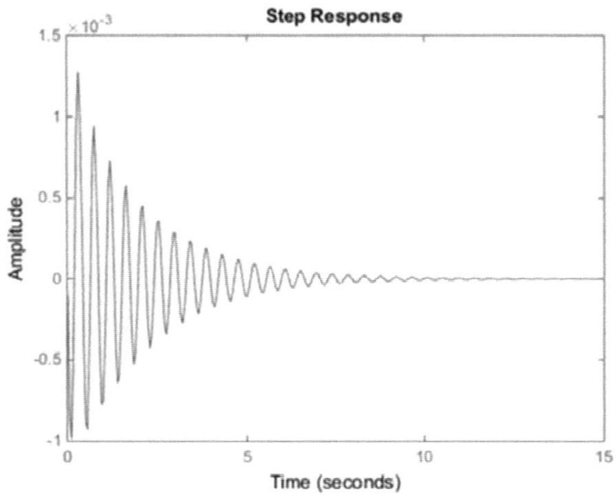

Figure 6. Closed-Loop Response with PID Controller for $c_1 = 400$ [N*s/m]

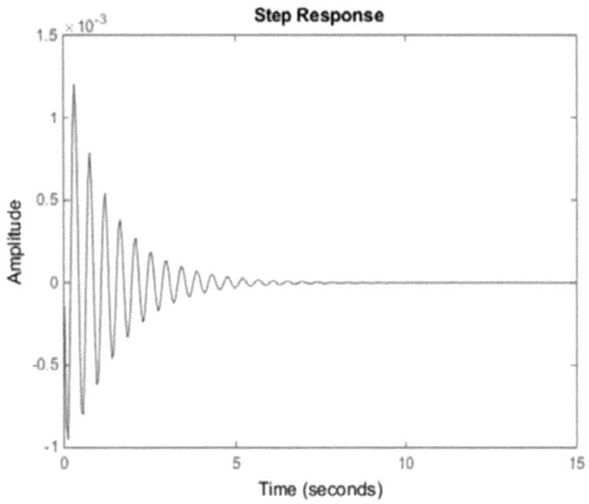

Figure 7. Closed-Loop Response with PID Controller for $c_1 = 600$ [N*s/m]

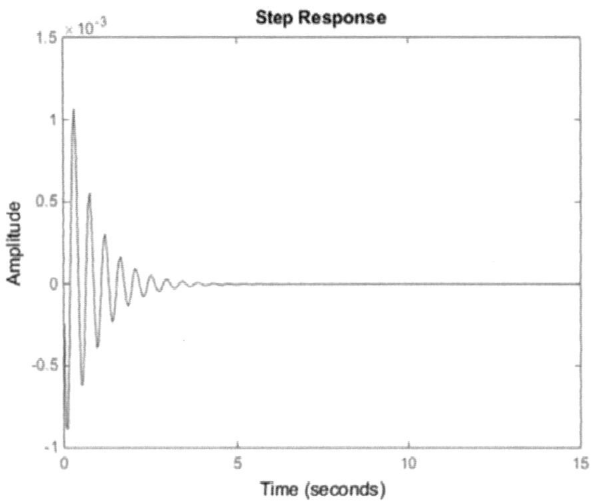

Figure 8. Closed-Loop Response with PID Controller for $c_1 = 1000$ [N*s/m]

6. Conclusions

Active suspension systems can maintain the necessary stability and comfort due to the adaptability of the suspension according to the driving conditions.

In the present paper we presented the main types of active suspensions, and the model shown showed the efficiency of the active suspension using the Matlab computing program.

The analysis of the three diagrams resulted in the following conclusions:

- the damping time was reduced by about 30%
- the mass distribution m1 has considerably decreased

References

[1] Youngjoo Cho, Byung Suk Song, Kyongsu Yi, *A Road-Adaptive Control Law for Semi-Active Suspensions,* KSME International Journal, Vol 13, No. 10, pp. 667 - 676, 1999.

[2] Hamid Reza Karimi, *A Semiactive Vibration Control Design for Suspension Systems with Mr Dampers.*

[3] Bart L. J. Gysen, Johannes J. H. Paulides, E.A. Lomonova, Jeroen L. G. Janssen, *Active Electromagnetic Suspension System for Improved Vehicle Dynamics*, IEEE Transactions On Vehicular Technology, Vol. 59, No. 3, March 2010.

[4] *Adaptive air suspension in the Audi A8*, Home study program 292.

[5] Guido P. A. Koch, *Adaptive Control of Mechatronic Vehicle Suspension Systems*, 2011.

[6] Tomoaki Mori, Itthisek Nilkhamhang, Akira Sano, *Adaptive semi-active control of suspension system with mr damper*, Yokohama 223-8522, Japan.

[7] Wajdi S. ABOUD, Sallehuddin Mohamed HARIS, Yuzita YAACOB, *Advances in the control of mechatronic suspension systems*, Aboud et al. / J Zhejiang Univ-Sci C (Comput & Electron) 2014 15(10):848-860.

[8] M.S.P Leegwater, *An active suspension system capable of economically leveling a car during cornering*, 2007.

[9] Ayman A. Aly, *Car Suspension Control Systems: Basic Principles*, International Journal Of Control, Automation And Systems Vol.1, No.1, January 2012.

[10] Wolf R Kruger Ondrej Vaculin, Martin Spieck, *Semi Active suspension systems for road and off road vehicles*, Mecca 4/2004.

[11] Hassan Elahi, Asif Israr, M. Zubair Khan, and Shamraiz Ahmad, *Robust Vehicle Suspension System by Converting Active & Passive Control of a Vehicle to Semi-Active Control System Analytically*, Journal of Automation and Control Engineering Vol. 4, No. 4, August 2016.

[12] Emanuele Guglielmino, Tudor Sireteanu, Charles W. Stammers, Gheorghe Ghita Marius Giuclea, *Semi-active Suspension Control*, 2008.

[13] Takayoshi Fujita, Takanori Fukao,Tomohiro Kinoshita,Noriaki Itagaki, *Semi-Active Suspension Improving Both Ride Comfort and Handling Feel*, 7th IFAC Symposium on Advances in Automotive Control The International Federation of Automatic Control September 4-7, 2013. Tokyo, Japan.

[14] X. D. Xue, K. W. E. Cheng, Z. Zhang, J. K. Lin, D. H. Wang, Y. J. Bao, M. K. Wong, and N. Cheung, *Study of Art of Automotive Active Suspensions*, July 2011.

[15] Sharad S Jadhav, Dr.S.S.Kulkarni, Prof. Amod .P.Shrotri, *Suspension System With Broad Classification And Various Models: A Review*, International journal of advances technology in engineering and science.

[16] Rajesh Rajamani, *Vehicle Dynamics and Control*, 2012.

[17] Pandrea, N., Popa D., Parlac S, *Modele pentru studiul vibratiilor automobilelor*, Ed. TIPARG, Pitesti, 2001.

IOP Conf. Series: Materials Science and Engineering **252** (2017) 012033 doi:10.1088/1757-899X/252/1/012033

Advanced Engineering Methods in Design and Education

Krasimir Kamenov, Antoaneta Dobreva, Vyarka Ronkova

University of Ruse, Ruse 7017, Studentska str. 8, Bulgaria

adobreva@uni-ruse.bg

Abstract. The paper presented deals with the challenges occurring during the education of the new digital generation of students in the area of engineering subjects. This new situation in teaching technologies imposes the obtaining of appropriate feedback from the students during the lectures and tutorials. The objective of the research is to investigate the impact of the application of video games, graphical presentations, animations, etc. The feedback received in such interactive way gives the opportunity to improve the teaching models and to increase the active participation of the students during the lectures and tutorials.

1. Introduction

The classic teaching methods of lecturing scientific study material consisting of theoretical knowledge, empirical relationships and formulas often turn to be a challenge for the mentality and perception potential of the university students. These factors usually lead to difficulties in assimilation of specific knowledge, which is based upon previous theoretical subjects. Besides, negative emotions to certain topics from the study syllabi occur.

Finding solutions for such complicated problems both from pedagogical and from methodological points of view is a real challenge for the academic staff responsible for the study process. This dynamic situation imposes changes in the teaching and lecturing methods applying new and modern technological solutions stimulated by the development of computer and information technologies.

The authors' team has experience in presenting and explaining academic study material, for example – application of multimedia educational systems, electronic laboratories and software products, [1]. The efforts indicated lead to significant results. Nevertheless, the process of improving the teaching methods is dynamic and it requires continuous changes and innovations.

2. Investigating options for using mobile applications in study process

Mobile applications have increasing influence upon the academic study process for students from different degree courses regardless for what kind of devices they are designed: iPhone, Windows or Android, [2, 4]. The university study process involving mobile devices includes three different aspects: focusing on the mobile device; focusing on the learning outside the lecture hall and focusing on the mobility of every educated and trained student.

The research presented treats another additional aspect: the usage of mobile applications in the process of lecturing and acquiring new study material. This option give the possibility to engage the attention of the students to a greater extend and to encourage their active involvement in the assimilation of the material delivered. The theoretical survey made by the authors' team established that mobile applications could be especially appropriate for achieving feedback from the students concerning the level of the study material acquired, [3, 5].

All students possess mobile devices and some of them – several such devices. The students follow with enormous interest the new information in their mobile devices. This circumstance creates preconditions for improving the level of their attention concentration applying mobile devices for test control.

3. Specific features of the test structure for mobile applications

Creating question for students educated is carried out with the help of logic, taking into consideration several factors. These factors influence the level of complexity of the test control. The most important factor is the specifics of the education in the academic degree program. Students from different degree programs within the Higher education system learn study material different in scope and in degree of difficulty. The aim is to select questions in such a way that the level of the study material acquired by the students is to be controlled definitely.

The number of the questions is different for each control test depending on the topic taught. Within a topic, the selection of the questions is random. The selection of questions is to be done through algorithm ensuring different combinations of possible questions and corresponding answers. This procedure avoids the frequent repetition of the same questions and prevents memorizing the correct answers by heart. It is preferable to project each question through multimedia tools for a strictly defined time period. After the time interval is over, the next question is to be presented. In order to ensure independence and equality of all students evaluated, the system does not give options for making corrections or answering the questions after the time is over.

Control tests in different subjects are elaborated: machine science, machine elements, engineering graphics, internal combustion engines, intermodal transportation operations, etc. The subjects from the area of engineering design and engineering graphics are fundamental. They are especially important because they are the first technical subjects for future engineer professionals. These subjects form the necessary base within the education and training of these students. For the first time the students face different issues of technical nature during acquiring knowledge and skills within these subjects. They learn to apply the accumulated theoretical knowledge by solving problems from the design engineering subjects. Besides, the students acquire experience and routine necessary for their future engineering practice. These fundamental subjects introduce a great part of the technical terminology and systematize the applied terms and notions of mechanical components.

Therefore, it is very important to evaluate whether the students comprehend, understand fully and remember these terms and notions. Besides, the lecturers are delivering and presenting relative complicated 3D models and 2D drawings, especially in accordance with students' level of training. A great number of difficulties occur when the students have to figure out the information on drawings and to understand its meaning. It is envisaged that the control tests for mobile applications will include two difficulty levels. The first level should involve relatively easy questions which should be visualized in the form of video games. This degree of difficulty is intended mainly for students with a low level of attention concentration. The second level of difficulty contains more complicated control questions. Their number is significantly smaller than the quantity of control questions from the first level. This level is intended for students who demonstrate interest not only in the study activities but also in scientific and research problems. It is foreseen that the test control for mobile applications is to be implemented at the end of the relevant tutorial. At the beginning of the test session several questions from the first level of difficulty are to be given to the students. After these questions are over, one or two questions of the second difficulty level are to be presented.

The aim of all test control questions is to receive feedback from the students and to examine the opportunities of mobile applications for accomplishing better communication between lecturers and students. Creating such options, the academic staff will be able to evaluate in details the degree assimilation of the study material and to increase the activity and the attention of the students to the training process. The final goal of the usage of mobile application for test control will be to help the students to understand and learn the greater and the more important part of the study material during the lecture or tutorial.

4. System for cross-platform mobile applications MATE

Based upon the theoretical research on the possibilities for using mobile applications in the academic study process and the analysis concerning the specific methodological requirements to the structure of the control tests for these applications, the authors' team created a system for test control MATE: Mobile Application Test Evaluation.

The system for control and assessment of the knowledge acquired by students during lectures and tutorials consists of the following components: a desktop or a laptop computer; a multimedia projector or a LCD television set; a server and a mobile device (a smart phone or a tablet). An application is to be installed on the lecturer's computer. The function of this application is to generate test questions on each topic or section of the study material delivered, included in the test system MATE, see Fig. 1.

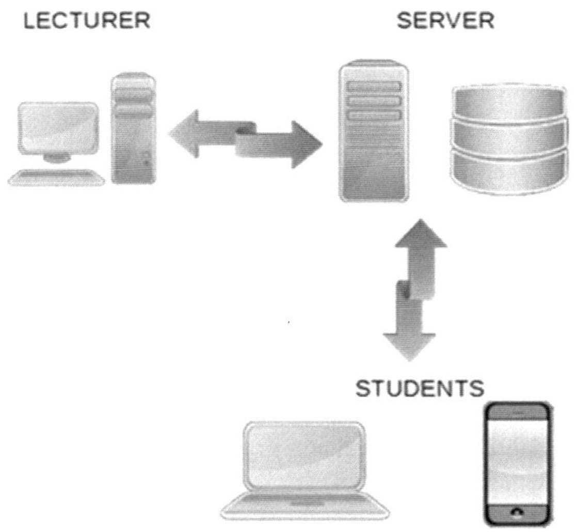

Figure 1. MATE design.

The diversity of mobile devices with various displays leads to a number of difficulties in visualizing the specific test questions and the possible answers. For example, graphical information cannot be represented well enough in a great number of mobile phone models. This graphical information is especially important because the students need to reflect on its meaning.

Therefore, the system MATE solves this problem by visualization tools which are equal for all students attending the tutorials. Most often, these tools are multimedia projector, interactive whiteboard or LCD television sets. The selection of these multimedia and visualization tools of the test questions is driven by the fact that the greatest part of the study material within general engineering and engineering design subjects is related to equations, diagrams, schemes, drawings, three dimensional models and pictures of real objects and to significantly less amount of text explanations. Consequently, the authors' team made the decisions that the small mobile phone displays are not suitable for visualizing the information and knowledge necessary for the aims of the study process and the research.

Fig. 2 shows one of the questions which is used during the education and training in the subject Engineering graphics. A plane is defined through two intersecting lines. Three points are designated on the drawing. The students have to decide for a certain period of time which point (among the 3 points) belongs to the preliminary given plane. The same graphical presentation can be used for a variety of questions about the appurtenance of a point to a line and a plane. Each question related to this drawing can be chosen on a random basis and in random order.

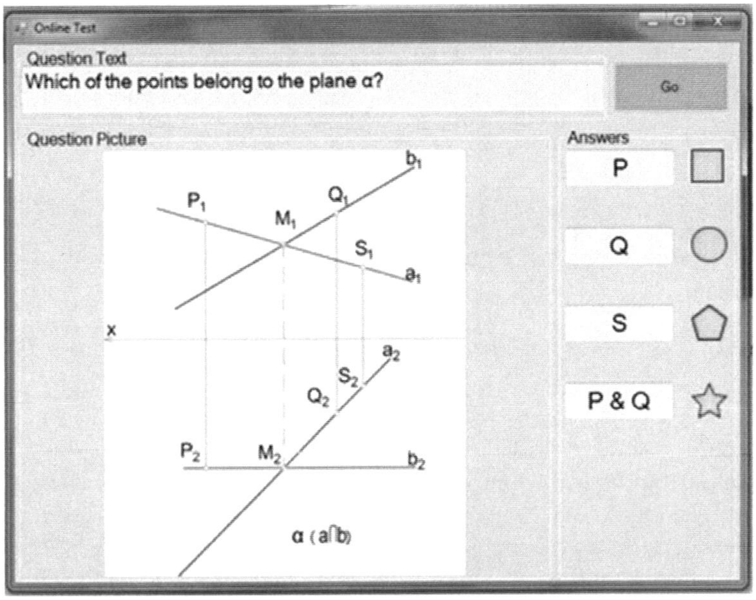

Figure 2. Typical MATE question.

The system MATE consists of two separate software products. The first one is installed on the lecturer's PC or laptop. The second component is installed on the mobile device. The computer application is developed on the language C# in two compiled versions. The first version is designed for Windows. The second one is intended for Linux. Due to this option, a great number of available computers and operating systems are covered. The application is also developed in portable version. Thus its usage on computers with administrator restrictions is possible. The application for the mobile device with operating system Android is designed on Java. The downloading and installation can be done in two possible ways: from FTP (File Transfer Protocol) server or with the help of Bluetooth.

The algorithm of the computer application consists of the following units: unit for the selection of a topic, unit for the selection of a question, unit for generating the possible answers. After executing the whole algorithm, a timer is activated. It defines the time for answering. After this time is over, it is impossible for the students to answer the corresponding question. The time for answering a test question is the same for all students participating in the test session. However, time periods can be different for different question. The algorithm of the application for the Android device includes communication with the SQL (Structured Query Language) server. Thereby, it reads the number of the possible answers and the maximum time necessary for the relevant response. After starting the process of answering the questions, the students are given the option to choose and confirm the correct answers. When the time for answering is over, the corrections become impossible.

The appearance for a mobile device showing the buttons for the selection of the true answer by the student is to be seen on Fig. 3. The database containing the subject topics, the possible questions and answers and the necessary time for each question is located on the SQL server. The server exchanges information with the lecturer's PC and also with the mobile devices of the students participating in the test control session.

The system MATE introduces a new center of attention – a focus on feedback from the students aiming to make some correction in implementing the lecture or the tutorial in order to increase the quality of the study process through better comprehensibility of the teaching material. In order to make this option possible several control tests during the lecture or tutorial are made.

Figure 3. Typical MATE answer display.

They do not take long time from the study process and at the same time the tests give valuable information to the lecturer. Another important advantage of application of the elaborated test control system MATE in real time is the optimal time use during the study process. The volume of study material delivered in not diminished and the quality of the study process ca be increased. This advantage of MATE complies with the circumstance that all students possess one or several mobile devices – smart phones, tablets and laptops. The application of MATE leads to another positive effect. The students show increased attention to the study material delivered because the answering of the questions imposes certain minimum of knowledge acquired during the relevant lecture or tutorial

5. Conclusions and future work

The elaboration and usage of the system MATE has contributed to the investigation of the opportunities of the mobile applications for performing better communication between lecturers and students. MATE facilitates the lecturer's evaluation of the level of the acquired material by the students and contributes to increasing the activity and the attention of the students during the study process. Due to this option the greater and the more essential part of the material delivered is to be understood during the lecture or the tutorial. Therefore, the quality of the study process is increased significantly. Applying MATE new academic teaching models for students from the engineering bachelor programs are created and tested. MATE has fulfilled the aims set: the attention concentration of the students has been increased during the study process and the motivation of students with significant interests in study and scientific activities has been improved.

References

[1] Dobreva, A., Haralanova, V. (2012). Measuring and Evaluation in Machine Science and Design Education, based upon Diagnostic Research. Procedia Social and Behavioral Sciences, WCLTA, Brussels, 163, 51–59.

[2] Strunk, W., Jr., & White, E. B. (1979). The elements of style (3rd ed.). New York: MacMillan.

[3] Mettam, G. R., & Adams, L. B. (1999). How to prepare an electronic version of your article. In B. S. Jones & R. Z. Smith (Eds.), Introduction to the electronic age (pp. 281–304). New York: E-Publishing Inc.

[4] Fachinger, J., den Exter, M., Grambow, B., Holgerson, S., Landesmann, C., Titov, M., et al. (2004). Behavior of spent HTR fuel elements in aquatic phases of repository host rock formations, 2nd International Topical Meeting on High Temperature Reactor Technology. Beijing, China, paper #B08.

[5] Fachinger, J. (2006). Behavior of HTR fuel elements in aquatic phases of repository host rock formations. Nuclear Engineering & Design, 236, 54.

IOP Conf. Series: Materials Science and Engineering **252** (2017) 012034 doi:10.1088/1757-899X/252/1/012034

Finite Elements Contact Modelling of Planetary Gear Trains

Svetlin Stoyanov, Vasko Dobrev, Antoaneta Dobreva

University of Ruse, Ruse 7017, Studentska str. 8, Bulgaria

adobreva@uni-ruse.bg

Abstract. The paper presents the application aspects of computer technologies concerning the process of creating theoretical contact models of planetary gear trains using Abaqus/Explicit. The necessary assumptions, constrains and specific features of these gear drives are discussed in details. The models created are appropriate and useful tools for computer simulation research of the dynamic behaviour of planetary gear trains.

1. Introduction

Planetary gear trains possess significant kinematic features – they can be uses as summing or differential mechanisms (operating as reducers or multiplier). Depending on the kinematic scheme they can reach a gear ratio up to 3000. Compared to conventional mechanical drives, the planetary gear trains are characterized with 2 to 4 times less weight and dimensions transmitting the same or greater torques. The power is transmitted through several symmetrically positioned satellite. Therefore, the power flow is distributed in several branches. This is the reason for diminishing the load of the meshing and of the rolling bearings and the efficiency coefficient reaches for some kinematic schemes the values of 98 %.

Due to these and some other advantages, planetary gear trains are used considerably in many industry branches. They are very attractive especially for aeronautical applications (turboprop engines, helicopter transmissions, etc.), for transmissions in automotive industry and national defense.

Very important problems for these drive trains are the inaccuracies which appear during the manufacturing of their machine elements: wheel with internal meshing, central pinion, satellites, carrier, housing details and also during the assembly process of the components. The distribution of the loading between the planetary gear train components becomes irregular. This specific feature leads to occurrence of vibration and noise. The reduction and control of noise and vibration is absolutely necessary because these phenomena lead to damages of the machine details and units. Besides, they have negative effects for the staff operating with the planetary transmissions. Therefore, one of the main goals of design professionals is to reduce of the level of noise and vibrations.

Another important machine design objective is the concept of reducing the period of design layout and calculations. The entire period of the implementation of new industry products includes the following stages: design layout, creating an experimental model, conducting functional and reliability tests, executing corrections if necessary and manufacturing the market products. A very important approach to reduce the design and manufacturing process is the application of different software products.

According to the experience of the authors' team, the software system Abaqus FEA is the most appropriate tool for diminishing significantly the necessary time period for the design process.

Abaqus/Standard product is applied mainly for operations with relative low speeds. Abaqus/Explicit can be used for applications with high speeds, in which non linear transitional areas dominate. This advantage ensures simulation with great deformations and nonlinear behavior of material including inertia effects as results of fast movements. Analyzing assembly units of solid bodies, the product can determine the speed and the acceleration of each detail, the forces and the reactions within the connections between the machine design components. The dynamic analysis of flexible components possesses also the advantages mentioned and besides, it provides the deflections and internal stresses within the details of the assembly unit. Because of these reasons, the authors' team has applied the software package Abaqus/Explicit.

The software system Abacus has been used for model discretization and analyses in a variety of scientific publications. The product Abaqus/CAE gives the ability to have two different seed steps for one and the same part – global and local. Therefore, the designer can mesh the entire part with a global step, which has a great value of the seed step. Together with this global step, the designer can mesh the contact lines and surfaces with a local step, which has a small value of the seed size. This procedure provides high accuracy of the model discretization with a minimal number of elements. Therefore, minimal computational cost could be ensured.

The objective of the research presented it to describe the process of creating the theoretical contact model considering the necessary assumptions, constrains and specific features of the planetary gear trains.

2. State of art concerning theoretical modelling of planetary gear trains

The advanced gear design is based upon the modern theory of gearing and upon the application of enhanced approaches of tooth contact analysis, application of Finite Element (FE) analysis for stress determination aiming to optimize gear tooth surfaces, providing high endurance, low levels of noise and vibration for gear drives. The FE models can incorporate different kinds of deformations and these models generated can be exported for main commercial FE analysis computer programs such as: ABAQUS, ANSYS, etc.

Finite Element (FE) simulation of gear contact with 3D models runs very well in ABAQUS software. There is a variety of research work already done on different kinds of analyzes with the help of ABAQUS. This software product was used for determination of transmission errors in [1, 2]. Special gear simulations were also presented in [3, 4].

In order to avoid unexpected gear box failure it is important to have in mind the effect of different parameters in gear contact analysis. Particular geometry issues of gears have been conducted: in [5] - for pin stiffness and misalignment; in [5, 6, 7] –for tooth wedging and bearing clearance nonlinearities, special regimes and eccentricities. The influence of the eccentricity on planetary gear dynamics is presented also in [5, 8]. Since contact forces in meshing gears are extremely difficult and expensive to measure, the simulation of these forces plays an important role. A typical gear pairing is selected and impacts between one tooth pair are investigated for different boundary and initial conditions with 3 different models: a multibody model, a fully nonlinear finite-element model and a modally reduced model in combination with a local force equations, in [9]. Several investigations were carried out in order to measure the contact forces of different types of gears in the field of FE contact stress analysis. The authors of [10], were working on the stress distribution in a planetary gear set by means of a 2D model with FE method created by ABAQUS software.

In the field of FE stress analysis, several investigations are especially important: A study described in [11] considers each gear as a deformable body and predicts load, stress and deformations. Based on these results, a deformable body analysis with a thin rim was carried out; An advanced nonlinear FE method used to simulate accurately gear contact behavior under real load conditions was developed in [12]. This method was successful in automotive transmission gear surface fatigue wear reduction; Stress analysis of a helical gear set with localized bearing contact was developed in [13]. Though FEA package, ABAQUS/Stand the gear stress distribution was shown. Furthermore, the paper presents some examples, which demonstrate the influences of the gear's design parameters, [13].

Based upon different approaches, interesting researches are implemented aiming to create and investigate FE gear train models. In [14], a complete model of planetary gear set is developed in FE model. In contact problems, the accuracy of contact stresses obtained is associated with the meshing of contact positions. Therefore, gear teeth are meshed more refined in the model presented in [14]. Abaqus/standard solver is used to calculate the contact stresses, by assumption of ignoring inertia effects in analysis.

A number of studies have been implemented on the dynamic behaviour of transmission systems and planetary gear trains. The authors of [15] presented unusual system behaviour observed in a lumped-parameter planetary gear model. The main focus of the investigation was on the phenomena that occur at especially high speeds. In [16] the mesh phase relations of general compound planetary gears were systematically studied. The results from this study have been used further by the authors for analytical multibody investigations on the static or dynamic response of general compound planetary gears. In [17] the dynamic response of an example planetary gear was studied including the nonlinear tooth behavior and its correlation with planet bearing forces. The major causes for the vibrations caused by gravity forces and the presence of clearance-type nonlinearities in the form of backlash and bearing clearance were analyzed. In [18] the dynamic responses of planetary gears were analysed when component gears had time-varying pressure angles and contact ratios caused by bearing deformations. In [19] an original lumped parameter model of planetary gears, which accounted for planet position errors and simulated their contribution to the static and dynamic load distributed amongst the planets, was presented. In [20] the three-dimensional nonlinear vibration of gear pairs was investigated where the nonlinearity was due to portions of gear teeth contact lines.

Although there is a great variety of literature sources in the area of theoretical modelling of planetary gear trains, they focused mainly on certain geometry and loading conditions and most of them do not discuss the influence of sliding, slippage and lubrication process. In this study, the model of planetary gear train was built considering these important features of the meshing contact. Therefore, the authors team is aiming to achieve another objective of the study presented: to serve as a useful source of reference for engineers in calculating and designing such gear systems.

3. Modelling of contact interaction

The software product Abaqus/Explicit provides two algorithms for modelling contact interactions: general contact algorithm and contact pair algorithm, [21].

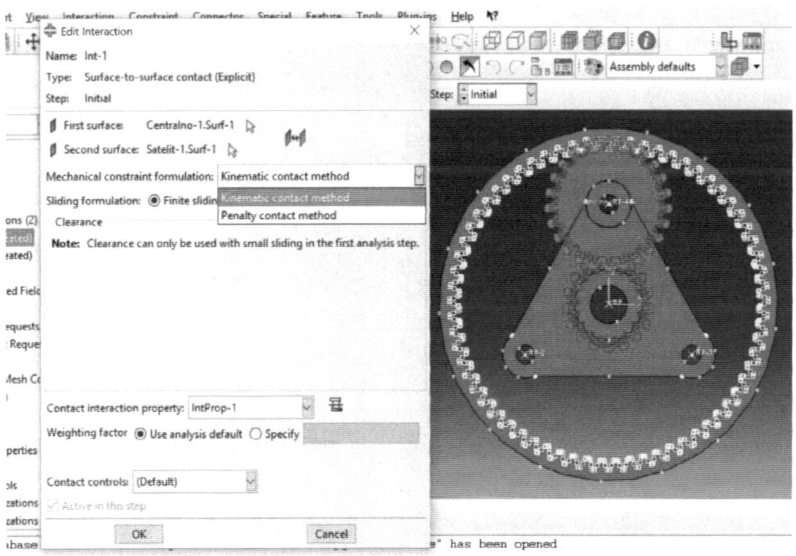

Figure 1. Selecting contact interaction properties.

The general contact algorithm allows simple contact definitions with a small number of restrictions concerning the types of surfaces involved. The contact pair algorithm has more restrictions considering the types of surfaces involved and it often requires a more complicated contact definition. However, the contact pair algorithm allows the investigation of interaction behaviors that are not available applying the general contact algorithm. Another advantage of Abaqus/Explicit is that the designer can combine these two algorithms in order to simulate contacts between rigid and deformable bodies or the contact of a body with itself. Besides, contacts with eroding bodies can be simulated.

4. Implementing Abaqus/Explicit for imposing contact constrains

The explicit algorithm uses two different methods to enforce contact constraints, (Fig.1): The kinematic contact algorithm. It uses a kinematic predictor/corrector contact algorithm in order to enforce strictly contact constraints. For example, no penetrations between the profiles of the meshing gears are to be allowed; The penalty contact algorithm. It has a weaker enforcement of contact constraints, but it allows operating with more general types of contact.

Concerning the contact pair algorithm, the explicit algorithm uses a kinematic contact formulation by default. This circumstance ensures precise compliances with the contact conditions with a help of the predictor/corrector method. At first, during the increment the algorithm operates under the assumption that contact does not occur. In case (at the end of the increment), there is an overclosure, the acceleration is modified to obtain a corrected configuration in which the contact constraints are enforced. The penalty method applies a force at the contact points which have penetrated across the target surface. This force aims to eliminate the penetration. The magnitude of the force necessary to prevent penetration is unknown (preliminary). It is necessary that the force should be great enough to push the contact surface back. At the same time, it should not have an extremely great magnitude in order to prevent the contact loss causing errors and instability. The positive aspect of the penalty method is its simplicity. The negative aspect is that the interaction steps end with a finite value of penetration, because some level of penetration is necessary for the contact force generation. The pressure – overclosure parameters are shown on Fig. 2a. The creating of a "hard" contact is presented. In this case, the resisting force is defined in the following way: if this force would be applied during the increment, it will cause the slave node to contact without penetration and without partial loss of contact with the master surface.

5. Defining sliding and slippage

The explicit algorithm uses two different sliding formulation – finite sliding or small sliding. The planetary gear train model created is based upon the option: finite sliding. This selection procedure is shown on Fig. 2b.

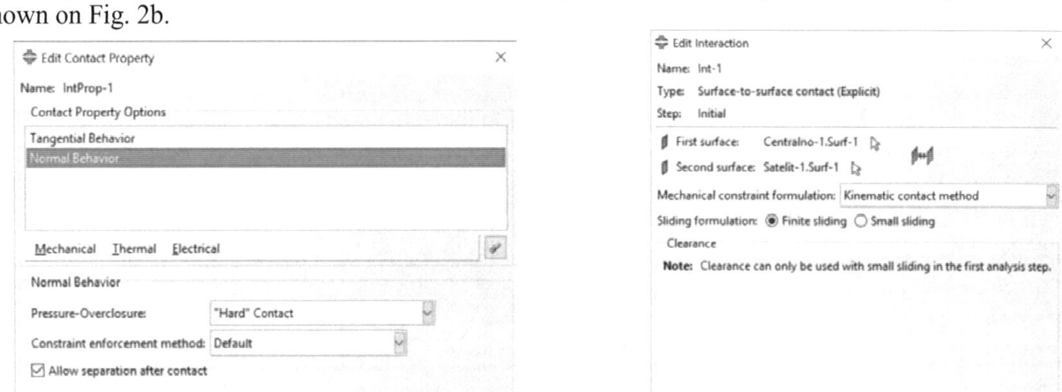

a b

Figure 2. Contact settings: a) Pressure/ Overclosure method; b) Sliding formulation – finite or small sliding.

6. Modelling the lubrication process in planetary gear trains

Damping is an energy dissipation which is proportional to velocity, i.e. viscous friction. The option to apply damping parameters concerning contacts or in connector sections can be used to simulate oil presence in gear trains, Fig. 3.

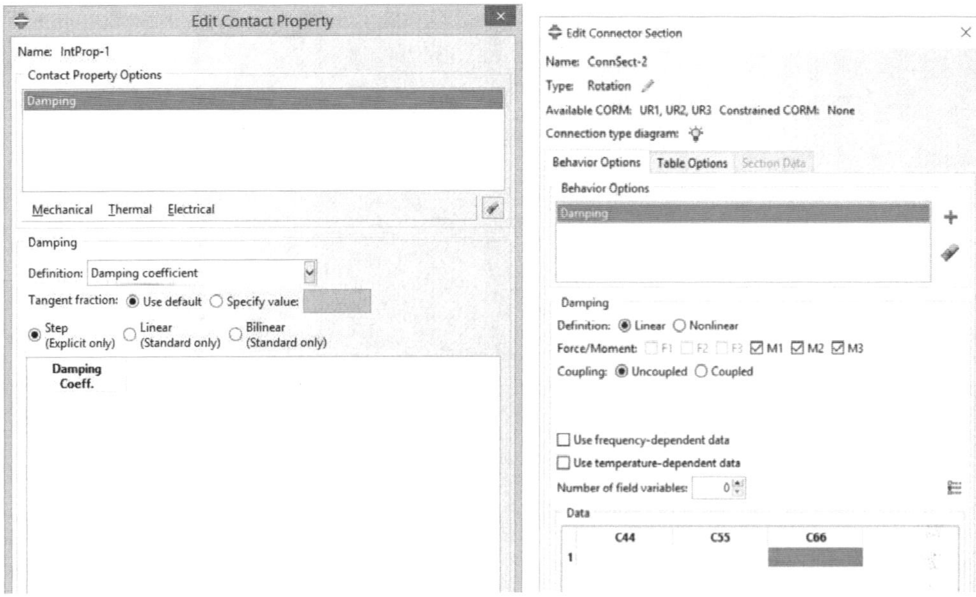

a b

Figure 3. Damping settings: a) contact damping settings; b) connector damping settings.

The damping effects can influence both the normal and tangential to the surfaces motions. In the normal direction, it is proportional to the relative velocity between the surfaces, [21]. In the tangential direction, it is proportional to the elastic slip rate associated with friction.

7. Application aspects of computer modelling of planetary gear trains

Several finite elements models of gear trains has been developed. Two of them are shown on Fig. 4 and Fig. 5, [22, 23]. They have been used for vibration characteristics determination of planetary gear trains.

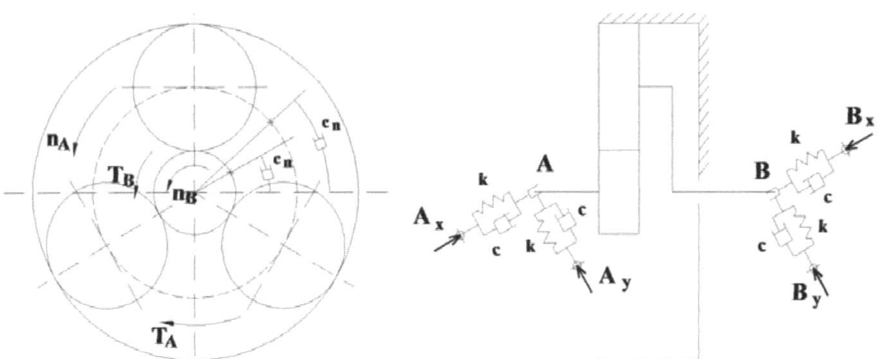

Figure 4. Mechanical model of a planetary gear train.

IOP Conf. Series: Materials Science and Engineering **252** (2017) 012034 doi:10.1088/1757-899X/252/1/012034

Concerning these models, the "hard" contact type. The viscous friction is taken into account trough the coefficient c_n, Fig. 4.

Significant part of the results related to the dynamic behavior of planetary gear trains and the analysis of the simulation research implemented are described in details in [22, 23].

Figure 5. Meshing of a planetary gear train model.

Based upon the theoretical FE model presented a variety of simulation experiments have been implemented. Some of the results are shown on Fig. 6 and Fig.7. Further results are shown and discussed in details in [24].

In order to dimension correctly the gear pairs in a planetary gear train concerning the criterion minimal vibrations, the obtained horizontal and vertical deviations of the axis of the central (sun) gear has to be limited. The determination of this deviation as a function of the time t is shown in a graphical way in Fig.6.

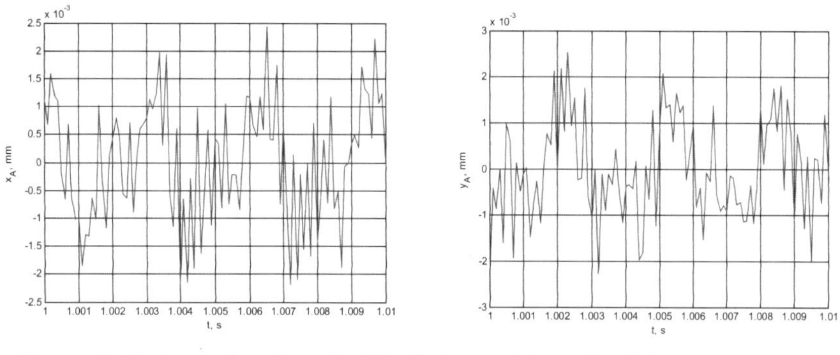

Figure 6. Horizontal and vertical deviations of the axis of the central gear

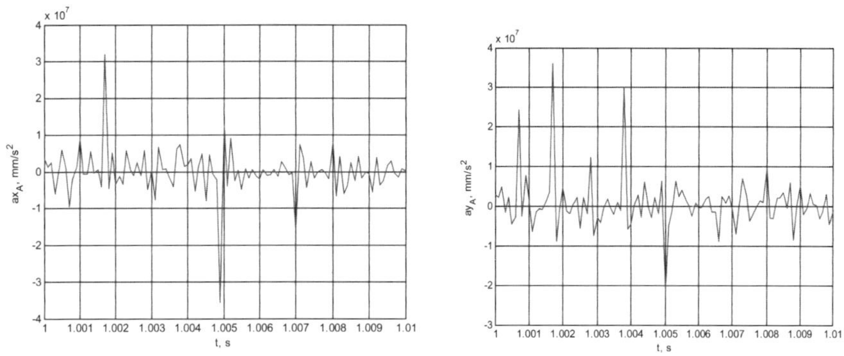

Figure 7. Speed of the horizontal and vertical vibrations of the axis of the central gear

The graphical results, which are shown on the diagram on Fig. 6 are obtained during the first investigated seconds of the t-interval. This approach gives the option to define precisely the special time interval, in which the maximal values of the horizontal and vertical deviations of the axis of the central gear occur. Based upon the experience of the authors' team, a decision was made to investigate the speed of the horizontal and the vertical vibrations (Fig.7) of the axis of the central (sun) gear. This precise visualization gives the opportunity to define accurately the special time interval, in which the maximal values of the speed of the horizontal vibrations of the axis of the central (sun) gear are to be observed.

8. Conclusions

The theoretical model creation considering the necessary assumptions, constrains and specific features of the planetary gear trains has led to the following conclusions:

1. The software Abaqus turns to be really effective enough for the process of contact modelling of planetary gear trains
2. Basic requirements concerning the design of planetary gear trains are taken in consideration
3. The models created are appropriate tools for simulation research of the dynamic behavior of planetary gear trains.

References

[1] Tharmakulasigam R 2009 Transmission Errors in Spur Gears *Brunel University Report*

[2] Mao K 2006 An Approach for Powertrain Gear Transmission Errors Prediction Using the Non-linear Finite Element Method *J. of Automotive Engineering* **220** pp. 1454 – 63

[3] Howard J 2004 The Torsional Stiffness of Involute Spur Gears *J. of Mech. Eng. Science* **218** pp 131 – 142

[4] Sommer A 2011 Vibration – based Health Monitoring of Multiple Stage Gear Train and Differential Planetary Transmission involving Teeth Damage and Backlash Nonlinearity *San Luis Obispo Report*

[5] Gu P 2013 On the Dynamic Simulation of Eccentricity Errors in Planetary Gears *Mechanisms and Machine Theory* pp 14-19

[6] Guo R 2010 Dynamic Modeling and Analysis of a Spur Planetary Gear Involving Tooth Wedging and Bearing Clearance Nonlinearity *European Journal of Mechanics A/Solids* **29** pp 1022 – 33

[7] Khabou N 2011 Study of a Spur Dynamic behavior in Transient regime *Mechanical Systems and Signal processing* **25** pp 3089 – 3101

[8] Gu P 2012 A Dynamic Model to Study the Influence of Planet Position Errors in Planetary Gears *J. of Sound and Vibration* pp 4554 – 74

[9] Ziegler P, Eberhard P and B Schweizer 2006 Simulation of Impacts in Gear Trains Using Different Approaches *Appl Mech* **76** pp 537–548

[10] Chen Z and Y Shao 2011 Dynamic simulation of spur gear with tooth root crack propagation along tooth width and crack depth *Engineering Failure Analysis* **18** pp 2149-64
[11] Kahraman A, Kharazi A and Umrani M 2003 A Deformable Body Dynamic Analysis of Planetary Gears with Thin Rims *J. of Sound and vibration* **262** pp 752-768
[12] Mao K 2007 Gear Tooth Contact Analysis and its Application in the Reduction of Fatigue Wear *Wear* **262** pp 1281-88
[13] Chen Y and Tsay C 2002 Stress Analysis of a Helical Gear Set with Localized Bearing Contact *Finite Element in Analysis and Design* **38** pp 707-723
[14] Mozafari H and Najafian S 2013 Stress Analysis and Study the Fracture Mechanics in Planetary Gear Set Using FEM *Acta Mechanica Slovaca* **17** pp 68 – 74
[15] Cooley C and Parker R 2013 Unusual gyroscopic system eigenvalue behavior in high-speed planetary gears *J. of Sound and Vibration* **332** pp 1820–28
[16] Guo Y and Parker R 2011 Analytical determination of mesh phase relations in general compound planetary gears *Mech. and Mach. Theory* **46** pp 1869–87
[17] Guo Y and Parker R 2010 Dynamic modeling and analysis of a spur planetary gear involving tooth wedging and bearing clearance nonlinearity *Eur. J. of Mechanics* **29** pp 1022–33
[18] Kim W, Lee J and Chung J 2012 Dynamic analysis for a planetary gear with time-varying pressure angles and contact ratios *J. of Sound and Vibration* **331** pp 883–901
[19] Gu X and Velex P 2012 A dynamic model to study the influence of planet position errors in planetary gears *J. of Sound and Vibration* **331** pp 4554–74
[20] Eritenel T and Parker R 2012 Three-dimensional nonlinear vibration of gear pairs, Journal of Sound and Vibration *J. of Sound and Vibration* **331** pp 3628–48
[21] Dassault Systemes 2014 *Abaqus 6.14 – Analysis user's guide*, **II**: Analysis
[22] Dobrev V, Stoyanov S, Dobreva A 2015 Design, Simulation and Modal Dynamics of Gears and Transmissions *International Conference on Gears* vol 3 (Munich) pp 695 - 707
[23] Stoyanov S 2015 Explicit dynamics of gear pair using finite element model *Scient. Proc. of University of Ruse* **54** pp. 67-71
[24] Stoyanov S, Dobrev V and Dobreva A 2016 Numerical Investigation of Planetary Gear Trains and Transmissions *5th Int. Conf. on Power Transmission BAPT (Ohrid)* vol 1, pp 155-162

Study on the Influence of the Convoy Rolling over Aerodynamic Resistance

D Iozsa[1], C Stan[2] and L Ilea[3]

Automotive Engineering Dept., POLITEHNICA University of Bucharest, 313 Splaiul Independentei, Bucharest, Romania

[1] daniel_iozsa@yahoo.com, [2] corraista@yahoo.com, [3] laurentiu.ilea@yahoo.com

Abstract. The aim of the study is to investigate how the aerodynamic resistance is influenced by the convoy rolling and to see how much this is possible by varying the distance between trucks. Then to see how the gains correlate with the position occupied by the truck in the convoy. The study starts from current research on the premises of running in convoy. Aerodynamic analysis was performed using software finite element of Computational Fluid Dynamics (CFD) type, where it was modeled the convoy rolling of a variable number of trucks. The number of trucks and the distance between them was varied in the model in order to acquire an understanding of the flow field around the trucks and how the distance between them can improve the aerodynamic parameters. The results are presented in the form of streamlines of the air, which indicates the air volume travel speed and direction and of the pressure distribution on the surface of the body. The most significant drop in pressure on the front surface was obtained for the second truck of the convoy, whereas for the following ones the reduction was less important. The participation in a convoy of more than two trucks is justified by the reduction of the whirls that appear and by the uniform air flow. The main advantage of running in convoy mode is to decrease aerodynamic resistance, with beneficial consequences on economic and ecological parameters. Continuing work from here on, it could be analyzed the impact of changing the distance between trucks on the aerodynamic coefficient. The results of CFD simulations need to be verified with experimental data, such as wind-tunnel test, to ensure reliability of the results.

1. Introduction

Transport is a vital part of each industry sector and our daily life. According to recent E.U. statistics, more than half of transport of goods is made by road (50.6%) [1], thus having a significant impact on the economy and environment. To reduce emissions footprint of heavy trucks, extensive research is necessary on all systems, in order to reduce energy consumption. This paper proposes to study the aerodynamic drag which is an important part of the total force that has to be exceeded by a moving vehicle [2].

To improve the aerodynamic performance of a vehicle body, studies are performed on air flow around the vehicle body and engine compartment in order to minimize the energy lost by vortices [3]. Studies performed at highway speed showed that aerodynamic losses can raise up to 55% [4]. Due to large projected area and massive shape, engineers are highly challenged to find ways of reducing aerodynamic drag. Thus, every solution can bring an important gain. Engineers started to conduct aerodynamic studies once the highway speed exceeded 80 km/h. Nowadays, studies are performed by research facilities all around the world. The result is a new design that is with approx. 40% aerodynamically improved as compared to models from mid-20th century.

2. Convoy concept

To keep reducing the aerodynamic drag, researchers are now focusing on the particular way of movement of trucks on the high-speed roads, one behind another. The term of "convoy" is now seen as a good way of improving the movement efficiency of goods by road. Experiments performed with CFD tools showed that reducing the distance between 2 or more moving trucks, can bring a significant reduction of drag for the vehicles included in the platoon, thus leading to energy save [5].

As seen in figure 1, the c_d value (drag coefficient -parameter that shows the aerodynamic performance) reduction for the second vehicle in the convoy depends on the distance between trucks. The c_d value for the third truck is even smaller, however, the c_d value difference between the second and the third truck is significantly smaller than the first and the second truck. This leads to the conclusion that by moving in a convoy, the total energy consumption is reduced and transport performance is increased.

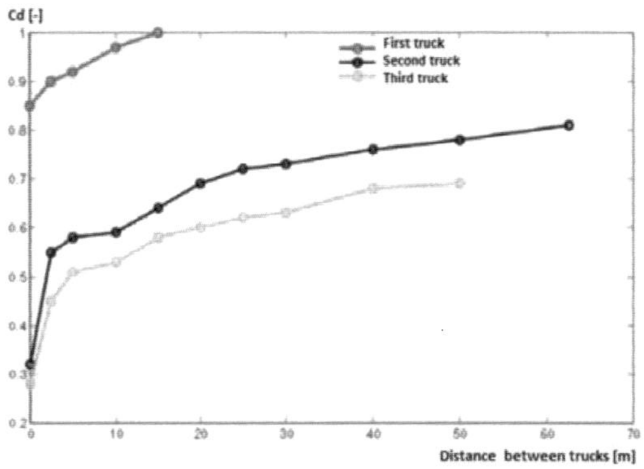

Figure 1. Influence of distance between vehicles on c_d, when moving in convoy [5]

The main constraint in reducing the distance between vehicles is technologic capacity of assuring safe travel condition – allowing all vehicles to stop at every moment, if necessary. Due to human factor, the minimum distance between vehicles to assure safe stop is not small enough to reduce the c_d value, at highway speed. To compensate this, introduction of autonomous functions are imperative. In a controlled environment, trucks equipped with all the necessary sensors and special trained drivers, managed to roll with a distance of just 3m between them (figure 2) [6]. Even if the test was not performed in heavy traffic conditions, this first step may be the down of autonomous driving trucks, where moving in convoy can become the general way of rolling on highways.

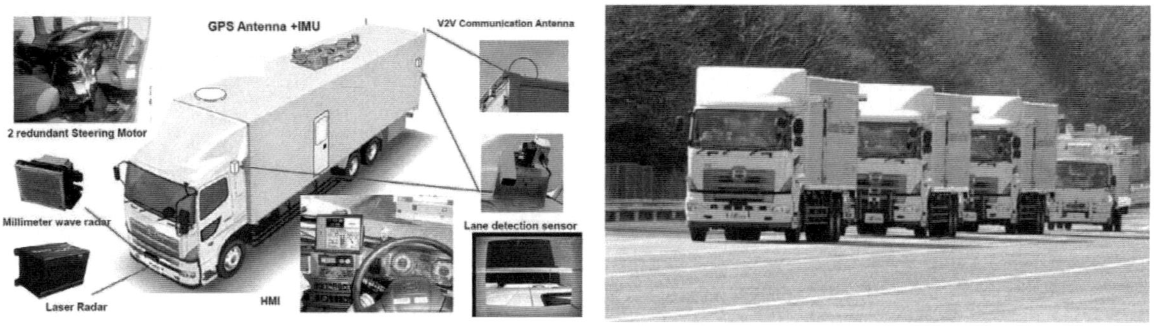

Figure 2. Trucks rolling in convoy, in controlled environment [6]

3. CFD analysis

The aerodynamic drag of trucks moving in convoy is studied in this paper by using finite element method and ANSYS software. The first step consists in defining the geometry of the 3D model (figure 3) - a rigid frame truck model with aerodynamic kit applied was chosen. The applied aerodynamic elements mimic the recent truck models and were validated in previous work.

The simulated air speed is 25 m/s. This value was chosen based on general speed limitation on EU roads, for this type of vehicles. Solutions that bring a gain at this speed will be amplified if the speed is increased.

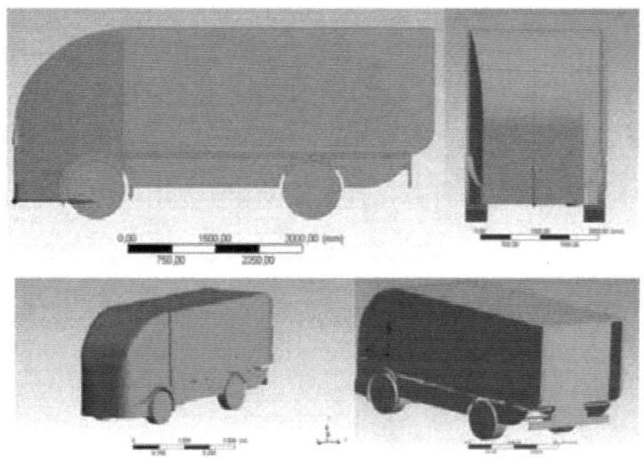

Figure 3. 3D model

The simulation consists in studying the air flow and the aerodynamic drag for a convoy of four identical trucks that follow each other at a three meters distance. This total number of trucks was chosen as a start point for the general picture of the air flow. The three meters distance between trucks was set as such because physical tests confirmed that present technology can sustain travelling in safe conditions at this distance.

This is a 3-parameters analysis: (i) vector analysis, (ii) air flow streamline and streamlines plotted on surface analysis and (iii) pressure plotted on surface analysis.

3.1. Vector analysis

In figure 4, vectors plotted on Y=0 plane show how the efforts are distributed near each truck of the convoy. It can be observed that the first truck encounters a higher resistance than the following ones (that travel in its wake).This may lead to the conclusion that the first truck will have a higher fuel consumption (similar to a truck that travels alone). The following trucks should have a reduced fuel consumption. Therefore, an improved average fuel consumption should be obtained leading also to a lower emission footprint - both environmental and financial advantages.

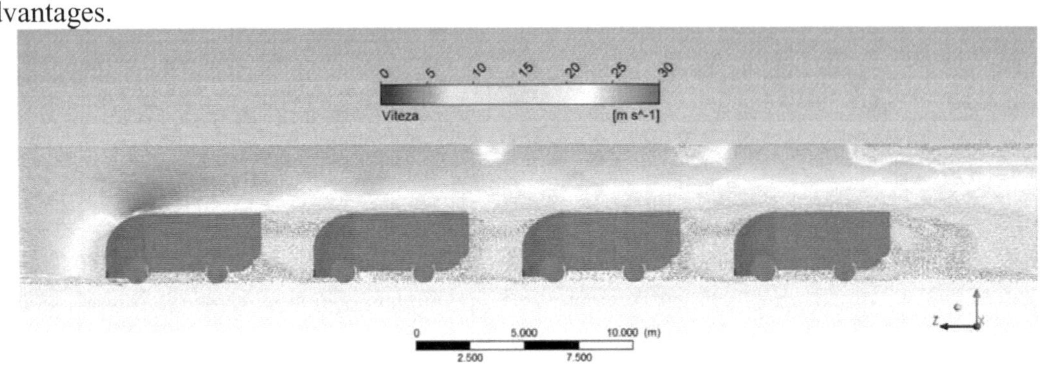

Figure 4. Air speed by vector at Y=0 for a 4 truck convoy, inlet air speed =25m/s

The wake between the trucks has similar shape, except the wake between first two trucks which does not follow yet the formed pattern between the other vehicles. This shows that the second truck has a significant advantage if part of the convoy. In going further, the aerodynamic resistance of the third and fourth trucks are very similar. Also, the picture shows how a stream of air with higher speed envelopes the convoy, thus reducing the aerodynamic drag.

3.2 Streamline analysis

The plotted streamlines (figure 5) show how the wake tends to have the same pattern. Therefore, the drag can be controlled easier by developing new aerodynamic parts for this rolling conditions. This picture also confirms the higher resistance encountered by the first truck.

The streamlines plotted on the surface of the trucks show how the 2nd and 3rd vehicle have very similar contour maps implying a similar air flow. It also can be seen that the second truck encounters a slightly higher turbulence than the following ones.

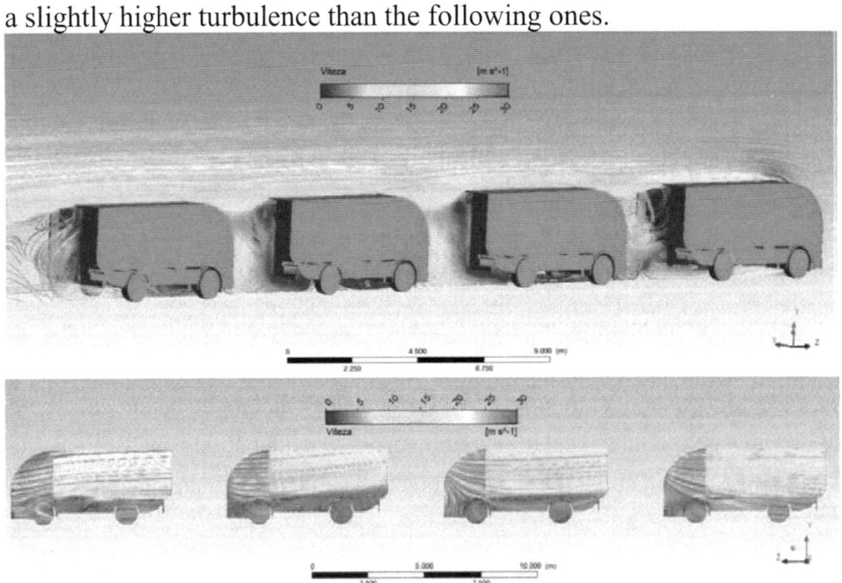

Figure 5. Air speed by streamline, isometric and Y=0 view for a 4 truck convoy, inlet air speed =25m/s

3.3 Pressure map

The air flow image (figure 6) is completed by plotting the pressure on the surface of each vehicle. By identifying the areas with different pressure values, two conclusions may be drawn in relation to: (1) the pressure distribution and (2) the area where the air detaches from the body, leading to vortex' birth. From the picture, it can be easily seen that the three trucks in the convoy have similar pressure patterns. By analyzing the pressure borders, we can identify areas of the body that may be improved in order to reduce energy losses – front part of cab, rear cab area.

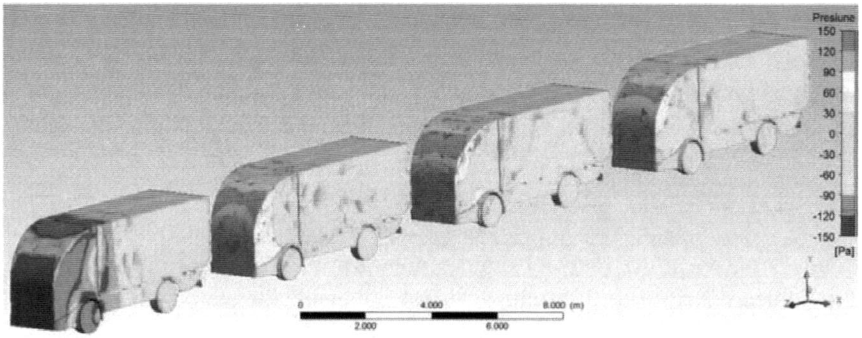

Figure 6. Pressure map on truck body for a 4 truck convoy, inlet air speed =25m/s

By computing the pressure on frontal panels we can obtain an overview of the resistance of each truck in the convoy and its advantage depending on its place in the convoy (figure 7). The image shows that the second truck encounters a pressure with almost 75% smaller than the convoy leader. For the following trucks, the pressure is still decreasing but with far less. It can be easily concluded that by reducing the aerodynamic resistance with convoy formation, the average fuel consumption is reduced. Considering the advantages brought by this solution it would be worth implementing the convoy formation in practice.

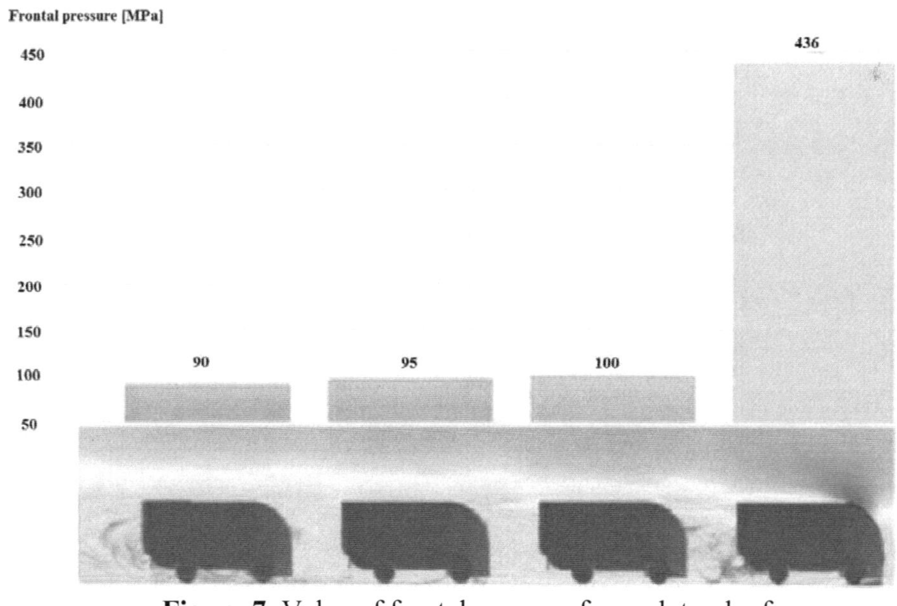

Figure 7. Value of frontal pressure for each truck of convoy

4. Conclusions

The perspective of new legislation on emission footprint (quite restrictive) and the competitive environment of business, drive engineers to come up with ideas that may seem eccentric. However, these are necessary if new regulatory framework should be met.

The convoy concept that was first inspired by nature wild life may bring a big advantage in fuel consumption by reducing the aerodynamic drag for those vehicles following the leader. The advantage is directly dependent on the distance between the vehicles. A distance of three meters between vehicles can reduce by 75% the aerodynamic resistance for the vehicles included in convoy.

A further study may focus on varying the distance between trucks in order to define a dependence equation between c_d and distance.

A direct consequence of traveling in a convoy is that, the trucks in the convoy (except the leader) may benefit from a reduction of engine cooling capacity. Due to reduced resistance, the engine can run at a lower rpm, which implies lower cooling requirements. Therefore, an optimization of cooling system should be performed. On the other hand, the lack of direct air on heat exchangers may lead to a higher energy consumption for cooling module. Most likely, the increased electric energy requirement will not cover all the aerodynamic advantage from convoy formation, so the balance should be still positive.

References

[1] Freight transport statistics - modal split, Eurostat, Statistics explained, May 2017, ec.europa.eu/eurostat/statistics-explained/

[2] Andreescu C *Dinamica autovehiculelor pe roti* vol. 1, Ed. Politehnica Press, 2010

[3] Iozsa M D *Caroserii pentru autovehicule rutiere*, Ed. Politehnica Press, 2016

[4] Patten J, McAuliffe B, Mayda W and Tanguay B *Review of aerodynamic drag reduction devices for heavy trucks and buses,* National Research Council Canada, 2012

[5] Lammert M, Duran A, Diez J, Burton K et al. *Effect of platooning on fuel consumption of class 8 vehicles over a range of speeds, following distances and mass,* Commercial vehicle, SAE, 2014

[6] Ashley S *Convoys of wireless-linked semi-autonomous vehicles could soon be hitting our roads, giving drivers a chance to put their feet up on the morning commute,* BBC Future, November 2014, www.bbc.com/future/

Research on optimization of test cycles for comfort to the special vehicles

Marian MITROI[1], **Anghel CHIRU**[2,1]
[1]"Ovidius" University of Constanta, Department of Mechanical Engineering, Bd. Mamaia 124, Romania
[2]"Transilvania" University of Brasov, Department of Mechanical Engineering, Bd. Eroilor 29, Romania
Corresponding autor: e-mail (ejju_marian@yahoo.com), (achiru@unitbv.ro).

Abstract. The comfort of vehicles, regardless of their type is represent a requirement to by fulfilled in the context of current technological developments special vehicles generally move under different soil, time, or season conditions, and the land in which the vehicles move is complex and varied in the physical structure. Due to the high level of involvement in the driveability, safety and comfort of automotive, suspension system is a key factor with major implications for vibration and noise, affecting the human body. The objective of the research is related to determining the test cycles of special vehicles that are approaching real situations, to determine the level of comfort. The evaluate of the degree of comfort will be realized on acceleration values recorded, especially the vertical ones that have the highest influence on the human body. Thus, in this way the tests can be established needed to determine the level of comfort required for each particular type of special vehicle. The utility of the test cycles to optimize comfort is given to the accurate identification of the specific test needs, depending on the each vehicle.
Keywords: comfort, optimizing, vibration.

1.Introduction

Subjectively speaking, the comfort of the vehicles is closely related to a series of factors that together generate the comfort and satisfaction of those inside it and have the effect of safely moving at a satisfactory speed without producing negative reactions by point of view psycho-physiological of the body or damage to the materials and the transported technique.

The feeling of comfort is a very complex ensemble of physical and psychological physiological responses, which appear in response to the multitude of factors that act upon it, in the time to travel with the vehicle. Vehicle displacements in rough terrain often cause human body disturbances due to preponderant shocks of over 2g, which cause substantial force loads, damage to the musculoskeletal structure. Their influence is more evident when produced at the natural resonance frequency of the human body - 5Hz. About the way how affect the comfort the unevenness of roads and road shocks, at the car seat level, a number of notable Huston & Zhao researches (2000) **[1]** were carried out. Nar et al (2008) **[2]** mentions that shifts by exposure for 1 hour to such vibrations cause chronic muscle fatigue and structural damage to the spine. Matsumoto & Kunimatsu (2010) **[3]** mentions also as that the human body has a high degree of sensitivity at certain frequency ranges: 4-6Hz in the vertical direction and 8-12Hz in the horizontal direction. The research in the comfort domain revealed that the most important factors that have direct influence on it are: suspensions, seats and its facilities, vibrations and noise produced during the journey.

Corroborating these aspects, we can understand the importance of establishing test cycles for vehicles to determine the degree of comfort provided by special vehicles, on different road categories. In this direction, the research objectives have been established to obtain significant data on the vibrations felt at the level of the special motor vehicle structure, following the approach of real and varied routes as a physical structure, as well as the establishment of reference cycles for identify the level of comfort. These cycles will be correlated with current comfort rules and requirements, based on the analysis of recorded data, to determine new, more realistic tests valid for this type of vehicles.

2. Establishing test cycles

During the research processes, design, development and operation of the vehicles, are highlighted the necessity and the importance of the tests.

These activities contribute in a definitive way to their further and permanent improvement. The test cycles (figure 1) are of particular importance and are carried out in order to verify that the technical and qualitative characteristics of the systems are exactly fulfilled, but also to check the optimizations that can be applied later. These include tests on different ground and climate conditions.

Figure1. Testing trails: a-highway road, b-paved road, c-road with unevenness, d-road with obstacles

Thus, by optimizing the comfort test cycles, all the qualitative spectrum is constantly improved, new technical and constructive solutions are identified.

Technically speaking, the main aspect of importance for vehicle comfort is the choice of component parts of the suspension system. This is a compromise between the maneuverability and comfort of the vehicle. Special vehicles (military) is traveling In very varied terrain, time or season, much more different than those predetermined in the homologation tests.

They are moving in terrains with a very complex physical structure and involves dynamic forces much higher than expected. Because of these situations, the vehicles must ensure good maneuverability and adequate comfort. The development of the research are considering to identify correctly the routes that correspond to the realities of the displacements to which the special vehicles are subjected, the analysis of the main aspect affecting the comfort factor and the subsequent establishment of some test cycles in order to determine the passengers' comfort, for these types of vehicles. The approach of land areas with a varied structure, under the conditions imposed by the specific actions, allows the correct identification of the values that affect the state of comfort. To establish the correct cycles, vertical accelerations are of particular importance due to the negative influence on the human body. Their amplitude value is directly proportional to the discomfort created. The choice of cycles is determined by the needs and demands of special vehicles during mission trips.

In the research process, a route was identified that could accommodate several areas of the comfort test, which corresponds to the actual displacements performed. The cycles chosen for the tests highlight the level of mechanical oscillations induced by the route.In the following image (figure 2) we can see the route and the topography of the terrain, during the researches.

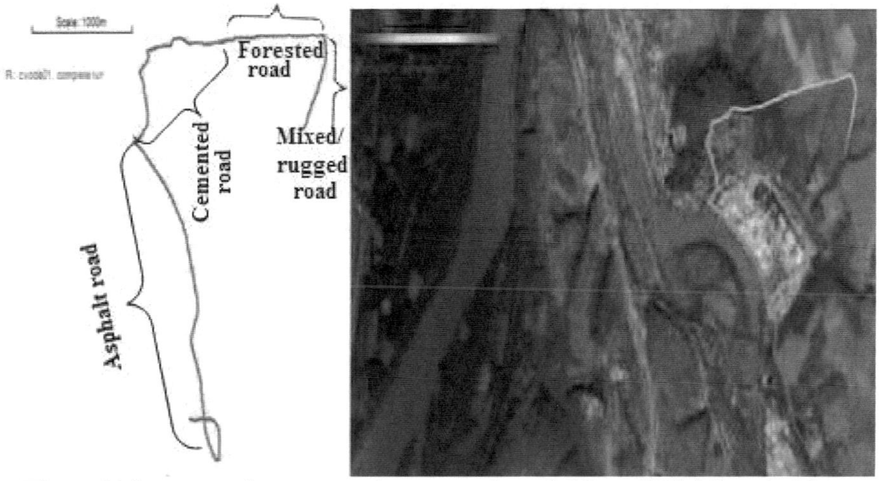

Figure2.The route where research was conducted: Vbox-GPS / Google Map

The established cycles follow the values of the vertical accelerations recorded on each part of the field, by moving the vehicles at speeds imposed by the created situation of the action, and maintaining the safety and the minimum stability of the vehicle. Thus, the amplitudes of recorded vertical oscillations characterize the level of comfort of the vehicle's personnel, by reference to ISO 2631-1 specifications **[4]**.

3. The making out research

For the development of the researches, an area has been established up, that comprising various sections of relief, as well as different physical structures of the terrain.

Thus, the area of Cernavoda met the requirements imposed by the research. The tests were carried out with a special VW T4 full-traction vehicle (figure 3), used by the Gendarmerie in specific tactical actions. In research development, has been foreseen that data acquisitions will follow the values of vertical accelerations by passing through a variety of terrain: asphalt with small irregularities, concrete-industrial, earth-forest, mixed-rugged.

Figure3. Special vehicle VW T4 in rugged terrain – Gendarmerie

The physical variety of the lands (figure 4) is given by the complexity of the travels of the special vehicles during the specific actions.

Figure4. The rugged trails in which special vehicles are moving

The equipment used for the data acquisition was established in two categories: the first category is allocated for the collection of data on the topography of the land, and the second one for the recording of the accelerations specific to the sections of the land in which the travels is performed. The equipment chosen for data acquisition about vibrations is produced by RACELOGIC Technology, respectively a Vbox module with GPS antenna and triaxial accelerometers type RLVBIMU 03. In accordance with the ISO 8608 **[5]** for a value of the land unevenness index w = 2, in the following (table 1) road classes and their specific values are presented, depending on the spectral density characteristic (PSD).

Table1. The road classification according to ISO 8608

Road Class	$G_H(\Omega_0)$ $[10^{-6}\ m^3/rad]$ ($\Omega_0 = 1 rad/m$)		
	Lower limit	**Upper limit**	**Geometric mean**
A	-	2	1
B	2	8	4
C	8	32	16
D	32	128	64
E	128	512	256
F	512	2,048	1,024
G	2,048	8,192	4,096
H	8,192	-	16,384

However, the value of the "w" index is not uniform for all road classes, in reality its value is up to w = 3,5. By changing its value, low frequencies generated by the road profile become more pronounced. This affects the more sensitive human body in this frequency range. At the same time, a small value of the index generates high frequencies, which affect the comfort and response of the vehicle suspension system **[6]**. The Power Spectral Density (PSD) factor reflects the vibration response of the vehicle and in fact represents the unevenness of the road, given by the planarity index "C" and the undulation "w". The estimation of parameters C and w is mentioned in the standards: ISO 8608 and EN 13036-5.

For a vehicle traveling on a rough road with a speed (v), the road non-uniformity relation (PSD) can be described by an equivalent mathematical function in the spatial domain G_H (Ω) in the time domain G_H (ω):

$$G_{H(\omega)} = G_{H(\Omega)/v} \tag{1}$$

where: ω = angular frequency [rad/s]; $\Omega = \omega v$.
Knowing that the frequency of excitation in the time domain is given by the relationship:

$$f = \frac{\omega}{2\pi} = \frac{v}{\lambda} \tag{2}$$

and that, vibrations generally propagate through the structure of vehicles with frequencies ranging from f = 0.5-15 Hz, and the human body is very sensitive on the intervals: f = 1-2 Hz; 4-8Hz; 10-12Hz, we can say that this range is covered by waves which satisfies the relationship:

$$0{,}5Hz \leq \frac{v}{\lambda} \geq 15Hz \tag{3}$$

During the research, the speed of movement showed a lot of variations but, according to the recorded data, for the areas of land of which particular interest, were registered the maximum travel speed V_{max} = 59 km/h and the average V_{mid} = 27,4 km/h. Reconstructing the relationship (3) we can find the value of the wavelength of the frequencies generated by the road in which the vehicle moved, on the vibration interval:

- for V_{max}: 59km/h

$$16{,}38\, m/s = 15Hz \cdot \lambda \Rightarrow 1{,}092m^{-1} \tag{4}$$
$$16{,}38\, m/s = 0{,}5Hz \cdot \lambda \Rightarrow 32{,}76m^{-1} \tag{5}$$

- for V_{mid}: 27,4km/h

$$7{,}61\, m/s = 15Hz \cdot \lambda \Rightarrow 0{,}507m^{-1} \tag{6}$$
$$7{,}61\, m/s = 0{,}5Hz \cdot \lambda \Rightarrow 15{,}22m^{-1} \tag{7}$$

On the basis of the values obtained, we can understand that for the sections of road where the travel speed was higher, the road roughness did not generate significant oscillations for the vehicle, so it did not create a high degree of discomfort. In comparison, the much more difficult route that generated totally different approach situations, the λ values were high, which shows a great discomfort. Analyzing the data from the following graph (figure 5) we can see the influence of the factor λ on the movement of the vehicle through the rough areas of the land where the travel was made.

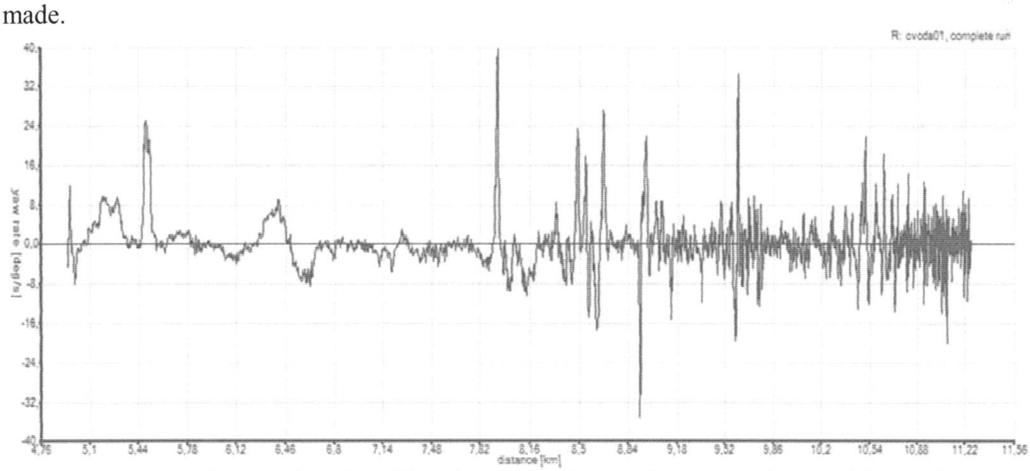

Figure5. The speed under which the vehicle it transits the roughness of the road

IOP Conf. Series: Materials Science and Engineering **252** (2017) 012036 doi:10.1088/1757-899X/252/1/012036

4. Results and data analysis

Road is structure with the most important for the vehicles, from a dynamic point of view.

Due to this, features such as vehicle viability and comfort are influenced by the road type and its irregularities as well as the speed of travel. Ensuring comfort regarding the vibration for the driver and transported persons depends on the entire shock and vibration damping system. Taking into account the profile of the routes in which the travel was made and their physical characteristics, we can understand that the vertical, but also the longitudinal or lateral accelerations were completely different. Also, in this respect, we can say that the accelerations have manifested differently from the front axle at the rear axle. The wheels, not go in aligned way on the unevenly road profile so the values on the right side of the vehicle are different from those on the left. In the following (fig. 6), can be observed along the route the values of longitudinal, lateral and vertical accelerations, in accordance with the altitude of the terrain.

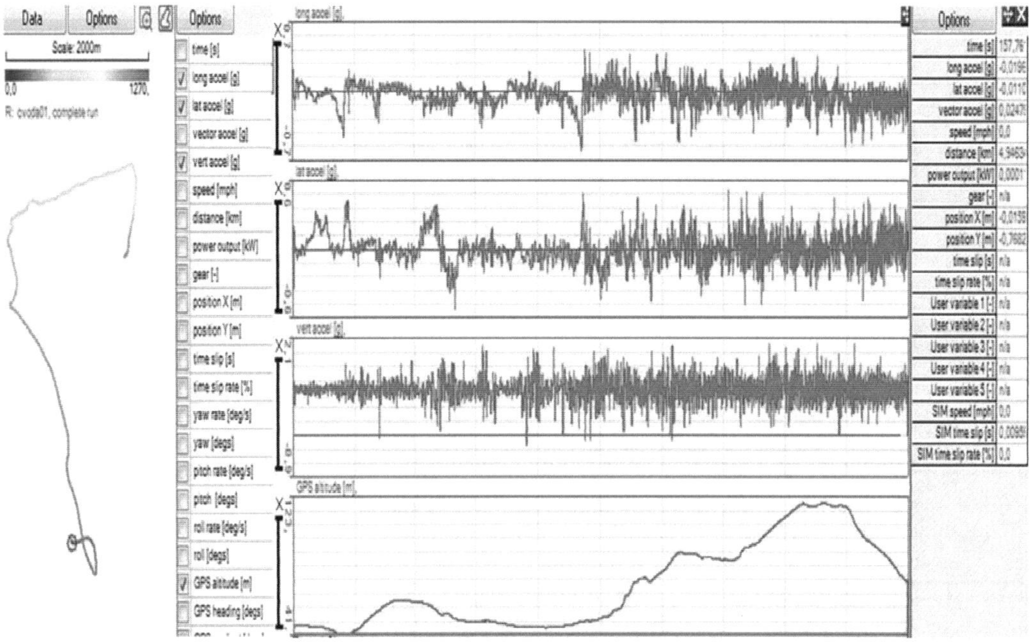

Figure6. The acceleration values recorded according to the altitude of the terrain

By analyzing the route, the vertical acceleration values of each section of the road were identified and compared. Each route, as shown in the following picture (figure 7), clearly revealed the discomfort created by the personnel through the recorded values.

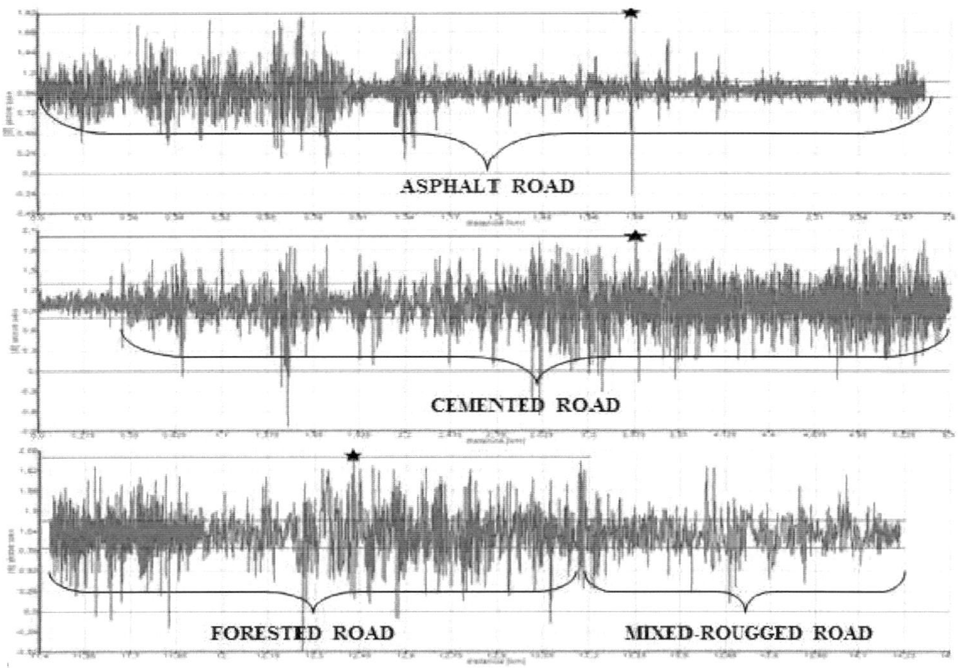

Figure7. Vertical (unfiltered) accelerations recorded on each section of road

The level of mechanical oscillations amplitude was more pronounced in the concrete road sector, due to the speed of the vehicle. High values also occurred in the forest and mixed-road sectors, but they were lower due to the limited speed of the vehicle during the movement of the vehicle on the route. Thus, the accelerations were predominantly in the range of 1.2-1.5 m/s^2 with multiple manifestations in the frequency range of 1.8 m/s^2. These values demonstrate a very high degree of discomfort and a risk to the health of the body. In the same way, the forestry section indicated many vertical acceleration values of 1.2 m/s^2, which is also a high level of discomfort. In this case the travel speed was lower due to the pronounced ground deviations. However, discomfort also manifested itself in this case. The asphalt road sector was of no particular importance, except for a small portion where the road was built by overlapping asphalt layers, which created non-uniform displacement.
In according with ISO 2631-1 (Annex C) and BS 6841 **[6]**, in the following table (table 2) can be observed weighted values of the accelerations and how their influence on the people comfort.

Table 2. Guidelines ISO 2631 and BS 6841 for vibration and human perception

Values *(m/s^2)*	**Perception**
<0.315	Not uncomfortable
0.315–0.63	A little uncomfortable
0.5–1.0	Fairly uncomfortable
0.8–1.6	Uncomfortable
1.25–2.0	Very uncomfortable
>2.5	Extremely uncomfortable

The ISO 2631-1 standard specifies that after the value 1,25m/s^2 RMS, status uncomfortable is increasing. Therefore, in the conditions of the route that was taken during the research driver has felt different conditions of discomfort.
The sensitivity of the body to vibration depends on the direction and the character of the oscillatory movement, but also on the subjective perceptions of each individual.

The oscillations produced by the vehicle during the travel represent a lot of movements, in different directions, having different frequencies and amplitudes, even in the case of simple sinusoidal excitations. In the following (figure 8), can also be observed the values of the recorded longitudinal / lateral accelerations depending on the terrain: asphalt-concrete-forest-mixed, the speed of the vehicle and the altitude of the terrain.

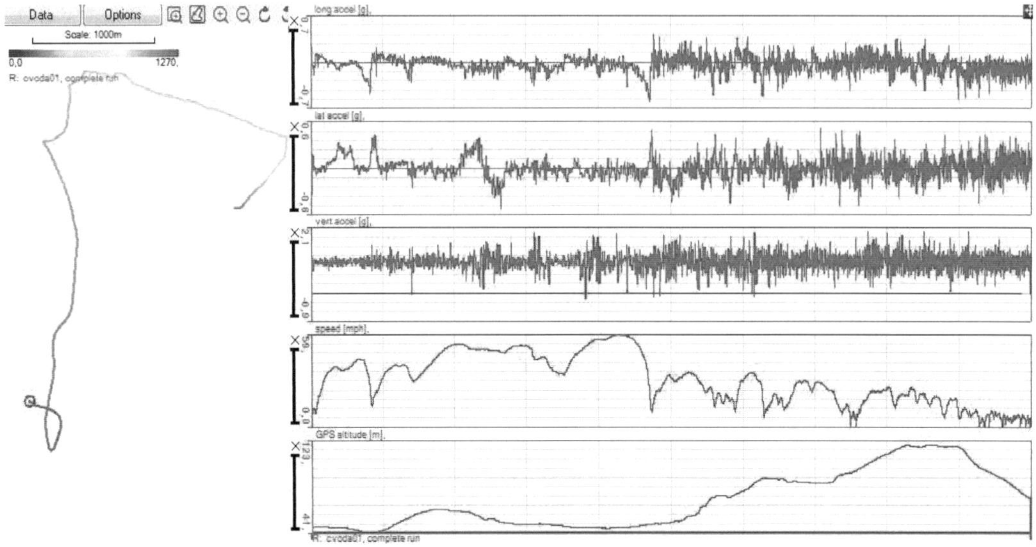

Figure8. Longitudinal / lateral accelerations recorded relative to the speed and altitude of the terrain

According to the above image (figure 8), we can understand that the altitude of the terrain created problems to travel with the speeding due to its physical characteristics.
Also, the structure of the ground through physical irregularities has limited the uniform movement of the vehicle, generating frequent crossings over obstacles, which has created a high level of discomfort to staff, especially through the generated shocks. The portion of concrete road and forest road indicated significant comfort problems through the high and constant values of vertical accelerations. In accordance with the international standard VDI 2057, the following table (table 3) presents the scale of subjective perceptions of comfort.

Table3. Thresholds of perception and subjective appreciation of comfort

Step	The threshold of perception "K"	Subjective appreciation
A	0,1	It does not feel
B	0,25	Appearthe first sensations
C	0,63	Minimum level of sensations
D	1,6	It feels good
E	4	He feels strong
F	10	Very strong level of perception
G	25	Very strong level of perception
H	63	Very strong level of perception

Factor K sets the human body's tolerance limits to periodic vibrations for the frequency range of 0.5-80 Hz. The VDI 2057-1 standard recommends in practice the summation of the squares of the partial comfort indices of the n different variations.

$$K = \sqrt{\sum_{i=1}^{n} k^2 i} \quad (8)$$

IOP Conf. Series: Materials Science and Engineering **252** (2017) 012036 doi:10.1088/1757-899X/252/1/012036

where: ki = partial comfort index; n = number of component vibrations.

The evolution of vertical accelerations recorded throughout the test cycles can be seen in the following figure (figure 9). Consistent with this, we can say that the reference values for determining the level of comfort are those included in the concrete and forest road section, which have accelerations predominantly above the value of the comfort threshold: 1.2 m/s^2.

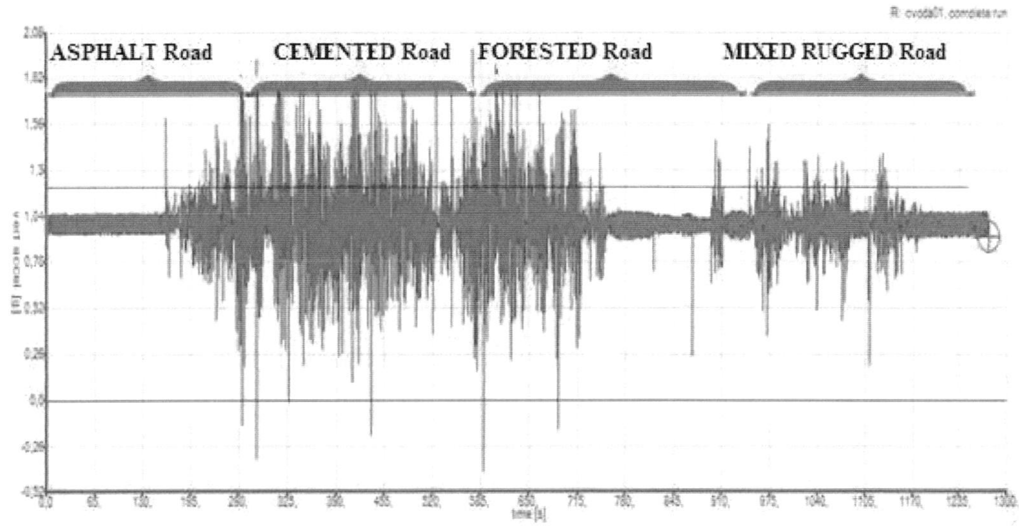

Figure9. Vertical accelerations recorded throughout the entire route

After result of the general analyzes we can see in the following table (table 4) the random vibration values recorded after filtration and mediation.

Table4. The values of random vibration recorded in the all sectors of road

Road tipe	Direction "x"		Direction "y"		Direction "z"	
	Peak acc. m/s^2	Peak acc. m/s^2	Peak acc. m/s^2	Low acc. m/s^2	Low acc. m/s^2	Low acc. m/s^2
Asphalt	0,26	0,1	0,49	0,1	1,56	1,04
Concrete	0,56	0,1	0,55	0,12	1,82	1,2
Forested	0,3	0,12	0,6	0,1	1,98	1,3
Mixed / stone	0,47	0,13	1,2	0,49	1,56	1,16

From the above we can understand that on some portions of the route the lateral and horizontal vibrations are slightly elevated, which signifies a feeling of inconvenience at the threshold of acceptance. However, due to the structure of the land, these types of vibrations were received by the structure of the vehicle by summing them up. So, the feeling of inconvenience due to vertical vibrations is was also amplified by these. To deepen the phenomenon and more accurately determine the degree of inconvenience, VDV $_{total}$ is the most appropriate index for assessment at the level of the seat structure. It can be established with the help of the following relationship, indicated by the international standard BS 6841:

$$VDV_{tota} = \left(VDV_{xs}^4 + VDV_{xy}^4 + VDV_{xz}^4 + VDV_{xb}^4\right)^{1/4} \quad (9)$$

where:

VDV^4xs, VDV^4xy, VDV^4xz, = vibrations on the three directions in the seat cushion;

VDV^4xb = longitudinal vibrations (x) received at the seat backrest.

Compared to the standard test methods, which, although providing for trails with different characteristics: bumps, inclinations, physical structure, the research presents more relevant aspects of the travels due to the values recorded on each part of the road. These include the real terrain and conditions in which the journeys are made, the speed of travel, aspects that are very different from those currently set as a reference for this category of motor vehicles.

The defining aspects of research are:

- travel speed - is on the limit of safety, all most of the time;
- the route followed - is the shortest way to reach the goal, regardless of structure and obstacles.

Its features are very different from the standard ones.

If for normal journeys within cities the methods established by the norms are fully satisfactory, in analyzing the degree of comfort of the staff in case of displacements in rugged areas, it is necessary to approach some more challenging routes as a structure and as a methodology of realization. The optimization of these comfort test cycles is necessary due to the infrastructure in the areas where journeys the special vehicle, but also because of the structure of the terrain.

5. Conclusions

Vehicle comfort is a very complex feature at present, due to the many factors involved in generating it.

Road roughness information is very important in assessing the vehicle's dynamic response, so in appreciating the comfort of the staff. For the research carried out are of importance, the specifications of the road classes on which the displacements (B-G) is performed.

By using the artificial road profile derived from PSD, we can study the vibration response of the vehicle and optimize the suspension parameters. The correct identification of some road sectors that corresponds to the actual movements of special vehicles makes it possible to understand correctly the phenomena that occur during the journey.

Determining accurately the acceleration values induced by routes allows the establishment of comfort limits by reference to existing international standards. Determining the degree of comfort in relation to the land being approached facilitates the optimization of the test cycles at the most realistic level of the trips.

The correct identification of the vibrations phenomena that occur at the level of the structure of special vehicles, during the travel to actions through very rough terrain and their reporting to the biological manifestation of the human body allows us to correctly understand the negative effects they produce.

Reference:

[1]. Huston D R, Zhao X., Whole body shock and vibration: Frequency and amplitude dependence of comfort, *Journal of Sound and Vibration*, **Vol.230**, 2000.

[2].Nar M J, Hosseini F. et al., - Index for Vehicle Acoustical Comfort Inside a Passenger Car, *Applied Acoustics* **Vol.69 (4),** 2008.

[3]. Matsumoto Y, Kunimatsu S., – Evaluation of Perception Thresholds of Horizontal Transient Vibration in Lateral Direction, *Proceedings of the 2010 Autum Meeting the Institute of Noise Control Engineering of Japan*, p.183-186, 2010.

[4]. ISO 2631-1 /1997 – Mechanical vibration and shock-Evaluation of human response to whole-body vibration. **Part 1**: General requirements.

[5].ISO 8608 /1995 – Mechanical vibration – Road surface profiles, Reporting of measured data.

[6].Türkay S, Akcay H., - Spectral Modeling of Longitudinal Road Profiles, **IEEE 28th** *Canadian Conference on Electrical and Computer Engineering (CCECE)*, 2015.

[7].BS 6841/1987- Guide to measurement and evaluation of human exposure to whole-body mechanical vibration and repeated shock, England.

[8].VDI 2057-1/2002-Human exposure to mechanical vibrations - Whole-body vibration. Germany.

An approach for multi-objective optimization of vehicle suspension system

D Koulocheris, G Papaioannou[1] and D Christodoulou

Vehicles Laboratory, School of Mechanical Engineering, National Technical University of Athens, 15780 Athens, Greece

[1]gpapaioan@central.ntua.gr

Abstract. In this paper, a half car model of with nonlinear suspension systems is selected in order to study the vertical vibrations and optimize its suspension system with respect to ride comfort and road holding. A road bump was used as road profile. At first, the optimization problem is solved with the use of Genetic Algorithms with respect to 6 optimization targets. Then the k – ε optimization method was implemented to locate one optimum solution. Furthermore, an alternative approach is presented in this work: the previous optimization targets are separated in main and supplementary ones, depending on their importance in the analysis. The supplementary targets are not crucial to the optimization but they could enhance the main objectives. Thus, the problem was solved again using Genetic Algorithms with respect to the 3 main targets of the optimization. Having obtained the Pareto set of solutions, the k – ε optimality method was implemented for the 3 main targets and the supplementary ones, evaluated by the simulation of the vehicle model. The results of both cases are presented and discussed in terms of convergence of the optimization and computational time. The optimum solutions acquired from both cases are compared based on performance metrics as well.

1. Introduction

The multibody dynamics have been used extensively by automotive industry so as to model and design the vehicle and its parts i.e. suspension system, chassis, tires. Depending on the complexity and the depth of each study, various models have been developed, so as to describe the dynamic behavior of the vehicle. These models vary from simple (Quarter Car Model [1]) to more complex and detailed ones (Half Car Model [2] or Full Car Model [3, 4]), simulating the tyre, the body and the suspension system of the vehicle. The latter is considered one of the most important subsystems of the vehicle, because not only it transmits the loads from the lower part of the vehicle to the upper, but also it ensures the comfort of the travel by absorbing any unwanted vibrations. Through the suspension system, various aspects of the dynamic behavior can be studied, with the main ones being the ride comfort of the passengers and the road holding of the vehicle [5]. The aforementioned characteristics depict one of the main conflicts in the automotive industry: the trade – off between the ride comfort and the handling of the vehicle. The suspension systems are simulated in multibody dynamics with springs and dampers. In literature, different systems can be found, depending on the needs of each study. The most common ones are the passive systems [6], but the need to adjust the characteristics of the vehicle in order to be able to cope with different road surfaces, has led to the development of the semi – active [7] and active suspension systems [4]. The main disadvantage of these suspension types is that they can be too complex and expensive at times. In order to achieve similar behaviour of the vehicle, certain nonlinearities can be introduced in the components of the passive suspension system. These nonlinearities were investigated in previous work [5], [8] and are discussed in this paper as well. Finally, in order to investigate how the passengers are affected by the driving experience various models have been introduced simulating also the body of the driver and the seats of the vehicle [9]. In this way, the vibrations that are transmitted to the driver can be studied in depth taking into consideration also ISO 2631 which illustrates methods to evaluate the human exposure to the vibrations induced by the road.

The rapid growth of modern optimization methods has been beneficial to the study of the suspension systems. Taking advantage of the versatility of the methods, as well as the advances in computer analysis, fast optimal solutions to the key parameters of the suspension systems are produced. Various optimization problems have been formed, involving either one optimization target (single – objective SOO) or multiple (multi – objective MOO). The MOO offers an insight of the conflicted targets that were mentioned before, whereas the SOO results in a much simpler and non-time consuming study. The most common multi – objective optimization method is the Pareto Front, where the different targets are separated throughout the optimization process, and the result of the method is a set of optimal solutions. The Pareto Front optimization method has been extensively used in the study of the dynamical behavior of the vehicle with the use of different targets including both the ride comfort of the passengers as well as the road holding of the vehicle [5]. Different characteristics of the performance of the vehicle have been used in order to achieve better optimization results such as the root mean square of the body acceleration, the variance of the tire forces [3] among others. The great variety in the possible objective functions often leads to an exaggeration in the selected number of targets, which increases the complexity of the problem in question.

As it was mentioned above, the result of the Pareto Front method is a set of optimal solution. Each one of these solutions is optimal in respect to the targets involved in the study. The selection of the optimum solution lies upon the designer, who is responsible of evaluating the solutions based on requested demands. This process can be tiresome and time – consuming, due to the potentially large number of optimal solutions and the complexity of the problem in question. In order to overcome this problem, various methods of solutions sorting have been developed. The aim of these methods is to sort the solutions, according to various indexes, aiming to produce one optimum solution or to decrease the number of optimal solutions significantly. Das (1999) introduced the k – optimality method which is based on the partial dominance between the solutions [10], and later Gobi et al. (2006, 2013) expanded this theory, launching the k – ε optimality method, which shows not only the dominance between different solutions but also the measure of their dominance [11].

The object of this paper is to optimize the parameters of the nonlinear suspension system of a heavy vehicle using the Pareto Front optimization method and extract one optimal solution based on the k – ε algorithm. In order to save computational time, a new approach is proposed, which involves the separation of the optimization targets to primary and supplementary ones, and it is described in the sections below. This paper is organized as follows: in Chapter 2 a brief description of the models used is provided, in Chapter 3 the road excitation applied is illustrated, in Chapter 4 the optimization methods are described in detail, while in Chapter 5 the optimization procedure and the results are highlighted and finally in Chapter 6 conclusions are presented.

2. Simulation Models

In the current study, the dynamic behavior of a heavy vehicle is investigated, as far as the vertical vibrations induced by the road surface are concerned. Thus, a half car model was chosen, including the front and rear axle of the vehicle allowing the pitch phenomena to be observed.

As shown in Figure 1, the vehicle model consists of three basic subsystems: the tires, the suspension systems and the body of the vehicle. The latter is considered as a rigid body of mass m, equal to the half of the total mass of the vehicle. The distance of the front and rear unsprung masses (m_F and m_R) from the center of mass is equal to a_F and a_R respectively. The front and rear tires are modelled as linear springs, which receive as input the irregularities of the road profile. The governing equations of the model were described in detail in previous work [5].

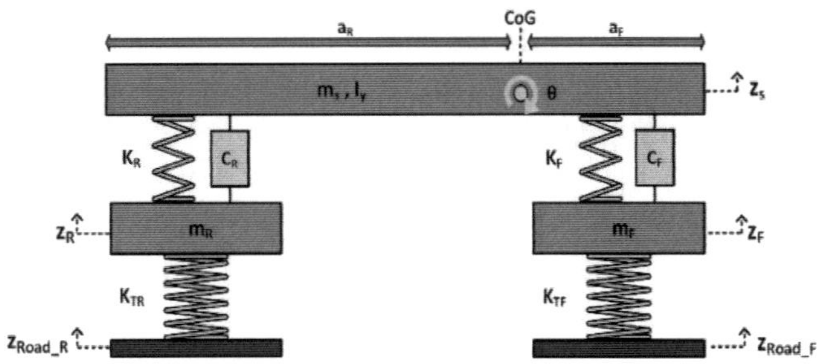

Figure 1. Half Car Model.

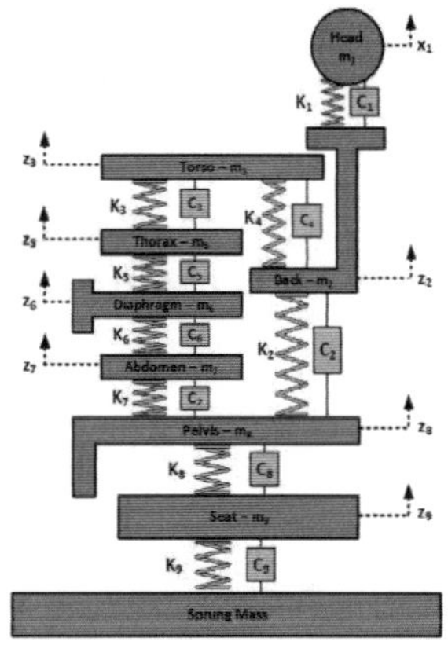

Figure 2. Seat – Driver Model.

Table 1. Nomenclature of vehicle parameters.

Parameters		**Subscripts**	
z	Vertical motion coordinate	**S**	Body
θ	Pitch motion coordinate	**F**	Front
z_{road}	Road excitation	**R**	Rear
m	Mass	***l***	Linear
C	Damper's coefficient	**n*l***	Non – Linear
K	Spring's stiffness coefficient	**T**	Tyre
a	Distance of the centre of the mass of the vehicle		

Table 2. Parameters of the Half Car Model.

Parameters	**Values**	**Parameters**	**Values**
m_s [kg]	2200	**a_F [m]**	1.61
I_z [kg·m^2]	1142	**a_R [m]**	1.67
m_F [kg]	50.00	**K_{TF} [N/m]**	$4.00 \cdot 10^5$
m_R [m]	100.00	**K_{TR} [N/m]**	$8.00 \cdot 10^5$

Ideally, a suspension system should adjust its characteristics in order to function properly under different road conditions, ensuring the comfort of the travel and the safety of the vehicle. This adjustment is achieved through either the use of controllers (active suspensions) or through added nonlinearities in the components of the suspension (springs and dampers). These nonlinearities are achieved by adding a nonlinear term in the suspension spring force which is described mathematically by equation (1) and represents a Duffing oscillator which could simulate with higher accuracy spring like variable coil spring or variable coil pitch.

$$F_{spring} = K_l \cdot x + K_{nl} \cdot x^3 \tag{1}$$

where x is the suspension travel. In Tables 1 and 2 the parameters of the model are explained and their values are given.

Additionally, the seat and the driver of a vehicle are modelled in order to investigate in depth the ride comfort of the passenger, as depicted in Figure 2. Different parts of the human body such as the pelvis, the diaphragm, the thorax etc. are described via several m – c – k subsystems. This model was based in the work of Abbas et al. (2010) [13], the equations of the model as well as the parameters of each body part are taken from this research paper.

Important quantities or metrics indicating the dynamical behavior of the vehicle model are the suspension travel and the tire deflection. The suspension travel of the suspension is the term $(z_S - z_F - a_F \cdot \theta)$ and $(z_S - z_R - a_R \cdot \theta)$ for the front and rear suspension respectively and the tire deflection is $(z_F - z_{Road_F})$ and $(z_R - z_{Road_R})$ for the front and rear tire respectively. Both quantities are key parameters in the optimization study that follows in the next chapters. Other important metrics of the dynamical behavior of the vehicle are the tire forces which are described in equation (2) and the ratio of the tire force to the static force between the tire and the ground [12]. If the value of this ratio exceeds 1, then loss of contact between the wheel and the ground is indicated.

Through the results of the passenger model, the ride comfort could be measured via the Vibration Dose Value (VDV) and the Crest Factor (CF) which are calculated through equations (3) and (4), respectively. These characteristics are proposed by the ISO 2631 standard which evaluates the human exposure to whole – body vibration.

$$VDV_{head} = \left(\int_0^T \left(a_{wh}(t)\right)^4 dt\right)^{\frac{1}{4}} \quad (3)$$

$$\mathrm{CF}_{\mathrm{head}} = \frac{\max(\mathrm{acc}_{\mathrm{head}})}{rms(acc_{head})} \quad (4)$$

3. Road excitation

One of the important aspects in the analysis of the dynamic behavior of a vehicle is the road profile that is used as an input function. Road profile generation is a strong asset for the researchers, as it enables them to test the vehicle under different road conditions. In this research, a road bump was generated as it is shown in Figure 3, using a sinusoidal function [7]. .

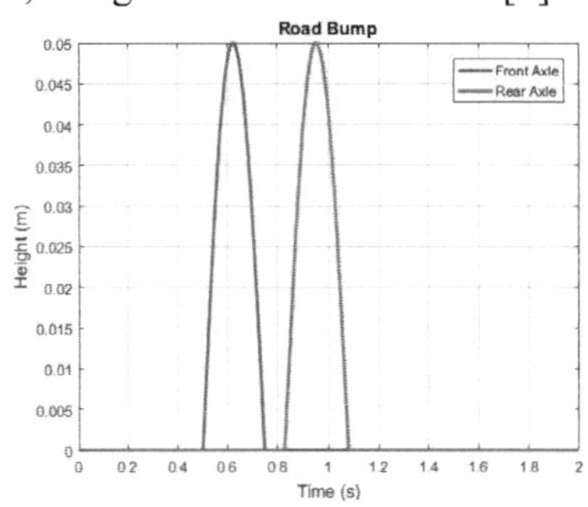

Figure 3. Road Bump

The geometrical aspects of the bump were chosen as follows: its height was set to 0.05m and its length was set to 2m. The velocity of the vehicle was considered constant, at 10 m/s. Finally, a time lag between the front and rear axles of the vehicle was applied. Specifically, the two wheels follow the same trajectory with a time delay due to the distance $a_F + a_R$ between them ($t_{distance} = \frac{a_F + a_R}{V}$), as it is shown in Figure 3.

4. Optimization

In this work, a multi – objective optimization approach was followed. The optimization problem is expressed as a problem of minimization of the objective functions, with the selected design variables being subjected to a number of inequality constraints [5].

In multi – objective optimization the objective function is not a scalar number, as in SOO problems, but a vector. Thus, in order to rank the different solutions, the concept of partial ordering is adopted. The Pareto dominance, which is used in this paper, is based on this concept. The solutions of the Pareto set are non – dominated by any other solution and are considered as equally optimal solutions [5]. When dealing with complex problems, the size of the Pareto set might be quite large, increasing significantly the workload so as to select one optimum solution based on this set. In order to bypass this problem, many algorithms have been developed. Their goal is to reduce the size of the Pareto set or even extract one optimum solution to the problem. Das (1998) introduced the k optimality method, which is based on the partial dominance of the Pareto solutions, locating solutions which achieve a better trade – off between the optimization targets and eventually "offer something more" than the other optimal solutions. The solutions are sorted using the k index, which is a measure of their optimality and can only acquire integer values [10]. Based on the aforementioned method, Levi et al (2005) introduced the k –∑ optimality method, which is implemented in this paper, which not only vets the as k-method, but also measures the entity of this variation. In k-e, the k index can receive not only integer values on contrary with k method, but also real values. In this way, a continuous degree of optimality is achieved. These two methods were described in detail in previous work [5].

5. Optimization procedure.

The objective of this report is the optimization of the dynamic behavior of a heavy vehicle under specific road conditions. In the multi – objective approach that was adopted, the design variables were chosen as the key parameters of the suspension systems, and are shown in equation (5).

$$\text{cdrhf mu` qh aklr<Z } {}_{k_E}\text{: B}_E\text{: J }{}_{k_Q}\text{:B}_Q\text{:J }{}_{nlk_E}\text{:J }{}_{nlk_Q}\backslash \quad (5)$$

Where K_F, K_{nlF}, K_R and K_{nlR} are the linear and nonlinear coefficients of the front and rear springs and C_F and C_R are the linear coefficients of the front and rear dampers. The upper and lower bounds of these design variables, were chosen based on previous work conducted by the authors and are shown in Table 3 [5, 8]. As far as the constraints of the problem are concerned, they were selected in regard to the ride comfort of the passengers and the nonlinear part of the suspension systems, demanding the rms of the sprung mass acceleration to be less than 0.8 m/s^2 and a 10 – 30% usage of the nonlinear part of the suspension ($\frac{K_{nl} \cdot x^3}{K_l \cdot x + K_{nl} \cdot x^3}$), based on equation 1) through all the simulation. [5].

Table 3. Upper and Lower Bounds of the design variables.

Design Variables	Bump	
	Lower Bound	**Upper Bound**
J $_{k_E}$+J $_{k_O}$ (N/m)	$3.2 \cdot 10^4$	$1.5 \cdot 10^5$
C_F, C_R (N/m·s)	$2.0 \cdot 10^3$	$1.0 \cdot 10^4$
J $_{nlk_E}$+J $_{nlk_O}$ (N/m^3)	$5 \cdot 10^5$	$3 \cdot 10^8$

The selection of the vector of the objective function is of vital importance to the optimization procedure. As it was mentioned above six optimization targets were selected (equations (6) – (11)): the variance of the acceleration of the body of the vehicle (f_1), the mean of the variances of the front and rear suspension system (f_2), the mean of the variance of the front and rear tire deflections (f_3), the vibration dose value of the head (f_4), the crest factor of the head (f_5) and the variance of the pitch angle (f_6). The targets f_1, f_4, f_5 are linked to the ride comfort of the passengers whereas the remaining three are involved with the road holding of the vehicle. In the first segment of the analysis, a multi – objective optimization problem was solved with all the aforementioned targets involved. In the latter part of the analysis, the targets were split into two groups: the main and the supplementary ones. In the first group, the variance of the body acceleration (f_1) and the mean variances of the suspension travels (f_2) and the tire deflections (f_3) were included. The remaining three targets were used as supplementary ones: the targets f_4 and f_5 enhance the ride comfort of the passenger whereas target f_6 is linked with the road holding of the vehicle. The Pareto optimization procedure was applied in the half – car model.

$$f_1 = \mathrm{var}(\mathrm{acc}_{\mathrm{sprung\ mass}}) \quad (6)$$

$$f_2 = 0.5 \cdot [\mathrm{var}(\mathrm{suspension\ travel}_F) + \mathrm{var}(\mathrm{suspension\ travel}_R)] \quad (7)$$

$$f_3 = 0.5 \cdot [\mathrm{var}(\mathrm{tire\ deflection}_F) + \mathrm{var}(\mathrm{tire\ deflection}_R)] \quad (8)$$

$$f_4 = \mathrm{VDV}_{\mathrm{head}} \quad (9)$$

$$f_5 = \mathrm{CF}_{\mathrm{head}} \quad (10)$$

$$f_6 = 10^4 \cdot \mathrm{var}(\theta) \quad (11)$$

6. Results

In this section the two resulting optimal set of solutions are presented. In both cases a threshold of 13% of the maximum value of every objective function was selected, based on previous work [5]. In Figure 4 the k - ∑ levels of optimal solutions are depicted for both cases. For the first case, which is presented in Figure 4 (a), the k – ∑ algorithm has detected one optimum solution of level 4, whereas for the second case (Figure 4 (b)) the algorithm has detected 3 solutions of level 4. As far as the second case is concerned, the solution with the maximum k value was selected in order to compare the optimum solutions of the two problems.

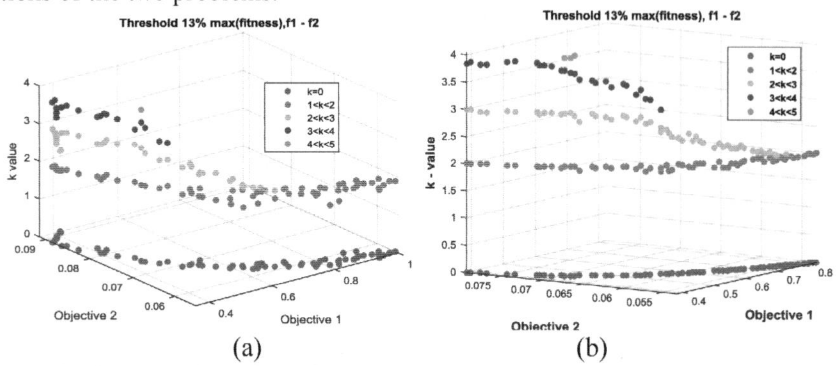

Figure 4. Optimal solutions $f_2=f_2(f_1,k)$, k – ε levels (a) 6 targets case, (b) 3+3 targets case

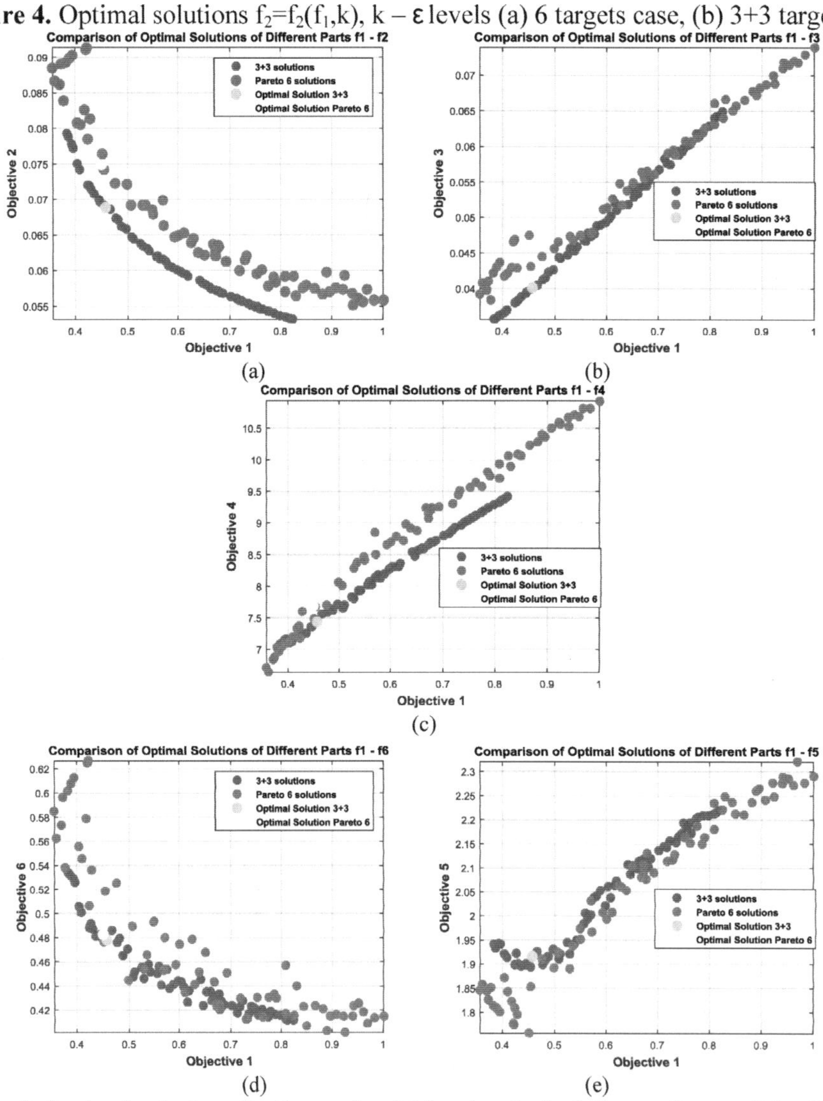

Figure 5. Optimal solutions for Pareto Set (a) k-e levels (b-f) Pareto fronts of the Objectives

Whereas, in Figure 5 the Pareto fronts for all targets in respect to the first objective are depicted. Both sets of optimal solutions are compared: with blue dots stand the solutions of the 3+3 optimization targets case whereas the red ones represent the solutions of the 6 optimization targets case. Furthermore, the optimum solutions of each set are depicted with different colour (green / yellow), based on the results of the $k-\varepsilon$ algorithm, which was applied in both cases. The results of both cases confirm the supplementary character of the 3 last target: Figure 5(a) and Figure 5(e) follow the same pattern where the same applies for the other three Figures (5(b) – 5(d)). Moreover, the conflicting relationship between the different optimization targets is highlighted: the targets f_2 and f_6 are in prone of the suspension travel of the vehicle, so if their values are decreased then the value of the f_1 target, which indicates ride comfort, increases (Figure 5(a) and 5(e)). On the other hand, targets f_4 and f_5 represent the ride comfort, so their values are proportional to the ones of the first objective target (Figure 5(c) and 5(d)). Additionally, the effect of f_3 to the ride comfort and the road holding is illustrated in Figure 5c, a subject which was investigated in depth in a previous work [8].

The optimum solutions of each set are compared in Table 4. In Table 3 the values of the optimization targets are presented. In both tables Solution 1 stands for the optimum solution of the 6 optimization targets case, whereas Solution 2 stands for the optimum solution of the 3+3 optimization targets case. The two solutions are similar, but the solution of the 3+3 optimization problem dominates in 4 out of 6 optimization targets.

Table 4. Comparison of the design variables and the optimization targets of 2 solutions.

Design Variables	Values		Objective	Values	
	Solution 1	Solution 2		Solution 1	Solution 2
K_{lf} (N/m)	70449	59645	f_1	0.450	0.456
C_f (Ns/m)	5322	5306	$f_2(\cdot 10^2)$	7.487	6.881
K_{lr} (N/m)	64000	46881	$f_3(\cdot 10^2)$	4.156	4.011
Cr (Ns/m)	3256	4675	f_4	7.680	7.430
K_{nlf}(N/m^3)	$2.88\cdot 10^7$	$2.11\cdot 10^7$	f_5	1.820	1.910
K_{nlr}(N/m^3)	$1.74\cdot 10^7$	$2.72\cdot 10^7$	f_6	0.482	0.478

An important aspect of this analysis concerns the dynamic behavior of the vehicle. In Table 4, the resulting values of the design variables are depicted. Comparing the values it is shown that, the solution of the 3+3 targets case has produced softer suspension systems compared to the 6 targets case, without neglecting the road holding. This is depicted also in the fact that the root mean square of the body acceleration is in the same range for both solutions (~0.67 m/s^2) without risking the road holding of the vehicle in any of the cases, and this is not only a success of the alternative approach in the optimization procedure but also the advantage of the non-linear suspension system, discussed in detail in previous work [5, 8]. Discussing the dynamical behavior of the vehicle, the solution of the 3+3 targets case produced almost equal values of the maximum tire force only in the front suspension system (3578N), while in the rear suspension system the value of the tire force is higher (3908 for solution 1 and 4805 for solution 2), due to the lower suspension travel of the rear suspension system. Moreover, the maximum absolute value of the η index are also in the same range for both solutions (~0.161 for the front suspension system, while for the rear the values are ranging from 0.171 – 0.212), way far from the limit of 1 which indicates the wheel lift off.

7. Conclusions

To sum up, this paper focuses on the comparison of two approaches for the solution of the same optimization problem, this of the optimum suspension design of a heavy vehicle, which involve different handling of the optimization targets. In the first approach, all the desired targets were introduced in the optimization process and then the resulting solutions were implemented in the $k-\varepsilon$ algorithm. In the second one, only the main targets of the study were introduced in the optimization algorithm and the rest of them were calculated later. Finally, the resulting set of solutions was implemented again in the $k-\varepsilon$ algorithm. The optimum solutions were compared in terms of the values of each target through figures, the design variables, and important metrics describing the dynamic behaviour of the vehicle.

As it was illustrated in previous chapters, the solutions and the Pareto front produced by the 3+3 targets case dominate the ones of the 6 targets case in terms of the values of the optimization targets as well as the dynamic behaviour of the vehicle is concerned. It is important to mention that by increasing the complexity of an optimization problem does not necessarily means that a better solution will be produced. As it was pointed out in this work, better results might be produced if the number of the main optimization targets is selected appropriately, and more focus is given in the selection of one optimum solution after the optimization process is finished by using as factors of the selection any other supplementary optimization targets. By adding more targets in the selection of the optimum solution, proved to be a more efficient process than adding them in the main optimization process. Apart from delivering a "better optimal" solution, is also less time consuming, due to the decreased complexity of the optimization process. Further work is in progress in order to extend the conclusions of this study and present them in more details.

References

[1] Mitra A C, Desai G J, Patwardhan S R, Shirke P H, Kurne W M and Banerjee N, 2016 "Optimization of Passive Vehicle Suspension System by Genetic Algorithm." *Procedia Engineering* **144** : 1158-66.

[2] Gadhvi B, Savsani V and Patel V, 2016, Multi-Objective Optimization of Vehicle Passive Suspension System Using NSGA-II, SPEA2 and PESA-II. *Procedia Technology*, **23,** 361-68.

[3] Shirahatti A, Prasad P S S, Panzade P and Kulkarni M M, 2008, "Optimal design of passenger car suspension for ride and road holding." *Journal of the Brazilian Society of Mechanical Sciences and Engineering* **30.1**: 66-76.

[4] Gobbi M, Levi F, and Mastinu G, 2006 "Multi-objective stochastic optimization of the suspension system of road vehicles", *Journal of sound and vibration*, **298.4**, 1055-72.

[5] Koulocheris D, Papaioannou G and Christodoulou D, 2017 "Optimal design solution among pareto alternatives for vehicle nonlinear suspension system", *Proceedings of 26th International Automotive Conference "Science and Motor Vehicles" (26th JUMV),* Beograd, Serbia, 2017.

[6] Alkhatib R, Jazar G N, and Golnaraghi M F, 2004, "Optimal design of passive linear suspension using genetic algorithm." *Journal of Sound and vibration* **275.3** : 665-91.

[7] Georgiou G, Verros G, and Natsiavas S, 2007. "Multi-objective optimization of quarter-car models with a passive or semi-active suspension system." *Vehicle System Dynamics* **45.1** : 77-92.

[8] Koulocheris D, Papaioannou G and Christodoulou D, 2016 "Assessment of the optimization procedure for the nonlinear suspension system of a heavy vehicle." *Mobility and Vehicle Mechanics* **42.2**: 17-35

[9] Gündoğdu Ö, 2016 "Optimal seat and suspension design for a quarter car with driver model using genetic algorithms." *International Journal of Industrial Ergonomics* **37.4** : 327-32.

[10] Das I, 1999 "A preference ordering among various Pareto optimal alternatives." *Structural optimization* **18.1**: 30-35.

[11] Levi F, Gobbi M, Farina M and Mastinu G. 2004, "Multi-objective design and selection of one single optimal solution." *International Mechanical Engineering Congress and Exposition. American Society of Mechanical Engineers (ASME 2004).*

[12] Ataei M, Asadi E, Goodarzi A, Khajepour A and Khamesee M B. 2017. "Multi-objective optimization of a hybrid electromagnetic suspension system for ride comfort, road holding and regenerated power". *Journal of Vibration and Control,23*(5), 782-93.

IOP Conf. Series: Materials Science and Engineering **252** (2017) 012038 doi:10.1088/1757-899X/252/1/012038

Experimental investigations of tungsten inert gas assisted friction stir welding of pure copper plates

M A Constantin[1,3], A Boşneag[1], E Nitu[2] and M Iordache[2]

[1] University Politehnica of Bucharest, Machine Building Technology Department, Splaiul Independenţei Street No. 313, Romania
[2] University of Pitesti, Manufacturing and Industrial Management Department, Târgul din Vale Street No.1, Romania

[3] constantin_marius.adrian@yahoo.com

Abstract. Welding copper and its alloys is usually difficult to join by conventional fusion welding processes because of high thermal diffusivity of the copper, alloying elements, necessity of using a shielding gas and a clean surface. To overcome this inconvenience, Friction Stir Welding (FSW), a solid state joining process that relies on frictional heating and plastic deformation, is used as a feasible welding process. In order to achieve an increased welding speed and a reduction in tool wear, this process is assisted by another one (WIG) which generates and adds heat to the process. The aim of this paper is to identify the influence of the additional heat on the process parameters and on the welding joint properties (distribution of the temperature, hardness and roughness). The research includes two experiments for the FSW process and one experiment for tungsten inert gas assisted FSW process. The outcomes of the investigation are compared and analysed for both welding variants. Adding a supplementary heat source, the plates are preheated and are obtain some advantages such as reduced forces used in process and FSW tool wear, faster and better plasticization of the material, increased welding speed and a proper weld quality.

1. Introduction

Friction Stir Welding is a solid state joining process that relies on frictional heating and plastic deformation realised at the interaction between a non-consumable welding tool that rotates on the contact surfaces of the workpieces. The welding tool is positioned vertically on the material and then moved at welding speed along the joint line. The plasticized material is transferred behind the tool, forming a welded joint, figure 1. FSW involves the joining of two metal pieces at the molecular level without melting. Due to its solid state nature FSW has many benefits over fusion welding techniques. The main advantage of this is its ability to weld alloys which are mostly non-weldable using fusion welding methods due to problems with oxidization, solidification, shrinkage, sensitivity to cracking and the resultant porosity problem [1].

The final disposal canister for spent nuclear fuel is the most critical application for FSW of copper. The other available welding method for sealing the corrosion barrier canister is electron beam welding (EBW). The microstructure and mechanical properties (especially tensile strength and elongation to fracture) of copper FSW welds are very similar to those of the hot worked base materials and superior as compared to EB welds [2].

The microhardness of FSW welded joints is the mechanical property most analyzed in the scientific papers. This is due, on the one hand, to the fact that microhardness depends on the microstructural changes, the grain size and the precipitation state generated by the welding process, and on the other hand, because there is a good correlation between the microhardness profile and the profile of the other mechanical properties (Rp0.2, Rm, A%, ...) [3], these correlation is making from the microhardness a marker of mechanical properties.

Research in this field has shown that FSW welding microhardness values are different depending on the nature of the joined materials. For cooper alloys' increasing the heat input leads to annealing softening and lower hardness. Lowering the heat input leads to refined grain size and eventually to equal or even higher hardness of the welds as compared to the base material [4 - 7].

The aim of this paper is to identify the influence of the additional heat generated from tungsten inert gas source on the process parameters, especially on welding speed, in order to identify the possibility of increasing the productivity of the process and on the welding joint properties (temperature distribution, hardness and roughness).

2. Experimental program

2.1. Experimental stand

The experimental stand includes a FSW welding machine, a TIG welding head, an orientation and fixing device, and various data acquisition, recording and monitoring systems, figure 2. Welding processes are monitored by analyzing the temperature and axial force recorded values throughout the process.

During the process, the temperature was measured by using a high-speed and high-sensitivity thermographic infrared camera (FLIR A40M). The used camera has a field of temperature measurement between -40 [^{0}C] and + 2000 [^{0}C]. The compressive force was measured with a mechanical device that has a fixed force transducer, type AM, with range between 0 [KN] and 20 [KN], assembled on the main spindle. The FSW machine and the information captured are extracted using a special soft.

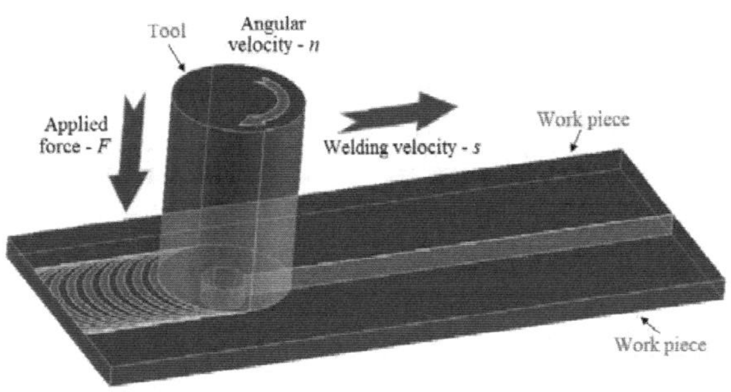

Figure 1. A schematic presentation of the friction stir butt welding process

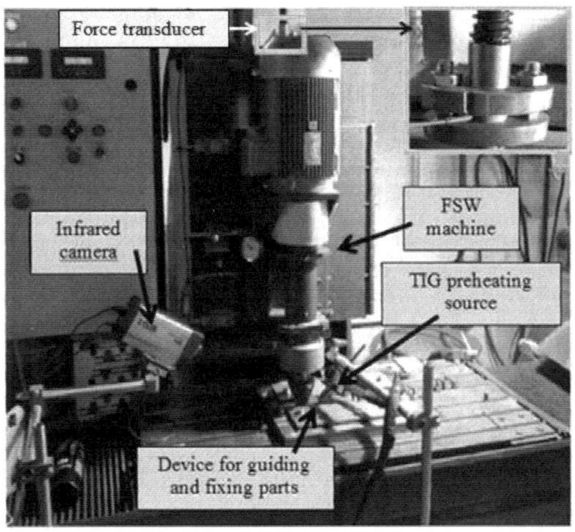

Figure 2. Experimental stand of TIG assisted friction stir welding process

2.2. Welded material

Copper has a face-centred cubic (fcc) crystal structure and it doesn't undergo phase changes after solidification. Melting point of copper is 1084 °C. Other properties of copper are including high thermal and electrical conductivity, formability, and corrosion resistance [8]. Because of high thermal conductivity, welding copper and its alloys is usually difficult to join by conventional fusion welding processes but this characteristic makes it feasible for solid state joining processes such as FSW.

The effective chemical composition of the Cu-DHP (phosphorus–deoxidized copper) sheet used in the experimental program is provided in the table 1 and the effective mechanical properties in the table 2.

Table 1. Chemical composition of Cu-DHP

Alloying element	Percentage [%]
Cu	99,9
P	0,015 - 0,04

Table 2. Mechanical characteristics of Cu-DHP

Name of the characteristic	Value
Tensile strength, Rm	260 MPa
Yield strength, Rp0,2	206 MPa
Vickers hardness, HV0,3	81 HV0,3
Elongation at break, A5	40 %

All the workpieces were cut by having the same rolling direction and more precisely with the side which the joint will be made parallel to the rolling direction of the initial blank, figure 3.

2.3. *Welding tool*

The welding tool used has a classical monoblock structure with conical pin and is made of **P20+S** (carbide of sintered tungsten), figure 3.

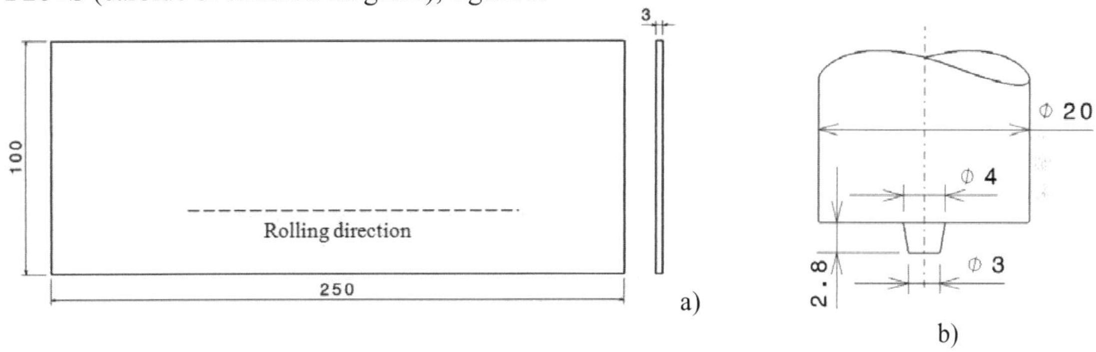

Figure 3. Geometry and dimensions of the workpieces (a) and of the welding tool (b)

2.4. *Experiments*

The experiments on FSW and TIG assisted FSW processes were developed for butt welding Cu-DHP sheets, table 3. The process' parameters which were varied are the rotational speed of the tool [rpm] and the welding speed [mm/min].

Table 3. Experimental plan of FSW and TIG assisted FSW processes

Exp. Nb.	Rotational speed [rpm]	Welding speed [mm/min]	TIG welding current [A]	Exp. Cod.
1	800	150	-	I.6
2	1200	150	-	I.3
3	1000	250	100	II.3

2.5. *Sampling of specimens*

From the welded joints, samples were cut for macroscopic and microscopic analysis, for measurement of microhardness and roughness, but also for traction and bending attempts. This paper presents and analyzes the results regarding the distribution of temperature, axial force, hardness and roughness values resulted from experiments. The relative positioning of the samples which microhardness and roughness were measured according to the geometry of the welding bead is shown in figure 4 and figure 5. These were cut from the beginning and from the end of the joint for highlight how microhardness and roughness are affected by process stability and local process parameters.

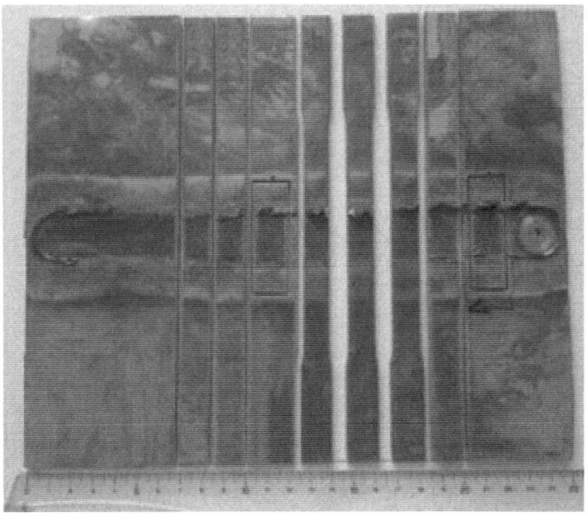

Figure 4. Photo with the samples cutted from the resulting welded joint

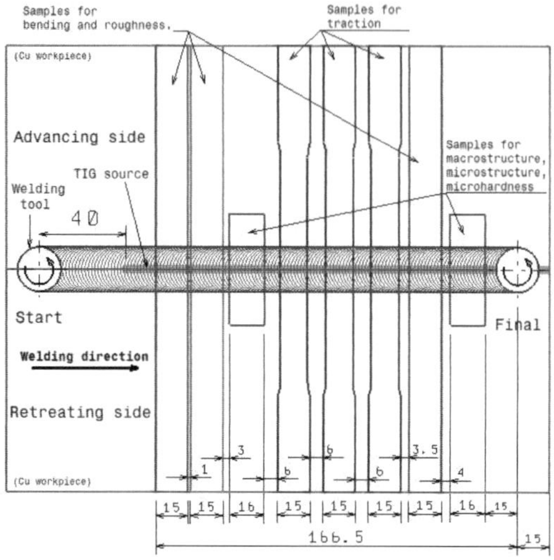

Figure 5. The relative position of the specimens used to determine microhardness and roughness

3. Results and discussions

3.1. Temperature and Plunging force analysis

In order to highlight the local conditions which the joining was made in the areas where the microhardness and roughness was determined, but also to highlight the moment from which the values of the compressive force become approximately constant during the experiments (and more precisely the moment when the process becomes stable), the two process parameters, the temperature and the axial force, were represented on the same diagram, having as background the specimen sampling scheme, figures 6 - 8.

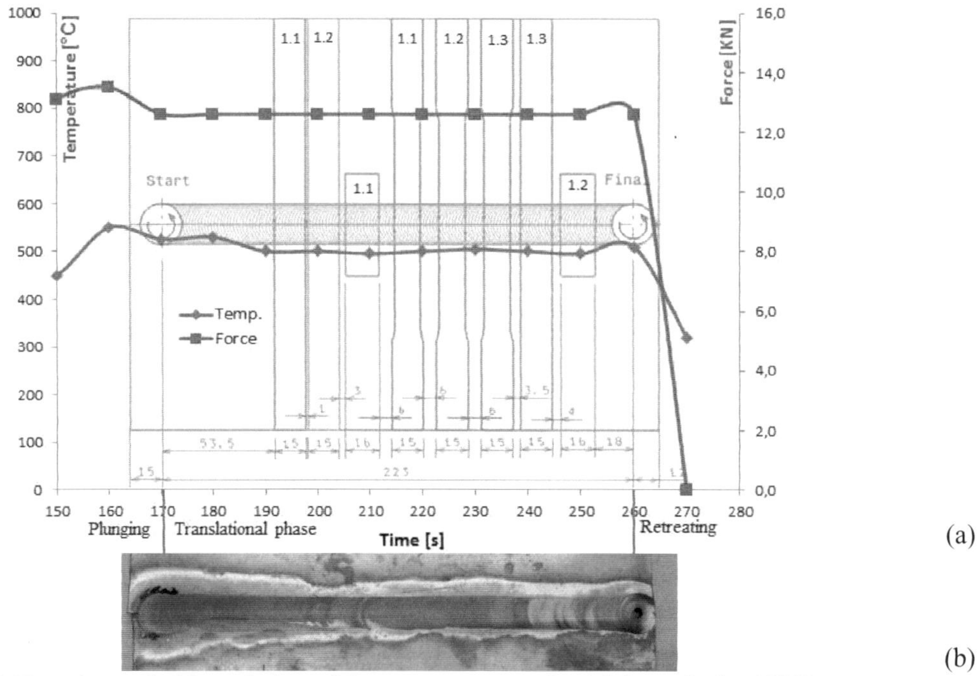

Figure 6. Experiment 1: (a) evolution of the temperature and axial force during FSW process; (b) welded joint.

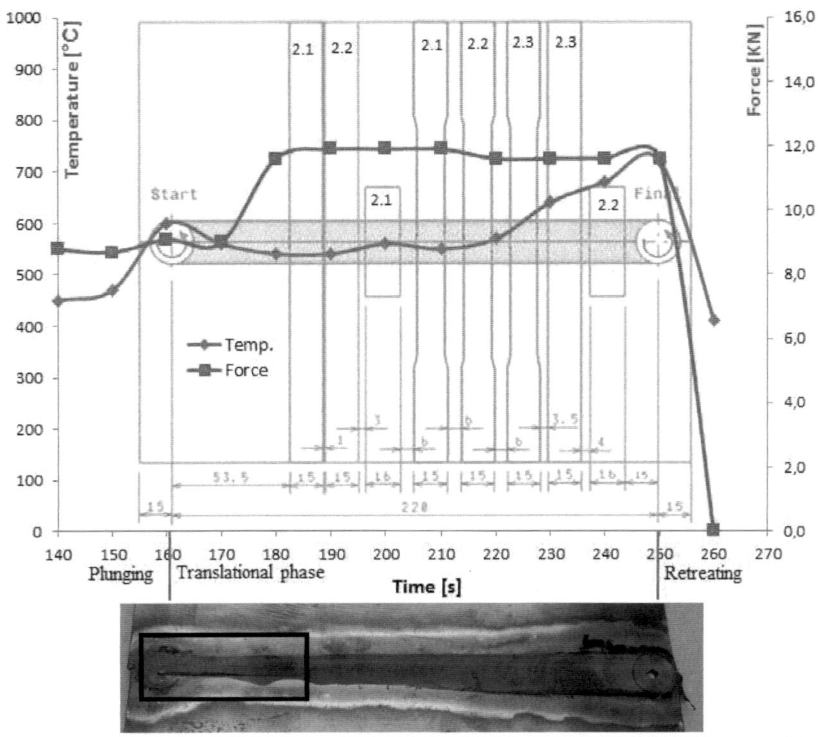

(a)

(b)

Figure 7. Experiment 2: (a) evolution of the temperature and axial force during FSW process; (b) welded joint.

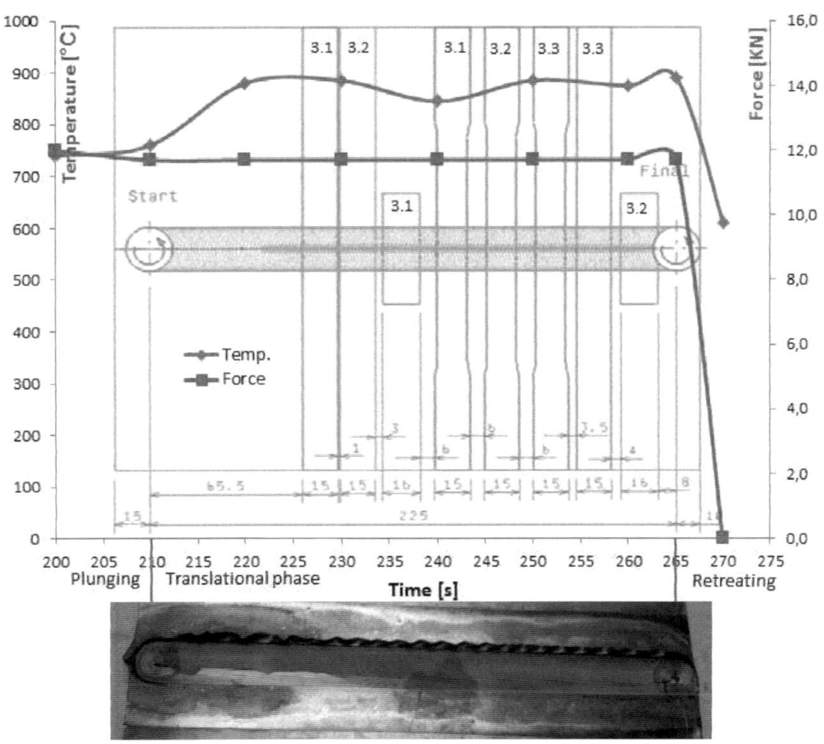

(a)

(b)

Figure 8. Experiment 3: (a) evolution of the temperature and axial force during FSW process; (b) welded joint.

The analysis of evolution for the two process parameters (temperature and axial force) from the previous diagrams, highlights the following aspects:

- Temperature in FSW process increases with increasing tool rotational speed. The values average temperature recorded during the three experiments were: 506 °C for experiment 1, 596 °C for experiment 2 and 860°C for experiment 3. The highest temperature was found, as expected, in TIG assisted FSW process.
- The compressive force decreases with increasing tool rotational speed (from about 13 kN in experiment 1 to about 12 kN in experiment 2). And also additional heat input (experiment 3) presenting a slight decrease in the value of the compressive force. Compressive force plays a very important role in development of FSW joints without defects, figure 7 show that the channel defect is eliminated only after increasing the compressive force.

3.2. Microhardness analysis

Microhardness profiles were measured over the transverse cross-section of Cu-DHP FSW welds with hot-rolled base material, figure 9. From each joint were extracted two samples, one from the beginning of the joint and the other from the end. These samples subsequently have been subjected to an automatic grinding and polishing.

The microhardness was measured with an electronic microhardness tester, type InnovaTest Falcon 500. The Vickers microhardness values of the welded specimens were determined under the load of 300 g during 10 s of penetration time.

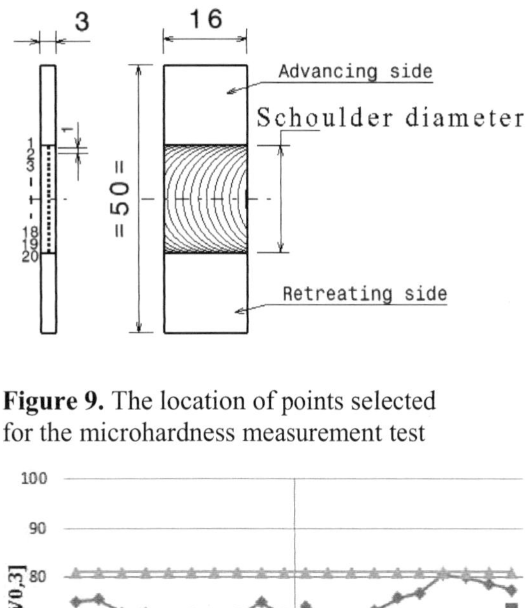

Figure 9. The location of points selected for the microhardness measurement test

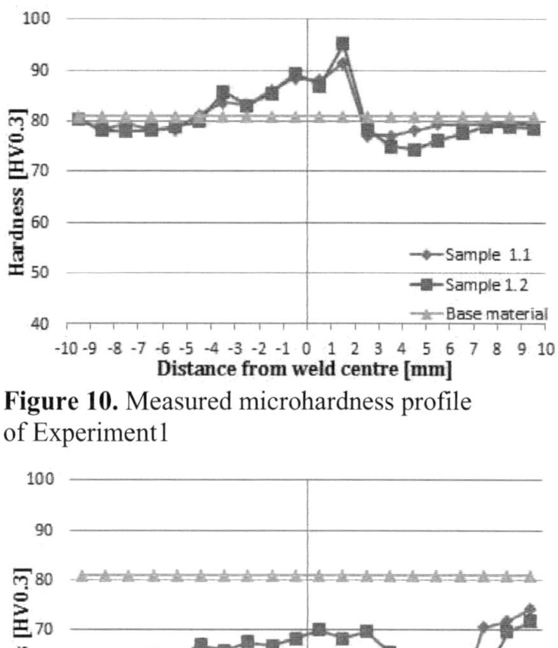

Figure 10. Measured microhardness profile of Experiment1

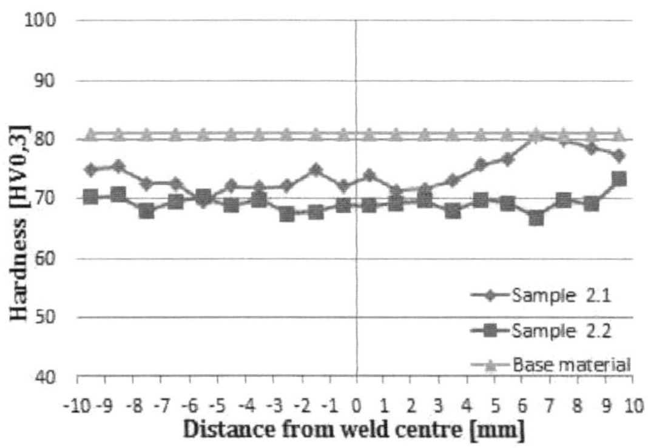

Figure 11. Measured microhardness profile of Experiment 2

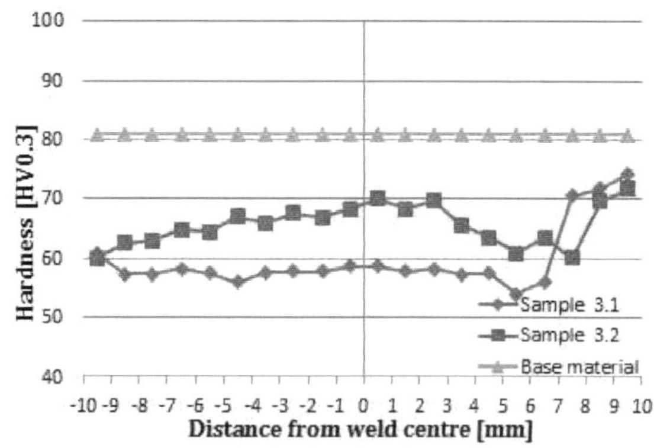

Figure 12. Measured microhardness profile of Experiment 3

Two competitive mechanisms are influencing the hardness of copper FSW welds: anneal softening and grain refinement. The weld hardness depends on the state in which the base material is found (annealed or work-hardened) as well as the heat input generated in the process. Heat input increases with increasing tool rotational speed and decreasing welding speed and leads to annealing softening and lower hardness, figures 11 and 12. So, the lowest values of the hardness were identified for the hybrid process in which the temperature recorded presents the highest values (experiment 3). Lowering the heat input leads to refined grain size and eventually to equal or even higher hardness of the welds as compared to the base material, figure 10.

Regarding the evolution of microhardness profile according to the local conditions registered on the samples which the determination was carried out, is ascertained the following:

- In experiment 1, there are no differences between the measured values on the sample taken at the beginning of the joint compared to the one taken at the end of this. This can be correlated with approximately equal values of the process temperature in the two areas;
- In experiment 2, the values measured on the sample taken at the beginning of the joint are slightly higher than those measured on the sample taken from the end of the joint. This is explained by the temperature differences in these two areas, lower in the beginning and higher to the end.
- In experiment 3, the values measured on the sample taken at the beginning of the joint are lower than those measured on the sample taken at the end of the joint. Although the temperature values in the two areas are approximately equal, these differences can be attributed to different cooling rates (at the end of the process, the additional TIG heat input being interrupted).

No rule of variation of the measured values on the advancing side compared to those measured on the tool retreating side has been identified.

3.3. Roughness analysis

Roughness measurement has been made on three samples for each welded joint, two from the beginning of the joint and the other from the end, figure 5. The roughness of samples was measured with the electronic roughness tester, type MarSurf PS 10. The measurements have been made in the same direction as the welding direction. For each sample there was made three measurements, the values shown in figure 13 representing their average.

Average roughness values show similar evolutions to those of microhardness, heat input also having a defining role in the variation of this parameter. So, with the temperature increase in the process, the material is more deformed and consequently has higher roughness. The smallest values is corresponding to the experiment 1 (with the lowest process temperature) and the highest values to the experiment 3, which benefits of additional TIG heat input.

Between the measured average values of roughness no dependence was identified according to the position of the sample.

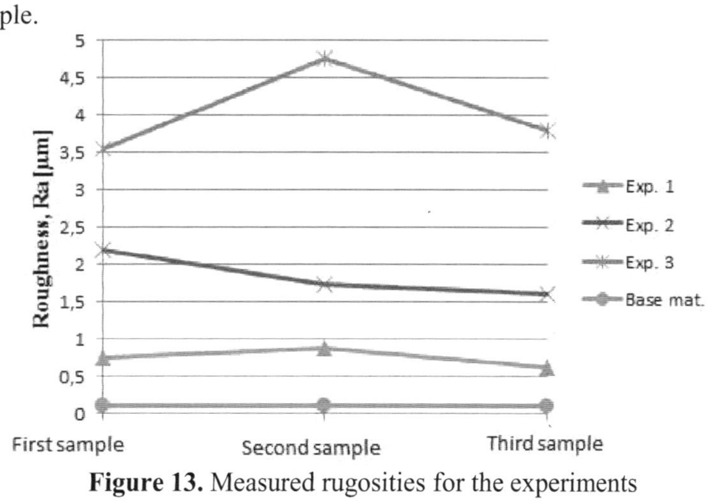

Figure 13. Measured rugosities for the experiments

4. Conclusions

As a result of the researches carried out and the analyzes presented above, the following main conclusions are presented:

1) The best properties of the welded joints of Cu made in these researches were recorded for experiment 1, using the smallest process parameter values (rotational speed & welding speed).

2) Adding a supplementary heat source, the plates are preheated and are obtained some advantages such as reduced forces used in process and FSW, tool wear, a faster and better plasticization of the material, increased welding speed at speeds up to 2 times higher than in FSW processes. However, in case of copper, this additional heat also brings some disadvantages to the properties of the welded joint, such as the decrease in the microhardness and the increase in the roughness of the material.

3) The increasing heat input leads to annealing softening and lower hardness. The lowering heat input leads to refined grain size and consequently to equal or even higher hardness of the welds as compared to the base material.

4) Also, heat input has a defining role in the variation of roughness. As the temperature rises, the material is more deformed and presents higher roughness.

In the next researches, this study will be extend to a larger set of parameters and it will be analyzed the cooling effect on properties of TIG assisted FSW joining for pure copper plates.

Acknowledgements

The work was done with the support of the National Institute for Welding and Material Testing, ISIM Timişoara.

References

[1] Dawes J and Thomas M 1996 *Friction Stir Process Welds Aluminium Alloys* Welding J. 31-45

[2] Savolainen K 2012 *Friction Stir Welding of Copper and Microstructure and Properties of the Welds* (Doctoral Thesis, Aalto University)

[3] Genevois C 2004 *Genèse des microstructures lors du soudage par friction malaxage d alliages d'aluminium de la série 2000 et 5000 et comportement mécanique résultant* (Doctoral Thesis, Institut National Polytechnique de Grenoble)

[4] Xie G M Ma Z Y and Geng L 2007 *Development of a Fine-Grained Microstructure and the Properties of a Nugget Zone in Friction Stir Welded Pure Copper* Scripta Materialia vol 57 73-76

[5] Surekha K and Els-Botes A 2011 *Development of High Strength, High Conductivity Copper by Friction Stir Processing* Materials and Design vol 32 911-916

[6] Cam G 2011 *Friction Stir Welded Structural Materials: Beyond Al-Alloys* International Materials Reviews vol 56 1-48

[7] Savolainen K Mononen J Saukkonen H and Koivula J 2004 *Friction Stir Weldability of Copper Alloys* 5th International Friction Stir Welding Conference (September 14-16) Metz France

[8] Won-Bae L and Seung-Boo J 2004 *The joint properties of copper by friction stir welding* Materials Letters vol 58 1041-1046

METHOD OF OPTIMIZING THE CONSTRUCTION OF MACHINING, ASSEMBLY AND CONTROL DEVICES

D.M. Iordache[1,2] , A.Costea[1], E.L. Niţu[1], A.D. Rizea[1] and A. Babă[1]

[1]University of Piteşti, Manufacturing and Industrial Management Department, Târgul din Vale Street No.1, Piteşti, Romania

[2] monica.iordache@upit.ro

Abstract. Industry dynamics, driven by economic and social requirements, must generate more interest in technological optimization, capable of ensuring a steady development of advanced technical means to equip machining processes. For these reasons, the development of tools, devices, work equipment and control, as well as the modernization of machine tools, is the certain solution to modernize production systems that require considerable time and effort. This type of approach is also related to our theoretical, experimental and industrial applications of recent years, presented in this paper, which have as main objectives the elaboration and use of mathematical models, new calculation methods, optimization algorithms, new processing and control methods, as well as some structures for the construction and configuration of technological equipment with a high level of performance and substantially reduced costs..

1. Introduction

Modern engineering methods and strategies require a permanent analysis of the machining phase from the moment the product is conceived and then at all stages of its existence.

The current trend in the field of machines construction's technology [1-4] promotes a new concept of multi-criteria optimization of manufacturing processes and equipment based on associated importance criteria, prioritized according to the objectives pursued.

The systemic approach to optimizing problems in the field of technological engineering is determined by the need to highlight the multitude of factors that influence it and their interdependencies.

The modernization of the manufacturing processes necessarily implies the development and evolution of the technological equipment for machining, assembly and control, through the development of new solutions in the field of orientation and fixing devices of workpiece [5-10].

In a previous paper [11], there was established a method to optimize the orientation and fixing system of workpieces for the construction of devices, which is a theoretical study on the generating and quotation bases for a particular technological operation.

The present paper is a continuation of the above-mentioned one, which proposes an organological study of constructive structures to establish the optimal solution for the construction, configuration and operation of devices.

2. Methodology to optimize structures for the construction of machining, assembly and control devices

Under current conditions the manufacture of a new product involves the design, execution and approval of technological equipment, especially of devices, at a very fast pace, so that the time and expenses for the technological preparation of the machining should be as small as possible, but to ensure the level of technical and economical performance imposed.

For machining and control devices, specialized papers [1,11,13] present the methodology to establish the optimal orientation and fixing scheme (SOF-O) expressed by information symbolization of the supports of the orientation and fixing system of the device.

The evolution of machining systems and technological equipment determined a multitude of constructive variants of devices that do not have total correspondence in the information symbolization, and for the analysis of technically possible orientation variants (VO-TP), only the orientation errors in the contact area between the workpiece and support were considered, without taking into account the errors caused by the elements of construction and operation of the support.

2.1. The design-optimization algorithm for the construction of orientation and fixing devices for workpiece

For a more rigorous assessment of the orientation errors in relation to the required machining or control precision and the establishment of rational technical solutions based on the calculation method for the errors of the construction orientation [3], there is proposed a design-optimization algorithm for the construction of the devices shown in Figure 1, following stages:

E_{01} – Knowing or determining SOF-O;
E_{02} – Coding construction variants of supports associated SOF-O;
E_{03} – Establishing the optimizing technical criterion;
E_{04} – Establishing the optimizing economical criterion;
E_{05} – Defining the optimizing method;
E_{06} – Formulating the centralizing table of variants and criteria analyzed;
E_{07} – Selecting the structure variants for the construction of supports;
E_{08} – Establishing the optimal variant to construct the device.

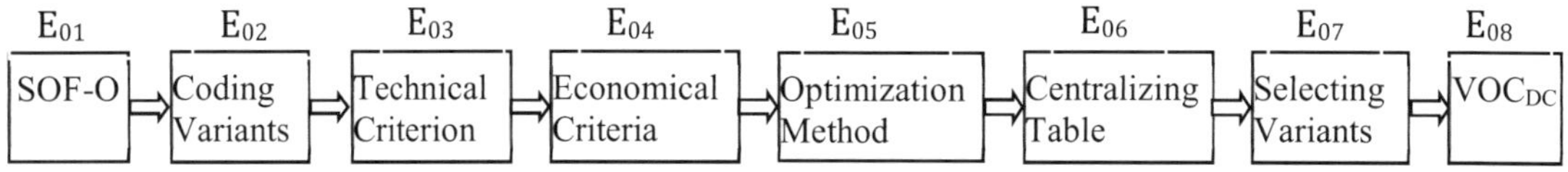

Figure 1. Algorithm to design-optimize devices

2.2. Optimizing the construction of devices for machining, assembly and control operations

Devices, as subsystems of the technological system, in the general context of the development and diversification of machines, tools, workpiece and drives, are made in a multitude of constructive variants, which create difficulties in their study and improvement.

For these reasons, the general design-optimization algorithm of the construction of devices with the eight stages of work was established.

E01 – Knowing or determining SOF-O

Initially, the quotas and conditions imposed on the analyzed operation or the optimal orientation and fixing scheme (SOF-O) are known as a combination of graphical signs in the information symbolization of the bases of orientation and fixing of the workpiece.

$$SOF\text{-}O = f(SI_o, SI_f) \tag{1}$$

where,

SOF-O – optimal orientation and fixing scheme for processing, assembly and control operations;
SI_o – information symbolization associated to the orientation and fixing bases and supports of device;
SI_f – information symbolization associated to the orientation and fixing bases and clamping elements of the device.

E02 Coding the construction variants of the orientation and fixing elements of the device

Constructive variants for structures, which are known, are defined and coded for each symbol element of the quotation bases BCI, BCII, BCIII, and fixing bases BFIV, BFV of the optimal orientation and fixing scheme set forth in [1], or new constructive variants to be achieved later can also be analyzed.

System of main orientation supports:

For BCI-SI_{OI}, structure variants are noted with X_{ij},
where i = 1, ..., m; j = 1, ..., n,

m – is the number corresponding to the quotation and fixing bases BCI, ..., BFV;

n – is the number of structure variants associated to each quotation and fixing base BCI, ..., BFV

In this case for BCI the structure variants are between X_{11}, ..., X_{1n}.

For BCII–SI_{OII}, structure variants are noted with X_{ij},
where i = 2; j = 1, ..., n, meaning X_{21} , ... , X_{2n} where n is the number of structures associated to symbol of quotation base BCII.

For BCIII–SI_{OIII}, structure variants are noted with X_{ij}, where i = 3; j = 1 ... n, meaning X_{31} , ... , X_{3n}, where n is the number of structures associated to symbol of quotation base BCIII.

Fixing system with additional supports:

For BFIV-SIf_{IV}, structure variants are noted with X_{ij}, where i = 4, j = 1 ... n, meaning X_{41} , ... , X_{4n}, where n is the number of structures associated to symbol of fixing base BFIV.

For BFV- SIf_{V}, structure variants are noted with X_{ij}, where i = 5, j = 1 ... n, meaning X_{51} , ... , X_{5n}, where n is the number of structures associated to symbol of fixing base BFV.

E03 – Establishing the technical optimizing criterion

For any construction, the most important technical selection criterion must be precision, which becomes eliminatory and requires the technically acceptable construction solution (SC-TA) to meet condition $\varepsilon_{c.\ s}^{o-p} < \varepsilon_{ad.\ s}^{0-p}$, where:

$\varepsilon_{c.\ s}^{o-p}$ - orientation-positioning error of the construction structure;

$\varepsilon_{ad.\ s}^{o-p}$ - admissible orientation-positioning error of the structure.

These two types of errors are calculated with the relations presented in the works [3,11] and are introduced in the centralizing table of variants and criteria of optimization.

For BCI and SI_{OI}, the established variants of constructive structures X_{ij} are analyzed as precision through the calculation of admissible orientation-positioning error of the structure ($\varepsilon_{ad.\ s}^{o-p}$), and of orientation-positioning errors of the structure construction ($\varepsilon_{c.\ s}^{o-p}$).

Value $\varepsilon_{ad.\ s}^{o-p}$, is obtained from relation:

$$\varepsilon_i = \sqrt{\varepsilon_o^2 + \varepsilon_f^2 + \varepsilon_d^2} \tag{2}$$

presented in [4], where:

ε_i – installation (positioning) error;

ε_o – orientation error;

ε_f - fixing error;

ε_d – error connected to the device construction.

Formula (2) is true if the three errors are orthogonal or they fulfil normal (Gaussian) distributions.

In the case of machining devices $\varepsilon_{iad} = (\frac{1}{2} ... \frac{1}{3})T_p$, where T_p – is the tolerance of the part at the quotation or condition analyzed.

For the construction of control devices $\varepsilon_{iad} = (\frac{1}{5} ... \frac{1}{10})T_p$

The precision condition is $\varepsilon_i \leq \varepsilon_{iad}$

From relation (2), the maximum value of error of the device construction is:

$$\varepsilon_{dmax} = \sqrt{(\frac{1}{2} ... \frac{1}{3})^2 T_p^2 - (\varepsilon_o^2 + \varepsilon_f^2)} \text{ , for processing devices} \tag{3}$$

$$\varepsilon_{dmax} = \sqrt{(\frac{1}{5} \cdots \frac{1}{10})^2 T_p^2 - (\varepsilon_o^2 + \varepsilon_f^2)} \text{ , for control devices} \quad (4)$$

Finally, it can be written:

$$\varepsilon_{ad.\ s}^{o-p} = \varepsilon_{dmax} \quad (5)$$

Value $\varepsilon_{c.\ s}^{o-p}$, is obtained from the analysis of errors produced by the construction and operation of elements composing each structure of variants with X_{ij} coded supports.

The calculated values of $\varepsilon_{c.\ s}^{o-p}$ and $\varepsilon_{ad.\ s}^{o-p}$ are actual values introduced in Table 1 for variant X_{ij}. Proceed similarly for BCII and SI_{OII}, BCIII and SI_{OIII}, BFIV and SI_{fIV}, BFV and SI_{fV}, calculate $\varepsilon_{ad.\ s}^{o-p}$ and $\varepsilon_{c.\ s}^{o-p}$ for each variant of structures X_{ij}, then introduce in Table 1.

E04 – Establishing economical optimization criteria

Establish the economical optimization criteria CE_1, CE_2, ..., CE_6, representative of the construction of the devices [3,11].

For the simplification of calculations, there are chosen only two economical selection criteria, cost (CE_3) and flexibility (CE_6), the actual values of these criteria are determined with relations from [13] and are introduced in Table 1.

E05 – Defining the optimization method

In order to determine the optimal solution of the variants of structures designed for the construction of the device, the global utility method [14] was customized for the analyzed case with the following stages:

- choose a variance interval $\{0, \ldots, 1\}$, considered utility, denoted by $u(v_i)$;
- the most favourable decisional consequence has value 1, noted with $u(v_i) = 1$, and the most unfavourable consequence has value zero, $u(v_i) = 0$;
- the other utilities will be calculated in stage E_{06} by linear interpolation with relation:

$$u_{ij} = \frac{a_{ij} - a_j^0}{a_j^1 - a_j^0} \quad (6)$$

where:

u_{ij} – utility of consequence i following criterion j;
a_{ij} - value of consequence i following criterion j;
a_j^0 – value of the most unfavourable consequence of criterion j;
a_j^1 – value of the most favourable consequence of criterion j.

This simple method makes it possible to optimize under certainty conditions and can be used in situations where decision criteria are equally important or decision criteria have different coefficients:

$$u_g(v_i) = \Sigma u_{ij} \cdot k_j u_{ij} , \quad (7)$$

where $u_g(v_i)$ global utility, and k_j, is the coefficient of importance of criterion j.

E06 – Formulating the centralizing table of constructive variants and optimization criteria

Table 1 centralizing the variants and optimization criteria with encoded variants X_{ij} of groups of structures BCI, ..., BFi, the optimization criteria (precision - C_{pr}, ..., costs C_c- CE_3, flexibility C_f- CE_6, ...), with their actual, admissible and critical value of coefficients of importance (CI) depending on how a variant is related to the criterion analyzed.

In case the interest is different for the decision criteria assign to each criterion a correction coefficient (priority) k_c, according to their importance.

Table 1. Centralizing table of constructive variants and optimization criteria

Nb.	Structure variants for BCI, ..., BFV	Precision criterion C_{pr}				Economical criteria – CE and Importance coefficients - CI						ΣCI	Decision
		$\varepsilon_{c.\ s}^{o-p}$	$\varepsilon_{ad.\ s}^{o-p}$	SC-TA		Cost criterion C_c - CE_3			Flexibility criterion C_f - CE_6				
BCI-SI_{OI}		Actual values		CI_{pr}	k_c^{pr}	Actual values	CI_c	k_c^c	Actual values	CI_f	k_c^f	(Max)	Optimal
1	X_{ij}	...	...	...	...	...	...	...	...	...	...	...	...
...	...	...	...	...	...	...	...	...	...	...	...	...	...
BCIII-SI_{OIII}													
...	X_{ij}	...	...	...	...	...	...	...	...	...	...	...	...
...	...	...	...	...	...	...	...	...	...	...	...	...	...
BFV-SIf_V													
...	X_{ij}	...	...	...	...	...	...	...	...	...	...	...	...
...	...	...	...	...	...	...	...	...	...	...	...	...	...

The actual and admissible values are calculated in E_{03} and E_{04}, and those of the coefficients of importance in this table are calculated for the optimization criteria established.

Precision criterion (C_{pr})

The value of the coefficients of importance (CI_{pr}) calculated for the precision criterion associated to optimization with the global utility method assumes the choice of a variation interval, which in the analyzed case can be between {0, ..., 1}. The most favourable consequence is assigned value 1 and the most unfavourable consequence value zero. The other utilities will be calculated by interpolation.

Given that $\varepsilon_{c.\ s}^{o-p}$ is in a reverse relation with the variance of the coefficients of importance (CI_{pr}), when one increases the other one decrease, a method of calculating by inverse interpolation is proposed with the following stages:

- Establish the value of interval (v_i) of structure X_{ij}:

$$v_i = \varepsilon_{ad.\ s}^{o-p} - \varepsilon_{c.\ s}^{o-p}, \tag{8}$$

- Determine the number of intervals (n_i) of the variation interval for the actual orientation-positioning errors of the structures:

$$n_i = \frac{\varepsilon_{ad.\ s}^{o-p} - \min \varepsilon_{c.\ s}^{o-p}}{v_i}, \tag{9}$$

- Calculate the coefficient of importance of an interval (CI_i):

$$CI_i = \frac{1}{n_i}, \tag{10}$$

- Determine the number of actual intervals (n_{ie}) of a structure $\varepsilon_{c.\ s}^{o-p}$,

$$n_{ie} = \frac{\varepsilon_{c.\ s}^{o-p}}{v_i}, \tag{11}$$

- Establish the value of apparent coefficient of importance (CI_a) of a structure X_{ij},

$$CI_a = n_{ie} \cdot CI_i, \tag{12}$$

Because the errors ($\varepsilon_{c.\ s}^{o-p}$) and coefficients of importance (CI) are in a reverse relation, the value of the coefficient of importance of the precision criterion (CI_{pr}) is calculated with relation:

$$CI_{pr} = 1 - CI_a, \tag{13}$$

Cost criterion (C_c-CE_3)
Calculate the actual cost of materials, workmanship and heat treatment or operating costs for each X_{ij} coded structure and introduce data in Table 1.

Flexibility criterion (C_f- CE_6)
Assess the degree of flexibility to adjust expressed in time (G_{AT}) with relation from [6]:

$$G_{AT} = e^{-(TR/TRO)}, \tag{14}$$

Where:
G_{AT} – degree of flexibility to adjust expressed in time;
TR – time to re-equip for the analyzed system;
TRO – time to re-equip for the comparison system.
For each type of X_{ij} structure, the degree of flexibility to adjust $G_{AT(Xij)}$ is calculated as the actual value and introduced in Table 1.

E07 Selecting the variants to construct the device
After filling the centralizing Table 1, the establishment of the optimal solution for the construction of the orientation and fixing device of the semi-finished product can be done by two methods.
Method 1
Sum up the coefficients of importance of the optimization criteria chosen for each analyzed X_{ij} variant and introduce them into the ΣCI column, then for each group of structures BCI, BCII, ..., BFi, locally optimal of the structure with the maximum value (max ΣCI) is identified.
The optimal construction variant of the device is a sum of structures identified as locally optimal.
The drawback of this method is that sometimes in the case of structures with insufficient precision there is a risk that they may be economically analyzed and lead to errors.
For this reason a new calculation method is proposed for establishing the optimal solution in which the variants of structures with insufficient precision ($\varepsilon_{c.\ s}^{o-p} > \varepsilon_{ad.\ s}^{o-p}$) are eliminated before the final phase of optimization.
Method 2
Write the relations for the quoting and fixing bases BCI, BCII, ... , BFi, as follows:

$$\mathrm{MaxV_{BCI}}[CI_{pr}^{X_{11}} \cdot k_c^{pr}(CI_c^{X_{11}} \cdot k_c^c + CI_f^{X_{11}} \cdot k_c^f); \ldots ; CI_{pr}^{X_{ij}} \cdot k_c^{pr}(CI_c^{X_{ij}} \cdot k_c^c + CI_f^{X_{ij}}\ k_c^f);\ \ldots \ldots ; CI_{pr}^{X_{mn}} \cdot k_c^{pr}(CI_c^{X_{mn}} \cdot k_c^c + CI_f^{X_{mn}} \cdot k_c^f)], \tag{15}$$

$$\mathrm{MaxV_{BCII}}[CI_{pr}^{X_{21}} \cdot k_c^{pr}(CI_c^{X_{21}} \cdot k_c^c + CI_f^{X_{21}} \cdot k_c^f); \ldots ; CI_{pr}^{X_{ij}} \cdot k_c^{pr}(CI_c^{X_{ij}} \cdot k_c^c + CI_f^{X_{ij}}\ k_c^f);\ \ldots ; CI_{pr}^{X_{mn}} \cdot k_c^{pr}(CI_c^{X_{mn}} \cdot k_c^c + CI_f^{X_{mn}} \cdot k_c^f)], \tag{16}$$

where:
$\mathrm{MaxV_{BCI}}$ - maximum value of coefficients of importance of structure variants for BCI;
$CI_{pr}^{X_{ij}}$ - coefficients of importance of the precision criterion for the structure variants X_{ij};
$CI_c^{X_{ij}}$ - coefficients of importance of the cost criterion for the structure variants X_{ij};
$CI_f^{X_{ij}}$ - coefficients of importance of the flexibility criterion for the structure variants X_{ij};
k_c^{pr} - correction coefficient (prioritizing) of the precision criterion;
k_c^c - correction coefficient (prioritizing) of the cost criterion;
k_c^f - correction coefficient (prioritizing) of the flexibility criterion.
Similarly write the calculating relations for BCIII, ..., BFi.
The optimal variant of device construction is a combination of structures which meet the maximum condition of values for BCI, BCII, ..., BFi.

E08 Establishing the optimal construction variant for the orientation and fixing device for workpiece

For method 1 to calculate the optimal variant of device construction, there are established the locally optimal structures with Max ΣCI of each group X_{ij} of BCI, BCII, ... , BFi, so the relation is:

$$VOC_D^1 = \text{OptimBCI} + \text{OptimBCII} + ... + \text{Optim BFi}\,, \tag{17}$$

For method 2 to calculate the optimal variant of device construction, relation is:

$$VOC_D^2 = \text{MaxV}_{BCI} + \text{MaxV}_{BCII} + ... + \text{MaxV}_{BFi}\,, \tag{18}$$

3. CONCLUSIONS

The modernization of the manufacturing processes is related to the promotion of new methods to design, implement and approve the technological equipment, firstly, of devices, which will certainly ensure a high level of technical and economical performances.

This paper presents an original approach to the design and optimization method and algorithm of device construction, the tabular formulation for the analysis of constructive variants and performance criteria, the inverse interpolation calculation method to determine the precision importance coefficient and the calculation method to establish the optimal variant of the device construction.

The use of these methods ensures the optimization of the devices for machining, assembly and control and can be applied with the necessary customizations to the construction of other types of devices, machines and machining systems.

References

[1] Brăgaru A, Picoş C and Ivan NV 1996 *Optimization of processes and technological equipment* (Bucarest: Didactic and Pedagogic Publishing Bucarest)

[2] Ivan NV, Păunescu T, Udroiu R, Ivan MC, Găvruş C and Pescaru R 2010 *Machine Building Technology. Theory and innovative approaches* (Braşov: Transilvania University of Braşov)

[3] Iordache DM, Costea A and Babă A 2016 *Calculation methods and mathematical models for design optimization devices* (Piteşti: University of Piteşti Publishing House)

[4] Qin GH and Zhang WH 2006 A mathematical approach to analysis and optimal design of a fixture locating scheme, *Int. J. Adv. Manuf. Tech.* **29** 349-359

[5] Păunescu T. A 2005 Monte Carlo approach of modular fixture systems locating accuracy. Part 1. The geometric modelling, *2nd Int. Conf. Mechanics and Machine Elements* (Sofia, Bulgaria)

[6] Păunescu T. A 2005 Monte Carlo approach of modular fixture systems locating accuracy. Part 2. The Monte Carlo simulation, *2nd Int. Conf. Mechanics and Machine Elements* (Sofia, Bulgaria)

[7] Zhu SW, Ding GF, Qin SF, Lei J, Li Z, Yan KY 2012 Integrated geometric error modeling, identification and compensation of CNC machine tools. *Int. J. Mach. Tools Manuf.* **52** 24–29

[8] Zhu S. W., Ding G. F., Ma S. W., Yan K. Y., Qin S. F. 2013 Workpiece locating error prediction and compensation in fixtures *Int. J. Adv. Manuf. Tech.* **67** 1423–1432

[9] Raghu A., Melkote S.N. 2003 Prediction of workpiece location error due to fixture geometric errors and fixture-workpiece compliance *Int. Mech. Eng. Congres* (Washington)

[10] Rong Y., Hu W. 2001 *Location error analysis and tolerance assigment computer aided design*, Int. J. Prog. Res. **39** 3529-3545

[11] Iordache DM, Rizea AD, Costea A, Niţu EL and Babă A 2017 Method for optimization of the orientation and fixing system of workpiece for the construction of control devices *MATEC Web of Conferences* **112** 06003 (2017)

[12] Vasii Roşculeţ S, Gojineţchi N, Andronic C, Şelariu M and Gherghel I 1982 Device design (Bucarest: Didactic and Pedagogic Publishing Bucarest)

[13] Abrudan I 1996 *Flexible Manufacturing Systems. Design and management concepts* (Cluj-Napoca: Dacia Publishing).

[14] Boldur Gh 1973 *The complex foundation of the economic decision-making process* (Bucarest: Scientific Publishing House).

DESIGN AND ANALYSIS OF A FUSED DEPOSITION MODELLING MANUFACTURED PART

Claudiu DIACONESCU[1], Stefan TABACU[2], Alexandru OLTEAN[3]
[1]ALSECA Engineering, Bucharest, Romania; [2]University of Pitesti, Pitesti, Argeş, Romania

claudiu.diaconescu@alseca.com

Abstract. Fused deposition modelling (FDM) technology was investigated in order to produce a part designed to support a test load without permanent deformation. A topology optimization script was developed in order to define the final shape of the part. The solver used for the numerical analysis is LS-Dyna (implicit formulation). The algorithm follows the element removal based on the stress state and it can be very simply implemented. This solution of developing a custom code was adopted because plastic strain state was monitored and this is not yet implemented in other topology optimization applications. The results of the traction test were assembled and investigated in order to outline some of the typical parameters used for mechanical characterization. As a general overview of the results it was found that the parts manufactured using a 45°/45° stacking configuration are stronger that the ones manufactured using 0°/90° stacking configuration.

1. Introduction

Rapid prototyping consists in technologies capable to manufacture complex part in a short period of time. Without considering the costs related to infrastructure the parts are less expensive compared to the ones manufactured using conventional technologies. Another important aspect is that parts with internal voids can be manufacture.

Among the 3D printing technologies plays an important role [1] and plays an important role. Fused deposition modelling (FDM) [2] is one technique of rapid prototyping based of 3D printing that uses an extruded plastic wire in order to form complex part. The materials used to form the parts are similar to some currently used in injection molding technology [3].

Ahn et al. [4] investigated ABS P400 material as used for injection moulded parts and for parts manufactured using Fused Deposition Modelling. The influence of the raster orientation and the gap between two adjacent paths were identified as the major parameters that influence the mechanical performances of the parts manufactured using Rapid Prototyping techniques. A set of rules are outlined to conclude their work and provides a valuable start point in designing parts to be manufactures using FDM.

Also the work of Dawoud et al. [5] is focused on study of the mechanical behaviour of ABS. The raster angle and the gap are also in this case the main parameters associated with the manufactures parts. The printing process parameters were not altered from one part to other. It was found that using a negative gap can increase both traction and flexural performances of the part (up to -12% in performance compared to injection moulded specimens). The differences are explained also by the structure of the material - injection moulding allows the formation of crystals while FDM results in an amorphous (yet less strong) structure.

Garg and Bhattacharya propose in their work [6] a numerical model that in capable to accurately describe by the mean of finite elements method the behave of structures under tensile loading. The method is applied to investigate specific specimens for tensile test. Results show a good agreement between the experiment and simulation results. The study completes the knowledge in the field showing that the strongest parts are obtained when the part is aligned with the building plane.

A study on the build parameters in case of fused deposition modelling developed by Griffiths et al. [7] points that the infill level and number of shells plays an important role when the improvement of mechanical properties of parts is addressed.

2. Investigation of the mechanical properties

In order to investigate the mechanical properties of ABS a number of samples were manufactured using MakerBoot Experimental 2X printer that used Fused Deposition Modelling technology for rapid prototyping. The specimens were tested in traction using a universal testing machine. The tests were performed according to the prescriptions of ASTM D 638.

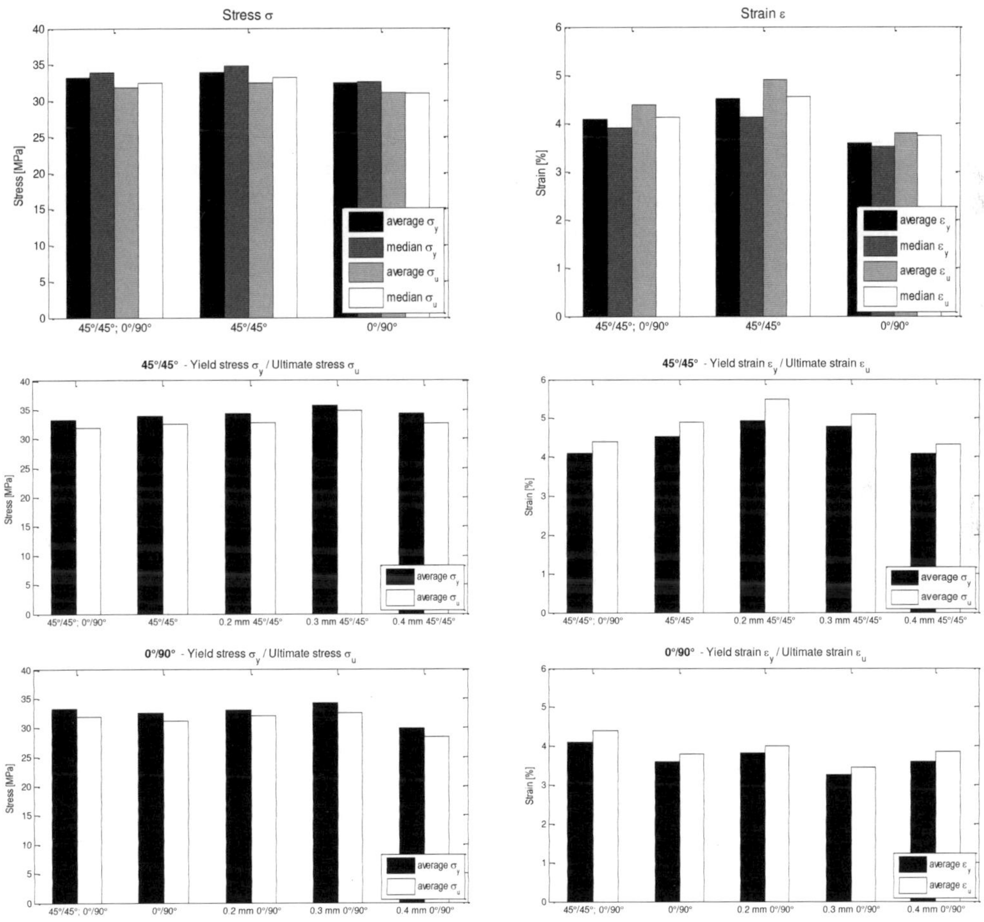

a) *b)*

Figure 1. Analysis results of the tensile tests.

a) overall results – stress yield/ultimate; b) overall results – strain yield / ultimate;

The results of the traction test were assembled and investigated in order to outline some of the typical parameters used for mechanical characterization. As a general overview of the results it was found, as expected and also supported by literature [5], that the parts manufactured using a 45°/45° stacking configuration are stronger [8] that the ones manufactured using 0°/90° stacking configuration (Figure 1 *a)*).

In case of a 45°/45° configuration, Figure 1 *c)* show that increasing the layer thickness the strength of the part increases. However the increase of the layer thickness results in a decrease of the strain as resulting from Figure 1 *d)*. A smaller layer thickness gives a more compact construction yet the raw material might be affected during extruding thus the mechanical properties altered. This gives the small decrease in strength. The higher value of the stress is recorded for a layer thickness of .3 mm close to the optimal value recommended for the 3D printer.

For the 0°/90° there is a similar pattern. The stress increases with the increase of the layer thickness (Figure 1 *e)*) while the strain decreases (Figure 1 *f)*)

The fractured sections of the parts are presented in Figure 2. Figure 2 *a)* presents the cross section of the 45°/45° sample with a layer thickness of 0.2. The section shows some small voids on the edge (contour section). The voids [9] can also be explained by the standard configuration with a null gap. Figure 2 *b)* presents the cross section of the 0°/90° sample with a layer thickness of 0.4. The previous observations are also valid.

a) 0.2 mm - 45°/45° *b)* 0.2 mm - 0°/90°

c) 0.4 mm - 45°/45° *d)* 0.4 mm - 0°/90°

Figure 2. SEM images of the fractured section.

In the case of 45°/45° parts there is a similar pattern for the fractured section as presented in Figure 2 *a)*, *c)*. It worth mentioning that there is predominant direction for the failure of the material. However the part with a layer thickness of 0.4 mm displays more voids. The stress is lower because the number of longitudinal filaments.

Figure 3 *a)* displays the fracture pattern for a part with 0.3 mm layer thickness and a stacking sequence of 45°/-45°. It shows that the 45° and -45° filaments fused together. Figure 3 *a)* displays the fracture pattern for a part with a stacking sequence of 0°/-90°. The layers fused together. The failure is produced staring from the transverse filament followed by the failure of the longitudinal one.

a) 0.3 mm - 45°/-45°

b) 0.3 mm - 0°/90°

Figure 3. SEM images of the fractured section (details for the 0.3 mm layer thickness).

In order to provide an input for the design of part, Figure 4 presents typical stress – strain curves for the ABS material. Data presented follow the results of the analysis presented in Figure 1.

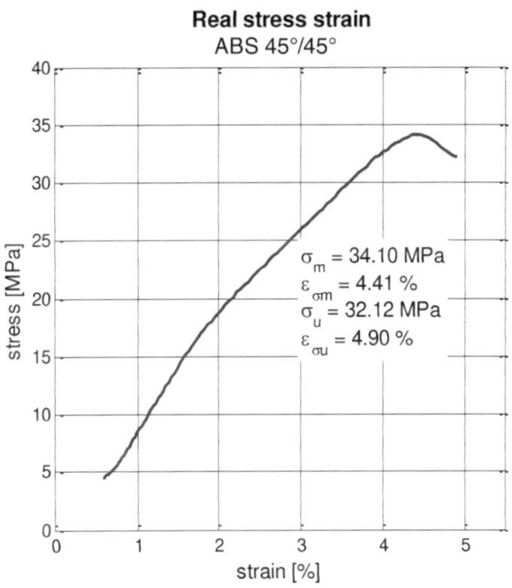

45°/-45° stacking sequence

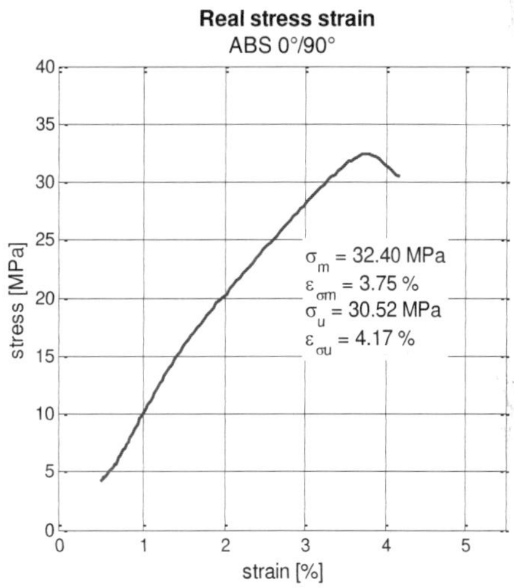

0°/90° stacking sequence

Figure 4. Stress strain data for ABS.

Although literature displays for ABS values of the Young Modulus of 1950-2000 MPa the present study reports values between 1000-1100 MPa. For design purposes of the parts a value of 1050 MPa was recommended. Poisson's coefficient must be considered according to literature thus a value of 0.36 can be used.

3. Part design

A sample part is designed in order to support a load of 1500 N without permanent deformation. The part is considered to be clamped at the bottom supports and the load is applied at mid-span.

IOP Conf. Series: Materials Science and Engineering **252** (2017) 012040 doi:10.1088/1757-899X/252/1/012040

A topology optimization [10–12] script was developed in order to define the final shape of the part. The solver used for the numerical analysis [6] is LS-Dyna (implicit formulation). The algorithm follows the element removal based on the stress state and it can be very simply implemented. This solution of developing a custom code was adopted because plastic strain state was monitored and this is not yet implemented in other topology optimization applications.

The input data are obtained from the AVS database by *DATABASE_AVSFLT card in an ASCII format. The values reported are stress tensor components and the plastic strain.

For each step a stress value is imposed thus the elements with an associated value below that minimum are removed from the model. The function giving the threshold value can be adapted to different situations by the selection of the exponent parameter $stpexp$. The exponent should be selected according to each application as a function of the model size, load configuration, load magnitude and support configuration. For the current model a "conservative" exponent was selected (5.0). Additional parameters like nodal or rigid displacement can be imposed as termination criteria.

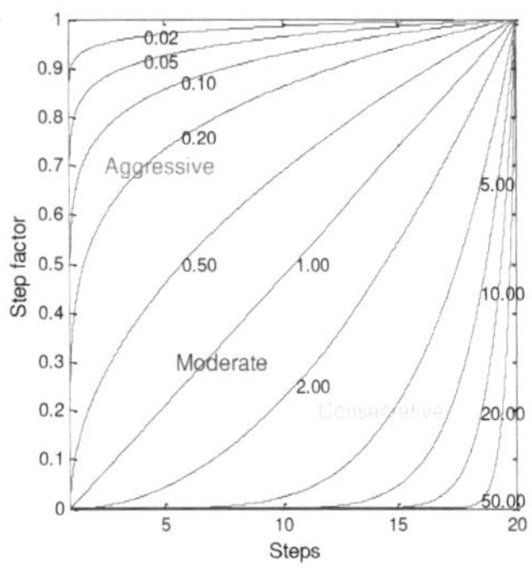

$$\sigma_c = \sigma_m \cdot \left(\frac{step}{steps}\right)^{stpexp}$$

User parameters:
σ_m - maximum stress state
$steps$ - number of steps for the optimization process
$stpexp$ - gives the current factor $\left(\frac{step}{steps}\right)^{stpexp}$

Figure 5. Optimization parameters.

The initial design domain (displayed in yellow) is presented in Figure 6 (step 1). It can be noticed that given the value of the $stpexp$ for the first steps the shape of the part doesn't change too much. The final steps defined the actual shape of the part. Considering 20 iterations the plastic strain limit was reached at iteration 15 thus the final shape of the part is the one obtained for step 14.

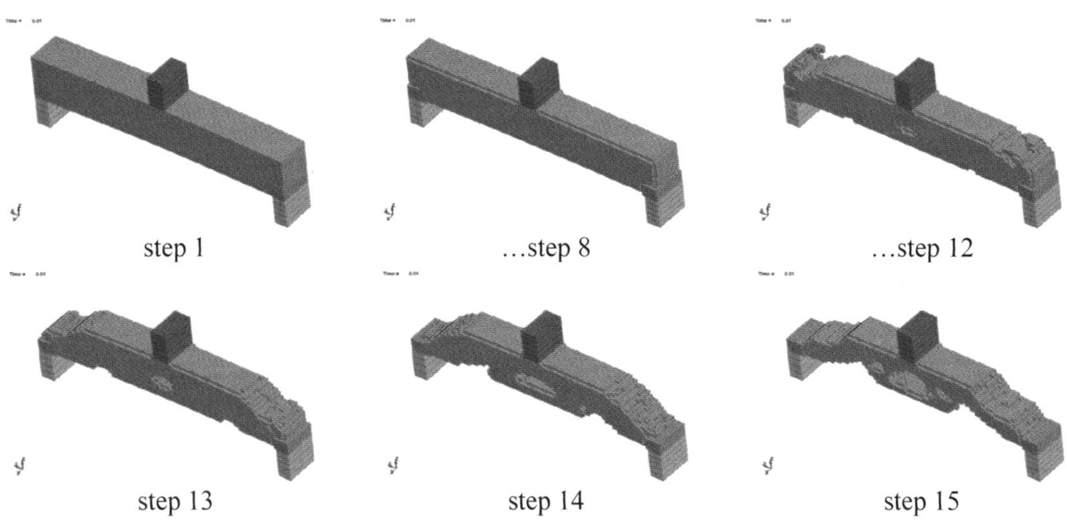

Figure 6. Optimization solution.

It should be mentioned that the section displayed in green and brown are add-on features intended to fix the part and to load it.

Considering the geometrical solution obtained after the optimization the part was printed using MakerBoot Experimental 2X printer. Figure 7 presents the outline of the part that it was designed following the optimization solution. Details on the printing pattern are provided for the narrow sections of the model.

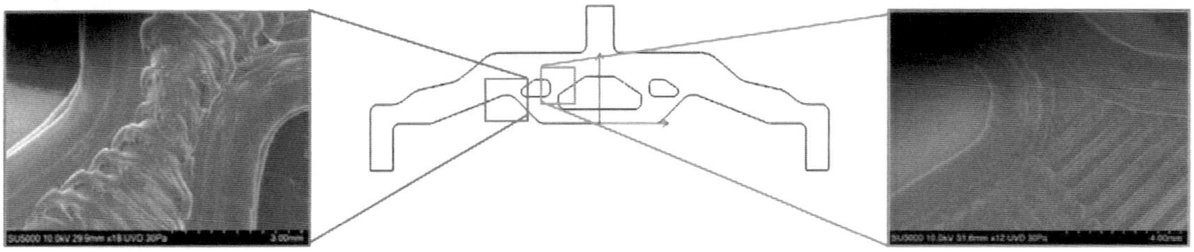

Figure 7. Outline of the part and details on the printing pattern.

4. Experimental investigation

The finished part was fitted on the universal testing machine using a custom 3 points bending fixture. The bottom left and right end were clamped while the load was applied on top of the part. The loading configuration if presented in Figure 8*a)*.

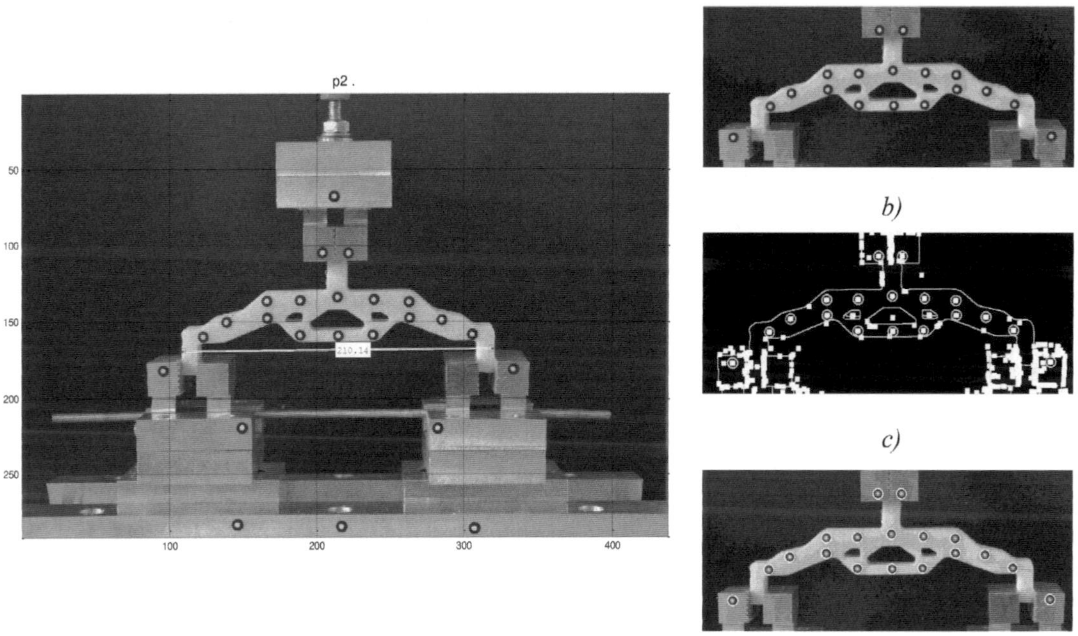

b)

c)

a) *d)*

Figure 8. Experimental analysis of the part.

a) loading configuration; *c)-d)* image processing procedure.

The top displacement was recorded using the machine equipment. In order to improve the reading of the results and to obtain results for a complex deformation analysis a number of high resolution markers were added on the part.

Using digital camera, images of the part during loading were captured. Figure 8*b)* presents the first step of the video analysis. The colour image was processed to grey. Using edge-detection the positions of the markers were identified as presented in Figure 8*c)*. Considering the circular shape the centre of each marker was identified (Figure 8*d)*) and the coordinates were saved to a data file.

The markers located at the bottom of the machine were used as a reference in order to eliminate the motion of the camera during shooting. Considering the first recorded frame fixed the subsequent positions of the markers located on the machine were corrected. This was the actual deformed shape of the part can be determined.

Figure 9 present the positions of the markers during loading process. The initial state, final state and intermediary positions can be identified. This way there is a good traceability of each point and the global deformation can be correlated with local values.

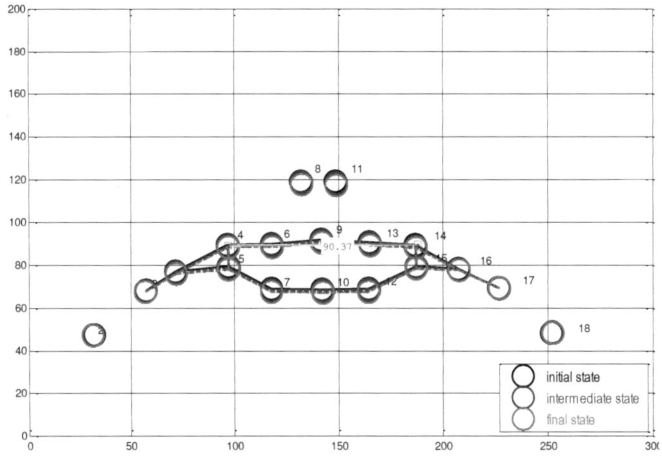

Figure 9. Deformation history of the part.

Figure 10 presents a comparison of the results as obtained from simulation and experiment. Overall there is a good agreement between the two sets of results. The nonlinear behaviour of the part as obtained from the experiment can be explained by the existence of friction between the mechanical parts.

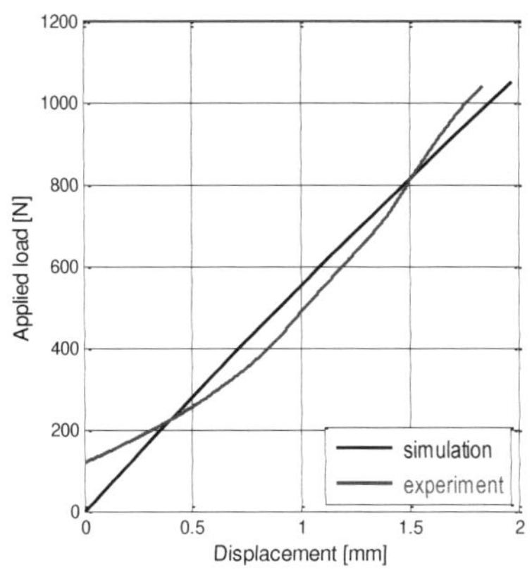

Figure 10. Displacement of the part at loading point.

The part is just a little bit stiffer than it was estimated by the model. This is considered acceptable as the numerical models should be considered conservative in order not to overestimate of to overcome some characteristics of the real model.

IOP Conf. Series: Materials Science and Engineering **252** (2017) 012040 doi:10.1088/1757-899X/252/1/012040

5. Conclusions

The present paper gives a brief description over the design and analysis process of a fused deposition modelling manufactured part.

The process debuted with the examination of the mechanical properties of the material. A number of specimens were manufactured and tested using a universal testing machine. Considering the number of parameters that can be controlled, the influences of stacking and layer thickness over the mechanical properties were investigated.

Once a global mechanical model was defined a part was constructed. The construction of the part is the results of an optimization process using numerical simulation by the finite elements method.

The part was finished using computer aided design and finally manufactured with a 3D printer.

In order to evaluate the performances a three-point bending experiment was developed. Image analysis was employed in order to enhance the results from the experimental analysis.

A good agreement was obtained between results from the numerical analysis and simulation. The small differences (showing a stiffer printed model) can be explained by the use of filleting radii that enhances flow stress. Nonlinearities can be explained by the existence of friction of the mechanical components of the testing equipment.

References

[1] Schniederjans D G 2017 Adoption of 3D-printing technologies in manufacturing: A survey analysis *Int. J. Prod. Econ.* **183** 287–98

[2] Thompson M K, Moroni G, Vaneker T, Fadel G, Campbell R I, Gibson I, Bernard A, Schulz J, Graf P, Ahuja B and Martina F 2016 Design for Additive Manufacturing: Trends, opportunities, considerations, and constraints *CIRP Ann. - Manuf. Technol.* **65** 737–60

[3] Carneiro O S, Silva A F and Gomes R 2015 Fused deposition modeling with polypropylene *Mater. Des.* **83** 768–76

[4] Ahn S, Montero M, Odell D, Roundy S and Wright P K 2002 Anisotropic material properties of fused deposition modeling ABS *Rapid Prototyp. J.* **8** 248–57

[5] Dawoud M, Taha I and Ebeid S J 2016 Mechanical behaviour of ABS: An experimental study using FDM and injection moulding techniques *J. Manuf. Process.* **21** 39–45

[6] Garg A and Bhattacharya A 2016 An Insight to the Failure of FDM Parts under Tensile Loading: Finite Element Analysis and Experimental Study *Int. J. Mech. Sci.*

[7] Griffiths C A, Howarth J, Rowbotham G D A and Rees A 2016 Effect of Build Parameters on Processing Efficiency and Material Performance in Fused Deposition Modelling *Procedia CIRP* **49** 28–32

[8] Zaldivar R J J, Witkin D B B, McLouth T, Patel D N N, Schmitt K and Nokes J P P 2017 Influence of processing and orientation print effects on the mechanical and thermal behavior of 3D-Printed ULTEM 9085 Material *Addit. Manuf.* **13** 71–80

[9] Mohamed O A, Masood S H and Bhowmik J L 2017 Characterization and dynamic mechanical analysis of PC-ABS material processed by fused deposition modelling: An investigation through I-optimal response surface methodology *Meas. J. Int. Meas. Confed.* **107** 128–41

[10] Gardan N and Schneider A 2015 Topological optimization of internal patterns and support in additive manufacturing *J. Manuf. Syst.* **37** 417–25

[11] Langelaar M 2016 Topology optimization of 3D self-supporting structures for additive manufacturing *Addit. Manuf.* **12** 60–70

[12] Rezaie R, Badrossamay M, Ghaie A and Moosavi H 2013 Topology optimization for fused deposition modeling process *Procedia CIRP* **6** 521–6

IOP Conf. Series: Materials Science and Engineering **252** (2017) 012041 doi:10.1088/1757-899X/252/1/012041

Friction Stir Welding of three dissimilar aluminium alloy used in aeronautics industry

A Boşneag [1,3], M A Constantin [1], E Niţu[2] and M Iordache[2]

[1] University Politehnica of Bucharest, Machine Building Technology Department,Splaiul Independenţei Street No. 313, Romania
[2] University of Pitesti, Manufacturing and Industrial Management Department, Târgul din Vale Street No.1, Romania

[3] ana.bosneag89@yahoo.com

Abstract. Friction Stir Welding (FSW) is an innovative solid-state joining process. This process was, in first time, develop to join the similar aluminum plates but now the technology can be used to weld a large area of materials similar or dissimilar. Taking these into account FSW process, for dissimilar materials are increasingly required, more than traditional arc welding, in industrial environment. More than that FSW is used in aeronautics industry because of very good result and very good weldability between aluminum alloy used at building of airplanes, where the body of airplane are 20% aluminum alloy and this percent can be increaser in future. In this paper is presented an experimental study which includes welding three dissimilar aluminum alloy, with different properties, used in aeronautics industry, this materials are: AA 2024, AA6061 and AA7075. After welding with different parameters, the welding join and welding process will be analyzed considering process temperature, process vertical force, and roughnessof welding seams, visual aspect and microhardness.

1. Introduction

Friction stir welding is a new welding technique invented for aluminum alloy by The Welding Institute - TWI, U.K., in 1991, but before that the principle of solid state welding process have emerged in 1956 in the Soviet Union. In last years this innovative welding process gaining wider industrial applicability like: shipbuilding and offshore, aerospace, aeronautics, automotive, railways, general fabrication, nuclear, military, robotics and computers [1]. Now, this process is developed form materials similar or dissimilar such as:cooper, brass, magnesium, titanium, steel [2], polymeric materials [3] or metal matrix composites (MMCs), such as Al2O3, SiC, Si3N4 or B4C [4].

Compared to other welding processes, the FSW has many advantages including the following:edge pieces do not have needed additional preparation,the procedure can be automated and performed in all positions,the welding procedure is perform without consumables, FSW can be used for alloys that cannot be welded with traditional method [5] and is termed "green technology" due to its energy efficiency and environmental friendliness [6]. On the other hand FSW process has several disadvantages such as: a great tool wear, weld speeds are slower, equipment is massive and expensive, friction stir welding for high melting temperature materials have limitations [7].

Considering that FSW process was invented to combine the aluminium alloy, until now the primary research and industrial interest has been to join aluminium alloys. Defect-free welds with good mechanical properties have been made for variety types of aluminium alloy even those previously thought to be impossible to weld, in thicknesses from less than 1[mm] to more than 35 [mm] [8].

IOP Conf. Series: Materials Science and Engineering **252** (2017) 012041 doi:10.1088/1757-899X/252/1/012041

The dissimilar welding of aluminium alloys has attracted more attention, since it offers an insight into many phenomena which were not clear during the friction stir welding of similar aluminium alloys. Many topics like variation of microhardness, material flow, material location, temperature distribution, vertical force distribution, residual stresses, and so forth, across the interface of the abutting materials and their consequent effect on the mechanical properties are of interest to many researchers [9–11].

Since the occurrence of this process, FSW has found great development and importance in the aeronautics industry, in especially for the welding parts of aluminium alloy like: AA2024, AA6061 and AA7075. Until now, this three alloys was weld and mixed together two by two, AA2024 with AA6061 [12], AA2024 with AA7075 [13] and AA6061 with AA7075 [14] with good result, but they have never been welded all three. This paper has purpose to analyse the behaviour of joints these three materials according to input parameters.

2. Experimental procedure

2.1. Base materials

In this experimental procedure was used three different aluminum alloy, welding together, one above the other. All of them are used in aeronautics industry, for construction of different parts of planes, this materials are: AA2024, AA6061 and AA7075.

Plates dimensions for this are 140 [mm] x250 [mm] with thickness equal with 2 [mm]. For all three, the effective chemical composition are listed in Table 1 and effective mechanical characteristics are listed in Table 2.

Table 1. Chemical composition of AA2024, AA6061 and AA7075 aluminium alloys

	Si	Fe	Cu	Mn	Mg	Cr	Zn	Ti	Ti+Zr
AA2024	0.10	0.11	4.40	0.47	1.50	0.01	0.14	0.04	0.05
AA6061	0.74	0.40	0.22	0.14	0.90	0.18	0.09	0.05	-
AA7075	0.05	0.10	1.60	0.05	2.70	0.19	5.80	0.05	0.01

Table 2. Mechanical characteristics for AA2024, AA6061 and AA7075 aluminium alloys

	UTS [MPa]	YS [MPa]	Elongation [%]
AA2024	464 to 466	344 to 348,8	17 to 18
AA6061	317 to 319	286 to 290	10 to 12
AA7075	593 to 594	531 to 532	11 to 12

2.2. Welding process

The dissimilar materials AA2024, AA6061 and AA7075 were joined according lower configuration, presented in Figure 1 and Figure 2. The position of this, in package, for welding are: AA6061 in upper position, AA7075 in meddle position and AA2024 in lower position.

All three plates were fixed with a special device, on the machine table, in welding time. The welding process was performed in three steps: (1) tool was vertically plunged through the upper plate (AA6061), the second plate (AA7075) and partially in third plate (AA2021), (2) the tool travel along the welding seam and joining this plates, (3) at the end, the tool was vertically extraction and on the following this, in materials remains a hole with a diameter equal to the diameter of the pin tool. The width difference, between welding plates, was compensated through three additional plates with dimensions 60x250 [mm] with thickness 2 [mm].

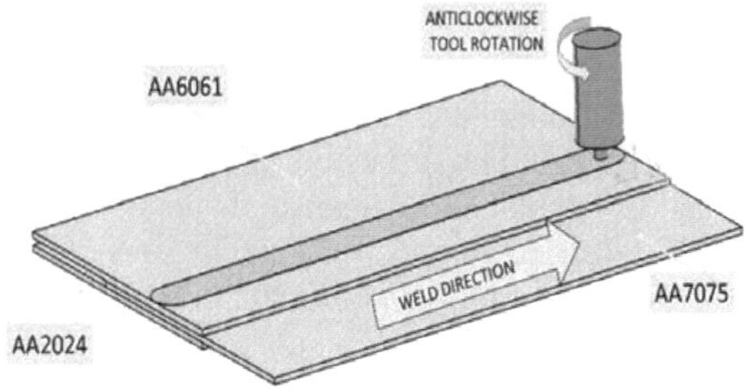

Figure 1. Friction Stir Welding process – isometric view

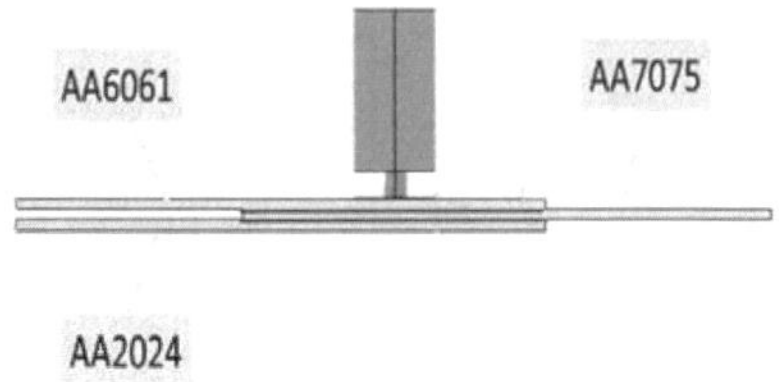

Figure 2. Friction Stir Welding process – front view

The welding tool, used to joint this three aluminium alloy, is a cylindrical tool, with the diameter of the shoulder equal with ϕ22 [mm] and the pin with shape threaded M6 and high equal with 5.3 [mm]. The tool material isP20+S (carbide of sintered tungsten), Figure 3. The joining by FSW process of the plates was performed with the anticlockwise tool rotation on the middle of the package upper described. The process performing are showed in Figure 4.

Figure 3. Welding tool

Figure 4. FSW process performing

2.3. Measurement of data during the process

During the friction stir welding process it was measure two output parameters: the temperatures and the vertical force exerted by the welding head. In process time, the temperature was measured using a high-speed and high-sensitivity thermographic infrared camera (FLIR A40M), the used camera have a field of temperature measurement between -40 [^{0}C] and + 2000 [^{0}C], this is presented in Figure 5. Measurement was made on the welding line, behind the welding tool shoulder, very close by them. The measure dates was extracted using ThermaCAM^{TM}Researcher specialized soft, and look like in Figure 6.

Figure 5. Thermographic infrared camera (FLIR A40M)

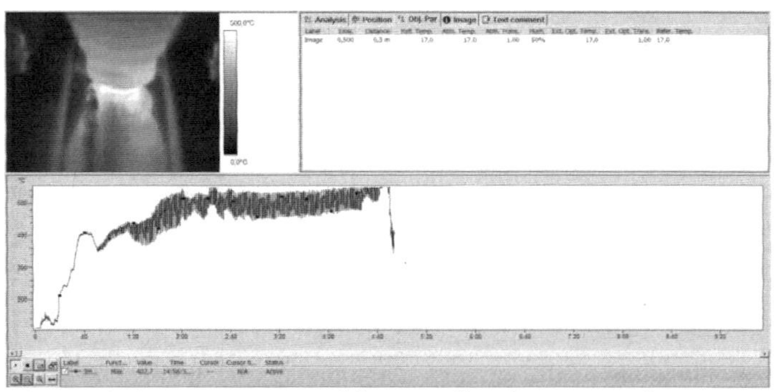

Figure 6. Temperature graphic recording in process time

IOP Conf. Series: Materials Science and Engineering **252** (2017) 012041 doi:10.1088/1757-899X/252/1/012041

Interaction between tool and plates generate a reaction force from the material, the most important component of this force for friction stir welding process are vertical force. This vertical force is necessary to accomplish effective welding and it represents the force of penetration of a tool in base material. The vertical force was measure with a mechanical device that has a fixed force transducer, type AM, with range between 0 [KN] and 20 [KN], mounted on main spindle the FSW machine and the information captured are extracted using a special soft. In Figure 7 is represented force traducer type AM mounted on the FSW machine and in Figure 8 is represented graphic with values of force recording in process time.

Figure 7. Force traducer type AM fixed on the welding machine

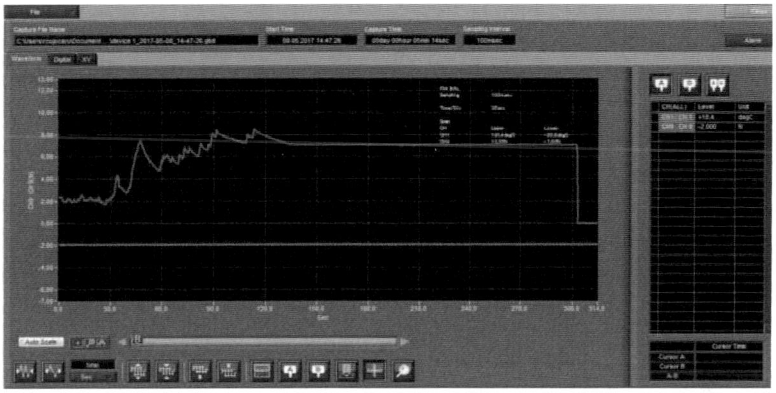

Figure 8. Vertical force graphic recording in process time

After welding process, on the samples, was measured roughness and microhardness. The roughness, was measured with the electronic roughness tester, type MarSurf PS 10, and microhardness was measured with electronic microhardness tester, type Innova Test Falcon 500.

2.4. Experimental plan

In friction stir welding, the input parameters used are: the rotation speed of the welding tool [rpm] and welding speed [mm/min]. In this experiments, the welding speed was keep the same for both attempts and rotation speed of the welding tool was modified in two situation (minimum and maximum) according with lower table.

Table 3. Experimental plan for FSW

Code of experiment	Welding speed [mm/min]	Rotation speed [rpm]
2.1	70	600
2.2	70	1400

3. Results and discussions

3.1. Temperature evolution analysis

Temperature evolution during FSW process for both two experimental situation is presented in Figure 9. In first part of graph, in front of the orange line is represented evolution of temperature in penetration and preheating phase and after orange line is represented evolution of temperature in welding process time on the length of the welding seam.

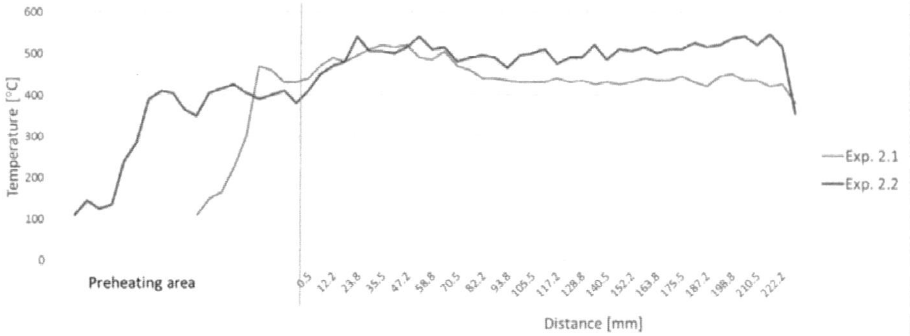

Figure 9. Evolution of temperature for experiments 2.1 and 2.2

The maximum temperature recorded in process time by thermographic infrared camera have very similar for the two experimental cases: for experiment. 2.1 is 520 [°C] and for experiment. 2.2 is 545 [°C], below the melting point. Difference between them being the position of the welding seams, for the experiment 2.1 the maximum temperature are at 35 mm and 50 mm from the start and for experiment 2.2 the maximum temperature are at 25 mm, 55 mm and at 220 mm from the start. The average temperature, in time of process, is bigger with 50°C for experiment 2.2 than experiment 2.1. For one is 450°C and for second is 500°C.

The differences between this two experiments show a small influences of rotation speed concerning temperature. Like example for experiment with bigger rotation speed (more than double) the average of temperature in process time was bigger, and, the same, the maxim point of temperature was bigger but, the difference is just by 50 [°C] for average value and 25 [°C] for maximum value.

3.2. Vertical force evolution analysis

Vertical force evolution during FSW process for both two experimental cases is presented in Figure 10. In first part of graph, in front of the orange line is represented evolution of vertical force in penetration and preheating phase and after orange line is represented evolution of vertical force in welding process time on the length of the welding seam.

The maximum vertical force recording in process time, is very different for preheating area, for experiment 2.1, the maximum value is 16 [KN], double versus experiment 2.2, where the maximum vertical force in preheating area is 8 [KN]. This means that for lower value for rotation speed, the vertical force transmitted by the machine is higher.

The vertical force was stabilised after 60 mm from the start, for experiment 2.1, and faster, at 35 mm from the start, for experiment 2.2. Differences exist also at average vertical force in welding time, for first, experiment 2.1, the average force is 8,8 [KN], and for experiment 2,2 the average is 7,2 [KN]. This means that welding machine is much more demanded for experiments with lower level used for rotation speed.

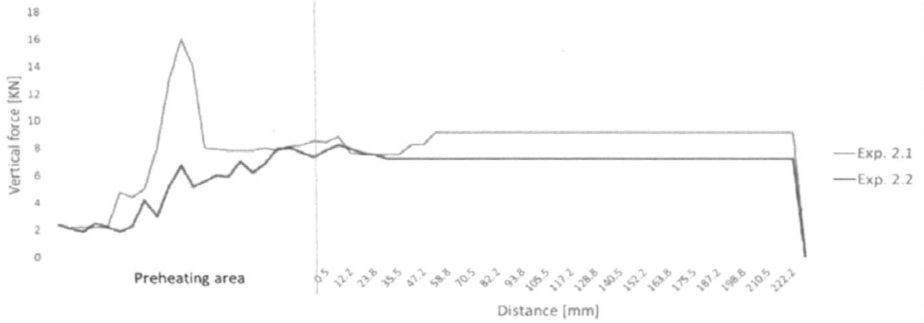

Figure 10. Evolution of vertical force for experiments 2.1 and 2.2

3.3. Roughness evolution analysis

The weld seams exhibit a satisfactory visual aspect, without hole or cracks, in both samples, the difference being due to the presence of burrs on the advancing side for experiment 2.2. Both of them are present it in Figure 11 and Figure 12. The causes for this can be the bigger temperature achieved during the process.

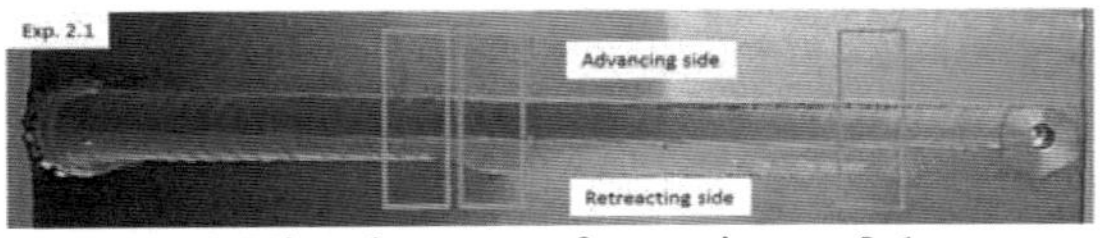

Figure 11. Visual aspect of experiment 2.1

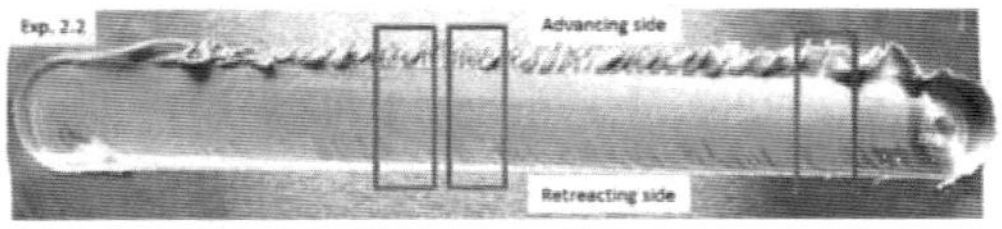

Figure 12. Visual aspect of experiment 2.2

The roughness was measure in three different area, on a sample located at 90 [mm] to the start point, on a sample located at 110 [mm] to the start point and on a sample located at 200 [mm] to the start point (blue/ red area marked in Figure. 11 and Figure. 12). Three measurements were made on each sample. The average value of the roughness, in all three samples are represented in Figure. 13. In lower graphic is shown a decrease of roughness value from the start welding seams to the end of welding seams. This evolution highlights a stabilization of the process during it is realization.

The difference between the values of roughness, for this two experiments, can be explained by rotation speed: for lower rotation speed the roughness is lower and for bigger rotation speed the roughness is bigger.

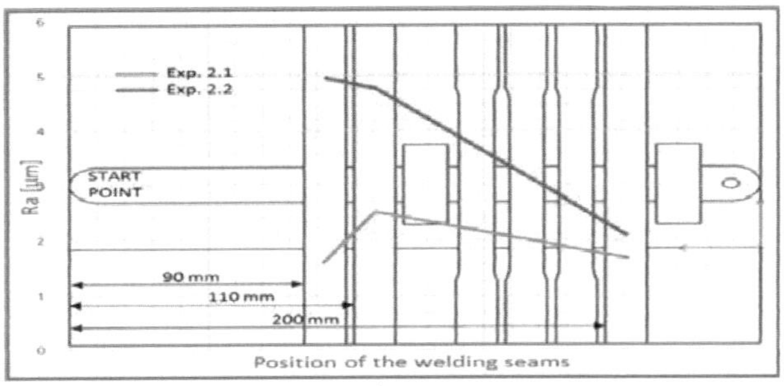

Figure 13. Evolution of roughness for experiments 2.1 and 2.2

3.4. Microhardness evolution analysis

The microhardness was measured in two area for both experiments, first on a sample located at 140 [mm] to the start of weld and second on a sample located at 230 [mm] to the start of weld. The microhardness (Vickers HV0,3) has been realized perpendicular on the tool movement direction along the welding elements, on three lines of depth, one for each material, at 1 [mm], 3 [mm] and 5 [mm] from the weld surface. On each line, microhardness was measured in 11 points, on 10 [mm] around the welding centre. The sketch with measured points are presented in Figure 14.

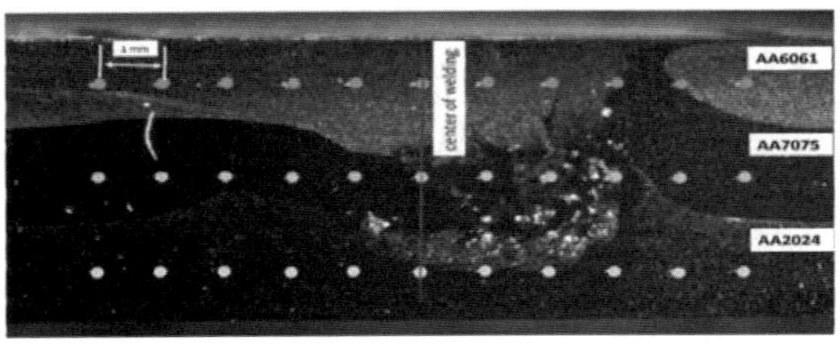

Figure 14. The place of the microhardness measurements points

The evolution of microhardness, for experiments 2.1 and 2.2, in two positions for every one of them are presented in Figure 15, Figure 16, Figure 17 and Figure 18 and average value for microhardness are presented in Table 4. For first experiments, 2.1, from first sample to second sample exist some differences. The value for average microhardness, are different, in meddle position the second is bigger wit 30 [HV0.3] than first. The second difference between them is represented by spread of values, for first, in AA7075 the minimum value of microhardness are spread from the centre with 4 [mm] in advancing side and with 1 [mm] in retracting side, for the second experiment spread are the next, 1 [mm] in advancing side and 1 [mm] in retracting side. This effect can be caused by stabilizing the FSW process.

Tabel 4. Average microhardness for experiment 2.1 and for experiment 2.2

Average microhardness [HV 0.3]	Experiment 2.1		Experiment 2.2	
	At 140 [mm] to the start	At 230 [mm] to the start	At 140 [mm] to the start	At 230 [mm] to the start
AA6061 upper position	44.7	41.5	58.0	59.4
AA7075 meddle position	91.5	121.2	124.0	131.4
AA2024 lower position	133.6	130.6	127.5	116.1

On the other hand, between experiment 2.1 and experiment 2.2 can be seen big differences. In first, the average of microhardness are bigger on the first two line with approximate 20 [HV 0.3] etch and smaller for lower line with approximate 10 [HV 0.3]. Another difference are between spread of value, for experiment 2.2, all big variations are after centre point on the retracting side between 1 [mm] and 5 [mm] or 0 [mm] and 4 [mm]. Third difference is represented by bigger variation of microhardness value for second experiment, this means a good blending of the three materials in welding process.

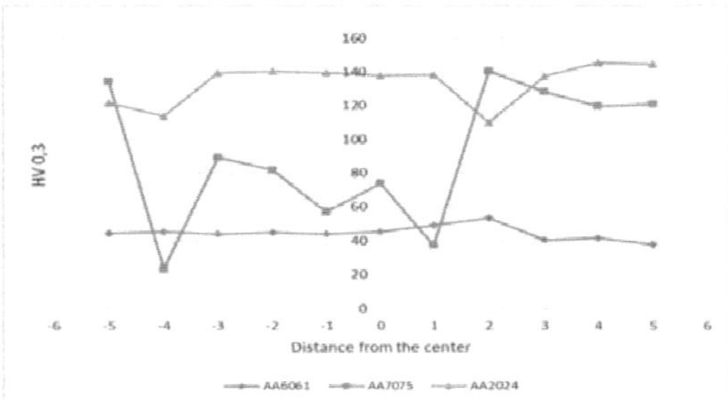

Figure 15. Evolution of microhardnessfor experiments 2.1 at 140[mm] to the start of weld

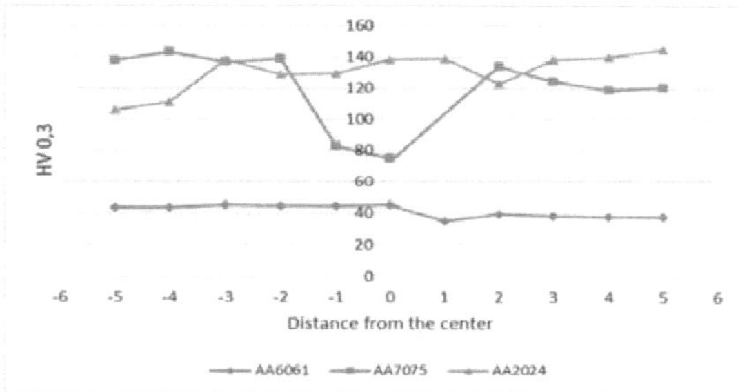

Figure 16. Evolution of microhardness for experiments 2.1 at 230[mm] to the start of weld

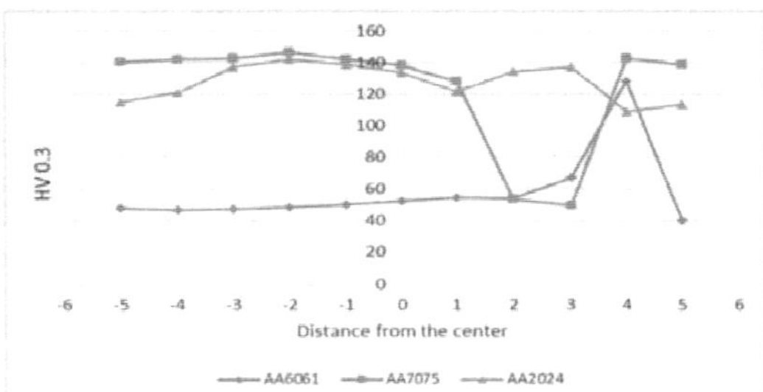

Figure 17. Evolution of microhardness for experiments 2.2 at 140[mm] to the start of weld

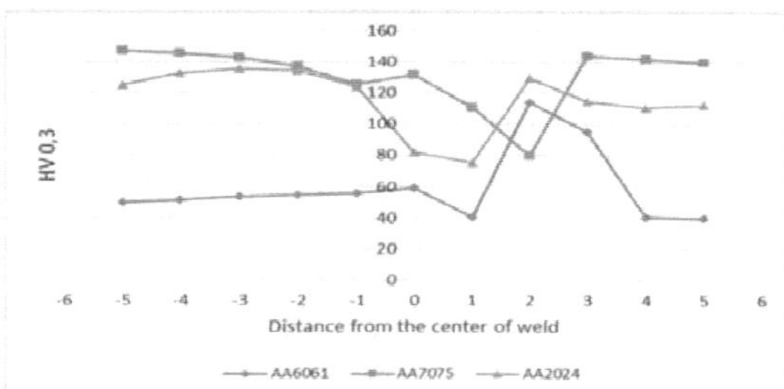

Figure 18. Evolution of microhardness for experiments 2.2 at 230[mm] to the start of weld

4. Conclusions

As a result of the researches carried out and the analysis presented above, the following main conclusions are drawn:

1. Welding of three dissimilar materials, is a new step to the new research topics.
2. In FSW process, the rotation speed is a very important parameter and have a big influence on the temperature of the process. In this study differences of 50 [°C] is given by the doubled value of rotation speed.

3. Vertical force is an important output parameter, when this parameter is bigger means that welding is much more demanded. In this study, was demonstrated that higher value for rotation speed improves (decreases) the vertical force.
4. The value of roughness is smaller when the rotation speed is smaller, and increases with increasing rotation speed. For both experiment the value of roughness decreased with the progress of the welding process, from the beginning of the seams to the end of this.
5. Evolution of microhardness was different for this two experiment. The good results, and the better shuffle of metals was identified in second experiment, where the rotation speed value was bigger.

In the future work this study will be extend to a large set of parameters, other variants to positioning this three materials and to analysis of macrostructure, microstructure, tensile test etc.

Acknowledgements
The work was done with the support of the National Institute for Welding and Material Testing, ISIM Timişoara.

References
[1] Suri A 2014 *An Improved FSW Tool for Joining Commercial Aluminum Plates,* Procedia Materials Science 6 1857-1864
[2] Leitao C M A 2013 *Influence of base material plastic properties and process parameters on friction stir weldability* (Universidade de Coimbra)
[3] Kiss Z, Czigany T 2007 *Applicability of friction stir welding in polymeric materials* Mech. Eng. 51(1) 15-18
[4] Minak G, Ceschini L, Boromei I, Ponte M 2010 *Fatigue properties of friction stir welded particulate reinforced aluminium matrix composites* Int. J. of Fatigue 32 218-226
[5] Kohn G, Greenberg Y, Makover I, Munitz A 2002 *Laser-Assisted Friction Stir Welding* American Welding Society 81 46–48
[6] Ni D R, Chen D L, Wang D, Xiao B L, Ma Z Y 2014 *Tensile properties and strain-hardening behaviour of friction stir welded SiCp/AA2009 composite joints* Mat. Science & Eng. A 608 1-10
[7] Campanelli S L, Casalino G, Casavola C, Moramarco V 2013 *Analysis and Comparison of Friction Stir Welding and Laser Assisted Friction Stir Welding of Aluminum Alloy* Materials 6 5923-5941
[8] Hatamleh O and DeWald A 2009 *An investigation of peening effects on the residual stresses in friction stir welded 2195 and 7075 aluminum alloy joints* J. of Mat. Proc. Tech. vol. 209, no. 10 4822-4829
[9] Lee C Y, Lee W B, Kim J W, Choi D H, Yeon Y M and Jung S B 2008 *Lap joint properties of FSWed dissimilar formed 5052 Al and 6061 Al alloys with different thickness* J. of Mat. Sci. vol. 43 no. 9 3296-3304
[10] Park J C and Kim S J 2010*The effect of traveling and rotation speeds on mechanical properties during friction stir welding of dissimilar Al alloys* Def. and Diff. Forum vol. 297-301 590-595
[11] Kwon Y J, and Son H J 2010*Effect of material locations on properties of friction stir welding joints of dissimilar aluminium alloys* Science and Technology of Welding and Joining vol. 15 no. 4 331–336
[12] Ouyang J H, Kovacevic R 2002 *Material flow and microstructure i the friction stir butt welds of the same and dissimilar aluminium alloys* Journal of Materials Engineer Perform 11, 51-63
[13] Song S W, Kim B C, Yoon T J, Kim N K, Kim I B, Kang C Y 2010 *Effect of welding parameters on weld formation and mechanical properties in dissimilar Al alloy joints by FSW* Materials Transactions 51, 1319-1325
[14] Guo J F, Chen H C, Sun C N, Bi G, Sun Z, Wei J 2014 *Friction stir welding of dissimilar materials between AA6061 and AA7075 Al alloys effects of parameters* Mat.&Design 56, 185-192

Wheel liner design for improved sound and structural performances

Alexandru OLTEAN[1], Claudiu DIACONESCU[1], Ştefan TABACU[2]

[1]ALSECA Engineering, Romania, [2]University of Pitesti, Romania

alexandru.oltean@alseca.com

Abstract. Vehicle noise is composed mainly of wheel-road noise and noise from the power unit. At low speeds power unit noise dominates while at high speeds wheel-road noise dominates as wheel-road noise level increases approximately logarithmically with speed. The wheel liner is designed as a component of the vehicle that has a multiple role. It has to prevent the dirt or water from the road surface that are engaged by the wheel to access the engine/front bay. Same time it has the important role to reduce perceived noised in the passenger's compartment that comes from the wheel-road interaction. Progress in plastic injection moulding technology allowed for new structures to be developed – nonwoven materials in combination with a PP based carrier structure which benefits from a cell structure caused by MuCell injection moulding. The results are light parts with increased sound absorption performances. An adapted combination of materials and production processes can provide the solution for stiff yet soundproofing structures valued for modern vehicles. Sound absorption characteristics of materials used for wheel liners applications were reported in this study. Different polypropylene and polyester fibre-based thermally bonded nonwovens varying in weight and thickness were investigated. Having as a background the performances of the nonwoven material the microcellular structure was part of the analysis. Acoustical absorptive behaviour was explained by analysing the results obtained using the impedance tube and correlating with the knowledge of materials structure.

1. Introduction.

Noise refers to the irregular and chaotic sound. With rapid development of modern industry and transportation, noise pollution has become increasingly prominent [1]. When it comes to transportation sector, the vehicle noise is composed mainly of wheel-road noise and noise from the power unit. At low speeds power unit noise dominates while at high speeds wheel-road noise dominates as wheel-road noise level increases approximately logarithmically with speed. Designed as a multi role barrier [2] the wheel liner is an important component of the vehicle. Initially designed to prevent the dirt or water from the road surface that are engaged by the wheel to access the engine/front bay now days the focus is on the important role to reduce perceived noised in the passenger's compartment that comes from the wheel-road interaction.

Progress in plastic injection moulding technology [3] allowed for new structures to be developed – nonwoven materials in combination with a PP based carrier structure which benefits from a cell structure caused by MuCell injection moulding.

In 2001, Trexel successfully developed a microcellular injection moulding process (MuCell) as an innovative process aimed to assist injection moulding [4]. By that time there have been developed studies on process, material, and technical developments. The process is still undergoing studies in order to materialize the full process potential [5–9]. Microcellular injection moulding [10] offers a number of advantages such as cost savings, weight reduction, ease in processing due to low viscosity, and dimensional accuracy [11]. Due to these advantages, the microcellular injection moulding process is used in many industries such as automotive, electrical goods, and home appliances. The process can be also applied to a wide range of thermoplastics [12].

Figure 1 presents a cross-sectional view of SEM micrograph of foamed specimens, allowing the skin layer and foam core layer to be identified.

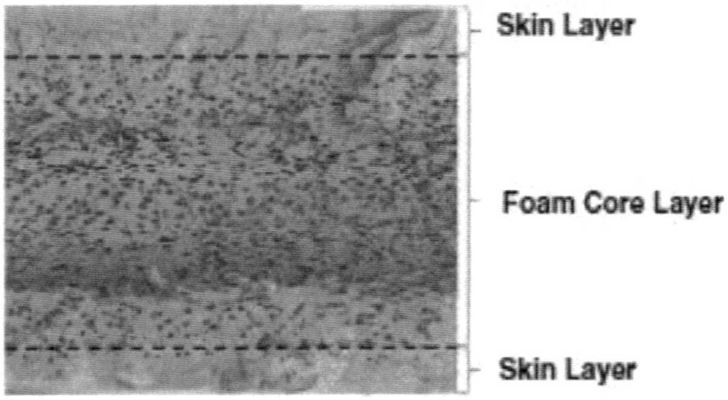

Figure 1. SEM micrograph of foamed specimens [13].

As it can be identified in Figure 1 the material has a multiple structure as a rigid skin layer and a porous core that may be addresses also as a sound absorbing material.

2. Experimental study

A number of specimens were investigated as components of existing solutions for vehicle wheel liners. Using specialized equipment the sound absorption performances were investigated.

Transmission loss is the number of sound decibels that are stopped by a structure at a given frequency or the accumulated decrease in acoustic intensity as an acoustic pressure wave propagates outwards from a source. As the acoustic wave propagates outwards from the source the intensity of the signal is reduced with increasing range due to spreading and attenuation.

A second parameter that is investigated is the Sound Absorption Coefficient that refers to the process by which a material, structure, or object takes in sound energy when sound waves are encountered. The measuring principles using the impedance tube are presented in Figure 2.

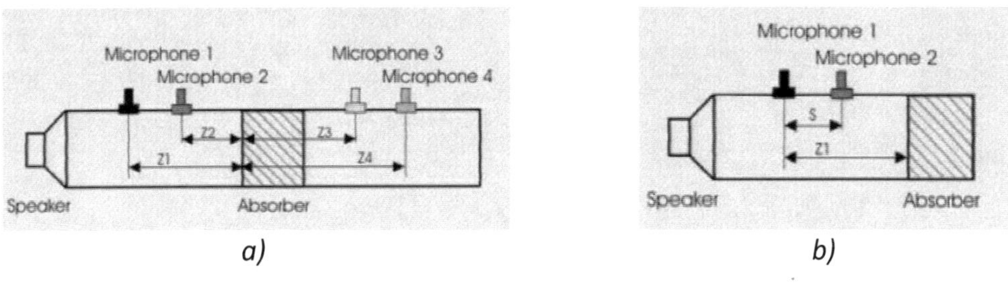

Figure 2. The measuring method.

a) method for Sound Absorption performance; *b)* method for Transmission Loss performance.

The layout of the measuring equipment comprising in the sound impedance tube, microphones and data acquisition system are presented in Figure 3.

Figure 3. Measuring equipment.

The samples collection is presented in Figure 4. It consists in four specimens of PP (polypropylene) based structures, four non-woven thermoformed parts and a specialized sound absorption material.

a) Plastic A
(1.8 mm)

b) Plastic / MuCell
(2.4 mm)

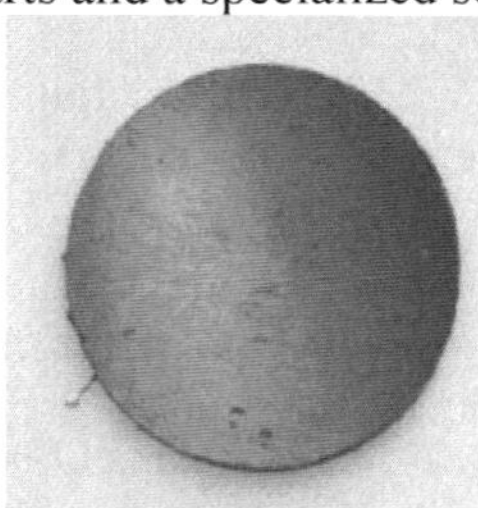

c) Plastic B
(2.0 mm)

d) Non-woven B
(800 g/sq.m.)

e) Non-woven A
(850 g/sq.m.)

f) Non-woven D
(1000 g/sq.m.)

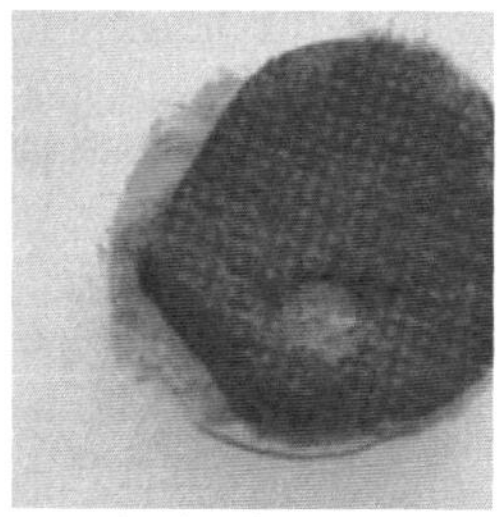

g) Plastic + 3M

Figure 4. Samples (plastic, non-woven and insulating).

3. Results and sound performance analysis

For a first set of related parts (PP manufactured solutions) the Transmission loss parameter is presented in Figure 5. The low frequencies are difficult to be absorbed and a more breathable material should provide a better performance. The results follow a similar pattern showing the lowest performance in the range of 1000 - 1600 Hz (low frequencies).

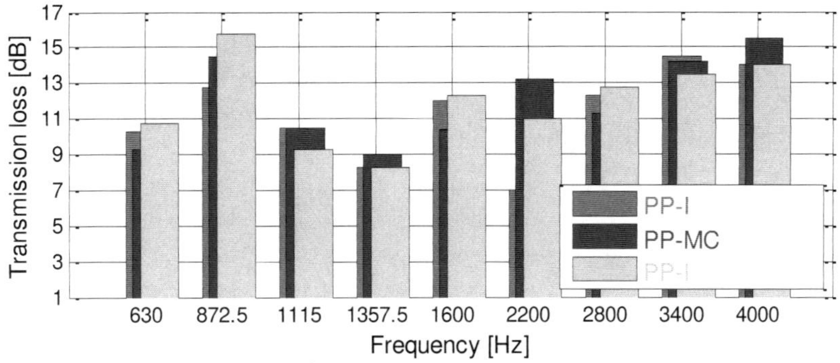

Figure 5. Transmission Loss performance for plastic specimens *(a)-c))*.

However it can be noticed that there is a good performance of the solution with a foam core (MuCell) compared to solid structures. Figure 6 presents the results obtained for non-woven samples. The raw material was thermoformed to specific thickness reaching the mentioned specific weight.

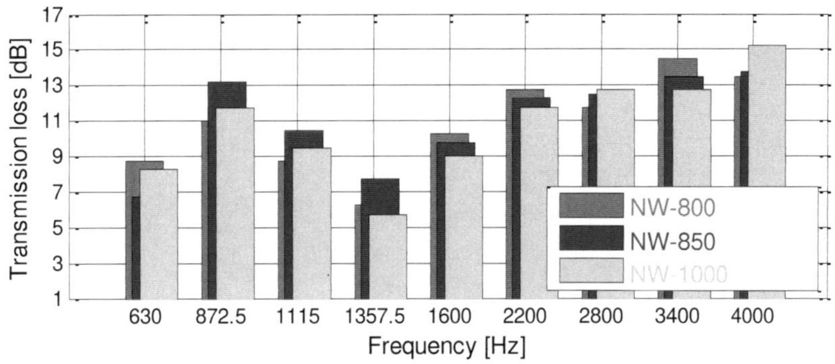

Figure 6. Transmission Loss performance for non-woven specimens *(d)-f))*.

An analysis of the above presented results shows that the performance suffers in the range of 1000-1600 Hz and the result might be related to the thickness of the part. Further studies will involve the analysis of the structures. It is possible for the results to the related with the thickness [14] of the part a mandatory input when the total weight of the structure is in the focus. Figure 7 presents the relative performance of the structures uses as a solution for material selection criteria. Results show that the MuCell structure has a good performance on the investigated frequency range.

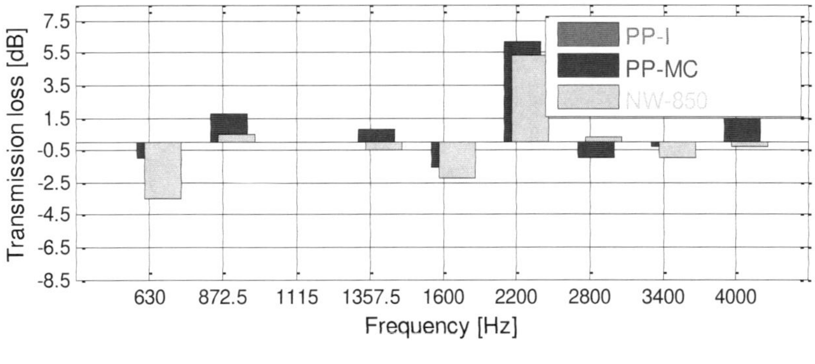

Figure 7. Transmission Loss relative performance for a specific product (wheel liner A). The results are referenced to the standard construction (PP) *(a),b),e))*.

A second solution was investigated involving parts manufacture using plastic, non-woven material and a combination of plastic and specialized insulator. Over all the solution of adding a specialized insulator provides the best results. This is according to method of soundproofing structures using a material to block and a material to absorb the sound [15].

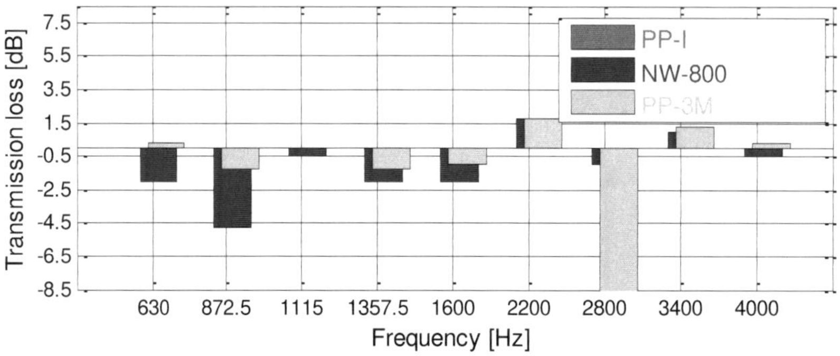

Figure 8. Transmission Loss relative performance for a specific product (wheel liner). The results are referenced to the standard construction (PP) (c),f),g)).

The relative performance analysis (Figure 10) shows that the use of specialized insulator provides a better performance in the range of low frequencies. This because that the sound is rather absorbed than reflected. The analysis is completed by the evaluation of the Sound Absorption Coefficient. At this point the performance of the foam core obtained by MuCell can be noticed.

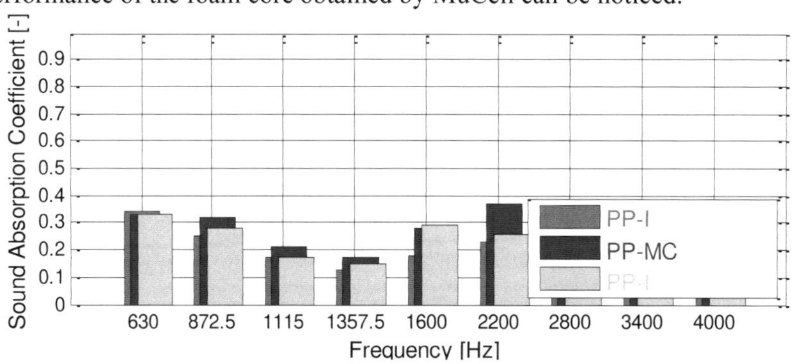

Figure 9. Sound Absorption performance for plastic specimens (*a),b),c)*).

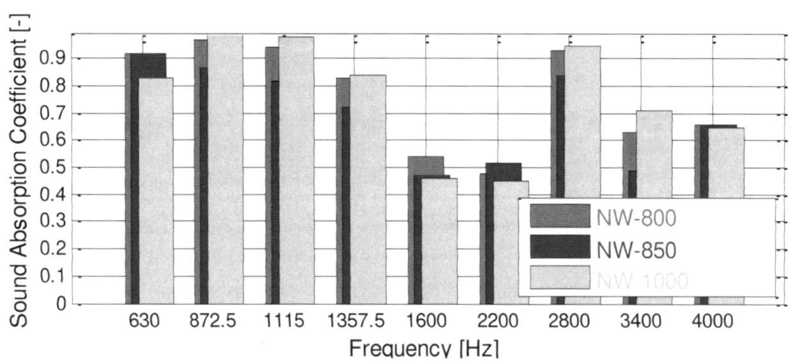

Figure 10. Sound Absorption performance for non-woven specimens (e*),f),g)*).

In order to look for a better solution Figure 11 provides the Sound Absorption Coefficient for solid plastic, MuCell structure and a non-woven.

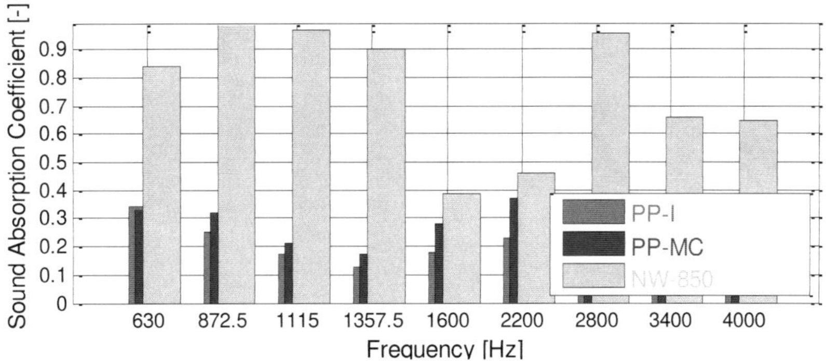

Figure 11. Sound Absorption performance for plastic and non-woven specimens applied to a given product solution (wheel liner A) ((*a),b),e)*).

Comparing the results Figure 7 and Figure 8 shows that a performance of -2.5÷ -3 dB is achievable by individual solution or by a combination [16].

4. Wheel liner study

Most efforts are paid in the direction of the vehicle body weight decrease and each individual component is investigated in order to achieve the maximum performance. Using non-woven that provides a good sound absorption is rather difficult to use because in order to obtain the required stiffness high forming thickness ratio are required. On the other hand the solid structures although individually show a good performance rather deflect the sound instead of capturing it.

A combination might be a good solution. A rigid frame provides the stiffness while the sound absorption material can be formed as a more breathable material.

In order to evaluate the performances of the structure a free-free eigenvalue analysis was performed. The recorded first two normal modes are reported in Figure 12.

a) *b)*

Figure 12. Normal model analysis. *a)* mode I, 15.9 Hz; *b)* mode II 19.2 Hz.

Following a design analysis of the structure the most reliable solution for a material replacement was defined. Figure 13 presents the model of the frame (solid frame) with the fixation points.

a) *b)*

Figure 13. Updated frame model. *a)* finite elements model; *b)* contour (outline).

Considering the patches of the thermoformed non-woven material the results recorded from the numerical analysis are presented in Figure 14.

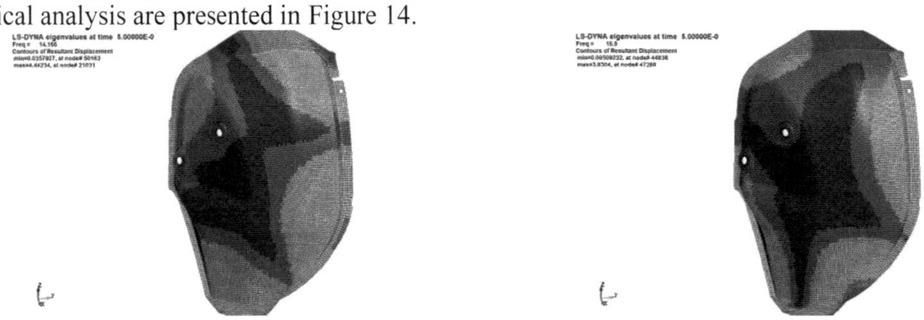

a) *b)*

Figure 14. Normal model analysis. *a)* mode I, 14.2 Hz; *b)* mode II 16.9 Hz.

The recorded values are 14.2 Hz (-10%) and 16.9 Hz (-12%) with no overlapping with the body in white natural frequencies [17].

Figure 15 presents the finished part of the hybrid wheel liner with a solid frame and non-woven sound absorbing patches.

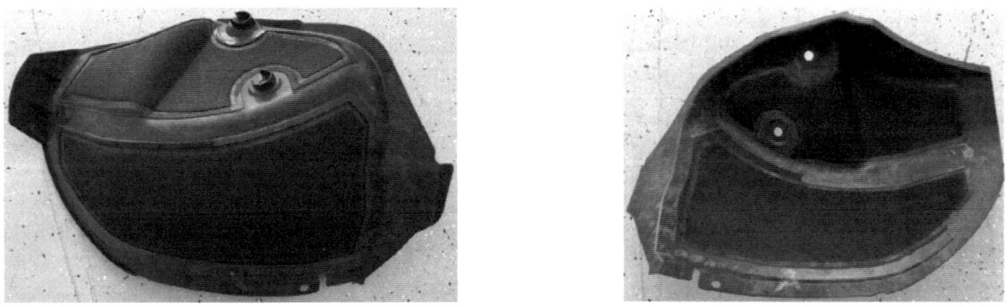

a) *b)*

Figure 15. Hybrid wheel liner. a) inner side; b) visible side.

5. Conclusions

The presents study was focused on the possibility of defining a sound absorbing structure for the wheel liner. It debuted with the analysis of sound transmission loss performance for a number of material specimens. The study proved the performances of the MuCell processed part with a porous core. However if the sound absorption coefficient has to be considered in order to evaluate the reflected sound the performance of non-woven materials has to be noticed. In this care a combination of the two materials may be a good solution as discussed and presented in our case study the resulted in the design of a hybrid wheel liner.

References

[1] Chen D, Li J and Ren J 2010 Study on sound absorption property of ramie fiber reinforced poly(l-lactic acid) composites: Morphology and properties *Compos. Part A Appl. Sci. Manuf.* **41** 1012–8
[2] Pieren R and Heutschi K 2015 Predicting sound absorption coefficients of lightweight multilayer curtains using the equivalent circuit method *Appl. Acoust.* **92** 27–41
[3] Lyu M Y and Choi T G 2015 Research trends in polymer materials for use in lightweight vehicles *Int. J. Precis. Eng. Manuf.* **16** 213–20
[4] Leicher S, Will J, Haugen H and Wintermantel E 2005 MuCell® technology for injection molding: A processing method for polyether-urethane scaffolds *J. Mater. Sci.* **40** 4613–8
[5] Chen S C, Li H M, Hwang S S and Wang H H 2008 Passive mold temperature control by a hybrid filming-microcellular injection molding processing *Int. Commun. Heat Mass Transf.* **35** 822–7
[6] Heim H-P and Tromm M 2015 General aspects of foam injection molding using local precision mold opening technology *Polymer (Guildf).* **56** 111–8
[7] Huang H X and Wang J K 2008 Equipment development and experimental investigation on the cellular structure of microcellular injection molded parts *Polym. Test.* **27** 513–9
[8] Tomasko D L, Burley A, Feng L, Yeh S K, Miyazono K, Nirmal-Kumar S, Kusaka I and Koelling K 2009 Development of CO2 for polymer foam applications *J. Supercrit. Fluids* **47** 493–9
[9] Annicchiarico D and Alcock J R 2014 Review of Factors That Affect Shrinkage of Moulded Part in Injection Moulding *Mater. Manuf. Process.* **6914** 140212072328009
[10] Saiz-Arroyo C, De Saja J A, Velasco J I and Rodríguez-Pérez M Á 2012 Moulded polypropylene foams produced using chemical or physical blowing agents: Structure-properties relationship *J. Mater. Sci.* **47** 5680–92
[11] Hwang S S, Hsu P P and Chiang C W 2008 Shrinkage study of textile roller molded by conventional/microcellular injection-molding process *Int. Commun. Heat Mass Transf.* **35** 735–43
[12] Xiao C L, Huang H X and Yang X 2016 Development and application of rapid thermal cycling molding with electric heating for improving surface quality of microcellular injection molded parts *Appl. Therm. Eng.* **100** 478–89
[13] Chen S C, Liao W H and Chien R Der 2012 Structure and mechanical properties of polystyrene foams made through microcellular injection molding via control mechanisms of gas counter pressure and mold temperature *Int. Commun. Heat Mass Transf.* **39** 1125–31
[14] Jahani D, Ameli A, Jung P U, Barzegari M R, Park C B and Naguib H 2014 Open-cell cavity-integrated injection-molded acoustic polypropylene foams *Mater. Des.* **53** 20–8
[15] Arunkumar M P, Jagadeesh M, Pitchaimani J, Gangadharan K V and Babu M C L 2016 Sound radiation and transmission loss characteristics of a honeycomb sandwich panel with composite facings: Effect of inherent material damping *J. Sound Vib.* **383** 221–32
[16] Wyerman B R and Jay G C 2007 Tire noise reduction with fiber exterior wheel arch liners *SAE Tech. Pap.*
[17] Yang Y, Zhao G, Ma D and Xu X 2011 Mode calculation and testing of a car body in white *Shock Vib.* **18** 289–98

IOP Conf. Series: Materials Science and Engineering **252** (2017) 012043 doi:10.1088/1757-899X/252/1/012043

A NEURAL NETWORKS APPLICATION FOR THE STUDY OF THE INFLUENCE OF TRANSPORT CONDITIONS ON THE WORKING PERFORMANCE

D-C ANGHEL[1], A ENE[2], C ŞTIRBU[3], G SICOE[4]

[1, 4]University of Piteşti, Manufacturing and Industrial Management Department, Târgul din Vale Street No.1, Piteşti, Romania
[2, 3]University of Piteşti, Department of Electronics, Computers, Communications and Electrical Engineering, Târgul din Vale Street No.1, Piteşti, Romania

Abstract. This paper presents a study about the factors that influence the working performances of workers in the automotive industry. These factors regard mainly the transportations conditions, taking into account the fact that a large number of workers live in places that are far away of the enterprise. The quantitative data obtained from this study will be generalized by using a neural network, software simulated. The neural network is able to estimate the performance of workers even for the combinations of input factors that had been not recorded by the study. The experimental data obtained from the study will be divided in two classes. The first class that contains approximately 80% of data will be used by the Java software for the training of the neural network. The weights resulted from the training process will be saved in a text file. The other class that contains the rest of the 20% of experimental data will be used to validate the neural network. The training and the validation of the networks are performed in a Java software (TrainAndValidate java class). We designed another java class, Test.java that will be used with new input data, for new situations. The experimental data collected from the study. The software that simulated the neural network. The software that estimates the working performance, when new situations are met. This application is useful for human resources department of an enterprise. The output results are not quantitative. They are qualitative (from low performance to high performance, divided in five classes).

1. Introduction

The relationships between job performance and job satisfaction, on one hand, and the transportations conditions and job performance, on the other hand, are the studies the most significant in industrial-organizational psychology. Landy [1] describe this relationship as the "Holy Grail" of industrial psychologists. The purpose of this article is to review the state of the literature concerning the relationship between: job performance and transportations conditions to study the mutual influences based on a theoretical model, using Artificial Neural Networks.
The article is organized into five major sections. First, we qualitatively review previous research on the job motivation-job satisfaction-job performance relationships. In this section, we recapitulate previous reviews of the literature and then consider various conceptualizations of the satisfaction-performance and transportation conditions and performance relationships. Second, we developed a theoretical model, a research approach and establish a list of evaluating criteria. Third, an Artificial Neural Networks are

designed based on the parameters. Fourthly, simulations with trained Artificial Neural Networks was made in order to study the relationships and the correlation between the parameters, for discrete values. Finally, we propose an improved model for the study and we provide some suggestions for future research.

2. Literature review

Job satisfaction is a positive emotional state resulting from the valuation of their work [2]. Optimistic attitude towards work leads to greater job satisfaction which in return perk up the performance of the individual [3]. Employee satisfaction is very important because discouraged workers are negatively affecting the desire to work [4]. „The neglect of job satisfaction reflects professional suspicion of what may be called subjective variables: variables that measure "what people say" rather than "what people do." Job satisfaction should be maximized, and the social reason for seeking to improve the quality of employment is to increase resultant satisfaction.
This must be treated not only as a valued goal, but also as a dynamic process occurring primarily as an individual phenomenon but having significant implications for societal adaptivity as well [5]. The relevant theory to work redesign was made it by Herzberg two-factor theory of satisfaction and motivation [6,7]. The theory proposes that the primary determinants of employee satisfaction are factors intrinsic (i.e., recognition, achievement, responsibility, advancement). This specifies that a job will enhance work motivation and satisfaction only to the degree that "motivators" are designed into the work itself.
Over the time different micro-models have been developed to study the relationships between satisfaction and performance. Our study was focused on the analysis of the following two models:

JOB SATISFACTION AND JOB PERFORMANCE ARE RECIPROCALLY

Five studies were investigated on the relationship of reciprocity between job satisfaction and job performance. Two of these studies have suggested that performance resulted because of satisfaction, but not and vice versa [8,9]. Sheridan and Slocum [10] gave partial support for a reciprocal relationship and Wanous [11] consider a reciprocal relationship that depends on satisfaction type, figure 1.

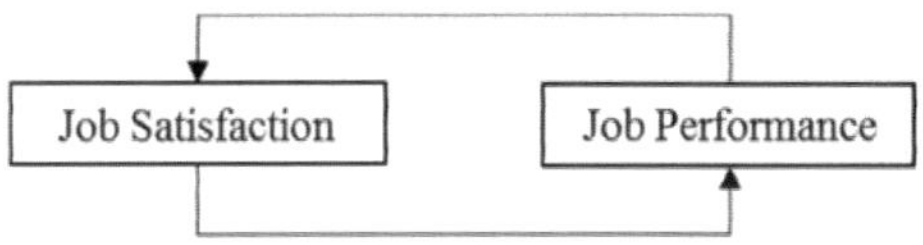

Figure 1. Job satisfaction and job performance are reciprocally

For extrinsic satisfaction result satisfaction - performance, while for intrinsic satisfaction the relation is performance - satisfaction.
In conclusion four of the five studies suggest a causal effect of job performance in job satisfaction, and two of the five suggesting a causal effect of job satisfaction on performance at work.

JOB SATISFACTION AND JOB PERFORMANCE IS MODERATED BY OTHER VARIABLES

Many studies have hypothesized that performance should affect satisfaction at work to the extent that people are compensated based on their performance.
The most frequently investigated moderator of the relationship satisfaction - performance is self-esteem. This mean that employees deemed to have value intrinsic rewards, such as the nature of work itself more than pay [12]. In this sense have been proposed a variety of relationships satisfaction - performance such as: job fit [13], need for achievement [14], career stage [15] etc., but very few of them have been tested, figure 2.

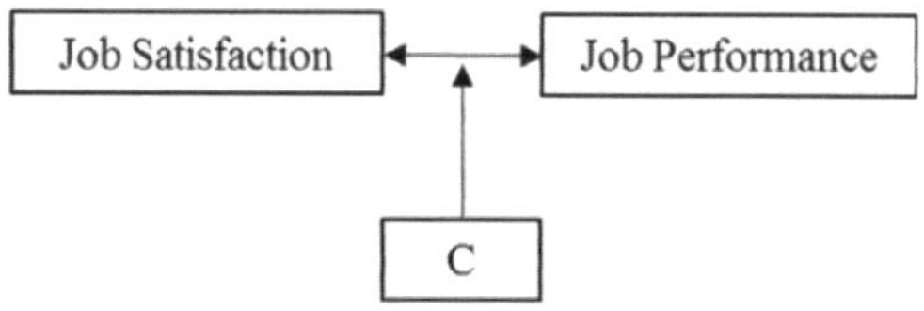

Figure 2. Job satisfaction and job performance moderated by other variables

TRANSPORTATIONS CONDITIONS AND JOB PERFORMANCE

In order to establish the transportation parameters, a study was carried out through discussions with the employees, with the employer and with the persons involved in the transport of employees at the workplace. So, the influence parameters are derived from:

- the duration of the journey;
- the posture of the person while traveling;
- the comfort;
- feeling fatigued etc.

ARTIFICIAL NEURAL NETWORKS

Feed forward neural network are intelligent devices that learn to solve problems in the same way people solve them: they learn using examples [16]. Feed forward neural networks are also called multi-layered perceptron networks (MLP). They are composed from layers of interconnected artificial neurons (perceptrons). Each neuron from a layer is connected with all the neurons from the next layer. Every connection between two neurons has associated a real number (positive or negative) called weight, which is a measure of the strength of the connection.
Neural networks are excellent tools for pattern recognition. They do not need to know a mathematical model for the object to be recognized, they learn through examples. They are excellent tools for the classification of objects being widely used in industrial, medical, and military applications. A study about the use of feed forward neural networks in ergonomics is presented in [17, 18]. In our application, a feed forward neural network is used to classify the input parameters in four classes: very good, … In order to classify objects, a feed forward neural network typically needs to have three layers of neurons: the input layer, the hidden layer and the output layer.

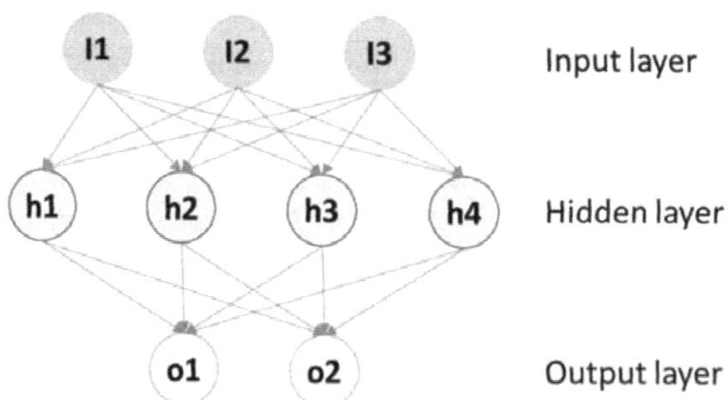

Figure 3. Example of an Artificial Neural Network

The examples (patterns) from which the network learns, are used to calculate the weights. In a network that has NI input neurons, NH hidden neurons and NO output neurons, there are NI*NH+NH*NO weights [16]. Using a single training example, starting from its inputs, we can we can propagate it through the network (forward propagation) and we can calculate the outputs of the network. So, we

obtain an equation that has as variables, the weights of the network. But we have many different training examples and so, we can obtain a system of equations that has the same number of equations as the number of weights. But we cannot solve exactly this equations system, because all the equations are nonlinear (because all artificial neurons from the hidden and from the output layer are nonlinear devices, having exponential transfer functions). The only way to find the weights is iteratively and in this manner, the backpropagation algorithm, calculates the weights.
A trained neural network can be used to classify new sets of input data. In order, not to repeat the training every time a network is used for a classification, the weights calculated with the backpropagation algorithm, must be saved in an external file. So, before using the network for classifications, the weights from the external file have to be loaded into the neural network.

3. Research model

The proposed model for this study, figure 4, is based on the model presented in the literature review. The hypothesis that we made involves the study of the direct influence of the Transportation Conditions on Job Performance. In order to determine these influences were used an Artificial Neural Network.

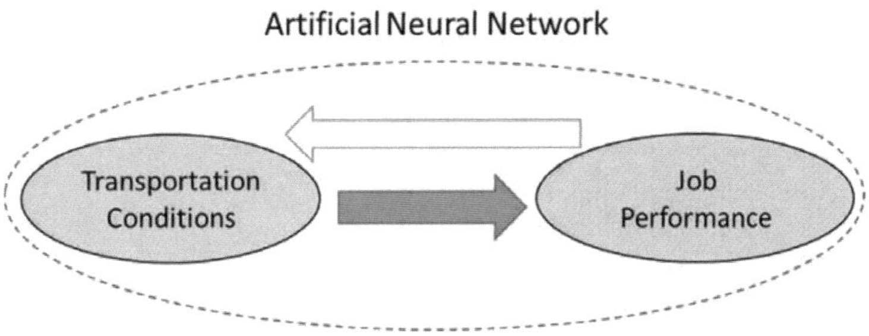

Figure 4. The research model based on the transportation conditions and job performance

The research approach used for this study is shown in figure 5.
That research approach starts with establishment of criteria for evaluation, data collection based on an investigation with a questionnaire and creating a database. An Artificial Neural Networks was designed based on the two parameters. The Artificial Neural Networks are trained, tested and validated. Application of trained Artificial Neural Networks on a new set of data in order to study the correlation between the three parameters.

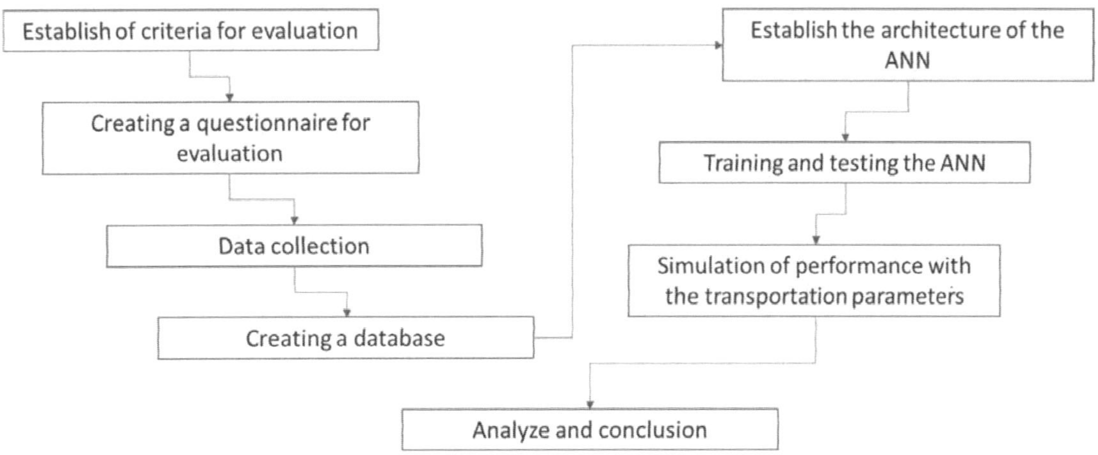

Figure 5. The research approach

The evaluation criteria of the parameters "Job Performance" and "Transportation condition" are presented in table 1.

Table 1. The evaluation criteria

Evaluation criteria of transport (Transportation condition parameters)	
TC1	The duration of the journey
TC2	The posture of the person while traveling
TC3	The presence / absence of hollow or road abnormalities
TC4	The thermal comfort
TC5	The acoustic comfort
TC6	The level of air pollution
TC7	The humidity level, especially during the heat cycle
TC8	The driving style of the driver
TC9	The number of stops on the route
TC10	Your profession requires physical effort
TC11	Your profession requires intellectual effort
TC12	Do you feel tired when you arrive at work
Evaluation criteria of performance of employees (Job Performance parameters)	
P1	Quality of work (accuracy)
P2	Labor efficiency (yield)
P3	Professional knowledge
P4	Responsibility

4. The software application and results

The software application is used for the estimation of the four parameters: P1, P2, P3, P4. It consists of three software programs. Two of them are used for the pre-processing of input data and the third is the one that simulates the neural network.
All the initial input data, obtained from the questionnaires, are stored in a text file. Each line has 16 integer values. The first 12 values represent the input parameters (TC1, TC2... TC12) and the last 4 values represent the output parameters P1, P2, P3, P4. For every parameter, there are possible only 4 values: -2, -1, 1 or 2.
Here is an example for a typical line from this file (file_1.txt):

1 2 1 1 2 1 1 1 1 -2 -2 -2 2 -1 -2 -2

We decided to use a separate neural network for the estimation of each output parameter, so we have a neural network for the estimation of parameter A, another neural network for the estimation of parameter B, etc.
For the estimation of parameter A, we use only the first 13 values (12 inputs and one output) from every line of file_1.txt:

1 2 1 1 2 1 1 1 1 -2 -2 -2   2

For the estimation of parameter B we use the first 12 input values, and the 14th:

1 2 1 1 2 1 1 1 1 -2 -2 -2   -1

In order to use this file for the training of the neural network, we had to pre-process it. The first pre-processing consisted in the elimination of duplicate lines, and this was done with the aid of a separate software program.
The algorithm used for the elimination of duplicate lines is the following:

Copy the first line from file_1.txt in the array of lines b[ ]
Set the current dimension of array b (nb), to 1
For i=2 to N repeat
Begin
  Read the current line crtLine from file_1.txt

IOP Conf. Series: Materials Science and Engineering **252** (2017) 012043 doi:10.1088/1757-899X/252/1/012043

Set flag areSimilar as false
For j=1 to nb repeat
Begin
if crtLine equals with the j line from array b than
set flag areSimilar as true
End for j
If flag areSimilar= false than copy crtLine in array b and increment nb
End for i
Copy all the array b[] in an output text file file_2.txt

The dimension of the obtained file (file_2.txt) is 28.
Now, all the lines from the file_2.txt are different.
This file file_2.txt was used as an input to another software program that we created in order to do the second pre-processing. For an input value (and there are 12 input values in each record) that can have 4 different values (-2, -1, 1, 2) we binary code it using two bits, in the following manner:

0 0 for -2
0 1 for -1
1 0 for 1
1 1 for 2.

In this way, each of the 12 input data is coded using 2 bits, so, we will use in the input layer of the neural network 24 neurons.
For the single output data from the current record that can also have 4 values (-2, -1, 1, 2), we code this using 4 binary values, in the following manner:

-2 : 1 0 0 0
-1: 0 1 0 0
1: 0 0 1 0
2: 0 0 0 1

In this way, the neural network will have 4 output neurons.
The third program simulates the neural network.
The architecture of the network results from the way we coded the data, and is:

24 – 14 -4

We used from the available 28 records, table 2, resulted from the pre-processing 23 records as patterns for the training process, and the other 5 records to test the accuracy of the network.

Table 2. The dataset used to train and test the network for the second experiment

Pattern no.		**1**	**2**	**3**	**4**	**5**	**6**		**28**
Input parameters Transportation Condition	TC1	1	1	2	1	1	1		1
	TC2	1	2	1	1	1	1		1
	TC3	1	1	2	1	2	1		1
	TC4	2	1	2	-1	2	1		1
	TC5	2	1	2	1	1	1		2
	TC6	1	1	1	-1	1	1		1
	TC7	-1	1	2	2	1	2		1
	TC8	-2	1	1	2	1	2	...	2
	TC9	1	1	1	-2	2	1		1
	TC10	1	-2	1	-1	1	2		2
	TC11	2	-2	1	1	1	1		1
	TC12	2	-2	1	-2	1	1		1
Output parameters Job Performance	P1	2	2	2	-1	2	2		1
	P2	2	-1	1	1	2	2		2
	P3	2	-2	2	1	1	1		1
	P4	2	-2	1	2	1	2		1

The error that we used for the training of the network is 1%.
The network learns all the training patterns in about 500 epochs. The algorithm used for the training (backpropagation algorithm) is the following:

Set E, the learning error (in our application is 1%)
Initialize all weights with small random values
Converged=false
While(not converged)
Begin
For all training patterns repeat
Begin
Forward propagate the inputs of the current pattern
Calculate the error of learning the current pattern
Backpropagate (modify all weights according to the delta rule)
End for
Calculate ERR, the error of learning all the training set of patterns
If(ERR<=E)converged=true
End while

After the network converges, all the calculated weights are saved in a text file.
In the testing phase, the input test data are forward propagated through the network, and the outputs are calculated.
Here are some experimental results obtained with this program, for two input test data:

First input test data:
First output: Ideal: 0 Real: 0.002556441418702507
Second output: Ideal: 0 Real: 0.007147458227179302
Third output: Ideal: 0 Real: 0.006617343440648846
Fourth output: Ideal: 1 Real: 0.9659198790230381
Second input test data:
First output: Ideal: 0 Real: 0.004862528784744236
Second output: Ideal: 0 Real: 0.0019752387364835675
Third output: Ideal: 0 Real: 0.0016502716524455837
Fourth output: Ideal: 1 Real: 0.995498458365162

It can be seen that the feed forward neural network correctly classifies the test data.
In the literature, we had identified the following factors of influence, in order to complete our model: time pressure (Bhagat, 1982), similarity in problem-solving styles (Goldsmith, McNeilly & Ross, 1989), affective disposition (Hochwarter, Perrewé, Ferris & Brymer, 1999).

5. Conclusions

This paper gives as a review and a study of the relationship between job performance and the transportation conditions.
Our study has found that the parameters are closely correlated.
We have decided do not use a classical approach because, in this case the number of employees involved in the study it would be great. By using ANN, the number of experiments it will be very low than because ANN are able to learn from examples and are able to make connections.
This is an obvious advantage because it is economical and once a network is trained, it can be used in various simulations.
This study allowed us to complete the initial research with other factors of influence.
In the future, we plan a more complex approach, so the results to be richer and the external influencing factors to be identified and characterized more precisely.

IOP Conf. Series: Materials Science and Engineering **252** (2017) 012043 doi:10.1088/1757-899X/252/1/012043

References

[1] Landy, F. J. (1989). Psychology of work behavior. Thomson Brooks/Cole Publishing Co.
[2] Locke, E. A. (1976). The nature and causes of job satisfaction. Handbook of industrial and organizational psychology, 1, 1297-1343.
[3] Linz, S. J. (2003). Job satisfaction among Russian workers. International Journal of Manpower, 24(6), 626-652.
[4] Nelson, N. (2006). A little appreciation can go a long way toward employee job satisfaction. Employment Relations Today, 33(1), 19-26.
[5] Misztal, A., Butlewski, M., Belu, N., Ionescu, L. M. (2014). Creating involvement of production workers by reliable technical maintenance. International Conference on Production Research - Regional Conference Africa, Europe and the Middle East (ICPR-AEM) / 3rd International Conference on Quality and Innovation in Engineering and Management (QIEM), 322-327.
[6] Herzberg, F., Mausner, B., & Snyderman, B. (1959). The motivation to work New York.
[7] Herzberg, F. I. (1966). Work and the nature of man.
[8] Bagozzi, R. P. (1980). Performance and satisfaction in an industrial sales force: An examination of their antecedents and simultaneity. The Journal of Marketing, 65-77.
[9] Siegel, J. P., & Bowen, D. (1971). Satisfaction and performance: Causal relationships and moderating effects. Journal of Vocational Behavior, 1(3), 263-269.
[10] Sheridan, J. E., & Slocum, J. W. (1975). The direction of the causal relationship between job satisfaction and work performance. Organizational Behavior and Human Performance, 14(2), 159-172.
[11] Wanous, J. P. (1974). A causal-correlational analysis of the job satisfaction and performance relationship. Journal of Applied Psychology, 59(2), 139.
[12] Jurgensen, C. E. (1978). Job preferences (What makes a job good or bad?). Journal of Applied psychology, 63(3), 267.
[13] Carlson, R. E. (1969). Degree of job fit as a moderator of the relationship between job performance and job satisfaction. Personnel Psychology, 22(2), 159-170.
[14] Steers, R. M. (1975). Effects of need for achievement on the job performance-job attitude relationship. Journal of Applied Psychology, 60(6), 678.
[15] Stumpf, S. A., & Rabinowitz, S. (1981). Career stage as a moderator of performance relationships with facets of job satisfaction and role perceptions. Journal of Vocational Behavior, 18(2), 202-218.
[16] Ene, A., Stirbu, ., "Retele neuronale. Teorie si aplicatii in Java", Ed. Univ. din Pitesti, 2008.
[17] Ene, A., & Anghel, D. C. (2013, June). A neural networks application in ergonomics. In Electronics, Computers and Artificial Intelligence (ECAI), 2013 International Conference on (pp. 1-4). IEEE.
[18] Anghel, D. C., Ene, A., & Belu, N. (2014). A matlab neural network application for the study of working conditions. Advanced Materials Research, 837, 310-315.

The potential of natural fibres for automotive sector - *review* -

MS Fogorasi[1] and I Barbu[2]

[1] Assoc.Prof. Ph.D, Aurel Vlaicu University, Faculty of Engineering, Arad, RO
[2] Prof. Ph.D, Aurel Vlaicu University, Faculty of Engineering, Arad, RO
E-mail:mfogorasi@yahoo.com, ionelbarbu@yahoo.com

Keywords: natural fibres, biocomposite, interfacial adhesion, mechanical properties, enzymatic modification, environmental impact

Abstract - Due to environment and sustainability regulation in the last decades considerable performance in green technology in the field of materials science through the development of biocomposites could be noticed. Thus, the development of high-performance materials produced from natural resources expands worldwide. This can be attributed mainly due to their assets compared to their synthetic contourparts like low cost, low weight, less damage to processing equipment, improved surface finish of moulded parts composite, good mechanical properties, biodegradability, abundant and renewable resources. Natural fibres are valuable and versatile resources with multiple advantages.

Nowadays all the more, the automotive industry is under increasing pressure to fulfill environmental and performance demands and higher fuel efficiency at competitive costs. Automakers recognize potential in biocomposites if these materials can offer the same performance as traditional composites but with lower weight. Additional they exhibit non-brittle fracture on impact, which is another significant requirement for automotive sector. Other drivers that scores for use of natural fibres reinforced polymer composites (NFRPC) in automotive applications imply reduced waste disposal, reduction of greenhouse gas emission and Life Cycle Consideration.

In spite of their benefits, the significant challenge for producers and supplier to handle with natural fiber reinforced polymer composites resides in large inconsistency of their properties. The chemical composition of vegetal fibres relies on several factors comprising fiber variety, time of harvesting, climatic history, soil characteristics and fibre processing technology. All these factors exert an influence on their final properties when used as reinforcements in biocomposite materials.

In this review a wide range of issues is addressed with special reference to mechanical properties of fibres, interface adhesion and environmental implication of NFRPC. The discussion on the cellulosic/lignocellulosic fibre properties is conducted in order to relate their chemical composition, microstructure and mechanical properties and to understand their *use* and *limits* as reinforcements in composite materials.

The variation within the mechanical properties of natural fibres is a challenge towards designing predictable components for industry since the engineers are accustomed to the precise and reproducible properties of synthetic fibres.

The hydrophilic nature of lignocellulosic fibres causes poor resistance to moisture and incompatibility to hydrophobic polymer matrix. As a consequence, this incompatibility causes a weak fibre/matrix interface, which consecutively leads to diminished mechanical properties of the biocomposites. Therefore, it is important to ensure a good adhesion between matrix and fibres to enhance the mechanical strength of NFRPC.

This study aims to provide an overview of the *greener surface treatments* without use of hazardous chemicals, with emphasize on the enzymatic surface modification of natural fibres. The effectiveness of the treatment on the mechanical properties of the resulting NFRPC is also reviewed.

Environmental impact of NFRPC is another important issue addressed in this review. A comparison of the environmental impacts between the NFRPC and SFRPC applied in the automobile sector, based on LCA studies, will be traced.

IOP Conf. Series: Materials Science and Engineering **252** (2017) 012044 doi:10.1088/1757-899X/252/1/012044

1. Introduction

Guidelines for the development of the 21 century generation of materials, products and processes are sustainability, industrial ecology, eco-efficiency and green chemistry.

We assist to an increasing pressure to develop novel bio- and other innovative technologies aiming to rescue extensive dependence on fossil fuel. Synthetic fibres like glass, carbon and aramid are well established reinforcements in polymer composites due to their high stiffness and strength properties [1].

Despite this, serious drawbacks such high processing cost, machine abrasion, high density, health hazards when inhaled and most significant, high energy consumption, non-biodegradability and -recyclability, coupled with severe environmental legislation, shifts attention from synthetic to natural fibres, from Synthetic Reinforced Polymer Composites to Bio- Composites or Natural Fiber Reinforced Polymer Composites (NFRPC) [2]. Besides, not all applications are addressed for structural or load –bearing goals, thus utilizing high- performance fillers for low and non-load bearing are over performance and not cost-effective [3]. Bio-composites defined as composite materials in which at least one of the component is derived from natural resources, can be classed in three subgroups [4]:

1. Petroleum derived polymers reinforced by natural fiber which are non-biodegradable
2. Biopolymers reinforced by natural fibers known as green composites (biodegradable)
3. Biopolymers reinforced synthetic fibers (i.e. glass, carbon etc.) (non-biodegradable).

On the other hand, various industrial application, including automotive, require characteristics that are not met by neat materials. Thus composite materials can be appropriate to fulfill the industrial requirements due to their proper mechanical properties through the synergetic action of the components, matrix and reinforcement, which support the load stress [5].

Under these circumstances, natural fiber composites (NFRPC) as environmentally appealing materials are emerging as a viable alternative to the glass-reinforced or carbon reinforced polymer composites. The final properties and performance of engineering products manufactured from NFRPC depend upon the characteristics of their individual constituents as well as their compatibility and matrix/filler interface.

Currently, NFRPC occupy several industrial area that proofs the potential of natural fibres as replacements for their synthetic counterparts for various application. The fields in which natural fibre composites are being broadly employed include the automotive and construction sectors [6]. Low density and satisfactory high specific properties of natural fibers are features that recommend NFPC for application in the transportation, particularly automotive industry. Attempts for their implication in automobile sector dates back in the 90s, when Mercedes -Benz manufactured door panels containing jute fibers. Flax has been the most relevant natural fibre for the German automotive industry for years, hemp are the second most important natural fibres [7].

Nowadays all the main international automotive manufacturers have adopted in their new models NFRPC for interior and exterior vehicle parts such door panels, package trays, trunk liners, seat backs, boot lining, hat rack, spare tyre lining, headliner panel, boot lining, noise insulation panels, moulded foot well linings, door trim, Interior door paneling, windshield, dashboard, back cushions, parcel shelf, insulation, rear storage shel, door trim panels (PP matrix with NF), engine or transmission cover side and back, spare wheel compartment cover, to name a few [6, 8 - 12].

This shift was prompted by the European Union End of Life Vehicle Directive (2020) imposing that 80wt% of a waste vehicle has to be reused, recovered or recycled [13].

Composite materials find their application also in aircraft, railway and truck industry [14]. Common applications of natural fibre reinforced composites include also household products and furniture, electrical and electronics, sports and leisure items [15] food industry [16] and medical applications [17].

Despite their numerous and manifold uses, there are still obstacles that are obstructing their growth development inside industrial area, particularly for structural purposes.

Apart from synthetic fibres, most natural fibres are relatively hydrophilic, have a rough surface and are physico-chemically heterogeneous. These characteristics exert a strong influence on the fibre-matrix interfacial adhesion in the composite. The poor interfacial bonding leads to defective mechanical properties of the biocomposite [18].
This paper reviews the latest trends in green surface treatments of natural fibers without use of hazardous chemical substances with focus on biotechnology as a viable alternative to the physico-chemical approaches. In this regard, the importance of surface modifications through enzymatic treatment and the resultant enhancement in the properties of the composites are discussed.
An exhaustive approach of the natural fiber structure, chemical composition and properties tries to underline the interdependence exiting between these factors. An understanding of the interconnection between the presence of different chemical constituents in the fiber and their influence on the fiber properties and later on upon the NFRPC, represent the key drive for a successfully enzymatic treatment and in turns improved properties of the composite materials for specific purposes. Environmental impact of NFRPC is another important topic addressed in this review.

2. Natural fibers

2.1.Microstructure and properties.
Natural fibres encloses all types of fibres that occur within nature, and are found in vegetables respectively plants (cellulose fibres), animals (protein fibres) and minerals (asbestos). A natural fibre may be further defined as a flexible material which large aspect ratio and high tensile strength.
Even if fibrous materials exist in abundance in nature, particularly cellulosic types like cotton, wood, grains, and straw, not all materials can be spooned into fibres for textile products or other industrial applications. Aside from economic perspectives, the appropriateness of a fibre for commercial purposes is driven by its attributes like length, strength, pliability, elasticity, abrasion resistance, absorbency mand surface properties.
Between the plant fibres, lignocellulosic fibres, especially bast fibres (flax, hemp, jute, kenaf, ramie) are by far the most prominent group of natural fibres used as substitute for the synthetic fibres in natural fiber composites in automotive field due to their specific strength, modulus characteristics and availability. Their high content of cellulose make them appropriate for stress- bearing applications [8]. So this review is confined only to plant-based fibres since lignocellulosic fibres accounts for the most technical application.
Lignocellulosic fibers might be considered themselves as composites material due to their cellular structures consisted of cells that contains crystalline and amorphous regions of cellulose linked together through an amorphous matrix of lignin and hemicellulose and pectin. Cellulose is the main constituent of plant fibers, although the percentage composition of pure cellulose, hemicellulose, lignin, pectin and other organic compounds such waxes, fatty acids, fats, pectins varies for different fibres [19].
The hemicellulose molecules are hydrogen bonded to cellulose and act as cementing matrix between the cellulose microfibrils, forming the cellulose – hemicellulose network, which is considered to be the main structural component of the fiber cell. The hydrophobic lignin grid activates as a binder agent of the cellulose/hemicellulose composite, being responsible for the rigidity and low elasticity [20]. In addition, the low elasticity is a result of a low value of microfibrilar angle of the fibers [21].
Furthermore, the cell wall in a fiber is not a homogenous membrane. Actually, each fiber has a complex, stratified structure comprising a thin primary wall surrounding a secondary wall. This secondary wall is an assembly of three layers and the thick middle layer is responsible for the mechanical properties of the fiber [20]. The middle layer has a complex structure composed by a series of helically winding microfibrils along the hollow fibre axis. Microfibrilar angle values differs from fiber to fiber and provide mechanical strength to the fiber. The spiral orientation of the microfibrils to the fiber axis increases the ductility of the fibers [22, 23].

Selected physical and tensile properties of various plant fibers and synthetic fibers are listed in tabel 1 [8, 24-26]. It can be seen from table 1 that glass fibers have higher tensile strength and higher Young's modulus than cellulose-based fibers. However, comparing the lower density values of the lignocellulosic fibres to the one of glass, it is obvious that the specific tensile strength and specific Young modulus of both types of fibres are comparable. This is an important aspect for automotive industry where weight reduction is a priority [19, 27].

Table 1. Mechanical properties of lignocellulosic fibers

Fibre	Density (g/cm^3)	Diameter (μm)	Tensile strength (MPa)	Stiffness/ Young's modulus (GPa)	Specific tensile strength (MPa/g/ cm^3)	Elongation at break (%)	Specific Young's modulus (GPa·cm^3/g)
Flax	1.38 - 1.52	5 - 600	343 - 1830	27 - 100	227 -1220	1.2 - 3.2	18 - 53
Kenaf	1.2	12 - 36	295 - 930	22 - 60	246 - 993	1.6 - 6.9	18 - 5 0
Jute	1.23 - 1.5	5 - 200	187 - 800	10 - 55	140 - 610	1.16 - 3.1	7.1 - 39
Hemp	1.35 - 1.51	10 - 500	550 -1110	30 - 70	210 - 740	1.6 - 4.5	20 - 47
Ramie	1.44 - 1.55	18 - 80	400 - 938	44 - 128	258 - 620	1.2 - 4	29 - 85
Coir	1.1 - 1.46	7 - 460	130 - 580	4 - 62	92 - 180	15 - 40	3.3 - 5.2
Sisal	1.2 - 1.5	7 - 200	468 - 855	9 - 28	55 - 610	1.9 - 7	6 - 20
E glass	2.5 - 2.55	15 - 25	200 - 3500	68.9 - 73	666 -1400	2.5 - 3	29

Lignocellulosic fibres proved to be an appropriate reinforcement for structural composites but they do expose some deficiencies emerging from their inherent nature. The tensile properties values comprised in Table 1 reveal a large variability even inside the same type of fibre. This variation can be ascribed to the innate scatter of the fibre properties and test methods applied for tensile properties measurement of single fibre [8, 27] Strength of vegetal fibres is controlled by a complexity of factors including: the ratio and perfection of crystalline cellulose in the fiber, fiber asymmetry (cellulose molecular chain length-to-width ratio), polydispersity and the cellulose microfibril angle measured to the fibre axis [8, 28].
Further, the length of natural fibre and orientation is an argument for their selection in composites for automotive application [29] since the use of long fibres in composites imparts improved mechanical properties and capability to adhesion over the matrix to the fiber [30].

2.2.Chemical composition of natural fibre.
The chemical constituents of the natural fibres such cellulose, hemicellulose and lignin exert a significant influence upon the overall properties of the fiber and thus upon their applications [19, 31]. As follows, hemicellulose is responsible for the biodegradation, moisture absorption, and thermal degradation of the fiber as it shows reduced level of resistance whereas lignin is thermally stable but responsible for the UV degradation.
Figure 1 exemplifies the interdependency of the natural fibres properties on their major chemical components [32].

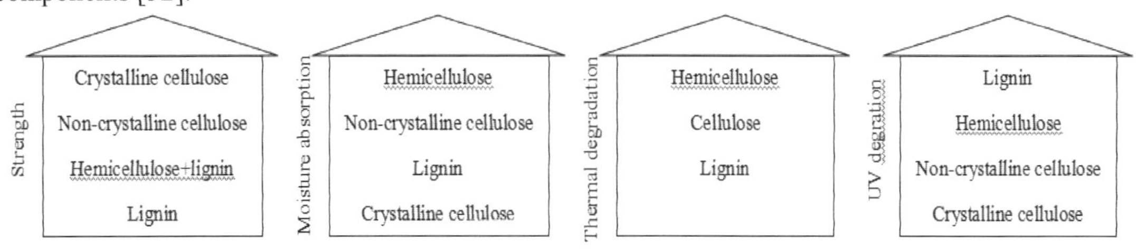

Figure 1. Influence of chemical components upon the properties of natural fibres

The percentage composition of each of these components varies on the type of fiber for different fiber As a general remark, increased cellulose content will increase tensile strength and Young's modulus of fibers.

3. Natural Fibres Reinforced Polymer Composites – Advantages and challenges

A significant application aria of natural fibres is their use as reinforcement for synthetic polymers in the automotive industry. Prior the use of a material in technical application, it must go through an extensive design process, which covers material selection, the generation of a part geometry and calculation of part performance. Since material costs usually exceed 50% of the global production costs, material selection is of utmost relevance, taking into consideration technical and economic requirements. Thus natural fibres will have to compete with materials that are appropriate for a specific technical application. Actually, the criteria for their selection relies in their performance in all aspects which implies a good understanding of their properties, the root cause for their weaknesses and improvement possibilities. The absence of reliable material data for selection and design is the major obstacle for natural-fibres-reinforced materials [33].

A satisfactory prediction of the response of a composite part involves an accurate prediction of the properties of the components (fibre, matrix and fillers), level of adhesion between them and the distribution of the constituents within the composite.

Since synthetic fibres and matrices are usually manufactured under controlled circumstances, their final chemical and physical characteristics are easier to predict. Thus, they can be tailored to have specific properties according to their final application. Unlike synthetic fibres, vegetable fibres, are deriving from nature. Thus, a comprehensive knowledge of their characteristics is more difficult to acquire [5].

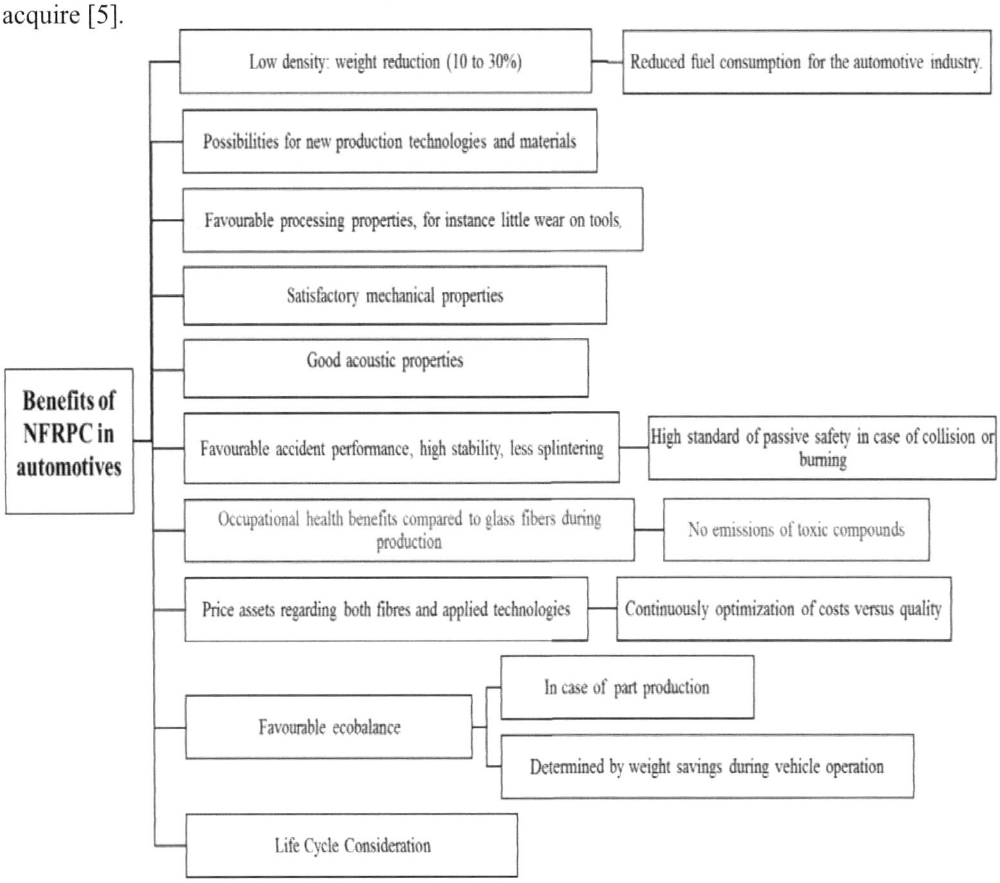

Figure 2. Main benefits of NFRPC in automotive

The prime driving forces for car –makers to replace Synthetic fibre reinforced polymer composites with NFRPC are depicted in fig. 2 [7].
In addition, several factors that are endorsing the application in the automotive sector refers to:

- Reduction of greenhouse gas emissions [34]
- Competitive pricing [11]
- Natural fibers are renewable resources
- Safety, Crashworthiness [35]
- Technical benefits (lower energy loss, better wear protection, expanded lifetime of tools
- Socio- economic benefits (expended usage of NFC in automotive will create new jobs in agriculture [36, 37].

Although these benefits are well known to the industry, natural fiber polymer composite application has been hindered by some drawbacks. Compared to their synthetic counterparts, following aspects have to be taken into consideration [38, 39]

- Structure is highly inhomogeneous
- Fibers discontinuity that causes lower performance of their composites
- Lower durability – possibility to improve considerable
- High moisture absorption which causes swelling
- Lower strength, especially impact strength
- Large scatter of the properties
- Limited thermal stability.

Parameters controlling the mechanical performance of NFRPC are extensively discussed in literature. The main factors in this regard include fibre selection (type, harvest time, extraction, fibre content), matrix selection, fibre dispersion, interfacial strength, composite manufacturing process and porosity [38].

4. Biotechnological procedures for interfacial adhesion improvement

The nature of the fiber-matrix interface exert a critical influence upon the overall properties of composite. The fiber-matrix adhesion and nature of the interphase control the stress transfer efficiency between fiber and matrix, stress distribution, mechanisms of damage propagation and accumulation. Besides the reinforcement of high-strength fibers, high mechanical performance of the composite material is expected in case of a strong fiber-matrix adhesion [40-42].
Natural fibres have some deficiencies due to their strong hydrophilic character which engenders two considerable restrains when used as reinforcement in biocomposites. These include: first, their accentuate sensitivity to water and moisture which provoke in composite a diminishing of mechanical properties upon moisture absorption; second, their incompatibility with the hydrophobic polymeric matrix used, which in turns leads to weak interfacial adhesion [43]. Thus, in several cases NFRPC do not exhibit the same grade of performances compared to the glass –fibers reinforced composites [44].
To overcome these major impediments associated with improper fiber performance, lignocellulosic fibers are subjected to various surface changes. Generally, the adhesion at the composite interface can be ascribed to the following major interactions: physical adhesion related to wettability and compatibility of the fibre - matrix complex, which are controlled by the surface energies of the materials, chemical bonding, and mechanical interlocking build up on the uneven fibre surface [40, 45].
Various modification strategies have been intensively discussed in literature. Processing and surface treatments to compose the interface in NFRPC can envisage either interface engineering trough matrix or fiber modification and the use of interface active additives [46]. Most applied approaches mentioned in literature for improved interface adhesion comprise physical and chemical modification [18, 20]. Chemical approaches are more discussed within the literature than physical with superior improvements obtained till now. Even though positive effects were attained on NFRPC, the applied physical and chemical methods boost the risk for fiber degradation and consecutively to augmentation in the processing cost. Furthermore, it is difficult to solve all the deficiencies associated with NFRPC

using one method [44]. A solution to improve interfacial adhesion in composite and to prevent fiber chain degradation in natural fibre is the use of biological agents, such as fungi and enzymes [47]. Currently, the application of biotechnology in the field of textile and natural fiber modification is well established [48, 49,50].

The major reason for embracing this technology relies in the high reaction specificity of the enzyme which leads to non-destructive transformation at the fiber surface. Besides, enzymes act under milder condition as compared to chemical methods, decreasing water and energy consumption, producing less toxic wastewaters and are biodegradable [51, 52].

Enzymes used for surface modification of natural fibres belong to the classes of hydrolysis and oxidoreductases. Enzymatic treatments proofed their effectiveness for lignocellulosic fibers in the retting stage [53] as well as in surface modification procedures [54]. Biotechnological retting using specific enzymes is a promising alternative of improving the traditional retting process due to greater control of processing parameters. Thus, the enzymatic retting preserve the natural fiber structure and result in superior mechanical properties compared with chemical retting, which damages structure and lessens quality [55].

This review is conceived with the biotechnological procedures for fiber surface modification of bast fibers, for improved interfacial strength. Some representative literature references are discussed in detail. There are several reports describing improvement of composite properties ascribed to enzymatic treatment of natural fibres. Z Saleem *et al.* [56] report the effect of pectinase treatment of hemp on mechanical properties of fibre reinforced polypropylene. Their results support evidence for an effective refinement of hemp fibres when treated with a commercial pectinase. Although the structural changes caused by the pectinase action altered the mechanical properties, the composite prepared using enzymatic treated hemp present improved tensile and flexural characteristics. Actually, the fibre refinement results in separation of technical fibres into smaller bundles and single fibre cells which in turns means an increasing in surface to mass ratio of the fibres and an enlarged interfacial area between fibre and matrix. Further studies reported that the enzymatic treatment of natural fibres leads to increased surface area and increased interfacial bonding coupled with increased impact properties [57].

Karaduman *et al* [58] also confirmed these findings by tracing a comparison between alkaline and enzyme-treated jute fibres on the mechanical properties of jute fiber-reinforced polyester composites. A set of various enzymes (pectinase, laccase, cellulase and xylanase), enzyme combinations and treatment time according to experimental design were employed in the study. Tensile and flexural properties measurement were carried out to evaluate the efficacy of the treatments. As a result of the enzymes action, pectin, hemicelluloses and lignin were removed from the fiber bundle interface. This action reduced the technical fiber diameter and hence increased the fiber aspect ratio which enables better fiber-matrix adhesion and improved mechanical properties of the composites [58].

Dong A *et al* [59] investigated the possibility to obtain an hydrophobic jute fabrics via the laccase-mediated grafting of octadecylamine (OA) on the lignin moieties. The hydrophobized jute was used as the reinforced material of the polypropylene (PP) matrix composites to improve the interfacial compatibility between the fiber reinforcements and the hydrophobic synthetic polymer. To evaluate the effectiveness of the enzymatic treatment on jute fabric/ PP composites, the tensile properties, dynamic mechanical performances and fracture surface of were analyzed [59]. Through laccase-mediated grafting of OA, modified jute gained excellent water repellency. As a consequence of the specificity of laccase to substrates and the surface modification of the enzymatic grafting no evident decrease of thermal properties was noticed. In addition, the tensile strength, tensile modulus as well as the breaking elongation of the hydrophobized jute/PP composites were enhanced. The treatment lead also to an improved interfacial adhesion between the jute reinforcement and the PP resin [59]. Laccases – assisted grafting process jute fibers using different chemical compounds proofed its efficiency, too [60, 61].

IOP Conf. Series: Materials Science and Engineering **252** (2017) 012044 doi:10.1088/1757-899X/252/1/012044

After the enzymatic graft modification the jute/PP composites present improved tensile and dynamic mechanical properties and good water absorption repellency. Besides, the hydrophobic modified jute represent a prime condition for good interfacial compatibility.
Improved surface properties of the natural fiber reinforcement can be achieved also through removal of amorphous constituents in order to gain as much cellulose as possible with higher degree of crystallinity. For this purpose the effect of various enzymatic systems on the surface chemical, morphological and thermal properties of bast fibres was studied [62]. Flax and hemp fibres were subjected to treatment with hemicellulases, pectinases and oxidoreductase. Removal of amorphous hemicellulose from the fibre surface and following exposure of the crystalline cellulose caused a lower contact angle for all the treated samples.
Other beneficial effects of enzymatic treatments are associated with a homogeneous cellulose surface for subsequent activation [63] and an improved thermal stability due to the removal of the thermally labile pectic and hemicellulosic fiber constituents. Compared to hemp fibres, flax is more susceptible to enzymatic degradation. The results confirm the enzymatic treatment as an inexpensive and environmentally appealing alternative to improve the surfaces of natural fibres for composite applications [62].

5. Environmental performance of NFRPC

In order to reduce environmental impact the automotive industry has embraced several directions: development of sophisticated emissions control technology that provide much cleaner vehicle with catalytic converters reducing smog-forming emissions for car; development of alternative fuels, hybrid technology [64]. Accordingly, considerable efforts are also directed towards improving the fuel efficiency of vehicles as required by global regulations. These target can be fulfill by weigh reduction because lower weight vehicles combine the use of less fuel with fewer carbon dioxide emissions [34]. Thus, NFRPC application in automotive has increased significantly mainly due to their ability to meet diverse requirements like significant *weight savings,* specific strength, stiffness and *environmental benefits.*
Natural fiber composites are claimed to offer multiple environmental advantages such as: reduced dependence on non-renewable energy/material sources, lower pollutant emissions, lower greenhouse gas emissions [65] enhanced energy recovery and end of life biodegradability of components [39].
However, all these assertions should be supported by quantitative analysis. In this regard, Joshia *et al.* analyzed three comparative Life Cycle assessment (LCA) studies to assess the overall environmental performances. Results reveal that in their specific applications, analyzed in the study, NFRPC composites are environmentally superior to glass fiber composites on most midpoint indicators [66].
Drivers of superior environmental performance depicted from the comparative analyses are as follows [66]:

- Less environmental impacts of natural fiber production compared to glass fiber production
- Substitution of matrix (synthetic polymer) by higher content of natural fiber in the composite
- Improvement of fuel efficiency and lower use phase emissions due to weight reduction
- Energy and carbon credits from end of life incineration of natural fibers [39].

Yet, these benefits are counter-balanced by the environmental burdens caused by the use of fertilizer in natural fibre cultivation. Thus, the high nitrate and phosphate emissions can lead to increased eutrophication in local water bodies. A reduced operating lifetime compared to the glass fibre composites affect the environmental superiority of natural fibre composites [66].
In their study, George M *et al.* highlight also the environmentally superiority of NFRPC compared to their synthetic counterparts. For this purpose, hemp fibres modified by chemical treatments for composite application were evaluated for their environmental impact. Two novel acid sulfonic procedures and conventional chemical methods were applied for surface fiber improvement. Previously, a comparison between the novel and established chemical procedures in terms of environmental burdens was traced. The new methods had a comparably lower environmental impact than the conventional ones. Furthermore, all the chemical methods proved to have environmental

lower footprint than the glass fiber reinforced polymer. Even with a chemical treatment the NFRPC were superior to glass fiber. These assumptions are based on LCA outcomes [67].

6. Conclusion

Fibres as reinforcement present some features that will interfere in the final properties of composites. Vegetal fibers, due to hydrophilic characteristic, are incompatible with most hydrophobic matrixes, weakening the interfacing of composite.

To overcome the hydrophilic/hydrophobic incompatibility, the enzymatic surface treatments envisage either the removal of amorphous constituents aiming to get as much cellulose as possible with higher degree of crystallinity or the mediation of a grafting process. The results obtained in several studies proved the potential of enzymatic surface treatment of natural fibres as an effective and environmentally friendly tool for natural composites with improved interfacial adhesion and mechanical properties.

Additional, studies assessing life cycle environmental performance of natural fiber composites and glass fiber reinforced composites considered for specific applications, reveal the environmentally superiority of natural fiber composites.

7. References

[1] Mohanty A K, Misra M and Drzal L T 2002 *J Polym Environ* **10** 1-2 pp 19-26
[2] Bledzki A K and Gassan J 1999 *Prog Polym Sci* **24** 2 pp 221-74
[3] Misnon M I, Islam M M, Epaarachchi J A and Lau K T 2014 *Materials and Design* **59** pp 359-68
[4] Al-Oqla F M and Omari M A 2017 ed Springer International Publishing AG pp 13-29
[5] Furtado S C R, Araújo A L, Silva A, Alves C and Ribeiro A M R 2014 *Int. J. Automotive* Composites **1** 1 pp 18-38
[6] Brett C S 2008 *Proceedings of the Symposium on Natural Fibres* (Rome), ftp://ftp.fao.org/docrep/fao/011/i0709e/i0709e10.pdf
[7] Karus M and Kaup M, Use of natural fibres in the German automotive industry Nova-Institut, Huerth, Germany http://www.internationalhempassociation.org/jiha/jiha6209.html
[8] Kozłowski R M 2012 Handbook of natural fibres: processing and applications vol 2 ed Woodhead Publishing Limited p 221
[9] FAO 2012 Unlocking the commercial potential of natural fibres: Food and Agriculture Organization of the United Nations, Division TaM (Rome) www.fao.org/.../natural-fibres/UnlockingFibrePotentialFinal
[10] Holbery J, Houston D 2006 *J Miner Met Mater Soc* **58** 11 pp 80–86
[11] Mohammed A L, Ansari M N M, Pua G, Mohammad J and Saiful Islam M 2015 *Int J Polym Sci* 243947
[12] Chandramohan D and Bharanichandar J 2013 *American Journal of Environmental Science* **9** 6 pp 494-504
[13] Directive 2000/53/EC on end of life vehicles: guidance document by the European Commission, 2005 Version 1 http://ec.europa.eu/environment/waste/pdf/guidance_doc.pdf
[14] Kozlowski R, J. Kozlowska J, M. Rawluk M and Barriga J 2004 *Nonlinear Optics, Quantum Optics* **31** pp 61-89 ed Old City Publishing Inc
[15] Ho M P, Wang H, Lee J H, Ho C K, Lau K T, Leng J and Hui D 2012 *Compos B Eng* **43** 8 pp 3549–62
[16] Majeed K, Jawaid M, Hassan A, Abu Bakar A, Abdul Khalil H, Salema A and Inuwa I 2013 *Mater Des* **46** pp 391-410
[17] Al-Oqla F M, Sapuan S M, Anwer T, Jawaid M and Hoque M 2015 *Synthetic Met* **206** pp 42-54
[18] Tran L Q N, Yuan X, Bhattacharyya D, Fuentes C A, Van Vuure A W and Verpoest I 2015 *Int J Mod Phys* **4**
[19] Nabi Saheb D and Jog J P 1999 *Adv Polym Tech* **18** 4 pp 351-63
[20] Jacob John M and Rajesh D. Anandjiwala R D 2008 *Polym Composite* pp 187-206
[21] Daescu C 1996 Natural fibrous materials ed Helicon (Timisoara) p 131
[22] *Textile Fibre* in Handbook of Textile Engineer 2002 ed AGIR (Bucureşti) pp 91-397
[23] Ray A K, Moandal S, Das S K, and Ramachandrarao P 2005 *J Mater Sci* **40** 5249
[24] Arpitha G R and B. Yogesha B 2017 *5th International on material processing and characterization – Materials Today: Proceedings* **4** pp 2755-60
[25] Everitt N M, Aboulkhair N T and Clifford M J 2013 *Conference papers in Materials Science* 141204 ed Hindawi Publishing Corporation

[26] Thangavel G, Mohanty S and Nayak S 2015 *Compos Part A-Appl S* **77** pp 1-25
[27] Fortea-Verdejo M, Bumbaris E, Burgstaller C, Bismarck A and Lee K Y 2017 *Int Mater Rev*
[28] Moharir A V 2000 *Indian J Fibre Text* **25** 1 pp 1-7
[29] Laine L and Rozite L 2010 State of the Art, Eco-efficient Composite Materials https://people.arcada.fi/~rene/biocomposites/lectures/STATE%20OF%20THE%20ART_material_examples.pdf
[30] Jeyanthi S and Janci Rani J 2012 *Int J Phys Sci* **7** 43 pp 5765-5771
[31] Gurunathan T, Mohanty S and Nayak S K 2015 *Composites Part A* **77** pp 1–25
[32] Sain M and Panthapulakkal S 2004 *Green Composites: Polymer Composites and the Environment* ed Woodhead (Cambridge) pp 181-88
[33] Baur E and Otremba F 2010 *Industrial Applications of Natural Fibres: Structure, Properties and Technical Applications* ed John Wiley & Sons pp 397- 407
[34] Lighter weight automotive parts, naturally, PolyOne, *www.materiautech.org/.../download/.../PolyOne_whitepaper*
[35] Bhardwaj B P 2014 The complete book on production of automobile components and allied products ed Niir Project Consultancy Services NPCS Edition pp 9-15
[36] Pickering K L 2008 Properties and performance of natural fibre composites ed Woodhead Publishing
[37] Vazguez A, Riccieri J and Carvalho L 1999 *Polym Composite* **20** 1 pp 29-37
[38] Pickering K L, Aruan Efendi M G and Le T M 2016 *Composites Part A* **83** pp 98-112
[39] Spārniņš E 2009 Mechanical properties of flax fibers and their composites Luleå University of Technology Sweden Ph.D Thesis
[40] Güven O, Monteiro S N, Moura E A B and Drelich J W 2016 *Polym Rev* **56** pp 702-36
[41] Faruk O, Bledzki A K, Fink H P and Sain M 2012 *Prog Polym Sci* **37** pp 1552-96
[42] Mohanty A K, Misra M and Drzal L T 2001 *Compos Interface* **8** pp 313-43
[43] Zafeiropoulos N 2011 Interface in natural fibre composites and their characterization in: *Interface Engineering of Natural Fibre Composites for Maximum Performance*, ed Woodhead Publishing pp 242-43
[44] Väisänen T, Das O and Tomppo L 2017 *J Clean Prod* 15 pp 582-96
[45] Ungureanu D, Ţăranu N, Lupăşteanu V, Roşu A R and Mihai P 2016 *Bul. Inst. Polit. Iaşi, Series Construction-Arhitecture* **62** 2 pp 37-45
[46] Rohit K and Dixit S 2016 *Polymers from Renewable Resources* **7** 2 pp 43-59
[47] Pickering K L, Li Y, Farrell R L and Lay M 2007 *J Biobased Mater Bio* **1** 1 pp 109-17
[48] Fogorasi M S 2003 Ph.D Thesis Technical University Gh Asachi Iassy
[49] Fogorasi M S, Puscas E L, Stanescu M D and Grigoriu A 2003 Enzyme influence in cotton dyeing DWI Report 127:B, 1-5 ISSN 0952301X
[50] Shahid M, Mohammad F, Chen G, Tang R C and Xing T 2016 *Green Chem* **8**
[51] Puscas E L, Stanescu M D, Fogorasi M S and Dalea V 2003 Sustainable development by environmental protection and textile biotechnology 1st edition ed Universitatii Aurel Vlaicu (Arad) pp 250- 324
[52] Stanescu M D and Fogorasi M S 2000 *Ind Textila* 51 pp 184-88
[53] Foulk J A, Rho D, Alcock M M, Ulven C A and Huo S 2011 *Adv Mater Sci Eng* 179023
[54] Kalia S, Thakur K, Celli A, Kiechel M A and Schauer C L 2013 *Journal of Environmental Chemical Engineering* **1** 3 pp 97-112
[55] Ramesh D, Ayre B G, Webber C L and D'Souza N A 2015 *Text Res J* **85** 19 pp 2059-70
[56] Saleem Z, Rennebaum H, Pudel F and Grimm E 2008 *Compos Sci Technol* **68** pp 471–76
[57] Bledzki K, Mamun A A Jaszkiewicz A and Erdmann K 2010 *Compos Sci Technol* **70** 5 pp 854-60
[58] Karaduman Y, Gokcan D and Onal L 2012 *J Compos Mater*
[59] Dong A, Wu1 H, Fan X, Wang Q, Yu1 Y and Cavaco-Paulo A 2016 *Express Polym Lett* **10** 5 pp 420–29
[60] Dong A, Yu Y, Yuan J, Wang Q and Fan X 2014 *Appl Surf Sci* **301** pp 418–27
[61] Ni X, Dong A, Fan X, Wang Q, Yu Y and Cavaco-Paulo A 2015 *Fiber Polym* (2015) **16** pp 2276-83
[62] George M, Mussone P G and Bressler D C 2014 *Ind Crop Prod* **53** pp 365-73
[63] Stuart T, Liu Q, Hughes M, McCall R D, Sharma H S S and Norton A 2006 *Compos Part A – Appl S* **37** 3 pp 393-404
[64] Wilson A 2017 *Reinforced Plastics* **61** 2
[65] Begum K and Islam M A 2013 *Research Journal of Engineering Sciences* **2** 4 pp 46-53
[66] Joshia S V, Drzalb L T, Mohantyb A K and Arorac S 2004 *Composites: Part A* **35** pp 371-76
[67] George M and Bressle D C 2017 *J Clean Prod* **149** pp 491-501

Improvement of the customer satisfaction through Quality Assurance Matrix and QC-Story methods: A case study from automotive industry

G M Sicoe[1], N Belu[1], N Rachieru[1] and E V Nicolae[2]

[1]Department of Manufacturing and Industrial Management, Faculty of Mechanics and Technology, University of Piteşti, 110040 Pitesti, Targu din Vale Street, no. 1, Argeş, Romania

[2]Automotive and Transport Department, Faculty of Mechanics and Technology, University of Piteşti 110040 Pitesti, Targu din Vale Street, no. 1, Argeş, Romania

Abstract. Presently, in the automotive industry, the tendency is to adapt permanently to the changes and introduce the market tendency in the new products that leads of the customer satisfaction. Many quality techniques were adopted in this field to continuous improvement of product and process quality and advantages were also gained. The present paper has focused on possibilities that offers the use of Quality Assurance Matrix (QAM) and Quality Control Story (QC Story) to provide largest protection against nonconformities in the production process, throughout a case study in the automotive industry. There is a direct relationship from the QAM to a QC Story analysis. The failures identified using QAM are treated with QC Story methodology. Using this methods, will help to decrease the PPM values and will increase the quality performance and the customer satisfaction.

1. Introduction

In automotive industry the development of new products, involve complex engineering processes subject to time pressures and fierce competition on the automotive market. The companies have to adapt their products precisely to customer needs, and therefore parameters of manufactured products are adjusted to the individual requirements of customers [1]. An important role in the development of such projects it is user of the modern quality tools in the development of new products. Existing tools and methods have become inadequate because they do not meet all additional requirements [2, 3]. It is therefore necessary to look for new solutions or improve old and trusted methods. In this context, the use of modern quality methods such as the Quality Assurance Matrix (QAM) and the Quality Control Story (QC Story) is opportune.

QAM and QC Story are tools of IATF 16949. In October 2016, the IATF will publish a revised automotive industry standard, and the first edition will be referred to as IATF 16949. This new standard will supersede and replace the current ISO/TS 16949:2009, defining the requirements of a quality management system for organizations in the automotive industry. IATF 16949 is aligned with and refers to the most recent version of ISO's quality management systems standard, ISO 9001:2015, fully respecting its structure and requirements. According IATF, 10.2.3 Problem solving: The organization shall have a documented process (es) for problem solving: c) root cause analysis, methodology used, analysis, and results. Where the customer has specific prescribed processes, tools, or systems for problem solving, the organization shall use those processes, tools, or systems unless otherwise approved by the customer [4]. In this paper, we present the solution of a quality problem by applying the QAM and QC Story methods for the manufacturing-assembly process of a subassembly-axle for vehicles. Using QAM we identify the failures, then these are treated with QC Story.

2. Quality Assurance Matrix and QC Story - Quality problem solving tools

2.1. Quality Assurance Matrix

Quality Assurance Matrix is a quality tool that is based on the principle that any failure (noncompliance) to a manufacturing process that affects a customer (who may be the next process or final customer) must imperatively be eradicated. The method ranks, for potential and existing defects, the reliability limits of the control systems in the manufacturing process, thus allowing the implementation a necessary corrective measures and achieving in this way a quality objectives [2,5]. Comparing to the Failure Mode and Effects Analysis (FMEA), it can be applicable in already implemented production [6]. QAM permit a periodic quantification of the level of production processes quality, verifying the reliability (in the sense of trust or even effectiveness) of means of control existing in the production process, the necessity and the possibility of implementation of anti-error systems (Poka-Yoke) or other nonconformities protection systems [5]. QAM is built on the Plan-Do-Check-Act (PDCA) cycle structure [5,7,8]: Plan - Establish quality goals and defining the technological process; Do - Making analysis; Check – Verifying the reliability of control set; Act – Implementation of the corrective actions for the operations that do not guarantee the quality level established.

2.2. QC Story

The QC Story originated in the Seisakusho Li Tianjin plant, which is located in Komatsu, Ishikawa in Japan. To improve the performance of the quality control circle - QCC, a structural process for the activity of the QCC is presented. The output of the QCC can be arranged and presented based on the process, namely, the QC Story [9]. The QC Story is a structural process for the presentation of QCC activities. It is frequently used in practical problem-solving procedures and methods [10]. For example, in the literature, QC Story methodology is used to improvement of Kaizen projects in local governments [11]. Also, Sha et al. [12,13] proposed a creative problem-solving quality control story (CPS-QC story) to improve the traditional quality control process. CPS-QC story consists of the combination of the traditional QC story and TRIZ.
In automotive industry, QC Story is a problem solving method based on considering the facts and data, without speculation, for a problem caused by several factors. It is applicable not only to quality problems, but also to problems of productivity, costs, logistics, energy, security etc [14]. Therefore, QC Story, which uses a different standard process and different tools, is applicable to different kinds of problems. The method is applied using nine steps and basic principles of quality by taking into account tools and techniques from various approaches based on the PDCA cycle structure, shown in table 1.

Table 1. QC Story Method steps.

PLAN	DO	CHECK	ACT
1. Choosing the subject	5. Analysis	7. Confirming the effects	8. Standardization
2. Explaining the reasons of the choice	6. Applying corrective actions		9. Synthesising and planning future actions
3. Understanding the current situation			
4. Choosing the targets			

The approach used by QC Story applies to solving problems both as a group and as individuals. Sometimes using the format is not essential but may be necessary at the beginning or to communicate the results.

IOP Conf. Series: Materials Science and Engineering **252** (2017) 012045 doi:10.1088/1757-899X/252/1/012045

3. Quality Assurance Matrix application

QAM application to analyse the manufacturing-assembly process of a subassembly-axle for vehicles is shown in table 4. The main steps of this analysis are to determine the failure modes, the parameters that may influence the occurrence of these failures and the proposed corrective measures to prevent the occurrence of these failures. All failures modes are not given in this table. The proposed measures presented in table 2 take into account the criticality (importance) of the failure for the client, as quoted in table 3 [5,7].

Table 2. Quotations of different levels for quality guarantee.

Quotation	Quality Guarantee
◉ 5 points	100% Automatic control; Impossibility of assembling or post machining; Interdiction / control Poka Yoke
○ 3 points	Warning Poka Yoke; Frequency Control (measurements); Manual control 100% in the line; Human control 100% with identification
△ 1 point	Frequency control; Human control 100% without identification; Periodically audit Human control 100% with or without identification

Table 3. The criticality of the failure for the client.

Degree	Client impact	Criteria for assessment	Quotation
A	Grave	Failure preventing use your product or creates a strong dissatisfaction, with repair request	5
B	Very embarrassing	Fault for the customer expresses dissatisfaction in a survey	3
C	Embarrassing	Discovered fault of the customer, but tolerated	1

Table 4. QAM for manufacturing-assembly process of a subassembly-axle for vehicles.

Failure modes	Parameters	Importance	Type of control	Control process inside the assembly workshop	Control process outside the assembly workshop
Oxidation of the part	Paint appearance nonconforming: stains, agglomerates, matte paint, orange peel, exfoliated paint	C	△	Visual control 100%	
Oxidation of the part	Contact areas on balance	C	△	Visual control 100%	Control in the workshop cataphoresis 1/week with calibre
Lack of articulation and incorrect orientation	Manual articulation load	A	◉	Present part in post ensuring the orientation of the part (the missing part and no advance of mobile arm in the work area prevent the table from rotating)	
Lack of security welding cords	Visual inspection and marking of security welding cords	C	△	Visual control and permanent marker marking 10%	
Diversity nonconforming part with the chosen gust	Axle type error	A	◉	100% automatic (read axle grading in the loading station)	

Table 4. *(continued)*

Failure modes	Parameters	Importance	Type of control	Control process inside the assembly workshop	Control process outside the assembly workshop
Pressing articulation nonconforming	Pressing effort nonconforming	A	○	Control 100% automatic	
Impossibility to fix the strings	Welding grids fillet nonconforming	C	○	Control 100% operator (screwdriver and screw)	
Sticking label product	Reference error	A	◉	100% automatic SIP MECA (in compliance with Data Matrix post 1)	
	Axle without sticker label	A	◉	100% automatic SIP MECA (in compliance with Data Matrix post 1)	Control 100% present label
Support shock absorber broken	Lack of welding grids	A	○	Control 100% present marked cords	Control 100% operator in the welding workshop
Mixed parts in box	Reference error	B	◉	100% automatic SIP MECA (in compliance with Data Matrix post 1)	
Impossibility to read serial number from sticker label	Duplicate sticker labels	B	○	Banning key removing labels manually	
Reference error Galia	Reference error	B	◉	Zipping stickers 100%,	

All proposed corrective actions aim to ensure a global quality assurance level of 94%. Applying these control measures as a result of QAM analysis did not ensure the protection of customers against the appearance of very embarrassing failures (B). After the start of production and delivery of parts to the customer there was a complaint from a customer who found labelling with the same series of two delivered axles.

Following the QAM analysis, the proposed measures for failure modes with impact on labelling were: 100% automatic SIP MECA (in compliance with Data Matrix post 1), banning key removing labels manually and zipping stickers 100%.

Since the compliant labelling measures initiated following the QAM analysis did not eliminate the possibility of this failure, it moved on to the use of the QC Story method which allows detailed analysis of all the potential causes of a failure and the establishment of corrective and preventive measures.

4. QC Story application

The steps of the QC Story methodology involve the following:

Step 1: Choosing the subject. Removing the failure: Edison axle double labelling, appeared on week 27 at the Hambach customer two axles reference 591R labelled with the same serial number, zero.

Step 2: Explaining the reasons of the choice. Given the company's quality policy by applying the QC Story method it is aimed to:

1. Eliminate the risk of the emergence of this failure at other customers.
2. Treat the problem quickly; deadline to solve the problem Week 28;
3. Reduce the costs generated by solving quality problems.

Step 3: Understanding the current situation. The current situation shown in table 5.

Step 4: Choosing the targets (objectives). Given the company's quality policy it was established that the goal of solving the problem would be Week 28.

IOP Conf. Series: Materials Science and Engineering **252** (2017) 012045 doi:10.1088/1757-899X/252/1/012045

Table 5. Current situation.

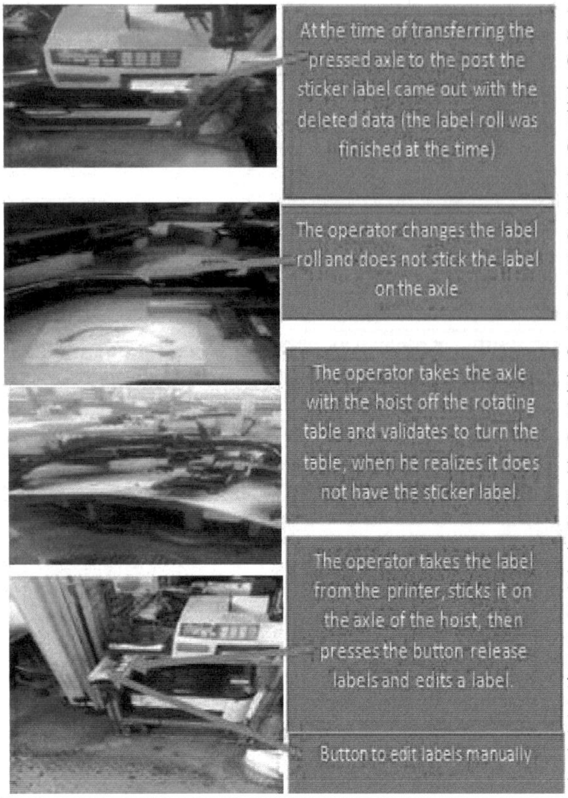

Personnel

a. The operator did not respect the Standard Operations Sheet – SOP- (took the piece off the presser.

b. The operator did not respect the Standard Operations Sheet (scanned the sticker label of the axle in box and not on the working device).

Method

c. There is no the Standard Operations Sheet roll labels and ribbon change.

d. There is no traceability for the last record and the first series after roll or ribbon change.

e. It is not mentioned in instruction sheet to clean the printer head.

f. Lack of key point in the Standard Operations Sheet (forbidden to apply sticker label with the same serial number on two different axles).

g. Lack of key point in the Standard Operations Sheet (forbidden to unload the part from the presser without the sticker label stuck and scanned).

Means

i. Printer (allows editing multiple labels with the same serial number).

j. Printer edits (illegible labels).

Management

k. Team leader did not explain to the operators what unitary traceability means and what role this traceability has.

Step 5: Analysis. The Ishikawa diagram has been used in order to establish the precise causes that led to the nonconformity, shown in figure 1.

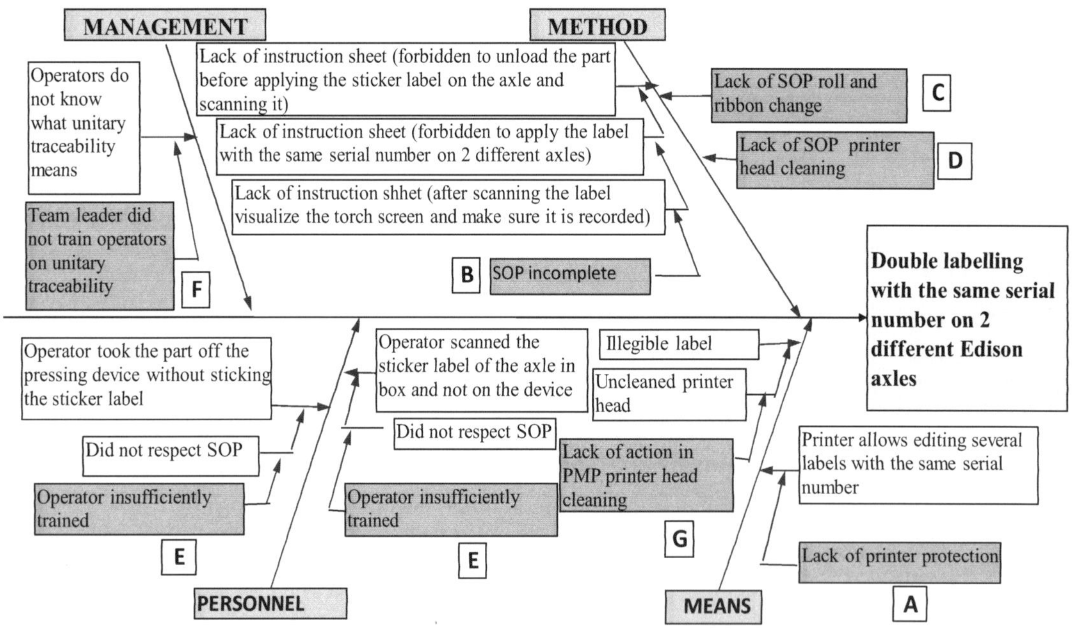

Figure 1. Ishikawa diagram for: double labelling with the same serial number on two different Edison axles.

Step 6: Applying corrective actions. The corrective actions and their efficiency – quality, cost and deadline are shown in table 6.

Table 6. Corrective actions.

No	Problems	Action	Efficiency				Responsible	Deadline
			Q	C	D	All		
A	Printer allows editing several labels with the same serial number	Blocking operator access to re-edit labels	9	6	6	324	Resp.1	W 28
B	Standard Operations Sheet incomplete	Introducing key point in the Standard Operations Sheet (forbidden to unload the part before applying the sticker label on the axle and scanning it)	6	6	6	216	Resp.2	W 28
		Introducing key point in the Standard Operations Sheet (forbidden to apply a label with the same serial number on two different axles)	6	6	6	216		
		Introducing key point in the Standard Operations Sheet (after scanning the label visualize the torch screen and make sure it is recorded)	6	6	6	216		
		Using it for the pressing articulation connecting rod post too	6	6	6	216		
C	Lack of the Standard Operations Sheet ribbon and sticker label change	Editing the Standard Operations Sheet ribbon change, sticker label and using on all posts	6	6	6	216	Resp.3	W 28
D	Lack of the Standard Operations Sheet printer head cleaning	Editing the Standard Operations Sheet printer head cleaning and using on all posts	6	6	3	54	Resp.4	W 28
E	Insufficiently trained operator	Training operator in three stages after the Standard Operations Sheet modified analysis, ribbon change, printer head cleaning	6	6	3	54	Resp.5	W 28
F	Team leader did not train operators on unitary traceability	Training operators on unitary traceability	6	6	3	54	Resp.6	W 28
G	Lack of action printer head cleaning	Introducing action in the instruction sheet and using it for all printer-equipped posts	6	6	3	54	Resp.7	W 28

Q = 9 – high quality; Q = 6 medium quality; Q = 3 minimum quality
C = 9 – high costs; C = 6 medium costs; C = 3 minimum costs
D = 9 – time > 5 days; D = 6 - 3 < time < 5 days; D = 3 – time <3 days

Step 7: Confirming the effects
In order to achieve this step the current situation is compared with the initial conditions, using graphs, Pareto diagram, figure 2.

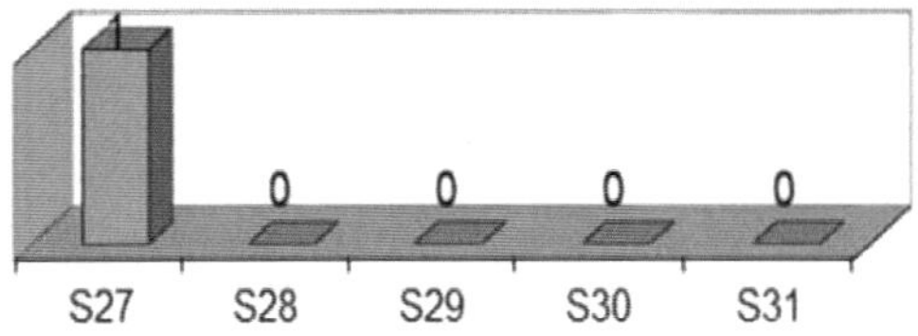

Figure 2. The number of cases occurred in workstation.

Step 8: Standardization
To avoid other similar cases the following standardizing actions were established:
1. Editing the Standard Operations Sheet and ribbon and sticker labels change.
2. Editing the Standard Operations Sheet and printer head cleaning.
3. Introducing in instruction sheet printer head cleaning.
4. Introducing key point in the Standard Operations Sheet (forbidden to unload the part before applying sticker labels on the axle and scanning it.
5. Introducing key point in the Standard Operations Sheet (forbidden to apply a label with the same serial number on two different axles).
6. Introducing key point in the Standard Operations Sheet (after scanning visualize the torch screen and make sure it is registered).

5. Conclusion

By applying the QC Story method, ten additional corrective measures have been established and implemented in relation to the control plan that emerged as a result of the QAM method, which led to the elimination of the failure: labelling with the same series two delivered axles.
The conclusion of this study confirms the complementarity of the two methods and the need for their separate application. QAM is applied in the process preparation stage and allows the identification of failure modes on the manufacturing flow and the parameters that influence these failure modes. The quantification of the impact of these failures on the customer allows establishing the type of control to ensure the delivery of compliant parts after each operation of the manufacturing stream. These control measures include frequency control for low-impact (embarrassing) failures on the customer up to a 100% automatic control or Poka Yoke system for major impact failures that prevent the operator from making mistakes (to perform a non-compliant operation).
By applying the QC Story method as a result of a failure with a serious impact on the customer, there was identified a series of seven nonconformities that could generate the failure: labelling with the same series two delivered axles, there were established nine corrective measures which were implemented to eliminate the possibility of occurrence of this failure. As well, the corrective actions (documentation preparation) were standardized in order to avoid the occurrence of failures.
In the situation of implementing of a new products line in a company from automotive industry, taking in consideration decorative parts, is necessary a good correlation between Quality Assurance Matrix and QC Story methods.

6. References

[1] Doshi J A, Desai D 2016 Overview of Automotive Core Tools: Applications and Benefits *J. Inst. Eng.* India Ser. **C** pp. 1-12
[2] Belu N, Misztal A, Ionescu L M 2016 Quality Assurance Matrix as the advanced, generation of, quality control *Proceedings of International Conference on Economics and Management Innovations (ICEMI 2016)* China pp 251 -256

[3] Misztal A, Bachorz S 2014 Quality planning of parts machine production based on housing of cylinder head milling machines *Applied Mechanics and Materials* **657** pp 986-990
[4] IATF 16949:2016, Automotive Quality Management Systems Standard- Quality management systems requirements for automotive production and relevant service parts organizations *International Automotive Task Force (IATF)*
[5] Stirbu L, Belu N, Bondoc M D 2012 Quality assurance matrix in automotive industry *Annals of the Oradea University. Fascicle of Management and Technological Engineering* **XI (XXI)** pp 5.117 – 21
[6] Hrbackova L 2016 Risk-based thinking in the production process using the methods of Quality Assurance Matrix and the FMEA process *Journal of Systems Integration* **1** pp 21-28
[7] Boroiu A, Grigore J, Bălteanu A, Bâldea M, Boroiu A A 2015 Possibilities to weighting the global indicators of quality determined with the Quality Assurance Matrix *The Scientific Bulletin of Valahia University – Materials and Mechanics* **10** pp 154 - 156
[8] Nicolae V, Boroiu A, Vieru I, Dobrescu I 2007 Utilization of Quality Assurance Matrix for Reducing the Non-Conformities within the Manufacturing Process *Volume of International Congress "Automobile, environment and farm machinery", AMMA 2007* pp. 103 - 105
[9] Sugiura Z, Shantian J 2003 QC Story Utilization Method, *Corporate Synergy Development* (Center Press, Taipei, Taiwan)
[10] Yang P J 2004 The Practical Manual on Enhancing the Problem Solving on Shop Floor: Problem-Solving QC-Story *Corporate Synergy Development* (Center Press Taipei Taiwan)
[11] Suárez-Barraza M F, Miguel-Dávila J A 2014 Assessing the design, management and improvement of Kaizen projects in local governments, *Business Process Management Journal*, **20** 3 pp. 392 – 411
[12] Sha D Y, Hsu S Y, Chen H Y 2016 Creative Problem-Solving QC Story Based on Systematic Innovation TRIZ *Journal of Quality* **23** pp 25 - 42
[13] Hsu S Y 2014 A new problem-solving procedure based on TRIZ methodology and QC story *Proceedings of the 5th International Asia Conference on Industrial Engineering and Management Innovation (IEMI 2014*) China pp 289 – 293
[14] Bidault F 2001 QC story, Manualul stagiarului **1.1**

A comparative critical study between FMEA and FTA risk analysis methods

G Cristea[1] and DM Constantinescu[1]

[1]University POLITEHNICA of Bucharest, Romania

E-mail: gabriel.cristea@upb.ro

Abstract. Today there is used an overwhelming number of different risk analyses techniques with acronyms such as: FMEA (Failure Modes and Effects Analysis) and its extension FMECA (Failure Mode, Effects, and Criticality Analysis), DRBFM (Design Review by Failure Mode), FTA (Fault Tree Analysis) and and its extension ETA (Event Tree Analysis), HAZOP (Hazard & Operability Studies), HACCP (Hazard Analysis and Critical Control Points) and What-if/Checklist. However, the most used analysis techniques in the mechanical and electrical industry are FMEA and FTA. In FMEA, which is an inductive method, information about the consequences and effects of the failures is usually collected through interviews with experienced people, and with different knowledge i.e., cross-functional groups. The FMEA is used to capture potential failures/risks & impacts and prioritize them on a numeric scale called Risk Priority Number (RPN) which ranges from 1 to 1000. FTA is a deductive method i.e., a general system state is decomposed into chains of more basic events of components. The logical interrelationship of how such basic events depend on and affect each other is often described analytically in a reliability structure which can be visualized as a tree. Both methods are very time-consuming to be applied thoroughly, and this is why it is oftenly not done so. As a consequence possible failure modes may not be identified. To address these shortcomings, it is proposed to use a combination of FTA and FMEA.

1. Introduction

The ability to predict the reliability of components and systems is important in technical design. Reliability prediction is designed to prevent malfunctions, for example through planned maintenance. The reliability of the individual components is superior to the corresponding system reliability [1].

Optimizing the reliability analysis process requires the use of appropriate quality control tools such as Failure Mode and Effect Analysis (FMEA) and Fault Tree Analysis (FTA). Through these analyzes it is possible to examine the critical parameters of the analyzed process.These are the two commonly used methods for failure analysis.

The first method is failure mode and effects analysis (FMEA). According to MIL-STD-1629 REV.A Standard [2] its definition is "The Failure Mode, Effects and Criticality Analysis (FMECA) is a reliability evaluation/design technique which examines the potential failure modes within a system and its equipment, in order to determine the effects on equipment and system performance. Each potential failure mode is classified according to its impact on mission success and personnel/equipment safety." It is a bottom-up method, starting at the component level, which is used to find failure modes and to identify their effects. By adding a criticality analysis, qualitative FMEA becomes quantitative FMECA (failure mode, effects and criticality analysis).

The second method for failure analysis is Fault Tree Analysis (FTA), which is a top-down method used to identify relationships between events, such as subsystem failures and their causes. According to SAE ARP4761 Standard [3] its definition is "Fault Tree Analysis (FTA) is a deductive failure analysis which focuses on one particular undesired event and provides a method for determining causes of this event. In other words, a Fault Tree Analysis is a "top-down" system evaluation procedure in which a qualitative model for a particular undesired event is formed and then evaluated."

Both methods are very time-consuming to be applied thoroughly, and this is why it is oftenly not done so.

2. Failure and risk analysis methods

Generally, risk represents the possibility of occurrence of some undesirable events that trigger different types of failures. Risk analysis is used to find the causes of failures and to prevent these failures from occurring in the future. The results of risk analysis can be used to optimize processes.

Risk analysis has two complementary branches:

1. Qualitative.
2. Quantitative.

The task of qualitative analysis is to identify the risk areas of a process, the types of risks and the factors that cause the risks. This is done by an expert through brainstorming. Quantitative analysis allows you to quantify the effect level for each type of risk.

2.1. Basic risk analysis methods

The basic methods for risk analysis are as follows:

1. Analogy.
2. Methods of expertise.
3. Statistical methods.
4. Modeling.

The analogy method focuses on examining analogies between data from a range of sources. Methods of expertise are used to collect the opinions of qualified specialists. A statistical approach to risk analysis uses different types of statistical methods for the processing of data obtained experimentally. The simulation is based on calculating different types of models and testing on these models in different situations.

Among the most commonly used risk analysis methodologies are [4]:

1. Failure modes and effects analysis (FMEA).
2. Fault tree analysis (FTA).
3. Structured What-If Technique (SWIFT).

2.2. Failure mode and effects analysis

FMEA (Failure Mode and Effects Analysis) is an inductive and bottom-up method to identify failure modes, effects and causes of technical systems [5]. The analysis starts at the component level where the possible component failure modes are identified and it is examined what the consequences are on a higher level. FMEA analysis typically involves a diverse team of people with different backgrounds (eg mechanical design, software, production, maintenance), as this increases the probability that all failures will be identified and the effects will be estimated correctly [6].

The FMEA can be extended to FMECA (Failure Mode, Effects and Criticality Analysis) by adding a criticality analysis. In this way, the purely qualitative FMEA can be made quantitative.

In FMECA, the criticality of each failure mode is quantified by the risk priority number (RPN). The risk priority number is the product of three indicators: (a) severity indicator (S), (b) occurrence indicator (O) and (c) detection indicator of a failure (D).

IOP Conf. Series: Materials Science and Engineering **252** (2017) 012046 doi:10.1088/1757-899X/252/1/012046

These indicators are usually rated on a scale from 1 to 10 [6]. Standards that provide guidelines for the assessment of each indicator are available, see, e.g., [2].

The severity (S) of a failure refers to the seriousness of the effect or impact of a certain failure. The severity indicator is rated from low impact (e.g., 1 on a scale 1 to 10) to very high impact (e.g., 10 on the same scale).

The occurrence (O) indicator of a failure refers to its failure frequency. The occurrence indicator should be rated from unlikely to occur to almost inevitable.

Finally, the detectability (D) indicator refers to the likelihood that the failure is not detected before it induces major subsequent effects (e.g., by means of process controls, procedures or operator detectability). The detectability indicator should be rated from almost sure detection to almost sure non-detection.

The risk priority number (RPN) represents the entire risk for the system user and serves as a decision criteria for the introduction of optimization actions.

As a matter of principle:

• The larger the RPN is, the greater the priority that the risk is lowered with the help of design and quality assuring actions;

• Likewise, individual values for S, O, and D that are greater than 8 should be more closely observed;

• The product O x D gives information concerning the remaining probability that defective undetected parts will reach the hands of the customer.

The risk assessment is conducted for actions, which have already been implemented. In order to lower the risk even more, additional actions are mostly required.

The observation of the absolute value of the risk priority number (the product S x O x D) is not always sufficient in many cases to find the starting points for optimization actions. Likewise, it is not reasonable to define a "fixed RPN" as a company wide action limit (e.g. optimization conducted for all RPN $\geq$ 250), since the circumstances of the assessment standards may differ for each FMEA and the observation of smaller risk priority numbers could be neglected.

Optimization actions are to be taken for high RPNs and high individual assessment values. Firstly, the calculated risk priority numbers are ranked according to their values. The optimization begins with the failure cause with the greatest RPN and should be ended at a certain lower limit (e.g. RPN = 125), depending on the scope of the analysis.

High individual assessment values must also be observed along with the RPN. A value of O > 8 means that a failure occurs most of the time. Naturally, this must be remedied. A severity value of S > 8 points to serious function impairment or to a serious safety risk. These cases must also be looked at more closely. Failures can be detected with difficulty for values of D > 8. For this reason the danger that these can reach the hands of the customer is increased.

2.3. *Fault tree analysis*

FTA is a very powerful systematic way which is widely used for estimating process quality. Starting from the top event, the fault-tree method uses a Boolean algebra and a deductive logical modeling to make a graphical representation of the relations among various failure events at different levels of the process.It enables the root causes of the failure events of a process to be found. This type of logic helps to establish a detailed scheme of relationships between the events in the process that can affect its quality.

The advantages of fault trees are as follows:

• Allows potential failure parts of the process to be seen in detail.

• Helps identify failures deductively.

• Enables a qualitative or quantitative analysis of the process to be made.

• The method can focus on individual parts of the process, and can extract specific failures.

• It clearly represents the behavior of the process.

The main advantage of the fault tree (in comparison with other methods) is that the analysis is limited to identify only those events of the process which lead to a specific process failure.

The disadvantages of fault trees are:

• Implementation of the method requires considerable inputs, because more process details lead to a geometric increase in the analyzed area, and the number of influencing events grows correspondingly.

• A fault tree is a Boolean logic diagram, which shows only two states: work and fail.

• It is difficult to estimate the state of partial failure of the process parts, because use of the method generally indicates that the process is either in good condition or in a faulty state.

• It requires a reliability specialist with deep knowledge of the process.

2.4. Combination of FMEA and FTA methods

Both methods are very time-consuming to be applied thoroughly, and this is why it is oftenly not done so. As a consequence possible failure modes may not be identified. To address these shortcomings, it is proposed to use a combination of FTA and FMEA.

Two options can be identified: (a) perform both an FMEA as well as an FTA separately or (b) use a mixed approach.

Concerning the first option, some authors [5] argue that this may expand the number of failure modes found due to the different starting points of both methods: bottom-up in FMEA, versus top-down in FTA. However, performing both analyses would be time consuming and may lead to a loss of focus on the most critical parts of the system, which the failure analysis typically aims to identify.

Concerning the second option, some authors [7] propose a mixed approach in which the FMEA is guided by a FTA. In their proposed approach, the analysis starts with the definition of a system failure event and continues with the construction of a fault tree for a particular system as a whole. Finally, each basic event is analyzed further with FMEA to identify the underlying failure modes of each component.

2.5. Variation mode and effects analysis

The ability to predict reliability of components and systems is important in engineering design. By predicting reliability it is possible to prevent failures. There are several methods for improving the prediction of reliability.

Today, many failures are caused by variations (strength, loads, manufacturing tolerances, etc.), resulting in expensive reclamations and dissatisfaction that may lead to loss of customers [8]. Therefore, a method called Variation Mode and Effect Analysis (VMEA), a relatively new methodology, which is a deductive method of identifying and managing sources of variation, has been developed. The VMEA method has been decomposed into three different levels [9]: Basic VMEA, Enhanced VMEA and Probabilistic VMEA.

1. Basic VMEA is used in the early design stages, where information about the variation is vague and the aim is to compare and evaluate different concepts.

2. Enhanced VMEA is used later in the design stages, where more information about the sources of variation is known.

3. Probabilistic VMEA is used in late design stages, where detailed and statistical information about sources of variation is available.

Product characteristics that are of particular interest from a variation standpoint are selected and are usually referred to as Key Product Characteristics (KPCs). The KPC can be compared to the top event in a fault tree. Each KPC is decomposed into a number of sub-KPCs that are affected by variations.

2.6. Shortcomings of presented methods

FMEA and FTA methods possess shortcomings. First, both methods distinguish only between two levels:

(1) system level and (2) component level. Second, the importance of the level of detail in a failure analysis is not treated with much emphasis. Third, the methods do not possess a feature which enables

to focus the analysis on the most critical elements of a machine. For this kind of analyses, one may want to concentrate on the most critical parts of the system and allocate more attention to these parts. This is related to the efficiency of the analysis, i.e. the required effort to perform the analysis versus the gained knowledge of failure behaviour.

3. Conclusions

Both FMEA and FTA can be used to identify the causes of failures. The result of this approach is a reliability analysis performed through the interaction between the FMEA and FTA reliability tools. Each of these risk analysis methods has advantages that allow for clearer investigation and observation of the process from different points of view.

The FMEA method is generally a library of all possible potential failures and their consequences, while the FTA allows for a detailed analysis of the logical and temporal relationships that lead to a failure taken from the top of the tree. Applying these two complementary methods provides deeper information than the separate application of the methods.

The FMEA method is often used to identify fault modes, to assess the consequences of specific malfunctions and to subjectively infer a risk and a priority number. The VMEA method is used instead to identify, evaluate, and manage unwanted variations to increase system reliability.

Acknowledgments

Mr. Gabriel Cristea acknowledges the PhD student scholarship given by the Ministry of National Education from Romania through the contract no. 06.16/2015 which made possible the present researches. He wishes to thank Professor Cristian Andreescu for the financial support given throughout this research.

References

[1] O'Connor P 2002 *Practical Reliability Engineering* ed John Wiley and Sons (Chichester) p 513
[2] MIL-STD-1629A, Military Standard: Procedures for Performing a Failure Mode, Effects and Criticality Analysis -http://everyspec.com/MIL-STD/MIL-STD-1600-1699/
[3] SAE ARP4761 Standard - Guidelines and Methods for Conducting the Safety Assessment Process on Civil Airborne Systems and Equipment - http://standards.sae.org/wip/arp4761a/
[4] Vose D 2008 *Risk Analysis: A Quantitative Guide* ed John Wiley and Sons (London) p 735
[5] Bertsche B 2008 *Reliability in Automotive and Mechanical Engineering: Determination of Component and System Reliability* ed Springer Berlin (Heidelberg) p 492
[6] Tinga T 2013 *Principles of Loads and Failure Mechanisms: Applications in Maintenance, Reliability and Design* ed Springer (London) p 307
[7] Yu S, Liu J, Yang Q and Pan M 2011 *Proc. of 9th Int. Conf. on Reliability, Maintainability and Safety* pp 954-60
[8] Johansson P 2006 Variation mode and effect analysis: A practical tool for quality improvement *Quality and Reliability Engineering International* **8** 865-76
[9] Johannesson P 2009 Variation mode and effect analysis: An application to fatigue life prediction *Quality and Reliability Engineering International* **2** 167-79

IOP Conf. Series: Materials Science and Engineering **252** (2017) 012047 doi:10.1088/1757-899X/252/1/012047

OPTIMIZING THE CONSTRUCTION OF DEVICES TO CONTROL INACCESIBLE SURFACES - CASE STUDY

E L Nițu[*1], A Costea[1], M D Iordache[1], A D Rizea[1] and Al Babă[1]

[1]University of Pitești, Manufacturing and Industrial Management Department, Târgul din Vale Street No.1, Pitești, Romania

Abstract. The modern concept for the evolution of manufacturing systems requires multi-criteria optimization of technological processes and equipments, prioritizing associated criteria according to their importance. Technological preparation of the manufacturing can be developed, depending on the volume of production, to the limit of favourable economical effects related to the recovery of the costs for the design and execution of the technological equipment. Devices, as subsystems of the technological system, in the general context of modernization and diversification of machines, tools, semi-finished products and drives, are made in a multitude of constructive variants, which in many cases do not allow their identification, study and improvement. This paper presents a case study in which the multi-criteria analysis of some structures, based on a general optimization method, of novelty character, is used in order to determine the optimal construction variant of a control device. The rational construction of the control device confirms that the optimization method and the proposed calculation methods are correct and determine a different system configuration, new features and functions, and a specific method of working to control inaccessible surfaces.

1. Introduction

The complex optimization of all activities at the design, manufacturing and operation stages [1-3] require an objective analysis of the technological processes in which the orientation and fixing device of the semi-finished products is an important component of the system which also includes the machine tool, the tool and the human factor.

The present work is a continuation of the previous one, which proposes an organological study of constructive variants for structures to establish the optimal solution for construction configuration, operation and working method of a new type of device designed to control the inaccessible surfaces of the parts to be measured.

2. Optimization methodology for the construction of control devices

The diversity of constructive variants for making devices, machines and manufacturing systems [1, 4] generates difficulties in establishing rational solutions to technological equipment, which can only be overcome by means of optimization methods.

2.1. Algorithm to design-optimize the construction of control devices

For a more rigorous analysis of the precision of machining or control required by the technical documentation and the establishment of well-grounded technical solutions, the previous paper

* eduard.nitu@upit.ro

IOP Conf. Series: Materials Science and Engineering **252** (2017) 012047 doi:10.1088/1757-899X/252/1/012047

proposed a design-optimization algorithm for the construction of devices, in general, a particular case in Figure 1 for the construction of devices to control inaccessible surfaces. The signification of optimizing stages (E_{01}, E_{02}, ..., E_{08}) is the same as in the previous paper.

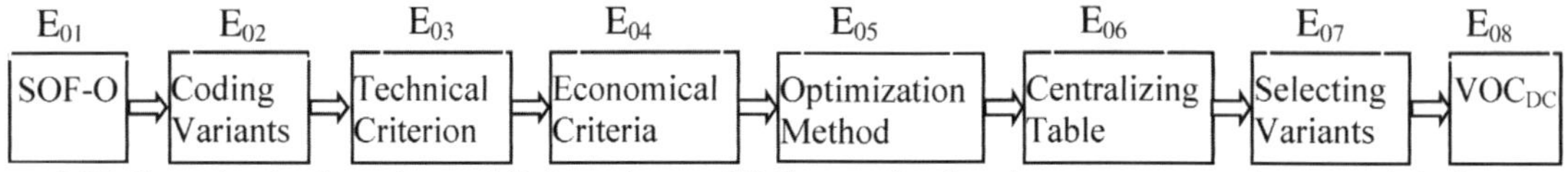

SOF-O: optimal orientation and fixing scheme; VOC_{DC}: optimal variant to construct the control device

Figure 1. Algorithm to design-optimize control devices

2.2. Optimizing the construction of devices to control inaccessible surfaces - case study

In many situations, especially in the case of subassemblies, certain surfaces cannot be controlled because they are in contact with the supports of the device, which make it impossible for the measuring elements to reach the work areas, and other surfaces have components assembled on them.

For this reason, there was defined the design-optimization algorithm presented in Figure 1, the working stages of this algorithm will be practiced on a case study of the water pump subassembly of the automotive engines for the construction of control devices.

E01 – Knowing or determining SOF-O

Initially, the quotas and conditions imposed on the analyzed control operation and the optimal orientation and fixing scheme (SOF-O) are known as a combination of graphical signs of the information symbolization of the orientation and fixing bases of the semi-finished products shown in Figure 2.

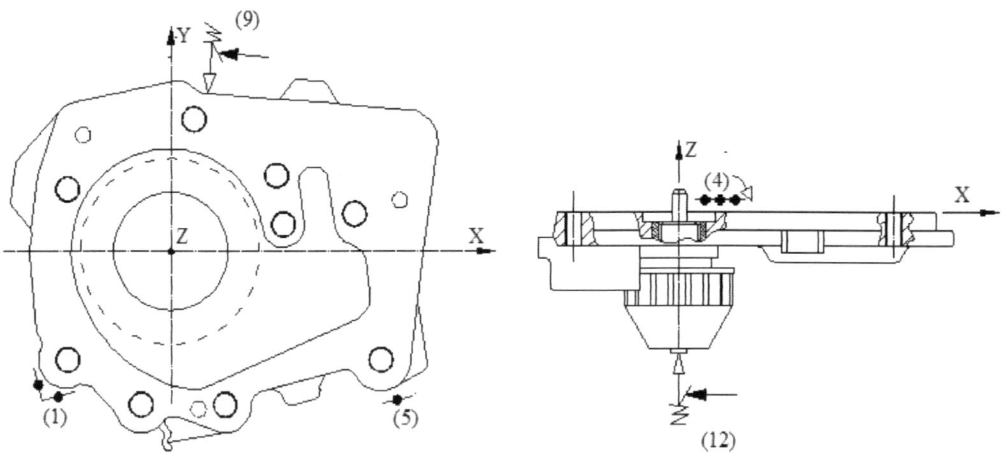

Figure 2. Operation scheme with SOF-O

SOF-O = (1) + (4) + (5) + (9) + (12)

The numbers of symbols in SOF-O have been preserved as in [5].

E02 Coding the construction variants of the orientation and fixing elements of the device

Constructive variants of structures, which are known, are defined and coded for each symbol element of quotation bases BCI, BCII, BCIII and fixing bases BFIV, BFV in the optimal orientation and fixing scheme set forth in [5], or new constructive variants to be achieved later can also be analyzed.

System of main orientation supports:

BCI - (1) X_{11} - short fixed prism mono-block;

X_{12} - short adjustable prism mono-block;
X_{13} - short fixed prism from two semi-prisms;
BCII - (4) X_{21}- planar flat base plate type mono-block;
X_{22} - planar flat base defined by three adjustable dowels with spherical contact heads;
X_{23} - planar flat base made of three adjustable dowels with planar oscillating contact surfaces, in modular version;
BCIII - (5) X_{31}- fixed planar support base (fixed dowel);
X_{32} - adjustable planar support base (adjustable dowel);

Fixing system with additional supports:

BFIV - (9) X41- additional support with self-positioning and subsequent blocking, as adjusting force;
BFV - (12) X51- additional support with self-positioning and subsequent blocking, as adjusting force and as main force, finally.

These additional supports (9, 12) will be analyzed in a single construction variant to simplify the presentation of this work, but these or other fixing components, bodies, guidance systems, or base plates may be analyzed by this method.

E03 – Establishing the optimizing technical criterion

The optimizing technical criterion is that of precision that becomes eliminatory and requires the technically acceptable constructive solution (SC-TA) to meet condition $\varepsilon_{c.\ s}^{o-p} < \varepsilon_{ad.\ s}^{0-p}$, where:

$\varepsilon_{c.\ s}^{o-p}$ - orientation-position error of the structure construction;
$\varepsilon_{ad.\ s}^{o-p}$ - orientation-position error admitted for the structure.

These two types of errors are calculated with relations presented in [4, 6] and are introduced in Table 1 centralizing the optimizing variants and criteria.

For BCI-SI$_{0I}$ (short fixed prism)

Calculation $\varepsilon_{ad.\ s}^{0-p}$

$$\varepsilon_i = \sqrt{\varepsilon_o^2 + \varepsilon_f^2 + \varepsilon_d^2} \tag{1}$$

where, ε_i,ε_o, ε_f, ε_d are defined in [4] and the previous paper.

The admissible installation error (ε_{iad})is calculated for the control device with relation:

$$\varepsilon_{iad} = (\frac{1}{5} \dots \frac{1}{10})T_p \tag{2}$$

where, $T_p = 0.080$, is the tolerance of the part at the quotation / condition analyzed.

The precision condition is $\varepsilon_i \leq \varepsilon_{iad}$.

$\varepsilon_o = 0$, because BC ≡ BO, and the orientation prism of the control device is identical to that of the processing operation.

ε_f = max. 0.010, due to the self-positioning and subsequent blocking fixing system.

From relation (1) it can be written:

$\varepsilon_d = \sqrt{\varepsilon_i^2 - (\varepsilon_o^2 + \varepsilon_f^2)}$ and: $\varepsilon_{dmax} = \sqrt{(\frac{1}{5} \cdot 0.080)^2 - (0^2 + 0.010^2)} = 0.012$

$\varepsilon_{ad.\ s}^{0-p} = \varepsilon_{dmax} = 0.012$

Calculation $\varepsilon_{c.\ s}^{0-p}$

X_{11} – short fixed prism mono-block (Figure 3): $\varepsilon_{c.\ s}^{0-p} = 0.010$, as execution and assembly precision;

X_{12} – short adjustable prism mono-block (Figure 4): $\varepsilon_{c.\ s}^{0-p} = 0$, as execution precision and possibility to adjust and block;

X_{13} – short fixed prism from two semi-prisms (Figure 5): $\varepsilon_{c.\ s}^{0-p} = 0.020$, as execution and assembly precision.

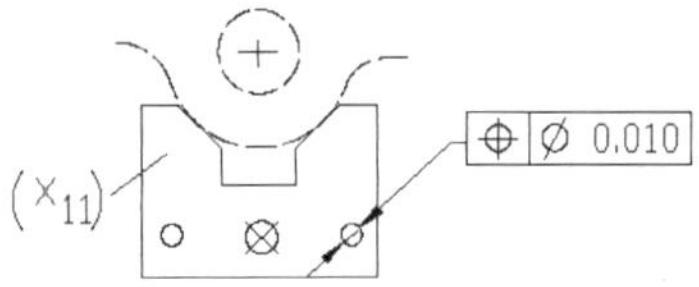

Figure 3. Short fixed prism mono-block

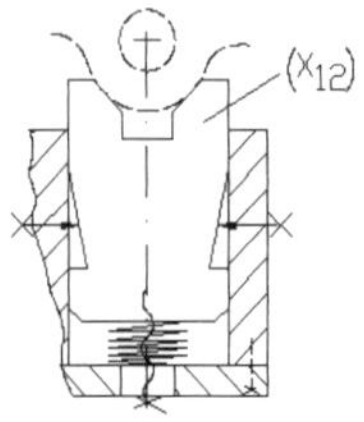

Figure 4. Short adjustable prism mono-block

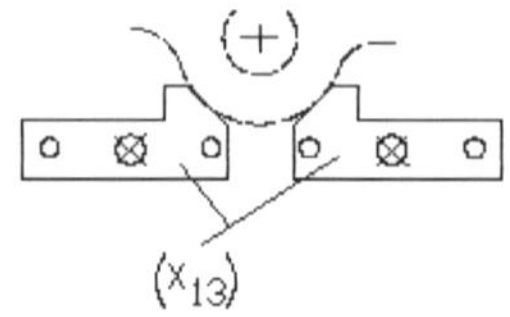

Figure 5. Short fixed prism from two semi-prisms

For BCII-SI$_{OII}$ (planar flat base)

Calculation $\varepsilon_{ad.\ s}^{0-p}$

There are used the relationships (1) and (2). From the part drawing [5]: $T_p = 0.100$; $\varepsilon_o = 0$ because BC ≡ BO; $\varepsilon_f = $ max. 0.010 due to the self-positioning and subsequent blocking fixing system.

$$\varepsilon_{dmax} = \sqrt{(\tfrac{1}{5}\cdot 0.100)^2 - (0^2 + 0.010^2)} = 0.017 \quad \Rightarrow \quad \varepsilon_{ad.\ s}^{0-p} = \varepsilon_{dmax} = 0.017$$

Calculation$\varepsilon_{c.\ s}^{0-p}$:

X_{21} – planar flat base mono-block:

$\varepsilon_{c.\ s}^{0-p} = 0.015 \ldots 0.020$, as deviation from of a planar surface

X_{22} – planar flat base defined by three adjustable dowels with spherical contact heads:

$\varepsilon_{c.\ s}^{0-p} = 0.010 \ldots 0.020$, due to the prints on the surface of the low-hardness part

X_{23} – planar flat base defined by three adjustable dowels with planar oscillating head:

$\varepsilon_{c.\ s}^{0-p} = 0$, planar contact surfaces of dowels do not produce any more local deformations

For BCIII-SI$_{OIII}$ (planar fixed restrained base)

Calculation $\varepsilon_{ad.\ s}^{0-p}$

There are used the relationships (1) and (2). From the part drawing [5]: $T_p = 0.080$; $\varepsilon_o = 0$ because on the control device the part is oriented on the same surface and support as for the processing operation; $\varepsilon_f = $ max. 0.010 due to the self-positioning and subsequent blocking fixing system.

It results: $\varepsilon_{ad.\ s}^{0-p} = 0.012$

Calculation $\varepsilon_{c.\ s}^{0-p}$:

X_{31} – narrow planar fixed base

$\varepsilon_{c.\ s}^{0-p} = 0.010$, as execution and assembly precision

X_{32} – narrow planar adjustable base.

$\varepsilon_{c.\ s}^{0-p} = 0$, because the execution errors (≈ 0.010) can be cancelled by adjustment.

For BFIV-SI$_{fIV}$ (pre-fixing system with self-positioning and subsequent blocking)

X_{41} – (a single constructive variant is analyzed to simplify calculations)

$\varepsilon_{ad.\ s}^{0-p} = 0.012$, like for the other types of structures.

$\varepsilon_{c.\ s}^{0-p} = 0$, due to the self-positioning and subsequent blocking fixing system.

For BFV-SI$_{fV}$ (pre-self-positioning and subsequent blocking fixing system)

X_{51} – (a single constructive variant is analyzed to simplify calculations)

$\varepsilon_{ad.\ s}^{0-p} = 0.012$, like for the other types of structures.

$\varepsilon_{c.\ s}^{0-p} = 0$, due to the self-positioning and subsequent blocking fixing system.

E04 – Establishing the economical optimization criteria

The six economical optimization criteria are established with the significance presented in the previous paper.

To simplify calculations, only two optimization criteria are chosen, cost (C_c) and flexibility (C_f).

Cost criterion (C_c - CE_3)

Calculate the actual cost of materials, workmanship and heat treatment costs or operating costs for each X_{ij} - coded structure and introduce data in Table 1. These values, in lei, are: X_{11}=346.5; X_{12}= 807; X_{13}=247.5; X_{21}=623; X_{22}=311.5; X_{23}^*=51.9; X_{31}=51.75; X_{32}= 12.3; X_{41}^*=200.62; X_{51}^*=200.62. Variants with sign * are modular structures.

Flexibility cost (C_f - CE_6)

Calculate the degree of flexibility to adjust in time (G_{AT}) with relation $G_{AT} = e^{-(TR/TRO)}$, where the terms have the signification of the previous paper and [7] and the values of time TR and TRO, in minutes, are:

BCI-SI$_{OI}$,X_{11} – (TR = 1920′; TRO = 1920′), $G_{AT}^{X_{11}} = e^{-(\frac{1920}{1920})} = \frac{1}{e^1} = 0.3678$

By proceeding similarly for the other structures as well it is obtained: $G_{AT}^{X_{12}}$= 0.1353, $G_{AT}^{X_{13}}$= 0.2865

BCII-SI$_{OII}$,$G_{AT}^{X_{21}}$ =0.3678, $G_{AT}^{X_{22}}$ = 0.4114, $G_{AT}^{X_{23}}$ = 0.9931

BCIII-SI$_{OIII}$,$G_{AT}^{X_{31}}$ =0.3678, $G_{AT}^{X_{32}}$ = 0.979,

BFIV-SI$_{fIV}$,$G_{AT}^{X_{41}}$ = 0.999

BFV-SI$_{fV}$, $G_{AT}^{X_{51}}$ = 0.999

E05 – Defining the optimization method

Determining the optimal solution of the structure variants for the construction of the device implies the customization of the global utility method in the analyzed case.

The calculation of the coefficients of importance for the precision criterion (CI_{pr}) is done by the inverse interpolation method, with relation $CI_{pr} = 1 - CI_a$, from the previous work, and their values are presented in E_{06} and introduced in Table 1.

The calculation of the coefficients of importance for the cost criterion(CI_c) *and of the coefficients of importance for the flexibility criterion* (CI_f) is based on the interpolation method for the minimization feature from [8] with relation:

$$CI_c^{X_{ij}} = \frac{\max(X_{ij}^c) - X_{ij}^c}{\max(X_{ij}^c) - \min(X_{ij}^c)} \quad (3)$$

$$CI_f^{X_{ij}} = \frac{X_{ij}^f - \min(X_{ij}^f)}{\max(X_{ij}^f) - \min(X_{ij}^f)} \quad (4)$$

where, i = 1, ... , m; j = 1, ... , n

$CI_c^{X_{ij}}$– coefficient of importance of cost for structure variant X_{ij};

$\max(X_{ij}^c)$ - maximum cost for structure variant X_{ij};

$\min(X_{ij}^c)$ - minimum cost for structure variant X_{ij};

X_{ij}^c - actual cost for structure variant X_{ij}.

$CI_f^{X_{ij}}$ – coefficient of importance of flexibility for structure variant X_{ij};

X_{ij}^f – actual flexibility for structure variant X_{ij};

$\min(X_{ij}^f)$– minimum flexibility for structure variant X_{ij};

$\max(X_{ij}^f)$ - maximum flexibility for structure variant X_{ij}.

Values $CI_c^{X_{ij}}$ and $CI_f^{X_{ij}}$ will be calculated in E_{06} and introduced in Table 1.

E06 – Formulating the centralizing table of constructive variants and optimization criteria

Table 1 with the X_{ij} variants of the five groups of structures BCI, ..., BFV, the three optimization criteria (C_{pr}, C_c-CE_3, C_f-CE_6), is built and completed with the actual and admissible values calculated in E_{03} and E_{04}, but also with the values of the coefficients of importance, as follows.

Coefficients of importance for the precision criterion(CI_{pr}) - Inverse Interpolation Calculation Method

The notes, actual and admissible values are those in E_{03}.

For X_{11}:

- Value of interval: $v_i = \varepsilon_{ad.\ s}^{o-p} - \varepsilon_{c.\ s}^{o-p} = 0.012 - 0.010 = 0.002$
- Number of intervals: $n_i = \frac{\varepsilon_{ad.\ s}^{o-p} - \min \varepsilon_{c.\ s}^{o-p}}{v_i} = \frac{0.012-0}{0.002} = 6$
- Coefficient of importance ofthe interval: $CI_i = \frac{1}{n_i} = \frac{1}{6} = 0.16$
- Number of affective intervals: $n_{ie} = \frac{\varepsilon_{c.\ s}^{o-p}}{v_i} = \frac{0.010}{0.002} = 5$
- Apparent importance coefficient: $CI_a = n_{ie} \cdot CI_i = 5 \cdot 0.16 = 0.8$
- Coefficient of importance of precision: $CI_{pr} = 1 - CI_a = 1 - 0.8 = 0.2 \quad \Rightarrow \quad CI_{pr}^{X_{11}} = 0.2$

Proceeding similarly for the other structure variants, it is obtained:

$CI_{pr}^{X_{12}} = 1$; $CI_{pr}^{X_{13}} = 0.020$; $CI_{pr}^{X_{21}} = 0$; $CI_{pr}^{X_{22}} = 0$; $CI_{pr}^{X_{23}} = 1$; $CI_{pr}^{X_{31}} = 0.2$; $CI_{pr}^{X_{32}} = 1$; $CI_{pr}^{X_{41}} = 1$; $CI_{pr}^{X_{51}} = 1$

Table 1. Centralizing table of constructive variants and optimization criteria

Nb.	Structure variants for BCI, ..., BFV	Precision criterion C_{pr}				Economical criteria – CE and Importance coefficients - CI						ΣCI	Decision
		$\varepsilon_{c.\ s}^{o-p}$	$\varepsilon_{ad.\ s}^{0-p}$	SC-TA		Cost criterion C_c - CE_3			Flexibility criterion C_f - CE_6				
BCI (1)		Actual values [mm]		CI_{pr}	k_c^{pr}	Actual val. [mm]	CI_c	k_c^c	Actual val. [mm]	CI_f	k_c^f	(Max)	Optimal
1	X_{11}	0.010	0.012	0.2	1	346.5	0.57	1.2	0.3678	0.26	1.3	1.222	Opt. BCI
2	X_{12}	0	0.012	1	1	807	0	1.2	0.1353	0	1.3	1	---
3	X_{13}	0.020	0.012	0	1	247.5	0.70	1.2	0.2865	0.17	1.3	1.061	---
BCII (4)													
4	X_{21}	0.020	0.017	0	1	623	0.23	1.2	0.3678	0.26	1.3	0.614	---
5	X_{22}	0.020	0.017	0	1	311.5	0.62	1.2	0.4114	0.31	1.3	1.147	---
6	X_{23}	0	0.017	1	1	51.9	0.95	1.2	0.9331	0.99	1.3	3.427	Opt. BCII
BCIII (5)													
7	X_{31}	0.010	0.012	0.2	1	51.75	0.95	1.2	0.3678	0.26	1.3	1.678	---
8	X_{32}	0	0.012	1	1	12.3	1	1.2	0.979	0.97	1.3	3.461	Opt. BCIII
BFIV													
9	X_{41}	0	0.012	1	1	200.62	0.76	1.2	0.999	1	1.3	3.212	Opt. BCIV
BFV													
10	X_{51}	0	0.012	1	1	200.62	0.76	1.2	0.999	1	1.3	3.212	Opt. BCV

Calculation of coefficients of importance for the cost criterion(CI_c)

By using the interpolation method for the minimizing feature values are determined with relation (3):

$$CI_c^{X_{11}} = \frac{807-346.5}{807-12.3} = 0.57; \quad CI_c^{X_{12}} = \frac{807-807}{807-12.3} = 0; \quad CI_c^{X_{13}} = \frac{807-247.5}{807-12.3} = 0.70$$

For the other structures, the values calculated similarly are:

$CI_c^{X_{21}} = 0.23$; $CI_c^{X_{22}} = 0.62$; $CI_c^{X_{23}} = 0.95$; $CI_c^{X_{31}} = 0.95$; $CI_c^{X_{32}} = 1$; $CI_c^{X_{41}} = 0.76$; $CI_c^{X_{51}} = 0.76$.

Calculation of coefficients of importance for the flexibility criterion (CI_f)

For CI_f it is considered that maximum flexibility has value 1, and the least flexible constructive variant has value zero. The coefficient of importance of flexibility CI_f is established for the maximizing feature with the relation from this paper (E_{05}) for X_{ij} analyzed structures:

$$CI_f^{X_{11}} = \frac{0.3678-0.1353}{0.999-0.1353} = 0.26$$

Similarly there are determined and introduced in Table 1 $CI_f^{X_{ij}}$ for the other X_{ij} structures, with the following values: $CI_f^{X_{12}}$= 0; $CI_f^{X_{13}}$= 0.17; $CI_f^{X_{21}}$= 0.26; $CI_f^{X_{22}}$= 0.31; $CI_f^{X_{23}}$= 0.99; $CI_f^{X_{31}}$= 0.26; $CI_f^{X_{32}}$= 0.97; $CI_f^{X_{41}}$= 1; $CI_f^{X_{51}}$= 1

In order to prioritize criteria, from an importance viewpoint, the correction coefficients for precision k_c^{pr}= 1, costs k_c^c= 1.2, and flexibility k_c^f= 1.3 are established.

E07 Selecting the construction variants of the device

Method 1: The optimal variant to construct the control device (VOC_{DC}^1) is a combination of structures which meet the condition locally optimal as max ΣCI of BCI, ..., BFV:

$\Sigma BCI^{X_{11}}$= $CI_{pr}\cdot k_c^{pr}$ + $CI_c\cdot k_c^c$+ $CI_f\cdot k_c^f$=0.2·1+0.57·1.2+0.26·1.3= 1.222 ⇒ $\Sigma BCI^{X_{11}}$ = 1.222

By proceeding similarly it is obtained:

$\Sigma BCI^{X_{12}}$ = 1, $\Sigma BCI^{X_{13}}$ = 1.061, $\Sigma BCII^{X_{21}}$= 0.614, $\Sigma BCII^{X_{22}}$ = 1.147, $\Sigma BCII^{X_{23}}$ = 3.427, $\Sigma BCIII^{X_{31}}$ = 1.678, $\Sigma BCIII^{X_{32}}$ =3.461, $\Sigma BCIV^{X_{41}}$ = 3.212, $\Sigma BCV^{X_{51}}$ = 3.212.

Method 2: The optimal variant of the control device (VOC_{DC}^2) is a combination of structures which meet condition $MaxV_{BCI}$, ..., $MaxV_{Fi}$ from the calculation relation (5):

$$\mathrm{MaxV_{BCI}}[CI_{pr}^{X_{11}}\cdot k_c^{pr}(CI_c^{X_{11}}\cdot k_c^c + CI_f^{X_{11}}\cdot k_c^f);\ldots;CI_{pr}^{X_{ij}}\cdot k_c^{pr}(CI_c^{X_{ij}}\cdot k_c^c + CI_f^{X_{ij}} k_c^f);\ldots;CI_{pr}^{X_{mn}}\cdot k_c^{pr}(CI_c^{X_{mn}}\cdot k_c^c + CI_f^{X_{mn}}\cdot k_c^f)] \quad (5)$$

$\mathrm{MaxV_{BCI}}[0.2\cdot 1(0.57\cdot 1.2+0.36\cdot 1.3); 1\cdot 1(0\cdot 1.2+0.13\cdot 1.3); 0\cdot 1(0.7\cdot 1.2+0.21\cdot 1.3)] \Rightarrow$

$\mathrm{MaxV_{BCI}}(0.221;0.169;0) \Rightarrow \quad \mathrm{MaxV_{BCI}} = 0.221$

By proceeding similarly for the other groups it is obtained:

$\mathrm{MaxV_{BCII}}$ = 1.47; $\mathrm{MaxV_{BCIII}}$ = 2.461; $\mathrm{MaxV_{BFIV}}$ = 2.199; $\mathrm{MaxV_{BFV}}$ = 2.199

E08 Establishing the optimal variants to construct the control device

Method 1:

$$VOC_{DC}^1 = \mathrm{OptimBCI + OptimBCII + OptimBCIII + OptimBFIV + OptimBFV} \quad (6)$$

It results: $VOC_{DC}^1 = X_{11} + X_{23} + X_{32} + X_{41} + X_{51}$

Method 2:

The optimal variant to construct the control device is a combination of structures which meet the condition of maximum of coefficients of importance from relation (7) of analyzed variants:

$$VOC_{DC}^2 = \mathrm{MaxV_{BCI} + MaxV_{BCII} + MaxV_{BCIII} + MaxV_{BFIV} + MaxV_{BFV}} \quad (7)$$

It results: $VOC_{DC}^2 = X_{11} + X_{23} + X_{32} + X_{41} + X_{51}$

3. Construction of the control device

For the construction of the control device and any other type of device, choose or design elements or subassemblies of the orientation and fixing system of the semi-finished product, bodies, connecting and guiding elements with features and functions identical to those of VOC_{DC}^2structures. The constructive solution can be made in the special variant or from modular elements depending on the size of the manufacturing series and the optimization criteria chosen. In this case study the structures X_{11}, X_{23}, X_{32}, X_{41}, X_{51} are found in the overall drawing of the device, in Figure 6.

The construction, configuration and working method with this device are atypical because some of the main and mandatory orientation elements do not allow the access of the control system to the surfaces to be measured.

For this reason it is necessary for the optimization algorithm to add, as a further conditioning, a planar pre-orientation base (11) and a planar substitution base (10) to the existing BCI, ..., BFV, and the sequence of work phase to be the following:

- Place the semi-finished product on the planar pre-orientation base of the support dowels 11, the short prism 1 and the support dowel 5 under its own weight and elastic tightening of the support 9 with self-positioning and subsequent blocking;

IOP Conf. Series: Materials Science and Engineering **252** (2017) 012047 doi:10.1088/1757-899X/252/1/012047

- The three elements 4 of the main orientation base are rotated and blocked in the working position;
- The movable element of the support 12 with self-positioning is brought into contact with the lower part of the semi-finished product, which rises from the planar pre-orientation base 11 and is placed on the main planar base 4, and then blocks;
- In contact with some surfaces of the part to be measured, located below the level of the main positioning base, the three self-positioning supports 10 and the conical movable elements, block afterwards, thus establishing a planar substitution base;
- The three movable elements 4 rotate to allow the access of the control system to check the quotations and measurement conditions, while the semi-finished product keeps its main orientation and fixing unaltered.

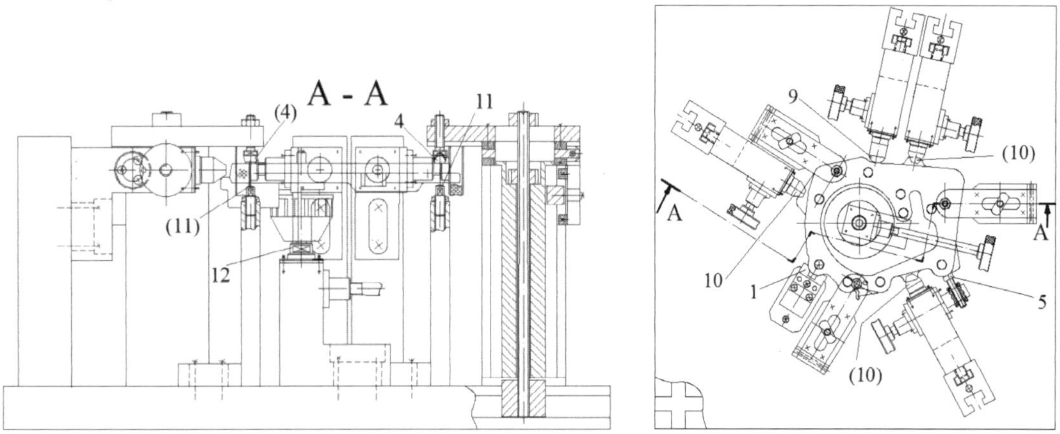

Figure 6. Overall drawing of the device to control inaccessible surfaces

4. Conclusions

The development of manufacturing systems through mathematical modelling, of construction and devices operation, of technological equipment in general is certainly very useful and much more economical than many costly and long lasting practical attempts to establish the optimal solution.

This paper presents a case study that confirms that not only the original optimization method but also the proposed calculation methods provide a reliable optimization solution for rational device construction.

References

[1] Brăgaru A, Picoş C and Ivan NV 1996 *Optimization of processes and technological equipment* (Bucarest: Didactic and Pedagogic Publishing Bucarest)

[2] Ivan NV, Păunescu T, Udroiu R, Ivan MC, Găvruş C and Pescaru R 2010 *Machine Building Technology. Theory and innovative approaches* (Braşov: Transilvania University of Braşov)

[3] Li B and Melkote SN 2001 *Int. J. Adv. Manuf. Tech.* **17** 104-113

[4] Vasii Roşculeţ S, Gojineţchi N, Andronic C, Şelariu M and Gherghel I 1982 Device design (Bucarest: Didactic and Pedagogic Publishing Bucarest)

[5] Iordache DM, Rizea AD, Costea A, Niţu EL and Babă A 2017 Method for optimization of the orientation and fixing system of workpiece for the construction of control devices *MATEC Web of Conferences* **112** 06003 (2017)

[6] Costea A and Rachieru N 2005 Flexibility and performance of processing equipment (Bucarest: BREN Publishing House)

[7] Abrudan I 1996 *Flexible Manufacturing Systems. Design and management concepts* (Cluj-Napoca: Dacia Publishing).

[8] Boldur Gh 1973 *The complex foundation of the economic decision-making process* (Bucarest: Scientific Publishing House)

Experimental and theoretical study of friction torque from radial ball bearings

Ionut Geonea[1], Nicolae Dumitru[2], Ilie Dumitru[3]
[1, 2, 3]University of Craiova, Faculty of Mechanics

igeonea@yahoo.com

Abstract. In this paper it is presented a numerical simulation and an experimental study of total friction torque from radial ball bearings. For this purpose it is conceived a virtual CAD model of the experimental test bench for bearing friction torque measurement. The virtual model it is used for numerical simulation in Adams software, that allows dynamic study of multi-body systems and in particularly with facility Adams Machinery of dynamic behavior of machine parts. It is manufactured an experimental prototype of the test bench for radial ball bearings friction torque measurement. In order to measure the friction torque of the tested bearings it is used an equal resistance elastic beam element, with strain gauge transducer to measure bending deformations. The actuation electric motor of the bench has the shaft mounted on two bearings and the motor housing is fixed to the free side of the elastic beam, which is bended by a force proportional with the total friction torque. The beam elastic element with strain gauge transducer is calibrated in order to measure the force occurred. Experimental determination of the friction torque is made for several progressive radial loads. It is established the correlation from the friction torque and bearing radial load. The bench allows testing of several types and dimensions of radial bearings, in order to establish the bearing durability and of total friction torque.

1. Introduction

In the literature are presented a large number of studies concerning the friction, lubrication and wear of materials. Test procedures for experimental measuring for the friction torque in rolling bearings are presented in [1]. For this purpose it is used a modified four ball machine, in order to test rolling bearings. It is monitored the friction torque and operating temperature.

Other research [2] examines the frictional power loss of a needle roller bearing lubricated with grease. Obtained results reveal that the test bearing has higher friction compared with bearings lubricated with conventional lubrication.

Low cost systems for the force and torque measurement, for wheel bearings are presented in [3]. Also, a study for the friction of a ball screw is presented in [4]. It is presented a theoretical model for the friction between balls of the screw. The study is useful because provides a theory support for reasonably reducing the screw ball friction.

Studies based on the lubrication theory are presented in [5]. It is studied the wear behavior, using a four ball wear tester. Tests are made with a low viscosity additivated mineral oil. Two types of balls are used, steel and ceramic and obtained results shows greater resistance of the ceramic balls.

The effects of roughness upon friction of the plastics, intended for bearing applications are studied in [6]. Obtained results shows that not exists an optimal roughness for minimum friction for polymers and the friction depends on the bulk properties of the polymer.

IOP Conf. Series: Materials Science and Engineering **252** (2017) 012048 doi:10.1088/1757-899X/252/1/012048

Aspects concerning automotive tribology are presented in [7]. In this research is presented an overview of various lubrication aspects, for typical power train, including engine, transmission and driveline. Also an aspect presented consist in the currents status and future trends in automotive lubricants. Other comparative studies on the tribological behavior of lubricants are presented in [8]. Test methods for engine lubricants are presented in ASTM (American Society for Testing and Materials) standards and also are available a large number of patents [9], concerning the bearing friction.

2. Theoretical considerations

The resistant torque that appears in roller bearings is produced by a complex of friction appearance ways. The rolling friction phenomena complexity is generated by a large number of factors that occurs simultaneously. The most important sources for bearing friction are: friction generated by the contact deformations, rolling friction on the contact surfaces, friction produced by the lubricants, sliding of bearing elements, friction from seals. For usual calculations the friction torque can be estimated, with sufficient precision, with relations obtained from experimental results. The equation (1) establishes the total friction torque (M_{tc}).

$$M_{tc} = M_f + M_l \tag{1}$$

where: M_l is the resistant torque produced by the fluid friction of bearing elements in contact with the lubricant; M_f is the resistant torque produced by the bearing load.

Equation (1) is used for bearings that operate to moderate speed and loads. In case of the ball bearings, used for high speed, when the frictions produced by the spin and gyroscopic motion are important, should be considered the friction torque produced by these motions.

The resistant torque (M_f) produced by the bearing load, is computed with the equation (2).

$$M_f = f_1 \cdot F \cdot d_m \tag{2}$$

where: F [N] is the bearing radial load; d_m is the bearing medium diameter, in [m].

For radial ball bearings the factor f_l is established with the relation:

$$f_1 = 0{,}0009 \cdot \left(\frac{P_0}{C_0} \right)^{0{,}55} \tag{3}$$

where: $P_0 = F_r$ [N]; C_0= 8000 N, for radial ball bearing type 6204.

In order to compute the resistant torque (M_l) produced by the fluid friction of bearing elements in contact with the lubricant are used equations (4) or (5):

$$M_l = f_0 \cdot (\nu \cdot n)^{\frac{2}{3}} \cdot d_m \cdot 10^3 \quad \text{[Nm], for } \nu \cdot n \geq 20 \cdot 10^{-3} \tag{4}$$

$$M_l = 16 \cdot f_0 \cdot d_m^3 \quad \text{[Nm], for } \nu \cdot n \leq 20 \cdot 10^{-2} \tag{5}$$

Where: ν is the kinematic viscosity of the lubricant, in [m^2/s]; n is the bearing operating speed, in [rpm]. For the considered case, f_0= 1,5~2 and ν = (25~33)$\cdot 10^{-6}$ [m^2/s].

3. Experimental test bench description

For experimental measurement of the bearing torque it is used the test bench shown in figure 1, where the tested bearing is mounted on the shaft (2) and it has oscillating exterior housing, as see in figure 1. The actuation is made with an electric motor, with the nominal speed of 1450 rpm.

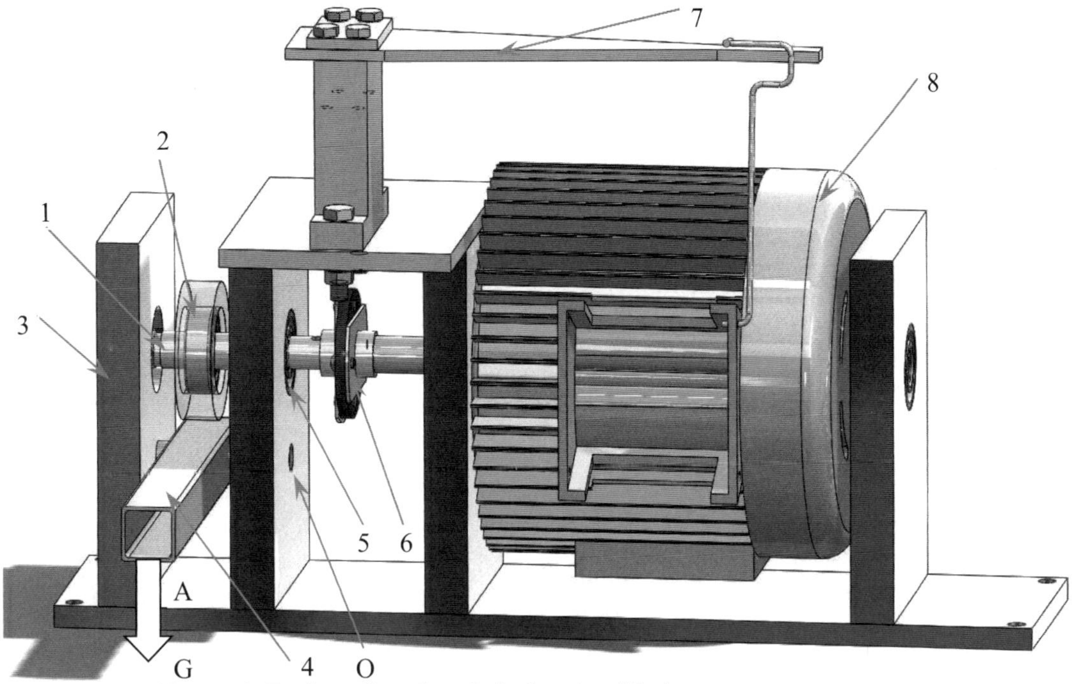

Figure 1. Design of test bench for bearing friction torque measurement.

Components of the test bench are: (1) shaft, (2) –tested bearing, mounted on oscillating housing; (3) – shaft bearings supports; (4) –lever to produce the radial load; (5) –shaft bearings; (6) – elastic coupling; (7) –beam with strain gauge transducers; (8) – electric motor.

Radial ball bearing (2) subjected to experimental tests is mounted on the shaft (1). The outer rig of the tested ball bearing is mounted to an oscillating cylindrical bush. The radial load of the bearing is created with the lever (4) that is articulated to point O (figure 1). To point A are added additional weights G (as see in Fig. 4) in order to create different radial loads for the bearing. Electric motor (8) is used to rotate the shaft (1). The shafts of tested bearing and electric motor are connected with an elastic coupling (6). Electric motor stator is fixed with a tie rod to the elastic beam (7) free side. The elastic beam is subjected to deformations proportional with the motor torque. In order to measure the beam bending force, produced by the motor torque are used the strain gauge transducers (10), as see in figure 2. With experimental calibration is established the dependency between elastic deformation versus bending force and motor torque, which is necessary to be determined.

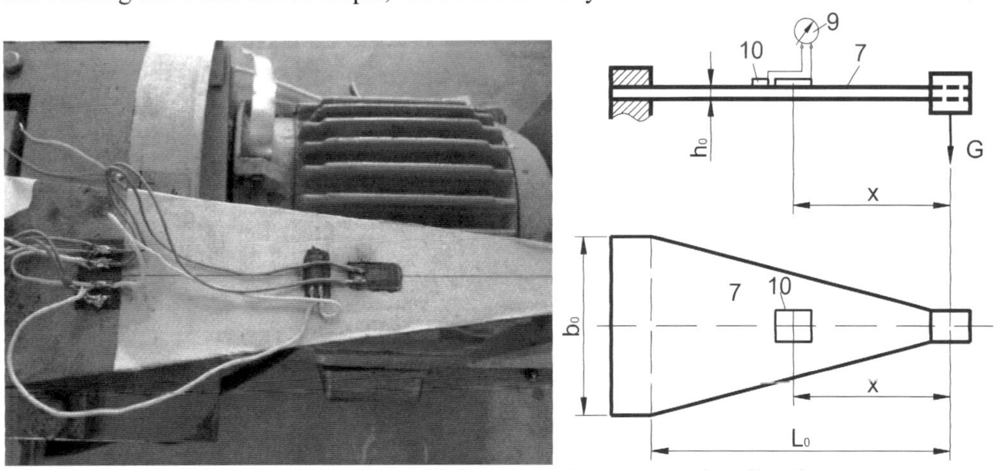

Figure 2. Equal strength beam used to measure bending force.

Bending stress that appears in the beam is computed with equation (6).

$$\sigma_j^{ex}(x) = E \cdot \varepsilon_j^{ex} \cdot 10^{-6} [MPa] \quad (6)$$

where: E is the Young's module for the beam material, in [MPa], ε_j^{ex} is the experimental measured deformation, in $[\mu m / m]$. For experimental measurement of the deformation it is used the MGCPlus acquisition system from Hottinger Baldwin Messtechnik. In order to evaluate the bending force, the beam transducer it is calibrated. The calibration diagram of the strain gauge transducer is presented in figure 3.

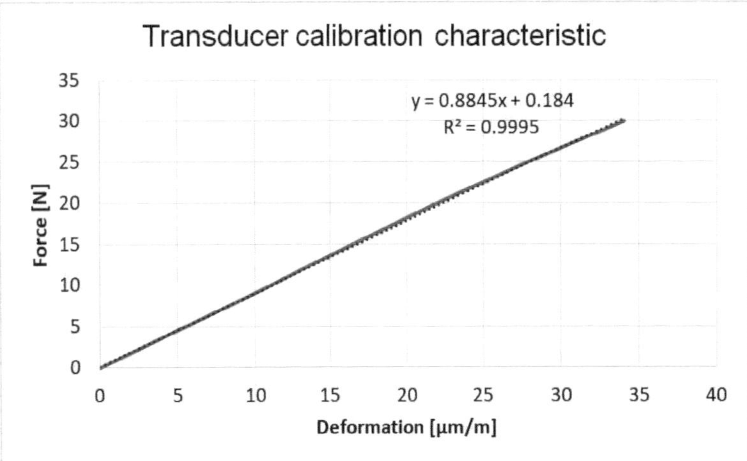

Figure 3. Transducer experimental calibration characteristic.

The experimental test bench for bearing friction torque measurement, developed in the Machine Design Laboratory, from Faculty of Mechanics, Craiova, is presented in figure 4. Also is shown the experimental acquisition system MGCPlus used to measure friction torque.

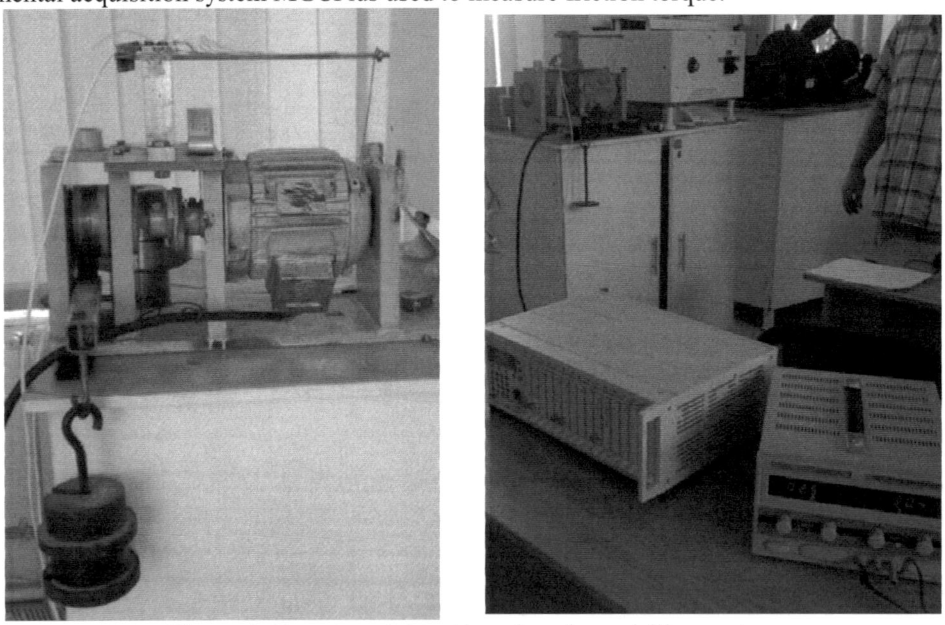

Figure 4. Experimental test bench and acquisition system.

Preliminary obtained results consist in the deformation of the elastic beam measurement. Originally acquired results, with the acquisition software Catman AP are presented in figure 5. The registration presented in figure 5 corresponds to the bearing function with zero radial loads. Tested radial ball bearing is type 6204, manufactured by SKF. To this operating regime the elastic beam deformation is stabilized to 12.5 μm/m. To this deformation corresponds a bending force of 11.06 N. Considering the distance between electric motor shaft and beam longitudinal center, as 71.3 mm, it results a friction measured torque by 788.57 Nmm.

The radial force in order to load the bearing is created by adding metal discs with 1.8 kg weight of the lever (4), as see in figure 4. Registered results of the beam deformations, by adding three supplementary loads are presented in figure 6. Obtained results, corresponding to radial load regime, shows an increase of the total friction torque.

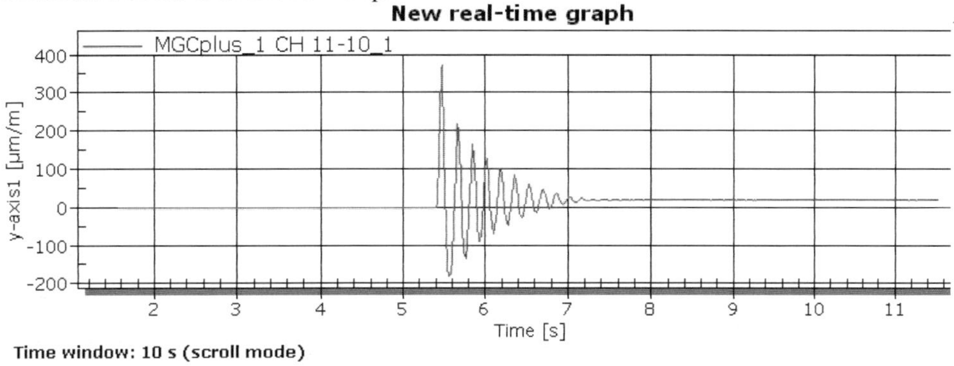

Figure 5. Experimental registered deformations of the elastic beam (bench start and no radial load).

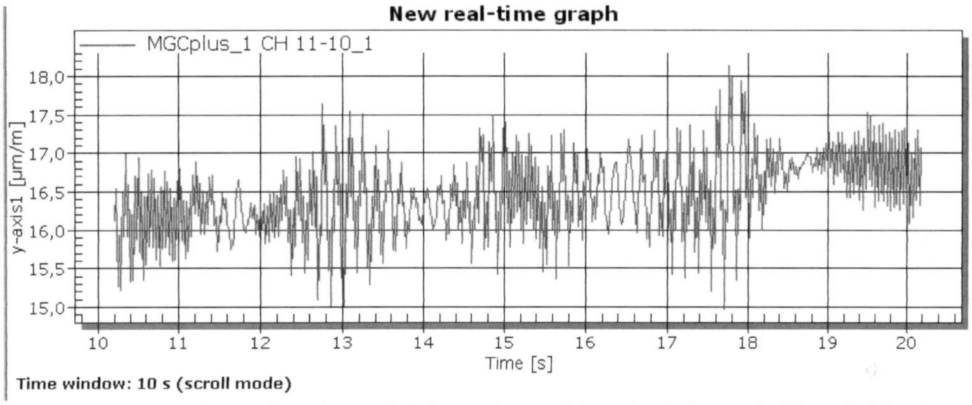

Figure 6. Experimental registered deformations of the elastic beam (with radial load).

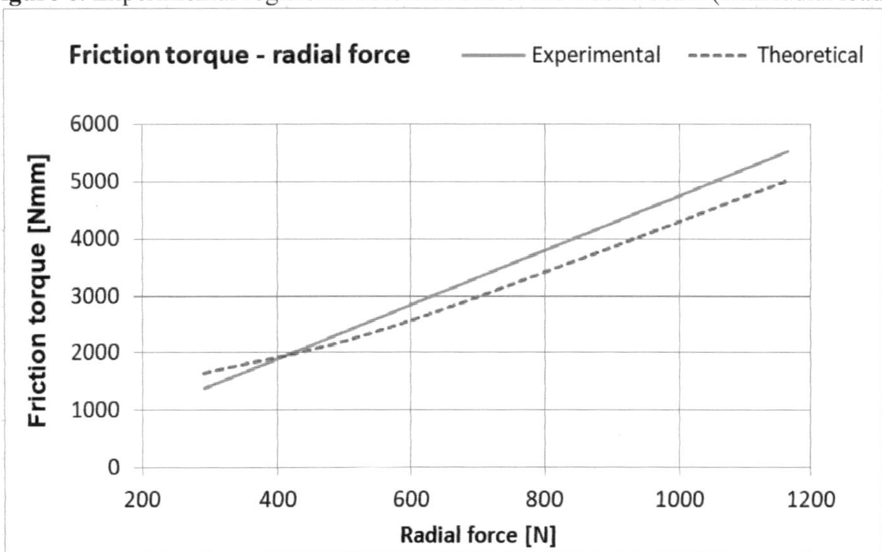

Figure 7. Experimental and theoretical variation of the bearing friction torque versus radial load.

Obtained theoretical dependence of the bearing friction torque versus radial load is presented in figure 7. Also, experimental obtained dependence of the bearing friction torque versus redial bearing load is presented in the same graphic, figure 7. It is observed a linear dependence, the friction torque increasing with the bearing radial load. Similar linear dependencies for the friction torque versus radial load are obtained by bearing manufacturer, to performed tests [10].

4. Numerical simulation in ADAMS of bearing test bench

ADAMS software offers the possibility with machinery plug-in to define a various number of machine elements. To design the dynamic model of the bearing test bench in ADAMS, for the kinematic elements have been specified the materials and are defined the revolute joints as ball bearings.
The ball bearing construction in Adams Machinery is specified as shown in figure 8.

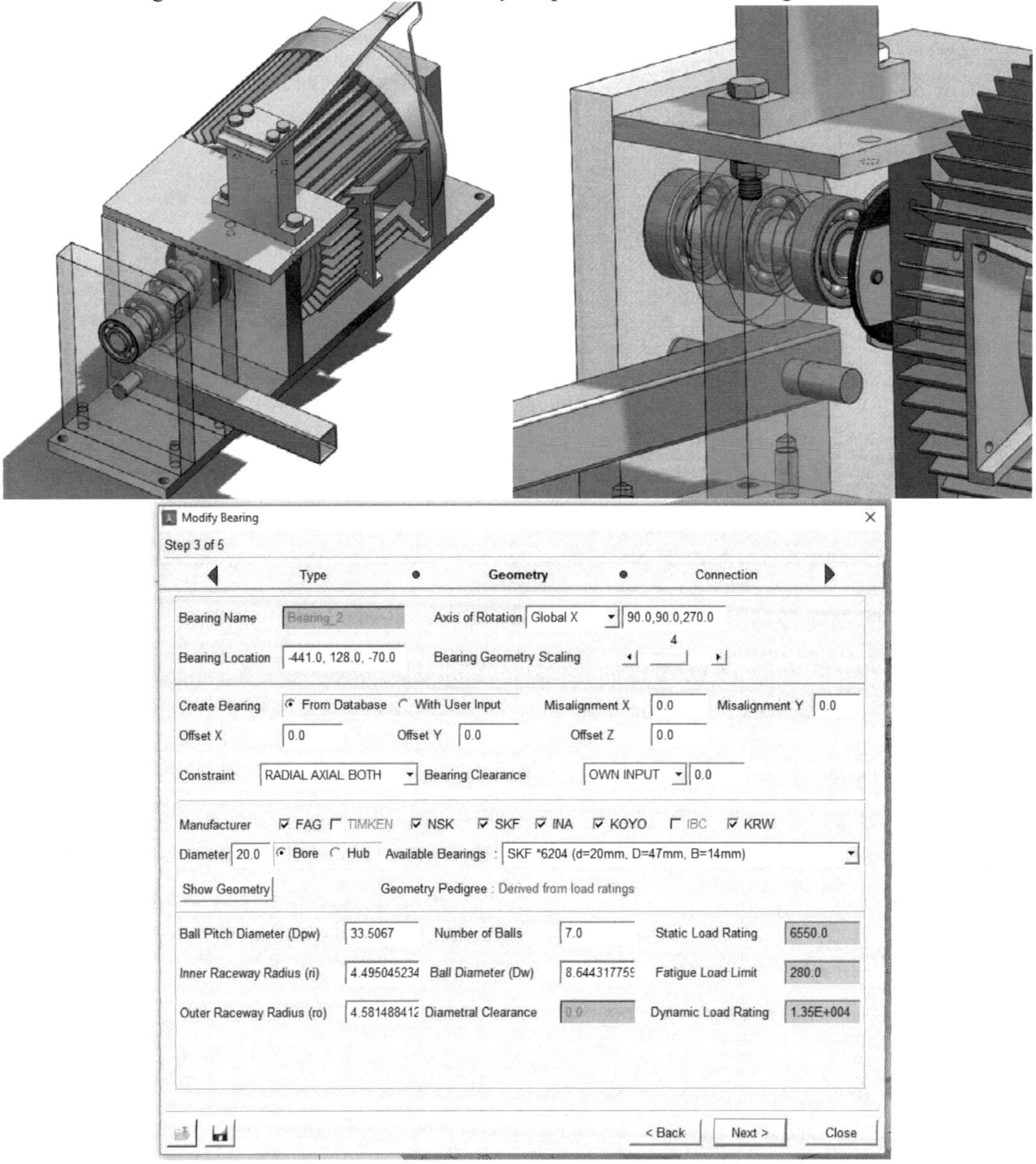

Figure 8. Definition of ball bearing using Adams Machinery feature.

The motion for the electric motor is defined to 157 rad/s. To obtain accurate results the shaft (1) of the test bench is considered as a deformable body. Based on Adams Machinery feature, the tested bearing life is 526 hours, as presented in figure 9.

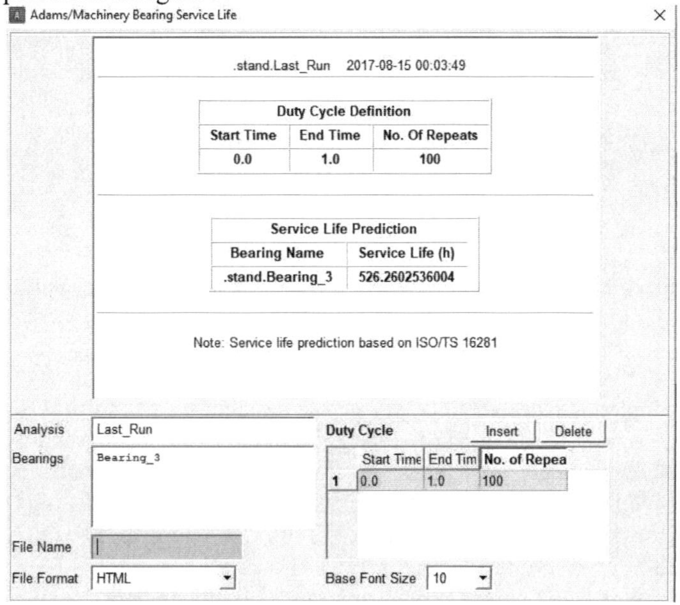

Figure 9. Bearing life report using Adams Machinery feature.

In figure 10, is presented the bearing center marker translational displacement upon X axis (axial). In figure 11, is presented the computed translational displacement of bearing center marker upon Y axis (radial). The axial displacement has small values (reaches 0.0015 mm), and the radial displacement is larger, reaching the value of 0.022 mm. Adams computed translational deformation of the shaft (1) center marker are shown in figure 12.

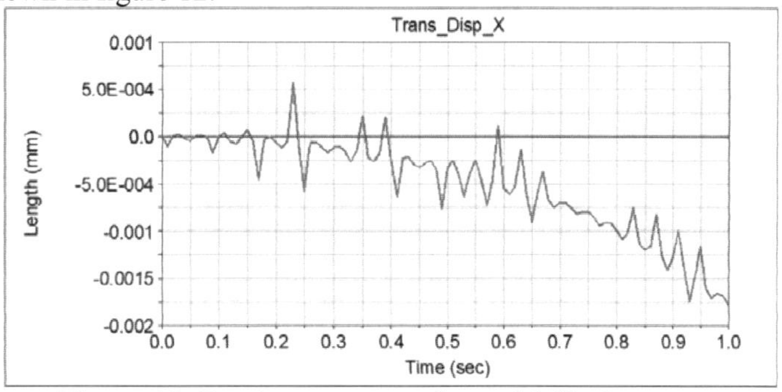

Figure 10. Bearing center marker axial displacement.

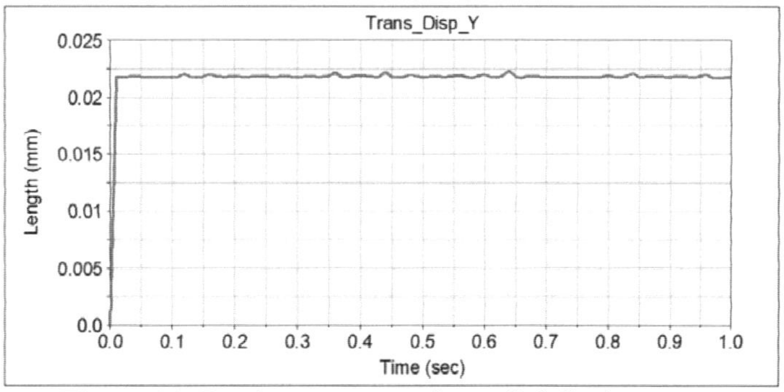

Figure 11. Bearing center marker radial displacement.

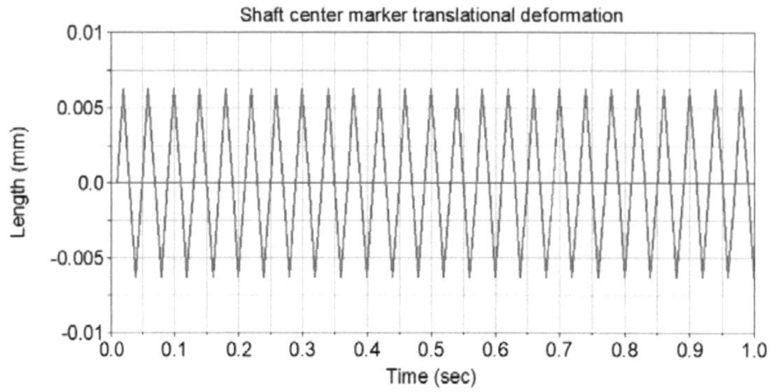

Figure 12. Shaft center marker vertical translational deformation.

5. Conclusion

In this paper it is presented the design of a test bench for bearing friction torque measurement. The ball bearing subjected to tests in this study is type SKF 6204. From theory and experimental tests are obtained similar linear dependencies between the friction torque and bearing radial load. A graphic comparison for the obtained theoretical and experimental obtained results is presented in figure 7. Theoretically, it is computed a total friction torque of 1647.96 Nmm for the radial load of 291 N and the torque increases to 5018.17 Nmm in case of 1164 N bearing radial load. Experimentally, at 291 N bearing radial load it is measured a total torque of 1383.22 Nmm and for the maximum radial load of 1164 N the value of the friction torque reaches 5532.88 Nmm. The numerical simulation of the test bench is performed with Adams software, considering as flexible body the shaft mounted on bearings. Adams machinery reports a bearing life of 526.26 hours, as presented in figure 9. Are computed in Adams and presented in the paper the bearing center marker translational displacements. Because of the radial load it is computed a translational of 0.022 mm for the marker attached to the bearing center. The shaft deformations shown in figure 12, have the maximum amplitude of 0.13 mm. Proposed test bench can be used to tests different types of radial bearings, with different lubricants.

6. References

[1] Cousseau T, Graça B, Campos A and Seabra J 2010 Experimental measuring procedure for the friction torque in rolling bearings *Lubr. Sci.* **22** pp 133–47

[2] Iqbal S, Al-Bender F, Croes J, Pluymers B. and Desmet W 2013 Frictional power loss in solid-grease-lubricated needle roller bearing *Lubr. Sci.* **25** pp 351–67

[3] Kwapisz D and Stephant J 2008 Instrumented bearing for force and moment measurements *IEEE 2008 Sensors*, pp 1480-83

[4] Nannan X and Wencheng T 2015 Modeling analysis and experimental study for the friction of a ball screw *Mech. and Mach. Theory* **87** pp 57-69

[5] Paleu V, Bercea I, Cretu S and Bercea M 2005 Lubricant oils additivated with polymers in EHD contacts: Part 2. Tests using a four-ball machine *Lubr. Sci.* **17** pp 173–84

[6] Quaglini V and Dubini P 2009 Influence of counterface roughness on friction properties of engineering plastics for bearing applications *Materials and Design* **30** pp 1650–58

[7] Tung S C and McMillan M L 2004 Automotive tribology overview of current advances and challenges for the future *Tribology International* **37** pp 517-36

[8] Pawlaka Z and Kaldonskic T 2009 A comparative study on the tribological behaviour of hexagonal boron nitride (h-BN) as lubricating micro-particles—An additive in porous sliding bearings for a car clutch *Wear* **267** pp 1198–202

[9] Bearing friction tester US Patent 4672838 A

[10] Yuchao W, Fuyan L, Huazhuang J and Wenming Y 2015 Investigation on frictional characteristic of deep-groove ball bearings subjected to radial loads *Advances in Mechanical Engineering* **7** pp1-12

IOP Conf. Series: Materials Science and Engineering **252** (2017) 012049 doi:10.1088/1757-899X/252/1/012049

Characterization of the dynamic behaviour of flax fibre reinforced composites using vibration measurements

Ali El-Hafidi; Papa Birame Gning; Benoit Piezel; Stéphane Fontaine

DRIVE EA1859, Université de Bourgogne Franche-Comté, 58027 Nevers, France

Corresponding author: Ali.el-hafidi@u-bourgogne.fr

Abstract. Experimental and numerical methods to identify the linear viscoelastic properties of flax fibre reinforced epoxy (FFRE) composite are presented in this study. The method relies on the evolution of storage modulus and loss factor as observed through the frequency response. Free-free symmetrically guided beams were excited on the dynamic range of 10 Hz to 4 kHz with a swept sine excitation focused around their first modes. A fractional derivative Zener model has been identified to predict the complex moduli. A modified ply constitutive law has been then implemented in a classical laminates theory calculation (CLT) routine.

1. Introduction

Structural parts of vehicles are submitted to dynamic loading. Excitations coming from the powertrain, road surfaces and aerodynamic flows cause mechanical vibrations. In order to improve the acoustical and vibrational comfort, natural fibre reinforced polymer composites, exhibiting interesting dissipative properties, can be used. Indeed, compared to conventional composite materials, the damping performance of flax fibre reinforced polymer (FFRP) can be respectively twice or three times higher than that of glass or carbon fibre reinforced polymer composites [1,2].

Damping in undamaged composite materials is induced by several microscopic level mechanisms, such as viscoelastic elongation of the matrix and/or fibres, friction between both components at their interfaces. Moreover, in the particular case of natural fibre reinforced polymer composites, the friction between fibres inside bundles can also increase this phenomenon.

In order to assess the damping within materials, vibration techniques have the advantage of rapidly exploring a wide range of frequencies (10 Hz to 1 kHz). Thus, many authors have studied the frequency dependence of composite materials [3,4,5,6]

The frequency dependence is a typical feature of viscoelastic materials. This dependence induces variations of the complex moduli when the frequency of excitation changes. In order to assess these variations, it is necessary to understand the material constitutive equations that relate the stresses to the strains with respect to time or frequency [7], with the help of the linear viscoelasticity theory. These relations are expressed by linear differential equations or convolution integrals. They have the main benefit of being expressed both in frequency and time domains.

The present study aims at presenting a method dedicated to the identification of the evolution of complex moduli of FFRP laminates on a large frequency band with the help of a fractional derivative Zener model. This identification has been done thanks to experimental tests on a specific device, between 10 Hz and 4 kHz.

2. Linear viscoelastic model

Materials submitted to mechanical loading store energy by elastic deformation. For purely elastic materials, this energy is totally and immediately returned during the unloading phase. However, in the case of viscoelastic materials, the energy stored is completely returned but with some delay due to inner rearrangements. For such materials the stress depends on the strain history [7]. Several models describing the behaviour of a viscoelastic material in the structural dynamic domain are available in the literature. Among them, the fractional derivative Zener model (Fig. 1) has been used by many authors to describe the viscoelastic behaviour. This model is mathematically described in Eq. 1, where $\sigma(t)$, $\varepsilon(t)$, τ, α are respectively the stress and strain as a function of time, the relaxation time and the α-order fractional derivative coefficient (0 < α < 1). $D^{\alpha}[\bullet]$ is an α-order fractional derivative operator. The mechanical stiffness E_0 and E_{∞} are respectively the dynamic modulus at very low frequency and at high frequency. The fractional derivative model is described in Fig. 1.

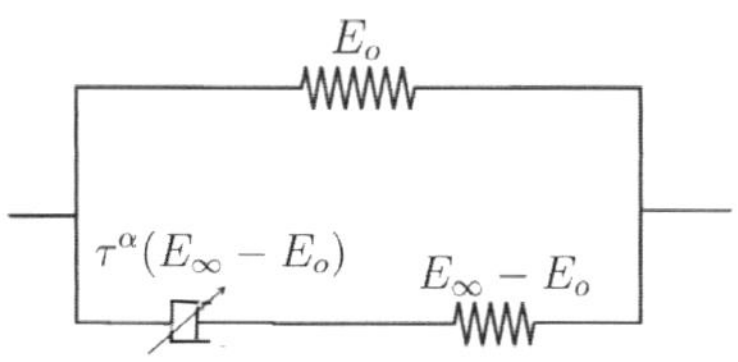

Figure 1. Fractional derivative Zener model.

$$\sigma(t)+\tau^{\alpha}D^{\alpha}[\sigma(t)]=E_0\varepsilon(t)+E_{\infty}\,\tau^{\alpha}D^{\alpha}[\varepsilon(t)] \quad (1)$$

In the case of a steady state harmonic excitation, the complex modulus (E^*)is derived from Eq. 1, where the fractional derivative operator $D^{\alpha}[\bullet]$ is replaced by a multiplication by $(j\omega)^{\alpha}$. j is the imaginary unit and $\omega_n=\omega\tau$.

$$E^*(\omega)=\frac{\sigma^*(\omega)}{\varepsilon^*(\omega)}=\frac{E_0+E_{\infty}\,(j\omega_n)^{\alpha}}{1+(j\omega_n)^{\alpha}} \quad (2)$$

Thus, it is possible to write Eq. 2 in the form of Eq. 3, in which $\eta(\omega)=E''(\omega)/E'(\omega)$ represents the material's loss factor. The real part $E'(\omega)$ of (E^*) represents the elastic behaviour of the tested material. It is the storage modulus while the imaginary part $E''(\omega)$ or loss modulus represents the viscous or damping behavior of the material.

$$E^*(\omega)=E'(\omega)+jE''(\omega)=E'(\omega)[1+j\eta(\omega)] \quad (3)$$

Using the constitutive equation (Eq. 1), $E'(\omega)$ and $\eta(\omega)$ can be expressed by Eqs. 4 and 5, where $c=E_{\infty}/E_0$. :

$$E'(\omega)=E_0\frac{1+(c+1)\cos(\alpha\pi/2)\omega_n^{\alpha}+c\omega_n^{2\alpha}}{1+2\cos(\alpha\pi/2)\omega_n^{\alpha}+\omega_n^{2\alpha}} \quad (4)$$

$$\eta(\omega)=\frac{(c-1)\sin(\alpha\pi/2)\,\omega_n^{\alpha}}{1+(c+1)\cos(\alpha\pi/2)\omega_n^{\alpha}+c\omega_n^{2\alpha}} \quad (5)$$

3. Classical laminate theory applied to viscoelasticity

With the help of the individual properties of each ply, the CLT allows one to compute the elastic properties of a stack of unidirectional (UD) composite plies (i.e. longitudinal and transverse storage moduli, Poisson's ratios and shear moduli). In order to establish a viscoelastic version of CLT, the so-called correspondence principle has been used. Initially introduced on homogeneous materials, this principle has been extended to heterogeneous and composite materials by Hashin [9]. In the case of a sinusoidal steady-state excitation, it is possible to measure and replace the elastic properties by the appropriate complex ones, i.e. the conversion of the elastic solutions to viscoelastic ones. Thus, the laminate's relations derived from the CLT can be used to predict the overall viscoelastic moduli of the laminate from the elementary ply properties. According to the elastic CLT method, the complex constitutive relation linking forces (N^*) and bending moments (M^*) to strains (ε^*) and membrane curvatures (κ^*) can be expressed by Eq. 6.

$$\begin{bmatrix} N^*(\omega) \\ M^*(\omega) \end{bmatrix} = \begin{bmatrix} A^*(\omega) & B^*(\omega) \\ B^*(\omega) & D^*(\omega) \end{bmatrix} \begin{bmatrix} \varepsilon^*(\omega) \\ \kappa^*(\omega) \end{bmatrix} \tag{6}$$

The complex laminate stiffness matrices $[ABD]^*$ are given by Eqs. 7 to 9, based on the complex reduced stiffness matrix $[\tilde{Q}^*_{ij}(\omega)]$, where h_n represents the position of the nth ply in the thickness of the laminate and N is the total number of plies (Fig. 2).

$$A^*_{ij} = A'_{ij} + jA''_{ij} = \sum_{n=1}^{N} [\tilde{Q}^*_{ij}]_n (h_n - h_{n-1}) \tag{7}$$

$$B^*_{ij} = B'_{ij} + jB''_{ij} = \frac{1}{2} \sum_{n=1}^{N} [\tilde{Q}^*_{ij}]_n (h_n^2 - h_{n-1}^2) \tag{8}$$

$$D^*_{ij} = D'_{ij} + jD''_{ij} = \frac{1}{3} \sum_{n=1}^{N} [\tilde{Q}^*_{ij}]_n (h_n^3 - h_{n-1}^3) \tag{9}$$

The nth layer in which plies are oriented with an angle (α_n) with respect to the loading direction, the coefficients $[\tilde{Q}^*_{ij}]_n$ are obtained from $[Q^*_{11}(\omega)]_n$ using Eqs. 10 to 15, where $p_n = \cos \alpha_n$ and $q_n = \sin \alpha_n$. The $[Q^*_{11}]_n$ are the complex reduced stiffness.

$$[\tilde{Q}^*_{11}]_n = [Q^*_{11}]_n p_n^4 + [Q^*_{22}]_n q_n^4 + 2([Q^*_{12}]_n + 2[Q^*_{66}]_n) p_n^2 q_n^2 \tag{10}$$

$$[\tilde{Q}^*_{12}]_n = ([Q^*_{11}]_n + [Q^*_{22}]_n - 4[Q^*_{66}]_n) p_n^2 q_n^2 + [Q^*_{12}]_n (p_n^4 + q_n^4) \tag{11}$$

$$[\tilde{Q}^*_{16}]_n = ([Q^*_{11}]_n - [Q^*_{12}]_n - 2[Q^*_{66}]_n) p_n^3 q_n + ([Q^*_{12}]_n - [Q^*_{22}]_n - 2[Q^*_{66}]_n) p_n q_n^3 \tag{12}$$

$$[\tilde{Q}^*_{22}]_n = [Q^*_{11}]_n q_n^4 + [Q^*_{22}]_n p_n^4 + 2([Q^*_{12}]_n + 2[Q^*_{66}]_n) p_n^2 q_n^2 \tag{13}$$

$$[\tilde{Q}^*_{26}]_n = ([Q^*_{11}]_n - [Q^*_{12}]_n - 2[Q^*_{66}]_n) p_n q_n^3 + ([Q^*_{12}]_n - [Q^*_{22}]_n - 2[Q^*_{66}]_n) p_n^3 q_n \tag{14}$$

$$[\tilde{Q}^*_{66}]_n = \left([Q^*_{11}]_n + [Q^*_{22}]_n - 2([Q^*_{16}]_n + [Q^*_{66}]_n)\right) p_n^2 q_n^2 + [Q^*_{66}]_n (p_n^4 + q_n^4) \tag{15}$$

For each of the n layers, the Q^*_{ij} are determined in the principal directions of the UD layer based on the engineer's moduli (Eqs. 16 to 19). E'_L and E'_T represent, respectively, the storage moduli in the fibres and the transverse directions of the elementary ply. η_L and η_T are longitudinal and transverse loss factors. G'_{LT} is the ply shear modulus. ν_{LT} and ν_{TL} are its Poisson's ratios.

$$Q^*_{11}(\omega) = \frac{E^*_L(\omega)}{1-\nu_{TL}\nu_{LT}} = \frac{E'_L(\omega)[1+j\eta_L(\omega)]}{1-\nu_{TL}\nu_{LT}} \tag{16}$$

$$Q^*_{22}(\omega) = \frac{E^*_T(\omega)}{1-\nu_{TL}\nu_{LT}} = \frac{E'_T(\omega)[1+j\eta_T(\omega)]}{1-\nu_{TL}\nu_{LT}} \tag{17}$$

$$Q_{12}^{*}(\omega) = \frac{\nu_{LT}E_T^{*}(\omega)}{1-\nu_{TL}\nu_{LT}} = \frac{\nu_{LT}E_T'(\omega)[1+j\eta_T(\omega)]}{1-\nu_{TL}\nu_{LT}} \tag{18}$$

$$Q_{66}^{*}(\omega) = G_{LT}^{*}(w) = G_{LT}'(w)[1+j\eta_{LT}(w)] \tag{19}$$

Thus, the effective bending stiffness and the loss factor of a composite cantilever beam are given by relations 20 and 21. D_{11}^{*-1} represents the first element of the inverse of matrix [D] defined by Eq. 20.

$$E^{*}(\omega) = \frac{12}{h^3 D_{11}^{*-1}(\omega)} = E' + jE'' \tag{20}$$

$$\eta(\omega) = \frac{E''(\omega)}{E'(\omega)} \tag{21}$$

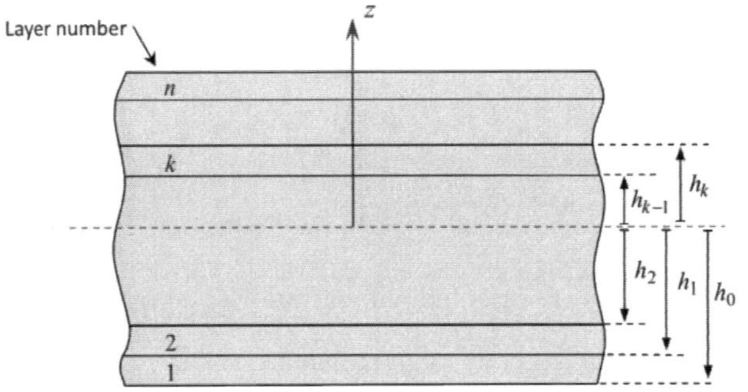

Figure 2. Laminate's Architecture representation

4. Experimental method

Materials and specimens

The FFRE specimens used were made of the Hermès variety of flax cultivated in northern France. A commercial dry roll of unidirectional (UD) flax fabric was supplied by BioRenforts, a local textile company. The fabric consisted of aligned flax UD fibres stitched by a cotton thread, without any treatment. The matrix was an epoxy system based on the resin SR 8200 with the SR 8205 hardener provided by SICOMIN. Fibre layers were first cut from the roll and manually impregnated with the liquid matrix. The plies were hand-laid, before being stacked under a pressure of 7 bars in a hydraulic press equipped with heating plates.

Slender composite specimens made of four layers, i.e. [0]4, [90]4, [±45]s and [±60]s with a thickness (t) of 1 mm, a width (b) of 10 mm and a free length (L) of 170 mm, were cut out from laminated plates with a high speed rotating grinding disk.

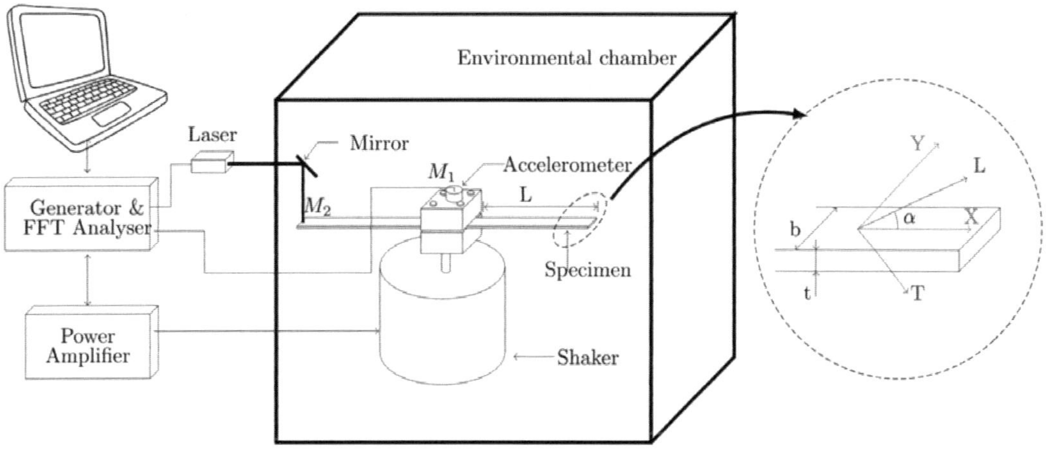

Figure 3. Schematic diagram of the specimen and test set-up.

The test apparatus is shown on Fig. 3. A PCB© accelerometer was fixed to the end of the shaft of the shaker, at point M_1, in order to measure the acceleration $(a(M_1))$ imposed to the sample. Then, the vibration velocity $(v(M_2))$ of the free end of the beam, at point M_2, was measured by a vibrometre sensor head Polytec-OFV-503 coupled to a controller unit Polytec-OFV-500, via a mirror inclined at 45° above M_2 allowing the laser signal to enter in the chamber through a small transparent window. A real-time signal analyser Pulse LabShop-B&K© gave the frequency response function $(H(\omega))$ between $v(M_2)$ and $a(M_1)$ (Eq. 22).

The storage modulus $(E'(\omega_i))$ is computed from the beam theory (Eq. 23). The loss factor $(\eta(\omega_i))$ has been calculated by the "-3 dB bandwidth" method, in the vicinity of ω_i^2.

$$(\omega) = \frac{v(M_2)}{a(M_1)} \tag{22}$$

$$E'(\omega_i) = \frac{12\rho L^4}{X_n^2 h^2}\omega_i^2 \tag{23}$$

5. Results and discussion

5.1. Material properties identification

The linear viscoelastic Zener's parameters (τ, α, E_0 and E_∞) have been identified for $[0]_4$ and $[90]_4$ using Eqs. 4 and 5 in an iterative procedure minimizing the error between the experimental data and the theoretical values. The results of the identification are given in Table 1.

Table 1. Optimal parameters of Zener model for E_L^* and E_T^* laminates.

	E_0 (GPa)	E_∞ (GPa)	τ (s)	α
$[0]_4$	19.2	20.7	1.59×10^{-5}	0.422
$[90]_4$	5.03	5.63	1.59×10^{-5}	0.422

Experimental data with standard deviations and Zener model curves for $[0]_4$ and $[90]_4$ laminates are plotted on Figure 4. The analytical calculation of the moduli and loss factors show good agreement with the test measurements, despite the inherent scattering of the properties of fibres. The measured moduli increased with the frequency of 10 and 17% on average, respectively for $[0]_4$ and $[90]_4$ laminates in the bandwidth (Figure 4a). The loss factors exhibited important increase with a non-linear trend of 22 and 20% on average, respectively for $[0]_4$ and $[90]_4$ samples (Figure 4b).. Conversely, η was dependent on frequency. Despite the inherent scattering of plant fibres, the analytical predictions show good agreement with the test measurements. Thus, the Zener model was assumed to be relevant.

5.2. Validation method

After the identification of the parameters of the Zener model (E_L, E_T, η_L and η_T) on $[0]_4$ and $[90]_4$ specimens presented above, the identification of the storage modulus and loss factor have been extended to additional $[\pm 60]_s$ $[\pm 45]_s$ laminates. The coefficients $[\tilde{Q}^*_{16}]$ were computed by the CLT, and used to determine the $[ABD]^*$ matrices. Inversion of these matrices gave the desired complex modulus. ν_{LT} was supposed to be frequency-independent and was measured at 0.43 by quasi-static tensile testing. ν_{TL} was determined using the conventional relationship $\nu_{TL} = \nu_{LT} \times E_T/E_L$. The frequency dependence of the shear modulus G_{LT} was similar to that of the complex modules insofar as it uses the fractional derivative Zener model. The model coefficients that minimize the difference between the theoretical values and the experimental data were estimated at ${G_{LT}}_0 = 0.15\ GPa$ and ${G_{LT}}_\infty = 1.5\ GPa$.

The comparison between the theoretical values and the experimental data for $[\pm 60]_s$ and $[\pm 45]_s$ laminates appears on Fig. 4.

As for $[0]_s$ and $[90]_s$ laminates, storage modulus for $[\pm 45]_s$ and $[\pm 60]_s$ laminates seems to depend slightly on the frequency (a). The difference between theoretical values and experimental data is relatively small for the $[\pm 60]_s$ laminate, but it is around 12% for $[\pm 45]_s$ at kHz. When the frequency increases, this difference tends to reduce to less than the experimental standard deviation.

Loss factor curves follow a non-linear law, with a stabilization beyond 1 kHz (b). Once again, the difference between the theoretical curves and experimental values is lower for $[\pm 60]_s$, but with increasing frequency, it remains within the standard deviation for $[\pm 45]_s$ laminate

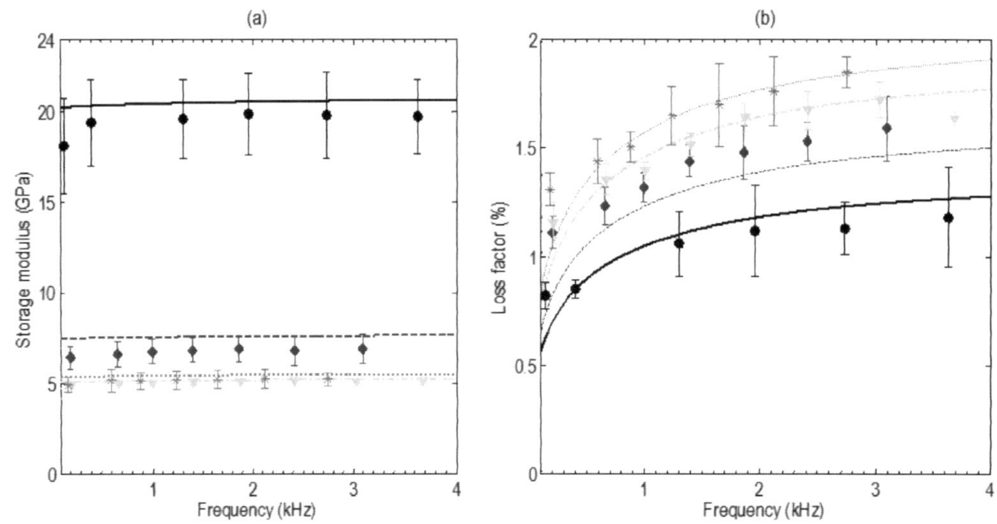

Figure 4. Experimental and analytic frequency dependence of E_L (a) and η (b) for $[0]_4$ (—: model, ●: experimental data), $[90]_4$ (---: model, *: experimental data), $[\pm 45]_s$ (·····:Model, ♦: experimental data) and $[\pm 60]_s$ (– · – : model, ▼ experimental data)

6. Results and discussion

An experimental vibration technique for fast identification of the dissipative properties of flax fibres reinforced epoxy composites has been presented. The characterization method allowed the measurement of Young's modulus and loss factor of beams in a large frequency band (10 Hz to 4 kHz).
Parameters of the fractional derivative Zener models identified in fibre and in-plane transverse directions of UD composites have been used to predict the frequency evolution of both moduli and loss factors. Based on the elastic-viscoelastic correspondence principle, CLT has been successfully used to predict the linear viscoelastic behaviour of different composite laminates. This prediction method was validated by comparison with the experimental results obtained for $[0]_4$, $[\pm 45]_s$, $[\pm 60]_s$ and $[90]_4$. laminates.

References

[1] F. Duc, P. E. Bourban and J. A. Månson, "The role of twist and crimp on the vibration behaviour of flax fibre composites," Composites Science and Technology, vol. 102, pp. 94-99, 2014.A reference
[2] S. Liang, P. B. Gning and L. Guillaumat, "A comparative study of fatigue behaviour of flax/epoxy and glass/epoxy composites," Composites Science and Technology, pp. 535-543, 2012.
[3] M. Berthelot, M. Assarar, Y. Sefrani and A. El Mahi, "Damping analysis of composite materials and structures," Composite Structures, vol. 85, no. 3, pp. 189-204, 2008.
[4] R. G. Ni, D. X. Lin and R. D. Adams, "The dynamic properties of carbon-glass fibre sandwich-laminated composites: theoretical, experimental and economic considerations," Composites, vol. 15, no. 4, pp. 297-304, 1984.
[5] A. B. Schultz and S. W. Tsai, "Measurements of complex dynamic moduli for laminated fiber-reinforced composites," Journal of Composite Materials, vol. 3, no. 3, pp. 434-443, 1969
[6] R. M. Crane and J. W. Gillespie, "Characterization of the vibration damping loss factor of glass and graphite fiber composites," Composites science and technology, vol. 40, no. 4, pp. 355-375, 1991.
[7] N. W. Tschoegl, The Phenomenological Theory of Linear Viscoelastic Behaviour: An Introduction, New York: Springer, 1989
[8] F. Duc, P. Bourban, C. Plummer and J.-A. Månson., "Damping of thermoset and thermoplastic flax fibre composites," Composites: Part A, vol. 64, pp. 115-123, 2014.
[9] Z. Hashin, "Complex moduli of viscoelastic composites fiber reinforced materials," International Journal of Solids and Structures, vol. 6, no. 6, pp. 797-807, 1970.

The transport forecast – an important stage of transport management

Vasile DRAGU[1], Oana DINU[1], Cristina OPREA[1], Eugenia Alina ROMAN[1]
[1]University Politehnica of Bucharest, Faculty of Transports, Transport, traffic and logistics Department, Splaiul Independenţei, No. 313, 060042, Bucharest, Romania

v_dragu@yahoo.com

Abstract. The transport system is a powerful system with varying loads in operation coming from changes in freight and passenger traffic in different time periods. The variations are due to the specific conditions of organization and development of socio-economic activities. The causes of varying loads can be included in three groups: economic, technical and organizational. The assessing of transport demand variability leads to proper forecast and development of the transport system, knowing that the market price is determined on equilibrium between supply and demand. The reduction of transport demand variability through different technical solutions, organizational, administrative, legislative leads to an increase in the efficiency and effectiveness of transport. The paper presents a new way of assessing the future needs of transport through dynamic series. Both researchers and practitioners in transport planning can benefit from the research results. This paper aims to analyze in an original approach how a good transport forecast can lead to a better management in transport, with significant effects on transport demand full meeting in quality terms. The case study shows how dynamic series of statistics can be used to identify the size of future demand addressed to the transport system.

1. Introduction

Transport management is differentiated from industrial management by the fact that the system is subject to variable demands over time which can often not be accurately predicted and as such, satisfying the demand for transport can be quantitatively and qualitatively affected.

The specificity of socio-economic activities leads to a demand for transport that has different time attributes and manifests itself differently in time and space. The non-uniformity of traffic over time has many causes. Traffic peaks in in freight transport can be identified in agricultural campaigns and traffic peaks in passenger transport can be identified during holidays, legal holidays, cultural, sporting, religious events or at different intervals of the day, etc.

Non-uniformity over time leads to difficulties in scheduling the park of vehicles to satisfy all transport demands. If the traffic would be uniform, forecasting and scheduling would be extremely easy for programmers and decision makers in transport. As this does not happen, it is attempted to optimize the programming so as to meet the schedule with a minimum vehicle park and a minimum number of board staff, taking into consideration some restrictions, both for vehicles and staff [1].

Non-uniformity over time can be *seasonal, daily and hourly* [2]. Ensuring traffic uniformity is based on the results obtained in previous periods, by pertinent traffic forecasts that are as accurate as possible to the values of the regression indicators.

To study the effects of the causes of the non-uniformity of transport, it is necessary to group them into categories. Starting from this criterion, the causes that produce variable exploitation demands were classified as economic, technical and organizational [3]. Economic causes have as a source the variation in the volume of traffic due to the increase in the income of the population and the share of free time, the seasonal nature of the holiday periods, or the change in the links between the travel's origin and destination areas. Technical causes have as a source the variations of traffic currents, both in size and rhythm, on shipment from point of origin or from intermodal nodes. Organizational causes result from non-uniformities due to discontinuous working regime of economical agents or due to interruptions in traffic.

Non-uniformity of freight and passenger transport influences all aspects of the foresight and planning activity of operators in the field. Above all, the unevenness of transport has a considerable influence on the level of the offered transport capacity level. The correct assessment of the size of the required transport capacity lies at the basis of the design of all the necessary constructions and facilities, and therefore the volume of investments and the deadlines for their recovery depend on it.

In order to solve the problems related to the size of the technical equipment of the transport operators and the improvement of the technologies, it is necessary to ensure the correlation between the tasks and the possibilities, namely between the effective and the maximum capacity, for which the total operating expenses are minimal.

On the transport market the level of traffic is set to the balance between transport offer and transport demand [4]. From the transport system – activity system interaction, the resulted traffic is characterized by the volume V_0 and the level of the quality of the insured services, t_0 (travel time), determined as solutions of the equilibrium between the offer (curve marked with O) and demand (curve marked with D), as shown in figure 1.

The decision to improve the transport system changes the transport offer from O_0 to O_1 (fig. 1), so there is an increase in the volume of traffic in the system $\Delta V = V_1 - V_0$, called *induced traffic*. Most often, the change of the offer from O_0 in O_1 leads to a service quality level t_1 lower than the anticipated t_0, in the hypothesis that traffic is maintained at the same level.

Improving the transport system by moving the offer curve from O_0 to O_1, stimulates the travel demand whose curve is transferred from D_0 to D_1 (fig. 1). The equilibrium traffic C2 corresponds to a travel time $t_2 < t_0$, which can stimulate the demand for transport. The new equilibrium (t_3, v_3) with $t_3 > t_0$ is no longer able to stimulate the further growth of transport demand.

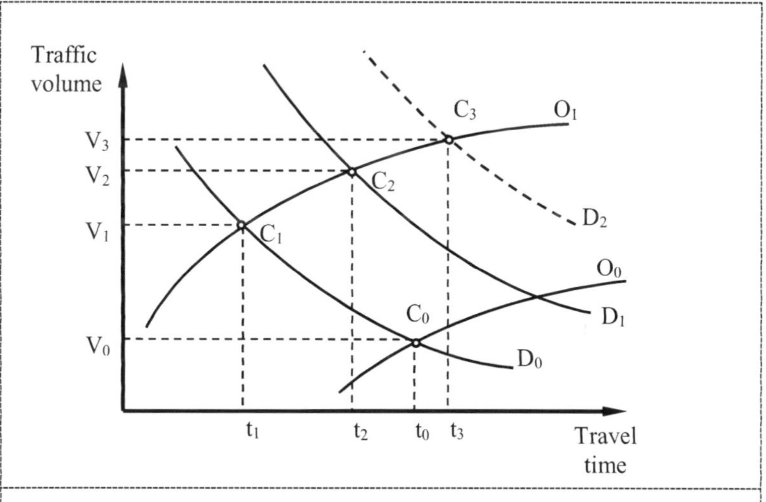

Figure 1. The dynamics of demand-offer equilibrium (source: [4])

From the situation presented in figure 1, there is a dynamic of the equilibrium between transport demand and transport offer, which has both positive and negative effects. The positive effects result from the continuous take-overs of new transport tasks, and the negative ones due to the externalities

caused by the worsening of the land use and all the undesirable effects that result from the proliferation of traffic [4, 5].
The transport and traffic forecast, as the initial stage of transport management, must be able to capture all these moments of changing the dynamic supply-demand equilibrium so that the transport system is prepared to successfully cope with all these dynamics so that all demands addressed to it to be satisfied in superior quality conditions.

2. Aspects of transport forecast

The forecast is defined as estimating the evolution of processes and future phenomena, the positive and negative effects they can generate on the managed system as well as the different strategies and scenarios of action for minimizing risks and maximizing the degree of achievement of the pursued objectives [6]. The forecast is made through diagnostic, forecasting and planning techniques.
The forecast models can be framed, as well as management types, over how long the concepts are achievable with great probability, in: strategic, tactical, and operative. Planning, as a management function (in English managerial literature, the foresight function is presented under the name of Planning), has been distorted in the centralized command economy, in which the essential relationships between the sectors of the economy and the functioning of the market was neglected, by neglecting the supply and demand law and by quasi-total elimination of the economic competition [7].
In the field of transport, abandoning planning was equivalent to limiting the state intervention in the field, which led to a fierce competition of the transport modes in both road and rail transport. The lack of regulation, and especially the lack of coherence in the applied pricing, has led from the complementarity of transport modes to competition, even where the areas of interest and competence clearly belong to one mode of transport.
The forecast, as a function of management, consists in the continuous attempt of building a more realistic picture of the future that corresponds to the achievement of the proposed objectives of the management. The degree to which the prefigured future becomes reality is the true confirmation of the quality of the exercise of the foresight function.
The foresight function in transport is equivalent to how the system needs to be prepared to meet the demands of the society in terms of quantity and quality. This is a problem of balance in the context of a permanent dynamic of supply and demand that must constantly adapt to the formulated demands. Knowing the socio-economic evolutions can lead to economic sizing of transport system's constructions and installations to satisfy the demand with minimal resource consumption. It is known that the greatest waste of resources is encountered in transportation, whether due to internal causes that are not sufficiently studied (management of empty vehicles, systemic incompatibilities, geographic separation requiring different transhipment / gauge, correlation of schedules, interdependence in traffic charts, etc.), or due to external causes and environmental influences (goods that are not on time, non-respecting the schedules, unfavourable weather conditions, quantities of goods not in line with initial planning etc.).
The necessary data in the forecast are data regarding the evolution of the system and the factors that determined it. The used methods are those specific to statistical analysis, in uncertainty conditions, which allow to establish probable areas of values of some indicators, based on dynamic series analysis, regression analysis etc. [7]. Unfortunately, there is no way to determine by objective calculation the type of curve that best approximates the dynamic series available [8]:
The choice of the type of function that most accurately describes the known values is made by the curve aspect obtained by graphic representation of the observed values taking the time on the abscissa. After choosing the function, its coefficients are calculated by the least squares method called the regression method.
The chosen theoretical curve best describes the observed values when the sum of squares of the differences between the observed values and the calculated theoretical values is minimal. But the sum is minimal when the first derivative is zero, so, by doing the partial derivatives with respect to each of the unknown parameters, a system of equations from which these parameters can be calculated is obtained. The right to use the least squares method is demonstrated by the principle of maximum verisimilitude from probability theory.

Thus, for the case of the linear distribution having the form $y = a_0 + a_1 t$, the value of the model coefficients is calculated with the relations [9, 10]:

$$a_0 = \frac{\sum_{i=1}^{m} y_i}{m} \text{ and } a_1 = \frac{\sum_{i=1}^{m} t_i y_i}{\sum_{i=1}^{m} t_i^2} \tag{1}$$

and for the case of the parabolic distribution having the form $y = a_0 + a_1 t + a_2 t^2$ the coefficients of the model are calculated with the relations:

$$a_0 = \frac{\sum_{i=1}^{m} t_i^4 \sum_{i=1}^{m} y_i - \sum_{i=1}^{m} t_i^2 \sum_{i=1}^{m} t_i^2 y_i}{m\sum_{i=1}^{m} t_i^4 - \left(\sum_{i=1}^{m} t_i^2\right)^2}; \quad a_2 = \frac{m\sum_{i=1}^{m} t_i^2 y_i - \sum_{i=1}^{m} t_i^2 \sum_{i=1}^{m} y_i}{m\sum_{i=1}^{m} t_i^4 - \left(\sum_{i=1}^{m} t_i^2\right)^2} \tag{2}$$

with the observation that the expression of the coefficient a_1 is the same as in the linear model and the relations are valid for the case where the origin of the time is chosen so that the relation $\sum_{i=1}^{m} t_i = 0$ is true.

The measure of the intensity of the link between the y_i and t_i is given by the correlation coefficient (r) for the linear model and by the correlation ratio (η) for the parabolic model.

$$r = \frac{m\sum_{i=1}^{m} t_i y_i}{\sqrt{m\sum_{i=1}^{m} t_i^2 \left[m\sum_{i=1}^{m} y_i^2 - \left(\sum_{i=1}^{m} y_i\right)^2\right]}}; \qquad \eta = \sqrt{1 - \frac{\sum_{i=1}^{m} (y_i - y^*)^2}{\sum_{i=1}^{m} (y_i - \bar{y})^2}} \tag{3}$$

where y_i represent the observed values; y^* – values calculated by the adjustment function; $\bar{y}$ – the average of observed values.

The exponential model can be written as a linear model by logarithm.

$$\ln y = \ln a + bt \tag{4}$$

We note $\ln y = Y$; $\ln a = A$; $b = B$ and we obtain:

$$Y = A + Bt \tag{5}$$

$\sum_{i=1}^{m} t_i = 0$ and the coefficients a and b are calculated with the expressions:

$$\ln a = \frac{\sum_{i=1}^{m} \ln y_i}{m} \Rightarrow a = e^{\frac{\sum_{i=1}^{m} \ln y_i}{m}}; \quad b = \frac{\sum_{i=1}^{m} t_i \ln y_i}{\sum_{i=1}^{m} t_i^2} \tag{6}$$

The quality of the exponential model adjustment is given, as in the parabolic model, by the correlation ratio η.

Another model used for forecasts is the logistic one also called the *Pearl Reed* curve. The logistic model has the expression [11, 12]:

$$y = \frac{a}{1 + be^{-ct}} \tag{7}$$

having the coefficients a, b and c fulfilling the condition $a, b, c > 0$. The value a is assimilated to the saturation limit.The model is used in the study of the evolution of the population, in estimating the demand for durable goods, in ecology, in vehicle fleet evolution, and in general, in modelling phenomena that have a saturation limit over time (fig. 2).
Figure 2 shows that the logistics function is always positive and monotonous ascending. Parameters a, b, and c of the logistic curve are calculated according to the rule from Table 1 and to the condition $y_1 < y_2 < y_3$, if three points in the equidistant dynamic series are known.

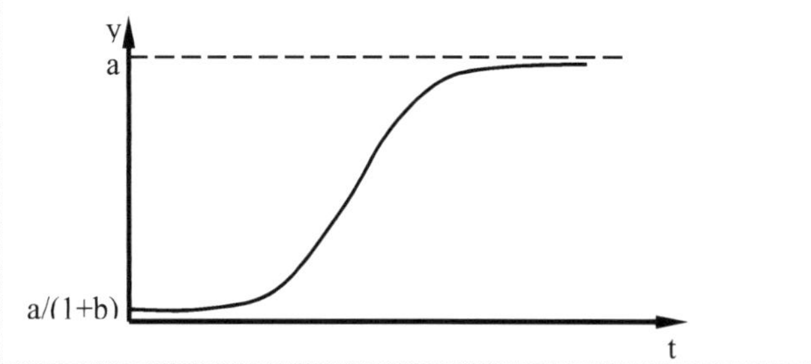

Figure 2. Variation of logistics function

In Table 1, t is the number of values in the dynamic series that separates two consecutive values as y_i. It is recommended that the y_i values to be chosen so that y_3 is the last value in the chronological series and y_1 is as close as possible in time to the first value of the series.

Table 1. The necessary conditions for determining the parameters of the logistic function

t	0	t	2t
y	y_1	y_2	y_3

The coefficients from the function of the forecasting model (a, b and c) are determined with the relations:

$$a = \frac{2y_1y_2y_3 - y_2^2(y_1 + y_3)}{y_1y_3 - y_2^2};\ b = \frac{a - y_1}{y_1};\ c = \frac{1}{t} ln \frac{y_2(a - y_1)}{y_1(a - y_2)} \tag{8}$$

Another model used for forecasts is the production function model. The most known model is the one proposed by R. Cobb and P. Douglas [13], which has the form:

$$Q = a\, M^{\alpha}\, F^{\beta}\, r \tag{9}$$

where: a, α and β are the parameters of the model; Q – production volume; M – the volume of labour; F – the volume of fixed assets; R – residue of estimation as an average of the differences between recorded and calculated values.
The calculation of function parameters is made by logging the expression that defines it and then by the least squares method the searched values are obtained.

$$\ln Q = \ln a + \alpha \ln M + \beta \ln F \tag{10}$$

We note: $ln Q = q$, $ln a = A$, $ln M = t$ şi $ln F = f$ (11)

Using the least squares method we obtain:

$$\begin{cases} \sum_{i=1}^{m} q_i = mA + \alpha \sum_{i=1}^{m} t_i + \beta \sum_{i=1}^{m} f_i \\ \sum_{i=1}^{m} t_i q_i = A \sum_{i=1}^{m} t_i + \alpha \sum_{i=1}^{m} t_i^2 + \beta \sum_{i=1}^{m} t_i f_i \\ \sum_{i=1}^{m} f_i q_i = A \sum_{i=1}^{m} f_i + \alpha \sum_{i=1}^{m} t_i f_i + \beta \sum_{i=1}^{m} f_i^2 \end{cases} \quad (12)$$

Evolution models based on extrapolation of the trend can be used for short and medium term forecasts. Each time it is mandatory to estimate the probable error by attaching to the forecast a confidence interval and a threshold of significance as to any other statistical approach.

3. Case study

The evolution in time of the volume of goods transported for a certain transport relationship is known (table 2).

Table 2. Evolution of the volume of transported goods

Year	2009	2010	2011	2012	2013	2014	2015	2016
Quantity × 10^3[t]	142	150	158	162	165	177	190	202

Considering that the evolution of the socio-economic factors will be preserved in the next forecasting period, the most appropriate expression should be established which will approximate in the future the present values of the volumes of transported goods. Parameters of linear, parabolic, exponential and logistic models are determined for this. These models were chosen because the time evolution of the data follows an ascending curve. The necessary elements for calculating the parameters of the linear and parabolic model are presented in Table 3.

For the linear model, a_0, a_1 and r are determined using the equations (1) and (3). The obtained values are: $a_0 = 168{,}25$; $a_1 = 4{,}05$ si $r = 0{,}980063$.

For the parabolic model a_0, a_1, a_2 and η are determined using the equations (1), (2) and (3). The obtained values are: $a_0 = 164{,}875$; $a_1 = 4{,}05$; $a_2 = 0{,}1607$; $\eta = 0{,}992347$. It is noticed that $\eta > r$ which leads to the conclusion that the parabolic model approximates better the given values.

Table 3 The required elements to calculate the parameters of the linear and parabolic model

t_i	y_i	t_i^2	$t_i y_i$	y_i^2	t_i^4	$t_i^2 y_i$	y^*	$(y_i - y^*)^2$	$(y_i - \bar{y})^2$
-7	142	49	-994	20164	2401	6958	144,40	5,76	689,06
-5	150	25	-750	22500	625	3750	148,64	1,84	333,06
-3	158	9	-474	24964	81	1422	154,17	14,66	105,06
-1	162	1	-162	26244	1	162	160,99	1,03	39,06
1	165	1	165	27225	1	165	169,09	16,69	10,56
3	177	9	531	31329	81	1593	178,47	2,16	76,56
5	190	25	950	36100	625	4750	189,14	0,74	473,06
7	202	49	1414	40804	2401	9898	201,10	0,81	1139,06
Σ	1346	168	680	229330	6216	28698	1346,00	43,69	2865,5

However, the linear model also offers a good approximation, since the value of the correlation coefficient is very close to the unit. The elements required for calculating the exponential model parameters are shown in Table 4.

Table 4 The elements required for calculating the exponential model parameters

t_i	y_i	t_i^2	$\ln y$	$t_i \ln y_i$	y_i^*	$y_i - y_i^*$	$(y_i - y_i^*)^2$	$y_i - \overline{y}$	$(y_i - \overline{y})^2$
-7	142	49	4,956	-34,691	141,074	0,926	0,858	-26,25	689,063
-5	150	25	5,011	-25,053	148,007	1,993	3,970	-18,25	333,063
-3	158	9	5,063	-15,188	155,282	2,718	7,387	-10,25	105,063
-1	162	1	5,088	-5,088	162,914	-0,914	0,836	-6,25	39,063
1	165	1	5,106	5,106	170,922	-5,922	35,066	-3,25	10,563
3	177	9	5,176	15,528	179,323	-2,323	5,394	8,75	76,563
5	190	25	5,247	26,235	188,136	1,864	3,473	21,75	473,063
7	202	49	5,308	37,158	197,383	4,617	21,313	33,75	1139,063
Σ	1346	168	40,954	4,008	1343,042	2,958	78,298	0,00	2865,500

For the exponential model, *a*, *b* and η are determined using the equations (6) and (3).
The values a = 166,87; b = 0,023857 si η = 0,986243 are obtained. The function is $y = a\,e^{bt} \Rightarrow y = 166{,}87 \cdot e^{0{,}024\,t}$.
The attempt to model the data in Table 2 through a logistic function fails because, for the coefficients in the function of the forecasting model (*a, b, c*), we obtain the values: a = 135,22; b = - 0,09854; c = - 0,20174, values that cannot represent a pertinent logistic function because both for *b* and for *c*, negative values have been obtained.
Table 5 shows the evolution of the park of vehicles in circulation for eight consecutive years.

Tabelul 5 The evolution of the park of vehicles

t_i	-1	0	1	2	3	4	5	6
y_i	70	76	84	92	107	116	130	144

Using the values from Table 5, where the origin of time was established, the parameters *a*, *b* and *c* of the logistic model are determined. The values that are chosen are y_3 = 144, y_2 = 107 și y_1 = 76, for t = 3. For the logistic model we determine *a*, *b* and *c* using the equation (8). The obtained values are: a = 350,028; b = 3,61 and c = 0,154. The logistics function is defined as follows: $y = \frac{350{,}028}{1 + 3{,}61\,e^{-0{,}154}}$

The asymptomatic tendency towards *a* value is demonstrated using the data listed in table 6.

Table 6 Future values of the series obtained with the logistics function

t_i	40	45	50	60	70	80	90	100
y_i	347,38	348,80	349,46	349,91	350,01	350,024	350,026	350,028

Table 6 shows that series values are asymptomatically approaching the value of *a* (*a* = 350,028) and beyond the value t = 70, the increases are insignificant.
Table 7 shows the differences between the recorded values and those approximated by the logistics function.

Table 7 The known values of the series determined using the logistics function

t_i	-1	0	1	2	3	4	5	6
y_i	70	76	84	92	107	116	130	144
y_i^{cal}	67,17	75,92	85,48	95,81	106,89	118,65	131,02	143,87
$\frac{y_i^{cal} - y_i}{y_i} \cdot 100$	−4,28	−0,11	1,76	4,14	−0,10	2,28	0,708	−0,09

Table 7 shows that the differences between the known values of the series and those approximated by the logistic function are between - 4,28% and -0,09% as negative values and 4,14% and 0,78% as positive values. So, both positive and negative differences are below the ± 5% threshold.

4. Conclusions

The relatively rapid variations of the demand, both in size and structure, addressed to the transport system, lead to the need of knowing with anticipation and security the future developments. Eliminating or reducing traffic non-uniformities is an essential task for transport planners and decision-makers from all levels involved in the socio-economic activity that generates travel needs. Through these actions, the temporal variations of the demand will be known and the transport system will be prepared to meet the demands in conditions of safety, security and quality.

The paper presented the most known predictive models (polynomial, exponential, logarithmic, trigonometric polynomial, Cobb Douglas and logistic model) and carried out a case study for real transport data.

The presented models are useful in forming an image of the future evolution of the systems, but the expected results must be carefully considered and analysed, considering that one of the conditions of application is the preservation of present and future socio-economic conditions. This is the main criticism of the prediction models - *the fact that the past is prolonging in the future* - which makes them vulnerable to changes in the external environment of the analysed system.

Unlike the exponential adjustment model, which grows infinitely, the logistic model has a finite growth. This curve is important because it models the transport growth, many of which have finite growths.

Example: increased traffic with a finite limit dictated by the capacity of existing or even designed arteries and parking facilities. The increase in the volume of goods transported in any of the transport systems also has a limit: the increase in vehicle park or, more recently, dependence of the speed of density in traffic flow theory. There is even a reverse logistic function that has asymptotes at the top and bottom, known as the speed, even in free flow mode and has finite values.

Acknowledgement

This work has been funded by University Politehnica of Bucharest, through the "Excellence Research Grants" Program, UPB – GEEX. Identifier: UPB–EXCELENȚĂ–2016 Research project title MODELING AND SIMULATION OF INTERMODAL TRANSPORT AND URBAN LOGISTICS, Contract number 346.

5. References

[1] Dragu V 2003 Management specific in transport, *Management in the 21st Century*, pp.188-193
[2] Șerban D 1988 *Rail traffic and transport coordination*, ed IPB (Bucharest) p 360
[3] Raicu S and Mașala Gh 1981 *Rail transport. Operation – Development - Efficiency*, ed. Stiintifica și enciclopedica, Bucharest p 317
[4] Raicu S 2007 *Transport systems*, ed. AGIR, Bucharest p 484
[5] Dekker S and Verhaeghe R A 2006 Modelling Approach for Integrated Planning of Port Capacity-Trade-Offs in Rotterdam Investment Planning, *Promet-Traffic&Transportation* vol. 18, no. 2, pp 53-58
[6] Tănăsuică I 1998 *Logistics and Transportation Engineering Management*, ed. Matrix Rom, Bucharest p 280
[7] Rădăceanu E 2008 *Successful management. Methods-Realities-Perspectives*, ed. AGIR, Bucharest p 352
[8] Giurgiu N et al. 1980 *Statistics* ed. Babes-Bolyai University, Cluj-Napoca p 274
[9] Dragu V 2004 *Techniques and models for estimating journeys in the urban transport system*, ed. BREN, Bucharest p 192
[10] Dragu V, Popa M, Stefanica C 2015 *Road traffic I*, ed. Printech, Bucharest p 127
[11] Todoran I 1989 *Possible answers - Correlation and forecast*, ed Dacia, Cluj Napoca p 322
[12] Neacșu G 2009 *Concepts and methods used in statistics*, ed. Universitara, Bucharest p 182
[13] Douglas P and Cobb R 1928 A Theory of Production, *American Review - Papers and Proceedings* vol 18, pp. 139 – 165

IOP Conf. Series: Materials Science and Engineering **252** (2017) 012051 doi:10.1088/1757-899X/252/1/012051

Smart Transportation CO_2 Emission Reduction Strategies

S Tarulescu[1], R Tarulescu[1], A Soica[1] and C I Leahu[1]
[1]Transilvania University of Brasov, Eroilor Boulevard Nr. 29, Romania

Abstract. Transport represents the sector with the fastest growing greenhouse gas emissions around the world. The main global objective is to reduce energy usage and associated greenhouse gas emissions from the transportation sector. For this study it was analyzed the road transportation system from Brasov Metropolitan area. The study was made for the transportation route that connects Ghimbav city to the main surrounding objectives. In this study ware considered four optimization measures: vehicle fleet renewal; building the detour belt for the city; road increasing the average travel speed; making bicycle lanes; and implementing an urban public transport system for Ghimbav. For each measure it was used a mathematical model to calculate the energy consumption and carbon emissions from the road transportation sector. After all four measures was analyzed is calculated the general energy consumption and CO_2 reduction if this are applied from year 2017 to 2020.

1. Introduction

The cities from Europe are centres of many activities: economic, social and innovation. Many of them face many challenges to their mobility systems: traffic jams, poor air quality, noise pollution, green house gases emissions and road accidents. All of these have significant negative impacts over the environment, human health and economic performance of cities. These problems will increase in intensity in the future, because, the cities continue to grow [1].

Urban transport systems are important elements of the European transport system and are a concern for the European Common Transport Policy. The 2011 Transport White Paper sets very ambitious targets to address these challenges, especially for Central European Countries. In this case, there is a wide range of instruments available to implement the policy efforts: efficient land usage, urban planning, pricing policy, infrastructure for non-motorised vehicles, charging/refuelling of clean vehicles to reduce emissions, 100% green transportation systems and infrastructure upgrades. These instruments should be part of a meaningful strategy in order to develop cost-effective interventions.

The target is the year 2030, in order to meet the Transport White Paper objectives for urban transportation [1].

The development of the roadmaps helps to recognise that different types of policy measures are appropriate for different types of cities. The analysis of scenarios and roadmaps help to assess the impact of the European urban mobility policies in the long term [1].

In Romania transport policy aims to align the national transport system continues to Community Transport Policy principles defined in the Transport White Paper (with corresponding updates) and the requirements of sustainable development of Romania [2].

For this study it was analyzed the road transportation system from Brasov Metropolitan area. The study was made for the transportation route that connects Ghimbav city to the main surrounding objectives. This region faces many mobility problems, because is situated in the centre of Romania, at the junction of several main transportation routes. This mobility problems affect Brasov city and all communities from the metropolitan zone. The city of Ghimbav is an important industrial, economic and residential city, despite its size (5200 inhabitants).

2. The analyzed area

For this study it was analyzed the transportation modes from Brasov Metropolitan Area, especially in the surroundings of Ghimbav city. The city of Ghimbav is located in the west of Brasov County, with the following geographic coordinates: 45 degrees 39 minutes wide. N; 25 degrees 36 minutes long. E. The road connections with these areas are insured by the national road network (DN1 and DN73B) and county road (DJ 103C).
This area is an important industrial zone, but also is an expanding residential zone. In the future in this area will be developed some important objectives, such as: International Airport Ghimbav-Brasov, Brasov-Comarnic highway, an agreement zone and more residential districts. The connections between the Ghimbav city centre and all this objectives will be made by road transportation (passengers and freights). In the figure 1 is presented the Ghimbav zone and its connections with actual and future objectives.

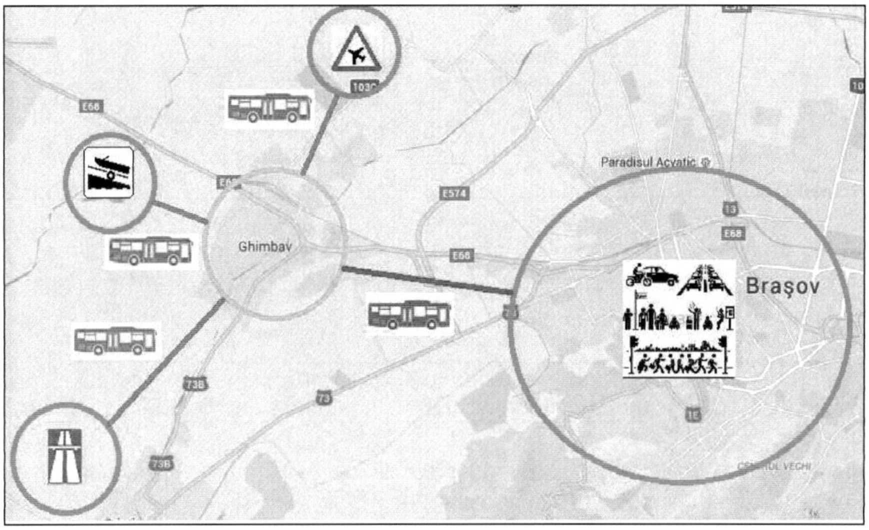

Figure 1. Analyzed area - Ghimbav city and its connections with actual and future objectives (Google Maps)

3. Energy consumption and CO_2 mitigation model for

3.1. Mathematical model for Ghimbav transportation system

For the city of Ghimbav it was made a prediction model in order to estimate the energy consumption diminish and the CO_2 amount resulted from road transportation. Three optimization measures ware considered: vehicle fleet renewal; making bicycle lanes; and green public transportation system implementation.
There were calculated the proportions of decrease for the following:

- Final energy consumption [MWh].
- Total CO_2 Emissions / CO_2 equivalent [6].

The study was start from initial data: Ghimbav vehicle fleet composition during 2008 - 2012; values expressed in tones of CO_2 produced in the 2008 -2012 period.
The prediction model uses traffic volumes data from 2010 - 2015 period in order to predict the traffic flows for the 2016 - 2020 period. Using existing data on vehicle fleet, the total energy consumption and CO_2 emissions for Ghimbav we can calculate the total energy and CO_2 emissions for 2016 - 2020

period. Following the mathematical modelling results the dependence of total energy consumption and amount of emitted CO_2 over the vehicle fleet evolution [6].

Without applying any measures, the energy consumption will increase in the following years. Equation (1) described by polynomial second degree regression curve, represents Energy consumption evolution up to 2020, for transportation sector (also in Figure 2):

$$X_{Energy} = 25{,}432 \cdot V_E^2 - 10217 \cdot V_E + 10^8 \quad (1)$$

Where: X_{Energy} is the approximate amount of the total energy consumed, and V_E is the number of etalon vehicles. For this function, the standard deviation (R^2) is 0.9963.

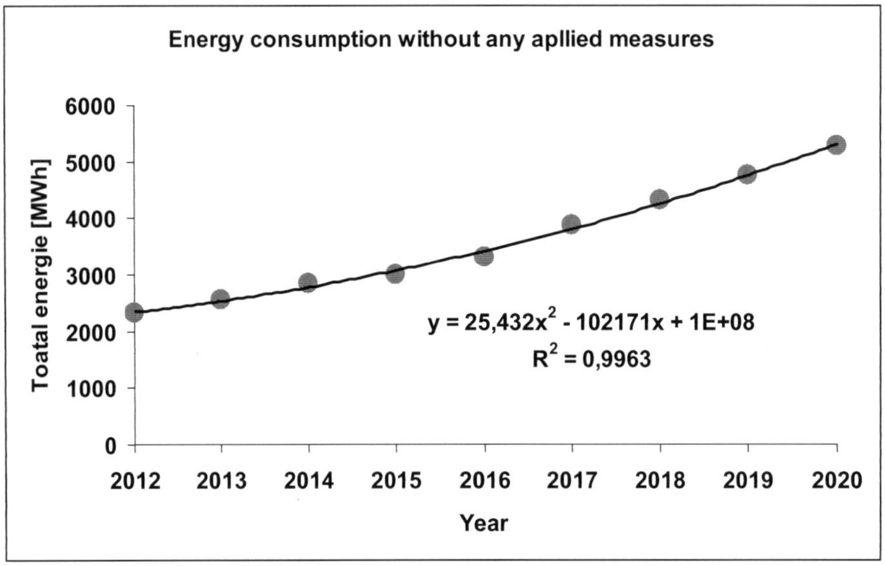

Figure 2. Energy consumption prevision for 2020 without any transportation applied measures

The CO_2 emission values will increase directly proportional with energy consumption. In order to decrease energy consumption and CO_2 emission values it is necessary to implement the measures mentioned above [8].

The standard emissions factors applied in this study are:

- 0.249 [tones CO_2 / MWh] - for spark ignition engines;
- 0.267 [tones CO_2 / MWh] - for compression ignition engines;
- 0.000 [tones CO_2 / MWh] - for biofuels [6].

For the transportation sector, data for each fuel and vehicle were calculated using the formula:

$$X_{UsedFuel} = D \cdot C_{average} \cdot Y \quad (2)$$

Where: $X_{UsedFuel}$ - Energy consumption for road transportation [kWh]; D - route distance [km]; $C_{average}$ - average fuel consumption [liters/km]; Y - conversion factor for each fuel [kWh/liters]. Y = 9.2 for gasoline and Y = 10 for Diesel [8].

3.2. Vehicle fleet renewal

This measure can only be applied to the municipal and local transport fleet. Using the data on the number of municipal vehicles fleet of Ghimbav city and from related fuel consumptions, result the following [6].

Fleet renewal for municipal fleet will lead to a consumption decrease for: from 10 liters/100 km to 8 liters/100 km for light vehicles and from 30 liters/ 00 km to 25 liters/100 km for heavy vehicles. After applying these measures will have a reduction: 448.97 [MWh] and 117,84 [tones CO_2].

3.3. Bicycle lanes implementation

Cycling has a major role for mobility quality in metropolitan areas. By using non-motored vehicles, congestions will be lighter and local air pollution and emissions of greenhouse gases will decrease. About 20% of car trips are less than 3 kilometres, a distance that can be easily made by the bike in less than 15 minutes [5]. The study is based on approximate traffic volumes decrease by creating bicycle lanes in Ghimbav city and on the main roads that connects Ghimbav with the stranding cities (Brasov, Codlea and Cristian).

Figure 3. Bicycle lanes implantation strategy: bicycle lane example (left); bicycle lanes strategy in Ghimbav city for 2020

According to the analyzed data in this study for bicycle lanes, resulted a decrease of 8% road traffic volumes after the construction of a bicycle lane in Ghimbav. After applying this percentage reduction for total vehicle we have a reduction of: 1015.28 [MWh] and 260,82 [tones CO_2] [6].

3.4. Implementing a green urban public transport system for Ghimbav

Public transportation systems, including electric buses, contributes to cleaner air and quality of life in cities and reduces green house gas emissions, as it uses up fewer resources and emits less CO_2 than other transportation modes. Compared to private motorised mobility, public transportation makes the most efficient and equitable use of scarce resources, such as fuel and public space. By occupying less space and generating fewer emissions per passenger, public transport enhances the quality of life in cities. The full potential can be exploited in the right conditions, when buses operate on reserved priority lanes and with prioritization measures [3].
Also, implementing a public transportation with electric buses will lead to a noise level diminish in city area. Road traffic noise may depend by several parameters: the traffic volume, the traffic flow typology, the vehicles typology, the road and pavement features, the speed. The influence of the speed on the noise is determined by two main factors: the interaction between tires and road and the power train noise, influenced by the engine speed. The power train noise will be significant decrease if electric buses are used [7].
For Ghimbav city was proposed two public buses routes. First route will cover the Ghimbav city centre, connections to schools, City Hall, commercial objectives and residential areas. The second route will cover residential areas, City Hall, industrial areas, Ghimbav Industrial Park, Aerotec an Airbus (in the perimeter of future International Airport) and commercial objectives.
For this paper it will be presented Route 1. The public transportation system parameters are presented in the following paragraphs:

- The total length of the route: 5,50 km.

- Total journey time: 30 minutes.
- Number of bus stations (stops) on the route: 11.
- The average bus stop time in stations: 1,2 minutes.
- Number of intersections (without priority) on the route: 7.
- The average stop time at intersections is 30 seconds.
- Buses speed limit on the route: 50 km/h.
- Cruising speed (average) of the route: 25 km/h.
- Peaks of passenger demand: Monday - Friday (working days): between 6:00 h – 10:00 h and 14:00 h – 18:00 h; Saturday: between 18:00 h – 22:00 h; Sunday: between 18:00 h – 22:00 h.
- Necessary buses: 3 peaces.

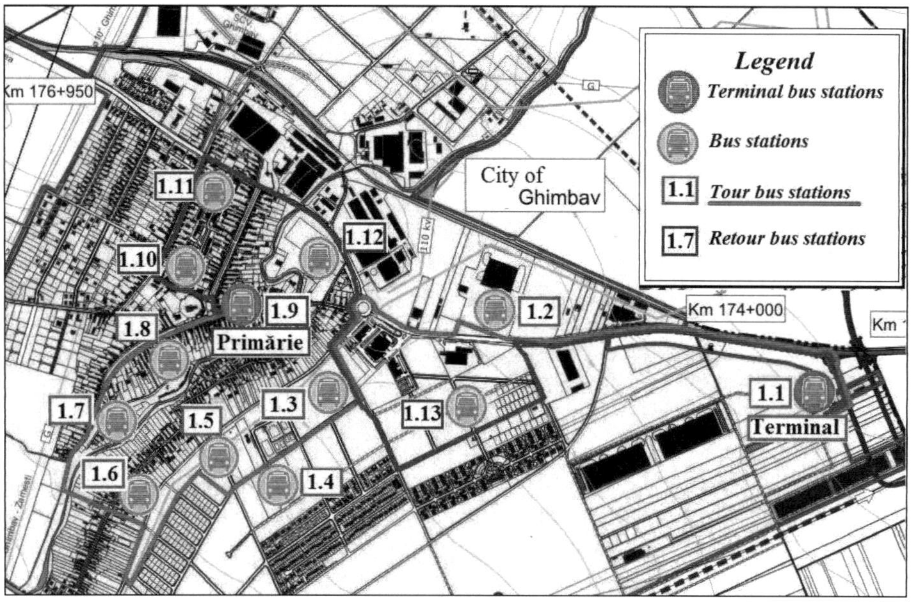

Figure 4. Electric public transportation system for Ghimbav - Route 1 presentation

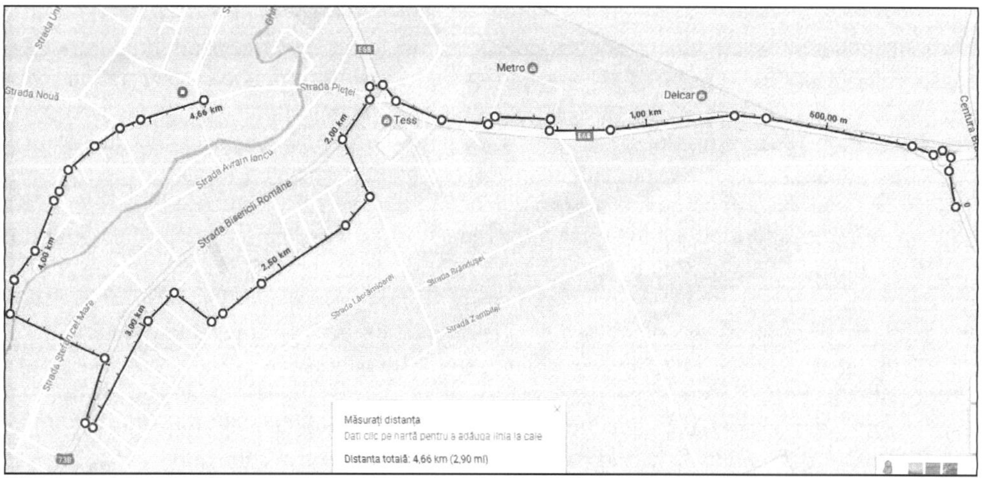

Figure 5. Electric bus Route 1 length (Google Maps)

The arrival frequency of bus stops in weekly hours is shown in the Table 1.

Table 1. Frequency buses arrival in station (hourly intervals)

Weekdays	*Hour interval*	*Time frame between consecutive arrivals*
Monday - Friday	22:00 h – 6:00 h	60 minutes
	10:00 h – 14:00 h şi 18:00 h – 22:00 h	20 minutes
	6:00 h – 10:00 h şi 14:00 h – 18:00 h	10 minutes
Saturday	22:00 h – 6:00 h	60 minutes
	6:00 h - 22:00 h	20 minutes
Sunday	22:00 h – 6:00 h	60 minutes
	6:00 h - 22:00 h	20 minutes

The transportation system will have an integrated management system in order to control very efficient the network. It can be used as an example the public transportation management system from Brasov. Necessary infrastructure for electric buses will be needed. Also, two terminal stations and 15 bus stations will be made.

Figure 6. The public transportation management system from RAT Brasov

The implementation of an urban public transport system serving the Ghimbav population will reduce the number of daily journeys made with personal cars in the following years as follows:

Table 2. The decrease percentage of local road traffic

Year	*2016*	*2017*	*2018*	*2019*	*2020*
The decrease percentage of local road traffic after public transportation implementation:	5 %	15 %	30 %	40 %	50 %

After applying this measure, we will have a reduction of: 1785,35 [MWh] and 458,64 [tones CO_2].

3.5. Estimated impact of analyzed measures in the future

In order to estimate the impact of implementing the proposed long-term measures, a calculation resource used in the European Union was used. This resource is Urban Transport Roadmaps framework (it is one of the tools used by the metropolitan development agencies). Using this framework, we can estimate the benefits of implementing the three measures for reducing energy consumption for Ghimbav, up to year 2030 (Figure 7, 8, 9 and 10) [4].

In Figure 7 is presented the evolution of carbon dioxide emissions after the implementation of the measures (reference year - 2017). On the left is anticipated the quantitative reduction of CO_2 from 2015 to 2030, after the implementation of the measures in 2017. On the right is the evolution of the CO_2 level produced by the studied region from 2015 to 2030, after applying the measures in 2017 (cumulated CO_2 level).

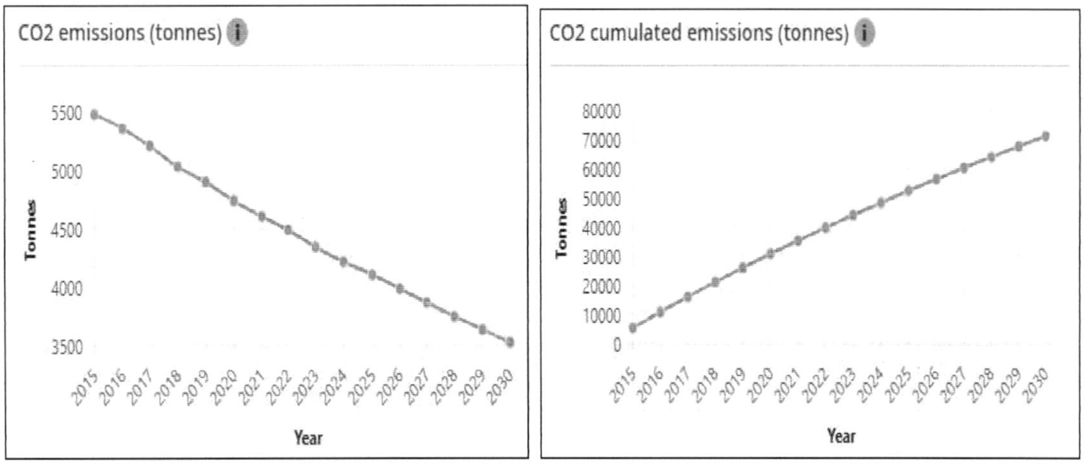

Figure 7. CO_2 emissions reduction after smart measures implementation [4]

In Figure 8 is presented the evolution of carbon monoxide pollution level after the implementation of the measures. On the left is anticipated the quantitative reduction of CO from 2015 to 2030, and on the right is the evolution of the CO level produced by the studied region, after applying the measures in 2017 (cumulated CO level).

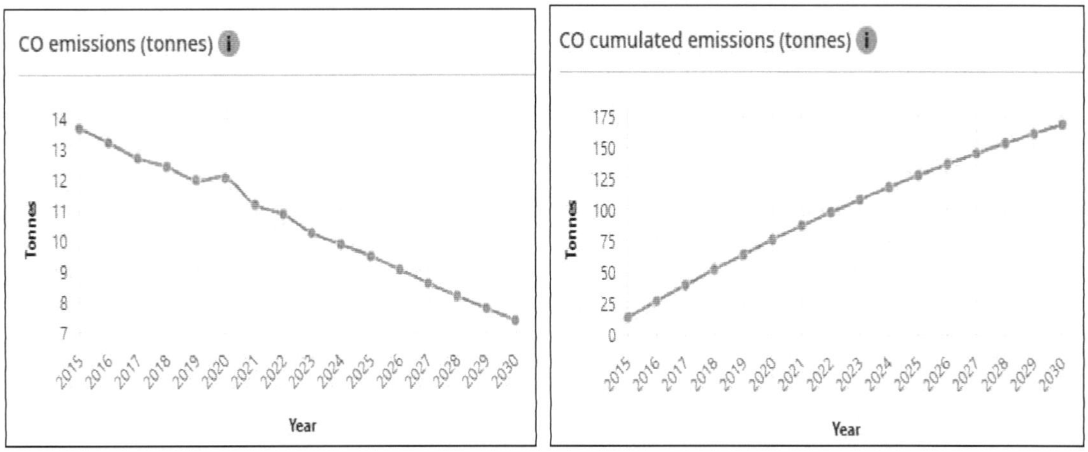

Figure 8. CO emissions reduction after smart measures implementation [4]

In Figure 9 is presented the evolution of particulate matter (PM) pollution level after the implementation of the measures. On the left is anticipated the quantitative reduction of PM from 2015 to 2030, and on the right is the evolution of the PM level produced by the studied region, after applying the measures in 2017 (cumulated PM level).

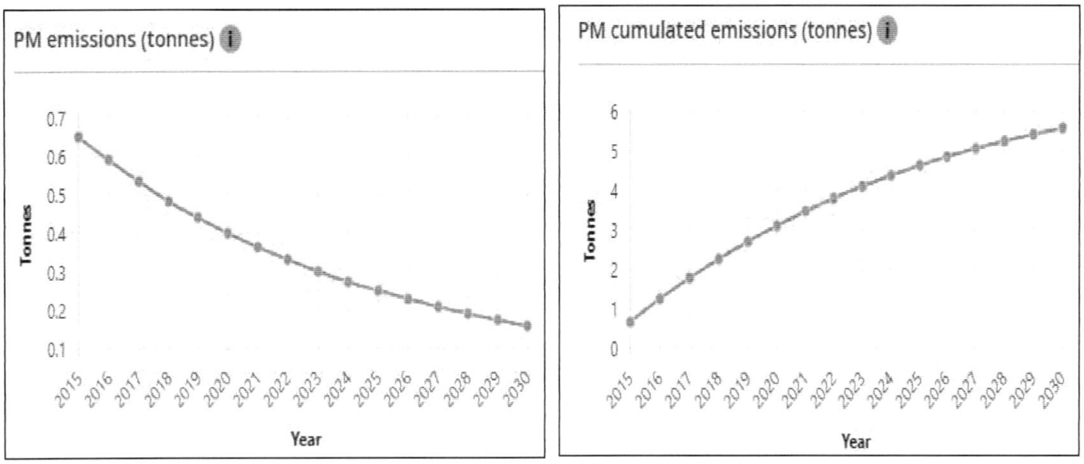

Figure 9. Particulate Matter (PM) emissions reduction after smart measures implementation [4]

The economy output after vehicle fleet renewal, making bicycle lanes and green public transportation system implementation is expressed in the Figure 10.

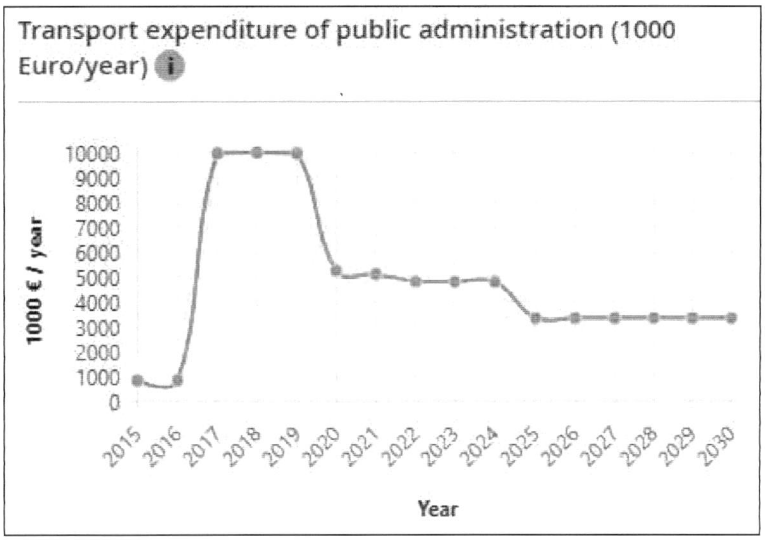

Figure 10. Transport expenditure of public administration after smart measure implementation for Ghimbav city [4]

4. Conclusions

For this study were analyzed three measures that can be adopted in order reduce energy consumption and CO_2 emissions: vehicle fleet renewal; making bicycle lanes; and green public transportation system implementation. The most effective measure, implementing a green urban public transport system for Ghimbav, will result in a decrease in energy consumption and CO_2 emissions by 19.71%. Making bicycle lanes will lead to lower energy consumption and CO_2 emissions by 11,21%. Vehicle fleet renewal strategy will lead to lower energy consumption and CO_2 emissions by 5%.

Using Urban Transport Roadmaps framework, we can estimate the benefits of implementing the three measures for reducing energy consumption for Ghimbav, up to year 2030. It can by noticed a substantial reduction of the CO_2, CO and PM emissions after implementation of the smart measures at

IOP Conf. Series: Materials Science and Engineering **252** (2017) 012051 doi:10.1088/1757-899X/252/1/012051

Ghimbav City. Also, the pollution level of the analyzed area will increase less after the implementation of these measures.

5. References

[1] C. de Stasio (TRT), D. Fiorello (TRT), F. Fermi (TRT), G. Hitchcock (Ricardo Energy & Environment), S. Kollamthodi (Ricard Energy & Environment), *Study On European Urban Transport Roadmaps 2030, Urban transport policy roadmaps*, European Platform on Sustainable Urban Mobility Plans, Ref: MOVE/C1/2013-188-2, March 2016.

[2] E.Vasile, M. Balan, I. Grabara, S. Balan, *Measures to Reduce Transportation Greenhouse Gas Emissions in Romania*, Polish Journal of Management Studies, 2012.

[3] The International Association of Public Transport (UITP), *Bus systems in Europe: Towards a higher quality of urban life and a reduction of pollutants and* CO_2 *emissions*, June 2015, Bruxelles.

[4] G. Hitchcock, *European Urban Transport Roadmaps to 2030, A policy support tool for small and medium size cities,* Ricardo-AEA, *http://www.urban-transport-roadmaps.eu/scenario/2bc5e0cf#*

[5] Protected Bicycle Lanes, *New York City Department of Transportation – Polly Trottenberg*, Commissioner September 2014.

[6] U.A.T. Ghimbav, *Action Plan for Sustainable Energy Of Ghimbav City*, *http://primaria-ghimbav.ro/docs/2016/2016-09-16-PAED-Ghimbav-2016.pdf,* September 2016.

[7] D. Covaciu, D. Florea, J. Timar, *Estimation of the noise level produced by road traffic in roundabouts*, Applied Acoustics, Volume 98, November 2015, Pages 43-51.

[8] S. Tarulescu Stelian, R. Tarulescu, *Urban Transportation Solutions for the* CO_2 *Emissions Reduction Contributions*, 12th International Congress of Automotive and Transport Engineering (CONAT), DOI: 10.1007/978-3-319-45447-4_49, ISBN:978-3-319-45447-4, 2016.

Acknowledgments
We hereby acknowledge the national project "*Introducere sistem de transport urban inteligent și ecologic în orașul Ghimbav. Trasee și infrastructură pentru transport electric*", 1755/16.02.2017 and Transilvania University of Brasov for providing the infrastructure used in this work.
Also, we hereby acknowledge the structural founds project PRO-DD (POS-CCE. O.2.2.1.. ID 123. SMIS 2637. ctr. No 11/2009) and Transilvania University of Brasov for providing the infrastructure used in this work.

IOP Conf. Series: Materials Science and Engineering **252** (2017) 012052 doi:10.1088/1757-899X/252/1/012052

Solutions to Improve Road Circulation in the Pitesti City Based on Analysis-Diagnostics of Road Traffic

A Vîlcan[1], E Neagu[2], H Badarau Suster[2] and A A Boroiu[2]

[1] AV Transport Planning srl, 9 Ionel Perlea Street, Apt 1, Sector 1, 010208, Bucharest, Romania

[2] Automotive and Transports Department, University of Pitesti, 1, Targul din Vale Street, 110040, Pitesti, Romania

E-mail: adrian@vilcan.com

Abstract. Road traffic congestion has become a daily phenomenon in the central area of Pitesti in the peak traffic periods. In order to achieve the mobility plan of Pitesti, an important stage is the diagnostic analysis of the road traffic. For this purpose, the urban road network was formalized through a graph containing the most important 40 intersections and traffic measurements were made at all these intersections and on the main roads connecting the peri-urban area. The data obtained by traffic macrosimulation confirmed the overloading of the street network during peak traffic hours and the analyzes made for various road traffic organization scenarios have shown that there are sustainable solutions for urban mobility only if the road network is fundamentally reconfigured (a belt outside the city and a median ring). Thus, the necessity of realizing the road passage in the Prundu neighbourhood and the finishing of the city belt by realizing the "detour West" of the city is argued. The importance of the work is that it brings scientific arguments for the realization of these road infrastructure projects, integrated in the urban mobility plan, which will base the development strategy of the Pitesti municipality.

1. Introduction

A comprehensive analysis of the traffic in the municipality of Pitesti is generated by the achievement of the *Plan of Sustainable Urban Mobility of the Municipality of Pitesti*, for which the transport model has been made and subjected to public debates [11]. The model is conceived for the base year 2017 and, depending on the scenarios applied, forecasts are made for 2023 and 2030.

The studies conducted have taken into account the existing road network, with the main arteries of entering the city (figure 1) and regulating the movement in the major road intersections (figure 2).

2. Developing the transport model for Pitesti city

A transport model allows the estimation of the way in which the movements of people and goods will respond, in time, to changes in what concerns:

• *the transport supply*: both services (introduction of new bus routes, increase or decrease in frequency, improvement of the services offered, changes in public transport prices) and the infrastructure (construction of a ring road, optimization of an intersection, construction of a new tram line, etc.);

• *the transport demand*: increase or decrease in population, change of its spatial distribution, change of the socio-economic (e.g. the degree of motorization) or demographic characteristics, growth or decline of economic activities, etc.

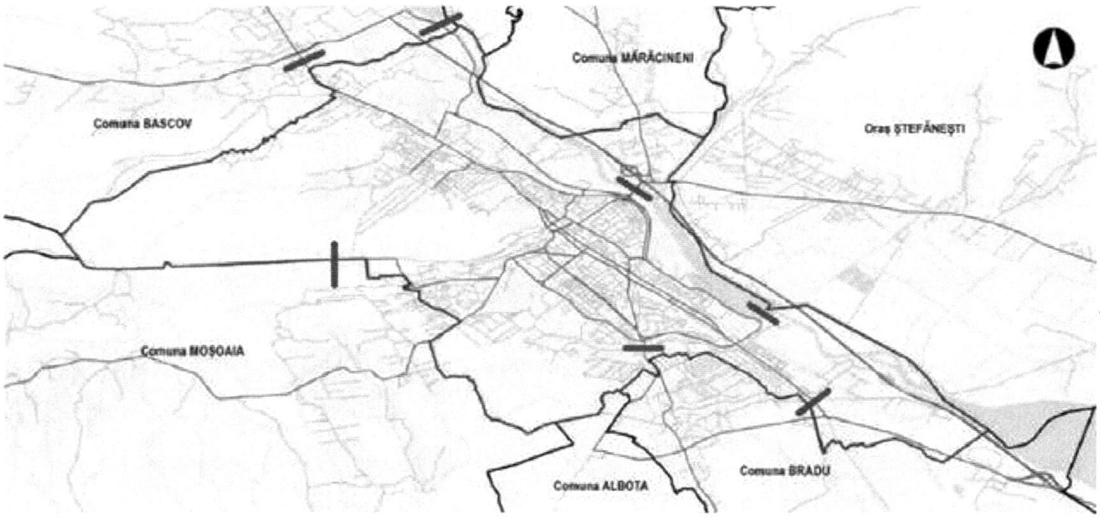

Figure 1. The road network in the municipality of Pitesti and the main arteries of entering the city.

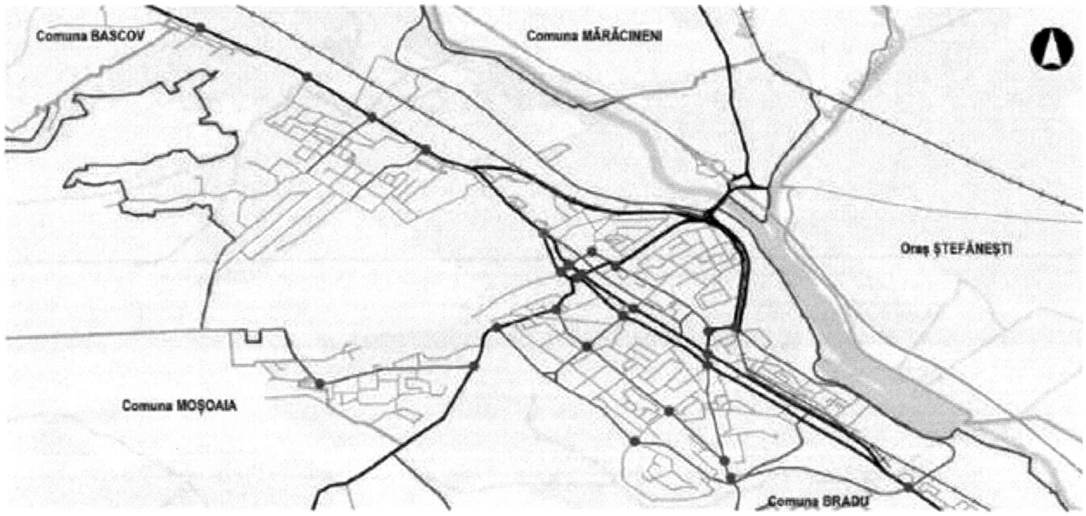

Figure 2. Roundabout intersections (blue) and traffic lights intersections (red) in the municipality of Pitesti.

An important component in the transport model approach is to select the most appropriate modelling software. PTV Vision software was chosen because it possesses the ability to satisfy all conditions studied, being widely used in Eastern Europe and featuring a wide range of applications in the region. In addition, PTV VISUM software has been used in recent years for most of the growth poles (Bucharest, Constanta, Brasov, etc.), so it is advantageous to perform the study with the same software [10].

2.1. *Data collection*

This first stage of creating the model was carried out for the following components:

2.1.1. Road network. An analysis of the network of roads and intersections in the city was performed with the purpose of defining the attributes of the road network from the transport model. These are the number of traffic lanes in each direction, the speed limit or the average speed, one-way or two-way road/street, lanes dedicated to public transport or priority lanes, parkings on the road, road signs, road markings, types of intersections and times of configuration/channelling/signalling.

2.1.2. The public transport system in the municipality of Pitesti. The transport network of the municipality of Pitesti has developed mostly in a linear direction following the North-South direction, as a consequence of the geographical location of the city (terraced parallel to the central axis), which did not allow the radial development of transport lines [1]. The bus network length is 41.07 kilometres, double track, and the length of the other tracks is 168.03 km, there being 100 bus stations; the 8 linebreaks are: Arpechim, Doja, Razboieni, Gara Sud, Trivale, Gavana, Alprom, Bascov (Peco Mol) [2].

2.1.3. Home interviews on population mobility. Interviews at home on population mobility have been carried out in accordance with the standard procedures normally used in transport studies. Family structure was identified, from the head of the family and continuing with all the members of the family. The gender, age, regardless of the subjects' attending some form of education or being employed, were recorded for each household member, as well as their holding a driving license, together with the total number and type of vehicles the respective household has at its disposal. This was followed by a record of the movements, performed separately for each member of the household; normally, the travels that had been made the previous day. To carry out a representative survey, there was established a percentage of 1.5% relative to the population of Pitesti, i.e. 2700 valid forms [7]. This component of the study describes the current situation of movement in Pitesti and represents the basis for modelling transport demand in the forecast component. This resulted in a series of useful data for traffic modelling, grouped as follows:

a) Socio-economic data:

• household size: about 75% of households are made up of between 2 and 4 persons, the average household size is 2.48 members;

• *economically active members*: about 22.0% of households have no economically active members, these households, most often, being households consisting solely of pensioners, on average, households have 1.50 economically active people;

• availability of vehicles: 39% of households do not have access to a private vehicle, 53% of households have access to a single private vehicle, the average of vehicle availability per household is 0.69 vehicles/household or 276 cars per 1000 people;

• access to public transport: access to public transport is good, with an average travel time on foot to the nearest public transport station of 7.35 minutes, 25% of households can access public transport in less than five minutes, 60% of households can access public transport into 5 - 10 minutes;

• average household income: 33% of households have an average monthly household income under 368 EUR, 41% of households have an average monthly household income between 368 and 690 EUR.

b) Personal information. For each household member there were obtained data on personal characteristics (age and sex, driving license, employment status situation, industrial category, workplace or education institution).

c) Information on movements. Travel data provides a wealth of data on the behaviour and habits of traveling for residents of Pitesti.

The following information is presented as information about a classical modelling of the transport process in four steps:

• generating movements: data on the number of movements (there was estimated a daily average of 1.35 trips per capita) and movement time (for the first movement, there is a distinct peak in the morning between 07:30 and 8:30);

• attracting movements;

• area distribution;

• modes of transport distribution: the main modes of transport are walking (29%), by private car (35%) and by bus (23%) - this equates to 88% of all journeys.

2.1.4. Traffic censuses in intersections. Traffic censuses were performed in 40 intersections (figure 3). These were conducted 10 hours a day, in the following time intervals: 6:00 to 11:00 and 2:00 p.m. to 7:00 p.m.

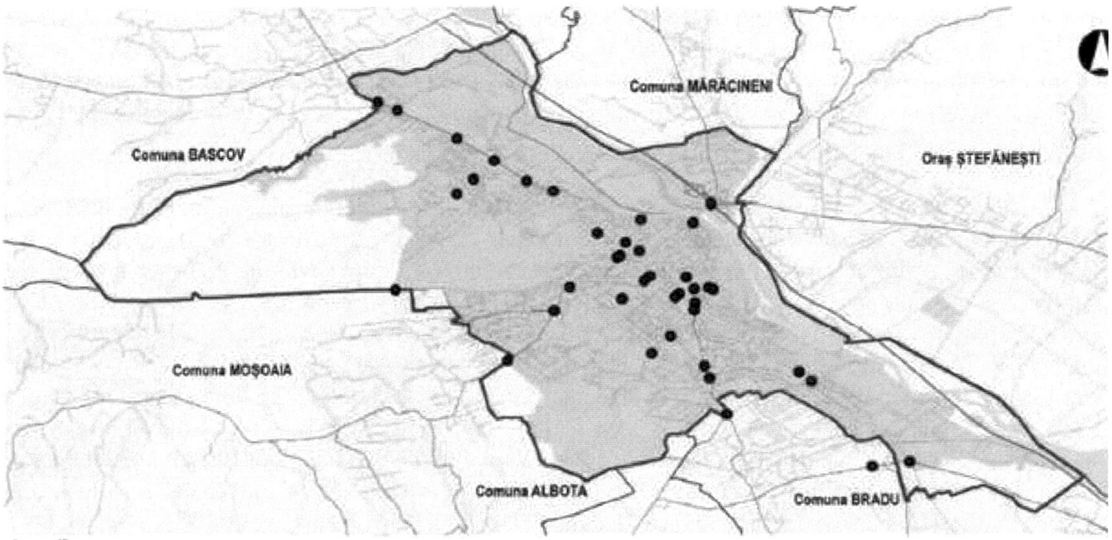

Figure 3. The 40 intersections in the municipality of Pitesti where traffic censuses were conducted.

The aggregated results for the 40 intersections reveal high levels of traffic flows measured in intervals of 15 minutes and two traffic peaks: a morning traffic peak (figure 4), between 07.30-08.30 a.m., and a second traffic peak in the afternoon, between 4.00 p.m. - 5.00 p.m. The figures below show the total traffic flows at the intersections that were the subject of the censuses.

2.1.5. Traffic censuses in section. The sections for which traffic censuses were conducted 24 hours a day are actually the most important intersections in the municipality of Pitesti, where cameras are located and the recorded images could be processed (DN65B - Bd. Petrochimistilor, Calea Craiovei - Basarabiei Str. - Vasile Parvan Str., Calea Dragasani - Bd. Libertatii, Bd. Nicolae Balcescu - Labusesti Str., Bd. Republicii - Bd. Eroilor, Armand Calinescu Str. - Bd. Eroilor – Egalitatii Str., Calea Bucuresti - Costache Negri Str. - Depozitelor Str.) [6]. The aggregated results for the seven sections where the traffic census was conducted show high levels of traffic flows measured in intervals of 15 minutes, and two traffic peaks (figure 5), just as in the case of the 40 intersections where traffic censuses were conducted manually by traffic observers.

2.1.6. Origin-Destination surveys in traffic. O-D surveys in traffic were conducted in seven locations outside the administrative boundaries of the city [7]. The reviewing and the interviews for the origin-destination surveys from the locations mentioned above were conducted between 06:00 a.m. - 9:00 p.m., i.e. for a period of about 15 hours. In this period there were counted about 50000 cars traveling to/from Pitesti. The main vehicle category was represented by the car, which counted for about 75% of all vehicles reviewed.

2.1.7. Reviewing public passenger transport. Studies of public transport were made to collect data on public buses in Pitesti during peak hours. There emerged certain characteristics of public transport such as information on the loading degree and the working hours of the buses. Reviewing transport lines was performed between 7:00 to 10:00 a.m. and 1:00 p.m. to 5:30 p.m. on a normal working day. After analyzing the data obtained the usage degree of public transport stations was determined, being noted that the largest number of passengers is concentrated in the centre, on the Bd. Republicii and Armand Calinescu Str. arteries, both in the morning and in the afternoon.

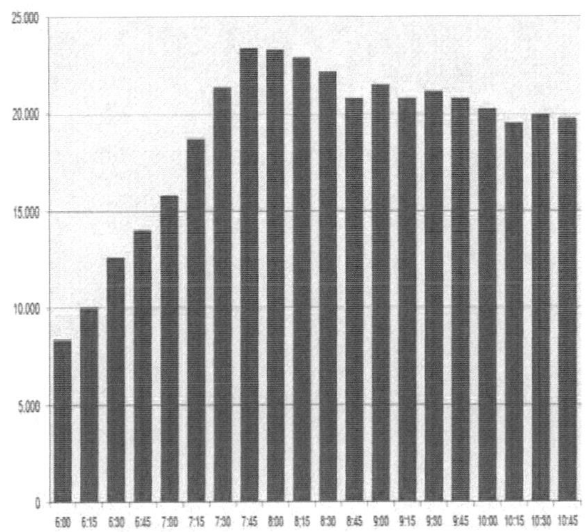

Figure 4. Traffic flows over intervals of 15 minutes, in the time interval of 06.00-11.00 a.m., cumulated for the 40 intersections.

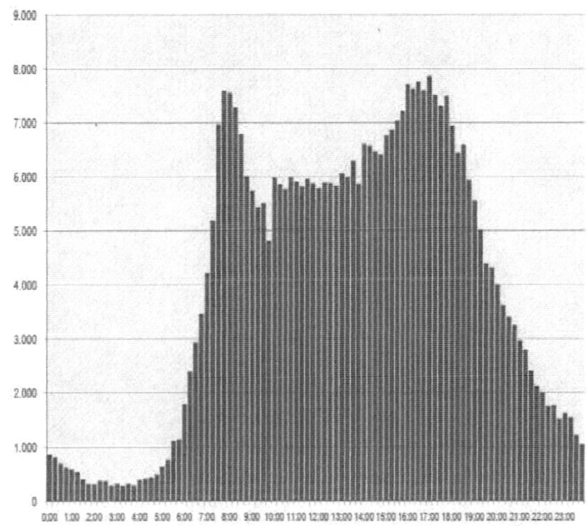

Figure 5. Traffic flows in intervals of 15 minutes, cumulative for the 7 sections monitored.

2.2. Evaluation of the transport demand

2.2.1. Definition of the zoning system. The zoning system needs to be defined so that the main daily models of inter-urban migration could be properly captured and modelled. This is especially important as cities begin to expand and the so-called satellite cities are developing towards their periphery.

The area specific to the transport model is an important element in the representation of demand for mobility, as movements are generated and attracted by the specific functions of each area. Thus, the zoning system is an integrated planning element, building at the same time an intrinsic link between urban features and mobility [4].

A condition for a transport model to be as close to reality as possible is to consider a large number of areas and the areas to be taken into account must be linked to the socio-economic data available and the purposes for which the model is developed.

The starting point in the zoning of the transport model was the delimitation of polling stations in view of organizing and conducting the 2016 elections, being later detailed in relation to their functional, socio-economic and demographic characteristics. The study area consists of 168 areas, of which 144 areas cover the municipality of Pitesti and 24 areas cover the territory of influence (figure 6).

The socio-economic data related to each area consist of the information regarding population (categories of age, occupation, income level, possession of a car or lack of possession of a car), the number of jobs by type of activity, places in education (educational establishments, number of places for students and pupils at the related educational establishments), income level and availability of using a car, economic activities (industry, services, trade, tourism, agriculture).

In the case of the large commercial or industrial centres, specific areas were considered within the transport model because of their specificity that leads to attracting significant flows of people and goods.

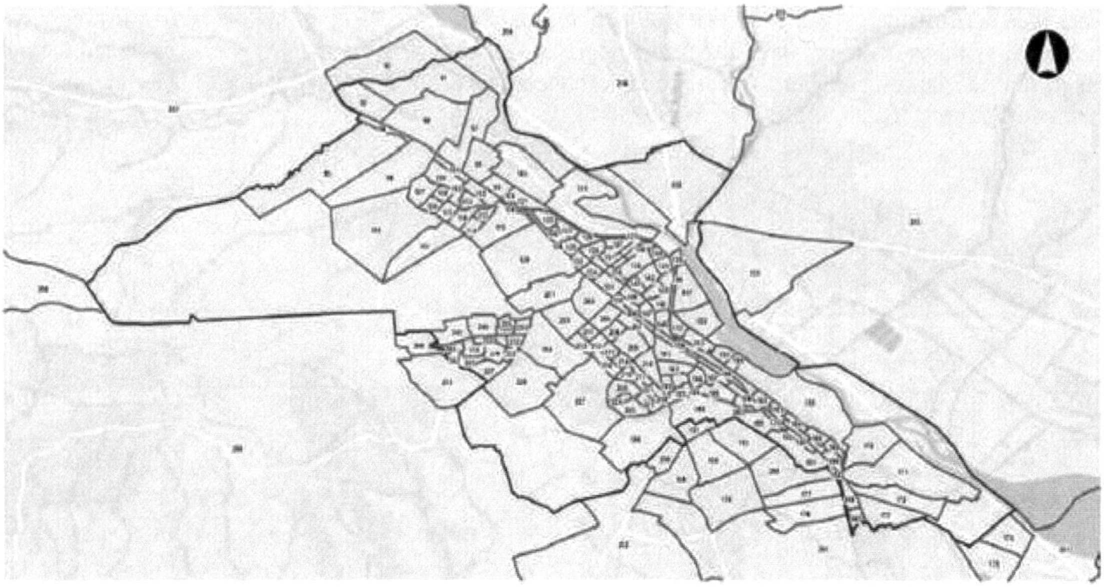

Figure 6. Zoning of the territory under study.

2.2.2. The a priori development of origin-destination matrices. The evaluation of the demand for passenger and freight transportation based on the data previously collected for morning peak hours (a.m.), afternoon peak hours (p.m.) and the daily annual average is an essential activity, the demand for passenger and freight transportation at the OD matrix level for the study area - a priori OD matrix, estimated directly based on home interviews, on the segments of mobility demand considered, is a major input for the calibration of the transport model. The informations taken into account for the development of the a priori matrices were information from home interviews and database of the 2012 General Transport Master Plan. The a priori OD matrices (observed) have been structured as a priori matrices for people movements and a priori O-D matrices for freight transport. The OD matrices for the base year 2017 were estimated based on a priori OD matrices, aggregated and converted into standard vehicles and based on the traffic census, the traffic flows being observed in equivalent standard vehicles and in vehicles related to public transport. There were taken into account techniques to estimate the matrices implemented in Visum. The information on the traffic flows observed was based on the traffic censuses carried out in the project, plus the automatic traffic counting conducted by CNAIR and the traffic censuses for national roads available for the project.

2.3. Data calibration and validation

2.3.1. Calibration of the urban transport model. Calibration is the iterative process whereby the model is revised repeatedly until it becomes stable and ensures the most faithful reproduction of the conditions of the reference year. The process involves refining (or finishing) in stages the network from the model in order to better display the existing conditions, such as the speed-flow relationships, the drive/steering constraints, the class in the network connection, the connectors, the traffic capacity, etc. After completion of the network model, the traffic volumes allocated are compared with the volumes observed or with the censuses conducted. The volumes of the application from the model are calibrated so that the flows modelled could be as close as possible to those reviewed either by manual manipulation of the matrix or automatically, by estimation of the matrix. The urban transport model calibration consists of calibrating the models specific to each of the stages in the process of modelling mobility [8].

An effective calibration of models is performed by estimating the parameters of the above equations for each individual model. The results of calibration within each model of generation and attraction of movements on travel purposes are evaluated by the statistical parameters of linear regression considered in the calibration process models. Thus, it is recommended that the standard deviation should be not less than 85% in most cases. The models of generating and attracting movements are important elements because they represent the link between the characteristics of the population and the functional characteristics of each area and the number of movements generated and attracted by it, being thus used for the forecast of movements generated and attracted by each area according to the new socio-economic and functional characteristics of the respective area. For the models of attracting and generating movements the following types of equations were considered:

$$G_i = a_i + b_i \cdot X_1 + c_i \cdot X_2 + d_i \cdot X_3 \tag{1}$$

$$A_j = a_j + b_j \cdot X_1 + c_j \cdot X_2 + d_j \cdot X_3 \tag{2}$$

where:

G_i – movements generated by the area „i";

A_j - movements generated by the area „j";

X_1, X_2, X_3 - socio-economic indicators for each area (population, active population, number of pupils, students, etc.);

a, b, c,... - parameters to be calibrated.

Within the models of generation and attraction of movements there are two categories of users: those who have access to a car and those who do not. So, finally, there were calibrated 69 models for generating traffic, 8 different models of distribution, 8 different models for slow movements, 8 different models for the Modal Split and for the level of motorization.

Models of distribution between areas carries out the distribution of movements generated and attracted by each area, between each pair of O-D areas, on purposes of travel, at peak morning (a.m.) and afternoon (p.m.) hours. The models for each peak hour represent distinct models. The gravitational model was considered when modelling the distribution of movements [5]. The gravitational model is based on the theory of gravity in Newtonian physics and requires that the number of movements between areas be directly related to the number of movements generated by each area and inversely related to the difficulty of making the movement. Mathematically, the gravitational model is defined as follows:

$$T_{ij} = G_i \cdot \frac{A_j \cdot F_{ij} \cdot K_{ij}}{\sum_{k=1}^{n} A_j \cdot F_{ij} \cdot K_{ij}} \tag{3}$$

where:

T_{ij} – number of movements from Area i to j.

G_i- number of movements generated from Area i.

A_j- number of movements attracted by Area j.

F_{ij}- impedance factors related to the separation between Areas i and j.

K_{ij} – the adjustment factor of distribution of optional movements for interchanges between Areas i and j.

n = number of areas (n = 136).

The correct development and calibration of the model of modal distribution provides the functionality of the model, giving information on the use of transport modes depending on the services offered by each mode. The model is based on the relative attractiveness of each mode to the other. In addition, this facilitates testing the operational and/or infrastructure improvements to each mode and allows the quantifying of their impact on traffic generated and specific to a mode. The complexity of the model increases as choices grow and travel parameters change. The main objective is to assess the impact of improving public transport systems (accessibility, rates, new modes, etc.).

Home interviews make available information on the choice of transport mode depending on the fact of owning a car or not. With an increase of the income and of the availability of the car, its use becomes more affordable to the members of the household, which leads to the performing of a larger number of movements by car. The choice of the transport modes considers the categories like slow modes, the choice of transport mode for those who do not have a car available, the choice of transport mode for those who have a car available. Most models based on choosing the model are based on formulation of the Logit model. Mathematically, the Logit model is defined by the formula [3]:

$$T_{ij,m} = T_{ij} \cdot \frac{e^{c \cdot U_{ij,m}}}{\sum_{k=1}^{n} e^{c \cdot U_{ij,k}}} = T_{ij} \cdot P_{ij,m} \quad (4)$$

where:

T_{ij} - Number of movements between the Areas i and j;

$T_{ij,m}$ _ - the number of movements made by mode of transport m;

c - the parameter of the procedure;

$U_{ij,m}$ – the generalized cost of the travel from i to j by mode m;

$P_{ij,m}$ _ - the proportion of movements made by mode of transport m.

This approach calculates the share of demand among different modes of transport competing for travel, based on a number of inputs that influence the overall cost of the travel (generalized cost of the travel). A first step is that in which a distinction is made between the slow modes (walking and cycling) and the motorized ones. To do this, the following type of model was used:

$$P_{ij,w} = \frac{1}{1+e^{a+b \cdot D_{ij}}} \quad (5)$$

where:

$P_{ij,w}$ _ - the share of movements performed by slow modes (walking and cycling), denoted by w from walk (go by foot);

$D_{ij,w}$ - distance between the Areas i and j

a, b – the parameters to be calibrated.

The model was calibrated using information from the interviews at home. In the second step_the modal distribution model is calibrated. This module models the distribution between motorized modes of transport, between car and public transport. The model makes the difference between users who have a car and those who do not.

Allocation on the network constitutes the final stage of the 4-stage determination of the transport demand and the associated traffic flows on the street network and those taken over by public transport of passengers. Traffic allocation is the process after which the amount of traffic on each segment of the transport network and on turns in intersections is determined. Any version of traffic allocation on network based on the principle of balance uses such functions as volume - delay both for the links and for the turns, thus impedance changes depending on the traffic. Within a model of urban transport, the volume-delay functions related to turns are very important because intersections affect the transport network performance more than links. For the purposes of allocation on the road transport network the Allocation on the Network with an Intersection Capacity Analysis (ICA) was employed, which takes into account the impedance of the intersections in a detailed manner. This procedure of allocation on the network includes allocating lanes and traffic light times. Furthermore, the interdependence of directions allowed in each traffic hub within the network was also considered.

2.3.2. Validation of the model of transport demand. The validation of the model of transport demand represents the stage of analyzing the results for each stage of the model with the data observed, in order to determine the degree of representation of reality. The validation of the model of transport demand is the stage that follows after the calibration of the model and consists mainly of comparing the outputs of the model with an independent set of data not used in the process of development of the model, in order to make sure that the model reflects the travels volumes and velocities observed across the entire area of the model, not just in the locations used for the development of the movement matrices. A suitable model for the goal pursued meets the standards required for both calibration and for validation. According to [9], for a model of transport to be considered valid four criteria should be met, by comparing the demand allocated, in more than 85% of cases in order to be considered accepted. The demand for private vehicles, passengers or freight should have a margin of 15% compared to the values observed. The flows on the windrows should have a margin of 5 % compared to the values observed. The GEH5 value (for individual flows GEH < 5, for total values for the windrow GEH < 4). The last criteria take into consideration was comparing travel times (times within a margin of error of 15% or of more than one minute).

Calibration of the transport model was satisfactory at this stage, the next step being to continue to examine in detail the transport network until the testing of scenarios. If certain aspects within the transport model are to be adjusted, this is done until the testing of scenarios.

2.4. *Model results*

The main results of the model are matrix on mode of transport and main OD relations, traffic flows on the network, network performance statistics. The figures below shows the flows for private and public transport traffic for the base year 2017 at peak a.m. hours (figure 7 and figure 8).

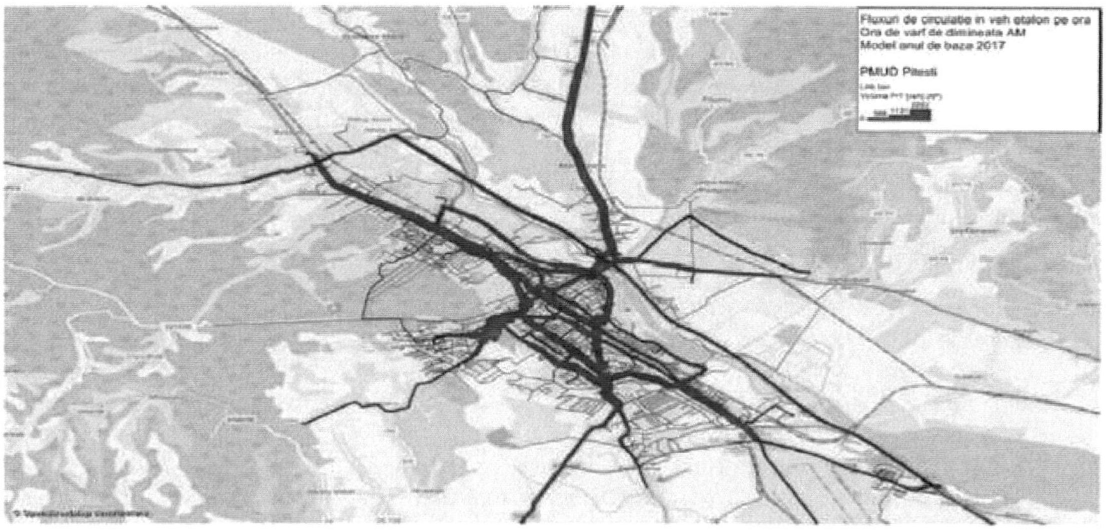

Figure 7. Hourly flows of road vehicles, in equivalent vehicles, for the morning peak.

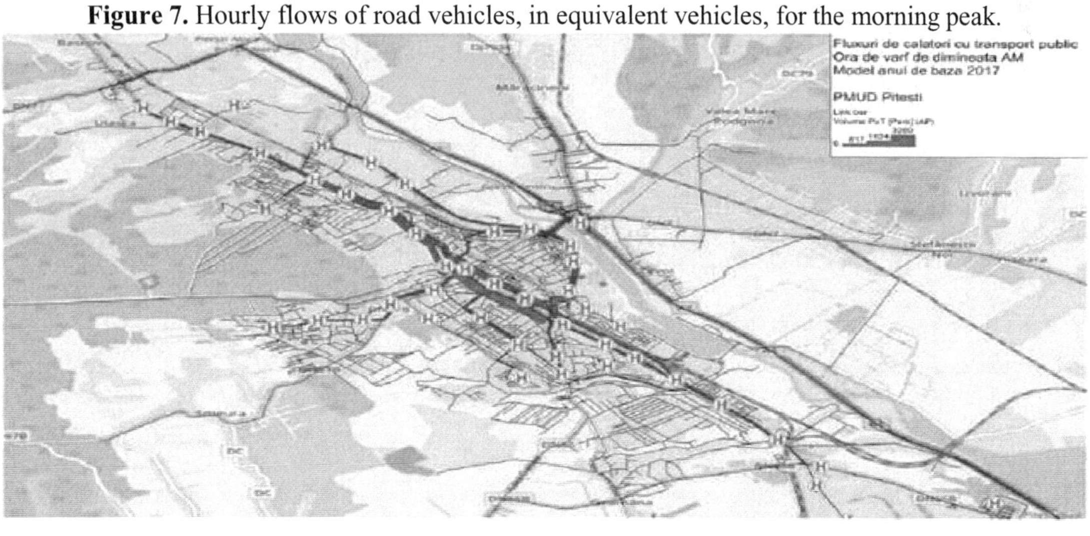

Figure 8. Hourly flows of passengers (public transport traffic) for the morning peak.

IOP Conf. Series: Materials Science and Engineering **252** (2017) 012052 doi:10.1088/1757-899X/252/1/012052

3. Conclusions. Forecasts. Suggested remedies

The increase or decrease of the population, of the level of motorization, of incomes, of tourism, of the residential development and of the GDP are the key factors that directly influence the increase/decrease of the number of persons' movements within the municipality of Pitesti, locally and regionally, and these aspects were taken into account in the travel demand forecasts for the 2023 and 2030 timeframes.

After the base year model is validated, a synthetic base model is also developed for forecasting purposes. The synthetic model for the base year is calibrated based on the relations between transport supply, demand and socio-economic parameters. This allows the socio-economic factors to be incorporated in the forecast, such as population, income and the increase of the number of car owners.

Two scenarios were evaluated to test the sensitivity of the model:

- scenario 1: A 10% increase in the cost of fuel for cars;
- scenario 2: A 10% increase in the cost of public transport tickets.

The data obtained by traffic macrosimulation confirmed the overloading of the street network during peak traffic hours and the analyses made by the macrosimulation of road traffic for various road traffic organization scenarios have shown that there are sustainable solutions for urban mobility only if the road network is fundamentally reconfigured. So, there are sustainable solutions only if the street network is bounded by a belt outside the city, which will ensure that there are no transit traffic and a median ring that will allow travel between different districts in the city to be made without penetration in the central area. Thus, the necessity of realizing the road passage in the Prundu neighbourhood and the finishing of the city belt by realizing the „detour West" of the city is argued.

The importance of the work is that it brings scientific arguments for the realization of these road infrastructure projects, integrated in the urban mobility plan, which will base the development strategy of the Pitesti municipality.

4. References

[1] Boroiu A 2009 *Transporturi de persoane* (Pitesti: University of Pitesti Publishing House)

[2] Boroiu A, Nicolae V, Ilie S, Zaharia C and Soare C 2008 *Proiectarea strategiei de transport public de persoane in judetul Arges pentru perioada 2008-2013* si *proiectarea programului de transport public de persoane in judetul Arges pentru perioada 2008-2011* Contract no 541/07.02.2008 Beneficiary Arges County Council.

[3] Boroiu A A and Neagu E 2015 *Trafic rutier si siguranta circulatiei rutiere. Aplicatii* (Pitesti: University of Pitesti Publishing House)

[4] Boroiu A A 2017 *Studii si cercetari privind reducerea poluarii fonice produse de autovehicule prin organizarea circulatiei rutiere*, PhD thesis, University of Pitesti

[5] Florea D, Cofaru C and Soica A 2000 – *Managementul traficului rutier* (Brasov: University Transilvania of Brasov Publishing House)

[6] Neagu E 2003 *Trafic rutier şi siguranta circulatiei* (Pitesti: University of Pitesti Publishing House)

[7] Nicolae V, Neagu E, Istrate M, Boroiu A, Vieru I, Badarau Suster H and Boroiu A A 2016 *Planul de Mobilitate Urbana Durabila al municipiului Pitesti. Culegerea şi prelucrarea datelor de mobilitate*, Subcontracting agreement no. 10581/22.08.2016, Beneficiary S.C. SYNERGESTICS CORPORATION SRL Bucharest

[8] Raicu S 2007 *Sisteme de transport* (Bucharest: AGIR Publishing House)

[9] JASPERS (Joint Assistance to Support Projects in European Region) 2014 JASPERS Appraisal Guidance (Transport) *The Use of Transport Models in Transport Planning and Project Appraisal*

[10] PTV Vision Software 2011 VISUM *User Manual* PTV AG Karlsruhe (Berlin: epubli GmbH)

[11] http://www.primariapitesti.ro *The Plan of Sustainable Urban Mobility of the Municipality of Pitesti*, Project in public debates

Computing Programs for Determining Traffic Flows from Roundabouts

A A Boroiu[1], I Tabacu[1], A Ene[2], E Neagu[1] and A Boroiu[1]

[1] Automotive and Transports Department, University of Pitesti, 1, Targul din Vale Street, 110040, Pitesti, Romania
[2] Electronics, Computers and Electrical Engineering Department, University of Pitesti, 1, Targul din Vale Street, 110040, Pitesti, Romania

E-mail: boroiu_alexandru@yahoo.com

Abstract. For modelling road traffic at the level of a road network it is necessary to specify the flows of all traffic currents at each intersection. These data can be obtained by direct measurements at the traffic light intersections, but in the case of a roundabout this is not possible directly and the literature as well as the traffic modelling software doesn't offer ways to solve this issue. Two sets of formulas are proposed by which all traffic flows from the roundabouts with 3 or 4 arms are calculated based on the streams that can be measured. The objective of this paper is to develop computational programs to operate with these formulas. For each of the two sets of analytical relations, a computational program was developed in the Java operating language. The obtained results fully confirm the applicability of the calculation programs. The final stage for capitalizing these programs will be to make them web pages in HTML format, so that they can be accessed and used on the Internet. The achievements presented in this paper are an important step to provide a necessary tool for traffic modelling because these computational programs can be easily integrated into specialized software.

1. Objectives of the paper

In recent years, a phenomenon of roundabout proliferation has been remarked, being noted that this type of intersection brings forward a number of advantages, especially to the specifics of urban traffic. As a result, there is constant concern to improve the service level of these intersections [3,7-8].

The methodology for determining the level of service for the roundabout intersections is shown in [11], a reference work that was recently reviewed (2015). The main mathematical models that evaluate the capacity of each arm of the intersection are based on the volume of the conflict flow [6,9-10]. Thus, ***the capacity of a roundabout intersection with a lane on the annular path*** is determined with the relation:

$$c_a = \frac{3600}{\cdot t_f} \cdot \frac{V_c \cdot e^{\frac{-V_c \cdot t_c}{3600}}}{1 - e^{\frac{-V_c \cdot t_f}{3600}}} \qquad (1)$$

where:

- c_a is the capacity of the arm (standard veh./hour);
- V_c is the conflict volume related to the arm (standard veh./hour);
- t_c is the critical access time from the arm to the intersection [sec];
- t_f is the time of tracking on the annular path [sec].

The conflict volume can be simply registered by a traffic operator (one operator for each arm of the intersection), but the methodology provides for a calculation algorithm by which this is established function of the current volumes of traffic at the intersection, as is the case, for example, of the conflict volume for arm b of a 4-arm roundabout intersection, presented below: adopting a trigonometrically increasing notation (b, b+1, b+2, b+3), this is given by the following analytical relation [1]:

$$V_c^b = V_{return}^{b+1} + (V_{left}^{b+2} + V_{return}^{b+2}) + (V_{forward}^{b+3} + V_{left}^{b+3} + V_{return}^{b+3}) \quad (2)$$

Accordingly, in the software programs dedicated to traffic modelling, the determination of the levels of service for roundabout intersections is based on analytical models in which the first step is the introduction, as input data, of the output of each traffic stream in the intersection or the introduction of entry outputs and mentioning the proportions which the traffic streams are divided into [4]. It is to be noted that in both cases there is no requirement for the measurement of flows of conflict and their introduction as input data, but rather for specifying the size of the traffic streams. But determining these traffic streams can only be achieved by following the route of each vehicle, and this would require recording the entire crossroad and then identifying the route of each individual vehicle – a method that involves a large amount of time dedicated to the further processing of the recordings.

As a consequence, the following question arises: *is it possible that, for an intersection where the traffic is regulated as a roundabout on the basis of the measured values (with fixed observers) for the flows of entry, flows of exit and flows of conflict (circular), one could determine the volume of traffic streams (figure 1)? Exact values would emerge from this analytical method.* To find the answer to this question, road circulation in the roundabout was formalized by noting the volume of traffic flows (number of standard vehicles in a particular timeframe) as similar as possible to the notations in specialised literature (written in Romanian or in a foreign language), as follows[11].:

- ingoing flows volume: Vi,m;
- outgoing flows volume: Ve,n;
- conflict (circular) flows volume: Vc,m

The index m represents the sequential number of the ingoing arm (origin), m = 1, 2, 3, 4, ..., and n is the sequential number of the outgoing arm (destination), n = 1 2, 3, 4 ....

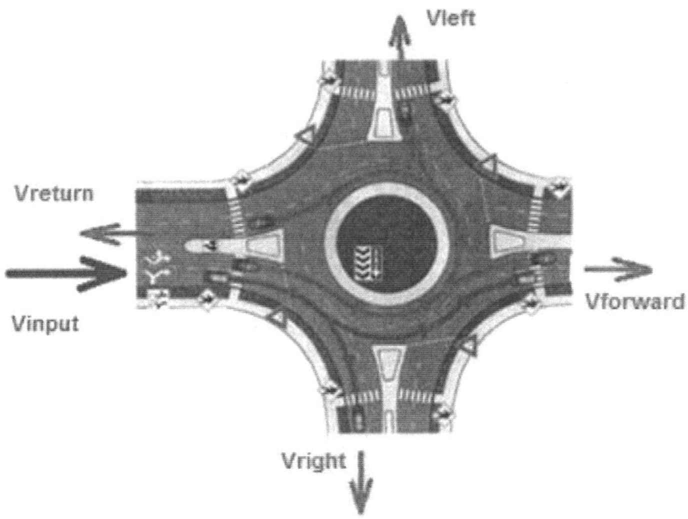

Figure 1. Traffic streams corresponding to an arm of the roundabout intersection [1].

The first three categories of flows are the flows that will be measured, so they represent the given data for the previously stated problem, and the last category of flows represents the values to be determined, therefore they constitute the unknown data of the problem. As a result, this issue will be analysed to discover if it can be solved, starting with the simplest roundabout intersection: the 3-arm roundabout. By cumulating the results derived from the theoretical research undertaken [1], carried out to discover whether it is possible to analytically determine traffic streams in roundabout intersections based on traffic measurements performed with fixed observers, the following results were obtained:

• In the case of the ***3-arm intersection***, one can formulate the analytical equations for the flows of return and flows of left turn as follows [2]:

$$V_{return} = V_{conflict}^{left\ arm} - (V_{entry}^{right\ arm.} - V_{right}^{right\ arm}) \quad (3)$$

$$V_{left} = V_{entry} - (V_{left} + V_{return}) \quad (4)$$

where:

- V_{entry}, $V_{right.}$, $V_{left.}$, V_{return} are the volumes of the flows related to the arm for which the calculation is performed (arm under analysis) – figure 2;
- $V_{conflict}^{left\ arm}$ is the conflict volume for the arm to the left of the arm under study;
- $V_{entry}^{right\ arm}$ is the entry volume for the arm to the right of the arm under study;
- $V_{right}^{right\ arm}$ is the volume of right-turn for the arm to the right of the arm under study.

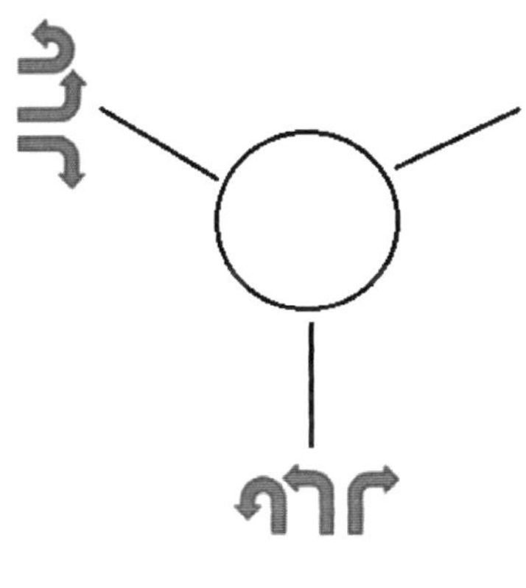

Figure 2. Traffic streams in a 3-arm roundabout intersection with return flows [1].

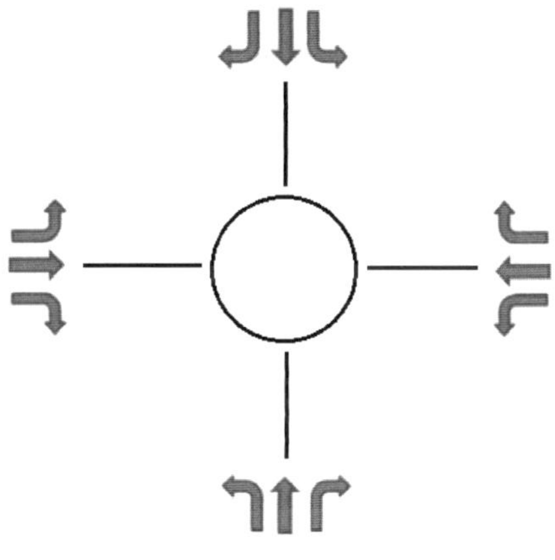

Figure 3. Traffic streams in a 4-arm roundabout intersection without return flows [1].

• In the case of the ***4-arm intersection***, it is possible to determine traffic streams only for the scenario „without return flows", the analytical relations for the flows forward and the left turn flows being [2]:

$$V_{left} = V_{conflict}^{forward\ arm} - (V_{entry}^{right\ arm} - V_{right}^{right\ arm}) \quad (5)$$

$$V_{forward} = V_{entry} - (V_{right} + V_{left}) \quad (6)$$

where:

• $V_{incoming}$, $V_{right.}$, $V_{left.}$, $V_{forward}$ are the volumes of the flows related to the arm for which the calculation is performed (arm under analysis) – figure 3;

- $V_{conflict}^{forward\ arm}$ is the conflict volume for the arm in front of the arm under analysis;
- $V_{entry}^{right\ arm}$ is the incoming volume for the arm to the left of the arm under analysis;
- $V_{right}^{right\ arm}$ is the right-turn volume for the arm to the right of the arm under analysis.

But a full capitalization of the analytical method discovered involves designing software programs in one of the dedicated programming languages, so the proposed analytical method could be used with ease – these representing ***the objectives of this paper***.

2. Development of computation programs for the analytical determination of traffic volumes at roundabout intersections

For each of the two cases that proved to be solvable by using data measured by fixed observers, a computation program was developed based on each set of analytical relations, a program that enables one to achieve, after introducing the data measured (volumes of entry, volumes of conflict and volumes of right turn), the values for the unknown flows on the annular path.

Thus, *two computation programs were developed in the Java programming language* (figure 4 and figure 5), with very user-friendly interfaces.

To meet any need, the programs also work when only partial results are necessary (a situation when only the necessary flows are to be measured).

If an analysis is required for all arms of the intersection, full measurement will be carried out.

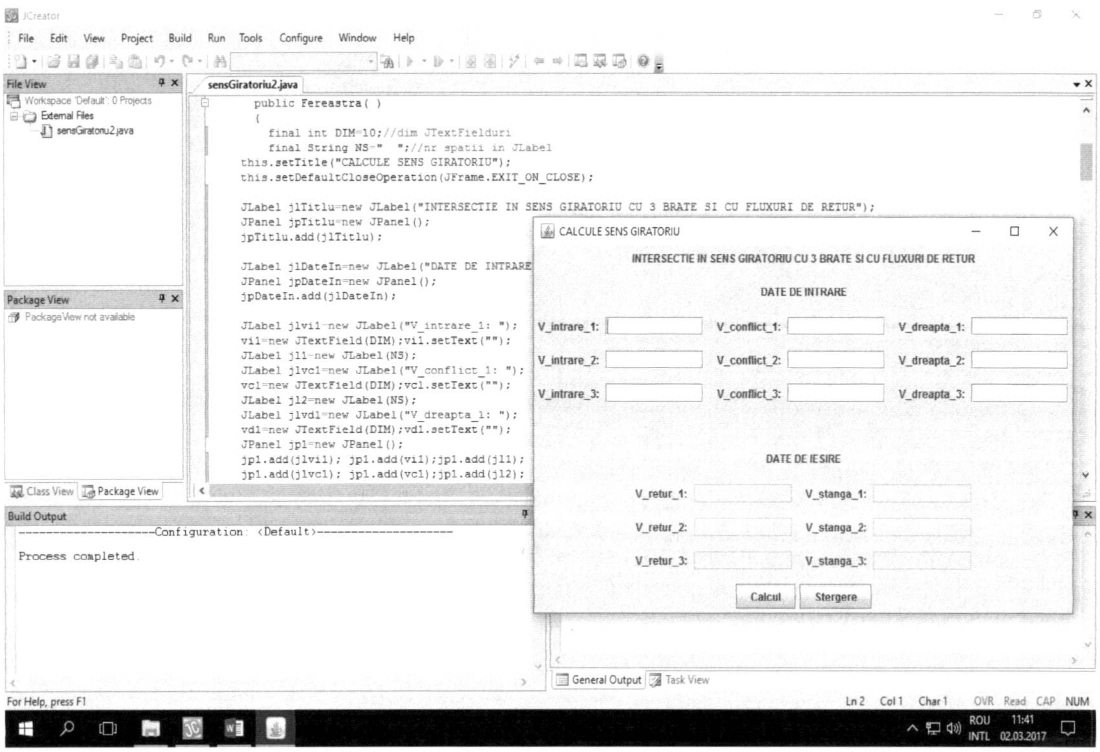

Figure 4. Developing the Java program for 3-arm roundabout intersections with return flows.

IOP Conf. Series: Materials Science and Engineering **252** (2017) 012053 doi:10.1088/1757-899X/252/1/012053

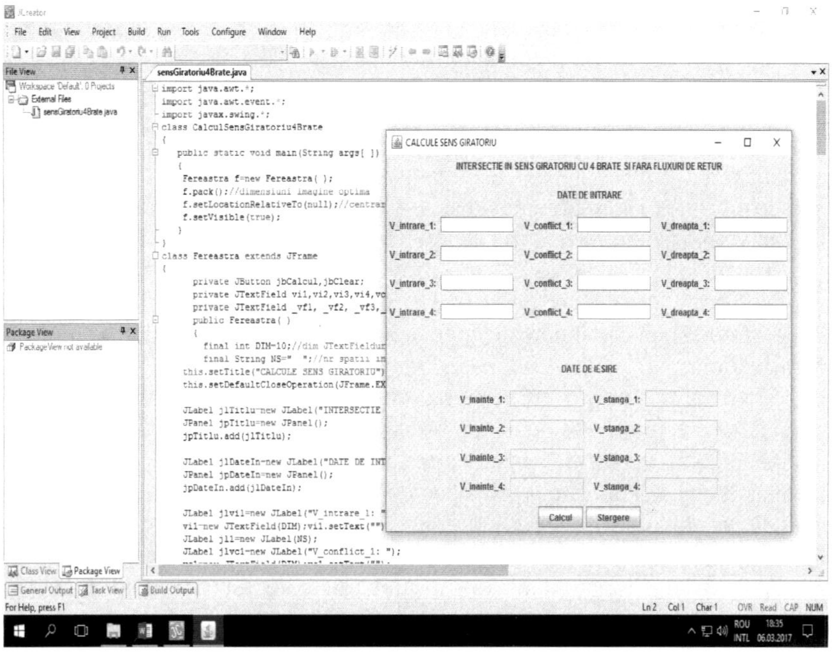

Figure 5. Developing the Java program for 4-arm roundabout intersections without return flows.

3. Results obtained by using the computation programs developed

With the help of the program developed for the 3-arm intersection, the data measured with fixed observers at the Podul Viilor roundabout intersection of Pitesti were processed - figure 6.

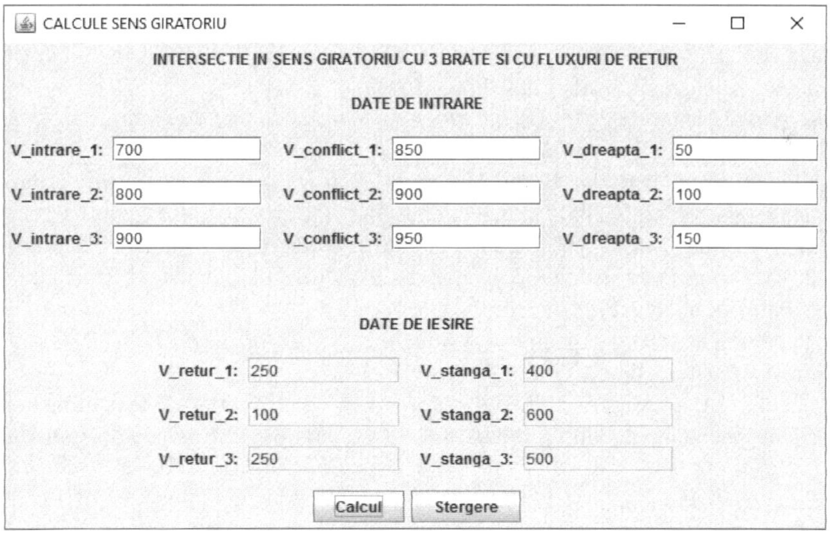

Figure 6. Traffic analysis for all arms with the Java program for the Podul Viilor 3-arm roundabout intersection with return flows.

Thus, it was noted that, on the Eastern arm, the return flow represents 70% of the incoming flow and it was possible to estimate that this flow, which increases the conflict flows for each of the other two arms, will be reduced substantially after the overpass artery road is put into service , so the traffic congestion reported in paper [1] will no longer take place.

The new measurements, made after the Calea Bucuresti overpass was put into service [12] confirmed the prediction.

4. Conclusions

It has been demonstrated that, for the 3-arm intersections with return flows and for the 4-arm intersections without return flows, the traffic streams may be fully known based on the traffic flows measured by fixed observers (measurement that are to mandatorily include the flows to the right) which is an absolutely unique and interesting achievement (the two sets of formulas represent personal contributions to traffic theory), because a knowledge of these traffic streams allows the reconsideration of traffic regulation in the area where the roundabout intersection is placed, for the purpose of traffic fluidization in the respective area, as well as in order to eliminate possible occurrences of congestion.

For each of the two cases that proved to be solvable with the help of the data measured by fixed observers, a computation program was developed based on each set of analytical relations, *in the Java programming language*, with very user-friendly interfaces.

To meet any need, the programs also work when only partial results are necessary (a situation when only the necessary flows are to be measured).

If an analysis is required for all arms of the intersection, full measurement will be carried out.

The use of these software programs allows formulating forecasts for road traffic in roundabout intersections.

The final step for the capitalization of these two programs will be to develop them as web pages [5], in HTML (HyperText Markup Language) format so that can be freely accessed and freely used on the Internet by any interested person.

5. References

[1] Boroiu A A 2017 *Studii si cercetari privind reducerea poluarii fonice produse de autovehicule prin organizarea circulatiei rutiere* PhD thesis, University of Pitesti

[2] Boroiu A A and Boroiu A 2016 Evaluarea curenţilor de trafic din intersecţiile în sens giratoriu pe baza fluxurilor măsurate în puncte fixe, *Scientific Conf. TRANSLU (Bucuresti), year XXI, Supliment* (Bucharest: AGIR Publishing House)

[3] Boroiu A A and Neagu E 2015 *Trafic rutier si siguranta circulatiei rutiere. Aplicatii* (Pitesti: University of Pitesti Publishing House)

[4] Boroiu A A, Tabacu, I, Parlac S and Neagu E 2016 Research on the correlation between the characteristics of traffic at intersections and the road noise level *Scientific Bulletin Automotive Series, year XXII, no. 24 (1)* (Pitesti: University of Pitesti Publishing House)

[5] Ene A 2015 *Programare orientată pe obiecte. Lucrari de laborator*, (Pitesti: University of Pitesti Publishing House)

[6] Kimber R 1980 The traffic capacity of roundabouts, *Transport and Road Research Laboratory Report LR942* (Crowthorne)

[7] Martin-Gasulla M, Garcia A and Tsui Moreno A 2016 Benefit Measurement of Metering Signals at Roundabouts with Unbalanced Flow: Patterns in Spain, *Transportation Research Record: Journal of the Transportation Research Board* 2585, p. 20-28

[8] Neagu E 2003 *Trafic rutier şi siguranta circulatiei* (Pitesti: University of Pitesti Publishing House)

[9] Steenbrink P A 1984 *Optimization of transport network* (New York: John Wiley & Sons)

[10] PTV Vision Software 2011 VISUM *User Manual* PTV AG Karlsruhe (Berlin: epubli GmbH)

[11] Transportation Research Board (TRB) of National Academies of Science Engineering Medicine 2010 *Highway Capacity Manual* (Washington: National Academies of Sciences)

[12] http://www.primariapitesti.ro *Proiect in dezbatere publica: Planul de Mobilitate Urbana Durabila al municipiului Pitesti*

Solutions to Improve Person Transport System in the Pitesti City by Analyzing Public Transport vs. Private Transport

Istrate Mihaela, Boroiu Alexandru, Nicolae Viorel, Vieru Ionel

University of Pitesti, Targu din Vale street, no.1, code 110040, Romania

mmihaela_1971@yahoo.com

Abstract. One of the major problems facing the Pitesti city is the road congestion that occurs in the central area of the city during the peak hours. With all the measures taken in recent years - the widening of road arteries, increasing the number of parking spaces, the creation of overground road passages - it is obvious that the problem can only be solved by a new philosophy regarding urban mobility: it is no longer possible to continue through solutions to increase the accessibility of the central area of the city, but it is necessary, on the contrary, to promote a policy of discouraging the penetration of vehicles in the city center, coupled with a policy of improving the connection between urban public transport and county public transport. This new approach is also proposed in the new Urban Mobility Plan of Pitesti city, under development. The most convincing argument for the necessity of this new orientation in the Pitesti city mobility plan is based on the analysis of the current situation of passenger transport on the territory of Pitesti city: the analysis of "public transport versus private transport" reveals a very low occupancy rate for cars and the fact that the road surface required for a passenger (the dynamic area) is much higher in the case of private transport than in the case of public transport. Measurements of passenger flows and vehicle flows on the 6 penetration ways in the city have been made and the calculations clearly demonstrate the benefits of an urban public transport system connected by "transshipment buses" to be made at the edge of the city, to the county public transport system. In terms of inter-county transport, it will continue to be connected to the urban public transport system by existing bus Station, within the city: South Bus Station and North Bus Station. The usefulness of the paper is that it identifies the solutions for sustainable mobility in Pitesti city and proposes concrete solutions for the development of the integrated passenger transport system.

1. Objectives of the paper

One of the major problems facing the city of Pitesti is the road congestion that occurs in the central area of the city during the peak hours of traffic [11, 12].

With all the measures taken in recent years - the widening of road arteries, increasing the number of parking spaces, the creation of overground road passages - it is obvious that the problem can only be solved by a new philosophy regarding urban mobility: it is no longer possible to continue through solutions to increase the accessibility of the central area of the city, but it is necessary, on the contrary, to promote a policy of discouraging the penetration of vehicles in the city center, coupled with a policy of improving the connection between urban public transport and county public transport.

These ideas have been promoted since 2008 [5, 6, 7] when it was proposed to extend the urban public transport network of Pitesti city to the periurbane area and to connect this extended local public

IOP Conf. Series: Materials Science and Engineering **252** (2017) 012054 doi:10.1088/1757-899X/252/1/012054

transport system with the county transport system by means of transhipment stations to be carried out at the edge of the future metropolitan area.

These proposals have not been put into operation, and recent research [4] has shown that road congestion has become a daily phenomenon at several points on the city street network.

The realization of the new sustainable mobility plan for Pitesti city [8] offered the opportunity for a detailed assessment of the demand for passenger transport within the territory of Piteşti city, so that the following objective could be set: evaluation of the demand for transport of persons and identifying the measures by which it can become more efficient in relation to traffic generated by means of transport - public transport versus private transport.

2. The researches and results

According to the official statement, "The Pitesti Municipality proposes the elaboration of an integrated urban mobility plan, which is a strategic document meant to plan in a sustainable way the fulfillment of the mobility needs of the people and of the economic sector in the city and its surroundings, for a harmonious socio-economic development and a better quality of life, today and in the future. This strategic document harnesses the existing planning practices for all modes of transport for people and goods and takes into account the principles of integration, socio-economic efficiency, participation and evaluation."[12].

The area of influence and the external area that affects the population mobility in the Pitesti city includes all the neighboring towns and villages (Ştefăneşti, Bradu, Albota, Moşoaia, Băbana, Bascov, Marăcineni) plus Mioveni, the industrial pole nearby, as shown in Figure 1 [12].

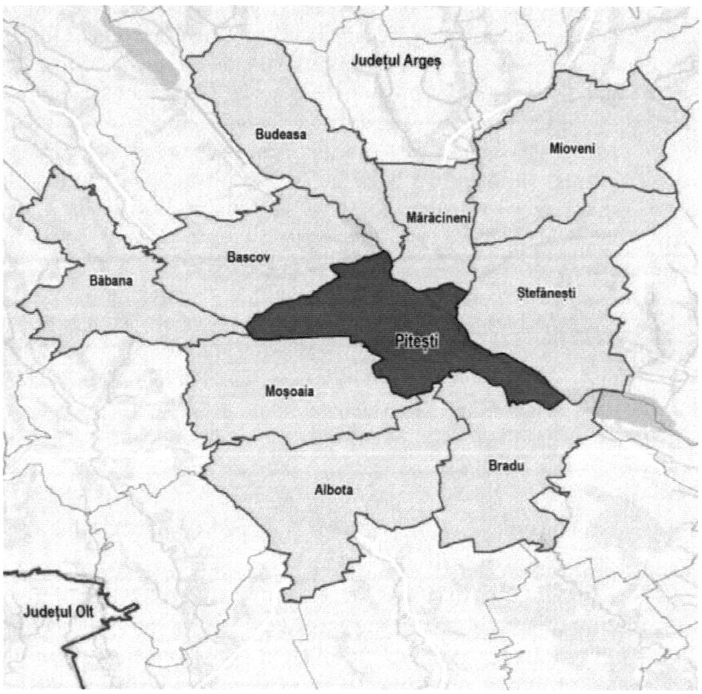

Figure 1. The influence area for the mobility of the people of Piteşti city.

In order to make the ex-post assessment of public transport demand within the influence zone, there were counted the travellers in time-travels within the six main inlets / outlets of the Pitesti city, for the connecting roads with the area of influence shown in Table 1, the survey points being highlighted in Figure 2.

Table 1. Recording stations for volume of passengers and the connecting roads with zone of influence.

Inlets / outlets	*Connecting roads with the Pitești city*
Viilor	DN73, DN7
Prundu	DN65B
Craiovei	DN65
Drăgășani	DN67 B
Băbana	DJ 703 E
Bascov	DN 7, DN7A

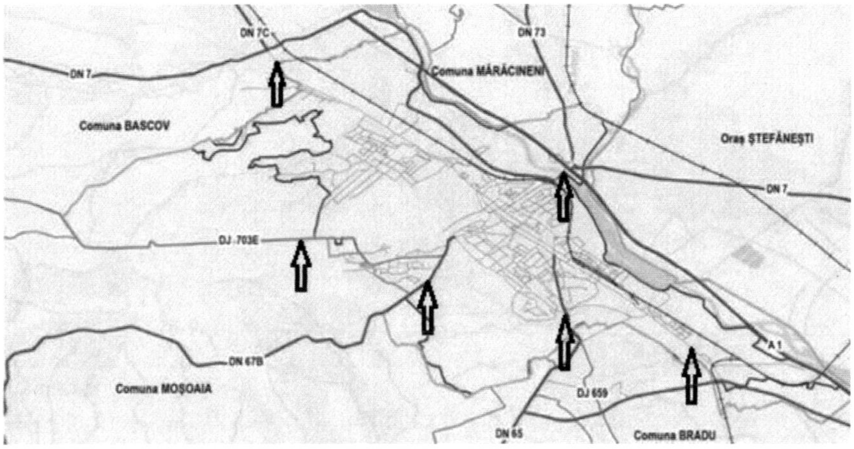

Figure 2. The six inlets / outlets of the Pitești city where the passengers in the buses were counted.

The records in the aforementioned locations took place between 06:00 and 22:00, over a period of 16 hours.

The situation with the number of passengers per hour, for each recording station and cumulated at the city level, is presented in Table 2 [8].

Table 2. Number of passengers recorded on time intervals.

Hourly Interval	Bascov		Babana		Dragasani		Prundu		Craiovei		Viilor		TOTAL	
	I	E	I	E	I	E	I	E	I	E	I	E	I	E
6.00-7.00	267	38	51	4	98	64	73	453	476	118	446	1171	1411	1848
7.00-8.00	236	127	47	26	64	85	93	115	613	189	747	702	1800	1244
8.00-9.00	113	72	47	9	128	64	73	95	381	179	775	258	1517	677
9.00-10.00	74	52	21	0	81	98	50	55	389	84	552	327	1167	616
10.00-11.00	38	39	0	4	59	75	64	43	30	47	213	430	404	638
11.00-12.00	107	97	12	21	59	40	34	55	268	149	353	535	833	897
12.00-13.00	129	143	26	12	44	88	51	38	234	183	256	557	740	1021
13.00-14.00	157	121	77	26	110	92	135	77	607	182	305	394	1391	892
14.00-15.00	132	206	4	51	82	58	90	455	132	201	672	801	1112	1772
15.00-16.00	256	105	21	30	92	160	479	93	147	478	1287	627	2282	1493
16.00-17.00	137	158	9	60	52	167	80	196	128	373	994	460	1400	1414
17.00-18.00	97	96	9	38	64	172	109	114	144	349	654	512	1077	1281
18.00-19.00	230	115	9	17	39	110	229	252	136	300	358	604	1001	1398
19.00-20.00	86	110	0	55	33	138	145	98	183	252	285	587	732	1240
20.00-21.00	73	110	0	21	6	77	91	17	64	187	214	317	448	729
21.00-22.00	50	40	4	0	59	33	9	69	55	9	140	122	317	273
TOTAL	**2182**	**1629**	**337**	**374**	**1070**	**1521**	**1805**	**2225**	**3987**	**3280**	**8251**	**8404**	**17632**	**17433**

IOP Conf. Series: Materials Science and Engineering **252** (2017) 012054 doi:10.1088/1757-899X/252/1/012054

Graphical presentation of the hourly distribution of passenger flows by public transport means shows that they all have a similar allotment, with maximum transport demand in the morning and afternoon (like in Figure 3 for Podul Viilor), and the same allure has the graph for the cumulative values at all those six inputs / outputs (Figure 4).

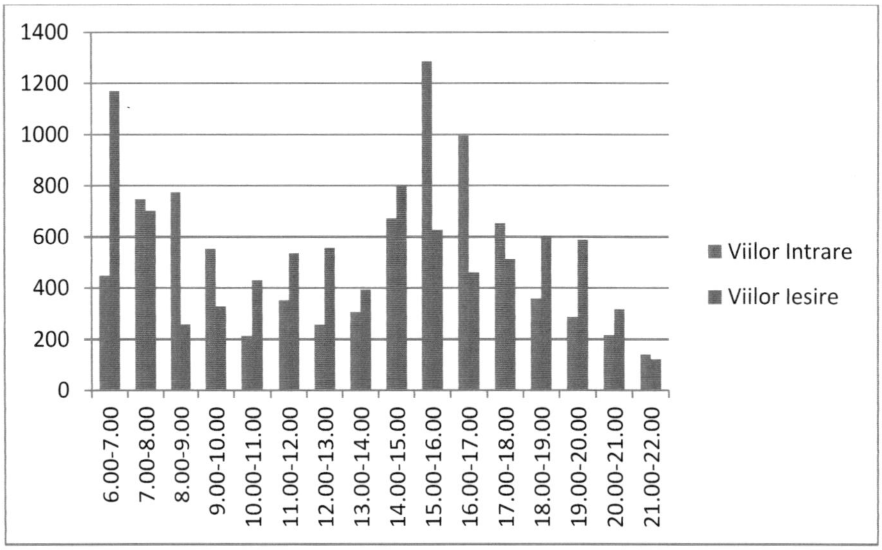

Figure 3. Hourly distribution of bus passenger flows at Podul Viilor recording station.

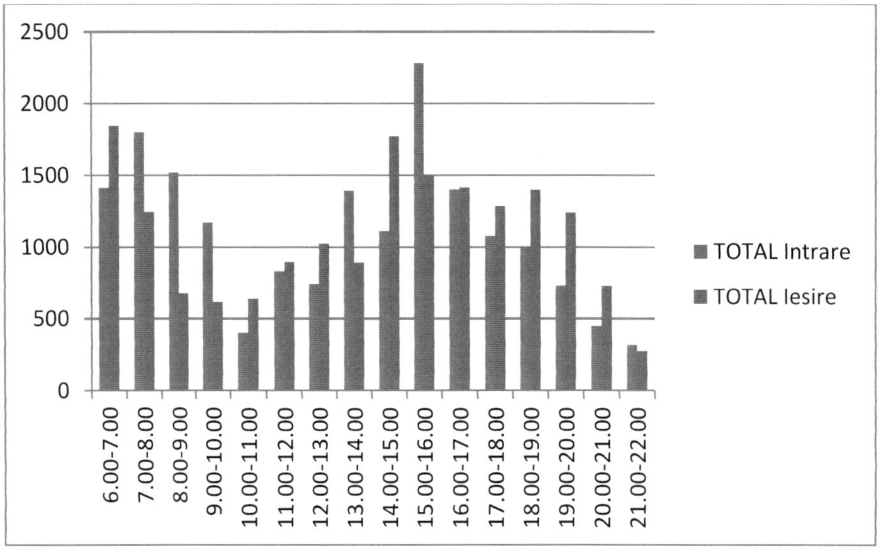

Figure 4. Hourly distribution of cumulative bus passenger flows for Piteşti city.

The largest passenger traffic flow was found at Podul Viilor (Figure 5), as expected, because this is the only main road connecting Piteşti city with important industrial destinations in the area of influence: Ştefăneşti and Mioveni.

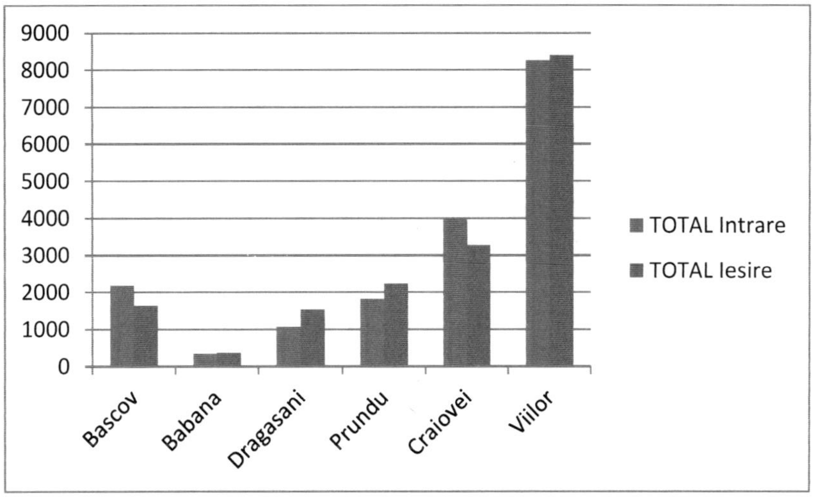

Figure 5. Daily distribution of passenger flows within the six inlets / outlets of the Pitești city.

During the same time period there were developed six O-D surveys (Origin-Destination) for all vehicles traveling between Pitesti city and its zone of influence [9, 12].

There have been established seven locations for the O-D surveys, creating a "belt" around Pitesti city, and thus all the traffic flows to / from the influence area of Pitesti city was recorded (Figure 6).

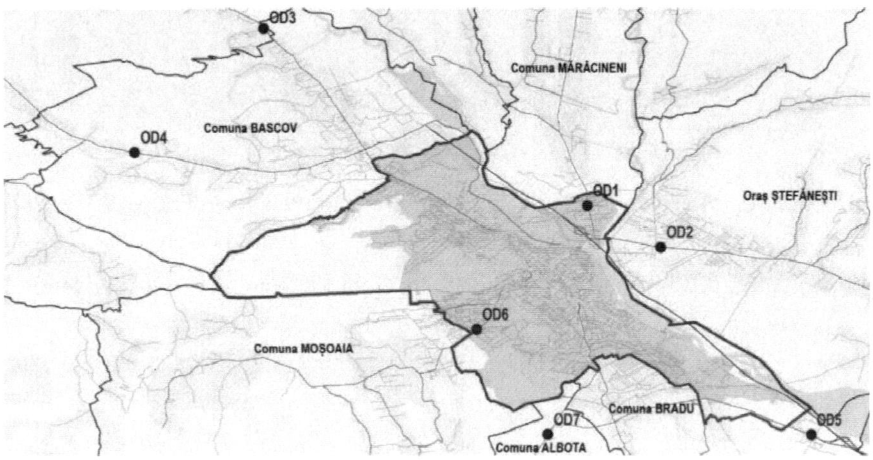

Figure 6. Location of the six recording points for the O-D surveys.

There were developed seven O-D surveys in the locations presented in Table 3 [12].

Table 3. Position of the seven recording points for the O-D surveys.

O-D survey point	*Position of the survey points on the connecting roads with the Pitești city*
OD 1	DN73 Calea Câmpulung, nearby Milestone 1 - Câmpulung 49 km
OD 2	DN7, at the IATSA Hotel
OD 3	DN7C, nearby Milestone 4 - Curtea de Argeș 28 km
OD 4	DN7, nearby Milestone 124 - Râmnicu Vâlcea 50 km
OD 5	DN65B, nearby Jupiter City Shopping Centre
OD 6	DN67B, nearby Capitol Hotel within Trivale residential quarter
OD 7	DN65B, nearby Tiriac Auto car dealer in Albota

For traffic flows, the following categories were recorded: bicycles, motorcycles, scooters, mopeds, passenger cars, taxis, minibuses, buses, vans, pick-ups, 2-axle trucks and derivatives, 3-axle or 4-axle trucks and derivatives, articulated vehicles (5 or more axles), special vehicles, tractors.

The O-D surveys and records in the aforementioned locations took place between 06:00 and 21:00, 15 hours.

The distribution of the vehicle volumes traveling to the areas of influence of the Pitesti city (and therefore of the passenger volumes using private transport or for own interest) recorded during 15 minutes intervals is represented in Figure 7 and it could be observed that it has a similar shape to the distribution of the passenger flows using public transport, which is in line with the demand for passenger transport in Pitesti city and its zones of influence [12].

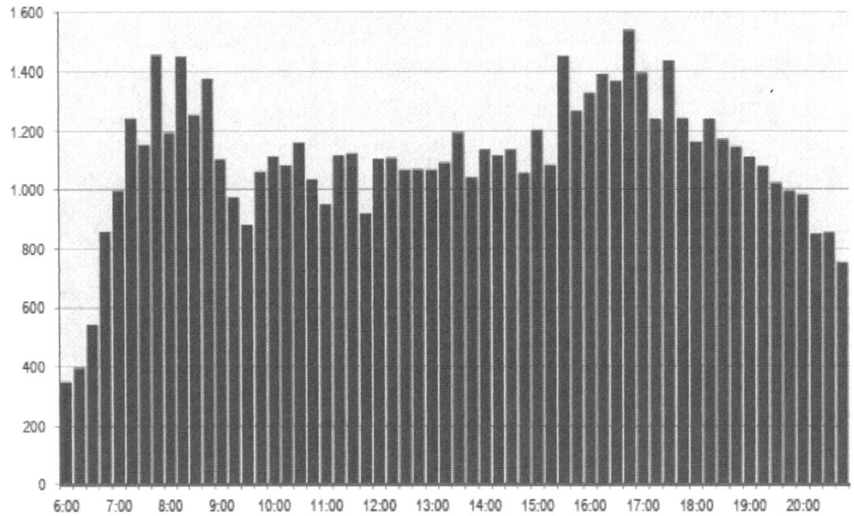

Figure 7. Distribution of vehicle flows during 15 minute intervals from Piteşti city to its zones of influence [12].

A similar distribution it was recorded for the traffic from the zones of influence to the Piteşti city (Figure 8).

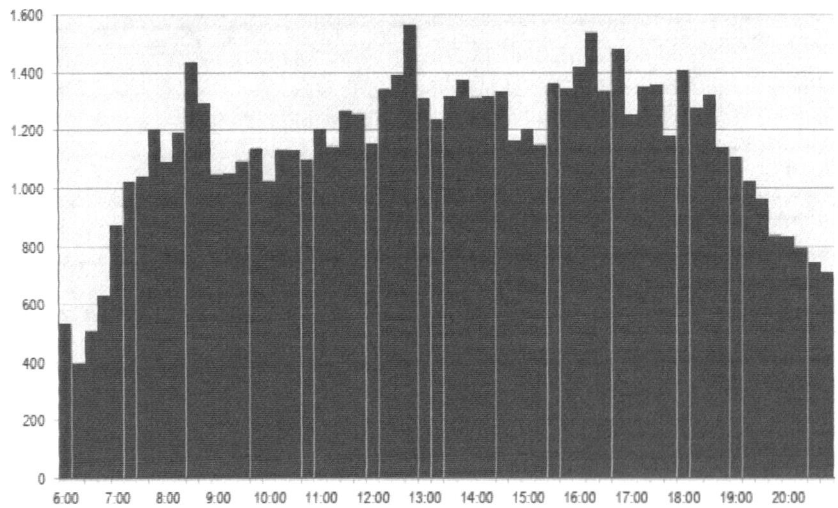

Figure 8. Distribution of vehicle flows during 15 minute intervals from the zones of influence to the Piteşti city [12].

Thus, it was found that 50000 vehicles go in and out from Pitesti city every day, out of which cars are around 38000 (about 75%).

From the distributions shown in Figures 7 and 8 it was found that, at the peak hours, the hourly traffic volume is about 5000 vehicles / hour, a very high value which can be taken into account in order to recommend to people the alternative of traveling using public transport.

Observations about the load of passenger cars revealed an average number of occupants of 1.25 passengers / car, which means that the daily volume of travellers with their own car is about 1.25 x 38000 = 47500 passengers / day / sense.

This magnitude can be analyzed in relation to the daily number of travellers with the public transport, which based on the data presented in Table 1 is about 17500 passengers / day / sense.

The most relevant analysis criterion is the dynamic area occupied for a passenger by the means of transport in the two cases - bus versus passenger car.

A first assessment can use the comparative information already known through numerous campaigns to encourage the use of public transport despite of passenger cars. Thus, a relevant picture is that presented in Figure 8, where it is highlighted that a bus carrying 75 passengers can replace 60 cars with an average load of 1.25 passengers per car, which would lead to a very large reduction of the occupied road surface, the source of information being an issue of the European Commission [13].

Figure 9. The road surface occupied by passenger cars or bus carrying 75 passengers [13].

The picture is relevant, but this comparison on the busy road surface can even be done based on an analytical calculation, for a legal speed of 50 km/h in the city.

Since each of the two types of means of transport (the bus or the passenger car) occupies the entire lane on which it moves, the problem is reduced to determining only the length of the area of the occupied dynamic surface, which is calculated from the condition of safety moving, represented by the condition to maintain a safe distance from the front vehicle which allows the driver to safely stop in the safe and comfortable conditions [3].

This means that when the front vehicle make an emergency braking (with the maximum brake deceleration), the vehicle of the traveler will brakes in comfort conditions (with a mean deceleration).

As a result, it could be considered the model in which the safety distance Lf is equal to the distance traveled in the braking process with a deceleration of 5 m/s2 for both types of transport [4]:

$$L_f = \frac{v^2}{2 \cdot d} = \frac{\left(\frac{50}{3.6}\right)^2}{2 \cdot 5} = 19.2\ m \qquad (1)$$

Thus, the length of the dynamic surface occupied by a passenger car (having a length L_{car} of about 4.5 m) will be:

$$L_{din,\ car} = L_{car} + L_f = 5.5 + 19.2 = 24.7\ m \tag{2}$$

For the bus (having a length L_{bus} of about 12 m), the length of the dynamic surface will be:

$$L_{din,\ bus} = L_{bus} + L_f = 12 + 19.2 = 31.2\ m \tag{3}$$

Based on these values, the ratio between the dynamic occupied area for transporting the same number of passengers (75) can be calculated for 60 passenger cars carrying 1.25 passengers/passenger car $S_{din,60car}$ compared to a medium-sized urban bus $S_{din,bus}$ with the following relation:

$$\frac{S_{din,60car}}{S_{din,bus}} = \frac{60 \cdot 24{,}7}{31{,}2} = 47.5 \tag{4}$$

This is a concrete value, which precisely characterizes the phrase "public transport versus private transport" in terms of the efficiency of the use of the road network in the passenger transport [1, 10].

It was found that in both transport directions (go in or out from Pitesti city), about 38000 passenger cars perform daily the transport of about 47500 persons in their own schedule and that for a person travelling with his own car it is necessary a road surface of 47.5 times higher than for travellling by public transport. These data are extremely useful in formulating proposals to improve the passenger transport in the considered area.

3. Conclusions

The researches carried out in the Pitesti city and its influence areas for the comparative analysis "public transport versus private transport" lead to the conclusion that, in order to improve the level of service of the road network in the Pitesti city, there are needed measures to encourage passengers to use public transport and, as a result, to give up their own passenger car journeys.

This can be achieved by extending the public transport network from the level of the Pitesti city to the level of the proximity influence area (the so-called peri-urban area), which can be legally constituted in the metropolitan area.

This will allow the development of an local public transport system extended throughout the area of influence so that it will be created more attractive public transport conditions for journeys that exceed the administrative limit of the Pitesti city.

These measures made to increase the attractiveness of public transport can be complemented by measures to discourage the use of passenger cars for own use in the central area of Pitesti city, the most easily applied measure being the rates increase in the public parking of the central area.

Regarding the access of cars in the central area of the Pitesti city, the strategy is: "not to meet the increasing demand for access of passenger cars in the central area, but to create conditions for changing the mode of transport at the borders of the metropolitan area - private transport / public transport".

This involves building car parks near the main entry streets in the metropolitan area, where travelers have easy access to the extensive local public transport so as to be motivated to travel further with public transport.

IOP Conf. Series: Materials Science and Engineering **252** (2017) 012054 doi:10.1088/1757-899X/252/1/012054

References

[1] Boroiu, A., *Transporturi de persoane*, Editura Universității din Pitești, 2009.

[2] Boroiu, A., Nicolae, V., Ilie, S., Zaharia, C., Soare, C. *Proiectarea strategiei de transport public de persoane in judetul Arges pentru perioada 2008-2013* si *proiectarea programului de transport public de persoane in judetul Arges pentru perioada 2008-2011*", Contract nr. 541/07.02.2008, Beneficiar: Consiliul Judetean Arges.

[3] Boroiu, A-A., Neagu, E., *Trafic rutier și siguranța circulației rutiere. Aplicații*, Editura Universității din Pitești, 2015.

[4] Boroiu, A-A, *Studii și cercetări privind reducerea poluării fonice produse de autovehicule prin organizarea circulației rutiere*, Teză de doctorat, Universitatea din Pitești, 2017.

[5] Bratu, M. - *Cercetări cu privire la creșterea performanțelor transportului public urban*, Teză de doctorat, Universitatea Politehnica din București, 2007

[6] Bratu, V., Boroiu, A., Tabacu, I. – *Researches regarding the improvement and the modernization on the public passengers transport in the peripheral area of Pitesti,* the 5th International Conference on Advanced Concepts in Mechanical Engineering, 14-15 June 2012, Bul. Inst. Polit. Iași, 2012

[7] Ciobâcă, A-M. - *Cercetări privind creșterea atractivității transportului public în marile aglomerări urbane*, Teză de doctorat, Universitatea Politehnica din București, 2015

[8] Nicolae V., Neagu, M., Istrate, M., Boroiu, A., Vieru, I., Șuster, H., Boroiu, A-A *Planul de Mobilitate Urbană Durabilă al municipiului Pitești. Culegerea și prelucrarea datelor de mobilitate*, Acord de subcontractare nr. 10581/22.08.2016, Beneficiar: S.C. SYNERGESTICS CORPORATION SRL București

[9] PTV AG (2010-1) - *PTV (Planung Transport Verkehr) Vision: Visum 11,5 - Basics*, Karlsruhe, 2010

[10] Steenbrink, P. - *Optimization of transport network*, Ed. John Wiley & Sons, New York, 1984

[11] Tabacu, I., Boroiu, A., Nicolae, V., Ivan, F. et al. - *Modernizarea transportului public de persoane în municipiul Pitești. Studiu de fezabilitate,* Contract de cercetare nr. 52230 / 10.11.2006, Beneficiar: Primăria municipiului Pitești

[12] http://www.primariapitesti.ro *Proiect în dezbatere publică: Planul de Mobilitate Urbană Durabilă al municipiului Pitești*

[13] http://www.autofreie-stadt.de

Proposal for the Use of a New Parameter in the Composite Road Traffic Model

A Boroiu, I Tabacu, A A Boroiu, E Neagu and I Vieru

Automotive and Transports Department, University of Pitesti, 1, Targul din Vale Street, 110040, Pitesti, Romania

E-mail: alexandru.boroiu@upit.ro

Abstract. Comparative analysis of the models used in the literature for the fundamental traffic diagram reveals that the improvement of these models can only be achieved by taking into account the vehicle tracking models. Given that it is of great interest to model the relation between density and speed, to identify the conditions in which maximum flow can be achieved with a high operating speed and safely. For the non-homogeneous traffic, met in our country due to the lack of highways, the model describing the minimum range of vehicles was developed based on the relation of the minimum distance between vehicles taking into account the difference between the vehicle braking modes. Finally it is explained the maximum volume of traffic according to a parameter that includes the decelerations of the two vehicles, which determined the proposed use of this new concept in road traffic theory. This has made it possible to conclude that in the case of non-homogeneous traffic the volume increases only to a certain speed value, then decreases, while in the case of homogeneous traffic the volume continues to increase to a limit value. The paper argues the necessity of using this new parameter for road traffic models.

1. Papers objectives

The comparative analysis of the various models used in specialised literature for the fundamental traffic diagram (single-regime models: Greenshields, Greenberg, Underwood, Northwestern, multi-regime models: Edie, linear-two regimes, modified Greenberg, linear-three regimes...) leads to the conclusion that the improvement of these models can only be achieved by taking into account the vehicle tracking models, which involves the microscopic modelling of road traffic flows [2].

Given that what is of great practical interest is the modelling of the relation between density and speed in order to identify the conditions in which maximum flow can be achieved (this is the interest of the road infrastructure designer), but also with an operating speed as high as possible (this is the interest of the user of the infrastructure, who wishes to reach his/her destination as soon as possible), and under safe circumstances (this is true for all parties involved), a detailed analysis of the most interesting cases has been carried out to obtain maximum traffic (to ensure high efficiency for road infrastructure) at high operating speeds (to ensure the economical use of vehicles) and under safe circumstances [5].

The analysis of traffic flows as a whole (macroscopically) also uses the parameters of the microscopic traffic flow - which describe the flow behaviour of the individual vehicle (singular vehicle speed) or of the pairs of vehicles inside the flow (the intervals between successive vehicles) [9].

Therefore, it is important for the vehicle range inside the flow to be able to be evaluated, a range that can be measured as time interval or space interval. The intervals between vehicles are primarily dependent on the flow velocity (they increase by the flow rate, for security reasons), but depend largely on the characteristics and condition of the track, these factors being the same for all road users. But, for the same conditions of speed and road, the intervals between vehicles are not constant since they depend on the other two factors in the traffic system - the driver (the human factor) and the vehicle (the technical factor), which causes a pronounced variability of the intervals between vehicles [7]. If, in what the technical factor is concerned, in the case of traffic on upgraded arteries, with many lanes in each traffic direction, the flows get homogenised (because vehicles with similar technical characteristics and dynamic performance will drive on the same lane), in the case of the vehicle drivers, their very different driving style is the main reason for the variability of the intervals between vehicles of the same type on the same lane.

As a result, since, within the actual flows, a constant interval between the vehicles cannot be maintained, the most suitable modelling should be based on a statistical approach, in view of the diversity of temper types, so it is necessary to evaluate the average of intervals [6]. Thus, the objective of this paper is to identify ways to improve the performance of traffic by analysing the intervals between successive vehicles.

2. Research undertaken

Empirical data are used to calculate the statistical indicators of vehicle intervals distribution within the flow, starting from the definition of the intervals depending on the measurement method. The interval between vehicles is an important quantitative parameter of traffic flows, by means of which performance of the road system or the level of road service can be measured, as well as the efficiency of traffic signals.

The time interval between vehicles is a measure of the time needed for the passage of two successive vehicles in the flow, through a designated benchmark on the roadside. The measure of the level of service of a road artery is closely linked to the time intervals between vehicles and can be expressed in two ways:

- the weight of the time interval in which a vehicle is forced to follow another vehicle on a road with two lanes;
- frequency of speed of travel adjustment manoeuvres a driver performs in order to keep a minimum time interval to the vehicle in front.

The movement of the two vehicles is shown in figure 1, where the following notations are used:

- l_{v1} – length of the vehicle in front (the first vehicle, the vehicle followed);
- d_{min} – space of the vehicle in the back (the second vehicle, the vehicle that is following) and the vehicle followed;
- s_0 – safety space;
- t_r – response time of the driver of the vehicle that is following;
- s_1 – the space covered by the vehicle in front in the time interval t_{f1}, braking with the deceleration a_{f1} to a halt;
- s_2 – the space covered by the vehicle in the back, made up of two components; the space covered at the speed v in the response time t_r, until the driver operates the braking system; the space covered in the time interval t_{f2} braking with the deceleration a_{f2} to a halt.

The analytical equation that indicates the minimum distance to be maintained between the two vehicles during movement at speed v is rendered in what follows:

$$l_v + d_{min} + s_1 = s_2 + l_v + s_0 \quad (1)$$

There thus ensues the value for the minimum distance d_{min} function of the parameters that define the movements of the two vehicles (speed v, driver's response time t_r, the decelerations of the two

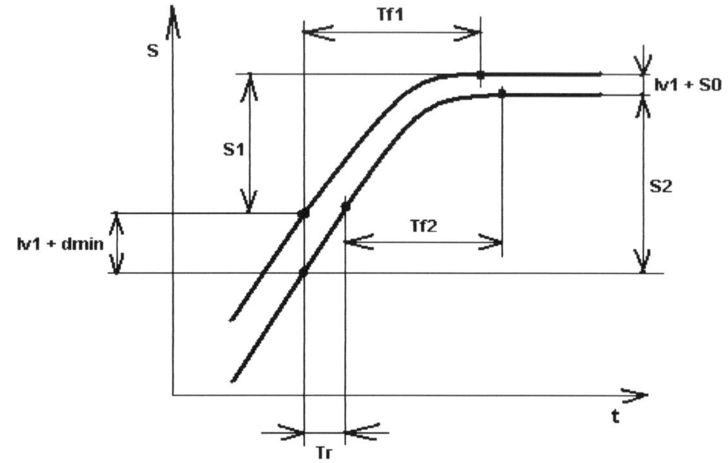

Figure 1. Graphical representation of the distance between two vehicles that follow each other if the first vehicle brakes to a halt [3].

vehicles, a_{f1} și a_{f2}, and the safety space s_0:

$$d_{min} = s_2 + s_0 - s_1 = \left(v \cdot t_r + \frac{v^2}{2a_{f2}}\right) + s_0 - \frac{v^2}{2a_{f1}} = v \cdot t_r + \frac{v^2}{2}\left(\frac{1}{a_{f2}} - \frac{1}{a_{f1}}\right) + s_0 \qquad (2)$$

There are two cases to be discussed for the minimum distance:

1 – homogeneous traffic: $a_{f1} = a_{f2}$ - the optimum case because the minimum distance between vehicles will have the lowest value:

$$d_{min} = v \cdot t_r + s_0 \qquad (3)$$

In this case the variability range of vehicles is given by the different options for drivers, depending on their physical and mental characteristics.

2 – non-homogeneous traffic: $a_{f1} \neq a_{f2}$

In this case, the vehicles have different braking capacities, and the succession of the various vehicles in the flow is random, so that the minimum interval between vehicles can be described through the equation that uses the absolute value for the difference ($1/a_{f1} - 1/a_{f2}$):

$$d_{min} = v \cdot t_r + \frac{v^2}{2}\left|\frac{1}{a_{f2}} - \frac{1}{a_{f1}}\right| + s_0 \qquad (4)$$

Thus, for two cars that follow each other and that are moving at high speed, there are three possible cases [8]:

a) both the vehicle followed and the vehicle that is following brake under emergency conditions – given the fact that the vehicle that is following brakes under emergency conditions, this is the case of *maximum safety* (take into account the fact that the vehicle followed will hit the brakes vigorously) and *discomfort* for the vehicle that is following (because it has to hit the brakes vigorously);

b) the vehicle followed brakes under emergency conditions and the vehicle that is following brakes under normal conditions – this is the case of *maximum safety* (take into account the fact that the vehicle followed will hit the brakes vigorously) and *comfort* for the vehicle that is following (which does not have to hit the brakes vigorously, it will brake under normal conditions);

c) both the vehicle followed and the vehicle that is following brake under normal conditions – this is the case of *reduced safety* (one does not take into account the fact that the vehicle that is following could brake vigorously, one takes into account only the variant in which it brakes under normal conditions) and *comfort* for the vehicle that is following (which does not have to hit the brakes vigorously, it will brake under normal conditions).

It can be noticed that, when the vehicles brake in the same way (with the same decelerations), the minimum required distance between them is the same regardless of the speed of travel. But to ensure safe travel in maximum comfort, the case that is considered to be appropriate is case b) *the vehicle followed brakes in emergency, and the vehicle that is following brakes under normal conditions*.

Of great practical interest, however, is the modelling of the relation between density and speed to identify conditions in which a maximum flow rate can be obtained. As a result, further on there will be examined the most interesting cases to obtain a high volume of traffic at high operating speed (to ensure economic use of the vehicle), under safe conditions. Thus, the most interesting case, not only from a theoretical point of view, is that providing the highest traffic volume possible, which is the one that allows that a minimum distance d_{min} be kept between vehicles (this is the case of **homogeneous traffic**, when the vehicles belong to the same category), a case in which the distance between the vehicles that can brake with similar decelerations is rendered through the equation (3), presented before. The analytical equation for the interval between two vehicles I_v that succeed one another during the movement at speed v is:

$$I_v = l_v + d_{min} = l_v + v \cdot t_r + s_0 \tag{5}$$

where l_v is the length of the vehicle. As a result, traffic density will be:

$$k\left[\frac{veh}{km}\right] = \frac{1000\ m}{l_v[m]+v\left[\frac{m}{s}\right]\cdot t_r[s]+s_{0[m]}} = \frac{1000}{l_v[m]+\frac{1}{3.6}v\left[\frac{km}{h}\right]\cdot t_r[s]+s_0[m]} \tag{6}$$

Thus, the maximum volume of homogenous traffic will be:

$$q\left[\frac{veh}{h}\right] = k\left[\frac{veh}{km}\right]\cdot v\left[\frac{km}{h}\right] = \frac{1000}{l_v[m]+\frac{1}{3.6}v\left[\frac{km}{h}\right]\cdot t_r[s]+s_0[m]}\cdot v\left[\frac{km}{h}\right] = \frac{3600}{\frac{l_v[m]+s_0[m]}{v\left[\frac{km}{h}\right]}+t_r[s]} \tag{7}$$

Analyzing the last equation, it is to be noted that the first order derivative of the function q (v) is always positive and continuously decreasing:

$$\frac{dq}{dv} = \frac{l_v+s_0}{v^2} \tag{8}$$

so it is not cancelled for any value of speed v, which means that, in this case (which is not just theoretical, but is rather frequently met with on highways, where it can be considered that each lane is populated by vehicles that are equipped with similar braking capacities) the maximum volume of traffic q continuously increases along with increasing the operation speed (from value zero, for $v \to 0$), but increasingly less, aiming asymptotically at (for $v \to \infty$) the value $3600/t_r[s]$ – figure 2. As, in the present, vehicles of the same type are comparable in braking performance, it appears that the only factor that makes a difference in terms of the required distance between vehicles is the drivers' behaviour, in terms of response time. The ideal situation is that where highways have three lanes, because trucks will drive on the first lane (which have a maximum speed limit), and cars will drive on the second and third lanes (on the second lane at an average speed, and on the third lane at a higher speed). This is how the „layering" of cars is performed on the second and third lanes, depending on the speed adopted and therefore the response time characterizing the drivers of the respective vehicles.

A second case is that of a ***non-homogeneous traffic*** (with various categories of vehicles - cars, trucks, buses ...), when the vehicle decelerations values are different, so there are situations when the vehicle followed brakes by a deceleration greater than the vehicle that is following [3]. The model for this second case is based on the relation that describes the minimum interval between vehicles, in view of the difference of braking modes of vehicles that are part of the road flow (different decelerations) - equation (4), presented above. The analytical equation for the interval between two vehicles that succeed one another during the movement at speed v is in this case:

$$I_v = l_v + d_{min} = l_v + v \cdot t_r + \frac{v^2}{2}\left|\frac{1}{a_{f2}} - \frac{1}{a_{f1}}\right| + s_0 \quad (9)$$

The traffic density will be:

$$k\left[\frac{veh}{km}\right] = \frac{1000\,m}{l_v[m] + v\left[\frac{m}{s}\right] \cdot t_r[s] + \frac{(\frac{v}{3.6})^2}{2}\left|\frac{1}{a_{f2}} - \frac{1}{a_{f1}}\right| + s_{0[m]}} \quad (10)$$

where speed v is expressed in [km/h]. As a result, the maximum volume of non-homogeneous traffic will be:

$$q\left[\frac{veh}{h}\right] = \frac{1000\text{ m}}{l_v[m] + \frac{1}{3.6} \cdot v\left[\frac{km}{h}\right] \cdot t_r[s] + \frac{(\frac{v}{3.6})^2}{2}\left|\frac{1}{a_{f2}} - \frac{1}{a_{f1}}\right| + s_{0[m]}} \cdot v\left[\frac{km}{h}\right] = \frac{1000}{\frac{l_v + s_0}{v} + \frac{1}{3.6} \cdot t_r + \frac{v}{2 \cdot 3.6^2} \cdot \frac{|a_{f1} - a_{f2}|}{a_{f1} \cdot a_{f2}}} \quad (11)$$

It is to be noted that, in this case, the first order derivative of the function q(v) is cancelled for a certain value of the speed, at which traffic volume records a maximum [3], a value that corresponds to a maximum traffic volume and is noted as v_M:

$$\frac{dq}{dv} = 0 => -\frac{(l_v + s_0)}{{v_M}^2} + \frac{1}{2 \cdot 3.6^2} \cdot \frac{|a_{f1} - a_{f2}|}{a_{f1} \cdot a_{f2}} = 0 \quad (12)$$

It results:

$$v_M\left[\frac{km}{h}\right] = 3.6 \cdot \sqrt{\frac{2 \cdot (l_v + s_0) \cdot a_{f1} \cdot a_{f2}}{|a_{f1} - a_{f2}|}} = 3.6 \cdot \sqrt{\frac{2 \cdot (l_v + s_0)}{\frac{|a_{f1} - a_{f2}|}{a_{f1} \cdot a_{f2}}}} = 3.6 \cdot \sqrt{\frac{2 \cdot (l_v + s_0)}{C}} \quad (13)$$

where, for simplicity reasons, the following notation was introduced:

$$C = \frac{|a_{f1} - a_{f2}|}{a_{f1} \cdot a_{f2}} \quad (14)$$

C is a coefficient that expresses the variability of braking decelerations for the vehicles in the continuous flow [4], a value inverse to deceleration: C $[1/(m/s^2)]$. Considering further on speed v expressed in [km/h] and the newly-adopted coefficient C (that can be referred to as a “deceleration variation coefficient”), the highest value for the maximum traffic volume will be:

$$q_M\left[\frac{veh}{h}\right] = \frac{1000}{\frac{l_v + s_0}{v_M} + \frac{1}{3.6} \cdot t_r + \frac{v_M}{2 \cdot 3.6^2} \cdot \frac{|a_{f1} - a_{f2}|}{a_{f1} \cdot a_{f2}}} = \frac{3600}{t_r + \sqrt{2 \cdot (l_v + s_0) \cdot \frac{|a_{f1} - a_{f2}|}{a_{f1} \cdot a_{f2}}}} = \frac{3600}{t_r + \sqrt{2 \cdot (l_v + s_0) \cdot C}} \quad (15)$$

The graph presenting the maximum traffic volume for the two cases is shown in figure 2 [1]. It is to be noticed that, in fact, the case of homogeneous traffic is a particular case of the model developed for non-homogeneous traffic: the case $a_{f1} = a_{f2}$, i.e. when there is no deceleration variability, so the deceleration variability coefficient is null: C = 0. For this model it is equally interesting to determine the traffic density that corresponds to the maximum traffic volume:

$$k_M\left[\frac{ve}{km}\right] = \frac{q_M\left[\frac{ve}{h}\right]}{vk_M\left[\frac{km}{h}\right]} = \frac{\frac{3600}{t_r + \sqrt{2 \cdot (l_v + s_0) \cdot C}}}{3.6 \cdot \sqrt{\frac{2 \cdot (l_v + s_0)}{C}}} = \frac{1000}{2 \cdot (l_v + s_0) + t_r \cdot \sqrt{\frac{2 \cdot (l_v + s_0)}{C}}} \quad (16)$$

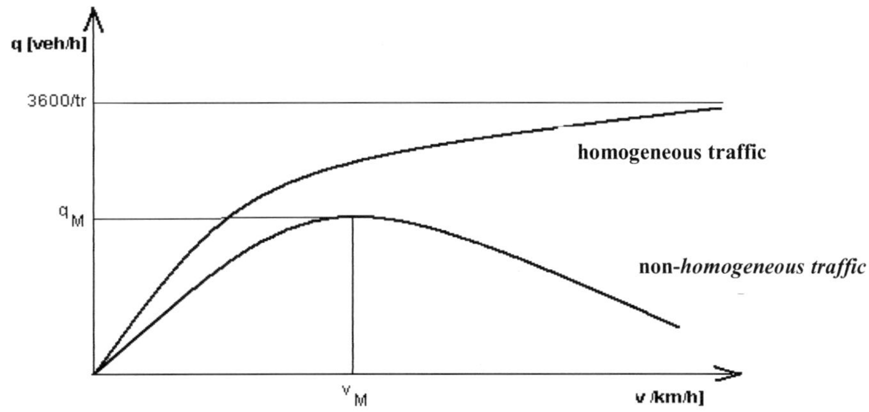

Figure 2. Relation maximum output - speed for homogeneous and non-homogeneous traffic.

One notes that for C = 0 (the case of homogeneous traffic), the density of maximum traffic tends to reach the value of zero, which seems surprising, but can be explained through the fact that for the finite value of maximum traffic (equal to $3600/t_r$) the speed tends to reach the infinite, and this can be justified mathematically only through the case of non-determination for the product $q_M \cdot k_M$.

As far as the deceleration variation coefficient C is concerned, it may be considered to have the following extreme values:

• for $a_{f1} = a_{f2}$, the minimum value: $C_{min} = 0$;

• for $a_{f1} \neq a_{f2}$, one may take into account a range of values starting from 3 m/s^2 to 9 m/s^2, resulting in the maximum value for C:

$$C_{max} = \frac{|a_{f1} - a_{f2}|}{a_{f1} \cdot a_{f2}} = \frac{|3-9|}{3 \cdot 9} = \frac{6}{27} \approx 0{,}2$$

Thus, using the model for the maximum flows of traffic obtained (equation 16) and following the highlighting of size C as a parameter, the variation of the maximum volume of traffic for the following values of deceleration variation coefficient was represented: C = 0; C = 0,05; C = 0,10; C = 0,20 and for the fixed values of the other parameters: l_v = 4,5 m; s_0 = 1,5 m; t_r = 1 s [3].

There were determined the highest values of maximum traffic output, q_M, and the speeds v_M at which these maximum values of traffic output can be obtained.

Thus, the maximum volume of traffic has been highlighted as a function of speed (in km/h), where the variation deceleration coefficient C is a parameter, through the equation:

$$q\left[\frac{veh}{h}\right] = \frac{1000}{\frac{l_v + s_0}{v} + \frac{1}{3.6} \cdot t_r + \frac{v}{2 \cdot 3.6^2} \cdot \frac{|a_{f1} - a_{f2}|}{a_{f1} \cdot a_{f2}}} = \frac{1000}{\frac{4.5+1.5}{v} + \frac{1}{3.6} \cdot 1 + \frac{v}{2 \cdot 3.6^2} \cdot C} = \frac{1000}{\frac{6}{v} + \frac{1}{3.6} + \frac{C}{25.92} \cdot v} \quad (17)$$

In order to represent the variation of the maximum volume of traffic function of the speed, for the five values of the parameter C an analytical calculation was carried out according to the equation above, using the Excel program for speed values from 0 to 150 km/h, at a pace of 10 km/h, there emerging the values from table 1.

In the table 1 there were identified and marked with bold characters the peaks of output for each case (corresponding to some of the values adopted at a pace of 10 km/h), but the exact values for the highest outputs, as well as for the corresponding appropriate speeds, were calculated analytically according to the analytical equations (13) and (15).

Table 1. The volumes of traffic function of speed for various values of C.

C [s²/m]	0.00	0.05	0.10	0.15	0.20
v[km/h]			**q [veh/h]**		
10	1139	1115	1091	1069	1047
20	1731	1622	1527	1442	1366
30	2093	1867	1685	**1535**	**1410**
40	2338	1980	**1718**	1517	1358
50	2514	2023	1693	1455	1276
60	2647	**2026**	1641	1379	1189
70	2751	2006	1578	1301	1107
80	2835	1972	1512	1226	1031
90	2903	1930	1446	1156	963
100	2961	1884	1382	1091	901
110	3009	1836	1322	1032	847
120	3051	1788	1265	978	798
130	3087	1740	1211	929	754
140	3119	1693	1162	884	714
150	**3147**	1647	1115	843	678

The exact calculation - achieved analytically according to these equations using Excel, for the 5 values of parameter C - led to the values recorded in table 2.

Table 2. The maximum volume of traffic and the corresponding speed, for different values of C.

C [s2/m]	0.00	0.05	0.10	0.15	0.20
v_M [km/h]	∞	55.8	39.4	32.2	27.9
q_M [veh/h]	3600	**2029**	**1718**	**1537**	**1412**

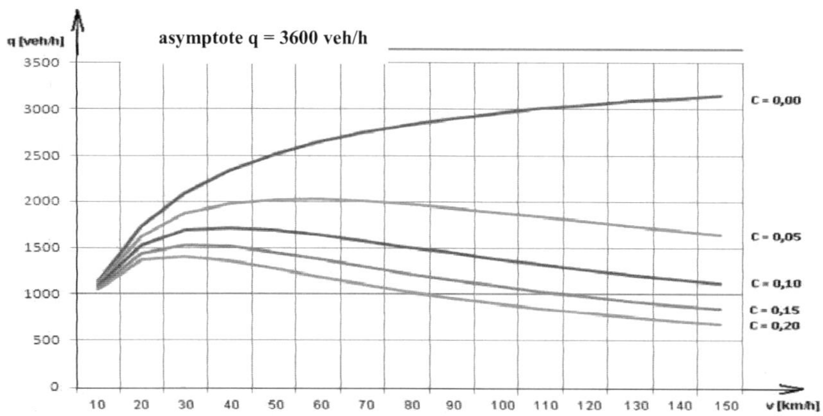

Figure 3. Variation of traffic volume with speed, for various values of C.

The graphical representation (figure 3) shows that, in the case of non-homogeneous traffic (when this parameter has non-zero values, for example = 0,05; 0,10; 0,15 or 0,20) the volume of traffic increases only up to a certain speed value (the smaller it is, the more non-homogeneous the traffic is), then decreases. On the other hand, in the case of homogeneous traffic (C = 0,00), the volume of traffic continues to increase at a constant speed, increasing to a high value and tending asymptomatically to a limit value.

3. Conclusions

It has been demonstrated that, in order to achieve a high volume of traffic, it is necessary to ensure conditions for maintaining a minimum distance between vehicles, and this can only be achieved in the case of *homogeneous traffic* (when the vehicles belong to the same category, therefore they have similar dynamic and braking performance). For this case (which can be met with on roads with restricted access for larger vehicles or on highways with several lanes in each direction, where traffic is naturally layered on lanes), the conclusion can be drawn that the only factor that makes the difference in what the distance between vehicles is concerned is the drivers' response time. For the case of *non-homogeneous traffic* (with a composite structure of vehicles, which have different braking and acceleration possibilities), frequently met with in our country due to the lack of highways, the mathematical model describing the minimum range of vehicles was developed based on the relation describing the minimum distance between vehicles taking into account the difference between the vehicle braking modes (different decelerations). Thus, the value for the minimum distance between two successive vehicles was defined according to the parameters defining the movements of the two vehicles (speed, driver response time, deceleration of the two vehicles and safety space) and, finally, the point is reached when the maximum volume of traffic is explained according to a parameter that includes the deceleration of the two vehicles, which determined the proposed use of this value concept as a new concept in road traffic theory: C = coefficient expressing the variability of braking deceleration for vehicles in the composite continuous flows [3], this coefficient having the equivalent of deceleration inverse: C [1/(m/s2)]. This has made it possible to conclude that, in the case of non-homogeneous traffic (when C has non-zero values) the volume of traffic increases only to a certain speed value (the smaller it is, the more non-homogeneous the traffic is), then decreases, while in the case of homogeneous traffic (C=0), the volume of traffic continues to increase at a constant speed, increasing to a high value and tending asymptomatically to a limit value that depends on the value of the drivers' response time.

4. References

[1] Akcelik R 2002 *Speed-flow Models for Uninterrupted Traffic Facilities* Technical Report (Greythorn Victoria: Akcelik & Associates Pty Ltd)

[2] Ammoser H 2014 *Fundamental diagram of traffic flow* (Dresden: Fakultät Verkehrswissenschaften)

[3] Boroiu A A 2017 *Studii si cercetari privind reducerea poluarii fonice produse de autovehicule prin organizarea circulatiei rutiere*, PhD thesis, University of Pitesti

[4] Boroiu A A and Neagu E 2015 *Trafic rutier si siguranta circulatiei rutiere. Aplicatii* (Pitesti: University of Pitesti Publishing House)

[5] Boroiu A A, Neagu E, Vieru I and Boroiu A 2014 Use of the Fundamental Diagram of Traffic Flow for the Organisation of Road Traffic During the Infrastructure Rehabilitation Works, *Proc. 3rd Int. Congress Science and Management of Automotive and Transportation Engineering (Craiova)* vol I (Craiova: Universitaria Publishing House) p 257-262

[6] Neagu E 2003 *Trafic rutier şi siguranta circulatiei* (Pitesti: University of Pitesti Publishing House)

[7] Olstam J J and Tapani A 2004 VTI meddelande 960 *A Comparison of Car-following Models*, (Linkoping: Swedish National Road and Transport Research Institute)

[8] Raicu S 2007 *Sisteme de transport* (Bucharest: AGIR Publishing House)

[9] Steenbrink P A 1984 *Optimization of transport network* (New York: John Wiley & Sons)

IOP Conf. Series: Materials Science and Engineering **252** (2017) 012056 doi:10.1088/1757-899X/252/1/012056

Comparative analysis of the performance of One-Way and Two-Way urban road networks

Carmen Gheorghe

Transilvania University of Brasov, Brasov, Romania

gheorghecarmen25@gmail.com

Abstract. The fact that the number of vehicles is increasing year after year represents a challenge in road traffic management because it is necessary to adjust the road traffic, in order to prevent any incidents, using mostly the same road infrastructure.

At this moment one-way road network provides efficient traffic flow for vehicles but it is not ideal for pedestrians. Therefore, a proper solution must be found and applied when and where it is necessary. Replacing one-way road network with two-way road network may be a viable solution especially if in the area is high pedestrian traffic.

The paper aims to highlight the influence of both, one-way and two-way urban road networks through an experimental research which was performed by using traffic data collected in the field. Each of the two scenarios analyzed were based on the same traffic data, the same geometrical conditions of the road (lane width, total road segment width, road slopes, total length of the road network) and also the same signaling conditions(signalised intersection or roundabout).

The analysis which involves two-way scenario reveals changes in the performance parameters like delay average, stops average, delay stop average and vehicle speed average.

Based on the values obtained, it was possible to perform a comparative analysis between the real, one-way, scenario and the theoretical, two-way, scenario.

1. Introduction

If you plan cities for vehicles and traffic, you will get vehicles and traffic. If you plan it for people and places you will get people and places [1]. If we take a look in the past we can observe that traffic planners and engineers considered that with the continued development of road infrastructure they will keep up with increasing demand for transport. The way the streets are organized influences daily mobility.

If we refer only to the movement of vehicles we can say that one-way streets bring intersections with greater capacity, less conflict points and higher travel speeds but two-way streets are better for downtown areas because it provides higher accessibility and helps the driver to take the shortest route to destination.

There are three basic elements that constitute each long-term planning process:

1. Prognosis of requesting a system for different levels of endowment;

2. Describe the economic, social and environmental changes that accompany the development of the system at the same levels of plant endowment;

3. An evaluation of the system in terms of investment and benefit for the various options considered.

In general, forecasts of population and economic activities are obtained by specialists in demographic and economic statistics outside the scope of transport studies. Numerous models have been developed to report changes in land use for independent variables such as:

- accessibility to the job;
- the share of available vacant land;
- the degree of land use;
- the use of land surfaces in different situations;
- the degree of occupation with different types of land use;
- the maximum value of journey times and distances in the study area;

From the human factor (pedestrian - user and non-user of the urban road network) point of view, two-way streets are a better option than one-way streets because the two-way streets limit the travel speeds and also allows installation of median separators so the pedestrians can cross the street more safely. In the case of one-way streets, the tendency is to eliminate the cross-walk in order to eliminate the conflict points. When cross-walks are eliminated an alternative must be found for pedestrians to cross and then, underground or over ground passages are implemented. In the big city centers there are many underground passages but these are not always safe for pedestrians, especially if they are crossed at late hours, if they are not properly illuminated, monitored and effectively guarded.

From the point of view of the development of the downtown area two-way road network is also preferred to one-way road network because the local businesses have more visibility.

The present paper aims to highlight the impact one-way and two-way road networks have on the road traffic performance.

2. One-Way and Two-Way road networks particularities

One-Way road networks have a number of characteristics that can be used as advantages and disadvantages. They are used often in comparative analyzes in order to help design an overview and are static concepts, based on the descriptive parameters of an area, over a period of time. The characteristics represent what already exist.

2.1. Advantages and disadvantages of One-Way road networks [5]

Advantages of One-Way road networks:

- Travel speed increases
- On long arteries the travel time decreases
- Signalized intersections are easier to coordinate
- The conflict points with the pedestrians decrease
- Making a left turn is no longer difficult

Disadvantages of One-Way road networks:

- Decreased road safety because of the accidents caused by the high travel speed
- Increased number of conflict points between vehicles when changing lanes
- Increased travel distance from origin to destination
- Many cross-walks are removed because the risk of injuries is higher. Pedestrians are forced to walk a longer distance in order to cross the street

2.2. Advantages and disadvantages of Two-Way road networks [5]

Advantages of Two-Way road networks:

- Road safety is improved by the decrease of the travel speed
- Travel time is shorter which leads to a smaller level of chemical pollution generated by fuel consumption
- Local businesses grow easier because of the higher visibility

IOP Conf. Series: Materials Science and Engineering **252** (2017) 012056 doi:10.1088/1757-899X/252/1/012056

- Travel distance from origin to destination is shorter
- There is a possibility to implement, for pedestrians, at cross-walks, refuges between the two ways
- The accessibility for vehicles is improved, allowing them to travel in both directions

Disadvantages of Two-Way road networks:

- Difficulties in coordinating intersections
- Travel speed is lower
- Many conflict points with pedestrians
- Many conflict points between vehicles when making a left turn

3. Performance measurement

For many years, traffic engineers were mandated to "move as much traffic as possible, as quickly as possible", often resulting in degradation of movement for other modes of travel. The unequivocal movement of the motor vehicle through a downtown network was of paramount concern; all other modes of travel took a back seat. Effectiveness of the network was measured by the amount of delay a motorist would encounter on a given street segment or intersection during either the morning or afternoon peak hours. [2].

Performance measurement focuses on identifying, tracking and communicating performance results with the help of performance indicators or parameters. It reveals, quantitatively, important aspects of the road traffic like: flow rate, speed, density.

3.1. Flow rate

Flow rate is a measure that quantify the amount of traffic passing a point on a lane during a giving time interval. As distinct from traffic volume which is the total number of vehicles that pass over a giving point during a given time interval, the flow rate is the equivalent hourly rate at which vehicles pass over a given point during a given time interval, but less than one hour. [3]

$$q = \frac{N}{t} \tag{1}$$

where,
q is the flow rate (veh/h)
n is the number of vehicles passing over a point
t is the time interval, usually one hour (h)

3.2. Average travel speed

Speed is defined as a rate of motion expressed as distance per unit of time. Average travel speed is computed by dividing the length of the street section or segment under consideration by the average travel time of the vehicles traversing it. [3]

$$s_a = \frac{L}{t_a} \tag{2}$$

where,
s_a is the average travel speed (km/h)
L is the length of the street (km)
t_a is the average travel time over L (h)

3.3. Density

Speed is defined as a rate of motion expressed as distance per unit of time. Average travel speed is computed by dividing the length of the street section or segment under consideration by the average travel time of the vehicles traversing it. [3]

IOP Conf. Series: Materials Science and Engineering **252** (2017) 012056 doi:10.1088/1757-899X/252/1/012056

$$D = \frac{q}{s_a} \tag{3}$$

where,
D is the density (veh/km)
q is the flow rate (veh/h)
s_a is the average travel speed (km/h)

The relation between flow and density, density and speed, speed and flow, can be represented with the help of some curves. They are referred to as the fundamental diagrams of traffic flow presented in Fig. 1.

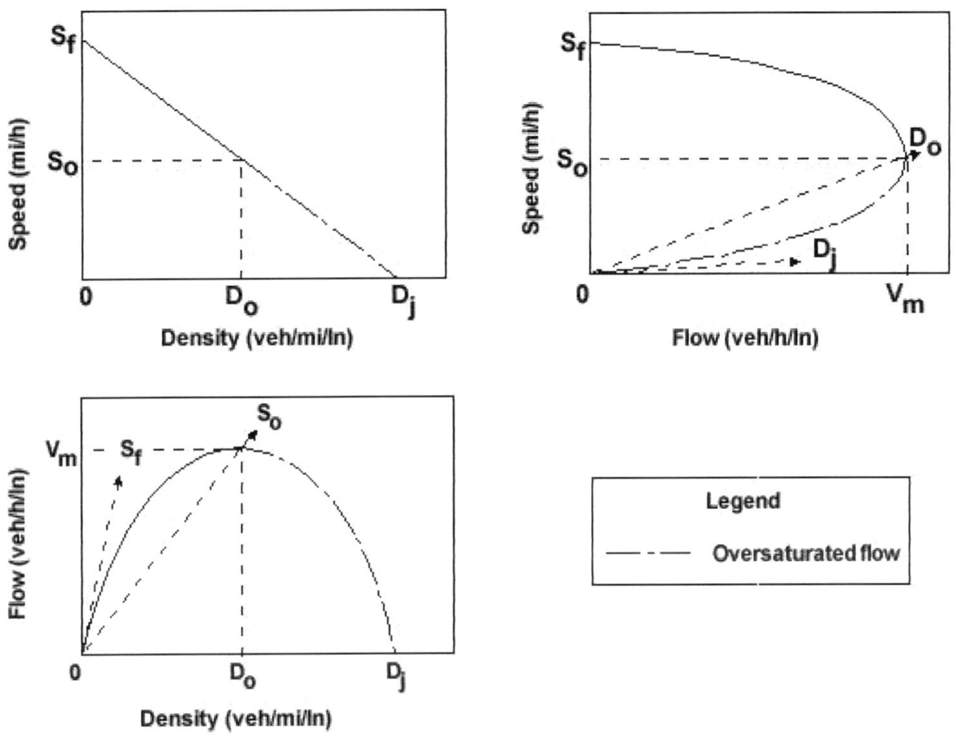

Fig. 1. Fundamental diagram of traffic flow [3]

On each of the diagrams above are presented traffic flow parameters as following: Sf is the free-flow speed, So is the optimum speed, Dj represents the jam density, Do is the optimum density and Vm represents the maximum flow.

The flow and density varies with time and location and similar to the flow-density relationship, speed will be maximum, referred to as the free flow speed, and when the density is maximum, the speed will be zero.

4. Analysis of the Civic Center Area of the Brasov City

The case study conducted on an area in the Civic Center of the Brasov City consists in comparing One-Way and Two-Way urban road networks. The first scenario consists in calculating the performance parameters for the real situation, Fig. 2, one-way road segments.

IOP Conf. Series: Materials Science and Engineering **252** (2017) 012056 doi:10.1088/1757-899X/252/1/012056

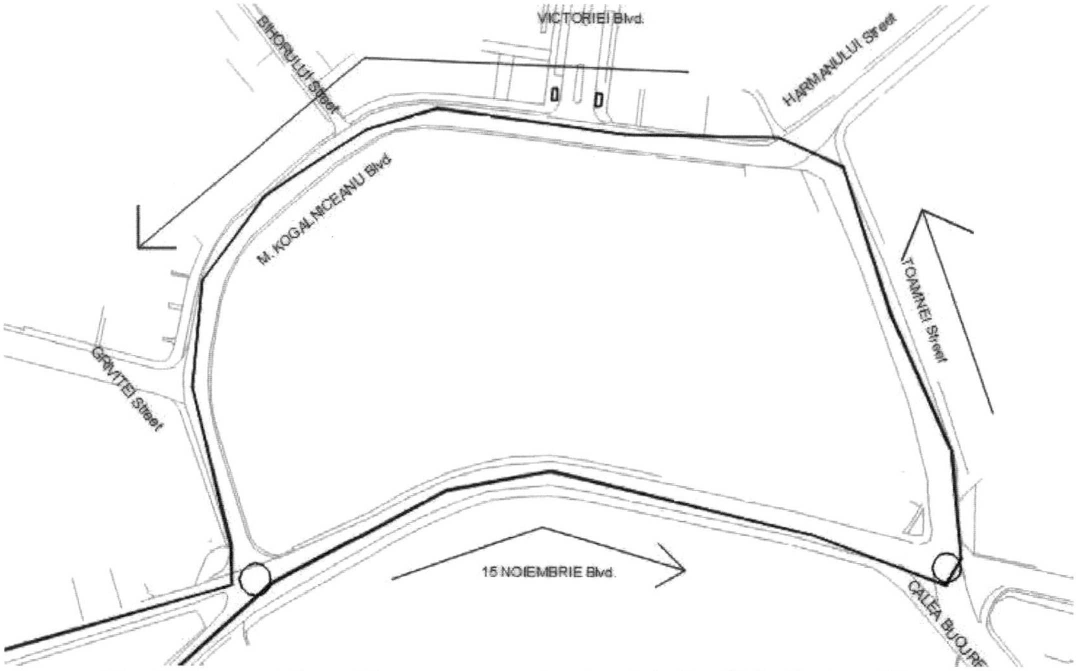

Fig. 2. Representation of the one-way road network in the Civic Center of Brasov City

The geometric and signaling conditions in the first scenario consist of two roundabouts and a traffic light coordinated intersection (Harman Street - Autumn Str.). Between the other intersections there is a continuous flow of traffic due to the six lanes used in one direction, only traffic signs being installed.

The second scenario, Fig. 3, consists in introducing two-way travel on all road segments.

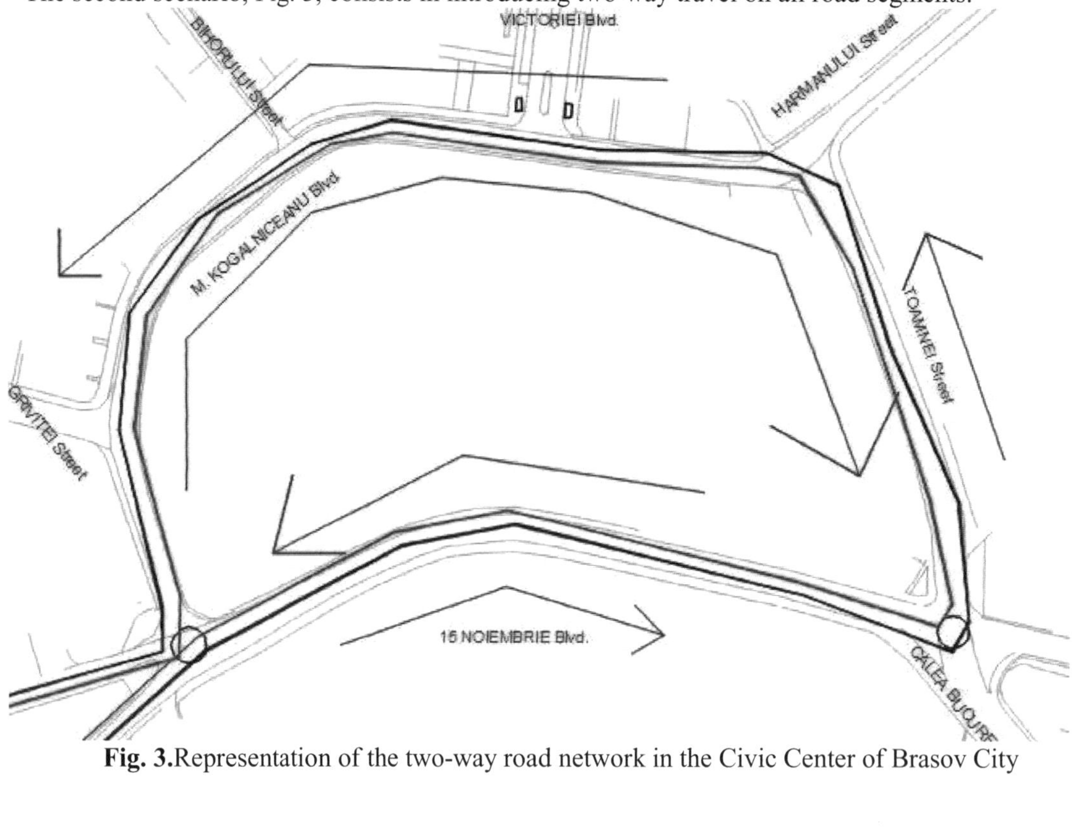

Fig. 3. Representation of the two-way road network in the Civic Center of Brasov City

In the second scenario the geometrical and signaling conditions were preserved, the only change being brought through systematization. Lanes 4, 5 and 6 were used for the opposite movement. However, it can be noted that, when introducing the two-way traffic, the signaling conditions for the following intersections:Kogalniceanu Blvd. - Victoriei Blvd., Bihor Str. and Grivitei Str. can not be preserved. It is necessary to introduce traffic light to improve traffic safety and to reduce points of conflict between vehicles. Also, a well-coordinated traffic light systemwould reduce to minimum the conflict points between vehicles and pedestrians, except for the Kogalniceanu Blvd. – Victoriei Blvd. intersection, where there is an underground passage for pedestrians.

4.1. The comparative analysis of the flow rate

In the analysis of the Civic Center area of Brasov City road network the traffic data were collected in 2014.[6] At the time of data collection, they were used for the elaboration of the bachelor project.

The first step consists in calculation of the flow rate, for each street that forms the Civic Center area, using equation 1. The number of vehicles used for the first way of circulation was collected in 2014 at the ingress of each intersection. For the second way of circulation the data used were estimated at 70% of each one-way lane.

The synthesis of results for the calculus of flow rates, for the two scenarios, is presented in Table 1.

Table 1. Results obtained for the flow rates calculation.

Segment name	Length (km)	*One-Way*	Two-Way
Toamnei Street	0,35 km	495 veh/h	421 veh/h
Kogălniceanu Blvd.	0,75 km	935 veh/h	795 veh/h
15 Noiembrie Blvd.	0,55 km	769 veh/h	653 veh/h

As can be seen in the table, in the second scenario, Two-Way, the flow rate is lower than in the first scenario and the explanation is that, in the Two-Way road network, the vehicles arrive at destination faster so the flow rate decreases. The destination is the place to which something is going and it could be a street, a place, a specific point.

4.2. The comparative analysis of the average travel speed

The second step consists in calculation of the average travel speed, for each street that forms the Civic Center area, using equation 2.

The synthesis of results for the calculus of average travel speed, for the two scenarios, is presented in Table 2.

Table 2. Results obtained for the average travel speed calculation

Segment name	One-Way	*Two-Way*
Toamnei Street	44 km/h	32 km/h
Kogălniceanu Blvd.	50 km/h	40 km/h
15 Noiembrie Blvd.	50 km/h	42 km/h

As can be seen in the table, in the second scenario, Two-Way, the average travel speed is lower than in the first scenario and the explanation is that the traffic flow is often interrupted, so when calculate the speed, the acceleration and deceleration at intersections are also introduced in the equation.

4.3. The comparative analysis of the density

The third step consists in calculation of the density, for each street that forms the Civic Center area, using equation 3.

The synthesis of results for the calculus of average travel speed, for the two scenarios, is presented in Table 3.

Table 3. Results obtained for the density calculation

Segment name	One-Way	*Two-Way*
Toamnei Street	11 veh/km	13 veh/km
Kogălniceanu Blvd.	19 veh/km	20 veh/km
15 Noiembrie Blvd.	15 veh/km	16 veh/km

It is worth pointing out that all densities presented in Table 3 are the number of vehicles in one lane per km and as can be seen the number is slightly higher in two-way road segments especially due to the lower speed. Even if the flow rate is lower in the second scenario, the lower average travel speed makes the density grow.

5. Conclusions

This paper was aimed to establish the influence of road network performance parameters. Based on the calculations and the results presented above, it is noted that two-way road networks are more efficient than one-way ones only if there is an effective signalling system so as to minimize the conflict points.

Speed, flow, and density are all related to each other. The relationships between speed and density are not difficult to observe in the field, while the effects of speed and density on flow are not quite as apparent and that is because the flow rate is measured as an hourly rate.

As the maximum density accepted is 100 vehicles per kilometre, if we report density to the measured length of the segment then it can be said that, in two-wayscenario the segments will have an optimum occupancy degree.

Even if the benefits of one-way road networks are evident, in this case study, due to the central position of the area and because the accessibility function has priority over the mobility function, the second scenario (two-way road network) is more appropriate.

References

[1] Kent, F.: Streets are people places https://www.pps.org/blog/transportationasplace/
[2] Walker, W., Kulash, W., McHugh, B.: Are We Strangling Ourselves on One-Way Networks? https://nacto.org/wp-content/uploads/2015/04/Are-We-Strangling-ourselves-on-one-way-networks_Walker.pdf
[3] Transportation Research Board: Highway Capacity Manual, 2000
[4] Hall, F.: Traffic stream characteristics. McMaster University, Department of Civil Engineering and Department of Geography
[5] Florea, D., Covaciu, D., Timar, J.: Analysis of the influence of One-Way Streets on the Urban Road Networks Connectivity. International Congress of Automotive and Transport Engineering, CONAT, 2016
[6] Gheorghe, C.: The modelling and microsimulation of road traffic in the Civic Center Area in Brasov City (using PTV VISSIM Software), Bachelor Project, *Transilvania* University of Brasov, 2014

IOP Conf. Series: Materials Science and Engineering **252** (2017) 012057 doi:10.1088/1757-899X/252/1/012057

Road Traffic Noise Pollution Analysis for Cernavoda City

L. Manea[1], A. Manea[1], D. Florea[2] and S. Tarulescu[2]
[1]Ovidius University of Constanta, Mamaia Boulevard Nr.124, Romania
[1]Transilvania University of Brasov, Eroilor Boulevard Nr. 29, Romania

Abstract. In the present paper was studied the noise pollution in Cernavodă city. The noise measurements were made for nine intersections from different city areas. Noise measurements were taken for three chosen routes with high population density, heavy traffic, commercial and residential buildings. Average, maximum and minimum values were collected and compared with standards. The impact of road traffic noise on the community depends on various factors such as road location and design, land use planning measures, building design, traffic composition, driver behaviour and the relief. In the study area 9 locations are identified to measure noise level. By using sound level meter noise levels are measured at different peak sessions i.e. morning, afternoon and evening. The presented values were collected for evening rush hour.

1. Introduction

Traffic noise is the most rigorous type of noise pollution. Traffic represents a serious problem because of inadequate urban planning of the cities. Residential areas, schools, commercial zones, hospitals, and other community buildings were built close to the main roads without buffer zones or adequate sound proofing. The problem has been compounded by increases in traffic volumes in the last years. This alarming increase in the volume of traffic is actually inversely related to the degradation of the environment. Sound pressure is a measure of the vibrations of air that makes up sound and these levels are measured on the logarithmic scale with units of decibel (dB) [7].

Global, or strategic, noise planning tries to prevent noise issues arising and to optimize the use of limited resources by mapping and managing the noise environment of a large area such as a city. Road traffic noise accounts for more than 90% of unacceptable noise levels (daytime LAeq > 65dB(A)) in Europe [6]. Environmental noise levels can either be assessed by measurement or by calculation. Although a measured level (i.e. the reading of some measurement device) tends to be more convincing to the public than a calculated level (i.e. the output of some computer program) the preferred method is very often the calculation. The reasons for this preference are: the laborious assessment of the exact situation of the source can be avoided; long term average weather conditions can be taken into account directly; noise levels can be assessed even for future developments; possibly disturbing effects of "background" noise can be avoided; detailed information on the contributions of different partial sources can be assessed more easily; different scenarios can easily be synthesized in the case of noise action planning [3], [5].

2. Noise measurement methodology

2.1. Theoretical elements

Environmental noise caused by road traffic is assessed for different purposes:

- to assess the acceptability of new building developments in noisy areas,
- to determine the nature and amount of noise abatement provisions, in case of excess of the limit value,
- to assess the acceptability of a new infrastructure in the neighbourhoods of residential areas,
- to assess the necessity of remedial action in the case of excessive noise in residential areas [1].

For noise measurements was used Sound Level Meter - Type 2250 from Bruel&Kjaêr. This sound meter has been developed specifically for measuring occupational, environmental and product noise, while complying fully with all the relevant national and international standards [2].

The tasks for environmental noise measurements are varied, so the instrument you pick for your measurements needs to be flexible, easy to configure, powerful and accurate. 2250 Light is built on the core platform of the award winning design of Type 2250. It borrows the robust construction, intuitive touch screen interface, and legendary Bruel&Kjaêr measurement accuracy [2].

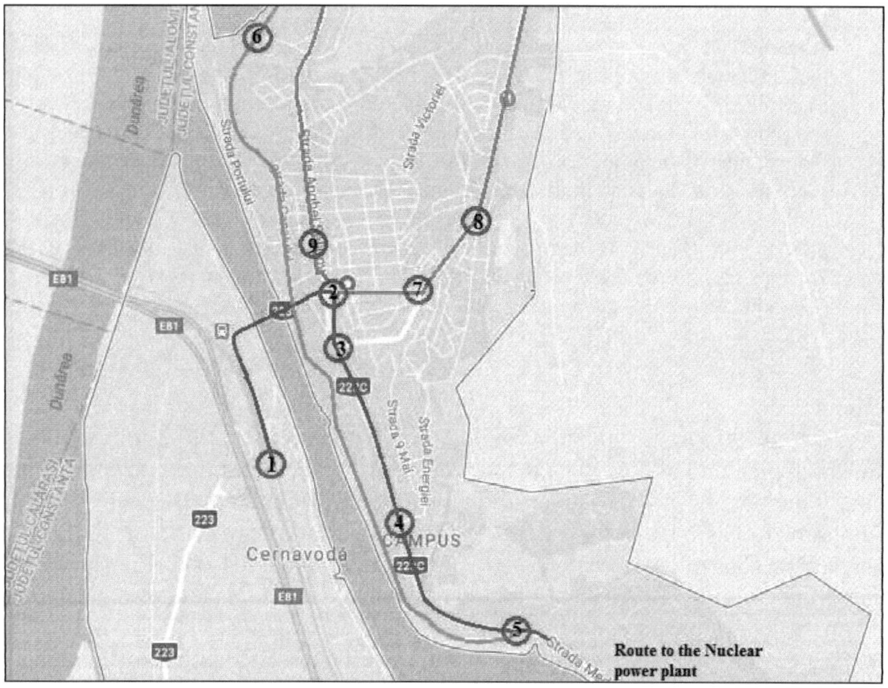

Figure 1. Analyzed area - Cernavodă road routes and studied intersections (Google Maps)

2.2. The analyzed area

In the Figure 1 are presented the analyzed routes from Cernavodă City. The studied intersections are numbered as follows: 1. Constanței Str. + DN22C Intersection; 2. Nicolae Titulescu Str. + Gării Str. + Crișan Str. Intersection; 3. Nicolae Titulescu Str. + Dacia Str. + Seimeni Str. + Medgidiei Str. Intersection; 4. Medgidiei Str. + Panait Cerna Str. Intersection; 5. Medgidiei Str. + Canalului Str. Intersection; 6. Canalului Str.+ Sălciei Str. Intersection; 7. 24 Ianuarie 1859 Str. + Seimeni Str. + Pictor N. Grigorescu Str. Intersection; 8 Seimeni Str. + Independenței Str. + Gheorghe Doja Str. Intersection; 9. Anghel Saligny Str. + Ovidiu Str. Intersection.

The analyzed route can be separated into three distinct routes: first route includes intersections no.1, 2, 3, 4 and 5 and is the route that connects Cernavodă city entrance with main road that leads to Nuclear power plant, characterized by wide roads with 2 lanes and high traffic flows. The second route includes intersections no. 2, 7, 8 and 9 and is a residential streets route, characterized by narrow roads with 1 lane and medium traffic flows. The third route includes intersections no. 5 and 6 and is the Cernavodă bypass road characterized by high heavy vehicles traffic.

3. Routes analysis

3.1. Routes measurement points

For each route ware selected noise measurement points relevant to the transited area. For residential route the measurement points ware selected in function of pedestrian zones and main socio-commercial objectives. For the main city route, the measurement points were selected in the junction areas, on pedestrian zones. For the Cernavodă bypass road, the measurement points were selected in residential areas and in junction areas.

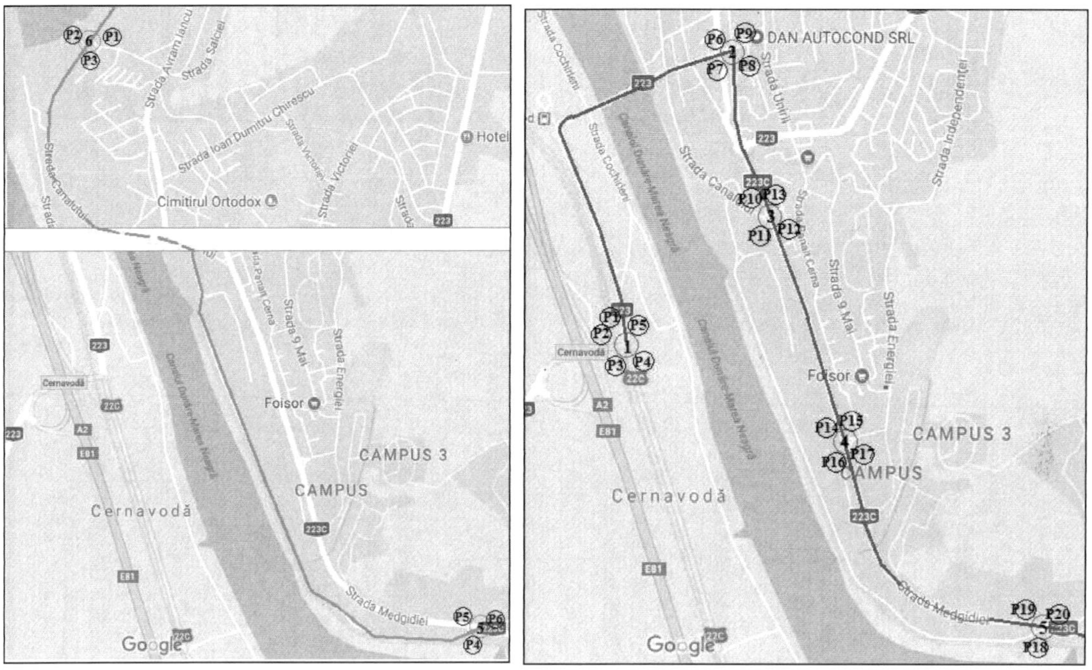

Figure 2. Noise measurement points for: Route 3 (orange - left) and Route 1 (violet- right) (Google)

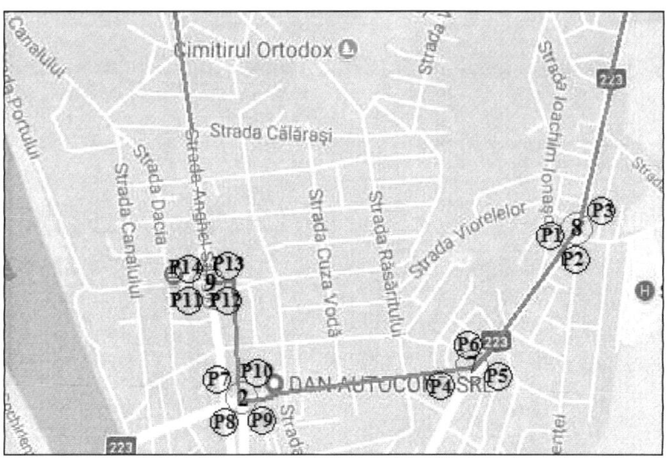

Figure 3. Noise measurement points for Route 2 (green) (Google Maps)

For each of the nine analyzed intersections were centralize the noise average and maximum values over the period of measurements, recorded for evening rush hour interval (15.45 – 16.45).
The noise level for each route (for all measurement points) are presented in the Figures 4, 5 and 6.

IOP Conf. Series: Materials Science and Engineering **252** (2017) 012057 doi:10.1088/1757-899X/252/1/012057

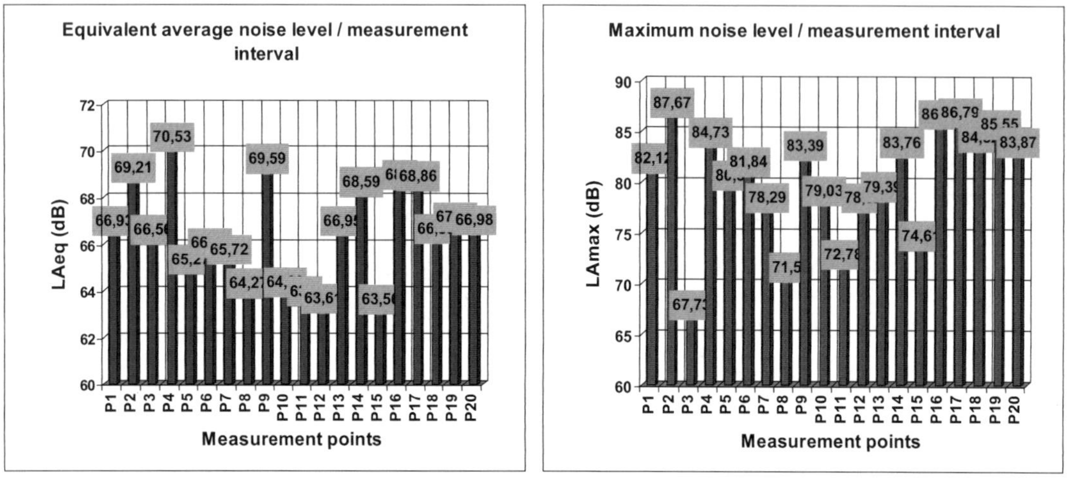

Figure 4. Noise level for Route 1 (LAeq values - left; LAmax values - right)

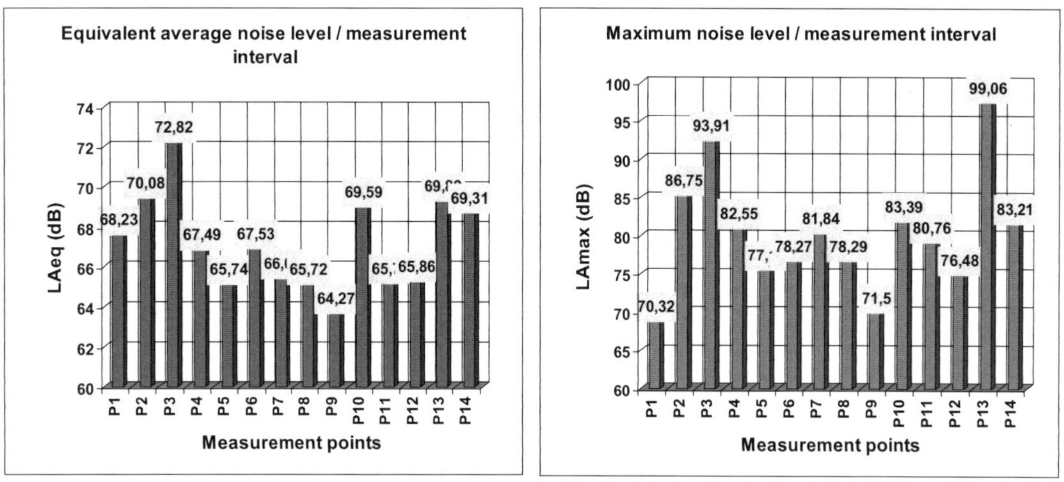

Figure 5. Noise level for Route 2 (LAeq values - left; LAmax values - right)

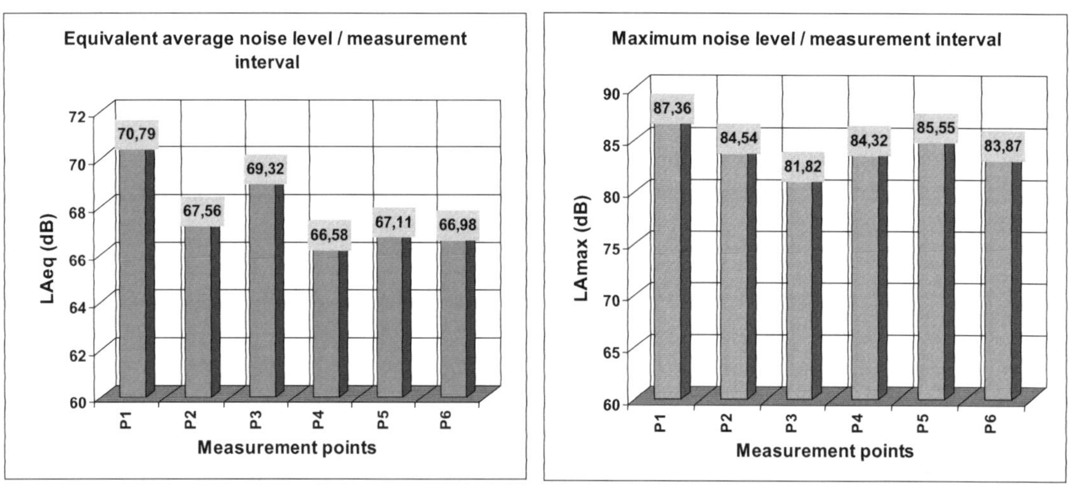

Figure 6. Noise level for Route 3 (LAeq values - left; LAmax values - right)

In the Table 1 are presented the traffic flows parameters for the nine analyzed intersections.

Table 1. Traffic flows parameters (number of vehicle for the evening rush hour and vehicles categories)

Intersection no.	*Light vehicles*	*Vans*	*Buses*	*Trucks*
		Route 1		
Intersection 1	325	63	5	66
Intersection 2	695	50	11	2
Intersection 3	837	99	26	1
Intersection 4	807	67	26	1
Intersection 5	656	71	30	6
		Route 2		
Intersection 2	695	50	11	2
Intersection 7	252	28	8	3
Intersection 8	360	38	2	1
Intersection 9	427	25	4	3
		Route 3		
Intersection 5	656	71	30	6
Intersection 6	198	21	2	27

3.2. Noise data analysis

The values for the three experimentally examined routes differ from one route to another. For the first route, corresponding to the area connecting the city main entrance with the most important local institution (Cernavodă Nuclear power plant), the measured average noise range varies between 63 and 70 decibels. The maximum noise level varies between 68 and 88 decibels. These values have been collected in the proximity of the road, in the intersections access and evacuation areas, where vehicles engines works transient. Another important factor that influences the noise levels is the composition of the traffic flow. Along with cars, buses equipped with compression ignition engines have an important share of influence. For this route we noticed a big buses number, because of the Nuclear power plant commuters. In the case of Intersection 1 - Constanței Str. + DN22C, we noticed a big number of trucks, that transit the national road DN22C, but not enter in the city. Also, for Intersection 2 - Nicolae Titulescu Str. + Gării Str. + Crișan Str. Intersection and Intersection 3 - Nicolae Titulescu Str. + Dacia Str. + Seimeni Str. + Medgidiei Str., the noise peaks ware registered in the proximity of metallic drainage channels.

Figure 7. High noise sources for Route 1 - metallic drainage channels on the main road (left) and traffic composition: commuting buses (right)

For the second route, corresponding to the residential areas from Cernavodă city, the average noise values vary between 64 and 73 decibels. In the case of the maximum noise values, that's vary between 70 and 99 decibels. Although, compared to the first route, the traffic flow is much smaller; a factor that increases the noise level is the relief of the region (hills), and the geometry of the streets (narrow streets with abrasive pavement). Also, the number of the vans, buses and trucks is smaller then Route 1. The main reason for the higher noise level comparing to the Route 1 is the streets declivity, that force the vehicles to use transient regimes and higher engine speeds.

Figure 8. High noise sources for Route 2 - road declivity (left) and the traffic composition - vans, trucks and old light vehicles (right)

For the third route, corresponding to the Cernavodă bypass road, the average noise values vary between 66 and 71 decibels. In the case of the maximum noise values, that's vary between 82 and 88 decibels. The high noise level, in this case, is produced by the heavy vehicles that use this route, for harbour and Dunăre-Black Sea Chanel activities.

IOP Conf. Series: Materials Science and Engineering 252 (2017) 012057 doi:10.1088/1757-899X/252/1/012057

Figure 9. High noise sources for Route 3 - traffic composition - vans and trucks

The European noise standard specify that the daytime average noise level LAeq < 70 dB(A) [4]. For Route 1 the highest average noise level was registered for intersection 1, that is outside the city. For Intersections 2, 3, 4 and 5 the average noise level is smaller than 70 dB. For Route 2 the highest average noise level was registered for intersection 8, that is situated on a high declivity. For Intersections 2, 7 and 9 the average noise level is smaller than 70 dB. For Route 3 the highest average noise level was registered in Intersection 6, because of the heavy vehicles activity.

4. Conclusions

The noise analysis of the Cernavodă city indicated that the noise levels in the region are escalating at a very fast rate with growing population, road infrastructure degradation and heavy vehicles and Diesel buses accumulation. Noise levels obtained at different locations of the city, in commercial, residential, and industrial areas are found to be exceeding the noise level /limits prescribed by the European standards.

This study was made in function of three main routes: first route connects Cernavodă city entrance with main road that leads to Nuclear power plant, characterized by wide roads with 2 lanes and high traffic flows; the second route is characterized by narrow roads, with high declivity, with 1 lane and medium traffic flows; the third route is the Cernavodă bypass road characterized by high heavy vehicles traffic.

It is also observed that noise level is closely related with the number and composition of road traffic. In this case, some measures should be taken to control the level of noise pollution in the city. To reduce noise pollution, several measures can be implemented [7]:

- infrastructure upgrades for most of the analyzed intersections, and also for all main city roads;
- Nuclear power plant transportation system improvement, and fleet renewal;
- Cernavodă bypass road rehabilitation, in order to increase its usage;
- traffic flow management optimisation in residential areas;
- plantation of trees and construction of sound barriers.

Noise pollution is becoming a major health concern with all of its potential biological and social effects (cardiovascular diseases, hearing problems, children performance at school and psychological problems for all age categories.

5. References

[1] A. Chiru, D. Covaciu, D. Florea, J. Timar, *Noise Mapping for Urban Road Traffic and its Effect on the Local Community*, The 3rd International Conference on Computational Mechanics and Virtual Engineering, 29 – 30 OCTOBER 2009, Brasov, Romania.

[2] Bruel&Kjaer, *Hand-held Analyzer 2250 Light*, Technical documentation, English BE 1766 –21, Denmark, October 2012.
[3] C. Cofaru, *Pollution legislation in road transportation*, Transilvania University of Brasov, 2002.
[4] Government Decision no. 674/28.06.2007. *Monitorul Oficial* (the official journal of Romania), no. 485/19.07.2007 (in Romanian).
[5] M. Deliu, Ghe. Deliu, *Urban Road Traffic Noise Mapping* RECENT Magazine, Vol. 9, no. 3(24), Brasov, November, 2008.
[6] N. Filip,, N. Cordos, I. Rus, *Urban Noise and Road Traffic*, Todesco Editor, Cluj-Napoca 2002, ISBN 973-8198-25-9.
[7] P. Balashanmugam, V. Nehrukumar, K. Balasubramaniyan, G. Balasubramanian, *Effect of Road Traffic Noise Pollution in Cuddalore Town: A Case Study*, International Journal of Innovative Research in Science, Engineering and Technology, Vol. 2, Issue 10, October 2013.

Acknowledgments
Authors wishing to acknowledge assistance or encouragement from colleagues, special work by technical staff and support from Cernavodă City hall.

IOP Conf. Series: Materials Science and Engineering **252** (2017) 012058 doi:10.1088/1757-899X/252/1/012058

Comparison of Different Battery Types for Electric Vehicles

C Iclodean[1], B Varga, N Burnete, D Cimerdean, B Jurchiş

Technical University of Cluj-Napoca, Romania, Muncii Bd. 103 - 105

[1] E-mail: calin.iclodean@auto.utcluj.ro

Abstract. Battery powered Electric Vehicles are starting to play a significant role in today's automotive industry. There are many types of batteries found in the construction of today's Electric Vehicles, being hard to decide which one fulfils best all the most important characteristics, from different viewpoints, such as energy storage efficiency, constructive characteristics, cost price, safety and utilization life. This study presents the autonomy of an Electric Vehicle that utilizes four different types of batteries: Lithium Ion (Li-Ion), Molten Salt ($Na\text{-}NiCl_2$), Nickel Metal Hydride (Ni-MH) and Lithium Sulphur (Li-S), all of them having the same electric energy storage capacity. The novelty of this scientific work is the implementation of four different types of batteries for Electric Vehicles on the same model to evaluate the vehicle's autonomy and the efficiency of these battery types on a driving cycle, in real time, digitized by computer simulation.

1. Introduction

Due to high demand of fossil fuels on the international markets together with the aggravation of environment problems caused by an increased number of internal combustion engine vehicles, there is an increased interest in the research and development of batteries used in electric and hybrid vehicles. These vehicles represent a solution for the future in road transportation field, taking into consideration the interest in reducing greenhouse gas emissions, as well as air and sound pollution [1, 2].

According to [3], transportation sector represents one of the main determinant factors of climatic changes, 23 % of the greenhouse gas from the atmosphere coming from this sector, being second in this classification after the industrial sector. Due to this reason, in 2015, "Paris declaration on Electro-Mobility and Climate Change and Call to Action" has been adopted. This declaration has as a main objective reducing global warming with more than 2 degrees. This goal is achievable if electric vehicles represent 35 % from the total number of vehicles sold until 2030.

In order to reach this target, a decrease in the acquisition price of the electric vehicles is mandatory until it reaches a level closer to that of the internal combustion engine vehicles. Nowadays, the most expensive part of an electric vehicle is the battery, which represents 25 … 50 % of the price of the electric vehicle, depending of the technology used [4, 5, 6].

As presented by [5, 7, 8], a decrease of battery cost is anticipated by 2025, reaching a price of 225 Euros/kWh, which will determine a significate decrease in the acquisition price of the electric vehicles, helping them reach a value closer to the price of internal combustion engine vehicles. As presented by [4], production costs of the Li-Ion batteries decreased with over 50 % from 2007 until 2014.

Figure 1 presents a comparative market price evaluation of different electric vehicles, depending on battery capacity:

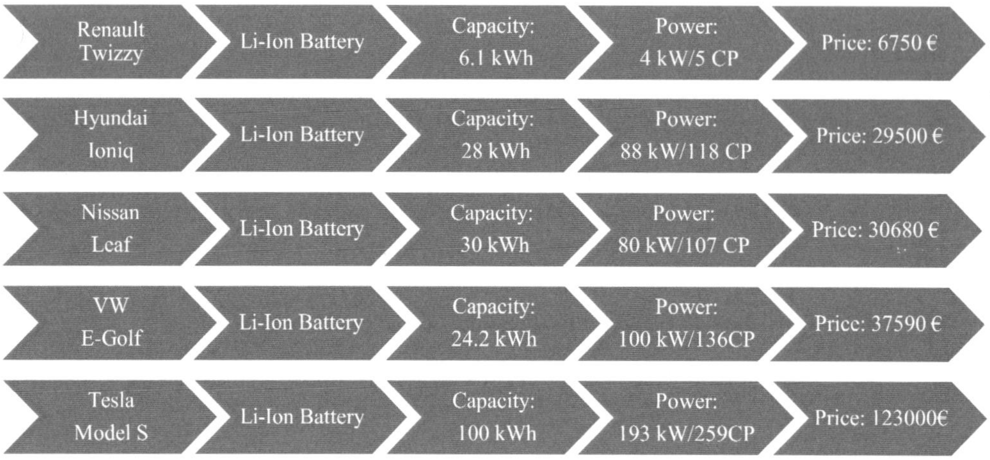

Figure 1. Comparative evaluation of different electric vehicle market cost.

Analyzing these types of electric vehicles presented above, according to the studies of [4, 5, 6] an approximate cost for one kWh energy is between 255 € (Nissan Leaf) and 388 € (VW E-Golf), cost calculated at a battery value of 25 % from the total vehicle amount.

As observed in figure 1, Li-Ion batteries are nowadays representing the most used technology in electric vehicles, both thanks to high energy density and increased power per mass battery unit, allowing the development of some types of batteries with reduced weight and dimensions at competitive prices. Trough studies carried out by [8, 9, 10, 11] Li-Ion batteries used in electric vehicle industry were studied, highlighting increased power (800 ... 2000 W/kg), and specific energy (100 ... 250 Wh/kg), in comparison with Ni-MH batteries.

In accordance with [12] studies, this technology represents the best "charge to weight" solution, fulfilling one of the most important conditions for a battery used in electric vehicle industry, which made an easily replacement of Ni-MH batteries by them. Another advantage is the lack of memory effect (to gradually lose the maximum energy capacity in case of repeatedly recharge, without being totally discharged), resulting an increased life cycle.

The disadvantage of Li-Ion batteries is represented by high developed operational temperature, which could affect energetic performances, among with lifetime and safety in exploitation [13]. This technology requires one management battery system in order to control and monitor internal cell temperature. Apart from the disadvantages caused by exploitation temperature, there are also problems related to high production costs [14], recycling capacity of batteries out of use [15] and recharging infrastructure [16].

At the beginning of 2000, Ni-MH batteries represented the most advanced technology used in hybrid and electric vehicles, being considered the first step to achieving the technology used today.

In comparison with the batteries used in those days, namely Ni-Cd and Lead-Acid batteries, Ni-MH technology was meeting the requirements imposed to batteries which were developed in order to be used in automotive industry. The advantages were: high energy density and power, allowing an autonomy of over 300 km while using batteries with 70 Wh/kg specific energy. Moreover, these batteries can be used with success in propulsion systems equipped with electric engines of 320 V AC, or 180 V DC, showing an increased lifecycle (until 80 % Depth of Discharge DOD). Other advantages were: the capability of using regenerative energy recovered from braking, recyclable materials were used in their development, excellent thermal properties (operating temperature starting from – 30 °C up until + 70 °C), battery charging and discharging safety, etc. [17, 18]. From a comparison made by [19] between Li-Ion and Ni-MH batteries, it is shown that, from a cell point of view, Li-Ion cells offer a specific energy increased by 20 %. But, if the whole battery system is taken into account, with the added mass of battery management system in the Li-Ion technology, it can be said that the management system of the Ni-MH battery is simpler, with reduced weight. If we compare the electric Nissan Leaf EV, equipped with a Li-Ion battery, having an autonomy of 160 km with the electric car of GM, EV1, equipped with Ni-MH battery, having an autonomy of 280 km, it results that Ni-MH batteries are more efficient, from an autonomy point of view. The disadvantages of Ni-MH batteries are increased weight and obsolete technology. Researches conducted in the last years in the field of electric vehicles battery technology have met an advancement by developing other battery types among the classic Ni-MH batteries.

Na-$NiCl_2$ batteries, also known as ZEBRA batteries, (Zeolite Battery Research Africa) were used for equipping some concept cars and buses used in urban public transportation [20]. These types of batteries are remarkable especially for their increased energy density (90 … 120 WH/kg), and also due to lower price, compared with other existing technologies. Other advantages are: overcharge and over discharge resistance, increased cycle life, and constructive robustness, which permits their utilization in harsh environments, their performances not being affected by low temperatures.

The big downsize of the Na-$NiCl_2$ batteries it's represented by increased internal operating temperatures (270 °C … 350 °C), continuous utilization of the electric vehicle being necessary in order to avoid freezing the battery electrolyte. In case the car is not used, maintaining the system at the operating temperature can be possible through an external heating system, which consumes 90 Wh power from the battery. In contrast, 12 to 15 hours are needed in order to defrost the battery and to bring it back to its functional parameters [21].

Although Li-Ion batteries are now in an advanced development stage, close to the theoretical values obtained through researches [22, 23], this type of battery cannot yet satisfy requests regarding especially the autonomy, making it necessary to find new technologies which can offer a high storage energy capacity, among with an extended lifespan [24]. Another type of technology, namely the Li-S battery, has showed a lot of interest from researchers mostly due to increased theoretical specific energy (2500 Wh/kg), among with its theoretical specific capacity (1672 mAh/kg), thus being considerate a strong competitor for the technology used today, representing a viable solution in the development of future electric vehicles. At the moment, due to reduced lifespan and energy retention capacity, the use of this battery type is limited [25, 26].

2. Materials and Methods

In this paper a virtual electric vehicle, based on the constructive parameters of the electric VW E-Golf was developed in AVL Cruise, in order to emphasize autonomy differences and to evaluate the performances of these four types of batteries. On this virtual model, all four battery types will be implemented and then details regarding vehicle autonomy will be studied trough computerized simulation.

2.1. Methodology of the simulation process

The steps used to define the algorithm of the computerized simulation process are presented in figure 2. The first phase consists in developing the model of the virtual electric vehicle in AVL Cruise.

AVL Cruise is a simulation package which has the possibility to simulate a large variety of vehicle propulsion systems, in order to study the evolution of the energy utilized in vehicle propulsion, starting from one energy source (electric battery) to one energy consumer (transmission system). A preprocessing tool is used to enter initial and input data, as well as technical characteristics of the new simulation model. After the assembly of the vehicle parts, along with supplementary systems, the software analyzes and calculates all processes requested during simulation time, with the help of all mathematical equations and computational algorithms used behind the Graphical User Interface [27].

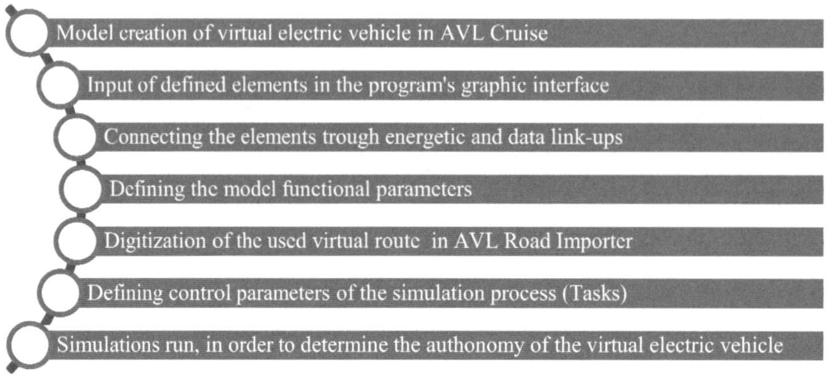

Figure 2. Steps of the computerized simulation process algorithm

2.2. Virtual model of the electric vehicle

In order to simulate the functioning of an electric vehicle, one virtual model of the VW E-Golf was created and developed in AVL Cruise (figure 3). Constructive data were taken from manufacturer's technical documentation in the interest of developing this virtual model [28, 29].

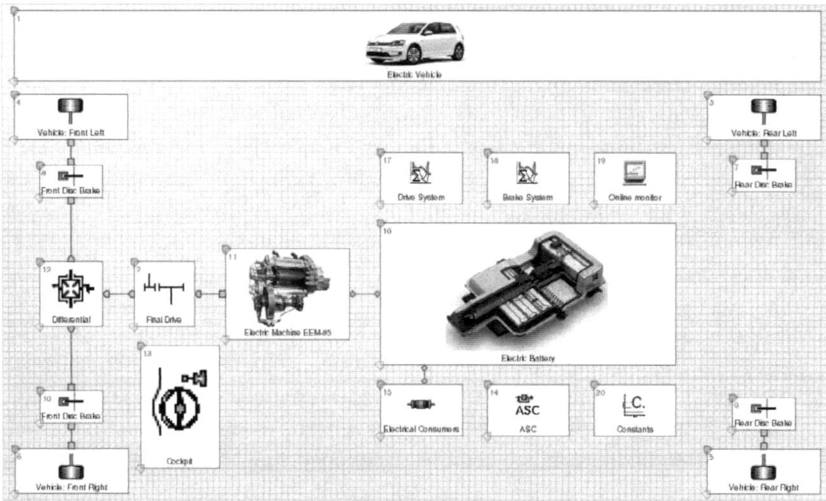

Figure 3. Virtual model of the studied electric vehicle in AVL Cruise

The elements of the model are the following: Electric Vehicle (1), Final Drive (2), Vehicle Wheel Rear Left (3), Vehicle Wheel Front Left (4), Vehicle Wheel Front Right (5), Vehicle Wheel Rear Right (6), Rear Disk Brake (7), Front Disk Brake (8), Rear Disk Brake (9), Front Disk Brake (10), Electric Machine (11), Differential (12), Cockpit (13), ASC Control (14), Electrical Consumers (15), Electric Battery (16), Drive System (17), Brake System (18), Online Monitor (19) and Constants (20). The informational connections are made through Data Bus Connection.

The Electric Vehicle component (1) contains general data, such as: constructive parameters, aerodynamic features, etc. [30, 31]. The Final Drive element (2) is used as a link in the transmission of the cinematic chain before the differential. The Wheel (3 ... 6) takes into consideration many variables and their influence and effect on the rolling state. All input data for all vehicle's wheels is identical. The Brake component (7 ... 10) defines the characteristics of the braking elements. The E-Machine component (11) defines general data of the electric machine (table 1). For calculating power losses, Maximum Power (Torque) (figure 4) and Efficiency Maps (figure 5) are defined [32].

Table 1. General data of the Electric Machine.

Name	Value	Unit
Type of Machine	PSM	-
Characteristic Maps and Curves	Overall	-
Nominal Voltage	323	V
Inertia Moment	1.0e-4	kg*m^2
Maximum Speed – motor related	12000	1/min
Maximum Power – motor related	85	kW
Maximum Torque – motor related	270	Nm
Maximum Speed – generator related	7000	1/min
Maximum Power – generator related	84	kW
Maximum Torque – generator related	264	Nm
Maximum Efficiency	92.5	%
Initial Temperature	20	° C

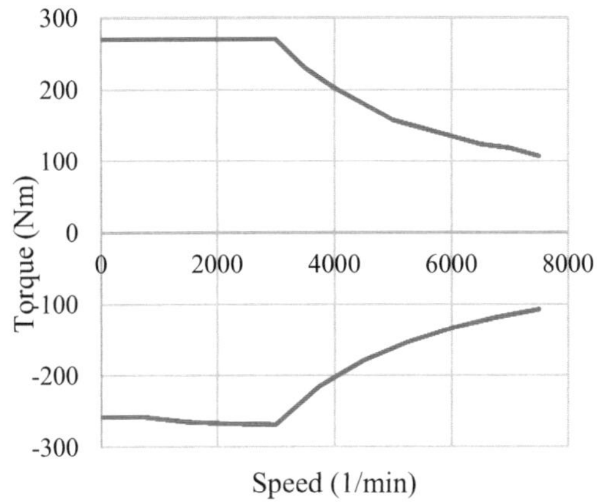

Figure 4. Electric Machine Torque Map

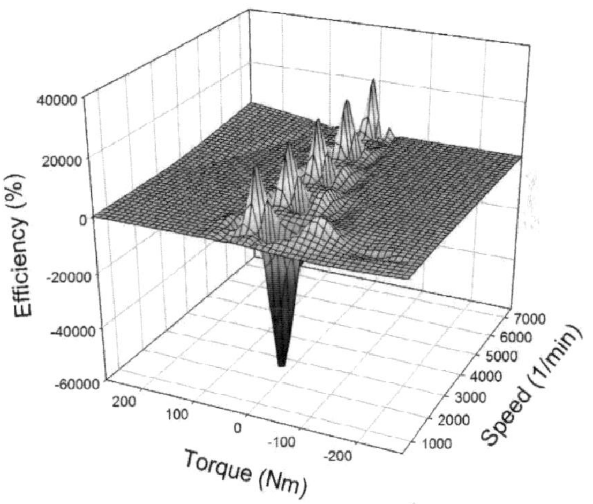

Figure 5. Electric Machine Efficiency Map

The Differential unit (12) compensates for discrepancies in the respective rotation rates of the drive wheels, allowing different wheel speed during cornering. The Cockpit (13) links the driver and the vehicle, connections being made only via Data Bus.

Anti-Slip Control component (14) controls the coefficient of friction of the single wheels, by checking the force transmission factor (ratio between the force which should be transmitted and maximum transmittable force) of all connected wheels (driven wheels).

The Electrical Consumers (15) are all the consumers of the electric vehicle network. In the modelling process, they are the ohmic resistors in the onboard network, which represents an electric current loss.

The Electric Battery component (16) simulates the battery used in the power supply of the electric vehicle. The electrical characteristics of all four battery types studied in this paper are presented in table 2 [32, 33, 34, 35, 36, 37, 38, 39, 40].

The Drive System (17) is a function defined by the user in C-Code, used to simulate the engine control unit, both in engine and generator mode. The Brake System (18) is a function defined by user in C-Code for controlling the conversion of the motor moment in braking pressure. The Monitor element (19) has the role of monitoring specific signals of the Data Bus Connection. The Constants component (20) allows the user to define all the constant values used by the elements of the electric vehicle trough Data Bus Connection [32].

Table 2. General parameters of the Electric Battery component.

Name	Value for Battery Type				Unit
	Li-Ion	Na-NiCl$_2$	Ni-MH	Li-S	
Maximum Charge	75	84	85	80	Ah
Nominal Voltage	323	289	288	305	V
Stored Energy	24.2	24.2	24.2	24.2	kWh
Maximum Voltage / Minimum Voltage	339 / 308	275 / 304	274 / 302	290 / 320	V
Initial Charge	100	100	100	100	%
Number of Cells per Cell-Row	12	12	20	26	-
Number of Cell-Row	17	30	20	1	-
Internal Resistance charge/discharge	1 / 1	1 / 1	1 / 1	1 / 1	Ω
Operating Temperature	33	270	36	30	°C
Specific Heat Transition	0.4	6	0.4	0.08	W/K
Specific Heat Capacity	795	950	677	1650	J/kg*K
Mass of Battery	318	457	534	173	kg
Battery Price	300	500	400	250	€

2.3. Digitized infrastructure of the virtual road

A Virtual Road is a representation obtained through computerized digitization of one real road sector, generated by AVL Road Importer, based on the coordinates taken from a software platform that shows in detail the terrestrial surface (Google Earth). It has an altitude profile which is punctually implemented with the GPS Prune application that uses GPS coordinates, which represent the Latitude, Longitude and Height of the reference points.

In order to evaluate the autonomy of the electric vehicle in AVL Cruise, the Nardo circuit from Italy was used as a virtual road (figure 6). This circuit with a length of 12.6 km is a testing track for vehicles and due to its circular geometry, it allows the displacement of the virtual vehicle in a closed loop until it reaches the maximum battery autonomy. Defining the geometry of the route was accomplished through a sequence of parametrized segments, which follow the altitude gradient of the real track.

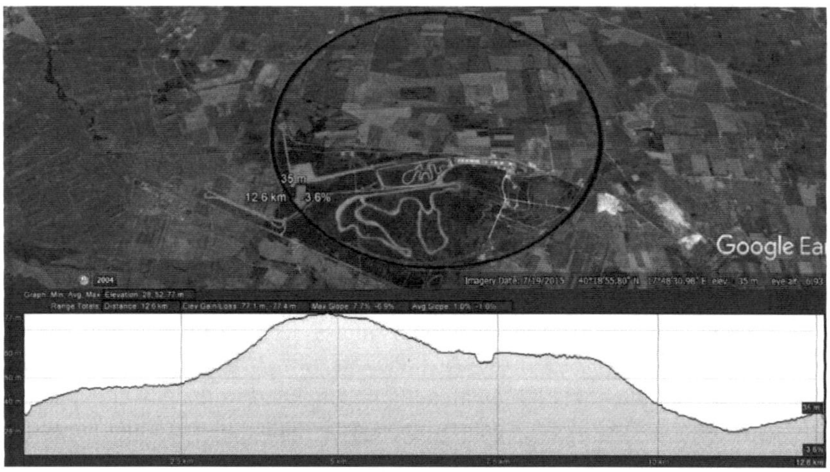

Figure 6. Nardo Ring, Nardo, Italy

In order to digitize the virtual track, the Keyhole Markup Language (KML) file from the Google Earth application was imported. This file contains all data used in the digitization of the road. They were introduced as initial data in the GPS Prune tool, which afterwards added punctually the altitude values for each GPS coordinates set, both for latitude and longitude.

The obtained virtual road was afterwards exported in a compatible format with AVL Cruise application, and was multiplied by 20 times in order to realize a closed loop circuit. This therefore was loaded in the Cycle Run section and afterwards used in the definition of the tasks of the simulations. The calculation tasks contain information regarding environment conditions, traveled distance, speed limits, driving time, behavior of the driver, etc. [27].

3. Results and Discussions

The results of the four battery types obtained from AVL Cruise simulations are presented in table 3.

Table 3. AVL Cruise simulations results.

Name	Value for Battery Type				Unit
	Li-Ion	Na-NiCl₂	Ni-MH	Li-S	
Electrical Consumption (VW Data)	12.7	-	-	-	kWh/100km
Electrical Consumption (Real Test Drive)	18.2	-	-	-	kWh/100km
Electrical Consumption (AVL Cruise)	14.7	12.6	15.8	17.2	kWh/100km
Electrical Vehicle Autonomy (VW Data)	190	-	-	-	km
Electrical Vehicle Autonomy (Real Test Drive)	130 - 190	-	-	-	km
Electrical Vehicle Autonomy (AVL Cruise)	165	192	153	140	km
Total Output Energy	6826	5927	7006	7979	kJ
Total Input Energy	184	222	184	159	kJ

The Electrical Consumption (kWh) results, presented in figure 7, were obtained on Nardo Ring (12.6 km), each simulation being done in identical conditions. After analyzing the results, it can be observed that the lowest energy consumption was achieved while using Na-NiCl2 batteries, which helped to develop a maximum autonomy of 192 km, with an average consumption of 12.6 kWh/100 km or 1.59 kWh/lap. Li-Ion batteries achieved a maximum autonomy of 165 km, with an average consumption of 14.7 kWh/100 km or 1.85 kWh/lap. Ni-MH batteries reached a maximum autonomy of 153 km, with an average consumption of 15.8 kWh/100 km, and 1.90 kWh for a lap. The maximum energy consumption was achieved while using Li-S batteries, which helped to develop a maximum autonomy of 140 km, with an average consumption of 17.2 kWh/100 km and 2.17 kWh/lap.

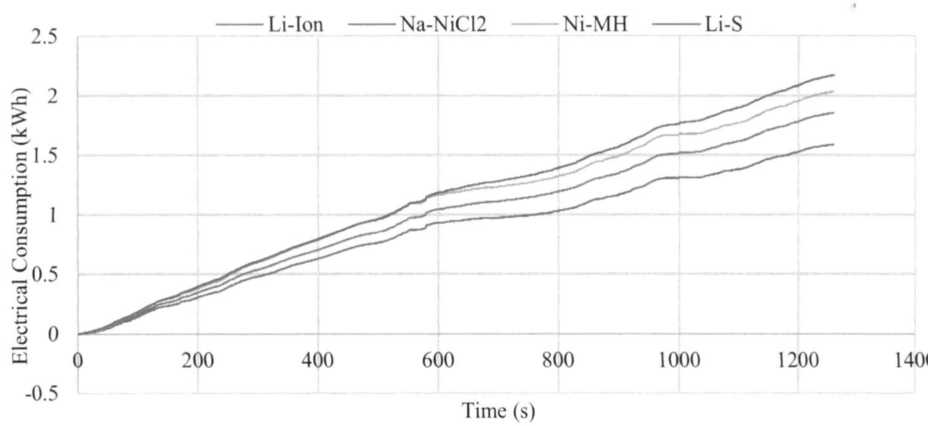

Figure 7. Electrical Consumption (Nardo Ring 12.6 km)

The obtained values are validated only for Li-Ion batteries, through official data received from the producer of the electric vehicle, and through data obtained from testing the vehicle in real traffic conditions. The results of the other battery types were obtained only through computerized simulation, of a model validated based on real data, without being validated through experimental measuring.

Total Output Energy (kJ) and Total Input Energy (kJ) results obtained for each simulation, with all the four battery types, are presented in figure 8. The results' analysis reveals that the recovered energy represents between 2 and 4 % of the total consumed energy. The total amount of energy obtained from recovery braking depends on the altitude profile, but also on the instant storage capacity of each battery type, which may lead to different results.

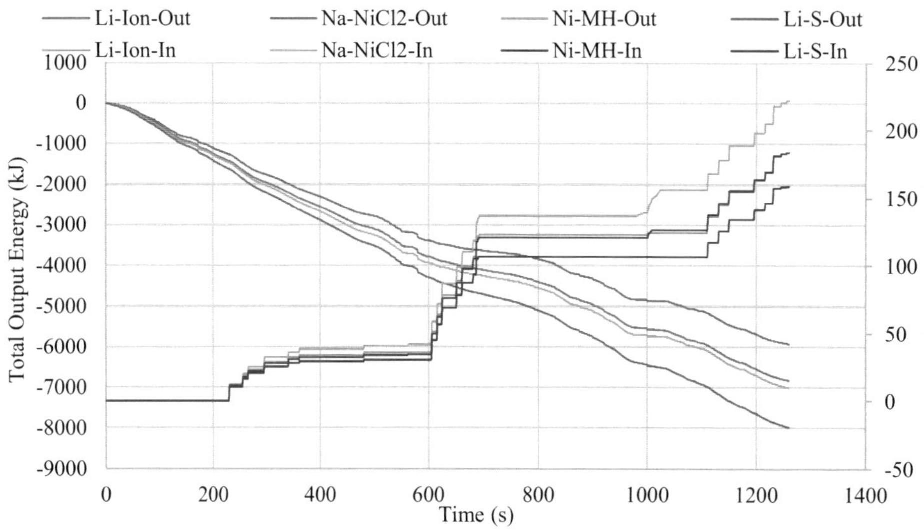

Figure 8. Total Output Energy and Total Input Energy

The Electrical Power (kW) results, obtained after each simulation, are presented in figure 9. The power developed by the electric engine (identical in all four cases) is similar for all four battery types, significant differences only occur while changing the operating mode of the electric engine depending on the altitude profile of the route. Compared to the power values developed by the electric engine powered by a Li-Ion battery, the Li-S batteries generate 0.5 % less power, $Na\text{-}NiCl_2$ generate 0.5 % more power, and Ni-MH batteries generate 1.3 % more power.

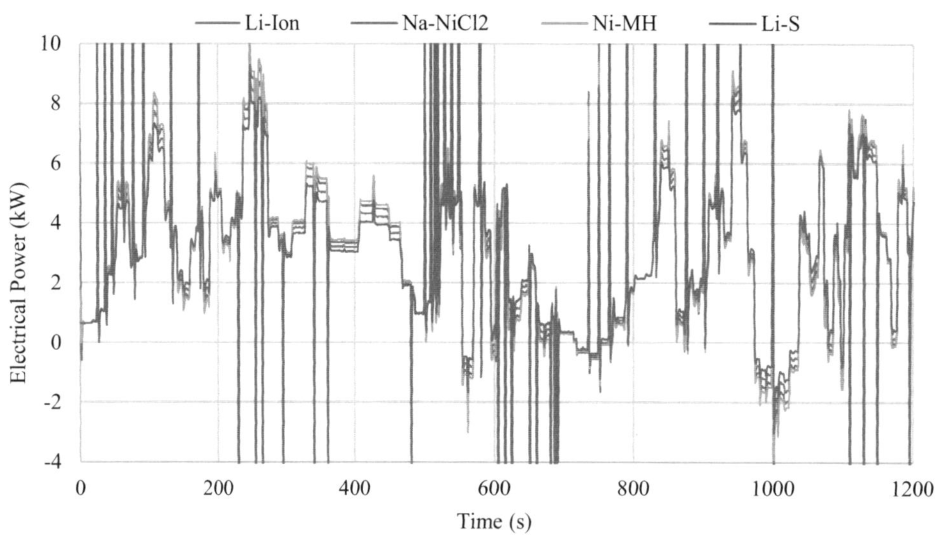

Figure 9. Electrical Power

4. Conclusions

Na-NiCl2 batteries have proven to be the best choice from an energy consumption point of view (12.6 kWh/100 km). Besides that, other important advantages are their low price, increased lifecycle or great functioning under normal parameters in harsh environments. One disadvantage of these batteries is increased operating temperature, which is causing the battery electrolyte to solidify if the vehicle is not used. That is why, it is necessary to have one external system which maintains the battery's operating temperature under functional parameters.

According to these conducted studies, it has been demonstrated that the highest energy consumption (17.2 kWh/100 km) is accomplished by Li-S batteries. However due to their low weight, increased energy storage capacity and low price, compared to other battery technologies, they might be one of the best solutions for systems with high energy storage capacity.

Ni-MH batteries, despite having a reasonable energy consumption, (15.7 kWh/100 km) they are inefficient, having an increased energy density and power, heavy weight as well as an outdated technology.

Nowadays, Li-Ion batteries have the biggest market segment in equipping electric vehicles. Moderate energy consumption (14.7 kWh/100 km), continuous decline of the cost price, advanced manufacturing technology, increased cycle life, low weight and high energy storage potential make Li-Ion batteries an optimal choice in this field. Their disadvantage is represented by high functioning temperatures, which may have negative effects on their energetic performances and lifecycle. All these represent risks regarding safe exploitation of the vehicle.

Acknowledgment

The simulations presented in the paper were done using the software AVL Cruise supported by AVL List GmbH, Graz, Austria.

References

[1] J. Brady, M. O'Mahony, 2011, *Travel to work in Dublin. The potential impacts of electric vehicles* (Elsevier: Transportation Research Part D: Transport and environment, vol 16, issue 2) p. 188
[2] T.R. Hawkins, B. Singh, 2012, *Comparative Environmental Life Cycle Assessment of Conventional and Electric Vehicles* (Journal of Industrial Ecology, vol 17, issue 1) p 53
[3] International Energy Agency, 2015, *Energy and climate change* (IEA Publishing)
[4] B. Nykvist, M. Nilsson, 2015, *Rapidly falling costs of battery packs for electric vehicles* (Nature Climate Change, vol 5) p 329
[5] B. Propfe, M. Redelbach, D.J. Santini, H. Friedric, 2012, *Cost analysis of Plug-in Hybrid Electric Vehicles including Maintenance & Repair Costs and Resale Values* (Conference EVS26)
[6] Z. Chen, N. Guo, X. Li, J. Shen, R. Xiao, S. Li, 2017, *Battery Pack Grouping and Capacity Improvement for Electric Vehicles Based on a Genetic Algorithm* (Energies, vol 10, issue 4)
[7] G.J. Offer, M. Contestabile, D.A. Howey, R. Clague, N.P. Brandon, 2011, *Techno-economic and behavioral analysis of battery electric, hydrogen fuel cell and hybrid vehicles in a future sustainable road transport system in the UK* (Energy Policy, vol 39, issue 4) p. 1939
[8] A. Sakti, J.J. Michalek, 2015, *A techno-economic analysis and optimization of Li-ion batteries for light-duty passenger vehicle electrification* (Journal of Power Sources, vol 273) p. 966
[9] C. Alaoui, 2013, *Solid-State Thermal Management for Lithium-Ion EV Batteries* (IEEE Transactions on Vehicular Technology, Vol 62, issue 1) p. 98
[10] L. Lu, X. Han, J. Li, J. Hua, M. Ouyang, 2012, *A review on the key issues for lithium-ion battery management in electric vehicles* (Journal of Power Sources, vol 226) p. 272
[11] S. J. Gerssen-Gondelach, A. P.C. Faaij, 2012, *Performance of batteries for electric vehicles on short and longer term* (Journal of Power Sources, vol 212) p. 111
[12] T. G. Goonan, 2012, *Lithium use in batteries* (U.S. Geological Survey, Circular 1371)
[13] D. Doughty, E.P. Roth, 2012, *A General Discussion of Li Ion Battery Safety* (The Electrochemical Society Interface, vol 21, no 2) p. 37

[14] A, Lajunen, J. Suomela, 2012, *Evaluation of Energy Storage System Requirements for Hybrid Mining Loaders* (IEEE Transactions on Vehicular Technology vol 61, issue 8) p. 3387
[15] L. Gaines, 2011, *The future of automotive lithium-ion battery recycling: Charting a sustainable course* (Sustainable Materials and Technologies, vol 1–2) p. 2
[16] O. Veneri, L. Ferraro, C. Capasso, D. Iannuzzi, 2012, *Charging infrastructures for EV: Overview of technologies and issues* (IEEE ESARS) p.1
[17] D. Linden, T.B. Reddy, 2002, *Handbook of batteries, third edition* (McGraw Hill)
[18] D. Linden, T.B. Reddy, 2010, *Linden's Handbook of batteries, fourth edition* (McGraw Hill)
[19] K. Young, C. Wang, L.Y. Wang, K. Strunz, 2013, *Electric Vehicle Battery Technologies* (Chapter: Electric Vehicle Integration into Modern Power Networks, Springer) p. 15
[20] T.M. O'Sullivan, C.M. Bingham, R.E. Clark, 2006, *Zebra battery technologies for all electric smart car* (SPEEDAM 2006) p. 6
[21] Meridian, 2005, *The Sodium Nickel Chloride "Zebra" Battery* (Meridian International Research)
[22] H. Wang, L.F. Cui, Y. Yang, H.S. Casalongue, J.T. Robinson, Y. Liang, Y. Cui, H. Dai, 2010, *Mn_3O_4 - Graphene Hybrid as a High-Capacity Anode Material for Lithium Ion Batteries* (Journal of the American Chemical Society, vol 132, issue 40) p. 13978
[23] A.D.W. Todd, P.P. Ferguson, M.D. Fleischauer, J.R. Dahn, 2010, Tin-based materials as negative electrodes for Li-ion batteries: Combinatorial approaches and mechanical methods (International Journal of Energy Research, vol 34) p. 535
[24] A.F. Burke, 2007, *Batteries and Ultracapacitors for Electric, Hybrid, and Fuel Cell Vehicles* (Proceedings of the IEEE vol 95, issue 4) p. 806
[25] D. Aurbach, E. Pollak, R. Elazari, G. Salitra, C.S. Kelley, J. Affinito, 2009, *On the Surface Chemical Aspects of Very High Energy Density, Rechargeable Li–Sulfur Batteries* (Journal of the Electrochemical Society, vol 156, issue 8) p. A694
[26] L. Chen, L.L. Shaw, 2014, *Recent advances in lithium-sulfur batteries* (Journal of Power Sources, Vol 267) p. 770
[27] B.O. Varga, F. Mariasiu, D. Moldovanu, C. Iclodean, 2015, *Electric and Plug-In Hybrid Vehicles: Advanced Simulation Methodologies* (Springer)
[28] Volkswagen, 2014, *VW e-Golf 2014 Brochure*, (Volkswagen AG, version: 03/2014)
[29] Volkswagen, 2015, *VW E-Golf Owner Manual*, (Volkswagen AG)
[30] A. Todorut, N. Cordos, I. Barabas, R.D. Muresan, M. Balcau, 2016, *Comparative study on the dynamic behavior in cornering from different classes of passenger cars, by experimental and simulation methods* (Acta Technica Napocensis, Series: Applied mathematics, mechanics and engineering, vol 59, issue 3) p. 285
[31] A. Todorut, N. Cordos, M.D. Burdea, M. Balcau, 2015, *The evaluation of normal load redistribution on the static axles and on the wheels, when the vehicle is in motion* (Acta Technica Napocensis, Series: Applied mathematics, mechanics and engineering, vol 58, issue 3) p. 349
[32] AVL, 2011, *Users Guide AVL Cruise Version 2011*, (AVL List GmbH, Edition 06/2011)
[33] A.A. Pesaran, M. Keyser, 2001, *Thermal Characteristics of Selected EV and HEV Batteries* (Sixteenth Annual Battery Conference on Applications and Advances) p. 219
[34] Daimler Benz Aktiengesellschaft, 1997, *The A-Class Electric Vehicle Powered by the ZEBRA Battery System* (Daimler Benz AG Kommunication)
[35] Eurobat, 2014, *A Review of Battery Technologies for Automotive Applications*, (EUROBAT)
[36] H.A. Kiehne, 2003, *Battery Technology Handbook, Second Edition* (Marcel Decker Inc.)
[37] J. Larminie, J. Lowry, 2004, *Electric Vehicle Technology Explained* (John Wiley & Sons Ltd.)
[38] Y.V. Mikhaylik, J. R. Akridge, 2003, *Low Temperature Performance of Li/S Batteries* (Journal of The Electrochemical Society, vol 150, issue 3) p. A306
[39] J.L. Sudworth, 2001, *The Sodium/Nickel chloride (ZEBRA) battery* (Journal of Power Sources, vol 100, issues 1–2) p. 149
[40] Z. Zhang, S.S. Zhang, 2015, *Rechargeable Batteries - Materials, Technologies and New Trends* (Springer)

Analysis of the economic and ecological performances in the transient regimes of the European driving cycle for a midsize SUV equipped with a DHEP, using the simulation platforms

Gheorghe Bancă[1], Florian Ivan[2], Daniel Iozsa[1], Valentin Nisulescu[1]

[1]POLITEHNICA University of Bucharest, Romania,
[2]University of Piteşti, Romania

banca_2004@yahoo.com

Abstract. Currently, the tendency of the car manufacturers is to continue the expansion of the global production of SUVs (Sport Utility Vehicle), while observing the requirements imposed by the new pollution standards by developing new technologies like DHEP (Diesel Hybrid Electric Powertrain). Experience has shown that the transient regimes are the most difficult to control from an economic and ecological perspective. As a result, this paper will highlight the behaviour of such engines that are provided in a middle class SUV (Sport Utility Vehicle), which operates in such states. We selected the transient regimes characteristic to the NMVEG (New Motor Vehicle Emissions Group) cycle. The investigations using the modelling platform AMESim allowed for rigorous interpretations for the 16 acceleration and 18 deceleration states. The results obtained from the simulation will be validated by experiments.

1. Introduction

The continuous increase of the vehicle fleet at a global level, corroborated with the climatic changes of the last decades led to the development of the various antipollution norms, with a strictness of the criteria, becoming harder and harder to be respected by the car manufacturers, norms based on the driving of a certain number of kilometres for a specific representative cycle like, for example, the NMVEG cycle, undertaken on the roller testing bench after a prior maceration at a temperature between 20 - 30 °C for the norms Euro 1(1992)…Euro 6b (2014).

The NMVEG cycle (Figure 1) is composed of an urban part, 4 cycles ECE or UDC (Urban Driving Cycle) – (Figure 2), a cycle EUDC (Extra Urban Driving Cycle) – (Figure 3) and is characterized by 3 types of regimes:

- idle speed regimes;
- transient regimes for acceleration/deceleration;
- stabilized regimes (constant speed sequences).

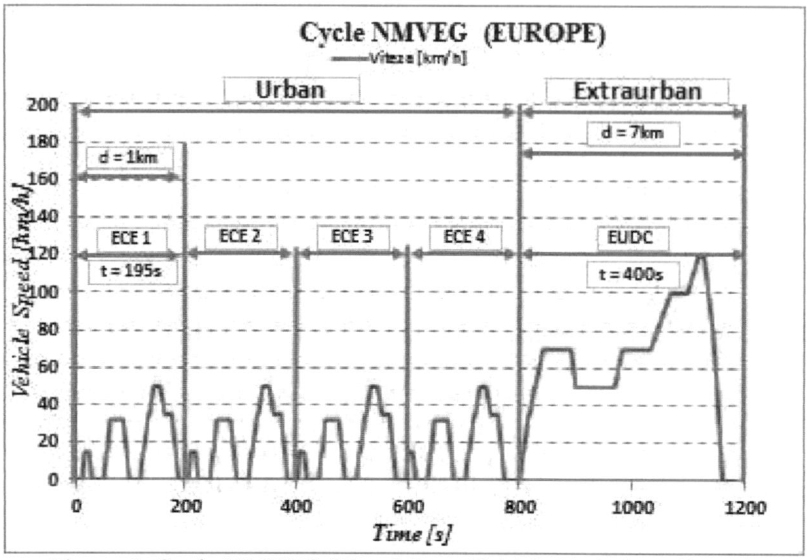

Figure 1. Cycle NMVEG (New Motor Vehicle Emissions Group).

The severity of the acceleration regimes specific for a testing cycle is given by a parameter called relative positive acceleration (RPA) in [m/s^2], having the formula [1]:

$$RPA = \left(\frac{1}{d_{tot}}\right)\sum_{i=1}^{n}\begin{cases} \frac{a_i v_i}{3.6} & if\ (a > 0) \\ 0 & (otherwise) \end{cases} \tag{1}$$

where d_{tot} is the total distance travelled, a_i the acceleration of the vehicle in [m/s^2] for time step *"i"* if $a_i > 0$ and v_i the vehicle speed in [m/s] for time step *"i"*.

In table 1 is presented a summary of the main characteristics of the NMVEG cycle:

Table 1. Main characteristics of the NMVEG cycle

Characteristics of the cycle	Unit	NMVEG
Total distance travelled	km	10.993
Time spent to idling speed	s	267
Average speed (with stop)	km/h	33.54
Average speed (without stop)	km/h	43.16
Average acceleration	m/s^2	0.59
Maximum acceleration	m/s^2	1.04
RPA ECE [1]	m/s^2	0.147
RPA EUDC [1]	m/s^2	0.0936

Noting with "*r*" - the idle speed regimes, with "*a*" - the acceleration transient regimes, with "*s*" - the stabilized regimes and with "*d*" - the deceleration transient regimes, like in figures 2 and 3, for the NMVEG cycle we obtain:

- (3x4+2) = 14 idle speed regimes *"r"*;
- (3x4+4) = 16 acceleration regimes *"a"*;
- (4x4+2) = 18 deceleration regimes *"d"*;
- (4x4+5) = 21 stabilized regimes, constant speed regimes *"s"*.

Analysing the transient regimes marked with *"a"* and *"d"*, we observe the following:

- the acceleration regimes, *"a"*, have a duration of 293s meaning 25% of the total duration of the cycle.
- the deceleration regimes, *"d"*, have a duration of 204s from the total of 1180 s, meaning 17% of the duration of the cycle.

This means that overall the transient regimes represent 42% of the duration of the cycle. As a result, the influence of these regimes on the economy and polluting emissions becomes significant, such influences being highlighted by this paper.

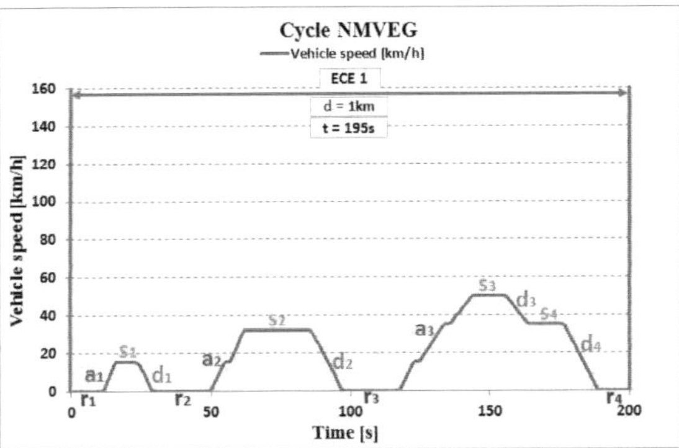

Figure 2. Cycle NMVEG ECE1

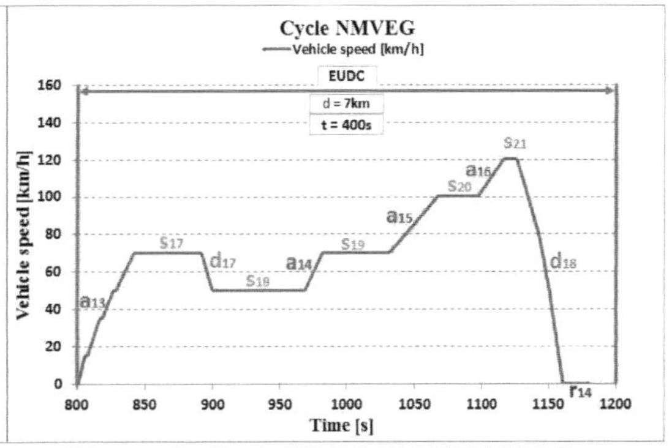

Figure 3. Cycle NMVEG ECE EUDC

2. Presentation of the model

The model is specific to a SUV from the class Renault DUSTER equipped with a propulsion system diesel hybrid electric parallel (figure 4), [2], [3]. The study was achieved with the multi-domain simulation platform, AMESim, designed to simulate three battery loading states (82%, 41% and 2%) in the transient regimes of the NMVEG cycle from which only the transient acceleration regimes were selected.

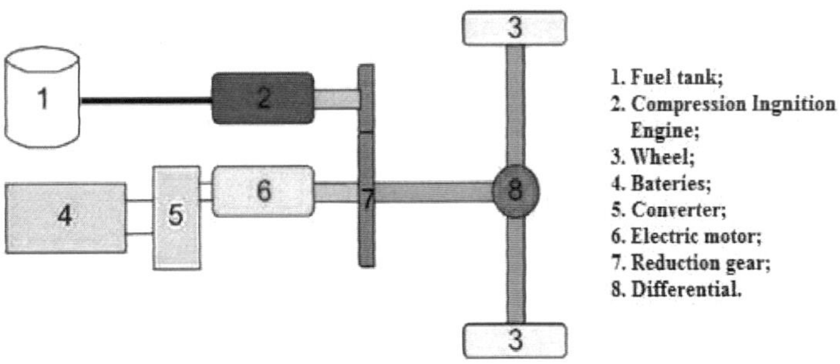

Figure 4. Structure of a DHEP

The AMESim model proposed in figure 4 is composed from 8 large components, as follows:

- a driver component which is used to control the vehicle on the cycle;
- a battery of accumulators with a voltage of 310 V;
- an electric machine (motor/power generator) of approx. 36 kW;

- a computer VCU-PH (Vehicle Control Unit – Parallel Hybrid) which manages the power and the operating mode for the combustion engine and the electric engine;
- compression combustion engine - Diesel (component based on schematics for torque, polluting emissions and fuel consumption identified on an ISO-N, for an engine of 1.5l dCI 81kW Euro 6);
- mechanical transmission with ratios parameterized according to the chosen vehicle;
- an exhaust line that includes the post-treatment systems for the exhaust gases specific to a Diesel vehicle (oxidation catalyst, NOx-trap, particle filter);
- a vehicle parametrized based on an SUV from the class Renault DUSTER.

Along with these components the model also presents elements like: type of cycle used for testing, two clutches, elements simulating the inertia and frictions and various sensors to recover information regarding the status, torque, temperatures and polluting emissions.

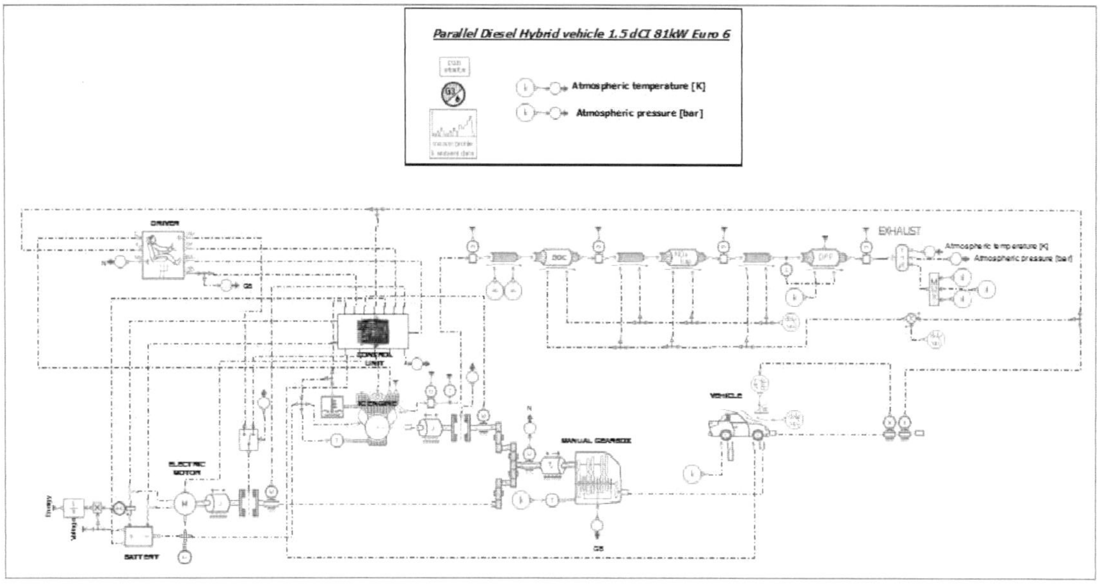

Figure 5. Model AMESim proposed for a vehicle equipped with DHEP

In the model used, a component with a major impact on the representativeness of the simulation from the economy and ecology perspective is the vehicle submodel (figure 6), [4], defined in AMESim as a vehicle with constant mass allowing the user to choose from two configurations for parameterization:

- road, in this configuration, the model is based on the rolling resistance of the vehicle, the aerodynamic resistance and the resistance of the slope, if present;
- roller testing bench, for this configuration, the user introduces the friction coefficients, similar to adapting the vehicle to the roller bench to reproduce the "curve of the road".

Also, depending on the longitudinal sliding of the tire, the user can chose between two configurations:

- no sliding between the tire and the carriageway, like in this paper;
- with sliding, the user defines the parameters of the longitudinal sliding of the tire.

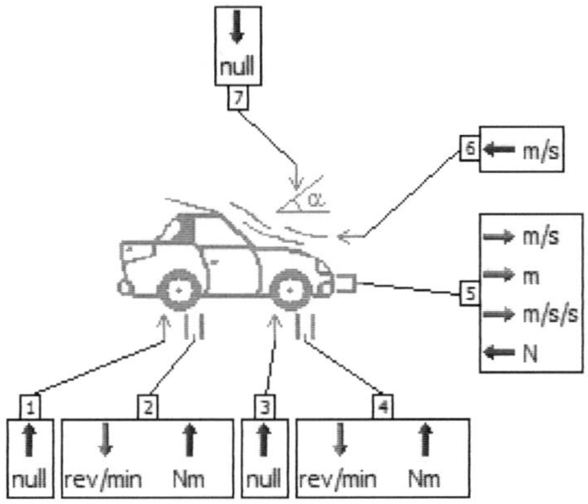

Figure 6. Vehicle Component

Returning to figure 6 (vehicle sub-model), the ports 1 and 3 are signal ports receiving the front and rear braking commands, the braking torque for each axle being determined by the input signal, the maximum braking torque and the distribution between the two axles, front and rear, and ports 2 and 4 are rotating mechanical ports that receive the driving torque and return the angular velocity for each axle.

The port 5, [4], being a mechanical translation port, receives the external forces applied to the car and returns its linear displacement, the speed, acceleration, and the ports 6 and 7 are signal ports receiving the wind velocity and the road slope.

Regarding the configuration chosen by the authors for the simulation presented in this paper, we chose the reproduction of the behavior of a SUV car from the Renault DUSTER class on the roller bench without longitudinal slide of the tire, using the following parameters: the distribution of the mass, characteristics of the brake, moment of inertia of the wheel plus the dimensions of the tire and the coefficients used to reproduce the "curve of the road".

For the calculation formulas used by the model for the development of the simulation we can list:

- wheel radius: $$R_R = \frac{1}{2} \cdot D_r + 0.01 \cdot H \cdot l \quad (2)$$

where D_r is the diameter of the wheel rim [m] , H the height of the tire [%] and l is the width of the tire [m].

- the total mass of the vehicle being considered and the effect of the moment of inertia of the wheel during the linear movement:

$$m_v = m + 4 \cdot \frac{J_R}{R_{R^2}} \quad (3)$$

in which m is the total weight of the vehicle in $[kg]$, J_R is the moment of inertia of the wheel $[kgm^2]$ and R_R is the radius of the wheel $[m]$.

In the configuration without the longitudinal sliding of the tire, the driving force shall be calculated by the formula:

$$F_m = \frac{T_2 + T_4}{R_d} \quad (4)$$

where T_2 and T_4 are the torque inputs for the ports 2 and 4 (for the two axles front and rear) $[Nm]$ and R_d is the dynamic radius of the wheel $[m]$.

For the case selecting the configuration with longitudinal slide of the tire, for each axis this slide A_L is considered in the submodel and is defined as:

$$A_L = 100 \cdot \frac{(R_d \cdot \omega_r \cdot \frac{\pi}{30} - v)}{v} \tag{5}$$

in which R_d is the dynamic radius of the wheel $[m]$, ω_r the angular velocity of the wheel $[rot/min]$ and v is the translational speed of the vehicle $[m/s]$.

The resistance forces simulated by the vehicle submodel are given by the following relationships:

- force given by the resistance of the slope $R_p[N]$:

$$R_p = m \cdot g \cdot \sin(\arctan(0.01 \cdot \alpha)) \tag{6}$$

where m is the total weight of the vehicle in $[kg]$, g the gravitational acceleration: g = 9,81 $[m/s^2]$ and α is the slope of the track [%].

- the aerodynamic resistance $R_a[N]$ is calculated using the following relationship:

$$R_a = \frac{1}{2} \cdot \rho_a \cdot C_x \cdot S \cdot (v + v_{v\hat{\imath}})^2 \tag{7}$$

in which ρ_a is the air density defined in the submodel $[kg/m^3]$, C_x is the aerodynamic coefficient [ ], S is the section of the vehicle $[m^2]$, v the speed of the vehicle $[m/s]$ and $v_{v\hat{\imath}}$ is the wind speed $[m/s]$ used as input for the port 6 (figure 6), [4].

- the force provided by the rolling resistance $R_r[N]$ provided by the formula:

$$R_r = m \cdot g \cdot (f + k \cdot v + wind \cdot v^2) \tag{8}$$

where f, k and $wind$ are defined by the user as constants, [4].

Then, taking into account all this forces, the resistive forces applied to the vehicle [5], [6], [7] and [8], are obtained from:

$$\sum R = R_r + R_p + R_a \tag{9}$$

Because the roller bench will have to reproduce all these resistive forces, the force of total resistances, [4], becomes:

$$\sum R = R_p + (a + b \cdot v + c \cdot v^2) \tag{10}$$

in which: a $[N]$, b $[N/(m/s)]$, $c[N/(m/s^2)]$ are the coefficients defined by the user as characteristic to the roller bench.

3. Results and discussions

The goal of the model used, described in chapter II, is to identify in the project development stage the behavior of the polluting emissions and the fuel consumption for the transient regimes for a SUV equipped with an powertrain diesel/hybrid for the 3 states of charging of the battery (SOC – State Of Charge), respectively for SOC = 82%, SOC = 41% and SOC = 2%.

Analyzing the global results of the instantaneous engine emissions, in the transient regimes for the whole NMVEG cycle (figures 7.a, 8.a, 9.a, 10.a, 11.a) we observe that they present similarities with the evolution of the engine torque provided by the internal combustion engine, figure 7, which indicates the presence or absence of the combustion process (disabling/enabled of the electric motor), and engine coolant temperature (figure 8).

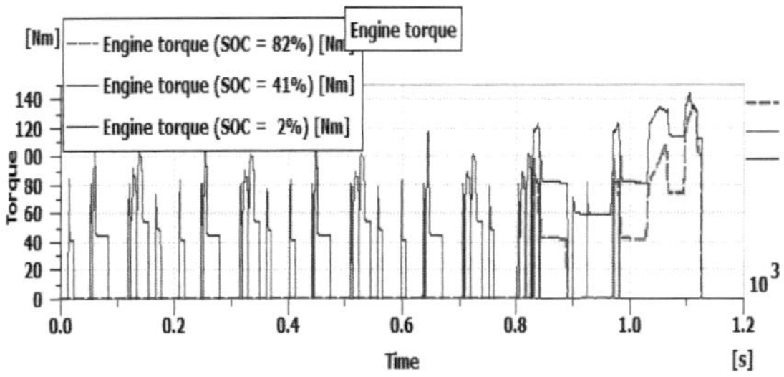

Figure 7. Engine torque on the NMVEG cycle

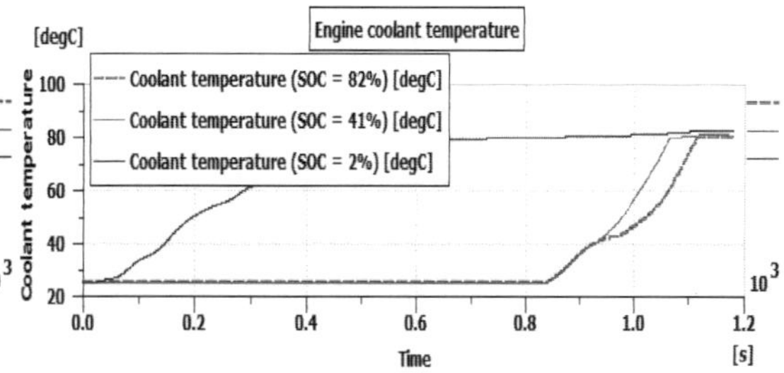

Figure 8. Coolant temperature on the NMVEG cycle

Realizing also the cumul of the emissions from the figures (9.a, 10.a, 11.a, 12.a, 13.a), by integrating each polluting emission depending on time for each individual transient acceleration regimes we obtain the figures (9.b, 10.b, 11.b, 12.b, 13.b). The decelerations were not considered because the engine polluting emissions are almost null during the deceleration, due to the fact that for the modern internal combustion engines the deceleration is done with injection cut during deceleration, and even with energy recovery for the hybrid powertrain.

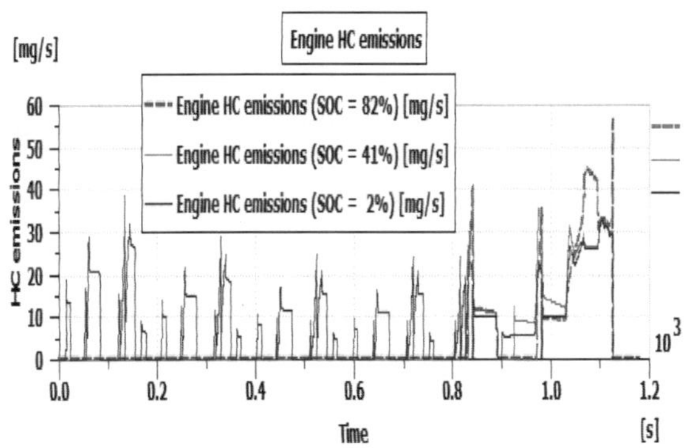

Figure 9.a. Instantaneous HC emissions

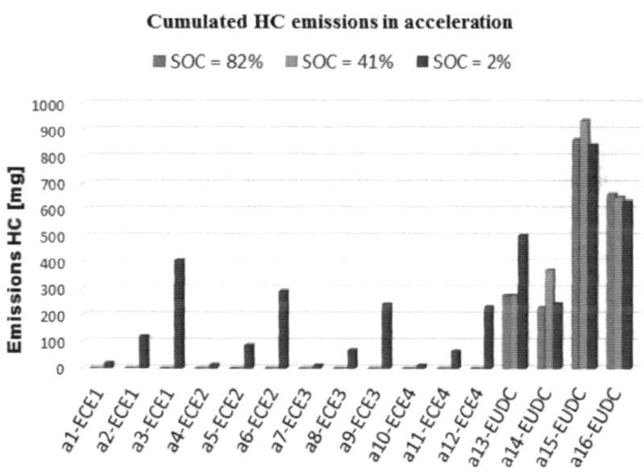

Figure 9.b. Cumulated HC emissions

It is known that the exit of the internal combustion engine from the thermal regime has negative consequences on both the fuel consumption and the pollutants [9], [10]. In the case of figure 9.b, the cumulated HC emissions present similarities compared to the CO emissions figure 10.b, when the engine is cold, the level of the HC emissions being up to 62% higher for the regime a1, 83% higher for the regime a2 and 75% for the regime a3.

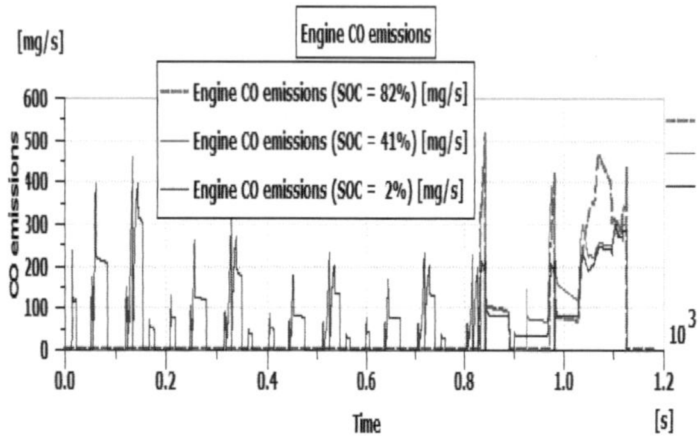

Figure 10.a. Instantaneous CO emissions

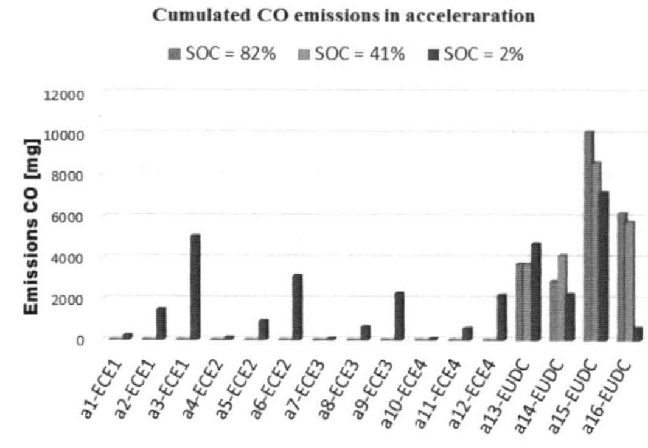

Figure 10.b. Cumulated CO emissions

For the configuration SOC = 2%, in figure 10.b is observed that the evolution of the cumulated CO emissions in the acceleration regimes for the 4 ECEs, presents an improvement depending on the evolution of the coolant temperature, figure 8. Thus, when the engine is cold (ECE 1), the CO emissions are up to 182% higher for the regime a1, 154% higher for the regime a2 and 130% for the regime a3.

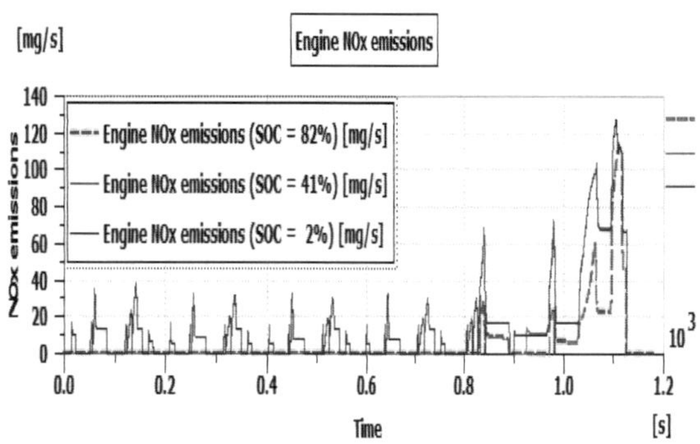

Figure 11.a. Instantaneous NOx emissions

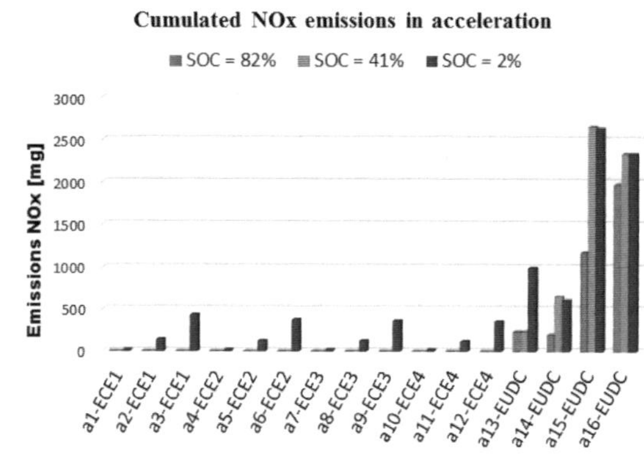

Figure 11.b. Cumulated NOx emissions

The cumulation of the NOx emissions for the transient acceleration regimes in figure 11.b, does not present a significant evolution for the first 4 ECEs. It can be observed, also, that the torque contribution provided by the electric motor in the case SOC = 82% leads to an improvement of up to 76% for the regime a13, 67% for the regime a14, 55% for the regime a15 and 15% for the regime a16. In the case SOC = 42% the improvement of the NOx emissions is of 76% for the regime a13 and only 8% for the regime a14 and 1% for the regime a15.

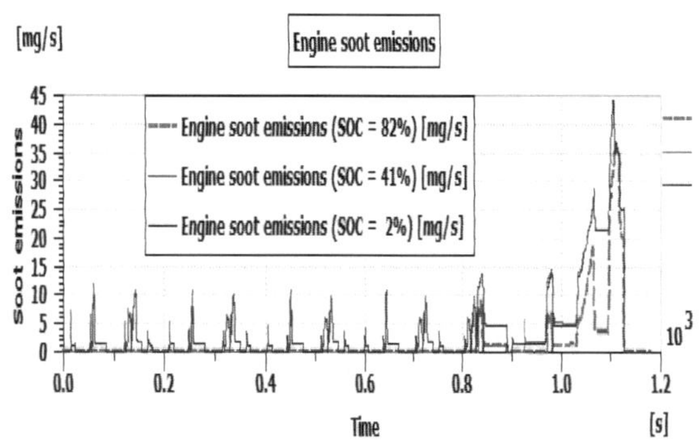

Figure 12.a. Instantaneous PM emissions

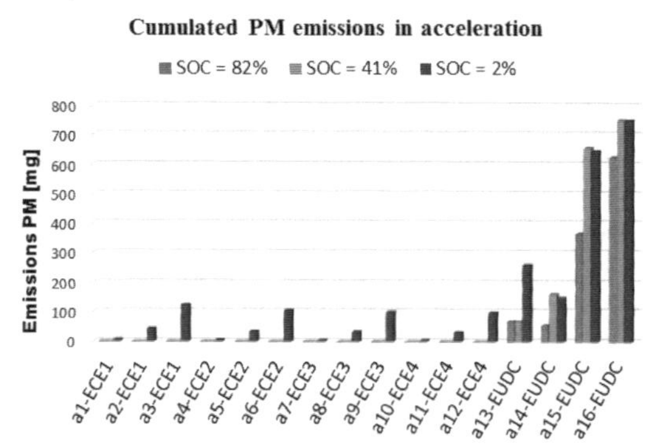

Figure 12.b. Cumulated PM emissions

When analyzing the cumulated emissions of mechanical particulates for the acceleration regimes, figure 12.b, is observed both an improvement of these emissions depending on the temperature increase for the engine coolant for the combustion engine and an improvement depending on the charging state of the battery. From this, was obtained a degrading of 48% for the regime a1, 35% for the regime a2 and 27% for the regime a3 for SOC = 2% as a result of the simulation of the engine start from cold on the cycle and an improvement of 74% for the regime a13, 63% for the regime a14, 43% for the regime a15 and 17% for the regime a16. In the case SOC = 41% we obtain a reduction of the emissions of 74% for the regime a13, 8% for the regime a14 and 2% for the regime a15.

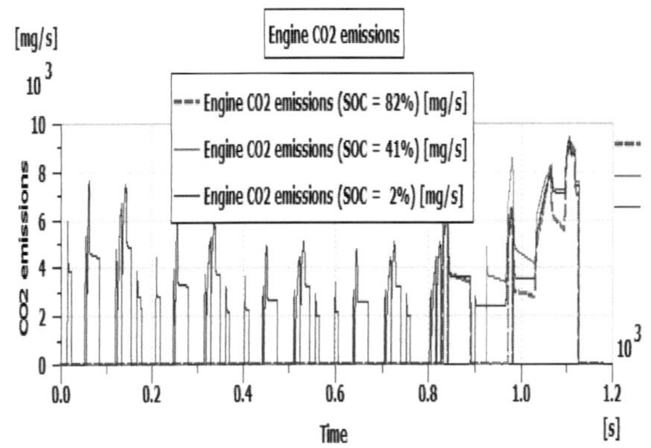

Figure 13.a. Instantaneous CO2 emissions

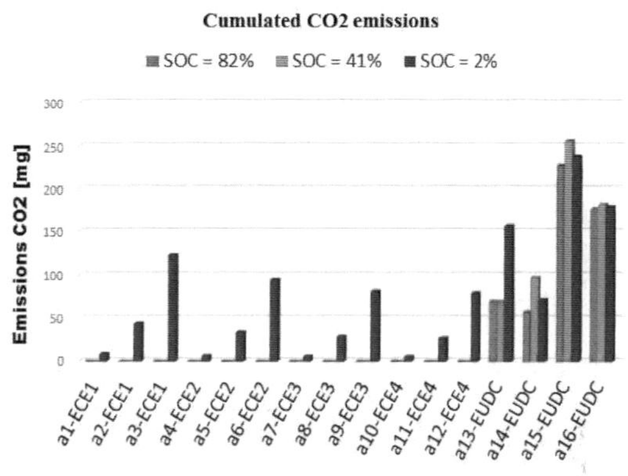

Figure 13.b. Cumulated CO2 emissions

For the analysis of the fuel consumption were used the summed CO2 emissions during the acceleration regims, figure 13.b. With regard to the 4 ECEs we can observe the same improvement tendency depending on the temperature of the cooling fluid. Thus, for SOC = 2% was obtained an increase of the CO2 emissions with 55% for a1, 63% for a2 and 56% for a3. On the EUDC sequence the results show a distinct reduction of the CO2 emissions, only for the case SOC = 82% thus obtaining an improvement of 56% for a13, 19% for a14, 5% for a15 and 1% for a16. In the case SOC = 41% was obtained an improvement of 56% only for a13, the other regimes presenting a degrading of 35% for a14, 8% for a15 and 2% for a16.

4. Conclusions

During the transient acceleration regimes corresponding to the 4 ECEs of the NMVEG cycle, urban use, the technology "Diesel Hybrid Total" without charging from the domestic electrical network can provide significant gains with regard to the polluting emissions HC/CO, NOx, PM and the fuel consumption. During the study were observed zero polluting emissions, zero CO2 emissions, for SOC = 82% respectively 41%.

During the urban use of a vehicle "Diesel Hybrid" with a discharged battery, in the case of the simulation conducted for SOC = 2%, for the transient acceleration regimes, the combustion engine being the main source of energy, the polluting emissions being generated exclusively by it, the most critical regimes being (a3, a6, a9 and a12), this means that a proper optimization of the calibration of the combustion engine is required.

In the case of using a vehicle with a discharged battery, the switching from an operating mode for the internal combustion engine to another (cold ~ 20 °C to hot ~85 °C) – figure 13, is accompanied by a reduction of the polluting emissions for the transient regimes as a result of the reduction of the internal friction of the engine. For the simulation conducted the polluting emissions were reduced by 100% to 150% depending on the pollutant, figures (7.b, 8.b, 9.b, 10.b, 11.b).

As a result of the activation of the internal combustion engine (vehicle speed > 55 km/h) for the extra urban use of the vehicle, with regard to the NOx and PM emissions, we observe that the SOC has a strong negative influence only for the regime a13 (first acceleration 0-70 km/h), the battery charging state depriving the internal combustion engine of the torque contribution of the electric motor.

References

[1] Barlow TJ, Latham S, McCrae IS and Boulter PG, June 2009, *A reference book of driving cycle for use in the measurement of the road vehicle emission*, ISBN 978-1-84608-816-2;
[2] Cezar Leandru Corneliu Babici, Iași 2012, *Control strategies for hybrid electric powertrain vehicles,* PhD thesis;
[3] Jefferson CM, Barnard RH, Wit press 2002, *Hybrid vehicle propulsion;*
[4] DRVVEH4A, constant load vehicle with or without slip, help AMESim Rev12;
[5] Aurel Stoicescu, Ion Tabacu, Marin Untaru, Gheorghe Pereș, Gheorghe Poțincu, București 1981, *Dynamics of vehicles on wheels;*
[6] Tiberiu-Nicolae Macarie, Pitești 2003, *Automotive Dynamics;*
[7] Wolf-Heinrich Hucho, 1987, *Aerodynamics of road vehicles;*
[8] Pierre Duysinx, Liege 2015, *Performances du vehicule;*
[9] Florian Ivan, Daniel Liță, Andrei Bușoi, 2014, *Methods and means of depollution of the engines for vehicles*, ISBN 978-606-25-0038-2;
[10] Rory Lewis, Jason Bruke, Costa Mesa 2000, *Detailed vehicle thermal system modeling in ADVISOR trough integration with FLOWMASTER2.*

IOP Conf. Series: Materials Science and Engineering **252** (2017) 012060 doi:10.1088/1757-899X/252/1/012060

Simulation of an electric vehicle model on the new WLTC test cycle using AVL CRUISE software

[1]Constantin Cristian CIOROIANU*, [1]Dănuţ Gabriel MARINESCU, [1]Adrian IORGA, [1]Adrian Răzvan SIBICEANU

[1] University of Piteşti, Str. Târgu din Vale No. 1, 110040 Piteşti, România

E-mail : constantin.cioroianu@yahoo.com, dan.marinescu@upit.ro, adyiorga@yahoo.com, sibiceanu_adrian_razvan@yahoo.com

Abstract. Nowadays, environmental pollution has become a general issue and the automotive industry is probably the most affected. The principal air-quality pollutant emissions from petrol, diesel and LPG engines are carbon dioxide, oxides of nitrogen, un-burnt hydrocarbons. Modern cars produce only quite small quantities of the air quality pollutants, but the emissions from large numbers of cars add to a significant air quality problem. Electric vehicles are an answer to this problem because they have absolutely no emissions. These vehicles have some major disadvantages regarding cost and range. In this paper, an electric vehicle model will be created in the AVL Cruise software. The constructed model is based on the existing Dacia Sandero. Also unlike the real car, the model presented has different characteristics since it is a full electric vehicle. It has an electric motor instead of the petrol engine and a battery pack placed in the trunk. The model will be simulated in order to obtain data regarding vehicle performance, energy consumption and range on the new WLTC test cycle. The obtained know-how will help on later improvements of the electric model regarding methods to increase the vehicle range on the new WLTC test cycle.

1.Introduction

One of the greatest challenges of recent years is air pollution caused by emissions of road vehicles. These emissions and the pollutants that they contain affect not only human health but also the air, water, soil, vegetation and fauna. In densely populated areas such as cities where there are many cars, the effect of pollution is even more drastic and air quality becomes a serious issue. Due to harsh regulations, modern cars have low emissions. This is not enough because the number of cars keeps growing from year to year. These tough emission regulations mean that automotive manufacturers have to invest more and more money on exhaust treatment systems, thus increasing the cost of the vehicle. Electric vehicles represent a viable solution to this issue, as they have no emissions and also they are very quiet. They could also help reduce noise levels found in big cities. The major disadvantages of electric vehicles are low range and high acquisition cost. In a battery electric vehicle (BEV) all energy to propel the vehicle is carried in batteries, which has major consequences for the vehicle design and performance[1]. Also, charging these batteries is a problem since it takes time. Most likely, EV drivers will use EV charging stations where they work[2], so this is a way to solve this problem. Developing these vehicles is an expensive and time consuming process and that's why computerized simulations are used in order to obtain results before carrying out experimental tests.

At the "Automotive Engineering" Research Center of the University of Piteşti there is the Sandero Electron concept. This concept was developed starting from a Dacia Sandero series model. Unlike the original model, the Electron concept is a full electric vehicle.

A globally harmonized standard for the determination of energy consumption and electric range from light duty vehicles (passenger and light commercial vehicles) is defined through the Worldwide harmonized Light vehicles Test Procedures (WLTP). The conditions regarding dynamometer tests and vehicle load were provided through a strict guidance from the WLTP test procedure with WLTC test cycle[3]. Since there is no information regarding the range and energy consumption of the Electron concept and because the WLTC will become a mandatory procedure for homologating vehicles, the decision was made to simulate a model of the Electron concept using AVL Cruise software. This software is an excellent tool to determine vehicle range and energy consumption.

2. Presentation of the Sandero Electron Concept

This vehicle was chosen as a base model because it is a low cost vehicle that can accommodate five passengers. It has a simple design that made the electric vehicle conversion a more easier process. It also provides enough space for the electric equipment; after the conversion the room for the passengers is not affected.
The Sandero Electron concept can be observed in figures 1 and 2 below.

Figure 1. Sandero Electron (front view)

Figure 2. Sandero Electron (rear view)

After the removal of all the elements that are specific to the petrol engine (engine, gearbox, fuel system and exhaust) the components for electric propulsion were installed. The liquid cooled electric motor is made by MES-DEA Switzerland and features two running modes: normal and boost. In normal mode the maximum power output is 18kW (24.5 HP) between 2850 – 9000 rot/min and maximum torque is 90 Nm at 2850 rot/min. In boost mode (short periods of time) the maximum power output is 31kW (42 HP) between 2850 – 9000 rot/min and maximum torque is 160 Nm at 1400 rot/min. The transmission features a fixed ratio reduction and has a shift lock for parking. The electronic management unit for the electric motor is TIM 400 (Traction Inverter Module), made by the same manufacturer as the motor. It is specially designed for electric and hybrid vehicles. The battery pack that empowers the electric system can be seen in figure 3 below.

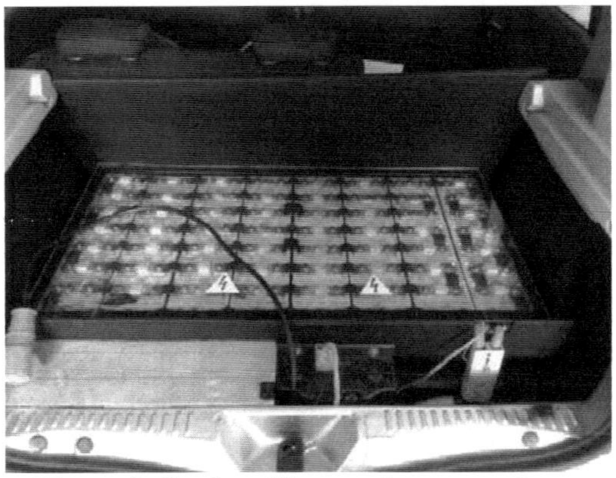

Figure 3. Sandero Electron battery pack

Battery data:

- Maximum charge: 60 Ah
- Nominal voltage: 3.2 V / cell
- Minimum voltage: 2.8V / cell
- Maximum voltage: 4V / cell
- Battery pack consists of 64 cells

The nominal voltage of the battery pack is 205V; minimum voltage is 179V and maximum voltage is 256V. The battery can be charged while driving, by regenerative braking or plug-in to the grid. Regenerative braking was implemented because stopping a 1500kg vehicle from 100km/h to zero speed dissipates about 0.16kWh of energy [$(1/2)MV^2$] in a few tens of meters. If this amount of energy is dissipated by coasting and only by drag forces (rolling resistance and aerodynamic drag) without braking, the vehicle will travel about 2 km [4].
From the vehicle brochure we find that the maximum weight (vehicle + load) is 1550 kg.
All these represent the input data for the simulation.

3. Simulation for the Sandero Electron model on WLTC test cycle in AVL Cruise

In order to achieve this simulation, the WLTC test cycle will be adapted to the Cruise software for the Electron model. This test cycle was chosen because it will become mandatory. Also in the WLTC significant higher velocities have to be driven for longer periods and the acceleration values are corresponding to real driving conditions [5]. Previously, vehicles were homologated using the NEDC but numerous complaints regarding optimistic numbers of fuel consumption lead to this change. It is estimated that, if in 2001 the difference between the theoretical and the real consumptions was 7 – 8%, 10 years after these differences are 3 – 4 times higher. For example, in 2013 was found a difference of 31% for private cars [6] between announced and real fuel consumptions. That's why the WLTC is used here for obtaining data that is as close to real running conditions as possible.
The first step is creating a model in AVL Cruise starting from the real Electron concept. The Cruise model can be seen in the following figure 4.

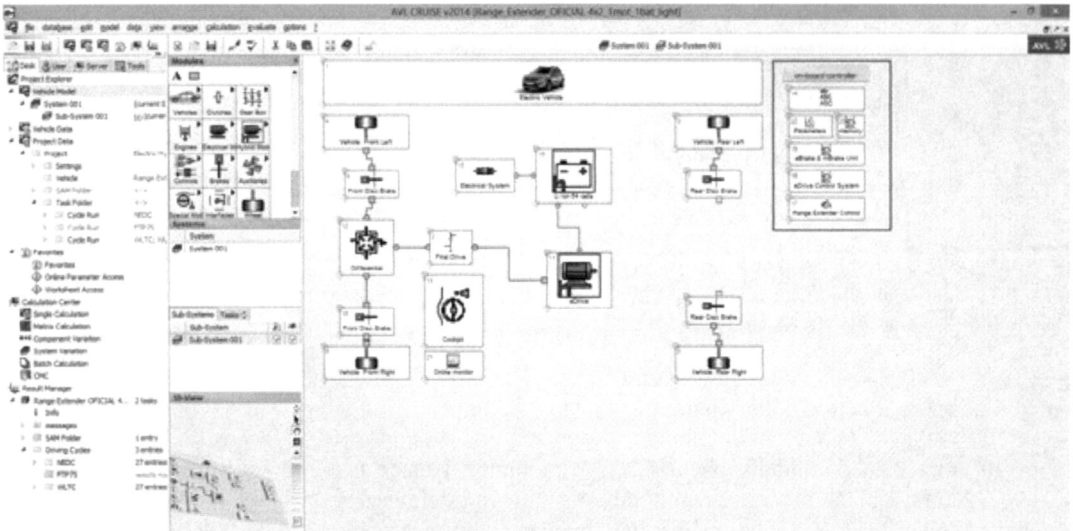

Figure 4. Sandero Electron model created in AVL Cruise (Source: AVL Cruise 2014)

The main characteristics of the WLTC test procedure are (data extracted from [6]):

- Total distance of 23.27 km
- Duration is 1800 s
- Mean velocity is 46.5 km
- Maximum velocity is 131.3 km/h
- 9 Number of stop phases

These characteristics are better shown in the following figure 5.

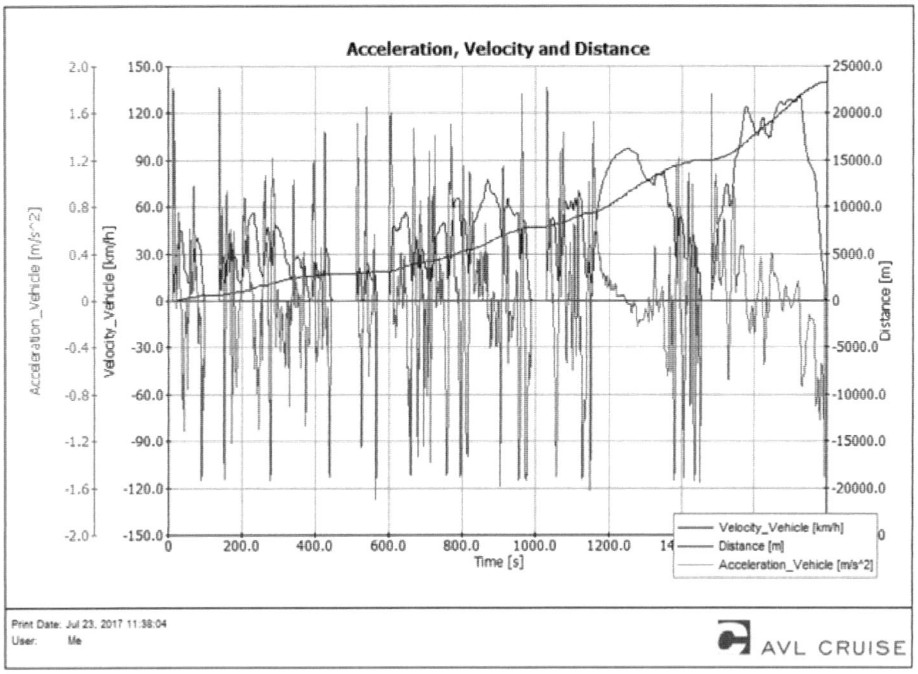

Figure 5. WLTC characteristics (Source: AVL Cruise 2014)

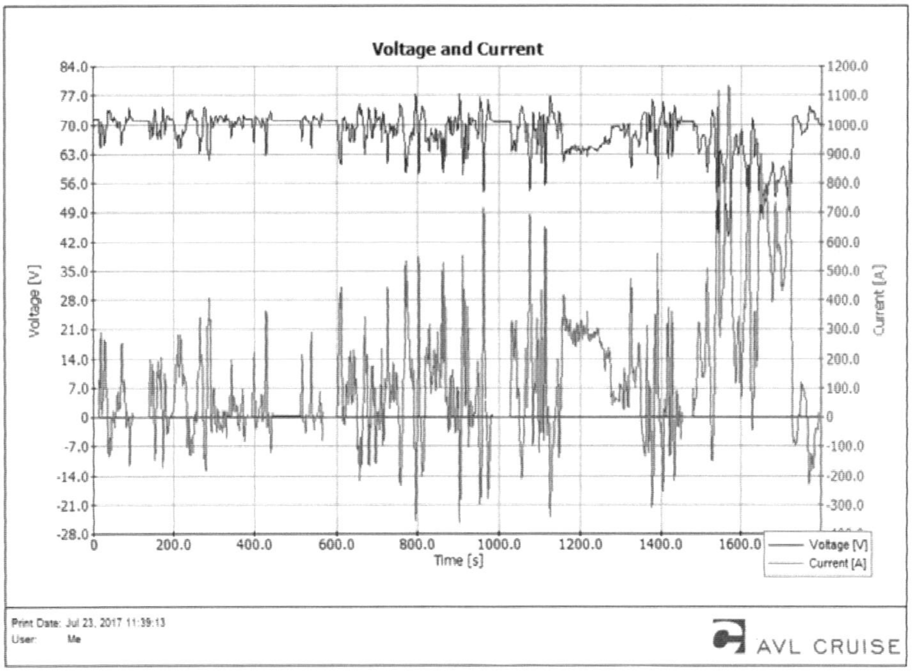

Figure 6. Voltage and current variations (Source: AVL Cruise 2014)

Figure 6 shows the variations of voltage and current. During acceleration phases, battery voltage drops because it is under load. As shown in the figure, during decelerating phases, the current has negative values. This happens because the vehicle has regenerative braking, the battery is being during these phases.

4. Results

After the WLTC simulation in AVL Cruise, the results are shown. In figure 7 the most important data is covered distance and the energy consumption achieved by the model.

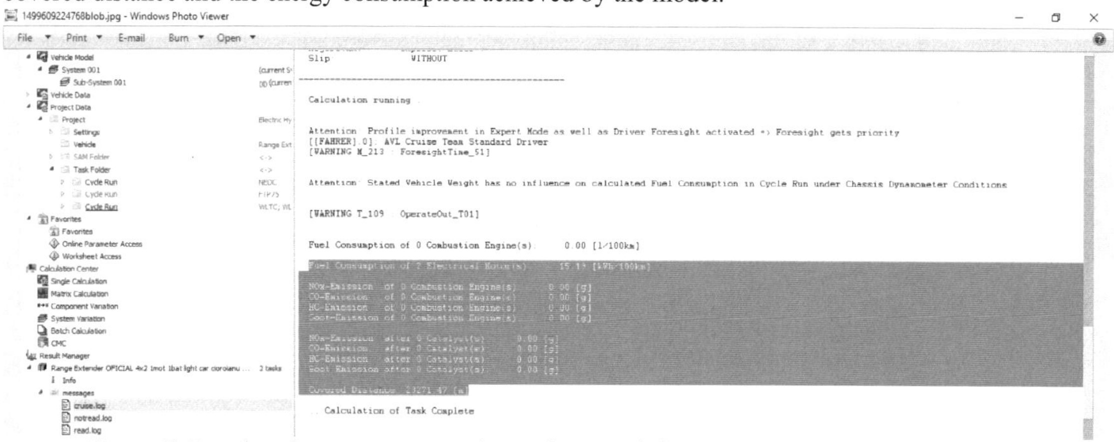

Figure 7. Results of energy consumption and covered distance (Source: AVL Cruise 2014)

The covered distance is 23271.47 m and the mean energy consumption is 15.19 kWh / 100 km. The energy consumption needed to perform a WLTC test cycle is 3.3538 kWh.

Knowing these consumption figures and the battery capacity, it is possible to determine the maximum range. Each cell has a maximum voltage of 4V and the battery pack features 64 cells. This means that the maximum voltage of the battery pack is the voltage of each cell times the number of cells. The maximum voltage of the battery pack is 256 V. Knowing the maximum charge of 60 Ah, it is possible to determine the power capacity of battery which is maximum voltage times the maximum charge. The power capacity of the battery pack is 15.36 kWh.

Considering a mean energy consumption of 15.19 kWh / 100 km and a maximum battery capacity of 15.36 kW, this leads to the maximum range of 101 km.

5. Conclusions

The use of computerized simulation with the help of AVL Cruise offers many advantages like reduced time and costs for vehicle development. It helps to refine the stage of the model at any step of the process. The obtained results show a relatively small range compared to other electric vehicles but the testing conditions were harsh: fully loaded vehicle and severe transitory regimes. Since the range is small, this vehicle model is clearly suitable for commuting or city driving. It is not suitable for driving on the highway, at high speeds. The model presented shows energy consumption of an electric vehicle and it is possible to determine the range afterwards. The obtained results will help on later improvements of the electric model, such as ways to increase the range by using a higher capacity battery or aerodynamic improvements.

IOP Conf. Series: Materials Science and Engineering **252** (2017) 012060 doi:10.1088/1757-899X/252/1/012060

References

1. I.J.M Besselink, P.F. van Oorschot, E. Meinders, H. Nijmeijer, *Design of an efficient, low weight battery electric vehicle based on a VW Lupo 3L,* Eindhoven University of Technology, 2010

2. D. Mayfield, *Sitting electric vehicle charging stations*, Editor Carlotta Collette, 2012

3. B. O. Varga, N. Burnete, C. Iclodean, Validation Procedure for Worldwide Harmonized Light Vehicles Test Cycle via Hardware in the Loop – Real Time Testing, CONAT Braşov, ISBN 978-3-319-45446-7, Springer Vol. 1, pp 128-135, 2016

4. M. Ehsani, Y. Gao, A. Emadi, Modern Electric, *Hybrid Electric and Fuel Cell Vehicles – Fundamentals, Theory and Design*, CRC Press, 2010, 411

5. M. Butsch, V. Rescigno, *Presentation of Tests with a Dynamometer Related to the Determination of Fuel Consumption in the NEDC and WLTC Driving Cycles and as Well Related to the Dynamic Behavior of Vehicles*, CONAT Braşov, ISBN 978-3-319-45446-7, Springer Vol. 1, pp 120-127, 2016

6. A.R. Sibiceanu, A. Iorga, V. Nicolae, F. Ivan, *Consideration on the Implications of the WLTC (Worldwide Harmonized Light-Duty Test Cycle) for a Middle Class Car*, CONAT Braşov ISSN 2069-0401, Transilvania University Press, pp 203-211, 2016

Dynamic behavior of the mechanical systems from the structure of a hybrid automobile

Popa Dinel, Tudor Irina, Stănescu Nicolae-Doru

University of Pitesti

dinel_popa@yahoo.com

Abstract. In introduction are presented solutions of planetary mechanisms that can be used in the construction of the hybrid automobiles where the thermal and electrical sources must be coupled. The systems have in their composition a planetary mechanism with two degrees of mobility at which are coupled a thermal engine, two revertible electrical machines, a gear transmission with four gears and a differential mechanism which transmits the motion at the driving wheels. For the study of the dynamical behavior, with numerical results, one designs such mechanisms, models the elements with solids in AutoCAD, and obtains the mechanical properties of the elements. Further on, we present and solve the equations of motion of a hybrid automotive for which one knows the dynamical parameters.

1. Introduction

In the last years all great constructors of automobiles promoted solutions of hybrid automotive as solution to reduce the noxae emitted by the thermal engines, especially in the urban environment.

More and more countries, by financial facilitations or traffic restrictions, encourages the production or buying of hybrid or electrical automotive.

No matter the type of the hybrid automotive: mini-hybrid, semi-hybrid or totally hybrid, it has in its composition a source of thermal power and a source of electrical power.

In [1] are described three types of coupling systems of these two sources of power: mechanical systems, hydraulic systems and electrical systems. In the same paper are also presented the differential equations of motions for a planetary mechanism used by Toyota Corp. in the series production on the hybrid automobile Prius, launched in fabrication in 1997.

The idea of using the planetary mechanism in the coupling of power sources was extended in [2] also for other configurations of mechanisms. Another four such mechanisms are analyzed from the kinematic point of view, mechanisms which can be used for the coupling of the power sources. Also in [2] is analyzed the stability of the five studied mechanical systems. Authors established the conditions that must be fulfilled by certain mechanical, kinematical and dynamical parameters in order to assure a stable motion. Modifying the configuration of the mechanism, one also modifies its stability of the motion.

In the present paper we purpose to study, by comparison, the dynamical behavior of a hybrid automobile which contains a planetary mechanism with four elements, in two configurations, mechanism which sums the power of two electrical machines and a thermal engine, and a mechanical transmission that transmits the motion to the wheel of the automotive.

2. Solution of analyzed planetary mechanisms

The planetary mechanisms are mechanisms with two degrees of freedom and they can couple a thermal engine and two electrical machines. The solution of mechanical systems used on Prius automobile is that in Fig. 1.

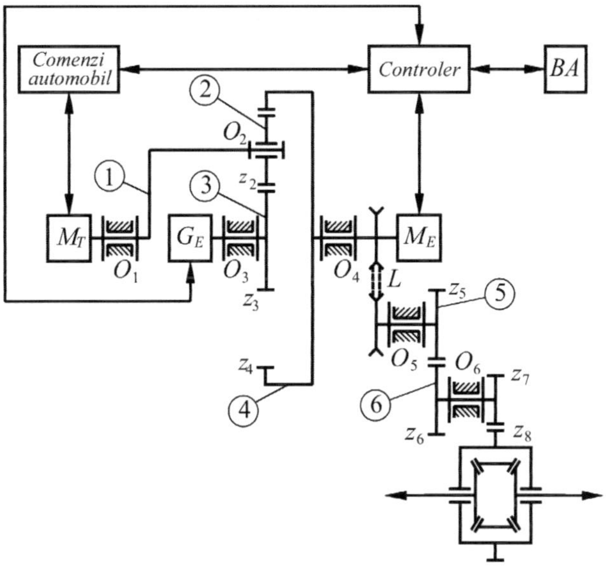

Figure 1. Mechanical system used at the Toyota Prius automobile.

In this configuration of planetary mechanism with four elements, the thermal engine is coupled at the port-satellite lever,1, the electrical machine GE is coupled at the solar wheel 3, and the electrical machine ME at the crown 4 of the mechanism. The motion transmits to the main transmission of the automobile through a chain transmission and through a gear with two cylindrical gears $z_5 - z_6$. The electrical machines ME and GE are used as motor or generator depending of the speed of the automobile and the electrical energy heaped in accumulator battery BAT. A controller controls the functioning modes, depending on the energy generated by the automobile.

In Fig. 2 we present another configuration of planetary mechanism with four elements.

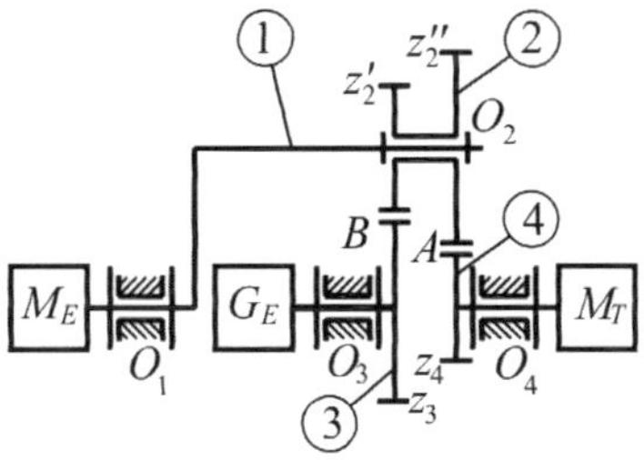

Figure 2. Planetary mechanism with four elements and double satellite.

Contrary to the solution analyzed in [3], where the thermal engine MT acts the port-satellite lever, in this configuration it acts upon the solar wheel 4. The electrical machine GE acts the solar wheel 3, and the electrical machine ME acts the port-satellite lever 1.

3. Kinematic analysis of the studied mechanisms

For the mechanism in Fig. 1 we denote the number of teeth of the gears 2, 3, 4 by z_2, z_3 and z_4, respectively. In the case of the mechanism in Fig. 2, we denote by z_2', z_2'' the number of teeth of the double satellite 2, and by z_3, z_4 the number of teeth of the gears 3 and 4.

In both cases, the absolute angular velocities of the elements 1, ..., 4 are denoted by ω_1, ..., ω_4.

Applying the Willis relations for both mechanisms, we successively obtain the relation in Table 1.

Table 1.

Kinematic relations for the mechanism in Fig. 1	Kinematic relations for the mechanism in Fig. 2
$\frac{\omega_3-\omega_1}{\omega_2-\omega_1} = -\frac{z_2}{z_3}$, $\frac{\omega_2-\omega_1}{\omega_4-\omega_1} = \frac{z_4}{z_2} = i_1$, $\frac{\omega_3-\omega_1}{\omega_4-\omega_1} = -\frac{z_4}{z_3} = i_2$	$\frac{\omega_3-\omega_1}{\omega_2-\omega_1} = -\frac{z_2'}{z_3}$, $\frac{\omega_2-\omega_1}{\omega_4-\omega_1} = -\frac{z_4}{z_2''} = i_1$, $\frac{\omega_3-\omega_1}{\omega_4-\omega_1} = \frac{z_4 z_2'}{z_3 z_2''} = i_2$
Notations	Notations
$i_1 = \frac{z_4}{z_2}$, $i_2 = -\frac{z_4}{z_3}$	$i_1 = -\frac{z_4}{z_2''}$, $i_2 = \frac{z_4 z_2'}{z_3 z_2''}$

$$\omega_2 = \omega_1(1-i_1) + \omega_4 i_1, \tag{1}$$

$$\omega_3 = \omega_1(1-i_2) + \omega_4 i_2, \tag{2}$$

Number of teeth of the gears $z_2 = 15$, $z_3 = 18$, $z_4 = 48$	Number of teeth of the gears $z_2' = 45$, $z_2'' = 30$, $z_3 = 20$, $z_4 = 35$
$i_1 = 3.2$; $i_2 = -2.6$;	$i_1 = -1.1$; $i_2 = 2.6$
$\omega_2 = -2.2\omega_1 + 3.2\omega_4$, $\omega_3 = 3.6\omega_1 - 2.6\omega_4$	$\omega_2 = 2.1\cdot_1 - 1.1\omega_4$, $\omega_3 = -1.6\omega_1 + 2.6\omega_4$

In the selection of the number of teeth of the gears in the two cases, we took into account that the ratio i_2, in modulus, has close values for both mechanisms in order to compare the results.

4. The equations of motion of the planetary mechanisms

Considering as known the configuration of the planetary mechanism, the number of teeth, the modulus and the thickness of the gears, one determines the masses m_1, m_2, m_3, m_4 of the elements and their moments of inertia J_1, J_2, J_3, J_4.

The expression of the mechanical power developed by the three motors reads

$$P = M_1\omega_1 + M_3\omega_3 + M_4\omega_4 \tag{3}$$

where M_1, M_3, M_4 are the torques developed by the three motors, function of the configuration of the planetary mechanism (Fig. 1 or Fig. 2).

Keeping into account the relation (2), we obtain the expression

$$P = [M_1 + M_3(1-i_2)]\omega_1 + [M_3 i_2 + M_4]\omega_4. \tag{4}$$

The generalized forces Q_1, Q_4 are obtained from the relation

$$\partial L = [M_1 + M_3(1-i_2)]\partial\theta_1 + [M_3 i_2 + M_4]\partial\theta_4, \tag{5}$$

from which it results

$$Q_1 = M_1 + M_3(1 - i_2), \tag{6}$$

$$Q_4 = M_3 i_2 + M_4. \tag{7}$$

To be out to obtain the Lagrange equations we start from the expression of the kinetic energy T,

$$T = \frac{1}{2}\left(J_1\omega_1^2 + m_2 R_2^2 \omega_1^2 + J_2\omega_2^2 + J_3\omega_3^2 + J_4\omega_4^2\right) \tag{8}$$

where R_2 is the radius at which there are the satellites 2.

The kinetic energy, keeping into account the relations (1) and (2), may be also written as

$$T = \frac{1}{2}\left(A_{11}\dot{\theta}_1^2 + A_{22}\dot{\theta}_4^2\right) + A_{12}\theta_1\theta_4 \tag{9}$$

where

$$A_{11} = J_1 + J_2(1 - i_1)^2 + J_3(1 - i_2)^2 + m_2 R_2^2, \tag{10}$$

$$A_{22} = J_2 i_1^2 + J_3 i_2^2 + J_4, \tag{11}$$

$$A_{12} = J_2(1 - i_1)i_1 + J_3(1 - i_2)i_2. \tag{12}$$

In the Lagrange equations

$$\frac{\mathrm{d}}{\mathrm{d}t}\left(\frac{\partial T}{\partial \dot{\theta}_1}\right) - \frac{\partial T}{\partial \theta_1} = Q_1, \quad \frac{\mathrm{d}}{\mathrm{d}t}\left(\frac{\partial T}{\partial \dot{\theta}_4}\right) - \frac{\partial T}{\partial \theta_4} = Q_4, \tag{13}$$

replacing the partial derivatives

$$\frac{\partial T}{\partial \dot{\theta}_1} = A_{11}\dot{\theta}_1 - A_{12}\dot{\theta}_4, \quad \frac{\partial T}{\partial \theta_1} = 0, \tag{14}$$

$$\frac{\partial T}{\partial \dot{\theta}_4} = A_{12}\dot{\theta}_1 + A_{22}\dot{\theta}_4, \quad \frac{\partial T}{\partial \theta_4} = 0, \tag{15}$$

we obtain the system of differential equations

$$A_{11}\ddot{\theta}_1 + A_{12}\ddot{\theta}_4 = M_1 + M_3(1 - i_2), \quad A_{12}\ddot{\theta}_1 + A_{22}\ddot{\theta}_4 = M_3 i_2 + M_4, \tag{16}$$

from which it results the solution

$$\ddot{\theta}_1 = \frac{[M_1 + M_3(1 - i_2)]A_{22} - [M_3 i_2 + M_4]A_{12}}{A_{11}A_{22} - A_{12}^2}, \quad \ddot{\theta}_4 = \frac{-[M_1 + M_3(1 - i_2)]A_{12} + [M_3 i_2 + M_4]A_{11}}{A_{11}A_{22} - A_{12}^2} \tag{17}$$

5. The equations of motion of the hybrid automobile

For the determination of the equations of motion of the hybrid automobile we proceed similarly to the determination of the equations of motion of the planetary mechanisms..

In this case the expression of the mechanical power is

$$P = M_1\omega_1 \pm M_3\omega_3 + M_4\omega_4 - P_R - P_a \tag{18}$$

where the power consumed by the rolling resistance P_R has the expression

$$P_R = f \cdot G_a \cdot v_a \tag{19}$$

while the power consumed by the air resistance P_a has the expression

$$P_a = \frac{1}{2}\rho c_x A v_a^2 \tag{20}$$

We used the classical notations: v_a – the speed of the automobile, G_a – the weight of the automobile, f – the coefficient of the rolling resistance, ρ – the density of the air, c_x – the coefficient of air resistance, A – the transversal area of the automobile.

The sign $\pm$ in front of the expression $M_3\omega_3$ was introduce because the electrical machine GE can be generator and then it consumes power (the sign –) or it can be motor and then it produces power (the sign +).

For the determination of the relation between the speed of the automobile and the angular velocity ω at the axle of the machine ME we start from the relation

$$\omega_r = \frac{\omega}{i_*} \tag{21}$$

where by ω_r we denoted the angular velocity of the wheel of automobile and by i_* the transmission ratio from the axle of the machine ME to the wheel of the automobile. Knowing the rolling radius r_r of the wheel of the automobile, expression (18) becomes

$$P = M_1\omega_1 \pm M_3\omega_3 + M_4\omega_4 - G_a f \frac{\omega}{i_*} r_r - \frac{1}{2}\rho c_x A \frac{\omega^3}{i_*^3} r_r^3 . \tag{22}$$

In the previous relation the angular velocity ω, for the mechanism in Fig. 1 is $\omega = \omega_4$, while for the mechanism in Fig. 2 it is $\omega = \omega_1$.

Keeping into account the relation (2), relation (22) becomes

$$P = [M_1 \pm M_3(1 - i_2)]\omega_1 + [\pm M_3 i_3 + M_4]\omega_4 - G_a f \frac{\omega}{i_*} r_r - \frac{1}{2}\rho c_x A \frac{\omega^3}{i_*^3} r_r^3 . \tag{23}$$

From the expression of the elementary work it results the expressions of the generalized forces in the cases of the two mechanisms. For the mechanism in Fig. 1 the generalized forces are

$$Q_1 = M_1 \pm M_3(1 - i_2), \tag{24}$$

$$Q_4 = \pm M_3 i_2 + M_4 - G_a f \frac{r_r}{i_*} - \frac{3}{2}\frac{\rho c_x A r_r^2}{i_*^2}\omega_4^2 , \tag{25}$$

while for the mechanism in Fig. 2 they are

$$Q_1 = M_1 \pm M_3(1 - i_2) - G_a f \frac{r_r}{i_*} - \frac{3}{2}\frac{\rho c_x A r_r^2}{i_*^2}\omega_1^2 , \tag{26}$$

$$Q_4 = \pm M_3 i_2 + M_4 . \tag{27}$$

The expression (8) of the kinetic energy completes with the kinetic energy of the automobile which, in a simplified model, is a mass m_a in translational motion, 2 driving wheels with the inertial moments J_{RM} and 2 free wheels with the moment of inertia J_{RL}. In the moments of inertia of the wheels we considered as included all the moments of inertia reduced to them (breaking system, the system of transmission of the rotational motion from the axle of ME to the wheels etc.). In this case the expression of the kinetic energy reads

$$T = \frac{1}{2}\left(J_1\omega_1^2 + m_2R_2^2\omega_1^2 + J_2\omega_2^2 + J_3\omega_3^2 + J_4\omega_4^2 + m_a\left(\frac{\omega}{i_*}\right)^2 r_r^2 + 2(J_{RM} + J_{RL})\left(\frac{\omega}{i_*}\right)^2 \right) \tag{28}$$

Using, for the mechanism in Fig. 1, the notations

$$A_{11} = J_1 + J_2(1 - i_1)^2 + J_3(1 - i_2)^2 + m_2R_2^2,\ A_{22} = J_2i_1^2 + J_3i_2^2 + J_4 + \frac{m_ar_r^2 + 2(J_{RM} + J_{RL})}{i_*^2},$$
$$A_{12} = J_2(1 - i_1)i_1 + J_3(1 - i_2)i_2 \tag{29}$$

and, for the mechanism in Fig. 2, the notations

$$A_{11} = J_1 + J_2(1 - i_1)^2 + J_3(1 - i_2)^2 + m_2R_2^2 + \frac{m_ar_r^2 + 2(J_{RM} + J_{RL})}{i_*^2},\ A_{22} = J_2i_1^2 + J_3i_2^2 + J_4,$$
$$A_{12} = J_2(1 - i_1)i_1 + J_3(1 - i_2)i_2, \tag{30}$$

we obtain the expression of the kinetic energy given by the relation (9).

From the Lagrange equations (13) it results, in a similar way, the system of differential equations (16) from which one obtains, for the mechanism in Fig. 1, the solution

$$\ddot{\theta}_1 = \frac{[M_1 \pm M_3(1 - i_2)]A_{22} - \left[\pm M_3i_2 + M_4 - \frac{G_afr_r}{i_*} - \frac{3\rho c_xAr_r^3}{2i_*^3}\dot{\theta}_4^2\right]A_{12}}{A_{11}A_{22} - A_{12}^2},$$
$$\ddot{\theta}_4 = \frac{-\left[M_1 \pm M_3(1 - i_2) - \frac{G_afr_r}{i_*} - \frac{3\rho c_xAr_r^3}{2i_*^3}\dot{\theta}_4^2\right]A_{12} + [\pm M_3i_2 + M_4]A_{11}}{A_{11}A_{22} - A_{12}^2}. \tag{31}$$

while for the mechanism in Fig. 2, the solution

$$\ddot{\theta}_1 = \frac{\left[M_1 \pm M_3(1 - i_2) - \frac{G_afr_r}{i_*} - \frac{3\rho c_xAr_r^3}{2i_*^3}\dot{\theta}_1^2\right]A_{22} - [\pm M_3i_2 + M_4]A_{12}}{A_{11}A_{22} - A_{12}^2},$$
$$\ddot{\theta}_4 = \frac{-[M_1 \pm M_3(1 - i_2)]A_{12} + \left[\pm M_3i_2 + M_4 - \frac{G_afr_r}{i_*} - \frac{3\rho c_xAr_r^3}{2i_*^3}\dot{\theta}_1^2\right]A_{11}}{A_{11}A_{22} - A_{12}^2}. \tag{32}$$

6. Integration of the equations of motion

For the integration of the equations of motion we need the mechanical properties of the elements. For this, we modeled with solids in AutoCAD the gears and the coupling elements. The gears were obtained with the AutoLisp functions described in [6]. All gears are manufactured from steel, have the same modulus $m = 2.5$ and the same thickness $b = 20\ \text{mm}$. Both mechanism have three satellites.

With the aid of the AutoCAD command MASSPROP one obtains the geometric properties from which one retains the mass of the element and the inertial moment relative to the axis OZ . In Table 2 are presented the obtain data.

Table 2.

Data for the mechanism in Fig. 1	Data for the mechanism in Fig. 2
$m_1 = 0.2564\,\text{kg}$, $J_1 = 1.372 \cdot 10^{-4}\,\text{kgm}^2$,	$m_1 = 0.4742\,\text{kg}$, $J_1 = 1.333 \cdot 10^{-3}\,\text{kgm}^2$,
$m_2 = 0.5169\,\text{kg}$, $J_2 = 9.7042 \cdot 10^{-4}\,\text{kgm}^2$,	$m_2 = 6.5688\,\text{kg}$, $J_2 = 5.043 \cdot 10^{-2}\,\text{kgm}^2$,
$m_3 = 0.2007\,\text{kg}$, $J_3 = 6.197 \cdot 10^{-5}\,\text{kgm}^2$,	$m_3 = 0.5468\,\text{kg}$, $J_3 = 1.396 \cdot 10^{-4}\,\text{kgm}^2$,
$m_4 = 1.3086\,\text{kg}$, $J_4 = 4.595 \cdot 10^{-3}\,\text{kgm}^2$.	$m_4 = 0.5468\,\text{kg}$, $J_4 = 8.777 \cdot 10^{-4}\,\text{kgm}^2$.

Further on, we need the characteristics of the motors as functions of the angular velocity.

The characteristic of the thermal engine is approximated with a function of second degree

$$M_{MT} = \begin{cases} 0 \text{ if } \omega < \omega_{10}, \\ M_{01} + a_1\omega + b_1\omega^2. \end{cases} \tag{33}$$

For the characteristics of the electrical motors we preferred ([2]) a function of first degree.

For the electrical machine ME the characteristic is

$$M_{ME} = \begin{cases} M_{02} \text{ if } \omega < \omega_{20}, \\ M_{02} + a_2\omega \text{ if } \omega > \omega_{20} \end{cases} \tag{34}$$

and for the electrical machine GE the characteristic is

$$M_{GE} = \begin{cases} M_{03} \text{ if } \omega < \omega_{30}, \\ M_{03} + a_3\omega \text{ if } \omega > \omega_{30}. \end{cases} \tag{35}$$

In the previous relations M_{0i}, ω_{i0}, a_i, with $i = 1, 2, 3$, and b_1 are constants which approximate the external characteristic given by the producer.

Considering an automobile with the mass $m_a = 2000\,\text{kg}$, with a coefficient of air resistance $c_x = 0.34$, a frontal area $A = 2.22\,\text{m}^2$, the coefficient of rolling resistance $f = 0.019$ and an efficiency of the transmission $\eta = 0.9$, it is necessary a power of approximate $90\,\text{kW}$ in order to reach a maximum speed of $180\,\text{km/h}$.

For the start-up and functioning in urban cycle one uses only the electric motor ME. Considering that at $2000\,\text{rot/min}$ ($\omega = 209.44\,\text{rad/s}$) for the angular speed of ME the automobile reaches a speed of $60\,\text{km/h}$ and the dynamic rolling radius has the value $r_r = 0.305\,\text{m}$, it results the value of the transmission ratio i_*

$$i_* = \frac{\omega}{\omega_r} = \frac{\omega}{\left(\frac{v_a}{r_r}\right)} = \frac{209.44}{\frac{16.666}{0.305}} = 3.83 \tag{36}$$

At maximum speed ME being directly coupled to the transmission will have the angular speed of $6000\,\text{rot/min}$.

Considering also that the thermal engine (a spark ignition one) has the angular speed of maximum power at $6000\,\text{rot/min}$, and for the reach of the maximum speed the energy is given by the thermal engine and by the two electrical machines feed from the accumulator battery, then from the relation

(2) it results that the angular speed of the electrical machine GE is equal to those of the two motors, $\omega_3 = \omega_1 = \omega_4$ no matter the configuration of the coupling mechanism (Figs. 1 or 2).

In the expression of the mechanical power (3), since the angular speeds are equal, it results

$$M = M_{MT} + M_{GE} + M_{ME} \tag{37}$$

If we choose the power of the thermal engine MT, $P_{MT} = 60\,\text{kW}$ at $6000\,\text{rot/min}$ the driving torque is $M_{MT} = 95.49\,\text{Nm}$. If in the expression (33) we consider three functioning points (95, 157.07), (110, 387.46) and (95.5, 628.31), then it results the expression of MT

$$M_{MT} = \begin{cases} 0 \text{ if } \omega < 100\,\text{rad/s}, \\ 68.588 + 0.210\omega - 0.00027\omega^2 \text{ if } \omega > 100\,\text{rad/s}. \end{cases} \tag{38}$$

Similarly we determine the external characteristic of the two electrical machines.

For ME we consider the functioning points (50, 209.44), (23.80, 628.32) and it results the characteristic

$$M_{ME} = \begin{cases} 63.10 \text{ if } \omega < 209.44\,\text{rad/s}, \\ 63.10 - 0.063\omega \text{ if } \omega > 209.44\,\text{rad/s}. \end{cases} \tag{39}$$

For GE we consider the functioning point (23.80, 628.32), (5, 1047.2) and it results

$$M_{GE} = \begin{cases} 52 \text{ if } \omega < 628.32\,\text{rad/s}, \\ 52 - 0.045\omega \text{ if } \omega > 628.32\,\text{rad/s}. \end{cases} \tag{40}$$

We will have to calculate also the constants given by the relations (29) for the mechanism in Fig. 1 and relations (30) for the mechanism in Fig. 2. In Table 3 are presented the obtained data.

Table 3.

Data for the mechanism in Fig. 1	Data for the mechanism in Fig. 2
$A_{11} = 0.006571$, $A_{22} = 12.889127$, $A_{12} = -0.007412$.	$A_{11} = 13.140103$, $A_{22} = 0.062842$, $A_{12} = -0.117074$.

The integration of the equations of motion is performed with the aid of the fourth order Runge Kutta method. One transforms the system of two second order differential equations in a system of four first order differential equations.

7. Conclusions

In the paper are analyzed, from the kinematic and dynamic point of view, the hybrid automobiles which have in their composition a planetary mechanical system with four components. The planetary mechanism with four elements that couples a thermal engine and two electrical machines is a mechanism with two degrees of mobility.

Comparing to the classic solution adopted by Toyota Corp. on model Prius we present a second solution of planetary mechanism (with four elements and double satellite) which can be used in coupling of three power sources.

From the performed kinematic analysis, by comparison to the two mechanisms, it resulted that both solutions of mechanisms are compatible.

For the study of the motion of the planetary mechanism one uses a system of two second order differential equations obtained by the aid of Lagrange equations.

Considering the mechanical power consumed by an automotive to beat the rolling and air resistances, proceeding in a similar way to that of obtaining the equations of motion for the planetary mechanism, it results a system of two second order differential equations which describe the motion of the hybrid automotive.

For the numerical integration of these differential equations one determined the mechanical properties of the elements, by modeling them with solids in AutoCAD, and the mechanical characteristics $M(\omega)$ of the three machines. The mechanical properties were determined in the case of the mechanism used by Toyota Corp. and in the case of the mechanism with double satellite.

For the determination of the powers of the thermal engine and electrical machines we considered an automobile with the mass of $2000\,\text{kg}$ that can move with a maximum speed of $180\,\text{km/h}$ .The solving of the equations of motion is performed by the fourth order Runge-Kutta method and by transformation of the system of two second order differential equations in a system of four first order differential equations.

In a future paper we will study the functioning modes of the hybrid automotive: the start-up in electric mode, light acceleration when the thermal engine has an angular speed greater than that of the electric motor, displacement with constant average speed, strong acceleration, displacement with constant high speed, recuperative breaking and reverse displacement.

Reference

[1] Popa D, Stănescu ND and Tudor I, 2013, The study systems used in coupling power source, *Scientific Bulletin Automotive series* **23** 60–7

[2] Popa D and Stănescu ND, 2015, Some observations concerning the stability of the mechanical systems used in the construction of the summing power planetary mechanisms, *RJAV* **XII**

[3] Pandrea N and Popa D, 2007, The coupling systems of the power sources, used at hybrid automobile, *The 2nd International Conference "Computational Mechanics and Virtual Engineering" COMEC 2007*, Brasov

[4] Molina S, 1997, Mechanism design. The Practical Kinematics and Dynamics of Machinery. *Ed. Pergamon*, Great Britain,

[5] Norton RL, 2003, Design of Machinery: An Introduction to the Synthesis and Analysis of Mechanisms and Machines, *McGraw-Hill.*

[6] Collins JA, Busby HR and Staab GH, 2003, Mechanical Design of Machine Elements and Machines: A Failure Prevention Perspective, 2nd Edition, *Wiley*, New York

[7] Dresig H, Holzwei, 2010, Dynamics of Machinery: Theory and Applications, *Springer* New York

[8] Mabie HH and Reinholtz CF, 1987, Mechanisms and Dynamics of Machinery, 4th Edition, *Wiley*, New York

[9] Russell K, Shen Q and Sodhi RS, 2015, Kinematics and Dynamics of Mechanical Systems: Implementation in MATLAB and Sim Mechanics, *CRC Press*, Boca Raton

IOP Conf. Series: Materials Science and Engineering **252** (2017) 012062 doi:10.1088/1757-899X/252/1/012062

A comparative study of planetary mechanisms usable for power sources coupling

Irina Tudor[1], Dinel Popa[1] and Nicolae–Doru Stănescu[1]

[1]University of Pitești, Târgul din Vale 1, Pitești 110040, Romania

dinel_popa@yahoo.com

Abstract. In the first part of the paper, a comparative study of the planetary mechanisms solutions which can be used in the coupling of power sources is made, this coming in addition to the previous works of the authors. It is analyzed the functioning mode of two planetary mechanisms used in different configurations and optimal variants are set.

1. Introduction

In the last decades, road traffic has increased considerably. Because motor vehicles are the main fossil fuel users and the source of greenhouse gases, it has been necessary to develop alternative methods of generating the energy required for propulsion. These methods and new technical solutions for the propulsion of motor vehicles are used to reduce the operating time of the thermal engine that could never reach a zero pollution level [5]. For this purpose, automotive engineers have conducted studies in the field of propulsion hybrid systems.

The hybrid vehicle is that vehicle that has at least two different energy sources that act as propulsion. The main elements of the propulsion system of the hybrid vehicle are the reversible energy storage system (for example an electric motor powered by a set of battery cells), the irreversible energy transformer (e.g. the internal combustion engine, whether it is a diesel engine or a spark-ignition engine) and a coupling system that connects the energy systems and conveys the movement to the wheels of the vehicle [1][6].

The coupling system can be classified into a hydraulic, electrical or mechanical coupling system.

This paper is dedicated to mechanical coupling systems used in the configuration of hybrid vehicle transmission, namely planetary mechanisms. These are mainly used by hybrid vehicle manufacturers due to their high efficiency, due to the fact that they are compact, they have a high-performance transmission ratio, a small size and a simple construction [2]. Planetary mechanisms are composed of central gear, crowns, planet carrier and one or more satellite pinions.

2. Mechanical systems for coupling power sources

The most commonly used planetary mechanism in power sources coupling is the power split mechanism with two degrees of freedom patented by Toyota (figure 1). The solution adopted by the Japanese manufacturer has changed to Hybrid Synergy Drive since 2004, when it started equipping vehicles outside Toyota brand, Lexus and Nissan vehicles [1].

The planetary mechanism components are the crown 2, the central gear 3, the satellite 4 and the planet carrier 1 [1][2][4]. For these, we have marked the numbers of teeth with z_2, z_3, z_4 and the absolute angular velocities with ω_1, ω_2, ω_3, ω_4.

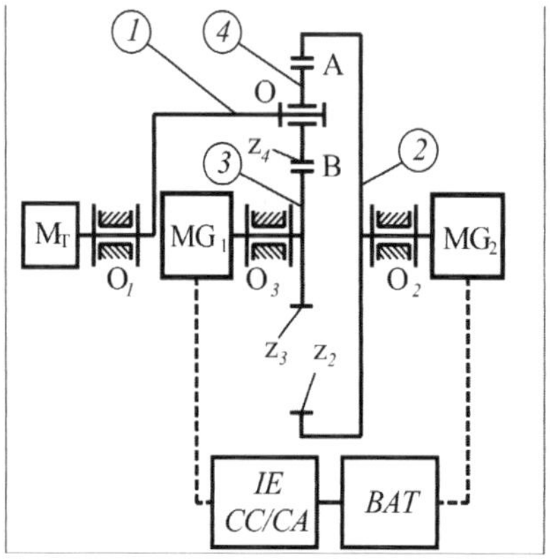

Figure 1. Power split, simple planetary mechanism– configuration Toyota Prius

Another solution that may be used for coupling power sources is by using planetary mechanisms with two degrees of freedom, double satellite and two eternal meshing [1][7] (figure 2a and b).

For these coupling systems we consider z_2, z_3, z_4', z_4'' the number of teeth of the gears and ω_1, ω_2, ω_3, ω_4 the absolute angular speeds of the elements 1, 2, 3, 4 [7] .

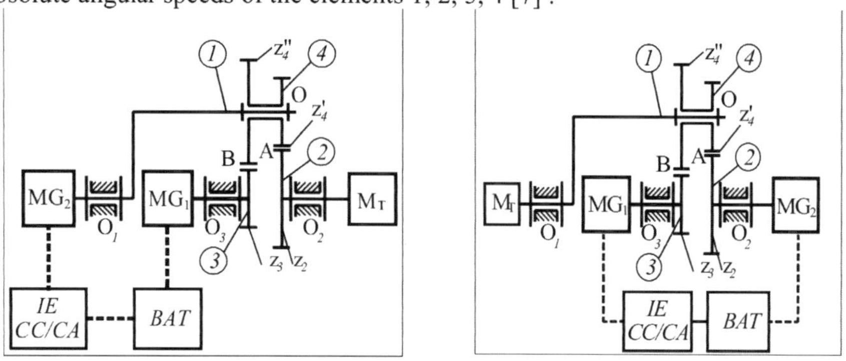

a) b)

Figure 2. Planetary mechanisms with double satellite and two eternal meshing.

We will apply the Willis method [3] in order to obtain kinematic relations [1] [8].

$$\frac{\omega_3-\omega_1}{\omega_4-\omega_1}=-\frac{z_4}{z_3}\,,\;\frac{\omega_4-\omega_1}{\omega_2-\omega_1}=\frac{z_2}{z_4}\,; \tag{1}$$

from which it is obtained :

$$\frac{\omega_3-\omega_1}{\omega_2-\omega_1}=-\frac{z_2}{z_3}\,. \tag{2}$$

If we denote the ratio:

$$i=-\frac{z_2}{z_3}; \tag{3}$$

by replacing into relations (2) it results:

$$\omega_3=\omega_1(1-i)+i\omega_2\,. \tag{4}$$

If we consider the ratio i_1 :

$$i_1=\frac{z_2}{z_4}\,; \tag{5}$$

from relation (1) and (5) we conclude:

$$\omega_4=\omega_1(1-i_1)+i_1\omega_2\,. \tag{6}$$

For the coupling systems in figure 2a and b, the ratios are [7]:

$$i_1=\frac{z_2}{z_4'}\,; \tag{7}$$

$$i = \frac{z_2 z_4^{''}}{z_3 z_4^{'}}.$$

By using the Willis method for the mechanisms in figure 2a and b one obtains:

$$\frac{\omega_4 - \omega_1}{\omega_2 - \omega_1} = -\frac{z_2}{z_4^{'}}, \frac{\omega_3 - \omega_1}{\omega_4 - \omega_1} = -\frac{z_4^{''}}{z_3}; \quad (8)$$

by replacing equation (7) into equation (8) we obtain the following equations [7]:

$$\omega_3 = \omega_1(1 - i) + i\omega_2;$$
$$\omega_4 = \omega_1(1 + i_1) + i_1\omega_2. \quad (9)$$

3. A comparative study of Toyota planetary mechanism solution versus the planetary mechanism with double satellite solution

The powertrain's operating mode (electric or hybrid propulsion) or the functioning regime of the two electric machines (engine or generator) can be explained by exemplifying the correlations of the rotations of the planetary mechanism elements (central gear, planet carrier and crown).

Next, we will compare mechanical systems solutions that use planetary mechanism with a double satellite and two eternal meshing gears with the solution used in the Prius configuration.

For the planetary mechanisms in the figure 1 and 2a, we consider:

MG1 – Alternator: ω_3, $\omega_3 = \omega_1(1 - i) + i\omega_2$;
MG2 – Electric motor: ω_2; (10)
MT - Thermal engine: ω_1.

For the configuration that has the planetary mechanism with a double satellite and the electric motor connected to the planet carrier, figure 2b, we note:

MG1 – Alternator: ω_3;
MG2 – Electric motor: ω_1; (11)
MT - Thermal engine: ω_2.

We will consider solution I, the configuration described in figure 2a, for $i = 2.5$, solution II the same configuration in figure 2a but for $i = 0.5$ and solution III represented in figure 1c and $i = 2.5$.

All these solutions will be compared to the Prius solution, represented in figure 1 for which it was considered $i = -2.5$. From relation (10) one obtains:

Planetary mechanism used in the Prius configuration ($i = -2.5$) – figure 1:	$\omega_3 = 3.5\omega_1 - 2.5\omega_2$;	(12)
Solution I - Planetary mechanism with double satellite ($i = 2.5$) - figure 2a:	$\omega_3 = -1.5\omega_1 + 2.5\omega_2$;	(13)
Solution II - Planetary mechanism with double satellite ($i = 0.5$) – figure 2a:	$\omega_3 = 0.5\omega_1 + 0.5\omega_2$;	(14)
Solution III - Planetary mechanism with double satellite ($i = 2.5$) - figure 2b:	$\omega_3 = -1.5\omega_1 + 2.5\omega_2$.	(15)

The rotations of the two motors (electric and thermal) and of the electric generator are determined with the relations:

Planetary mechanism used in the Prius configuration ($i = -2.5$) – figure 1:	$n_3 = 3.5n_1 - 2.5n_2$ - alternator; $n_1 = (n_3 + 2.5n_2)/3.5$ - thermal engine; $n_2 = (3.5 \cdot n_1 - n_3)/2.5$ - electric motor;	(16)
Solution I - Planetary mechanism with double satellite ($i = 2.5$) - figure 2a:	$n_3 = -1.5n_1 + 2.5n_2$ - alternator; $n_1 = (-n_3 + 2.5n_2)/1.5$ - thermal engine; $n_2 = (1.5 \cdot n_1 + n_3)/2.5$ - electric motor;	(17)
Solution II - Planetary mechanism with double satellite ($i = 0.5$) – figure 2a:	$n_3 = 0.5n_1 + 0.5n_2$ - alternator; $n_1 = (n_3 - 0.5n_2)/0.5$ - thermal engine; $n_2 = (-0.5 \cdot n_1 + n_3)/0.5$ - electric motor;	(18)
Solution III - Planetary mechanism with double satellite ($i = 2.5$) - figure 2b:	$n_3 = -1.5n_1 + 2.5n_2$ - alternator; $n_1 = (-n_3 + 2.5n_2)/1.5$ - electric motor;	(19)

$n_2 = (1.5 \cdot n_1 + n_3)/2.5$ - thermal engine.

Next, we describe the functioning modes for these kinematic solutions in different driving stages.

3.1. Starting out the vehicle (electric mode)

After starting, if the driver presses the accelerator pedal, the vehicle starts to move, being propelled by MG2 that will be operating in the electric motor mode. This will determine MG1 to rotate in the opposite direction of MG2, and because the thermal engine is shut down, this will immobilize the planet carrier of the planetary mechanism. By using the equations (10) and (12), (13), (14), (15), one obtains:

Planetary mechanism used in the Prius configuration ($i = -2.5$) – figure 1:	$\omega_1 = 0, \omega_3 = -2.5\omega_2;$	(20)
Solution I - Planetary mechanism with double satellite ($i = 2.5$) - figure 2a:	$\omega_1 = 0, \omega_3 = 2.5\omega_2;$	(21)
Solution II - Planetary mechanism with double satellite ($i = 0.5$) – figure 2a:	$\omega_1 = 0, \omega_3 = 0.5\omega_2;$	(22)
Solution III - Planetary mechanism with double satellite ($i = 2.5$) - figure 2b:	$\omega_2 = 0, \omega_3 = -1.5\omega_1.$	(23)

3.2. Engine starting

In electric power mode, if the vehicle's speed exceeds a certain limit (about 50 km / h), the thermal engine is started with the help of MG1. The inverter will power MG1 that will function in motor mode. The direction of rotation of MG1 and MG2 will be the same which will drive the thermal engine.

The equations (16) - (19) lead to:

Planetary mechanism used in the Prius configuration ($i = -2.5$) – figure 1:	$n_2 = 2000[rot/min]$, $n_1 = (1000 + 2.5 \cdot 2000)/3.5 = 1714[rot/min];$	(24)
Solution I - Planetary mechanism with double satellite ($i = 2.5$) - figure 2a:	$n_2 = 2000[rot/min]$, $n_1 = (-1000 + 2.5 \cdot 2000)/1.5 = 2666[rot/min];$	(25)
Solution II - Planetary mechanism with double satellite ($i = 0.5$) – figure 2a:	$n_2 = 2000[rot/min], n_3 = 500[rot/min]$, $n_1 = (500 - 0.5 \cdot 2000)/0.5 = -1000[rot/min];$ $n_3 = 1000[rot/min]$, $n_1 = (1000 - 0.5 \cdot 2000)/0.5 = 0[rot/min];$ $n_3 = 3000[rot/min]$, $n_1 = (3000 - 0.5 \cdot 2000)/0.5 = 1000[rot/min]$ – the engine will initially spin in one direction and then change its direction of rotation. This shortcoming can be remedied with a directional coupling.	(26)
Solution III - Planetary mechanism with double satellite ($i = 2.5$) - figure 2b:	$n_1 = 2000[rot/min]$, $n_2 = (1.5 \cdot 2000 + 1000)/2.5 = 1600[rot/min].$	(27)

3.3. Light acceleration

After starting the thermal engine if the driver desires a light acceleration of the vehicle, the thermal engine will increase its speed above the value of MG2. In this case MG1 will function in the electric generator mode. The electric energy produced by MG1 is used for MG2.

Planetary mechanism used in the Prius configuration ($i = -2.5$) – figure 1:	$n_2 = 2000[rot/min], n_1 = 2500[rot/min]$, $n_3 = 3.5 \cdot 2500 - 2.5 \cdot 2000 = 3780[rot/min];$	(28)

$$n_2 = 2000[rot/min], n_1 = 2500[rot/min],$$

Solution I - Planetary mechanism with double satellite (i = 2.5) - figure 2a: (29)

$$n_3 = -1.5 \cdot 2500 + 2.5 \cdot 2000 = 1250[rot/min]$$
– the alternator does not charge;

Solution II - Planetary mechanism with double satellite (i = 0.5) – figure 2a:
$$n_2 = 2000[rot/min], n_1 = 2500[rot/min], \quad n_3 = 0.5 \cdot 2000 + 0.5 \cdot 2500 = 2250[rot/min]; \tag{30}$$

Solution III - Planetary mechanism with double satellite (i = 2.5) - figure 2b:
$$n_1 = 2000[rot/min],\ n_2 = 2500[rot/min], \quad n_3 = -1.5 \cdot 2000 + 2.5 \cdot 2500 = 3250[rot/min] \tag{31}$$
- the alternator is charging.

According to the result obtained with relation (29) it is observed that the alternator does not charge in the case of a light acceleration when it is used the planetary mechanism with double satellite presented in the configuration in figure 2a, for i = 2.5. The solution III is solving this drawback in this driving stage.

3.4. *Light load cruise – light constant speed*

In this driving stage, the vehicle is propelled in hybrid mode, by the internal combustion engine and the electric motor MG2. The MG1 generator, produces electric energy that it is supplied to MG2.

Planetary mechanism used in the Prius configuration (i = −2.5) – figure 1:
$$n_2 = 3000[rot/min], n_1 = 3000[rot/min], \quad n_3 = 3.5 \cdot 3000 - 2.5 \cdot 3000 = 3000[rot/min]; \tag{32}$$

Solution I - Planetary mechanism with double satellite (i = 2.5) - figure 2a:
$$n_2 = 3000[rot/min], n_1 = 3000[rot/min], \quad n_3 = -1.5 \cdot 3000 + 2.5 \cdot 3000 = 3000[rot/min]; \tag{33}$$

Solution II - Planetary mechanism with double satellite (i = 0.5) – figure 2a:
$$n_2 = 3000[rot/min], n_1 = 3000[rot/min], \quad n_3 = 0.5 \cdot 3000 + 0.5 \cdot 3000 = 3000[rot/min]; \tag{34}$$

Solution III - Planetary mechanism with double satellite (i = 2.5) - figure 2b:
$$n_1 = 3000[rot/min], n_2 = 3000[rot/min], \quad n_3 = -1.5 \cdot 3000 + 2.5 \cdot 3000 = 3000[rot/min]; \tag{35}$$

3.5. *Full acceleration*

If a full acceleration of the vehicle is desired, the thermal engine speed increases. The vehicle is powered in hybrid mode (electric and thermal). MG1 will produce electric power to power MG2. In addition, in order to get a maximum torque from the MG2, the battery will additionally supply the required power [1].

Planetary mechanism used in the Prius configuration (i = −2.5) – figure 1:
$$n_2 = 3000[rot/min], n_1 = 5000[rot/min], \quad n_3 = 3.5 \cdot 5000 - 2.5 \cdot 3000 = 10000[rot/min]; \tag{36}$$

Solution I - Planetary mechanism with double satellite (i = 2.5) - figure 2a:
$$n_2 = 3000[rot/min], n_1 = 5000[rot/min], \quad n_3 = -1.5 \cdot 5000 + 2.5 \cdot 3000 = 0[rot/min \tag{37}$$
– the alternator does not charge;

Solution II - Planetary mechanism with double satellite (i = 0.5) – figure 2a:
$$n_2 = 3000[rot/min], n_1 = 5000[rot/min], \quad n_3 = 0.5 \cdot 5000 + 0.5 \cdot 3000 = 4000[rot/min]; \tag{38}$$

Solution III - Planetary mechanism with double satellite (i = 2.5) - figure 2b:
$$n_1 = 3000[rot/min], n_2 = 5000[rot/min], \quad n_3 = -1.5 \cdot 3000 + 2.5 \cdot 5000 = 8000[rot/min] \tag{39}$$
– the alternator charges.

It can be observed that even in the case of a full acceleration, according to the relation (37), the alternator does not charge in the configuration shown in solution II. Solution III is however a favourable option this time as well.

3.6. Full load cruise – constant speed

For a given speed of the thermal engine, in order to increase the speed of MG2, MG1 will be locked (zero speed). In this driving stage, the vehicle is propelled in hybrid mode, MG2 being powered by the battery pack.

Planetary mechanism used in the Prius configuration ($i = -2.5$) – figure 1:	$n_2 = 6000[rot/min], n_3 = 0[rot/min],$ $n_1 = 2.5 \cdot 6000/3.5 = 4286[rot/min];$	(40)
Solution I - Planetary mechanism with double satellite ($i = 2.5$) - figure 2a:	$n_2 = 6000[rot/min], n_3 = 0[rot/min],$ $n_1 = 2.5 \cdot 6000/1.5 = 10000[rot/min]$ – the thermal engine runs at high speeds;	(41)
Solution II - Planetary mechanism with double satellite ($i = 0.5$) – figure 2a:	$n_2 = 6000[rot/min], n_3 = 0[rot/min],$ $n_1 = -n_2 = 6000[rot/min]$ – the thermal motor turns backwards; We can choose the solution: $n_3 = 6000,$ $n_1 = (n_3 - 0.5n_2)/0.5,$ $n_1 = (6000 - 0{,}5 \cdot 6000)/0.5 =$ $6000[rot/min]$ – a directional coupling is needed, since it starts at negative speeds;	(42)
Solution III - Planetary mechanism with double satellite ($i = 2.5$) - figure 2b:	$n_1 = 6000[rot/min], n_3 = 0[rot/min],$ $n_2 = 1.5 \cdot 6000/2.5 = 3600[rot/min]$ – the thermal engine rpm's are normal.	(43)

For solution I this is an unfavourable situation. The thermal engine's rotations per minute are very high when driving with a constant high speed. In the case of solution II, we can see that the result suggests that the thermal engine rotates in reverse direction, which suggests that this solution is not viable one.

3.7. Maximum speed

For the same speed of the thermal engine, MG2 will have a higher speed if MG1 is in engine mode and the rotational direction is reversed. The vehicle is propelled hybrid mode (electrical and thermal), the battery pack powering both MG2 and MG1 electric motors.

Planetary mechanism used in the Prius configuration ($i = -2.5$) – figure 1:	$n_1 = 6000[rot/min], n_3 = -3000[rot/min],$ $n_2 = (3.5 \cdot 6000 + 3000)/3.5 = 6857[rot/min];$	(44)
Solution I - Planetary mechanism with double satellite ($i = 2.5$) - figure 2a:	$n_1 = 6000[rot/min], n_3 = 3000[rot/min],$ $n_2 = (1.5 \cdot 6000 + 3000)/1.5 = 8000[rot/min]$ –it is required an alternator with a higher power;	(45)
Solution II - Planetary mechanism with double satellite ($i = 0.5$) – figure 2a:	$n_1 = 6000[rot/min], n_3 = 9000[rot/min],$ $n_2 = (-0.5 \cdot 6000 + 9000)/0.5 = 6000[rot/min]$ - alternator with high speed;	(46)
Solution III - Planetary mechanism with double satellite ($i = 2.5$) - figure 2b:	$n_2 = 6000[rot/min], n_3 = 3000[rot/min],$ $n_1 = (3000 + 2.5 \cdot 6000)/1.5 = 8000[rot/min]$ – it is not necessary to reverse the alternator's direction.	(47)

In order to be able to meet the power requirements, it is necessary that the hybrid vehicle equipped with a planetary mechanism with double satellite from the solution I, to be equipped with a high power alternator. In the case of solution II it is necessary to use a high speed alternator.

3.8. Deceleration - regenerative braking

When the driver presses the brake pedal to reduce the speed of the vehicle, the thermal engine is switched off. In this mode, the MG2 will operate in electric generator mode, being driven by the vehicle's wheels (motor brake) [1]. The energy produced by MG2 is used to charge the battery pack. MG1 will not be powered, and will rotate in reverse direction to MG2 due to the zero speed of the thermal engine.

Planetary mechanism used in the Prius configuration ($i = -2.5$) – figure 1:	$n_1 = 0\ [rot/min], n_3 = -2.5n_2;$	(48)
Solution I - Planetary mechanism with double satellite ($i = 2.5$) - figure 2a:	$n_1 = 0\ [rot/min], n_3 = 2.5n_2;$	(49)
Solution II - Planetary mechanism with double satellite ($i = 0.5$) – figure 2a:	$n_1 = 0\ [rot/min],\ n_3 = 0.5n_2$ - The energy recovered after braking the vehicle is very low;	(50)
Solution III - Planetary mechanism with double satellite ($i = 2.5$) - figure 2b:	$n_2 = 0\ [rot/min], n_3 = -1.5n_1\,.$	(51)

3.9. Reverse
In this driving stage the vehicle is powered exclusively electric, MG2 being in electric motor mode but its rotational direction being opposite from the "Drive" mode. The thermal engine is turned off and MG1 will rotate freely. The electric energy for MG2 is supplied by the battery pack.

4. Conclusions
This paper reviews the general principles and features of hybrid propulsion systems, focusing on the mechanical systems used for coupling power sources.

The kinematic analysis made for the Prius planetary mechanism as well as for the planetary mechanism with a double satellite has helped determine the functioning modes of the hybrid solutions. Following the analysis of the functioning modes of the studied solutions, it can be concluded that only two of them can equip hybrid vehicle. One of these is, of course, the Prius mechanism, but also the planetary mechanism with a double satellite for which the electric motor is connected to planet carrier.

This solution, referred to in the paper as solution III, solves the drawbacks encountered in the use of solution I. For solution I, the planet carrier is connected to the thermal engine this time and is not a favourable solution to be used in the hybrid vehicle configurations.

Analyzing the functioning mode for the solutions with planetary mechanism, were determined the optimal constructions which can be used to become powertrain configuration for hybrid vehicles.

It is necessary to know the functioning characteristics in order to determine the dynamic performance of hybrid vehicle in road traffic but also to construct the driving cycles.

By using the explanations presented in point 3, of this current paper, we plan to determine the driving cycles for the two chosen architectures in a forthcoming paper.

By determining these driving cycles we will be able to compare the performances produced in each of these driving stages that will occur in real life driving traffic conditions.

References
[1] Popa D, Stănescu N, and Tudor I, 2013, The Study Of The Systems Used For The Power Sources' Coupling, *Scie. B.:- Automot. Ser.* **23 A,** pp 60- 67,

[2] Tudor I, Florescu B and Tecau F, Analiza cinematică şi dinamică a transmisiilor planetare cu două grade de mobilitate utilizate în cuplarea surselor de putere 2015, *Scie. B.:- Automot. Ser.*,

[3] Pandrea N and Popa D, 2000, Mecanisme. Teorie şi aplicaţii CAD, ed. Tehnică, Bucureşti,

[4] Bostan I, Dulgheru V and Grigoraş S, Transmisii planetare precesionale şi armonice, 1997, ed Atlas Tehnică Bucureşti, ed. Tehnică UTM,

[5] Oprean I, Automobilul modern. Cerinţe, restricţii, soluţii, 2003, ed. Academiei Române, Bucureşti,

[6] Marinescu D, Sisteme neconvenţionale de propulsie şi transport, 2003 ed. Universităţii din Piteşti,
[7] Pandrea N and Popa D, 2007, The coupling systems of the power sources, used at hybrid automobile, *The 2nd Int. Conf. "Comput Mech. and V. Eng." COMEC*, p 537,
[8] Popa D and Stănescu N, Experimental Model Used in the Study of Power Sources Coupling, *Ad. in Prod., Auto. and Transp. Sys.* 2015, p 288.

IOP Conf. Series: Materials Science and Engineering **252** (2017) 012063 doi:10.1088/1757-899X/252/1/012063

Determination of the needed power of an electric motor on the basis of acceleration time of the electric car

M Sapundzhiev[1], **I Evtimov**[2] **and R Ivanov**[2]

[1]Branch Silistra, University of Ruse, 8, Studentska str. 7017 Ruse, Bulgaria
[2]Department Engines and Vehicles, University of Ruse, 8, Studentska str. 7017 Ruse, Bulgaria
milenvs@abv.bg

Abstract. The paper presents an upgraded methodology for determination of the electric motor power considering the time for acceleration. The influence of the speed factor of electric motor on the value of needed power at same acceleration time is studied. Some calculations on the basis of real vehicle were made. The numeric and graphical results are given. They show a decrease of needed power with the increase of the speed factor of motor, because the high speed factor allows the use of a larger range of the characteristic with the maximum power of the motor. An experimental verification of methodology was done.

1. Introduction

When designing an electric car, the needed power of the electric motor is selected primarily based on the specified maximum speed [1, 7 and 9] but this criterion does not guarantee its acceleration to a certain speed at a set time, as well as a good inserting in the traffic.

This approach typically does not take into account the energy losses for the operation of the ancillary systems of the electric vehicle and the resulting power is considerably less than what is actually needed [2, 3, 4 and 5].

The classic methodology for selecting an electric motor does not take into account the intensity of acceleration, which is also important in the calculation of the electric vehicle traction [6, 8 and 9].

When a more powerful electric motor is used, the maximum speed is determined by the maximum rotation speed of the electric motor [1, 6 and 9].

The problem concerning determination of needed power is also actual in the field of hybrid and converted vehicles [3, 11 and 12]. Usually they resolve that problem using complex modeling and simulation tools [4, 7 and 10].

The main goal of the presented paper is to develop a relatively simple methodology for determination of the needed power of electric motor, based on time for acceleration. Other goals are investigation the influence of different factors on value of needed power and experimental verification the methodology.

2. Method of determination the needed power of an electric motor according to a given acceleration time

The existing [6], rather as an idea, approach to determining the power needed to accelerate the electric vehicle to a specified speed at a given time t_a is based on the following relation

$$t_a = \int_0^{V_b} \frac{\frac{G}{g}\delta_a}{\frac{P}{V_b} - G\,f_f - \frac{1}{2}k_в\,S\,V_b^2} dV + \int_{V_b}^{V_f} \frac{\frac{G}{g}\delta_a}{\frac{P}{V_f} - G\,f_f - \frac{1}{2}k_в\,S\,V_f^2} dV, \quad (1)$$

where V_f is the final speed, *km/h*;
V_b – the base speed, *km/h*;
f_f – the rolling resistance coefficient;
δ_a – the coefficient of influence of the rotating masses of the car;
g – the Earth acceleration.
G– the weight of the electric car, *kN*;
$k_в = c_x\,\rho$ – the coefficient of air resistance of the electric car;
c_x – the drag coefficient of the electric car;
ρ – the air density;
S– the front area of the electric car, m^2;
P – the power of the electric motor, *kW*;

The right side of the equation contains two parts. The idea may be illustrated in figure 1 - the change in torque and power of the electric motor vs the speed of the electric car. The first part of the equation (1) refers to the speed section from *0* to V_b, and the second from V_b to V_f.

The figure depicts the operation of the electric motor when driving an electric car with a transmission with constant gear ratio - starting and accelerating to speed V_f. If a gearbox with manual gearshift is used to drive the wheels, the interruption of the power flow can be compensated by the higher torque and the higher acceleration at the low gears.

In fact, it is difficult to find an analytical solution to equation (1), so [6] proposes a number of assumptions and simplifications.

The resistive forces acting on the electric car in acceleration mode on a straight, horizontal road are rolling resistance, air resistance and acceleration resistance.

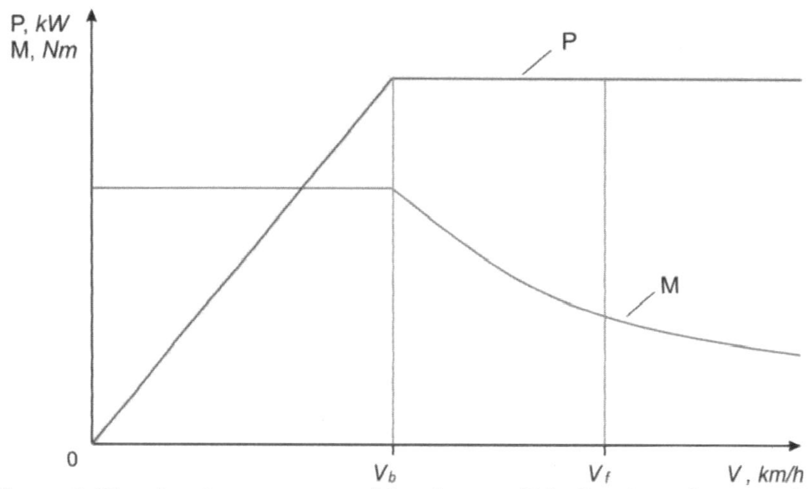

Figure 1. The electric motor torque M and power P in function of car speed *V*

Assuming that the power for overcome the rolling P_f and air resistances $P_в$ are not dependent on the curve of the motor torque but only on the speed of motion, then P_f and $P_в$ can be presented separately and after integration [6] are expressed by the equations:

– the power to overcome rolling resistance

$$P_f = \frac{2}{3}\, G\, f_f\, V_f\, ; \tag{2}$$

– the power to overcome air resistance

$$P_в = \frac{1}{5}\, k_в\, S\, V_f^2\, . \tag{3}$$

The coefficients 2/3 and 1/5 are a result of the integration and used simplifications [6]. After P_f and $P_в$ are presented separately, in equation (1) remains the acceleration time of the electric vehicle, expressed by relation of the speed and the mass to the power.

After integration, the acceleration time t_a of the electric car can be represented as

$$t_a = \frac{\delta_a\, G}{2\, g\, P_a}(V_f^2 + V_b^2) \tag{4}$$

After remaking of (4), the average power needed during the acceleration time of the electric vehicle is equal to

$$P_a = \frac{\delta_a\, G}{2\, g\, t_a}(V_f^2 + V_b^2) \tag{5}$$

Equation (5) expresses approximately the needed power P_a according to the intensity of the electric car acceleration, due to the different character in the two speed sections of the electric motor characteristic.

After replacing with (2), (3) and (5) in equation (1) it becomes as follows

$$P = \frac{\delta_a\, G}{2\, t_a\, g}\left(V_f^2 + V_b^2\right) + \frac{2}{3} G\, f_f V_f + \frac{1}{5} k_в\, S\, V_f^2 . \tag{6}$$

Relation (6) allows, though not very accurately, to determine the power needed of the electric motor based on a given acceleration time.

Figure 2 shows a theoretical comparison of the characteristics of electric motors of equal power which V_b is reached at different speeds. It can be seen that with increasing speed, the share distribution of resistance power is changed - the power needed to overcome rolling and air resistance increases and the acceleration power is reduced. It is also clear that the electric motor with lower base speed V_{b1} has a higher power reserve for acceleration from 0 to the moment when the speed is equal to V_{b2}.

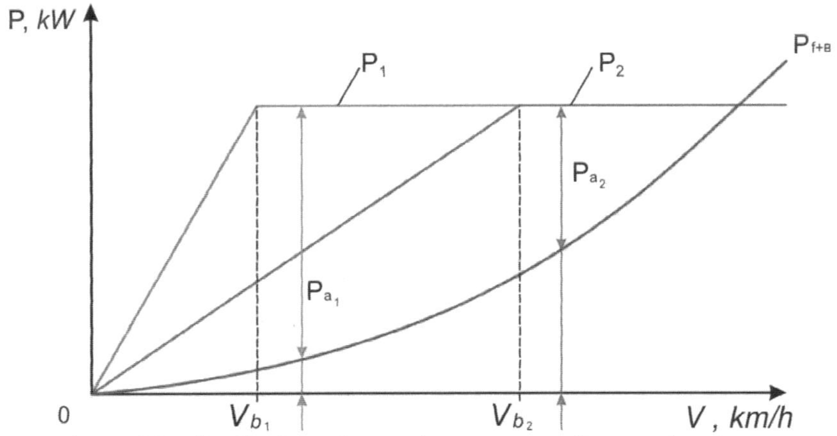

Figure 2. Comparison of the distribution of car resistance power for two electric motors with different base speeds: $V_{b1}$and V_{b2} – the base speeds of two motors; P_1 and P_2 – maximal power of motors; P_{f+B}– full resistance power of the car; P_{a1} and P_{a2} power for acceleration

Using the described approach to determine the needed power of an electric motor based on the acceleration time, the calculations were done for an electric car with the following parameters: G = 15,89 *kN*; f_0 = 0,012; $k_в$ = 0,0002 kNs^2/m^4; S = 2,35 m^2. The power needed for acceleration from 0 to V_f = *100 km/h*at different values of t_a is calculated. Several variants of different times (t_a = 5 – 30 *s*) and different base angular speed of electric motor (ω_b= 200 … 335*rad/s)* are considered. The results of the calculations are presented in table 1 and figure 3.

From the calculated results, it can be seen that as the acceleration time of the electric car increases, the needed power of the electric motor decreases. The same results are represented on figure 3 in function of speed factor V_f/V_b .

Table 1. Needed power of the electric motor at different values of acceleration time t_a of electric car to100 *km/h*and different values of base speed of motor ω_b

$P\omega_b$	$t_a = 5s$	$t_a = 10s$	$t_a = 15s$	$t_a = 20s$	$t_a = 25s$	$t_a = 30s$
P_{200}	185,3 *kW*	94,4 *kW*	64,1 *kW*	48,9 *kW*	39,9 *kW*	33,8 *kW*
P_{220}	195,3 *kW*	99,4 *kW*	67,5 *kW*	51,5 *kW*	41,9 *kW*	35,5 *kW*
P_{240}	206,3 *kW*	104,9 *kW*	71,1 *kW*	54,2 *kW*	44,1 *kW*	37,3 *kW*
P_{260}	218,3 *kW*	110,9 *kW*	75,1 *kW*	57,2 *kW*	46,5 *kW*	39,3 *kW*
P_{280}	231,2 *kW*	117,3 *kW*	79,4 *kW*	60,4 *kW*	49,0 *kW*	41,5 *kW*
P_{300}	245,0 *kW*	124,3 *kW*	84,0 *kW*	63,9 *kW*	51,8 *kW*	43,8 *kW*
P_{320}	259,8 *kW*	131,7 *kW*	89,0 *kW*	67,6 *kW*	54,8 *kW*	46,2 *kW*
P_{335}	271,6 *kW*	137,5 *kW*	92,9 *kW*	70,5 *kW*	57,1 *kW*	48,2 *kW*

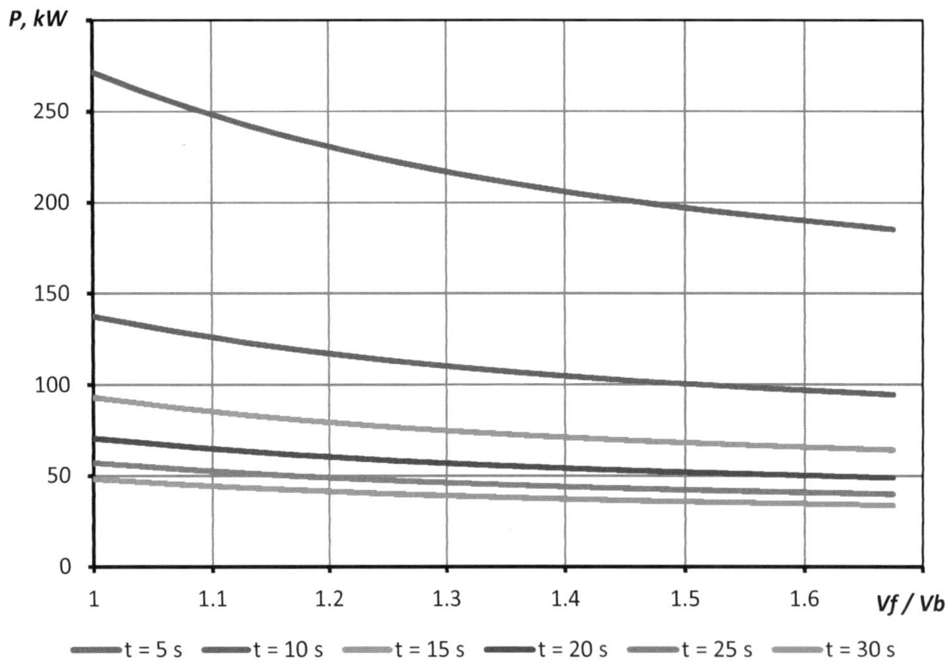

Figure 3. Needed power *P* of the electric motor vs the speed factor V_f / V_b

IOP Conf. Series: Materials Science and Engineering **252** (2017) 012063 doi:10.1088/1757-899X/252/1/012063

Decreasing of the needed power is also observed when V_b is reached at a lower angular speed. Therefore, the relationship between the two V_f / V_b speeds, or the so called speed factor, has a significant impact on the power needed by the electric motor.

At the same maximum speed, electric motors with a higher speed factor have better acceleration performance. The high V_f / V_b allows the use of a larger range of the characteristic with the maximum power of the motor (see figure 2).

Using this upgraded approach to determine the needed power for a given acceleration time, provides the possibility for a more reasoned determination of the necessary parameters of the motor - power, maximum and basic angular speed.

3. Experimental verification

Several experiments were done for verification the theoretical calculations made following the above described methodology. A converted car Citroen Berlingo to electric was used. It has the same parameters as used in calculations. Figure 4 shows the structure of the converted electric vehicle and figure 5 - the electric car in the test area.

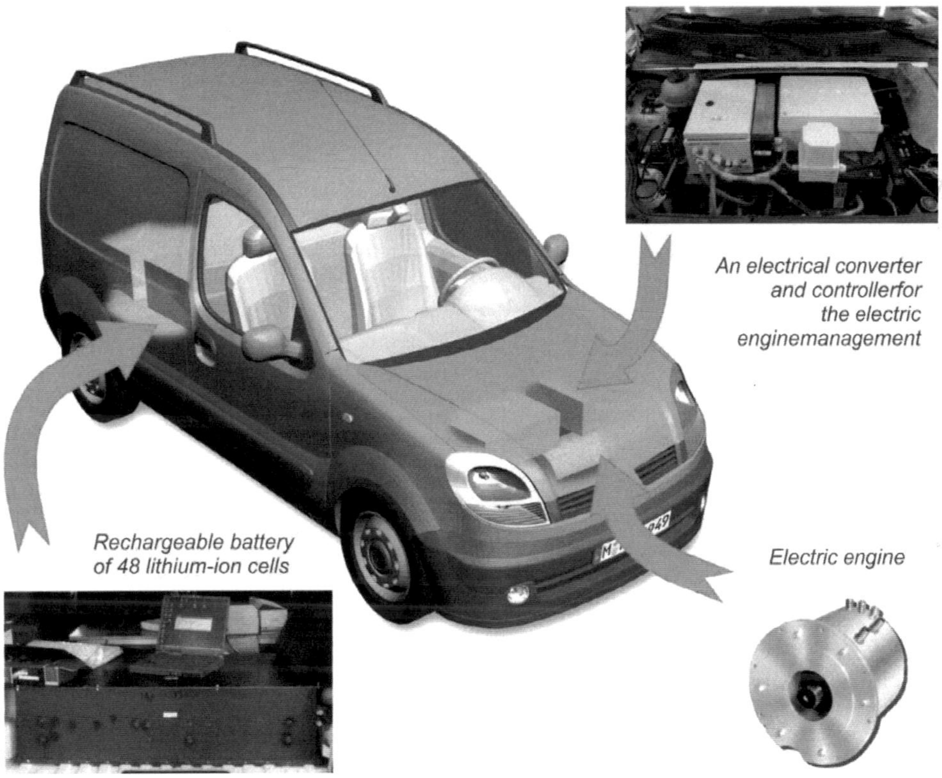

Figure 4. Structure of the converted electric vehicle on the base of Citroen Berlingo

The electric parameters during tests were registered by a multi-channel measuring and recording system, and then analyzed by a computer programme. The frequency of measurements was 0,1s for electric quantities. The accuracy for the voltage and the current is 0,1%

For the speed and time measurement a "fifth" wheel system of Peiseler was used. The accuracy for the distance is 0,2%. A transducer with 500 impulses for a rotation is built in the "fifth" wheel.

Figure 5. The converted vehicle, equipped with "fifth" wheel during the tests

In the tests, the highest registered peak power of the electric motor was in the range of 47 - 51 *kW*.

Using the described method of theoretical determination of the needed power on the base of acceleration time, the result was P = 44,3*kW* at the experimentally obtained acceleration time of 29s.The theoretical result is 6 - 13% less than the highest registered power in the tests. Taking into account the complex of road test and simplifications in theoretical methodology, the accuracy if satisfied.

4. Conclusions

1. The upgraded approach to determine the needed power of the electric motor based on the acceleration time of the electric car allows a more reasoned selection of the electric motor and the necessary parameters such as maximum power, maximum and basic angular speed.

2. The increase in the speed factor reduces the necessary power, at equal acceleration times of the electric vehicle. The effect of the base speed, as an electric motor's characteristic, on the acceleration properties of the electric car was investigated. Higher speed factor allows the use of a larger range of the motor characteristic with maximum power. At the same maximum speed, electric motors with lower base speed have better accelerating properties, because a larger relative share of the power is used for acceleration of the electric car.

References

[1] Evtimov I and Ivanov R 2016 *Electric Vehicles* (Ruse: Print center of University of Ruse) p. 232 ISBN 978-954-712-682-4

[2] Evtimov I and R. Ivanov 2009 Comparative Assessment of the Efficiency of Electric Vehicles and Conventional Cars. *Proc.International Scientific Conference BULTRANS-2009* (Sozopol) p. 94.

[3] Jia-Shiun Chen 2015 Energy Efficiency Comparison between Hydraulic Hybrid and Hybrid Electric Vehicles, *Energies* 2015, **8** doi:10.3390/en8064697 ISSN 1996-1073, p.4697-4723

[4] Madarás J 2012 Mathematica lmodeling of the electric drivetrain of the sports car *J. Electrotehnica (aveliable at*https://www.scribd.com/document/135212049/Mathematical-Modeling-of-the-Electric-Drive-Train-of-the-Sports-Car)

[5] Larminie J and Lowry J 2003 *Electric Vehicle Technology Explained* (John Wiley & Sons) ISBN 0-470-85163-5

[6] Ehsani M, Gao Y and Emadi A 2010 *Modern Electric, Hybrid Electric, and Fuel Cell Vehicles* (CRC Press) ISBN 978-1-4200-5398-2

[7] Juraj M. *Mathematical modeling of the electric drive train of the sports car*

[8] Wang Yan, Yin Tianming and Yin Haochun *Electric Vehicles - The Benefits and Barriers*. ISBN 978-953-307-287-6

[9] Besselink I.J.M., P.F. van Oorschot, E. Meinders, and H. Nijmeijer 2010 Design of an efficient, low weight battery electric vehicle based on a VW Lupo 3L. *25th International Battery, Hybrid and Fuel Cell Electric Vehicle Symposium* (China, Shenzen) p. 32-41

[10] Erik Schaltz *Electric Vehicles - Modelling and Simulations*, ISBN 978-953-307-477-1

[11] Marinescu D.G., Tabacu, I. and Nae C. 2009 Research Regarding the Development of an Electric Vehicle. *International Conference on Ecological Vehicles & Renewable Energies*, (Monaco).

[12] Marinescu D.G., Tabacu I., Tabacu S., Serban Fl., Nicolae V, and Vieru I. 2012 A Plug-in Hybrid Diesel-Electric Vehicle. *EVS26 -Worldwide International Battery, Hybrid and Fuel Cell Electric Vehicle Symposium & Exhibition* (Los Angeles, California)

IOP Conf. Series: Materials Science and Engineering **252** (2017) 012064 doi:10.1088/1757-899X/252/1/012064

Design-Parameters Setup for Power-Split Dual-Regime IVT

Ion Preda, Gheorghe Ciolan, Dinu Covaciu

Transilvania University of Brasov, Romania

{pion, cgicu, dinu.covaciu}@unitbv.ro

Abstract. To analyze the working possibilities of power-split infinitely variable transmissions (IVTs) it is necessary to follow a systematic approach. The method proposed in this paper consists of generating a block diagram of the transmission and then, based on this diagram, to derive the kinematics and dynamics equations of the transmission. For an actual numerical case, the derived equations are used to find characteristic values of the transmission components (gear and chain drives, planetary units) necessary to calculate the speed ratios, the speeds, torques and powers acting on the shafts and coupling (control) elements, and even to estimate the overall efficiency of the transmission.

1. Introduction

Although known for some time, the *continuously variable transmissions* (CVTs) began to be used in mass-produced automotive-vehicle only in the second half of the 20th century. As their name indicates, these transmissions have the ability to change the speed ratio (transmission ratio) continuously, [1], [2], [3], versus the conventional transmissions where speeds are changed in steps, and therefore discontinuously. The part of the transmission which allows a continuous gear-ratio change is named variator and represents the core element of any CVT.

The *variator* is a convertor (transformer) of rotational mechanical energy. It receives at its input shaft a power flow with certain angular speed and moment (torque) and delivers at its output shaft another power flow with possible different speed and moment. Thus, the variator is able to change, in an automatic or controlled way, its speed ratio (transmission ratio), while the output torque will be equal to the product of input torque, transmission ratio and efficiency.

Considering the working principle (the type of energy used to modify the parameters of the power between input and output), friction, hydrodynamic, hydrostatic and electric variators can be identified.

While the acronym CVTs refers mainly to transmissions employing mechanical friction variators, the aspects presented in this paper are also applicable to continuously variable transmissions with hydrodynamic, hydrostatic or electric variators.

Nowadays, the mechanical friction CVTs have important market share of the total mechanical transmissions, with Japan reaching about 40% of the total. A steady increase of CVT use has occurred during the last decades [4].

The main factors responsible of the rapid growth of CVT use in automobiles are as follows:

- the increased demand for environment protection and fuel saving by the automotive industry;
- the recent improvements in steel and traction oils;
- the advances in electronic control systems;

- the emergence of new and innovative ideas on new transmissions or hybrid propulsion systems layouts.

The present work is an attempt to help in the respect of this last item, i.e. to validate the viability of a new transmission layout and to permit the adoption of the primary kinematic and dynamic parameters of the design.

To analyze the working possibilities of such CVT it is necessary to follow a systematic approach. The method proposed in this paper consists in realizing first a block diagram of the transmission and then, based on that diagram, to derive the governing equations of the transmission kinematics and dynamics functioning. On the base of some adopted design data, the next step consists in the use of the derived equations to find characteristic values of the transmission components (gear and chain drives, planetary units), necessary to start the design. Finally, the functional equations will be used to calculate the speed ratios, the speeds, torques and powers acting on the shafts and coupling (control) elements, and even to estimate the overall efficiency of the transmission.

2. Components of the Multi-Regime CVTs

In some cases, a hydrostatic or an electric variator can solely represent the entire CVT, because:

- the variator input (respectively the pump or the generator) can be connected at the engine and the variator output (the hydrostatic or electric motor) can drive the wheels directly;
- the variator permits to reverse the vehicle direction of movement (both the forward and reverse driving are allowed);
- the variator is able to adapt enough the vehicle speed- and tractive force, ensuring inclusively the vehicle start-up, in order to cover all driving conditions; which means that the CVT represented by a hydrostatic or an electric variator is in fact an infinitely variable transmission (IVT).

Because the others batch-produced variator-types (the hydrodynamic torque convertors and the mechanical friction variators) are not able to invert the driving direction or to ensure a sufficient spread of the speed ratios, in addition to the variator, there is the need for other functional modules to obtain a CVT [4]. These are driving invertor, start-up device, gear or chain drives to adapt the speed, a parking brake, and a control system to change the speed ratio, together with a hydraulic power source for lubrication, cooling and actuation. Moreover, if the drive axle is placed on the same side of the vehicle as the engine, the CVT will also include the axle subassemblies i.e. the final drive and the differential. Consequently, the resultant CVT can become quite complex.

2.1. The variator

Over time, numerous CVT designs, with different type of variators, have been introduced in motor vehicle production. Some electro-mechanical or hydro-mechanical transmissions for heavy-vehicles contain a variator which is an electric generator-motor assembly [1] or a hydrodynamic torque convertor [1], [5]. A successful CVT with a hydrostatic variator used in tractor is presented in reference [6].

The friction variators (traction drives) used today in motor-vehicles are generally of two types [1], [2], [3]: pulleys variators and toroidal variators.

At the pulleys variators the axes of the input and output shafts are parallel and the two shafts are rotating in the same direction [7], [8]. At the toroidal variators the input and output shafts are coaxial and the two shafts may rotate either in the opposite direction (single-roller full-toroidal [9], or single-roller semi-toroidal [10]) or in the same direction in other designs (double-roller full-toroidal [11] or VariGlide [12]). The keeping or inversion of the rotation direction at the output shaft with respect to the input shaft is indicated on the symbols presented in the **Table 1** through a plus (+) or minus (-) sign.

The efficiency of the previously presented variators depends on many factors, from which the most important are the instantaneous speed ratio and input torque [1], [4].

Table 1. Suggested symbols to be used in the construction of the CVTs block diagrams

Parallel axes variator (pulleys variator)	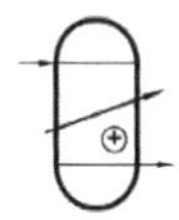	Coaxial variator (toroidal variator)	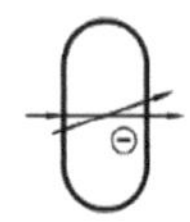		
Electric or hydrostatic variator	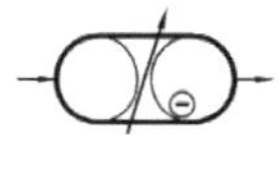	Electric generator or hydrostatic pump	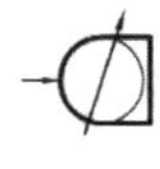	Electric or hydrostatic motor	
Gear drive	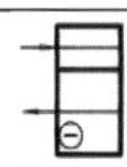	Chain drive	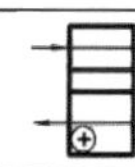	Clutch or brake	

2.2. The mechanical drives with constant speed ratio

The devices used to reduce or increase the speed with a fix step are gear- or chain drives (**Table 1**). While the chain drives are maintaining the same direction of rotation of the input and output shafts (indicated by the plus sign), the spur gears will invert the rotation in the case of exterior meshing (-), but will keep the rotation way in the case of interior meshing (+).

2.3. The control devices

The clutches or brakes used to change the CVT working regime will be represented in the block diagrams by a switch sign (**Table 1**). If the switch is on (clutch or brake engaged) the torque will be transmitted. If the switch is off (clutch or brake disengaged) the torque will be interrupted.

2.4. The planetary units

A *planetary unit* (epicyclic train) is a mechanism (generally with gears or sometimes with revolution bodies) having three coaxial shafts and two degrees of freedom. Combining two or more such elementary planetary mechanisms, a planetary (epicyclic) *gearset* will be obtained which may have three or more degrees of freedom.

Further, the three central elements (shafts) of the planetary unit will be indicated by the indices h, x and y (where h indicates the planet carrier and x and y the other shafts).

The next equation due to Willis [1], [13],

$$\omega_x - i_0\,\omega_y - (1 - i_h)\omega_h = 0 \qquad \text{(equivalent with: } \frac{\omega_x - \omega_h}{\omega_y - \omega_h} = i_0 = (-1)^p\,\frac{z_y}{z_x}\text{)} \tag{1}$$

will be used in the study of the planetary unit kinematics. Here p is the number of planet wheels groups (normally, p=1 or p=2), z_x and z_y are the teeth numbers of the gear wheels x and y respectively, and i_0 is known as the *internal speed ratio* of the planetary unit (the speed ratio from the wheel x to the wheel y when the carrier is blocked: ω_h=0).

Assuming negligible friction and stationary working regime (constant speeds for all three shafts) [1], the dynamics of the planetary unit will follow equations (2) and (3) below:

$$M_x + M_y + M_z = 0 \tag{2}$$

which is obtained applying the principle of moment equilibrium for the whole planetary unit, and

$$i_h\,M_x + M_y = 0 \qquad \text{(equivalent with: } \frac{M_y}{M_x} = -i_h = (-1)^{p-1}\,\frac{z_y}{z_x}\text{)} \tag{3}$$

which is derived applying the principle of moment equilibrium for the planet wheels.

The most widely used planetary units in the constructions of automotive transmissions consist of a sun wheel (s), a ring wheel (r), a planet carrier (h) and more planet wheels. Noting with $k = \frac{z_y}{z_x} = |i_h|$

the positive ratio of the outer and inner wheels teeth-numbers, from the equations (1), (2) and (3) one can obtain particular equations for the most common planetary units (**Table 2**).

Table 2. Characteristics of the most used planetary units in the CVTs design

PU type	*PU kin. sch.*	*PU block diag.*	i_h	*Torque ratio*
Single-planet epicyclic unit (spur gears)	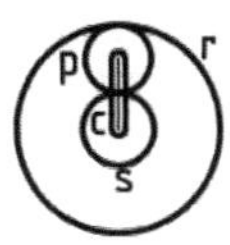	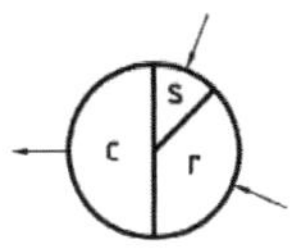	$-\frac{z_r}{z_s}$	$\frac{M_r}{M_s} = k = \frac{z_r}{z_s}$
Double-planet epicyclic unit (spur gears)	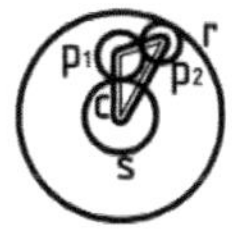	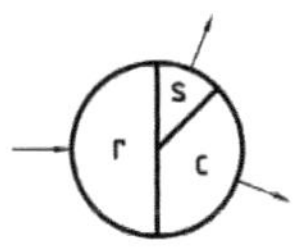	$+\frac{z_r}{z_s}$	$\frac{M_c}{M_s} = 1 + k = 1 + \frac{z_r}{z_s}$
Single-planet epicyclic unit (bevel gears)	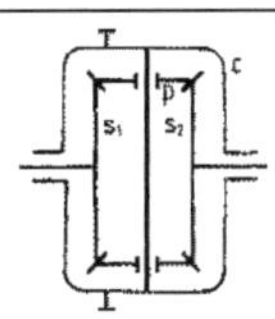	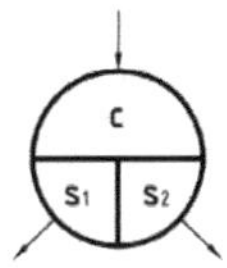	-1	$\frac{M_{s2}}{M_{s1}} = k = 1$

The moments applied to the planetary-unit shafts are considered positive for the driving ones, and negative for the resistant ones (in fulfilment of equation 2). Due to its three shafts, in a planetary unit the magnitude (absolute value) of one moment will be always equal with the sum of the other two moment magnitudes. In the symbols presented in the third column of the **Table 2**, the area associated with one central element (carrier *c*, sun *s* and ring *r*) is equal with the summed areas of the other two. Also, arrows will be used to indicate the input and output moments (the orientation of the power flows).

Some friction variators, as Dana's VariGlide toroidal variator [12], can be realized (and used) in two configurations: as a normal variator, with one degree of freedom, or as a planetary unit, with two degree of freedom. In this second case, the design includes three rotating shafts (the balls-axes support is rotating also). Thus, this mechanical device can have two inputs and one output for the torque or vice-versa, one input and two outputs. For that reason, the variator behaves similarly to a planetary mechanism (the internal speed ratio i_h of which may be changed, unlike the gear planetary mechanism which maintains constant ratio). This constructive solution was therefore named CVP (Continuously Variable Planetary) variator.

3. Multi-Regime Power-Split CVTs

Due to their possibility to generate more speed ratio or to join or split power flows, the planetary units are widely included in the construction of mechanical transmissions, CVTs being no exception. For example, the CVTs with a single power flow use commonly a planetary unit as *inversor* [4], allowing not only the vehicle forward-driving, but also the reverse driving, generally with a smaller speed. Its control devices (normally a clutch and a brake) may be used also as *starting device*.

A *power-split* configuration of a CVT realizes firstly the splitting of the power flow in two or more mechanical paths and finally the re-joining of these flows at the transmission output. Such a design has several important advantages:

- bigger CVT power input for the same variator power loading capability;
- bigger overall efficiency of the CVT;
- the possibility to obtain, at a certain variator speed ratio, a particular functioning situation named geared-neutral condition.

IOP Conf. Series: Materials Science and Engineering **252** (2017) 012064 doi:10.1088/1757-899X/252/1/012064

At the *geared-neutral* (also named powered-neutral), the transmission is engaged and ensures an infinity speed-ratio, at which the vehicle speed remains zero, while the engine is running under certain load, determined by external forces acting on the vehicle (like grade and aerodynamic forces). That means a continuous and smooth speed change it is possible from forward driving, passing through stop, to reverse driving (and vice-versa) – the vehicle can shuttle (can switch its movement between forward and reverse) without any actuation of the CVT clutch or brake [14].

To obtain the "geared-neutral", the planetary unit must be placed at the transmission output (as summation device). If the planetary unit is placed at the input, an opposed functional case can be obtained: speed ratios with very small absolute values (inclusive zero). Named „geared zero" [15], this case corresponds to the working situation in which the transmission is engaged (under power), the vehicle is moving, but the engine is stopped. That situation may be used to stop and to restart the engine during driving, for example at the hybrid propulsion systems (the start/stop function).

A *multi-regime* transmission is able to change its power flows structure by engaging or disengaging, in an appropriate mode, some control elements i.e. clutches, brakes, dog clutches or one-way clutches. In the case of such CVT, the variator needs to pass ("sweep") more times through its speed-ratio range in order to cover the larger range necessary for the transmission (for example, two times for a two-regime CVT, as is presented in **Figure 1**, bottom side).

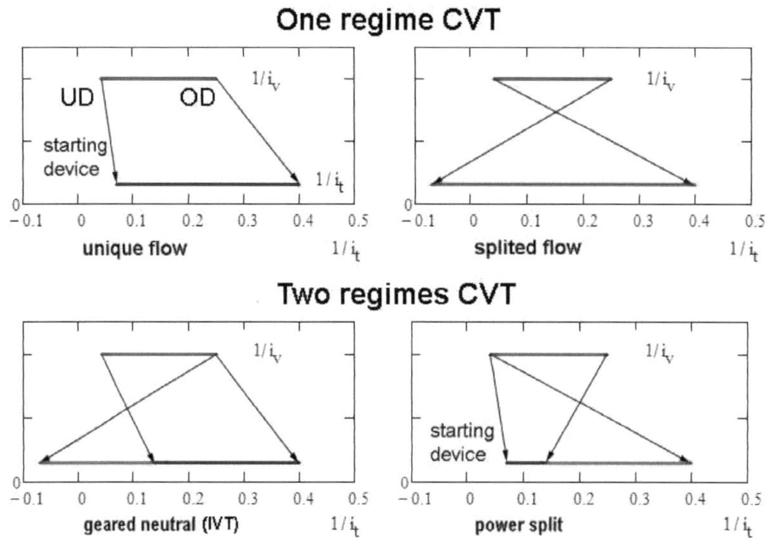

Fig. 1 Possible transposition of the variator speed ratio into the CVT overall speed ratio [14]

The variant of two-regimes CVT with power-split realized in the high-range speed ratios (**Figure 1**, bottom right) has the advantage of a higher efficiency on the most driving time (at medium and high vehicle speeds), but the necessity of a starting device still exists. The other variant, with the power-split attained at low vehicle speeds (**Figure 1**, bottom left), eliminates the need of a starting device and ensures excellent drivability in difficult terrains, but with a penalty of the vehicle overall fuel efficiency.

Despite of their added complexity, IVT with three or even four regimes are studied [16], because the power-split architecture may have key benefits, providing an increase in tractive effort simultaneously with higher efficiency and durability.

4. Requirements for the Multi-Regime CVTs

Even if currently there are only few power-split CVTs implemented in motor-vehicle series-production, a large number of such solutions (based on many variator types and planetary units) have been studied by researchers [10], [9], [15], [17], [18], [19].

The packaging, kinematic, dynamic and control possibilities of the adopted variator types will influence the layout of the transmission and, consequently, its suitability for different vehicle categories (front-wheel drive, rear-wheel drive or all-wheel drive vehicles). For example, the variator positions of the shafts or number of working cavities will impose different implementations of the other transmission components. The variator speed-ratio spread, capable torque, efficiency variations or reaction time are also very important aspects of the possible power-split designs.

The diminish of the number, together with the dimensional minimization and functional optimization of the control elements (couplings, brakes, dog clutches or one-way clutches) are also major concerns of the design engineers.

Even aspects which seem to be of minor importance are studied, as the avoidance (as much as possible) of the situations in which the variator shafts may stop rotating or the rotation way is reversed.

The existent variators present different control possibilities: in some designs the speed ratio it is controlled, while in others the torque is controlled, which makes one variator or the other more suitable for a specific application or driving situation. For the friction variators, a second action of control is oriented towards the contact force or towards the slip between the discs and rollers. All these aspects represent big problems for the engineers.

5. Example of CVT Design Parameters Setup

A constructive solution of a double-roller full-toroidal IVT, suggested by the Australian company Ultimate Transmissions on its website [11], will be analyzed to setup a hypothetic transmission.

In the left side of the **Figure 2** it is presented the kinematic scheme drawn by the authors. That was the base for the construction of the power-flows diagram, shown in the same figure.

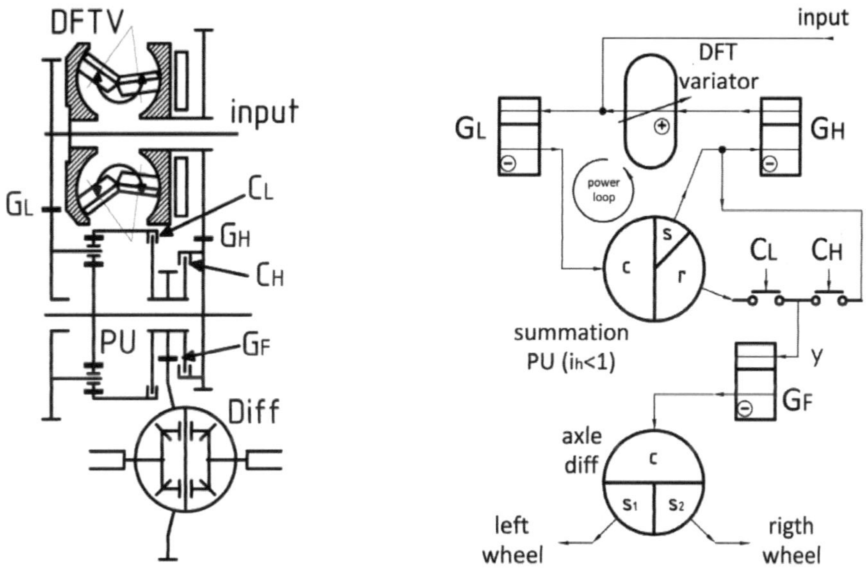

Fig. 2 Kinematic scheme (left) and power-flows diagram (right) of a dual-mode power-split IVT, based on double-roller full-toroidal variator

The power flow splits at the level of the input shaft. The two flows are joined again at the summation planetary unit. That has a single group of planet wheels.

Three exterior-meshing spur gears (G_L, G_H and G_F) and the bevel-gear axle-differential are also necessary for this transmission. Two clutches (C_L and C_H) ensure the two working regimes.

Obtained when C_L is off (disengaged) and C_H is on (engaged), the *High* regime is characterized by a single power flow and is used for medium and high vehicle speeds. The power passes through the input shaft, the dual-roller full-toroidal (DFT) variator, the gears G_H and G_F (the final drive) and has the output to the axle differential.

In the *Low* regime, obtained when C_L is on and C_H is off, the power has two different paths. That regime should ensure *geared neutral* and low speed driving, both in forward and reverse.

A third situation is also possible: a true neutral, when both clutches are disengaged. The situation when both clutches are engaged is not permitted.

The first ratio-sweep moves from the variator high gear (overdrive) to low gear (the IVT *Low* regime), and then, after the switchover which disconnects the planetary gear set, the variator moves from low to high.

The first aspect to be verified is if the vehicle moves forward in the CVT *High* regime. Considering that the engine is mounted in the vehicle right side, the (+) sign of the variator (caused by the double roller) and the two (-) signs indicate that the vehicle drive-wheels will rotate in the same way as the engine. That means the condition is ensured.

5.1. The initial data

Before starting the full transmission set-up, some initial values must be known.

From the calculation of the vehicle dynamical performances, the next transmission overall speed-ratios are already established: i_{Tv}=3.64, the ratio at which the vehicle will obtain its maximal speed; i_{To}=1.952, the minimum ratio for forward driving (overdrive, necessary for cruising with the legal speed on high-ways, at low engine RPM); i_{Tr}=-6.25, the ratio for the maximal speed in reverse.

Also, there are known the variator minimum and maximum ratios: i_{Vo}=0.427 (overdrive) and i_{Vu}=2.343 (underdrive) [11]. From these, the variator ratio spread (or domain or range) D_V results as

$$D_V = \frac{i_{Vu}}{i_{Vo}} = 5.487 \tag{4}$$

and that allows to find the transmission ratio at which the two regimes are changed (switched over)

$$i_{Tc} = i_{To} D_V = 10.711. \tag{5}$$

Another initial calculation permits to find the variator ratio at which the vehicle reaches its maximal speed:

$$i_{Vv} = \frac{i_{Tv}}{i_{To}} i_{Vo} = 0.796. \tag{6}$$

5.2. The transmission ratios

Studying the power-flows diagram from **Figure 2**, it can be observed the possibility to write three equations (two equations for the speed ratios in the transmission *High* and *Low* regimes and the equation of Willis for the summation planetary unit). Because the number of design unknowns is four (the intern ratio of the planetary unit i_0 and the speed ratios i_L, i_H, i_F of the gears), that means it is necessary to adopt one value. In this case, the ratio of the final drive is adopted, i_F=-3.76, to maintain a certain compatibility with the conventional transaxles of the vehicle.

Because the overall transmission ratio in the *High* regime is

$$i_T = i_V i_H i_F, \tag{7}$$

in the particular case of the minimum ratio for forward driving i_{To}, when the variator works with its highest ratio i_{Vo}, it results the ratio of the gear G_H:

$$i_H = \frac{i_{To}}{i_{Vo} i_F} = -1.216, \tag{8}$$

In the transmission *Low* regime, the overall ratio is

$$i_T = i_{PS} i_F, \tag{9}$$

where $i_{PS} = \frac{\omega_{in}}{\omega_r}$ is the speed ratio of the power-split part.

Observing from the **Figure 2** that

$$\omega_c = \frac{\omega_{in}}{i_L}; \qquad \omega_s = \frac{\omega_{in}}{i_V\, i_H} \tag{10}$$

and introducing in the equation of Willis

$$\omega_s - i_0\, \omega_r - (1 - i_0)\omega_c = 0 \tag{11}$$

it results

$$i_{PS} = \frac{i_0}{\frac{1}{i_V\, i_H} - \frac{1-i_0}{i_L}}. \tag{12}$$

where the ratios i_0 and i_L, are unknown.

Particularizing for the extreme ratios of the *Low* regime, corresponding to the variator ratios i_{Vo} and i_{Vu}, it obtains

$$\frac{i_0}{\frac{1}{i_{Vo}\, i_H} - \frac{1-i_0}{i_L}} = \frac{i_{Tr}}{i_F} = i_{PSr}=1.662; \qquad \frac{i_0}{\frac{1}{i_{Vu}\, i_H} - \frac{1-i_0}{i_L}} = \frac{i_{Tc}}{i_F} = i_{PSc}=-2.849. \tag{13}$$

Solving this last system of two equations, it obtain the internal ratio of the planetary unit,

$$i_0 = \frac{\frac{1}{i_H}\left(\frac{1}{i_{Vo}} - \frac{1}{i_{Vu}}\right)}{\frac{1}{i_{PSr}} - \frac{1}{i_{PSc}}} = -1.653, \tag{14}$$

and the ratio of the gear G_L,

$$i_L = \frac{1-i_0}{\frac{1}{i_{Vu}\, i_H} - \frac{i_0}{i_{PSc}}} = -2.849. \tag{15}$$

The last element needed to finalize the setup of the IVT main parameters is the variator ratio at which the *geared-neutral* will be obtained. For that, the ratio i_{PS} must have an infinite value, which is possible only if the denominator in the equation (12) will be equal with zero. Putting this condition, the *geared-neutral* variator ratio is obtained:

$$i_{Vgn} = \frac{i_L}{i_H(1-i_0)} =0.883 \tag{16}$$

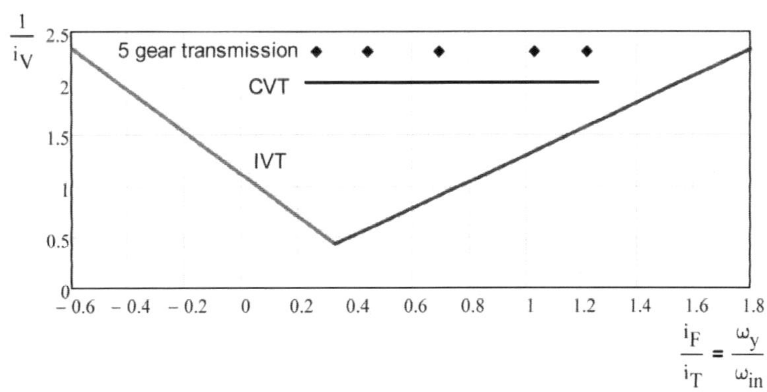

Fig. 3 Ratio spread of the IVT, compared with common CVT- and 5-gear manual transmissions

A graphical representation of the resultant IVT kinematic characteristics (eliminating the contribution of the final drive ratio i_F) is presented in **Figure 3**. Here it can be seen how the two regimes are distributed and also the important advantage of a larger ratio domain, if one compare the

IVT with a 5-speed gearbox and with a normal CVT (excluding also the influence of the final drive ratio).

5.3. The shaft torques

The calculation of the torques (torsional moments) acting on the IVT shafts are obtained starting from the schemes in **Figure** 2 and on the equations in the first line of **Table** 2.

Noting the torques acting on the input shafts of the variator and gears with M_V, M_H, M_L respectively, and the torques acting on the shafts of the planetary unit with M_c, M_s, M_r, a system of six equations may be written:

$$M_{in} + M_L + M_V = 0, \qquad M_c = M_L i_L \eta_L \qquad M_H = M_V i_V \eta_V^q$$
$$M_s = M_H i_H \eta_H^q \qquad M_c + M_s + M_r = 0 \qquad M_r = -i_0 \eta_0^s M_s \tag{17}$$

In these equations, the values η_V, η_H, η_L and η_0 represent respectively the efficiencies of the variator, gears G_H and G_L and planetary unit (when only the frictions proportional with the input torques are considered). Also, the exponent $q=\pm1$ will consider the direction of the power flow in the branch of the variator ($q=+1$ for normal flow; $q=-1$ for inversed flow, when the power loop appears), while the exponent $s=\pm1$ keep account of the direction of power flow through the planetary unit ($s=+1$ if the sun wheel rotates faster as the ring wheel i.e. $|\omega_s|>|\omega_r|$).

Solving the equations system (17), the torques M_V, M_H, M_L, M_c, M_s, M_r can be expressed as functions of the input torque M_{in}. For example, the variator input torque and the output torque of the transmission power-split part are:

$$M_V = M_{in}\frac{i_L\eta_L i_v \eta_V^q}{den} \qquad M_r = M_{in}\frac{-i_0\eta_0^s i_L\eta_L i_v\eta_V^q i_H\eta_H^q}{den} = M_V(-i_0\eta_0^s i_H\eta_H^q) \tag{18}$$

where the denominator is

$$\text{den} = (1 - i_0\eta_0^s) i_v \eta_V^q i_H \eta_H^q - i_L\eta_L \tag{19}$$

Using the equations (18), in the **Figure** 4 it is presented how many times the variator input torque and the output torque of the transmission power-split part (the ring wheel torque) are increased with respect to the transmission input torque. As can be seen, at the variator ratio $i_{Vgn}=0.883$ (geared neutral), the gain is theoretically infinite. Negative values are obtained for $i_V< i_{Vgn}$. That means also that, while the vehicle moves in reverse, the power flow in the variator branch is normal, not inversed as is presented in the **Figure** 2 (no power loop in this case).

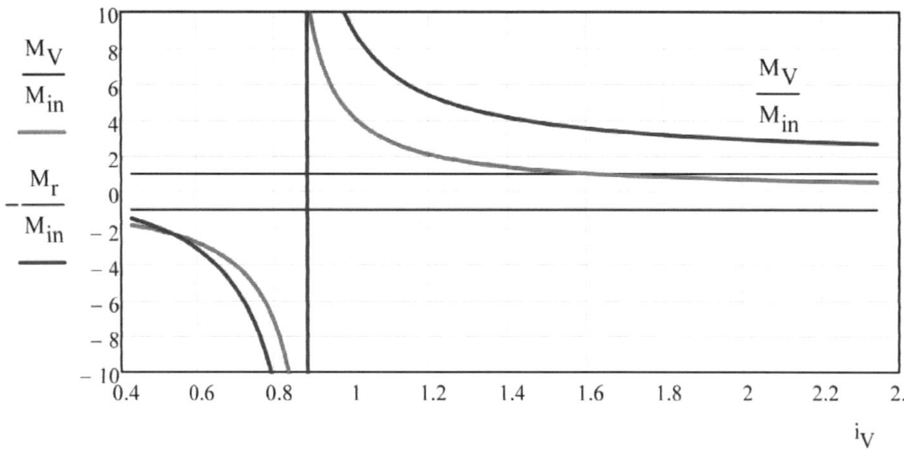

Fig. 4 Torques amplification vs. variator ratio (IVT “Low” regime)

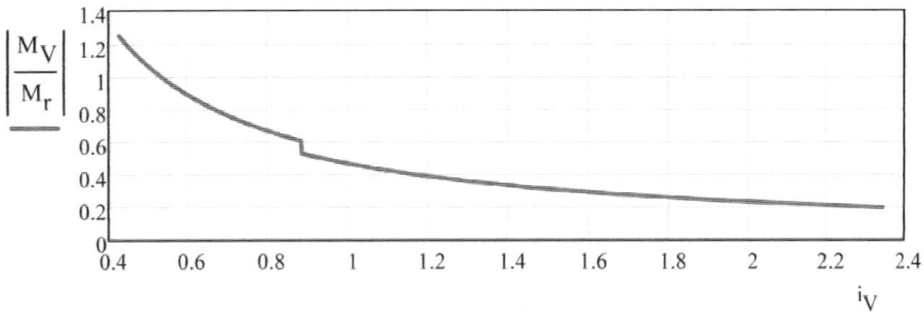

Fig. 5 Ratio of the variator and ring wheel torques vs. variator ratio (IVT "Low" regime)

In the **Figure** 5 it can be observed how big the variator input torque M_V is if compared with the output torque M_r of the transmission power-split part. As can be seen, the variator torque exceeds the output torque only when the vehicle moves fast in reverse. The discontinuity of the curve is explained by the change of the power flow in the variator branch.

5.4. The efficiency

The efficiency of the IVT will be calculated in different ways for the two working regimes. While in the *High* regime (with a single power path) the overall efficiency is simply

$$\eta_T = \eta_V \eta_H \eta_F \tag{20}$$

in the *Low* regime, characterized by two power paths, the overall efficiency is

$$\eta_T = \eta_{PS} \eta_F = \frac{M_r}{M_{in}} \frac{\omega_r}{\omega_{in}} \eta_F = \frac{M_r}{M_{in}} \frac{1}{i_{PS}} \eta_F \tag{21}$$

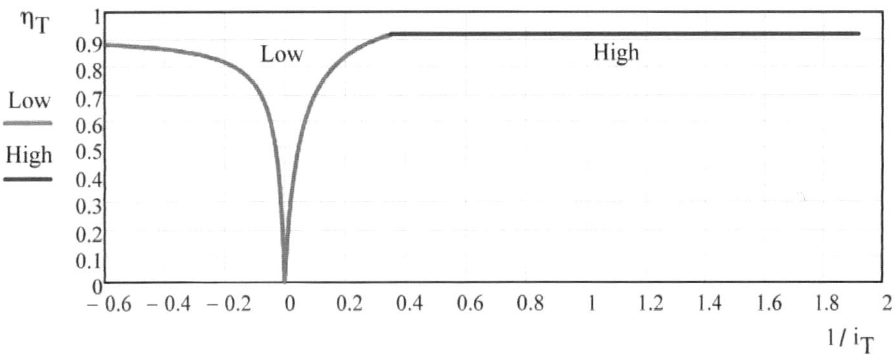

Fig. 6 Calculated overall efficiency of the IVT

In the **Figure** 6 it is represented the variation of the theoretical efficiency of the IVT for the entire domain of the transmission ratios. The calculation was performed considering the next constant values for the component efficiencies: η_V=0.955 [11], η_H=η_L=0.98 [1] and η_0=0.965 [1], [13].

In this hypothetical case of IVT, the power split layout of the *Low* regime ensures an efficiency at reduced (forward and reverse) vehicle speeds which is superior to the vehicles equipped with a gearbox or a normal CVT using as starting device a friction clutch or even hydrodynamic torque convertor. This conclusion was also presented in reference [14].

6. Conclusions

The graphical representations of the power flows with changing structures in complex CVTs are useful in a better understanding and analysis of the behavior of the transmission. In the present paper, the authors have shown how such diagrams allow a simplification of the writing of the functional equations.

On the base of these equations it is then possible to setup the main characteristics of the transmission components. A practical example illustrating the procedure was also shown in the last part of the paper.

The method presented here can be useful also for the analysis of epicyclic automatic transmissions and even for hybrid propulsion systems that use planetary units.

References

[1] Ciolan, G., Preda, I. and Pereş, G. Cutii de viteze pentru automobile. Editura Didactica si Pedagogica, Bucharest, 1998.
[2] Naunheimer, H. Bertsche, B. Ryborz, J. Novak, W. Automotive Transmissions – Fundamentals, Selection, Design and Application. Second edition. Springer, 2011.
[3] Fischer, R., Küçükay, F., Jürgens, G., Najork, R. and Pollak, B. The Automotive Transmission Book. Springer, 2015.
[4] Preda, I., Ciolan, G., Diaconescu, E. and Cristea, D. Transmisii mecanice cu fricţiune pentru autovehicule (Friction mechanical transmissions for automotive vehicles). Ed. Universitatii din Pitesti, 2012.
[5] *** Automatic transmissions for buses. DIWA Transmission Type D506. Voith, 2010.
[6] *** Fendt 200 Vario Tractor Brochure. Fendt, 2010.
[7] van der Sluis, F. Fuel consumption potential of the pushbelt CVT. 5th International CTI-Symposium "Innovative Automotive Transmissions", 2006, Berlin, Germany.
[8] Englisch, A. Faust, H. Homm, M. Lauinger, C. Vornehm, M. CVT Without Limits – Components for Commercial Vehicle Transmissions. Luk Colloquium, 2006.
[9] Brockbank, C. Heumann, H. Delivery of IVT for a 5 Litre SUV: Addressing the Concerns of Geared Neutral. www.torotrak.com, 2002.
[10] Miyata, S. Machida, H. Development of the Half-Toroidal CVT POWERTOROS Unit. (3) Development of the Power-Split System. NSK Technical Journal Motion & Control No.11, October 2001.
[11] *** Double Roller Full Toroidal Variator based CVT. UltimateTransmissions, 2011.
[12] McIndoe, G. Multi-Mode Passenger Car Transmission Concepts featuring a VariGlide® CVT. CTImag - 5th edition, December 2015, pp. 21-24.
[13] Ciobotaru, T., Frunzeti, D. and Jaentschi, L. Method for analysing multi-path power flow transmissions, 2009.
[14] Preda, I. Theoretical Aspects on the Multi-Regime Friction CVTs' Dynamical Behavior. International Journal of Engineering, Annals of Engineering Faculty Hunedoara, Tome XI, Fascicule 3, Year 2013, pp. 255-260.
[15] Lauinger, C. Vornehm, M. Englisch, A. 500 Nm CVT – LuK Components in Power Split. LuK Symposium 2002, pp.75-87.
[16] Brockbank, C. Multi regime full-toroidal IVT for high torque and Commercial Vehicle applications. 9th International CTI-Symposium, Detroit, 2010.
[17] Shinojima, T. Toyoda, T. Miyata, S. Imanishi, T. Inoue, E. Machida, H. Development of the Next-Generation Half-Toroidal CVT with Geared Neutral and Power-Split Systems for 450 N-m Engines. SAE 2004-34-2878.
[18] Resch, R. Leistungsverzweigte Mehrbereichs-fahrantriebe mit Kettenwandlern. Ph.D. Thesis, Technischen Universitaet Müuenchen, Muenchen, 2004.
[19] Fuchs, R. Hasuda, Y. James, I. Modeling, Simulation and Validation for the Control Development of a Full-Toroidal IVT. www.torotrak.com, 2005.

IOP Conf. Series: Materials Science and Engineering **252** (2017) 012065 doi:10.1088/1757-899X/252/1/012065

Study of emissions for a compression ignition engine fueled with a mix of DME and diesel

[1]Jurchiş, Bogdan, [1]Burnete, Nicolae, [1]Iclodean, Călin, [1]Burnete, Nicolae Vlad
Technical University of Cluj-Napoca, Romania, Muncii Bd. 103 - 105

[1] E-mail: calin.iclodean@auto.utcluj.ro

Abstract. Currently, there is a growing demand for diesel engines, primarily due to the relatively low fuel consumption compared to spark-ignition engines. However, these engines have a great disadvantage in terms of pollution because they produce solid particles that ultimately form particulate matter (PM), which has harmful effects on human health and also on the environment. The toxic emissions from the diesel engine exhaust, like particulate matter (PM) and NO_x, generated by the combustion of fossil fuels, lead to the necessity to develop green fuels which on one hand should be obtained from regenerative resources and on the other hand less polluting. In this paper, the authors focused on the amount of emissions produced by a diesel engine when running with a fuel mixture consisting of diesel and DME. Dimethyl ether (DME) is developed mainly by converting natural gas or biomass to synthesis gas (syngas). It is an extremely attractive resource for the future used in the transport industry, given that it can be obtained at low costs from renewable resources. Using DME mixed with diesel for the combustion process, besides the fact that it produces less smoke, the emission levels of particulate matter is reduced compared to diesel and in some situations, NOx emissions may decrease. DME has a high enough cetane number to perform well as a compression-ignition fuel but due to the poor lubrication and viscosity, it is difficult to be used as the main fuel for combustion

1. Introduction

More stringent global pollution rules force the automotive industry to adopt various solutions to limit pollutant emissions below the regulated level. To meet this goal, car manufacturers have had to find all sorts of solutions to be able to reduce vehicle emissions, starting from fitting them with diesel particulate filters and sophisticated exhaust after-treatment systems, up to increase the production of hybrid or fully electric vehicles. This option makes the purchase price of the car to be higher due to high production costs.

Due to the fact that The European Union set a target to EU countries to use at least 10% of their transport fuels from renewable sources by 2020, many studies and research have carried out on the use of less polluting fuels for so-called green fuels due to lower emissions than conventional fuels [1].

DME is a gaseous ether at atmospheric pressure and room temperature, with a reduced carbon to hydrogen ratio having the chemical formula CH3-O-CH3. The storage, handling and transport requirements for DME are similar to LPG because of their similar properties. At atmospheric conditions of 0.1 MPa and 298 K, DME is an invisible gas. By increasing the pressure to 0.5 MPa at standard temperature it passes from the vapor phase to the liquid phase. DME gas is denser than dry air, and the density of DME in liquid phase is 660 kg/m^3 at normal temperature and pressure [2][3][4]. One of the great advantages of this fuel is that it can be obtained from a multitude of raw materials such as coal, natural gas or biomass [6][4]. In this paper, we will focus on the potential of using Dimethyl ether

IOP Conf. Series: Materials Science and Engineering **252** (2017) 012065 doi:10.1088/1757-899X/252/1/012065

(DME) in different proportions in a blend with diesel, this paper focuses on the emission trend for particulate matter and NOx.

2. DME Properties

In his chemical structure, DME has no direct C-C bond and has a higher oxygen content than other fossil fuels, having as well good evaporation properties [5]. DME cetane number is between 55-60 being higher than conventional diesel which has a cetane value between 40-55 (table 1), thus making it suitable for diesel engines. Due to the high oxygen content of approximately 34.8%, DME supports the combustion process and plays a significant role in reducing emissions. The high cetane number of DME makes to have a lower auto-ignition temperature compared to pure diesel fuel (figure 1) [2][3][5][7][8].

Table 1. Physical and chemical properties of DME and diesel [2]

Property	DME	Diesel
Chemical formula	CH_3OCH_3	C_8 to C_{25}
Molecular weight	46.07	96 ~
Vapor pressure at 20 ^{0}C [bar]	5.1	< 0.01
Boiling temperature [^{0}C]	-25	150 - 380
Liquid density at 20 ^{0}C [kg/m^3]	660	800 - 840
Liquid viscosity at 25 ^{0}C [kg/ms]	0.12-0.15	2 -4
Gas specific gravity [vs air]	1.59	-
Lower heating value [MJ/kg]	28.43	42.5
Cetane number	55 - 60	40 - 55
Stoichiometric A/F ratio [kg/kg]	9.0	14.6
Enthalpy of vaporization at NTP[a] [kJ/kg]	460 [-20 ^{0}C]	250[a]

DME has low toxicity, there is no significant risk if it is inhaled, therefore, it does not have toxic effects for normal use and, at the same time, contact with the skin has no adverse effects[2][9]. In terms of LHV (lower heating value), DME has a LHV of 28.3 [MJ / kg] compared to diesel which has 42.5 [MJ / kg] [2]. This may be a disadvantage due to the fact that more fuel should be injected per cycle. In

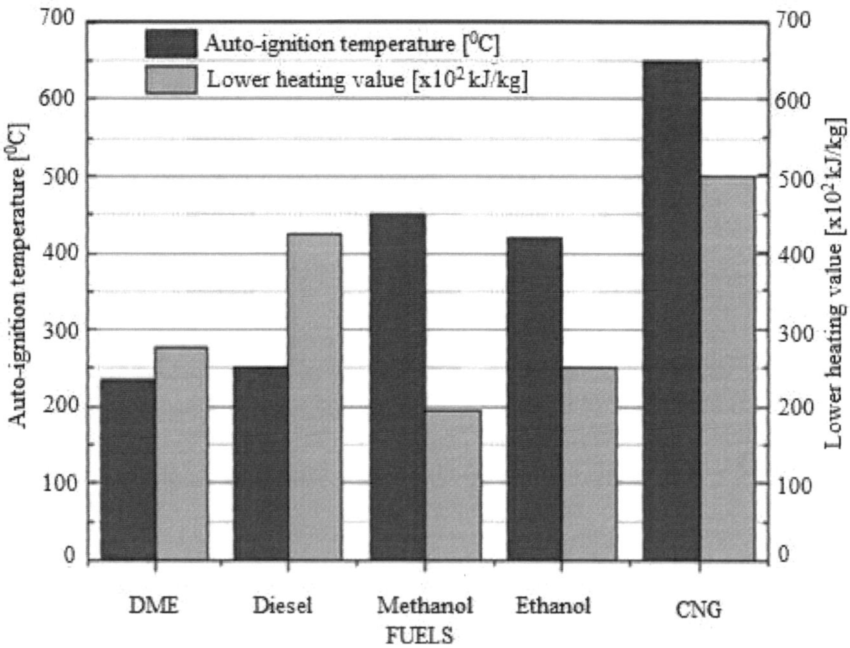

Figure 1 Auto-ignition temperature and heating values of various fuels [2]

terms of existing drawbacks in the application of DME as fuel for compression-ignition engines, is that DME has low lubrication characteristics and low viscosity compared to diesel, which causes fuel leakage and a bigger wear on moving parts of the injection system, thus requiring the use of additives to prevent fuel leakage and surface wear [2][4][5][10][11].

2.1. DME Production

Normally the production of dimethyl ether includes two step process [4] :

- Conversion of syngas to methanol
- Dehydration of methanol to dimethyl ether

The chemical reactions described by Troy et al [4] for DME formation are :

- Methanol synthesis :

$$CO + 2H_2 \rightarrow CH_3OH \qquad (1)$$

- Methanol dehydration :

$$2CH_3OH \rightarrow CH_3OCH_3 + H_2O \qquad (2)$$

- Water-gas shift :

$$H_2O + CO \rightarrow H_2 + CO_2 \qquad (3)$$

- Net Reaction :

$$3H_2 + 3CO \rightarrow CH_3OCH_3 + CO_2 \qquad (4)$$

2.2 DME Impact on Pollutant Emissions

Research on emissions produced by DME shows a downward trend in pollutant emissions especially in the case of particulate matter produced by the engine. These studies which have been taken into account, treats the case of the use of pure DME, as well as the case of mixing DME and diesel. Troy et al [4] present that the advantages of dimethyl ether over conventional diesel include reduced emissions of NOx, hydrocarbons and carbon monoxide and that DME does not produce soot (figure 2). According to

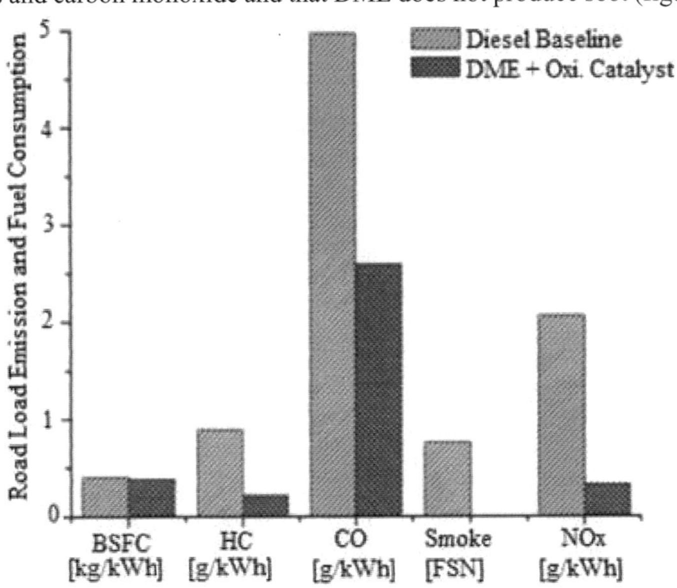

Figure 2 Road load test data comparing engine emissions using diesel and neat DME [4][18]

Ohno et al [12], they point out that the DME combustion characteristics include a similar thermal efficiency with diesel fuel, but with lower NOx emissions and without soot emissions.

A study of a DME blend and diesel was made by Wang et al[13] and they highlighted that the impact on emissions varies depending on engine operating conditions, at a high engine load, it reduces the smoke emissions around 58% to 68% for a mixture of DME and diesel compared to pure diesel. An interesting fact they describe is by adding DME, NOx emissions decrease a little, while CO emissions and unburned HC emissions increase under most operating conditions. Park et al [14] have shown the advantage of oxygen content in fuel in terms of lowering soot emissions.

There are some papers showing that NOx emissions increase if DME is added in a blend with diesel. Kim et al [15] reported higher NOx emissions with DME compared to diesel for all injection strategy. Park et al [16] submit that the emissions of NOx is higher for DME compared to diesel and the application of EGR is necessary to reduce the NO_X emissions during DME combustion.

3. Methodology

For this study the diesel particulate filter DPF was modeled using AVL BOOST. This way it was possible to consider correlations and connections between the amount of soot, the type of DPF channel structure and the type of fuel used for combustion. By using AVL BOOST simulation the authors were able to highlight the emission trend for a fuel mixture of diesel and DME in different proportions. Based on the obtained data it was possible to study an injection strategy to improve the emissions[17].

3.1. Engine Model

The model for this study is based on a single cylinder research diesel engine, namely the AVL 5402 of the Testecocel laboratory, from The Department of Automotive Engineering and Transports, Faculty of Mechanics of Technical University of Cluj Napoca. The technical data of the engine was used for the development of simulation model. The diagrams and the experimental data have been extracted from the testbed measurements and stored in the post processing software AVL CONCERTO. The Cylinder of the model is connected with the Engine element and it defines the type of engine used, operating speeds, moments of inertia and the brake mean effective pressure (BMEP). The combustion model used for simulation was the AVL MCC Model (Mixing Controlled Combustion), a model that predicts the rate of heat release (ROHR), NOx and accumulated Soot emissions based on the quantity of fuel injected in the cylinder and the turbulent kinetic energy introduced by the injection process. To calculate the injection rate of the combustion process, the normalized rate of injection (ROI) was used. This determines the fuel injection rate with respect to time [17].

The simulation model was built based on the single engine Cylinder Research Engine 5402 with the following technical data :

Table 2. 5402 Engine Technical Data

Parameter	Value	M.U.
Number of cylinders	1	[-]
Bore x Stroke	85 x 90	[mm]
Compression ratio	17.1	[-]
Maximum power	6	[kW]
Rated speed	4200	[min^{-1}]
Combustion system	4 valve	[-]
Displacement	510.7	[cm^3]

Based on the data from table 2, a simulation model was created in AVL boost. The simulation model in AVL BOOST was built by adding the constructive elements presented in the table above, respecting the technical data of the AVL 5402 single-cylinder engine. The simulation model presented in figure 3 contains the following elements: SB1; SB2 - Boundary elements, MP1; MP2 - measuring points, C1 - cylinder, CAT1 - catalyst, PF1 - particulate filter.

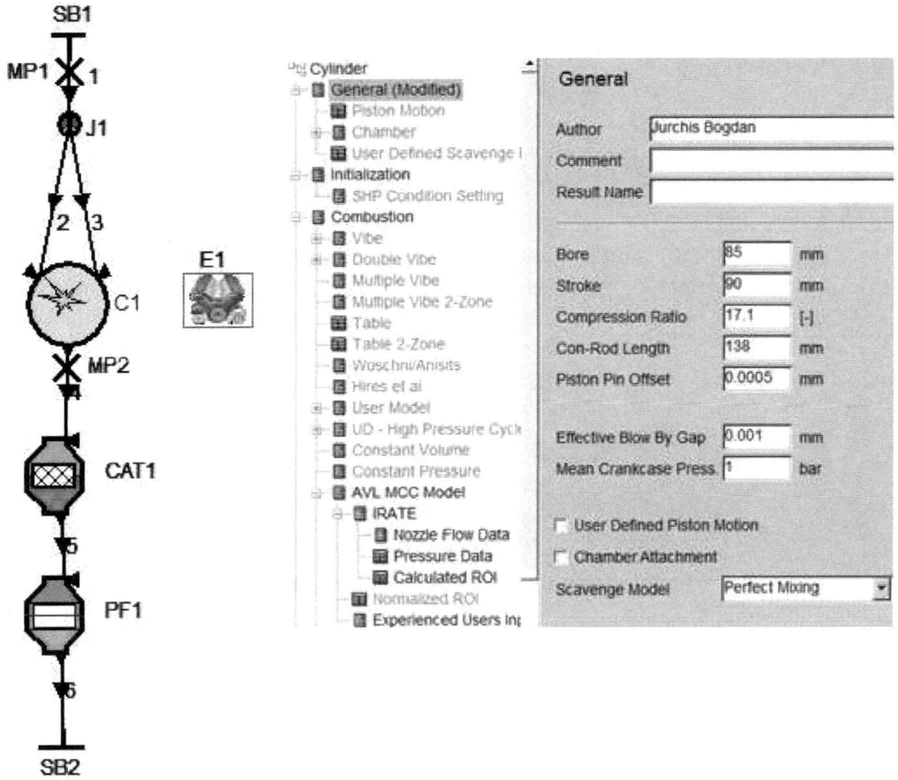

Figure 3 The AVL BOOST simulation model

3.2 Model Validation

The model was validated by comparing the pressure curves at 2000 rpm between the simulation and the experimental case, for the validation model we used pure diesel as fuel (figure 4). Based on the above

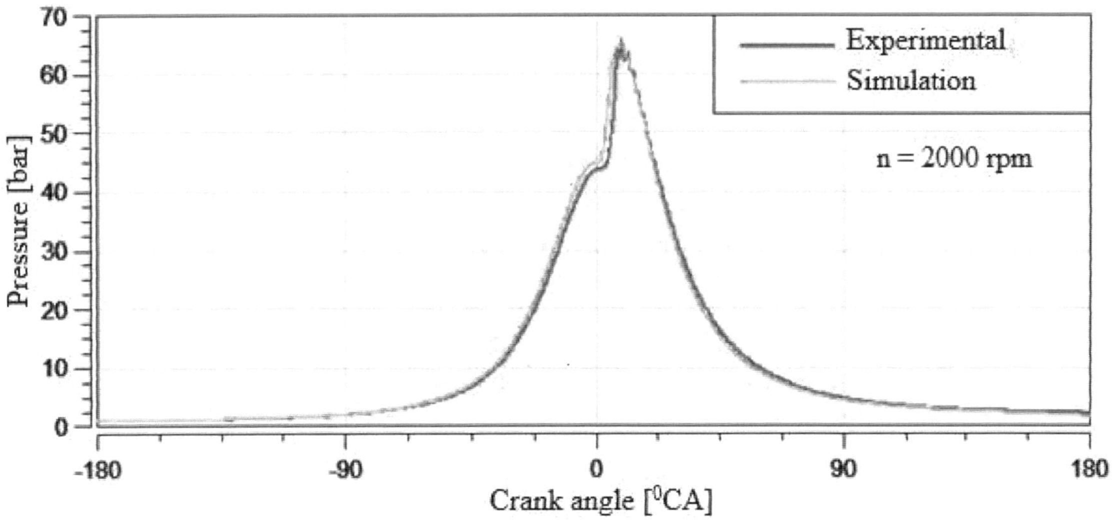

Figure 4 Model validation at 2000 min^{-1} fueled with pure diesel

chart it was considered that the simulation data is in accord with the experimental data and the simulation model can be used to highlight the NOx and PM emissions.

4. Results and discussions

In this paper, the tests that have been carried out have revealed that DME is a fuel that can be taken into account in terms of less polluting fuels. Tests have shown that adding DME favors burning due to increased oxygen content, producing less smoke after burning. The proportion we used in this paper was 10% or 20% DME in a blend with pure diesel at different engine speeds the emissions of soot were reduced by the addition of DME for all engine speeds as shown in the figure 5 at 1500 min^{-1}, figure 6 at 2000 min^{-1} and figure 7 at 2500 min^{-1}. Comparing cases between DME10 and DME20 showed that increasing the proportion of DME reduces the soot accumulation for all engine speed as it can be observed in figure 5, figure 6 and figure 7. From here it can be concluded that DME is a sootless fuel which can be used in a blend with diesel to reduce particulate matter. The change in injection rate did not yield any notable results in terms of particulate emissions.

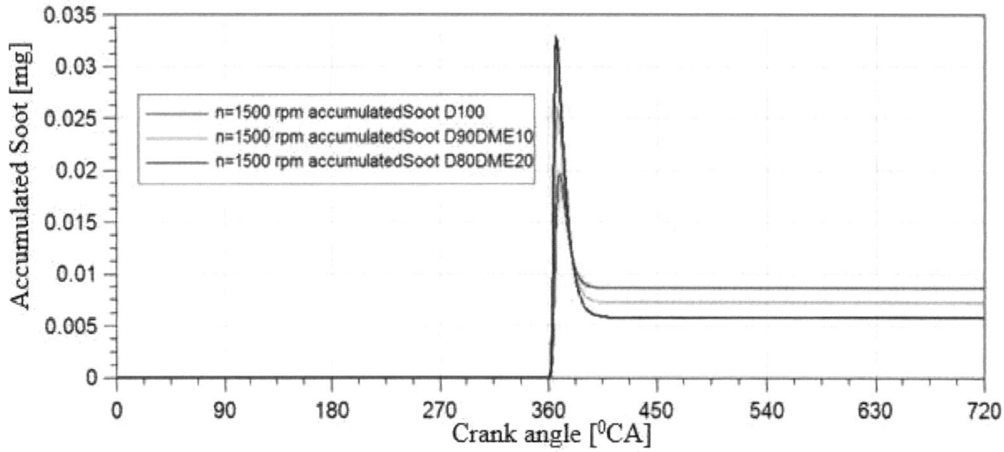

Figure 5 Soot accumulation for n=1500 min^{-1}

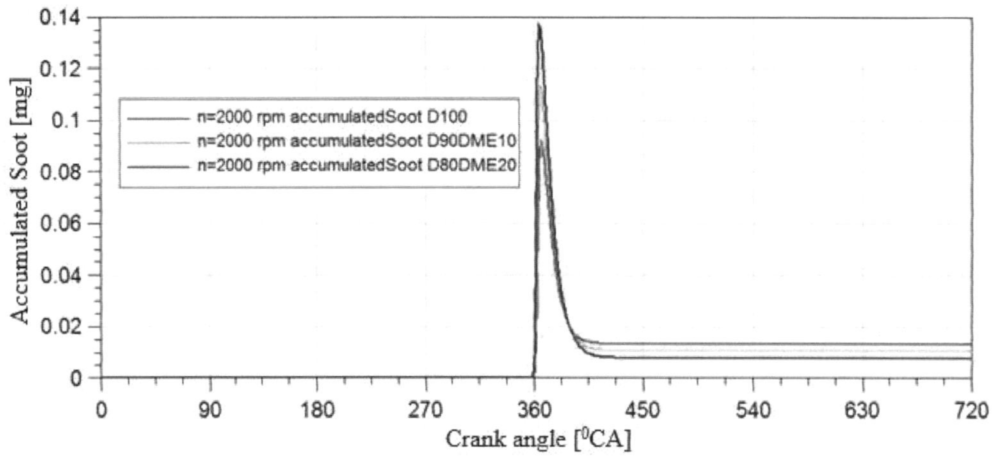

Figure 6 Soot accumulation for n=2000 min^{-1}

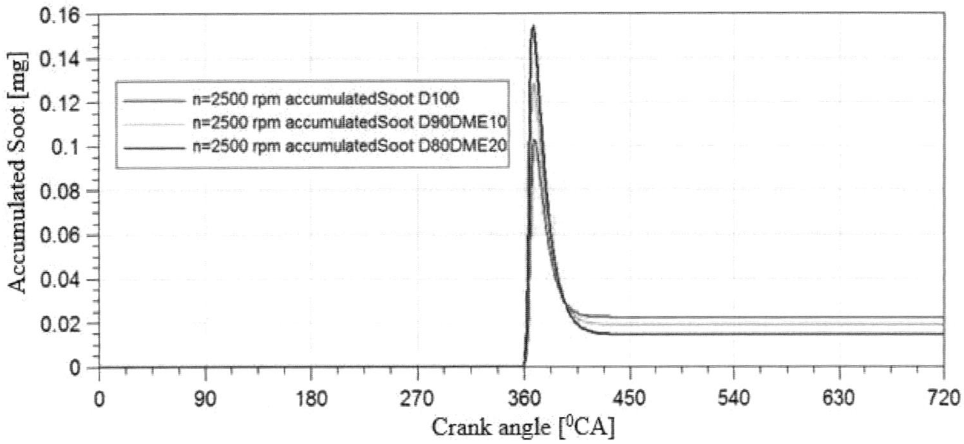

Figure 7 Soot accumulation for n=2500 min^{-1}

Another analysis that has been carried out in this paper is about the NOx emissions produced when operating with a mixture of diesel and DME. The amount of NOx produced was directly proportional to the amount of DME injected, as can be seen in the figure 8, figure 9, and figure 10 for all engine speeds.

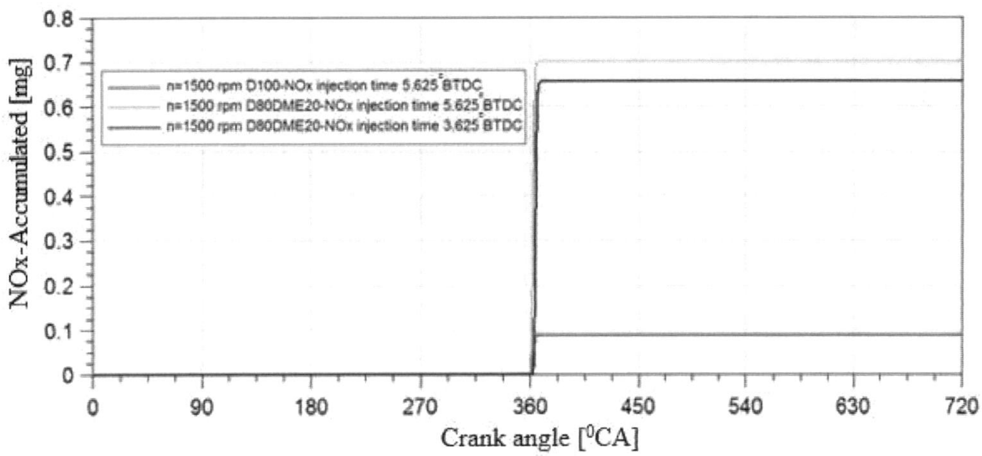

Figure 8 NOx emissions for n=1500 min^{-1}

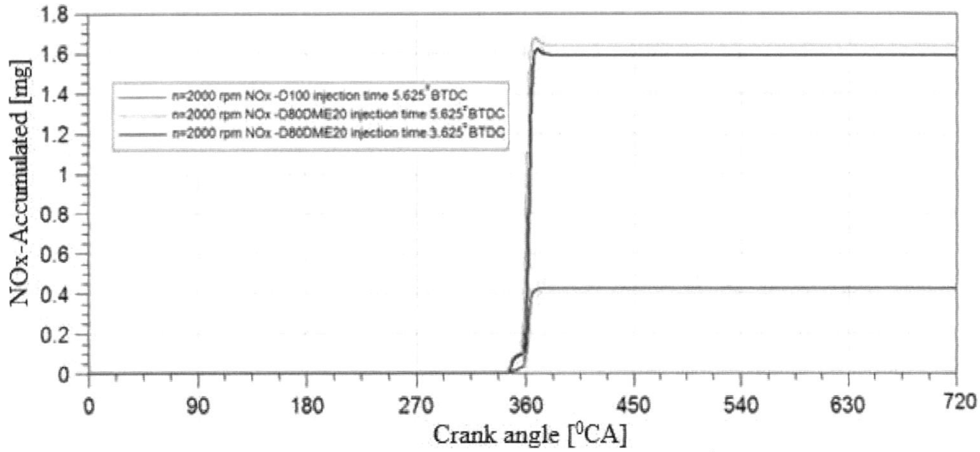

Figure 9 NOx emissions for n=2000 min^{-1}

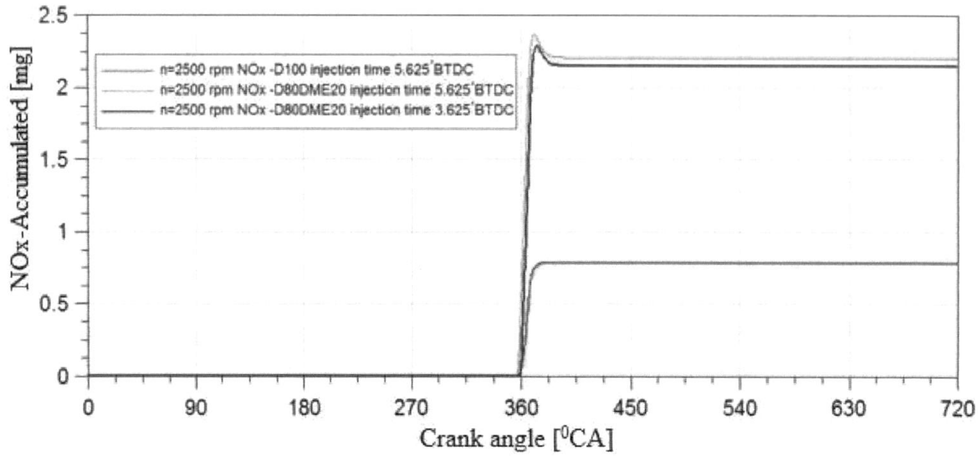

Figure 10 NOx emissions for n=2500 min^{-1}

Tests have shown that NOx emissions have increased slightly compared to pure diesel, and this trend has been kept for all engine speeds, DME20 producing more NOx than DME10. After many simulations it resulted that modifying the injection rate from 5,6250 BTDC to 3,6250 BTDC, brings a NOx reduction but not enough comparable to diesel so that another solutions for lowering NOx have to be considered by adding DME to the diesel.

5. Conclusions

The purpose of this work was to track pollutant emissions of soot and NOx by adding different proportions of DME mixed with diesel.

General conclusions that can be drawn from the tests are:

- DME is a fuel that improves the burning process by producing less soot due to its oxygen content
- Soot emissions decreases if the amount of DME is increased for all engine speeds
- NOx emissions are higher by adding DME compared to diesel for all engine speeds
- It has been noticed that the change in the injection pressure has reduced the NOx emission but not enough compared to diesel, so exhaust gas after-treatment systems such as the gas recirculation valve should be used to engines that are fueled by a mix of DME and diesel

References

[1] ***http://ec.europa.eu/energy/en/topics/renewable-energy
[2] Park S H, Lee C S 2014 Energ Convers Manage **86** 848–863
[3] Arcoumanis C, Bae C, Crookes R, Kinoshita E 2008 Fuel **87** 1014–1030
[4] Semelsberger T A, Borup R L, Greene H L 2006 J Power Sources **156** 497–511
[5] Thomas G, Feng B, Veeraragavan A, Cleary M J, Drinnan N 2014 Fuel Process Technol **119** 286–304
[6] Azizi Z, Rezaeimanesh M, Tohidian T, Rahimpour M R 2014 Chem Eng Process **82** 150–172
[7] Kim H J, Park S H 2016 Fuel **182** 541–549
[8] Zhu Z, Li DK, Liu J, Wei YJ, Liu SH 2012 Appl Therm Eng **35** 9–14
[9] Bîldea C S, Győrgy R, Brunchi C C, Kiss A A 2017 Comput Chem Eng **105** 142-151
[10] Jang J, Lee Y, Cho C, Woo Y, Bae C 2013 Fuel **113** 617–624
[11] Ying W, Longbao Z 2008 Appl Therm Eng **28** 1589–1595
[12] Ohno Y, Inoue N, Ogawa T, Ono M, Shikada T, Hayashi H 2001 NKK Technical Review **85**

23–28
[13] Ying W, Genbao L, Wei Z, Longbao Z 2008 Fuel Process Technol **89** 1272–1280
[14] Park W, Park S, Reitz R D, Kurtz E 2017 Combust Flame **180** 276–283
[15] Kim M Y, Yoon S H, Ryu B W, Lee C S 2008 Fuel **87** 2779–2786
[16] Park S H, Lee C S 2013 Prog Energ Combust **39** 147–168
[17] Iclodean C D, 2013 Studii şi cercetări privind influenţa parametrilor gestionaţi de catre unitatea electronică de control asupra motoarelor cu ardere internă care funcţionează cu combustibili alternativi (doctoral thesis). Universitatea Tehnică din Cluj-Napoca
[18] ***http://www.ascension-publishing.com/BIZ/DMEoverview.pdf.

IOP Conf. Series: Materials Science and Engineering **252** (2017) 012066 doi:10.1088/1757-899X/252/1/012066

Influences of diesel pilot injection on ethanol autoignition - a numerical analysis

N V Burnete[1], N Burnete[1], B Jurchis[1] and C Iclodean[1]

[1] Technical Univeristy of Cluj-Napoca, Faculty of Mechanics, Department of Automotive Engineering and Transports, Str. Muncii 103-105, Cluj-Napoca, Romania

E-mail: nicolae.vlad.burnete@auto.utcluj.ro

Abstract. The aim of this study is to highlight the influences of the diesel pilot quantity as well as the timing on the autoignition of ethanol and the pollutant emissions resulting from the combustion process. The combustion concept presented in this paper requires the injection of a small quantity of diesel fuel in order to create the required autoignition conditions for ethanol. The combustion of the diesel droplets injected in the combustion chamber lead to the creation of high temperature locations that favour the autoignition of ethanol. However, due to the high vaporization enthalpy and the better distribution inside the combustion chamber of ethanol, the peak temperature values are reduced. Due to the lower temperature values and the high burning velocity of ethanol (combined with the fact that there are multiple ignition sources) the conditions required for the formation of nitric oxides are not achieved anymore, thus leading to significantly lower NOx emissions. This way the benefits of the Diesel engine and of the constant volume combustion are combined to enable a more efficient and environmentally friendly combustion process.

1. Introduction

In light of the current efforts to clean the air in cities all around Europe by banning cars with diesel engines due to their high levels of particulate matter (PM) and nitric oxides (NO_x), the current study proposes a solution that could reinvent the Diesel process with the aid of renewable fuels, in this case bioethanol [1]. Even if the current solution implies the use of a small quantity of diesel – that anyway represents under 10% of the total amount of fuel supplied to the cylinder – for future solution the diesel fuel could be replaced with renewable solutions like dimethylether, which has been proved by many researchers to be a viable solution for diesel [2-5]. A survey of the available literature has showed that there are numerous studies about the use of ethanol fumigation [6-9] and diesel-ethanol blends [10-15], but very few studies that cover such a process that involves a separate direct injection of both fuels (diesel and ethanol), in which the diesel pilot injection is used to create the required conditions for the autoignition of a main ethanol injection, which is believed to be due to its higher complexity. The aim of this solution is to reduce the NO_x and soot emissions, while taking advantage of the higher efficiency of the Diesel process. If one takes into account that the current solutions to address these pollutants are selective catalytic reduction (SCR) and/or NO_x traps for nitric oxides and diesel particulate filters (DPF) for the PM, which are actually aftertreatment solutions, the solution presented above becomes a very attractive alternative because it affects the combustion itself. The current paper is continuation of a previous study in which the solution was compared with pure diesel combustion [1]. The main limitation of this study is that, currently, there is no experimental test bench available to back up the simulation results, but this is being addressed by the authors with the development of such a test bench.

IOP Conf. Series: Materials Science and Engineering **252** (2017) 012066 doi:10.1088/1757-899X/252/1/012066

2. Methodology

The study of the proposed combustion process involving the autoignition of a main ethanol injection with the aid of a diesel pilot injection required the development of a specific simulation strategy which was presented in [1]. This strategy involved the following steps: a) Run the simulation with diesel fuel to obtain the reference case; b) Run a second simulation with diesel fuel up to the main injection event; c) Stop the simulation just before the main injection and change the fuel from diesel to ethanol; d) Run the simulation from the point of main injection. The sector simulation model is based on a single cylinder research engine (SCRE) AVL 5402 with the following characteristics:

Table 1 Technical data of the engine

	Value	M.U.
Bore x Stroke	85 x 90	[mm] x [mm]
Compression ratio	17.1	[-]
Displacement	510.7	[cm^3]
Intake	1 Tangential port 1 Swirl port	
Power (naturally aspirated)	6	[kW]
Injection system	BOSCH Common Rail	
Injector (injection holes x diameter)	8 x 0.12	[-] x [mm]
Injection pressure	80	[MPa]

Following the development of the simulation model an extensive validation for diesel and a blend of diesel and ethanol was done for different speeds and loads, which is presented in [1]. To improve accuracy of the input data a more detailed simulation model was also used [16, 17]. The experimental data required for the testing was gathered using the equipment in the Laboratory of research, testing and homologation of internal combustion engines (TESTECOCEL) of the Department of Automotive and Transport within the Technical University of Cluj-Napoca [18]. The set-up of this laboratory for the SCRE used for this study is presented in Figure 1.

Figure 1. TESTECOCEL Laboratory set-up for SCRE AVL 5402

In the previous studies the injection timing and pilot quantities were the same as for the conventional Diesel combustion. To further the knowledge regarding the proposed combustion process, for this study the pilot injection parameters quantity and timing where changed to identify their influence on the combustion of the main ethanol injection. The influence was evaluated with respect to the temporal evolution of in-cylinder pressure and temperature, rate of heat release, accumulated heat released, ignition delay and pollutant emissions (NO and soot), as well as to the spatial evolution of the equivalence ratio, temperature and NO. The studied cases are presented in table 2. The engine speed was the same in all cases: 1500 [min^{-1}].

Table 2 Description of the simulated cases

D100_E100 case	Pilot		Main	
	Quantity [mg]	Timing [°CA] bTDC	Quantity [mg]	Timing [°CA] bTDC
0.5mg_24mg	0.5	21	24	0,6
0.5mg_24mg_4degCA	0.5	25		
1mg_24mg	1	21		
1mg_24mg_4degCA	1	25		
2mg_24mg	2	21		
2mg_24mg_4degCA	2	25		

The limitations of the study arise from the employed simulation strategy, which introduces an error in the total amount of energy and in the absolute value of pollutant emissions. However, a closer analysis of the results revealed that this error is lower than 3% in all cases [1].

3. Results and discussions

3.1. Quantity

For a better understanding of the phenomenon the analysis has been split in two parts: temporal evolution – where the parameters of interest are analyzed based on their evolution during the cycle – and spatial evolution – where cuts inside the cylinder at different crank angles are done to show the distribution of fuel, temperature and NO.

3.1.1. Temporal evolution. The temporal analysis of the results showed that the highest pressure was obtained in the case D100_E100_2mg_24mg (Figure 2), which was to be expected, since in this case the total fuel quantity is the highest. Due to the high diesel pilot quantity there is also a high pressure and temperature (Figure 3) increase, which leads to favourable autoignition conditions and a sufficient amount of energy for the main ethanol injection. As a result the rate of heat release shows lower peak values and a less steep increase compared to the other two cases.

The second highest pressure was obtained in the D100_E100_0.5mg_24mg. This is due to the fact that the lower pilot quantity leads to a larger ignition delay of the second stage of ethanol combustion. The heat released by the 0.5 [mg] pilot is enough to lead to the autoignition of the first ethanol drops, but, as the main injection of ethanol continues, the energy required for combustion is competing with the energy required for the evaporation of the fuel. This then leads, earlier than in the D100_E100_1mg_24mg case, to a stagnation of rate of heat release, until the ethanol is sufficiently mixed with air and heat is released again. Due to the fact that during this second stage of combustion the burnt quantity is larger in the D100_E100_0.5mg_24mg case than in the D100_E100_1mg_24mg case and due to the high burning velocity of ethanol, there is a rapid pressure increase that leads to a higher peak pressure. In the D100_E100_1mg_24mg case the stagnation is a bit more retarded and, as a result, the amount of fuel that is burnt in the second stage is smaller and the combustion evolves with an inferior speed. The phenomenon is also visible in the temperature curves (Figure 2). It must be noted that the maximum mean temperature in the cylinder corresponds to the amount of energy supplied to the cycle with the highest value for D100_E100_2mg_24mg and the lowest value for D100_E100_0.5mg_24mg.

A closer analysis also shows that, in the case D100_E100_2mg_24mg the combustion takes place in a single stage and not in two stages like in the other two case with 0.5 and 1 [mg] diesel pilot. The less violent combustion would also lead to lower mechanical stress of the engine components. However, in order to have a complete understanding, this would require additional experimental results.

The temporal development of combustion determines the end products of the process. Since the parameters of the pilot injection have a strong influence on the combustion process, the emissions vary significantly (Figure 4).

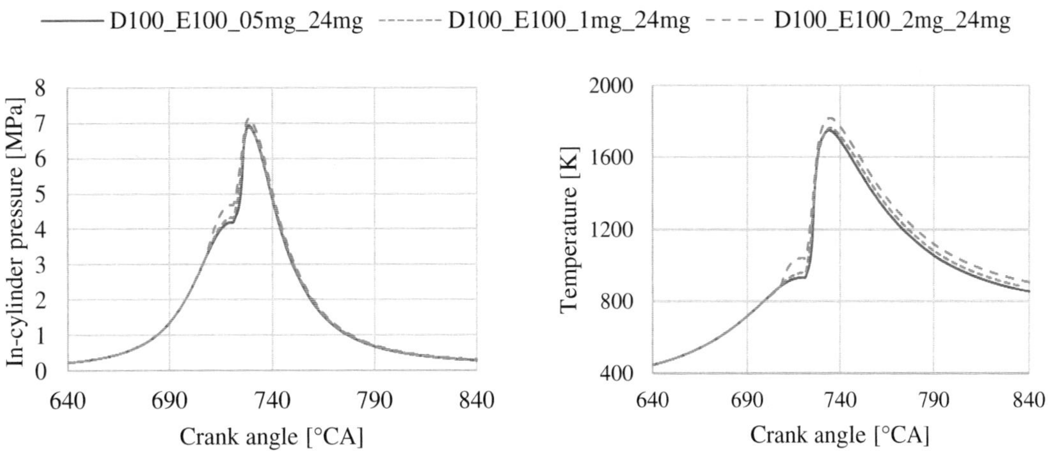

Figure 2. In-cylinder pressure and temperature evolution for a pilot advance of 21 [°CA]

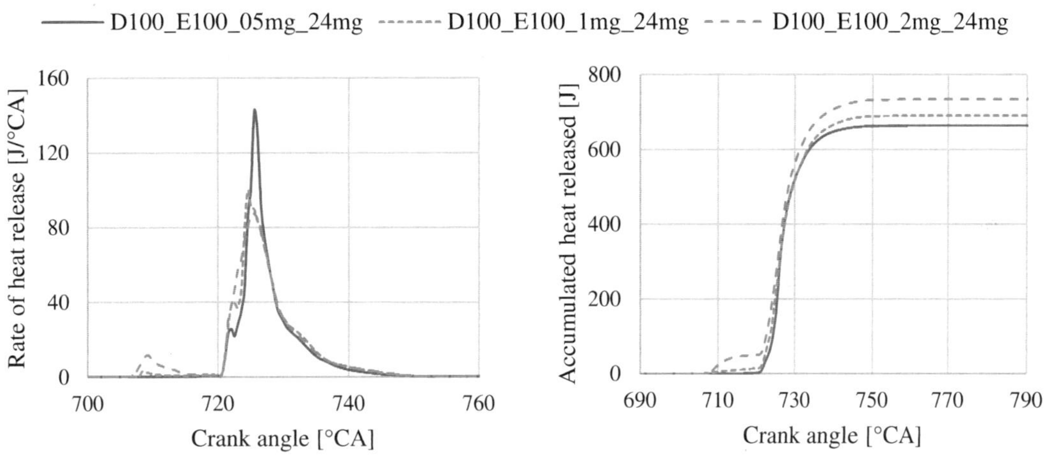

Figure 3. The rate of heat release and the accumulated heat released for a pilot advance of 21 [°CA]

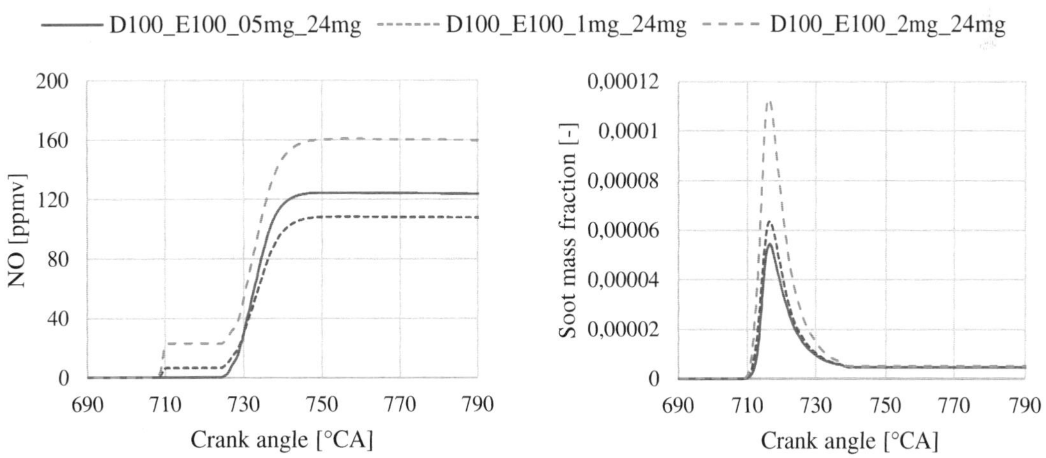

Figure 4. The temporal evolution of NO and soot for a pilot advance of 21 [°CA]

The analysis of NO showed that the highest values are obtained in the D100_E100_2mg_24mg, which is a direct result of the higher diesel quantity. The more interesting aspect is that the lowest values were obtained in the D100_E100_1mg_24mg case. The larger values in the D100_E100_0.5mg_24mg case are however a result of the larger ignition delay and of the larger quantity burnt in the second stage of the process. With respect to soot, the influence of the pilot quantity on the end values is very small.

3.1.2. Spatial evolution. In Figure 5 is presented the spatial distribution of the fuel (equivalence ratio), temperature and NO at 723, 730 and 740 [°CA] for the three cases with a pilot advance of 21 [°CA].

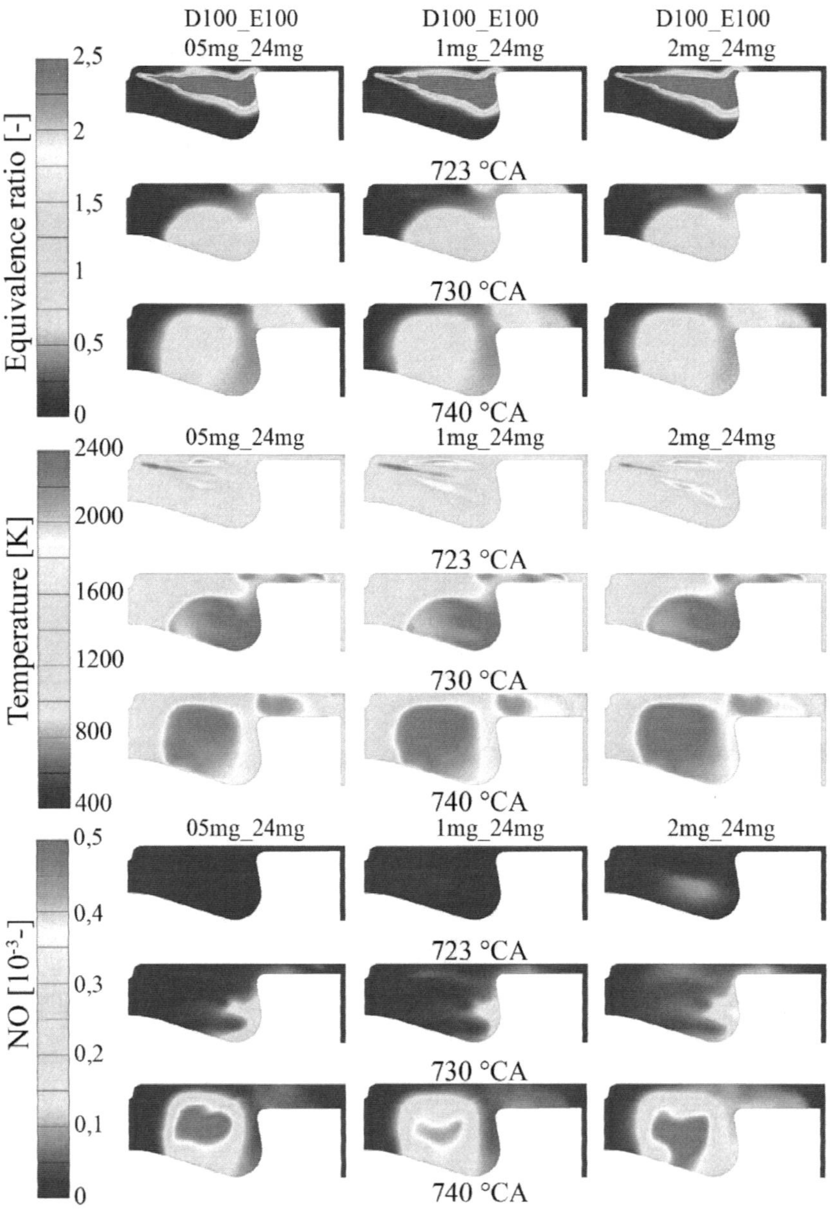

Figure 5. Equivalence ratio, temperature and NO spatial distribution at different crank angles for a pilot advance of 21 [°CA]

For the D100_E100_2mg_24mg case higher NO values (see Figure 5) are to be expected due to the higher amount of diesel fuel that leads to a higher pressure and higher local temperatures in the combustion chamber. The other two cases require a more detailed analysis of the images combined with the temporal evolution of the variables of interest. The point of interest for these two cases is around 730 [°CA] because this were the NO values in the D100_E100_05mg_24mg case overcome the values of the D100_E100_1mg_24mg case. The combined analysis shows that the larger amount of fuel that burns in the second stage of the D100_E100_05mg_24mg case leads to higher local temperatures (see Figure 5 for 730 [°CA]) that then lead to higher NO formation rates.

3.2. Timing

The second part of this study was the effect of the pilot timing on the variables of interest. In this sense, the pilot diesel injection advance was increased by 4 [°CA] for all three quantities.

3.2.1. Temporal evolution. Compared to the initial cases with an advance of 21 [°CA], when the advance is increased to 25 [°CA], the main difference that can be noticed is for the D100_E100_05mg_24mg_4degCA – where 4degCA designates the additional 4 [°CA] compared to the first three cases. Between the other two cases D100_E100_1mg_24mg_4degCA and D100_E100_2mg_24mg_4degCA one can observe similar differences as presented at 3.1.1.
By increasing the advance, in case D100_E100_05mg_24mg_4degCA one can notice a very late combustion which would seriously impair the operation of the real engine. In this case the combustion starts at 746.5 [°CA], which can be seen in the cylinder pressure and temperature as well in the rate of heat release. A closer analysis of this revealed that the combination of small pilot quantity, increased ignition delay and charge movement leads to a much to lean mixture that cannot ignite. The injection of ethanol aids in the formation of a rich enough mixture, but the high enthalpy of vaporization further increases the ignition delay. When it finally starts, the combustion evolves with a very low heat release rate and as a result the accumulated heat released is much lower than in the other cases. Also, the temperature is much too low and therefore, the end NO values are very low.

When comparing the values of the variables of interest of the cases with 21 [°CA] advance with those of the cases with 25 [°CA] there are only very small differences with respect to in-cylinder pressure and temperature. The increase pilot injection advance of the D100_E100_1mg_24mg_4degCA case revealed a single stage combustion with a lower peak value compared to the D100_E100_1mg_24mg case, which can be seen in the rate of heat release. The end values of NO and soot are somewhat higher in this case, but the difference is negligible.

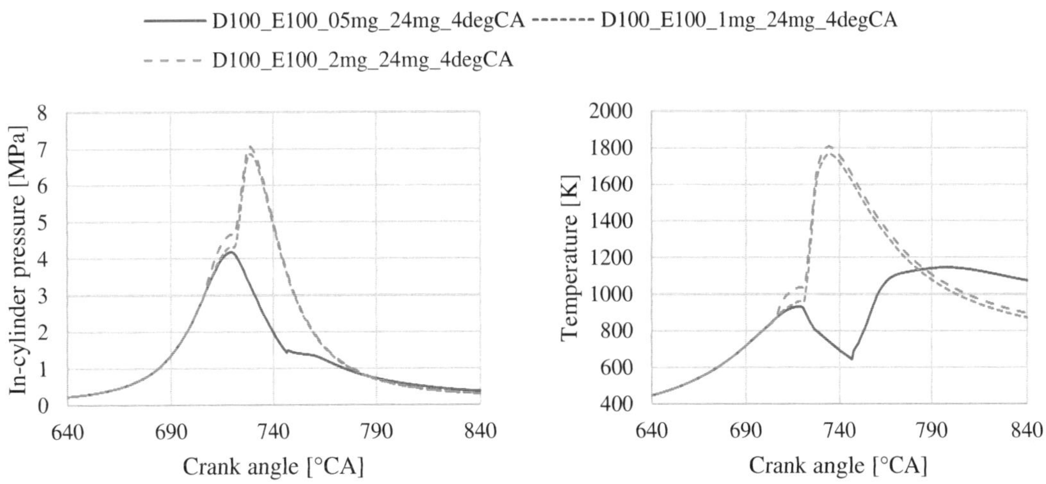

Figure 6. In-cylinder pressure and temperature evolution for a pilot advance of 25 [°CA]

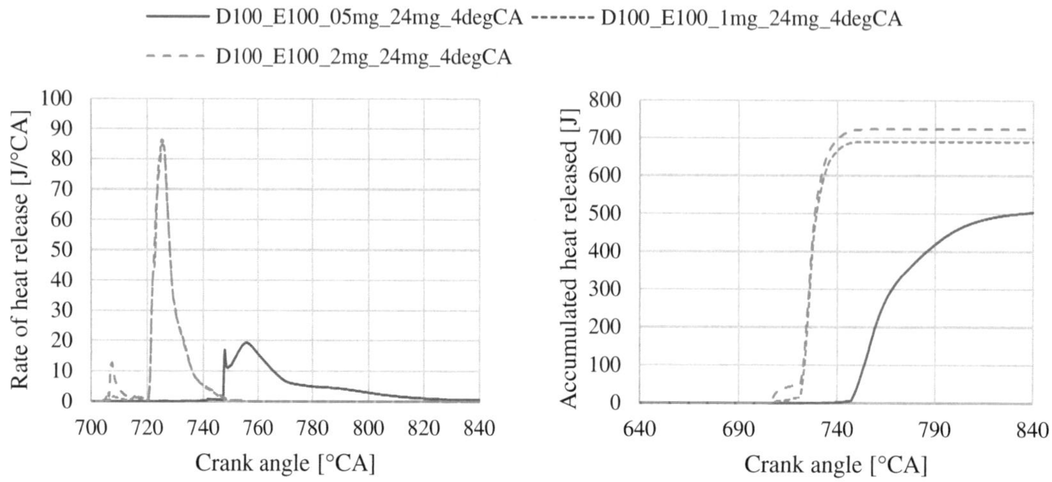

Figure 7. The rate of heat release and the accumulated heat released for a pilot advance of 25 [°CA]

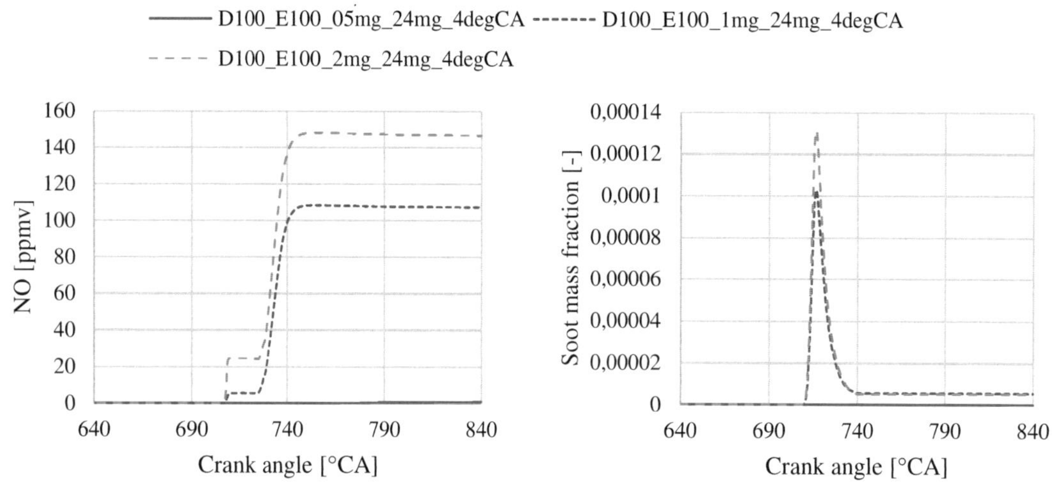

Figure 8. The temporal evolution of NO and soot for a pilot advance of 25 [°CA]

Between the D100_E100_2mg_24mg_4degCA case and the D100_E100_2mg_24mg case the trends are similar with one difference: the NO values decrease by approximately 8% when the pilot advance is increased, which is a result of the lower ignition delay of the main injection of ethanol.

3.2.2. Spatial evolution. The analysis of the spatial distribution of fuel, temperature and NO shows that there is a negligible influence of the increased pilot injection advance in the cases with 1 and 2 [mg] of diesel pilot quantity. However, it needs to be mentioned that the lower NO values obtained in the D100_E100_2mg_24mg_4degCA case compared to the D100_E100_2mg_24mg case are a result of the fewer points with temperatures, thus affecting the nitric oxides formation rates.

As highlighted above, for the D100_E100_05mg_24mg_4degCA, there is no combustion of the diesel pilot which can be observed in the temperature values inside the combustion chamber. As a result, the main ethanol injection is more compact and does not impinge on the piston wall like in the other cases. Also, by combining the cooling effect of the fuel vaporization and the increase of the combustion chamber volume, the ignition delay is further increased. This results in a very late combustion with a very low efficiency that would not be desired in the real engine. Due to the very low local temperatures the NO emissions that result are negligible.

As in the cases with 21 [°CA] and as it was highlighted at 3.1.3, the highest NO concentrations are obtained in the case with 2 [mg] diesel pilot quantity. Even if the combustion of ethanol produces very similar amounts of NO in both cases, D100_E100_1mg_24mg_4degCA and D100_E100_2mg_24mg_4degCA, the higher values of nitric oxide resulting from the combustion of 2 [mg] of diesel lead to a higher end value.

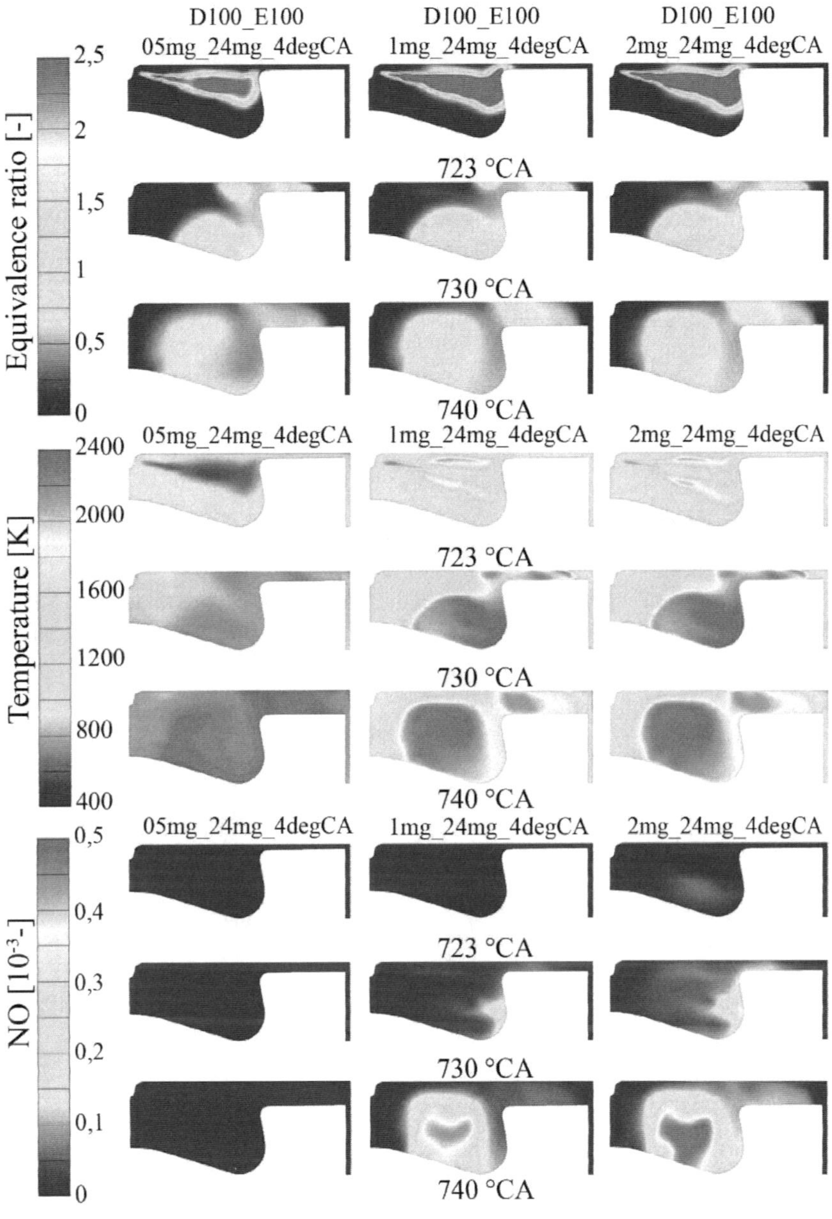

Figure 9. Equivalence ratio, temperature and NO spatial distribution at different crank angles a pilot advance of 25 [°CA]

4. Conclusions

This study comes to further the knowledge and understanding of a newly proposed combustion process that involves a pilot diesel injection and a main ethanol injection. The idea is to use combine the higher efficiency of the Diesel process with a more clean combustion of a renewable fuel with the aim of considerably reducing the NO emissions. In the sense mentioned above three diesel pilot quantities – 0.5, 1 and 2 [mg] – with two different timings – 21 and 25 [°CA] – were studied with the aid of simulation. Based on the obtained results the following main conclusions can be drawn:

- for an advance of 21 [°CA] the a pilot quantity of 0.5 and 1 [mg] leads to a two stage combustion of ethanol; a pilot quantity of 2 [mg] leads to a single stage combustion;
- increasing the advance to 25 [°CA] causes a single stage combustion in the case with 1 [mg] of diesel pilot as well;
- and advance of 25 [°CA] combined with a high ignition delay leads to too lean conditions that cause the pilot injection to not burn; as a result, the ignition delay of the main ethanol injection is very high and the combustion takes place with a very poor efficiency;
- the lowest NO emissions were obtained for a diesel pilot quantity of 1 [mg].

Based on the above analysis it was concluded that for this case, with a main injection quantity of 24 [mg] of ethanol the optimum diesel pilot quantity is 1 [mg]. Also, this shows that there is a strong correlation between the diesel pilot quantity, diesel pilot advance – by using the pilot injection, the required autoignition conditions of ethanol are created – and the combustion of the main ethanol injection.

ACKNOWLEDGEMENT
The results presented in this paper were obtained with the support of the Technical University of Cluj-Napoca through the research Contract no. 2014/12.07.2017, Internal Competition CICDI-2017.
This work would have not been possible without the hardware and software of the AVL LIST GmbH Company, available as part of the partnership between AVL LIST GmbH Company and the Technical University of Cluj-Napoca.

References

[1] Burnete N V 2017 *Therm Sci* **21** 451-63
[2] Kim H J and Park S H 2016 *Fuel* **182** 541-9
[3] Lamani V T, Yadav A K and Narayanappa K G 2017 *Environ Scie Pollut R* **24** 15500-9
[4] Lee U, Han J, Wang M, Ward J, Hicks E, Goodwin D, Boudreaux R, Hanarp P, Salsing H, Desai P, Varenne E, Klintbom P, Willems W, Winkler S L, Maas H, De Kleine R, Hansen J, Shim T and Furusjo E 2016 *Sae International Journal of Fuels and Lubricants* **9** 546-57
[5] Thomas G, Feng B, Veeraragavan A, Cleary M J and Drinnan N 2014 *Fuel Process Technol* **119** 286-304
[6] Mariasiu F, Burnete N V, Moldovanu D, Varga B O, Iclodean C and Kocsis L 2015 *Therm Sci* **19** 1931-41
[7] Hansdah D and Murugan S 2014 *Fuel* **130** 324–33
[8] López A F, Cadrazco M, Agudelo A F, Corredor L A, Vélez J A and Agudelo J R 2015 *Fuel* **153** 483–91
[9] Chauhan B S, Kumar N, Pal S S and Du Jun Y 2011 *Energy* **36** 1030–38
[10] Burnete N V, Filip N and Barabás I 2015 *Rom J Automot Eng* **21** 89-106
[12] Rakopoulos D C, Rakopoulos C D, Kakaras E C and Giakoumis E G 2008 *Energ Convers Manage*, **49** 3155–62
[13] Armas O, Mata C and Martinez-Martinez S 2012 *Int J Engine Res*, **13** 417–28
[14] Hansen A C, Zhang Q and Lyne P W L 2005 *Bioresource Technol*, **96** 277–85
[15] Torres-Jimenez E, Jerman M S, Gregorc A, Lisec I, Dorado M P and Kegl B 2011 *Fuel* **90** 795–802
[16] Moldovanu D and Burnete N 2013 *Therm Sci* **17** 195-203
[17] Burnete N V, Iclodean C, Moldovanu D and Filip N 2017 *Conat 2016: International Congress of Automotive and Transport Engineering* 379-86
[18] Burnete N V, Iclodean C, Popescu G and Filip N 2017 *Conat 2016: International Congress of Automotive and Transport Engineering* 387-94

The comparative study of vehicles engine in different functioning situations

M Nicolau[1], C I Pantece[1], G I Dorobantu[2], V J Radulescu[3] and I Copae[1]
[1]Military Technical Academy, Romania
[2]Debt Management Expert, Romania
[3]Metrorex S.A., Romania
Corresponding author's email: **nicolaumarin@yahoo.com**

Abstract: The paper presents a comparative study of a Diesel engine functioning in different situations and at different partial loads and engine speeds. There are shown graphs of some experimental results, obtained by using concepts and algorithms from mathematical statistics. It is presented a study of functional factors influence on engine dynamics and fuel saving, by using correlation analysis, variance analysis, information theory and sensitivity analysis. There are highlighted characteristics of vehicle driving in urban and extra-urban environment.

1. Introduction

The study of engine dynamical behavior has developed new possibilities due to electronic control of engine parameters. The in-depth of phenomenas that appear during functioning can be studied based on informations sent from on-board computer, which implies a systemic and interdisciplinary handling of modern engine dynamics, approach resulted from the engine structural and of control complexity. In view of all this, the paper performs an analysis of experimental data acquired from embedded sensors which were stored by the on-board computer; the analysis of experimental data offers the possibility to establish engine dynamics and fuel saving performances.

2. Experimental research

Experimental researches were conducted with a Ford Focus automobile equipped with a supercharged Diesel engine and a "common rail" supply unit. For the acquisition of engine and vehicle functional parameters were used the Ford's FoCOM software and interface [1, 2]. The tests allowed the acquisition and storage of a certain number of experimental data, whereat are added those obtained by calculus. From all data were kept the 80 most significant experimental samples, which marked a normal driving; from these, 40 samples represent urban driving, and the other 40, extra-urban driving. One sample contains 4000 values for each parameter, the time period is of 400 s; results a measuring rate of 10 values/s.

3. Data statistical processing

Follow-up are presented some of the results, which contain instantaneous values of functional parameters, as it is presented in figure 1 for vehicle driving speed V and in figure 2 for engine speed n. Likewise, by using first order statistical characteristics, in graphs were are also presented minimal, average and maximal values of vehicle speed and engine speed, in case of urban and extra-urban driving. Graphs from figure 1 and figure 2 show that in case of extra-urban environment the values of vehicle speed and engine speed are higher than in urban environment, which was expected.

For example, from figure 1a results that in case of urban environment all 160000 values are in the range 0-70 km/h; instead, from figure 1b it can be observed that in extra-urban environment 86.8% of all values are in a superior interval, meaning in the range 70-130 km/h.

Similarly, from figure 2 results that in urban environment 34.6% of values are in the range 1500-2500 rev/min, and in case of extra-urban environment there are 85.6% of values within the interval.

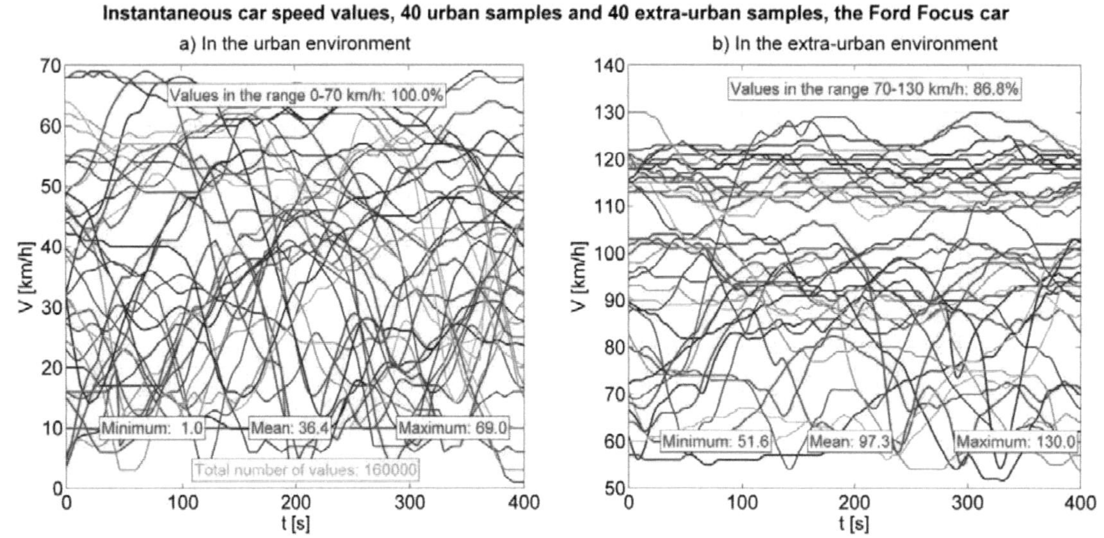

Figure 1.

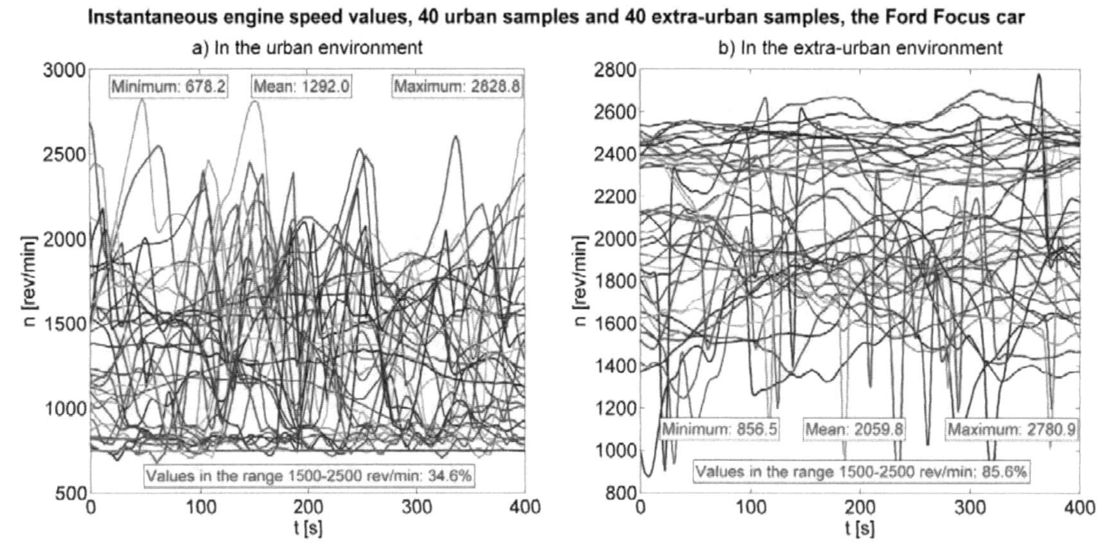

Figure 2.

In figure 3 are presented instantaneous, minimal, maximal and average values of throttle shutter's position *p*, stated as a percentage of the maximum opening. As it can be seen from graphs, throttle shutter's position has also higher values in extra-urban environment relative to urban environment.

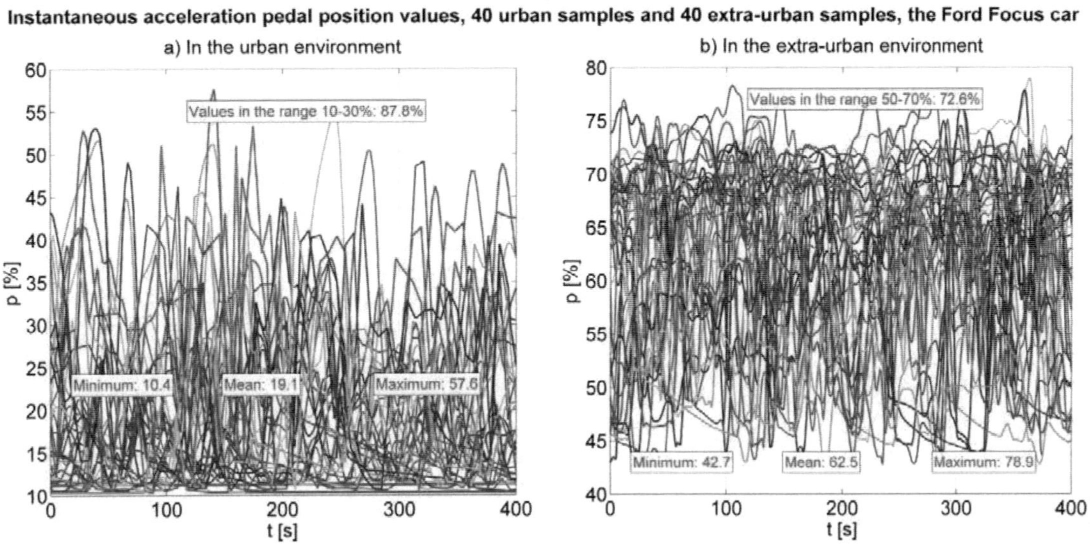

Figure 3.

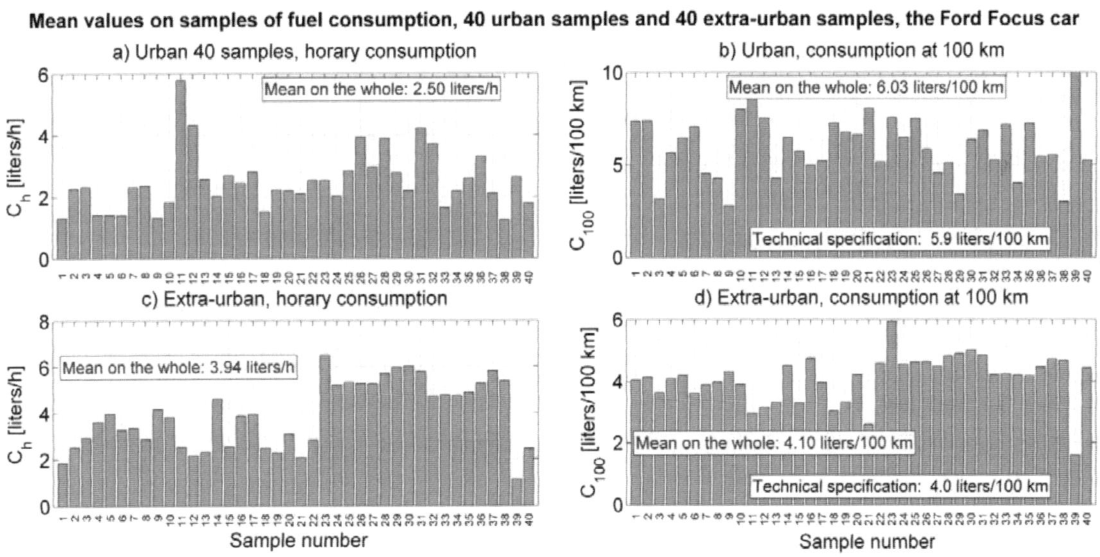

Figure 4.

In figure 4 are shown average values of fuel consumption; graphs from the left illustrate engine hourly fuel consumption C_h, and those from the right contain fuel consumption at 100 km C_{100}. From these graphs results a higher hourly consumption in case of urban driving, and a fuel consumption at 100 km lower in case of extra-urban driving.

The explanation results from formula [3]:

$$C_{100} = 100 C_h / V \tag{1}$$

Indeed, even if hourly consumption is higher for extra-urban driving, fuel consumption at 100 km is lower due to higher speeds in this case (the effect of speed is predominant).

In figure 4 are shown the values of fuel consumption at 100 km, from vehicle's technical specifications 5.9 liters/100 km for urban driving and 4 liters/100 km for extra-urban driving.

Graphs from figure 5 present average values of the quantity of recirculated gas γ and of rotational angle of turbine vanes α. From these graphs it can be observed that for all samples, the two sizes are smaller in case of extra-urban driving.

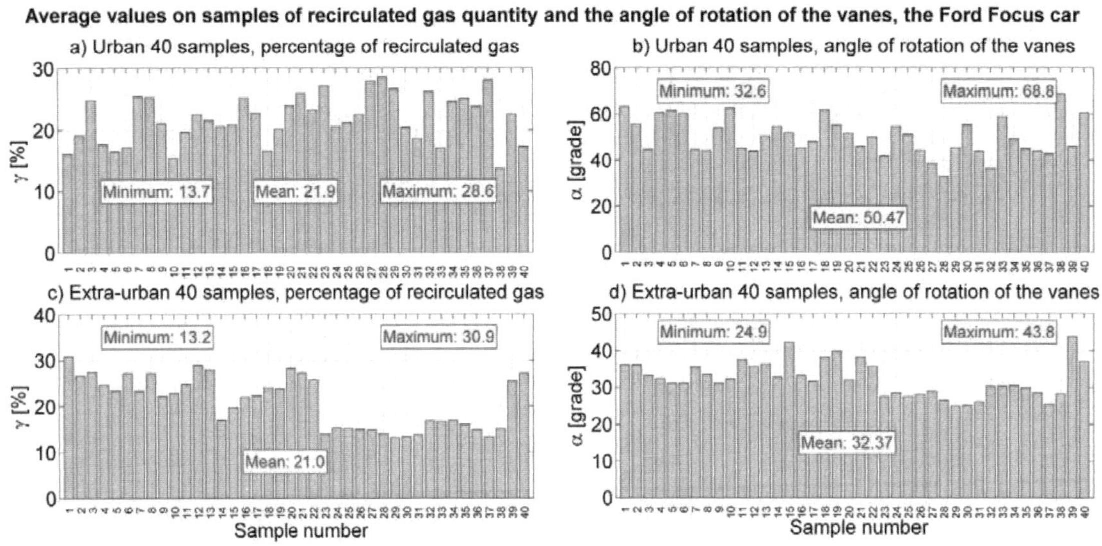

Figure 5.

In figure 6 are presented the histograms of engine torque M_e and engine power P_e, in case of urban and extra-urban driving. Upper graphs show a placement of the two sizes closer to lower values in case of urban driving; in lower graphs the sizes are closer to average values in case of extra-urban driving.

Likewise, from figure 6 it can be seen that the two sizes are subject of normal distribution, which confirms that engine torque motor from figure 7 with the values of repartition function $F(x)$ where is also applied the Smirnov-Kolmogorov test, noted S-K [3, 4], as well as from figure 8 where are shown the values for density of probability $f(x)$. As it can be seen from figure 7, the experimental curve is different from the ideal one and the data are not subject of normal distribution.

Likewise, as it can be seen from figure 8, curve functions $f(x)$ do not have a single maximum placed at the average value M_{em}, meaning the so called Gauss bell; in the lower part of graphs from figure 8 it is shown the placement of experimental values in order to present a sugestive image for density of probability.

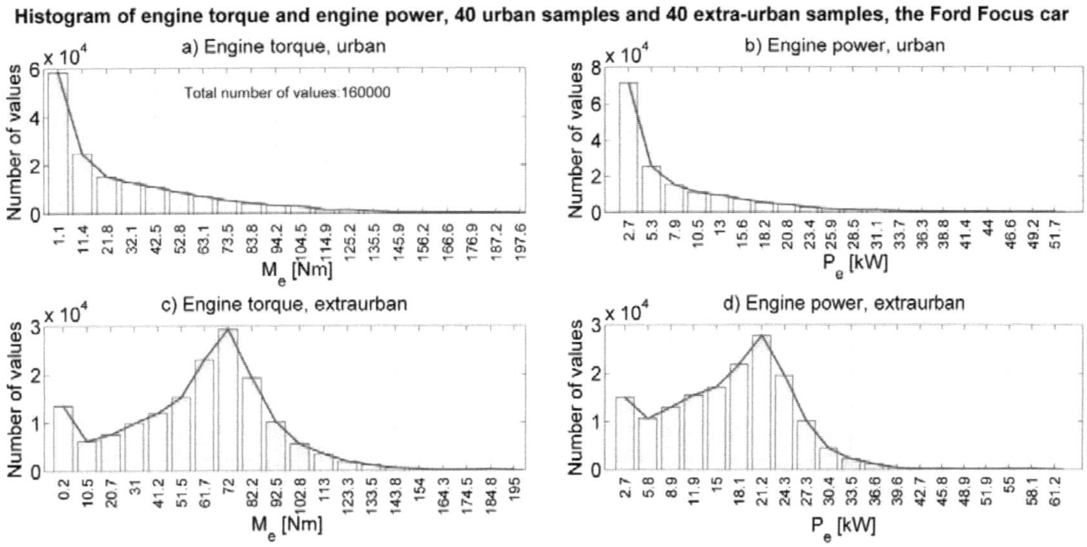

Figure 6.

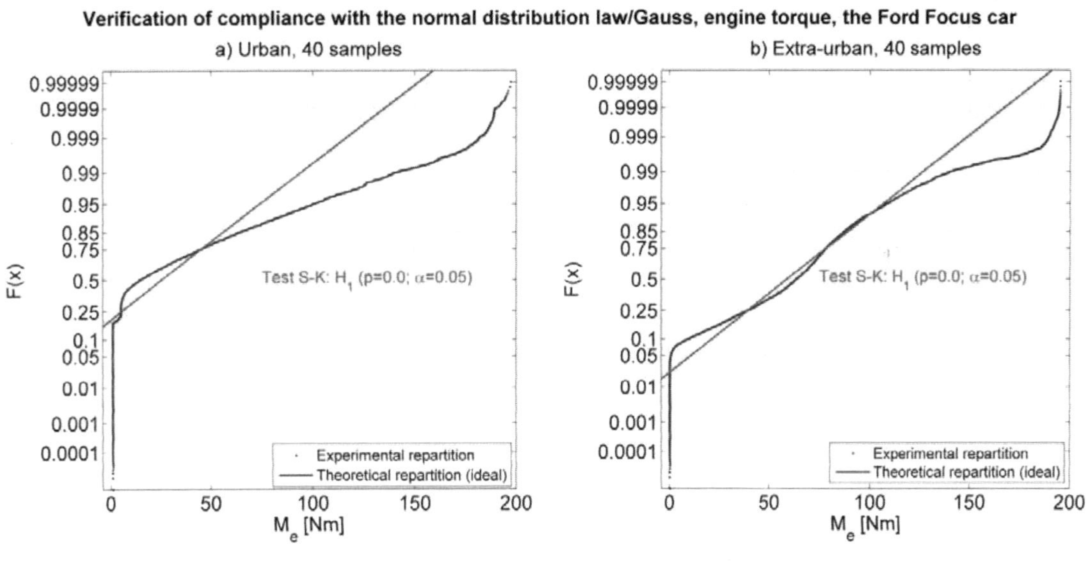

Figure 7.

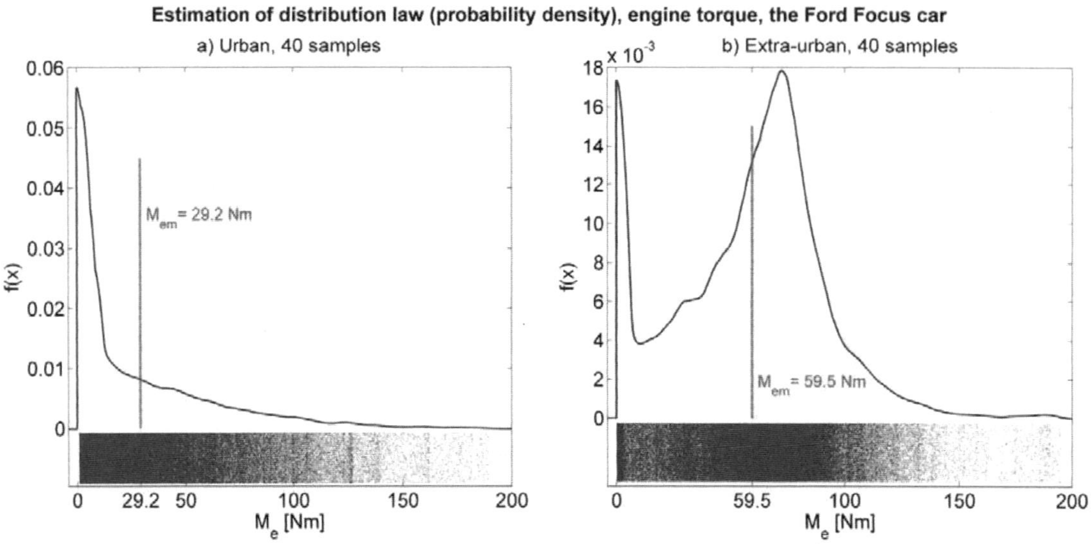

Figure 8.

4. Some factors influence on engine functioning

The paper contains a study of main factors influence on sizes that define engine dynamics and fuel saving performances. For a quantitative highlighting of targeted sizes on engine functioning there can be used correlation analysis, variance analysis, information theory and sensitivity analysis.

4.1. Correlation analysis. In order to highlight if functional sizes are independent or not and to establish the dependency between sizes (linear or non-linear), it is frequently used the correlation coefficient ρ, defined by formula [3,4]:

$$\rho_{xy} = R_{xy}(0) / \sqrt{R_{xx}(0)R_{yy}(0)} \tag{2}$$

with values $\rho \in [-1;1]$, a perfect linear dependency if $\rho^2 = 1$. If ρ=1 then there is perfect direct linear dependency, and if ρ=-1 there is perfect indirect linear dependency; if $0 < \rho \leq 1$ there is a direct dependency, and if $-1 \leq \rho < 0$ there is an indirect dependency. Finally, if ρ=0, the two sizes *x* and *y* are independent. So, if ρ^2 is far away of unitary value (without reaching the null value), the highest the non-linearity.

In addition, in formula (2), at numerator is the intercorrelation function in the origin of discrete time, meaning for τ=0, and under the radical are intercorrelation functions for τ=0.

For example, in figure 9 are presented values of correlation coefficient in case of urban driving; factorial sizes (influence factors) are fuel pressure from the common rail p_c, intake air presssure p_a, throttle shutter's position *p* (engine load) and engine speed *n*, and resultative size is engine torque M_e.

As it can be seen from graphs, there is no experimental sample at which the correlation coefficient is nul and because all sizes are not independent, which was functionally expected. Likewise, because for all samples the correlation coefficients are subunitary, it can be concluded that between sizes there are nonlinear dependencies; the most pronounced non-linear dependency is between engine torque and engine speed, the correlation coefficients having the lowest values, including the mean for all samples ρ=0.351.

This final aspect leads to the conclusion that engine functioning must be described by non-linear mathematical models in order to ensure high precission.

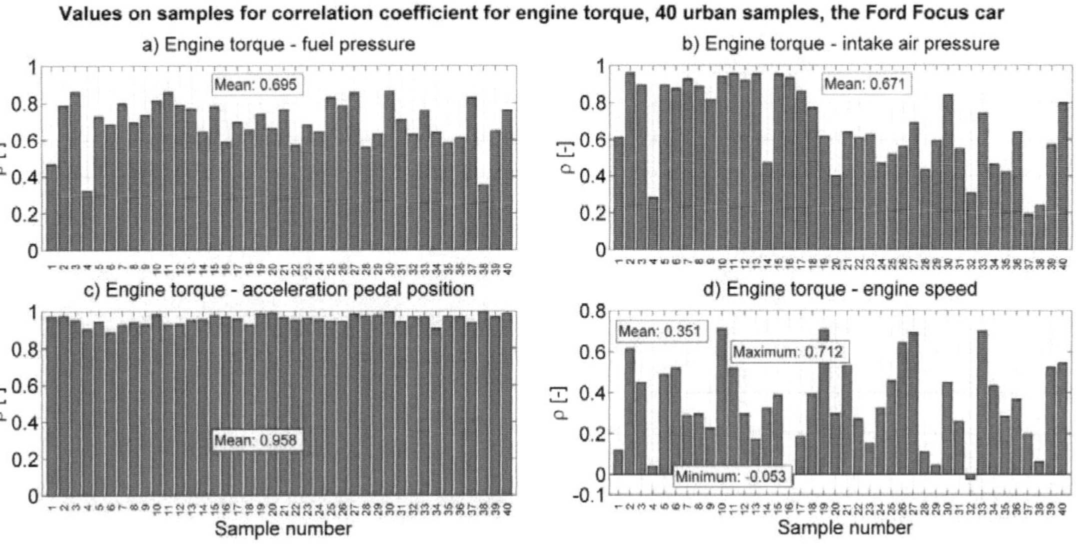

Figure 9.

4.2. Variance analysis. The study of some factors influence also uses variance analysis (ANOVA – **AN**alyse **O**f **V**ariance, MANOVA – **M**ultivariate **AN**alyse **O**f **VA**riance); dispersion has an important role in the analysis of factors influence on a dynamic process. English matematician and statistician Ronald Fisher, the founder of variance analysis, proved that by estimating variance of a characteristic influenced by a factor, and by removing that influence and comparing the two variances, there can be obtained quantitative informations regarding the influence.

In conclusion, variance analysis consists of comparing two types of variances, factorial and residual. If factorial variance is higher than the residual one, the respective factor has a sensible influence on the targeted process. On the other hand, if factorial variance (individual and of interaction with another factor) is lower that the residual one, the factor has a neglectable influence on targeted process. Practically, this comparison can be made by establishing percentage contribution of each factor and residual in case of total variance [3, 5, 6].

In figure 10 are presented results after applying generalized MANOVA algorithm (there are considered the targeted sizes and their interactions), by the study of throttle shutter's position influence p and engine speedt n on engine torque M_e. From figure 10 it can be seen that in both cases, variance of engine speed is higher that the residual one (89.2%, respective 79.5%) as well as the variance of throttle shutter's position (9.9%, respective 14.7%). So, in both cases, throttle shutter's position (considered the engine load) has the highest influence on engine torque variance. From graphs it can also be observed a lower influence on engine torque variance given by the interaction between throttle shutter's position and engine speed.

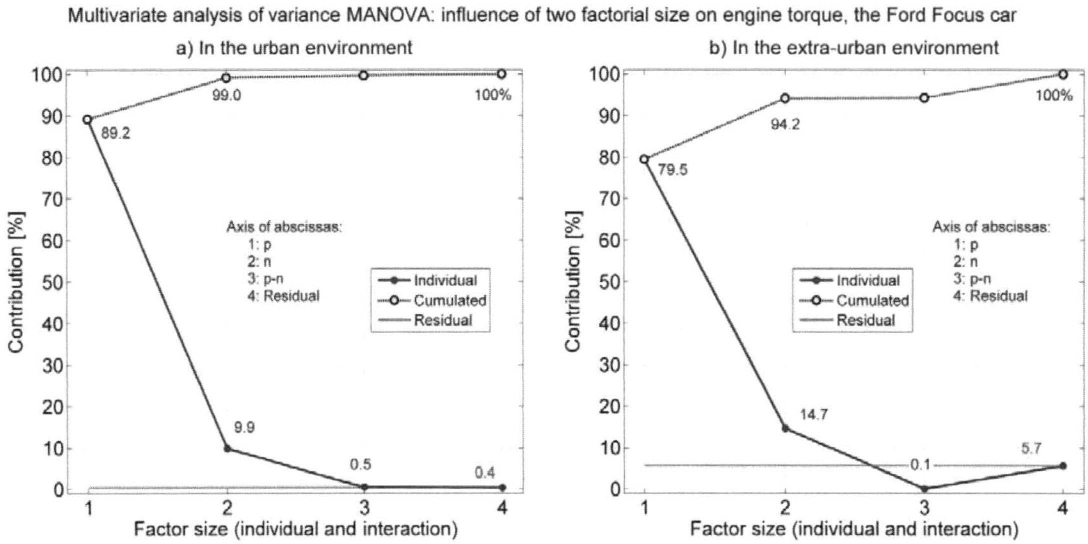

Figure 10.

4.3. Informational analysis. The study of factors influence also uses informational teory, which allows establishing significant parameters that define engine functioning. Informational analysis is based on two main concepts from information theory: entropy and mutual information [3, 7, 8, 9, 10, 11]. Mutual information represents a concept regarding the quantitative estimation of uncertainty attenuation, meaning the increase in prediction. The higher the values of mutual information, the lower the uncertainties, meaning higher predictions. Considering this, mutual information represents a basic concept for the study of systems dynamics and represents a measure of interdependencies between variables. Therefore, in establishing mathematical models must be chosen those variables characterized by the highest mutual informations, which assures the highest predictions; those are called significant variables, attached to the concept of relevance. Considering all reasons mentioned above, information theory represents a generalisation of classical theory, and mutual information represents a measure of relevance.

For example, in figure 11 is presented the graph containing results of informational analysis in case of engine torque as a resultant parameter (placed in upper part) and by considering 6 factorial sizes (influence factors): engine load (through throttle shutter's position p), engine speed n, fuel presure from the common rail p_c, intake air pressure p_a, quantity of recirculated gas γ and rotational angle of turbine vanes α.

In graphs' joints are shown the values of entropy H. It can be observed that the highest entropy is owned by engine speed, of 7.1 bits, and the lowest belongs to intake air pressure: H=2.8 bits; according to those mentioned it results that the usage of engine speed in dynamic calculus has the effect of most reduced prediction. On graphs' bend are written the values of mutual information between two sizes I. It can be seen that the first two relevant variables (with the highest influence on engine torque) are throttle shutter's position (a mutual information with engine torque of 1.165 bits) and fuel pressure (I=0.485 bits).

In figure 11 are also shown mutual informations between the 6 factorial sizes; it can be observed that that the highest mutual information is between fuel pressure and rotational angle of vanes, of 2.213 bits, which also exceeds those mentioned above; this aspect confirms the neccessity that engine's functional study also target the influence between different factors, not only the influence of factors on resulting parameter.

Similarly, it is made an analysis in case of extra-urban driving, the first two significant sizes being the same as in case of urban driving.

It should also be observed that variance analysis targets only parameters variance, and informational analysis targets the entire parameter, not only the variance.

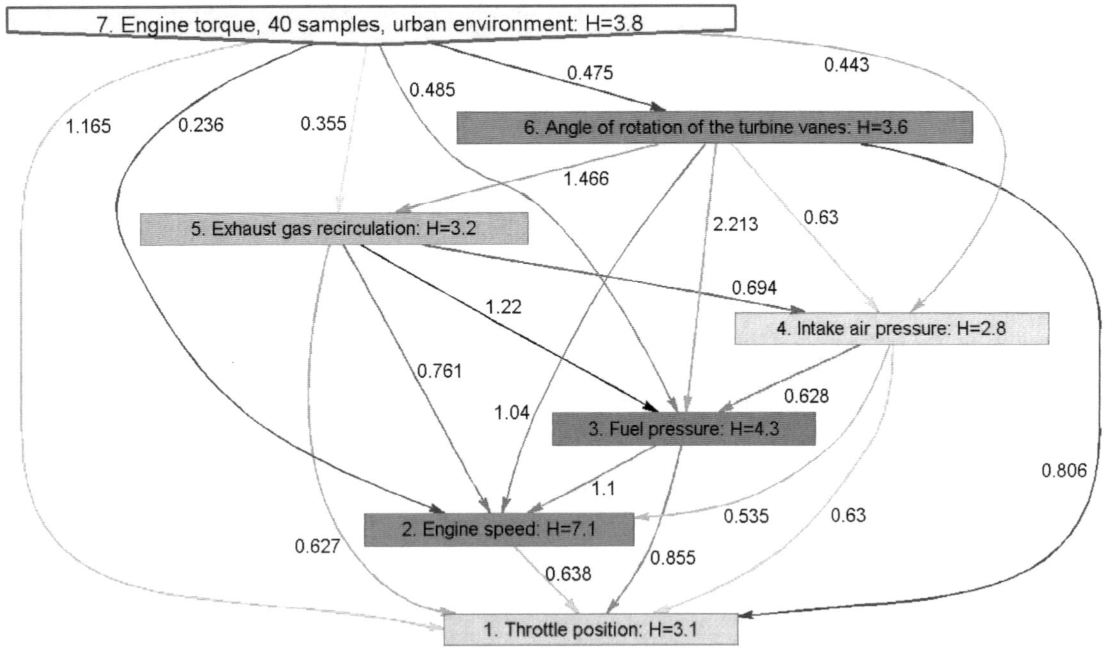

Figure 11.

5. Engine energy efficiency

For a detailed study of engine functioning it should be targeted its energy efficiency, meaning both engine dynamics and fuel saving [3].

For the study of energy efficiency can be established different assessing criteria, which can show the intake of energy in order to obtain certain dynamical performances.

For example, in figure 12 is presented the fuel consumption, expressed in mililiters, needed to obtain an engine power of 1 kW and an engine torque of 1 Nm. In figure 13 are shown mean values of the ratio between vehicle kinetic energy (meaning the energy used for actual driving) and the energy introduced with the fuel.

From figure 12 it can be seen that in order to obtain a unity of power and torque, it is consumed a higher quantity in case of urban driving; so, energy efficiency is higher in case of extra-urban driving.

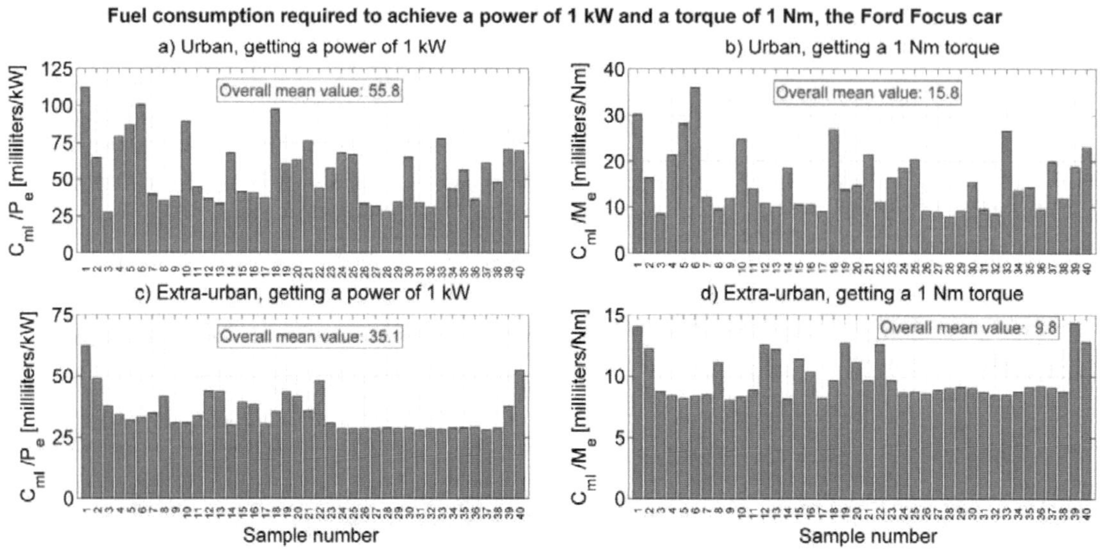

Figure 12.

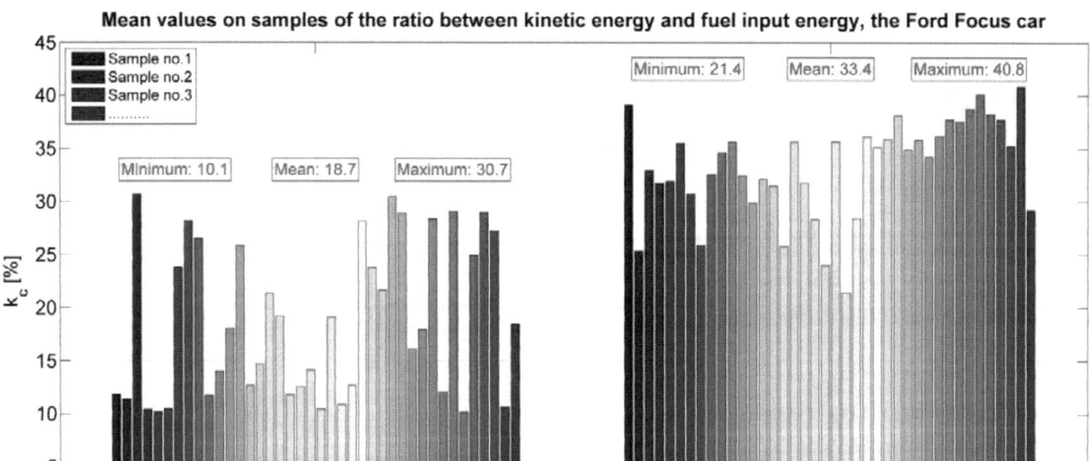

Figure 13.

In figure 13, the ratio between vehicle kinetic energy W_{cin} and energy introduced by fuel W_i is established with ratio:

$$k_c = W_{cin} / W_i \cdot 100\ [\%] \tag{3}$$

where:

$$W_{cin} = m_a v_a^2 / 2; \quad W_i = C_h Q_i S / V \tag{4}$$

and in which there are noted: m_a and v_a (V) – autovehicle weight and speed; S – distance covered; Q_i – fuel inferior calorific power; C_h – fuel hourly consumption.

Graph from figure 13 shows that for the 40 experimental samples in urban environment, from the energy introduced by fuel only 10.1%-30.7% it is used for actual driving; the mean of energy used at all 40 samples is 18.7%. Similarly, from figure 13 results that for all 40 samples in extra-urban driving, from the energy introduced by fuel only 21.4%-40.8% it is used for actual driving; the mean of energy used at all 40 samples is 33.4%. In conclusion, even for this assessment criterion, energy efficiency is higher in case of extra-urban driving. It must be mentioned that the values presented are also confirmed by specialty literature from autovehicle field, proving so its poor energy efficiency.

6. Conclusions

The equipment of current automobiles with sensors and actuators embedded from fabrication and with an on-board computer, allows conducting complex studies, which go deep in the core of highly rapid dynamic processes that are specific to engine functioning.

Assessment criteria for engine fuel saving, currently used in specialty literature, do not satisfy the requests of a study regarding energy efficiency, that is why there must be established criteria specific to this issue.

The study of factors influence on engine functioning must also take into account interactions between factors, but also the fact that factorial sizes vary simultaneously during driving, these two are the main differencies from the classical study presented in specialty literature.

7. References

[1] *** Technical note of Ford Focus car, with 1.6 TDCi engine, 66 kW (90 CP)
[2] *** Diagnostics software FoCOM, http://www.obdtester.com/focom
[3] Copae I, Lespezeanu I, 2006, Cazacu C, Dinamica autovehiculelor, *Editura ERICOM*, Bucureşti
[4] Hyötyniemi H, 1998, Multivariate Statistical Methods in Systems Engineering. Report 112, Helsinki University of Technology
[5] Carey G, 1998 , Multivariate Analysis of Variance (MANOVA). Colorado State University
[6] Langsrud Ø, 2003, MANOVA for Matlab, Division of Statistics, Departament of Mathematics, University of Oslo, Norway
[7] Gray R, 2007, Entropy and information theory, *Stanford University*, New York
[8] Hoenselaar A, 2007, A mutual information as a measure of relevance in neural coding. Thesis. Providence, Rhode Island
[9] MacKay D, 2003, Information theory, inference, and learning algorithms. *Cambridge University Press*
[10] Nemenman I, 2002, Information theory, multivariate dependence, and genetic network inference, University of California
[11] Norris M, 2003, Redundancy through time, University of Plymouth, Plymouth UK

Experimental and numerical assessment of ignition delay period for pure diesel and biodiesel B20

Mohanad Aldhaidhawi [*,1,2], **Marek Brabec** [4,5], **Miron Lucian**[1], **Radu Chiriac** [1,3], **Viorel Bădescu**[1]

[1] University Politehnica of Bucharest, Faculty of Mechanical Engineering, Spl. Independentei 313, Sector 6, 060042, Bucharest, Romania

[2]Al-Furat Al-Awsat Technical University, Technical Institute of Kufa, Iraq

[3] Associated Member to EA7341 CMGPCE of Conservatoire National des Arts et Metiers, Paris, France

[4] Institute of Computer Science of the ASCR, Pod Vodarenskou vezi 2, 182 07 Prague 8, Czech Republic

[5] Czech Institute of Informatics, Robotics, and Cybernetics, Czech Technical University in Prague, Zikova 4, 166 36 Prague 6, Czech Republic

Abstract. The ignition delay period for a compression ignition engine fueled alternatively with pure diesel and with biodiesel B20 has been experimentally and numerically investigated. The engine was operated under full load conditions for two speeds, 1400 rpm speed for maximum brake torque and 2400 rpm speed for maximum brake power. Different parameters suggested as important to define the start of combustion have been considered before the acceptance of a certain evaluation technique of ignition delay. Correlations between these parameters were analyzed and concluded about the best method to identify the start of combustion. The experimental results were further compared with the ignition delay predicted by some correlations. The results showed that the determined ignition delays are in good agreement with those of the Arrhenius type expressions for pure diesel fuel, while for biodiesel B20 the correlation results are significantly different than the experimental results.

Keyword: Ignition delay, biodiesel B20, diesel engine

1. Introduction

Presently, there is an increasing interest in biodiesel behavior as fuel in compression ignition engines due to many environmental and economic benefits. Biodiesel is composed of fatty acids produced from vegetable oil or animal fat via chemical processes called transesterification. Biodiesel has higher cetane number, higher oxygen content, lower aromatics and is free of sulphur content. It is contributing as an improver of the combustion process and reduce the exhaust gas emissions [2-3]. Biodiesel is safe to store and handle due to its high flash point and lower volatility with respect to diesel fuel. Since biodiesel has a higher bulk modulus and higher density, these characteristics lead to an advance in the start of fuel injection [4-8]. The higher cetane number of biodiesel is expected to reduce the ignition delay period resulting in emissions reductions [9,10]. Ignition delay is the period between the start of fuel injection (SOI) and the start of combustion (SOC) [11]. The start of fuel injection is easy to identify because is related to the time when the injector needle lifts off while the start of combustion is more difficult to be defined. There are several methods currently used for the measurement of the start of combustion such as an abrupt change in the cylinder pressure gradient, the combustion of a certain mass of fuel, the change of slope in the heat release profile, light emission detected by a photocell diode and cylinder temperature rise due to combustion [11]. An abrupt change in the cylinder pressure gradient and change of slope in the heat release profile methods are widely used to determine the start of combustion in the diesel engines. However, selecting the suitable technique to identify the start of combustion is still an important question in the ignition delay studies. In previous studies, such as that of Alkhulaifi and Hamdalla [12] the start of combustion was determined based on the change of gradient in the heat release rate is determined by the minimum value on the curve. Rodriguez et al. [13] used the maximum value of the first derivative of in-cylinder pressure as a technique to identify the start of combustion. Assanis et al. [14] concluded that the second derivative of in-cylinder pressure is a suitable method for determining the start of combustion in compression ignition engines compared to the other techniques. Katrasnik et al. [15] determination the start of combustion (SOC) based on the maximum of the third order derivative of the cylinder pressure with respect to the crank angle. Thoo et al. [16] proposed the point at which the net heat-release rate first reaches 2 J/degree, this value was set above the maximum heat-release rate of the motored engine cycles with respect to the crank angle as the appropriate criterion for SOC detection. In Theobald [17] and Dhole et al. [18] the point at which the heat release rate is zero was considered as the initiation of combustion.

The main objectives of this study are to test different parameters suggested as important to define the start of combustion in diesel engine and to choose the best assessment evaluation technique. The experimentally determination of ignition delay period by conducting a series of test runs on a tractor diesel engine under full load operating conditions and engine speeds 1400 rpm and 2400 rpm fueled with pure diesel and biodiesel B20 and comparing it with those obtained from different correlations suggested in previous studies represent the work developed by the research team. This work emphasis some experimental techniques used for ignition delay determination and the differences which occur by comparison with different theoretical correlations.

2. Experimental setup

A tractor diesel engine, four- cylinder, four-stroke, naturally-aspirated, DI was used. Several types of equipment were installed on the test engine to measure the engine performance and combustion characteristic as shown in figure 1. The main engine specifications were listed in table 1. Two AVL GM 12 D pressure transducers with a sensitivity of 15.76 pC/bar were used to measures the in-cylinder pressure rate, and one AVL QL21D pressure transducer with a sensitivity of 2.5 pC /bar was used to measures the high-pressure line values. The pressure transducers measurements will be used later to identify the start of injection and the start of combustion. The fuel system was adapted for multiple fuels operation, allowing the engine to be alternatively fueled with pure diesel and biodiesel B20. The experiment was conducted under full load operating conditions at engine speeds 1400 rpm (speed of maximum brake torque) and 2400 rpm (speed of maximum effective power). The properties of the test fuels pure diesel, biodiesel B20 and B100 (as a reference) were measured in two different places. The cetane number and cold filter plugging point were tested in the laboratories of Rompetrol oil company, located in Constanta city, Romania. The fuels density, viscosity, flash point, pour point and cloud point were tested in the chemical laboratory (LAMC), Bucharest, Romania. The test fuels properties with the test method are listed in table 2.

Figure 1. Schematic layout of the test bench.

Table 1. The main engine specifications.

Tractor Diesel engine	4-stroke
No. of cylinders	4 in line, vertical
Bore x stroke (mm)	102 x 115
Displacement (cm^3)	3759
Fueling system	Direct injection (DI)
Maximum brake torque (Nm) at 1400 rpm	228
Rated power (kW) at 2400 rpm	50
Compression ratio	17.5:1

Table 2. The test fuel properties.

Properties	**Pure diesel**	**B20**	**B100**	**Test method**
Density (g/cm^3) @ 20°C	0.820	0.8565	0.886	SR EN ASO 3675
Viscosity (cSt)@ 20°C	2.53	5.12	8.06	SR EN ASO 3104
Flash point (°C)	58.5	85	184	SR 5489
Cetane Number	51.1	55.5	-	EN ISO 516598
Cold filter plugging point (°C)	-24	-18	-	SR EN 116 2016
Cloud point (°C)	-16	-10	-4	SR EN 23015
Pour point (°C)	0	-7	-18	SR 13552
Lower Heating Value (MJ/kg)	41.87	40.59	37.4	ASTM D 240

3. Simulation Procedures

Due to advances in computational hardware, AVL Workspace has become an option to studies the engine performance, combustion process and emissions formation in a wide variety of engines at different operating conditions fueled with different standard fuels such as diesel, gasoline, hydrogen… etc. However, the biodiesel B20 properties were obtained as optional fuel in the AVL Boost program by the authors using the AVL procedures and data bases. Two models were created to understand the combustion process, one in AVL Boost and the other in AVL Fire.

3.1. AVL Boost Simulation Model

The engine model was built by using the AVL Boost 2016 simulation code to estimate the performance, combustion characteristic, and emissions. The model calibration for the combustion chamber process initially was selected based on documentation (AVL Boost theory and AVL Boost users guide v2016) [19]. The calibration parameters were adjusted according to the model results to reach a good agreement with experimental data. All engine components, such as the intake manifolds, air filter, cylinder geometry, system boundaries, exhaust pipe, and catalyst geometry were based on the real values taken from the test engine and linked together by pipes and implemented in the Boost interface as shown figure 2. The start and the rate of fuel injection, the air mass flow, and the fuel mass flow rate were experimentally measured at specified operating conditions to implement them in the AVL Boost before running the program. The AVL- MCC combustion model and the Woschni 1990 heat transfer model were chosen.

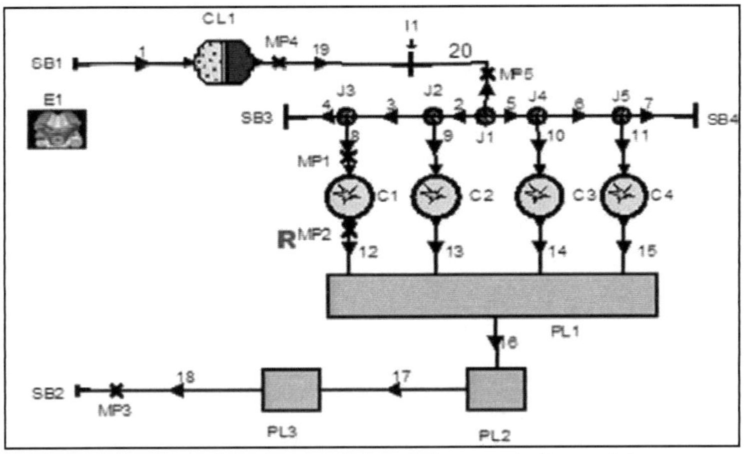

Figure 2. Schematic of the engine symbolic model.

3.2. AVL Fire Model

The simulation model was created by using the AVL Fire 2016 simulation. The 2D Sketcher software from AVL Fire was used to draw the piston bowl shape, and ESED was used for injector shape, block structure, and selections. The importance of the piston shape is due to the need for controlling the flow of fuel and air inside the cylinder together with injector characteristics. Piston bowl dimensions were considered according to the official work drawing of the part obtaining thus a cross-sectional area of the piston, as it is presented in figure 3. The injector characteristics were considered as five holes with 0.000239 m nozzle diameter with a spray angle of 160 degrees. Computational meshes were created by using the parameters entered in the block structure section. An average cell size of 0.001m was chosen in order to reduce the computational time. A compensation volume was adopted to take into account the pressure losses alongside the piston rings, thus obtaining a dynamic compression ratio equal to 16.4. The resulted mesh is seen in figure 4. AVL Fire ESED uses complex mathematical models with the aim of simulating the fuel spray development, vaporization, combustion and pollutant emissions formation specific to the direct injection diesel engine. The calculations were made within a range of -80/+80 crank angle degree (°CA) specific to top dead center (TDC) using initial values as cylinder pressure measured data and calculated bulk temperature obtained from AVL Boost. The ECFM-3 combustion model was used as it is specific to the diesel combustion process because it best correlates with our experimental data. The autoignition was controlled using the table autoignition model which utilizes detailed chemical kinetics. The Wave Break-up model was used for fuel jet spray dispersion as it increases the accuracy of the calculation. These models were selected and implemented in the ESED software with the aim of obtaining good accuracy while minimizing computational time.

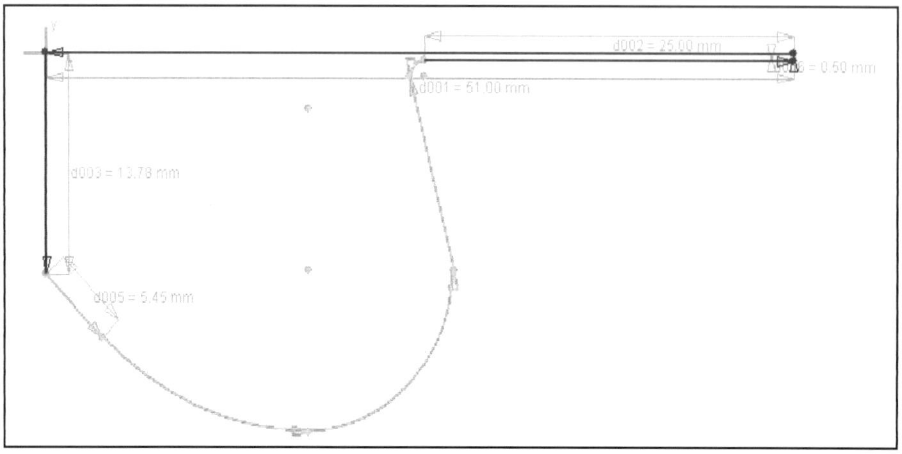

Figure 3. Cross-sectional area of the piston.

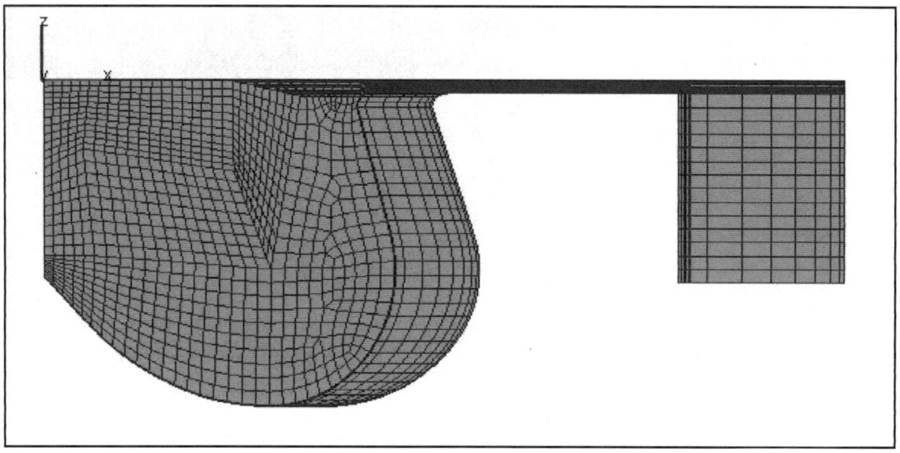

Figure 4. Meshes generated of a piston.

4. Theoretical Approach

The mean value of the cylinder pressure at all operating conditions for both test fuels was obtained from the experimental work. The mean cylinder gas temperature was obtained from the numerical simulation using AVL Boost 2016 code. The mean cylinder pressure and mean gas temperature during the ignition delay period were taken and implemented in several correlations collected from previous studies in order to evaluate the ignition delay period. The correlations were used to estimate the ignition delay for pure diesel fuel as following and to perform comparisons with ignition delays experimentally determined:

- Correlation proposed by Assanis [14]

$$\tau_{D10} = 2.4 \frac{\exp(\frac{2100}{T})}{p^{1.02} \, . \varphi^{0.2}} \tag{1}$$

- Correlation proposed by Watson [20]

$$\tau_{D100} = 3.45 \frac{\exp(\frac{2100}{T})}{p^{1.02}} \tag{2}$$

- Correlation proposed by Rodriguez et al. [13]

$$\tau_{D100} = \frac{\exp\left(\frac{950}{T}\right)}{p^{0.24} \cdot \varphi^{0.04}} \tag{3}$$

- Correlation proposed by El-Kasaby et al. [21]

$$\tau_{D100} = 26.06 \frac{\exp(\frac{1038}{T})}{p^{1.21} \cdot \varphi^{1.36}} \tag{4}$$

Only one correlation proposed in [21] was used to estimate the ignition delay period for biodiesel B20. The results obtained from this equation were compared with experimental data at both engine speeds under full load conditions:

$$\tau_{B20} = 74.32\, p^{-1.32}\ \ \varphi^{-1.39} \exp(\frac{1022}{T}) \tag{5}$$

Here: τ-ignition delay period (ms), T-temperature (K), p- pressure (bar), φ-equivalence ratio.

5. Results and Discussions

In this part of the study, the results from the simulation study are first compared to the experimental results to examine the usefulness of the models. In the second part, the start of combustion is estimated by using different methods. In the third one, the effect of diesel and biodiesel blend B20 on the ignition delay period is presented.

5.1. Cylinder Pressure

The variations of cylinder pressure with respect to the crank angle position, experimental and simulation (obtained from AVL Boost and AVL Fire) for diesel fuel at different engine speeds and full load operation conditions are given in figure 5. From this figure, a good agreement between experimental trace as average over 200 consecutive cycles and the simulations results can be observed. The relative deviation between the experimental and numerical for maximum cylinder pressure obtained from AVL Boost and AVL Fire for diesel fuel at engine speed of 2400 rpm is 1.108% and 0.95%, respectively. Similar good agreements between simulation and experiments have been obtained for all operating conditions specified in this study.

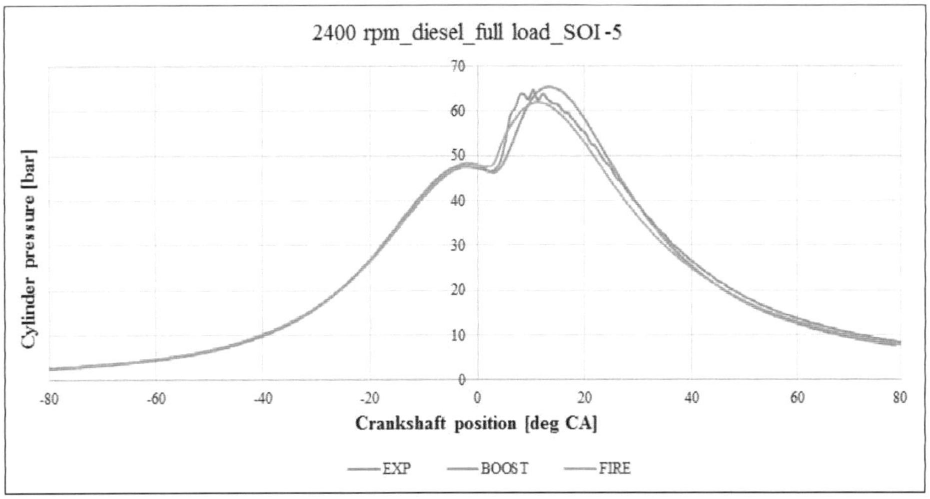

Figure 5. Comparison between experimental and simulation pressure traces for full load, 2400 rpm speed, diesel fuel.

5.2. Estimate the Start of Combustion (SOC)

There are several technical methods in the literature that were used to identify the start of combustion in compression ignition engines. The start of combustion was determined from the cylinder pressure and rate of heat release curves by using several evaluation techniques before the acceptance of a certain evaluation that was used to estimate the ignition delay period. The needle lift, the rate of heat release and the cylinder pressure results were used based on the average of 200 successive cycles. The large number of cycles was necessary to avoid the random deviations in measurements. To explore the SOC from the cylinder pressure trace and rate of heat release curve, a model for smoothing data was needed. Therefore, the raw data were smoothed via penalized spline GAM model (mgcv library in R), the fitted curve differentiated numerically via Deriv function from Pracma library (with step size 0.1). The raw cylinder pressure of diesel engine at 2400 rpm under full load conditions fueled with pure diesel fuel for 200 successive cycles is given in figure 6. The first and second derivatives result of smoothed cylinder pressure for all cycles were determined by this model and it is shown in figure 7. Figure 8 presents the start of combustion identified from the maximum value of the first, second, third and fourth order derivative (D1f, D2f, D3f, and D4f, respectively) of the cylinder pressure (as an example for one cycle). As shown in this figure, there is some deviation in SOC estimated from different derivatives. However, the second order derivative was more sensitive to identification of the start of combustion in comparison with the other derivatives. This finding is consistent with the results reported in the study [14].

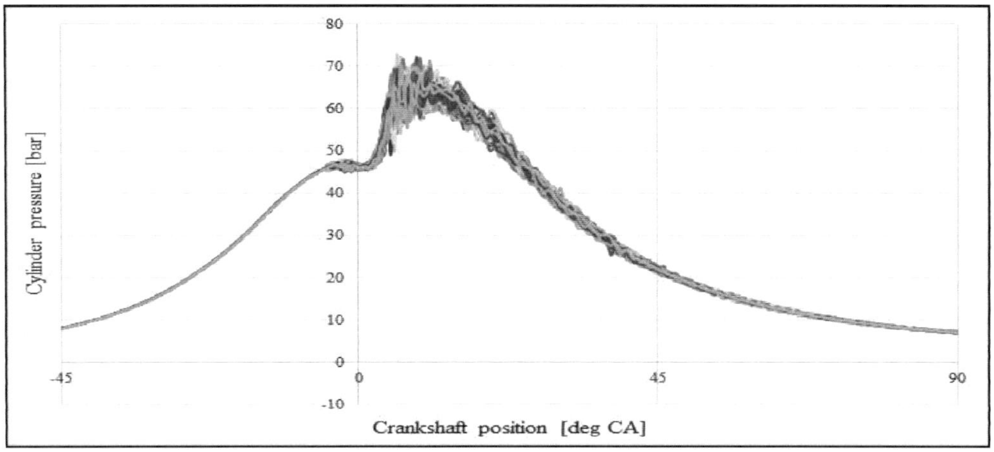

Figure 6. Cylinder pressure vs. crankshaft position at 2400 rpm, pure diesel fuel for 200 cycles.

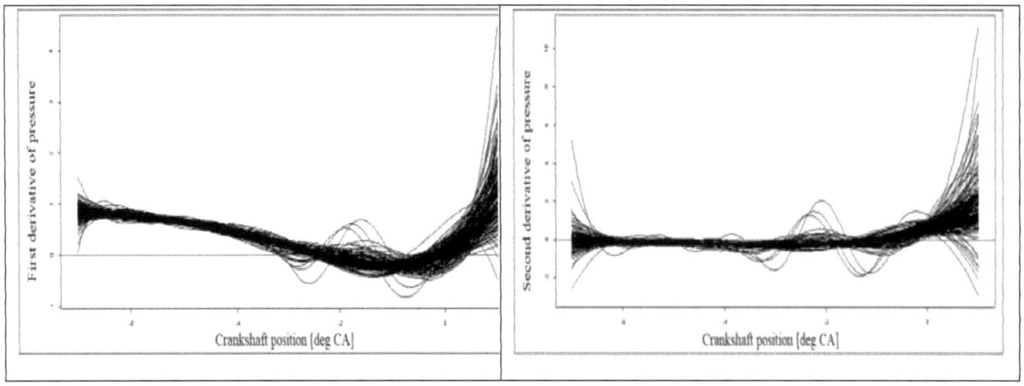

Figure 7. First and second order derivative of smoothed cylinder pressure at 2400 rpm, pure diesel fuel for 200 cycles.

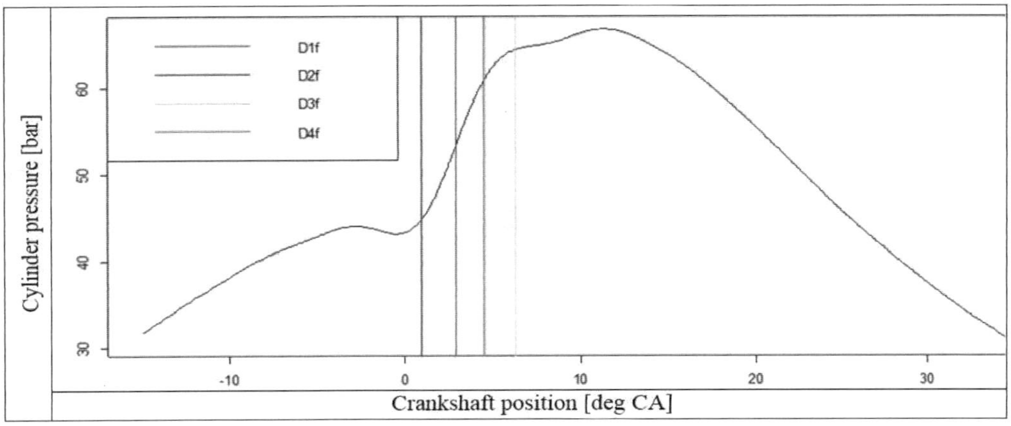

Figure 8. Crankshaft position of SOC identified by using first (Df1), second (Df2), third (Df3) and fourth (Df4) order derivative of cylinder pressure at 2400 rpm under full load fueled with pure diesel fuel for 200 cycles.

There is another method that was covered in this study to identify the start of combustion from the rate of heat release curve. The rate of heat release curves for 200 cycles for engine fueled with diesel fuel and running in the same conditions (full load and engine speed 2400 rpm) is given in figure 9. Figure 10 presents the smoothed of the rate of heat release. Here, the start of combustion SOC was considered as the root of the rate of heat release ($dQ_f/d\varphi = 0$) (see figure 11). Figure 12 presents the distributions of the start of combustion that were determined from the second order derivative of the cylinder pressure (the most sensitive way to identify the SOC compared to other derivatives according to our findings) and from the root of the rate of heat release. As shown in this figure, the root of the rate of heat release method appeared more sensitive for identification of the SOC than the second order derivative method, due to normal distributions of SOC with this method. However, the same steps listed above were followed to determine the SOC inside the cylinders when using pure diesel and biodiesel B20 at different engine speeds under full operating conditions. The same conclusion was obtained, that the root of the rate of heat release method ($dQ_f/d\varphi = 0$) is the proper technique to be used to identify the SOC. Therefore, this method will be considered in the determination of SOC in our calculation to predict the ignition delay period.

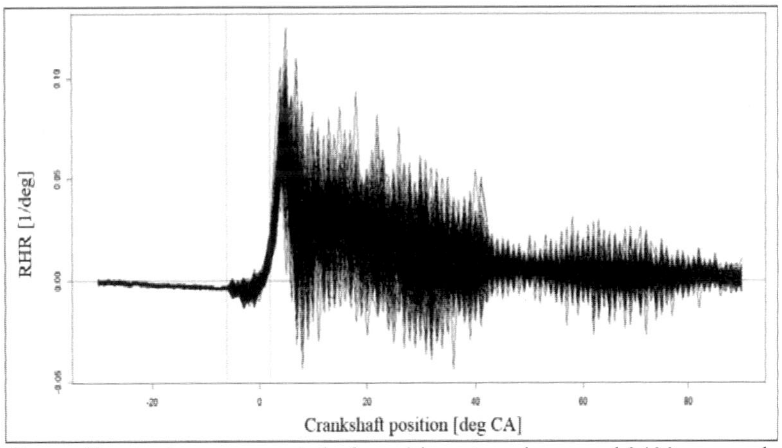

Figure 9. Rate of heat release vs crankshaft position at engine speed 2400 rpm under full load fueled with pure diesel fuel for 200 cycles raw data.

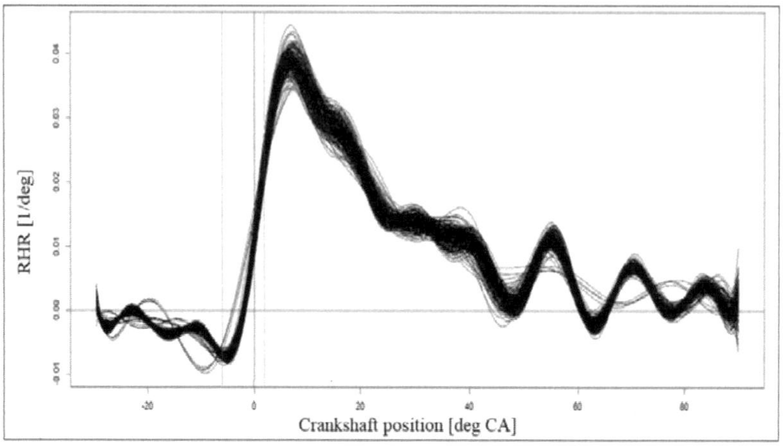

Figure 10. Rate of heat release smoothed curves vs at engine speed 2400 rpm at full load fueled with pure diesel fuel for 200 cycles.

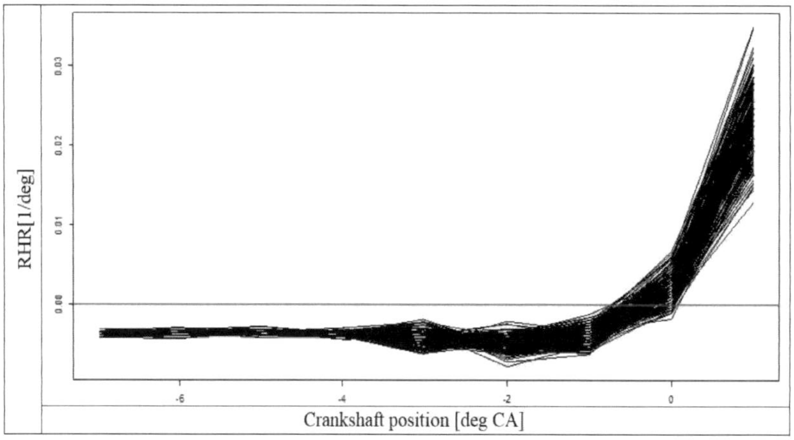

Figure 11. Crankshaft position of SOC estimated by using zero root of rate of heat release at 2400 rpm, full load fueled with pure diesel fuel for 200 cycles.

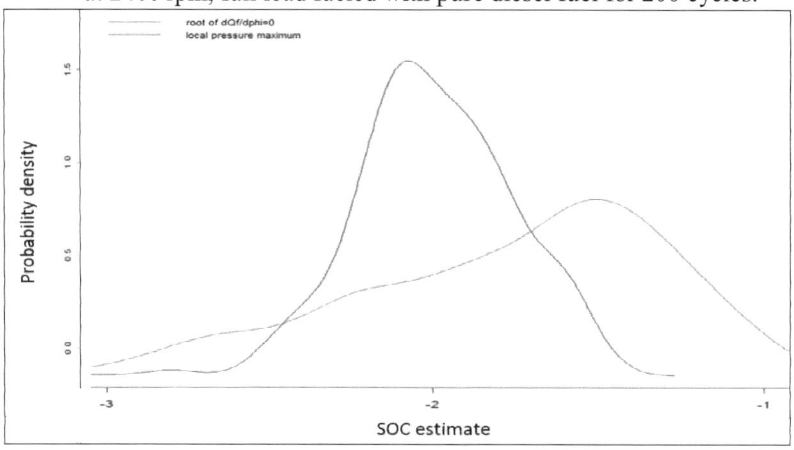

Figure 12. SOC distribution estimated from the rate of heat release (zero root) and from second order derivative of cylinder pressure at 2400 rpm under full load fueled with pure diesel fuel for 200 cycles.

5.3. Ignition delay period

Ignition delay period (ID) as mention before, is the duration between the start of fuel injection SOI and the start of combustion SOC [11]. In this study, the start of injection was taken as the time when the injector needle lifts off (threshold value was equal to 0.002 mm) while the start of combustion corresponds to the zero root of the heat release rate. The variations of needle lift versus the crankshaft position for diesel fuel at the engine speed of 2400 rpm and full load operating conditions for 200 cycles are presented in figure 13. The experimental tests were repeated three times in different days trying to keep the same ambient initial conditions (temperature 35°C and barometric pressure 1bar in the test cell) to verify the consistency of experimental data. At engine speed of 2400 rpm the start of injection for the average of cycles was 5 degrees before top dead center (BTDC).

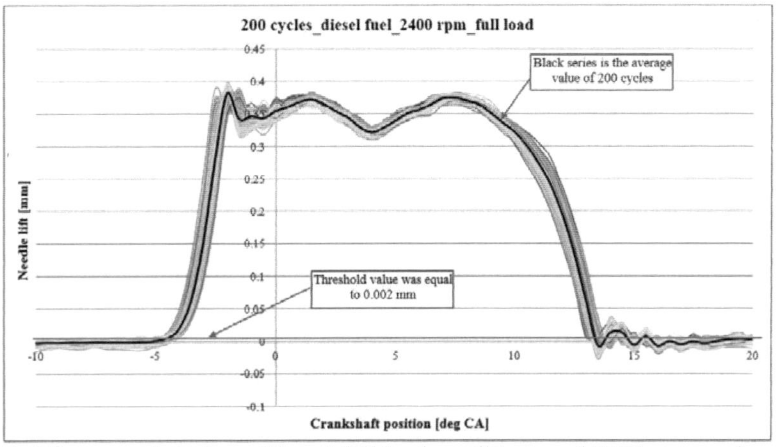

Figure 13. Needle lift vs. crankshaft position at full load conditions and 2400 rpm engine speed, the black series is the average value comprised in the envelope of 200 cycles, the SOI was considered at the threshold value equals to 0.002 mm.

The start of combustion (SOC) was placed at the zero root for the rate of heat release characteristic profile for tested regime, as presented in figure 14. Thus, the start of combustion point (average of cycles) was identified at 0.375 degree after top dead center (ATDC) at 2400 rpm speed. These values (SOI and SOC) for both engine operating conditions were taken into consideration to calculate the ignition delay period.

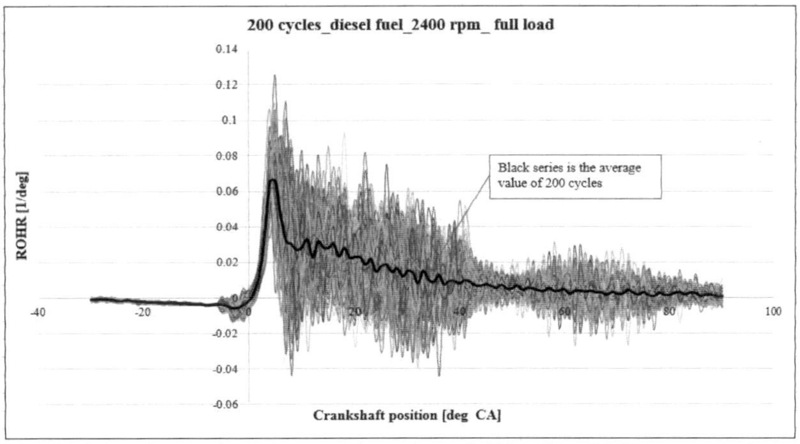

Figure 14. Rates of relative heat release vs. crank angle at full load and 2400 rpm engine speed, the black series is the average value comprised in the envelope of 200 cycles.

The mean cylinder pressure and temperature during the ignition delay period were used to estimate the ignition delay from different correlations. The results obtained from these correlations were compared to the experimental data. The ignition delay period for pure diesel and biodiesel B20 fuels, calculated from different correlations, are listed in table 3. Regarding diesel fuel, there are some deviations between the results that were obtained from these correlations with the experimental data. However, the correlation proposed by Assanis [14] is the closest to the experimental results, with relative deviations of 13% and 14.5% at engine speeds 1400 rpm and 2400 rpm, respectively. Concerning biodiesel B20, there is a significant difference between the experimental results and the results that were obtained from the correlation previously proposed by Al-Kasaby [21], with relative deviations of up to 75% and 80.54% at engine speeds 1400 rpm and 2400 rpm, respectively. The ignition delay period for biodiesel B20 was shorter than that of pure diesel fuel at the same operating conditions. This behavior may be due to the fact that the biodiesel has higher cetane number compared to pure diesel fuel.

Table 3. Ignition delay estimated from different correlations

For pure diesel fuel											
Engine speed	SOI	SOC	Duration	ID EXP	P	T	Phi	Assanis [14]	Watson [20]	Rodriguez [13]	EL-Kasaby [21]
[rpm]	[deg CA]	[deg CA]	[deg CA]	[ms]	[bar]	[K]	[-]	[ms]	[ms]	[ms]	[ms]
1400	-6	-0.33	5.67	0.394	44.39	994.5	0.64	0.453	0.595	1.065	1.38
2400	-5	0.375	5.37	0.37	46.93	1008.4	0.52	0.433	0.546	1.046	1.69
For biodiesel B20 fuel											
	SOI	SOC	Duration	ID EXP	P	T	Phi				EL-Kasaby [21]
1400	-6	-0.25	5.75	0.39	41.61	957.15	0.64	-	-	-	1.56
2400	-5	0.125	5.125	0.356	44.7	990.38	0.52	-	-	-	1.83

6. Conclusions

The ignition delay period for a compression ignition engine fueled with pure diesel fuel and biodiesel B20 at engine speeds 1400 rpm and 2400 rpm under full load operating conditions were experimentally determined. The experimental results were compared to that obtained from different correlations proposed in several studies by different authors. The main conclusions of this study are the following:

- The root of the rate heat release is a sensitive and proper method to identify the start of combustion compared to other methods that were considered.
- The correlation proposed by Assanis [14] seems to the closest to the experimental results for pure diesel fuel at tested operating conditions with relative deviations lower than 15%.
- The ignition delay period for B20 was shorter than that of pure diesel at same operating conditions.
- The correlation proposed by El- Kasaby [21] to predict the ignition delay period for biodiesel B20 is far than experimental results.
- This work marks a starting point which has to be followed by the efforts to investigate the combustion process of the diesel engine fueled with pure diesel fuel and biodiesel B20 at other engine speeds and full load condition.

Acknowledgements
The authors of this paper acknowledge the AVL Advanced Simulation Technologies team for the support they have offered in performing the simulation part of this work. One of the authors (Mohanad Aldhaidhawi) thanks the Iraqi government for its financial support.

References
[1] Liaquat A, Masjuki H, Kalam M, Rizwanul I, Hazrat M, Varman M, Mofijur M, Shahabuddin M, 2013 Effect of coconut biodiesel blended fuels on engine performance and emission characteristics *Procedia Engineering* **56** pp 583-590.
[2] Naik S, Goud V, Rout P and Dalai A, 2010 Production of first and second generation biofuels: A comprehensive review *Renewable and Sustainable Energy Reviews* **14** pp 578-597.
[3] Mofijur M, Atabani A, Masjuki H, Kalam M and Masum B, 2013 A study on the effects of promising edible and non-edible biodiesel feedstock on engine performance and emissions production: A comparative evaluation *Renewable and Sustainable Energy Reviews* **23** pp 391-404.
[4] Khalid A, Syamim M, Mustaffa N, Sapit A, Zaman I, Manshoor B, Samion S, 2014 Experimental investigations on the use of preheated biodiesel as fuel in various load conditions of diesel engine *Australian Journal of Basic and Applied Sciences* **8** pp 423-430.
[5] Lin B, Huang J, Huang D, 2009 Experimental study of the effects of vegetable oil methyl ester on DI Diesel engine performance characteristics and pollutant emissions *Fuel* **88** pp 1779-1785.
[6] Ozturk E, 2015 Performance, emissions, combustion and injection characteristics of a Diesel engine fueled with canola oil–hazelnut soap stock biodiesel mixture *Fuel Processing Technology* **129** pp 183-191.
[7] Harch C, Rasul M, Hassan N, Bhuiya M, 2014 Modelling of engine performance fueled with second generation biodiesel *Procedia Engineering* **90** pp 459-465.
[8] Dunn R, Shockley M, Bagby M, 1996 Improving the low-temperature properties of alternative Diesel fuels: Vegetable oil-derived methyl esters *Journal of the American Oil Chemists Society* **73** pp1719-1728.
[9] Grau B, Bernat E, Antoni R, Jordi-Roger R, 2010 Small-scale production of straight vegetable oil from rapeseed and its use as biofuel in the Spanish Territory *Energy Policy* **38** pp189-196.
[10] Korakianitis T, Namasivayam A, Crookes R, 2011 Diesel and rapeseed methyl ester (RME) pilot fuels for hydrogen and natural gas dual-fuel combustion in compression–ignition engines *Fuel* **90** pp 2384-2395.
[11] Heywood J B, 1988 *Internal Combustion Engines Fundamentals*, McGraw-Hill Inc. USA, ISBN 0-07-028637-X, 481pp.
[12] Alkhulaifi K, Hamdalla M, 2011 Ignition delay correlation for a direct injection diesel engine fuelled with automotive diesel and water diesel emulsion *International Journal of Chemical, Molecular, Nuclear, Materials and Metallurgical Engineering* **5** pp10-15.
[13] Rodriguez R. Sierens S, Verhelst S, 2011 Ignition delay in a palm oil and rapeseed oil biodiesel fueled engine and predicate correlation for the ignition delay period *Fuel* **90** pp 766-772.
[14] Assanis D, Filipi Z, Fiveland S, Simiris M 2003 A predictive ignition delay correlation under steady-state and transient operation of a direct injection diesel engine *Journal Engineering Gas Turbines Power* **125** pp 450-7.
[15] Katrasnik T, Opresnik S, Trenc F, Skific N, 2005 A new criterion to determine the start of combustion in diesel engine *Journal. Eng. Gas Turbines Power* **4** pp 928-933.
[16] Thoo W, Kevric A, Ng H, Gan S, Shayler P, Rocca A, 2014 Characterization of ignition delay period for a compression ignition engine operating on blended mixtures of diesel and gasoline *Applied thermal engineering* **66** pp 55-64.
[17] Theobald M A, 1986 *A numerical simulation of diesel autoignition* University of Texas, Austin, USA, PhD thesis.
[18] Dhole A, Yarasu R, Lata D, 2016 Investigations on the combustion duration and ignition delay period of a dual fuel diesel engine with hydrogen and producer gas as secondary fuels *Applied Thermal Engineering* **107** pp 524-532.
[19] AVL Boost theory and AVL Boost users guide 2016, https://www.avl.com/ro/boost.
[20] Watson N, Pilley A, Marzouk M 1980 A combustion correlation for diesel engine simulation. *SAE Technical* Paper 800029, doi:10.4271/800029.
[21] El-Kasaby M, Nemit-Allah M, 2013 Experimental investigations of ignition delay period and performance of a diesel engine operated with Jatropha oil Biodiesel *Alexandria Engineering Journal* **52** pp141-149.

Experimental Study on the Distillation Capacity of Alcohol-Gasoline Blends

C Stan[1] C Andreescu[2] A Dobre[3] and D Iozsa[4],
POLITEHNICA University of Bucharest, Romania

[1] corraista@yahoo.com, [2]andreescu.cristian@gmail.com, [3]alexandru.c.dobre@gmail.com, [4]daniel_iozsa@yahoo.com

Abstract. The paper objective is to highlight the consequences of adding different alcohols in gasoline on the distillation characteristics of these blends. Changes of the distillation parameters (ti, t10, t50, t90, tf, E70, E100, E150) have been evaluated and, also, the evolution trends of the distillation curves for different alcohol added in mixture with the gasoline have been estimated. Several types of gasoline sold on the market and methanol, ethanol, i-propanol and butanol were used during the experiments and the corresponding distillation curves have been analyzed. The alcohol fraction in mixtures varied between 5 and 20%. Double blends with alcohol added in gasoline and triple blends with two alcohols added in gasoline were used. The comparison of the distillation curves of the mixtures was done with respect to that of pure gasoline. It was specified how the values of the distillation parameters, E70, E100 and E150, were set within the limits of EN 228. The distillation was made on 100 ml of fuel and the measurements were made on each 10 ml of fuel transformed into vapor state and then condensed. The influence of the alcohols present in these mixtures was manifested by the changes in the shape of the distillation curve. The butanol influence on the distillation temperatures was found lower than that of ethanol, because the physicochemical properties of the butanol are closer to those of gasoline. The molecules of alcohols actively interact with the fractions of gasolines, their combination leading to a conjugate effect and to a modifying the distillation parameters values. The variation of these parameters depends on the alcohol fraction in the mixture.

1. Introduction

The current resources of fossil fuels are limited; they are estimated to end in about 45 years. There is a need to look for alternative fuel sources, especially as the world's population will grow and fuel consumption will increase. At the same time, the pollution level of the planet has increased significantly during the recent years, and both the European and world level are pursuing the implementation of increasingly severe environmental standards. An important contribution to the global level of pollutants emitted into the atmosphere is represented by the pollutant emissions resulting from the combustion of fuels in the internal combustion engines (i.c.e.). That is why they were limited for car manufacturers by regional regulations. The transport sector is ranked third by the European Union in terms of greenhouse gas emissions. In this context, the use of alcohols as independent fuel or mixed with fossil fuels represents a direction of great interest for the fuel industry [1]. Another argument in using alcohols is that some of them can be obtained from the fermentation of

biomass that is a renewable source. It also leads to a decrease in dependence on oil-holding countries and changes geo-political strategies.
It has been admitted since back the '80s that the methanol and ethanol are accepted, technically speaking, for use on i.c.e. and, in time, researches has focused on other alcohols such as propanol, butanol, pentanol. A history of the use of gasoline-alcohols mixtures is presented in [2].
Some researches aimed at studying combustible mixtures with large shares of alcohols (5%-85%) in mixture [2, 3], others investigated narrower registers (5%-20%) [4, 5, 6].
In the gasoline sold on the Romanian market, the share of alcohols is restricted to a few percentages, alongside other oxygenated compounds. Standard EN 228 makes the following specifications found in the blending formula from the technical specifications of gasoline: it limits the oxygenated compounds content to 2.7% m/m, methanol to 3% (v/v), ethanol to 5% (v/v) and it requires a minimum of 4.5% (v/v) biofuel.
The alcohols used in mixture with the gasoline [3,7] have different properties from those of it. In table 1, compared to gasoline, the most important characteristics of alcohols that were used in this study are presented.
The high octane number of alcohols leads to the improvement of the ON of the gasoline to which they are added. This way higher compression ratios or larger spark advances can be used, which leads to higher engine efficiency [8]. Oxygen in the alcohols' molecule reduces the minimum air for complete combustion and allows changing the ignition limits of the mixture.
The high latent heat of vaporization of alcohols leads to the reduction of the temperature at the end of the intake process, allowing the increase of the amount of substance evolving during the cycle in the cylinder. The effective power downward trend due to the lower calorific power of alcohols can be recovered by increasing the fuel consumption. This will require an enlargement of the fuel tank. The vapor pressure of alcohols; is lower than that of gasoline; consequently, it makes more difficult the cold start of the engine, a disadvantage that demands the adoption of some specific constructive solutions. The higher the percentage of added alcohol, the more extensive constructive adaptations are.
FFV-Flexible Fuel Vehicles are vehicles tailored to various amounts of alcohols [9], capable to regulate the engine operation in correlation with the measured alcohol content. Thus, many vehicles have been equipped with engines that can operate with gasoline and ethanol from 0% to 85% without constructive changes [10].
Although the percentage of ethanol may vary over a wide range, the best results for the exploitation characteristics have been reported for percentages up to 20%.
In terms of pollutant emissions [2, 11], when using mixtures of gasoline and alcohols, HC and CO emissions decrease while for NOx and volatile organic compounds involved in smog formation increases have been reported [7,12 13].
Regarding the fuel consumption, although oxygenated fuels have a lower calorific power than gasoline, some experiments showed similar consumption expressed in l / km and usually lower when expressed in units of energy (MJ / km).
The benefits of using alcohols in mixture with gasoline are maximum around the percentage of 20%. The corrosive effects of alcohols also lead to limitations of the percentage added to the blend [4,5,6,14].
The fractional distillation is based on heating the liquid to the boiling point when it vaporizes and then condensing it back into the liquid state. In the case of gasoline, which consists of hundreds of fractions, each fraction has its boiling point, and thus a temperature range for boiling is obtained for gasoline. The distillation curve expresses the percentage of fuel that distils up to a certain temperature. Thus, the first distil the light fractions, followed by the medium ones, then the heavy ones. By its specific parameters (ti,t10,t50,t50,tf, E70,E100, E150, total distilled percentage and the amount of residue), the distillation curve provides information on fuel volatility that affects the exploitation, transport and storage characteristics.
The parameter ti - the initial distillation temperature and t10 - the temperature at which 10% of the fuel volume distils characterize the cold start engine's capability and behavior. The T50 has an influence on

engine warm-up time and on its operating stability. The parameter T90 is related to the combustion characteristics of fuel, influencing the complete burning, the fuel consumption, the engine wear, and deposits inside the engine.

E70, E100 and E150 represent the percentage of liquid that distilled up to 70, 100 and 150 ° C respectively [5]. Tf-final distillation temperature is the maximum temperature recorded during the distillation process.

The addition of alcohols in gasoline leads to the modification of some characteristic parameters of the distillation curve of the mixture. The study aimed to highlight the influences of different alcohols (M-methanol, E-ethanol, iP-isopropanol and B-butanol) added in varying proportions to the distillation characteristics of the mixtures.

The quality of the produced and traded gasolines is controlled by standards such as EN 228, standard which compels the distillation parameters (t10, t50, t50, tf, E70, E100, E150) within certain limits.

2. Materials and methods

There have been used the ON 95 pure gasoline sold on the market and blends of gasoline with methanol, ethanol, i-propanol and butanol, the alcohols being added in proportions ranging from 5 to 20% (v / v). There have been used double mixtures, gasoline + an alcohol and triple mixtures gasoline -80% (v/v) + two types of alcohols (E+iP, E+B and iP+B) 20% (v/v), meaning 10% (v/v) for each of the two alcohols.

The fractional distillation was carried out under ASTM 86 conditions. Samples of 100 ml of gasoline and gasoline-alcohol blend were used. Heating was done at a rate of 5ml /min. Several determinations were performed for the same type of sample (Figure 1) and the average values of the measurements were used.

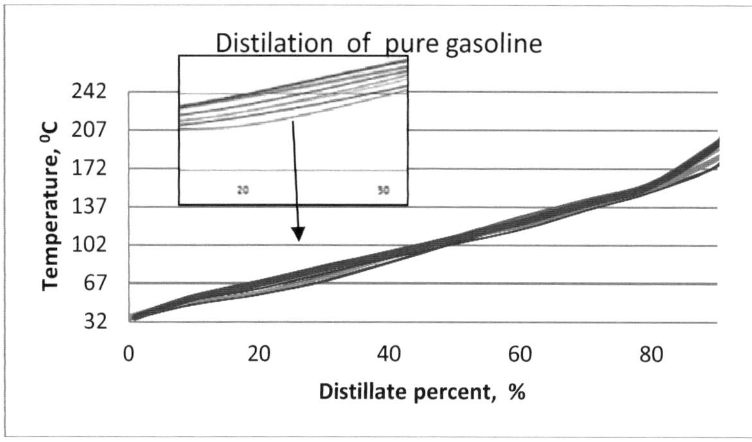

Figure 1. Several distillation curves for pure gasoline

3. Results and comments

The values obtained for the characteristic parameters of the fractional distillation are shown in table 1. The obtained distillation curves are shown in figure 2. The addition of alcohols led to changes in the distillation curves; a plateau area shows up close to the "steam point" of the alcohol.

In the case of addition of 20% methanol to the mixture (M20), the plateau area in the distillation curve is below the M100 pure methanol distillation point (64,5^{0}C). Lower values are obtained for the distillation temperature starting from more than 10°C below the methanol distillation temperature. The size of this range extends over a longer, almost double range from the 20 percent added methanol, which is higher than the volumetric proportion of methanol in the blend. This is due to the interaction of methanol molecules with mild HC, leading to the formation of azeotropes, with synergistic effects of increasing volatility [16].

Table 1. The distillation parameters of several alcohol-gasoline blends

Parameters	Gasoline Free alcohol	Gasoline blended with 20% alcohol						
		M20	E20	iP20	B20	E10+ iP10	E10+ B10	iP10+ B10
ti, °C	39	40	40	40	39	38	40	38
t10	53	53	54	57	60,5	55	57	55
t50	100,5	68	73	78	102,5	74	82	90
t90	169,5	168	158	153,5	163,5	152	159	158
tf	190	188	170	182	195	179	188	183
Distillate %v/v	95	95	93	96,5	97	95	97	97
Residuum %v/v	2	2	1	1	1,5	1,5	1,5	1,5
E70, %v/v	23,5	51,5	39,5	30	18	38	29	35
E100 %v/v	50	56	76	63	49	68	60	65
E150 %v/v	85,5	79	87	78	85	80	85	86,2

M=methanol, E=ethanol, iP=i-Propanol, B=butanol
M20=20% methanol +80% gasoline; E20=20%ethanol+80%gasoline;...
E10+ iP =10%ethanol+10%i-propanol+80%gasoline;...

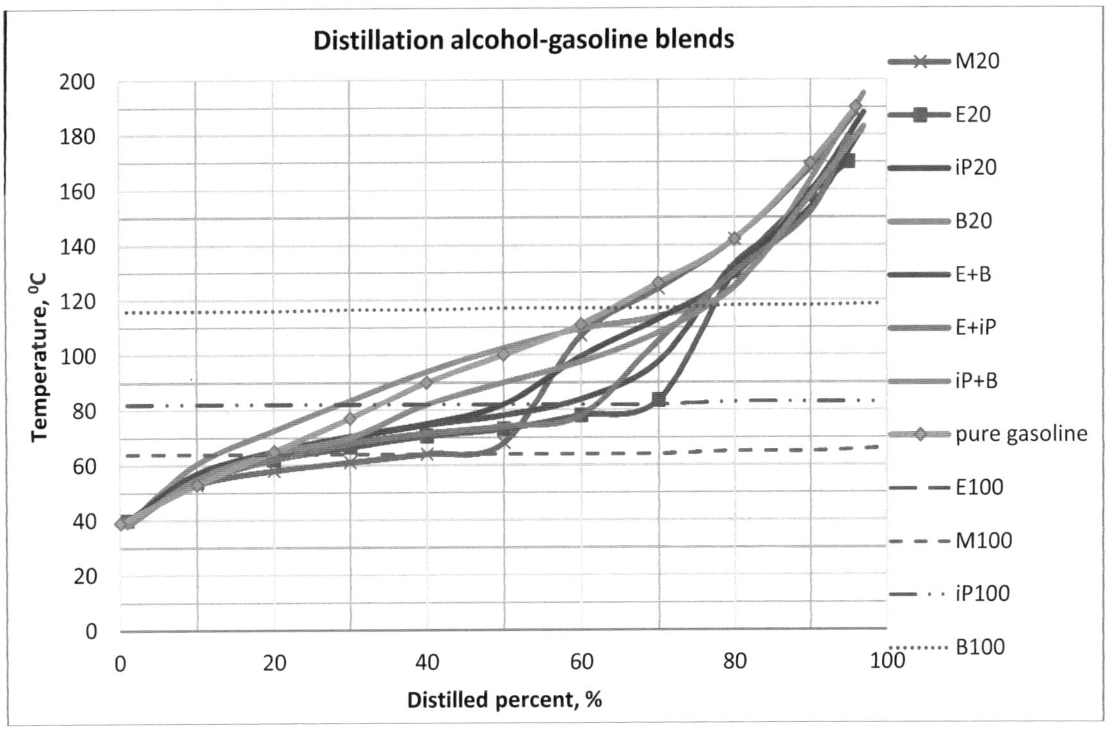

Figure 2. Distillation curves for pure gasoline, pure alcohols and alcohol gasoline blends

The M20 mixture records an E70 value (51° C) above the limit set by EN 228 (20⁰-50⁰C) and values that meet the requirements of the E100 (46⁰-71⁰C) and E150 (> 75⁰C) standards (Figure 3) [17].

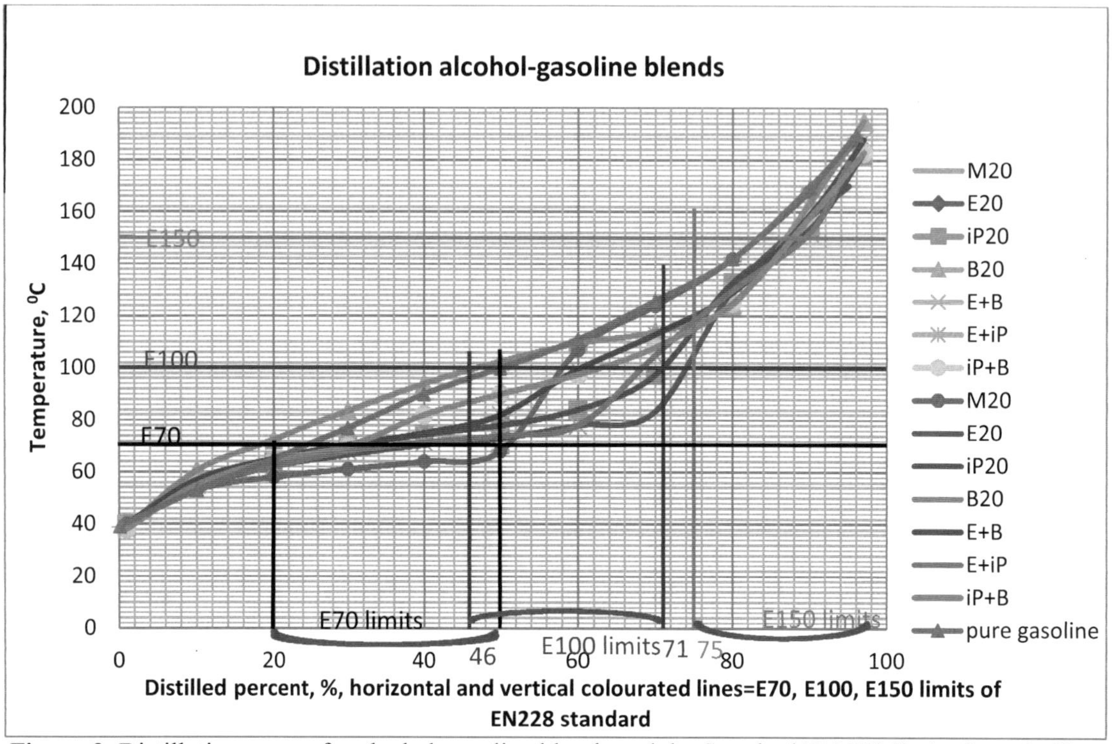

Figure 3. Distillation curves for alcohol-gasoline blends and the Standard EN 228 limits for volatility.

The horizontal and vertical colored lines represent the E70, E100, E150 limits of EN228 standard. The minimum value for t50 (68⁰C) and the maximum value of t90 (189⁰C) belongs to M20.

The E20 mixture also shows a distillation curve of a higher stretch on the axis of the distillate than methanol. The plateau is manifested at a higher temperature than the methanol mixture, justified by the ethanol distillation point (78⁰C) higher than that of methanol. For ethanol, the distillation curve shows the highest deviation from the pure gasoline distillation curve.

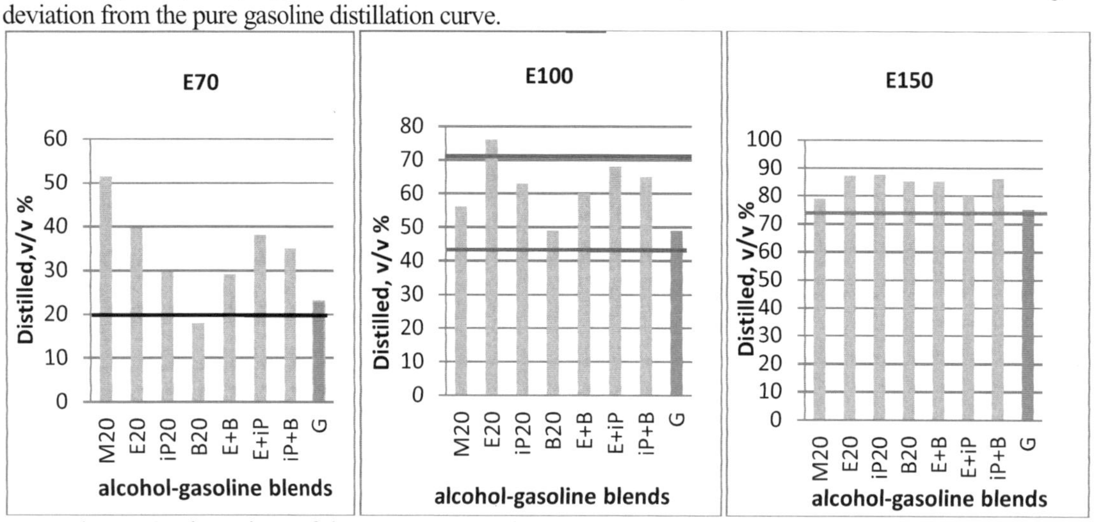

Figure 4. The values of the E70, E100 and E150 parameters for the alcohol gasoline blends
The horizontal colored lines = the limits of the EN 228 Standard

As in the case of M20, the plateau of distillation curve of the mixture E20 is manifested at a lower temperature than the distillation of ethanol. Near the 110^{0}C temperature, the distillation curve of the mixture with methanol returns to the characteristic values of the pure gasoline distillation curve.
In this case also, the interaction of ethanol molecules with light and medium fractions of gasoline and the formation of azeotropes take place. The E20 blend has E70 and E150 within European limits, but it exceeds the recommended range for the E100 (46^0-71^{0}C, Figure 3), showing greater volatility (74^{0}C, Figure 4).
The mixture with i-propanol, iP20, meets the volatility requirements for the E70, E100, E150. The degree of variation of the mixture distillation curve compared to that of the gasoline is lower than that that of M20 and E20.
The B20 distillation curve deviates from that of pure gasoline in both light and heavy hydrocarbons. In the first case, it has higher values, while in the second case its values are lower. The interaction of butanol with light HC molecules raises the B20 distillation temperature curve as against to the gasoline's while with the heavy HC the butanol causes the distillation temperature curve to decrease. The distillates of B20 blends are situated within the E100 and E150 volatility limits, but they are below the lower limit of the recommended range for E70 (Figure 4).
Notable for triple blends with two alcohols is that they all fit into the recommendations of the EN 228 Volatility Regulation. The deviations of the distillation curves are greatly attenuated. The largest change in the distillation curve compared to pure gasoline is shown in the E + iP mixture while in the case of iP + B mixture the deviation is the smallest. The initial distillation temperatures have very little variation from that of pure gasoline. The temperature t10 recorded the lowest value for the M20 mixture and the highest for B20.
T50 temperature has the lowest value for M20, followed by E20 and it is the highest for B20. T50 for the triple E + iP mixture is only 1 degree above the E20 mix.

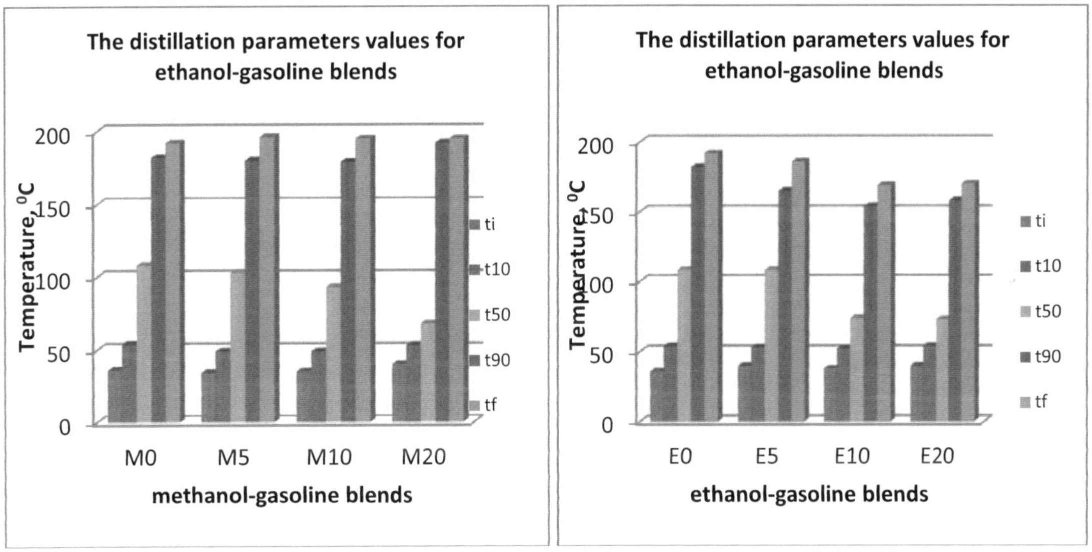

Figure 5. The variation of the distillation parameters ti,t10,t50,t90,tf, for the methanol-gasoline and ethanol-gasoline blends with the percentage of alcohol.

The temperature t90 has the lowest value for the triple E + iP mixture, very close to the iP20, and the highest value for B20, justified by the highest distillation point of pure butanol alcohols.
The distillations parameters ti,t10,t50,t90,tf, for each alcohol-gasoline mixture varies with the percentage of alcohol. Figure 5 presents the variation of the distillations parameters ti,t10,t50,t90,tf, for the methanol-gasoline and ethanol-gasoline blends with respect to the percentage of alcohol. In the case of methanol, there is an influence on t10 and t90 and, in the case of ethanol, it is recorded on t50, t90 and tf.

4. Conclusions

The double blends of pure gasoline with alcohols have modified the distillation curves compared to pure gasoline, the shape of the curve presents a plateau, whose size is proportional but greater than the percentage of added alcohol. The largest deviation was recorded for the E20 mixture.

E70 for blend B20 does not fall within the limits required by EN 228, being below them.

Mix E20 has a value for E100 that exceeds the maximum value allowed. The M20 mixture showed a higher volatility of the light fractions, the value of the E70 parameter exceeded the upper limit mentioned in the EN 228 standard.

In the case of triple blends with two alcohols, the distillation curves deviated less from those of pure gasoline. In all cases the characteristic values of the volatility of the triple blends were within the ranges specified in the Regulation.

5. References

[1] BP Statistical Review of World Energy June 2015

[2] Gravalos I Moshou D Gialamas T Xyradakis P Kateris D Tsiropoulos Z 2011 *Performance and Emission Characteristics of Spark Ignition Engine Fuelled with Ethanol and Methanol Gasoline Blended Fuels* Alternative Fuel book

[3] Andersen V F Anderson J E Wallington T J Mueller S A and Nielsen O J 2010 *Distillation Curves for Alcohol−Gasoline Blends* Energy Fuels 24 (4) pp 2683–2691

[4] Beverly L Smith and Thomas Bruno J *2007 Improvements in the Measurement of Distillation Curves Application to Gasoline and Gasoline + Methanol Mixtures* Ind. Eng. Chem. Res. 46 (1) pp 297–309

[5] Tarek M Aboul F Mazen O A and Ashour I 2017 *An Experimental Study on the Influence of Ethanol and Automotive Gasoline Blends* Journal of Petroleum & Environmental Biotechnology

[6] Belincanta J Alchorne J A Teixeira da Silva M 2016 *The Brazilian Experience with Ethanol Fuel* Brazilian Journal of Chemical Engineering vol.33 no.4

[7] Gautam M Martin D W Carder D 2000 *Emissions characteristics of higher alcohol/gasoline blends* Proceedings of the Institution of Mechanical Engineers Part A Journal of Power and Energy

[8] Ishibashi Y Asai M 1996 *Improving the Exhaust Emission of Two-Stroke Engines by Applying the Activated Radical Combustion* SAE 960742

[9] Hönig V Orsák M. Pexa M and Linhart Z 2015 *The distillation characteristics of automotive gasoline containing biobutanol, bioethanol and the influence of the oxygenates* Agronomy Research 13 (2), pp 558–567

[10] Cofaru C Sierens R Florea D Sandu V Lihtenchi M Verhelst S Chiru A Novac A Vasilovici N Bejan C Ispas N 2007 *Transport şi ingineria mediului* Brasov

[11] Doğan B Erol D Yaman H Kodanli E 2017 *The effect of ethanol-gasoline blends on performance and exhaust emissions of a spark ignition engine through exergy analysis* Applied Thermal Engineering Volume 120 pp 433-443

[12] Yoon S H et al 2009 *Effect of bioethanol as an alternative fuel on the emissions reduction characteristics and combustion stability in a spark ignition engine* Proceedings of the Institution of Mechanical Engineers Part D Journal of Automobile Engineering vol 223 pp 941-951

[13] Najafia G Ghobadiana B Tavakolia T Buttsworthb D R Yusafb T F Faizollahnejada M 2009 *Performance and exhaust emissions of a gasoline engine with ethanol blended gasoline fuels using artificial neural network* Applied Energy Volume 86 pp 630-639

[14] Hönig V Orsák M Pexa M and Linhart Z 2015 *The distillation characteristics of automotive gasoline containing biobutanol, bioethanol and the influence of the oxygenates* Agronomy Research 13(2) pp 558–567

[15] Kesin A Guru M 2011 M *Unleaded Gasoline on Exhaust and Noise Emissions of a Spark Ignition Engine* Energy Sources Journal Vol 33

[16] Hadler A Ott L Bruno T 2009 *Study of azeotropic mixtures with the advanced distillation curve approach* Fluid Phase Equilibria Volume 28 pp 49-59

[17] European Standard EN 228.

Research on cylinder processes of gasoline homogenous charge compression ignition (HCCI) engine

Corneliu Cofaru

Transilvania University of Brasov, 29, Eroilor Blvd. 500036 Romania

Abstract. This paper is designed to develop a HCCI engine starting from a spark ignition engine platform. The engine test was a single cylinder, four strokes provided with carburetor. The results of experimental research on this version were used as a baseline for the next phase of the work. After that, the engine was modified for a HCCI configuration, the carburetor was replaced by a direct fuel injection system in order to control precisely the fuel mass per cycle taking into account the measured intake air-mass. To ensure that the air – fuel mixture auto ignite, the compression ratio was increased from 9.7 to 11.5. The combustion process in HCCI regime is governed by chemical kinetics of mixture of air-fuel, rein ducted or trapped exhaust gases and fresh charge. To modify the quantities of trapped burnt gases, the exchange gas system was changed from fixed timing to variable valve timing. To analyze the processes taking place in the HCCI engine and synthesizing a control system, a model of the system which takes into account the engine configuration and operational parameters are needed. The cylinder processes were simulated on virtual model. The experimental research works were focused on determining the parameters which control the combustion timing of HCCI engine to obtain the best energetic and ecologic parameters.

1. Introduction

World urban areas face a number of major environmental challenges regarding the transport sector, their scale and intensity can vary, thus, in this case, a common set of issues can be identified. They are related to poor air quality, traffic volumes and congestion, high levels of ambient noise, lack of recreational areas and other factors of this kind. In the same time, these environmental challenges constitute a major problem and have significant impacts on human health, environment and economic performance.

Transport sector is important for the wealth of human societies and quality of life, providing access to jobs, services, education and leisure while creating the conditions to support economic growth, moving goods and people. This is why transport is an instrument in achieving some of the sustainable development goals, such as improving urban and rural access, improving safety and reducing air pollution.

Transport has also negative effects, such as pollution, including carbon dioxide emissions and accidents. In addition, transport with conventional power trains is also responsible for small amounts of other GHG emissions including methane, nitrous oxide and fluorinated gases , as well as acoustic noise and significant local air pollution from SOx, NOx, carbon monoxide, volatile organic compounds (VOC), unburnt hydrocarbons and particulates, all of them affecting human health and environment. The emissions' increasing was observed for both categories: passenger transport and freight transport. The values of this phenomenon are mainly due to an ever growing transport demands, characterized by large increases in passenger kilometers and tone kilometers. As for the passenger road transport, a relative decrease in the use of public transport is also noteworthy, even if the high level of efficiency on the passenger cars has not been sufficient to counteract this trend. As for the freight road transport, an increased share of road freight transport as opposed to other transport modes supplements the increased transport demand for goods.

There are some ways to achieve the tasks regarding diminishing fuel consumption, pollutants and CO_2 emissions by: homogenous charge compression ignition engine (HCCI) represents a recent combustion technology that combines the main advantages of the gasoline engines (the homogeneous charge) with the main advantages of the diesel engines (higher compression ratios which determinate higher efficiency) determining the potential to reduce emissions while increasing fuel economy. HCCI could give 50% improvement in engine efficiency at part-load compared to spark ignition engines and 30% compared to compression ignition engines.

An important advantage is the low temperature combustion that offers higher thermal efficiency and lower emissions of Nitrogen Oxides (NOx) and Particulate Matter (PM) in comparison with spark ignition and diesel engines. For such engines, a major problem is to control the start of the combustion because the air and fuel need to be mixed homogeneously before the starts of combustion and, the air-fuel mixture auto ignites due to the temperature increasing at the end of compression stroke.

2. Experimental setup

A single cylinder four-stroke engine was used to collect the experimental data in this research. The engine having the specifications presented in Table 1 was connected to a SIEMENS 1PH6167 4NG00 dynamometer, Figure 1.

Table 1. Engine's specifications

Engine type	SI	HCCI	[-]
No.of cylinders	1	1	[-]
Fuel	Gasoline	Gasoline	[-]
Displacement	652	652	[cm^3]
Compression Ratio	9.7:1	11.5:1	[-]
Bore	100	100	[mm]
Stroke	83	83	[mm]
Con-rod length	149	149	[mm]
Valve position	DOHC	DOHC -variable	[-]
Intake Valve Opening	17° BTDC	variable	
Intake Valve Closing	45° ABDC	variable	
Exhaust Valve Opening	47° BBDC	variable	
Exhaust Valve Closing	15° ATDC	variable	
No. of valves	4	4	[-]
Fuelling system	Carburettor	Direct injection	[-]
Injection Pressure	-	10	[MPa]
Lubrication	Dry Sump Lubrication by Oil Pump, Replaceable Oil	Dry Sump Lubrication by Oil Pump, Replaceable Oil	[-]
Cooling	Liquid Cooled	Liquid Cooled	[-]

During the tests, one could register data regarding: ambient, intake and exhaust pressures, injection and cylinder pressure, oil pressure, ambient, cylinder, oil, water temperatures, torque, the valves timing and pollutant emissions (HC, CO, NOx).

The standardized equipments were used for tests.[1,2,3,4,5,6,7]. The inlet air has the possibility to be conditioned in conformity with research requirements to obtain the control of combustion timing. The inlet and exhaust pressure was measured through KISTLER 6041A and KISTLER 4075A10 transducers. The fuel consumption was measured by using a fuel mass flow meter and the air flow was measured by using laminar anemometer.

IOP Conf. Series: Materials Science and Engineering **252** (2017) 012070 doi:10.1088/1757-899X/252/1/012070

Figure 1. The test engine on the research banch

The cylinder's pressure data were obtained with a KISTLER 6041A piezoelectric transducer. The information regarding crankshaft's position was obtained by using a transducer type ROD 426A.

The engine test incorporated a full electronic and manual control.

The calibration tool was also used to record engine's control parameters such as the intake manifold pressure, air flow rate, injection timing, fuel injection pressure, injection duration, equivalence ratio, etc.

The equipment used for pollutant measurements is presented in Table 2.

Table 2. Equipment for pollutants measurement

Equipment type	*Pollutant*
NDIR Pierburg	O_2/N_2
NDIR Pierburg	CO, CO_2
CLD Pierburg	NO/NO_x
HFID AVL	HC

The HCCI test engine was realized starting from a SI engine equipped with a carburetor as a platform. In order to obtain a HCCI operation cycle some modifications were implemented to the engine. The compression ratio was increased from 9.7:1 to 11.5:1, the fuelling system was changed, replacing the carburetor by direct injection system provided with an electromagnetic actuating injector. The injector was fitted by replacing the spark plug.

To modify the quality of cylinder's mixture in order to control the combustion process, the engine was provided with a variable gas distribution using a camshaft phaser system, as it is shown in Figure 2. The original camshafts were replaced by others camshafts with new cams profile. These cams provide a short valves opening period and a lower valve lifts to offer a safety clearance (valve-piston) when the valve timings are changed during the research work.

The thermodynamic analysis of combustion process was realized using the "BremoXP" soft-ware.

The engine tests matrix is presented in Table 3.

3. Experimental results and discussions

The main problem of the HCCI is to control the start of ignition, for this reason, the cylinder's charge needs to have a high temperature and sufficient dilution at the start of the compression process.

In cylinder, the air fuel mixture temperature must be at a high level to initiate and to sustain the chemical reactions to determine auto-ignition process and combustion process. The air-fuel mixture dilution is necessary to control the rates of heat release reactions.

Figure 2. The engine cam phaser

Table 3 Engine testing matrix

Parameter	Values				
EVC (CA)	630	644	646	649	-
IVC (CA)	204	210	215	-	-
IA (CA)	325	335	345	350	-
ID (μs)	2400	2500	2600	2700	-
N (rpm)	1900	2000	2100	2200	2300

For this research, the burnt gases meeting both requirements were used, to rise the mixture temperature and to realize the mixture's dilution, exploiting their effects: heating, dilution, heat capacity and chemical activity..

The analytical and experimental studies showed that in gasoline HCCI engine the ignition is dominated by the charge heating effects instead ,the combustion duration is dominated by the dilution and heat capacity effects of mixtures.

The heat release is influenced by a combination of charge heating effect and by the dilution and heat capacity effects. Taking in account these influences on ignition and combustion processes in gasoline HCCI, it is important to manage the using of burnt gases, because it is advantageous to use burnt gases at as high temperature as possible to determine auto ignition of air fuel mixtures especially at low loads. For this reason residual gas trapping was chosen, by using a negative valve overlap strategy (NVO). The great amount of residual gas is trapped in cylinder when the exhaust valve is closed early during the exhaust stroke. In order to keep the trapped burnt gases in cylinder and to avoid the back flow of these gases into the inlet manifold, the intake valve is opened later, after TDC at the beginning of the inlet stroke. Applying this strategy, a higher charge temperatures are obtained and it made possible to extend the combustion at low loads, but at high speeds and loads, the ignition can be also advanced and the results being a very fast pressure's increase.[8,9,10]

In order to use the trapped residual gas as to obtain a HCCI combustion, the set up of the valves timings was due to realize a negative valve overlap. During the engine test to determine the influence of quantity of burnt gases on combustion process, the engine speed (n), start of injection (IA), fuel injected mass (DI) and the intake valve timing were kept constant. The exhaust valve closing timing varied between 639 and 649 CA, these points are placed during the exhaust stroke. This means that the exhaust valves closes earlier than the piston reaches the TDC, on these conditions, the burnt gases are compressed after the closing of the exhaust valves, the consequence being the rising in the cylinder's pressures. Note that the exhaust valves open later than on the usual recommended values for the spark ignition engine

In order to avoid the back flow of the burnt gases into the inlet system, the intake valve will be opened later during the inlet stroke.

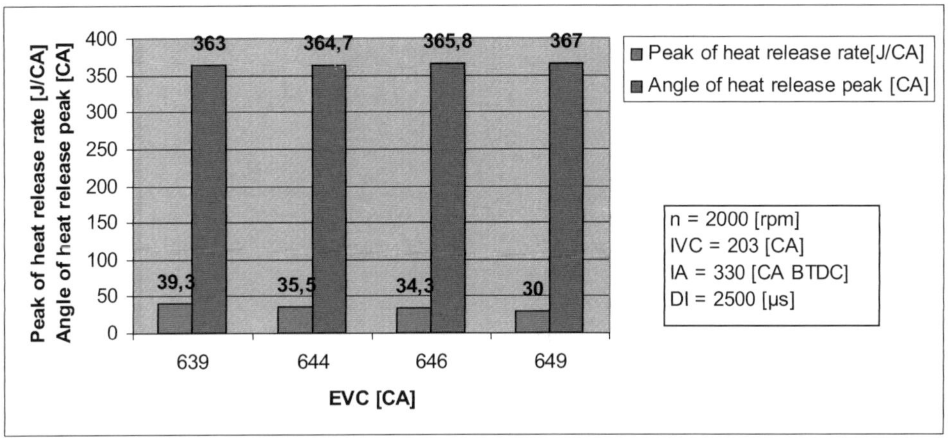

Figure 3. Infuence of exhaust valve closing (EVC) on heat release peak and angle of heat release peak.

Analyzing the cylinder's processes determined by changing the EVO values, the increased amount of burnt gases trapped in cylinder by earlier closing of exhaust valve can be noted and it determines a vigorous combustion which is placed near the TDC, the peak of rate heat release reaches 39,3 [J/CA] at EVO=639 [CA], retarding the closing of the exhaust valve at 649 [CA], the peak of heat release value is 30 [CA], representing a decrease of - 23,7 %. Diminishing the quantity of burnt gases in the cylinder's mixture determinates the moving of combustion in the expansion stroke, The angle of heat release rate peak moves from 363 to 367 [CA] (Figure 3).

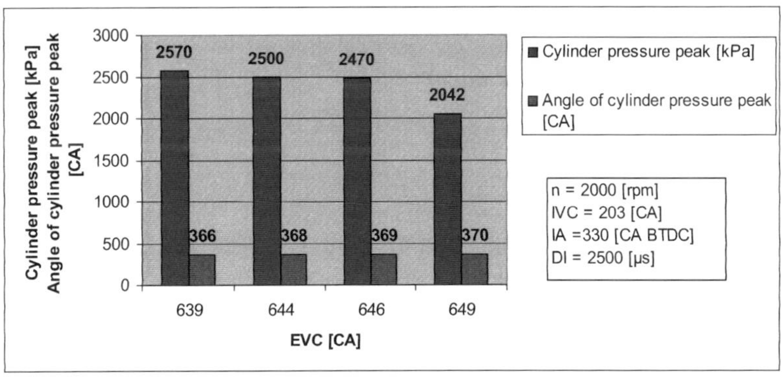

Figure 4 Infuence of exhaust valve's closing (EVC) on cylinder's pressure peak and angle of cylinder's pressure peak.

The improving of combustion process by keeping in cylinder a higher quantities of burnt gases are sustained by the cylinder's pressure and temperatures, the peak values following in the same way the rate of heat release, decreasing from 2570 [kPa] to 2042 [kPa] for the cylinder's pressure and decreasing from 1462 [K] to 1375[K] for temperatures (Figure 4 and Figure 5).The quantity of burnt gases kept in cylinder determines the oxygen concentration influencing the gaseous pollutants formation, the HC and CO increases with rising of burnt gases quantities, instead the NOx rises with the decreasing of burnt gases quantities because the mixtures is provided with more oxygen contents.

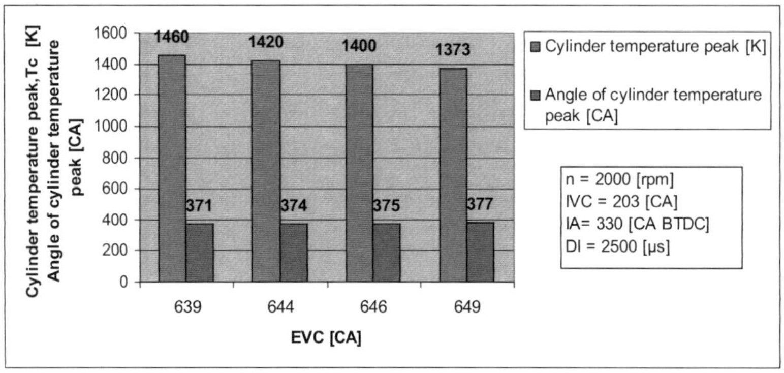

Figure 5. Infuence of exhaust valve's closing (EVC) on cylinder's temperature peak and angle of cylinder's temperature peak.

Regarding the intake process, the inlet valve's opening timing is placed later during the intake stroke to avoid the back flow of the cylinder's burnt gases; there is no influence of mixture compositions. The research was focused on the influences of the intake valve's timings upon cylinder processes: mixture quality and combustion. By the intake valve's closing time, the quality of the cylinder charge can be calibrated, in way of maximization or of reduction. The influence of the intake valve's closing was realized for 204, 210 and 215 [CA], maintaining constant the following parameters: EVC at 649 [CA], the injection advances at 330 [CA], the injection duration ID=2550 [µs] and the engine speed n = 2000 [rpm], in these cases, the quatities of burnt gases and fuel remainig constant in the cylinder.

By changing of values of IVC, the intake valve's opening timing, determining the modification of the fresh air content in the cylinder's mixture was also modified, in fact, the quality of mixtures are changed. Increasing the values of retarding the inlet valve's closing from 204 [CA] to210 and 215 [CA], the data of experimental research show that the peak of heat release rate decreases from 29,8 [J/CA] at IVC = 204 [CA] to 25,1[J/CA] at IVC = 210 [CA] (- 15,8%) and to 23,8 [J/CA] at IVC = 215 [CA] (- 20,1%),

The peaks of heat release rate are placed at the same crank angle. The peak of cylinder's temperature decreases from: 1377 [K] to 1347 [K] and, 1320 [K] for established IVC values. The pollutant emissions follow the values of cylinder's temperatures, HC increases from 650 [ppm] to 681 [ppm] and 775 [ppm], CO increases from 800 [ppm] to 1000 [ppm] and 1445 [ppm], and NOx decreases from 17,5 [ppm] to 15,2 [ppm] and 15,6 [ppm].

During the experimental tests, an important step was to determine the influence of the fuel injection advance to obtain the best point of start of injection (SOI) to have an HCCI engine. For this analyse, an engine operation point characterized by the following parameters was established: engine speed n =2000 [rpm], intake valve's closing timing IVC = 203 [CA], exhaust valve's closing timing EVC = 644 [CA], injection duration ID = 2500 [µs] in order to keep the cylinder's content constant at the moment of the fuel injection. The advance of start of the injection points (SOI) chosen were : 350 [CA], 345 [CA], 334 [CA] and 325 [CA], representing 10 [CA], 15 [CA], 26 [CA] and 35 [CA] on the engine's cycle.

The point (SOI) was placed during the intake stroke after the exhaust valve had been closed, but before the intake valve had been opened. The fuel was injected into hot burnt gases where it evaporated and mixed with them. After the opening of the intake valve, the burnt gases and the evaporated fuel mixed with air, and this mixture was homogenized during the intake and compression strokes. By decreasing of the fuel injection advance from 350 [CA] to 325 [CA] the peak of heat release rate decreased from 31,3 [J/CA] to 25,5 [J/CA], and combustion processes had a light moving in expansion stroke, influencing the peak of cylinder's temperature which decreased from 1373 [K] to 1327 [K]. The level of cylinder's temperature influenced the pollutants emissions , HC and CO emissions increased with decreasing o fuel injection advances, instead NOx decreased from 14,4 [ppm] at IA= 350 [CA] to 8,6 [ppm] at 325 [CA], representing - 40,28%.

The influence of engine's speed on the cylinder's processes was realized for 1900, 2000, 2100, 2200 and 2300 rpm, keeping constant the following parameters: intake valve's closing IVC = 203 [CA], exhaust valve's closing EVC = 644 [CA], fuel injection advance IA = 350 [CA] BTDC, fuel injection duration ID = 2500 [μs] which takes into account the constant content of the cylinder.

Table 4. Influence of engine's speeds on combustion process and emissions

Parameter / Engine speed (rpm)	Peak of heat release rate (J/CA)	HC (ppm)	CO (ppm)	NOx (ppm)
1900	23,5	847	1453	5,7
2000	25,0	767	1120	9,0
2100	27,5	727	1027	12,0
2200	29,0	680	907	15,0
2300	31,0	607	1057	17,2

On the one hand, the engine's speeds have an important inflence on the heat exhange process which decreases at the increasing of engine's speed, on the other hand, the homogenization process of cylinder's content is intensified. By increasing of engine's speed, the combustion process is improved, represented by an increased of heat release peak from 23,5 [J/CA] to 31,0[J/CA] representing +21,6%.

The influence of fuel quantity injected on cylinder's processes was realised by modyfing the fuel injection duration, the used values being 2400 [μs], 2500 [μs], 2600 [μs] and 2700 [μs], the engine's settings being: n =2000 [rpm], intake valve closing IVC = 203 [CA], exhaust valve's closing EVC = 644 [CA], fuel injection advance IA = 330 [CA] BTDC. In fact, by the injected fuel quantity, the energy released by the engine can matching with the demand to overcome the resistance, on this condition, the duration of injection can be considered as the engine's load.

The research regarding the developments of HCCI engine demonstrates the capability of solution regarding lowering the fuel consumption, pollutant emissions, especially that of NOx (Table 5).

Table 5. Comparison between SI and HCCI engines' performances (n=2000 [rpm], iemp=0,2 [MPa])

Parameter / Engine type	Specific fuel consumption (g/kWh)	HC (g/kWh)	CO (g/kWh)	NOx (g/kWh)
SI	340	2,9	35	0,84
HCCI	254	3,1	7,5	0,04

4. Conclusions

During the experimental research, the negative overlap strategy to control the engine's processes by residual gas trapped in cylinder was adopted. The amount of trapped burnt gases has a huge influence on the combustion process.The amount of residual gas trapped in the cylinder was changed by modifying the exhaust valve's closing timing earlier during the exhaust stroke. Closing the exhaust valve earlier during the exhaust strokes, the burnt gases are than compressed and this process determines the increasing of the pumping work.

The intake valve's timings were chosen in order to keep the trapped burnt gases in the cylinder and to avoid the back flow of these gases into inlet manifold; the intake valve is opened later, after the TDC at the beginning of the inlet stroke. The closing of the intake valve was adopted after the TDC to maintain the volumetric efficiency in an acceptable range. For this reason, the intake and exhaust valves timing needs to be tuned and depends on the engine's load and speed. The engine's load was modified by changing the injection duration. The increasing of quantities of injected fuels per engine's cycles determines the rising of the heat release rate.

The obtaining of homogenous mixture in HCCI needs to select carefully the advance of injection (start of injection). Within the experimental tests, the point (SOI) was placed during the intake stroke after the exhaust valve had been closed, but before the intake valve had been opened.

References

[1] Cosgarea R, Aleonte M, CofaruC 2011 The influence of the valve lift strategies on the combustion characteristics of a homogeneous charge compression ignition engine model. *International Journal of Mechanics* Issue 3, Volume 5, pp 191-201

[2] Cosgarea R, Aleonte M, Cofaru C 2011 The Influence of the Internal Exhaust Gas Recirculation (EGR) On the Pumping Losses. In: *Bulletin of the Transilvania University of Brasov* • Vol. 4 (53) No.1 – 2011 Series I, ISSN 2065-2119 (Print), ISSN 2065-2127 (CD-ROM) pp. 1-7

[3] Cosgarea R, Cofaru C, Aleonte M, Sandu G 2011An approach for modelling the homogeneous charge compression ignition (hcci) combustion using the vibe functions. *Proceedings of the 10th International Automotive Congress "Automotive and Environment" CAR 2011* Universitatea din Pitesti , ISSN 1453-1100

[4] Cosgarea,R. 2011 Cercetari privind utilizarea de sisteme avansate de ardere prin folosirea amestecurilor omogene/partial omogene la motoarele cu aprindere prin comprimare. *PhD Thesis*, Universitatea Transilvania din Brasov

[5] Sandu G, Cofaru C, Cosgarea R, Jelenschi L, Aleonte M L 2011 The impact of the valve variable timing upon the pumping work in the internal combustion engine. In: *Proceedings of the 10th International Automotive Congress "Automotive and Environment" CAR 2011* Universitatea din Pitesti , ISSN 1453-1100

[6] Maxim V Subbotin, Karl Lukas Knierim, Sungbae Park, Aleksandar Kojic, Jasim Ahmed 2008 Modeling and Control of a Two Stroke HCCI Engine. 2008 *American Control Conference* Westin Seattle Hotel, Seattle, Washington, USA, June 11-13,. ISBN978-1-4244-2079-7

[7] Mehran Bidarvatan, Vishal Thakkar, and Mahdi Shahbakhti 2014 Grey-box Modeling and Control of HCCI Engine Emissions. 2014 *American Control Conference (ACC)* June 4-6, Portland, Oregon, USA. ISBN 978-1-4799-3271-9

[8] Nick J. Killingsworth, Salvador M. Aceves, Daniel L. Flowers, Francisco Espinosa-Loza, and Miroslav Krstic 2009 HCCI Engine Combustion-Timing Control: Optimizing Gains and Fuel Consumption ViaExtremum Seeking. *IEEE Transactions on control systems technology*, vol. 17, no. 6, november, ISSN 1063-6536

[9] Nuraini A. Aziz, Najihah A. Rahman and Othman Innayatullah:2013 Control of Homogeneous Charge Compression Ignition (HCCI) Engine Through Stochastic Variables. *World Applied Sciences Journal* 27 (3): 359-366, ISSN 1818-4952

[10] Zhao H 2007 HCCI and CAI engines for the automotive industry, Cambridge edition: *Woodhead Publishing Limited.*

IOP Conf. Series: Materials Science and Engineering **252** (2017) 012071 doi:10.1088/1757-899X/252/1/012071

On the assessment of performance and emissions characteristics of a SI engine provided with a laser ignition system

A. Birtas[1,*], N. Boicea[1], F. Draghici[2], R. Chiriac[3,4,*], G. Croitoru[5], M. Dinca[6], T. Dascalu[5] and N. Pavel[5,*]

[1] Renault Technologie Roumanie, North Gate Business Center, B-dul Pipera, Nr. 2/III, Voluntari, Ilfov, 077190, Romania. Email: adrian.birtas@renault.com
[2] University Politehnica of Bucharest, Faculty of Electronics, Telecommunications and Information Technology, Spl. Independentei 313, Sector 6, 060042, Bucharest, Romania.
[3] University Politehnica of Bucharest, Faculty of Mechanical Engineering, Spl. Independentei 313, Sector 6, 060042, Bucharest, Romania. Email: radu.chiriac@pub.ro
[4] Associated member to EA7341 CMGPCE of Conservatoire National des Arts et Metiers, Paris, France
[5] National Institute for Laser, Plasma and Radiation Physics-INFLPR, Laboratory of Solid-State Quantum Electronics, Atomistilor 409, Magurele 077125, Ilfov, Romania. Email: nicolaie.pavel@inflpr.ro
[6] University of Bucharest, Faculty of Physics, Magurele 077125, Ilfov, Romania
[*]Authors to whom any correspondence could be addressed.

Abstract. Performance and exhaust emissions of spark ignition engines are strongly dependent on the development of the combustion process. Controlling this process in order to improve the performance and to reduce emissions by ensuring rapid and robust combustion depends on how ignition stage is achieved. An ignition system that seems to be able for providing such an enhanced combustion process is that based on plasma generation using a Q-switched solid state laser that delivers pulses with high peak power (of MW-order level). The laser-spark devices used in the present investigations were realized using compact diffusion-bonded Nd:YAG/Cr^{4+}:YAG ceramic media. The laser igniter was designed, integrated and built to resemble a classical spark plug and therefore it could be mounted directly on the cylinder head of a passenger car engine. In this study are reported the results obtained using such ignition system provided for a K7M 710 engine currently produced by Renault-Dacia, where the standard calibrations were changed towards the lean mixtures combustion zone. Results regarding the performance, the exhaust emissions and the combustion characteristics in optimized spark timing conditions, which demonstrate the potential of such an innovative ignition system, are presented.

1. Introduction

The actual internal combustion engines which have been developed in time either as spark ignition engines, or compression ignition engines will continue to dominate for the next decades not only the transport sector but also some parts of industrial and agricultural domains. Heat in the spark engines is released by expanding as turbulent flames over premixed homogeneous fuel/air mixtures in the case of multipoint fuel injection, or over nonhomogeneous mixtures in the case of gasoline direct injection. In both cases a controlled ignition is produced by the electrical system of the engine at a certain moment of the operating cycle and at a specific location of the combustion chamber where the spark plug is located.

Engines efficiency, performance and emissions will be continuously improved and even small enhancements, perhaps previously not considered worth pursuing, will become now important. In such context, traditional combustion problems as ignition and early flame development will acquire renewed importance. Ignitability of lean mixtures and stability of combustion in lean and diluted (high EGR) mixtures in spark ignited engines will represent the key issues to be solved in the next combustion technologies applied for heat engines [1].

A novel possible technological solution that seems to represent a promise future development of gasoline engines is the gasoline direct injection compression ignition. This idea was initially proposed to improve the problems of other advanced combustion concepts which were associated with the use of different new petroleum based fuels or alternative fuels. In this case, the major challenge is that the fuel must be injected early in the compression stroke when pressure and temperature are low and the suitable conditions for fast and robust ignition are poor. This challenge is associated too with the gasoline's very low Cetane Number (CN) which is usually estimated to be no higher than about CN15 [2] that implies longer ignition delays.

To overcome these drawbacks many researches have been dedicated in the last decades to the use of enhanced ignition systems like Radio Frequency Ignition or Laser Ignition (LI) systems. For ensure faster flame initiation in the region of the combustion chamber, where the mixtures are within the flammability limits, promoting thus a rapid combustion over very lean mixtures in the whole combustion chamber with substantial benefits on fuel efficiency and emissions, extremely high-power ignition sources have been tested. LI has been demonstrated to offer some advantages in comparison with classical spark plug ignition in terms of improved efficiency and reduced variability [3-5]. For the first time, such kind of ignition was reported in 1978, when a single cylinder internal combustion engine was ignited by a CO_2 laser [6]. Advanced Q-switched Nd:YAG lasers were used to ignite a four-cylinder engine in 2008 [7]. In those experiments the laser beams were directed to and then focused into the engine cylinders by common optics (lenses and mirrors).

Consequently to these researches, spark-plug like Nd:YAG/Cr^{4+}:YAG lasers that were side pumped [8] or longitudinally pumped [9-11] and that delivered one beam [8,9] or multi-beam output [10,11] were demonstrated to be viable for use in engines. Finally, as result of many years of research, the world first gasoline engine vehicle that was ignited only by laser igniters was presented in 2013 by T. Taira et al. [12].

All these efforts are nowadays dedicated to the improvement of LI systems in terms of their reliability, miniaturization, and better operation in correlation with combustion chamber architecture, fuel distribution, inner flow field, turbulence intensity which characterize the diverse and variable passenger car engine working conditions [13]. Moreover, new studies concerning the influence of delivered energy on a local or on multiple ignition points, the requirements concerning time and space distribution associated with the use of alternative fuels with low carbon content are under progress [14-18].

The current paper represents an extension of the previous researches that have been performed from some years at INFLPR and RTR concerning the possibility to use of a LI system to an automobile engine [19,20]. It consists in a theoretical and experimental study regarding the influence of the location of ignition point inside combustion chamber and ignition event on operating cycle expressed in terms of engine operating stability and combustion analysis.

2. Experimental Details

The spark ignition engine was a conventional multipoint fuel injection car engine with 4 cylinders with 1598 cm^3 total displaced volume and rated power of 64 kW at 5500 rpm. The engine was provided with open ECU Continental EMS 3132 and connected with an ECU control system ETAS-INCA v2.1. The experimental test bed was correspondingly equipped for the control of the engine operation and the collection of all the relevant parameters characterizing engine's operating conditions. The engine was coupled with a Froude Consine AG 250-UK eddy current dynamometer for control and the measurement of its torque and speed. Test bed control system is TEXCEL v6 having installed the corresponding software Froude for engine dyno-control and for low speed parameters variation data acquisition on 16 channels. The exhaust emissions (CO_2, CO, NO_x THC)

IOP Conf. Series: Materials Science and Engineering **252** (2017) 012071 doi:10.1088/1757-899X/252/1/012071

were measured and the relative air/fuel ratio λ was calculated by a gas analyzer Horiba Exsa -1500L. The engine fuel consumption was measured with a dynamic fuel meter AVL 733S coupled with a fuel temperature conditioner AVL 753C. A data acquisition system for high speed variation parameters as cylinder pressures, electric discharge parameters, injectors needle lift, type AVL Indimodul 621 for 8 channels was coupled with a micro-IFEM AVL system. Two cylinder-pressure sensors AVL ZF43 were installed in cylinder 1 and 4 and one AVL GU21D in the cylinder 1 by crossing the cylinder head. The crank angle encoder AVL 365 connected to the AVL Indi module was used for generating the marker and trigger signals necessary for data recording (figure 1).

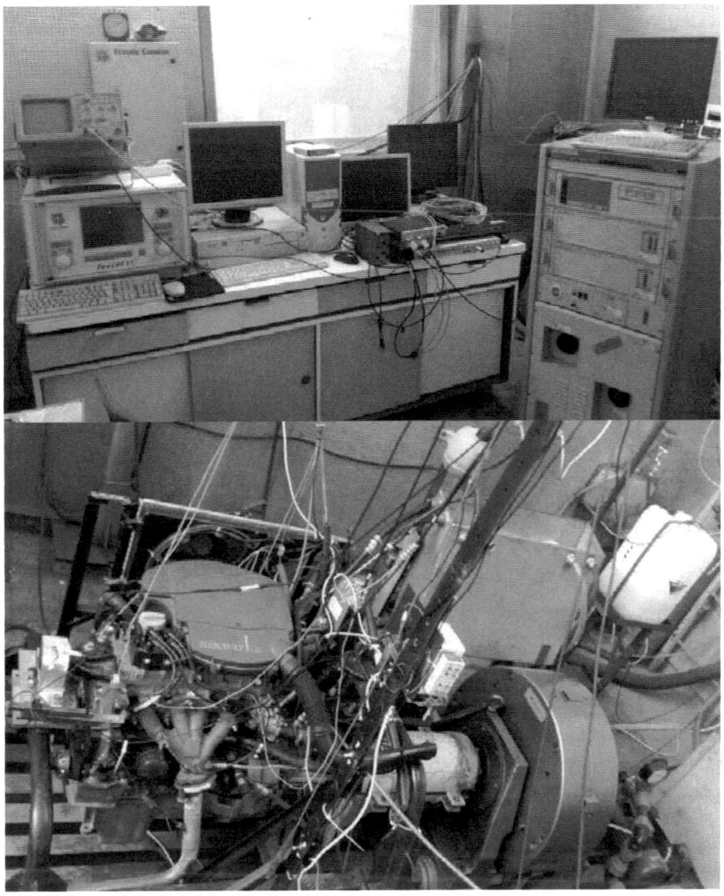

Figure 1. The experimental test bed.

For each tested operation point, pressure traces for 200 consecutive cycles were registered. The operation condition was light load, at 2000 rpm and 2 bar brake mean effective pressure (BMEP), considered as representative for the city traffic. The load was identified by the pressure existing in the inlet manifold equal to 430 mbar for light load. Final data comparisons concerning the effect of using different ignition systems at various λ and engine speed and load, were made for optimized ignition timings (minimum advance for best torque, MBT).

The LI system was developed at the National Institute for Laser, Plasma and Radiation Physics, Magurele, Romania. It consisted of an integrated four-laser sparks device that was triggered by the electronically control unit of the car. In general (or usual operation mode), each laser spark delivered pulses with energy up to E_p= 4.0 mJ and pulse duration around 0.8 ns; the repetition rate could be increased from few up to 60 Hz. In addition, by increasing the pump-pulse duration at a defined repetition rate, each laser spark could emit multiple pulses (i.e. two, three or even four pulses), realizing the so-called burst-pulse operation. More information about the design and performances of such a laser spark can be found in Ref. [19].

3. Simulation Details

The numerical model was built using the AVL BOOST 2016 simulation codes to estimate the engine performance, efficiency, combustion characteristics and exhaust gas emissions. The engine calibration parameters and cylinder processes' simulation were described by using codes v2016 (AVL BOOST Theory and AVL BOOST Users Guide) [21]. Hence, the engine components, such as the intake and exhaust manifolds, cylinder geometry, air filter or the system boundaries and catalyst were implemented in the BOOST interface based on the real values which were taken from the engine installed on the test bed and used in this study. All the engine components are linked together by pipes and its symbolic model is shown figure 2.

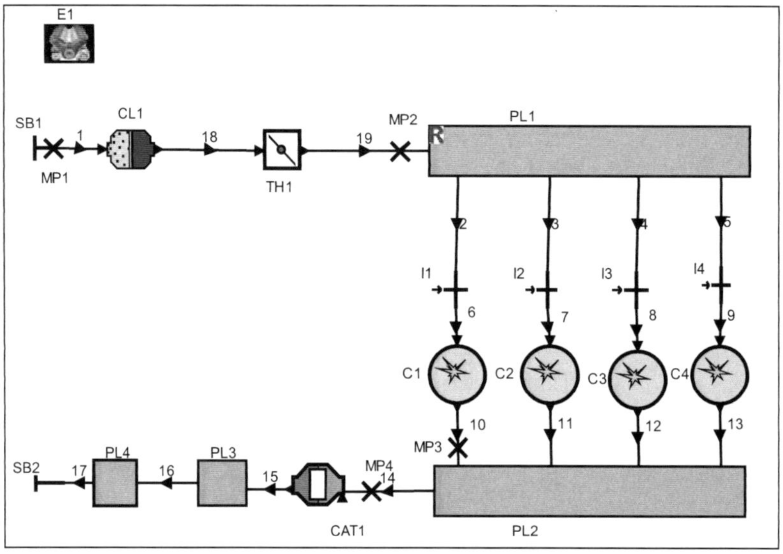

Figure 2. AVL BOOST numerical model.

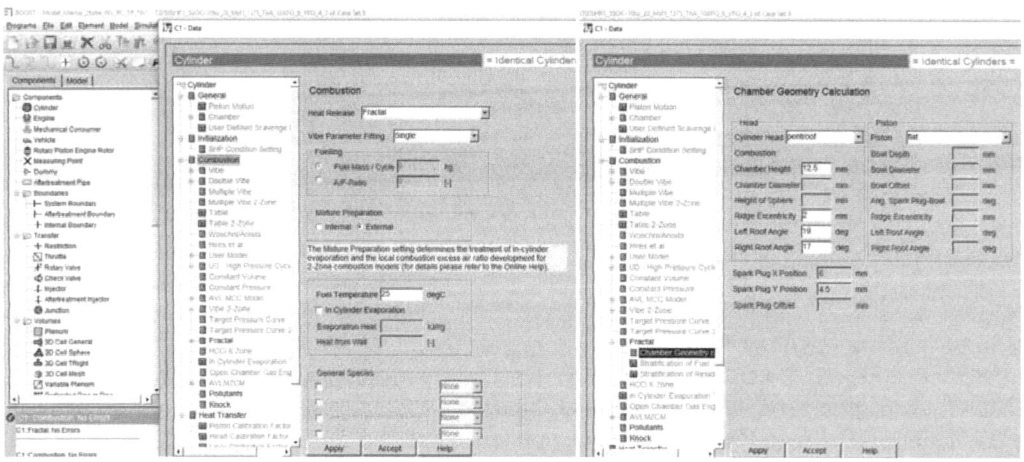

Figure 3. The Fractal combustion model characteristics.

An assessment of the possible influence of the location of ignition point inside the combustion chamber was performed by considering the Fractal combustion model (figure 3) that was selected from the library of the BOOST code. The combustion chamber shape manufactured in the cylinder head is Pentroof, while the flat piston top crown was adopted having the same dimensions to the real combustion chamber and piston. In this model the position (X,Y) of the flame kernel is considered as variable being initially connected for model calibration with the spark plug location. The distance to the combustion chamber center is related to the focal length of the lens that is used to focus the laser beam and is calculated based on current values of (X,Y) which were selected in different steps. This distance started with 10 mm equivalent to the original position of the spark plug and goes down to 0 mm, equivalent to the combustion chamber center.

4. Experimental results

The results obtained show a better engine stability when running with LI system, compared with the conventional spark plug ignition system. Reduced coefficients of variability (COV) were obtained for different engine characteristics for combustion as well as for smooth engine operation.

A smaller COV, in terms of maximum in-cylinder pressure for 200 consecutive cycles, was obtained for the LI system. Thus, the COV coefficient was improved by 11% for the LI operation in comparison with that of the ignition by a conventional spark plug (figure 4.) In addition, the difference between the COV of these two ignition systems in terms of angle of maximum pressure is 2.2 points of COV (figure 5).

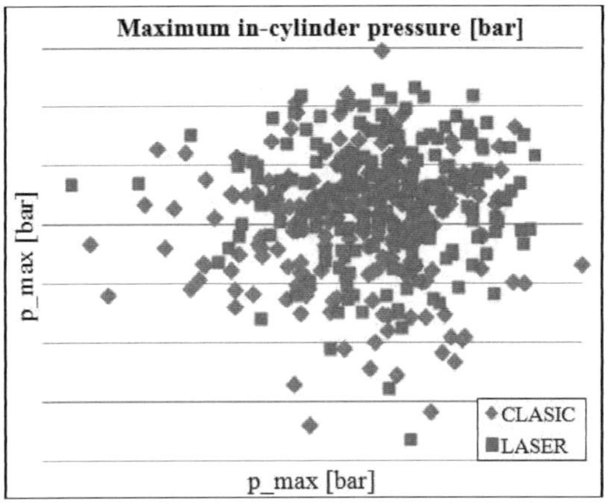

Figure 4. The variation of maximum in-cylinder pressure (p_max) [bar] for 200 consecutive cycles at 2000 rpm and 2 bar of BMEP (graph n/n+1).

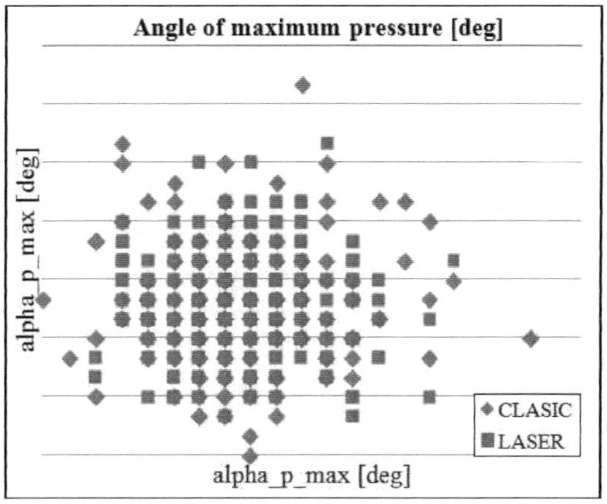

Figure 5. The variation of maximum in-cylinder pressure angle (alpha_p_max) [deg] for 200 consecutive cycles at 2000 rpm and 2 bar of BMEP (graph n/n+1).

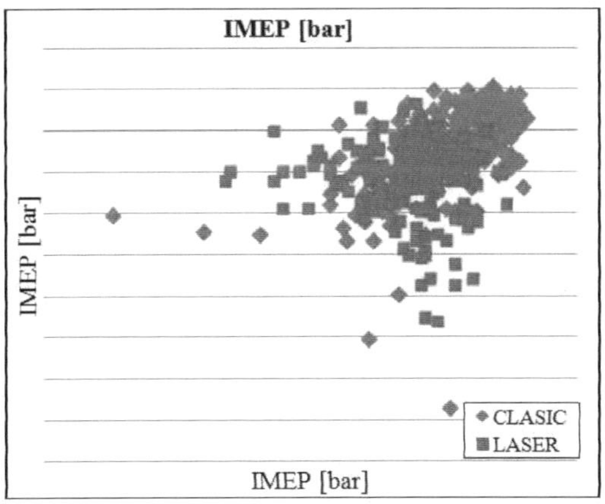

Figure 6. The variation of Indicated Mean Effective Pressure (IMEP) [bar] for 200 consecutive cycles at 2000 rpm and 2 bar of BMEP (graph n/n+1).

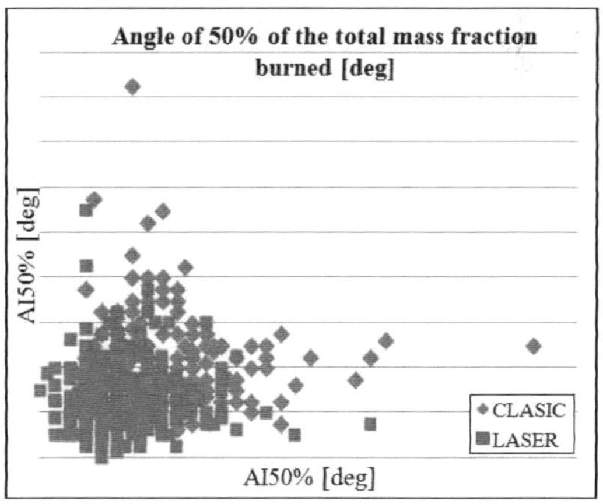

Figure 7. The angle when 50% of the total mass fraction is burned (AI50%) [deg] for 200 consecutive cycles at 2000 rpm and 2 bar of BMEP (graph n/n+1).

Also, the COV calculated for all 200 consecutive cycles in terms of indicated mean effective pressure (IMEP) [bar] was smaller (by 15.8%) for the LI system compared with the conventional one. (figure 6). Another important combustion characteristic analyzed was the variability of the angle where 50% of the total mass fraction was burned (AI50%), measured in crank angle degrees. A better stability for LI was measured with a smaller COV by 6.1% compared with the classical ignition system (figure 7).

In all above figures representing the stability of engine operation (figure 4 to figure 7) more disperse point repartition can be seen for the classical spark plug ignition system compared with LI system. These promising results in terms of stability when using a LI system support the extension of investigation in the operating areas of the engine in which the stability coefficients are high.

5. Simulation results

The numerical simulation model offered the base to analyse the variation trend of main characteristics of engine efficiency, combustion and exhaust emissions, when modifying the position of the ignition point inside the combustion chamber. Simulations for five different focal lengths were made for three different spark timings. An analysis for lean mixtures (λ=1.2 [-]) was made, highlighting the more significant advantages of this engine operation zone. We would like to mention that the above mentioned focusing positions can be very easily obtained be changing the lens that is used to focus the laser beam inside the cylinder.

In figure 8, the variation of brake specific fuel consumption (BSFC) [g/kWh] is presented. Higher drops of percentages for BSFC, up to 7.5%, is expected for all spark timings from the leaner mixture (λ=1.2) compared with the stoichiometric engine functioning condition (λ=1) (table 1).

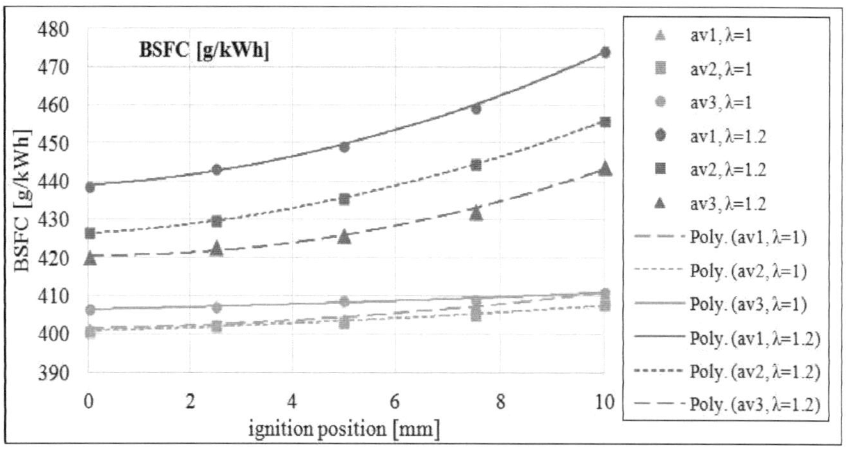

Figure 8. Brake Specific Fuel Consumption (BSCF) [g/kWh] depending on ignition position for λ=1 and λ=1.2 at 2000 rpm.

The total combustion duration characteristics from figure 9 shows, at λ=1, small variations depending on the ignition position. The most important combustion duration reduction was 35% for λ=1.2 at the 3rd spark advance investigated (av3) (table 1).

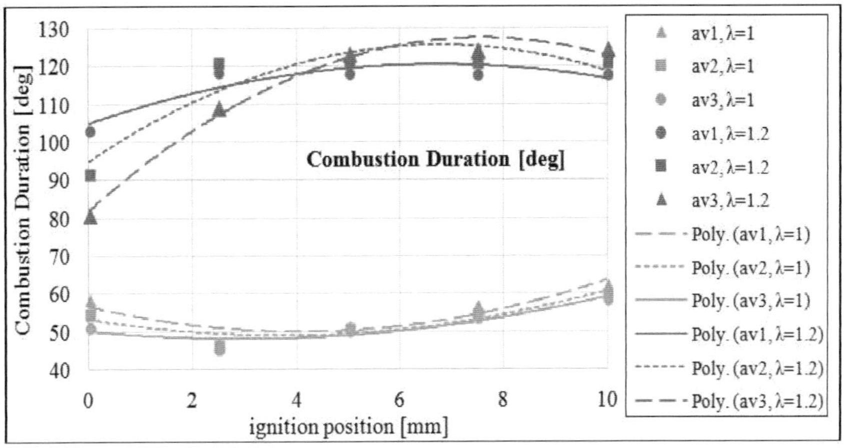

Figure 9. Combustion duration [deg] depending on ignition position for λ=1 and λ=1.2 at 2000 rpm.

The major disadvantage when changing the ignition position was obtained for the nitrogen oxides emissions (NO_x). This increase in NO_x occurs as the peak fire temperature rise along with the central position of the flame kernel (figure 10). The greater actual increase was for λ=1 and the 3rd spark advance (av3), of 2 g/kWh between the initial point and the center of the combustion chamber (table 1). Smaller disadvantages were obtained for λ=1.2 ; the biggest increase of 0.9 g/kWh was at the 3rd spark advance (av3). This drawback could be mitigating by using high diluted EGR mixtures together with an enhanced ignition system like LI.

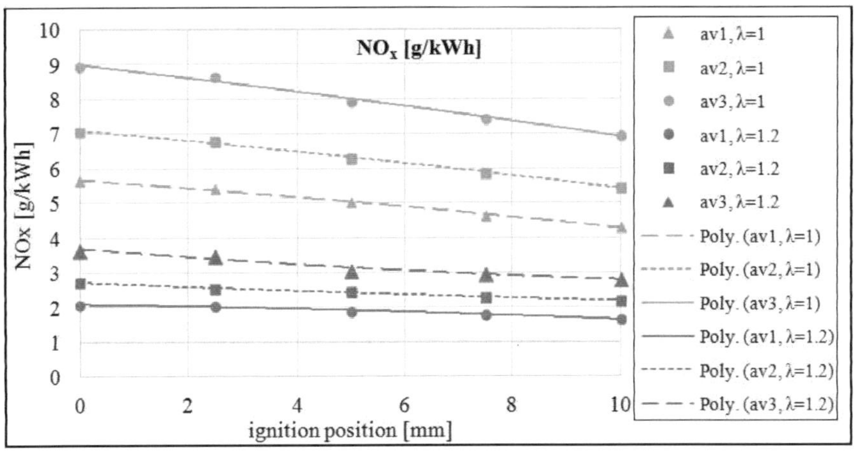

Figure 10. Emissions of nitrogen oxides NOx [g/kWh] depending on ignition position for λ=1 and λ=1.2 at 2000 rpm.

Table 1. Percentages of variation between conventional position (10 mm) and centered position (0 mm) for the main characteristics.

Spark timing [deg]	av1, λ=1	av2, λ=1	av3, λ=1	av1, λ=1.2	av2, λ=1.2	av3, λ=1.2
BSFC [%]	-2.2	-1.7	-1.1	-7.5	-6.4	-5.3
Combustion Duration [%]	-6.1	-8.7	-12.1	-12.6	-24.4	-35.0
NO_x [%]	+31.1	+29.9	+28.7	+25.0	+25.1	+31.0

6. Conclusions

A petrol multipoint fuel injection engine equipped successively with a spark ignition and a LI system was operated for a specified condition of speed 2000 rpm and 2 bar brake mean effective pressure. In the same time, a set of simulations were performed for the same engine and same condition but with different locations of ignition point inside the combustion chamber. It was made an analysis in terms of performance, efficiency and combustion characteristics for which the main findings can be summarized as follows:

- Theoretical investigations shown the possibility to improve efficiency by 2% when changing the ignition point location from the actual position of the spark plug to the center of the combustion chamber;
- This evolution seems to be similar when the stoichiometric mixture that is characteristic for the normal operation of the engine is replaced by lean mixture;
- The theoretical results obtained for stoichiometric mixture were in good agreement with an overall accuracy of 3% by the experimental data measured on the test bed;
- The extension towards the lean mixture associated with the use of an enhanced ignition system as LI can improve efficiency and engine operation stability by reducing cycle-by-cycle variability.

These investigations will continue for lean mixtures but covering also other engine operating conditions.

References

[1] Kalghatgi G T 2015 *Proceedings of the Combustion Institute* **35** 101.
[2] Concawe Fuels and Emissions Management Group 2014 *Exploring a Gasoline Compression Ignition (GCI) engine concept* Report no. 13/14.
[3] Rooney P D 1994 *Opt. Eng.* **33** 510.
[4] Tauer J, H. Kofler H and E. Wintner E 2010 *Laser & Photon. Rev.* **4** 99.
[5] Dearden G and Shenton T 2013 *Opt. Express* **21** A1113.
[6] Dale J D, Smy P R and Clements R M 1978 *SAE International* paper 780329; DOI: 10.4271/780329.
[7] Mullett J, Dickinson P, Shenton A, Dearden G and Watkins K G 2008 *SAE International* paper 2008-01-0470; DOI: 10.4271/2008-01-0470.
[8] Kroupa G, Franz G and Winkelhofer E 2009 *Opt. Eng.* **48** 014202.
[9] Tsunekane M, Inohara T, Ando A, Kido N, Kanehara K and Taira T 2010 *IEEE J. Quantum Electron.* **46** 277.
[10] Pavel N, Tsunekane M and Taira T 2011 *Opt. Express* **19** 9378.
[11] Dascalu T, Croitoru G, Grigore O and Pavel N 2016 *Photonics Research* **4** 267.
[12] Taira T, Morishima S, Kanehara K, Taguchi N, Sugiura A and Tsunekane M 2013 *1st Laser Ignition Conference (LIC'13),* Yokohama, Japan, April 23-25, paper LIC3-1.
[13] Grzeszik R 2017 in *Laser Ignition Conference 2017*, OSA Technical Digest (online) (Optical Society of America, 2017), paper LFA3.1.
[14] Saito T 2017 in *Laser Ignition Conference 2017*, OSA Technical Digest (online) (Optical Society of America, 2017), paper LWA4.2.
[15] Hayashi J, Nakatsuka N, Morimoto I and Akamatsu F 2017 in *Laser Ignition Conference 2017*, OSA Technical Digest (online) (Optical Society of America, 2017), paper LWA4.4.
[16] Gupta S, Almansour B, Vasu S, Wång Q, Leeuwen R V and Bihari B 2017 in *Laser Ignition Conference 2017*, OSA Technical Digest (online) (Optical Society of America, 2017), paper LThA3.1.
[17] Agarwal A K 2017 in *Laser Ignition Conference 2017*, OSA Technical Digest (online) (Optical Society of America, 2017), paper LThA3.3.
[18] Prasad R K and Agarwal A K 2017 in *Laser Ignition Conference 2017*, OSA Technical Digest (online) (Optical Society of America, 2017), paper LWA5.7.
[19] Pavel N, Dascalu T, Salamu G, Dinca M, Boicea N and Birtas A 2015 *Opt. Express* **23** 33028.
[20] Pavel N, Birtas A, Croitoru G, Dinca M, Boicea N and T. Dascalu T 2017 in *Laser Ignition Conference 2017*, OSA Technical Digest (online) (Optical Society of America, 2017), paper LWA4.3.
[21] AVL BOOST 2016 Theory and Usersguide documentation.

Acknowledgments

The authors from INFLPR acknowledges a grant of Ministry of Research and Innovation, CNCS - UEFISCDI, project number 137/2017, PN-III-P4-ID-PCE-2016-0332, within PNCDI III. The authors thanks the AVL Advanced Simulation Technologies team for the significant support that was offered during performing simulation investigations. Special acknowledgements are addressed to Renault Technologie Roumanie for their technical support offered as improvements of the experimental test bed.

Contribution of developing advanced engineering methods in interdisciplinary studying the piston rings from 1.6 spark ignited Ford engine at Technical University of Cluj-Napoca

[1]Cherecheş, Ioan - Aurel; [1]Borzan, Adela - Ioana; [1]Băldean, Doru - Laurean

[1]Technical University of Cluj-Napoca, Romania

aurel.chereches@auto.utcluj.ro doru.baldean@auto.utcluj.ro

Abstract. Study of construction and wearing process in the case of piston-rings and other significant components from internal combustion engines leads at any time to creative and useful optimizing ideas, both in designing and manufacturing phases. Main objective of the present paper is to realize an interdisciplinary research using advanced methods in piston-rings evaluation of a common vehicle on the streets which is Ford Focus FYDD. Specific objectives are a theoretical study of the idea for advanced analysis method in piston-rings evaluation and an applied research developed in at Technical University from Cluj-Napoca with the motor vehicle caught in the repairing process.

1. Introduction

The expansion of modern engines would have been unfeasible without advanced lubricant chemistry and proper lubricant formulation. Introduction of diamond like carbon (DLC) coatings opens further possibilities in improving performance of engine and transmission components, which cannot longer be achieved only by lubricant design. DLC coatings show extremely good promise for several applications in automotive components as they exhibit excellent tribological properties [1].

Despite extensive research into alternative methods, the internal combustion engine is expected to remain as the primary source of vehicular propulsion for the foreseeable future. There are still significant opportunities for improving fuel efficiency, thus directly reducing the harmful emissions. Consequently, mitigation of thermal and frictional losses has gradually become a priority. The piston-cylinder system accounts for the major share of all the losses as well as the emissions. Therefore, the need for an integrated approach, particularly of a predictive nature is essential. The cylinder liner temperature is critical in mitigating power loss as well as reducing Hydrocarbon (HC) and Nitrogen Oxide (NOx) emissions from the compression ring – cylinder liner conjunction [2].

The past work of the authors on this subject developed experimental testing procedures on the specific task of piston-ring technical state determination in the case of spark ignited engine from Ford Focus 1.6 16v Turnier, code DNW 74 kW. Piston-rings construction and operation, as well as its specific working-surface configuration, materials, dimensions and lubrication procedure, in various thermo-mechanical loading regimes, are essential in influencing the engine's life-expectance. In this paper are presented the piston-rings construction and their measured dimensions at the end of exploitation period of a spark ignited engine from Ford Focus. The importance for monitoring the piston-ring dimensions at different locations and directions consists in the possibility of controlling their wear-ness [3].

The vehicle and main components considered for this study are shown in the Figure 1.

Figure 1. The studied vehicle type model Ford Focus with spark ignited engine 1600 dm^3 FYDD [3]

Figure 2 shows the main features of piston rings which are installed on the piston assembly in order to isolate the combustion chamber from the oil sump.

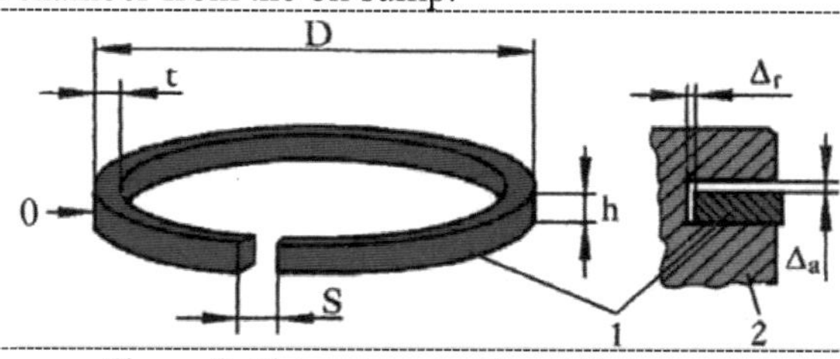

Figure 2. Piston ring significant features [4].
0-operational interface; 1-compression piston ring; 2-piston; D-exterior diameter; S-piston ring clearance; h-piston ring height; t-piston ring thickness; Δa-axial clearance between piston and ring; Δr-radial clearance between piston and ring.

There is an opportunity to develop the studies around piston rings in order to better understand the specific phenomenology of their wear process. This opportunity resides in the common use and availability of selected car in this study, as well as of its common engine. The operation of exchanging the piston rings is often applicable to engines after few hundred thousand kilometers in use or even earlier if the operating regime is higher (in the range of full loads).

By knowing the wear process specific evolution and by understanding the mechanical efforts applied on the surfaces and parts of the piston ring from the analyzed engine of Ford Focus it may be assessed the technical state of the entire engine or even the whole motor-car.

The second compression piston ring was took into detailed consideration in order to highlight the wear process, the critical wear area, to obtain some insight related to the working and wear evolution, as well as digital inspection of the wear parameters.

In figure 3 is shown the simplified model of the second compression piston ring from the analyzed Ford engine.

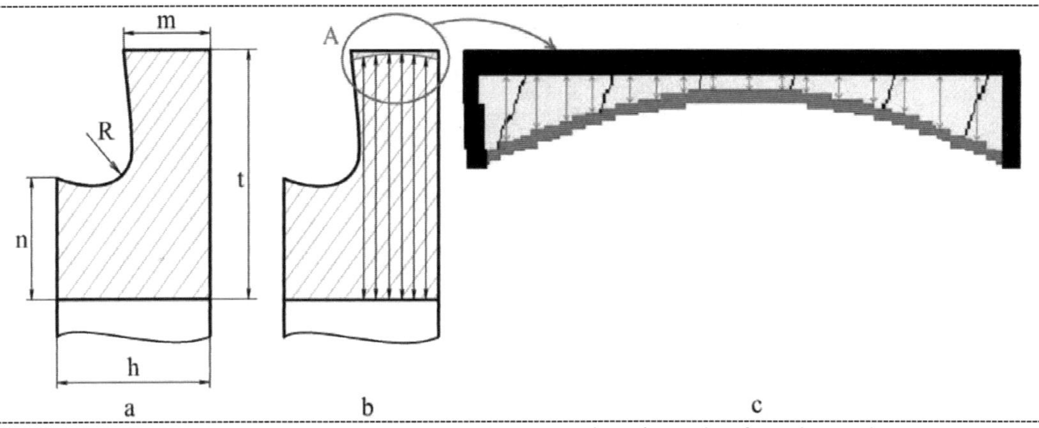

Figure 3. Simplified section cut through second compression piston ring from the Ford Focus FYDD engine.
a-dimensions; b-analyzed wear area; c-main wear zone; A-critical area detail.

The present paper shows the possibility of implementing an advanced method of digital evaluation and inspection of automotive mechanical components in order to improve the data acquisition and development process, as well as design of those parts.

2. Methodology and materials

2.1. Methodology

By analyzing the construction, considering the materials and design which were inspected or determined in the study, respectively by using the computer aided scanning and evaluation, it may be understood and interpreted the wear-ness level, as well as its evolution peculiarities. The present paper points out an engineering method for computer aided piston-ring evaluation and analyze in the field of physical wear-ness concerning the spark ignited engine in service. The researched and evaluated case is a real Ford Focus FYDD spark ignited engine, and was brought to be analyzed with precise methods by use of latest technology. An innovative method of advanced engineering analyze was put at work, using Digimizer application in order to inspect the piston ring surface. The correlation between computer processing power and piston - ring evaluation effort is mentioned.

First, there are some basic measurements of piston rings. Secondly, image capturing on test bed. Thirdly, software analyze of wear zone with Digimizer.

2.2. Materials

The present paper takes into close consideration a motored vehicle available at the Technical University from Cluj-Napoca in order to outline the possibility of digitizing and inspecting very closely the wear area. There were used devices and apparatus such as cameras and graphical processing units in order to improve data analyze from the piston ring cut section.

The graphic and image recording camera was placed close to the piston ring supporting bench. Both the camera and test bench were kept steady still while the proper image was taken and properly transferred to processing unit to further develop the analysis and wear inspection.

Using the Digimizer application (often and commonly used till now in medicine and biology) in automotive field is already an innovation by its own, but the present work takes the analyzing process to a new level by making progressive steps concerning the wear process evaluation and wear area inspection with mechanical, digital and graphic tools.

Figure 4 shows the some of the significant components that are considered in first stage of practical experimentation due to the necessity of starting with measured data.

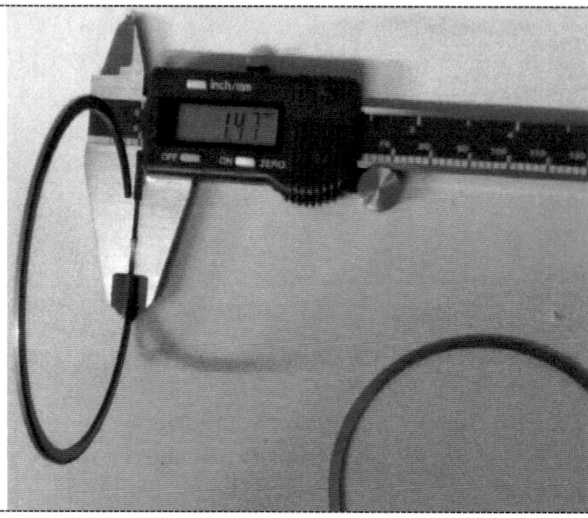

Figure 4. The first stage of measurement.

2.3. Mathematics and calculus

When the rings are installed on the piston and in operation there are some peculiar efforts and specific forces that applies to them which have to be considered in order to determine the causes of their degradation and lack of performance in operation.

The main effort in a piston ring is bending, when the maximum stresses appear in the external fiber, which determines the maximum bending torque during the operation time through middle fiber radius (r_o) and the surface pressure [5] which influences the dynamic rim wear in engine's operation:

$$M_{max} = 1.742 \cdot p_{em} \cdot r_o^2, \tag{1}$$

where: M_{max} is maximum torque generating the bending effort, in Nm; p_{em} – medium elastic pressure, in bar; r_o – middle fiber radius, in m.

Calculating the elastic pressures (that press the piston ring on the cylinder influencing the wear force and the rim slit) with the following mathematical model [5]:

$$p_{em} = 0.15E \frac{S_o / t}{\left(\frac{D}{t} - 1\right)^3 \frac{D}{t}}, \tag{2}$$

may be determined the basic bending effort in operation as follows:

$$\sigma_{f\,max} = \frac{M_{max}}{W} = 1.242 \frac{E \cdot S_o / t}{\pi D \left(\frac{D}{t} - 1\right)^3}, \tag{3}$$

where: W resistance modulus of the cut section; S_o – clearance between ring ends when free (unrestraint); t – radial thickness of piston ring; D – cylinder diameter.

Shear stress which is generated when installing the piston ring between the piston and the cylinder does not depends on the ring height, but make a significant difference in rim wear evolution.

When installed inside the engine cylinder the piston ring takes a stress given by the following relation [5]:

$$\sigma_{m\,max} = \frac{4 \cdot E \cdot \left[1 - \frac{S_o / t}{2{,}804 \cdot \pi}\right]}{\pi \frac{D}{t} \left(\frac{D}{t} - 1{,}4\right)^3}. \tag{4}$$

3. Experimental setup and protocols

3.1. Experimental setup protocol for testing and measurement

In order to achieve the desired and targeted applied results and to develop some interesting conclusions in relation with the practical measurements and Digimizer mapping of wear area in an effort of implementing the advanced and innovative engineering method in studying piston ring post-operation state and phenomena in a specialized research I.C.E. Laboratory of the Technical University from Cluj-Napoca there was designed and implemented a test bed represented simplified in figure 5 and also was configured a digital processing station for software applied research as in figure 6.

Figure 5 shows the basic model in the image capturing process when implementing the fundamental idea of the present work.

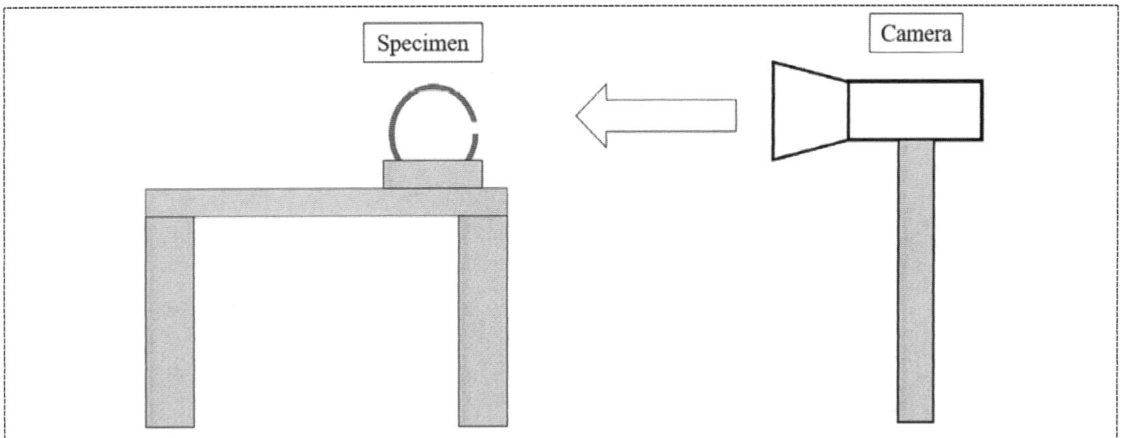

Figure 5. The simplified model of camera placement in front of the supporting bench with the studied piston ring.

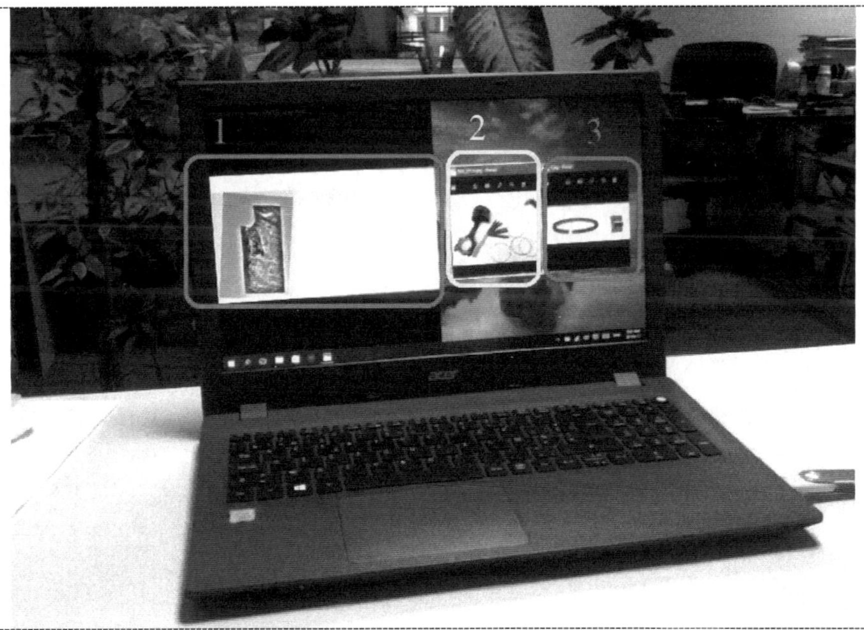

Figure 6. Post-operation processing unit used for data analyze and information storage.
1-Digimizer interface; 2-obtained data through practical measurements; 3-CAD for modeling the piston ring.

3.2. Experimental results and observations

Wear area of piston ring end parts may be determined specifically in each particular case after the engine was dismantled and components were separated and measured in order to get precise measured data necessary for digital processing phase. After measuring the piston rings in the present case and after taking the adequate image capture with the image recording camera (as it is shown in figure 5) a specialized software is used in order to determine the precise wear area according to the figure 7 and figure 8.

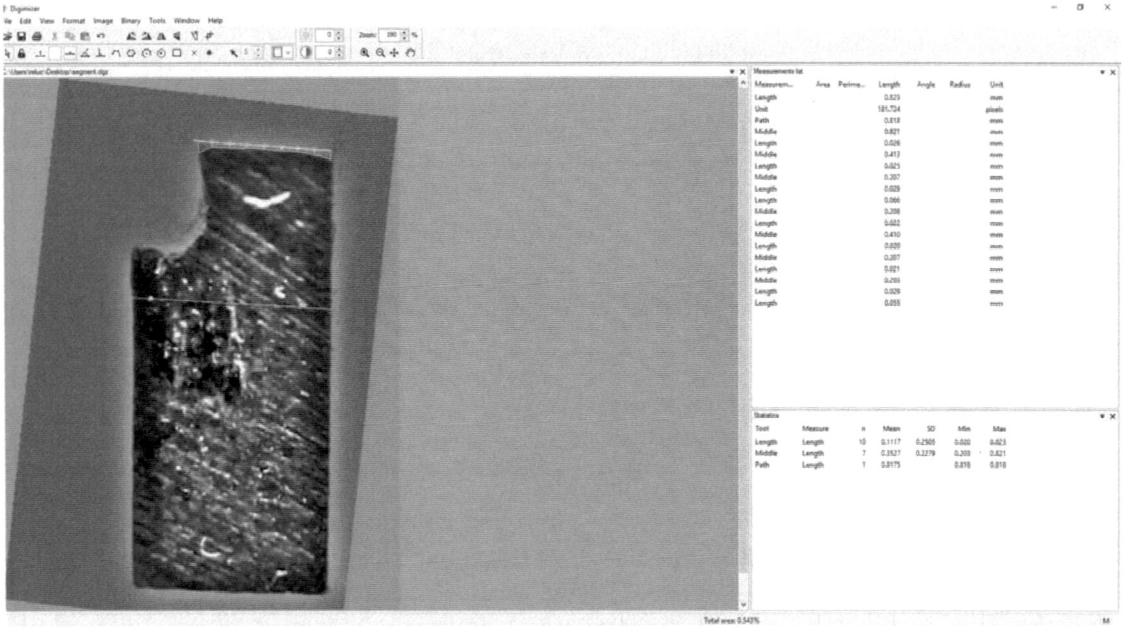

Figure 7. Digimizer interface with measurements and data concerning wear area of piston ring.

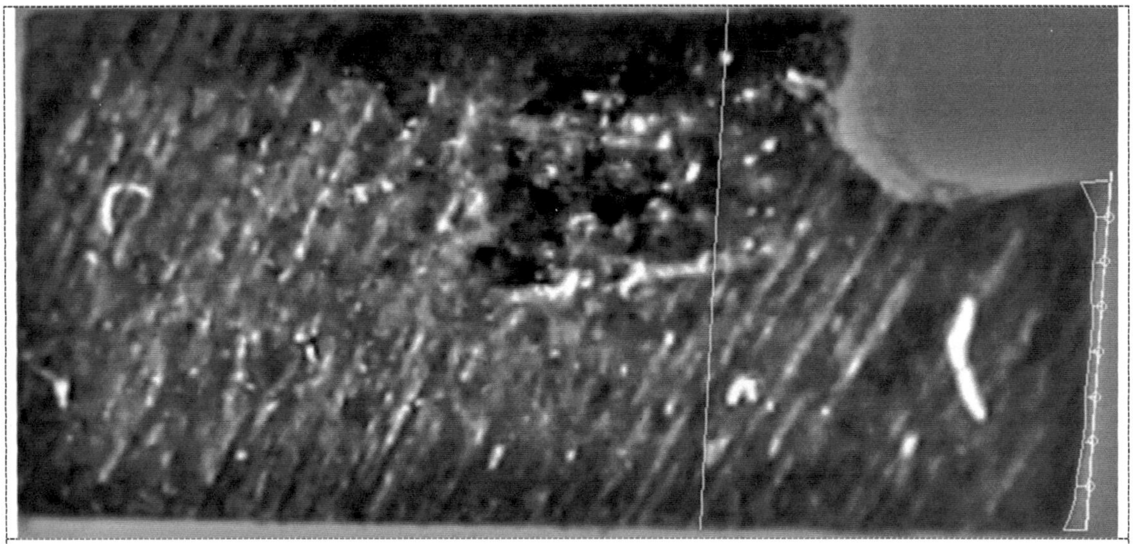

Figure 8. Detailed image capture of piston ring section cut with Digimizer advanced method of critical wear area determination

By defining up a computer assisted graphic content analysis model and determining the primary dimensions of wear area which may give consistent information about the piston ring damage, as well as precise data about the surface deposits and post operation state there were obtained the points and measurements represented schematically in figure 9, being caused by the oscillating movement of the piston ring. The vehicle technical specs are considered in simplified form as presented in table 1.

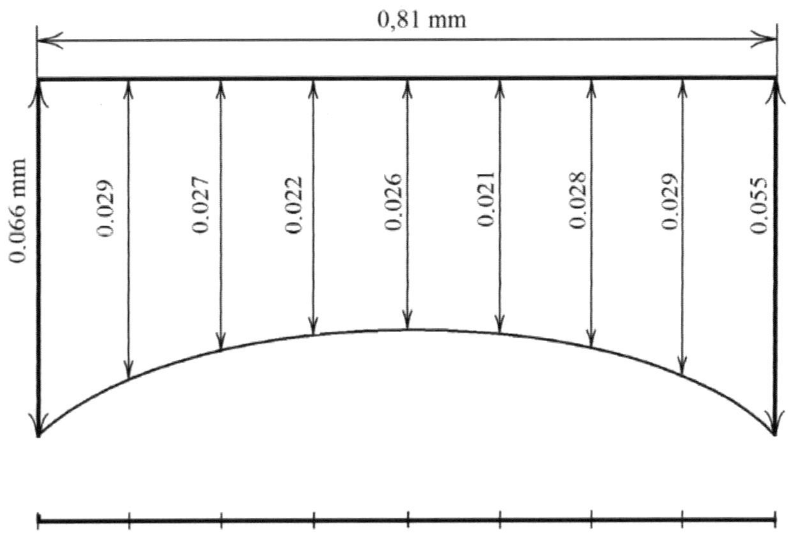

Figure 9. Measuring points and results on the simplified model wear area obtained in Digimizer application through software graphic content advanced analysis.

In a detailed analysis of the wear area of the end part from compression piston ring has a particular shape influenced both by the elastic pressure and the translation, which may be considered in order to formulate a comprehensive theory concerning the wear process in engine operation and to extend the knowledge on the other piston rings.

Table 1. Technical specs of tested motor-car.

Parameter	**Actual Value**
Manufacturer	Ford
Model	Focus
Type	FYDD
Engine displacement	1600 cm^3
Fuel type	Petrol
Number of valves	16
Code	DNW
Nominal power	74 kW
Diagnosis interface	OBD 2

3.3. Limitations of the proposed study
The present study is limited to the implementing of the new engineering interdisciplinary method between computer aided scanning evaluation for piston rings and automotive classic applications, but it may be used in other systems as well, for analyzing the engine components and other systems.

3.4. The novelty of the achievement
The authors are experienced in solid state phenomena, finite element method, manufacturing engineering, rapid prototyping, alternative technologies in manufacturing and computer aided

technologies. The consideration of piston - rings evaluation aspects in real conditions presents a development of research capacities and interdisciplinary study. Even if there were also studied in the previous works some aspects related to Ford Focus FYDD engine, this is the most complete research of many advanced aspects that define piston - ring evaluation and advanced engineering computational methods based on today processing power available at Technical University from Cluj-Napoca.

In figure 10 is presented the virtual model of the piston ring which was considered in the present analyze in order to point out the critical wear location.

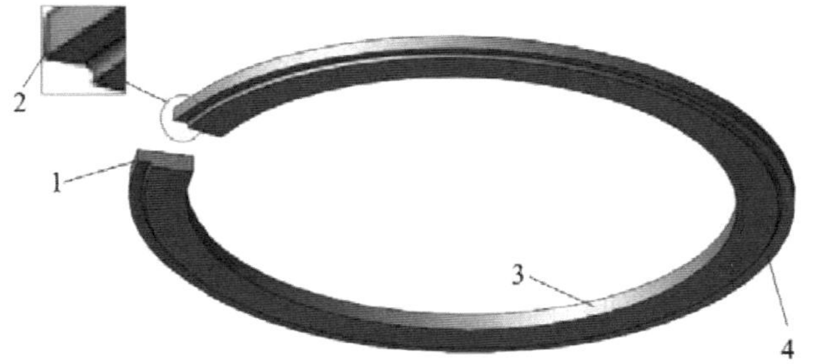

Figure 10. Virtual model of the piston ring that was studied and analyzed with Digimizer and by applying the modern advanced methods.

1-cut section end of piston ring; 2-detail with the critical wear area; 3-interior surface; 4-cut canal for oil scraper.

4. Conclusions

Defining the analysis methods in automotive and internal combustion engines field is close related to the applied science research. Piston - ring parameters are not interacting alone as sole factors in the cylinder and combustion chamber assembly. A piston – ring should be considered in relation with the complex assembly consisting in the cylinder liner, piston itself, con-rod and cylinder-head, which is a thermo-dynamic environment. The causes that determined the rim profile presented in Figure 9 consist in cyclic dynamic stress and high combustion pressures. The wear aspects of the piston-rings have a major influence in the operation or control of the internal combustion engines, but in the same time they lead to a more profound financial impact when it comes to maintenance and repairing processes. In conclusion, the results facilitate the development of an easy-to-use method of wear determination and its causality. Further developments are necessary concerning piston-ring wear and tear-off analysis, rotated with 90°, as well as numerical modeling with finite elements method. Innovative capabilities are very important in order to stimulate and generate other studies of the problem concerning the wear process and critical area in the case of piston rings and engine components.

References

[1] Mobarak H M et al 2014 *Effect of DLC coating on tribological behavior of cylinder liner-piston ring material combination when lubricated with Jatropha oil* Procedia Engineering 90 pp 733-739 www.sciencedirect.com

[2] Rahmani R et al 2017 *The effect of cylinder liner operating temperature on frictional loss and engine emissions in piston ring conjunction* Applied Energy 191 (1 April 2017) pp 568-581

[3] Chereches A Băldean D L 2017 *Contributions to experimental research of piston-rings wearness from Ford Focus FYDD 1.6 S.I.E.* Știință și Inginerie 31 85 http://stiintasiinginerie.ro/31-85-contributii-la-cercetarea-experimentala-a-uzurii-segmentilor-de-piston-de-la-m-a-s-fydd-ford-focus-1-6-l/

[4] Burnete N et al 2008 *Motoare Diesel şi biocombustibili pentru transportul urban* ISBN 978-973-713-217-8 Ed Mediamira Cluj-Napoca

[5] Bățagă N et al 1995 *Motoare cu ardere internă* ISBN 973-30-4922-0 E.D.P. Buc. pp 184-188

[6] *** 2015 *Digimizer image analysis software manual* http://www.digimizer.com/

Theoretical model for determination of the spark ignition engine thermo-gasodynamic parameters on various functional conditions

V Vornicu[1], T Ulian[1], E Rakosi[1], Gh Manolache[1] and L Gaiginschi[1]

[1] Technical University "Gheorghe Asachi" of Iasi, Faculty of Mechanical Engineering, 700050, Romania

E-mail: vladd.alex@yahoo.ro

Abstract. In this paper, an analysis and a model of the factors influencing the filling and compression of the fresh load is presented, in case of a spark ignition engine. Equally, the combustion and expansion processes of the engine are taken into account. All these processes are modelled together with the engine load changing, for different engine speed values.

1. Objective

This paper starts from the observation that the technical literature does not approach the theoretically car engine behaviour in different operating modes.

For this reason, it is not possible to determine the characteristic energy parameters, in particular, the power and fuel consumption, on a diversity of regimes, imposed by the automobile operation. In this context, the authors are proposing an original calculation model. This model, developed in the present paper, addresses to the specific processes for the spark ignition engine, in the context in which they are determining the evolution of the engine operating cycle assembly and subsequently its performance. Also, the related model envisages different engine operating regimes. It introduces changes of the main parameters when the engine operating mode is being changed.

2. Methodology

In this paper, an analysis and a model of the factors influencing the filling and compression of the fresh charge is presented, in case of a spark ignition engine. Equally, the combustion and expansion processes of the engine are taken into account. All these processes are modelled together with the engine load changing, for different engine speed values. The difficulty of the problem occurs especially for this engine type because the load setting is obtained by changing the throttle position [1] [2]. This adjustment significantly influences the pressure at the end of the intake process, thereby cylinder filling and engine performance. For the development of this theoretical model, it is necessary to be introduced some basic assumptions. It is mainly assumed that changing the functional mode will have an impact on the cylinder fill efficiency, in case of a spark ignition engine, being affected exclusively by the modification of the engine load. In the same time, within the developed model a certain variation of the exhaust gas coefficient in the cylinder engine, depending on its load, is adopted [3] [4].

3. Results

This paper, through the developed model, leads to the achievement of functions that emphasise pressure and temperature variations at the end of each cycle of the engine, in the same time with the engine load changing, for different engine speed values in the operating range.

The expression for the intake pressure, dependent on the engine load, can be written as the following:

$$p_{a(x)} = \Phi_0 \cdot (1-\varepsilon^{-1}) \cdot (1-\varphi_{pu}+\gamma_r) \cdot \begin{cases} -56 \cdot \chi^2 + 95 \cdot \chi + 299 \\ -28 \cdot \chi^2 + 48 \cdot \chi + 312 \\ -167 \cdot \chi^2 + 284 \cdot \chi + 263 \\ -278 \cdot \chi^2 + 473 \cdot \chi + 225 \end{cases} \cdot \begin{cases} 0{,}2 \cdot \chi + 0{,}56, & n = 2 \cdot 10^3 \\ 0{,}368 \cdot \chi + 0{,}406, & n = 3 \cdot 10^3 \\ 0{,}475 \cdot \chi + 0{,}325, & n = 4 \cdot 10^3 \\ 0{,}69 \cdot \chi + 0{,}17, & n = 5 \cdot 10^3 \end{cases} [rpm] \quad (1)$$

Also, this model has developed the expression for intake temperature parameter determination:

$$T_{a(x)} = \begin{cases} -56 \cdot \chi^2 + 95 \cdot \chi + 299, & n = 2 \cdot 10^3 \\ -28 \cdot \chi^2 + 48 \cdot \chi + 311, & n = 3 \cdot 10^3 \\ -167 \cdot \chi^2 + 284 \cdot \chi + 263, & n = 4 \cdot 10^3 \\ -278 \cdot \chi^2 + 473 \cdot \chi + 225, & n = 5 \cdot 10^3 \end{cases} [rpm] \quad (2)$$

The expression for the compression pressure, dependent on the engine load, can be written also:

$$p_{c(\chi)} = \Phi_0 \cdot (1-\varepsilon^{-1}) \cdot (1-\varphi_{pu}+\gamma_r) \cdot \begin{Bmatrix} -278 \cdot \chi^2 + 473 \cdot \chi + 225 \\ -167 \cdot \chi^2 + 284 \cdot \chi + 263 \\ -28 \cdot \chi^2 + 48 \cdot \chi + 311 \\ -56 \cdot \chi^2 + 95 \cdot \chi + 299 \end{Bmatrix} \cdot \begin{cases} 0{,}2 \cdot \chi + 0{,}56 \\ 0{,}368 \cdot \chi + 0{,}406 \\ 0{,}475 \cdot \chi + 0{,}325 \\ 0{,}69 \cdot \chi + 0{,}17 \end{cases} \cdot \begin{cases} \varepsilon^{1,3775}, & n = 2 \cdot 10^3 \\ \varepsilon^{1,3883}, & n = 3 \cdot 10^3 \\ \varepsilon^{1,3937}, & n = 4 \cdot 10^3 \\ \varepsilon^{1,397}, & n = 5 \cdot 10^3 \end{cases} [rpm] \quad (3)$$

Regarding the temperature parameter for the compression stroke, it can be determined by using the following formula, dependent on the engine load too:

$$T_c = \begin{Bmatrix} -56 \cdot \chi^2 + 95 \cdot \chi + 299 \\ -28 \cdot \chi^2 + 48 \cdot \chi + 311 \\ -167 \cdot \chi^2 + 284 \cdot \chi + 263 \\ -278 \cdot \chi^2 + 473 \cdot \chi + 225 \end{Bmatrix} \cdot \begin{Bmatrix} \varepsilon^{0,3775}, & n = 2 \cdot 10^3 \\ \varepsilon^{0.3883}, & n = 3 \cdot 10^3 \\ \varepsilon^{0,3937}, & n = 4 \cdot 10^3 \\ \varepsilon^{0.397}, & n = 5 \cdot 10^3 \end{Bmatrix} [K] \quad (4)$$

There was also developed an expression which can be used in order to find the evolution of the combustion process, dependent on the engine load, for four different engine speed values:

$$p_z = \begin{cases} -114.3 \cdot \chi^3 + 225.5 \cdot \chi^2 - 150 \cdot \chi + 88.5, & n = 2 \cdot 10^3\ [rpm] \\ 4.1 \cdot \chi^3 - 14.7 \cdot \chi^2 + 19.4 \cdot \chi + 47, & n = 3 \cdot 10^3\ [rpm] \\ -13 \cdot \chi^3 - 8.6 \cdot \chi^2 + 64.7 \cdot \chi + 31.3, & n = 4 \cdot 10^3\ [rpm] \\ 0.75 \cdot \chi^3 - 51.9 \cdot \chi^2 + 129 \cdot \chi + 7, & n = 5 \cdot 10^3\ [rpm] \end{cases} \quad (5)$$

The same considerate for the parameter of temperature:

$$T_z = \begin{cases} -189910 \cdot \chi^6 + 671518 \cdot \chi^5 - 935209 \cdot \chi^4 + 643787 \cdot \chi^3 - 224873 \cdot \chi^2 + 36937 \cdot \chi + 529, & n = 2 \cdot 10^3\ [rpm] \\ -13246 \cdot \chi^4 + 33828 \cdot \chi^3 - 30833 \cdot \chi^2 + 11857 \cdot \chi + 1280, & n = 3 \cdot 10^3\ [rpm] \\ -1868.3 \cdot \chi^4 + 4856.5 \cdot \chi^3 - 4241.5 \cdot \chi^2 + 1436.9 \cdot \chi + 2695, & n = 4 \cdot 10^3\ [rpm] \\ -17404 \cdot \chi^4 + 44889 \cdot \chi^3 - 40680 \cdot \chi^2 + 14959 \cdot \chi + 1069, & n = 5 \cdot 10^3\ [rpm] \end{cases} \quad (6)$$

For the power process, the authors has developed the folowing expressions:

$$p_d = \begin{cases} -5.8 \cdot \chi^3 + 11.4 \cdot \chi^2 - 7.6 \cdot \chi + 4.5, & n = 2 \cdot 10^3\ [rpm] \\ 0.22 \cdot \chi^3 - 0.78 \cdot \chi^2 + \chi + 2.5, & n = 3 \cdot 10^3\ [rpm] \\ -0.7 \cdot \chi^3 - 0.47 \cdot \chi^2 + 3{,}53 \cdot \chi + 1.7, & n = 4 \cdot 10^3\ [rpm] \\ 0.04 \cdot \chi^3 - 2.88 \cdot \chi^2 + 7.14 \cdot \chi + 0.4, & n = 5 \cdot 10^3\ [rpm] \end{cases} \quad (7)$$

$$T_d = \begin{cases} -97971 \cdot \chi^6 + 346423 \cdot \chi^5 - 482456 \cdot \chi^4 + 332117 \cdot \chi^3 - 116007 \cdot \chi^2 + 19055 \cdot \chi + 273, & n = 2 \cdot 10^3\ [rpm] \\ -7186 \cdot \chi^4 + 18352 \cdot \chi^3 - 16727 \cdot \chi^2 + 6432 \cdot \chi + 695, & n = 3 \cdot 10^3\ [rpm] \\ -1039{,}4 \cdot \chi^4 + 2702 \cdot \chi^3 - 2360 \cdot \chi^2 + 800 \cdot \chi + 1500, & n = 4 \cdot 10^3\ [rpm] \\ 14439 \cdot \chi^6 - 32479 \cdot \chi^5 + 9555 \cdot \chi^4 + 28990 \cdot \chi^3 - 29619 \cdot \chi^2 + 10247 \cdot \chi + 477, & n = 5 \cdot 10^3\ [rpm] \end{cases} \quad (8)$$

The main results obtained derive from the simulation performed on the modification of these defining parameters. These parameters modification is expressed and analysed through graphs which are illustrating the variation of them.

Simulation of the engine cycle at 2000 [rpm] and various loads is illustrated in figure 1 (p-V):

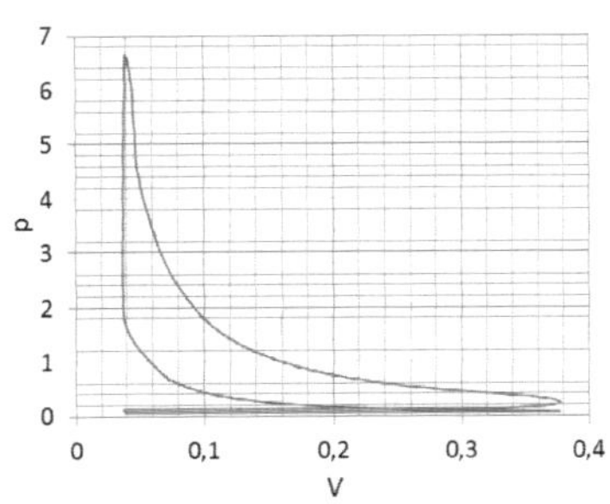

Engine load 20%

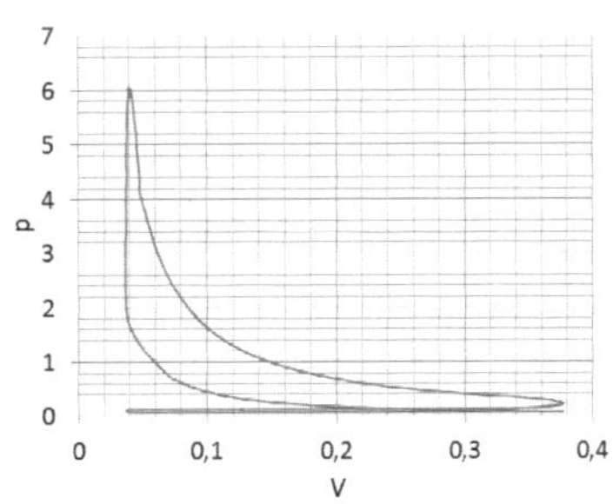

Engine load 30%

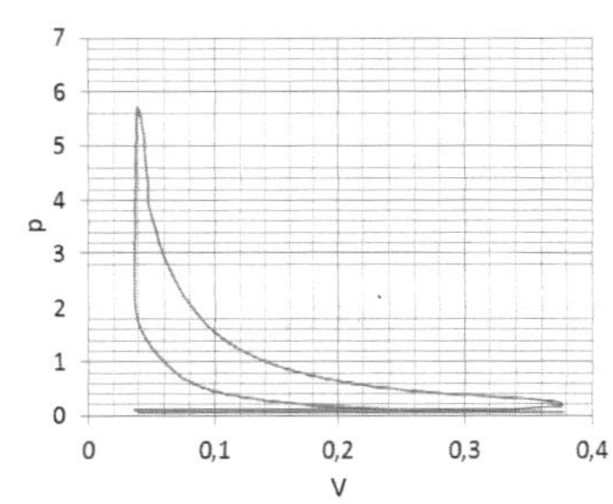

Engine load 40%

IOP Conf. Series: Materials Science and Engineering **252** (2017) 012073 doi:10.1088/1757-899X/252/1/012073

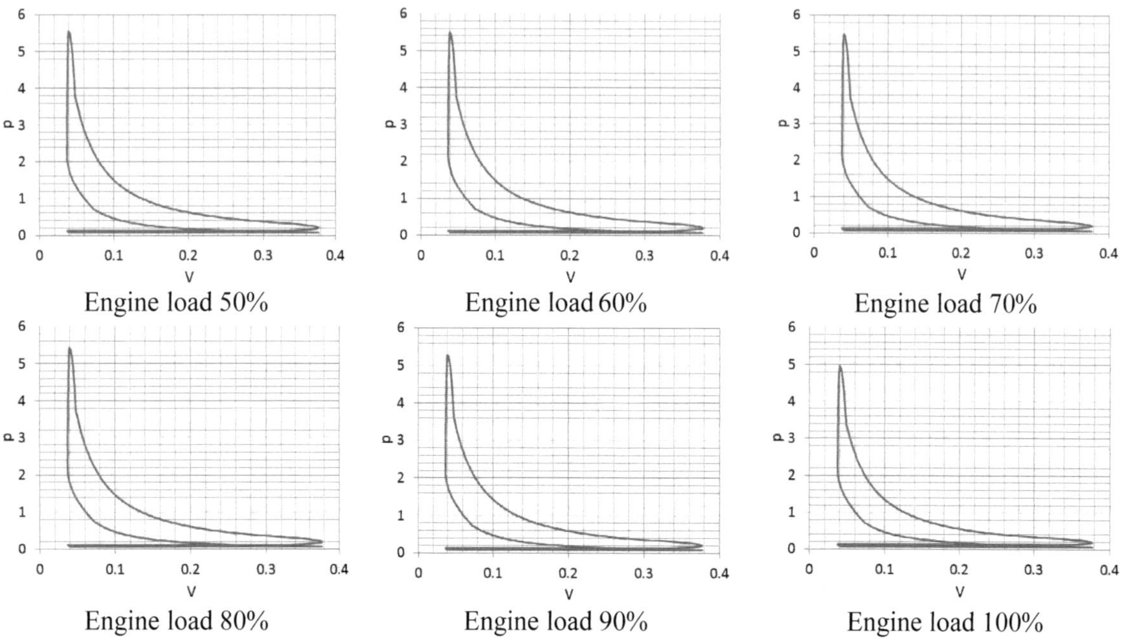

Figure 1: Engine cycle evolutions (p-V) at 2000 [rpm]

The figures 2, 3 and 4 (p-V) illustrates a simulation of the engine cycle evolutions at 3000, 4000 and 5000 [rpm] at various loads.

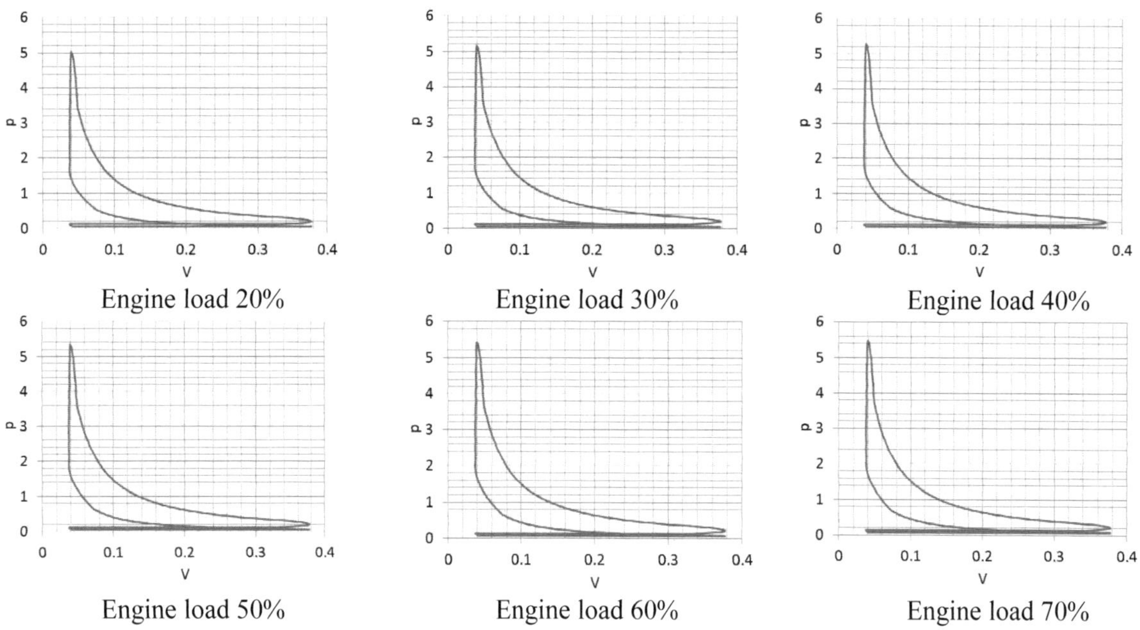

IOP Conf. Series: Materials Science and Engineering **252** (2017) 012073 doi:10.1088/1757-899X/252/1/012073

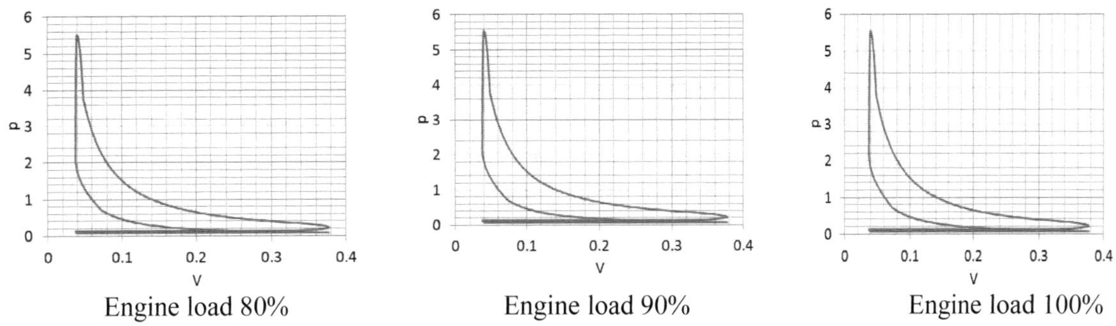

Figure 2: Engine cycle evolutions (p-V) at 3000 [rpm]

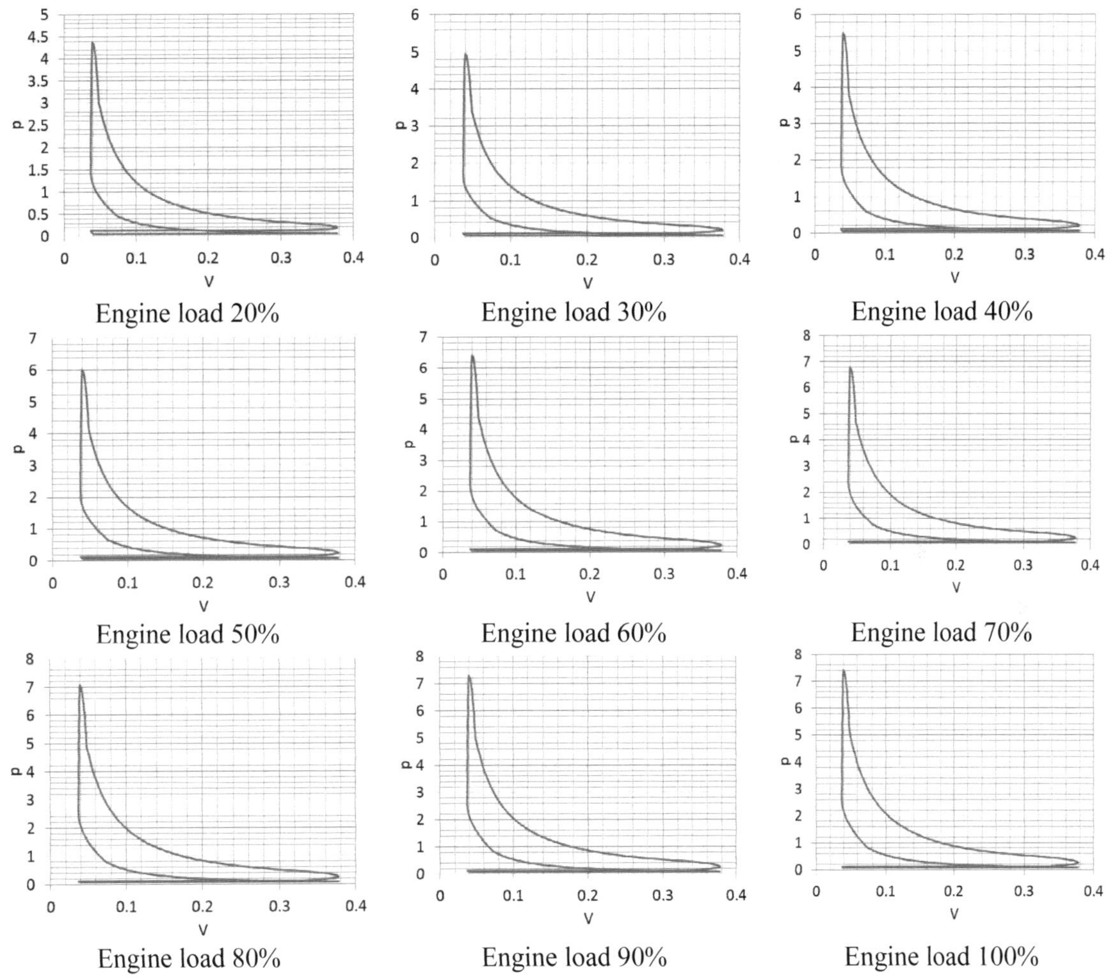

Figure 3: Engine cycle evolutions (p-V) at 4000 [rpm]

IOP Conf. Series: Materials Science and Engineering **252** (2017) 012073 doi:10.1088/1757-899X/252/1/012073

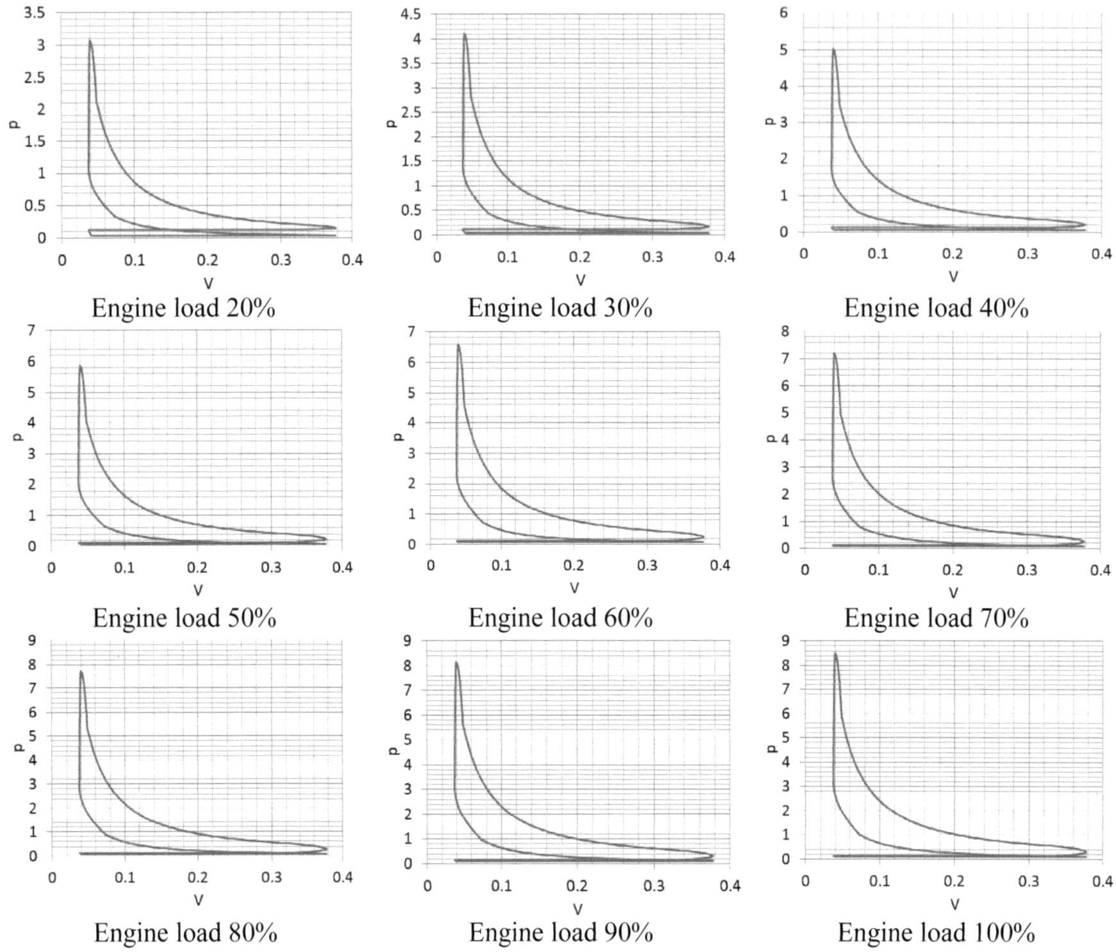

Figure 4: Engine cycle evolutions (p-V) at 5000 [rpm]

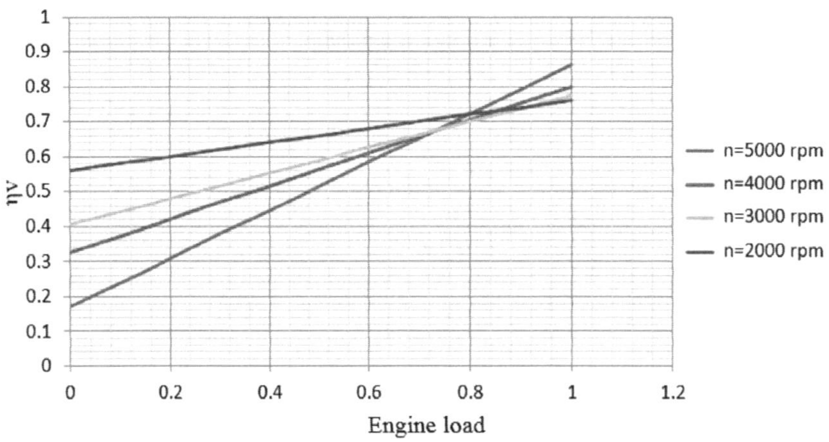

Figure 5: Dependence between filling coefficient and engine load

Regarding the dependence between the inlet pressure and engine load calculation, a linear function is introduced in the model which shows, for spark-ignition engines, the filling coefficient variation in case of engine load modification, for different engine speeds.
Establishment of this linear variation was made by analysing a data series from the literature [4] [5] [6]. The representation of this variation, for an engine speed interval which ensures a steady engine running is showed in figure 5, as a family of lines with different tendencies.
Analysis of this different tendencies highlights that for low engine loads, the filling coefficient has high values in low engine speeds domain. By this way, the lower rate of airflow contributes to lowering the gaso-dynamic losses. Increased efficiency of the filling in the low engine speeds domain is also favoured by increased of the filling time.
In case of high engine loads, the gaso-dynamic resistance is decreasing, irrespective of engine speed value, which leads to increased filling coefficient for the high engine speeds domain, when the inertia of the intake air flow is higher, encouraging in this way the filling.
This somewhat particular variation of the filling coefficient in the low-speed engine domain, which takes into account the experimental data available for low- and medium-spark ignition engines, influences the performance of the engine.
As can be seen from the graphical evolutions presented in figure 1, the developed model leads to values of the mean indicated pressure in accordance with the particular variation of the filling coefficient at the 2000 [rpm] engine speed.
On the other hand, for determining the mean indicated pressure of the engine cycle, p_i, this theoretical model does not take into account the change of the excess air coefficient values when changing the engine load. Thus, in real cases, the influence of the variation of the filling coefficient on the behavior of the engine is diminished.
This limitation of the theoretical model is diminished in the medium-speed and high-speed engine domains, the graphical evolutions in figures 2, 3 and 4 being consistent with the actual behavior of the propulsion engine in various regimes.
The proposed theoretical model establishes also a variation of the exhaust gases coefficient, depending on the engine load. Together with this, a new value of the constants that intervene in the calculation expression of the compression and expansion average polytropic exponent is proposed, dependent on the engine speed.

4. Conclusions

Due to the complex modification of the parameters involved in the calculation of the operating cycle, as well as the lower influence of the engine speed on certain indices, the model mainly takes into account the variation of some factors of influence, variation determined mainly by the engine load and not by the engine speed.

It is considered that the theoretical study and the presented model from this paper are useful in the interaction between the engine and the car, which can be improved through a better estimation of propulsion system performance.

Compared to other papers, developed by authors on this topic, this paper fully addresses the processes from the engine, integrating the burning and expansion processes.

These results lead to the idea of developing and perfecting the model by introducing several factors of influence, thus creating the premises of an experimental validation.

5. References

[1] Abd Alla G.H. 2002 Computer simulation of a four stroke spark ignition engine Energy Conversion and Management vol. 43 pp 1043-1061

[2] Bayraktar H. Durgun O. 2003 Mathematical modelling of spark ignition engine cycles Energy Sources vol. 25 pp 651-666

[3] Cengel Y A Boles M A 2002 Thermodynamics: An engineering approach, Fourth Edition, McGraw-Hill Book Co.

[4] Heisler H 1995 Advanced Engine Technology, SAE International

[5] Heywood B J 1988 Internal Combustion Engine Fundamentals McGraw – Hill Series in Mechanical Engineering Library of Congress Cataloging-in-Publication Data

[6] Kirkpatrick A T Ferguson C R. 2001 Internal Combustion Engines Applied Thermosciences John Wiley & Sons Inc.

ELEMENTS OF CAM'S SYNTHESIS FOR A DISTRIBUTION MECHANISM FOR THE MILLER-ATKINSON CYCLE

[1]Dragomir Ionuţ, [1]Mănescu, Bogdan, [2]Stănescu, Nicolae–Doru, [2]Pandrea, Nicolae, [2]Clenci, Adrian, [2]Popa, Dinel

[1]AKKA ROMSERV Bucharest, Romania, [2]University of Piteşti, Piteşti, Romania

s_doru@yahoo.com

Abstract. In this paper we present the synthesis of the cam for a distribution mechanism used for Miller-Atkinson cycle and described in previous works. The input data contain the angles at which the valve opens and closes, respectively, the dimensions of the main parts, the maximum displacement of the valve. and the law of motion of the valve. The authors consider that the law of motion of the valve is described by numerical values, even if in the simulation process they use analytical formulae for this displacement. The head of the valve is considered to be circular. A particular case of flat head of the valve is also described. The relative displacement of the lever and the valve is discussed in two hypotheses: symmetric position and asymmetric position of the contact point with respect to the axis of the valve. The contact between the head of the valve is assured by a cylindrical roller. The contact between the cam and the lever is considered with a roller tappet. In all cases the cam is obtained by numerical synthesis using a convenient angular degree. A great attention is paid to the convexity of the cam. The resulted cam is transformed in a convex one using the Jarvis March. The authors also perform a comparison between the theoretical profile of the cam and the profile of the convex cam. The reduced velocity and acceleration are obtained in all cases. Some aspects of the wear are discussed using the relative linear and angular velocities. The paper closes with a paragraph of conclusions.

1. Introduction

Some advantages of the Miller-Atkinson [1-8] cycle are given in [9, 10] and we will not repeat the presentation here.

Some experiments using intake valve closure are described in [11] leading to the increase of indicated thermal efficiency. For a diesel engine [12] the authors study the influence of Miller cycle and variable geometry turbocharger on combustion and emissions in steady and transient operation. An excellent review exploring the critical problems, challenges and prospective for the Miller-Atkinson cycle is given in [13]. Comparison between the Miller cycle and five stroke cycle is analyzed in [14] highlighting the advantage of the first at low loads in terms of fuel conversion efficiency. The application of the Miller cycle in the case of a hydrogen engine is presented in [15]. Gonca [16] performs a comparison between different types of engines from the point of view of the thermal efficiency, dimensionless power output and dimensionless power density. The authors [17] investigate the effects of compression ratio, intake valve closing retardation angle, and engine speed on the fuel consumption performance and power performance of the Miller cycle engine. Thermodynamic analysis of Otto engines working after Miller cycle is performed in [18], while the influence of engine design and operating parameters for a diesel engine working after the same cycle is analyzed in [19].

2. The mechanism

The mechanism in the asymmetric displacement of the lever (asymmetric case) is drawn in Fig. 1 and consists in the bar OB of length equal to l_1 and having at its end a roll of radius R_2. In the initial position, the angle between the bar and the horizontal direction is β_0 and it is considered as known. The roll of radius R_2 is in contact, at any moment, with the head of the valve, head considered as an arc of radius R_1. The center of the arc is always situated at the distance d, considered known, from the vertical axis.

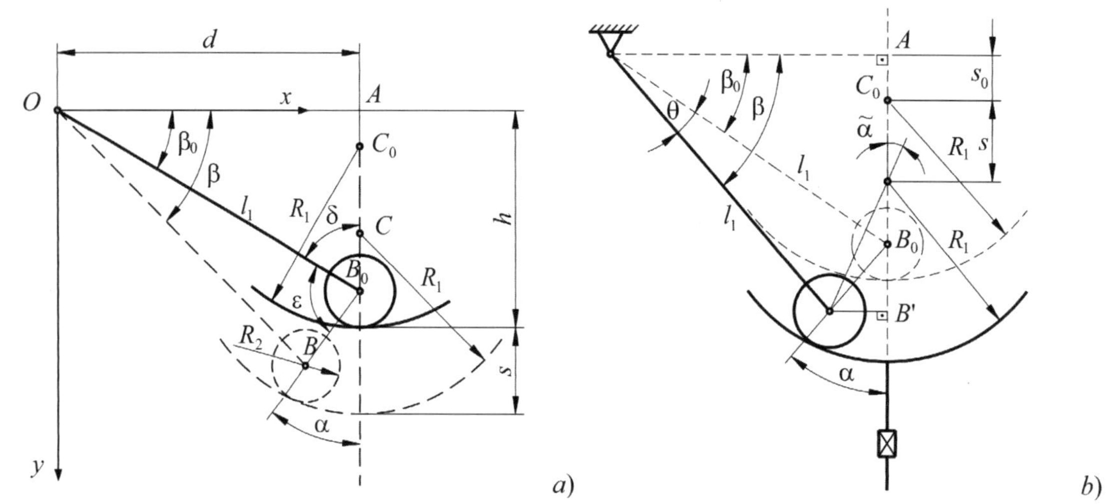

Figure 1. The mechanism in the asymmetric case: *a*) analytic approach; *b*) geometric approach.

The valve moves in the vertical direction with the displacement s leading to the motion of the point C (the center of the arc) from the point C_0 (the initial position) to the point C (the current position). The distance between the point C and the vertical axis Oy remains equal to d.

The bar (lever) OB of length l_1 rotates around the point O such that the roll of radius R_2 remains all the time tangent to the arc of radius R_1. The new angle between the lever and the horizontal direction is β.

In an analytical approach the equation which offers the angle β reads [9]

$$\begin{aligned} &\left\{2x_C(l_1\sin\beta - y_C)^2 + (l_1\cos\beta - x_C)\left[C_1 - 2y_C(l_1\sin\beta - y_C)\right]\right\}^2 \\ &\quad - \left(l_1^2 + x_C^2 + y_C^2 - 2l_1x_C\cos\beta - 2l_1y_C\sin\beta\right) \\ &\times \left\{4x_C^2(l_1\sin\beta - y_C)^2 + \left[C_1 - 2y_C(l_1\sin\beta - y_C)\right]^2 - 4R_1^2(l_1\sin\beta - y_C)^2\right\} = 0, \end{aligned} \tag{1}$$

where

$$C_1 = R_1^2 - R_2^2 + l_1^2 - \left(x_C^2 + y_C^2\right), \tag{2}$$

$$x_C = d,\ y_C = h + s - R_1. \tag{3}$$

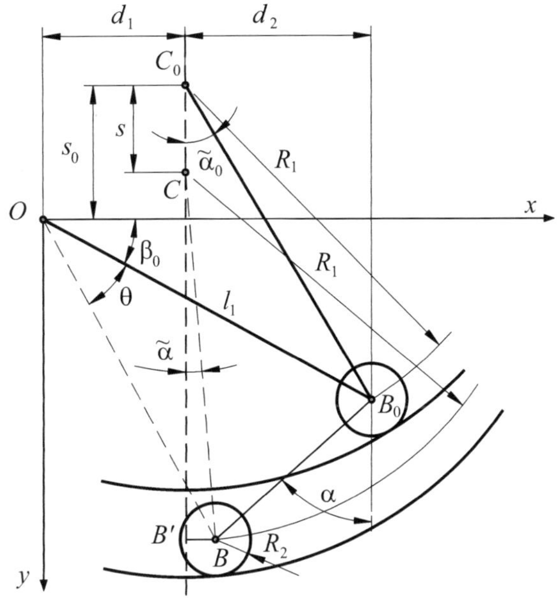

Figure 2. The mechanism in the symmetric case.

The geometric approach of the same case [10] offers the equation

$$-s + l_1[\sin(\theta + \beta_0) - \sin\beta_0] + (R_1 - R_2)(1 - \cos\tilde{\alpha}) = 0, \tag{4}$$

where

$$\tilde{\alpha} = \arcsin\left(\frac{l_1[\cos\beta_0 - \cos(\beta_0 + \theta)]}{R_1 - R_2}\right). \tag{5}$$

For the symmetric case (Fig. 2) we may write the following relations

$$\beta_0 = \arccos\left(\frac{d_1 + d_2}{R_1}\right), \tag{6}$$

$$\tilde{\alpha}_0 = \arcsin\left(\frac{d_2}{R_1}\right), \tag{7}$$

$$\alpha = \beta_0 + \frac{\theta}{2}, \tag{8}$$

$$B_0B = \tilde{d} = 2R_1\sin\left(\frac{\theta}{2}\right), \tag{9}$$

$$\sin\tilde{\alpha} = \frac{d_2 - \tilde{d}\sin\alpha}{R_1}, \tag{10}$$

$$\tilde{\alpha} = \arcsin\left(\frac{d_2 - 2R\sin\alpha\sin\left(\frac{\theta}{2}\right)}{R_1}\right), \tag{11}$$

$$CB' = R_1\cos\tilde{\alpha}, \tag{12}$$

$$C_0B' = R_1\cos\tilde{\alpha}_0 + \tilde{d}\cos\alpha, \tag{13}$$

$$s = C_0B' - CB', \tag{14}$$

$$s = R_1\cos\tilde{\alpha}_0 - R_1\cos\tilde{\alpha} + 2R_1\cos\alpha\sin\left(\frac{\theta}{2}\right). \tag{15}$$

The symmetry implies

$$\tilde{\alpha}_{final} = \tilde{\alpha}_f = -\alpha_0, \tag{16}$$

wherefrom

$$\sin\tilde{\alpha}_f = \frac{d_2 - 2R_1\sin\alpha\sin\left(\frac{\theta}{2}\right)}{R_1} = -\sin\alpha_0, \tag{17}$$

$$d_2 - 2R_1\sin\left(\beta_0 + \frac{\theta}{2}\right)\sin\left(\frac{\theta}{2}\right) = -d_2, \tag{18}$$

$$2d_2 = R_1[\cos\beta_0 - \cos(\beta_0 + \theta)], \tag{19}$$

$$\theta_f = \arccos\left(\frac{R_1\cos\beta_0 - 2d_2}{R_1}\right) - \beta_0, \tag{20}$$

$$\theta_f = \arccos\left(\frac{d_1 - d_2}{R_1}\right) - \beta_0. \tag{21}$$

The calculation would be complete if we would know the distance d_2.

Assuming that the maximum displacement of the valve s_{max} is known, for the calculation of the distance d_2 we proceed as follows:

1. we consider $d_2 = kd_1$, with $k = 0.002$;
2. we calculate in order:

– β_0 with relation (6),
– $\tilde{\alpha}_0$ with relation (7),
– θ_f with relation (21),
– α with relation (8),
– s with relation (15),
– the deviation Δ with the formula

$$\Delta = s - s_{\max} ; \tag{22}$$

3. if $|\Delta| > \varepsilon$, where ε is the a priori imposed error, we consider a new value for d_2,

$$d_2 \mapsto d_2 + kd_1 \tag{23}$$

and repeat the step 2; otherwise the distance d_2 was determined.

3. Synthesis of the cam

For the repose position (Fig. 3, *a*)) we have

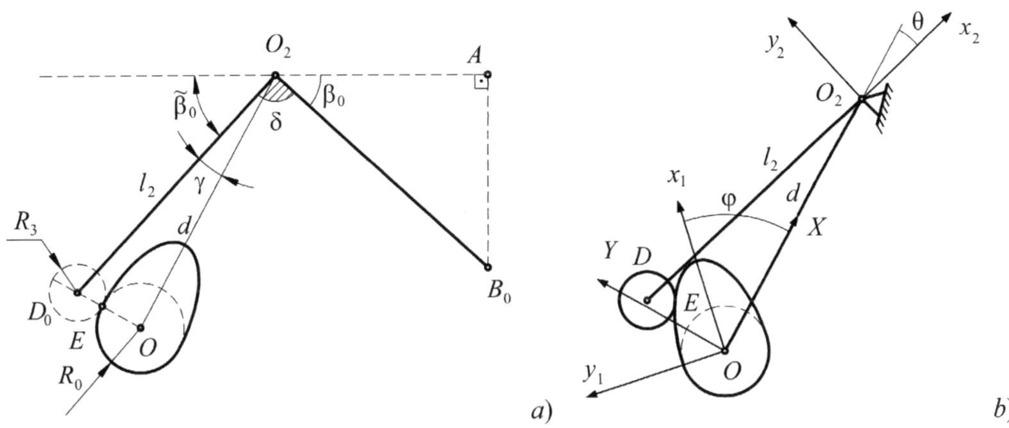

Figure 3. Synthesis of cam: *a*) repose position; *b*) current position

$$\tilde{\beta}_0 = \pi - \beta_0 - \delta_0, \tag{24}$$

$$\gamma = \arccos\left(\frac{l_2^2 + d^2 - (R_0 + R_3)^2}{2l_2 d}\right). \tag{25}$$

Following the steps described in [10], the parametric equations of the cam (Fig. 3, *b*)) are

$$x_1 = d\cos\varphi - l_2\cos(\theta + \gamma + \varphi) - R_3\sin(\theta + \gamma + \varphi + \lambda), \tag{26}$$

$$y_1 = -d\sin\varphi + l_2\sin(\theta + \gamma + \varphi) - R_3\cos(\theta + \gamma + \varphi + \lambda), \tag{27}$$

where the parameter λ is given by

$$\tan\lambda = \frac{d\cos(\theta + \gamma) - l_2\left(1 + \frac{d\theta}{d\varphi}\right)}{d\sin(\theta + \gamma)}. \tag{28}$$

4. Numerical analysis

We consider the case defined by $l_1 = 0.01612\,\text{m}$, $R_1 = 0.025\,\text{m}$, $R_2 = 0.010\,\text{m}$, $\beta_0 = 15^0$, $s_{\max} = 0.00935\,\text{m}$, $R_0 = 0.018\,\text{m}$, $d = 0.042\,\text{m}$, $l_2 = 0.060\,\text{m}$, $R_3 = 0.01867\,\text{m}$, $d_1 = 0.010\,\text{m}$, the angles $\varphi_1 = 92^0$ (when the valve starts to open), $\varphi_2 = 161^0$ (when the valve is completely open), $\varphi_3 = 180^0$ (when the valve starts to close), and $\varphi_4 = 244^0$ (when the valve is completely closed).

Two laws of motion are considered for the valve displacement

$$s_1(\varphi) = \begin{cases} 0 \text{ for } \varphi \in [0, \varphi_1), \\ s_{\max} \sin^2\left(\dfrac{90(\varphi - \varphi_1)}{\varphi_2 - \varphi_1}\right) \text{ for } \varphi \in [\varphi_1, \varphi_2], \\ s_{\max} \text{ for } \varphi \in (\varphi_2, \varphi_3), \\ s_{\max} \sin^2\left(\dfrac{90(\varphi + \varphi_4 - 2\varphi_3)}{\varphi_4 - \varphi_3}\right) \text{ for } \varphi \in [\varphi_3, \varphi_4], \\ 0 \text{ for } \varphi \in (\varphi_4, 360] \end{cases} \quad (29)$$

and

$$s_1(\varphi) = \begin{cases} 0 \text{ for } \varphi \in [0, \varphi_1), \\ s_{\max} \dfrac{(\varphi - \varphi_1)^2 (\varphi - 2\varphi_2 + \varphi_1)^2}{(\varphi_2 - \varphi_1)^4} \text{ for } \varphi \in [\varphi_1, \varphi_2], \\ s_{\max} \text{ for } \varphi \in (\varphi_2, \varphi_3), \\ s_{\max} \dfrac{(\varphi - \varphi_4)^2 (\varphi - 2\varphi_3 + \varphi_4)^2}{(\varphi_3 - \varphi_4)^4} \text{ for } \varphi \in [\varphi_3, \varphi_4], \\ 0 \text{ for } \varphi \in (\varphi_4, 360]. \end{cases} \quad (30)$$

The angle of rotation for the lever is determined from equation (1) in the asymmetric case.

For the symmetric case, the angle θ is obtained from the relation (15)

$$\sin\left(\frac{\theta}{2}\right) = \frac{s - R_1 \cos\tilde{\alpha}_0 - R_1 \cos\tilde{\alpha}}{2R_1 \cos\alpha}. \quad (31)$$

The resulted cams are presented in Figs. 4 and 5. The convex cams are obtained after the application of the Jarvis March. The original cams are drawn in blue, and the convex ones in red.

a) *b)*

Figure 4. The cam for the first law of motion: a) the asymmetric case; b) the symmetric case.

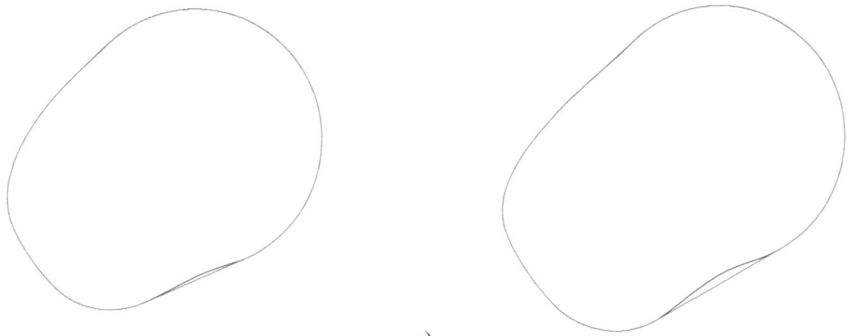

a) *b)*

Figure 5. The cam for the second law of motion: a) the asymmetric case; b) the symmetric case.

It is easy to observe that the symmetric case leads to a cam with bigger dimension no matter which law of motion is considered. Moreover, the obtained cam is not a convex one in all situation. The symmetric case leads to a more concave cam and the Jarvis March is constrained to avoid more characteristic points of the cam.

The reduced angular velocity and acceleration are given by

$$\left.\frac{d\theta}{d\varphi}\right|_{\varphi=\varphi_i} = \frac{\theta_{i+1} - \theta_{i-1}}{2\Delta\varphi}, \tag{32}$$

$$\left.\frac{d^2\theta}{d\varphi^2}\right|_{\varphi=\varphi_i} = \frac{\theta_{i+1} - 2\theta_i + \theta_{i-1}}{(\Delta\varphi)^2}. \tag{33}$$

The angular step $\Delta\varphi = 1^0 = \frac{\pi}{180}$ rad. The reduced angular velocity is a measure of the wear.

The diagrams are presented in Figs. 6–9.

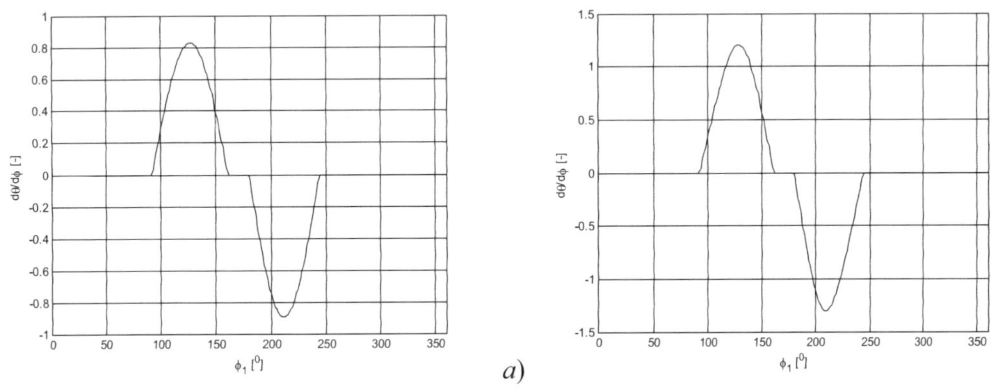

a) *b)*

Figure 6. The reduced angular velocity $\frac{d\theta}{d\varphi} = \frac{d\theta}{d\varphi}(\varphi)$ using the first law of motion for: *a)* the asymmetric case; *b)* the symmetric case.

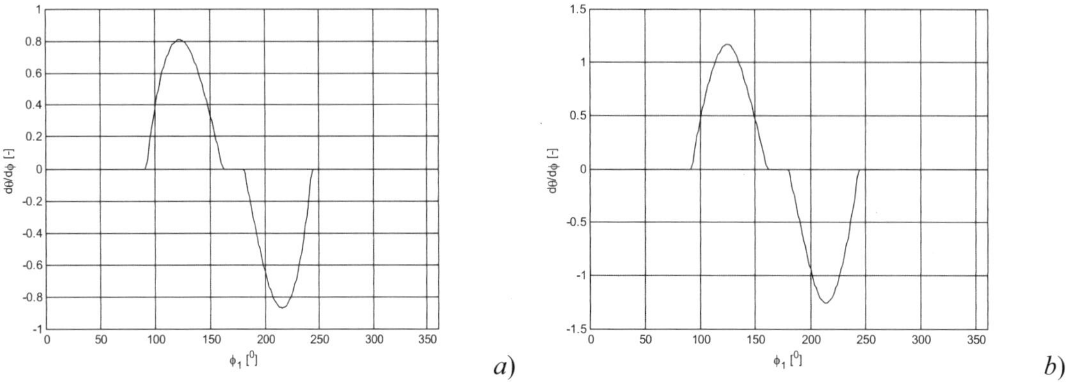

Figure 7. The reduced angular velocity $\frac{d\theta}{d\varphi} = \frac{d\theta}{d\varphi}(\varphi)$ using the second law of motion for: *a*) the asymmetric case; *b*) the symmetric case.

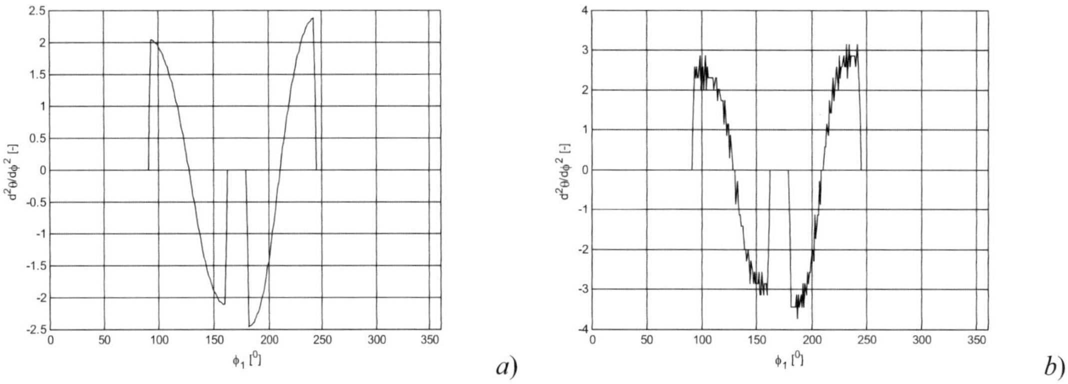

Figure 8. The reduced angular acceleration $\frac{d^2\theta}{d\varphi^2} = \frac{d^2\theta}{d\varphi^2}(\varphi)$ using the first law of motion for: *a*) the asymmetric case; *b*) the symmetric case.

5. Planar head of valve

For the case of the planar head of valve (Fig. 10), some simple geometric considerations lead to the following expression for the angle θ

$$\theta = \beta - \beta_0 = \arcsin\left(\frac{s + l_1 \sin\beta_0}{l_1}\right) - \beta_0 \tag{34}$$

The reduced angular velocity and acceleration may be obtained by direct calculation

$$\frac{d\theta}{d\varphi} = \frac{d\theta}{ds}\frac{ds}{d\varphi} = \frac{1}{\sqrt{1 - \left(\frac{s + l_1 \sin\beta_0}{l_1}\right)^2}}\frac{ds}{d\varphi}. \tag{35}$$

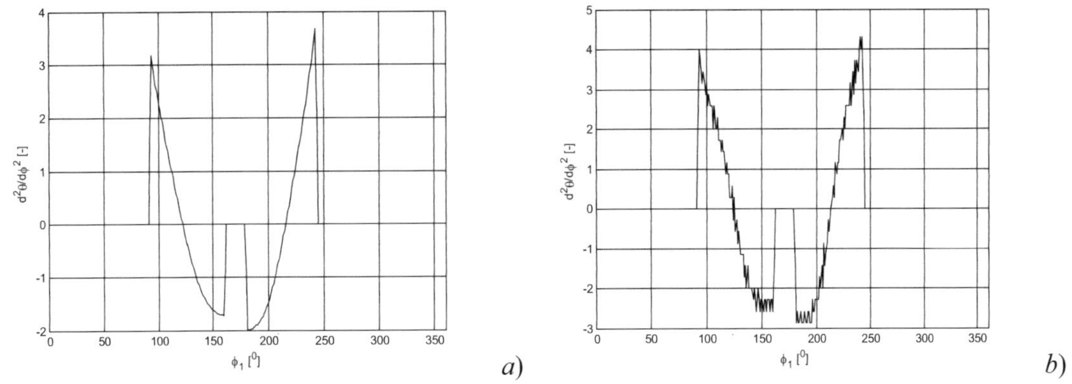

Figure 9. The reduced angular acceleration $\frac{d^2\theta}{d\varphi^2} = \frac{d^2\theta}{d\varphi^2}(\varphi)$ using the second law of motion for: *a*) the asymmetric case; *b*) the symmetric case.

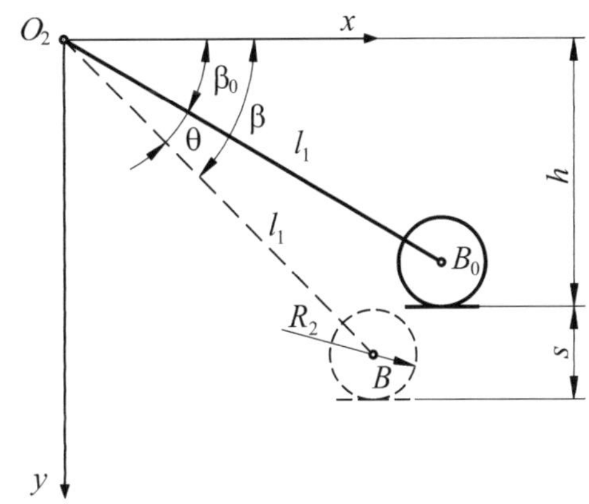

Figure 10. The case of the flat head of valve.

The cams obtained for the two laws of motion are captured in Fig. 11.

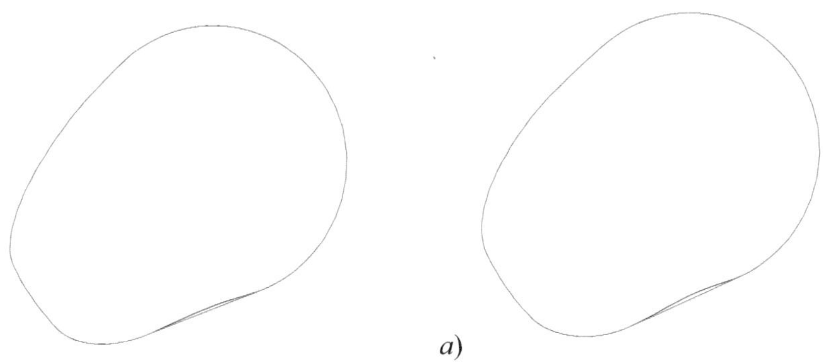

Figure 11. The cam for the flat head of valve: a) the first law of motion; b) the second law of motion.

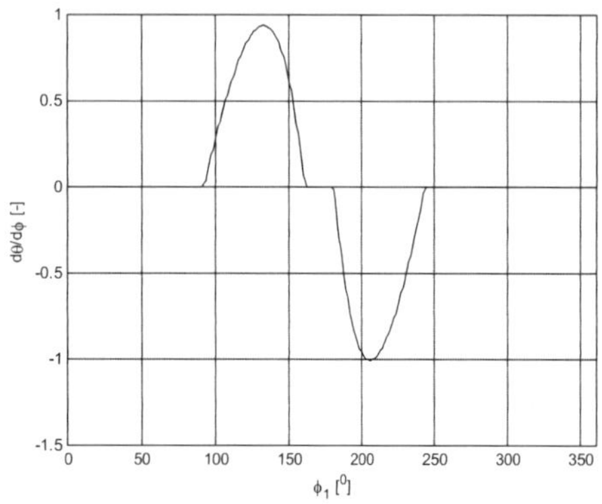

a)

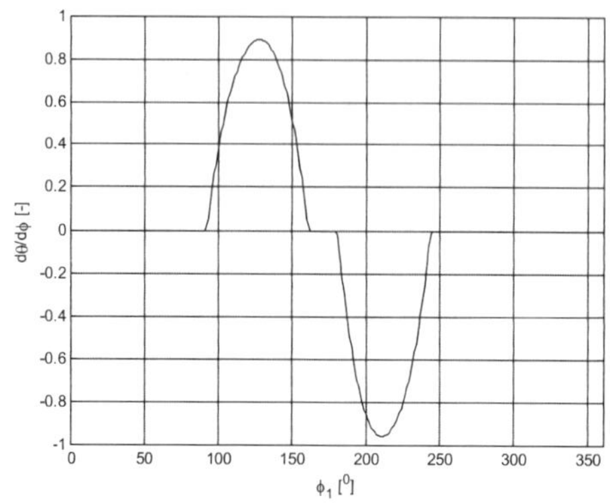

b)

Figure 12. The reduced angular velocity $\frac{d\theta}{d\varphi} = \frac{d\theta}{d\varphi}(\varphi)$ using: *a)* the first law of motion; *b)* the second law of motion.

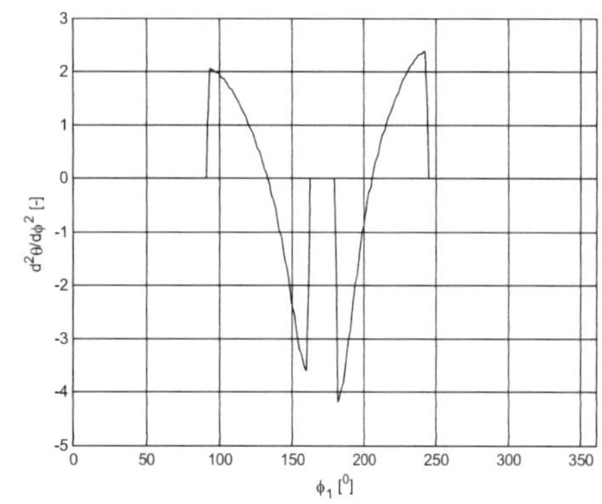

a)

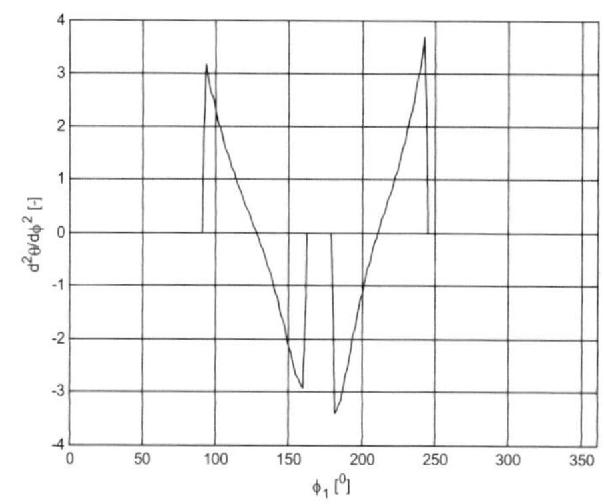

b)

Figure 13. The reduced angular acceleration $\frac{d^2\theta}{d\varphi^2} = \frac{d^2\theta}{d\varphi^2}(\varphi)$ using: *a)* the first law of motion; *b)* the second law of motion.

The corresponding diagrams are given in Fig. 12–13. The reader may observe the increase of the smoothness of the variation diagrams for the reduced angular velocity and acceleration.

6. Conclusions

From the previous diagrams we may conclude that, in the numerical case considered in this paper, the asymmetric case leads to a more smooth variation of the reduced angular velocity and acceleration and to cams with smaller dimensions than those obtained in the symmetric case. The symmetric case offers smaller bending moments acting on the valve comparing to the asymmetric case. The best situation is in the case of the flat head of valve.

Some future studies must be performed in order to obtain more smooth diagrams of variation for the reduced velocity and acceleration either by studying the influence of different parameters, or by considering more laws of motion for the valve. A great attention must be paid to the error in the law of motion caused by the Jarvis March which changes the profile of the cam. The best situation is that in which the obtained cam is a convex one. These aspects will be discussed in our next papers.

References

[1] Gonca G, Sahin B, Parlak A, Ust Y, Ayhan V, Cesur I and Boru, B, 2015, Theoretical and experimental investigation of the Miller cycle diesel engine in terms of performance and emission parameters, *Applied Energy* **138** 2015 11–20
[2] Martins MES and Lanzanova TDM, 2015, Full-load Miller cycle with ethanol and EGR: Potential benefits and challenges, *Applied Thermal Engineering* 274–85
[3] Ebrahimi R, 2011, Effects of mean piston speed, equivalence ratio and cylinder wall temperature on performance of an Atkinson engine, *Mathematical and Computer Modelling* **53** 1289–97
[4] Zammit JP, McGhee MJ, Shayler PJ, Lawa T and Pegg I, 2015, The effects of early inlet valve closing and cylinder disablement on fuel economy and emissions of a direct injection diesel engine, *Energy* **79** 100–10
[5] Murata Y, Kusaka J, Odaka M, Daisho Y, Kawano D, Suzuki H, Ishii H and Goto Y, 2006, Achievement of medium engine speed and load premixed diesel combustion with variable valve timing, *SAE paper 2006-01-0203*
[6] De Ojeda W, 2010, Effect of variable valve timing on diesel combustion characteristics, *SAE paper 2010-01-1124*
[7] Kéromnčs A, Delaporte B, Schmitz G and Le Moyne L, 2014, Development and validation of a 5 stroke engine for range extenders application, *Energy Conversion and Management* **82** 259–67.
[8] Zhu S, Deng K, Liu S and Qu S, 2015, Comparative analysis and evaluation of turbocharged Dual and Miller cycles under different operating conditions, *Energy* **93** 75–87
[9] Dragomir I, Mănescu B and Stănescu ND, 2017, The Analysis of a Distribution Mechanism for the Miller-Atkinson Cycle, *AVMS 2017*, Timişoara
[10] Dragomir I, Mănescu B, Stănescu ND and Pandrea N, 2017, Synthesis of the cam and determination of the reduced angular velocity and acceleration of the lever of a distribution mechanism for a Miller – Atkinson cycle, *ICMSAV XLI*, Cluj-Napoca
[11] Yan B, Wang H, Zheng Z, Qin Y and Yao M, 2017, The effects of LIVC Miller cycle on the combustion characteristics and thermal efficiency in a stoichiometric operation natural gas engine with EGR, *Applied Thermal Engineering* **122** 439–50
[12] Wu B, Zhan Q, Yu X, Lv G, Nie X and Liu S, 2017, Effects of Miller cycle and variable geometry turbocharger on combustion and emissions in steady and transient cold process, *Applied Thermal Engineering* **118** 621–9
[13] Zhao J, Research and application of over-expansion cycle (Atkinson and Miller) engines – A review, 2017, *Applied Energy* **185** 300–19
[14] Li T, Wang B and Zheng B, 2016, A comparison between Miller and five-stroke cycles for enabling deeply downsized, highly boosted, spark-ignition engines with ultra expansion, *Energy Conversion and Management* **123** 140–52
[15] Luo Q and Sun B, 2016, Effect of the Miller cycle on the performance of turbocharged hydrogen internal combustion engines, *Energy Conversion and Management* **123** 209–17
[16] Gonca G, 2016, Comparative performance analyses of irreversible OMCE (Otto Miller cycle engine)-DiMCE (Diesel Miller cycle engine)-DMCE (Dual Miller cycle engine), *Energy* **109** 152–9
[17] Wang Y, Zu B, Xu Y, Wang Z and Liu J, 2016, Performance analysis of a Miller cycle engine by an indirect analysis method with sparking and knock in consideration, *Energy Conversion and Management* **119** 316–26
[18] Dobrucali E, 2016, The effects of the engine design and running parameters on the performance of a Otto–Miller Cycle engine, *Energy*, **103** 119–26
[19] Gonca G and Sahin B, 2016, The influences of the engine design and operating parameters on the performance of a turbocharged and steam injected diesel engine running with the Miller cycle, *Applied Mathematical Modelling*, **40** 3764–82.

IOP Conf. Series: Materials Science and Engineering **252** (2017) 012075 doi:10.1088/1757-899X/252/1/012075

ASPECTS IN THE SYNTHESIS OF A VARIABLE COMPRESSION RATIO MECHANISM

[1] Mănescu, Bogdan, [1] Dragomir Ionuţ, [2]Stănescu Nicolae–Doru, [2]Pandrea Nicolae, [2]Clenci Adrian, [2]Popa Dinel

[1]AKKA ROMSERV Bucharest, Romania, [2]University of Piteşti, Piteşti, Romania

s_doru@yahoo.com

Abstract. The mechanism considered in this paper is a VCR one consisting in a crank, a shaft, an intermediate triangular element and a control lever and it was described in previous papers of the authors. The authors start from a classical crank-shaft mechanism for which the extreme positions are known. The first stage of the synthesis consists in determination of the constraint function which has to be fulfilled by the new mechanism so that the extreme position remain unchanged. The new step consists in imposing new conditions for the mechanism so that some new positions have to be obtained. The main hypothesis is that the positions of different characteristic points of the mechanism may be written as continuous functions of the certain input data. Due to aspect, the synthesis of the mechanism implies continuous variations of the output data and, consequently, there exists at least one solution for the synthesis process. In each case the authors determine the extreme positions of the piston. These extreme positions are also continuous functions of the input data. Two main approaches are discussed in the paper. One approach consists in the exact determination of the solution using a numerical procedure. The second one is an approximate one and consists in the determination of an approximate solution of the synthesis and verifying the deviation of this solution. For the new mechanism obtained by synthesis the authors determined the reduced velocities and accelerations of different characteristic elements. Some aspects of the wear are discussed with the aid of the reduced relative velocities.

1. Introduction

Some of the references [1–11] are commented in our previous works [12, 13] and will be not discussed again here.

In [14] the author studies the performance and emissions for a four strokes one cylinder diesel engine at different compression ratio. A study consisting in the kinematic and dynamic analysis and finite element analysis for an engine mechanism is performed in [15]. The influence of Miller cycle and variable geometry turbocharger on combustion and emissions are investigated in [16]. Performances of a special four-stroke engine with variable stroke-length and compression ratio are discussed by numerical simulation in [17]. Constructive aspects concerning the variable compression mechanisms are discussed in [18]. Tsai [19] and Yan [20] developed a methodology for the investigation of the opportunities for the development of variable compression engines. There exist more than one hundred different variable compression ratio engines designs. Some of these designs are rejected by different authors at the moments when they wrote their studies based on some more or less particular criteria. In time, some of these rejected designs are successfully implemented by different constructors of automobiles.

Excellent introduction and presentation on the variable compression mechanisms for engines can be found in [21] and [22] and the references within them. An overview for the variable compression ration is realized in [23].

2. Geometric approach

The mechanism (Fig. 1) was described in our previous works [12, 13] and we will not repeat here.

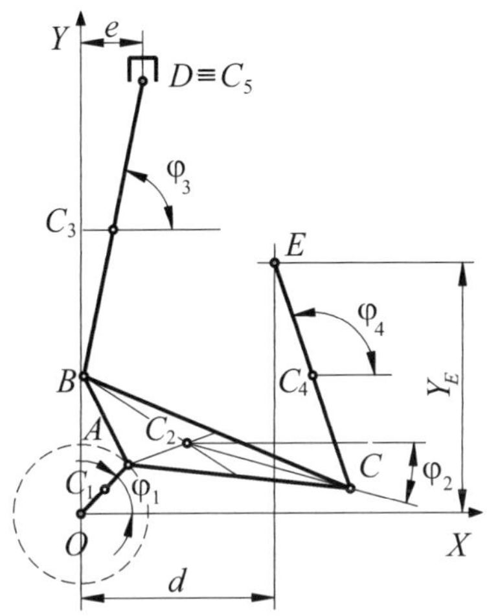

Figure 1. The mechanism.

The coordinates of the point A are

$$x_A = OA\cos\varphi_1,\ y_A = OA\sin\varphi_1. \tag{1}$$

The point C is obtained as the intersection of the circles with the centers at the points A and E, and radii AC and CE, respectively, that is, its coordinates fulfill the system

$$(x_C - x_A)^2 + (y_C - y_A)^2 = (AC)^2,\ (x_C - x_E)^2 + (y_C - y_E)^2 = (CE)^2. \tag{2}$$

Analogically, the following systems may be written for the points B and D

$$(x_B - x_A)^2 + (y_B - y_A)^2 = (AB)^2,\ (x_B - x_C)^2 + (y_B - y_C)^2 = (BC)^2, \tag{3}$$

$$(x_D - x_B)^2 + (y_D - y_B)^2 = (BD)^2,\ x_D = e, \tag{4}$$

where the coordinates of the point E are $x_E = d$ and $y_E = Y_E$.

Generally speaking, the intersection of two circles of equations

$$(x - x_1)^2 + (y - y_1)^2 = R_1^2,\ (x - x_2)^2 + (y - y_2)^2 = R_2^2 \tag{5}$$

is given by

$$x = \frac{-\beta \pm \sqrt{\beta^2 - \alpha\gamma}}{\alpha},\ y = Ax + b, \tag{6}$$

where

$$A = -\frac{x_2 - x_1}{y_2 - y_1},\ B = \frac{R_1^2 - R_2^2 + \left(x_2^2 - x_1^2\right) + \left(y_2^2 - y_1^2\right)}{2\left(y_2 - y_1\right)},\ y_1 \neq y_2, \tag{7}$$

$$\alpha = 1 + A^2,\ \beta = A\left(B - y_1\right) - x_1,\ \gamma = x_1^2 + \left(B - y_1\right)^2 - R_1^2 \tag{8}$$

Similarly, the intersection between a circle and a straight line is obtained as the solution of the following system

$$\left(x - x_1\right)^2 + \left(y - y_1\right)^2 = R_1^2,\ x = k, \tag{9}$$

wherefrom

$$y = y_1 \pm \sqrt{R_1^2 - \left(k - x_1\right)^2}. \tag{10}$$

It results that the coordinate y_D is a function of the angle φ_1 and, possibly, the coordinate y_E if this coordinate is a function of time.

Further on, we will consider that the coordinate y_E is fixed.

We may write the following sequence

$$\begin{aligned} y_D &= y_D\left(x_B, y_B\right) = y_D\left(x_B\left(x_C, y_C\right), y_B\left(x_C, y_C\right)\right) \\ &= y_D\left(x_B\left(x_C\left(x_A, y_A\right), y_C\left(x_A, y_A\right)\right), y_B\left(x_C\left(x_A, y_A\right), y_C\left(x_A, y_A\right)\right)\right). \end{aligned} \tag{11}$$

The extreme positions of the piston are the solutions of the equation $\frac{\mathrm{d}y_D}{\mathrm{d}\varphi_1} = 0$.

On the other hand

$$y_D = y_B \pm \sqrt{(BD)^2 - \left(d - x_B\right)^2}, \tag{12}$$

$$\frac{\mathrm{d}y_D}{\mathrm{d}\varphi_1} = \frac{\partial y_D}{\partial y_B}\frac{\partial y_B}{\partial \varphi_1} + \frac{\partial y_D}{\partial x_B}\frac{\partial x_B}{\partial \varphi_1}, \tag{13}$$

$$\frac{\partial y_D}{\partial y_B} = 1,\ \frac{\partial y_D}{\partial x_B} = \mp \frac{d - x_B}{\sqrt{(BD)^2 - \left(d - x_B\right)^2}} \tag{14}$$

and so on.

Using this approach, one has to solve a nonlinear equation in the unknown φ_1.

3. Multibody approach

The mechanism has five elements, the crank OA, the triangular shell ABC, the elements BD and CE, and the piston.

Denoting by $\left[\mathbf{A}_i\right]$, $i = \overline{1,5}$, the rotational matrices

$$\left[\mathbf{A}_i\right] = \begin{bmatrix} \cos\varphi_i & -\sin\varphi_i \\ \sin\varphi_i & \cos\varphi_i \end{bmatrix}, \tag{15}$$

we may write the relation

$$\begin{bmatrix} X_P \\ Y_P \end{bmatrix} = \begin{bmatrix} X_{C_i} \\ Y_{C_i} \end{bmatrix} + [\mathbf{A}_i] \begin{bmatrix} x_P^{(i)} \\ y_P^{(i)} \end{bmatrix}, \tag{16}$$

where P is a generic point, X_P, Y_P are the coordinates of the point P relative to the fixed reference system, X_{C_i}, Y_{C_i} are the coordinates of the center of mass C_i of the element i relative to the fixed reference frame, $x_P^{(i)}$, $y_P^{(i)}$ are the coordinates of the point P relative to the mobile reference system rigidly attached to the element i, while φ_i is the angle between the axis $C_i x_i$ and the axis OX.

The first two constraint functions are given by the coordinates of the point C_1

$$f_1(\{\mathbf{q}\}) = X_{C_1} - \frac{OA}{2}\cos\varphi_1 = 0, \tag{17}$$

$$f_2(\{\mathbf{q}\}) = Y_{C_1} - \frac{OA}{2}\sin\varphi_1 = 0, \tag{18}$$

where $\{\mathbf{q}\}$ is the column matrix

$$\{\mathbf{q}\} = \begin{bmatrix} X_{C_1} & Y_{C_1} & \varphi_1 & \dots & X_{C_5} & Y_{C_5} & \varphi_5 \end{bmatrix}^{\mathrm{T}}. \tag{19}$$

We denote by θ_1 and θ_2 the angles BC_2C and AC_2C, and by m_a, m_b, m_c the lengths of the medians of the triangle ABC.

The coordinates of the points A, B and C relative to the mobile reference system $C_2x_2y_2$ are

$$x_C^{(2)} = \frac{2}{3}m_c,\ y_C^{(2)} = 0, \tag{20}$$

$$x_B^{(2)} = \frac{2}{3}m_b\cos\theta_1,\ y_B^{(2)} = \frac{2}{3}m_b\sin\theta_1, \tag{21}$$

$$x_A^{(2)} = \frac{2}{3}m_a\cos\theta_2,\ y_B^{(2)} = \frac{2}{3}m_a\sin\theta_2, \tag{22}$$

where

$$m_a^2 = \frac{2\left[(AC)^2 + (AB)^2\right] - (BC)^2}{4},\ m_b^2 = \frac{2\left[(AB)^2 + (BC)^2\right] - (AC)^2}{4}, \\ m_c^2 = \frac{2\left[(BC)^2 + (AC)^2\right] - (AB)^2}{4}, \tag{23}$$

$$\theta_1 = \arccos\left(\frac{\left(\frac{2}{3}m_c\right)^2 + \left(\frac{2}{3}m_b\right)^2 - (BC)^2}{2\times\frac{2}{3}m_c\times\frac{2}{3}m_b}\right),\ \theta_2 = \arccos\left(\frac{\left(\frac{2}{3}m_a\right)^2 + \left(\frac{2}{3}m_c\right)^2 - (AC)^2}{2\times\frac{2}{3}m_a\times\frac{2}{3}m_c}\right). \tag{24}$$

Point A is situated on the bar OA and on the triangle ABC. It results the relations

$$\begin{bmatrix} X_A \\ Y_A \end{bmatrix} = \begin{bmatrix} X_{C_1} + \dfrac{OA}{2}\cos\varphi_1 \\ Y_{C_1} + \dfrac{OA}{2}\sin\varphi_1 \end{bmatrix} = \begin{bmatrix} X_{C_2} + x_A^{(2)}\cos\varphi_2 - y_A^{(2)}\sin\varphi_2 \\ Y_{C_2} + x_A^{(2)}\sin\varphi_2 + y_A^{(2)}\cos\varphi_2 \end{bmatrix} \quad (25)$$

and the constraint functions

$$f_3(\{\mathbf{q}\}) = X_{C_1} + \frac{OA}{2}\cos\varphi_1 - X_{C_2} - x_A^{(2)}\cos\varphi_2 + y_A^{(2)}\sin\varphi_2 = 0, \quad (26)$$

$$f_4(\{\mathbf{q}\}) = Y_{C_1} + \frac{OA}{2}\sin\varphi_1 - Y_{C_2} - x_A^{(2)}\sin\varphi_2 - y_A^{(2)}\cos\varphi_2 = 0. \quad (27)$$

Analogically we may write the relations

$$\begin{bmatrix} X_B \\ Y_B \end{bmatrix} = \begin{bmatrix} X_{C_2} + x_B^{(2)}\cos\varphi_2 - y_B^{(2)}\sin\varphi_2 \\ Y_{C_2} + x_B^{(2)}\sin\varphi_2 + y_B^{(2)}\cos\varphi_2 \end{bmatrix} = \begin{bmatrix} X_{C_3} - \dfrac{BD}{2}\cos\varphi_3 \\ Y_{C_3} - \dfrac{BD}{2}\sin\varphi_3 \end{bmatrix}, \quad (28)$$

$$\begin{bmatrix} X_C \\ Y_C \end{bmatrix} = \begin{bmatrix} X_{C_2} + x_C^{(2)}\cos\varphi_2 - y_C^{(2)}\sin\varphi_2 \\ Y_{C_2} + x_C^{(2)}\sin\varphi_2 + y_C^{(2)}\cos\varphi_2 \end{bmatrix} = \begin{bmatrix} X_{C_4} - \dfrac{CE}{2}\cos\varphi_4 \\ Y_{C_4} - \dfrac{CE}{2}\sin\varphi_4 \end{bmatrix}, \quad (29)$$

$$\begin{bmatrix} X_E \\ Y_E \end{bmatrix} = \begin{bmatrix} d \\ (Y_E)_0 \end{bmatrix} = \begin{bmatrix} X_{C_4} + \dfrac{CE}{2}\cos\varphi_4 \\ Y_{C_4} + \dfrac{CE}{2}\sin\varphi_4 \end{bmatrix}, \quad (30)$$

$$\begin{bmatrix} X_D \\ Y_D \end{bmatrix} = \begin{bmatrix} X_{C_3} + \dfrac{CD}{2}\cos\varphi_3 \\ Y_{C_3} + \dfrac{CD}{2}\sin\varphi_3 \end{bmatrix} = \begin{bmatrix} X_{C_5} \\ Y_{C_5} \end{bmatrix}, \quad (31)$$

$$X_{C_5} = e \quad (32)$$

$$\varphi_5 = 0 \quad (33)$$

and the corresponding constraint functions.

We thus obtain 14 constraint functions $f_i(\{\mathbf{q}\})$, $i = \overline{1,14}$, and, consequently, the mechanism has one degree of freedom.

The Lagrange function reads

$$F(\{\mathbf{q}\}, \lambda_1, \ldots, \lambda_{14}) = Y_D + \sum_{i=1}^{14} \lambda_i f_i(\{\mathbf{q}\}) = Y_{C_5} + \sum_{i=1}^{14} \lambda_i f_i(\{\mathbf{q}\}). \quad (34)$$

One has to solve the following nonlinear system of 29 equations with 29 unknowns (X_{C_1}, Y_{C_1}, φ_1, ..., X_{C_5}, Y_{C_5}, φ_5, λ_1, ..., λ_{14})

$$\frac{\partial F}{\partial q_j} = 0, \ \frac{\partial F}{\partial \lambda_i} = 0, \ i = \overline{1,14}, \ j = \overline{1,15} \tag{35}$$

4. Numerical approach

We start with the initial value $(\varphi_1)_0 = 0$ and consider a small enough incremental step $\Delta\varphi_1$. For each value $\varphi_1 = (\varphi_1)_0 + i\Delta\varphi_1$, $0 \le i \le \frac{2\pi}{\Delta\varphi_1}$, i integer, one calculates the value $(Y_D)_i$. From this sequence one determines the minimum $(Y_D)_{\min} = (Y_D)_k$ and maximum $(Y_D)_{\max} = (Y_D)_l$ values.

Let us consider now the minimum value $(Y_D)_{\min} = (Y_D)_k$. We take into account the interval $[(Y_D)_{k-1}, (Y_D)_{k+1}]$ and use a numerical method to obtain the value $(Y_D)_{\min}$ with the aid of the theory developed in paragraphs 2 and 3.

Another approach assumes another incremental step $\Delta\varphi_1^* << \Delta\varphi_1$ on the interval $[(Y_D)_{k-1}, (Y_D)_{k+1}]$.

No matter which approach is used, the procedure continues until the difference between of $(Y_D)_{\min}$ at two successive iterations is smaller than an imposed error ε.

The same way of calculation may be used for the determination of $(Y_D)_{\max}$.

5. The crank-shaft mechanism

The crank-shaft mechanism is drawn in Fig. 2. One may write the relation

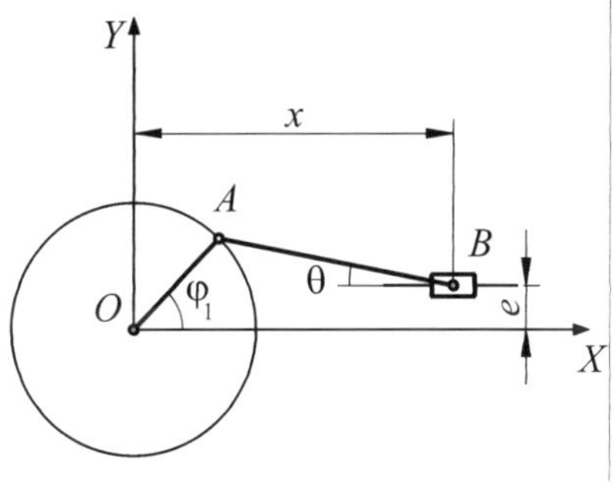

Figure 2. The crank-shaft mechanism.

$$OA \sin \varphi_1 = AB \sin \theta + e, \tag{36}$$

wherefrom

$$\theta = \arcsin\left(\frac{OA \sin \varphi_1 - e}{AB}\right), \tag{37}$$

$$x = OA \cos \varphi_1 + AB \cos \theta = OA \cos \varphi_1 + AB \cos \arcsin\left(\frac{OA \sin \varphi_1 - e}{AB}\right). \tag{38}$$

The extreme positions of the piston are obtained from the equation

$$\frac{dx}{d\varphi_1} = -2OA \sin \varphi_1 + \frac{eAB}{\sqrt{(AB)^2 - (OA \sin \varphi_1 - e)^2}} = 0. \tag{39}$$

IOP Conf. Series: Materials Science and Engineering **252** (2017) 012075 doi:10.1088/1757-899X/252/1/012075

The solutions of the equation (39) are

$$\tan\varphi_1^{(1)} = \frac{e}{OA + AB},\ \sin\varphi_1^{(1)} = \frac{e}{\sqrt{e^2 + (OA + AB)^2}},\ \cos\varphi_1^{(1)} = \frac{OA + AB}{\sqrt{e^2 + (OA + AB)^2}}, \tag{40}$$

$$\tan\varphi_1^{(2)} = \frac{e}{AB - OA},\ \sin\varphi_1^{(2)} = \frac{e}{\sqrt{e^2 + (AB - OA)^2}},\ \cos\varphi_1^{(2)} = \frac{AB - OA}{\sqrt{e^2 + (AB - OA)^2}}, \tag{41}$$

for which we get $x_{max} = x\left(\varphi_1^{(1)}\right)$ and $x_{min} = x\left(\varphi_1^{(2)}\right)$.

Imposing x_{max} and x_{min}, it results a system of two equations with the unknown e, OA and AB. Usually, the system may have an infinite number of solutions, if there is no condition for the three unknown parameters'

6. Synthesis of the mechanism

The mechanism is defined by the following dimensions OA, AB, BC, AC, CE, BD, e, d and $(Y_E)_0$, that is, a total of nine parameters.

Some geometric constraints may be written

$$AB + BC > AC,\ AC + BC > AB,\ AB + AC > BC, \tag{42}$$

$$Y_D \le \sqrt{(OA + AB + BD)^2 - e^2}, \tag{43}$$

$$Y_D \le \sqrt{(BD - OA - AB)^2 - e^2}, \tag{44}$$

$$d \le OA + AC + CE,\ d \ge -(OA + AC + CE), \tag{45}$$

$$d > |CE - OA - AC|, \tag{46}$$

at which one may add the conditions for the existence of the radicals.

Unfortunately, as we will see in paragraph 8, a mechanism that fulfills the geometric constraints may lead to a non-smooth motion, that is, the geometric constraints are only necessary conditions for a variable compression mechanism of this type, but they are not sufficient conditions too.

In addition, there is a particular case of the mechanism, when the angle BAC is equal to $\frac{\pi}{2}$ [24], which assures the same extreme positions of the piston, no mater the value of Y_E.

7. Numerical study

Let us consider the crank-shaft mechanism in Fig. 2 for which $e = 0\,\text{mm}$, $OA = 71.1\,\text{mm}$, $AB = 128.7\,\text{mm}$. These values assures the displacements $x_{max} = 199.8\,\text{mm}$ and $x_{min} = 57.6\,\text{mm}$.

We want to obtain the variable compression mechanism described in Fig. 1 knowing that $AB \in [42,44]\,\text{mm}$, $BC \in [127,129]\,\text{mm}$, $CA \in [98,100]\,\text{mm}$, $CE \in [102,104]\,\text{mm}$, $OA \in [29,31]\,\text{mm}$, $BD \in [129,131]\,\text{mm}$, $d \in [85,87]\,\text{mm}$, $Y_E = 108\,\text{mm}$, $e = 0\,\text{mm}$, $(Y_D)_{max} = x_{max} \pm 0.2\,\text{mm}$, $(Y_D)_{min} = x_{min} \pm 0.2\,\text{mm}$. In addition, if $Y_E = 105\,\text{mm}$, then $(Y_D)_{max} = 200.5 \pm 0.2\,\text{mm}$ and $(Y_D)_{min} = 57.7 \pm 0.2\,\text{mm}$.

Obviously, the problem may have more than one solution. Considering an incremental step for the distances equal to $0.0005\,\text{mm}$ and an incremental step for the angle $\Delta\varphi_1 = 1^0$, one obtains a number of 257 solutions. Reducing the tolerance from 0.2 mm to 0.1 mm, the number of solutions decreases to 36. These solutions are given in Table 1.

Table 1. The solutions.

Sol. #	AB	BC	CA	CE	OA	BD	d	$(Y_D)_{\max 1}$	$(Y_D)_{\min 1}$	$(Y_D)_{\max 2}$	$(Y_D)_{\min 2}$
1.	0.0420	0.1275	0.0990	0.1025	0.0310	0.1300	0.0855	0.1998	0.0576	0.2004	0.0578
2.	0.0420	0.1275	0.0990	0.1025	0.0310	0.1300	0.0860	0.1999	0.0576	0.2004	0.0578
3.	0.0420	0.1275	0.0990	0.1025	0.0310	0.1300	0.0865	0.1999	0.0576	0.2004	0.0578
4.	0.0420	0.1275	0.0990	0.1025	0.0310	0.1300	0.0870	0.1999	0.0576	0.2004	0.0578
5.	0.0425	0.1275	0.0990	0.1020	0.0305	0.1300	0.0850	0.1998	0.0575	0.2004	0.0577
6.	0.0425	0.1275	0.0990	0.1020	0.0305	0.1300	0.0855	0.1999	0.0575	0.2004	0.0577
7.	0.0425	0.1275	0.0990	0.1020	0.0305	0.1300	0.0860	0.1999	0.0575	0.2004	0.0577
8.	0.0425	0.1275	0.0990	0.1020	0.0305	0.1300	0.0865	0.1999	0.0575	0.2004	0.0577
9.	0.0425	0.1275	0.0990	0.1020	0.0305	0.1300	0.0870	0.1999	0.0575	0.2004	0.0577
10.	0.0425	0.1280	0.0995	0.1020	0.0305	0.1300	0.0870	0.1998	0.0575	0.2004	0.0577
11.	0.0425	0.1280	0.0995	0.1025	0.0305	0.1300	0.0850	0.1999	0.0576	0.2005	0.0578
12.	0.0425	0.1285	0.1000	0.1025	0.0305	0.1300	0.0850	0.1998	0.0576	0.2004	0.0578
13.	0.0425	0.1285	0.1000	0.1025	0.0305	0.1300	0.0855	0.1999	0.0578	0.2004	0.0578
14.	0.0425	0.1285	0.1000	0.1025	0.0305	0.1300	0.0860	0.1999	0.0576	0.2004	0.0578
15.	0.0425	0.1285	0.1000	0.1025	0.0305	0.1300	0.0865	0.1999	0.0576	0.2004	0.0578
16.	0.0425	0.1285	0.1000	0.1025	0.0305	0.1300	0.0870	0.1999	0.0576	0.2004	0.0578
17.	0.0430	0.1285	0.1000	0.1020	0.0300	0.1300	0.0850	0.1999	0.0575	0.2004	0.0577
18.	0.0430	0.1285	0.1000	0.1020	0.0300	0.1300	0.0855	0.1999	0.0575	0.2004	0.0577
19.	0.0430	0.1285	0.1000	0.1020	0.0300	0.1300	0.0860	0.1999	0.0575	0.2004	0.0577
20.	0.0430	0.1285	0.1000	0.1020	0.0300	0.1300	0.0865	0.1999	0.0575	0.2004	0.0577
21.	0.0435	0.1280	0.0990	0.1025	0.0295	0.1300	0.0855	0.1998	0.0576	0.2004	0.0578
22.	0.0435	0.1280	0.990	0.1025	0.0295	0.1300	0.0860	0.1998	0.0576	0.2004	0.0578
23.	0.0435	0.1280	0.990	0.1025	0.0295	0.1300	0.0865	0.1999	0.0576	0.2004	0.0578
24.	0.0435	0.1280	0.0990	0.1025	0.0295	0.1300	0.0870	0.1999	0.0576	0.2004	0.0578
25.	0.0440	0.1280	0.0990	0.1020	0.0290	0.1300	0.0850	0.1998	0.0575	0.2004	0.0577
26.	0.0440	0.1280	0.0990	0.1020	0.0290	0.1300	0.0855	0.1998	0.0575	0.2004	0.0577
27.	0.0440	0.1280	0.0990	0.1020	0.0290	0.1300	0.0860	0.1999	0.0575	0.2004	0.0577
28.	0.0440	0.1280	0.0990	0.1020	0.0290	0.1300	0.0865	0.1999	0.0575	0.2004	0.0577
29.	0.0440	0.1280	0.0990	0.1020	0.0290	0.1300	0.0870	0.1999	0.0575	0.2004	0.0577
30.	0.0440	0.1285	0.0995	0.1025	0.0290	0.1300	0.0850	0.1999	0.0575	0.2004	0.0577
31.	0.0440	0.1285	0.0995	0.1025	0.0290	0.1300	0.0855	0.1999	0.0575	0.2005	0.0577
32.	0.0440	0.1290	0.1000	0.1025	0.0290	0.1300	0.0850	0.1998	0.0576	0.2004	0.0577
33.	0.0440	0.1290	0.1000	0.1025	0.0290	0.1300	0.0855	0.1998	0.0576	0.2004	0.0577
34.	0.0440	0.1290	0.1000	0.1025	0.0290	0.1300	0.0860	0.1999	0.0576	0.2004	0.0577
35.	0.0440	0.1290	0.1000	0.1025	0.0290	0.1300	0.865	0.1999	0.0576	0.2004	0.0577
36.	0.0440	0.1290	0.1000	0.1025	0.0290	0.1300	0.0870	0.1999	0.0576	0.2004	0.0577

The reader may observe that all solutions are characterized by $BD = 0.1300\,\text{m}$, that is, the synthesis may lead to some imposed values for certain parameters,

Defining the norm of a solution by the expression

$$norm(sol) = \left[(Y_D)_{\max}\big|_{Y_E=0.108} - 199.8\right]^2 + \left[(Y_D)_{\min}\big|_{Y_E=0.108} - 57.6\right]^2 + \left[(Y_D)_{\max}\big|_{Y_E=0.105} - 200.5\right]^2 + \left[(Y_D)_{\min}\big|_{Y_E=0.105} - 57.7\right]^2, \tag{47}$$

IOP Conf. Series: Materials Science and Engineering **252** (2017) 012075 doi:10.1088/1757-899X/252/1/012075

one may choose the best solution from the previous table (the most closed to the imposed values). In this case the optimum solution is characterized by $AB = 0.0440\ \text{m}$, $BC = 0.129\ \text{m}$, $CA = 0.1000\ \text{m}$, $CE = 0.1025\ \text{m}$, $OA = 0.0290\ \text{m}$, $BD = 0.1300$, $d = 0.0860$, for which if $Y_E = 0.108\ \text{m}$, then $(Y_D)_{\max} = 0.1999\ \text{m}$, $(Y_D)_{\min} = 0.0576\ \text{mm}$, and if $Y_E = 0.105\ \text{m}$, then $(Y_D)_{\max} = 0.2004\ \text{m}$ and $(Y_D)_{\min} = 0.0577\ \text{m}$.

8. Reduced velocities and accelerations

We have to determine the derivatives $\frac{\mathrm{d}Y_D}{\mathrm{d}\varphi_1}$ and $\frac{\mathrm{d}^2 Y_D}{\mathrm{d}\varphi_1^2}$. These derivatives will be calculated by a numerical approach. Considering as known the elements of the mechanism and the incremental step $\Delta\varphi_1$, one may use the formulae

$$\left.\frac{\mathrm{d}Y_D}{\mathrm{d}\varphi_1}\right|_{\varphi_1=\varphi_{1i}} = \frac{(Y_D)_{i+1} - (Y_D)_i}{2\Delta\varphi_1}, \tag{48}$$

$$\left.\frac{\mathrm{d}^2 Y_D}{\mathrm{d}\varphi_1}\right|_{\varphi_1=\varphi_{1i}} = \frac{(Y_D)_{i+1} - 2(Y_D)_i + (Y_D)_{i-1}}{(\Delta\varphi_1)^2}. \tag{49}$$

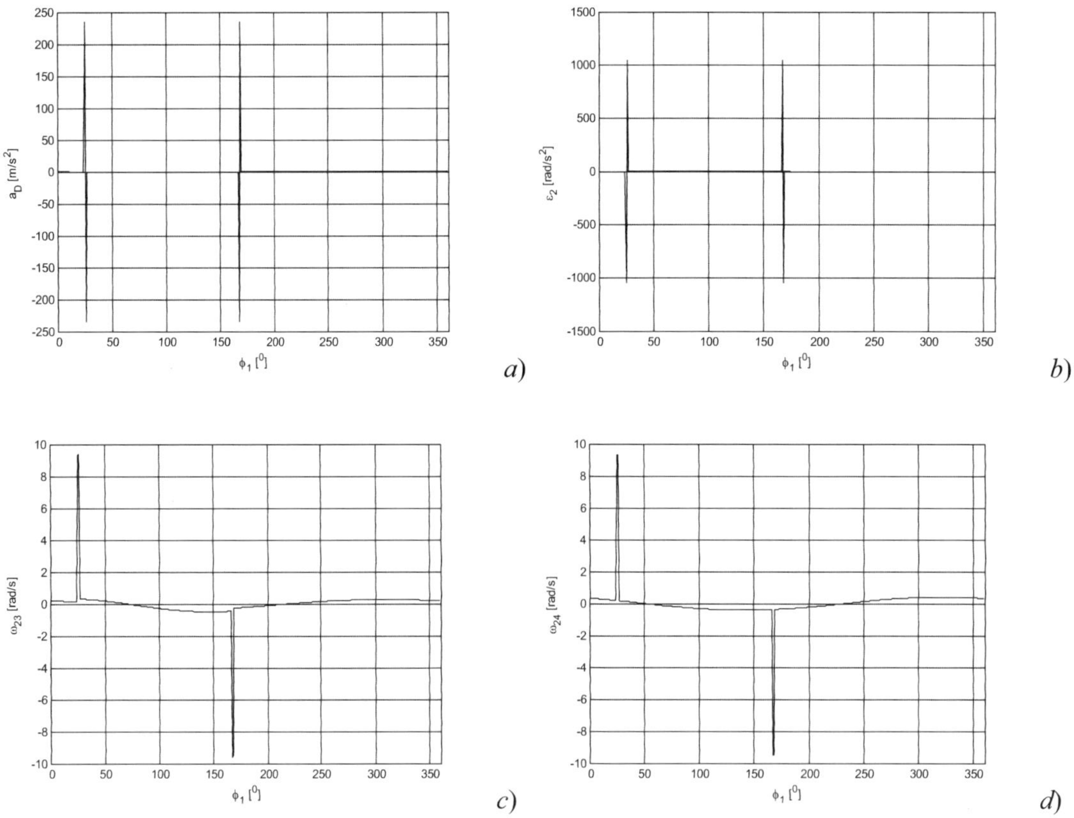

Fig. 3. Variation of: *a)* acceleration of point A; *b)* angular acceleration of the element 2; relative angular velocity between the elements *c*) 2 and 3; *d*) 2 and 4.

Similarly, we may determine $\varphi_2 = \varphi_2(\varphi_1)$, $\varphi_3 = \varphi_3(\varphi_1)$, $\varphi_4 = \varphi_4(\varphi_1)$ and the positions of the points C_i as functions of the angle φ_1. With the aid of these values we may determine the reduced linear and angular velocities and accelerations of the characteristic elements and points.

The reduced relative velocities are a good measure of the wear in the kinematic pairs.

Some corresponding diagrams are given in the Fig. 3. The diagrams are drawn considering that the element OA has a rotational motion with constant angular velocity $\omega_1 = 1\,\text{rad/s}$. The parameters are as follows: $AB = 0.043\,\text{m}$, $BC = 0.128\,\text{m}$, $CA = 0.099\,\text{m}$, $CE = 0.103\,\text{m}$, $OA = 0.030\,\text{m}$, $BD = 0.130\,\text{m}$, $d = 0.086\,\text{m}$, $e = 0\,\text{m}$, $(Y_E)_0 = 0.108$, the incremental angular step being $\Delta\varphi_1 = 1^0$.

The reader may easily observe the non-smooth of the diagrams. There exist two jumps of great values for each characteristic parameter. This jumps can be a criterion to accept or reject the mechanism. More studies must be performed in this direction.

9. Conclusions

The synthesis of such a variable compression mechanism is a difficult task and one has to consider two main problems:

– the geometric constraints which imposes some relations between the lengths of the elements;
– the smoothness of the motion of the resulted mechanism.

First problem may be divided into two parts. The first one considers only the general geometric constraints between the lengths of the elements, while the second one selects from the possible solutions only those ones which satisfy some constraints regarding the extreme positions of the piston. Some other constraints may take into account the space occupied by the mechanism (see [13] for a tiller curve, which imposes a certain free space around the mechanism).

Unfortunately, the second problem can not be solved by analytical methods. The only solution can be given by numerical simulations. An idea could be to construct the diagrams of variation of the measures of the jumps function of different parameters. In this way, one may vary the dimensions of the elements in order to obtain the minimum jump. This idea must be correlated with the diagrams for the variation of the extreme positions of the piston as function of the dimensions of the elements (some aspects concerning the extreme positions of the piston may be found in [13]).

In this study we did mot take into account the displacement of the control lever CE from an initial to a final position in a period of time. The consideration of this displacement leads to at least three problems: the existence of the solution for any intermediate position of the control lever, the smoothness of the law of motion, and the transitory vibrations of the mechanism between the initial and the final status. Some aspects of the last problem for some imposed laws of motion of the control lever are given in [12].

References

[1] Freudenstein F and Maki ER, 1983, Development of an optimum variable-stroke internal combustion engine mechanism from the viewpoint of kinematic structure, *Journal of Mechanisms, Transmissions and Automation in Design* **105** 259–66

[2] Hoeltgebaum T, Simoni R and Martins D, 2016, Reconfigurability of engines: A kinematic approach to variable compression ratio engines, *Mechanism and Machine Theory* **96** 308-22

[3] Hara V, Pandrea N, Popa D, Stan M and Boncea S, 1995, Motoare termice adaptative, Editura Universității din Pitești

[4] Clenci A, 2003, Autoturism echipat cu motor cu raport de comprimare variabil, Raport de cercetare

[5] Miller I, 2001, SAAB Variable Compression Motor, available online at www.me.udel.edu /meeg425/SaabVarComp.doc, 2001.

[6] Mahlesh PJ and Aparna VK, Variable compression ratio engine – A review of future power plant for automobile, *International Journal of Mechanical Engineering Research and Development (IJMERD)*, **2**

[7] Teodorczyk A, 1995, Variable compression ratio engine VR/LE concept, published on Institute of Heat Engineering Warsaw
[8] Published at http://freerepublic.com/ focus/f-news/1818771/posts by Red Badger in 2007.
[9] SAE article http://articles.sae.org/6043/ published in 2009 by John Kendall.
[10] SAE article http://articles.sae.org/15040/ published in 2016 by John Kendall.
[11] Nagarajaa S, Sooryaprakashb K and Sudhakaran R, 2015, Investigate the Effect of Compression Ratio over the Performance and Emission Characteristics of Variable Compression Ratio Engine Fueled with Preheated Palm Oil -Diesel Blends, *Procedia Earth and Planetary Science*
[12] Mănescu B, Dragomir I and Stănescu ND, 2017, The transitory vibrations for a variable compression ratio mechanism, *AVMS 2017*, Timişoara
[13] Mănescu B, Dragomir I, Stănescu ND and Pandrea N, 2017, Study of the influence of geometric parameters on the displacement of piston and compression ratio for a variable compression ratio mechanism, *ICMSAV XLI*, Cluj-Napoca
[14] Sivaramakrishnan K, 2017, Investigation on performance and emission characteristics of a variable compression multi fuel engine fuelled with Karanja biodiesel–diesel blend, *Egyptian Journal of Petroleum*, in press
[15] Satyanarayana K, Viswanath AK, Hanumanta Rao TV and Uma Maheswara Rao SV, 2017, Investigation of the Stresses Induced in Crank Shaft AISI E4340 Forged Steel, *Procedia Engineering* **173** 1672–7
[16] Wu B, Zhan Q, Yu X, Lv G, Nie X and Liu S, 2017, Effects of Miller cycle and variable geometry turbocharger on combustion and emissions in steady and transient cold process, *Applied Thermal Engineering* **118** 621–9
[17] Yamin JAA and Dado MH, 2004, Performance simulation of a four-stroke engine with variable stroke-length and compression ratio, *Applied Energy* **77** 447–63
[18] Eriksson L and Nielsen I, 2014, Modeling and control of engines and drivelines, *John Wiley & Sons*, London
[19] Tsai LW, 2000, Mechanism design: Enumeration of kinematic structures according to function, *CRC Press*, London
[20] Yan HS, 1998, Creative design of mechanical devices, *Springer*
[21] Nilsson Y, 2007, Modelling for fuel optimal control of a variable compression engine, *Linköping University*
[22] Podevin P, Clenci A, 2008, Moteurs à taux de compression variable, BM 2525, *Editions Techniques de l'Ingénieur*, ISSN 1762-8768
[23] Shaik A, Moorthi SV and Rudramoorthy, 2007, Variable compression ratio engine: a future power plant for automobiles – an overview, *Proc. IMechE Part D: J. Automobile Engineering* **221** 1159–68
[24] Pandrea N, 2017, Private communications

IOP Conf. Series: Materials Science and Engineering **252** (2017) 012076 doi:10.1088/1757-899X/252/1/012076

The physically-mathematically model for studying the thermal regime of the spark ignition engine concerning the filling coefficient

T Ulian[1], V Vornicu[1], E Rakosi[1], S Talif[1], I Agape[1]

[1]"Gheorghe Asachi" Technical University of Iasi, Faculty of Mechanical Engineering, 700050, Romania

uliantudor@yahoo.fr, vladd.alex@yahoo.com, edwardrakosi@yahoo.com, tsorinel@yahoo.com, agapeiulian@gmail.com

Abstract. The study of the development conditions of the intake process from spark plug engine, especially from thermal point of view, highlight complex influences. These influences affect in a decisive manner the energy performance of the engine. In this paper, the authors had done a study and an analysis of the aspects for the purpose to develop a physically-mathematically model which ensures the highlighting of the thermal regime influences of the spark plug engine concerning the efficiency of the filling.

1. Methodology

The study of the engine's thermal regime concerning the filling coefficient indicates that the heating of
The study of the engine's thermal regime concerning the filling coefficient indicates that the heating of
the fresh charge affects the filling by reducing the filling coefficient values.

This analyze shows that exist a complex mechanism by which the fresh charge, being at the beginning of the intake process at the ambient temperature, come into contact with the warm engine pieces and it is mixed with the residual gases from the combustion chamber [1]. These processes lead to an increase in the initial temperature of the fresh charge and to a significant reduction in the filling coefficient [3].

The modification of the state parameter changes of fresh charge from cylinder can be quantified on analytical way through the theoretical model developed by the authors. In this model was expressed as a distinct function the dependence of the filling degree on the fresh charge temperature at the end of the intake process. In the same idea, another distinct function was introduced to express the dependence of the filling degree on the increase in the temperature of the fresh charge at the contact with the warm surfaces of the engine during the intake process.

Considering that the heat flow Q is generated in the combustion chamber, it propagates to the intake channel through different areas of the cylinder head [2]. For the purpose of analyzing the phenomenon, a theoretical plan model is used to simulate the heat transmission through the walls of the cylinder, bounded by the control volume as shown in figure 1.

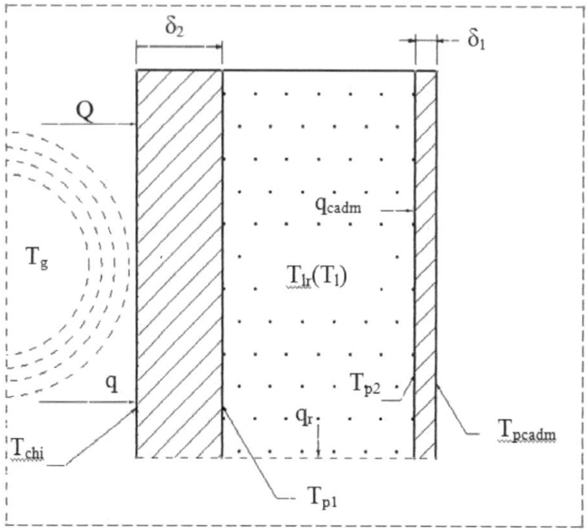

Figure 1. Theoretical plan model used to simulate the heat transmission

The equations for the successive transmission of the heat flux density, q, from the combustion chamber through the liquid cooling jacket, lead to the basic relations:

$$q = \alpha_g \cdot (T_g - T_{chi}) \quad (1)$$

$$q = \frac{\lambda_{Al}}{\delta_1} \cdot (T_{chi} - T_{p1}) \quad (2)$$

$$q = \alpha_l \cdot (T_{p1} - T_l) \quad (3)$$

where:

α_g – convective heat transfer coefficient from the combustion gases to the inner wall of the cylinder head;
Tg – mean cycle temperature of combustion gases;
T_{chi} – the temperature of the combustion chamber surface in the engine cylinder head;
T_{pl} – the surface temperature of the combustion chamber wall in contact with the coolant;
T_l – the liquid temperature of the cylinder head without heat release to the outside;
T_{lr} – the temperature of the cooling liquid with heat release to the outside environment;
α_l – convective thermal transfer coefficient from the cylinder head walls to the coolant;
δ_1 – the average wall thickness of the combustion chamber;
δ_2 – the average wall thickness of the intake channel in the cylinder head;
λ_{Al} – the conductive heat transfer coefficient of the aluminium alloy of the cylinder head.

Relationship (3) is valid in the situation $q_r = 0$, as a particular case which is not found in practice in a real motor. Within this model, taking into account the value of the engine efficiencies, the Relationship (3) is valid in qr = 0, as a particular case that is not practicable in a real motor. Within this model, taking into account the value of engine efficiency, the following situation is also interesting:

$q_r > 0,\ q_r >> q_{cadm}$

The balance of heat flux densities is:

$$q = q_r + q_{cadm} \quad (4)$$

which by replacement becomes:

$$q = \alpha_l \cdot (T_{p1} - T_{lr}) = \alpha_{lpr} \cdot (T_{lr} - T_{aer}) + q_{cadm} \quad (5)$$

where:

T_{aer} – average outdoor air temperature;
q_r – the density of the heat flow yielded by the coolant;

IOP Conf. Series: Materials Science and Engineering **252** (2017) 012076 doi:10.1088/1757-899X/252/1/012076

q_{cadm} – the density of the heat flow transmitted to the intake manifold in the cylinder head;
α_{lpr} – the convective heat exchange coefficient from the cooling liquid to the reduced isothermal equivalent wall, which releases heat to the outside environment.
This gives the expression of the heat flux density that reaches the intake channel in the cylinder depending on the temperature of the coolant, T_{lr}:

$$q_{cadm} = f(T_{lr}) \quad (6)$$

Taking into account the dependence of the temperature of the combustion chamber internal surface from inside the cylinder head on the coolant temperature, experimentally deducted by [6], the authors proposed an analytical expression. This is concretized into a function of the type:

$$T_{chi} = A \cdot T_{lr} + B \quad (7)$$

The variation of which is illustrated in figure 2.

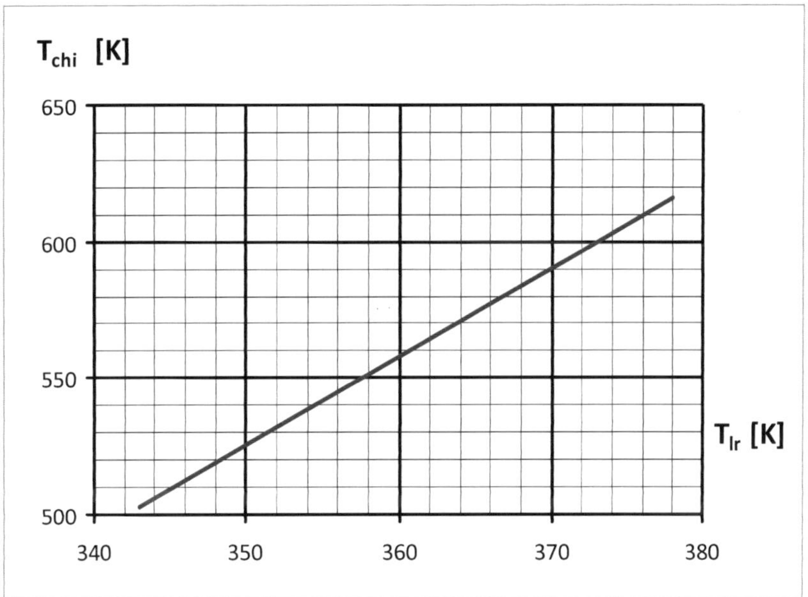

Figure 2. Temperature of the combustion chamber internal surface variation

This function provides a more general character for the proposed mathematical model. Determining the values of coefficients A and B of this function obtains: A = 3.2333 and B = -606.0333. Through development and grouping we obtain successively for q_{cadm}, the expressions:

$$q_{cadm} = \alpha_l \cdot A \cdot T_{lr} + \alpha_l \cdot B - \alpha_l \cdot \frac{\delta_1}{\lambda_{Al}} \cdot \alpha_g \cdot T_g + \alpha_l \cdot \alpha_g \cdot \frac{\delta_1}{\lambda_{Al}} \cdot A \cdot T_{lr} + \alpha_l \cdot \alpha_g \cdot \frac{\delta_1}{\lambda_{Al}} \cdot B + \alpha_{lpr} \cdot T_{aer} - \alpha_l \cdot T_{lr} - \alpha_{lpr} \cdot T_{lr} \quad (8)$$

respectively

$$q_{cadm} = \left[\alpha_l \cdot A \cdot \left(1 + \alpha_g \cdot \frac{\delta_1}{\lambda_{Al}}\right) - \left(\alpha_l + \alpha_{lpr}\right)\right] \cdot T_{lr} - \alpha_l \cdot \alpha_g \cdot \cdot \frac{\delta_1}{\lambda_{Al}} \cdot \left(T_g - B\right) + \alpha_{lpr} \cdot T_{aer} + \alpha_l \cdot B \quad (9)$$

Assimilating the inlet channel in the engine head with a tubular flow path having a linear and a curved portion, a convective heat transfer between the wall channel with the T_{pcadm} temperature and the fresh charge is considered [4]. The heat transfer from the intake channel wall leads to an average increase in the temperature of the fresh charge with the ΔT value. This temperature rise, ΔT, leads, as has been shown above, to a thermal loss at intake due to contact with the hot surfaces of the engine, especially with the inner surface of the intake channel. Thus, the mixture will heat from the temperature T_0 to the temperature $T_0'=T_0+\Delta T$.

As it is known, the particularities of heat transfer by convection influence in this case also the coefficient of convection α which, essentially, quantifies the set of factors defining the process. These factors are mainly identified with the type of motion of the mixture as well as by the shape and orientation of the heat exchange surface.

For the further development of the proposed model, which is in this case the forced convection and the turbulent flow of the mixture through the tubular flow through the cylinder head, a reference relationship, recommended by [5], is used. This relationship is valid in the following areas:

$$7 \cdot 10^3 \leq Re \leq 2 \cdot 10^5 \tag{10}$$

and

$$10^{-1} \leq Pr \leq 5 \cdot 10^2 \tag{11}$$

and materializes as follows:

$$St = \frac{Nu}{Pe} = \frac{Nu}{Re \cdot Pr} = \frac{\frac{\xi}{8}}{1+\Gamma \cdot Re^{-\frac{1}{8}} \cdot (Pr-1)} \tag{12}$$

In this relation Γ represents a constant dependent on the value of the Pr criterion.

Equally, the factor ξ which represents the coefficient of flow resistance of the fresh charge in to the channel occurs in this relation.

The heat exchange inside the cylinder head intake path is characterized by the equation:

$$q_{cadm} = \alpha_{pcadm} \cdot (T_{pcadm} - T_0') \tag{13}$$

In this way, for the fresh charge, we get the expression of the temperature increase due to contact with the warm surfaces of the intake channel from inside the cylinder head, ΔT, depending on the coolant head temperature, T_{lr}.

Due to the reduced wall thickness of the intake channel we have: $T_{pcadm}=T_{p2}$ and $\Delta T(T_{lr})$ becomes:

$$\Delta T(T_{lr}) = T_{p2} - \left[\frac{\alpha_l}{\alpha_m} \cdot A\left(1 + \alpha_g \frac{\delta_1}{\lambda_{Al}}\right) - \frac{\alpha_l+\alpha_{lpr}}{\alpha_m}\right] \cdot T_{lr} - \frac{\alpha_l \cdot \alpha_g}{\alpha_m} \cdot \frac{\delta_1}{\lambda_{Al}} \cdot (T_g - B) + \frac{\alpha_{lpr}}{\alpha_m} \cdot T_{aer} + \\ + \frac{\alpha_l}{\alpha_m} \cdot B - T_0 \tag{14}$$

Finally, the function that allows the calculation of the filling coefficient in dependence on the change in the coolant temperature from inside the cylinder head in accordance with the principle of zonal cooling of the engine (the zonal thermal regime of the engine) is obtained.

$$\eta_v(T_{lr}) = c_p \cdot c_{Tm} \cdot c_\varepsilon \cdot c_{up_2}^{-1} \cdot c_{es} \cdot \left\{T_{p2} - \left[\frac{\alpha_l}{\alpha_m} \cdot A\left(1 + \alpha_g \frac{\delta_1}{\lambda_{Al}}\right) - \frac{\alpha_l+\alpha_{lpr}}{\alpha_m}\right] \cdot T_{lr} - \frac{\alpha_l \cdot \alpha_g}{\alpha_m} \cdot \frac{\delta_1}{\lambda_{Al}}(T_g - \\ B) + \frac{\alpha_{lpr}}{\alpha_m} \cdot T_{aer} + \frac{\alpha_l}{\alpha_m} \cdot B - T_0\right\}^{-1} \tag{15}$$

in which:

$c_p = \frac{p_a}{p_0}$ – pressures constant;

$c_{Tm} = T_0$ – temperatures constant;

$c_\varepsilon = \frac{\varepsilon}{\varepsilon-1}$ – constructive engine constant;

$c_{up_2}^{-1} = 1 - \varphi_{pu} + \gamma_r$ – the second constant that influences the process of filling the cylinder with the fresh charge;

$c_{es} = 1 + \gamma_r$ – the constant that characterized the exhaust gases from the cylinder.

Considering the transfer of heat from the combustion chamber to the coolant and then from it to the intake channel in the cylinder, the proposed model finally determines the dependence of the filling coefficient on the engine's thermal regime through the coolant temperature. In order to attenuate the phenomenon of reducing the filling coefficient value and implicitly the performance of the engine with the heating of the fresh load, with the proposed model, optimization of the engine's thermal regime can be made.

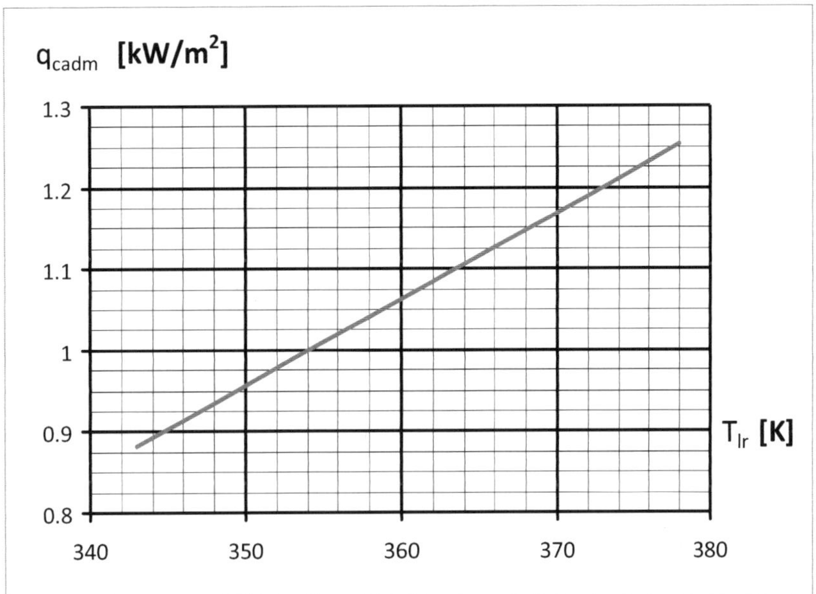

Figure 3. Variation of the heat exchange inside the cylinder head intake path

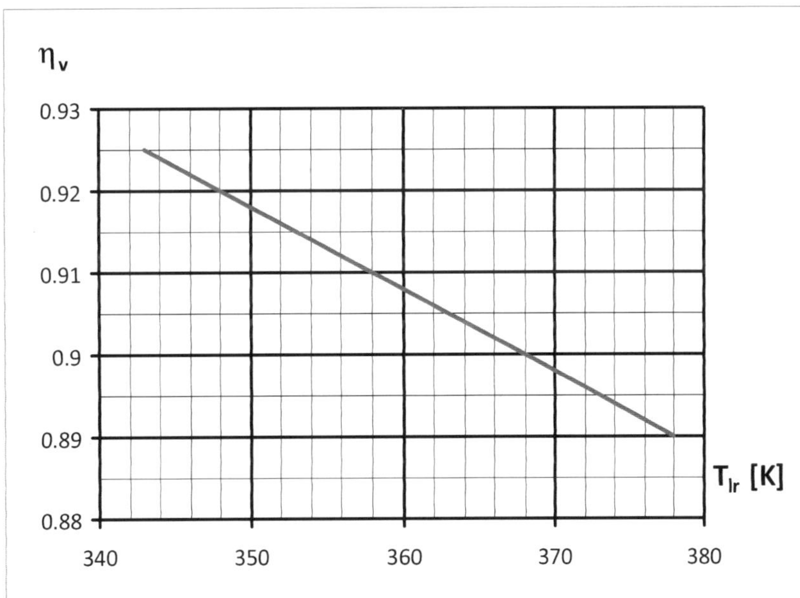

Figure 4. Filling coefficient dependence on the coolant temperature

IOP Conf. Series: Materials Science and Engineering **252** (2017) 012076 doi:10.1088/1757-899X/252/1/012076

2. Results

The synthetically presented physico - mathematical model assures first of all the emphasis of the heat exchange inside the cylinder head intake path dependence on the variation of the coolant temperature. Thus, the establishment and the graphical representation of this dependence, shown in figure 3, indicate a correlation between coolant temperature and the heat exchange inside the cylinder head, thus highlighting the advantages of a zonal thermal engine.

On the other hand, the model proposed by the authors highlights the improvement of the cylinder filling coefficient by reducing the cylinder head thermal regime. This dependence reveals a satisfactory increase of the filling coefficient, as shown in figure 4, in the conditions of an average decrease in the thermal regime of the engine.

3. Conclusions

Through the proposed model in this study it is realizing an analysis of the influence of the engine thermal regime on the cylinder filling phenomenon, based on the variation of the filling coefficient.

An original physico-mathematical model is proposed and developed to ensure that this influence is established. The model provides significant results that can be used to improve and to optimize the performance of the spark ignition engine.

In this idea, compared to previous studies realized by the authors, this paper contains a physico-mathematical model that is more complex, in which intervene the thermal regime of the engine.

We noted that the theoretical model proposed by the authors in this paper does not include the engine load modification.

4. References

[1] Allen D J Lasecki M P 2001 *Thermal management evolution and controlled coolant flow* SAE paper 2001–01-1732

[2] Alkidas A C 1980 *Heat transfer characteristics of a spark – ignition engine J Heat transfer* vol. 102(2) pp 189–193

[3] Eberth J F Wagner J R Afshar B A. and Foster R C 2004 *Modelling and Validation of Automotive Smart Thermal Management System Architectures* SAE paper 2004-01-0048

[4] Incropera F P and DeWitt D P 2002 *Introduction to Heat Transfer* John Wiley & Sons Inc.

[5] Popa B Aradau D Biris I Iosifescu C 1986 *Thermo Technician's Engineering Handbook* vol 1 Technical Publishing House Bucharest

[6] Popa B. Bataga N Madarasan T Marinescu M 1978 *Thermal Stress in Machinebuilding* Technical Publishing House Bucharest

IOP Conf. Series: Materials Science and Engineering **252** (2017) 012077 doi:10.1088/1757-899X/252/1/012077

Key fuel properties and engine performances of diesel-ethanol blends, using tetrahydrofuran as surfactant additive

A Molea[1]*, P Visuian[1], I Barabás[1], R C Suciu[2], N V Burnete[1]

[1]Technical University of Cluj-Napoca, Faculty of Mechanics, Department of Automotive Engineering and Transports, Cluj-Napoca, Romania
[2]National Institute for Research and Development of Isotopic and Molecular Technologies, Cluj-Napoca, Romania

andreia.molea@auto.utcluj.ro

Abstract. In this paper there were presented researches related to preparation and characterization of physicochemical properties of diesel-ethanol blends stabilized with tetrahydrofuran as surfactant, in order to be used as fuels in compression ignition engines. The main spray characteristics and engine performances of these blends were evaluated by using AVL Fire software. In the first stage of the studies, commercial diesel was mixed with ethanol, in different concentrations (between 2% and 15% v/v), followed by the addition of tetrahydrofuran (THF) until the blends were miscible, i.e. the blends were stabilized. The experiments were done at room temperature (22 ^{0}C). The obtained blends were characterized in order to determine the chemical composition and physicochemical properties, i.e. density, kinematic viscosity, surface tension. UV-Vis spectroscopy was utilized in order to determine a semi-quantitative evaluation regarding the chemical composition of the prepared blends and chemical interaction between diesel, ethanol and THF. Based on the determined properties, the fuel spray characteristics, engine performances and emission characteristics were evaluated by simulation using the AVL Fire software. The obtained results regarding physicochemical properties of blends were compared with diesel. Some improvements were observed when operating with the prepared blends compared to diesel with respect to engine performances and emission characteristics. Based on physicochemical evaluation and computer simulation, it was demonstrated that diesel-ethanol-tetrahydrofuran blends can be used as alternative fuel in compression ignition engines.

1. Introduction

Diesel fuel is used in internal combustion engines in many fields which include: automotive, engineering machinery, marine propulsion [1]. The consumption of fossil fuels leads to various types of air pollutants including nitrogen oxides (NO_x), particulate matter (PM), carbon monoxide (CO) and other harmful substances. Moreover, around 60% of greenhouse gas emissions such as carbon dioxide (CO_2) are derived from the direct combustion of fuels [1,2]. Therefore, arises the need to replace part of the diesel fuel used currently with other alternatives. One of these alternatives is ethanol which can be manufactured from bio/renewable materials and by a process that would make it bio-ethanol. This compound can be made from many raw materials such as: sugar cane, molasses, waste biomass materials, corn, barley, sugar beets, cellulose [2,3,4]. Ethanol has been used before especially in combination with gasoline, but it has been tested with diesel fuel also and the blends obtained were technically acceptable for compression ignition engines [3,4,5].

In order to test the feasibility of using the diesel-ethanol blends in compression ignition engines it is important to compare the physical properties of the blends and to see if they are comparable and within standards that allow them to be used in these engines. Such properties include, but are not limited to: stability, density, viscosity, lubricity, flash point, safety [4]. These properties vary with the quality of the components used, for example the water content of the ethanol or the aromatic compound content of the diesel fuel [2]. Therefore it is important to describe the substances used because the results are only available for these specific concentrations.

Studies have shown that diesel-ethanol blends, that contain no more than 15% (v/v) ethanol, are a solution to reduce the quantity of fossil-fuel used by internal combustion engines, and these blends show acceptable properties that fulfill the standards required for diesel fuel. They are stable at medium temperatures, but for lower temperatures, in order to maintain the stability, additives need to be added [1,3,4,6]. The majority of these additives are surfactants that help reduce the surface tension between the two substances, making them more miscible. Surfactants that have been used before by other researchers include: biodiesel, stearyl alcohol (saturated), 1-hexanol, oleyl alcohol (unsaturated), methyl oleate (unsaturated with ester group) [1,3,7]. In order to improve the effectiveness of these substances some additional additives can be added. Due to their role and the quantity in which they are added, this beeing lower compared to the surfactants, they are called co-surfactants. Some of these are: 1-butanol, 1-octanol, 1-decanol [7].

2. Materials and methods

2.1. Preparation of alternative fuel blends

In order to prepare the diesel-ethanol blends, commercial winter grade diesel fuel, without biodiesel content, was supplied by Petrom (Cluj-Napoca, Romania) which presents the qualitative parameters required by European Standard EN 590. Ethanol (99.2% purity, water content: max. 0.8%) and tetrahydrofuran (99.5% purity, water content: max. 0.05%) reagents were supplied by Nordic Invest.

In the first stage of the studies, five diesel-ethanol blends were prepared by mixing ethanol (EtOH) with diesel in proportion between 2%(v/v) and 15% (v/v), under magnetic stirring condition (table 2). After the binary blends were prepared, tetrahydrofuran (THF) was added as surfactant, in order to obtain stability at room temperature (22 ^{0}C), i.e. diesel to become miscible with ethanol.

2.2. Physicochemical characterization of blends

Density and kinematic viscosity of diesel-ethanol stabilized with tetrahydrofuran blends were measured using an automated SVM 3000 Anton Paar rotational Stabinger viscometer, equipped with vibrating U-tube densimeter and Peltier elements to thermoregulate the sample. Surface tension was determined using stalagmometric method. In order to assess the hazard classification of prepared blends, flash point was measured using an automated Pensky Martens closed cup flash point analyzer HFP 339.

Semi-quantitative analysis for the determination of chemical composition and/or the interaction between components was made using UV-Vis spectroscopy and spectrofluorimetry (PL), respectively. UV–Vis absorption spectra of pure components and blends were taken on a JASCO V-550 spectrometer, in a wavelength range of 200–900 nm. Photoluminescence measurements were carried out using an Able &Jasco FP 6500 spectrofluorometer with a xenon lamp of 150 W as the excitation source and a photomultiplier tube (PMT) as detector. For comparison, all emission spectra were measured at room temperature with the same instrumental parameters using an excitation at 389 nm specific to diesel and emission originated from the luminescent centres at 423 nm. All spectra were registered exactly in the same conditions: excitation/reflection into 3 mL of liquid sample with identical incidence intensity, the excitation slit (3 mm) and emission slit (5 mm).

2.3. AVL Fire simulation procedure

The model for engine simulation was developed in the AVL ESE Diesel simulation environment [8]. Based on the AVL 5402 type single cylinder engine, the main parameters of the model are presented in table 1. Since this engine has a symmetrical combustion chamber, only a 45° sector having a single injection hole was modelled (figure 1). The validation of the simulation model is described in detail in [9].

Table 1. The main engine parameters

Parameter	Value
Bore x Stroke (mm x mm)	90 x 85
Compression ratio	17.1
Speed (min^{-1})	2500
Number of holes, diameter and position (– x mm x °)	8 × 0.12 × 162°

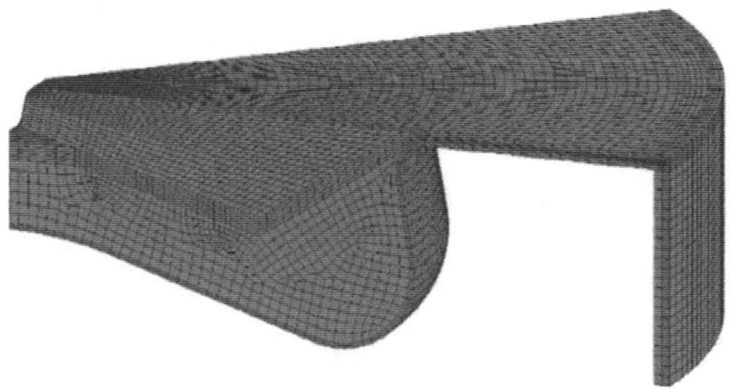

Figure 1. Simulation mesh (45° engine sector) [9].

3. Results and discussion

3.1. Stability of prepared blends

Room temperature stabile blends were obtained by using tetrahydrofuran with concentration between 0.6% (v/v) for diesel-2% (v/v) ethanol blend and 7% (v/v) for diesel-15% (v/v) ethanol blend, respectively. The total volume proportions of blends were detailed in table 2.

Table 2. Ratio of diesel, ethanol and tetrahydrofuran used for preparation of blends.

Sample	Diesel % v/v	Ethanol % v/v	Tetrahydrofuran % v/v
Diesel	100	-	-
2% EtOH	97.4	2	0.6
5% EtOH	92.7	4.9	2.4
8% EtOH	87.6	7.6	4.8
10% EtOH	84.8	9.4	5.8
15% EtOH	79	14	7.0
EtOH	-	100	-
THF	-	-	100

The stability of the blends was monitored for two months, until the prepared blends were considered homogenous. Therefore, in the first day of stability study, 1 mL THF was added into the 2% and 5% ethanol blends and 5 mL THF into the 8% to 15% ethanol blends and mixed using a magnetical stirrer, followed by keeping at room temperature for 3-4 days in order to observe if the phase's separation has occurred. According with the observation it was noted that, after two weeks, by adding a total percent volume of 0.6% and 2.4% THF, respectively, the blends which contain 2% and 5% ethanol were stabilized. For blends which contain over 8% ethanol, stability has been achieved after two months, after total percent volume of 4.8%, 5.8% and 7% THF, respectively, was added. In figure 2, the ternary stability diagram of diesel-ethanol-tetrahydrofuran blends, at room temperature can be seen.

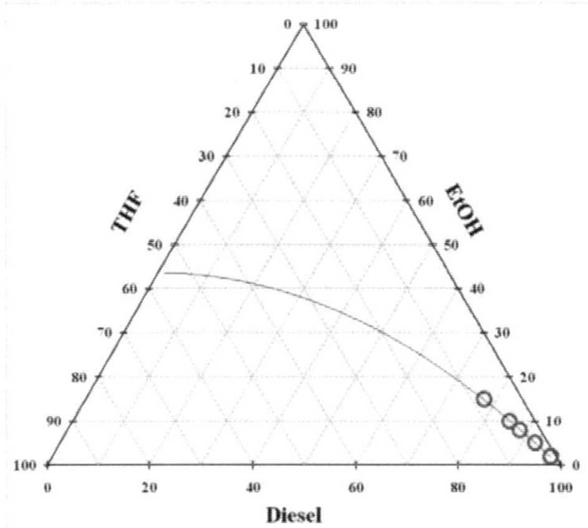

Figure 2. Stability diagram of diesel-ethanol-tetrahydrofuran ternary blends.

Since ethanol and diesel present different structure and chemical characteristics, in time, the blends become unstable, leading to the separation of components. Immiscibility of ethanol into diesel fuel is caused by water content from ethanol and aromatic constituents from diesel fuel. By adding tetrahydrofuran as surfactant, this acts as a bridging agent through molecular compatibility and bonding to produce a homogenous blend [3,10].

3.2. *Physicochemical properties*

Figure 3 depicts the variation of density with temperature, while figure 4 shows the density at 15 ^{0}C of diesel fuel, diesel-ethanol blends stabilized with THF, ethanol and THF pure components. It can be observed that for all the samples the density decreases linearly with temperature. Also, the blends' density at 15 ^{0}C are smaller compared to diesel fuel (0.8416 g/cm^3) which is attributed to the lower density of ethanol component, even if the density of THF is higher than superior limit for diesel density. However, the density at 15 ^{0}C of the blends are acceptable for the EN 590 standard; the inferior limit for diesel density is 0.8200 g/cm^3. The obtained results are in accordance with the results presented by other researchers [10,11].

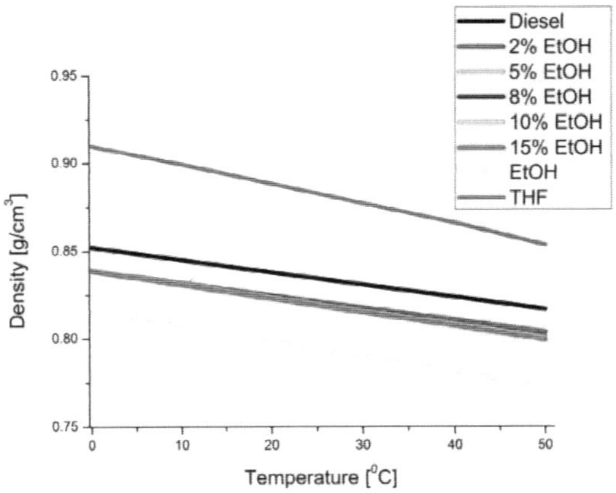

Figure 3. Density vs. temperature of pure components and blends.

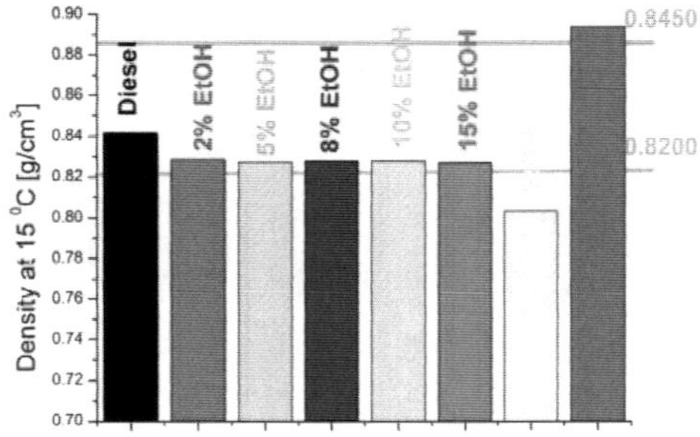

Figure 4. Density at 15 ^{0}C of pure components and blends.

The kinematic viscosity measurements vs. temperature of pure components and blends are presented in figure 5 and it can be seen that THF presents small variations of viscosity with temperature, with a viscosity of around 0.62 mm^2/s. Also, ethanol presents low kinematic viscosity with variations between 2.55 mm^2/s at 0 ^{0}C and 0.92 mm^2/s at 50 ^{0}C. According to the obtained data, blends which contained 2% and 5% ethanol (before adding THF) present a kinematic viscosity at 40 ^{0}C higher than the minimum level of viscosity for diesel fuel (2 mm^2/s) which can be considered an acceptable limit for a standard diesel fuel (figure 6). However, the increase of ethanol concentration to 15% and the addition of the needed THF for stabilization, leads to a decrease of viscosity, value which exceeds the required minimum for diesel fuel. A low viscosity could cause more fuel pump leakage that would limit the amount of fuel delivered during high load conditions. A lower viscosity would also cause the pulverized fuel droplets to have a smaller dimension which would lead, for the same amount of volume injected, to an increase of surface area of the fuel droplet which would influence the evaporation characteristics [3,4].

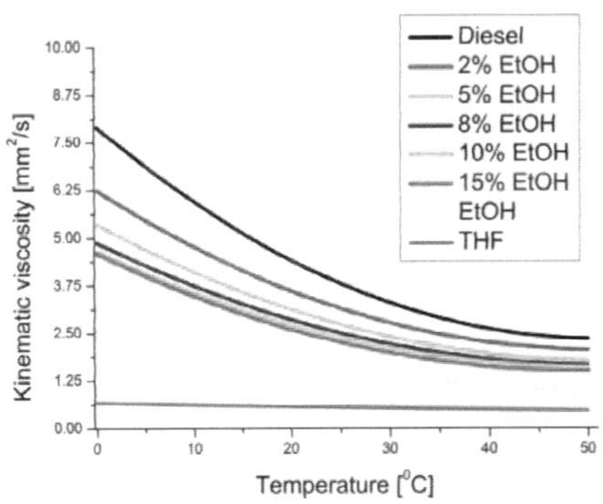

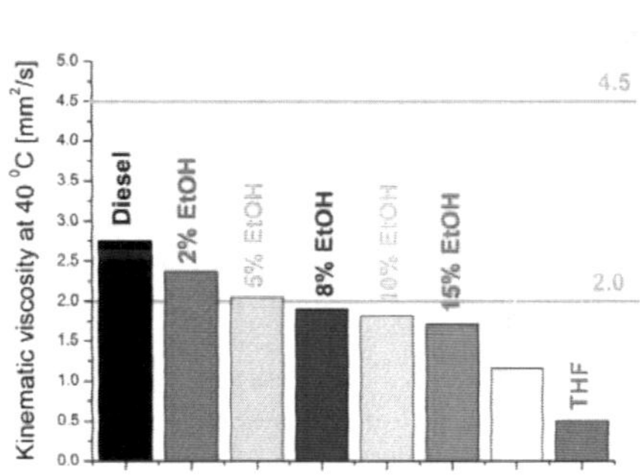

Figure 5. Kinematic viscosity vs. temperature of pure components and blends.

Figure 6. Kinematic viscosity at 40 ^{0}C of pure components and blends.

Surface tension of diesel fuel plays an important role in the fuel's combustion, affecting its initial stage, i.e. fuel atomization and spray characteristics. In figure 7 it can be seen that the surface tension of diesel is specific to a winter grade diesel (0.0309 N/m). For small concentrations of ethanol it can be observed that the value of surface tension is close to that of the diesel fuel (0.0306 N/m), but by increasing the ethanol concentration over 5%,the surface tension tends to decrease to 0.0279 N/m. However, the surface tension of the blends falls within the limits presented in the literature data [2,12].

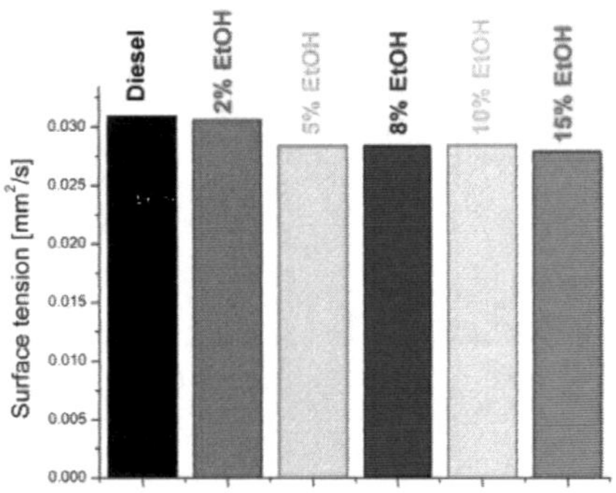

Figure 7. Surface tension at 20 ^{0}C of pure components and blends.

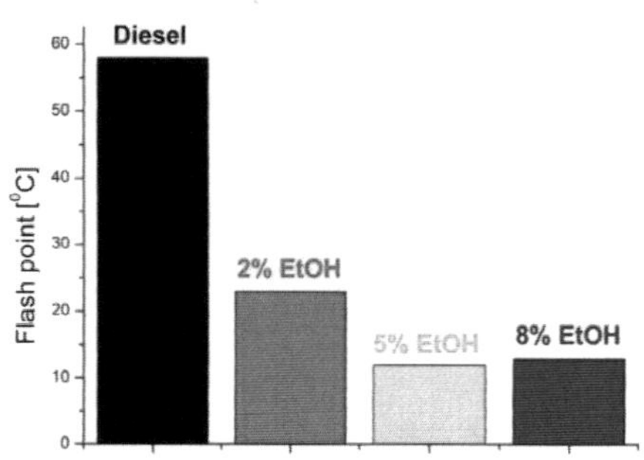

Figure 8. Flash point of diesel and blends which contains 2%, 5% and 8% ethanol.

Even if the density, viscosity and surface tension properties of blends which contain ethanol in concentrations lower that 8% have good characteristic for utilization as fuel in diesel engines, the safety properties need to be considered for shipping and storage classification and precautions should be taken in handling and transporting of the prepared blends. Therefore, figure 8 emphasizes that, due to the low flash point of both ethanol and tetrahydrofuran (i.e. 17 ^{0}C and -14 ^{0}C, respectively), the flash point of blends dramatically decrease even at low concentration of the components. As a result, the transportation, handling and storage of the blends require additional safety measures in order to avoid ignition and explosions. Flash point of diesel-ethanol blends was discussed by other researchers and our results are in accordance with the literature data [3,10,13].

Using UV-visible analysis it was observed that there are no chemical interactions between diesel and ethanol or tetrahydrofuran compounds due to the fact that,in the spectra, there are no new absorption bands, other than diesel characteristic absorption bands (figure 9).

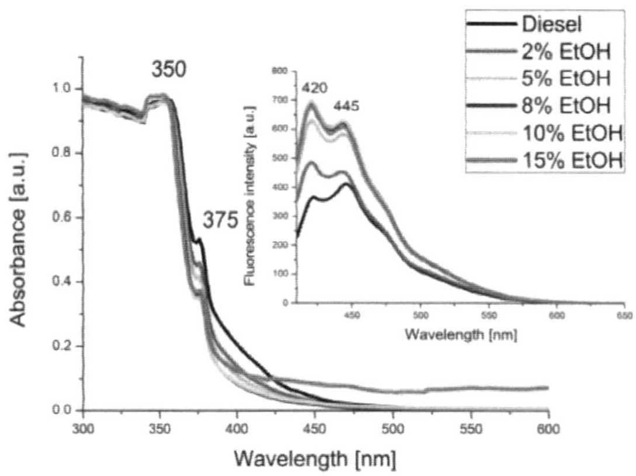

Figure 9. UV-Vis spectra of diesel and blends; Insert: PL spectra of diesel and blends.

The absorption bands at 350 nm and 375 nm wavelengths and emission bands from 420 nm and 445 nm identified upon excitation at 389 nm (insert figure 9), respectively, can be attributed with polycyclic aromatic compounds [14,15] from diesel structure. However, in the emission spectra it can be observed that the band from 420 nm increases while the band from 445 nm decreases, respectively, with increasing of ethanol and THF concentrations. Due to complex chemical structure of diesel fuel which is predominant formed from non-polar n-alkanes and moderately polar aromatic hydrocarbons, while ethanol present an highly polar character thru hydroxyl group, aromatic structures from diesel may interact with polar hydroxyl group of the ethanol and form transient dipoles or London dispersion forces which are a kind of intermolecular forces weaker than the hydrogen bond, leading to a stable and miscible bend [16,17]. At high concentration of ethanol, THF acts as a surface-active agent, leading to the formation of micelles which consists of polar heads and non-polar tails, forming a bridge between diesel and ethanol leading to formation of a miscible phase [3,10].

3.3. Simulation results

The variations of cylinder pressures for all fuels are presented in figure 10. In the cases of the blends, the ignition delays slightly increased (from 0.2 °CA – 2% EtOH to 0.7°CA – 15% EtOH) because of the low cetane number of ethanol and THF. Due to the diminished caloric values of the blends, the maximum values of in-cylinder pressures decreased with the increase in ethanol and THF contents (between 0.5% – 2% EtOH and 1.2 % – 15% EtOH). Same trends were observed in cases of the in-cylinder temperatures (figure 11). Consequently, the engine power is mildly reduced.

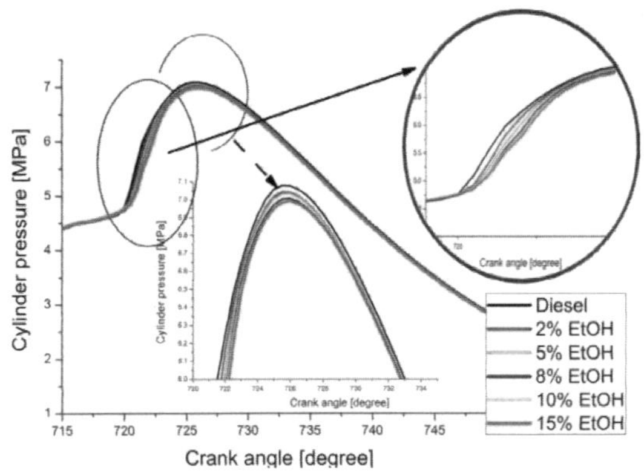

Figure 10. Cylinder pressure as a function of crank angle.

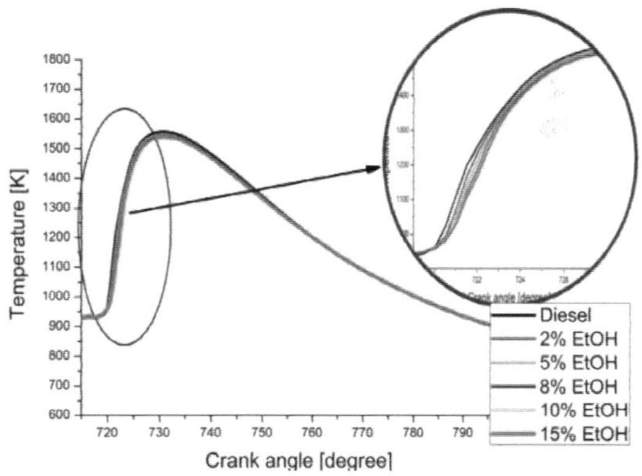

Figure 11. Cylinder temperature as a function of crank angle.

Except for CO emissions which increased significantly, the engine operation with the blends yielded better emission parameters in comparison to the diesel operation, as it can be observed from the figures 12 to 15. Comparing with diesel emissions, at the final of simulation, a decrease of NO emissions between 9% and 10.4% was observed (figure 12), with the increase of ethanol concentration (for the blends which contain 2% and 15%), while soot presents a drastic decrease from 15.5% up to 44%, for the blends mentioned above (figure 13). Also, a small decrease of CO_2 emissions (figure 14) of the blends was observed (from 0.46% specific to the blends which contain 2% ethanol to 1.28% in case of the blend with 15% ethanol content). However, a significant increase in CO emissions can be observed for the blends (up to 14%) comparative with diesel CO emissions (figure 15).

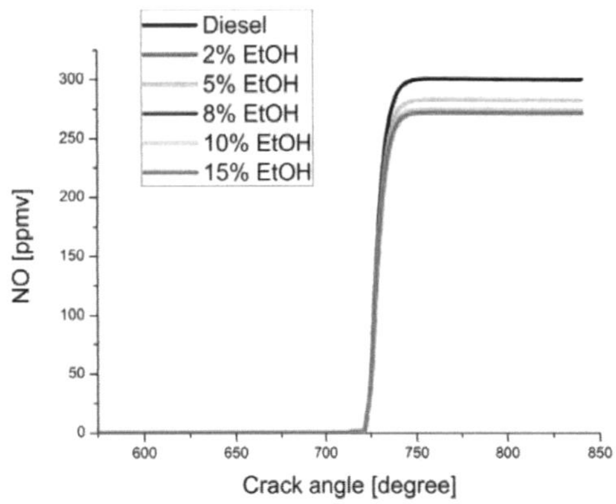

Figure 12. Variation of the NO concentration as a function of crank angle.

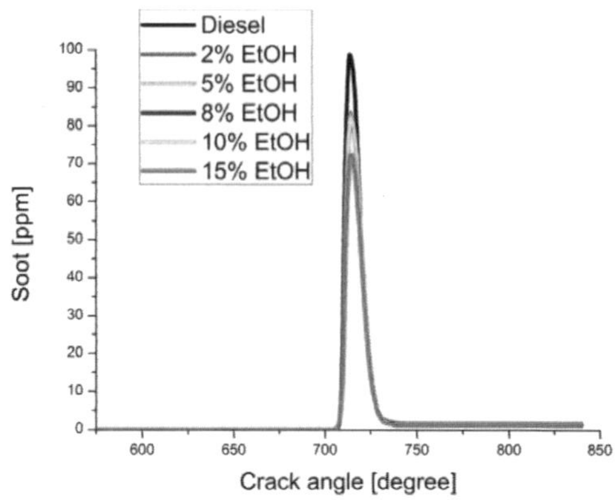

Figure 13. Evolution of the soot concentration as a function of crank angle.

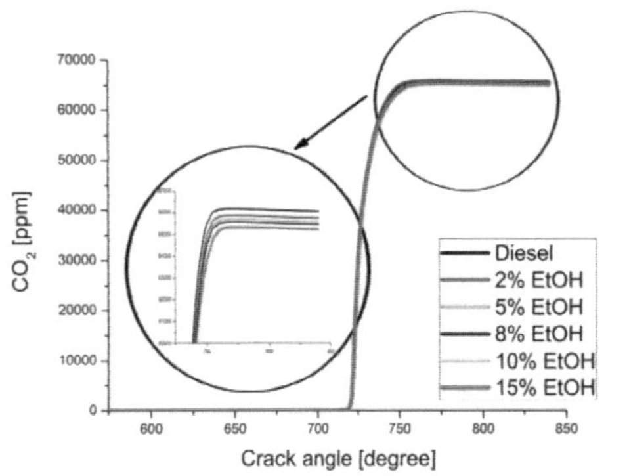

Figure 14. Variation of the CO_2 concentration as a function of crank angle.

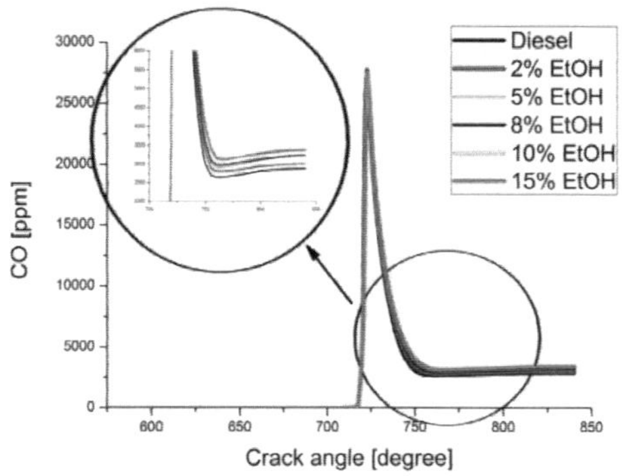

Figure 15. Variation of the CO concentration as a function of crank angle.

These improvements can be explained by lower combustion temperatures and the intrinsic oxygen content of the studied blends, which contributes to improving the combustion process. The CO emissions increased because of the prolonged ignition delay which reduced the allowable combustion time.

4. Conclusions

The studies presented emphasised the possible utilization of diesel-ethanol blends as alternative fuel in internal combustion engines, in order to decrease both utilization of fossil fuels and engine emissions, respectively. The blends were prepared using ethanol between up to 15% v/v concentration and based on stability studies it was observed that at low concentration of ethanol (under 5% v/v), the blends were stabilized at 22°C using 2.4 % v/v THF, in two weeks while at high concentration of ethanol, diesel and ethanol become miscible and stabile only by adding up to 7% v/v THF.

Based on physicochemical characterisation of the diesel-ethanol-THF blends it was observed that density values are within the EN 590 limits, while the kinematic viscosity values of the blends which contain up to 5% v/v ethanol are located near the inferior limit of the standard, but for the blends with above 5% v/v ethanol, the kinematic viscosity values are under the inferior limit. Surface tension values slightly decrease with the increase of the ethanol and THF concentration, but overall these three parameters of the blends have good limit values for utilization as fuels in diesel engines. However, due to the low flash point of both ethanol and THF, the blends present a flash point lower than 25°C which will require additional safety measures. Also, using VU-Visible spectroscopy it was emphasized that there are no chemical interactions between diesel and ethanol or tetrahydrofuran compounds and THF acts as a surface-active agent, forming a bridge between diesel and ethanol leading to formation of a miscible phase.

The engine simulation shows that the engine operation with the blends yielded slightly lower in-cylinder pressures and temperatures, but better emission parameters in comparison to the diesel operation.

References

[1] Liu H, Hu B and Jin C 2016 *Fuel* **184** 440-8
[2] Li D-g, Zhen H, Xingcai L, Wu-gao Z and Jian-guang Y 2005 *Renew. Energ.* **30** 967–76
[3] Shahir S A, Masjuki H H, Kalam M A, Imran A, Rizwanul Fattah I M and Sanjid A 2014 *Renew. Sust. Energ. Rev.* **32** 379–95
[4] Hansen A C, Zhang Q and Lyne P W L 2005 *Bioresour. Technol.* **96** 277–85
[5] Torres-Jimenez E, Jerman M S, Gregorc A, Lisec I, Dorado M P and Kegl B 2011 *Fuel* **90** 795-802
[6] Lapuerta M, Armas O and Garcia-Contreras R 2007 *Fuel* **86** 1351–7
[7] Arpornpong N, Attaphong C, Charoensaeng A, Sabatini D A and Khaodhiar S *Fuel* **132** 101-6
[8] AVL List GmbH2014 *AVL Fire Manual. Graz* (Austria)
[9] Burnete N V, Iclodean C, Moldovanu D and Filip N 2017 *CONAT 2016 International Congress of Automotive and Transport Engineering*, eds. Chiru A and Ispas N (Springer International Publishing Switzerland) 379
[10] Kwanchareon P, Luengnaruemitchai A and Jai-In S 2007 *Fuel* **86** 1053–61
[11] Guarieiro L L N, Figueiredo de Souza A, Torres E A and B de Andrade J 2009 *Atmos. Environ.* **43** 2754–61
[12] Xingcai L, Zhen H,Wu-gao Z and Li D-g 2004 *Combust. Sci. Technol.* **176** 1309-29
[13] Burnete N V, Filip N and Barabás I 2015 *Ro. J. A. E.* **21** 89-106
[14] Burkert A, Paa W, Reimert M, Klinkov K and Eigenbrod C 2013 *Fuel* **111**384–92
[15] Lapuerta M, Rodríguez-Fernandez J, Fernandez-Rodríguez D and Patino-Camino R 2017 *Fuel* **199** 332–38
[16] Lapuerta M, Rodriguez-Fernandez J, Garcia-Contreras R and Bogarra M 2015 *Fuel* **139** 171–79
[17] Bejaoui S, Mercier X, Desgroux P and Therssen E 2014 *Combust. Flame* **161** 2479-91

A COMPARATIVE STUDY OF CYCLE VARIABILITY OF LASER PLUG IGNITION VS CLASSICAL SPARK PLUG IGNITION IN COMBUSTION ENGINES

Bogdan Done
Renault Technologie Roumanie, University Politehnica of Bucharest,
Departmentof Thermotechnics, Engines,
Thermal Equipment and Refrigeration Installations, Bucharest, Romania

Abstract. Over the past 30 years numerous studies and laboratory experiments have researched the use of laser energy to ignite gas and fuel-air mixtures. The actual implementation of this laser application has still to be fully achieved in a commercial automotive application. Laser Plug Ignition as a replacement for Spark Plug Ignition in the internal combustion engines of automotive vehicles, offers several potential benefits such as extending lean burn capability, reducing the cyclic variability between combustion cycles and decreasing the total amount of ignition costs, and implicitly weight and energy requirements. The paper presents preliminary results of cycle variability study carried on a SI Engine equipped with laser Plug Ignition system. Versus classic ignition system, the use of the laser Plug Ignition system assures the reduction of the combustion process variability, reflected in the lower values of the coefficient of variability evaluated for indicated mean effective pressure, maximum pressure, maximum pressure angle and maximum pressure rise rate. The laser plug ignition system was mounted on an experimental spark ignition engine and tested at the regime of 90% load and 2800 rev/min, at dosage of λ=1.1. Compared to conventional spark plug, laser ignition assures the efficiency at lean dosage.

Keywords: laser, cycle variability, spark ignition engine, combustion and performance.

1. INTRODUCTION

In the current global context of severe restrictions regarding the limits of the level of pollutant emissions and greenhouse gases produced by automotive internal combustion engines and in view of efficiency improvement, researchers have been focusing on the use of new technologies regarding combustion process control, [1], [11].

Consequently, pollution reduction and fuel efficiency can be controlled at primary level in the engine in-cylinder, with direct control on ignition and combustion processes. Such a new technology may be represented by the Laser Plug Ignition system (LPI), also known as Laser Ignition system (LI), [2]. In order to take place, the process of LI requires two basic steps: spark formation (generally limited by breakdown intensity) and subsequent ignition (generally limited by a 'minimum ignition energy' or MIE), [1]. For example, it is possible either to deliver sufficient energy for ignition but with insufficient intensity (i.e. no spark forms), or to form a spark but with insufficient energy for combustion.There are four well-known main mechanisms in use by which laser radiation can ignite combustible gas mixtures [2] and which are constantly evolving: I. Thermal initiation (TI); II. Non-resonant breakdown (NRB); III. Resonant breakdown (RB); IV. Photo-chemical ignition (PCI).

The most widely studied LI mechanism is NRB. It is similar to conventional electric SI as it produces plasma that emits light, heat and a shockwave, [5], [6], [7]. However, laser-induced sparks are generally smaller in size, shorter in duration and have higher temperatures [5], [13]. Another important issue is the reduction of cycle variability in engine operation, with benefits on efficiency and emissions, [10], [13]. Basically, the phenomena of cycle dispersion are the results of the variations of the combustion process, produced by imperfect mixing of in-cylinder fill in terms of homogeneity, by the phenomena produced in the formation of the plasma channel between the spark plug electrodes, by heat transfer from the flame core to spark plug electrodes, by convective heat transfer from the developed nucleus to the mass of initial mixture and by variation of the engine in-cylinder turbulence, [10], [14], [15], [16]. Cycle variability is strongly influenced by dosage, local air-fuel ratio and by the in-cylinder turbulent velocity field, [10], [12], [13].

Researchers from the National Institute for Laser, Plasma and Radiation Physics (INFLPR) and Renault Technologie Roumanie (RTR), Bucharest, Romania, have presented comparative results regarding the operation of an automobile engine that was ignited with classical spark plugs but also with laser spark, [4]. In the case of the K7M 812k engine, at a speed of 1500 rpm, the coefficient of variation $(COV)_{Pmax}$ decreases by 15% and the $(COV)_{IMEP}$ improvement was in the range of 18.5% (at 920-mbar load) to 22.6% (at 880-mbar load), [4]. The researchers noticed that the cyclic variability of an engine is improved at both high speed and load regimes, showing a lower influence of LPI on the coefficients of variability, expected in these conditions [4]. At 2000 rpm speed regime and high 920-mbar load, small differences between $(COV)_{Pmax}$ and $(COV)_{IMEP}$ for classic and LI ignition systems were noticed, [4]. Also, the results indicate a higher stability of the car engine that was operated at medium speeds by LPI, resulting in reduced noise, vibrations and mechanical stress, [4].

Mullett and Dearden [5], studied the LI system performance and cycle variability on a Ford Zetec engine at the regime of 1500 rpm and 36° before Top Dead Centre (TDC), [5]. Mullett calculated, for different values of laser energy in-cylinder, between 4...92 mJ, the ratio between the COV of IMEP determinant for the LI system and for SI system, [5].

The ratio of COV values for IMEP evaluated for Laser Plug Ignition system and for classic ignition system continuously decreased with the increase of in-cylinder laser energy, [8]. Regarding the combustion stability, for stoichiometric operation, $\lambda = 1$, the LI system was found to outperform the SI system in terms of reduced $(COV)_{IMEP}$ [8]. Dickinson compares the values of $(COV)_{IMEP}$ over a wide range of ignition angles at 1500 rpm, 2.62 bar brake mean effective pressure (BMEP), and analyses the effect of load on $(COV)_{IMEP}$ when operating at 1500 rpm at minimum advance for best torque (MBT), [8].

Shenton, Mullett and Dearden show that LI system improves the combustion stability, explained by measured values of $(COV)_{IMEP}$, [8] and with proper control, these improvements can enable engines to be run under leaner conditions, with higher EGR concentrations, or at lower idle speeds without increasing the noise, vibration and harshness characteristics of a vehicle. LI gives significantly shorter plasma duration compared to SI, [8]. With the recent development of higher average power and higher pulse frequency lasers, it is expected that a multi-strike LI system and associated combustion control can reduce the probability of misfires under high levels of dilution. The prospects for LI are also interesting from a control perspective, from optical sensing of the in-cylinder combustion made possible through self-cleaning (SC) of the laser beam pathway, to the array of possible ignition activation and control mechanisms, [8]. It is anticipated that, combined with the capability to control the

IOP Conf. Series: Materials Science and Engineering **252** (2017) 012078 doi:10.1088/1757-899X/252/1/012078

ignition location and timing, this will play a significant role in the optimization of future engines by dynamic feedback control, [7]. In case of new ignition system use, like Laser Plug Ignition system, a study of cycle variability for the main parameters that characterize the engine running is imminent.

2. EXPERIMENTAL INVESTIGATION

The experimental research was developed on an experimental single cylinder SI engine, equipped with laser Plug Ignition. The operating regime was 2800 rev/min, 90 % load. This operating regime, defined by the fact that speed is close to the maximum torque speed regime, is often used in exploitation and presents interest for investigation. The load is reduced at 90% in order to assure acceptable mechanical stress of the engine and does not affect the mechanical structure and function of the laser spark plug during this preliminary investigation. The experimental engine was mounted on the test bed adequate instrumented for the experimental investigations carried, its schema being presented in Figure 1.

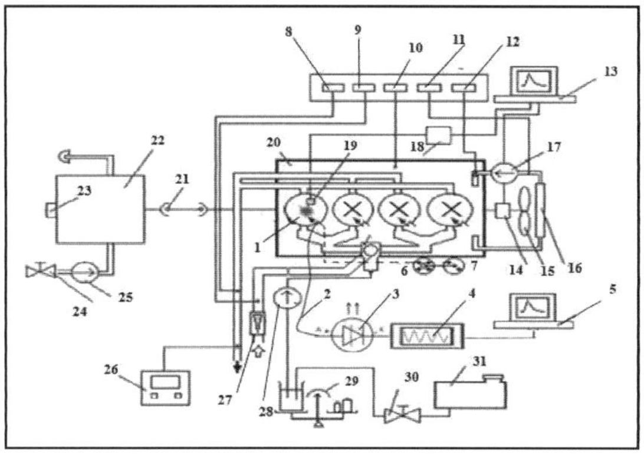

Figure 1: Experimental test bed schema

Figure 2: Compared LPI and SPI

The test bed: 1- laser plug ignition, 2 - optical fibre, 3- laser diode, 4 - laser power supply, 5 - PC with soft laser, 6,7 - the ensemble breaker distributor (cam with one corner), 8 - inlet air temperature measurement indicator, 9 - exhaust gas temperature measurement indicator, 10 - engine oil temperature measurement indicator, 11- engine oil pressure measurement indicator,12 -cooling liquid temperature measurement indicator,13 - PC equipped with AVL acquisition board, 14 - crank angle encoder, 15 - cooling fan, 16 - cooler,17 - engine water pump, 18 - Kistler charge amplifier , 19 - piezoelectric Kistler pressure transducer, 20 - spark plug ignition, 21- coupling, 22 - Schönebeck B4 hydraulic dynamometer, 23 - mechanical snuff speed, 24 - air flow meter, 25 - hydraulic dynamometer water pump, 26 - AVL DiCom Analyzer 4000, 27 - air flow meter, 28 - gasoline fuel pump, 29 - gravimetric fuel flow meter ,30 gasoline consumption tap, 31 - tank.

The laser spark used in the experiments was provided by INFLPR, Laboratory of Solid-State Quantum Electronics, Magurele, Romania. The photo in Figure 2 shows a laser spark plug compared with a classical spark plug. The plasma induced in air by optical breakdown is visible. The laser medium was a Nd:YAG/Cr^{4+}:YAG ceramic structure (Baikowski Co., Japan) that consisted of a 8.0-mm long, 1.0-at.% Nd:YAG ceramic, optically-bonded to a Cr^{4+}:YAG ceramic with saturable absorption (SA) [4], [9]. The initial transmission of Cr^{4+}: YAG SA was around 40%. Monolithic configuration of the resonator was obtained by coating the high reflectivity mirror at lasing wavelength, λ_{em}= 1.06 μm on the Nd:YAG free side and the outcoupling mirror with reflectivity R= 50% at λ_{em} on the Cr^{4+}:YAG free surface. The Nd:YAG side was coated for high transmission (T> 0.98) at the pump wavelength, λ_p= 807 nm. The optical pump was performed with a fiber-coupled diode laser (JOLD-120-QPXF-2P, Jenoptik, Germany) that was operated in quasi continuous-wave mode; the pump pulse duration was 250 μs and repetition rates up to 100 Hz were used. Typically, the laser yielded pulses with energy of 3.8 mJ at 1.06 μm for the pump with pulses of ~35 mJ at 807 nm; the laser pulse duration was around 1 ns. Cyclical variability is evaluated mainly by the variation of pressure differences, which are reflected in the calculated values of coefficients of cyclical variability, [10], [13].The cycle variability can be characterized by coefficients of a cylinder pressure variation. The intensity of the cycle variability phenomena is defined by the coefficient of cycle variability, [10], [13]. For "n" consecutive cycles, if is considered a normal distribution of the deviation probabilities, the squared average deviation can be calculated and the cycle variability coefficient is defined as:

$$(COV)a_i = \frac{\sqrt{\dfrac{\sum_{i=1}^{n}\left(a_i - \dfrac{\sum_{i=1}^{n} a_i}{n}\right)^2}{n-1}}}{\sum_{i=1}^{n} a_i} \cdot 100\%$$

where n is the number of cycles, a is the parameter of which variability is studiedand is defined for indicated mean effective pressure IMEP, maximum pressure pmax, maximum pressure rise rate (dp/dα)max and the angle where maximum pressure occurs, αpmax in the cycle number "i".

The way of cycle variability evaluation for regimes with spark timing closer to the value of spark timing for maximum torque brake (MTB) the coefficient of variation (COV) of maximum pressure is suitable, [10], [13]. When the maximum pressure occurs, the COV of maximum pressure angle is used for characterization of the combustion cycle variability during the initial phase of combustion [10], [13].

The variation of the IMEP, appreciated by $(COV)_{IMEP}$, is the most suitable instrument to define the engine respond to the combustion process variability. From this point of view, the limit value of $(COV)_{IMEP}$ defines practically the limit of mixture leaning, [10], [13].This cycle coefficient can also indicate the variability of flame development during the initial phase of combustion [10], [11].A higher combustion velocity reduces the influence of turbulence and reduces the cycle variability [10, 11].

The quality of the in-cylinder mixture influences the combustion process through chemical reaction speed, with a maximum in the area of rich dosage. As a result, the initial and final phases of the combustion process have minimal duration at the dosage for which the chemical reaction speeds are maximum, λ=1.1 [11], [12]. At the mixture leaning, the duration of those two phases increase and the total combustion duration also increases. The normal automotive engine manoeuvrability is assured if the coefficients values are fewer than 10% [10], [13].

IOP Conf. Series: Materials Science and Engineering **252** (2017) 012078 doi:10.1088/1757-899X/252/1/012078

3. RESULTS

The experimental research was carried on the SI engine firstly equipped with Spark Plug Ignition system (SPI), defined as reference, and secondly for the engine equipped with laser Plug Ignition system (LPI). The operating regime was 2800 rev/min, 90 % load and air-fuel ratio λ=1.1. For the investigated regime, the data were measured with a resolution of 1 CAD and recorded with the AVL data acquisition system. Analysis of the consecutive pressure diagrams, the cycle variability coefficients were calculated for IMEP, maximum pressure, maximum pressure rise rate and angle of maximum pressure. COV values are presented in the following figures.

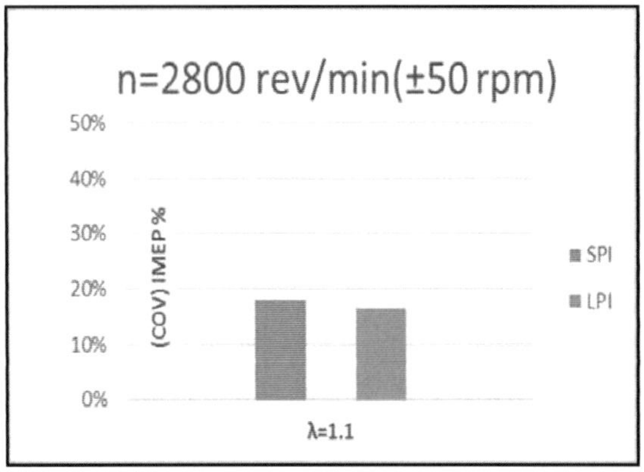

Figure 3: The $(COV)_{IMEP}$ evaluated for SPI and LPI systems

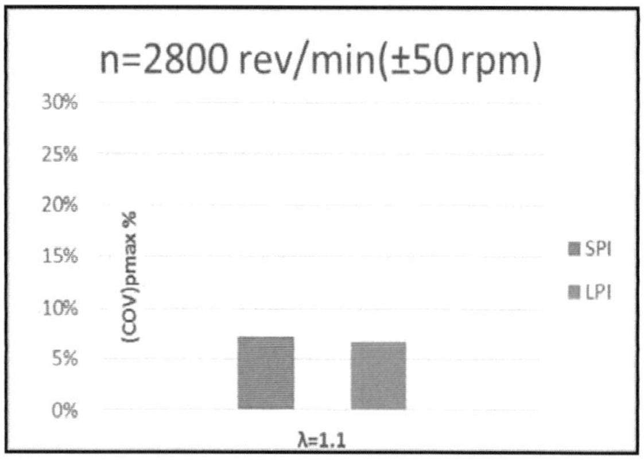

Figure 4: The (COV)pmax evaluated for SPI and LPI

The values of COV calculated for IMEP for Spark Plug Ignition (SPI) and for laser Plug Ignition (LPI) are presented in the Figure 3.

The value of COV of IMEP decreases with 1.8 % at the use of LPI, fact that shows an improvement of the combustion stability at the use of laser Plug Ignition system versus classic ignition system.

A lower value of $(COV)_{IMEP}$, registered at the LPI use, as figure shows, indicates a better engine respond at the variability of the combustion process, at λ=1.1. Also, the reduced value of the COV for indicated mean effective pressure shows a much lower variability of the flame development into the initial phase of the combustion process when the LPI is used comparative to the spark plug system.

The values of COV calculated for maximum pressure for Spark Plague Ignition (SPI) and for laser Plug Ignition (LPI) are presented in Figure 4. The value of COV of maximum pressure decreases from 7.2% down to 6,7 %. The variability coefficient improves its value with almost 0.5 % when the laser Plug Ignition system is used comparative to the classic spark ignition system. The decrease of COV for maximum pressure is correlated with the variation tendency registered for the COV of IMEP.

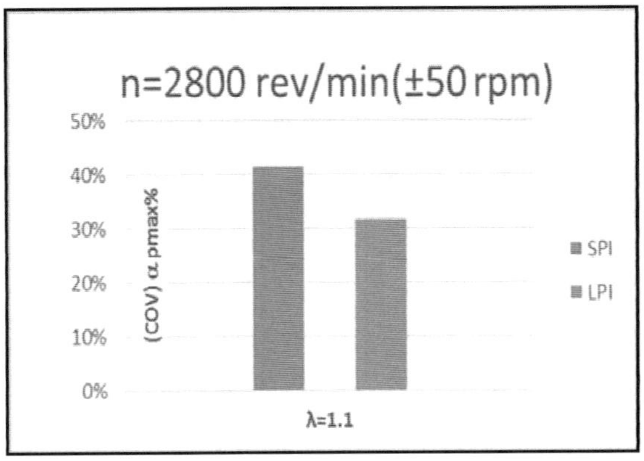

Figure 5: The (COV) $_{\alpha pmax}$ evaluated for SPI and LPI systems

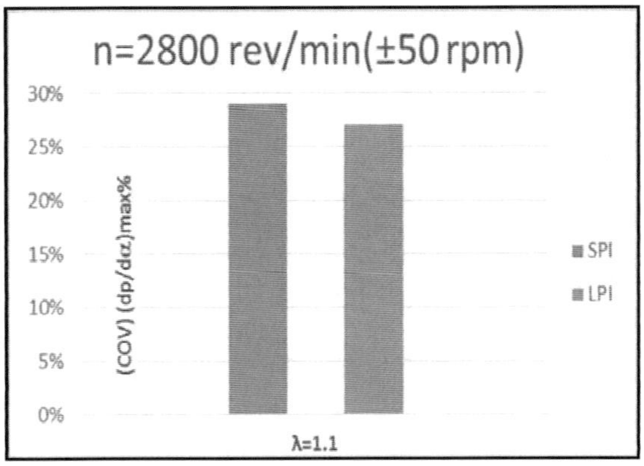

Figure 6: The (COV) $_{(dp/d\alpha)max}$ evaluated for SPI and LPI

The COV of maximum pressure angle, illustrated in Figure 5, defined by the angle when maximum pressure occurs per cycle, decreases from 41.5 % value registered for SPI down to 31.6% for LPI. The decrease of the COV of maximum pressure angle, with 9.9% at LASER Plug Ignition system use, reflects a lower cycle variability of the combustion process registered during the initial phase of the combustion; this fact is correlated with the variation tendency registered also for COV of IMEP, Figure 3.
The COV of maximum pressure rise rate is presented in Figure 6. The calculated values for the LPI system of the COV of maximum pressure rise rate decreases with almost ~ 2% comparative to the values registered for the classic ignition system with spark plugs. The decrease of the COV for $(dp/d\alpha)_{max}$ appears in correlation with the reduction of the other COV values calculated for IMEP, maximum pressure and angle of maximum pressure.

4. CONCLUSIONS

Regarding the experimental research of a new laser Plug Ignition system used on a SIE, the main conclusions of the cycle variability study can be formulated as it follows:
The values of COV calculated for IMEP for laser Plug Ignition system (LPI) decrease with 1.8%, fact that shows an improvement of the combustion stability at the use of laser Plug Ignition system versus classic ignition system.
Due to a lower value of $(COV)_{IMEP}$, registered when the laser Plug Ignition system is used, indicates a better engine respond at the variability of the combustion process, for $\lambda=1.1$. Moreover, a much lower variability of the flame development during the initial phase of the combustion process when the LPI is used comparative to spark plug system.
The values of $(COV)_{pmax}$ for laser Plug Ignition system (LPI) decrease from 7.2% down to 6.7%, the variability coefficient improves its value with almost 0.5%. The $(COV)_{pmax}$ decrease is in correlation with the variation tendency registered for $(COV)_{IMEP}$.
The decrease of the COV of maximum pressure angle, $(COV)_{\alpha pmax}$ with 9.9% when using laser Plug Ignition system, reflects a lower cycle variability of the combustion process registered during the initial phase of the combustion; this fact is correlated with the variation tendency registered for $(COV)_{IMEP}$.

The values of COV for maximum pressure rise rate, $(COV)_{(dp/d\alpha)max}$, decreases with almost ~2% % for LPI system use comparative to the values registered for the classic ignition system with spark plugs. The decreases are in correlation with the reduction of COV values calculated for IMEP, maximum pressure and angle of maximum pressure.
The improved values of the cycle variability coefficients registered for laser ignition versus classic ignition system show a good perspective for further experimental investigations carried on other engine operating regimes.

Acknowledgements

The experiments were performed with a laser spark device developed at National Institute for laser, Plasma and Radiation Physics, Laboratory of Solid-State Quantum Electronics, Magurele, Ilfov, 077125, Romania.The authors would like to thank to Mr. Pavel Nicolaie, Mr. Dinca Mihai and Mrs. Croitoru Gabriela for their help and assistance during the experiments.

References

[1]. *P. D. Ronney,* "laser versus conventional ignition of flames," Opt. Eng. 33(2), 510–522 (1994).
[2]. *C. Morgan,* "laser -Induced Breakdown of Gases," Rep. Prog. Phys. 38(5), 621–665 (1975).
[3]. *S. S. Vorontsov, V. N. Zudov, P. K. Tretyakov, and A. V. Tupikin,* "Peculiarities of the ignition of propane-air premixed flows by CO2 laser radiation," Thermophys. Aeromech. 13(4), 615–621 (2006).
[4]. *N. Pavel, T. Dascalu, G. Salamu, M. Dinca, N. Boicea, and A. Birtas*, "Ignition of an automobile engine by high-peak power Nd:YAG/Cr^{4+}:YAG laser -spark devices," Opt. Express 23(26), 33028-33037 (2015).
[5]. *G. Dearden and T. Shenton*, "laser ignited engines: progress, challenges and prospects," Opt. Express 21(S6 Suppl 6), A1113–A1125 (2013).
[6]. *Y. L. Chen, J. W. L. Lewis, and C. Parigger,* "Spatial & temporal profiles of pulsed laser -induced air plasma emissions," J. Quantitative Spectrosc. Radiative Transf. 67(2), 91–103 (2000).
[7]. *J.D.Mullett,*"Laser-Induced Systemsfor Gasoline Automotive Engines," PhD Thesis, University of Liverpool (2009).
[8]. *P. B. Dickinson, A. T. Shenton, J. D. Mullett, G. Dearden, and A. Scarisbrick,* "Prospects for laser ignition in gasoline engine control," 10th Int. Symp. on Advanced Vehicle Control (AVEC10), 22–26 (2010).
[9]. *T. Dascalu, G. Salamu, O. Sandu, M. Dinca, and N. Pavel,* "Scaling and passively Q-switch operation of a Nd:YAG laser pumped laterally through a YAG prism," Opt. & laser Techn. 67, 164-168 (2015).
[10]. *Heywood, B.* Internal Combustion Engine Fundamentals. New York; Mcgraw-Hill Book Company; 1988.
[11]. *Negurescu, N. Pana, C. Popa, M. G.* Internal Combustion Engines. Processes. Bucharest; Matrixrom, 2009.
[12]. *Frank, R. Heywood, J.B.* The Effect of Fuel Caracteristics on Combustion in a Spark-Ignited Direct-Injection Engine. SAE 902063.
[13]. *Gaiginschi, R.* Increasing of Running Cycle Stability of the Four Stroke Spark Ignition Engine Operating with Lean Mixtures. Doctoral Thesis; Institute Politehnic of Bucharest;1976.
[14]. *Sullivan, P., Ancimer, R., and Wallace, J.* (1999), "Turbulence averaging within spark ignition engines", Experiments in Fluids, Vol.27, No.1, pp.92-101.
[15]. Baby X, Dupont A, Ahmed A, Deslandes W, Charnay G and Michard M. A New Methodology to Analyze Cycle-to-Cycle Aerodynamic Variations. SAE 2002-01-2837. 2002.
[16]. *Ghandhi JB, Herold RE, Shakal JS and Strand TE.* Time-Resolved Particle Image Velocimetry Measurements in an Internal Combustion Engine. SAE Paper 2005-01-3868. 2005.

LPG AS A FUEL FOR DIESEL ENGINES-EXPERIMENTAL INVESTIGATIONS

Nikolaos Cristian Nutu[1], Constantin Pana[2], Niculae Negurescu[3], Alexandru Cernat[4] and Ionel Mirica[5]

[1]Lecturer, Rutier Vehicles Department, University Politehnica of Bucharest, Romania
[2]Professor, Thermotechnics, Thermal Engines and Equipments, Refrigeration Instalations Department, University Politehnica of Bucharest, Romania
[3]Professor, Thermotechnics, Thermal Engines and Equipments, Refrigeration Instalations Department, University Politehnica of Bucharest, Romania
[4]Lecturer, Thermotechnics, Thermal Engines and Equipments, Refrigeration Instalations Department, University Politehnica of Bucharest, Romania
[4]PhD Student, Thermotechnics, Thermal Engines and Equipments, Refrigeration Instalations Department, University Politehnica of Bucharest, Romania

e-mail: cristi_cmt@yahoo.com

Abstract. The main objective of the paper is to reduce the pollutant emissions of a compression ignition engine, fuelling the engine with liquefied petroleum gas (LPG), aiming to maintain the energetic performances of the engine. To optimise the engine operation a corelation between the substitute ratio of the diesel fuel with LPG and the adjustments for the investigated regimens must be made in order to limit the maximum pressure and smoke level, knock and rough engine functioning, fuel consumption and the level of the pollutant emissions.

The test bed situated in the Thermotechnics, Engines, Thermal Equipments and Refrigeration Instalations Department was adapted to be fuelled with liquefied petroleum gas. A conventional LPG fuelling instalation was adopted, consisting of a LPG tank, a vaporiser, conections between the tank and the vaporiser and a valve to adjust the gaseous fuel flow. Using the diesel-gas methode, in the intake manifold of the engine is injected LPG in gaseous aggregation state and the airr-LPG homogeneous mixture is ignited from the flame appeared in the diesel fuel sprays. To maintain the engine power at the same level like in the standard case of fuelling only with diesel fuel, for each investigated operate regimen the diesel fuel dose was reduced, being energetically substituted with LPG. The engine used for experimental investigations is a turbocharged truck diesel engine with a 10.34 dm^3 displacement. The investigated working regimen was 40% load and 1750 rpm and the energetic substitute ratios of the diesel fuel with LPG was situated between [0-25%].

1. INTRODUCTION

Liquefied petroleum gas is usually a mixture of two hydrocarbons, propane and butane, in different ratios depending on season. Because of its good burning properties and price liquefied petroleum gas is a very good choice to fuel a diesel engine.

Several liquefied petroleum gas properties are presented in the table 1, compared with the diesel fuel properties.

Table 1. Liquefied petroleum gas properties, comparative with diesel fuel properties [1].

Properties	Diesel fuel	Propane	Butane
Density [kg/m³]	800-840	503	500
Self-ignition [°C]	355	481	544
Stoichiometric A/F ratio [kg/kg]	15	15.71	15.49
Lowe r heating value[MJ/kg]	42.5	46.34	45.55
Cetane number	40-55	-2	-2
Flame temperature	2054	1900	-

The density of liquid LPG is lower, 503 kg/m³ for LPG and 800-840 kg/m³ for diesel fuel [1], leading to a lower fuel autonomy.

LPG needs a lower quantity of heat to vaporise 420 kJ/kg to 465 kJ/kg for diesel fuel [1], allowing to consume less local heat.

The LPG self-ignition temperature is higher than the diesel fuel self-ignition temperature, 481 °C – propane, 544 °C –butane, 355°C diesel fuel [1] and this combined with a very low cetane number gives LPG very poor self-ignition properties. Therefore, fuelling a diesel engine with LPG requires the use of specific methods.

The flame temperature of LPG lower than the diesel fuel flame temperature leads to an important reduction in nitrogen oxides emissions level.

The higher LPG lower heating value leads to an increase in the amount of heat released during combustion for the same fuel quantity.

Fuelling a diesel engine with LPG require specific methods. In this paper, the authors chose the Diesel-Gas method, which consists of gaseous LPG injection in the intake manifold of the engine.

The diesel gas method has been applied by Tariq Miqdam in [2] with good results as far as the brake specific energetic consumption and pollutant emissions are concerned. In the paper [3] the author fuelled with LPG a car diesel engine, experiments leading in a reduction of the vehicle operating price. Qi et. al. in the work [4] experimented the direct injection of a LPG-diesel fuel mixture with different proportions: 0, 10, 20, 30, 40 %, leading to a decrease in the pollutant emissions of the engine. In [5] the authors decreased the level of the nitrogen oxides emission fuelling the diesel engine with LPG but the level of unburned hydrocarbons increased. The authors managed to reduce the unburned HC emission level using a glow plug [5]. A decrease of unburned hydrocarbons emissions level was achieved by Pali Rosha et al. in [6], by LPG fuelling a single cylinder diesel engine and using exhaust gas recirculation. The level of nitrogen oxides emissions and carbon dioxide emission decreased compared to standard diesel engine fuelling case, while emissions of unburned hydrocarbons and carbon monoxide increased, especially in the partial load regimens. In order to reduce these emissions, the authors used a percentage of 16% recycled exhaust gas. Another example of a liquefied petroleum gas diesel engine fuelling is presented in the paper [7]. The authors used liquefied petroleum gas injection into the intake manifold of the engine and rapeseed oil for the pilot injection. The maximum degree of

IOP Conf. Series: Materials Science and Engineering **252** (2017) 012079 doi:10.1088/1757-899X/252/1/012079

substitution used of rapeseed oil with LPG was 60 %, where the engine cyclic variability was within acceptable limits.

This paper presents results of experimental investigations carried out on a truck compression ignition engine fueled with liquefied petroleum gas using the diesel-gas method.

2. Experimental study

The experimental study was carried out on a compression ignition engine type Roman D2156 MTN 8, with 6 cylinders in line. The main specification and performances of the engine are presented in the table 2 [8].

The test bed consists of: the Roman D2156 MTN 8 diesel engine, Hofman eddy current dyno, AVL data aquisition system, Kistler piezoelectric pressure transducer, AVL Dicom 4000 gas analyser and opacimeter, Optimass masic fuel flow meter, Meriam volumic air flow meter, thermocouples and thermoresistences for temperature measuring, gravimetric system for diesel fuel consumption measuring, gas leak detector. All the equipments were calibrated prior to measurements. The investigated regimen was 40% load and 1750 rpm. The Diesel-Gas method consists in gaseous LPG injection in the intake manifold of the engine. Therefore, the homogeneous mixture of air-LPG is ignited by the flame which appears in the diesel fuel jet.

Table 2. Specifications and performances of the engine D 2156 MTN 8 [8].

Number of cylinders	6
Bore [mm]	121
Stroke [mm]	150
Displacement [L]	10.34
Compression ratio	17
Rated power [kW]	188
Maximum torque [Nm]	900
Admision type	turbocharged

First time was determined the reference, fuelling the engine only with diesel fuel, then the diesel fuel was partially substituted with liquefied petroleum gas, concerning to maintain the same engine power like in the standard case of fuelling with diesel fuel. Therefore, for each substitute ratio investigated, the diesel fuel cycle dose was reduced and the LPG cycle dose was increased. The energetic substitute ratio was calculated with the following relation:

$$x_c = \frac{m_{LPG} H_{i_{LPG}}}{m_{LPG} H_{i_{LPG}} + m_{dieselfuel} H_{i_{dieselfuel}}} \quad (1)$$

where: m_{LPG}- the LPG dose;

$m_{dieselfuel}$-the diesel fuel dose;

H_i- the caloric heating value.

The investigated energetic substitute ratios of the diesel fuel with LPG was situated between [0-25] %.

3. Results and discussions

Because of the burning process intensification due to the presence of LPG-air mixture in the combustion chamber the in-cylinder pressure increased for all the investigated substitution ratios of diesel fuel with LPG. The measured in-cylinder pressure for the investigated cases is presented in the figure 1.

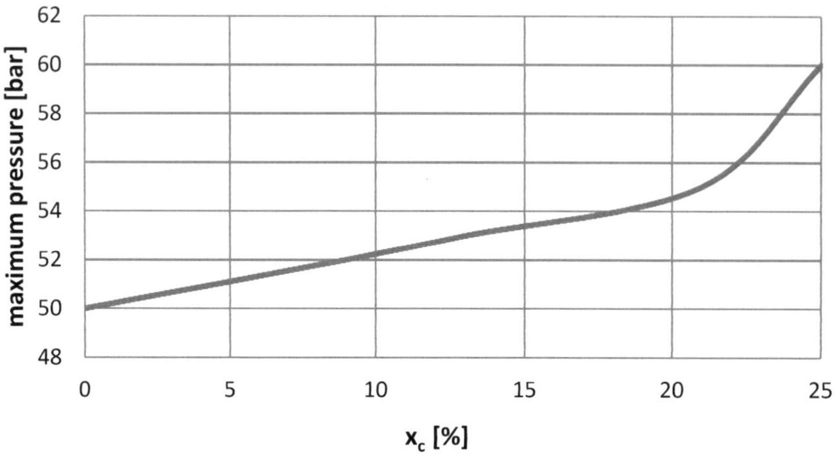

Fig. 1. The maximum pressure inside the cylinder versus the substitute ratio.

Because of a higher flame speed in the homogeneous mixture or air-LPG The maximum rate or pressure rise increased for all the investigated cases. The figure 2 presents the maximum rate of pressure rise for the investigated cases.

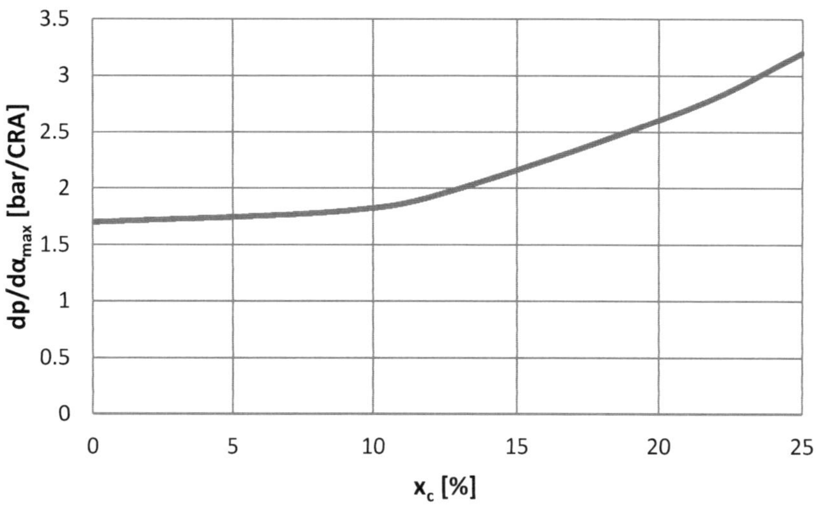

Fig. 2. The maximum rate of pressure rise versus the substitute ratio.

The nitrogen oxides emission level decreased for all the investigated substitute ratios of diesel fuel with LPG because liquefied petroleum gas has a lower flame temperature than diesel fuel. The nitrogen oxides emission variation is presented in the figure 3.

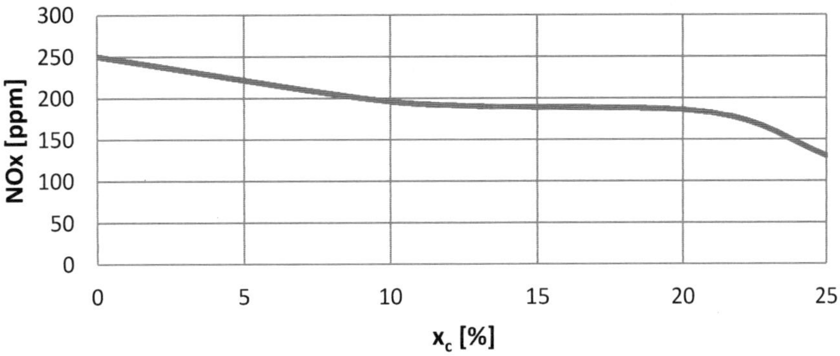

Fig. 3. The nitrogen oxides emission versus the substitute ratio.

The smoke emission level decreased for all the investigated substitute ratios of diesel fuel with LPG because when LPG is present in the combustion chamber the burning rate of diffusive mixtures decreases and the burning rate of preformed mixtures increase. The figure 4 presents the measured smoke emission level, evaluated by the coefficient of absorption k.

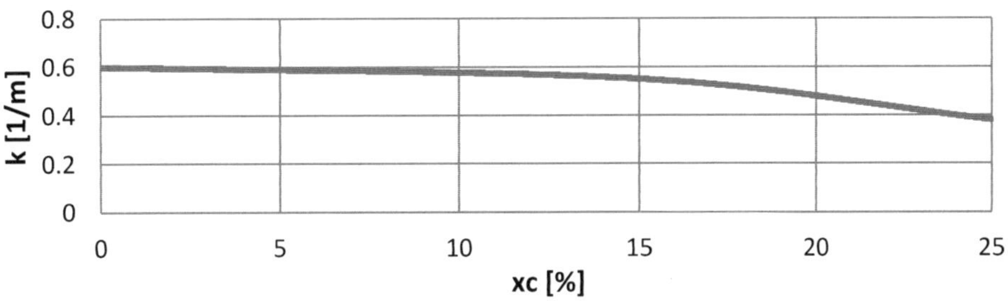

Fig. 4. The smoke emission level versus the substitute ratio.

The brake specific energetic consumption decreased for the substitute ratios of diesel fuel with LPG. Figure 5 presents the brake specific energetic consumption versus the substitute ratio.

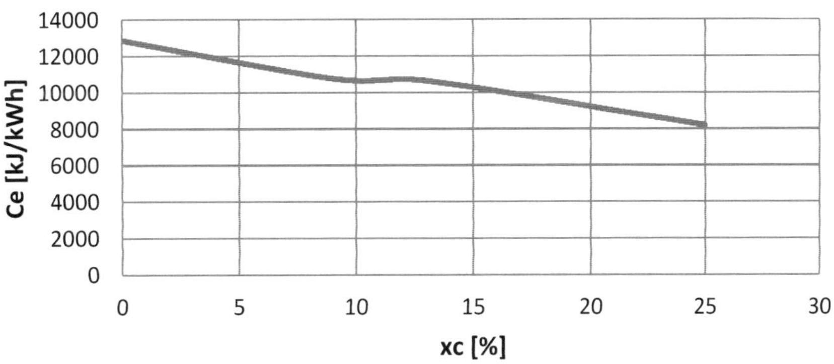

Fig. 5. The energetic specific fuel consumption versus the substitute ratio.

4. Conclusions

At the LPG engine fuelling were observed the following:

1. The brake specific energetic consumption decreased with ~20% when the substitute ratio was x_c=25.
2. The nitrogen oxides emission decreased with ~25% for the substitute ratio x_c=25.

3. The smoke emission level decreased because when LPG is present in the combustion chamber the burning rate of diffusive mixtures decreases.
4. The maximum pressure and the maximum rate of pressure rise increased in the case of LPG fuelling.

5. Acknowledgements

The authors would like to thank to AVL List GmbH Graz, Austria, for providing the possibility to use the research equipment.
This work has been funded by University Politehnica of Bucharest, through the "Excellence Research Grants" Program, UPB – GEX 2017. Identifier: UPB- GEX2017, Ctr. No. /2017.

6. References

[1]. Popa, M G, Neguresc u, N, Pana, C, Motoare Diesel, 2003, *Matrix ROM*, Bucuresti.
[2]. Miqdam, T, Exhaust analysis and performance of a single cylinder diesel engine run on dual fuels mode, 2011, *Journal of Engineering*, **Nr. 4**, Vol. 17.
[3]. Michal, J. et al, Analysis of engine parameters at using diesel-lpg and diesel-cng mixture in compression-ignition engine, *Acta Universitatis Agriculturae Et Silviculturae Mendelianae Brunensis*, 2014, vol. 62**, nr. 1**.
[4]. Qi, D. H., Bian, Y. Z., Ma, Z.Y., Zhang, C. H., Liu, S. Q., Combustion and exhaust emission characteristics of a compression ignition engine using liquefied petroleum gas–diesel blended fuel, *Journal of Energy Conversion Managemen*t, 2007, vol. 48, **no. 2**, pp 500-509.
[5]. Vijayabalan, P., Nagarajan, G., Performance, Emission and Combustion of LPG Diesel Dual Fuel Engine using Glow Plug, *Jordan Journal of Mechanical and Industrial Engineering*, 2009, Volume 3, **Number 2,** ISSN 1995-6665 Pages 105 – 110.
[6]. Rosha, P., Bharj, R.S., Gill, K., Performance and emission characteristics of Diesel+LPG dual fuel engine with exhaust gas recirculation, *International Journal of Science, Engineering and Technology Research (IJSETR)*, October 2014, Volume 3, **Issue 10**.
[7]. Tira, H.S, Herreros, J.M., Tsolakis, A., Wyszynski, M.L., Characteristics of LPG-diesel dual fuelled engine operated with rapeseed methyl ester and gas-to-liquid diesel fuels, *Energy*, 2012, **Volume 47**, 620-629.
[8]. Abăităncei, D., Motoare pentru autovehicule si tractoare fabricate in Romania, *Alma*, 2011, Craiova.

Injection system used into SI engines for complete combustion and reduction of exhaust emissions in the case of alcohol and petrol alcohol mixtures feed

N Ispas[1*], C Cofaru[2] and M Aleonte[3]

[1,2]Automotive and Transportation Department, Transilvania University of Brasov, Romania
[3] Karlsruhe Institute of Technology (KIT) – Institut für Kolbenmaschinen (IFKM), Karlsrue, Germany

*inicu@unitbv.ro

Abstract. Internal combustion engines still play a major role in today transportation but increasing the fuel efficiency and decreasing chemical emissions remain a great goal of the researchers. Direct injection and air assisted injection system can improve combustion and can reduce the concentration of the exhaust gas pollutes. Advanced air-to-fuel and combustion air-to-fuel injection system for mixtures, derivatives and alcohol gasoline blends represent a major asset in reducing pollutant emissions and controlling combustion processes in spark-ignition engines.
The use of these biofuel and biofuel blending systems for gasoline results in better control of spark ignition engine processes, making combustion as complete as possible, as well as lower levels of concentrations of pollutants in exhaust gases.
The main purpose of this paper was to provide most suitable tools for ensure the proven increase in the efficiency of spark ignition engines, making them more environmentally friendly.
The conclusions of the paper allow to highlight the paths leading to a better use of alcohols (biofuels) in internal combustion engines of modern transport units.

1. Introduction

The transport sector is almost completely dependent on oil and its derivatives both in the United States and in Europe, this sector accounts for about one third of total energy consumption and about 30% of CO_2 emissions [1]. The transport sector will contribute, following an EU-wide forecast, with an increase of about 90% in CO_2 emissions [1]. The increase in oil demand for transport in China, India and other Asian countries has led to rising prices and boosted the production of oil substitutes. Finding alternative solutions is a key issue and biofuels should be the easiest as alternative fuel because no significant changes are required in infrastructure or in existing vehicles and engines.
Currently, most transport fuels are derived from oil, with more subsequent negative attributes in terms of fuel and carbon emissions. Consequently, there has been renewed interest in the production and use of biofuels, such as biodiesel and bioethanol, which is produced from vegetable or animal oils, as well as bioethanol produced by fermentation. Biofuels are renewable and can be produced from agricultural products with the potential to reduce dependence on fossil fuels. Biofuels have low carbon emissions because they are produced in the short-term carbon cycle and their burning is as much atmosphere in the atmosphere as the one absorbed by these plants in the atmosphere. Therefore, unlike the burning of

fossil fuels, combustion of biofuels has the potential to be carbon neutral. Ethanol can reach 96% as volume purity by distillation and is clear as water, this purity being sufficient to burn ethanol. Ethanol burns cleaner than many other fuels and produces low CO_2. When burning is complete, the exhaust products are just CO_2 and water. That's why ethanol is a favourite for ecological transport schemes and is used in many sustainable transport systems. However, ethanol readily reacts or dissolves some tires and plastics and cannot be used in engines that do not have new material changes. In addition, ethanol has a higher-octane rating than regular gasoline, requiring changes in compression ratio or spark advance to achieve maximum performance [2]. Alcohols such as ethanol or methanol have been and are still used in pure state or mixed in high proportion in gasoline for fuelled spark ignition four-stroke engines and spark ignition two-stroke engines. Ethanol and its oxygenated derivatives can be used in low proportions as additive fuel or gasoline additives. This mode of use is rational because the advantages and constraints are no longer the same as when used in high proportions in mixtures. Two-stroke engines produce more power and are more compact than the four-stroke engines, and they are lightweight and less costly. On the other hand, some fuel gets wasted in a two-stroke engine, decreasing its efficiency.

In the case of two-stroke engines, for every two strokes of the piston inside the cylinder, one power stroke is produced. In four-stroke engines, power is produced once during four strokes of the piston. For the same size engine, the power produced by the two-stroke engine is more that the four-stroke engine. Ideally the power produced by the two-stroke engine is double that of the four-stroke engine, but in actual practice it is only about 30% more than four-stroke engine. Also, since the power produced by the two-stroke engine is higher, these engines are small and compact in size [2]. One disadvantage that applies to both diesel and petrol two-stroke engines is the extensive cooling and lubricating requirements of the two-stroke engines. Since in two-stroke engines power stroke is produced after every stroke, a large amount of heat is generated within them. To reduce the temperature of the engine and keep the moving parts well-lubricated, good lubrication and cooling systems for the engine are required. For reasons relating to the simple construction of the two-stroke engine cylinder head due to the lack of valve distribution system, the easy positioning of the spark plug and the air-assisted ethanol injector was one of the advantages of choosing this type of engine for the experimental research described in this paper.

2. Experimental research

A single cylinder two stroke engine was used to acquire the experimental data in this research (Table 1). The engine having the specification presented in Table 1was connected to an Eddy Current Dynamometer type Schenk W40.

Table 1. Experimental engine characteristics

Engine type	*Parameter*
Two stroke,	
Spark ignition engine (SI)	
No of cylinders	1
Compression ratio	8:1, 9:1
Swept volume (cm^3)	70.7
Bore (mm)	50
Stroke (mm)	36
Connecting rod length (mm)	75
Fuel delivery	Carburetor, Direct injection air assisted
Fuels	E0; E85, E100
	Crankcase scavenging
	Schnürle full loop
Engine cooling	Air

First fuelling system was by the carburettor, used with optimal settings for ignition and air-fuel ratio. The second fuel delivery system was an air-assisted direct injection system. In the Figure 1. is shown the two-stroke engine equipped with the injection system. Details of air-assisted injector are showed in Figure 2.

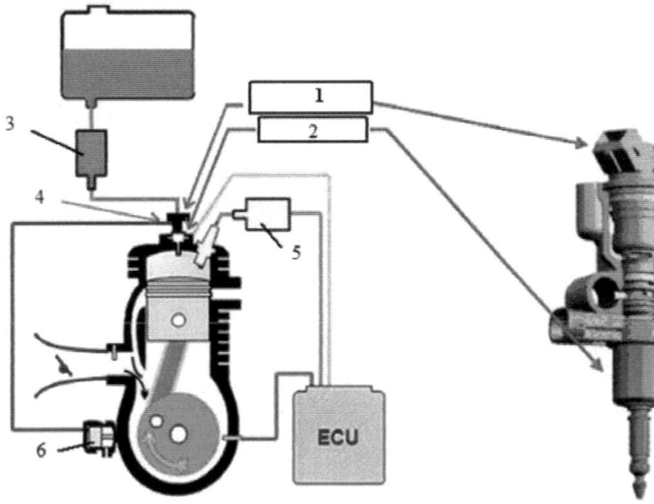

Figure 1. The two-stroke engine used in experimental and air-assisted direct injection system, where: 1- Fuel injector; 2 - Air-fuel mixture injector; 3 - Fuel pump; 4 - Pressure regulator; 5 – Ignition system; 6 – Air compressor; ECU – Electronic control unit.

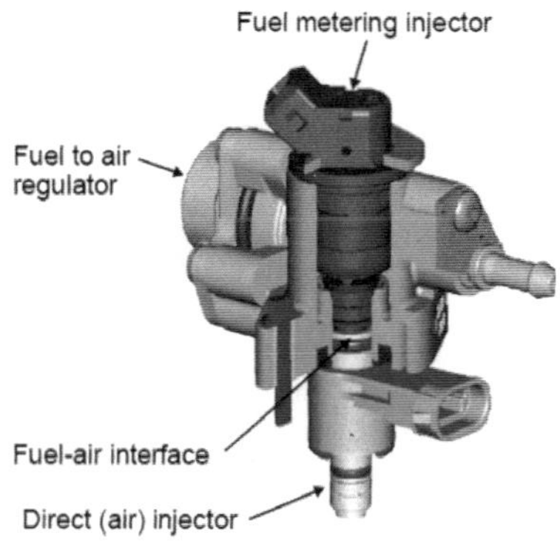

Figure 2. Details of air-fuel direct injection injector

The air-assisted injector operates in the following manner: the fuel pumped into the air-fuel mixture formation chamber contained in the air- assisted DI system had a pressure of 0.6 – 0.7 MPa. The air pressure was induced into the air-fuel mixture formation chamber in the air assisted direct injection system with a pressure of 0.45 – 0.55 MPa. In order to have a constant compressed air pressure, an air regulator was mounted on the air inlet system. The main advantage of air-assisted fuel injector is a good atomization of fuel; droplets' size having less than 60 SMD on the condition of air-assist pressure. In case of a direct fuel injection, the crank train lubrication and durability remain a problem to be solved; on this condition, the lubricant oil needs to be added to the intake air charge. The engine lubrication was made by fitting an oil pump as the oil is not mixed with the fuel, the pump delivering the oil into the intake air flow. The oil pump frequencies were controlled by taking in account the engine operating parameters.

3. Results and discussion

For two compression ratio values, using different fuel, the engine operating characteristics were determined according to the speed at full engine load. In Figure 3. engine torque (T), engine power (P), fuel consumption (FC) and brake specific fuel consumption (BSFC) for the two-stroke engine with spark ignition and air-assisted direct injection, at a compression ratio (CR) equal to 8 : 1 (petrol fuel) are shown.

In Figure 4. engine torque (T), engine power (P), fuel consumption (FC) and brake specific fuel consumption (BSFC) for the two-stroke engine with spark ignition and air-assisted direct injection, at a compression ratio (CR) equal to 9 : 1 (gasoline fuel) are shows.

As can see in Figure 5. the power and the motor torque in the version equipped with a compression ratio CR = 9 have a higher value than the variant equipped with CR = 8. At the speed of 5500 [rot / Min] for CR = 9, the power and the moment represent 106% for both cases of the value for CR = 8. At the speed of 6500 [rot / min] they become equal, because at the speed of 7500 [rpm] we have an increase of 2% and 3% respectively.

For fuel consumption at 5500 [rpm] for CR = 9 we have a 2% decrease and for a specific consumption of 5%. At 6500 [rpm] we have a decrease in hourly consumption and specifically by 2% and 4%, respectively. In this treatment, we considered the reference values obtained for the pneumatically assisted pneumatically-injected air-fuel mixture system and CR = 8.

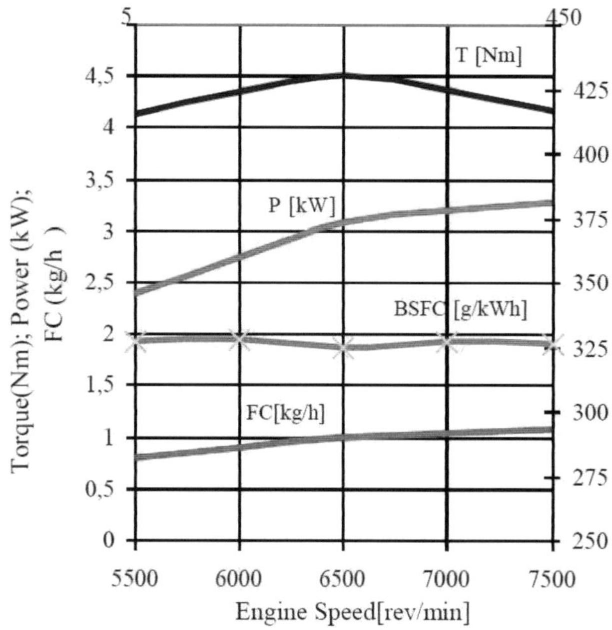

Figure 3. Engine speed operating characteristic at full load (CR = 8 : 1, petrol fuel).

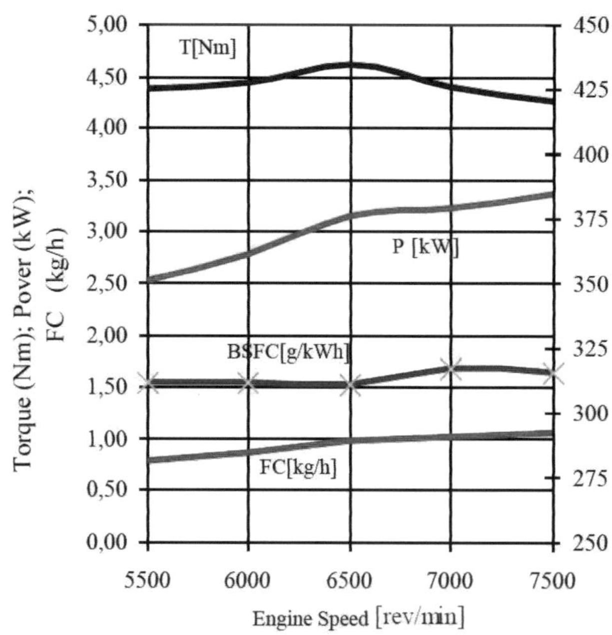

Figure 4. Engine speed operating characteristic at full load (CR = 9 : 1, petrol fuel).

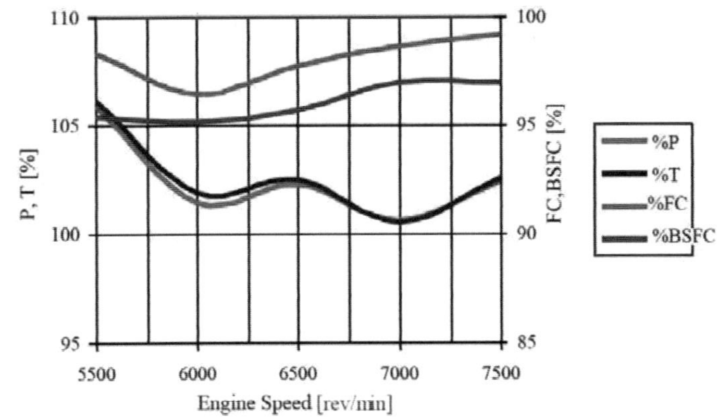

Figure 5. Comparison between engine parameters for CR = 8 : 1 and CR = 9 : 1 (gasoline fuel)

In Figure 6. engine torque (T), engine power (P), fuel consumption (FC) and brake specific fuel consumption (BSFC) for the two-stroke engine with spark ignition and air-assisted direct injection, at a compression ratio (CR) equal to 8 : 1 (E85 fuel – 85% Ethanol mixed with 15% gasoline, density = 781 [kg/m3) are shows.

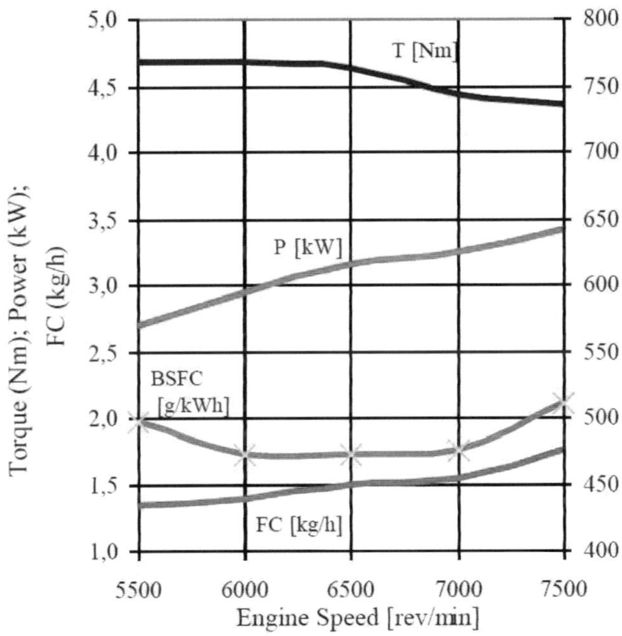

Figure 6. Engine speed operating characteristic at full load (CR = 8 : 1, E85 fuel).

In Figure 7. engine torque (T), engine power (P), fuel consumption (FC) and brake specific fuel consumption (BSFC) for the two-stroke engine with spark ignition and air-assisted direct injection, at a compression ratio (CR) equal to 9 : 1 (E85 fuel) are presented.

From Figure 7, for pneumatically assisted direct injection we can see that the engine power and torque, when operating with a compression ratio CR = 9, they have a higher value than the equipping option with CR = 8. Evolution of values can be see the in the Figure 8. Engine torque and power values reported at CR = 8 are equal for the given speeds. At the speed of 5500 [rpm] for CR = 9, the power and the engine torque These are maintained at 103% and at 7500 [rpm] their value being 102%. In the case of

FC at 5500 [rpm] for ε = 9 we have a decrease of 9% for the BSFC of 10%. At 6500 [rpm] we have one decrease in hourly and specific consumption by 2% and 5%, respectively.

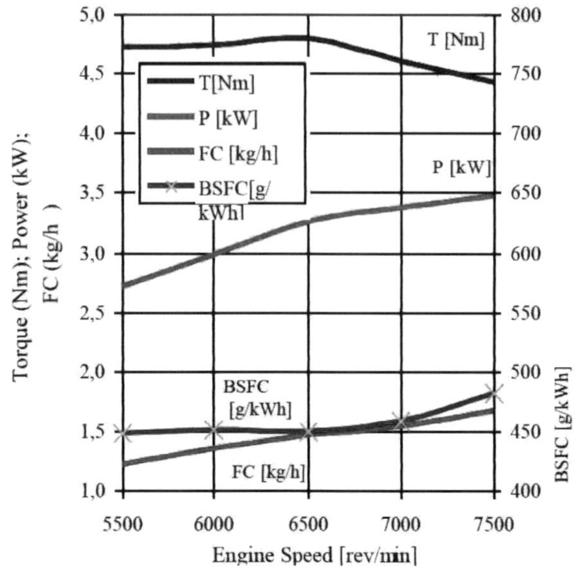

Figure 7. Engine speed operating characteristic at full load (CR = 9 : 1, E85 fuel).

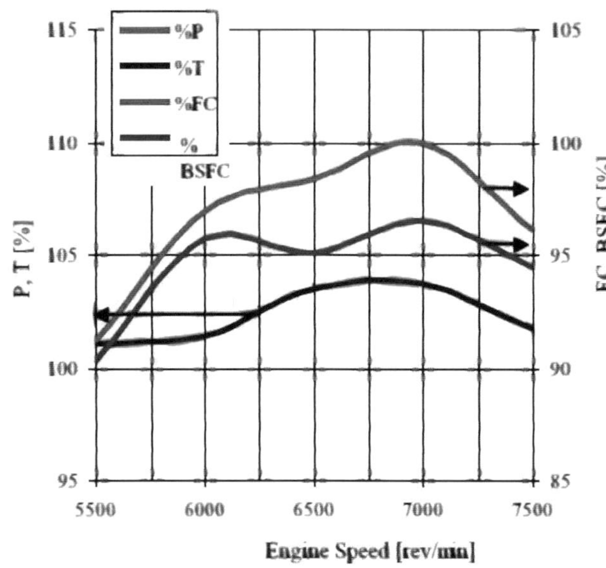

Figure 8 Comparison between engine parameters for CR = 8 : 1 and CR = 9 : 1 E85).

In figure 9. we can observe the evolutions of the engine torque (T), of the engine power (P), fuel consumption (FC) and brake specific fuel consumption (BSFC) for the engine with air assisted direct injection using E100 and operating with a compression ratio of 9:1.

From Figure 10, for air assisted direct injection, we can see that the engine power and torque, when operating with a compression ratio CR = 9, they have a higher value than the variant operating with CR = 8. The evolution of the values can see both in Figure 11 At the speed of 5500 [rpm] for CR = 9, power and torque represent 105%, for both cases, of the value for ε = 8. At speed of 6500 [rpm] these are maintained at 103% and at 7500 [rpm] their value being 102%. In case of fuel consumption at 5500 rpm, for ε = 9 we have a decrease of 3%, and for the specific fuel consumption of 8%.

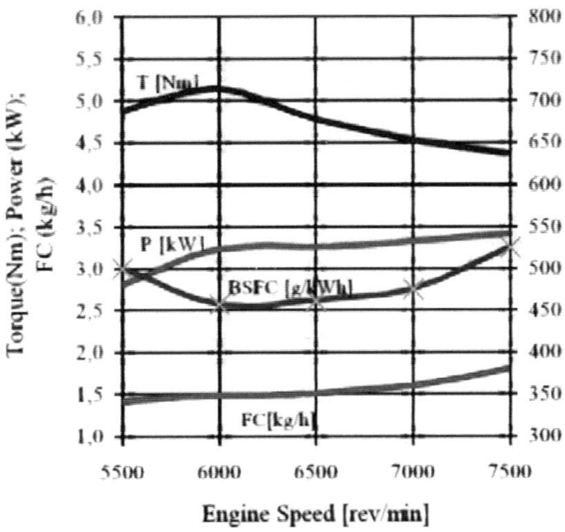

Figure 9. Engine speed operating characteristic at full load (CR = 8 : 1, E100 fuel).

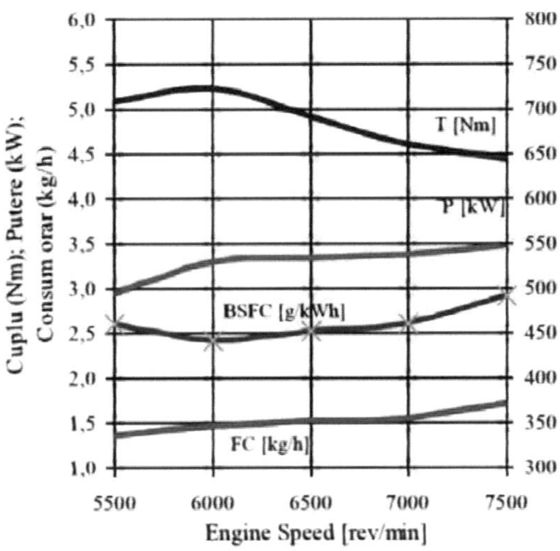

Figure 10. Engine speed operating characteristic at full load (CR = 9 : 1, E100 fuel).

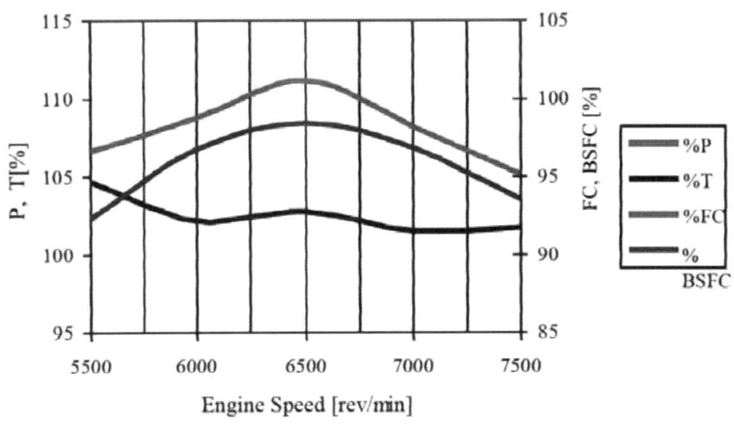

Figure 11. Comparison between engine parameters for CR = 8 : 1 and CR = 9 : 1, E100 fuel.

In Figure 12. shows the evolution of the heat release and heat release rate for engine fueled by E85 and air assisted injection system and a compression ratio CR = 8.
In Figure 13. shows the evolution of the heat release and heat release rate for engine fueled by E85 and air assisted injection system and a compression ratio CR = 9.

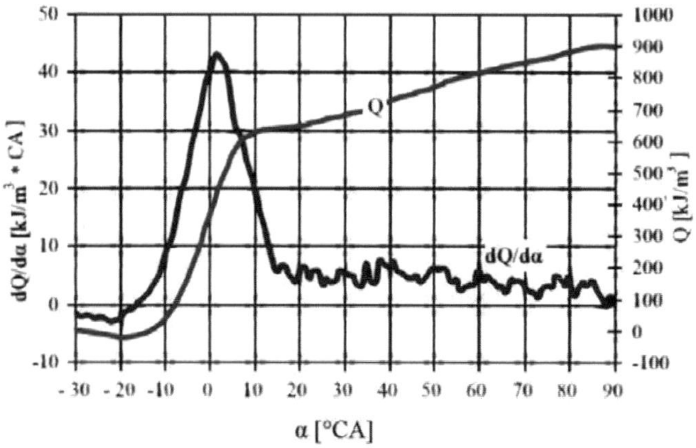

Figure 12. Heat release and heat release rate for engine running at maximum torque (CR = 8 : 1, E85 fuel).

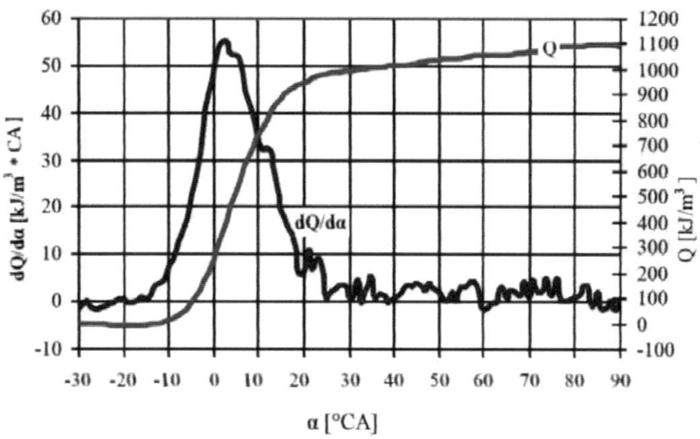

Figure 13. Heat release and heat release rate for engine running at maximum torque (CR = 8 : 1, E85 fuel).

The duration of the burning process for E100, E85, E20 and E0 is 15, 16, 18 and 25 ms. In the Figure 14. are showed the CO and CO_2 emission for all used fuelling system.

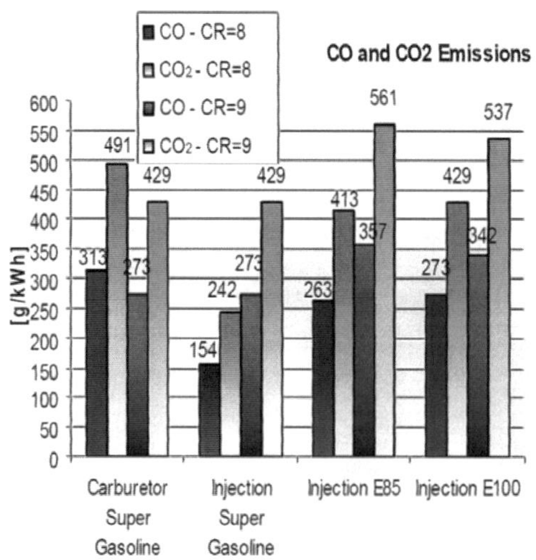

Figure 14. The CO and CO_2 emission for all used fuelling system.

The equipment needed for emissions measurements are presented in Table 2.

Table 2. Equipment for emissions measurement

Equipment type	Emissions
Hartmann & Braun Magnos 16 (%Vol)	O2/N2
Hartmann & Braun Uras 14 (%Vol)	CO2
Hartmann & Braun Uras 14 (%Vol)	CO
ECO Physics CLD 700 EL (ppm)	NO/NOx
Testa FID HC (ppm)	HC

4. Conclusions

In this paper, for fuelling systems, was study the direct injection-air-assisted system in order to exploit the vaporization properties of the fuels used. Air-assisted fuel injectors are used to permit the fuel to enter the combustion chamber in small droplets so, the fuel could atomize quickly and mix with the freshly scavenged air. It lessens the effects of charge and exhaust-gas mixing, significantly reduces short-circuiting, and offers precise air/fuel ratio control. Depending on the injection timing the homogenous mixture or stratified mixture may be organized in the combustion chamber. For early injection timing, when the intake ports are open, the air and the fuel have time to homogenize; simultaneously, a part of the mixture can escape through the exhaust ports. On the contrary, when the injection timing is set after the exhaust ports have been closed, there is no fuel lost to the exhaust. This second case may be able to promote mixture's stratified charge combustion at part loads, which greatly improves combustion stability especially at high levels of residual gas trapped.

One main change comprises the location of the Strata™ air-assisted fuel injector in the cylinder to realize a direct fuel injection. Air-assisted direct injection is applied to a range of 2-stroke engines, from 50 to 500 cc/cylinder displacements. The air-assisted injector operates in the following manner: the fuel pumped into the air-fuel mixture formation chamber contained in the air- assisted DI system had a pressure of 0.6 – 0.7 MPa. The air pressure was induced into the air-fuel mixture formation chamber in the air assisted direct injection system with a pressure of 0.45 – 0.55 MPa. In order to have a constant compressed air pressure, an air regulator was mounted on the air inlet system. The main advantage of air-assisted fuel injector is a good atomization of fuel; droplets' size having less than 60 SMD on the condition of air-assist pressure. The use of these injection system in experimental research with two-stroke engines is motived by the following reasons [3,4]:

1. Two-stroke engines produce more power and are more compact than the four-stroke engines, and they are lightweight and less costly.
2. In the case of two-stroke engines, for every two strokes of the piston inside the cylinder, one power stroke is produced. In four-stroke engines, power is produced once during four strokes of the piston.
3. For the same size engine, the power produced by the two-stroke engine is more that the four-stroke engine.
4. Ideally the power produced by the two-stroke engine is double that of the four-stroke engine, but in actual practice it is only about 30% more than four-stroke engine.
5. Since the power produced by the two-stroke engine is higher, these engines are small and compact in size.
6. The times of injection of the air-fuel mixture, their control, determine the operating parameters of this system as well as the parameters of the combustion processes and the pollutant emissions caused by the combustion of fuels tested on the two-stroke spark ignition engine
7. The use of several types of fuels to carry out combustion processes, super gasoline, gasoline blended with alcohols and pure alcohol led to the control parameters of the direct injection feed system. By controlling this power system, optimum operating parameters were obtained at the different operating modes of the two-stroke spark ignition engine.

Compared to the carburettor feed system, the main advantages of air-assisted injection system are:

- Reducing the amount of fuel reaching the surface of the engine cylinder wall;
- Improving atomization of the fuel;
- Increased control flexibility in air-fuel mixture formation, which in turn facilitates: low-emission pollutants at cold and hot start, reduction of transient emissions;
- Provide better stability of the combustion process;
- Provide fast and reliable cold start.

Following experimental research on flame propagation in the constant volume combustion chamber using the E0, E20, E85 and E100 fuels, the following aspects were found [4]:
- The maximum pressure has been reached for the E100 fuel;
- Flame diameter for E100, E85, E20 and E0 is 4.80, 3.81, 3.65, and 2.88 cm respectively;

Among the main issues of using alcohol as a fuel in spark ignition engines can be listed:
- The tendency to reduce effective power at a constant injection rate of alcohols due to their lower calorific power compared to gasoline (when burning methanol, an amount of energy is released by about 50% less than in the case of the burning of a quantity Equivalent to gasoline, nd burning ethanol results in only 66% of the energy released from the combustion of petrol) [4], [5]
- The presence of oxygen in the molecular structure of alcohols ensures, on the other hand, the reduction of the oxygen requirement for combustion, so that overall the calorific value of the fuel-air mixture, relative to the volume of the mixture, is slightly modified (methanol requires 44% Less air for combustion than gasoline, and ethanol - only 61% of the air required to burn petrol) [9]-[10]
-Therefore, it is possible to maintain unchanged power of the engine with a given cylinder, by increasing the fuel flow accordingly (to maintain the range of the vehicle, the capacity of the fuel tank must also be increased); [6] [7]
- The difficulty of cold start, caused by the low pressure of the vapors at low temperatures; In the case of the use of pure alcohols; Cold start can be solved by using auxiliary fuels (gasoline or liquefied petroleum gas) or by spraying (methanol requires 3.7 times more heat and ethanol - 2.6 times compared to gasoline); The tendency of worsening of evaporation in The intake system for carburettor engines, determined by high alcohol vaporising temperatures and requiring redesign of the intake system; [8], [9]
- The tendency to increase the incidence of incidents in the hot engine due to the formation of vapor plugs and the emission of alcohol (the boiling point of alcohols being low in gasoline); [9], [10]
- Unfavourable lubricating qualities, caused by the low viscosity of alcohols and directly affecting the friction torques, primarily at the pump level and in the high-pressure section of the feed system [10].

Even though one kg of gasoline contains about 850 grams of carbon and one kilogram of E85 contains only 570 grams of carbon dioxide (520 grams for E100), examining the Figure 13. shows an increase in CO_2 hourly emission, relative to the engine power, for the engine supply of E85 and E100 respectively, higher for CR = 9: 1 than 8 :1.

The main cause of this evolution is the drop in calorific power for E85 (29.29 MJ / kg fuel) and E100 (26.78 MJ / kg fuel) compared to the calorific power value of gasoline (43.5 MJ / kg of fuel).

Even when the United States came out of the Paris Accord, Europe and other countries (Japan, South Korea, Australia and others), continued efforts to reduce carbon dioxide emissions from human activities. Respecting the sustained natural cycle of ethanol production from the vegetal mass, using ethanol as a fuel in transport unit engines can help reduce substantial the carbon dioxide concentration in the atmosphere.

References

[1] Ispas N, Cofaru C, Aleonte M Carbon dioxide reduction by improving combustion of ethanol or petrol-ethanol mixtures into 2 stroke engines, 28th International Symposium on Transport Phenomena, University of Peradeniya, Sri Lanka, 22 – 24 September 2017.
[2] Cofaru C, Ispas N, Aleonte M Reduction emissions by optimizing the cylinder processes using ethanol and gasoline-ethanol blend into SI 2-stroke engine, 28th International Symposium on Transport Phenomena, University of Peradeniya, Sri Lanka, 22 – 24 September 2017.

[3] Ispas N and Nastasoiu M CO_2 Emission Determination in Accord with European Regulation for Old and Today Cars Powered by Diesel Engines, The Romanian Journal of Automotive Engineering, pp 13 – 18, RoJAE 22(1) 1 – 52, ISSN 1842 – 4074, 2016.
[4] Aleonte M Researches on using advanced air fuel mixture systems and the combustion by supplying with alcohols and their derivates and alcohols, derivates and gasoline's mixtures for a Spark Ignition Engine, PhD Thesis, Brasov, 2011.
[5] Ishibashi Y, Asai M, Improving the Exhaust Emission of Two-Stroke Engines by Applying the Activated Radical Combustion, SAE 960742, 1996.
[6] Jin, S. H., et al., An experimental study of the spray from an air-assisted direct fuel injector," Proceedings of the Institution of Mechanical Engineers Part D-Journal of Automobile Engineering, vol. 222, pp. 183-194, Oct 2008.
[7] Keskin, A The Influence of Ethanol-Gasoline Blends on Spark Ignition Engine Vibration Characteristics and Noise Emissions, Energy Sources Part a-Recovery Utilization and Environmental Effects, vol. 32, pp. 151-160, 2010.
[8] Lin, W. Y., et al., "Effect of Ethanol-Gasoline Blends on Small Engine Generator Energy Efficiency and Exhaust Emission," Journal of the Air & Waste Management Association, vol. 60, pp. 142-148, Feb 2010.
[9] Mittal, M. et al., Burn Rate Analysis of an Ethanol-Gasoline, Dual Fueled, Spark Ignition Engine. New York: Amer Soc Mechanical Engineers, 2009.
[10] Spicher U, Methoden zur Analyse der motorischen Verbrennung"," Institut für Kolbenmaschinen (IFKM) Universität Karlsruhe, 2006.

IOP Conf. Series: Materials Science and Engineering **252** (2017) 012081 doi:10.1088/1757-899X/252/1/012081

EVOLUTION OF THE PRESSURE WAVE SUPERCHARGER CONCEPT

[1]Costiuc, Iuliana*, [1]Chiru, Anghel

[1]Transilvania University of Braşov, Romania
*costiuciuliana@unitbv.ro

Abstract. Born more than a century ago, the concept of exploiting the pressure wave phenomenon has evolved with rather small steps, experiencing an accelerated progress over the past decades. This paper aims an overview on the researchers' results over time regarding the pressure wave technology and its applications, pointing out on the internal combustion engine's supercharging application.

This review complements the past reports on the subject, presenting the evolution of the concept and technology, as well as the researcher's efforts on solving the specific shortcomings of this pressure wave technology. Undoubtedly, the pressure wave rotors have been a research goal over the years. At first, most of the researches were experimental and the theoretical calculations required to improve the technology were too arduous. Recently, new computer software dedicated to accurate simulation of the processes governing the wave rotor operation, altogether with modern experimental measurement instruments and well-developed diagnostic techniques have opened wide possibilities to innovate the pressure wave supercharging technology.

This paper also highlights the challenges that specialists still have to overcome and aspects to become future preoccupations and research directions.

Keywords – Wave rotors, pressure wave supercharging, internal combustion engines, shock waves.

1. Introduction

The internal combustion engine (ICE) has a long history, evolving from its strange ancient "relatives", such as the rudimentary internal combustion piston engine using the gunpowder as fuel, suggested by Jean de Hautefeuille in 1678 [1], to the contemporary highly-technologized engines. In more than three centuries, our society became dependent on the internal combustion engines, the thermal engines being used now for transport, trade and energy generation. Indeed, without the internal combustion engine, the world, as we know it today, would be quite a different place!

The first engines, as we know them today, developed in the mid-nineteenth century and were immediately used for transport. The development of the internal combustion engine helped initially to liberate manual labor, especially through physical force; later it made possible the developing of the aviation industry and other forms of transport and also helped to revolutionize the generation of electricity. From small, useful equipment near the house to large machines, from small sailing crafts to marine or aircraft engines, the internal combustion engine is indispensable in our activities and lives.

A constant concern for engineers and scientists was the improvement of the engine's design, dimensions, performances, and, moreover, emissions in the latter days. The major research directions can be resumed as follows: improving the overall efficiency and performance, conservation of energy and heat recovery, reducing the fuel consumption and lowering the engine's emissions.

A highly prevalent manner to increase the power output and the engine's efficiency is supercharging, i.e. rising the intake manifold pressure by producing considerable boost for the inlet fresh mixture or combustion air. This is achieved by using superchargers or turbochargers, the equipment positioned within the intake manifold system.

The idea of supercharging is almost as old as that of the internal combustion engine itself. In 1885 Gottlieb Daimler received a patent describing the technique of using a gear-driven pump to force air into an internal combustion engine. In 1896, Rudolf Diesel tested the effectiveness of pre-compressed combustion air introduced into the admission of his engine. Around 1900, almost simultaneously, Sir Dugald Clark, Sulzer and Renault discovered that increasing artificially the volume of air charge entering an engine, substantially more power is produced. Right around the same time, Rateau developed the centrifugal compressor. In 1912-1915, Dr. Alfred Büchi developed the first exhaust gas-driven charger, and proposed the first turbocharged diesel engine, but his ideas were not well received at that moment. Only in 1925, Büchi achied a power output increase of more than 40% and demonstrated the advantages of using the exhaust gas turbocharging [2]. This initiated the introduction of supercharging of internal combustion engines into the automotive industry.

In fact, turbocharging of automotive engines began in 1938 with truck engines, built by the company Swiss Machine Works Saurer. The World War II and the need for transportation power especialy for the military purposes accelerated the development of the charging technology. The most stiking application of turbocharging was to war planes engines [2].

Supercharging is about raising the engine's charge density - air or fuel-air mixture - before filling the cylinders. In this way, higher mass of charge can be compressed in each cylinder, containing more oxygen available for combustion than conventional method of natural intake of fresh charge. As a result, combustion will be more effective, thereby raising the generated power.

Increasing the pressure intake air is performed usually in a compressor that can be driven directly from the crankshaft (supercharger) or by a turbine fixedly connected to the compressor shaft (turbocharger), turbine driven in turn by the engine's exhaust gases.

A particular type of "compressor" is the pressure wave device - known as "wave rotor" or "pressure wave supercharger" (PWS) - that use also the energy of the combustion gases to induce forced air into the admission manifold, but rely on the pressure waves action inside narrow channels. The fast response on the engine performance for any of the engine speeds makes the PWS a good option for supercharging the IC engines for road vehicles.

2. PWS Operating Principles

PW superchargers transfer energy from the combustion gases to the air intake by means pressure waves. The PWS basically consists of a rotor, also called "cell wheel" in which are machined longitudinal narrow channels, positioned radially on one or two rows (Fig.1). The channels are filled with air at one end and with exhaust gases at the other end, putting the two fluids into direct contact and interaction. The rotation of the cell wheel, together with a certain positioning of the air and gases inlet and outlet ports generate boost. In short terms, the exhaust gases produce shockwaves that expands within the channels and compress the intake fresh air.

The rotor is mounted inside a cylindrical casing, in between two static end plates provided with the inlet and outlet passages (ports) of air and gases, respectively.

IOP Conf. Series: Materials Science and Engineering **252** (2017) 012081 doi:10.1088/1757-899X/252/1/012081

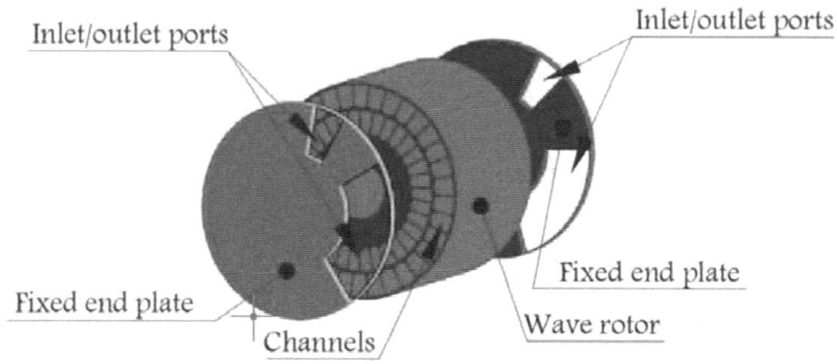

Figure 1 – PWS construction elements [4]

The PWS inlet working fluids are: the high pressure gases coming from the exhaust manifold of the engine (HPG) and the low pressure fresh air (LPA), while the outlet fluids are: the low pressure gases (LPG) evacuated through the engine exhaust system and the high pressure air (HPA) delivered into the admission manifold towards the cylinders (Fig.2).

The number and location of the ports, as well as the thermal energy source have made the distinction between various applications of these devices, serving different purposes. For instance, the four ports configuration was used for supercharging the internal combustion engines, three ports design was applied to pressure separators or pressure equalizers, while rotors with two, four, five or nine ports were integrated into gas turbines applications [5].

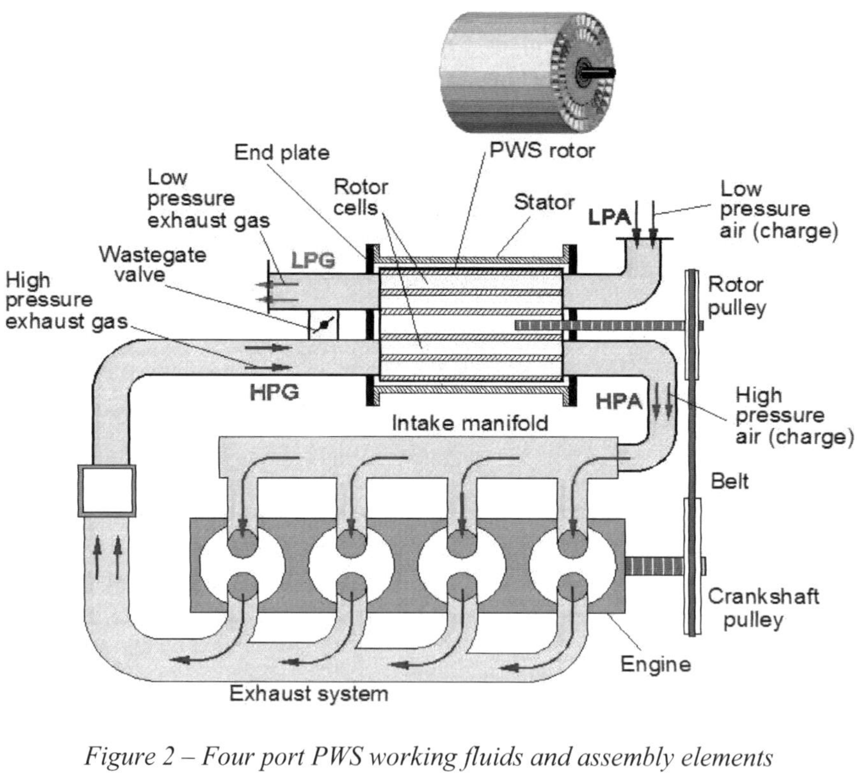

Figure 2 – Four port PWS working fluids and assembly elements

3. Pressure wave machines' history.

The concept of supercharging, i.e. supplying an internal combustion engine with pressurized air, dates back to early last century. The oscillatory and thermodynamic phenomena last since forever on Earth, but scientists have deciphered these processes in relatively recent years and continue to study them. The turbulent movement of particles, the steady or non-steady fluids flow, as well as the oscillatory phenomena has shown their potential more since the beginning of the 20th century. However, the laborious and detailed calculations necessary to study the non-steady phenomenon inside fluids have hindered progress to be made in this direction.

The beneficiaries of these complex and interconnected processes are the internal combustion engines and other machines that can record improved performances whether these phenomena are known and accurately described.

Basically, the concept that describes the operating principles of machines using the non-steady flow is the transfer of energy by means of pressure waves. The wave pressure devices have some common characteristic elements: a rotor with longitudinally shaped narrow channels within. The ports machined into the end fixed plates control the working fluids flow from the manifold pipes towards the pressure wave channels and vice-versa. The energy transfer is achieved through the direct interaction between fluids by using the non-steady pressure waves.

Unlike the steady-flow turbomachines that compress or expand the working fluids, the wave rotor performs both compression and expansion in the same component [5]. Despite the advantages of the rotor wave machines comparing to turbomachines, such as the rapid response, the low speed, the simple geometry translated into low fabrication costs, the reduced erosion of the channels and the possibility of good self-cooling of the rotor, the pressure wave machines raised some challenges that limited the large commercial implementation of these devices.

These difficulties were of mechanical nature - like sealing and thermal - and of theoretical nature - understanding and defining the complex non-steady flow phenomenon was the main factor that hampered the pressure wave machines development. During the last century, the efforts to improve the geometry and functioning features of the rotor wave, accordingly to a certain application, were sustained by the continuous progress made in solving the equations describing the inner nonlinear phenomena and in implementing the new techniques and technologies. Also, a continuous impetus for achieving energy efficiency, for withdrawing the oldest technology and for finding new, innovative and economic technical solutions, have stimulated researchers' interest in the pressure wave technology.

Shock wave engines with internal combustion, rotary thermal separators, such as wave rotor refrigerators, the shock tunnels, pulse detonation engines, wave rotors and its best known device, the Comprex, which was developed as a replacement for conventional turbochargers, etc., are some applications using the pressure wave compression and unsteady-flow phenomena.

The first pressure exchanger presentation dates from 1906 [8] when Knauff patented a cell drum called 'semi-static pressure exchanger" (SSPE) because its pressure characteristics are almost entirely independent of its speed [9]. This device was initially described as a rotor with curved rotor blades and inclined stator nozzles that provided output shaft power, therefore showing a pressure exchange engine, using a steady-flow energy transfer by mixing compressible or incompressible fluids [9]. In 1913 a German engineer, Burghard patented a device resembling with the actual cell wheel. His device was a rotating drum with axial channels shaped on the periphery, as a continuous source of pressurized air [7]. As the knowledge on the unsteady flow domain was insufficient, his device was never developed. In 1928 another SSPE was proposed by Lebre [10] for a refrigerating unit, both with a geometry implying long narrow channels.

Also, around 1928, Burghard recognized the process of unsteady-flow transfer in compressible fluids using the pressure wave process, and patented a more simpler device [11 in 9] known as 'dynamic' pressure exchangers (DPE) to distinguish it from previous "static pressure exchangers". Inside DPE the pressure waves are used in both compression and expansion processes eventuating within the rotor channels. This gave the designation of these devices as "wave rotors".

Inspired by Burghard patent, researchers have sought for decades solutions for the development of the wave rotors concept, being limited by the difficulties mainly related to poor understanding of unsteady-flow processes.

Some inventions related to the wave phenomenon were patented in the same period in different countries. In 1933 Michael Kadenacy (UK) tuned the engine exhaust pipes using the pressure wave effect, later called "Kadenacy effect",, obtaining considerable power output, but in a narrow range of speeds. Around 1938, Johann Wydler (USA) designed a rotating device that used the exhaust gases energy for supercharging a reciprocating engine. Prof. Arthur R. Kantrowitz at Cornell University designed a device working with high pressure ratio compression and expansion waves. His efforts were thwarted by the mechanical difficulties occurred [13]. However, at Cornell Aeronautical Laboratory CAL the non-steady flow concept was developed over the years, having an important contribution in developing of energy exchangers for gas turbine and various stationary power applications [12 in 5].

During the World War II, in 1940, Claude Seippel of the Brown Boveri Co., trying to apply the Lebre principle to a heat pump, acknowledged that the pressure waves can efficiently transmit energy, i.e. the expansion of a gas towards another gas to be compressed [13], both gases being in direct contact. The first machine using the wave rotor concept was implemented in Switzerland by Seippel as a high-pressure stage for a gas-turbine locomotive engine.

Seippel coined the term "COMPREX" due to COMPression - EXpansion processes that take place within the rotor channels.

The rotor had 30 channels shaped within, rotating with 6000rpm. Two ports on each side of the rotor allowed the passing of gas and air. The pressure ratio was initially 3:1 with a global efficiency of 69%, as reported in the tests performed in 1941-1943 [5]. According to Seippel's patents [14-17], a power boost of 80% and a 25% increase in performance were expected. Even though the first rotor worked satisfactory, proving the potential of pressure waves in transferring energy, when installed on an engine, it showed that improvement in designing and matching need to be done. The device was replaced by a heat exchanger which yielded a somewhat higher overall thermal efficiency. No further development work is known to have been made [9].

However, the work of these researchers outlined the idea of using the pressure wave rotor for turbocharging diesel engines. First attempt in implementing this new concept and to gather enough data to establish a correlation between measured performance and theory was made by ITE Circuit Breaker Co., Philadelphia, USA, under the supervision of Prof. Arthur R. Kantrowitz and support of the Bureau of Aeronautics. The first units were designed and built starting with 1949. In 1951 came the first encouraging results. In 1954 were run the first tests that proved that pressure wave superchargers can be used for supercharging the diesel engines. In 1955 a cooperative program between ITE Circuit Breaker Co. and Caterpillar Tractor Co. started, the result being a small PWS prototype that incorporated all the experience gained. ITE continued to perform tests until 1957 when started the tests on vehicle [13], using the so-called device COMPREX as a diesel-engine supercharger. The early version of the supercharger did not yield sufficient manifold pressure at very low engine speed, at which the clutch is engaged [9]. However, the tests proved that the wave rotor is a simple device, it can deliver high air density over a wide speed range and it allows rapid load changes with no lag or smoke [13].

A cooperative program between ITE and BBC Brown Boveri & Co. started in 1955 as a result of the promising results obtained. BBC, as a turbochargers producer, continued to develop pressure-wave superchargers for diesel engines, in a partnership with the Swiss Federal Institute of Technology (ETH, Zurich). They succeeded to make cycle modifications to overcome the deficiency of the early version of superchargers [9].

Around the same time, the British company Power Jets Ltd. started to work on multiple wave rotor applications, initially on IC supercharging but later on gas turbines, pressure equalizers, air cycle refrigeration devices. For instance, the Hungarian engineer Jendrassik worked at Power Jets Ltd. to develop the rotor wave as a high pressure topping stage for aircraft engines applications [5].

Also, in the mid 50's, a new rotor geometry was designed by Pearson at Ruston-Hornsby Turbine Company in UK, producer of diesel engines and gas turbines. The device called the Pearson Rotor used its helical channels to change the path of waves and to produce shaft work the same way a conventional gas turbine would. The rotor was 23 cm diameter and 7.6 cm length, and, despite its reduced dimensions, it realized a single cycle per each rotation [19 in 5]. The Pearson Rotor was known as the wave turbine engine and worked successfully in a wide range of operating speeds (3000–18,000 rpm), producing up to 26 kW at its design-point [5]. The Pearson Rotor encountered a series of difficulties, such as sealing against leakage and incomplete scavenging. Unfortunately, the motor was destroyed due to excessive speed and its further development ended abruptly. The Pearson rotor represented a notable design of a device made for producing significant power output, being in the same time a successful pressure exchanger.

In late 1955 at Cornell Aeronautical Laboratory, Inc., Buffalo, New York was initiated a study of a high-stagnation-temperature testing unit, that put the premises of developing a device based on shock-tubes principles, which could produce a continuous stream of high-temperature, high-velocity air, device called the CAL Wave Superheater [20]. The validated results of the small scale Wave Superheater led to the development of the 9000°R Wave Superheater, including an operational Wave Superheater Hypersonic Tunnel. The working principle involves the use of a group of shock tubes mounted on the periphery of a rotating drum, to generate high temperature, high density and high velocity air. It requires a controlled flow of pre-heated helium driver gas for the hypersonic wind tunnel and a controlled flow of pre-heated charge air (which has to be superheated prior to expansion to more than 4000 K and at a pressure of 120 atm), a controlled flow of pre-heated prime helium gas and a controlled flow of coolant helium gas [20]. The CAL Wave Superheater remained as a remarkable proof of the capabilities of wave rotor devices.

Klapproth (General Electric Company) and Goldstein (NASA), in the late 1950s, initiated conceptual and feasibility studies of combustion within rotating devices [22-24 in 21], with application on turbojets, gas turbines and ramjets. Between 1956 and 1963 General Electric Co. (GE) conducted investigations on pressure-exchangers and shock wave combustors. In Ohio, Klapproth & co. focused on designing and testing experimental four-port pressure-exchange wave engines, one with straight and one with curved channels [21], a wave engine using air-gap seals. Some difficulties, such as inaccurate flow calculations and a lack of attention given to the inner reflected waves resulted in a lower power output as was predicted [36 in 5]. However, the Klapproth rotor clearly proved the possibility of complete energy transfer within a wave rotor.

Therewith, in California, GE focused on designing a prototype of a shock wave engine having curved channels and the combustion process inside the rotor [21]. This configuration eliminates the external combustion chamber, resulting a lower weight and a compact size. Between 1956 and 1958 the experience gained was applied in fabrication of the first internal combustion wave rotor. As described by Weber [25], the only test performed on the prototype lasted about 20 seconds because the thermal expansion of the rotor exceeded the gap between the rotor and the end plates, thus, the rotor seized between the end casings. The test pointed on the difficulty in controlling the clearance between the end plates and the rotor because the thermal expansion phenomena. Despite the leaks, an overall pressure ratio of the wave rotor was achieved of 1.2 to 1.3, corresponding to global temperature ratios of 1.9 - 2.6 measured between inlet and outlet ports of low pressure [5]. In 1963 the GE company interrupted the studies of the wave rotor, reportedly due to changes in business strategy [26].

D. B. Spalding of Imperial College London continued the theoretical and experimental work of Jendrassik from Power Jets Ltd., elaborating the first wave rotor calculation methods that took into account the heat and friction losses and their effect. The numerical model provided solutions free from instabilities and improbabilities [28].

An interest in researching the wave rotor have shown, also in the mid-60s', Rolls-Royce company in the UK, by cooperating with Brown Boveri & Co. and with Berchtold of ETH Zurich și Spalding of Imperial College London as consultants [30 in 5]. Their efforts were focused on developing a wave rotor as a topping stage for a small helicopter engine, based on a reverse-flow configuration rotor incorporated into a single turbine engine. The difficulties in designing and quickly performing changes, as well as other challenges, like leakage, start-up, bearing durability, fuel control, affected the device performance. The program was canceled in 1972 against the background of financial difficulties [31].

In 1971 the first COMPREX prototype was installed on a truck engine by the company Valmet Tractors from Finland, while around the same year Mercedes-Benz started testing the COMPREX for their passenger diesel cars. Diesel engines' supercharging using pressure wave rotors became of big interest starting with 1974 when car manufacturers like Opel, Ford, Mercedes-Benz, Volvo, Peugeot and Ferrari focused their attention on COMPREX supercharging. The first success that became a limited edition was the 2.3 liter supercharged engine on the Opel Senator model.

The first extended application of Comprex® on vehicles was in 1987, when ABB (the new company formed by merging ASEA with Brown Bovery, Co.) sold the Comprex activity to Mazda. The new Mazda 626 Capella model [32] (Fig.3) was equiped with a 2.0 liter engine supercharged with Comprex. Mazda sold more than 150,000 units and is considered one of the most successful commercial applications of PW technology. ABB continued to use the pressure wave technology for gas turbine applications [5].

Figure 3 - Mazda 626 2.0L Diesel (under-hood view)

In 1994, when Mazda was taken over by Ford, the PWS series production was cancelled. Later, all the rights were sold to Caterpillar that, in collaboration with the new founded company Comprex AG (started by former BBC's employees) used the PWS technology to charge diesel engines in order to decrease the NO_x emissions [53]. In 1998 Comprex AG was taken-over by the swiss company Swissauto WENKO AG.

At the end of 70s', Mathematical Science Northwest Inc. with the support of DOE and DARPA, investigated the application of wave machines as a top stage for a gas turbine coal-derived fuels; as a high temperature air compressor for a power plant based on burning coal; and as a "dirty" gas expander or air compressor for coal burning on pressurized, fluidized bed power plants. The work implied the developing of an experimental wave rotor, with 100 channels, 45 cm diameter and 40 cm length, four ports and two additional small ports (provided for more uniform flow through ports). The calculations of energy exchanger performance were made using a computer program, FLOW, which was developed for modelling the one-dimensional unsteady flows. FLOW included also the modelling of wall friction and heat transfer processes [33]. The experimental results were very well correlated to the numerical computational ones, for a large range of operating conditions. The FLOW software used algorithms that took into account the real conditions and configurations: heat transfer, viscosity, influence of speed, port configuration and channel geometry, flow leakage [5]. Their influence on the wave rotor performance was analyzed for on-design, as well as off-design conditions. Configuration changes during the tests included variation of the clearance between rotor and end walls, configuration of the inlet and outlet ports, and increasing the area of the main driven stream outlet port. Operating parameters that were varied were: the driver and driven gas inlet pressures, driven gas outlet pressure, rotor speed, and flow rates through the port [33].

The first test conclusion, reported by Thayer, was that decreasing the clearance between the rotor and stationary port faces reduced the leakage and substantially improved the efficiency. Therefore, the leakage was proven to be a key problem for efficient operation of a pressure wave rotor. Also, the experiments showed that efficiency was relatively constant at pressure ratios less than approximately 0.8 and it dropped quite rapidly at higher pressure ratios, due to a decrease in mass flow rate through the compressed air outlet port. Another conclusion validated by tests was that a significant increase in efficiency can be achieved if the width of the exhaust gas outlet manifold was increased. The data obtained from tests on MSNW wave rotor and the computed analysis led to a much thorough understanding of the operation of real energy exchangers and the wave mechanisms [33].

The flow modelling was probably the most critical step in developing an accurate fluid dynamic representation of the dominant unsteady flow processes and the principal losses, many of which are two- or three-dimensional in nature. Efforts on numerical simulation were reported also during the conference hosted by the Naval Postgraduate School, Monterey, California in March, 1985, the ONR/NAVAIR Wave Rotor Research and Technology Workshop. According to Atul Mathur [34], an interest in developing and understanding of the basic flow processes and potential applications of pressure exchangers or wave engines prompted the initiation of a research effort at the Turbopropulsion Laboratory of the Naval Postgraduate School. The research directions were: development of an unsteady flow code, together with numerical studies and experimental work.

Mathur introduced a numerical one-dimensional code, using the Random Choice Method, based on the solution of Riemann problems. The method solves the governing Euler equations in one-dimension by solving a sequence of adjacent Riemann problems, specified as the initial conditions for each succeeding time step [34]. The code proved to be useful for more rapid computational calculations of the unsteady flow process inside the wave rotor and of some preliminary design features. However, it was not recommended for describing the real flow processes that involve friction and heat losses, or for multi-dimensional flows [35]. The results of the uni-dimensional code were used in developing of new software, called ENGINE, for performance calculations of jet engines [5].

A bi-dimensional code, based on Godunov method, was developed by S. Eidelman, showing that it is essential to take into account the gradual opening of the passages when designing a wave machine, for a proper timing, number and geometry of passages, and also because of the losses occurring due to mixing and wave reflections [5, 37].

Also, around 1985, at Cornell University, researches on wave rotor were resumed; the direction approached being mainly on new analytical methods and concepts of three-ports rotor wave, five-ports rotors, double-wave rotors, etc. [5]. The idea of increasing the pressure ratio and thus the efficiency by using a compound unit, including of two or more wave rotors, one being the supercharger for the other, was suggested first by Muller in 1954 [38 in 9].

As synthesized in [5], during the 90's other researchers became interested in investigating and developing numerical methods to describe processes inside pressure wave machines: Eldin et al. of University Wuppertal, Germany (numerical method based on characteristics theory), Piechna et al. of Warsaw University of Technology (one- and two-dimensional numerical models), Oguri et al. of Sophia University Japan, Guzzella et al. at ETH Switzerland (a control-oriented model for pressure wave devices used for engine supercharging, with accent on transient exhaust gas recirculation modeling), as well as a diesel NOx emissions investigation under Comprex® supercharging performed in Turkey.

Since the 90's, significant progress was made in computer and automation fields, therefore, analytical as well as experimental researches got a tremendous help, becoming easier to perform, to model, simulate and to process the data results. A significant number of studies revealed a real potential of pressure wave devices together with the improvements that need to be done in order to obtain higher efficiency or overall performance.

The National Aeronautics and Spatial Administration, NASA, was interested in studying the wave rotor combustors for improvements of aircraft propulsion systems performance. Daniel E. Paxson at NASA Glenn Research Center developed a wave rotor simulation cod, i.e. a quasi-one-dimensional, time-accurate, reactive, CFD Euler solver. The model was experimentally validated and was able to picture the cycle complex dynamic gas processes, to calculate the wave rotor geometry and to allow many cycle modifications in order to study their effect on the on-design and off-design performance [39, 40 in 21]. The code is recognized as a general analysis tool for wave rotors [21].

Comprex® developed by BBC in mass production implied also solving the shortcomings resulted, such as leakage, noise and rotor unequal heating: the rotor was mounted within a pressurized casing to limit the leakage and was produced by materials with low thermal expansion coefficient [36 in 5]. The performance was also improved by covering a wide engine speed range through pockets shaped within the end plates in order to control the reflected waves. The result was a reliable product for its purpose: ICE supercharging. There were produced 8 models with different geometrical dimensions, from CX-65 to CX-125, the numbers in the model's name representing the diameter (in mm) of the wave rotor. In Table 1 are presented the main features of the production Comprex models (except CX-65 and CX-125) [61], according to the dimensions represented in Fig. 4:

Comprex model	**Comprex lenghth L [mm]**	**External diameter D5 [mm]**	**LPA D1 [mm]**	**HPA D2 [mm]**	**HPG D3 [mm]**	**LPG D4 [mm]**	**G [kg]**	**Effective pressure of ICE [kW]**	**Maximum Comprex speed [rpm]**
CX – 71	274	113	49	36	39	50	4.9	25…40	23900
CX – 78	303	118	55	36	43	56	5.4	30…48	21800
CX – 85	310	126	62	36	47	61	6.2	35…57	20000
CX – 93	335	129	64	46	52	67	7.2	40…70	18300
CX -102	364	143	72	46	57	73	8.4	50…83	16700
CX - 112	396	150	80	46	62	80	11.6	60…100	15200

Table 1 – Main features of Comprex models (see Fig.4)

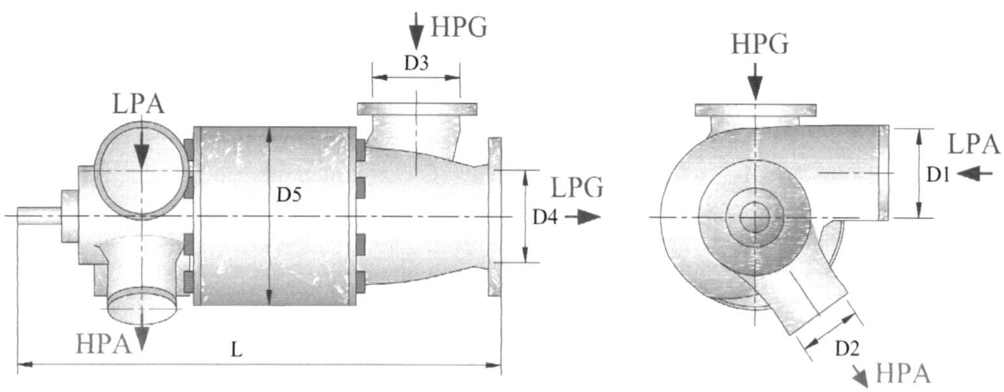

Figure 4 – Geometrical dimensions of Comprex models

Even though the Comprex® had shown a huge potential, it could not overpower the conventional turbocharger. In the early 90's, Swissauto WENKO AG in cooperation with two Swiss companies: BRM Design and Esoro developed for Greenpeace a novelty project, SmILE (Small, Intelligent, Light and Efficient), which had the targets to double the fuel mileage of a production car (Renault Twingo), keeping the same main features: performance, capacity of transport, safety and comfort. SmiLE was the first street legal car that attained consumption values of 2.3 - 2.5 Liters per 100km during test drives in real traffic conditions (3.4 l/100 km for NEDC), confirmed by the German TüV [45]. With the new designed engine, 360cc, the SAVE concept engine, supercharged with a PWS, emissions were lower than the standards for Euro III, while the overall weight of the car was reduced by 150 kg.

The three-cylinder engine was optimized according to the SAVE approach, being provided with a regular throttle valve to control the load for low torques, variable gas pockets to control high loads and an additional valve to control the scavenge air flow mass leaving the PWS. The pressure ratio of the PWS was controlled by a waste-gate valve. In the SAVE concept, an unusual element is boosting pressure with a PWS, while this was only used for diesel engines. Initially, the SAVE engine was supposed to work with a TC or a MC, but after six years of unsuccessful attempts (noise and comfort problems with the MC, unsatisfactory pressure ratios, high temperature, efficiency and high shaft speeds when using the TC, mainly because of unusual behavior on small displacement engines, under 1 liter), a PWS was chosen to increase the intake air pressure. Wenko improved over three years the PWS for the SmiLE project, increasing the pressure ratios at low speeds (from 1.2 bar to 2.4 at 2000rpm), lowering the noise level over the entire operation range by 8-15dB(A), as well as improving the compression efficiency at medium and high engine speeds [52].

SmiLE was the first success of downsizing/supercharging applied technology, proving the advantages of the PWS:

- high boost pressures at low speeds (an average value of 2 bar attained at engine speed of 1700 rpm);
- extremely fast response times (under 10ms);
- no significant problems when over-speeding, or due to mechanical or thermal stresses at high pressure levels;
- good noise reduction with the asymmetrical multi-cell version (the SmiLE vehicle was provided with a single absorption muffler that kept the noise unde the Swiss regulations value of 69.5dB).

From 1998 to 2008, Swissauto optimized the concept, presenting in 2008 the VW Golf 5 demonstrator car, equipped with 1.0 liter engine supercharged with the new Hyprex®, with no other changes made to the vehicle. The original Golf 5, with a NA double displacement engine of 2.0 liter, was compared with the highly supercharged demonstrator car, showing that downsizing with factor 2 and using the Hyprex® PWS bring significant improvement in the overall vehicle performance. Values of 210 Nm torque were obtained at 1400 rpm and an output power of 110 kW at 5000 rpm. More, acceleration and march-through were improved, the fuel consumption was reduced [45].

The Hyprex®, is driven by an electrical motor controlled by the engine's Electronic Control Unit (ECU). This new supercharger is not provided with pockets in its stator, it regulates the boost pressure and the scavenging process by means of a Gas Pocket Valve (GPV). This valve is also controlled by the ECU. As the Hyprex® is designed to work on small gasoline engines, for which the boost pressure as well as the EGR rate have a significant influence upon the engine's performance and emissions, the GPV together with the ECU have an important role [6]. Swissauto WENKO AG drew up all the patent applications regarding the relation of pressure wave charging for spark-ignition engine use [45]. The difficulties occurred in developing the pressure wave charger for gasoline engines were related to the appropriate adjustment of the charger to the wide operation map of these engines. Hyprex was further developed to our days because of its thermodynamic advantages, reliability and its direct responding speed, making it ideal for charging small displacement gasoline engines. The benefits of Hyprex imply boost response time and increased pressure ratios, lower specific fuel consumption and reduced noise and emissions [45].

In 2010, Swissauto Engineering SA renewed the Hyprex patent for the next 10 years, being known at the moment as the only company currently producing PWSs for small gasoline engines. Unfortunately, in 2015 the company was translated to Swissauto Technology AG that, in January 2017, was in full process of liquidation.

Since the beginning of the 90s', as reported by [5], a number of universities oriented their researches towards the wave rotor technology and its applications. University of Florida proposed numerical and analytical methods to investigate certain wave rotor geometry variations and its design. After 2000, at University of Tokyo, Nagashima and co. developed 1-D and 2-D numerical models that can simulate the flow behavior inside the rotor channels; also, the group investigated the rotor wave application in ultra-micro gas turbines [46, 47 in 5]. Since 2002, Michigan State University has been preoccupied by pressure wave technology application, such as wave rotor micro-turbines, refrigeration cycles using R718, ultra-micro-gas-turbines, wave rotors with radial flux (in collaboration with Warsaw University of Technology, Poland). Also, Michigan State University together with Warsaw University of Technology developed the FLUENT software [48-51] used for the investigation of the gaso-dynamic phenomena inside the rotors with axial and radial fluxes.

From 2014 to 2017 an interest on developing small scaled wave-rotors as study model, in order to improve power output values for aircraft propulsion, have shown Mataczynski et al. from Air Force Laboratory, D. Paxson from NASA, Hoke et al. from Innovative Scientific Solutions Co. In 2014, tests and flow calculations were performed on 16 inch length wave rotors working with a two-stroke internal combustion engine, proving that the PWS is capable of providing up to 50% increase in the intake mass flow, thus boosting the inlet pressure by up to 135%. With these values, the aircraft engine should work as it is close to the ground level, producing the designed power output [41]. In 2015, the same team validated the small scaled wave rotor results [42]. In 2016, representatives of the three institutions made improvements in design and performances of the small PWS and tests over a large range of operation conditions were performed. The changes and experiments provided valuable data for validation of the design and improvement of the simulation codes [43]. The study on small scaled superchargers described their performances, while the experimental results have shown the capability of the wave rotors, aiming the use of PWS in small aircraft propulsion [44].

4. PWS Performance Features. Comparison TC - PWS – MC - NA.

As depicted in [53], in terms of efficiency, PWSs showed a better downscaling behavior than turbochargers (TC) and mechanical superchargers (MC). In a PWS the energy is transmitted instantaneously and directly from one fluid to the other, therefore, no "turbo lag" can occur, as it does in a TC, caused by inertia, leakage, and local losses. This results in a good driveability for the PW supercharged engine, i.e. a fast dynamic torque response (fast, steady, and smooth rising of engine's torque) to the driver's request. Exhaust gas recirculation (EGR) can also occur within a PWS because of direct contact of fresh air and exhaust gas. In spark ignition engines, EGR has to be reduced, while in compression ignition engines EGR has to be enabled. EGR may be an advantage if it is controllable. Modern versions of PWS are provided with a controlled valve to solve the EGR problem. Regarding the noise, caused by the pressure waves in the exhaust manifold, tests have demonstrated a low noise level for PWS, triggering a smaller muffler built into the exhaust manifold. In opposition, the space for PWS installation is approximately twice as that of a turbocharger. Also, larger exhaust pipes are necessary to reduce the oscillatory pulses caused by the valves [54, 55 in 53]; this induces increased heat transfer and thus a longer time for the torque to be build-up. In order to reduce pulsations, a three-way catalytic converter can be installed between the engine's exhaust valves and the PWS. This solution is difficult to implement because of the small available space, especially for the earlier models of PWS which were belt-driven. The modern PWS are either driven by an electric motor or simply free-running [53].

In the 1980s, the cost for manufacturing a PWS was approximatively twice as that of a TC, mainly because the rotor was made by expensive materials and because the TC was already in serial production. Also, considering the system complexity due to back-pressure effect, as well as EGR control, the effort for producing engines functioning with PWS might be higher than for the turbocharged ones.

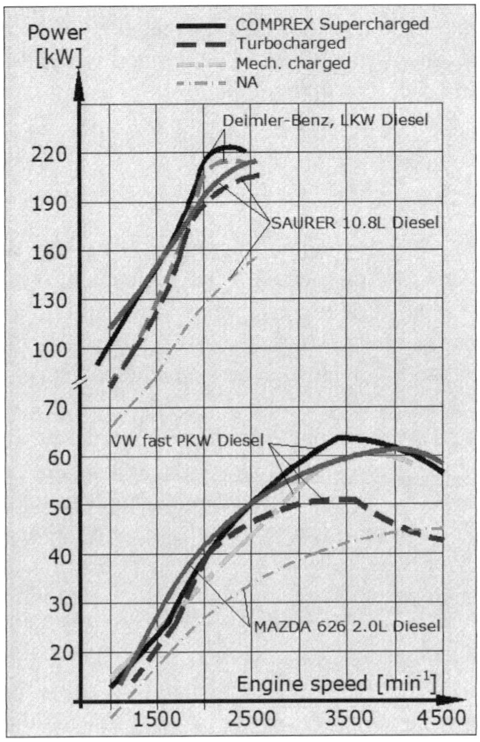

A benchmarking between the *operational behavior* of engine chargers, i.e. pressure wave superchargers (PWS), turbochargers (TC), mechanical compressors (MC) or naturally aspirated engines (NA) reveal significant differences in terms of efficiency, power, torque, specific fuel consumption. According to the results presented by Wiedemann and Rhode in [57], Tatsutomi et al. in [58], Wunsch in [59] and Hiereth in [60], collated in Fig.5, Fig.6 and Fig.7, the Comprex supercharger register a higher adiabatic efficiency than other conventional chargers in normal operating conditions. The high efficiency of PWS comes from the internal cooling effect of fresh air sweeping the channels, while in the TC or MC the rise in heat lowers the efficiency. Comprex is also experiencing losses caused by the bearings friction and windage, but they are negligible. Also, the power lost by driving the rotor is very low, about 0,5% of total load [56], and no work is lost for air compression.

Figure 5 – Comparative power output for passenger car engines (LKW) and truck engines (PKW).

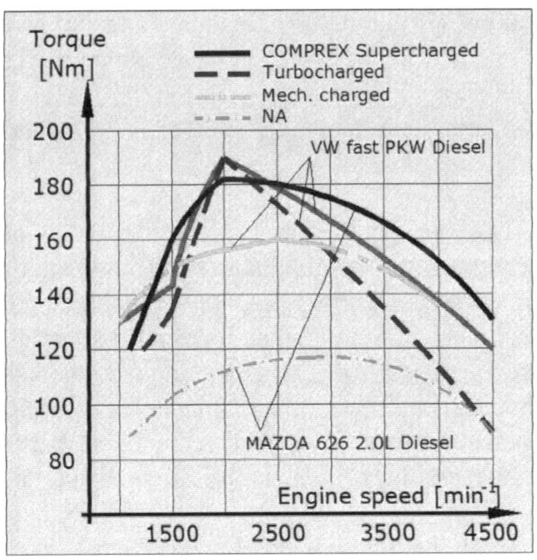

Figure 6 – Torque variation for passenger car engine

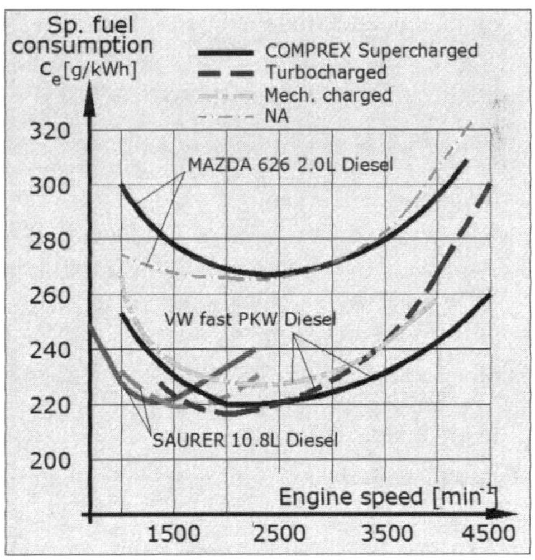

Figure 7 – Specific fuel consumption for PWS, TC, MC and NA engines

The mechanical supercharger gives an approximately equal torque curve over the entire rotational speed range as in the case of a large aspirated motor [57], similar with the NA engine. When comparing the TC with the PWS for passenger car engines, it is obvious that a very high torque was achieved at about 2000 rpm, whilst the maximum power is achieved over 3000 rpm, but for the TC the curve has a smoother slope, which keeps the peak value of the power closer to the peak torque value at the optimum engine speed, than for the PWS.

In the matter of fuel specific consumption for passenger cars (PKW), the NA engine is more effective at low loads in comparison to PWS system, while for the truck (LKW) engines the pressure wave supercharging is better than turbocharging. Therefore, the engine should be operated in bypass or combined mode according to the optimum speed range, in order to remain in the lower consumption zone. However, at high loads and real traffic conditions, the PWS has real consumption advantages.

5. Future research targets.

PWS has proved to be a solution with high potential to ensure efficient engine charging, being able to fulfill all main functions in the simplest way and with the shortest time delay. It has the capability to be tuned with the internal combustion engine over its full range of operating speeds. The solution of driving the rotor with an additional electric motor, operating at speeds that can be adjusted in a dynamic mode, accordingly to the waves' behavior inside the rotor channels, is an interesting and rewarding one. Researches in this field showed promising results.

In order to increase the efficiency of the load exchange and to reduce the intensity of the noise produced during operation, a future solution is that of introducing a device to ensure the adjustment of the angular positions of the windows (pockets) in the front covers. Their rotation can be achieved with a step-by-step motor and the command by a specialized electronic unit. Decreasing the rotor noise can be realized, e.g. by shielding the supercharger housing.

The connection of the pressure wave supercharger to the engine intake and exhaust manifolds with shorter piping and its positioning in the car body compartment are specific study subjects for each engine.

Also, providing pressure transducers (e.g. piezoelectric) on the PWS equipment, capable of ensuring the assessment of the air and flue gas pressures in dynamic mode, should ensure the necessary information for the rotor speed controlling and the front cap position settings.

Finally, the cost of PWS manufacturing is a permanent problem, but the new and modern technologies and materials will reduce the total costs of the PWS, increasing, in consequence, the PWS position within the automotive industry.

6. Conclusions.

The pressure wave supercharging concept, as well as the general wave technology, as shown in the section above, has sparked since 1906 the interest of researchers and manufacturers. In more than a century, thanks to the development and implementation of new technologies, new reliable high-strength and high-temperature resistant materials and nevertheless of the computational performances, workable wave machineries could be realized and improved. Comparing to other engineering and production fields, the wave technology registered relatively slow progress, though considerable results were achieved especially on steady-flow turbo-machineries, despite the well-recognized better efficiency of non-steady flow machines using the shock compression, process more efficient than the isentropic one.

Pressure wave superchargers offer advantages that cannot be overpassed, but still have some disadvantages that have to be overcome. Therefore, researchers have a wide range of possibilities to improve the technology, by theoretical studies, as well as by testing, observing, measuring and implementing innovative ideas. The complex phenomena occurring inside the wave rotor cells challenged the researchers more than a century - and still does – to find solutions for the theoretical equations that describe the fluid behavior, as well as geometrical adjustments, new configurations or new and reliable materials for the pressure wave supercharger elements. The results will thus highlight the main advantages, pointing on the most appropriate applications. Further theoretical and experimental studies will open the path for new large-scale industry-wide applications of the pressure wave technology and, undoubtedly, for promoting pressure wave superchargers for road vehicles applications.

References

[1] http://wiki.vintagemachinery.org/A-Brief-History-of-the-Internal-Combustion-Engine.ashx

[2] Petitt, J., "Sport Compact Turbos & Blowers", CarTech Inc, ISBN 1884089887, 2004

[3] Heisler, H. "Advances Engine Technology", SAE International, 1995, ISBN 1 56091 734 2

[4] Costiuc, I., Chiru, A., "Thermodynamic Process Modeling in Pressure Wave Superchargers", The International Congress of Automotive and Transport Engineering, ISSN 2069-0401, 2016, republished in "Ingineria Automobilului", No.43, 2017

[5] Akbari, P., Nalim, R., Mueller, N. – A Review of Wave Rotor Technology and its Applications. Journal of Engineering for Gas Turbines and Power, Vol.128, Oct.2006.

[6] Weber, F., "Mean Value Modelling of Wave Pressure Supercharger Including Exhaust Gas Recirculation Effects". Swiss Federal Institute of Technology: Zürich, Switzerland, 2001

[7] Pranav, A.S., Wave Rotor Test Rig Design Procedure for Gas Turbine Enhancement, Michigan State University. Department of Mechanical Engineering, 2008

[8] Knauff, R., 1906, "Converting Pressures of Liberated Gas Energy into Mechanical Work," British Patent 2818

[9] Azoury, P.H. – An Introduction to the Dynamic Pressure Exchanger, *Proceedings of the Institution of Mechanical Engineers,* Vol 180, Issue 1, pp. 451 – 480, 1965-66

[10] Lebre, A. F. British Patent No. 290 669, 1928

[11] Burghard, H. German Patent No. 485 386,1928

[12] Weatherston, R.C., Hertzberg, A. – The Energy Exchanger, A New Concept for High-Efficiency Gas Turbine Cycles", ASME Journal For Engineering For Power, 1966

[13] Berchtold, M., "The Comprex Diesel Supercharger", SAE Paper 590001, Vol.67, 1959

[14] Seippel, C., Swiss Patent 225426, 1940

[15] Seippel, C., 1942, Swiss Patent 229280

[16] Seippel, C., 1946, "Pressure Exchanger," U.S. Patent 2399394

[17] Seippel, C., 1949, "Gas Turbine Installation," U.S. Patent 2461186

[18] Chiru A. – Researches regarding supercharging of Diesel engines. PhD thesis, Universitatea "Transilvania", Braşov 1986.

[19] Pearson, R. D. - "Pressure Exchangers and Pressure Exchange Engines," in The Thermodynamics and Gas Dynamics of Internal Combustion Engines Vol. 1, R. Benson, ed., Oxford University Press, Chap. 16, pp. 903–940, 1982

[20] Carpenter, J.E., "Wave superheater hypersonic tunnel. Semi-annual Report", Cornell University Library, Report No. AD-1345-W-7, 1961-62

[21] Akbari, P., & Nalim, R. (2009). Review of recent developments in wave rotor combustion technology. Journal of Propulsion and Power, 25(4), 833-844.

[22] Schmidt, F. A. F., "Mechanically Controlled Multistage Combustion Chambers for Gas-Impulse-Type Engines and Improved Discharged Control Therefor," U.S. Patent No. 2, 937, 498, 1960.

[23] Spalding, D. B., "Pressure Exchangers," U.S. Patent No. 2, 940, 656, 1960.

[24] Spalding, D. B., "Combustion Ignition Pressure Exchanger," U.S. Patent No. 3, 003, 315, 1961.

[25] Weber, H.E. – Schock Wave Engine Design, John Wiley & Sons, Inc., N.Y., 1995, ISBN 0-471-59724-4

[26] Mathur, A., "A Brief Review of the GE Wave Engine Program (1958-1963)", Proceedings ONR/NAVAIR Wave Rotor Research and Technology Workshop, Naval Postgraduate School, Monterey, CA, Report NPS-67-85-008, pp. 171, 1985

[27] Leahu, C.I., Abaitancei, H., Radu, S., "Drive with rotative speed independent from the engine, of the pressure wave compressors", Recent Journal, martie 2013, vol. 14, no. 37, pp. 29-35.

[28] Azoury, P.H. – Engineering Applications of Unsteady Fluid Flow, John Wiley & Sons, N.Y., 1992, ISBN 978-0471929680

[29] Heywood, J.: Internal Combustion Engine Fundamentals, McGraw-Hill International Editions, I.Title, II. Series, 1988.

[30] Kentfield, J.A.C. – Wave Rotors and Highlights of Their Development, AIAA Paper 98-3248

[31] Moritz, R. "Rolls-Royce Stuy of Wave Rotors (1965-1970)", Proceedings ONR/NAVAIR Wave Rotor Research and Technology Workshop, Naval Postgraduate School, Monterey, CA, Report NPS-67-85- 008, pp. 116-124, 1985

[32] Zehnder, G., Mayer, A., and Mathews, L., "The Free Running Comprex®," SAE Paper 890452, 1989

[33] Thayer, W. J., "The MSNW Energy Exchanger Research Program," Proceedings ONR/NAVAIR Wave Rotor Research and Technology Workshop, Naval Postgraduate School, Monterey, CA, Report NPS-67-85-008, pp. 85– 116, 1985

[34] Mathur, A., "Design and Experimental Verification of Wave Rotor Cycles," Proceedings ONR/NAVAIR Wave Rotor Research and Technology Workshop, Naval Postgraduate School, Monterey, CA, Report NPS-67-85- 008, pp. 215–228, 1985

[35] Mathur, A., "Wave Rotor Research: A Computer Code for Preliminary Design of Wave Diagrams," Naval Postgraduate School, Monterey, CA, Report NPS67-85-006CR, 1985

[36] Taussig, R. T., and Hertzberg, A., "Wave Rotors for Turbomachinery", Winter Annual Meeting of the ASME, Machinery for Direct Fluid-Fluid Energy Exchange, edited by J. F. Sladky, AD-07, pp. 1–7., 1984

[37] Eidelman, S., "Gradual Opening of Rectangular and Skewed Wave Rotor Passages," Proceedings ONR/NAVAIR Wave Rotor Research and Technology Workshop, Naval Postgraduate School, Monterey, CA, Report NPS- 67-85-008, pp. 229–249, 1985

[38] Muller M.A. German Patent No.916 607 and No.924 845, 1954

[39] Paxson, D. E., and Perkins, H. D., "Thermal Load Considerations for Detonative Combustion-Based Gas Turbine Engines," AIAA Paper 2004-3396, 2004.

[40] Paxson, D.E., "A Performance Map for Ideal Air-Breathing Pulse Detonation Engines," AIAA paper 2001- 3465, July, 2001, also, AIAA Journal of Propulsion and Power

[41] Mataczynski,M. R., Polanka,M.D., Hoke, J., Paxson, D.E., "Design and Implementation of Waverotor Technology for Small-Bore Two-Stroke Internal Combustion Engines", 52nd Aerospace Sciences Meeting, AIAA SciTech Forum, (AIAA 2014-0528)

[42] Mataczynski,M. R., Hoke, J., Polanka,M.D., Nees, J.B., Paxson, D.E., "Validation and Simulation of a Small-Scale Pressure Wave Supercharger", 53th Aerospace Sciences Meeting, AIAA SciTech Forum, (AIAA 2015-1345)

[43] Mataczynski,M. R., Hoke, J., Polanka,M.D., Paxson, D.E., "Experimental Performance of a Small Scale Pressure Wave Supercharger", 54th Aerospace Sciences Meeting, AIAA SciTech Forum, (AIAA 2016-0768)

[44] Mataczynski,M. R., Paxson, D.E., Hoke, J., Schauer,F., "Design and Testing of a Small Pressure Wave Supercharger for an Industrial Diesel Engine", 55th Aerospace Sciences Meeting, AIAA SciTech Forum, (AIAA 2017-1624)

[45] http://www.swissauto.com/

[46] Okamoto, K., Nagashima, T., and Yamaguchi, K., 2003, "Introductory Inves-tigation of Micro Wave Rotor," ASME Paper IGTC03-FR-302, Japan.

[47] Okamoto, K., Nagashima, T., and Yamaguchi, K., "Design and Perfor-mance of a Micro Wave Rotor," 17th International Symposium on Airbreath-ing Engines, Paper ISABE-2005-1270, 2005

[48] Piechna, J., Akbari, P., Iancu, F., and Müller, N., "Radial-Flow Wave Rotor Concepts, Unconventional Designs and Applications," ASME Paper IMECE2004-59022, 2004

[49] Piechna, J., "The Micro Jet Wave Engine Idea," 2nd International Symposium on Innovative Aerial/Space Flyer Systems, Tokyo, 2005

[50] Frackowiak, M., Iancu, F., Potrzebowski, A., Akbari, P., Müller, N., and Piechna, J., "Numerical Simulation of Unsteady-Flow Processes in Wave Rotors," ASME Paper IMECE2004-60973, 2004

[51] Iancu, F., "Integration of a Wave Rotor to an Ultra-Micro Gas Turbine UμGT" Ph.D. thesis, Michigan State University, E. Lansing, MI, 2005.

[52] Guzella, L., Martin, R., "Das SAVE-Motorkonzept", Ch. "Neue Motoren", MTZ Motortechnische Zeitschrift 59, No.10/1998.

[53] Spring, P., "Modeling and Control of Pressure-Wave Supercharged Engine Systems", PhD Diss. ETH No. 16490, 2006.

[54] Endres, H., "Comprex - Aufladung schnellaufender direkteinspritzender PKW - Dieselmotoren", Diss. der Technischen Hochschule Aachen, Deutschland, 1985.

[55] Mayer, A., Kirchhofer, H., "The Comprex-Pressure Wave Supercharger in Passenger Car Applications", International Journal of Vehicle Design, Vol. 6, No. 1, 1985.

[56] Walzer, P., Emmenthal, K.-D., Rottenkolbe, P. – Aufladesysteme fur PKW – Antrieb, Automobil Industrie, 1989.

[57] Wiedemann, B., Rhode, W. – Aufladetechnische Konferenz Lehrstuhl f. Angewandte Thermodynamik RWTH, LAT-FEV , Aachen, 1984.

[58] Tatsutomi, Y., Yoshizu, K., Komagamine, M. – "Der Dieselmotor mit Comprex-Aufladung fur den Mazda 626" – Motortechnishe Zeitschrift 51, 1990.

[59] Wunsch, A. – "Aufladung von Fahrzeugdieselmotoren mit dem Abgasturbolader und mit der Druckwellenmaschine Comprex" - Motortechnishe Zeitschrift 31, 1990.

[60] Hiereth, H. – "Eignungsabschatzung neuer Aufladesystemefur Fahrzeugmotoren" - Aufladetechnische Konferenz Lehrstuhl f. Angewandte Thermodynamik RWTH, LAT-FEV, Aachen, 1984.

[61] Leahu, C., "Optimisation of Compression Ignition Engine using Supercharging Systems", PhD Diss, University Transilvania Brasov, 2011.

Surge detection on an automotive turbocharger during transient phases

M Deligant[1], A Danlos[2*], P Podevin[2], A Clenci[2,3], S Guilain[4]
[1]DynFluid Laboratory, EA92, Arts et Métiers ParisTech, CNAM, France
[2]Laboratoire de Chimie Moléculaire, Génie des Procédés Chimiques et Energétique (CMGPCE), EA7341, CNAM, France
[3]Department Automobiles and Transport, University of Pitesti, Romania
[4]Renault SA, France

*amelie.danlos@lecnam.net

Abstract. The surge limit on automotive turbocharger needs to be avoided to prevent operations with pressure and mass flow oscillations. Mild surge is accompanied by noise which is disturbing. Deep surge can cause significant loss of engine power and severe drivability issues. It is necessary to know the stationary limit in order to match a turbocharger with an engine, ensuring enough surge margin. However, this choice does not guarantee surge free operation during transient functioning. In this paper, the surge onset of a compressor while closing a downstream valve is studied. Various tests have been carried out varying the closing time, the position of the initial operating point and the volume of the circuit. The inlet and outlet signals of physical parameters are analyzed with spectral and temporal methods in order to define the instant of the surge occurrence.

1. Introduction

Turbochargers are usually used to increase the specific performances of internal combustion engines, which in the last years was used mainly to replace relatively large naturally aspirated engines with smaller ones without sacrificing the peak performance. This combination of downsizing and uploading an engine is known since long time, but the conversion in an acceptable way of this potential to practice is very challenging. Actually, for a low mass flow rate combined with a high load, the turbochargers may be submitted to instabilities. These instabilities can induce mechanical failures of the turbocharger components and affect engine functioning. As turbochargers manufacturers give only the characteristics measured in steady state operation, to prevent all problems in a vehicle, it is necessary to understand phenomena responsible for the oscillating flow inside the turbocharger, which is called surge. This surge appearance is studied since last three decades [1-4].

The surge phenomenon can also appear for high altitude operation [5] or for acceleration and deceleration [6]. In order to use turbochargers in a larger operating range, it is necessary to determine precisely the surge limit and to identify mild surge and deep surge [7-9]. Many authors have discussed different temporal or spectral methods to obtain a criterion allowing to detect the surge appearance on a compressor map. Classical method uses the detection of surge noise, which is a non-sufficient criterion because of the arbitrary feeling of the operator [10]. Sinusoidal signals of pressure ratio or mass flow rate characterize the deep surge and may be a better criterion to identify the surge limit [11,12]. Other methods as the analysis of the standard deviation of temperature signals [13] or frequency domain analysis [14,15] have also been applied on automotive turbocharger performance study. The criterion to determine the surge limit is crucial because it has been observed that the turbocharger changes its operating range when the upstream or downstream circuit is different [6,10,16,17]. This implies that manufacturers' compressor map given from a test bench is changed when the turbocharger is placed inside the vehicle.

The study presented in this paper deals with the appearance of surge phenomenon for transient operating conditions of an automotive turbocharger. The criterion for detecting the surge limit is discussed and compared with other methods in order to observe which physical variable is interesting to detect this limit.

2. Experimental set-up

2.1. Test bench description

In our laboratory, we use dry compressed air to set up the characteristic curves of turbocharger. An advantage is to do experiments close to adiabatic conditions which allows to calculate the power demand of compressor easily by applying first law of thermodynamic. The adiabatic hypothesis is not applicable in some case (ex: low speed). Previous works have been presented on this subject [18,19]. Nevertheless, accuracy is always higher than experiments conduct on hot test bench.

A scheme of the test bench is given in Figure 1. The turbine is fed with dry compressed cold air and a valve is used to adjust the speed of the turbocharger to the requirement. Another valve on the compressor downstream of the circuit is used to adjust the resistance of the circuit. By acting on these two valves, an iso-speed can be relatively easily achieved.

An additional ¼ ball valve is used for surge experiments.

The center housing is fed by the lubricating unit with SAE 5-30W oil. Oil temperature and pressure are adjustable respectively from 20 to 120 °C and from 0.5 to 4 bar.

Following measurements are done:

- Compressor and turbine upstream-downstream pressure: with strain gauge transducers
- Compressor and turbine upstream-downstream temperature: usual platinum resistance thermometers have been replaced by K thermocouple of 0.5mm for transient experiments
- Rotational Speed: detection of the influence of metal on inductance parameters
- Compressor air flow: thermal mass flow meter
- Turbine air flow: thermal mass flow meter

For transient experiments, a Constant Temperature Anemometry (CTA) system is linked to a hot film in order to measure instantaneous air-flow and a 0.5 K thermocouple in front of compressor wheel has been added.

All the sensor signals are converted to 0-10 voltage and sent to a USB data acquisition device. The device is controlled by Labview software. For surge measurements, 5120 measurements were done at a scanning speed of 1024 Hz, which corresponds to 5s duration. Data are stored in a file for further processing.

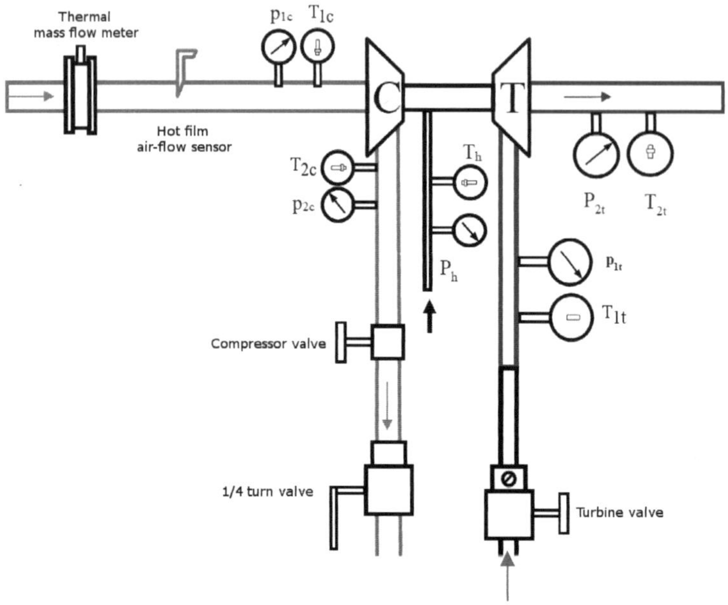

Figure 1: Experimental setup

2.2. Experimentations

This study consists in analysing the transient behaviour of the compressor while entering into surge due to a sudden closing of the downstream valve. Future experiments are planned to simulate surge behaviour on an automotive engine. The turbocharger is Honeywell-Garrett GT15 model equipped with a variable nozzle turbine (VNT). The control device of the guide vanes is set at a position chosen by the operator, in order to allow a quick acceleration of the turbocharger. The position of the turbine valve is adjusted to reach the chosen rotational speed, typically 150,000 rpm.

The initial compressor operating point is defined by the position of the valve at the compressor outlet. When the operating point is stabilized, the ¼ turn valve is manually suddenly closed (a mechanical stop is used to adjust the valve end position). Then the valve is reopened to avoid continued surge operation. The circuit downstream of the compressor is cylindrical with a diameter of 50 mm and a length of about 1 m up to the quick closing valve. In order to test the effect of a modification of the upstream or downstream circuit, a capacity of 0.5 litre can be inserted between the outlet of the compressor and the control valve. This makes it possible to increase the surge intensity and to decrease the resonance frequency. For each test, the acquisition of the data is started approximately one second before valve closing.

In this paper, results of 4 tests with initial rotational speed of 150,000 rpm are presented:

- Test 1: No capacity, initial operating point far from the surge limit
- Test 2: No capacity, initial operating point close to the surge limit
- Test 3: 0.5L capacity, initial operating point far from the surge limit
- Test 4: 0.5L capacity, initial operating point close to the surge limit

2.3. Signal processing

As there is no trigger to start recording the data, the beginning of the records is defined *a posteriori*. The instant of the beginning of the acceleration is defined by analysing the signal of the mass flow rate. Starting from a stabilized operating point, the beginning of the acceleration can be identified as the beginning of the mass flow rate decrease. For example, with the first experiment, the analysis of the signal in Figure 2 shows the beginning of the acceleration at t=1.5s. In order to easily compare the results of different experiments, the beginning of all records is modified to be at 0.5s before the acceleration.

The mass flow rate is measured using a hot film sensor which is very sensitive and has small response time. A correlation has been experimentally set between velocity measured by the hot film sensor and mass flow measured by the thermal mass flow meter. In case of deep surge, reverse flow can occur. The hot film sensor is not able to provide the direction of the flow. In this study, this problem is solved using the information provided by the inlet pressure sensor which sign is the opposite of the flow rate. So, the sign of the inlet pressure is extracted and used to correct the sign of the hot film sensor measurements. The obtained signal is then filtered with a low pass filter with a cut-off frequency of 100 Hz (see Figure 3).

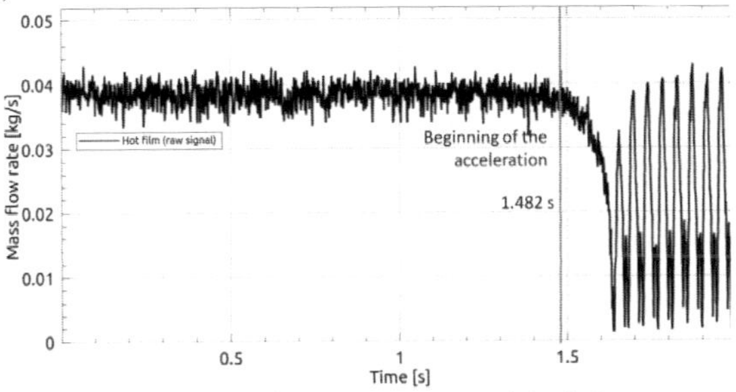

Figure 2: Mass flow rate versus original time

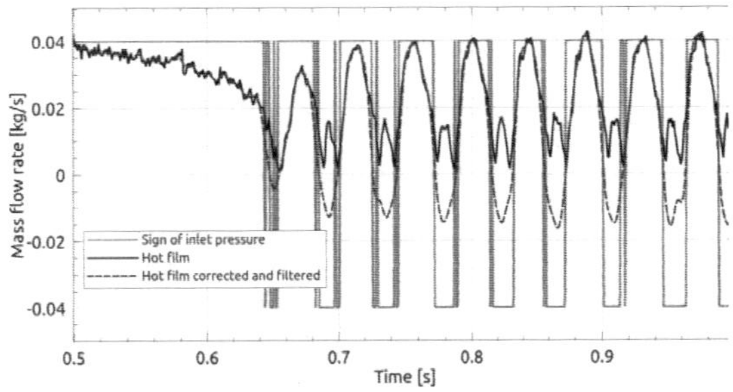

Figure 3: Processing of the hot film signal to get instantaneous mass flow rate

3. Results and discussion

Figure 4 and Figure 5 show the evolution of rotational speed. All tests start at the same rotational speed of 150,000 rpm. Test 1 and test 3 with a mass flow far from the surge line need more power than test 2 and test 4. When the downstream valve is closed, the rotational speed increases due to the reduction of the compressor power while the turbine inlet pressure remains constant. The higher the initial turbine power, the greater the acceleration of rotational speed.

It can also be noticed that the transition to surge appears at the same time (about 0.6s) whether the initial operating point is far or close to the surge limit.

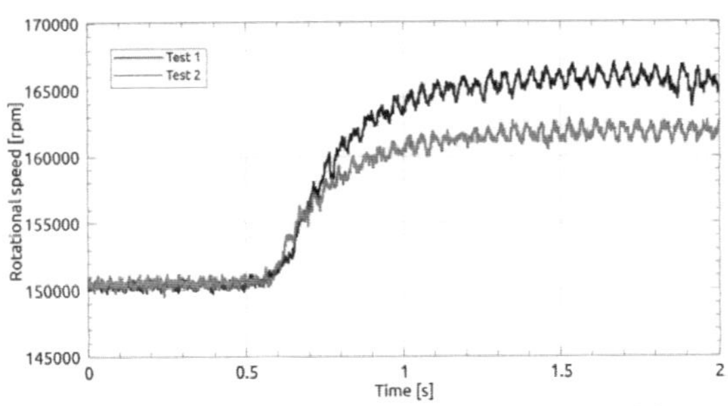

Figure 4: Rotational speed versus time (tests 1 and 2)

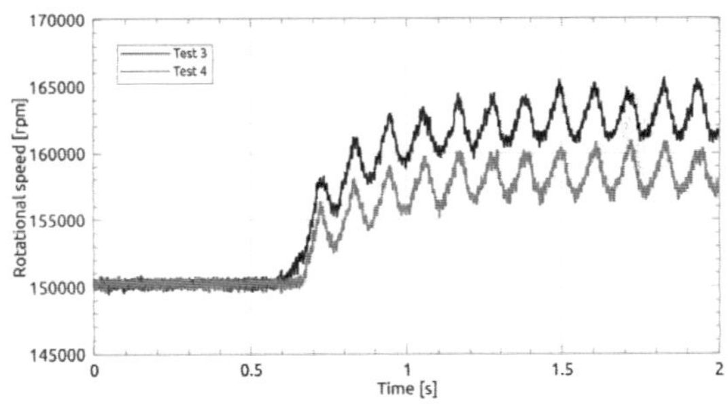

Figure 5: Rotational speed versus time (tests 3 and 4)

Figure 6, Figure 8 and Figure 10 present the temporal signal of pressures, mass flow and temperatures for test 1. On the pressure signals, it can clearly be seen that the pressures increase between 0.5s to 0.7s. Then, both inlet pressure and outlet pressure seem to stall before they start oscillating. The same behaviour can be observed on the mass flow rate signal. This is also confirmed by the observation of the temperature on front of the wheel, which starts increasing. This indicates the presence of flow reversing in front of the wheel which warm up the air at the inlet of the compressor.

Figure 7, Figure 9 and Figure 11 show similar results for test 3. The reduction of the frequency of the oscillation can be easily noticed by comparison with result of test 1. Fluctuations also appear on the temperature measurement.

For all tests, the temperature in front of the compressor wheel has a short response time. It increases quickly after the pressures stall. Using the ratio of the inlet temperature and the temperature in front of the wheel could help predicting the surge limit.

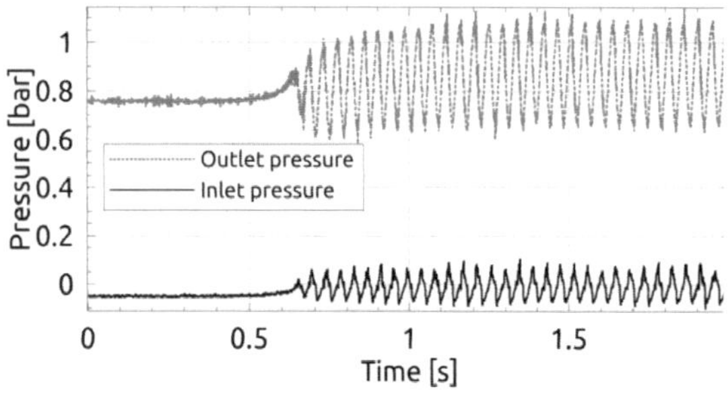

Figure 6: Pressures versus time (test 1)

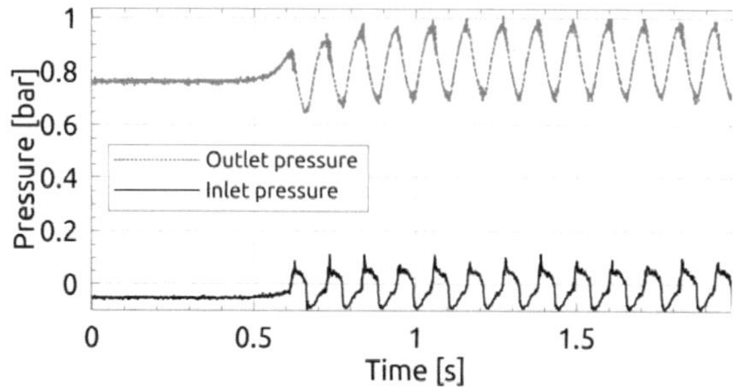

Figure 7: Pressures versus time (test 3)

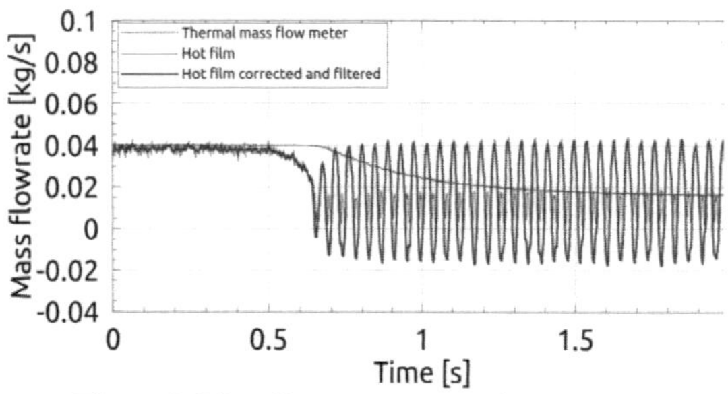

Figure 8: Mass flow rates versus time (test 1)

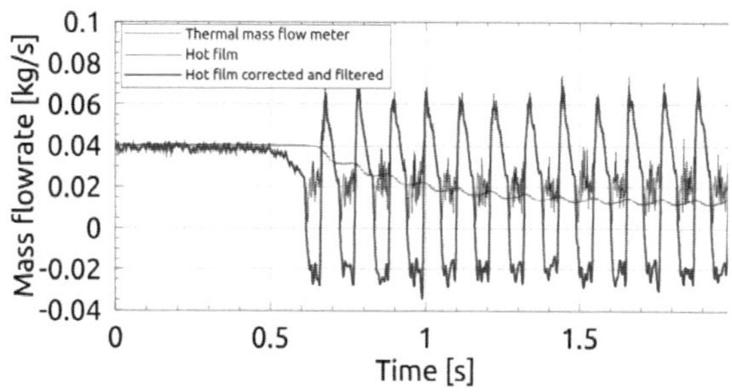

Figure 9: Mass flow rates versus time (test 3)

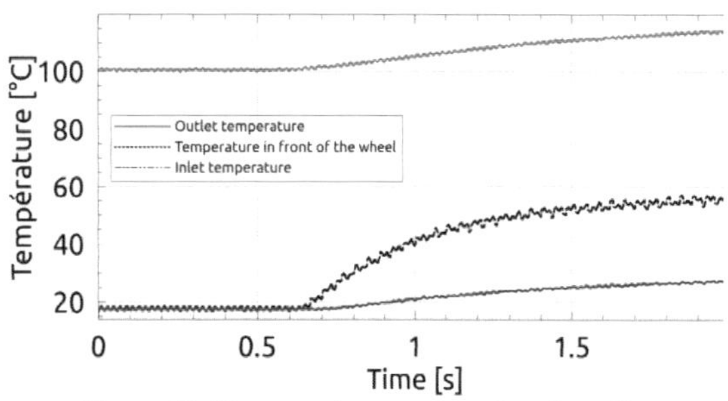

Figure 10: Temperatures versus time (test 1)

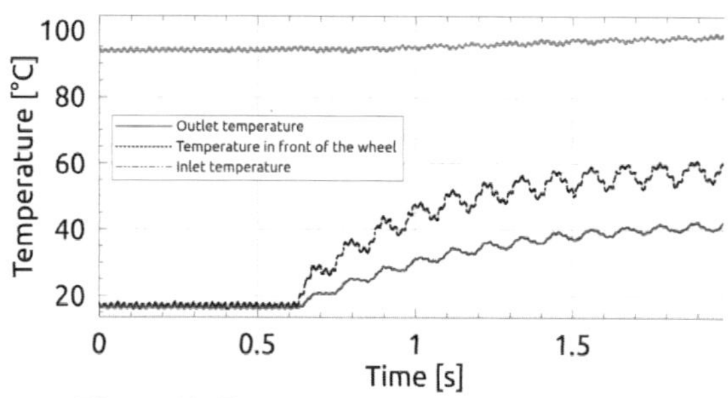

Figure 11: Temperatures versus time (test 3)

Figure 12 and Figure 13 present the amplitude spectrum of the pressures and mass flow signals. Figure 14 and Figure 15 present the amplitude spectrum of the temperatures signals. The FFT (Fast Fourier Transform) is applied on the portion of the signal where there are steady oscillations, after the appearance of surge. For the test carried out without the 0.5 L volume, the frequency of the pressures and flow signals is 23Hz. With the 0.5 L volume added on the outlet pipe, the frequency of the signals is 9 Hz. These frequencies can also be found in the spectrum of the temperature signals. We can also observe in the temperature signals several peaks and specifically one at 51Hz. Results are only presented for tests 1 and 3. Similar results are obtained for tests 2 and 4.

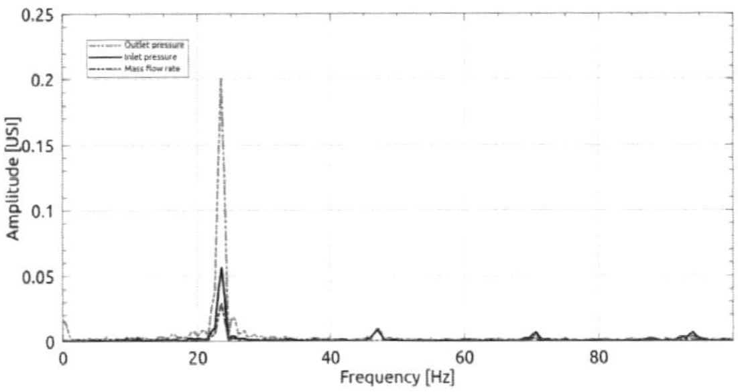

Figure 12: Spectra of pressure and flow rate signals (test 1)

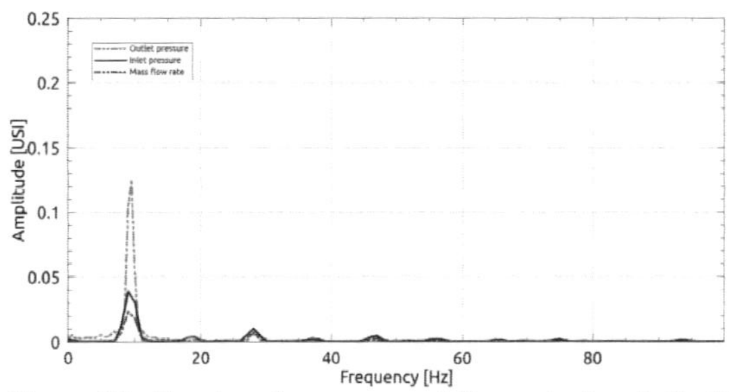

Figure 13 : Spectra of pressure and flow rate signals (test 3)

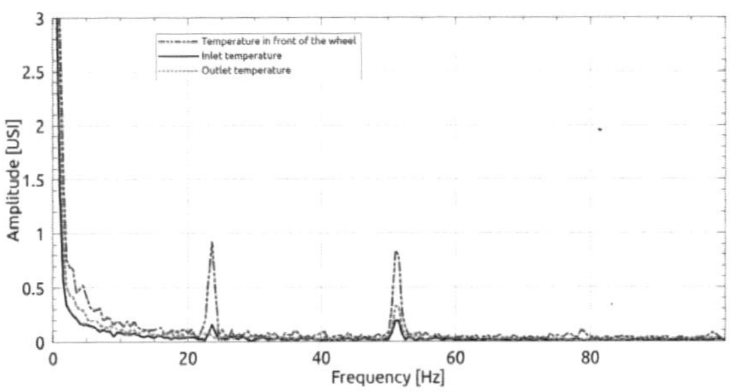

Figure 14: Spectra of temperature signals (test 1)

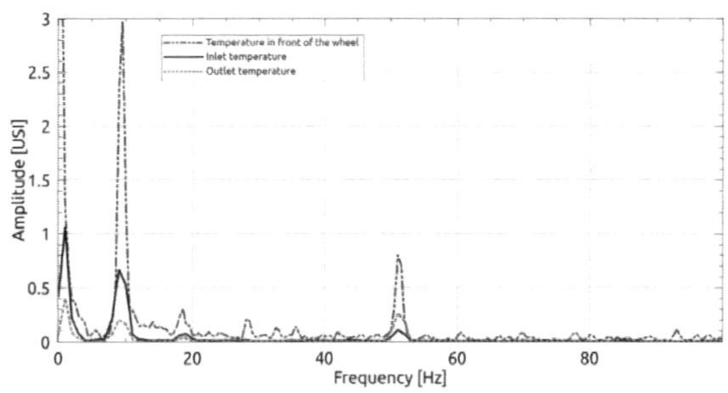

Figure 15: Spectra of temperature signals (test 3)

Figure 16 to Figure 19 present the evolution of pressure ratio versus air flow for the different test starting from initial point. The pressure ratio is filtered using a low pass filter with a cut-off frequency of 100 Hz. The effect of the capacity is obvious on the amplitude of the air flow, but without incidence on the intensity of the pressure ratio.

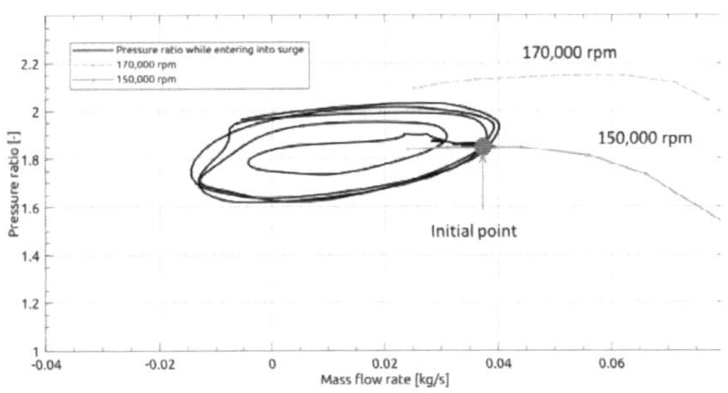

Figure 16: Surge loop (test 1)

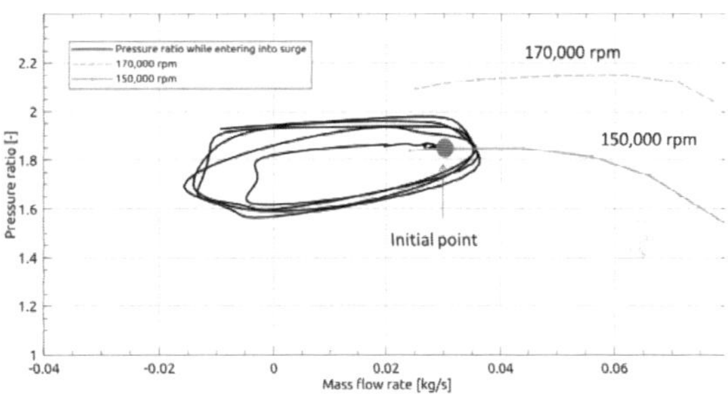

Figure 17: Surge loop (Test 2)

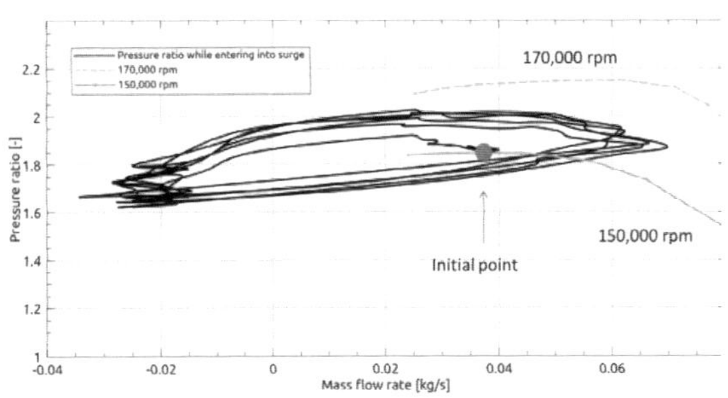

Figure 18: Surge loop (test 3)

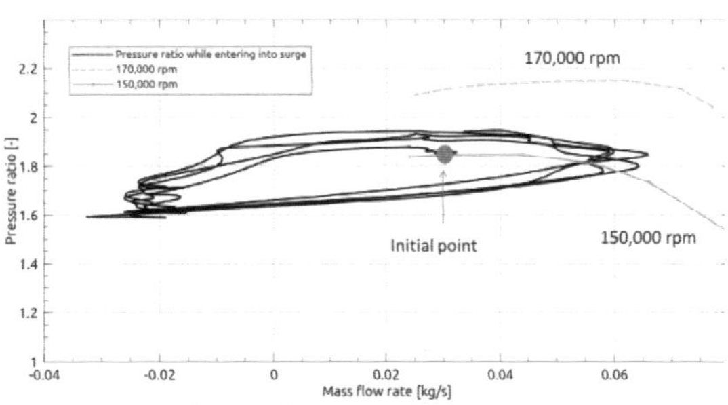

Figure 19: Surge loop (test 4)

4. Conclusion

Here has been presented a part of our experiments. Complementary tests of suddenly closing the ball valve further from surge point (Mass flow rate = 0.052 kg/s at 150000 rpm) conduct to the same result. Surge appears about 0.2s after turbocharger acceleration. This short delay is mainly due to the fact that in our test procedure the power given by the turbine is almost constant. Then later apparition of surge can be expected if we are able to reduce rapidly the turbine power as for example closing the VNT blades (or opening). Control of VNT is expected for further experiments combine also with control of compressor vane closing.

Concerning experimental results, it has been observed high variations of turbocharger speed in deep surge and even more with a capacity at the compressor outlet. Capacity can be assimilated to a charged air-cooler system. This capacity allows to observe the effect of the downstream and upstream circuit of the turbocharger. As manufacturers give performances of the turbocharger according to test bench, we can see that the operating range of this turbocharger changes when it is used in a vehicle. This phenomenon is confirmed by pressure variation at turbine inlet. Some calculation must be done to link variation of turbocharger speed to torque variation. An optical measurement of turbocharger speed will be added to get more accurate values.

During surge, fluctuation of temperatures has been noticed at main surge frequency. A frequency of 50Hz can also be observed. It seems that this one is due to the conditioning of the signal by the converter which transforms the low voltage of the thermocouple from 0 to 10V signal.
It is planned to replace these converters by others, dedicated to very high speed measurements.

Other experiments have been done on a compressor prototype, defining mild surge by acoustic detection with the operator's ear and deep surge by the occurrence of the surge frequency by a spectral analysis with FFT. Acoustic detection is rather subjective but we have linked this detection with temperature measurements. The same procedure has been applied for deep surge. Results are encouraging and offer a more precise and objective limit.

Surge limit remains difficult to set. Based on our experiments transient measurements look like a good opportunity to define this limit. Surge clearly appears for example by analysing the compressor pressure inlet. Nevertheless, it can be noticed close to this point relatively high variations of air flow which could make difficulties to define precisely the point in terms of air flow and pressure. It could be interesting to carry out investigations for this solution.

In order to obtain a more precise detection method of the surge limit, other signal processing techniques will be tested on pressure, mass-flow rate or temperature signals. Wavelet analysis or Short Fourier Transform could be more efficient to detect the surge appearance than FFT because it allows to analyse signals in very small temporal windows. The Stockwell transform is another way to obtain better temporal and amplitude resolutions of the surge detection.

Finally, modifying the test bench is expected to get closer to a real automotive engine operate range.

References

[1] Semlitsch B, Mihaescu M 2016 Flow phenomena leading to surge in a centrifugal compressor *Energy* **103** 572.
[2] Guillou E, Gancedo M, DiMicco RG, Gutmark E, Hellström F, Fuchs L and Mohamed A 2010 Surge characteristics in a ported shroud compressor using PIV measurements and large eddy simulation *9th International conference on turbochargers and turbocharging-institution of mechanical engineers, combustion engines and fuels group* 161.
[3] El Kamel C, Toussaint M, Lefebvre A, Marchal A and Descombes G 2011 Experimental study of the surge behaviour of a centrifugal compressor *13rd European Automotive Congress* Valencia, Spain
[4] Toussaint M, El Kamel C, Aumoite P 2010 Experimental study of the compressor surge *ISROMAC 13 -The 13th International Symposium on Transport Phenomena and Dynamics of Rotating Machinery* Honolulu, Hawaïi.

[5] Wang X, Ge Y, Yu L, Feng X 2013 Effects of altitude on the thermal efficiency of a heavy-duty diesel engine *Energy* **59** 543
[6] Galindo J, Tiseira A, Navarro R, Tari D, Meano C.M. 2017 Effect of the inlet geometry on performance, surge margin and noise emission of an automotive turbocharger compressor *Applied Thermal Engineering* **110** 875.
[7] Greitzer E.M, Moore F.K 1985 A theory of post-stall transients in axial compression systems: part II: application *ASME Paper GT-1985-172*.
[8] Camp T.R and Day I.J 1998 A study of spike and modal stall phenomena in a low-speed axial compressor *ASME Journal of Turbomachinery* **120** 393.
[9] Garnier V.H, Epstein A.H and Greitzer E.M 1991 Rotating waves as a stall inception indication in axial compressors *ASME Journal of Turbomachinery* **113** 290.
[10] Marelli S, Carraro C, Moggia S and Capobianco M. 2016 Effect of circuit geometry on steady flow performance of an automotive turbocharger compressor *Energy Procedia* **101** 630.
[11] Bontempo R., Cardone M., Manna M and Vorraro G 2017 A statistical approach to the analysis of the surge phenomenon *Energy* **124** 502.
[12] Andersen J, Lindström F and Westin F 2008 Surge definitions for radial compressors in automotive turbochargers *SAE Int J Engines* **1** 218.
[13] Liu A.X. and Zheng X.Q. 2013 Methods of surge point judgment for compressor experiments *Experimental Thermal and Fluid Science* **51** 204.
[14] Toussaint M, Danlos A and Descombes G. 2014 Analyse du fonctionnement en pompage d'un compresseur centrifuge refoulant à travers une vanne rotative *COFRET'14*, Paris, France.
[15] Galindo J, Serrano J.R, Guardiola C and Cervello C. 2006 Surge limit definition in a specific test bench for the characterization of automotive turbochargers *Experimental Thermal and Fluid Science* **30** 449.
[16] Marelli S, Carraro C, Moggia S and Capobianco M 2016 Effect of circuit geometry on steady flow performance of an automotive turbocharger compressor *Energy Procedia* **101** 630.
[17] Toussaint M, El Kamel C. 2012 Influence of the Network Geometry on the Surge Characteristics of a Centrifugal *Compressor ISROMAC 14 -The 14th International Symposium on Transport Phenomena,* Honolulu, United-States
[18] Deligant M. 2011 Caractérisation numérique et expérimentale des performances d'un turbocompresseur automobile aux bas régimes de rotation. Phd Thesis, Université Paris 6.
[19] Podevin P, Deligant M, Descombes G and Clenci A. 2011 Turbocharger characteristics at low speed and experimental measurements of friction losses *International Congress Automotive and Environment CAR 2011*, Pitesti, Romania.

IOP Conf. Series: Materials Science and Engineering **252** (2017) 012083 doi:10.1088/1757-899X/252/1/012083

Development of a test method for distillation of diesel-biodiesel-alcohols mixtures at reduced pressure

R Niculescu[1], I Iosub[3], A Clenci[1,2], C Zaharia[1], V Iorga-Simăn[1]

[1]University of Pitești, Dep. Automobiles and Transport, 1, Tg. din Vale Street, Pitești, Romania
[2]Le Cnam Paris, 292, rue St. Martin, Paris, France
[3]University of Pitești, Dep. of Natural Science, 1, Tg. din Vale Street, Pitești, Romania

E-mail: rodica.niculescu@upit.ro

Abstract. Increased environmental awareness and depletion of fossil petroleum resources are driving the automotive industry to seek out and use alternative fuels. For instance, the biofuel is a major renewable energy source to supplement declining fossil fuel resources. The addition of alcohols like methanol and ethanol is practical in biodiesel blends due to its miscibility with the pure biodiesel. Alcohols also improve physico-chemical properties of biodiesel blends, which lead to improved combustion efficiency. Proper volatility of fuels is critical to the operation of internal combustion engines with respect to both performance and emissions. Volatility may be characterised by various measurements, the most common of which are vapour pressure, distillation and the vapour/liquid ratio. The presence of ethanol or other oxygenates may affect these properties and, as a result, performance and emissions, as well. However, in the case of diesel-biodiesel-alcohols mixtures, the variance of component volatility makes difficult the analysis of the overall volatility.
Thus, the paper presents an experimental method of distilling diesel-biodiesel-alcohols mixtures by adjusting the boiler pressure of an i-Fischer Dist equipment.

1. Introduction

From the road transport standpoint, today, there are three major problems facing humanity: (1) concerns about greenhouse gas (GHG) emissions and global climate change; (2) a desire for renewable/sustainable energy sources; and (3) an interest in developing domestic and more secure fuel supplies, in other words, energy security. These issues have led to intense debate within international organizations and among political leaders on the impacts of the increased use of biofuel. [1]

The EU objective is to achieve an overall 80%–95% reduction in CO2 emissions by the year 2050, with respect to the 1990 level. Decarbonization of transport and the substitution of oil as transport fuel therefore have both the same time horizon of 2050. Thus, the European "Renewable Energy Directive" 2009/28/EC includes a binding target of 10 percent renewable fuels in transport in 2020 [2- 6].

S. K. Thangavelu et al. in [7] support the idea the most eminent alternative fuels for replacing fossil fuels in internal combustion engines are biofuels (biodiesel and bioethanol).

Also, Mohd Hafizil Mat Yasin et al. concluded in their study [8] that a lot of different researchers around the world revealed that biodiesel is proven as a substitute for mineral diesel and the use of additives such as alcohol is a solution to achieve objectives like reducing the emission significantly.

Ethanol and methanol has higher oxygen in basis (approximately 35% and 30%) as compared to mineral fuels that help internal combustion (IC) engine to achieve higher complete combustion. More oxygen in fuels means more complete combustion to be achieved [9 – 14].

The use of alcohol additives include methanol and ethanol are very practical in the biodiesel blends due to its miscibility with the pure biodiesel. Alcohol additives are very helpful to reduce the viscosity and density of the biodiesel which is higher compared to standard mineral diesel [9].

However, Obed M. Ali et al. say in [15] that these additives have a relatively low flash point and auto-ignition temperature compare to diesel and biodiesel fuel, and it may lead to noticeable increase in engine cycle-to-cycle variations when exceeds a certain limit. These variations might lead to both a reduction of engine output power and higher emissions; so, it is necessary to develop effective strategies to control the optimum additive ratio through gaining a better understanding of the different parameters that influence the overall combustion process.

An experiment was carried out by H.G. Howa et al. [16] on a common rail direct injection (DI) diesel engine with diesel, biodiesel and bioethanol fuel blends (B20E5) on engine performance, emissions and combustion characteristics at different engine load operations. They concluded that, the addition of higher oxygen content of ethanol as additive for biodiesel-diesel blend can be applied in compression ignition (CI) engine without any engine modifications.

Concerning the reduction of greenhouse gas emissions ethanol has proved to have a significant potential from 21% (in the case of using wheat as feedstock) to 75% (in case of sugar cane ethanol) or even to 87% (in case of wheat straw) [6].

Methanol is considered to be one of the favourable fuels for engines. Methanol has been studied in spark ignition (SI) engine application. Due to its physicochemical properties (lower cetane number, higher latent heat, and other) it is more difficult to fuel CI engines, sustain Wang L-J et al. in [11], citing a number of authors who investigated engine combustion characteristics using methanol/diesel blends. Methanol is also gaining interest as a renewable fuel, since its production derived from gasification of waste biomass followed by catalytic synthesis at high pressure provides both high yield (45–55% in weight) and energy efficiency (70–75%). In terms of cost and yield, Juan Hernandez et al. claim in this paper that methanol is better compared to ethanol. [6].

M. Mofijur et al. in their critically reviews show the benefits of using diesel-biodiesel-alcohols (ethanol) such as: reduce NO_X and HC significantly, the peaks of smoke emissions were reduced, PM emission reduced, lowered particle number concentration and particulate mass emission. [4]

The massive use of alcohol-diesel blends is limited by the following factors highlighted by Juan Hernandez et al. in [6]: a) some key properties of the alcohol blends like blending stability, lubricity, viscosity and cold filter plugging point affect the injection system; b) the low cetane number that affects the combustion process is an important limiting factor; c) the fuel storage and handling difficulties derived from the high volatility of alcohols, in this respect, different technical modifications in the fuelling system are necessary, also, to avoid vapour locks there are necessary special systems for recovering of fuel vapours, when increase the alcohol amount.

It is very known that crude oil and the various petroleum fractions and products derived from it are consist of a complex mixture of various components, mostly hydrocarbons with various volatility characteristics. Volatility is defined as the tendency of a material to change from a liquid state to gaseous state. When dealing with petroleum products, the principal and significant volatility characteristics are distillation, vapor pressure, and flammability.

Evaporation characteristics influence the spray structure and affect the preparation of air–fuel mixture under different engine load conditions. Changes of the Reid vapor pressure and distillation properties do not show a linear correlation as a function of the alcohol blend ratio [17].

As we said before, transportation fuels consist of a large number of species that belong to different families of organic compounds. Thus, the usage of detailed evaporation models is restricted by these species and demands an alternative approach. Existing computational approaches introduce challenges in generalizing to realistic fuel blends due to assumptions that are introduced to describe evaporation and composition. Govindaraju et al. in their paper [18], present a review of the concerns of some

IOP Conf. Series: Materials Science and Engineering **252** (2017) 012083 doi:10.1088/1757-899X/252/1/012083

researchers regarding the modeling of evaporation characteristics of transport fuels mixtures. They also propose a model for assessing the evaporation characteristics of multicomponent fuels. Information about physical properties of each compound in the fuel is required to perform calculations using the evaporation model. It is important to note that these physical properties are functions of pressure and temperature and should be determined under standard conditions.

So, for the application of such models it takes complete research infrastructure and a lot of work.

Considering the above, we assert, the determination of experimental volatility characteristics is an alternative option that requires less effort.

However, in the case of diesel-biodiesel-alcohols blends, the variance of component volatility makes it difficult to analyse overall volatility.

In this paper, we have conducted a study in order to develop a method for distillation of diesel-biodiesel-alcohols mixtures at reduced pressure.

2. Equipment description

The i-Fischer Dist D-1160 C distillation system used in our research (figure.1) is produced by Fischer Engineering, Germany and is a fully automatic vacuum distillation apparatus working according to ASTM D-1160. It can be used for the distillation of petroleum products, lubricating oils, palm oil, mixtures such as biodiesel-diesel, at atmospheric pressure and vacuum. It can also be used in an extended range - the system can be operated under atmospheric and vacuum conditions, e.g. pre-distillation at atmospheric pressure and further vacuum distillation, equally, according to ASTM D-1160. Table 1 shows the technical data of the equipment

Table 1. Technical data

Operating parameter	Value
Temperature	max. 400 ^{0}C
Boiling range	max. 600 ^{0}C (equivalent to atmospheric pressure)
Pressure range	0.1 ... 760 mm Hg (Torr)
Standard work pressure	1 mmHg (Torr)
Balloon size	500 ml distillation
Sample	200 ml
Ambient temperature	max. 30 ^{0}C
Power supply	230 V, 50 Hz

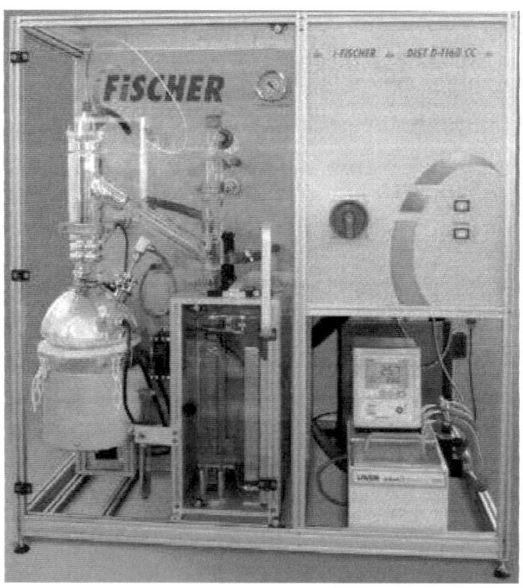

Figure 1. Vacuum distillation unit - i-Fischer Dist D-1160 C

3. D-1160 C Test Method description

For the first time, Distillation Measurement at Reduced Pressure has been described in 1938 by Fenske in a review of that time. This have evolved in a number of standard test methods to determine the distillation characteristics of petroleum products and fractions that would decompose if distilled at atmospheric pressure. ASTM D1160 was first published in 1951. The original version of this test method is similar to the current version, although this standard has undergone revision over time [19].

This boiling range, can be used in engineering calculations to design distillation equipment, to prepare appropriate blends for industrial purposes, to determine compliance with regulatory rules, to determine the suitability of the product as feed to a refining process, or for a host of other purposes. Many engineering design correlations have been developed on data by this test method [20].

ASTM D1160 "Standard Test Method for Distillation of Petroleum Products at Reduced Pressure" is a true analytical method the determination, at reduced pressures, of the range of boiling points for petroleum products that can be partially or completely vaporized at a maximum liquid temperature of 400 °C. D1160 is operated at a fixed distillation rate and for this, the rate of heating is adjusted, so that the distillate is recovered at a uniform rate of 6 to 8 mL/min. Moreover, interruption of the distillation is not permitted; the maximum end point is essentially dictated by the chosen operating pressure and the requirement that the liquid temperature shall not exceed 400 °C. As for any distillation method, the most important parameters having the largest impact on accuracy and precision are vapor temperature and operating pressure. For ASTM D1160, in addition, accurate volume measurement is also an important factor [21].

Temperature. The system must produce readings with an accuracy of ±0.5 °C over the range to 400 °C and have a response time of less than 200 s. The location of the vapour temperature sensor is extremely critical. The type of this sensor is very important, too. The sensor must to be able to respond sufficiently rapidly to changes in temperature such that no significant error due to lag is introduced in a rapidly rising temperature curve. This is very important when working with low pressures and the heat content of the vapours is minimal. However, in ASTM D1160, neither the mass, surface area, nor the heat transfer characteristics of the sensor is defined [21].

Distillation Pressure. Pressure is the second most important parameter to take care of, to assure reasonable reproducibility. D1160 test method is rather vague about the pressure to be applied. The sample is distilled at an accurately controlled pressure between 0.13 and 6.7 kPa (1 and 50 mm Hg). The pressure level also affects the temperature lag due to a rapidly rising vapor temperature in relation to sensor response time, thus introducing more "noise" in precision variance [21].

Volume Measurement. The boiling curve produced by ASTM D1160 is a plot of vapor temperature (AET atmospheric equivalent temperature) versus volume recovered. Volume is a function of temperature. Therefore, it is important that the intake volume measured (200 mL) has the same temperature as the distillate collected in the receiver. More, the accurate calibration and reading of the distillate receiver have a significant bearing on precision of ASTM D1160 results [21].

Accuracy. Distillation curves by D1160 will NOT match when they are obtained under different pressure conditions, or with different temperature measuring devices, or using D86 (atmospheric distillation method) [21].

4. Results and discussion

The mixtures for which a distillation method at reduced pressure was created are: B20M10 (diesel 70% + biodiesel 20% + methanol 10%) și B20E10 (diesel 70% + biodiesel 20% + ethanol 10%). The fuel samples were prepared as blended fuels: 70% diesel + 20% biodiesel + 10% ethanol / methanol, by volume, and strongly mixed manually for about 5 minutes before analyzing.

As mentioned in the previous paragraph, the distillation curves by D1160, obtained under different pressure conditions, will NOT match. So, to be able to compare the volatility of the two mixtures it is necessary that the determination is made in identical conditions. Table 2 presents these identical measurement parameters. As far as it was possible, we took care not to get out of the limits of the method. Figure 2 shows the notations of the set parameters.

IOP Conf. Series: Materials Science and Engineering **252** (2017) 012083 doi:10.1088/1757-899X/252/1/012083

Table 2. Set parameters of the method

Set Parameters	Symbol	UM	Value	Observation
Vacuum set point	Vacuum	[torr]	600	
Minimum temperature for the thermostatic bath (TH1)	T_{min}	[°C]	40	
Maximum temperature for the thermostatic bath (TH1)	T_{max}	[°C]	90	
Differential temperature between the head temperature (ACT*) and the temperature of the thermostatic bath (TH1)	T_{delta}	[°C]	120	According to this preset the temperature of the thermostatic bath is controlled.
Distillation rate	-	[ml/min]	7	According to this value the sump heater is controlled
Temperature of pre-heating the charge	$T_{preheating}$	[°C]	50	After starting a run, the charge is pre-heated until this temperature is reached
Time to reach the vacuum	Vacuum - Timeout	[sec]	1800	If the vacuum is not reached within this time, distillation is stopped
Maximum temperature difference between the sump heater $T_{fl\text{-}heat}$ and the charge T_{flask} during heating up phase	$dT_{max.\text{-}heat\text{-}flask}$	[°C]	200	The sump heater will be automatically switched off if the value is above this limit
The heating constant power of the sump heater	Start power	%	20	
Power after trigger		%	20	After the $T_{trigger}$ is reached the sump heater is heating with this power
The temperature of the receiver to have the exact volume and weight.	$T_{receiver}$	[°C]	60	After entering the density and the corresponding temperature of the receiver the weight of the charge with 200 ml get to calculate.
Gradient to detect $T_{trigger}$	$T_{trigger}$ - gradient	[°C/sec]	0.2	If the rise of the temperature of the vapours $T_{trigger}$ is more than this value per second the trigger is detected

**ACT – Actual Temperature (vapor temperature measured at operating pressure)*

To choose the parameters of the method, it was necessary to consider the very different volatility of the blends components. Applying the method with the determined parameters resulted in the distillation curves presented in fig. 3 and 4. As seen, three measurements were made for each type of fuel. The results are a clear indication of the repeatability of the experiment.

We wanted to verify the composition of the samples on Mid - FTIR Spectrometer, but difficulties were encountered. The equipment manufacturer's explanations were: *"The alcohols interfere with the FAME determination for FAME concentrations higher than about 6.5 vol %. So, in this case, it seems to be an interaction between FAME and alcohols such that this region is distorted, and then the normal calibration for FAME is not valid anymore. So, a new method is needed depending on the alcohol concentration, and that would have a completely new way to calculate the FAME concentration".*

We had to make this point because on the distillation curves obtained (fig. 3 and 4), the plateau believed to be the boiling area of the alcohol is more than 10 percent recovered. This could also be a cause of a volumetric measurement error.

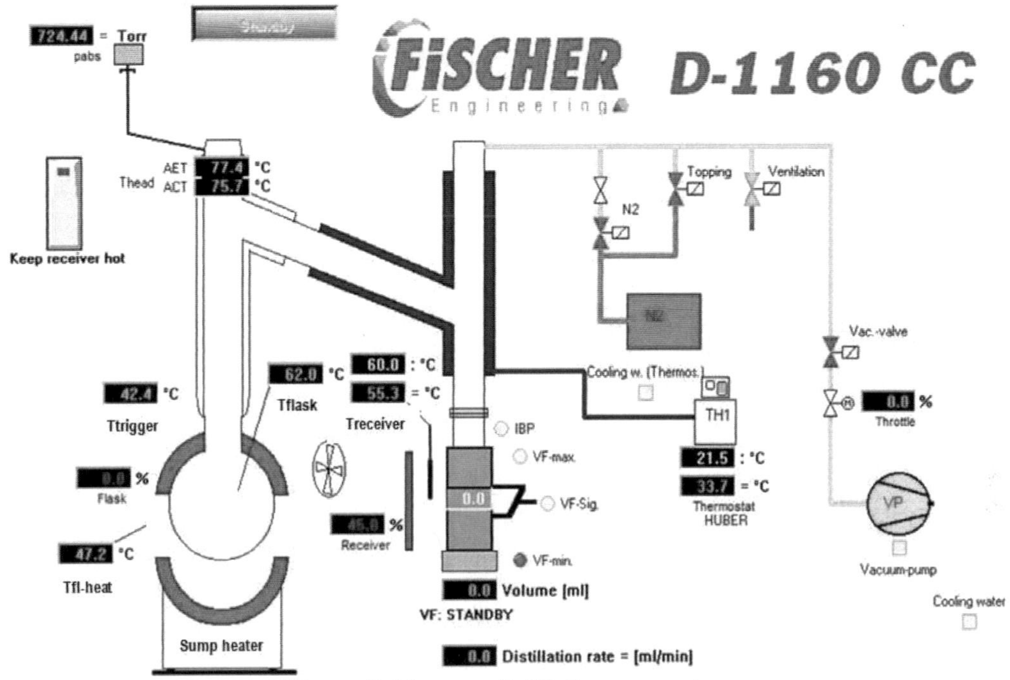

Figure 2. Vacuum distillation parameters

There could be other causes we need to talk about. It is noticed that the stability of samples is affected by three factors mainly: temperature, water content and initial alcohol content.

Concerning the stability of the B20M10 mixture, the following remarks are given:

- biodiesel is relatively difficult to be fully described regarding its composition;
- it contains mainly a mixture of esters (obtained by a transesterification reaction);
- in non-polar compounds, it is relatively stable;
- with the addition of relatively polar compounds such as the methanol (with a molecular dipole moment of 1.7 Debye) or the ethanol (with a molecular dipole moment of 1.66 Debye) interactions and decomposition can occur;
- thus, mixtures of alcohols and fatty acids can be formed;
- the steady state of esters is adversely affected by methanol and by traces of water, (see the Le Chatelier principle);
- moreover, methanol and ethanol may have a relative reactivity in the presence of certain unsaturated component from biodiesel;
- in conclusion, the final composition of the mixture may be affected.

Concerning the stability of the B20E10 mixture, the following remarks are provided in the addition of what was given above:

- the main drawback is that ethanol is immiscible with diesel fuel over a wide range of temperatures, leading to phase separation [22];
- but, esters, biodiesel could be an additive that help to increase the stability of ethanol-diesel blends [9,23];

- so, ethanol dissolves without problem in B20.

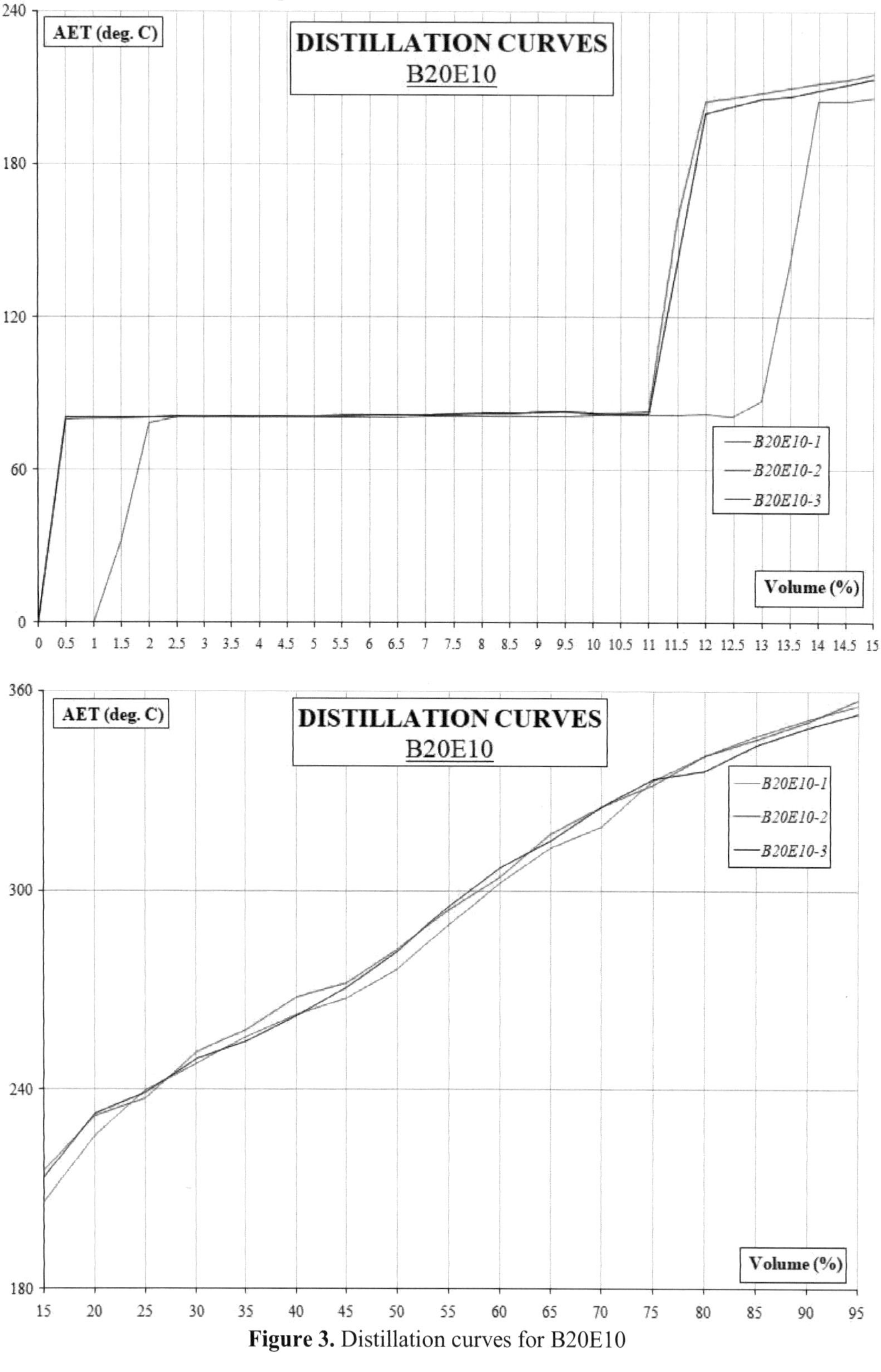

Figure 3. Distillation curves for B20E10

IOP Conf. Series: Materials Science and Engineering **252** (2017) 012083 doi:10.1088/1757-899X/252/1/012083

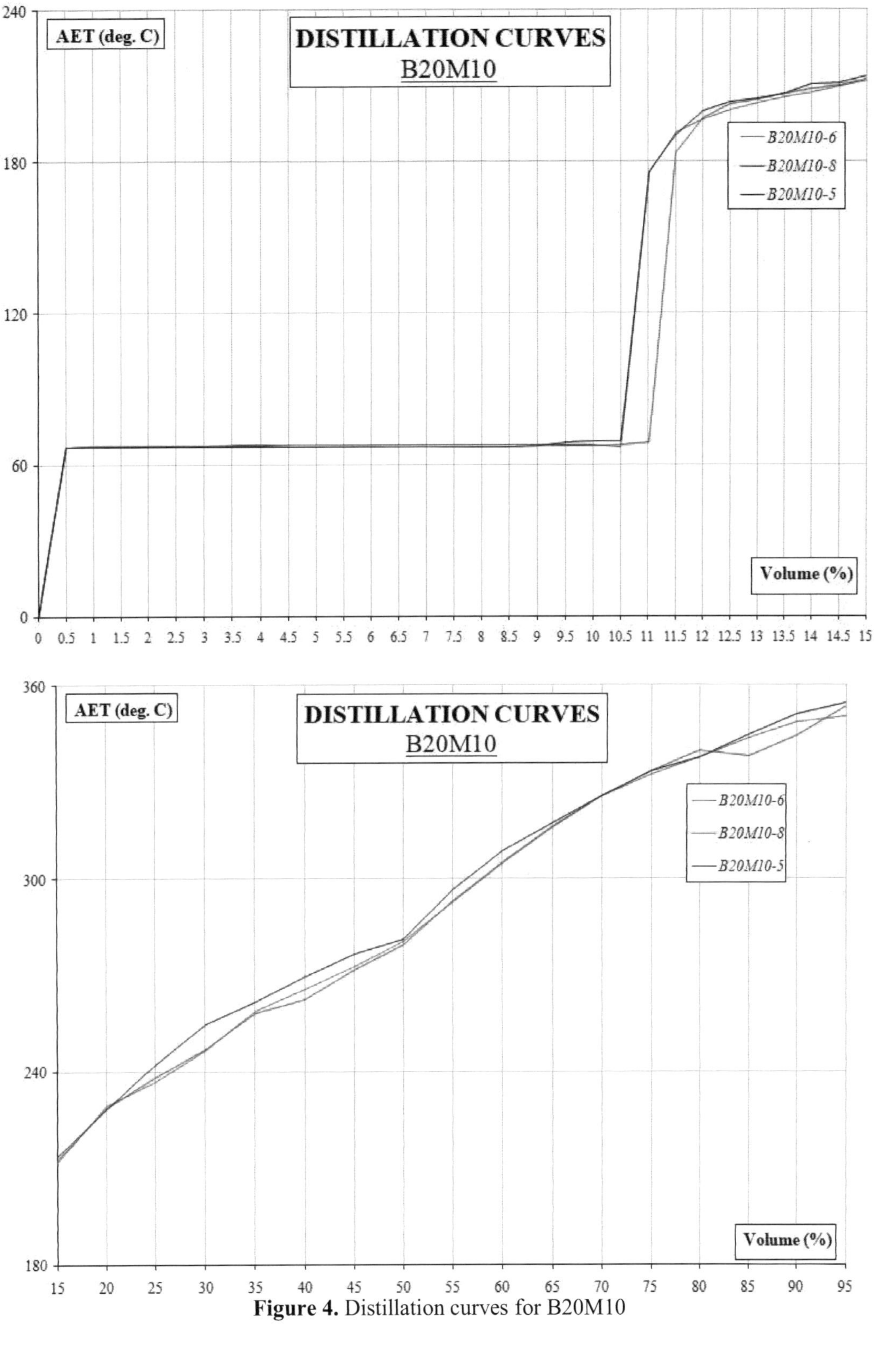

Figure 4. Distillation curves for B20M10

However, in fig. 3 it is observed that the curve B20E10-1 is slightly more distant than the other two. This could be explained by the fact that it was the first determination made with this method, and the device responded slower, given the conditions between different parameters (tab.2).

Finally, it can be seen that, in both cases, the values obtained are very close. This justifies saying that the parameters set for determining the distillation curve are appropriate.

5. Summary and conclusions

General results concluded from this study can be summarized as follows:

- The significance of using alternative fuels can be attributed to the following aspects: (1) pursuing energy sustainability through the extended usage of those alternative fuels derived from renewable energy sources; (2) improving engine efficiency and engine-out emissions; (3) relieving the unbalanced usage of fossil fuels.
- The use of alcohol additives (e.g., methanol and ethanol) in the biodiesel blends is very practical due to its miscibility with the pure biodiesel. Equally, they reduce viscosity and density of the resulted blend.
- Biodiesel is relatively difficult to be fully described regarding its composition. With the addition of relatively polar compounds such as methanol or ethanol, interactions and decomposition can occur. Moreover, methanol and ethanol could have a relative reactivity in the presence of certain unsaturated component from biodiesel.
- Determination of physico-chemical properties of these mixtures could be difficult and new analyzing methods are required.
- Evaporation characteristics influence the spray structure and affect the preparation of air–fuel mixture. Therefore, the more accurate knowledge of the fuel volatility property is necessary.
- For the distillation of substances with high boiling temperatures at atmospheric pressure, like biodiesel, distillation at low pressures is needed because high temperatures may change its chemical structure.
- ASTM D1160 is a true analytical method of determination at reduced pressures of the range of boiling points for petroleum products that can be partially or completely vaporized at a maximum liquid temperature of 400°C.
- Distillation curves by D1160, obtained under different pressure conditions, will NOT match!
- To be able to compare the volatility of two or more different fuels it is necessary that the determination is made in identical conditions. This results in the need to find appropriate analysis parameters.
- The method presented in this paper can be applied successfully for the two blends: B20M10 (diesel 70% + biodiesel 20% + methanol 10%) şi B20E10 (diesel 70% + biodiesel 20% + ethanol 10%).
- In the future, analyzes should be made in order to see if this method is applicable for higher percentage of methanol or ethanol, but also in case of other alcohols such as, butanol and isobutanol.

Acknowledgements

The authors are grateful to Compact Industrial company for its help with using the equipment to conduct experiments and for mediating communication with the equipment's manufacturer.

Author Contributions: The authors have equal contribution to the work presented in this article.

Abbreviations

ACT	Actual Temperature (vapor temperature measured at operating pressure)	FTIR	Fourier Transform Infrared Spectroscopy
AET	Atmospheric Equivalent Temperature	GHG	GreenHouse Gas
ASTM	American Society for Testing and Materials	HC	Hydrocarbons
B20	Biodiesel20% + Diesel80% blend	IC	Internal Combustion
B20E5	Diesel 75% Biodiesel 20% Ethanol 5% Blend	NO_x	Nitric Oxides
B20E10	Diesel 70% Biodiesel 20% Ethanol 10% Blend	PM	Particulate Matter
B20M10	Diesel 70% Biodiesel 20% Methanol 10% Blend	SI	Spark Ignition
CI	Compression Ignition	T10/T50/T90(°C)	Temperatures at which 10/50/90 % of the fuel boiled
CO_2	Carbon Dioxide	TH1	Thermostatic Bath
DI	Direct Injection	$T_{fl\text{-}heat}$	Sump Heater Temperature
EC	European Commission	T_{flask}	Charge Temperature
EU	European Union	$T_{trigger}$	Temperature of The Vapours
FAME	Fatty Acid Methyl Ester		

6. References

[1] Niculescu R, Clenci A, Iorga-Simăn V and Zaharia C 2016 Review on the use of bioethanol /bio methanol – gasoline blends in spark ignition engine *SCIENTIFIC BULLETIN Automotive Series University of Pitesti* **26** 6-19

[2] Clenci A, Niculescu R, Danlos A, Iorga-Siman V and Trica A Impact of Biodiesel Blends and Di-Ethyl-Ether on the Cold Starting Performance of a Compression Ignition Engine 2016 *Energies* **9** 284

[3] Renzi M, Bietresato M and Mazzetto F 2016 An experimental evaluation of the performance of a SI internal combustion engine for agricultural purposes fuelled with different bioethanol blends *Energy* **115** 1069-1080

[4] Mofijur M, Rasul M.G and Hyde J 2015 Recent Developments on Internal Combustion Engine Performance and Emissions Fuelled With Biodiesel-Diesel-Ethanol Blends *Procedia Engineering* **105** 658 – 664

[5] Tutak W, Jamrozik A, Pyrc M and Sobiepański M 2016 Investigation on combustion process and emissions characteristic in direct injection diesel engine powered by wet ethanol using blend mode *Fuel Processing Technology* **149** 86–95

[6] Hernández J, Lapuerta M, García-Contreras R and Agudelo J 2016 Modelling of evaporative losses in n-alcohol/diesel fuel blends *Applied Thermal Engineering* **102** 302–310

[7] Thangavelu S. K, Ahmed A. S and Ani F. N 2016 Review on bioethanol as alternative fuel for spark ignition engines *Renewable and Sustainable Energy Reviews* **56** 820–835

[8] Yasin M. H. M, Mamat R, Yusop A F, Rahim R, Aziz A and Shahb L A 2013 Fuel Physical Characteristics of Biodiesel Blend Fuels with Alcohol as Additives *Civil and Chemical Engineering, Procedia Engineering* **53** 701 – 706

[9] Yasin M. H. M, Mamat R, Yusop A. F, Aziz A, Najafi G 2015 Comparative study on biodiesel-methanol-diesel low proportion blends operating with a diesel engine *Energy Procedia* **75** 10 – 16

[10] Li W, Ren Y, Wang X-B, Miao H, Jiang D-M and Huang Z-H 2008 Combustion characteristics of a compression ignition engine fuelled with diesel–ethanol blends *JAUTO496 F IMechE Proc. IMechE* **222** (Part D: J. Automobile Engineering)

[11] Wang L-J, Song R-Z, Zou H-B, Liu S-H and Zhou L-B 2008 Study on combustion characteristics of a methanol–diesel dual-fuel compression ignition engine *JAUTO656 F Proc. IMechE* **222** (Part D: J. Automobile Engineering)

[12] Hasimoglu C 2008 Exhaust emission characteristics of a low-heat-rejection diesel engine fuelled with 10 percent ethanol and 90 percent diesel fuel mixture *JAUTO633 F IMechE Proc. IMechE* **222** (Part D: J. Automobile Engineering)

[13] Datta Bharadwaz Y, Govinda Rao B, Dharma Rao V and Anusha C 2016 Improvement of biodiesel methanol blends performance in a variable compression ratio engine, using response surface methodology *Alexandria Engineering Journal* **55** 1201–1209

[14] Umar Soni D and Gupta R 2016 Numerical investigation of emission reduction techniques applied on methanol blended diesel engine *Alexandria Engineering Journal* **55** 1867–1879

[15] Ali O M, Abdullah N R, Mamat R and Abdullah A A 2015 Comparison Of The Effect Of Different Alcohol Additives With Blended Fuel On Cyclic Variation In Diesel Engine *Energy Procedia* **75** 2357 – 2362

[16] How H G, Masjuki H H,. Kalam M A and Teoh Y H 2014 Engine performance, emission and combustion characteristics of a common-rail diesel engine fuelled with bioethanol as a fuel additive in coconut oil biodiesel blends *Energy Procedia* **61** 1655 – 1659 (ICAE2014)

[17] Choongsik B and Jaeheun K 2016 Alternative fuels for internal combustion engines *Proceedings of the Combustion Institute* **000** 1–25

[18] Govindaraju P and Ihme M 2016 Group contribution method for multicomponent evaporation with application to transportation fuels *International Journal of Heat and Mass Transfer* **102** 833–845

[19] Kishore Nadkarni R A Guide to Astm Test Methods for the Analysis of Petroleum Products and Lubricants, Second Edition. ASTM Stock No. MNL44–2nd ASTM International 100 Barr Harbor Drive PO Box C700 West Conshohocken, PA 19428-2959

[20] Designation: D 1160 – 06 Standard Test Method for Distillation of Petroleum Products at Reduced Pressure

[21] Montemayor R G Distillation and Vapor Pressure Measurement in Petroleum Products, , editor, ASTM International, 100 Barr Harbor Drive, PO Box C700, West Conshohocken, PA 19428–2959 Printed in the U.S.A. ASTM Stock Number: MNL51

[22] Lapuerta M, Armas O and Garcia-Contreras R 2007 Stability of diesel–bioethanol blends for use in diesel engines *Fuel* **86** 1351–1357

[23] Burnete N.V, Filip N and Barabas I 2015 Diesel-Ethanol Blends and their use in Diesel Engines *Ingineria automobilului*, **36** 12-20

IOP Conf. Series: Materials Science and Engineering **252** (2017) 012084 doi:10.1088/1757-899X/252/1/012084

Assessment of real driving emissions via portable emission measurement system

*A Clenci[1,2], V Sălan[3], R Niculescu[1], V Iorga-Simăn[1], C Zaharia[1],

[1]University of Pitești, 1, Tg. din Vale street, Pitești, 110040, Romania,
[2]Le Cnam Paris, 292, St. Martin street, Paris, 75003, France,
[3]Renault Technologie Roumanie, Voluntari, Pipera, 077190, Romania

E-mail: *adrian.clenci@upit.ro

Abstract. The European Commission approved a so-called Real Driving Emission (RDE) test in response to the criticisms to the current driving cycle used at chassis dyno for homologation purpose (NEDC): it is considered outdated and misleading since air pollutants in real driving conditions are considerably higher than the certification thresholds. So, what's at stake is the air quality which degraded continuously despite the ever-increasing severity of the regulations during the last almost three decades. Thus, from September 2017, the RDE test will become part of the type approval process for all cars sold in Europe. As its name points out, it will include "real world driving" using a portable emissions measurement system (PEMS). The paper presents the RDE features (PEMS mounting, testing environment, boundary conditions, driving dynamics) and presents a case study on the influence of the driving style upon the tail-pipe emissions under the RDE testing. The results presented in the paper issued from the existing cooperation on this topic between University of Pitesti and Renault Technologie Roumanie

1. Introduction and literature review

Despite the ever-increasing severity of the European regulations regarding automotive tail-pipe pollution and CO_2 emission, real driving emissions (RDE) barely changed over the last almost three decades [1]. One reason is that till now, the European passenger car certification procedure was only about testing the vehicle in lab condition using a non-representative driving cycle. Indeed, the New European Driving Cycle (NEDC) which was long used for the homologation purpose, was finally declared outdated and misleading since it consists in a driving style rather smooth (moderate and uniform accelerations and decelerations, constant cruising speeds and many idling events), which isn't to be met in real roads. Moreover, NEDC doesn't account neither for other factors related to the real world such as: the use of air conditioning, vehicle accessories, reduced tyre pressure, [2]. Thus, these result in fuel consumption, CO_2 emissions and pollution which are non-representative of actual real-world performance. There are many studies raising the problem of certification procedure's representativeness to the real-world driving [3–8]. In fact, the automotive manufacturers are simply developing technical solutions and testing them to the letter of the law. Few of them step further because of the involved costs, which could be a commercial disadvantage on the market. Consequently, this raises questions on the real progress of EU regarding its commitment to achieve at least 20% reduction of greenhouse gas (GHG) emissions by 2020 (compared to 1990 levels) [9].

However, the problem of GHG and pollution is a complex one since, currently, there is a lack of homogeneity in worldwide type-approval procedures: each major geographic area with its own style of regulations as if the Earth atmosphere shouldn't be treated as a whole, belonging to each and every one of us. On this matter, however, progress was made as in 2009 the World Forum for the Harmonization of Vehicle Regulations of the United Nations Economic Commission for Europe (UNECE), through its working party on pollution and energy (GRPE), launched a project with the aim to develop a worldwide harmonized light duty test cycle (WLTC) and test procedure (WLTP) [10]. Two working groups were established; the first group in charge of the development of harmonized cycle (DHC) and the second group working on development of test procedures (DTP). The Joint Research Centre (JRC) of the EC has been deeply involved in both groups, participating to the design, validation and revision of the new test cycle and in developing, testing and adjusting the new test procedures [7]. Figure 1 shows a comparison with former homologation driving cycle, NEDC. A quick analysis reveals that WLTC is longer in time and distance and features a more dynamic speed profile, which means it is more representative for the real world. So, now that WLTP exists, what remains to be done is that it is undertaken at global scale.

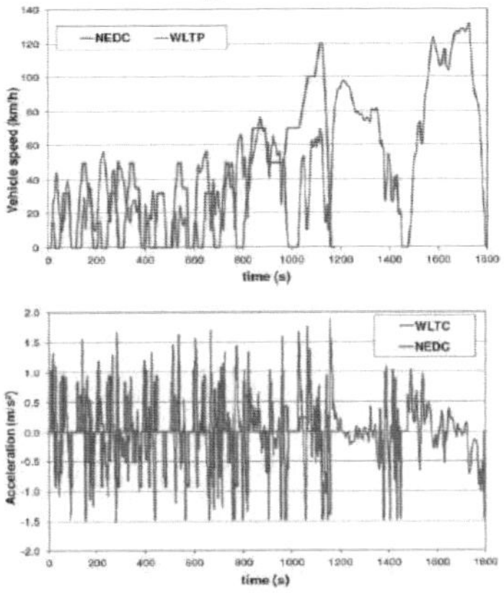

Criteria	NEDC	WLTC
Distance [km]	11.013	23.141
Duration [s]	1180	1800
Average speed [km/h]	33.6	46.3
Max. speed [km/h]	120	131
Max. acceleration [m/s^2]	1.04	1.88
Max. deceleration [m/s^2]	-1.39	-1.52
Standstill time [s]	280	227
Standstill proportion [%]	23.7	12.6
Constant cruise driving proportion [%]	39.3	3.4
Acceleration & deceleration proportion [%]	36	84

Figure 1. NEDC vs. WLTC

Paper [7] dealt with the effect of WLTC on CO_2 and regulated pollution gases with respect to the ones issued from NEDC. The study was performed on Euro 4, 5 and 6 production vehicles. Thus, surprisingly, CO_2 emission in WLTC is comparable with the one from NEDC (a bit lower in some cases, a bit higher in other cases). The authors explained it as follows: firstly, although WLTC is more demanding, however, the cold start emissions have a smaller impact due to its longer duration (the estimated contribution of cold start emission over NEDC is between 50-60% higher than for WLTC); secondly, because of the higher dynamic of WLTC, the engine is run at higher loads, thus resulting in higher average efficiency. On the matter of regulated pollutants, the same authors drew the following conclusions: "for those pollutants whose emissions are mainly due to the cold start phase of the cycle (e.g. THC for gasoline and diesel vehicles and CO for diesel vehicles), moving from NEDC to WLTC brings an improvement due to the lower weight of the cold start emissions on the whole cycle. On the other hand, the pollutants that are emitted also during high engine load conditions (e.g. NOx on diesel vehicles and CO on gasoline vehicles), the change from NEDC to WLTC can lead to substantial emission increase". Thus, the introduction of WLTC is expected to reduce the existing gap between overall emissions obtained from NEDC and the RDE [6,7].

However, in order to deal with high on-road emissions from passenger cars, the European Commission created also the "working group for the development of a RDE test procedure for light-duty vehicles (RDE-LDV)" in 2011. The RDE-LDV is chaired by the EC and supported by the JRC of the EC and technical experts from the member states, industry and non-governmental organizations (NGO) [1]. As explained in [1], the RDE test procedure has been, for practical reasons, split into several packages. The **first RDE package** was voted positively by the Member States in the Technical Committee of Motor Vehicles (TCMV) in May 2015 and has entered into force since March 2016 (see the EU regulation 2016/427, [11]). In the RDE procedure, pollutant emissions will be measured by portable emission measuring systems (PEMS) that will be attached to the car while driving in real conditions on road. This means that the car will be driven outside and on a real road featuring random acceleration and deceleration patterns. This package deals with the technical requirements for PEMS testing. In an initial phase that started in 2016, the RDE testing will only be done for monitoring purposes, without an impact on the actual type approval, which will continue to be delivered based on the laboratory measurements. On 28th of October 2015, Member States meeting in the TCMV voted by a large majority for **the second RDE package**, which was necessary for the RDE tests to have an actual impact on type approvals issued by national authorities (see the EU regulation 2016/646, [12]). From 1st of September 2017, the new RDE tests will determine whether a new car model can be put on the market. For that, a conformity factor (CF) is set for all new vehicles at 2.1 for NO_x and at 1.5 for PN. A second stage CF is equal to 1.0, with a margin parameter that considers the additional measurement uncertainties introduced by the PEMS equipment. The margin parameter is currently set at 0.5. The latter is subject to an annual review and shall be revised as a result of the improved quality of the PEMS procedure or technical progress. **The third RDE package** will define a procedure for the measurement of the particulates and include the effect of vehicle cold starts into the RDE testing. In addition, the Commission will oblige manufacturers to publish the CFs of an individual vehicle in its certificate of conformity (CoC). In this way, the consumer will have full transparency on the real emission performance. In **the fourth RDE package,** the EC will define the rules for independent RDE testing of vehicles being in-service, including the regulatory consequences in case of non-conformity.

On the matter of in-use regulation, it has to be said that they first appeared in US. Having in mind the need to be sure that emission standards are maintained throughout the course of the vehicle's useful lifetime, the US Environmental Protection Agency (EPA) and California Air Resources Board (CARB) have implemented in 2005 a series of regulations to control NOx and PM emissions from heavy-duty diesel trucks during in-use conditions [13]. One of the most important US regulations with respect to controlling in-use emissions is Not-To-Exceed (NTE) standard, which sets limits for gaseous and PM pollutants emitted during operation in a defined portion of the engine map and specifies the protocols required to make those measurements. Emission results from this test procedure must be less than or equal to 1.25 times the pre-existing Federal Test Procedure (FTP) standards. The NTE test and associated emission limits are effective starting with 2007 model year [14].

This being the context, the paper deals with RDE features (PEMS mounting, testing environment, boundary conditions, driving dynamics) and presents a case study on the influence of the driving style upon the tail-pipe emissions under the RDE testing. The results presented in the paper issued from the existing cooperation on this topic between University of Pitesti and Renault Technologie Roumanie.

2. Real driving emissions testing. An extract from the regulations

According to Commission regulation (EU) 2016/427 [11], the RDE means "the emissions of a vehicle under its normal conditions of use". Moreover, in the same regulation it is stated that throughout its normal life, the emissions of a vehicle type-approved according to Regulation (EC) 2007/715 [15] determined and emitted at a RDE test performed in accordance to the current requirements [11], shall not be higher than the following NTE values:

$$NTE_{pollutant} = CF_{pollutant} \cdot TF(p_1, p_2, \dots) \cdot EURO6, \qquad (1)$$

where EURO6 is the applicable limit from the regulation [11] and $TF(p_1, p_2, \dots)$ is a transfer function depending on different parameters (e.g. ambient temperature and humidity, etc).

Amongst the tail-pipe emissions, only CO_2, CO, NO, NO_2 and PN are mandatory to measure in RDE testing. HC measurement by FID was not considered because of safety reasons. However, currently, CFs are imposed only for NOx and PN (see table 1).

Concerning the boundary conditions (vehicle payload and test mass; ambient conditions; dynamic conditions; vehicle condition and operation), trip and operational requirements, they all are stated in [15]. Subsequently, a briefing of all these is given. First, table 1 presents a synopsis of the RDE features. In this table:

- Euro 6d-TEMP is the full Euro6 emission requirements assessed on regulatory lab test cycle and RDE testing against temporary CF, [15];
- according to the regulation, by way of derogation the lower temperature for moderate conditions shall be greater or equal to 3°C and the lower temperature for extended conditions shall be greater or equal to -2°C between the start of the application of binding NTE emission limits as defined in section

2.1 and until five years after the dates given in paragraphs 4 and 5 of Article 10 of Regulation (EC) No 715/2007, [15].

- as mentioned in [11], the cold start period covers the first 5 minutes after initial start of the combustion engine. If the coolant temperature can be reliably determined, this period ends once the coolant has reached 70 °C for the first time but no later than 5 min after initial engine start.
- if during a particular time interval during the RDE test, the ambient conditions are extended, the emissions during this period shall be divided by a value *ext* before being evaluated for compliance with the requirements [11].

Table 1. RDE features according to the regulations, [11,15]

Test	Application date	CF NO_x	CF PN	Moderate conditions		Extended Conditions		Cold start?
				Altitude [m]	Temperature [°C]	Altitude [m]	Temperature [°C]	
RDE monitoring	April 2016	w/o	w/o	0..700	0..30	700..1300	-7..0 or 30..35	No *Recorded but excluded from the emissions evaluation*
Euro 6d-TEMP	September 2017	2.1	1.5		3..30		-2..0 or 30..35	Yes
Euro 6d	January 2020	1.5	1.5		0..30		-7..0 or 30..35	Yes

The RDE test also means requirements for the trip to be used to measure the emissions of a vehicle under its normal conditions of use. Table 2 shows a synopsis of the RDE trip requirements, which consists in 3 parts that must be run continuously. Rural operation may be interrupted by short periods of urban operation when driving through urban areas. Highway operation may be also interrupted by short periods of urban or rural operation. Certainly, local speed limits remain in force at a PEMS test, though their violations per se do not invalidate the results of a PEMS test. Urban operation must contain several stop periods of min. 10 s whose share out of the urban time duration must be at account for 6-30%. The inclusion of one excessively long stop period that individually comprises >80 % of the total stop time of urban operation shall be avoided.

Table 2. RDE trip requirements, [11,12]

Detail \ Type	Urban	Rural or Extra-urban	Motorway or High-way
Trip sequence	1	2	3
Share [%] ± 10%	~ 34 *(but in any case, > 29 %)*	~ 33	~ 33
Speeds [km/h]	< 60	60..90	90..145 *(> 100 for at least 5 min)*
Average speed [km/h]	15..40	-	-
Stop periods share [%]	6..30	-	-
One stop period [s]	≥ 10	-	-
Min. distance of each trip [km]	16		
Trip duration [min]	90..120		
Difference in altitude between start and stop points [m]	≤ 100		

According to [11], the RDE test for the passenger cars or light-duty vehicles is to be carried out with a PEMS composed of the following components: gas analyzers for CO_2, CO, NO_2, NO concentration in the exhaust gas (in ppm or %); PN counter from the exhaust gas; exhaust flow mass or volume (EFM, EFV); global positioning system (GPS) to determine the position, altitude and speed of the vehicle; weather station for ambient temperature and relative humidity (RH); an engine control unit (ECU) connection for the acquisition of engine related parameters; an energy source independent of the vehicle to power the whole PEMS. Figure 2 shows the whole layout of the PEMS belonging to University of Pitesti, which was used for this study.

IOP Conf. Series: Materials Science and Engineering **252** (2017) 012084 doi:10.1088/1757-899X/252/1/012084

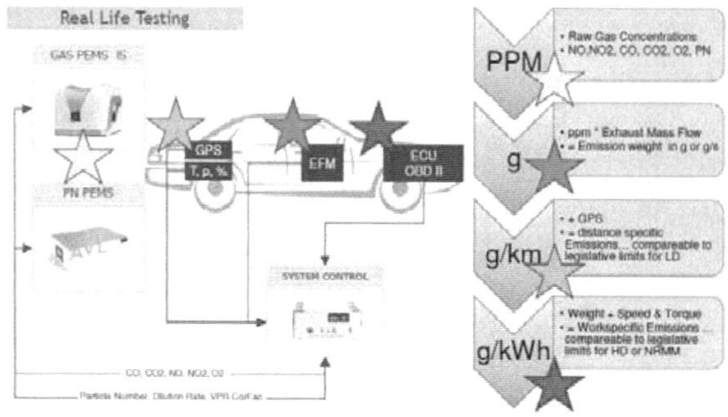

Figure 2. AVL's M.O.V.E. PEMS for RDE testing

In accordance to what figure 2 indicates, below are given the relations needed to transform the emissions from concentration (% or ppm) in distance-specific (DS) emissions, i.e., (g/km) or (#/km):

$$CO_2\left[\frac{g}{km}\right] = 10^3 \cdot \frac{CO_2[\%] \cdot 10^{-2} \cdot EFM\left[\frac{kg}{h}\right]}{v\left[\frac{km}{h}\right]} \tag{2}$$

$$CO\left[\frac{g}{km}\right] = 10^3 \cdot \frac{CO[\%] \cdot 10^{-2} \cdot EFM\left[\frac{kg}{h}\right]}{v\left[\frac{km}{h}\right]} \tag{3}$$

$$NO + NO_2 = NO_x\left[\frac{g}{km}\right] = 10^3 \cdot \frac{NO_x[ppm] \cdot 10^{-6} \cdot EFM\left[\frac{kg}{h}\right]}{v\left[\frac{km}{h}\right]} \tag{4}$$

$$PN\left[\frac{\#}{km}\right] = \frac{PN\left[\frac{\#}{m^3}\right] \cdot EFV\left[\frac{m^3}{h}\right]}{v\left[\frac{km}{h}\right]} \tag{5}$$

The techniques of measuring for each gas analyzer are usually non-dispersive ultraviolet (NDUV) for NO and NO_2 and non-dispersive infrared (NDIR) for CO and CO_2. RDE test parameters provided by the PEMS are acquired at a constant frequency of 1 Hz or higher. The installation of the PEMS shall be done in a way to influence the vehicle emissions and performance to the minimum extent possible. Usually, the PEMS may be mounted either outside the vehicle on a bike carrier attached on a usual hook at the vehicle or inside the vehicle (figure 3).

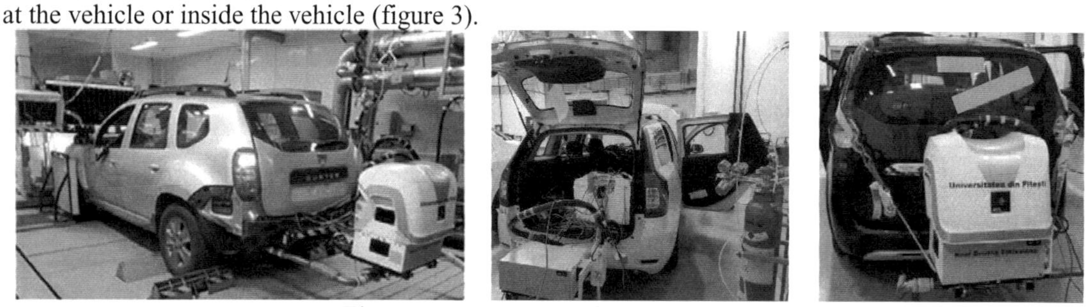

Figure 3. PEMS mounting on vehicles

Every RDE test consists of the following:

- pre-test procedure:
  - *leak-check, i.e., when the probe is disconnected from the exhaust system with the end plugged and with the analyser pump switched on, after an initial stabilisation period, all flow meters shall read nearly zero in the absence of a leak; if not, the sampling lines shall be checked and the fault corrected,*
  - *starting and stabilising the PEMS,*
  - *calibration of the gas analysers, i.e., zero and span calibration adjustments of the analysers shall be performed using calibration gases; the calibration gases shall be chosen to match the range of emissions concentrations expected during the emissions test.*

- emissions test procedure:
  - *sampling, measurement and recording of parameters shall begin prior to the start of the engine and continue throughout the on-road test of the vehicle,*
  - *the end of the test is reached when the vehicle completed the trip and the engine is switched off. The data recording shall continue until the response time of the sampling systems has elapsed.*
- post-test procedure:
  - *recalibration of gas analysers, i.e., the zero and span of the analysers of gaseous components shall be checked again by using calibration gases identical to the ones applied when performing the pre-test procedure; the purpose is to evaluate the analyser response drift compared to the pre-test calibration.*

The linearity of gas analysers, PN counter, flow-measuring instruments is to be performed at least every 3 months and shall comply with the requirements presented in [11]. Equally, it is recommended to validate the installed PEMS once for each PEMS-vehicle combination either before the test or after the completion of an on-road test. This is to be conducted on a chassis dynamometer using the WLTP and having to respect the permissible tolerances given in [11]. A validation procedure for the EFM is also provided in [11].

After establishing the validation of an RDE test according to criteria presented above (measuring and trip criteria), there are still two other conditions to be fulfilled in order to obtain a final validation of the RDE test: driving dynamics via data pre- and post-processing. The latter is either EMROAD moving average windows (MAW) or CLEAR power binning, [11,12]. On the first way to check the driving dynamics, some helpful references may be found in [16], which are also explained in [12]:

- relative positive acceleration per speed bin used to distinguish the soft driving:

$$RPA_k\left[\frac{m}{s^2},\frac{kW\cdot s}{kg\cdot km}\right]=\frac{\sum_j\left[\Delta t\cdot\left(v\cdot a_{pos}\right)_{j,k}\right]}{\sum_i d_{i,k}}, j=1\ to\ M_k, i=1\ to\ N_k, k=u,r,m, \tag{6}$$

where:
RPA_k is the relative positive acceleration for urban (*u*), rural (*r*) and motorway (*m*),
Δt, time difference equal to 1 second,
M_k the sample number for *u*, *r* and *m*,
N_k the total sample number for *u*, *r* and *m*.

- product between actual vehicle speed and positive acceleration greater than 0.1 m/s^2 in time step *j*, considering the *u*, *r* and *m* shares

$$\left(v\cdot a_{pos}\right)_{j,k}\left[\frac{m^2}{s^3},\frac{W}{kg}\right] \tag{7}$$

- 95th percentile of the product between actual vehicle speed and positive acceleration greater than 0.1 m/s^2 in time step *j*, considering the *u*, *r* and *m* shares

$$\left(v\cdot a_{pos}\right)_{k\_[95]}\left[\frac{m^2}{s^3},\frac{W}{kg}\right]. \tag{8}$$

The severity of the driving is judged by using $\left(v\cdot a_{pos}\right)_{k_[95]}$parameter, while the soft driving is distinguished through RPA_k:

$$\text{If}\quad \begin{aligned}&\bar{v}_k\le 74.6\frac{km}{h}\ and\ \left(v\cdot a_{pos}\right)_{k\_[95]}>(0.136\cdot\bar{v}_k+14.44)\ \text{or}\\ &\bar{v}_k>74.6\frac{km}{h}\ and\ \left(v\cdot a_{pos}\right)_{k\_[95]}>(0.0742\cdot\bar{v}_k+18.966)\end{aligned} \tag{9}$$

are fulfilled then the trip is invalid.

$$\text{If}\quad \begin{aligned}&\bar{v}_k\le 94.05\frac{km}{h}\ and\ RPA_k<(-0.0016\cdot\bar{v}_k+0.1755)\\ &\bar{v}_k>94.05\frac{km}{h}\ and\ RPA_k<0.025\end{aligned} \tag{10}$$

are fulfilled then the trip is invalid.

According to [3,4,11], the EMROAD validation based on the MAW method provides an insight on the RDE occurring during the test at a given scale. The test developed by JRC is divided in sub-sections (windows) and the subsequent statistical treatment aims at identifying which windows are suitable to assess the vehicle RDE performance. The mass emissions are not calculated for the complete data set, but for sub-sections of the complete data set whose length is being determined to match the CO_2 mass emitted by the vehicle over the reference laboratory cycle. The moving average calculations are conducted with a time increment corresponding to the data sampling frequency. Based on this method, the RDE test is validated if it is complete (i.e., it comprises at least 15 % of urban, rural and motorway windows, out of the total number of windows) and normal (i.e., at least 50 % of the urban, rural and motorway windows are within the primary tolerance defined for a characteristic curve).

The other method, which could be used for the validation is the so-called CLEAR or power binning developed by TU Graz. It is based on a normalization to a standardized wheel power frequency (SPF) distribution [3,4,12]. Based on this method, the RDE test is validated if the power distribution for the urban and total trip in each class falls within the tolerances given by [12].

3. Real driving emissions testing. Case study

For a better understanding of the RDE requirements and of the consequences which come with its implementation, our endeavour started with the challenge of finding the right trip. Figure 4 shows the features of a validated trip, which is currently used in our study.

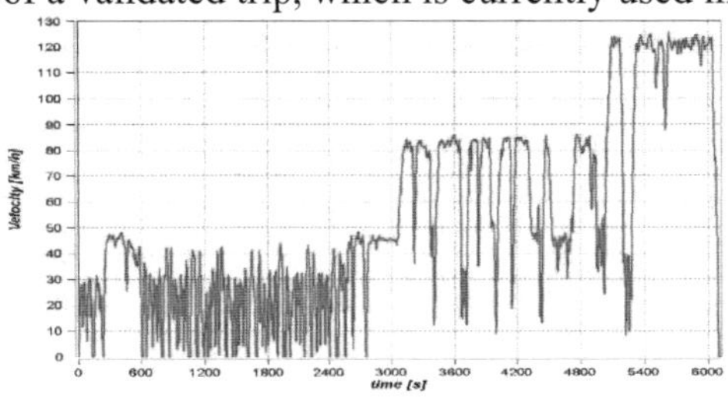

Total Trip Requirements	pass/fail
Trip Shares	
Urban 34% +10% and >=29%	pass
Rural 33% +-10%	pass
Motorway 33% +-10%	pass
Minimum Distance	
Urban 16 km	pass
Rural 16 km	pass
Motorway 16 km	pass
Total Trip Duration 90 -120 min	pass

Figure 4. RDE trip

Figure 5 pictures the operating points of a gasoline production engine during running in NEDC (in green), WLTC (in blue) and RDE (in red) conditions, as well as the speed profiles for each of the 3 tests.

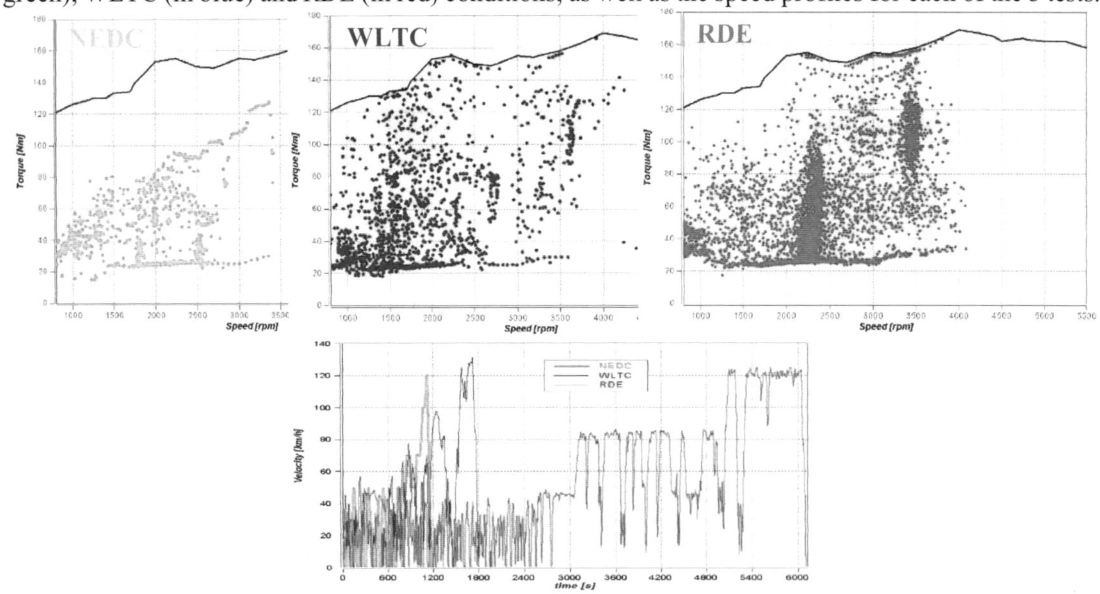

Figure 5. Engine's operating points and speed profiles for NEDC, WLTC and RDE

Comparing the 3 pictures, it becomes obvious that the type approval process by using WLTC and RDE is about addressing in a detailed way far more operating points of the engine. This requires the development of engines able to operate cleanly under all driving conditions. Thus, the engines which will comply with this far more severe type approval process must be adapted due to the extension of their operating range. Therefore, as indicated in [17], mixture enrichment usually performed in high loads and speeds in order to protect the exhaust valves, exhaust manifold, turbocharger (if any) and catalyst against temperature overload will be at issue. Equally, it is questionable to which extent the scavenging through valve overlap for the increase of low-end torque and transient response will still be feasible. Therefore, this must be considered early in the development phase in order to choose the right technical definition for the future engines (e.g. the integration of the exhaust manifold into the cylinder head for being possible the cooling of the exhaust gas).

Furthermore, to have a better understanding of different influences upon the RDE, the following plan was undertaken:

1. influence of the driving style, ("sport-driving vs. "eco-driving") in "quasi-iso-conditions", i.e., with the same driver, the same vehicle, during the same season and by using the same trip;
2. influence of the trip in "quasi-iso-conditions", i.e., when using the same driver with the "same" driving style, the same vehicle and during the same season.
3. influence of the ambient conditions or of the season (winter vs. summer) in "quasi-iso-conditions", i.e., when using the same driver with the "same" driving style, the same vehicle and the same trip.

Out of these three, only the first will be presented subsequently. Regarding the third objective, currently, a testing campaign carried out in the chassis dyno test cell is in progress. The objective is to see the effect of the ambient temperature upon the PEMS performance. Consequently, we plan a follow-up article to present the results obtained with the above methodology.

Coming back to the effect of the driving style, the vehicle which was used featured a manual transmission, so, for the "eco-driving" style, the pilot respected the recommendations of the gear-shift indicator (GSI). In short, it's about changing the gears not higher than 2500 rpm. Conversely, for the "sport-driving" style, the pilot changed the gears at about 3200 rpm.

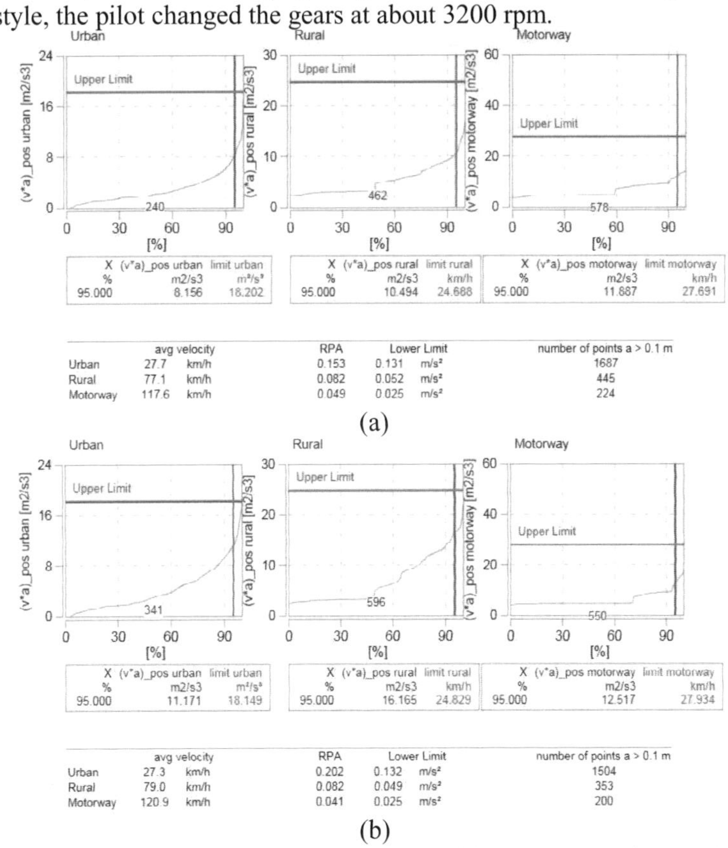

Figure 6. Parameters used to judge the driver's "aggressiveness"
(a) "eco-driving, (b) "sport-driving"

Figure 6 presents the main features of these two driving styles, judged via the parameters defined in relations (6), (7) and (8). The "aggressiveness" of the driver in the second case is obvious: in each of the three parts of the RDE trip (urban, rural and motorway), the $(v \cdot a_{pos})$ is higher and still, below the regulated limit. The same observation stands for the *RPA,* as well.

Two RDE trips per driving style were performed and then, the tail-pipe emissions results were averaged and analyzed comparatively. The ambient conditions while performing these four tests were close: ~ 22°C & ~ 23% RH for "eco-driving" and ~ 24°C & 22% RH for "sport-driving". Thus, it can be said the tests were performed in "quasi-iso-conditions", i.e., with the same driver, the same vehicle, very close ambient conditions and by using the same trip. The impact of the driving style while fulfilling all the RDE trip criteria upon the tail-pipe emissions is presented in figure 7.

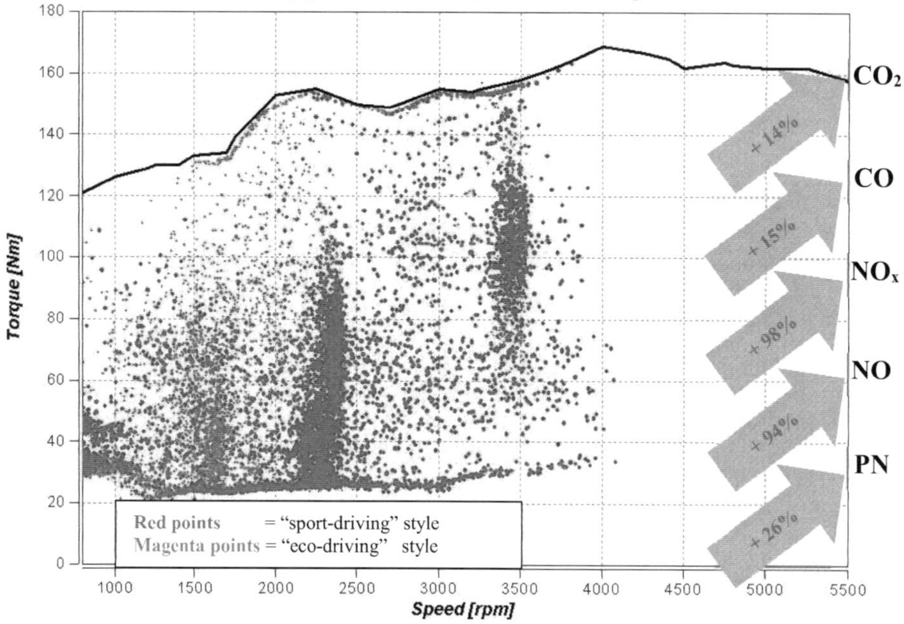

Figure 7. Driver's aggressiveness effect on tailpipe emissions

The increase in tail-pipe emissions while using the "sport-driving" are within expectations and of course are related with the engine's operation in higher loads and speeds. Thus, one may see that especially, NO_x and PN are strongly affected when driving more aggressive. Consequently, driving behavior can be the object of an additional measure to be taken to reduce air pollution from road vehicles.

4. Conclusions

Road mobility is the engine of economic development and is one of the fundamental aspects of social cohesion because it allows everyone the freedom of movement, along with the advantages inherent to it: easy access to various jobs, to public services, vacations/leisure etc. Even though it still represents the most spread energy source in the field of road mobility, the combustion engine faces environmental issues. Furthermore, despite the ever-increasing severity of the European regulations regarding automotive tail-pipe emissions, the population is exposed to air pollutants concentrations above the accepted limited. Consequently, the RDE testing was added to the type approval procedure for reasons of achieving a better representativeness to the real-world driving. Lab testing for type approval is and will be wanted because of the standardization and repeatability. Thus, both parts of the type approval process (lab testing + RDE) aim at achieving a real reduction in tail-pipe emissions. Nevertheless, it needs to be underlined, that even if vehicle emissions standards are a very powerful tool to reduce air pollution from road vehicles, they are not the only tool available for meeting air quality objectives. Additional measures which address traffic volumes, fuels and driving behavior are also necessary.

On the one hand, our study showed that RDE means new technical challenges for engine and after-treatment technology developers. Nonetheless, despite the inherent efforts of the manufacturers needed to cope with, it will also contribute to giving the combustion engines a more positive environmental image. On the other hand, it also presented a case study on the assessment of the driving style effect upon the emissions while performing RDE tests. As expected, results showed a significant increase in tail-pipe emissions while adopting a more aggressive driving style, which support the above statement regarding the addition measures which could be taken to meet the air quality objectives.

5. References

[1] European Commission 2016 Air pollutants from road transport *RDE Packag.*
[2] UNECE 2015 Uniform Provisions Concerning the Approval of Vehicles with Regard to the Emission of Pollutants According to Engine Fuel Requirements *Regul. No. 83 – Revis. 5, Geneva, Switz.*
[3] Badur J, Kohle F and Schmidt H Real Driving Emissions (RDE) – status of discussion and practical implementation *37th International Vienna Motor Symposium* ed VDI&OEVK (Vienna)
[4] Vlachos T, Bonnel P, Weiss M, Barouch G, Francesco R, Pablo M-V and Adolfo P 2016 The Euro 6 Real - Driving Emissions (RDE) procedure for light - duty vehicles : Effectiveness and practical aspects *37th Int. Vienna Mot. Symp.* 1–20
[5] Fontaras G and Dilara P 2012 The evolution of European passenger car characteristics 2000-2010 and its effects on real-world CO 2 emissions and CO 2 reduction policy *Energy Policy* **49** 719–30
[6] Pavlovic J, Marotta A and Ciuffo B 2016 CO2 emissions and energy demands of vehicles tested under the NEDC and the new WLTP type approval test procedures *Appl. Energy* **177** 661–70
[7] Marotta A, Pavlovic J, Ciuffo B, Serra S and Fontaras G 2015 Gaseous Emissions from Light-Duty Vehicles: Moving from NEDC to the New WLTP Test Procedure *Environ. Sci. Technol.* **49** 8315–22
[8] Marotta A, Pavlovic J, Serra S, Anagnostopoulos K, Ciuffo B, Fontaras G, Tsiakmakis S and Zacharof N 2015 A Comparison of Gaseous Emissions from Light-Duty Vehicles under the NEDC and the WLTP test procedures 13p
[9] European Commission 2007 *EU action against climate change—Leading global action to 2020 and beyond* (European Commission)
[10] UNECE 2015 Worldwide Harmonized Light Vehicles Test Procedure *Glob. Tech. Regul. No. 15, Geneva Switz.*
[11] European Commission 2016 Emissions from light passenger and commercial vehicles (Euro 6). RDE test procedure *Off. J. Eur. Union, Comm. Regul. 2016/427*
[12] Europe Commission 2016 Emissions from light passenger and commercial vehicles (Euro 6) *Off. J. Eur. Union, Comm. Regul. 2016/646*
[13] US Federal Register 2005 Control of emissions of air pollution from new motor vehicles: in-use testing for heavy-duty diesel engines and vehicles *Fed. Regist. vol. 70 No. 113*
[14] US Federal Register 2007 Not to Exceed test procedures *40 CFR 86.1370*
[15] European Commission 2007 On type approval of motor vehicles with respect to emissions from light pas senger and commercial vehicles *Off. J. Eur. Union, Regul. No 715/2007*
[16] Ericsson E 1996 How to measure driving patterns – a study of methodology (in Swedish). *Bull. 137. Dep. Traffic Planning. Lund Inst. Technol. Lund Univ.*
[17] Baumgarten H, Scharf J, Thewes M, Uhlmann T, Balazs A and Bohmer M 2016 Simulation - Based Development Methodology for Future Emission Legislation *37th International Vienna Motor Symposium* (Vienna) pp 1–23

Nomenclature. Abbreviations and notations

CARB	California Air Resources Board	LDV	Light-duty vehicles
CF	Conformity factor	MAW	Moving average windows
CO	Carbon monoxide	NDIR	Non-dispersive infrared
CO_2	Carbon dioxide	NDUV	Non-dispersive ultraviolet
CoC	Certificate of conformity	NEDC	New European Driving Cycle
DS	Distance-specific	NGO	Non-governmental organisations
EC	European Committee	NOx	Nitric oxides
ECU	Engine control unit	NTE	Not-To-Exceed
EFM/V	Exhaust Flow Mass/Volume	PEMS	Portable emission measuring systems
EPA	US Environmental Protection Agency	RDE	Real driving emissions
EU	European Union	RH	Relative humidity
FID	Flame ionization detection	RPA	Relative positive acceleration
FTP	Federal Test Procedure	SPF	Standardised power frequency
GHG	Greenhouse gas	TCMV	Technical Committee of Motor Vehicles
GPS	Global positioning system	THC	Total unburnt hydrocarbon
GSI	Gear shift indicator	UNECE	United Nations Economic Commission for Europe
HC	Unburnt hydrocarbons	WLTC	Worldwide harmonized light duty test cycle
JRC	Joint Research Centre		

Acknowledgment
The equipment which was used to carry out this joint study was acquired by University of Pitesti under the frame of POS CCE 221 program. The authors are, therefore, grateful to the Romanian Authority for Scientific Research and Innovation (ANCSI) for the financial grant received in 2015.

Disclaimer
The views expressed here belong purely to the authors and may not be regarded as official position of their institutions.

IOP Conf. Series: Materials Science and Engineering **252** (2017) 012085 doi:10.1088/1757-899X/252/1/012085

Analysis of performance and emissions of diesel engine using sunflower biodiesel

Dragos Tutunea, Ilie Dumitru

University of Craiova, Faculty of Mechanics

dragostutunea@yahoo.com

Abstract. The world consumption of fossil fuels is increasing rapidly and it affects the environment by green house gases causing health hazards. Biodiesel is emerging as an important promising alternative energy resource which can be used to reduce or even replace the usage of petroleum. Since is mainly derived from vegetable oil or animal fats can be produce for large scale by local farmers offering a great choice. However the extensive utilization of the biofuels can lead to shortages in the food chain. This paper analyzed the sunflower methyl ester (SFME) and its blends as an alternate source of fuel for diesel engines. Biodiesel was prepared from sunflower oil in laboratory in a small biodiesel installation (30L) by base transesterification. A 4 cylinder Deutz F4L912 diesel engine was used to perform the tests on various blends of sunflower biodiesel. The emissions of CO, HC were lower than diesel fuel for all blends tested. The NOx emissions were higher due to the high volatility and high viscosity of biodiesel.

1. Introduction

In the past the use of petroleum has led to atmospheric pollution and the global warming. The increase in human population worldwide and industrialization led to depletion of fossil fuel [1]. Gasoline and diesel-driven automobiles are the principal sources of greenhouse gas (GHG) emissions today [2]. Replacing the fossil fuels with biofuels seems to be a viable option because of the scarcity of petroleum reserves and continuously increasing stronger emission regulations [3]. However the fossil fuel cannot be replace in totality especially in transport and power generation because these represent the backbone for the economic growth in a country [4]. Biodiesel can be produced from various feedstock's (vegetable oils: soyabean, sunflower oil, rapeseed/canola, palm, cottonseed, coconut, peanut, pongamia, karanja, neem; animal fats: usually tallow; or waste oil such as frying oils from food industry) [5]. Due to their high viscosity and density these oils cannot be used in the engines until the glycerol component is removed by a reaction of transesterification [1]. The thermo-physical properties considered important to biofuels are density, viscosity, calorific value, cetane number, flash points, cloud and pour points. Also in several researches has been reported the properties of biofuels depends even by the chemical composition and fatty acid contents. All biofuels must be in the limits specified by ASTM D6751 and EN 14214 standards. High viscosity of biodiesel represents a problem because is directly linked with the injection system and affects fluidity and atomization causing carbon deposits and incomplete combustion in the engine [6]. Raheman and Phadatare [7] investigated the performance of karanja methyl ester and various blends (B20, B40, B60 and B80) and found a reduction of CO emissions by 73-94% compared to fossil diesel. Bhupendra Singh et al. [8] tested Karanja biodiesel at different proportions 5%, 10%, 20%, 30% and 100% to evaluate performance and emissions with diesel fuel. NOX emissions were higher for all the blends and the brake thermal efficiency was lower than neat diesel. Labeckas and Slavinskas [9] tested rapeseed oil methyl esters to evaluate performance and exhaust emissions in direct injection diesel engines and found a reduction in the emissions of HC, CO and CO_2. Mohamed F. Al Dawody et al. [10] tested on single cylinder diesel engine different blends of a soybean methyl ester and found a reduction up to 48.23% and an increase in brake specific fuel consumption with 14.65% compared with fossil diesel. The CO emission decreased with 11.36% for B20 and with 41.7% for B100. The NOX emissions were reported higher for all the blends of soybean methyl ester. In a study conducted by Pradhan et al. [11] in a 4 stroke single cylinder diesel engine fueled with biodiesel from waste mustard oil reported a reduction with 7.78% in brake thermal efficiency for B10 blend, higher Brake Specific Fuel Consumption (BFSC) for all biodiesel blends and an increase in NOx emissions. M. Habibullah et al. [12] investigated the effects of two biodiesel blends palm and coconut (PB30, CB30 and PB15CB15) on a single-cylinder, four-stroke, and direct-injection diesel engine under a full load and varying speed conditions. In engine performance PB30, CB30, and PB15CB15 engine brake power was lower by 3.92%, 4.71%, and 4.10%, whereas BSFC values were higher by 8.55–9.03% than diesel fuel. The NOx emissions were increased by 3.13–5.67% for all the tested biodiesel blends. M. Mofijur et al. [13] have reported on the performance and emission characteristics of Moringa oleifera biodiesel (B10 and B20) tested in a multi-cylinder diesel engine at various speeds and full load conditions. They reported that biodiesel blends reduced brake power and increased brake specific fuel consumption. In engine emissions CO was reduced by 10.60-22.93%, HC by 9.21- 23.68% and the NOx increased by 8.46-18.56% compared to diesel fuel. The present study focuses in evaluating the possibility of sunflower biodiesel to become an alternative source for diesel fuel in the existing engine without any modification.

2. Materials and methods

2.1. Production of biodiesel
The biodiesel was produced by alkali catalyzed transesterification process from sunflower oil, which was extracted from sunflower seeds. In this process methanol and a base catalyst, NaOH with a purity of 99% was used. Initially, the catalyst is dissolved into methanol and heated to 55°C. The agitation was kept constant at 450 rpm. The reaction time was of 60 minutes. The product separated by gravity into two layers namely ester layer and glycerol layer. The ester layer contains mainly methyl ester and methanol and the glycerol layer contains mainly glycerol and methanol. The mixture was allowed settling for 48 h to allow the separation of biodiesel and glycerin. The biodiesel was then washed with water to remove impurities and heated at 110°C to remove the remaining traces of the catalyst and methanol [14] (Fig.1).

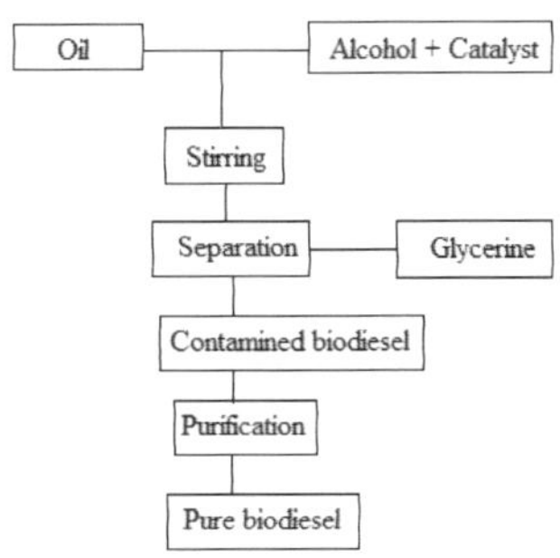

Fig. 1 Biodiesel production scheme

2.2. Characterization of biodiesel
The biodiesel prepared from sunflower oil (SFME) was characterized and tested for the compliance with EN standards. Biodiesel blends of B10 (10% biodiesel, 90% diesel), B20 (20% biodiesel, 80%

diesel), B30 (30% biodiesel, 70% diesel), B40 (40% biodiesel, 60% diesel) and B100 (pure biodiesel) were prepared on volume basis. The physical and chemical properties of SME are listed in Table 1.

Table 1 The physical and chemical properties of blends of SFME and Diesel EN14214

Property	Method	Specifications	Reference diesel	B10	B20	B30	B40	B100
Flash point °C	EN ISO 3679	min. 120	68	75	86	97	110	183
Kinematic viscosity 40°C, mm^2/s	EN ISO 3104	3.5-5	3.18	3.21	3.26	3.31	3.35	4,7
Density 15°C, kg/m^3	EN ISO 3675	860-900	832	835	839	843	848	868
Calorific value MJ/kg	-	-	44.8	43.6	42.2	41.7	40.3	33.5

2.3. Engine stand

Experiments were conducted on a four cylinder diesel engine with direct injection. The specifications of the engine for the current study are listed in Table 2 and a schematic diagram of the engine with instrumentations is shown in Fig.2.

Table 2 Specifications of Deutz F4L912 diesel engine

Number of cylinders	4
Bore/stroke (mm)	100/120
Displacement (l)	3.770
Compression ratio	19
Maximum rated speed (rpm)	2500
Mean piston speed (m/s)	10
Power (kW) @ 2350rpm	51
Mean effective pressure (bar)	6.9
Maximum torque (Nm) @ 1450rpm	238
Minimum idle speed (rpm)	650

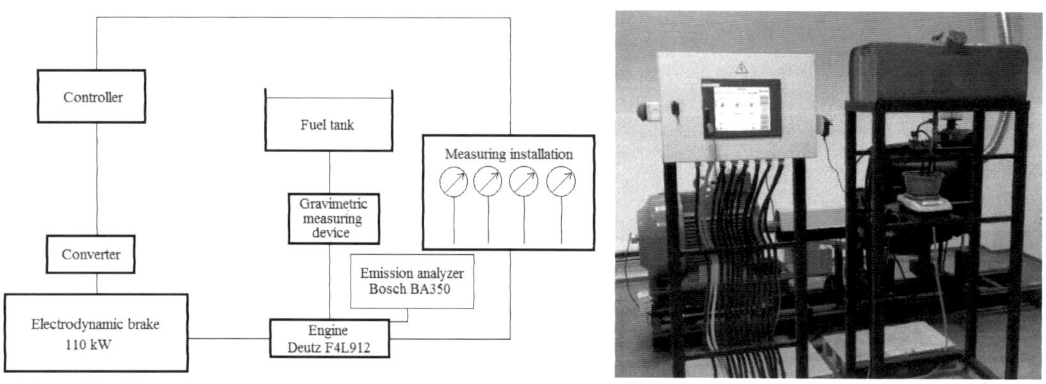

Fig.2 Schematic diagram of experimental setup

The engine stand is mounted on a chassis which is connected to electronic display and fuel reservoir. Components of testing stand are fuel feeding installation, controller and measuring equipment. Mechanical connection between engine and motor is supported by a shaft fitted with protective shields. Feeding system has the role to ensure sufficient supply of fuel to the engine by using a fuel tank which stays on electronic balance to measure the fuel consumption. The electric brake is a 110 kW three phase electric motor controlled by the inverter and ordered by the process computer. The feeding system consists of a rack with fuel tank and has separation tap. Controller is equipped with a touch screen which is used to control the engine. The precision of measuring system is given in Table 3. A BEA350 gas analyzer was used to measure engine emissions.

Table 3 Precision of parameters by measuring system

Parameter	Precision value
Speed (rpm)	0…6000±2%
Torque (Nm)	0…1000±2%
Fuel consumption (kg/h)	0…50±2%
Intake air temperature (°C)	0...50 ±5%
Inlet pressure (kPa)	-50…300±5%
Exhaust temperature (°C)	0…800±5

3. Results and discussion

EGT (exhaust gas temperature) is an indicator of the heat release rate of the fuels tested during combustion period. In generally the combustion in the case of biodiesel is improved due to the presence of excess oxygen in the fuel itself. It is observed that EGT increase with the increase in load for all blends of SFME (Fig.3). These findings are in accordance with several researches; Godiganur et al. [15] on a six cylinder turbocharged diesel engine fueled with mahua biodiesel blends observed an increase in EGT with the increase in engine load (about 12% higher than diesel in the case of B100); Datta et al. [16] on a double cylinder, four stroke, diesel engine fueled with jatropha biodiesel blends observed an increase in EGT due to the higher flame temperature of biodiesel.

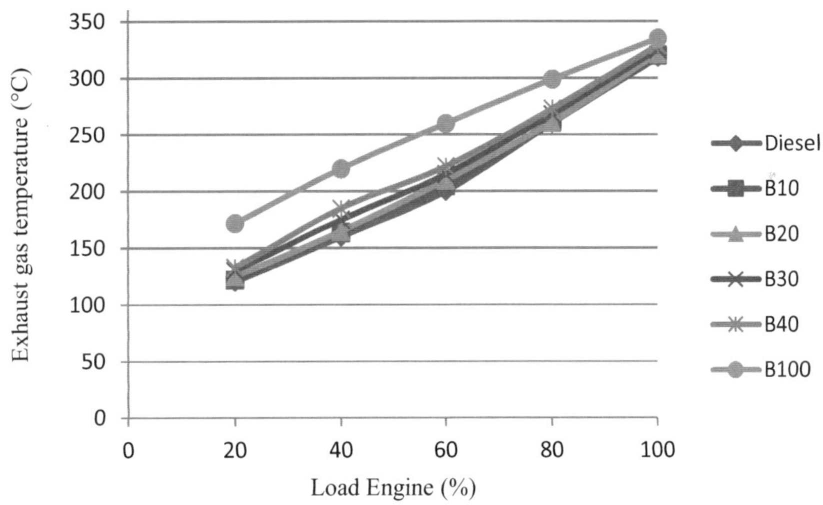

Fig.3 Variation in exhaust gas temperature with load

Carbon monoxide (CO) in the engine is produced by incomplete combustion of the fuel. It is observed that CO emissions decrease at lower loads due to the presence of oxygen (almost 11%) at lower loads and then increase sharply for all fuels tested (Fig.4). Other researchers found similar results; Nabi et al. [17] on a single cylinder fueled with cotton biodiesel the CO emissions were reduced with 24% in comparison with diesel; Çelikten et al. [18] on a four cylinder diesel engine fueled with soybean biodiesel found a decrease of 28% of CO emissions; Anand et al. [19] observed a reduction with 46.5% of CO emissions at higher load for karanja biodiesel.

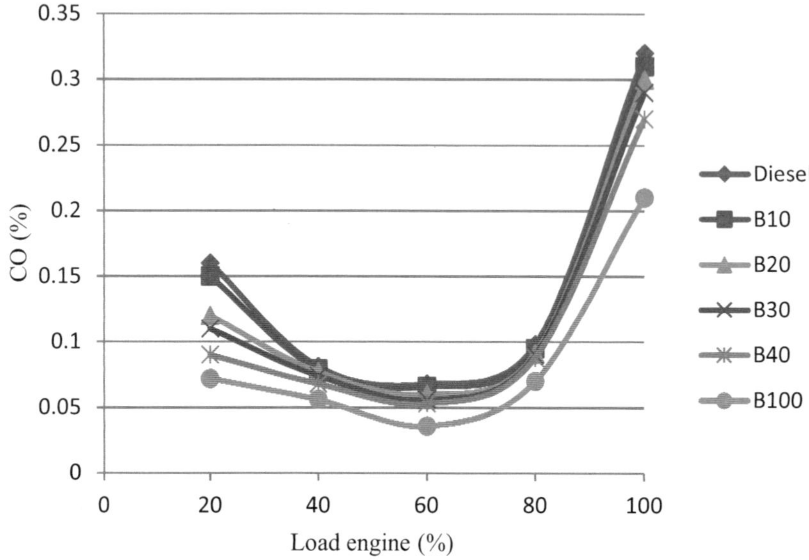

Fig.4 Variation in CO emission with load

Complete combustion inside the combustion chamber increase carbon dioxide (CO_2) emissions. It is observed that CO_2 emissions increase at all the loads for the tested fuels (Fig.5). Huzayyin et al. [20] found a increase in CO_2 emissions for jojoba biodiesel blends for all engine loads. Fontaras et al. [21] for blends of soybean the CO_2 emissions increase by 14% for B100 and 9% for B50. Aydin and İlkiliç [22] found a reduction by 16% of CO_2 emissions of rapeseed biodiesel compared with diesel.

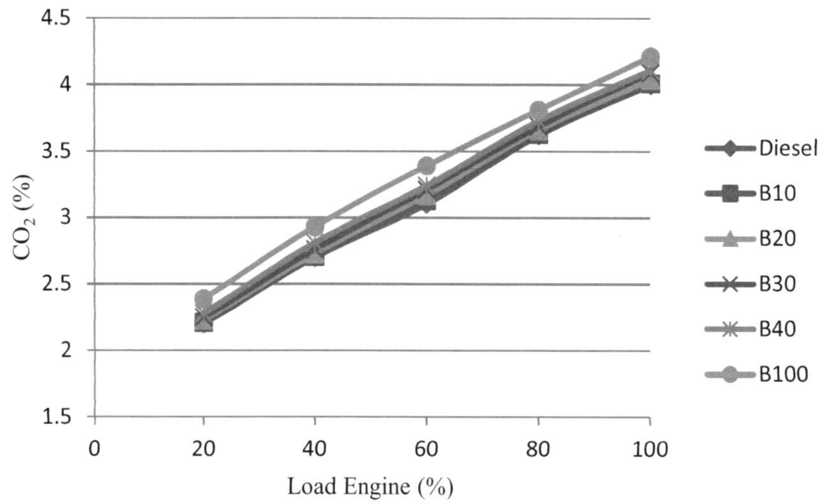

Fig.5 Variation in CO_2 emission with load

In general the emission of hydrocarbons (HC) depends mainly by compositions and combustion characteristics of the fuels tested. If combustion is improved the HC emissions decrease and vice versa. Because of the high content of oxygen in the biodiesel it is expected that HC emission will decrease for blends of SFME and diesel. It is observed that the decrease of HC emissions depends of the percentage of biodiesel in the blend (Fig.6). Sahoo et al. [23] found a reduction in HC emissions by 20.64%, 20.73% and 6.75% using biodiesel of karanja, jathropa and polanga. Tsolakis et al. [24] observed a reduction of nearly 50% for rapeseed biodiesel compared with low sulfur diesel.

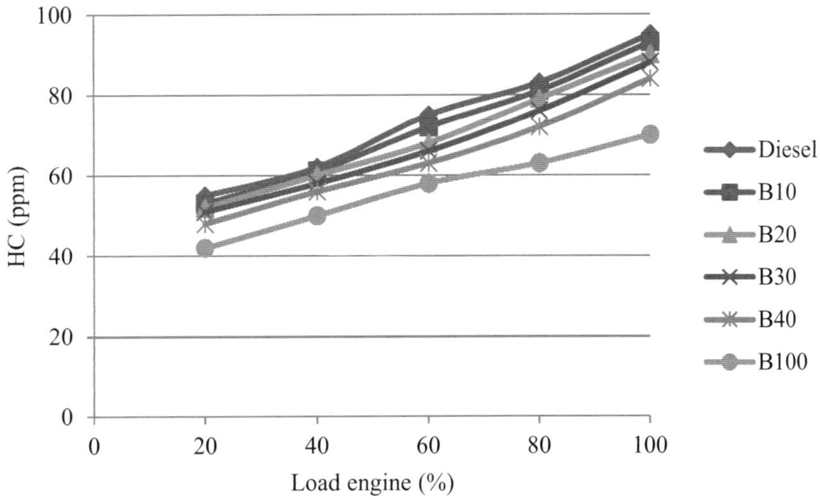

Fig.6 Variation in HC emission with load

Nitric oxide (NO) and nitrogen dioxide (NO_2) are formed in the process of oxidation of the nitrogen during combustion and depend of the combustion temperature and oxygen content. The tests show an increase in NOx emissions with the increase in engine load for all blends (Fig.7). It is found highest for SFME because of high oxygen content which results in complete combustion causing high combustion temperature. Gumus M. [25] found an increase of NOx emissions for apricot seed kernel oil methyl ester by 10% compared to diesel fuel. Aydin and İlkiliç [22] reported an increase of of NOx emissions 16.7% with B20 blend and 11.8% with B100 for biodiesel of rapeseed.

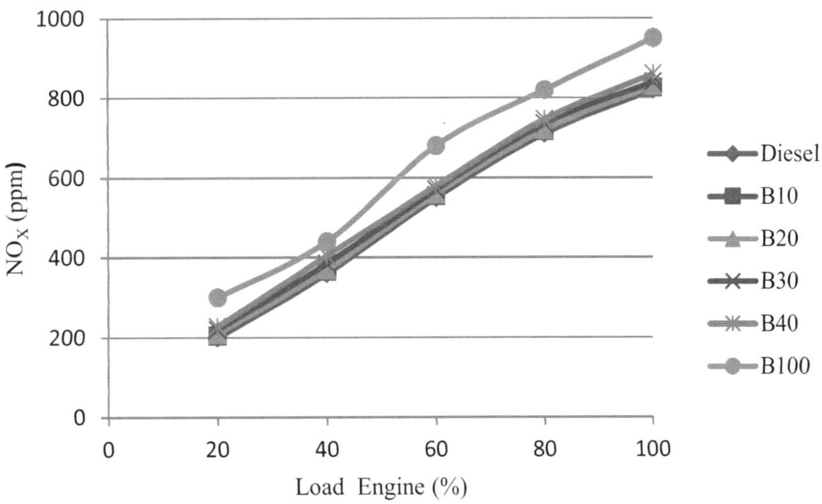

Fig.7 Variation in NO_X emission with load

IOP Conf. Series: Materials Science and Engineering **252** (2017) 012085 doi:10.1088/1757-899X/252/1/012085

Conclusion

The 4 cylinder diesel engine runs successfully during tests on biodiesel of sunflower oil and its blends. The blends were characterized for their various physical chemical properties. The EGT was found highest for pure biodiesel. The availability of extra oxygen enhanced the temperature of combustion resulting in higher temperature for all the blends tested. CO and HC emissions are highest for the diesel fuel and lower for the blends of SFME. The NO_X were found highest for pure SFME and its blends because of high volatility, low heat content and high viscosity compared with diesel fuel.

References

[1] Jayashri N. Nair, Ajay Kumar Kaviti, Arun Kumar Daram, 2016, *Analysis of performance and emission on compression ignition engine fuelled with blends of neem biodiesel*, Egyptian Journal of Petroleum.

[2] Puneet Verma, Mahendra Pal Sharma, Gaurav Dwivedi, 2016, *Potential use of eucalyptus biodiesel in compressed ignition engine*, Egyptian Journal of Petroleum 25, pp 91–95.

[3] Ayatallah Gharehghani, Mostafa Mirsalim, Reza Hosseini, 2017, Effects of waste fish oil biodiesel on diesel engine combustion characteristics and emission, Renewable Energy 101 pp 930-936.

[4] Varatharaju Perumal and Ilangkumaran M., 2017, *Experimental analysis of engine performance, combustion and emission using pongamia biodiesel as fuel in CI engine*, Energy 129 pp 228- 236

[5] Dwivedi G., Sharma M.P., 2015, *Investigation and Improvement in Cold Flow Properties of Pongamia Biodiesel*, Waste Biomass Valorization 6 (1) pp 73–79.

[6] Ambarish Datta, Bijan Kumar Mandal, 2016, *A comprehensive review of biodiesel as an alternative fuel for compression ignition engine*, Renewable and Sustainable Energy Reviews 57 pp 799–821.

[7] Raheman H., Phadatare A.G., 2004, Diesel engine emissions and performance from blends of karanja methyl ester and diesel, Biomass Bioenerg. 27 pp 393-397.

[8] Bhupendra Singh Chauhan, Naveen Kumar, Haeng Muk Cho, Hee Chang Lim, Energy 56 (2013) pp 1–7.

[9] Labeckas, G. and Slavinskas, S., 2006, *The effect of rapeseed oil methyl ester on direct injection diesel engine performance and exhaust emissions*. Energy Conversion Management, 47, pp. 1954–1967.

[10] Mohamed F. Al Dawody and S.K. Bhatti, 2014, *Experimental and Computational investigations for Combustion, Performance and Emission Parameters of a Diesel Engine Fueled with Soybean Biodiesel-Diesel Blend*, Energy Procedia 52 pp 421 – 430, Alternative Energy in Developing Countries and Emerging Economies

[11] Piasy Pradhan, Sourojeet Chakraborty, Rajat Chakraborty, 2016, *Optimization of infrared radiated fast and energy-efficient biodiesel production from waste mustard oil catalyzed by Amberlyst 15: Engine performance and emission quality assessments*, Fuel 173 pp. 60–68.

[12] Habibullah M., Masjuki H.H., Kalam M.A., Rizwanul Fattah I.M., Ashraful A.M., Mobarak H.M., 2014 *Biodiesel production and performance evaluation of coconut, palm and their combined blend with diesel in a single-cylinder diesel engine*, Energy Conversion and Management 87 pp 250–257.

[13] Mofijur M., Masjuki H.H., Kalam M.A., Atabani A.E., Arbab M.I., Cheng S.F., Gouk S.W., 2014, *Properties and use of Moringa oleifera biodiesel and diesel fuel blends in a multi-cylinder diesel engine*, Energy Conversion and Management 82 pp 169–176.

[14] Tutunea D., Bica M., Dumitru I., Dima A., 2014, *Experimental researches on biodiesel properties*, SMAT 2014 3rd International Congress Science and Management of Automotive and Transportation Engineering 23rd – 25rd of October, Craiova, Tome I, ISBN 978-606-14-0864-1, Universitaria, pp. 295-300.

[15] Godiganur S, Murthy CHS, Reddy RP., 2009, *6BTA 5.9 G2-1 Cummins engine performance*

and emission tests using methyl ester mahua (Madhuca indica) oil/diesel blends. Renew Energy 34 pp 2172–7.

[16] Datta A., Palit S, Mandal BK., 2014, *An experimental study on the performance and emission characteristics of a CI engine fuelled with Jatropha biodiesel and its blends with diesel*. J Mech Sci Technol 28(5) pp1961–6.

[17] Nabi M.N., Rahman M.M., Akhter M.S., 2009, *Biodiesel from cotton seed oil and its effect on engine performance and exhaust emissions*, Appl Therm Eng 29 pp 2265–70.

[18] Çelikten I., Koca A., Arslan M.A., 2010, *Comparison of performance and emissions of diesel fuel, rapeseed and soybean oil methyl esters injected at different pressures*. Renew Energy 35 pp 814–20.

[19] Anand K., Sharma RP, Mehta PS. Experimental investigations on combustion, performance and emissions characteristics of neat karanji biodiesel and its methanol blend in a diesel engine. Biomass Bioenergy 2011;35:533–41.

[20] Huzayyin A.S., Bawady A.H., Rady M.A., Dawood A., 2004, *Experimental evaluation of diesel engine performance and emission using blends of jojoba oil and diesel fuel*. Energy Convers Manag 45 pp 2093–112.

[21] Fontaras G., Karavalakis G., Kousoulidou M., Tzamkiozis T., Ntziachristos L., Bakeas E., Stournas S., Samaras Z., 2009, *Effects of biodiesel on passenger car fuel consumption, regulated and non-regulated pollutant emissions over legislated and real-world driving cycles*. Fuel 88 pp 1608–17.

[22] Aydin H., İlkiliç C., 2011, *Exhaust emissions of a CI engine operated with biodiesel from rapeseed oil*. Energy Sources Part A 33 pp1523–31.

[23] Sahoo P.K., Das L.M., Babu M.K.G., Arora P., Singh V.P., Kumar N.R., Varyani T.S., 2009, *Comparative evaluation of performance and emission characteristics of jatropha, karanja and polanga based biodiesel as fuel in a tractor engine*. Fuel 88 pp1698–707.

[24] Tsolakis A., Megaritis A., Wyszynski M.L., Theinnoi K., 2207, *Engine performance and emissions of a diesel engine operating on diesel-RME (rapeseed methyl ester) blends with EGR (exhaust gas recirculation)*. Energy 32 pp 2072–80.

[25] Gumus M., Kasifoglu S., 2010, *Performance and emission evaluation of a compres- sion ignition engine using a biodiesel (apricot seed kernel oil methyl ester) and its blends with diesel fuel*. Biomass Bioenergy 34 pp 134–9.

IOP Conf. Series: Materials Science and Engineering **252** (2017) 012086 doi:10.1088/1757-899X/252/1/012086

Evaluation of performance and emissions characteristics of methanol blend (gasohol) in a naturally aspirated spark ignition engine

Dima Alexandru, Dumitru Ilie, Tutunea Dragos

University of Craiova, Faculty of Mechanics

alexandru1dima@yahoo.com

Abstract. Alternative fuels for use in internal combustion engines have become recently in attention due the strict regulations regarding the environmental protection, emissions and to reduce the dependency of the fossil fuels. One choice is the use of methanol as it can be produce from renewable sources and blended with gasoline in any proportion. The aim of this study is to compare the effects of methanol – gasoline blends regarding performance, combustion and emission characteristics with gasoline. Five different blends M5, M10, M15, M20 and M25 were tested in a single cylinder spark ignition engine typically used in scooters applications. The experimental results in engine performance show a decrease of torque and power up to 10 %and in emissions characteristics a CO, CO2, HC. It can be concluded that gasohol is viable option to be used in gasoline engines to replace partially the fossil fuel.

1. Introduction

In the past decades global concerns have been rising to find solutions to the depletion of fossil fuels and to search for greener combustible fuels to be used in internal combustion engines. These problems have lead to the research of several possible renewable fuels like bioalcohols, vegetable oils and biodiesel to reduce the air pollution [1,2]. Among a variety of alcohols methanol, ethanol and butanol present the highest potential for use in transportation sector mainly because are cheap and have properties similar to gasoline [3,4]. Alcohols have a higher octane ratings, higher latent heat of vaporization and oxygen in their molecular structure leading to higher efficiency of combustion and low emissions [5]. In time researchers used M15 and M85 (USA in the early 1980s) while in China was used ''low-content methanol gasoline" called gasohol with a methanol proportion under 50%. Gasohol was used directly in the engines without adjustment of engine configuration and recalibration of ECU. Siwale et al. [7] studied a blend of methanol-nbutanol-gasoline and found that the emission of NOx were higher that gasoline with the increase in spark timing. Canakci et al. [8] found that methanol–gasoline fuel blends reduce the emissions of CO and HC at the vehicle speed of 80 km/h. Wai et al. [9] studied the effect of methanol blends in internal combustion engines and found for M85 a reduction of emissions of CO and NOx by 23% and 80%. Hu et al. [10] studied the methanol addition in gasoline and found that methanol reduce the phase of rapid burning and also advances the start of combustion. The maximum pressure of the blends was higher comparative with gasoline. Bardaie et al [11] found for pure methanol (M100) a loss o power of 4–5% comparative with gasoline. Biligin and Sezer [12] observed the emissions characteristics and performance for methanol blends and found a maximum brake mean effective pressure for M5. Dai et al. [13] found that for M15 the evaporative THC emissions increased by 63% and carbonyls increased by 19% comparative with gasoline. Palmer [14] found that 10% addition of alcohol to gasoline reduce the CO emissions by 30%. Yanju et al. [15] found the maximum pressure increase with methanol addition in the blend with the increase by 4-5° CA in ignition delay and by 5-6° CA in combustion duration for M85 compared to gasoline. Agarwal et al. [16] study the regulated and unregulated emissions on medium-duty transportation MPFI SI engine fuelled with M5 and M15 and found a reduction in emissions of benzene, iso-butylene, acetaldehyde, formic acid and nitrogen dioxide. Abu-Zaid et al. [17] studied the performance of a M3, M6, M9, M12 and M15 and found a maximum power output for M15 fuel blend. Tiegang Hu et al. [18] investigated on a three-cylinder engine the combustion and emission characteristics of methanol blends. The engine combustion was found better with the methanol addition into gasoline. During cold start the emissions of HC are reduced with 40% at 5 °C and 30% at 15 °C, CO with 70% for M30. In this study the effects of blending methanol and gasoline in a mono cylindrical engine are analyzed regarding emissions levels and performance.

2. Methods and materials

2.1. Fuels

In this study methanol was purchased from a regional supplier. Conventional gasoline and methanol – gasoline blends with five different substitution ratios of 5%, 10%, 15%, 20 % and 25% were prepared based on volume percentage basis and mechanical mixing. Fuel properties influence engine operations and performance. Physical properties of methanol vis-à-vis gasoline are shown in Table 1.

Table 1 Comparison of fuel properties [19]

Property	Gasoline	Methanol
Molecular formula	$C_4 - C_{12}$	CH_3OH
Molecular weight	95-120	32
Oxygen content (%)	0	50%
Density (kg/m^3)	740	792
LHV (Mj/m^3)	44,3	20
Octane number	>90	111
Auto-ignition temperature (°C)	228-470	465
Stoichiometric A/F ratio	14.8	6.47
Latent heat (kJ/kg)	305	1103
LHV of stoi-mixture (MJ/m^3)	3810	3906

2.2. Engine stand

The engine tests were performed at the Faculty of Mechanics, University of Craiova using a single cylinder gasoline engine (Table 2). It is used an air cooled single cylinder 4 stroke engine with a 72 x 60 bore and stroke values and a compression ratio of 9,8:1, which recommends the engine as suited for the experiment [20] The engine was warmed up until it reached steady condition with gasoline and afterwards was fuelled with alcohol gasoline blend. After that, the engine was left to run with the gasohol until it reached steady state condition. For each test the engine needs to run again with gasoline to eliminate remaining gasohol from de fuel line system. In addition each test was repeated for at least three times to increase the accuracy of the experimental data recorded. The schematic of the experimental setup is shown in Fig. 1. A dynamic torque transducer for measuring the torque was fitted between the shaft that transmits the motion from the engine to the cinematic chain that leads to the braking system. A Stargas 898 gas analyzer was used to measure the gas emissions of the engine and engine parameters with compatibility of other devices or with a PC offering real time information as R.P.M., temperature of oil engine.

Table 2 Engine specifications

Model	Honda CN 250
Engine type - Number of cylinders	Single cylinder, four-stroke
Engine size - Displacement - Engine capacity	244.00 ccm (14.28 cubic inches)
Maximum power - Output - Horsepower	17.00 HP (12.4 kW) @ 7000 rpm
Maximum torque	21.00 Nm (2.1 kgf-m or 15.5 ft.lbs) @ 5000 rpm

Fig.1 Engine stand
1 – engine; 2 – soft coupling; 3 – bearings; 4 – torque transducer; 5 – bearings; 6 – soft coupling; 7 – disc brake; 8 – brake disc cooler.

3. Results and conclusions

The present section shows the results obtained from the experiment conducted to analyze the behaviour of naturally aspirated spark ignited engine fuelled with gasoline and 5 different blends of 5,10,15,20 and 25 vol. % of methanol in gasoline, used as M5, M10, M15, M20, M25 further.

3.1. Engine performance

The energy performance of the engine concerning the torque and power are summarized in Fig. 2. and Fig. 3.

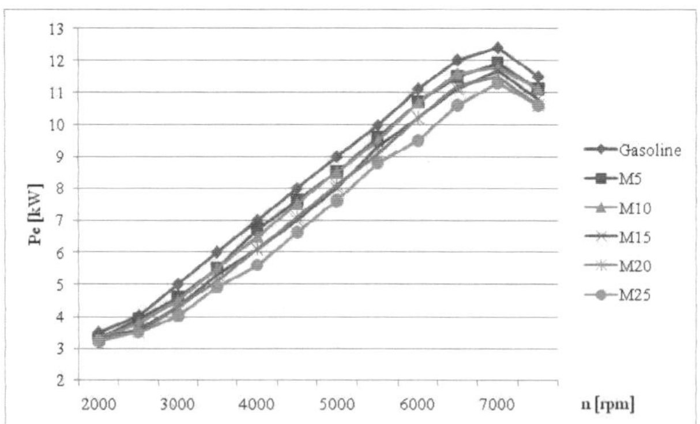

Fig. 2 Evolution of engine power for gasoline and methanol blends

IOP Conf. Series: Materials Science and Engineering **252** (2017) 012086 doi:10.1088/1757-899X/252/1/012086

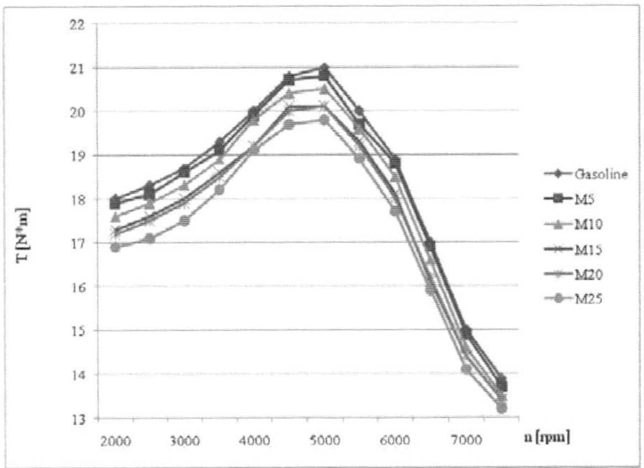

Fig. 3 Torque variation related to engine speed

The variation of engine torque and power is dependent by the quantity of methanol in the gasoline mixture. Increasing the fraction of methanol results in a decrease of performance, meaning lower value for the parameters examined previous at different speeds and loads of the engine. S. Liu et al. [21].

3.2. *Exhaust emissions*

The following figures, Fig. 4 – Fig. 6, show the variation of pollutant emissions versus the load of the engine for different engine speeds of 3600 and 4600 rpm, starting from where the load will be applied.

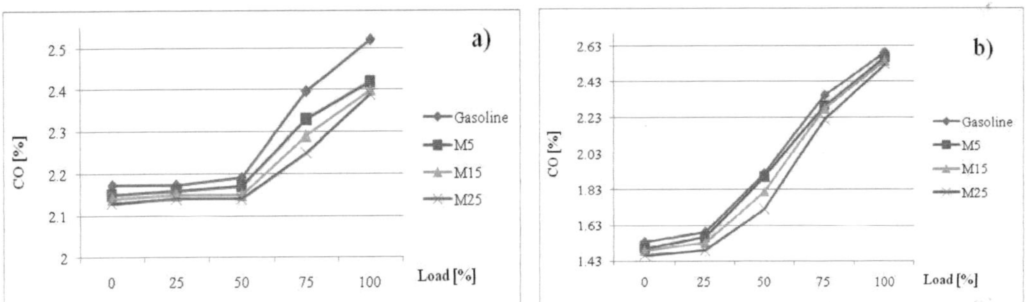

Fig. 4 Methanol influence on CO emissions, a) 3600 rpm, b) 4600 rpm

As it can be seen, the values of CO decrease once the percentage of methanol is increased. The effect is generated by the quantity of oxygen contained by the alcohol and by the carbon, that is found in greater proportion in gasoline rather than in methanol composition, [22].

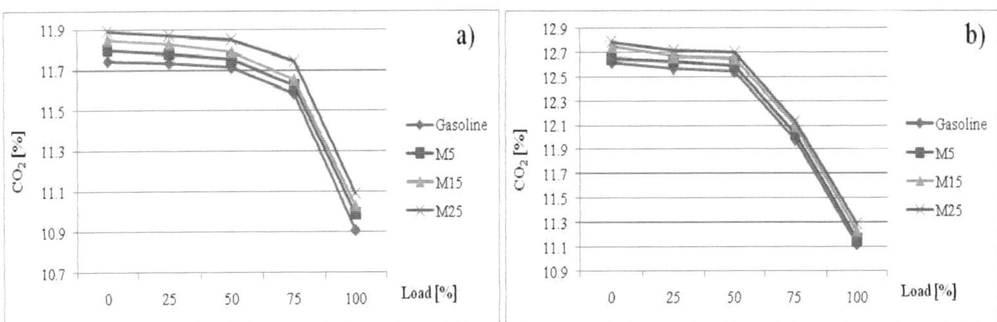

Fig. 5 Methanol influence on CO_2 emissions, a) 3600 rpm, b) 4600 rpm

Fig. 5 indicates the evolution of CO_2 emissions for two different engine speeds starting from 0 to 100 % load. It can be seen an increasing trend for the alcohol gasoline blends. At a high engine speed the emission value have a great value do to the quantity of air and fuel that increase the rate of complete combustion. Similar results were recorded with differences up to 13% higher CO_2 emissions for methanol-gasoline blends [23]

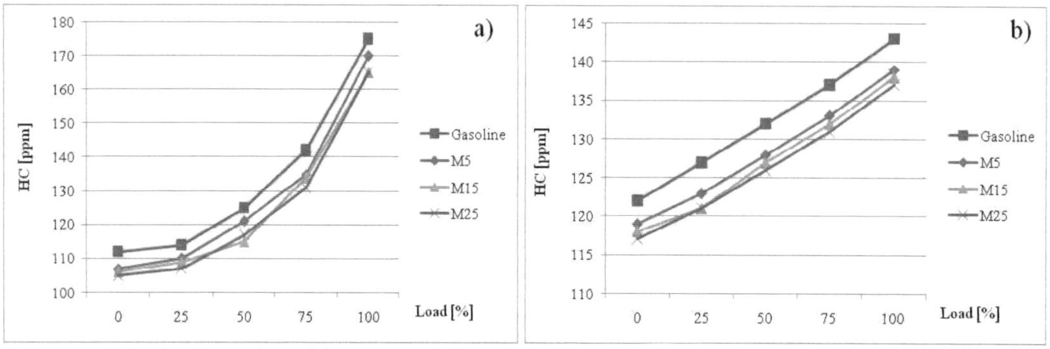

Fig. 6 Methanol influence on HC emissions, a) 3600 rpm, b) 4600 rpm

Fig. 6 depicts the evolution of HC emissions, which can rise rapidly due to incomplete combustion or heat losses in the cylinder. Increase of the HC emissions can be caused also by the existence of regional lean and rich mixtures in the combustion chamber. Experiments show a decreased value of the emissions when the gasoline is mixed with methanol, influenced by the carbon to hydrogen fuel ratio which is lower than in the case of gasoline, [24].

3.3. Conclusions

The purpose of present paper was to compare the performance and the emissions characteristics of the gasohol with the performance parameters obtained from the case of using gasoline as single fuel. The results showed in the above figures can sustain the possibility of usage of methanol blends as fuel for gasoline engines and also the idea of searching variations with other alcohols in order to obtain an alternative that could also satisfy the output given by the decrease of the energy performances. The results also show a decrease of the CO and CO_2 emissions and an equivalent variation of the HC emissions, which proves that the combination of these two could offer a viable solution.

References

[1] M. Eyidogan, et al., Impact of alcohol–gasoline fuel blends on the performance and combustion characteristics of an SI engine, Fuel 89 (10) (2010) 2713–2720.

[2] Costagliola MA, De Simio L, Iannaccone S, Prati MV. Combustion efficiency and engine out emissions of a S.I. engine fueled with alcohol/gasoline blends. Appl Energy 2013;111:1162–71.

[3] M. Eyidogan, A. Ozsezen,M. Canakci, A. Turkcan, Impact of alcohol–gasoline fuel blends on the performance and combustion characteristics of an SI engine, Fuel (2010) 1–7.

[4] Avinash Kumar Agarwal, Himanshu Karare, Atul Dhar, Combustion, performance, emissions and particulate characterization of a methanol–gasoline blend (gasohol) fuelled medium duty spark ignition transportation engine, Fuel Processing Technology 121 (2014) 16–24.

[5] B. Celik, B. Ozdalyan, F. Alkan, The use of pure methanol as fuel at high compression ratio in a single cylinder gasoline engine, Fuel 90 (2011) 1591–1598.

[6] Xin Wanga, Yunshan Ge, Linlin Liu, Zihang Peng , Lijun Hao, Hang Yin, Yan Ding, Junfang Wang, Evaluation on toxic reduction and fuel economy of a gasoline direct injection- (GDI-) powered passenger car fueled with methanol–gasoline blends with various substitution ratios, Applied Energy 157 (2015) 134–143

[7] Siwale L, Kristof L, Bereczky A, Mbarawa M, Kolesnikov A. Performance, combustion and emission characteristics of n-butanol additive in methanol– gasoline blend fired in a naturally-aspirated spark ignition engine. Fuel Process Technol 2014;118:318–26.

[8] Canakci M, Ozsezen AN, Alptekin E, Eyidogan M. Impact of alcohol–gasoline fuel blends on the exhaust emission of an SI engine. Renew Energy 2013;52:111–7.

[9] J.Wei,H.Liu,S.Li, R. Yang, J.Liu,Y.Wang, Effects of methanol/gasoline eblends on a spark ignition engine performance and emissions, Energy & Fuels 22 (2) (2008) 1254–1269.

[10] T. Hu, Y.Wei, S. Liu, L. Zhou, Improvement of spark-ignition (SI) engine combustion and emission during cold start, fueled with methanol/gasoline blends, Energy & Fuels 21 (2001) 171–175.

[11] Z. Bardaie, R. Janius, Conversion of spark-ignition engine for alcohol usage comparative performance, Agricultural Mechanization in Asia, Africa, and Latin America 15 (2) (1984) 31–40.

[12] A. Bilgin, I. Sezer, Effects of methanol addition to gasoline on the performance and fuel cost of a spark ignition engine, Energy & Fuels 22 (2008) 2782–2788.

[13] Dai, P.P., Ge, Y.S., Lin, Y.M., Su, S., Liang, B., 2013. Investigation on characteristics of exhaust and evaporative emissions from passenger cars fueled with gasoline/ methanol blends. Fuel 113, 10e16.

[14] F. Palmer, Vehicle Performance of Gasoline Containing Oxygenates, MI Mechanical, 1986. 33–46.

[15] W. Yanju, L. Shenghua, L. Hongsong, Y. Rui, L. Jie, W. Ying, Effects of methanol/ gasoline blends on a spark ignition engine performance and emissions, Energy & Fuels 22 (2008) 1254–1259.

[16] Avinash Kumar Agarwal, Pravesh Chandra Shukla, Jai Gopal Gupta, Chetankumar Patel, Rajesh Kumar Prasad, Nikhil Sharma, Unregulated emissions from a gasohol (E5, E15, M5, and M15) fuelled spark ignition engine

[17] Abu-Zaid M, Badran O, Yamin J. "Effect of Methanol Addition on the Performance of Spark Ignition Engines", Energy Fuels, 2004, Vol.18, pp.312–315.

[18] Tiegang Hu, Yanjv Wei, Shenghua Liu, Improvement of Spark Ignition (SI) Engine Combustion and Emission during Cold Start, Fueled with Methanol/Gasoline Blends, Energy & Fuels , 21(1), 171-175, 2007

[19] Vancoillie J, Demuynck J, Sileghem L, Van De Ginste M, Verhelst S, Brabant L, et al. The potential of methanol as a fuel for flex-fuel and dedicated sparkignition engines. Appl Energy 2013;102:140–9.

[20] M. Balki, C. Sayin, The effect of compression ratio on the performance, emissions and combustion of an SI (spark ignition) engine fueled with pure ethanol, methanol and unleaded gasoline, Energy 71 (2014) 194-201

[21] Liu Shenghua, Eddy R. Cuty Clemente, Hu Tiegang *, Wei Yanjv, Study of spark ignition engine fueled with methanol/gasoline fuel blends, Applied Thermal Engineering 27 (2007) 1904–1910.

[22] S. Babazadeh Shayan, S. M. Seyedpour, F. Ommi, S. H. Moosavy, M. Alizadeh, Impact of Methanol–Gasoline Fuel Blends on the Performance and Exhaust Emissions of a SI Engine, International Journal of Automotive Engineering, Vol. 1, Number 3, July 2011

[23] A. Elfasakhany, Investigations on the effects of ethanolemethanolegasoline blends in a spark-ignition engine: Performance and emissions analysis, Engineering Science and Technology, an International Journal 18 (2015) 713-719

[24] L. Siwale, L. Kristóf, A. Bereczky, M. Mbarawa, A. Kolesnikov, Performance, combustion and emission characteristics of n-butanol additive in methanol–gasoline blend fired in a naturally-aspirated spark ignition engine, Fuel Processing Technology 118 (2014) 318–326

IOP Conf. Series: Materials Science and Engineering **252** (2017) 012087 doi:10.1088/1757-899X/252/1/012087

The study of the energetic performance in test cycles

Cristian RENTEA, Alexandra TUCA, Mircea OPREAN, Marius BATAUS

Automotive Engineering Department, University Politehnica of Bucharest, 313 Spl. Independentei, Bucharest, Romania

mvbataus@yahoo.com

Abstract. One of the most important subsystems of modern passenger cars is the transmission. This paper aims to investigate the global performances of a modern transmission in different test cycles including the newly introduced WLTC (Worldwide Harmonized Light Vehicles Test Cycle). The study is done using a complex model developed in a performant simulation environment. Transmission efficiency calculation is emphasized, the efficiency being considered variable depending on engine torque, engine speed and gear ratio. The main important parameters (vehicle speed fluctuation, overall transmission efficiency, fuel consumption ratio) needed to compare test cycles and the transmissions performance are determined.

1. Introduction

Some of the main concerns for automotive manufacturers is to respect the mandatory emission reduction targets for new cars imposed by EU legislation. The fleet average to be achieved by all new cars by 2020-2021 is 95 g CO_2 / km, this means a fuel consumption of around 4.1 l/100 km of petrol or 3.6 l/100 km of diesel [1]. This target is based on NDEC (New European Driving Cycle) and is shown in [2] that this will translate in a WLTC (Worldwide Harmonized Light Vehicles Test Cycle) based target of 100 g CO_2 / km. The WLTC test cycle has as main objective the determination of the fuel consumption and pollutant emissions, close to the real driving. Unlike NEDC cycle, the WLTC test cycle is defined for several vehicle categories depending on the power/mass ratio and maximum speed declared by the manufacturer.

For emission prediction of newly developed vehicles the industry is heavily relying on modeling and simulation. Having such high targets it is essential to carefully choose the simplification hypothesis applied. One common simplification is the use of a constant transmission efficiency for every gear.

The transmission efficiency is dependent on several factors such as: transmitted torque, primary shaft speed, selected gear (or gear ratio), lubricant temperature, the level and quality of the lubricant and the constructive features of the transmission [3]. Therefore on different test cycles is functioning with different global efficiency. This paper aims to determine the effect of considering in simulations a variable transmission efficiency over the use of a constant value both at constant speed and in different test cycles. The efficiency is considered variable depending on transmitted moment, engine speed and coupled gear.

In order to simulate the performance, fuel consumption and emissions of a vehicle one can use LMS Imagine.Lab AMESim which is a programming environment for systems' modelling and simulation which is adequate for such studies [4], [5]. One of the main advantages is the availability of a library of components which can allow complex studies in most of the technical domains.

The fuel consumption in determined for a representative passenger car at constant velocity and for the following test cycles: US 06, Japanese 10-15, HWFET (Highway Fuel Economy Test Cycle), NEDC and WLTC. The vehicle speed fluctuation, overall transmission efficiency and fuel consumption ratio are determined and used to compare test cycles and the transmission performances.

2. Model development

Figure 1 presents the model that was developed in AMESim for the purpose of this study. The automobile's modelling for studying energetic performances needs the use of three libraries: IFP Drive, Mechanical and Signal & Control.

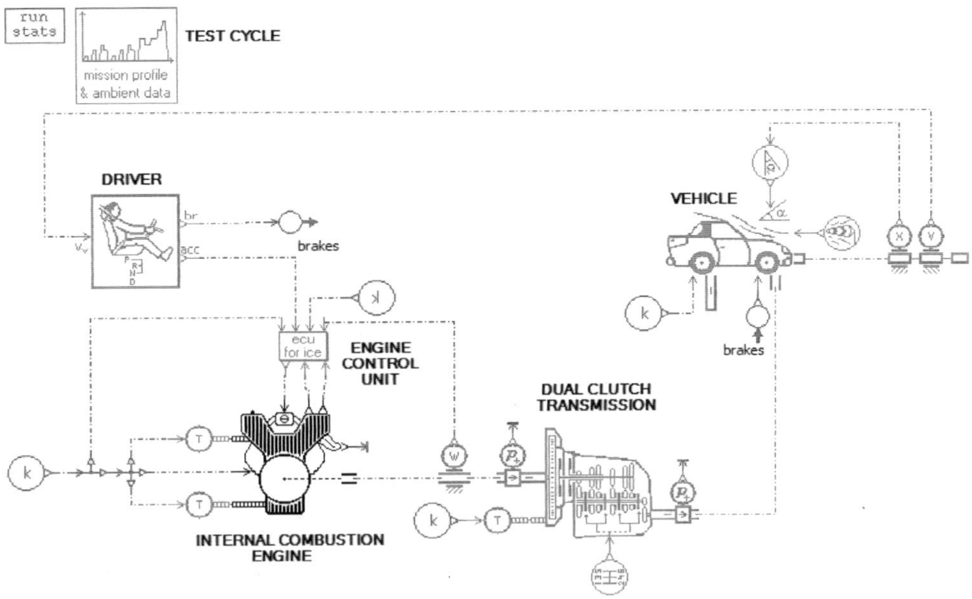

Figure 1. The vehicle model for energetic performances studies

The vehicle/powertrain model is a dynamic one and is composed by the following important submodels from IFP Drive library: internal combustion engine (DRVICE01D), engine control unit (DRVVECU1A), dual clutch transmission (DRVDCT01) and 1-D vehicle (DRVVEH4A). A driver submodel (DRVDRVA00A) is also needed to follow the velocity profile introduced with the mission profile and ambient data submodel (DRVMP2A). Because a single gear change law is used (for minimum fuel consumption) there is no need for a transmission electronic control unit and the mission profile submodel is used for computing the current gear. In this preliminary study the engine and transmission temperatures are considered constant and equal with the nominal ones.

The study is done for a middle class passenger car equipped with a 1.4 l turbocharged diesel engine and a 6 gears DCT (dual clutch transmission).

3. Transmission efficiency

In a general study on vehicles' fuel consumption [6], the powered axle efficiency is considered to be equal to the average value of 0.952, and the efficiency of the gearbox (η_g) is determined in each gear (j) using the relation (1).

$$\eta_{gj} = \frac{1}{1+a_j} - \frac{T_{0j}}{(1+a_j)T} \quad (1)$$

where: T_{0j} – resistant torque, a_j – coefficient determined experimentally and T – input torque.

The values obtained for the maximum input torque considering the given gearbox are centralized in table 1.

Taking into consideration the fact that will be performed also simulations where the variation of the transmission efficiency will be dependent on the transmitted torque, engine speed and coupled gear, the relation (2) is used, [6].

$$\eta_t = \eta_{tm} - \frac{R_0 \cdot r_r}{i_t \cdot T} \tag{2}$$

where: η_{tm} – transmission efficiency with regards to the torque dependent losses; i_t – total transmission ratio; T – input torque; r_r – rolling radius; R_0 – equivalent resistance to idling losses in the transmission.

Table 1. Efficiency values in each gear for maximum input torque

Gear	T_{0j} [Nm]	a_j [-]	η_{gj}
1	0.5	0.060	0.941
2	0.5	0.056	0.945
3	0.5	0.047	0.953
4	0.5	0.035	0.964
5	0.5	0.02	0.978
6	0.5	0	0.998

The transmission efficiency afferent to torque dependent losses is determined as follow:

$$\eta_{tm} = \eta_c^{N_c} \cdot \eta_{co}^{N_{co}} \cdot \eta_{ca}^{N_{ca}} \cdot \eta_{rul}^{N_{rul}} \tag{3}$$

where: η_c –efficiency of cylindrical gear; η_{co} – efficiency of conical gear; η_{ca} – efficiency of a universal joint; η_{rul} – efficiency of a rolling bearing; N_c – the number of pairs of cylindrical gears engaged under load; N_{co} – the number of pairs of conical gears engaged under load; N_{ca} – the number of universal joints; N_{rul} – the number of rolling bearings.

For the given transmission structure we have $N_c = 2$ and $N_{rul} = 7$. Considering from [6] the values $\eta_c = 0.97$ and $\eta_{rul} = 0.999$ will result $\eta_{tm} = 0.934$.

The resistance equivalent to idling losses in the transmission depends on the velocity of the vehicle, expressed as linear function, as shown in relation (4), [6].

$$R_0 = R_{00} + a_1 \cdot v \tag{4}$$

where: R_{00} – low speed transmission resistance; a_1 – coefficient of velocity influence; v – vehicle velocity.

From relations (2), (4) and considering the dependency relation between transmission input speed and vehicle velocity results:

$$\eta_t = \eta_{tm} - \left[\left(R_{00} + a_1 \cdot \frac{\pi \cdot n_m}{30} \cdot \frac{r_r}{i_t}\right) \cdot r_r\right] \cdot (i_t \cdot T)^{-1} \tag{5}$$

For front wheel drive with the engine of the vehicle transversely arranged we have: R_{00}=10 N and $a_1 = 0.3\ \mathrm{N} \cdot \frac{\mathrm{s}}{\mathrm{m}}$, [6].

Figure 2 depicts the variation of transmission efficiency for gears 1 and 6, depending on the engine torque for two values of the engine speed. The transmission efficiency increases with the increasing engine torque for a constant speed value. It can be observed that for gear 6 the transmission efficiency has higher values and, when the engine torque increases, a bigger variation of the efficiency does not appear as it does for gear 1.

Figure 3 underlines the variation in transmission efficiency for gears 1 and 6, depending on the engine speed for two values of the engine torque. Transmission efficiency varies slightly with the increasing engine speed when the engine is working under full load.

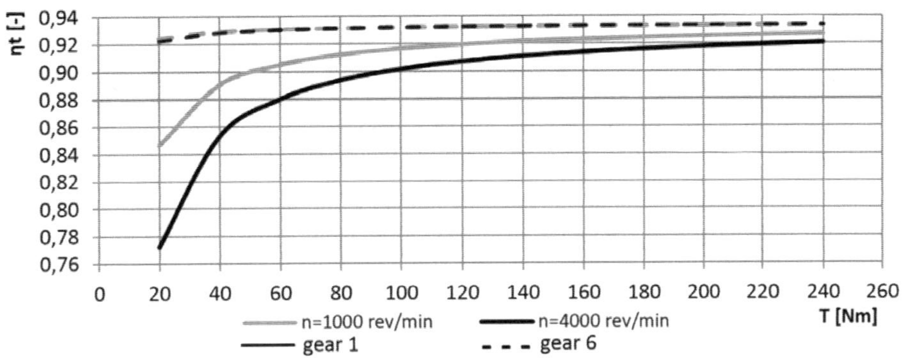

Figure 2: Variation in transmission efficiency depending on the engine torque

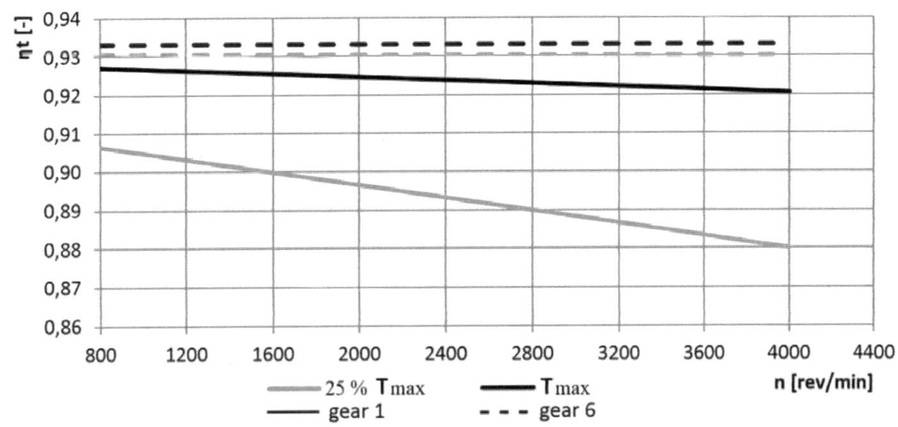

Figure 3: Variation in transmission efficiency depending on the engine speed

4. Vehicle Speed Fluctuation

In order to quantify the fluctuation of vehicle speed in a test cycle, a specific parameter called VSF (Vehicle Speed Fluctuation) is defined in [7] as follow:

$$VSF = \frac{\sigma}{m} \cdot 100 \; [\%] \tag{6}$$

where: σ – standard deviation of vehicle speed; m – average vehicle speed.

Table 2 summarize the VSF values computed from the cycles velocity profiles ([8], [9]) using the relations for normal distribution from [10].

Table 2. Values of the VSF for considered test cycles

Cycle	m [km/h]	σ [km/h]	VSF [%]
US 06	83.4	34.2	41
10-15	33.6	18.2	54.1
HWFET	78.2	15	19.2
NEDC	44.6	28.I	63
WLTC	53.3	33.5	62.8

Figure 4 presents normal speed distribution and frequency of different speed values with a multiple of 10 km/h for the WLTC.

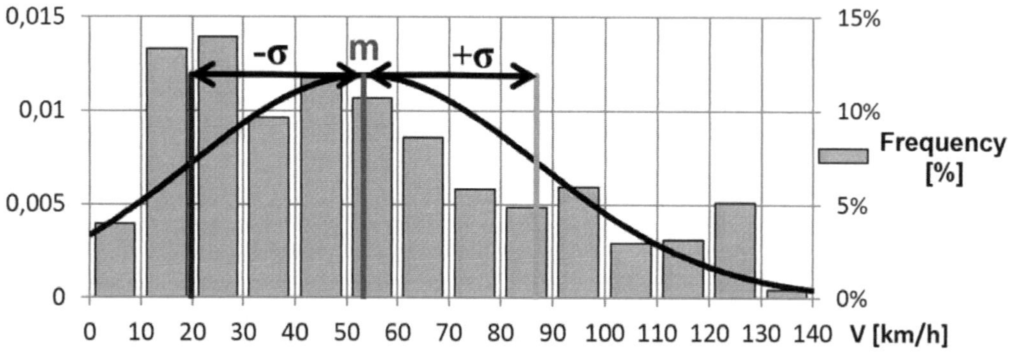

Figure 4. Normal speed distribution for the WLTC

For a speed of between 120 and 130 km/h the occurrence frequency is approximately 5%, which shows that the gearbox upper gears are used more often than in the NEDC.

5. Results

5.1 Transmission performance in test cycles

Using two power transducers to acquire the transmission input (P_i) and output (P_o) powers, the average transmission efficiency (including clutch losses) η_{t_cycle} is calculated over the entire cycle with relation (7), and the values obtained are centralized in table 3. Only the traction regime is considered (positive power).

$$\eta_{t\_cycle} = \frac{\int P_o\,dt}{\int P_i\,dt} \cdot 100\ [\%] \quad (7)$$

Table 3. Overall transmission efficiency

No.	Cycle	η_{t_cycle} [%]	
		For η_t constant	For η_t variable
1	US 06	90.9	89.5
2	10-15	81.2	82.1
3	HWFET	91.1	89.5
4	NEDC	84.9	84.7
5	WLTC	90.3	89.4

One can observe an absolute difference between 0.2 % and 1.4 % between the values obtained with the two methods. In terms of CO_2 emissions the maximum difference is of 2 g CO_2 / km (a 1.3 % error).

For the cycle 10-15, the lower gear shifts are used, so the global transmission efficiency in this cycle is lower than that obtained in the other cycles. The absolute difference between maximum and minimum global efficiency is 9.9 % for η_t constant and 7.4% η_t variable.

Figure 5 presents the variations of transmission input and output power and of transmission efficiency in WLTC. For engine brake and idle regimes the calculated values are replaced by 1. It can be seen that for short periods the transmission efficiency decreases dramatically attaining a minimum of 35%.

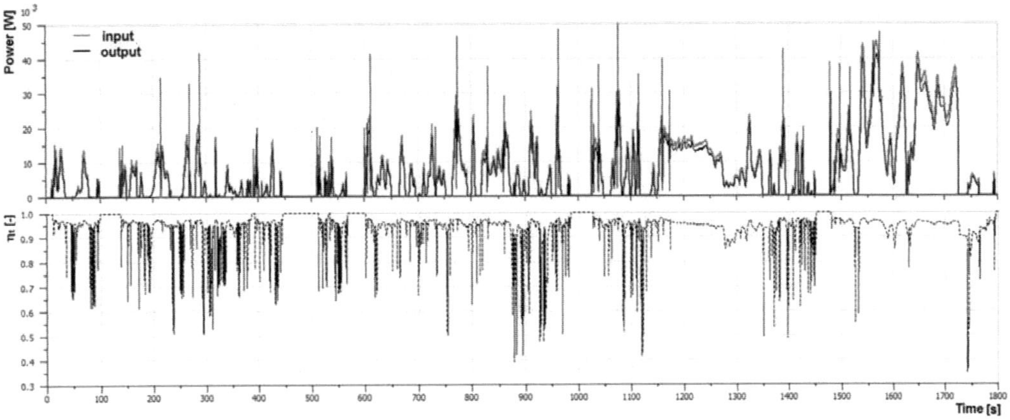

Figure 5. Variation of transmission efficiency in WLTC

5.2 Fuel consumption

In figure 6 is shown fuel consumption when driving at constant speed considering constant and variable transmission efficiency. Here, the speed at which the minimum fuel consumption is reached is around 53 km/h, which is the average speed corresponding to the WLTC cycle.

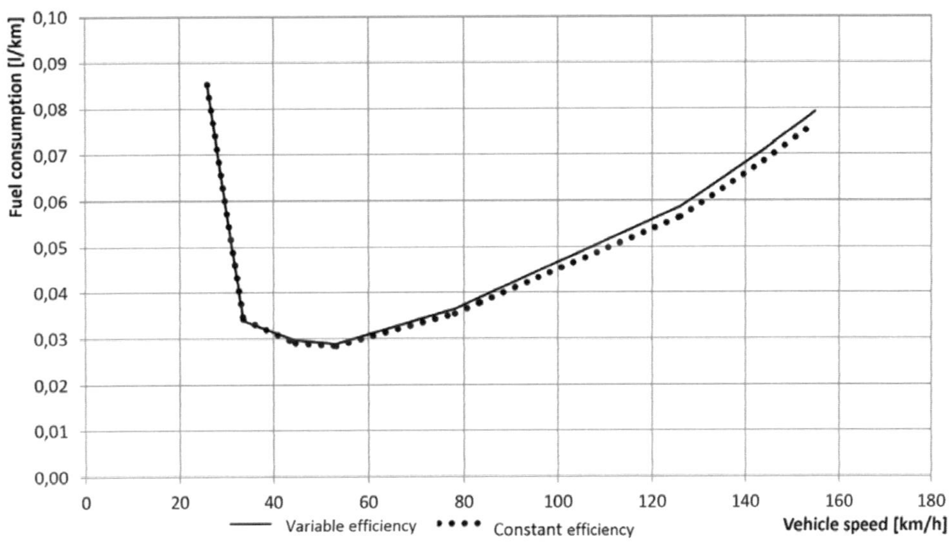

Figure 6. Fuel consumption at constant speed

In figure 7 the fuel consumption obtained at constant speed is compared with the fuel consumption obtained in the considered test cycles (using cycle average speed as abscissa). As expected, important difference can be observed regardless of the cycle average speed.

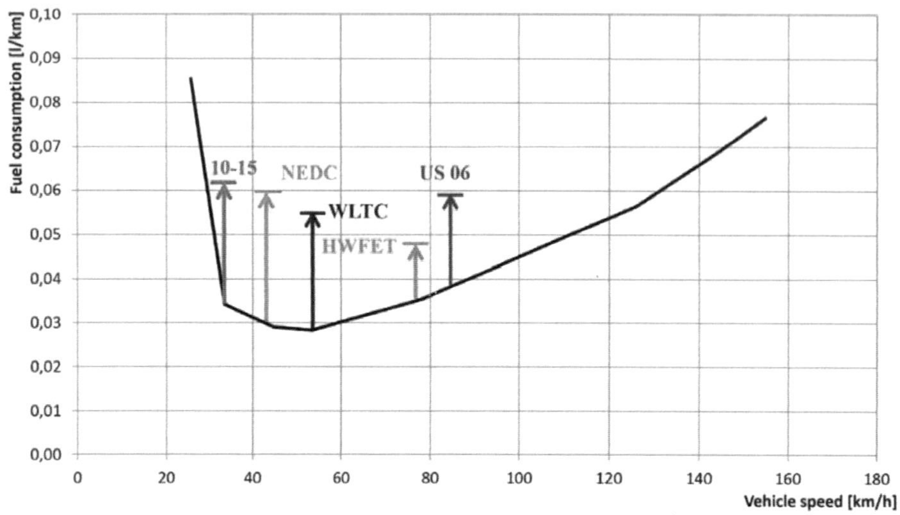

Figure 7. Fuel consumption obtained for constant efficiency

To compare the influence on fuel consumption of different test cycles we use a parameter called FCV (Fuel Consumption Ratio) defined in [7] as follow:

$$FCR = \frac{FC_{cycle}}{FC_{ct.speed}} \cdot 100\ [\%] \qquad (8)$$

where: FC_{cycle} – fuel consumption obtained in the test cycle; $FC_{ct.speed}$ – fuel consumption obtained at constant speed movement equal to the average speed of cycle;

In figure 8 is shown the correlation between FCR and VSF. A regression line having seat as intercept (imposed point) the corresponding point of the constant speed movement (0, 100) has a coefficient of determination $R^2 = 0.976$.

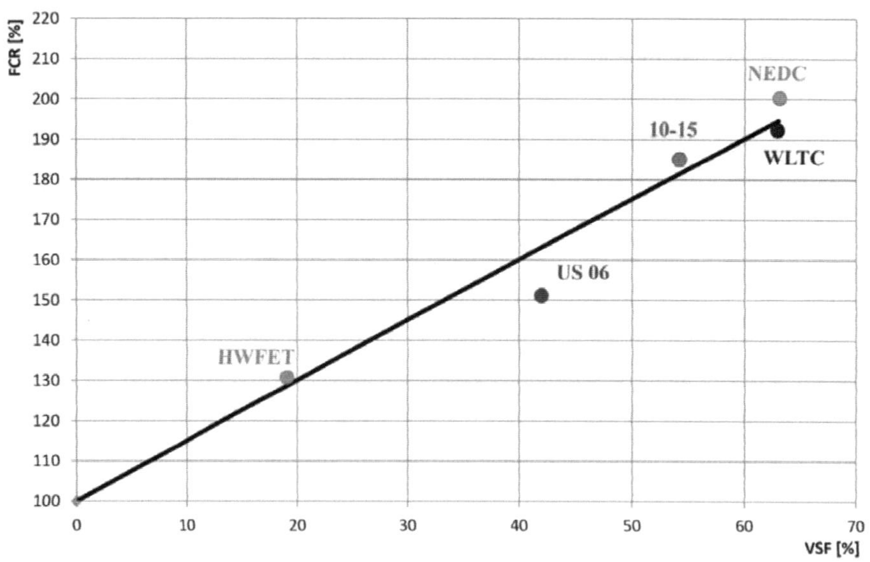

Figure 8. Correlation between FCR and VSF

6. Conclusions

When the engine and transmission temperatures are considered constant, considering only the variation of transmission efficiency with the gear can be sufficient regardless of the test cycle used. The maximum error in terms of CO_2 emission for the given application was of 1.3 % when compared to the values computed using variable transmission efficiency with gear, input torque and input speed.

In terms of CO_2 emissions the type of test cycle used has a big influence whit a maximum of 33 % relative difference between the minimum (in HWFET) and maximum (in 10-15). When compared using a relative parameter (FCR) the most disadvantageous cycles are NEDC and WLTC, the cycles with the highest VFS (around 63%). The linear correlation between FCR and VSF observed in [7] is also obtained for the current application and for the WLTC.

It is important to extend the study for a petrol engine application and to consider variable temperatures of engine and transmission.

Acknowledgements. This work has been funded by University Politehnica of Bucharest, through the "Excellence Research Grants" Program, UPB – GEX. Identifier: UPB–EXCELENȚĂ–2016, Research project: Use of scaled vehicles for automotive dynamics studies, Contract number: 46 (acronym: VEhRDI).

References

[1] Kohler, H. and Steiner, R. 2014 Efficient and Connected Powertrain - Challenges and Potentials *13th International CTI Symposium* (Berlin: Car Training Institute)

[2] Mock, P., Kühlwein, J., Tietge, U., Franco, V., Bandivadekar, A. and German, J. 2014 The WLTP: How a new test procedure for cars will affect fuel consumption values in the EU *International Council on Clean Transportation* pp 173–214

[3] Naunheimer, H., Bertsche, B., Ryborz, J. and Novak, W. 2011 *Automotive Transmissions. Fundamentals, Selection, Design and Application* (Springer-Verlag Berlin Heidelberg)

[4] Bățăuș, M.V., Dragne, F. D., Maciac, A. N., Oprean, I. M., Vasiliu, N. 2009 Models of Automotive Transmissions for Fuel Consumption Studies, Proceedings of Advanced Transmission for Low CO2 Vehicles (Paris: Institut Français du Pétrole)

[5] Bățăuș, M. and Vasiliu, N. 2012 Modelling of a Dual Clutch Transmission for Real Time Simulation, *UPB Sci. Bull.* D **74** 2 p 251

[6] Stoicescu, A. 2007 *Proiectarea performanțelor de tracțiune și de consum ale automobilelor* (Bucharest: Editura Tehnică)

[7] Oota, Y. 2013 The advantage of CVT in the real word *12th International CTI Symposium* (Berlin: Car Training Institute)

[8] Tutuianu, M., Marotta, A., Steven, H., Ericsson, E., Haniu, T., Ichikawa, N. and Ishii, H. 2013 Development of a Worldwide harmonized Light duty driving Test Cycle (WLTC). Technical Report (UNECE/WP. 29/GRPE/WLTP-IG DHC subgroup)

[9] Makarchuk, D., Kreicbergs, J., Grislis, A. and Gailis, M. 2015 Analysis of energies and speed profiles of driving cycles for fuel consumption measurements *14th International Scientific Conference on Engineering for Rural Development* (Jelgava) pp 265-271

[10] Forbes, C., Evans, M., Hastings, N. and Peacock, B. 2011 *Statistical Distributions* (John Wiley & Sons)

Comparative analysis by simulation for behavior of a spark-ignition engine fueled with gasoline and LPG in the transient regimes

Valentin Nisulescu[1], Florian Ivan[2], Daniel Iozsa[1], Gheorghe Banca[1]

[1]Polytechnic University of Bucharest, Romania,
[2]University of Pitesti, Romania

nisulescu.valentin@yahoo.com

Abstract. It is known that current vehicles must meet stringent demands on pollution limits but also must meet and the dynamical and economical performances. In this context the transient regimes are those affecting this performances, in this paper are presenting the results of the simulations for these regimes using a vehicle powered with two energy sources gasoline and LPG. Have been selected the transient regimes characteristic for NMVEG cycle (New Motor Vehicle Emissions Group). The simulation is performed using AMESim platform and the results have allowed meticulous interpretations for the 16 regimes of acceleration. The results obtained from the simulation will be validated experimentally.

1. Introduction

As a result of the continuous increase number of vehicles equipped with thermal engines at European and Global level has determined that the analysis of the quality of testing and the approval of pollutant emissions is currently a very important issue in the sense of realistic reproduction of the driving mode of the vehicle in relation to the current test cycles imposed [1]. All these changes regarding testing and approval become very difficult to achieve by vehicle manufacturers with the need for additional components mounted on motor vehicles to meet the legal requirements, components that increase manufacturing costs and depend on them the sale price of the vehicle [10], [6].
In this paper is carried out an analysis of pollutant emissions and fuel consumption using the speed profile of the European testing cycle. Were determined by simulation the differences of fuel consumption and pollutant emissions in the transient regimes for a vehicle equipped with spark ignition engine using double energy source, gasoline and liquefied petroleum gas. The simulation was realized using the software AMESim "Advanced Modeling Environment for Simulation of engineering Systems".

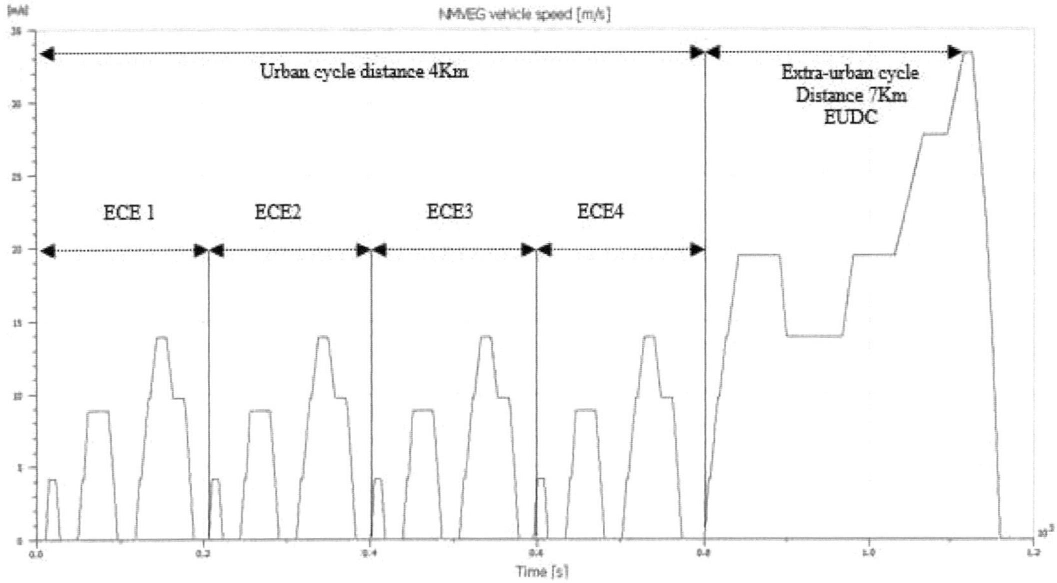

Figure 1. European testing cycle of the pollutant emissions NMVEG

As shown in figure 1 the current test cycle is characterized by a distance of 11,007 km to be performed by the vehicle in the speed conditions imposed, acceleration and deceleration conditions over a period of 1180 seconds [2].

The NMVEG cycle has two test areas Urban Driving Cycle (UDC) and Extra Urban Driving Cycle (EUDC). It can be easily seen that the urban area is realized between 0 km and 4 km while the extra-urban area is specific to the distance of 7 km to 11 km. For the urban area 4 identical speed profiles are encountered and each profile having a distance of 1 kilometer while for the extra-urban area there are 3 speed profiles.

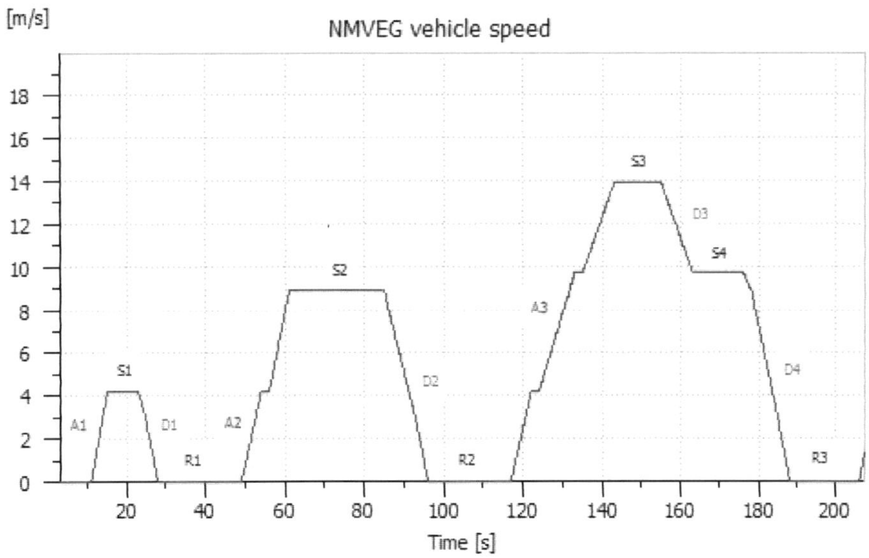

Figure 2. NMVEG Cycle Urban Zone

In figure 2 is indicated one of the four speed profiles. They were noted with R- idle regimes, with A- acceleration regimes, S-stabilized regimes and D - deceleration modes.

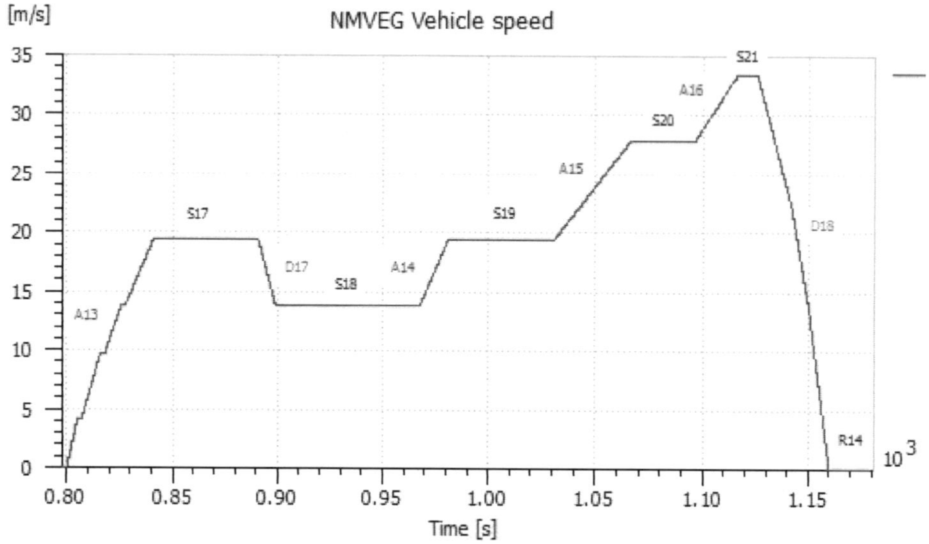

Figure 3. NMVEG Cycle extra-urban zone

In figure 3 is indicated the speed profile for the extra-urban area, the notations are identical to those in figure 2. Thus, considering all three figures 1, 2 and 3 we can easily determine that the regimes acceleration, deceleration, idle speed and stabilized can be encountered on the NMVEG cycle.
Their number has the following weight: 14 idle speed regimes, 16 acceleration regimes, 18 deceleration regimes, 21 constant speeds or stabilized regimes. Considering these aspects we find that the transient regimes, respectively the acceleration and deceleration regimes have together 497s of the total duration of the 1180s test and 42% of the cycle run. The transient acceleration regimes have a weight of 25% and the deceleration regimes of 17%. Values thus indicate the importance of these regimes to pollutant emissions and fuel consumption per cycle, hence concluding that pollutant emissions analysis on these regimes is justified [8].

2. Presentation of the model

As specified above, as the processing and simulation software was employed the program AMESim. The program is run for a vehicle of middle class equipped with a spark-ignition engine and manual gearbox with ratios parameterized according to the model of the chosen vehicle. The modeling is based on internal schematics for torque, polluting emissions and fuel consumption identified based on an actuation field for an engine of 1.6L MPI 75 kW.
Were also defined for model the following components:

- A component that simulates the driver component employed to ensure the control of the vehicle while running the NMVEG cycle. (fig.5)
- A component that simulates the exhaust gas treatment system specific for a spark ignition engine respectively three-way catalyst.
- A component that simulates the engine control unit.

The components listed above represent the main input values for the model, but at the same time it also uses other subcomponents, which provide information regarding the moment of inertia, engine torque, coupling stages, temperatures, masses of polluting emissions, which are the basis of the internal

calculation algorithms [5]. The model employed is intended to identify, based on the internal mathematical algorithms, the differences of performance, economic and ecological based on the consecrated formulas, differences shown in agreement with the running conditions imposed by the European cycle NMVEG for the two power sources.
The two types of fuel are characterized by different values of density and calorific power [9].

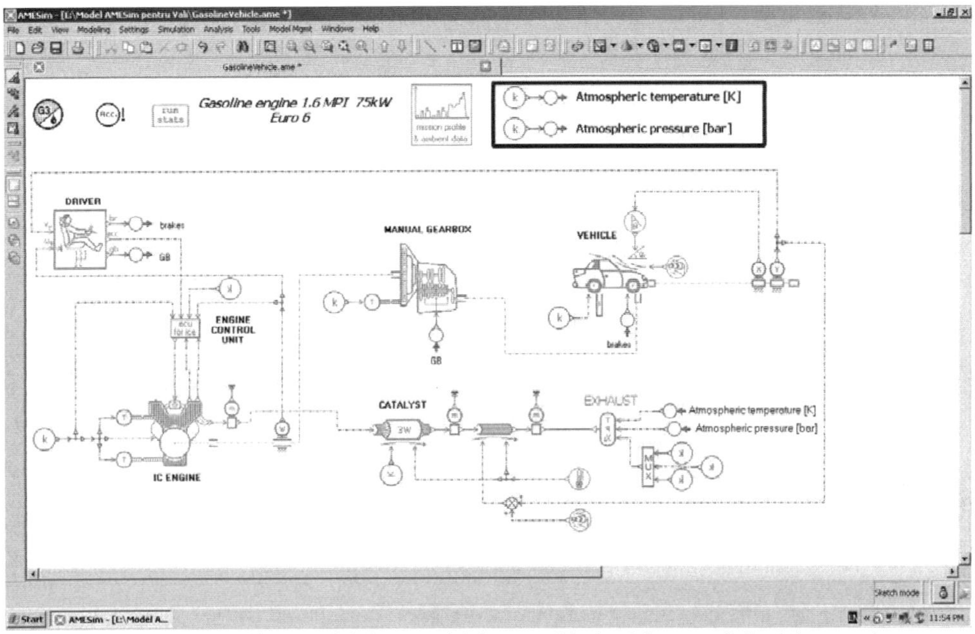

Figure 4. AMESim model for a vehicle provided with a spark ignition engine

Regarding the transient regimes only the acceleration regimes have been studied in the present paper, conscious that these are the major generators of pollutant emissions [3]. The transient deceleration regimes are characterized by a lack of injection which ultimately indicates very low and insignificant pollutant emissions to be taken into account [7].
A component in achieving the model in terms of vehicle speed control is the driver. This is described in figure 5 and he makes the accelerating, braking and gearshift as outputs commands [2].

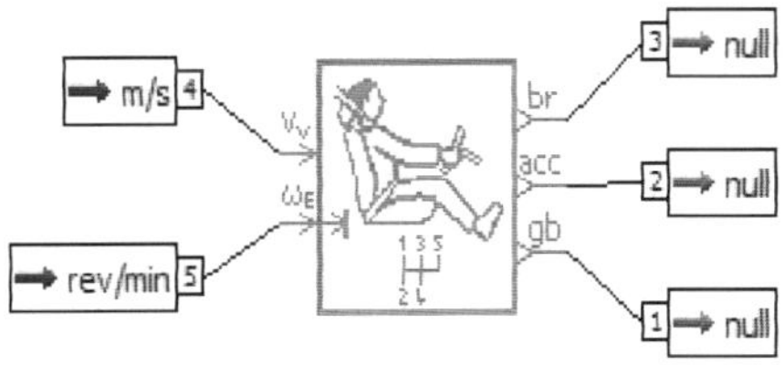

Figure 5. Driver component

The driver is one of the most important elements taken into account in the case of the test cycle because in reality the behavior of drivers on the road is different. Compared with gasoline test in the case of liquefied petroleum gas to respect the cycle speed condition it is necessary to press supplementary the accelerator pedal [4]. All these factors make the driver the "target component" because the results of pollutant emissions can be severely affected by its behavior. This is the reason why the driver component was selected to be detailed.

IOP Conf. Series: Materials Science and Engineering **252** (2017) 012088 doi:10.1088/1757-899X/252/1/012088

The control of the vehicle on the cycle for simulating accelerations and brakes is based on the use of the following relationships:

- The vehicle speed control error *err* is calculated with relation 1:

$$err = Vcont - Vveh \tag{1}$$

where *Vcont* is the control speed [m/s]; *Vveh* vehicle speed [m/s];

- For acceleration control *acc* [m/s^2] is used the relation 2:

$$acc = GPacc\cdot err + GIacc\cdot\int err\cdot dt + GAacc\cdot dvcontAnt \tag{2}$$

in which *GPacc* is proportional gain for acceleration control loop [m/s^2]; *err* vehicle speed control error [m/s]; *GIacc* integral gain for acceleration control loop [m/s^2]; *dt* time [s]; *GAacc* anticipative gain for acceleration control loop [m/s^2]; *dvcontAnt* derivative of future speed control [m/s];

- Therefore the derivative of control for the future speed $dvcontAnt$ takes the form from relation 3:

$$dvcontAnt = \frac{vcontAnt - vcont}{advAnt} \tag{3}$$

where $vcontAnt$ is speed control at a certain time [m/s]; $vcont$ the control speed [m/s]; $advAnt$ advance time for control anticipation[s];

- The plottable variable titled "integral part of the acceleration control loop" *Iacc* is defined as follows:

$$Iacc = GIacc\cdot\int err\cdot dt \tag{4}$$

in which *GIacc* is integral gain for acceleration control loop [m/s^2]; *dt* time [s];

- For braking control *brak* is used the relation 5:

$$brak = -GPbr\cdot err - GIbr\cdot\int err\cdot dt - GAbr\cdot dvcontAnt \tag{5}$$

where *GPbr* is proportional gain for braking control loop [m/s^2]; *err* vehicle speed control error [m/s]; *GIbr* integral gain for braking control loop [m/s^2]; *dt* time [s]; *GAbr* anticipative gain for braking control loop [m/s^2]; *dvcontAnt* derivative of future speed control [m/s];

- The plottable variable titled "integral part of the braking control loop" *Ibr* is defined as follows:

$$Ibr = -GIbr\cdot\int err\cdot dt \tag{6}$$

in which *GIbr* is integral gain for braking control loop [m/s^2]; *dt* time [s];

3. Results and discussions

As a result of the modeling have been realized graphs for all pollutants CO, CO_2, HC and NOx where may be observed their variations in relation with the idle speed, acceleration, deceleration and stabilized regimes. Also on the graphs are indicated the pollutants obtained by simulation for both energy sources, can be easily drawn a conclusion on the order of magnitude that characterize each source of energy. Regarding the acceleration regimes which have been analyzed in this paper the values obtained are marked in the tables and it is possible to observe for each transient acceleration regime from 1 to 16 the cumulative instantaneous emissions.

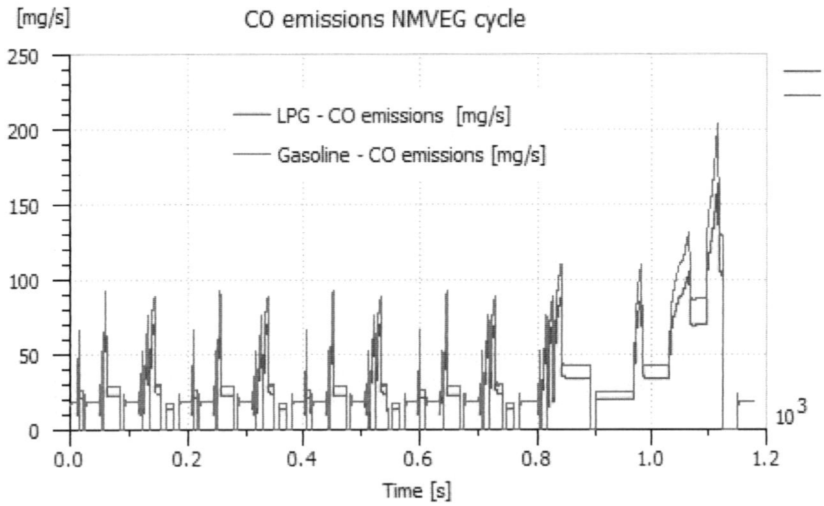

Figure 6. Instantaneous CO emissions of engine

In figure 6 is indicated the evolution of CO emissions at run to the European Cycle using the two energy sources. Regarding the first four parts of the cycle respectively the ECE urban cycle because they are identical at the speed profile roughly the same CO emissions were obtained. The precise values obtained for each acceleration regime specific to each fuel are shown in the table below.

Table 1. Cumulating of CO emissions in the acceleration regimes

Regime	CO emissions gasoline [mg]	CO emissions LPG [mg]
A1	133.9	106.5
A2	487.2	388.5
A3	1202.9	960.6
A4	133.9	106.5
A5	487.2	388.5
A6	1189.9	949.8
A7	139.9	111.3
A8	487.2	388.5
A9	1201	958.7
A10	133.9	106.5
A11	487.2	388.5
A12	1211.6	967.6
A13	2423.1	1936.5
A14	1166.9	930.6
A15	3701.4	2960.2
A16	3255.4	2606.5

In table 1 are displayed estimated values for the cumulating of CO emissions in the transient acceleration regimes for the two energy sources. An improvement of the CO emissions is observed in the case of the LPG energy source of about 20% for acceleration regimes from 1A to A16.

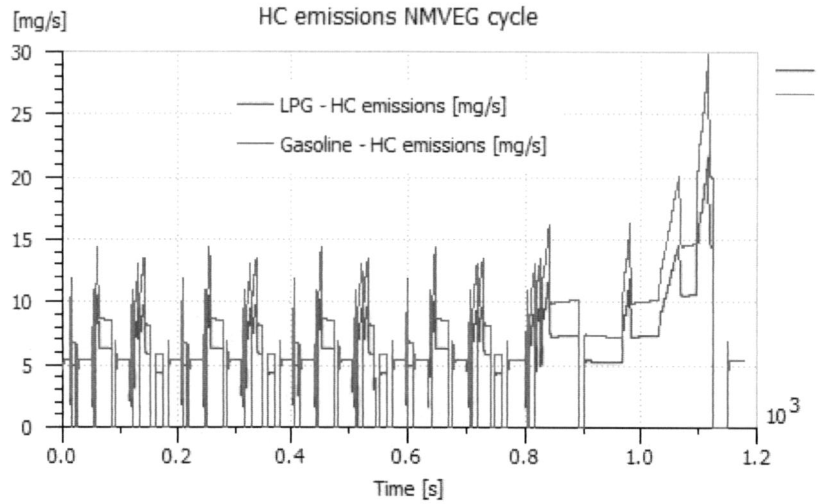

Figure 7. Instantaneous HC emissions of engine

In figure 7 is indicated the evolution of HC emissions at run to the European Cycle using the two energy sources. Regarding the first four parts of the cycle respectively the ECE urban cycle because they are identical at the speed profile roughly the same HC emissions were obtained. The precise values obtained for each acceleration regime specific to each fuel are shown in the table below.

Table 2. Cumulating of HC emissions in the acceleration regimes

Regime	HC emissions gasoline [mg]	HC emissions LPG [mg]
A1	29.1	21.2
A2	95.7	69.8
A3	233.7	169.8
A4	29.1	21.2
A5	95.7	69.8
A6	233.7	169.8
A7	29.7	21.7
A8	95.7	69.8
A9	233.7	169.8
A10	29.1	21.2
A11	95.7	69.8
A12	233.7	169.8
A13	421.9	306.3
A14	171.9	124.3
A15	562.4	407.6
A16	490.7	356.2

In table 2 are displayed estimated values for the cumulating of HC emissions in the transient acceleration regimes for the two energy sources. Here is also an improvement of HC emissions in the case of the LPG energy source of about 27% for acceleration regimes from 1A to A16.

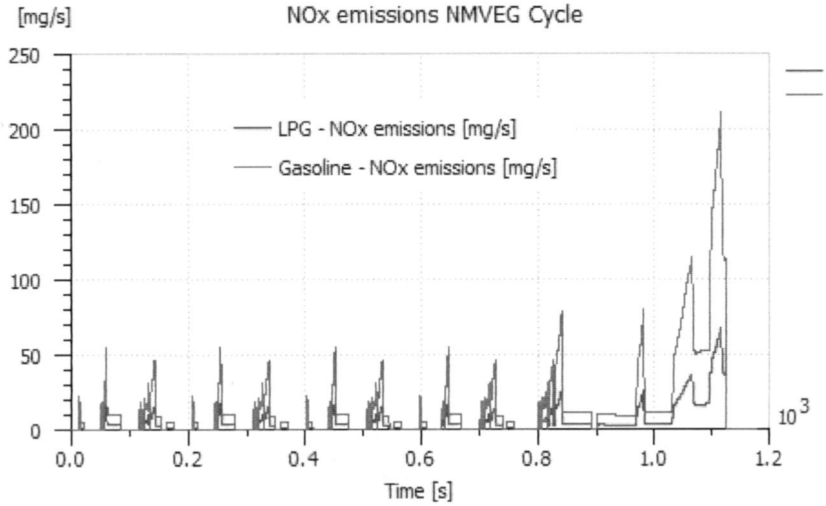

Figure 8. Instantaneous NOx emissions of engine

In figure 8 is indicated the evolution of NOx emissions at run to the European Cycle using the two energy sources. Regarding the first four parts of the cycle respectively the ECE urban cycle because they are identical at the speed profile roughly the same NOx emissions were obtained. The precise values obtained for each acceleration regime specific to each fuel are shown in the table below.

Table 3. Cumulating of NOx emissions in the acceleration regimes

Regime	NOx emissions Gasoline [mg]	NOx emissions LPG [mg]
A1	55.4	17.5
A2	223.4	70.9
A3	491.9	156.8
A4	55.4	17.5
A5	223.4	70.9
A6	491.9	156.8
A7	55.4	17.5
A8	223.4	70.9
A9	491.9	156.8
A10	55.4	17.5
A11	223.4	70.9
A12	491.9	156.8
A13	1204.9	384.5
A14	692.4	220.5
A15	2670.8	853.9
A16	3273.0	1047.9

In table 3 are displayed estimated values for the cumulating of NOx emissions in the transient acceleration regimes for the two energy sources.

Here is also an improvement of N0x emissions in the case of the LPG energy source of about 68% for acceleration regimes from 1A to A16.

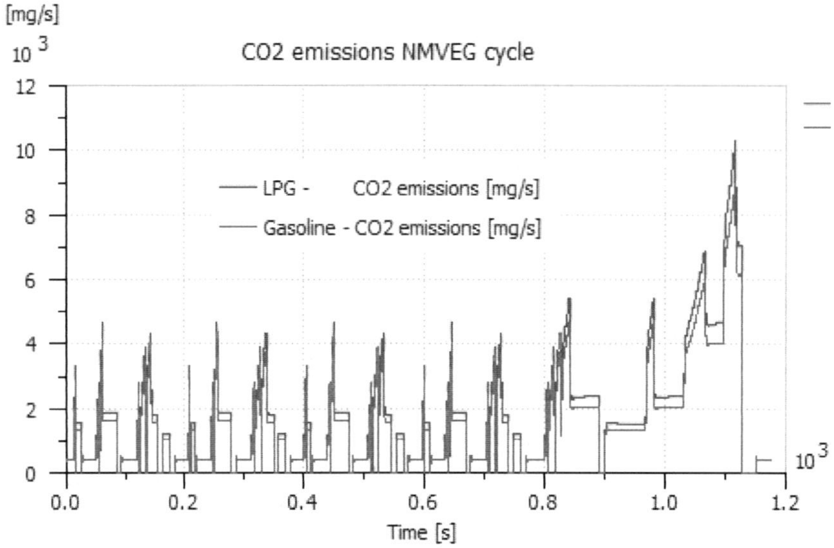

Figure 9. Instantaneous CO_2 emissions of engine

In figure 9 is indicated the evolution of CO_2 emissions at run to the European Cycle using the two energy sources. Regarding the first four parts of the cycle respectively the ECE urban cycle because they are identical at the speed profile roughly the same CO_2 emissions were obtained. The precise values obtained for each acceleration regime specific to each fuel are shown in the table below.

Table 4. Cumulating of CO_2 emissions in the acceleration regimes

Regime	CO_2 emissions Gasoline [g]	CO_2 emissions LPG [g]
A1	6.39	7.22
A2	22.73	25.91
A3	55.11	62.99
A4	6.39	7.22
A5	22.73	25.91
A6	55.11	62.99
A7	6.39	7.22
A8	22.73	25.91
A9	55.11	62.99
A10	6.39	7.22
A11	22.73	25.91
A12	55.11	62.99
A13	107.63	123.39
A14	48.80	55.91
A15	161.99	186.13
A16	149.09	171.40

In table 4 are displayed estimated values for the cumulating of CO_2 emissions in the transient acceleration regimes for the two energy sources. Unlike all other pollutants there is an increase of CO_2 emissions in the case of the LPG energy source of about 14% for the acceleration regimes from 1A to A16 which eventually indicates the difference in fuel consumption between the two fuels.

4. Conclusions

1. Regarding the values obtained by simulation for the cumulating of pollutant emissions in the transient acceleration regimes for the two energy sources resulted that during the LPG operation the emissions are reduced compared to the gasoline operation as follows: CO with about 20%, HC with about 27%, NOx with about 68%.

2. In the case of CO_2 emissions is an increase in the LPG mode with about 14% for the acceleration regimes from A1 to A16 which eventually indicates the difference in fuel consumption between the two fuels.

3. Given the prior conclusions can be easily stated that the liquefied petroleum gas represents a very good and current alternative power source for the spark ignition engines considering the indicators for economy and pollution. In this conditions presents an increasing interest for the car manufacturers, the number of vehicles provided from the factory with LPG systems is continuously expanding, having a very large coverage among the manufacturers.

References:

[1] Pundlik Nivrutti Patil, Dheeraj Sheshrao Deshmukh, Jitendra Gulabrao Patil *(ICGTETM-2016"A Review on Liquid Petroleum Gas as a Fuel in Spark Ignition Engine Application* "International Conference on Global Trends in Engineering, Technology and Management) ISSN: 2231-5381, pp 124-128.
[2] Help AMESim Rev12.
[3] BAUKAL, C. E. JR. 2000. *Heat Transfer in Industrial Combustion.* CRC, pp 65-117.
[4] HEYWOOD, J. B. 1988. *"Internal combustion engine fundamentals"*,New York: McGraw-Hill.
[5] Rory Lewis, Jason Bruke, Costa Mesa 2000, *Detailed vehicle thermal system modeling in ADVISOR trough integration with FLOWMASTER2.*
[6] J. W. Lee, H. S. Do, S. I. Kweon, K. K. Park, and J. H. Hong. 2010. *Effect of various LPG supply systems on exhaust particle emission in spark-ignited combustion engine.* International Journal of Automotive Technology Volume 11, Issue 6, pp 793–800
[7] Bo-yan Xu, Xiang-long Liu, Long-long Jiang, Juan Xu. 2017 *Simulation of mixed gas formation for a spray-wall complex guided LPG direct injection engine.* International Journal of Automotive Technology Volume 18, Issue 3, pp 489–497
[8] Florian Ivan, Daniel Lita, Andrei Busoi ,2014. "*Methods and means of depollution of the engines for vehicles* "ISBN 978-606-25-0038-2
[9] Thirumal Mamidi, Dr. J.G. Suryawnshi 2012 *"Investigations on S.I. Engine Using Liquefied Petroleum Gas (LPG) As an Alternative Fuel,"* vol. 2, no, ISSN: 2248-9622, pp. 362- 367.
[10] K.S. Shankar, P.Mohanan, 2011 "*The effect of LPG injection on combustion, performance and emission characteristics of a MPFI engine.*" International Journal of Energy and Environment, 2(4), pp761-770.

IOP Conf. Series: Materials Science and Engineering **252** (2017) 012089 doi:10.1088/1757-899X/252/1/012089

NEW TURBO COMPOUND SYSTEMS IN AUTOMOTIVE INDUSTRY FOR INTERNAL COMBUSTION ENGINE TO RECOVER ENERGY

R Chiriac[1], A Chiru[2] and O Condrea[3]

[1] Phd. Student, Faculty of Mechanical Engineering, Department of Automotive and Transport Engineering, "Transilvania" University, Romania
[2] Prof. Dr. Ing., Director of the Research Department, Faculty of Mechanical Engineering, "Transilvania" University, Romania
[3] Phd. Student, Faculty of Mechanical Engineering, Department of Automotive and Transport Engineering, "Transilvania" University, Romania

Abstract. The large amount of heat is scattered in the internal combustion engine through exhaust gas, coolant, convective and radiant heat transfer. Of all these residual heat sources, exhaust gases have the potential to recover using various modern heat recovery techniques. Waste heat recovery from an engine could directly reduce fuel consumption, increase available electrical power and improve overall system efficiency and if it would be used a turbochargers that can also produce energy. This solution is called turbo aggregation and has other ways to develop it in other areas of research like the electrical field. [1-3]

1. Introduction

The design of road transport engines is even more focused on reducing emissions. Recently, greenhouse gas commitments have brought new technological challenges. Exhaust gas recovery has great potential, given the amount of mechanical or electrical work that could be generated on board. The paper considers the recovery that could be obtained from exhaust gases that expand them into an additional turbine (turbo-mixing).

Over the years, the road transport sector has faced a period of strong technological change to reduce harmful emissions and, more recently, CO2 emissions, for example fuel consumption. The real technological surprise was that the emission reduction was achieved without losing the expected mechanical performance of the engine (torque, speed, driving fun, etc.). The results obtained were excellent: over the last two decades, the specific power (per unit of swept volume) increased by a factor of 1.5, while the emission level was 10 times. The EURO limits have progressively (and significantly) reduced steps, producing cleaner and more powerful engines. [4]

One of today's biggest challenges for reducing CO2 emissions is energy recovery, a key sector due to quantitative importance: the exhaust gases of an internal combustion engine have an energy content equal to about one-third of the chemical energy of the fuel. [5-6]

The first type of recovery seems to be more technologically simpler than the second one, which introduces a clear complexity when handled on board. The first recovery, which has the nature of direct recovery, is addressed in the literature as a turbo blend and has been extensively treated. Many manufacturers (John Deere, Volvo, Caterpillar), especially heavy engines, have already applied turbocharging, but have been very focused on the possibility of extra-mechanical power. Recently, in the literature, the possibility of having an energy recovery due to the turbo mix in the form of electricity has gained interest. [7]

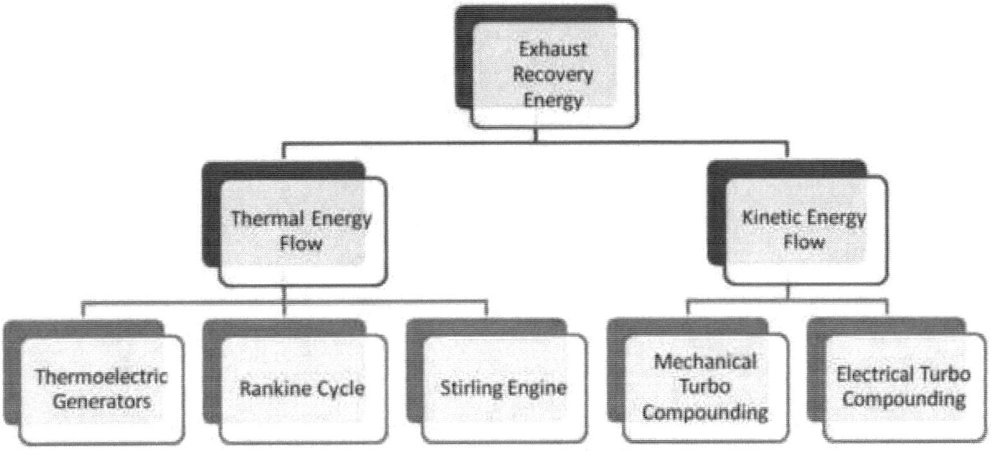

Figure 1. Main Methods of Exhaust Energy Recovery

A lot of fruitful efforts have been made to recover the energy from the exhaust gases. Figure 1 shows the main methods that have been practiced and applied to recover the energy from the exhaust gases. The most dominant technologies for the recovery of exhaust energy are the thermoelectric generators that convert heat directly into electricity, the implementation of the Rankine cycle through an expander, the Stirling engine, the mechanical combination with turbo and the blending with electric turbo compressors. [8]

Turbo compounding is the process of using an exhaust gas turbocharger to provide extra power to the crankshaft by means of a gear or hydrodynamic coupling or to power a generator that distributes power through an electronic power module. Technology is not new - the aerospace industry introduced the concept in the 1940s - but an article from the car and the driver says it could be "the next big thing in energy recovery" for automotive engines. Turbo compounding is a way to take advantage of the lost exhaust gas to provide more power to the engine. In the table above, the top right compressor (blue arrows) is fed by the turbine wheel to the left (red arrows) to send the improved intake air to the engine. However, the turbine sends exhaust gases to the left, where another turbine connected to a shaft drives a gearbox, essentially a high-speed transmission, eventually connected to the lower left crankshaft. [9]

2. Objective

Over time, it has been considered that the turbocharger has a greater potential to be developed in the field of research, such as mechanical, thermal and electric fields. But for these improvements we need to develop a constructive solution. Hybrid turbo compressors have dual purposes, such as, for example, the take-off of exhaust gas from the engine and also power generation by building the extended shaft to accommodate the alternator at the blower end to produce ecological electric power through the power Exhaust gases. The hybrid system can generate high energy efficiency, so high-speed rotation of the shaft and low weight of the parts, can produce no more than 5,50 KW without any problems. It also has an efficient construction component with a cooling system that is designed for self-cooling.

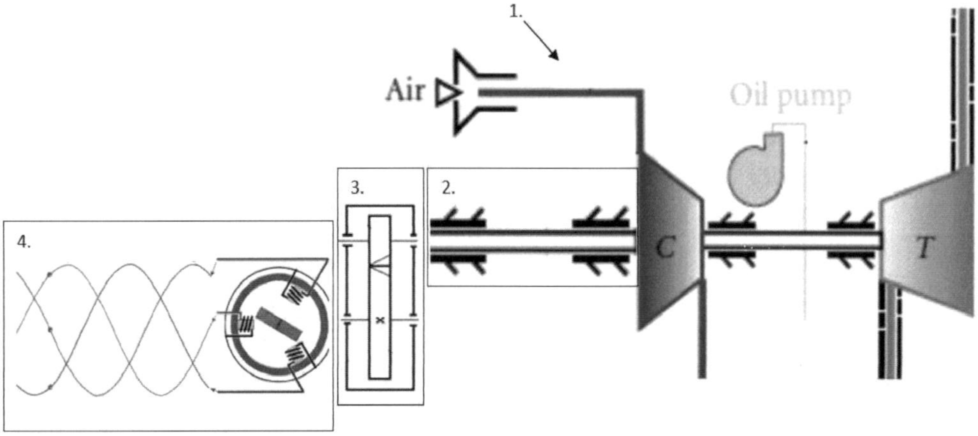

Figure 2. Schematic hybrid turbocharger solution

In figure 2- Schematic hybrid turbocharger solution, we can see that there are four groups of components which together form the proposed solution, namely: 1 - represented by the classic turbocharger (C-Compressor, T-Turbine, Oil pump), 2 - represented by external shaft, 3 – gearbox with cylindrical gear wheel, 4 - alternator / power generator. Also in the constructive schema can be seen in the marked area with 1, air inlet path labeled "Air". [10] The one-stage gear is optional but when it is used as a transmission solution is designed to reduce the generator rotational speed by a ratio 1/4 in relation with to the rotational speed which the turbocharger operates.

3. Methodology

The configurations that we are studying new turbo compound system in automotive industry for internal combustion engines to recover the energy from the engine and also a part of it to transform it into green energy for the automobile. The basics equations consist of energy, impulse and mass conservation, resolved along the average flow line. The mass and energy equations are solved for each volume and the impulse equation is resolved for each volume boundary.

Equation (1): Mass continuity equation

$$\frac{dm}{dt} = \sum\nolimits_{boundaries} m_{flux} \quad (1)$$

Equation (2): Conservation of energy equation

$$\frac{d(me)}{dt} = p\frac{dV}{dt} + \sum\nolimits_{boundaries} m_{flux} H - h_{flux} A\left(T_{gas} - T_{wall}\right) (2)$$

Equation (3): Engine power

$$P_{engine} = BME \quad {}_D\eta_{cycle} \; (3)$$

The values recorded for the energy balance are:

Tabel 1. Indicative values for the thermal flows heat balance

Engine type	$\dot{Q}_e$[%]	$\dot{Q}_{ch}$[%]	$\dot{Q}_w$[%]	$\dot{Q}_n$[%]	$\dot{Q}_r$[%]
Spark ignition engine	21 . . . 29	14 . . . 28	28 . . . 48	1 . . . 40	8 . . . 19
Diesel engine	28 . . . 41	16 . . . 33	24 . . . 44	1 . . . 4	8 . . . 16

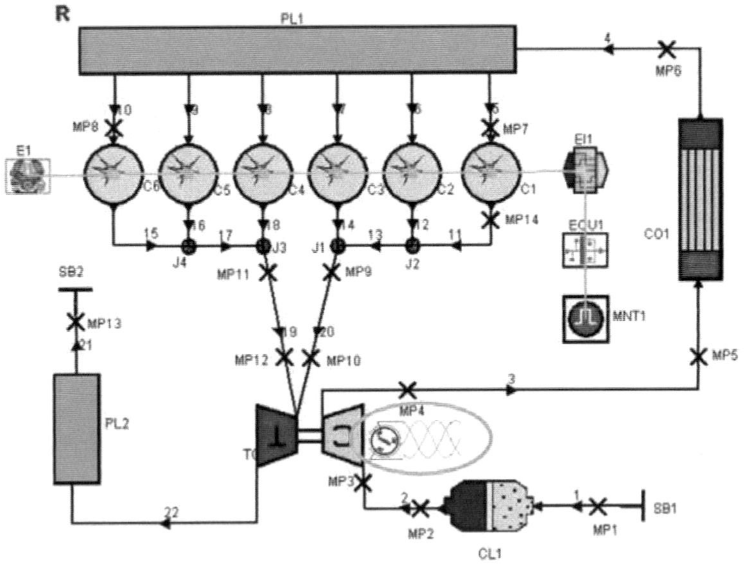

Figure 3. AVL model of the original engine

Figure 3 shows the AVL BOOST model of the original engine system that is with a turbocharger with a generator attached . The main goal of simulating the original engine is to have a reference engine performance and after that a reference of a engine simulation with a hybrid turbocharger. Which can then be compared with composite turbo engines. All the parameters set will be in accordance with the manufacturer's datasheet. [11]

4. Results

The external balance equation takes the following form:

Equation (4): Engine heat balance

$$\dot{Q} = \dot{Q}_e + \dot{Q}_{ch} + \dot{Q}_w + \dot{Q}_n + \dot{Q}_r \ (4)$$

where:
$\dot{Q}$ – the heat flow obtained by burning the fuel consumed by the engine (available energy); $\dot{Q}_e$ – useful heat; $\dot{Q}_{ch}$ – the heat flow lost through the cooling fluid; $\dot{Q}_w$ – the heat flow lost through the combustion gases along the evacuation; $\dot{Q}_n$ – the heat loss due to incomplete chemical burn; $\dot{Q}_r$ – the residue of the heat balance. [12]
Therefore the:

Equation (5): Useful heat

$$\dot{Q}_e = P_e \ (5)$$

P_e- engine power

IOP Conf. Series: Materials Science and Engineering **252** (2017) 012089 doi:10.1088/1757-899X/252/1/012089

Equation (6): Heat flow lost through the cooling fluid

$$\dot{Q}_{ch} = \dot{m}_{water} c(t_e - t_i) \ (6)$$

$\dot{m}_{wather}$- cooling water flow
c - average specific heat of water
t_e, t_i - inlet and outlet engine water temperatures

Equation (7): Heat flow lost through the combustion gases along the evacuation

$$\dot{Q}_w = \dot{m}_g c_{pmg} t_{ge} - \dot{m}_{air} c_{pma} t_a \ (7)$$

$\dot{m}_g$- the exhaust gas flow rate
$\dot{m}_{air}$ - air flow rate
c_{pmg}, c_{pma} - specific exhaust gases and air
t_{ge} - gas temperature in the outlet
t_a- inlet air temperature

Equation (8): Heat loss due to incomplete chemical burn

$$\dot{Q}_{ch} = \sum V_i Q_l \ (8)$$

V_i - the amount of uncombined components in the exhaust gas
Q_l - lower combustion heat of the gaseous components in the exhaust gas

Equation (9): Residue of the heat balance

$$\dot{Q}_r = \dot{Q} - (\dot{Q}_e + \dot{Q}_{ch} + \dot{Q}_w + \dot{Q}_n) \ (9)$$

After calculation, it can be estimated that 30% of engine exhaust gases can be recovered and from this proportion it can be produce further on green energy with the hybrid turbocharger. There are several solutions to recover some of it. A technical option is the one proposed in the paper. It is found in various other similar variants on existing engines like that for turbocompound engines but the principal of the hybrid turbocharger working is related with the electrical generators used in the marine industry. It can be seen in figure 4 an example of hybrid turbocharger used in the marine industry produced from Calnetix Company.

Figure 4. Hybrid Turbocharger- Calnetix product [13]

With the AVL Software it will be evaluate the energy limits of the system that it is proposed. This is partly done on the basis of the expertise, assimilated in the laboratories of the Research and Development Center - High Tech Products for Automotives – "Transilvania" University. Further results will be developed and simulated in the virtual environment and presented.

5. Limitations of the proposed study (if any)

Regarding the constructive part, the requirements for setting up hybrid turbocharger are these three basic things:
1- Conventional turbocharger with extended shaft to accommodate alternator at blower end,
2- A specially designed, very compact alternator to run at very high speed of around 100.000- 200.000 rpm.,
3- A cooling system for alternator, as heat generation will be more due to its compact size for given rpm (optional).
The rotation may vary due to the turbine's variable geometry on the turbine side with several running domains. The turbo compact has two parts: the turbine, where the flue gas is and the spiral compress with the fresh air. The two parts in turn are coupled by a shaft. The developed electric current has three possibilities:
1- Storage in the main battery of the vehicle, extending its lifetime,
2- Redirect to peripheral computer consumers,
3- Storage and redirection according to the central control computer.
Other studies will be presented. [14-15]

6. Conclusions

Turbochargers are used by Diesel engine manufacturers and a growing number of gasoline engines, especially in terms of size reduction. This component must be well understood and designed as a simulation because it is widely used at every stage of development. Indeed, cost and development time must be reduced to meet both customer wishes and stricter emissions standards. The current turbocharger simulation codes are mostly based on the search tables (airflow and air mass) provided by manufacturers. This papers aim is to present the new turbo compound system in automotive industry for internal combustion engines to recover therefore the hybrid turbocharger solution is a reliable, practical and very efficient solution in many ways: production costs, maintenance, reduced geometric geometry in terms of its applicability. From an ecological point of view: it reduces the consumption of the main battery of the vehicle and produces green electricity. From a constructive and economical point of view, production, installation and maintenance costs are very low and the product has a long life span in relation of its applicability. [15]

7. References

[1] F Millo and F Mallamo 2006 The Potential of Electric Exhaust Gas Turbocharging for HD Diesel Engines *SAE Technical Paper*
[2] S Nadaf and P Gangavati 2014 A review on waste heat recovery and utilization from diesel engines *Int. J. of Adv. Eng. Tech.*
[3] T Yamagami and N Oda 2009 Current Possibilities of Thermoelectric Technology Relative to Fuel Economy *SAE Technical Paper*
[4] International Council on Clean Transportation 2011 Global Passenger Vehicles Program Global Comparison of Light Duty Vehicle Fuel Economy GHG Emissions Standards
[5] F Jianqin and L Jingping 2011 A Study on the Prospect of Engine Exhaust Gas Energy International Conference on Electric Information and Control Engineering (Wuhan China)
[6] R Toom 2007 Waste heat regeneration system for internal combustion engines Messe Stuttgart
[7] A Patterson and R J and Tett J McGuire 2009 Exhaust heat recovery using electro-turbo generators *SAE Paper*
[8] D Hountalas and V Lamaris 2009 Recovering Energy from the Diesel Engine Exhaust Using Mechanical and Electrical Turbocompounding *SAE Technical Paper*

[9] D Hountalas and C O Katsanos 2007 Recovering energy from the diesel engine exhaust using mechanical and electrical turbo compounding *SAE Paper*
[10] J Seuma 2012 Impact of Turbocharger Non-Adiabatic Operation on Engine Volumetric Efficiency and Turbo Lag Int. J. of Rot. Mac. p 11
[11] A Teo 2015 Effectivess of series and parallel turbo compounding on turbocharged diesel engine *J. of Mech. Eng. and Sci.* pp. 1448
[12] M Luft 2012 Losses of IC engine a chance for electrical energy recuperation, Scientific Journals, Maritime University of Szczecin Poland
[13] www.calnetix.com
[14] N PERROT 2017 Experimental Study of Centrifugal Compressor Speed Lines Extrapolation for Automotive Turbochargers *SAE Paper*
[15] R Chiriac 2017 New constructive solutions for hybrid turbochargers- as electrical energy generator for increasing the green supply of the vehicule *International Conference for Doctoral Students* (Brasov: "Transilvania" University)

Acknowledgments
I would like to thank for the cooperation to the company Garrett, turbocharger producers, for support and interest.

Study on the combustion process in a modern diesel engine controlled by pre-injection strategy

P Punov [1,*], N Milkov [1], C Perilhon [2], P Podevin [2] and T Evtimov [1]

[1] Technical University of Sofia, Department of Internal Combustion Engines, Automobiles and Transport, Kl. Ohridski blvd. 8, 1756 Sofia, Bulgaria,
[2] Conservatoire National des Arts et Métiers, CMGPCE, 292 rue Saint Martin, 75003 Paris, France
*plamen_punov@tu-sofia.bg

Abstract. The paper aims to study the combustion process in a modern diesel engine over the engine operating map. In order to study the rate of heat release (ROHR), an automotive diesel engine was experimentally tested using the injection parameters factory defined. The experimental test was conducted over the engine operating map as the engine speed was limited to 2400 rpm. Then, an engine simulation model was developed in AVL Boost. By means of that model the ROHR was estimated and approximated by means of double Vibe function. In all engine operating points we found two peaks at the ROHR. The first is a result of the pilot injection as the second corresponds to the main injection. There was not found an overlap between both peaks. It was found that the first peak of ROHR occurs closely before top dead center (BTDC) at partial load than full load. The ROHR peak as a result of main injection begins from 4°BTDC to 18°ATDC. It starts earlier with increasing engine speed and load. The combustion duration varies from 30 °CA to 70 °CA. In order to verify the results pressure curve was estimated by means of defined Vibe function parameters and combustion duration. As a result, we observed small deviation between measured and simulated pressure curves.

1. Introduction

In direct injection diesel engines for passenger cars the pre-injection strategy is widely use in order to control the pressure rise in the combustion chamber, respectively the noise produced by engine [1]. That strategy requires precise control of the pilot injection quantity and timing as well as the timing of the main injection over the engine operating map [2, 3]. Thus, the variation of injection parameters causes difficulties in prediction the rate of heat release (ROHR) in numerical simulation of direct injection diesel engines. Jaipuria and Lakshminarayanan proposed a one-dimensional model for the prediction of the ROHR in common-rail direct-injection diesel engines with pilot, main, and post injections [4]. The results were found to be closed to experimental data.

Breuer in [5] studied double Vibe function parameters on a DI diesel engine using diesel fuel, rape oil and rape methyl ester. In this paper a "logarithmic anamorphosis" was used in order to define the six parameters of the function. Nikolov presented numerical study using a single Vibe function [6]. The variation of the Vibe function shape parameter leads to variation of the combustion beginning in order to achieve the maximum indicated mean effective pressure (IMEP). The double Vibe function parameters have been experimentally determined in [7]. The engine was tested with biodiesel blend. In this study the shape parameter of the first Vibe function varied within the range of 2.9 to 3.1 while the shape of the second Vibe function ranged from 0.7 to 1.6. Most of the studies are related to determination the double Vibe function parameters in diesel engine with single injection. For that reason, this paper aims to study the combustion process and the ROHR in direct injection diesel engine adapted to operate with pilot injection strategy.

2. Experimental set-up

In order to study experimentally the combustion process in a modern diesel engine an engine test bench was used. On that test bench the engine is coupled to a hydraulic engine brake which is manually controlled. An advanced engine management system was used in order to control the injection process as well as for data monitoring. This system is based on National Instruments real-time controller while the software is developed in LabVIEW. The combustion process was studied by means of in-cylinder pressure measurement. A piezo quartz pressure sensor was installed in the engine cylinder head. The indicated pressure was measured and registered by means of AVL indicating system. The scheme of the test facility is presented in Fig. 1.

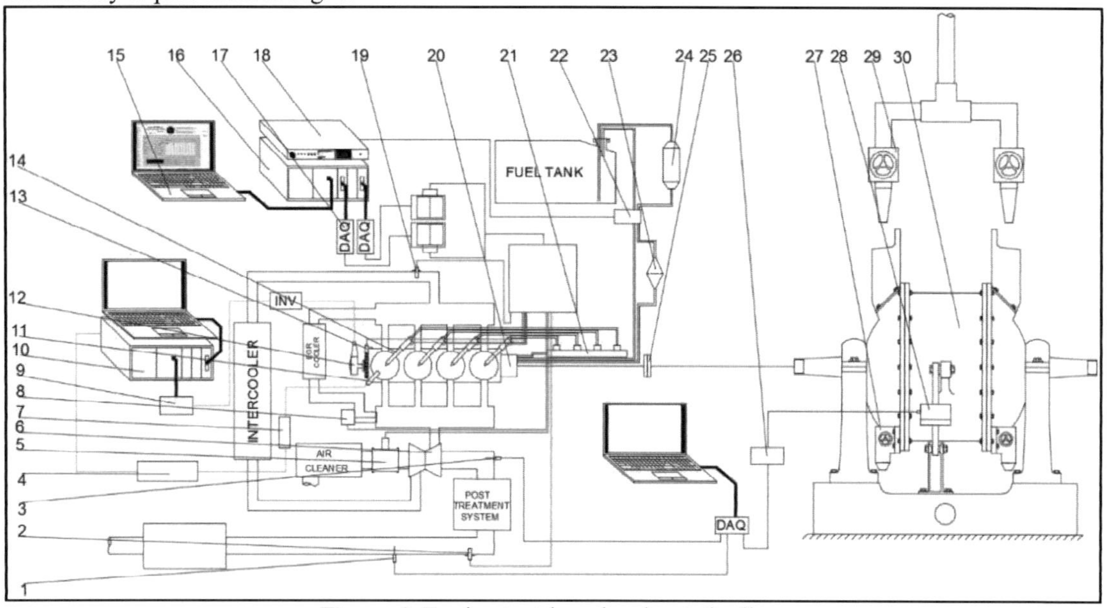

Figure 1. Engine test-bench schematic diagram

2.1. Studied engine

The engine under study is a direct injection diesel engine with displacement of 1998 cm^3. This engine is equipped with a Common-Rail system for direct injection of the fuel in the cylinder, turbocharging system, exhaust gas recirculation system (EGR), catalytic converter and etc. The rated engine power is 101 kW at 4000 min^{-1} and the maximum torque is 320 Nm at 2000 min^{-1}. The engine is equipped with a variable geometry turbocharger. The boost pressure is limited to 1.3 bar. The Common rail system of the engine is developed by Delphi. The maximum injection pressure is limited to 1600 bar. The main parameters of the engine are shown in Table 1.

Table 1. Engine main parameters

Engine type	HDI
Number of cylinders	4
Displacement	1998 cm^3
Bore	85 mm
Stroke	88 mm
Compression ratio	17,6
Valves per cylinder	4

2.2. Engine management system

The engine management system is based on National Instruments real-time controller. The hardware of the system consists of: PXI chassis - NI PXI-1031, Real-Time Embedded Controller - NI PXI-8106 RT, FPGA module - NI PXI-7813R, two R Series Expansion Chassis - NI CRIO 9151 as well as the modules developed by Drivven. In our project we used four modules: DI Driver – 2pcs, Low Side, AD Combo and O_2 sensor module. All of these modules are based on NI C series interface.

The entire system provides very large functionality for real-time control and monitoring of the engine control parameters such as: injection process control with up to five separate injections per cycle, injection pressure control, boost pressure control, exhaust gas recirculation (EGR) control, closed-loop control of injection by means of wide band oxygen sensor in exhaust gases and etc.

A LabVIEW project was developed in order to determine the control logic of the management system. The real-time data was visualized by means of the software CalVIEW developed by Drivven. The front panel of the software is presented in Fig. 2. This software was also used to set the operating parameters in real-time.

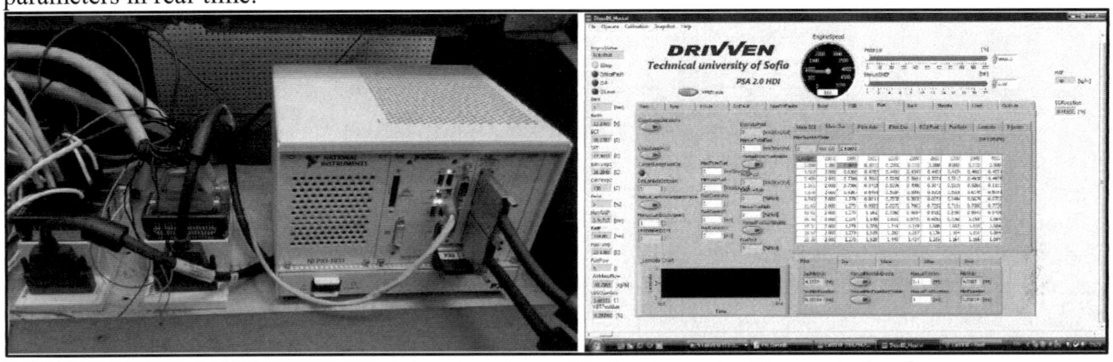

Figure 2. Engine management system with front panel of the control software

2.3. Cylinder pressure measurement

The measurement line consists of: In-cylinder pressure sensor AVL GH13P, Angle encoder AVL 364C, In-line converter PCB Piezotronics 422F01, Signal conditioner PCB Piezotronics F482B05 and indicating instrumentation AVL Indiskop 647. The pressure sensor is used in combination with a glow-plug adaptor AG04 a nearly flush mounted solution for diesel engines. It allows measurements without pipe oscillations. The measuring range of the sensor is up to 250 bar. In the measurement line an in-line converter and a signal conditioner were used. It provides signal converting to voltage without changing the signal level (converting 1pC to 1mV). The indicating system AVL Indiskop 647 was equipped with a two channel voltage amplifier 3065-A01.

3. Simulating model

In order to study the combustion process and to determine the ROHR a simulating model of the engine was developed by means of advanced software AVL BOOST. The block diagram of the model is shown in Fig. 3. It is a combination between 0D an 1D physical models developed for each specific element in the program. Our model consists of following elements: cylinders, plenums, pipes, air cooler, turbocharger, boundaries and the main engine element (E1).

Into the element called "cylinder" were defined geometrical parameters as well as the combustion model, heat transfer model and valve train mechanism by valve port specification, lift curve, flow coefficients and etc. The mathematical background is based on the first thermodynamics law applied to open control volume. The same 0D mathematical model is used in the plenums. AVL BOOST offers several approaches to modelling the turbocharger. In our case a simplified model was used that needs of target compressor and turbine pressure ratio as well as the constant isentropic and mechanical

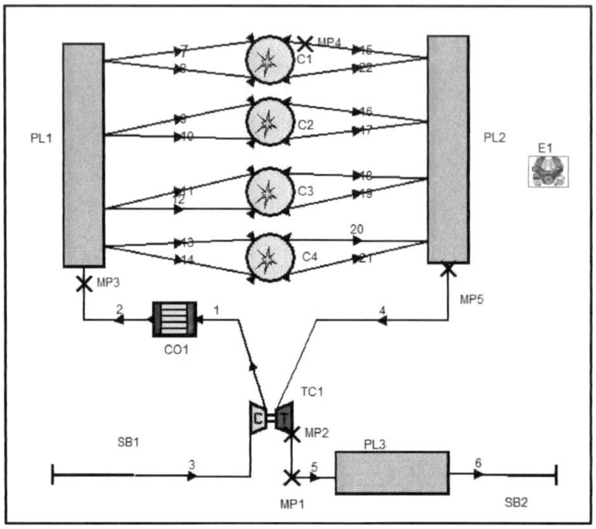

Figure 3. Engine simulating model in AVL BOOST

efficiency. The „Pipe" element connects the other elements of this model and represents the pipes of the manifolds. For each pipe were defined geometrical parameters such as the length and diameter. Also the friction coefficient and heat transfer coefficient between the wall and fluid were defined. A 1D model (Euler equation system) is used in order to define the fluid parameters along the pipe lines. The simulating conditions, friction losses and engine operating parameters were defined in the main "Engine" element.

In our study the simulating model was used in order to define the ROHR based on the experimental pressure curves. In this case a combustion model called "Target pressure curve" was chosen. Thus, the pressure curve was an input parameter in this study. Then, the ROHR was approximated by means of double Vibe function for each operating point. In order to assess this approach, estimated pressure curves were compared to these obtained experimentally.

The double Vibe function is defined as a sum of two simple Vibe functions in a correlation as follows:

$$\frac{dX}{d\varphi} = a_1.\frac{6{,}9}{\varphi_{z1}}.\left(\frac{\varphi_1}{\varphi_{z1}}\right)^{m_1}.(m_1+1)e^{\left(-6{,}9.\frac{\varphi_1}{\varphi_{z1}}\right)^{(m_1+1)}} + a_2.\frac{6{,}9}{\varphi_{z2}}.\left(\frac{\varphi_2}{\varphi_{z2}}\right)^{m_2}.(m_2+1)e^{\left(-6{,}9.\frac{\varphi_2}{\varphi_{z2}}\right)^{(m_2+1)}}$$

where:

$$a_1 + a_2 = 1$$

4. Results and discussion

4.1. Experimental results

The experimental study was carried-out over the engine operating range from1400 min^{-1} to 2400 min^{-1}, by an interval of 200 min^{-1}, and the engine load varies from idle to full load. This study was limited to that operating points due to the fact that the hydraulic brake is manually controlled which does not allow to keep the constant engine speed over 2400 min^{-1}. The results concerning the in-cylinder pressure variation are presented in Fig. 4. The maximum measured pressure at this engine speed of 1400 min^{-1} is 130.8 bar at BMEP of 15.2 bar. This value was measured at an angle position of the crankshaft 13 °CA after TDC. At engine speed of 1800 min^{-1}, the pressure reaches a maximum value of 168.9 bar at BMEP of 19.5 bar. This value is measured at an angle position of the crankshaft 11 °CA after TDC. At engine

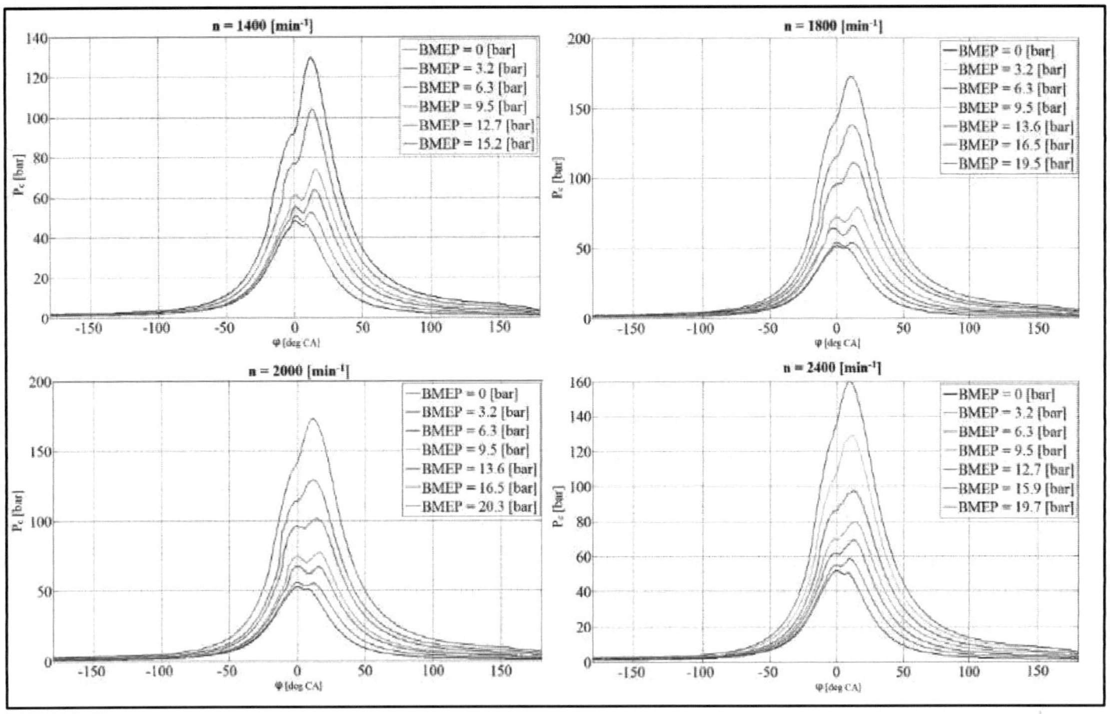

Figure 4. In-cylinder pressure measured experimentally

speed of 2000 min^{-1}, the maximum cylinder pressure is 172.5 bar, this value is measured at an angle position of the crankshaft 11 °CA after TDC. At engine speed of 2400 min^{-1}, the pressure reaches a maximum value of 159.9 bar at BMEP of 19.7 bar. This value is measured at an angle position of the

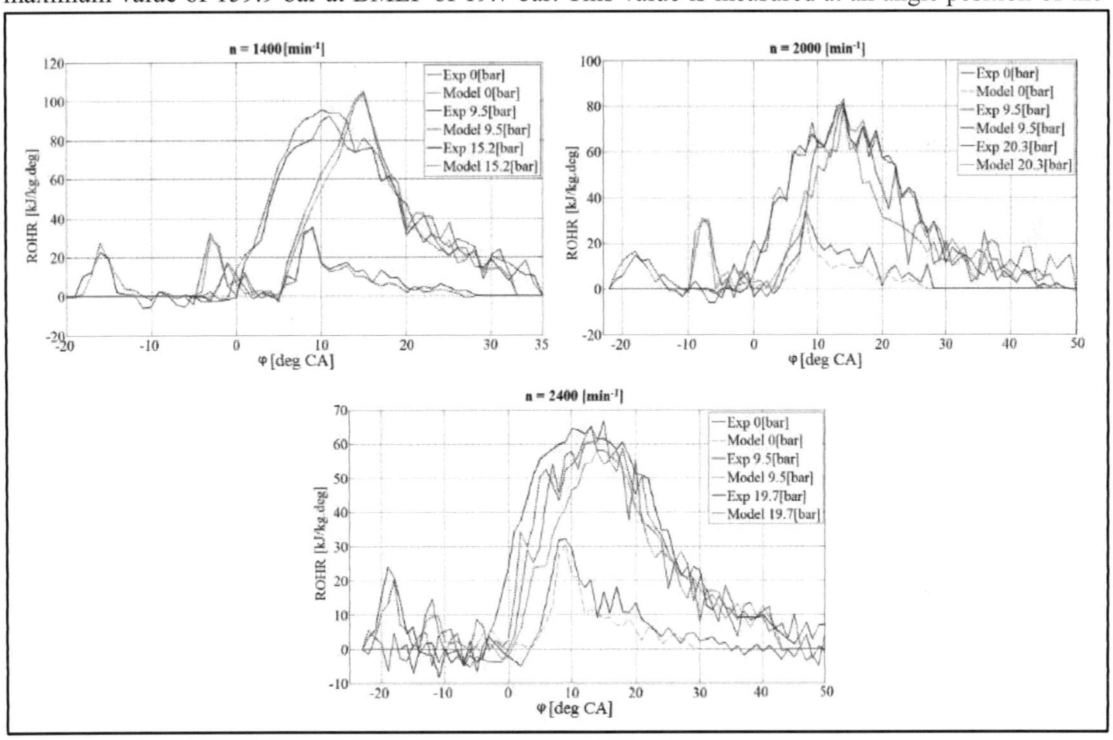

Figure 5. Rate of heat release

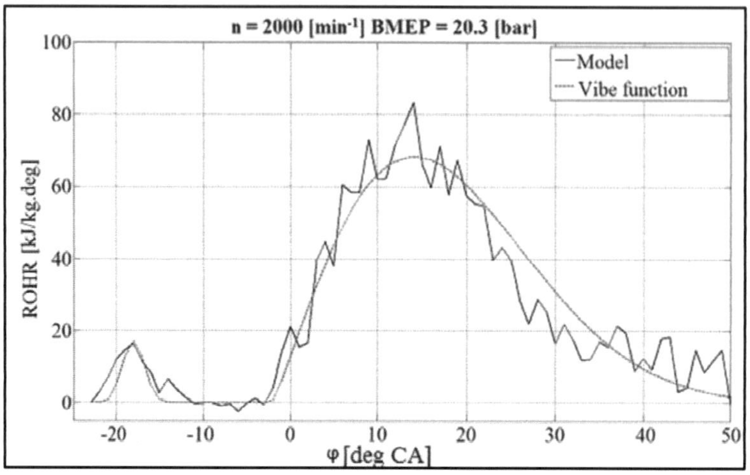

Figure 6. Approximated ROHR by double Vibe function

crankshaft 10 °CA after TDC. At the all studied operating points the pilot injection strategy is applied. Thus, it was observed slightly increased in-cylinder pressure before TDC due to the pre-injected fuel which causes higher pressure value at TDC than compression pressure. The results revealed that the maximum in-cylinder pressure occurs at the operating point corresponding to the maximum engine torque at 2000 min^{-1}. By increasing the engine speed over 2000 min^{-1} the maximum pressure slightly decreases.

4.2. Rate of heat release

The ROHR was estimated both by indicating instrument AVL Indiskop 647 and simulating model using the measured pressure curves and combustion algorithm "Target pressure curve". The results for engine speed of 1400, 2000 and 2400 min^{-1} are presented in Fig. 5. It was observed small deviation between curves estimated by both methods for all operating points. Dash line presented the results estimated by the engine model while the solid line was used in order to present the value estimated by Indiscop 647.

The results presented in Fig. 5 revealed that pilot injection strategy reduces the maximum value of the ROHR during the premixed combustion period. For each operating point we found two peaks at the ROHR. The first one is in result of the pilot injection while the second function is due to the main injection. There was not found an overlap between the peaks. It was observed slightly increasing ROHR after the start of main combustion without extremely high value due to the relatively small ignition delay period. The higher values for the ROHR were observed at low engine speed due to the lower mixing rate. The maximum value of 102 kJ/kg.deg is observed at BMEP of 9.5 bar. At full engine load the maximum of the ROHR is lower as it accounts to 94 kJ/kg.deg. At 2000 min^{-1} maximum value of ROHR as a result of the main injection was 90 kJ/kg.deg at 17 °CA after TDC.

4.3. ROHR approximation by double Vibe function

Due to the fact that the ROHR in diesel engine combustion simulation is easy to be defined by Vibe function or double Vibe function than real ROHR estimated by means of pressure curve we approximated the real ROHR by double Vibe function. It was done for each of studied engine operating points by defining the Vibe function parameters: shape parameter (m_1 and m_2), combustion duration (φ_{z1} and φ_{z2}) and the start of combustion. On the basis of obtained parameters as well as the combustion and mixing phenomena we defined the Vibe function parameters at whole engine operating range. Thus, we can study numerically the engine performances as well as the pollutant formation by means of our simulating model. Fig. 6 shows both ROHR estimated on the bases of pressure curve and approximated by double Vibe function. It concerns engine operating point at 2000 min^{-1} and full load (engine maximum torque). In order to validate the engine simulation model, we compared the experimentally

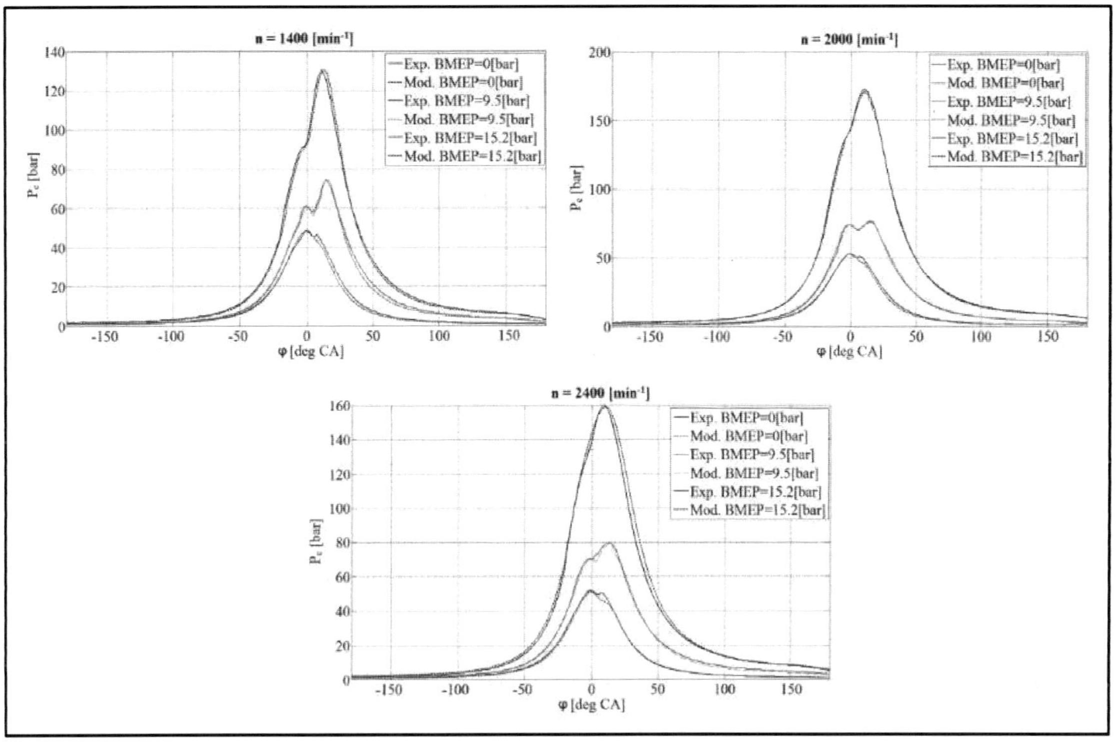

Figure 7. In-cylinder pressure model validation

measured cylinder pressure and those calculated by the model using the rate of heat release calculated by double Vibe function. The results are presented in Fig. 7. At 1400 min^{-1} we observed maximum

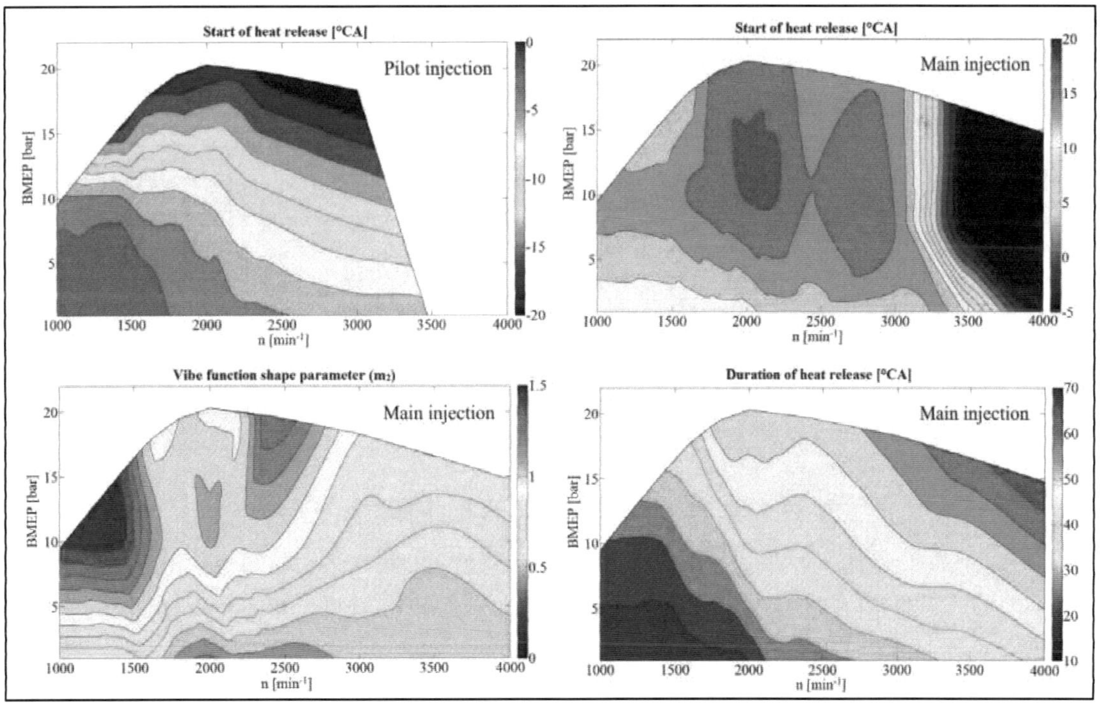

Figure 8. Parameters of the Vibe function

deviation between experimental and simulated values of 7 %. In absolute value it accounts to 3.25 bar at idle. At full load the maximum deviation was 6.3 %. At 2000 min^{-1} the maximum deviation of 7.4 % was estimated at idle. At other operating points the deviation is lower than 1.9%. At 2400 min^{-1} it was found the deviation of 9.1 % concerning operating point at idle and 8.9 % at full load.

The parameters of the Vibe function in the engine operating range outside the experimental range from 1400 min^{-1} to 2400 min^{-1} were determined as it was followed the trend of parameters variation taking into consideration the injection parameters and operating conditions as well. However, the engine operates with not pre-injection in the operating range above 3500 min^{-1}, thus a single Vibe function is used. In the zone without pilot injection, the Vibe function parameters were determined based on commonly used values adapted to direct injection diesel engines. Over the non-experimentally studied operating range, the validation of the model was carried-out by means of the maximum value of in-cylinder pressure. Therefore, the Vibe function parameters were determined over the engine operating range and the values are presented in Fig. 8. The meaning of colour fields is presented at the top of each diagram. Vibe shape parameter and duration concerning the pilot injection are not presented here due to the fact that the values are relatively constant over the whole operating range. The shape parameters were constantly equal to 2.5 and the duration ranged from 4 to 6 °CA.

5. Conclusions

A comprehensive analysis of the combustion process in a modern direct injection diesel engine is presented in this study. In order to study the ROHR an experimental test was conducted as the in-cylinder pressure was measured. Then, the ROHR was approximated by means of engine simulating model developed in advanced simulation product AVL BOOST. A double Vibe function was used in order to approximate the ROHR over the engine operating range when the pre-injection strategy is applied. Outside this area, a single Vibe function was used. The Vibe function parameters was determined and verified on the bases of estimated ROHR within the operating range from 1400 min^{-1} to 2400 min^{-1}. However, the Vibe parameters over 2400 min^{-1} was also determined follows the function variation trends and the injection parameters. The engine simulation model using double Vibe function was verified as the simulated in-cylinder pressure was compared to the measurement values. The maximum deviation was found to be 9.1 %. The results revealed that the Vibe shape parameters concerning the main injection varies from 0.6 to 1.5 over the engine map while the combustion duration of the main injection varies from 30 °CA to 70 °CA.

References

[1] Lakshminarayanan P A and Aghav Y V 2010 *Modelling Diesel Combustion*: Springer

[2] Punov P 2011 Research influence of multi-pulse pilot injection on combustion heat release and combustion proces in modern diesel engines In: *BulTrans,* Sozopol, Bulgaria pp 201-204

[3] Punov P and Evtimov T 2015 Combustion optimization in a modern diesel engine by means of pre-injection strategy *Machines, Technologies, Materials* **IX** pp 41-44

[4] Jaipuria A and Lakshminarayanan P A 2011 Prediction of the Rate of Heat Release of Mixing-Controlled Combustion in a Common-Rail Engine with Pilot and Post Injections *Proceedings of the Institution of Mechanical Engineers, Part D: Journal of Automobile Engineering* **225** pp 246-259

[5] Breuer C 1995 The influence of fuel properties on the heat release in DI-diesel engines *Fuel* **74** pp 1767-1771

[6] Nikolov V 2014 A simulation study of the influence of the relative heat release during combustion on the indicator diagram of a diesel engine *International Journal of Research in Engineering and Technology* **03** pp 175-179

[7] Pešić R B, Davinić A L, Taranović D S, Miloradović D M and Petković S D 2010 Experimental determination of double vibe function parameters in diesel engines with biodiesel *Thermal Science* **14** pp 197-208

Statistical models of petrol engines vehicles dynamics

C O ILIE[1,2], M MARINESCU[1], O ALEXA[1], R VILĂU[1] and D GROSU[1]

[1] Dept of Military Automotive Engineering and Transportation, Military Technical Academy - Bd. G Cosbuc nr. 39-49, sect 5, Bucharest, Romania

ovidiuilie66@yahoo.com

Abstract. This paper focuses on studying statistical models of vehicles dynamics. It was design and perform a one year testing program. There were used many same type cars with gasoline engines and different mileage. Experimental data were collected of onboard sensors and those on the engine test stand. A database containing data of 64th tests was created. Several mathematical modelling were developed using database and the system identification method. Each modelling is a SISO or a MISO linear predictive ARMAX (**A**uto**R**egressive–**M**oving-**A**verage with e**X**ogenous inputs) model. It represents a differential equation with constant coefficients. It were made 64th equations for each dependency like engine torque as output and engine's load and intake manifold pressure, as inputs. There were obtained strings with 64 values for each type of model. The final models were obtained using average values of the coefficients. The accuracy of models was assessed.

1. Introduction

In order to study any system, it is necessary to model it. The purpose of modeling is to obtain relevant conclusions of the original, based on model study. In the case of automotive systems, the model will be a mathematical one and describes their operation under the conditions or absence of disturbances imposed by the engineer. Mathematical model is theoretically elaborated and finalized on the basis of the data obtained from the experimental tests [5,6].

More and more restrictive pollution rules, traffic safety requirements and the need for effective diagnosis have led to the emergence and proliferation of electronic control systems onboard modern cars. All of these systems have reference databases containing records of the parameters concerned. At the vehicle level there is a large amount of data that needs to be processed in an efficient way. Thus, the need to use statistical methods and models arises [3,4].

Statistical models are obtained on the basis of large amounts of data. To achieve the goal, we needed to develop a testing program of vehicles dynamics. It aimed to capture a large range of operating regimes. There were used 14th same type cars with gasoline engines and different mileage. Vehicles have traveled over a year on runways like mosaic tiles, asphalt or pavement and in various weather conditions, snow, sleet, rain or sunny and warm weather. Driving style varied between normal and sport ones. Some of the engine parameters like power, consumption and torque were measured on test bench, but most data were recorded of on board sensors using a specialized tester with a reading frequency of 10 values per second. The result of the experimental research program is a set of 64th tests. A test contains 16 time-series of the measured parameters. Each time-series has 256 values. Data were saved in a database and became the reference for the modeling [1,7,8].

2. Model used

"Determining the best model in the set, guided by the data is the System Identification method." It is used parametric models having vectors as arguments. It is well known that the model of the vector ξ is noted $M(\xi)$. If the mathematical model of the process is " parameterized " by the vector ξ, the problem of the identification resides in determining or assessing the model's parameters on the basis of the experimental data of the input and output variables of the system (parameters identification). In other words, the analytical expression of the mathematical model is known, but the values of the coefficients (parameters) are not known. Model validation is done using different criteria. One of these is the one based on the least squares method. Thus, the objective function that is minimized is:

$$V(\theta, z) = \frac{1}{n}\sum_{i=1}^{n}[y(t) - \dot{y}(rI\theta)]^2, \quad (1)$$

where n is the number of values of the experimental series data, $y(t)$ is the experimental series, $\dot{y}(rI\theta)$ is estimated series, obtained by identification and z is an expression $z(t)=[y(t)u(t)]$ that encompasses y(t) as output sixe and $u(t)$ as input size. Both series are vectors with discrete values like experimental data are [2,1].

The general form of the model is:

$$A(q)y(t) = \frac{B(q)}{F(q)}u(t - nk) + \frac{C(q)}{D(q)}e(t) \quad (2)$$

where $y(t)$ is the system's output, $x(t)$ is the system's input, $e(t)$ is the noise and t is the independent variable (actually the time, usually given in discrete domain). The number of the delaying elements along the system's input-output chain is nk. The polynomials featuring the model are given by:

$$A(q) = 1 + a_1q^{-1} + a_2q^{-2} + \cdots + a_{na}q^{-na} \quad (3)$$

$$B(q) = b_1 + b_2q^{-1} + b_3q^{-2} + \cdots + b_{nb}q^{-nb+1} \quad (4)$$

$$C(q) = 1 + c_1q^{-1} + c_2q^{-2} + \cdots + c_{nc}q^{-nc} \quad (5)$$

$$D(q) = 1 + d_1q^{-1} + d_2q^{-2} + \cdots + d_{nd}q^{-nd} \quad (6)$$

$$F(q) = 1 + f_1q^{-1} + f_2q^{-2} + \cdots + f_{nf}q^{-nf} \quad (7)$$

where q is known as the delay operator and is given like:

$$q^{-i}u(t) = u(t - i), \quad (8)$$

while na, nb, nc, nd and nf are the polynomial's orders. Model described by equation (1) is a linear prediction if its polynomial coefficients are constant.

Particular models have been developed from the generalized model. Each of these is intended to model certain categories of processes. The vehicle taken as a whole can be considered an automatic system. The most proven model for automated systems is ARMAX (**A**uto**R**egressive–**M**oving-**A**verage with e**X**ogenous inputs). Its mathematical equation is:

$$A(q)y(t) = B(q)u(t - nk) + C(q)e(t) \quad (9)$$

and it is featured by the following conditions:

$$\begin{cases} nd = nf = 0 \\ D(q) = F(q) = 1 \end{cases} \quad (10)$$

The major advantage is the freedom in describing the properties of the disturbance term. It adds flexibility to that by describing the equation error as a moving average of white noise. This gives the model:

$$\begin{aligned} y(t) &+ a_1y(t-1) + \cdots + a_{na}y(t - na) \\ &= b_1u(t-1) + \cdots + b_{nb}u(t - nb) + e(t) + c_1e(t-1) + \cdots + c_{nc}e(t - nc) \end{aligned} \quad (11)$$

3. Modelling vehicles dynamics

The mostly used checking method is the "predicted value method". This is a simulation by which a process is reproduced in identical conditions by a certain number of times. Calculations require:

- the experimental series of input and output sizes;
- a mathematical model which is in this case ARMAX;
- the prediction horizon k that determines how many times the process is reproduced. In this study it is equal with 1. [9,10]

One model studied was a SISO one with vehicle speed as output and engine's load-ξ as input $V=f(\xi)$. Figure 1 presents the prediction and residual of test I20n.

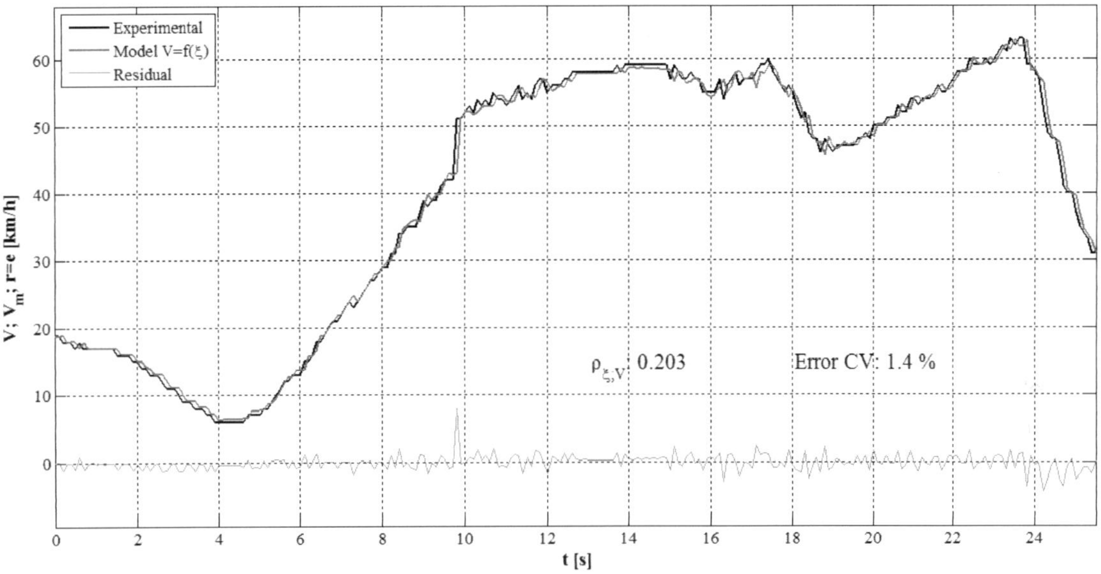

Figure 1. Prediction of model $V=f(\xi)$ for test I20n

Although, the linear dependence between the two vectors is reduced, the correlation coefficient being $\xi_{\xi,V}$ = 0,203, the error is low, 1,4%. [11] This is calculated by the formula (12):

$$Error\ CV = \frac{CV(y)-CV(ypred)}{CV(y)} * 100\ [\%]\ , \tag{12}$$

$y(t)$ is the experimental series, $ypred(t)$ is the prediction result and CV the coefficient of variation. The equation with differences obtained by applying the identification method is for I20n test:

$$V(k)=0,101V(k-1)+0,867V(k-2)+0,0592\ Q(k-1)-0,00173\ \xi(k-2)-e(k)+0,84e(k-1), \tag{13}$$

where k is the number of value in the series. Other models were engine torque M_e or hourly consumption C_h as outputs and engine load ξ as input. The equations with differences for each of the two modeling for I20n test are:

$$M_e\ (k)=1,01M_e\ (k-1)+0,0126M_e(k-2)+2,18\ \xi(k-1)-2,19\ \xi(k-2)+e(k)+0,71e(k-1) \tag{14}$$

$$C_h\ (k)=0,876C_h(k-1)+0,007C_h\ (k-2)+0,130\xi(k-1)-0,14\xi(k-2)+e(k)+0,432e(k-1) \tag{15}$$

The correlation coefficients are 0,992 and 0,872 and the errors are 0,71% and 0,42%.

4. Statistical models of vehicles dynamics'

Statistical models are derived from statistical data and statistical tools. In this case, the equation of each model is obtained using the average values of coefficients obtained by modeling the 64 tests.

Taking into account the $C_h=f(\xi)$ dependence, the following results were obtained:

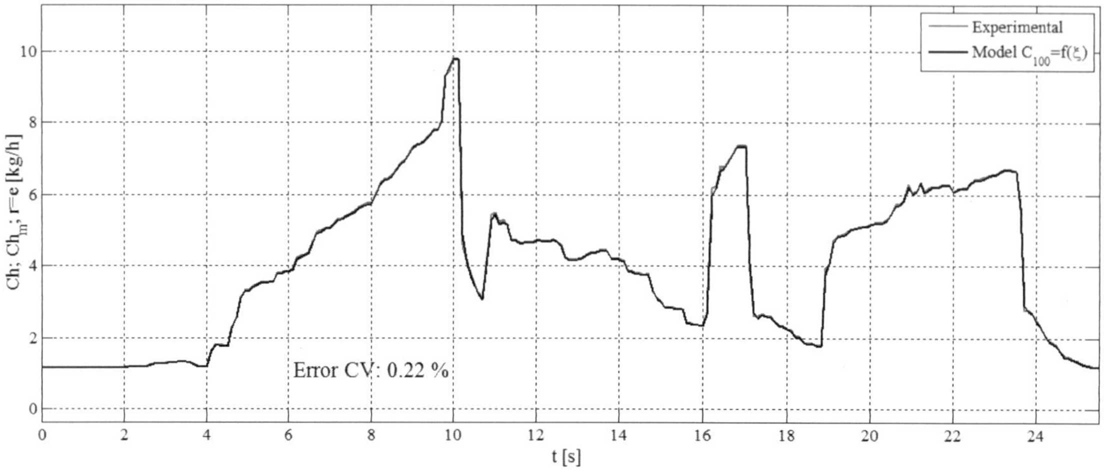

Figure 2. Prediction of $C_h=f(\xi)$ model with average values of coefficients for test I20n

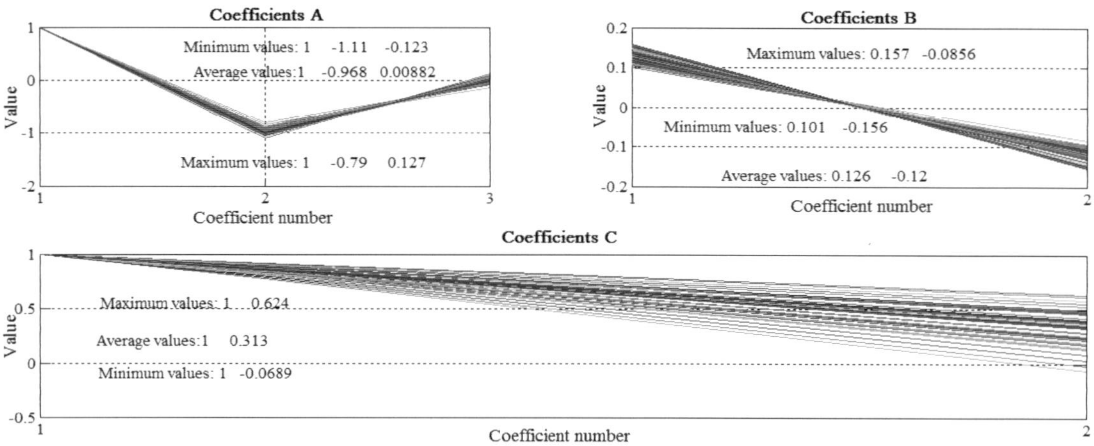

Figure 3. Coefficients values of all 64 predictions of $C_h=f(\xi)$ model

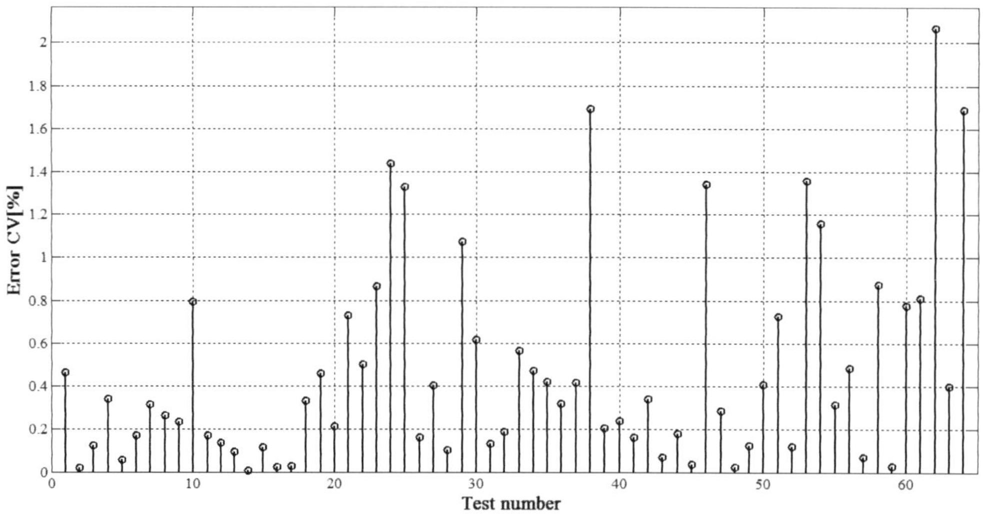

Figure 4. Errors of modeling $C_h=f(\xi)$ with average values of coefficients

Figure 4 shows o lower error in the case of test I20n for model generated with average coefficients. This is not true for all tests, but errors are lower than 3%, which is a reasonable limit of accuracy. Average coefficients values of this model are presented in the figure 3. Also, there are figured all coefficients values.

The equations with differences with average coefficients for the three models are:

$$C_h(k)=0,968C_h(k-1)+0,0082\,C_h(k-2)+0,126\,\xi(k-1)-0,12\,\xi(k-2)+e(k)+0,31e(k-1) \quad (16)$$

$$V(k)=0,899V(k-1)+0,087V(k-2)+0,0117\,\xi(k-1)-0,00975\,\xi(k-2)+e(k)-0,136e(k-1) \quad (17)$$

$$M_e(k)=0,969M_e(k-1)+0,00659M_e(k-2)+2,07\,\xi(k-1)-2,03\,\xi(k-2)+e(k)-0,2131e(k-1) \quad (18)$$

The modeling process can be developed with MISO (Multiple Inputs - Single Output). Engine angular speed n is added as a second input. New models are: $V=f(\xi,n)$, $M_e=f(\xi,n)$ and $C_h=f(\xi,n)$.

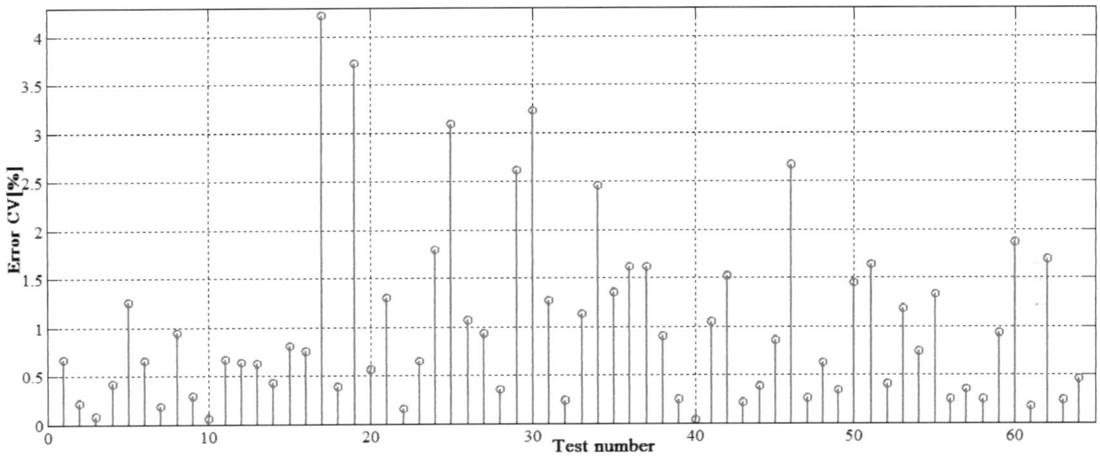

Figure 5. Errors of modeling $M_e=f(\xi,n)$with average values of coefficients

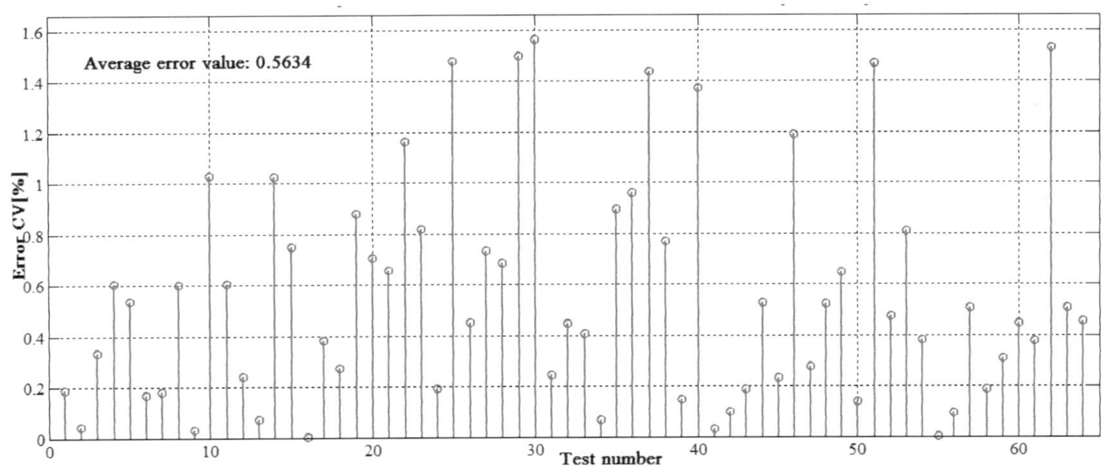

Figure 6. Errors of modeling $C_h=f(\xi,n)$with average values of coefficients

By comparing the results in Figures 4 and 6, it can be observed that the precision of the models increases with the increase of the number of entries. The equations with differences with average coefficients for the three MISO models with two inputs are:

$$C_h(k)=0,977C_h(k-1)+0,00173C_h(k-2)+...$$
$$+0,126\xi(k-1)-0,122\,\xi(k-2)+0,0009n(k-1)-0,0009n(k-2)+e(k)+0,308e(k-1) \quad (19)$$

$$V(k)=0,802V(k-1)+0,164V(k-2)+...$$
$$+0,0103\,\xi(k-1)+0,00303\xi(k-2)+0,00372n(k-1)-0,00325n(k-2)+e(k)-0,0832e(k-1) \quad (20)$$

$$M_e(k)=0,958M_e(k-1)+0,000918M_e(k-2)+...$$
$$+2,07\xi(k-1)-2,\xi(k-2)-0,00347n(k-1)+0,00373n(k-2)+e(k)+0,18e(k-1) \quad (21)$$

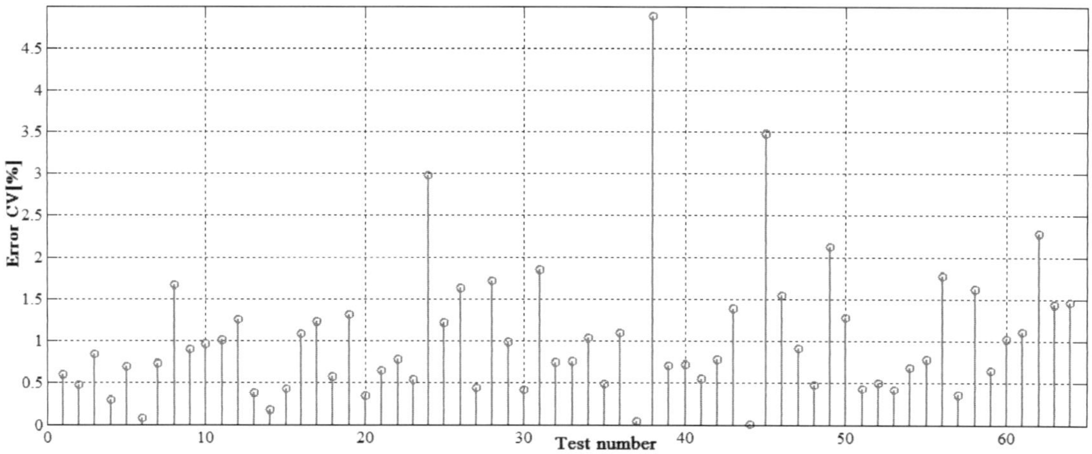

Figure 7. Errors of modeling $V=f(\xi,n)$with average values of coefficients

Maximum error of all two inputs MISO models and test is lower than 5%. This demonstrates that the models are well-designed and can be used successfully in analyzing the operation of gasoline powered vehicles.

5. Conclusions

Modelling vehicle dynamics using single or multiple linear prediction has good results. It is a useful, helpful and replicable process, which can be integrated in a multitude of experiments/designs. Using average values of coefficients produces models with very good accuracy; the error finding is less than 5%. However, it requires a great volume of accurate experimental data. Using parameters with significant correlation between them creates accurate models. The more factorial parameters are involved, the higher the accuracy of the model we get. Another observation is that the accuracy of the models created, and that of the data collected was directly proportional with the number of tests run.

References

[1] C O Ilie, Statistical Modeling of Vehicle Dynamics, *Military Technical Academy*, Bucharest, 2008, pp.7-19 ISBN 10 973-640-111-1, ISBN 13 978-973-640-111-4.

[2] L Ljung, 1987 System Identification. Department of Electrical Engineering, Linkoping University, Sweden

[3] I Lespezeanu, F M Militaru, O Alexa, C O Ilie, M Marinescu 2016 "Aspects Regarding the Influence of Lambda Control System Faults on Pollutant Emissions of Spark Ignition Engines", ICOME 2015, Craiova, Romania, *Applied Mechanics and Materials, Vol. 823, pp. 291-296*

[4] I. Lespezeanu, F M Militaru, O Alexa, C O Ilie, M Marinescu 2016 "On-Board Malfunction Simulations on Vehicles that Are Equipped with Electronic Control Systems", ICOME 2015, Craiova, Romania, *Applied Mechanics and Materials, Vol. 823, pp. 211-216*

IOP Conf. Series: Materials Science and Engineering **252** (2017) 012091 doi:10.1088/1757-899X/252/1/012091

[5] F Popescu, F Enache 2013 Training of RBF Neural Networks: a comparative overview, "Mircea cel Batran" Naval Academy Publishing House, *Scientific Bulletin*, Constanţa, Romania

[6] R M Stoica, V J Radulescu, D Neagu, C Trocan and I Copae 2016 - The analysis of experimental data obtained from automotives tests - *Proceedings of The 7th International Conference on Advanced Concepts in Mechanical Engineering*, ACME 2016, Iași, June 2016, ISBN: 978-1-5108-2999-2, ISSN: 1757-8981, Vol 147, pg.950

[7] M Marinescu, C O Ilie, R Vilău, A Octavian, D Voicu 2016 - Malfunction detection using spectral analysis for automotive braking system , 11th International Conference on COMMUNICATIONS (COMM), June 09-11, Bucharest, Romania, Conference Proceedings, AGIR Publishing House, pp. 330-336

[8] B Laszlo, D Voicu 2016 "Developing a measuring system able to determine the mechanical stress on a rectangular cross – section bar", *MTA Review,* Vol. XXVI, No. 1, pp. 47-52

[9] M Truţă, M Marinescu and V Vînturiş 2016 - Influence of the drawbar pull over the power flux within the automotive transmissions - in 7th International Conference on Advanced Concepts in Mechanical Engineering, IOP Conf. Series: *Materials Science and Engineering* 147 (2016) 012131 doi:10.1088/1757-899X/147/1/012131

[10] M. Trută, M. Marinescu, V. Vînturiş 2013 – Multi-Spectral Analysis of the Self Generated Torque's Signal within a 4×4 Automotive Driveline - in *COMEC Proceedings* 2013, pp. 252-257, Brasov, Romania, 2013

[11] Dobre A., A. DOBRE, A. HADĂR, D. VASILIU 2017 - Correlation Between Comfort and Energy Consumption for Automotive Suspensions", *U.P.B. Sci. Bull.*, Series D, Vol. 79, Iss. 1, ISSN 1454-2358, pp. 119-126, Publishing House Politehnica Press.

Biofuel effect on flame propagation and soot formation in a DISI engine

A Irimescu, S S Merola*, S Di Iorio, B M Vaglieco

Istituto Motori – CNR, via G. Marconi – 80125 Napoli (Italy)
*Corresponding author: s.merola@im.cnr.it

Abstract. The use of biofuels, especially in transportation and industrial processes, is seen as one of the most effective solutions to promote the reduction of greenhouse gases and pollutant emissions, as well as to lighten the dependence from petro-fuel producers. Biofuels are defined as a wide range of energy sources derived from biomass. In this category, alcohols produced through fermentation, such as ethanol and butanol, are considered some of the most suitable alternatives for transportation purposes. The benefits of bio-ethanol addition to gasoline have always been recognized for practical reasons. Apart from the variety of sources which it can be produced from, ethanol can raise the octane rating, given its improved anti-knock characteristics, allowing the use of higher compression ratios and higher thermal efficiency. However, ethanol's high latent heat of vaporization can cause problems during cold-start due to poor evaporation. On the other hand, in hot climates ethanol fuelling can result in adverse effects such as vapour lock. Butanol can be considered as an emergent alternative fuel. Normal butanol has several well-known advantages when compared to ethanol, including increased energy content, greater miscibility with transportation fuels, and lower propensity for water absorption. Despite of these pros, the costs of n-butanol production are higher due to lower yields compared to ethanol. Moreover, vaporization remains a critical aspect of this biofuel. Understanding the effect of biofuels on in-cylinder combustion processes is a key-point for the optimization of fuel flexibility and achieving lower CO2 emissions. To this aim, a combined thermodynamic and optical investigation was performed on a direct injection spark ignition engine fuelled with ethanol, butanol and gasoline. Fuels were compared by fixing the injection and spark ignition strategies. Thermodynamic measurements were coupled with optical investigations based on cycle resolved flame visualization. Optimized procedures of image processing were applied to follow the evolution of the flame front in terms of morphological parameters and to evaluate the local distribution of diffusive flames induced by oxidation of fuel deposits during late combustion. These data were correlated with exhaust gas measurements. The experiments confirmed that the chemical-physical specifications of the tested fuels strongly influenced the temporal and spatial evolution of the flame front. Moreover, different distributions and intensities of diffusive flames were observed. These results demonstrated the effect of the fuel on the deposits amount and distribution in the combustion chamber, at fixed operative conditions.

Keywords: Direct Injection, Spark Ignition Engine, Ethanol, Butanol, Flame Development

1. Introduction

The use of fossil fuels dominated the internal combustion engines field for about two centuries. Alternative propulsion and energy concepts (electric, wind, solar, and geothermal) are not yet at the stage to significantly impact petroleum use, mainly because none of them match simultaneously the balance between cost, energy density, and efficiency with respect to transportation and energy production of 'classical' liquid fuels. Renewable biofuels derived from bio-feedstock have confirmed significant potential to reduce the consumption of petroleum-based energy feedstock [1]. Ethanol is currently widely used for this purpose after blending with gasoline up to 85%$_{vol}$. Recently, n-butanol is also becoming a viable option, either as a neat fuel or an additive [2]. Both alcohols can be considered as renewable fuels if they are derived from fermentation processes of biomass. Normal butanol has several well-known advantages when compared to ethanol, including its higher energy content, increased compatibility with fuel systems, and lower propensity for water absorption. It may be blended with gasoline in any concentration, and its comparatively higher energy content would require a lower volumetric concentration compared to an equivalent ethanol-petroleum-derived fuel blend, for similar performance. On the other hand, costs and yield are still more advantageous for ethanol.

The use of bio-alcohol fuels in direct injection (DI) spark ignition (SI) engines has been widely investigated, but most of the works are focused on the analysis of performance and emissions under different operative conditions [3][4][5]. In general, DISI power units emit higher concentrations of particulate matter (PM), if compared to (PFI) engines. Several works confirmed that direct injection of the fuel into the combustion chamber induced incomplete evaporation and determined the impingement of droplets onto piston and cylinder walls, especially during cold start [4]. These lead to locally rich, diffusion-governed liquid fuel combustion or pyrolysis that is prone to PM formation [6][7]. As aforementioned, blending bio-alcohols (ethanol or butanol) with gasoline allows a decrease of petroleum consumption and fossil-based CO_2 emissions. On the other hand, they were found to feature an adverse influence on spray and evaporation characteristics, thus worsening PM emissions [4]. Conversely, if proper evaporation can be ensured, bio-alcohols addition can also reduce the effects of this pollutant category and therefore potentially serve as enablers for DI technology [8][9][10]. For these reasons, a detailed analysis of ethanol and butanol combustion with respect to commercial gasoline, can contribute to better understand the potential of bio-alcohols as replacement for fossil fuels in DISI engines.

This paper discusses the effect of fuel injection timing of n-butanol, ethanol and gasoline on the combustion process in a DISI engine under different settings of injection phasing in wide open throttle (WOT) conditions, with stoichiometric air-fuel ratio. Particular interest is devoted to the diffusive flames induced by oxidation of fuel deposits within the combustion chamber. Based on cycle-resolved imaging data, the morphological parameters of these flames were investigated and the fuel injection strategies are also analysed with respect to opacity at the exhaust emissions.

2. Experimental apparatus and measurement procedures

All tests were performed in an optically accessible spark ignition wall guided DISI engine with 79 mm bore and 81.3 mm stroke. More details on the experimental apparatus can be found in [9][10]. The injection pressure was fixed at 100 bar; the duration of injection was set in order to operate in stoichiometric condition. Apart from the two alcohols, gasoline was used as reference fuel. Compared to gasoline, injection duration was increased by a factor of 1.21 for butanol and 1.51 for ethanol, so as to compensate for the alcohols' lower stoichiometric air-fuel ratio. Mixture strength was monitored with an accuracy of ±1% through measurements with a wide band oxygen gas sensor fitted at the exhaust. The engine was operated at 2000 rpm with wide open throttle (WOT). Spark timing was fixed at 15 crank angle degrees (CAD) before top dead centre (BTDC) so that fluid dynamics was roughly the same at the instant of ignition; it should be noted that this setting was close to be point of maximum brake torque (MBT). Start of injection (SOI) was set at three different values, at 340, 300 and 260 CAD BTDC. In-cylinder pressure was measured with crank angle resolution of 0.2 CAD using a flush-installed piezo-electric transducer that provided an accuracy of ±1%. Exhaust gas concentrations of pollutant species were determined using a gas analyser based on the electrochemical method for nitrogen oxide (NO_x) emissions; the nondispersive infrared (NDIR) measurement principle for carbon monoxide (CO) and unburned hydrocarbons (HC). Accuracy was within 3%, and the resolution of the readings was 1 ppm for the first and third species and 0.01% for the other one. Smoke was measured by a part flow opacimeter with 0.01% resolution.

Flame front propagation and the evolution of the fuel deposits oxidation were investigated by a CMOS high-speed camera (Optronis CamRecord 5000) that featured a 16 μm x 16 μm pixel area that ensured good sensitivity in the visible wavelength range. It was coupled with a 50 mm focus Nikon lens (f/1.8D) and worked in full chip configuration (512x512 pixel), with an acquisition speed of 5000 fps and an exposure time of 200 μs. Synchronization of various control triggers for ignition, injection and camera was achieved using the optical encoder mounted on the crankshaft as an external clock connected to an AVL Engine Timing Unit. The f/stop of the lens was fixed at 5.6; in this way, the optical set-up gave a resolution of 190 μm per pixel, with a signal to noise ratio sufficient to apply image processing for the analysis of flame front propagation during the early combustion stages. The spark ignited flame evolution and fuel deposits distribution were analysed by a custom methodology developed by NI Vision Assistant software already used in previous works [9][10].

3. Results and discussion

Given the focus of the study, aimed at understanding fuel deposits characteristics and their distribution inside the combustion chamber, only a brief analysis of the in-cylinder pressure related data was performed. To this end, data recorded during 200 consecutive cycles was averaged and global parameters were calculated (i.e. indicated mean effective pressure IMEP and its coefficient of variation COV).

Overall engine performance (evaluated through the IMEP) was found to be roughly the same (within ±2% of the reference value recorded for gasoline), for the injection phasing of 300 CAD BTDC optimized for gasoline (Fig.1). Indeed, this setting provided the best compromise between jet penetration and intake air velocity, that ensured good mixing and improved fuel evaporation [9][10]. It is interesting to note that ethanol provided an increase of around 2% in IMEP compared to gasoline, even though it featured high latent heat of evaporation; of course, it is also characterized by increased laminar flame speed, but the dominating effect should be that of air-fuel mixture preparation. Most likely, a balance between the two properties was the reason behind the observed trend; nonetheless, saturation pressure is another important factor that needs to be taken into account when performing such evaluations. As a general observation, retarding or delaying injection timing from its optimized case of SOI 300 resulted in a drop in engine performance, more so for the alcohols. This result already hints to fuel evaporation issues; the COV data suggests more favourable charge stratification with early injection (SOI 340) compared to late fuel delivery (SOI 260). This can be related to the mechanism of jet impingement on the piston crown and cylinder liner, which resulted in 'richer' mixture close to the centre of the combustion chamber with the SOI 340 setting. With late injection, more fuel impinged on the liner, thus causing 'lean' combustion during the initial stages of combustion and 'rich' operation after flame front propagation. Overall the effect was practically insignificant for gasoline, but quite evident for the alcohols, more so for butanol.

These results and related mechanisms are in line with recorded exhaust gas concentrations of CO, HC and NO_x, as well as smoke opacity (Fig. 2). Carbon monoxide emissions were found to be comparable for all cases, suggesting an overall stoichiometric combustion evolution. HC concentrations confirm this observation, with slightly lower values for the alcohols and minor advantage for the late injection case. Measured NO_x values also hint towards extensive charge stratification with ethanol and butanol fuelling, both in the SOI340 and SOI260 cases; this needs to be evaluated in the context of inherently decreased NO formation for the alcohols, due to their lower flame temperature. The most interesting result is the significant increase in smoke opacity when employing early injection timing. This is directly related to increased fuel impingement on the piston crown for SOI 340. As expected, ethanol resulted in the lowest overall recorded opacity, even though it features the highest latent heat of evaporation. The fact that butanol fuelling with early injection promoted soot formation emphasizes its poor evaporation in the SOI 340 case, as well as the importance of saturation pressure as an important fuel property (given that it is the lowest for this alcohol).

IOP Conf. Series: Materials Science and Engineering **252** (2017) 012092 doi:10.1088/1757-899X/252/1/012092

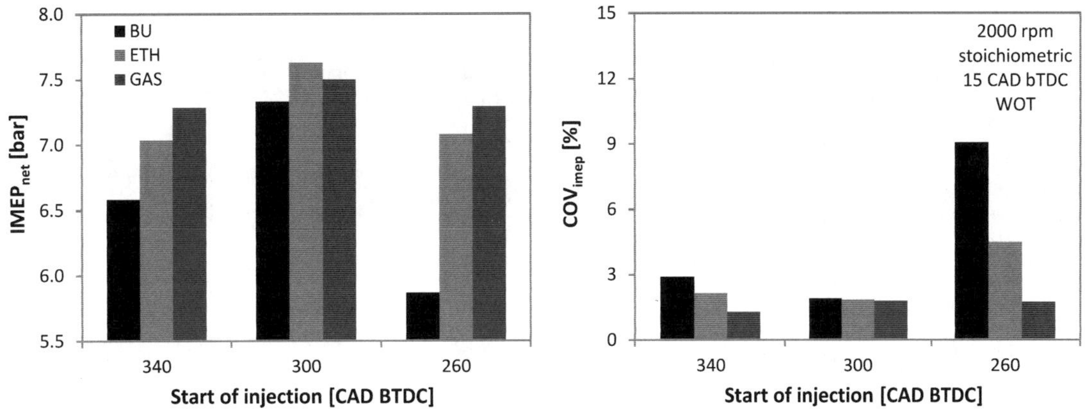

Figure 1. Overall engine performance indicators for butanol (BU), ethanol (ETH) and gasoline (GAS).

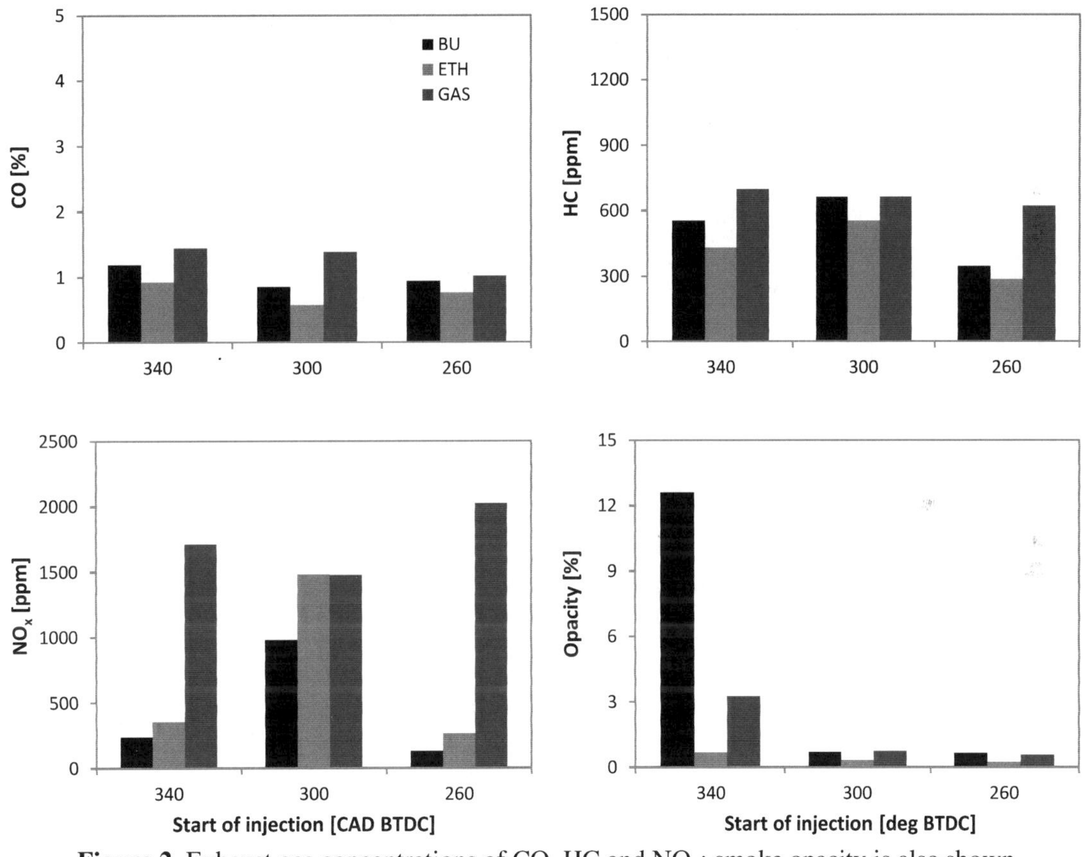

Figure 2. Exhaust gas concentrations of CO, HC and NO_x; smoke opacity is also shown.

In order to understand the effect of the fuel injection strategy on flame characteristics, cycle-resolved digital imaging was performed in all the selected operative conditions. Fig. 3 shows selections of images detected for gasoline at fixed SOI 340 CAD BTDC in the early and late combustion stages, respectively. The continuous lines represent the contour of the spark ignited flame (Fig. 3, top) and that of fuel deposits oxidation sites (Fig. 4, bottom) obtained through image processing. Sequences refer to the engine cycle with the pressure curve closest to the average trace. The frames are labelled by the crank angle degree after the start of spark (CAD ASOS). The intake valve side corresponds to the bottom of the images.

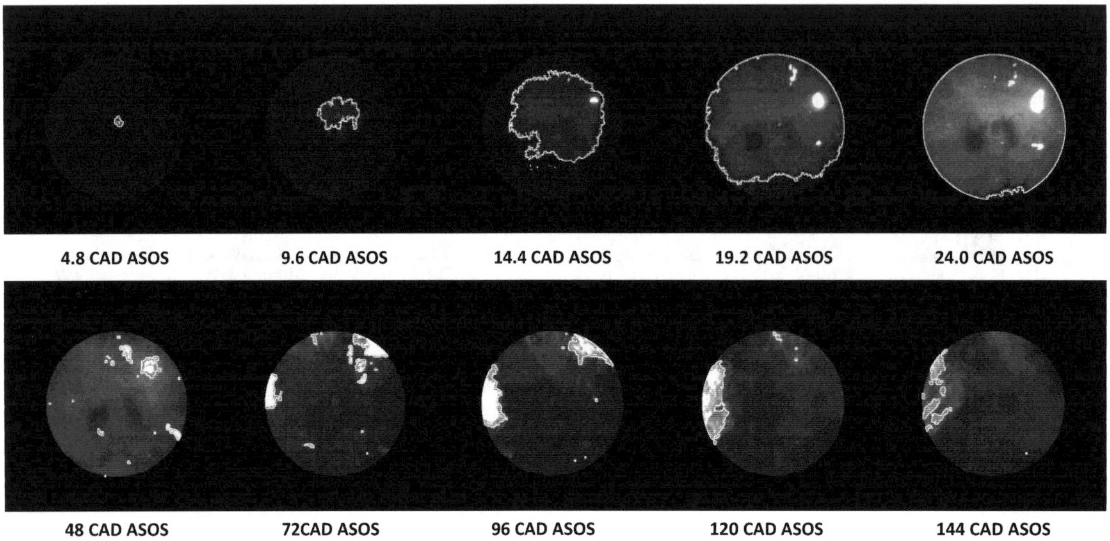

Figure 3. Flame evolution during early (top) and late (bottom) combustion for gasoline, SOI 340.

As observed in other works [9][10][11], the flame front initiated in the region of the spark plug and evolved roughly in the same way until reaching the optical limit. Faster flame development was observed along the direction towards the exhaust valves. Even with highly homogenous mixtures, the effect can be in part related to the gradient in the local gas temperature from the intake and exhaust valves, as well as due to tumble motion. Localized spots and small regions with high luminosity present in the burned gas, as well as the persistence of strong luminous flames during late combustion, well after the normal spark ignited flame propagated, demonstrate the occurrence of fuel deposits burning phenomena. These locally rich regions were determined by the impingement of the fuel jets on the piston surface. Previous investigations correlated the presence of such fuel deposits to a reduction in flame speed and caused an asymmetry in the flame front shape [9][10]. The custom procedures of image processing allowed the analysis of flame front propagation from kernel inception up to the optical limit, as well as the spatial distribution of the flames induced by fuel deposits. Figure 4 shows the evolution of the flame morphological parameters, evaluated for the selected fuels and SOIs; these traces correspond to the average-like engine cycles in each case. The normalized flame area traces (i.e. referenced to the bore cross-section) emphasize 'slower' combustion when advancing or retarding injection from its optimized setting of 300 CAD BTDC (Fig. 4, top). This effect is much more dominant for the alcohols, with butanol showing the highest sensitivity to changes in the injection strategy. These evolutions correlate well with the observed changes in engine performance and relative influence of each fuel. Heywood factor data (Fig. 4, middle), calculated as the ratio between the flame front perimeter and that of an equivalent circle with the same area, suggests quite circular flames for gasoline fuelling, with reduced influence of injection phasing. With butanol and ethanol, on the other hand, more distorted flames were observed, which correlates well with charge stratification; even for the optimized case of SOI 300 the alcohols featured overall slightly higher Heywood factors compared to gasoline. The fact that this parameter increased for advanced or retarded injection, hints towards more extensive fuel evaporation issues and related decrease in engine performance when using ethanol, and even more so for butanol. An interesting observation is that flame displacement was quite different for the three injection settings (Fig. 4, bottom; displacement was taken as positive along the direction from the intake to the exhaust valves, i.e. upwards when looking at Fig. 3 and 5). For the optimized case of SOI 300, reduced displacement was noted; with

IOP Conf. Series: Materials Science and Engineering **252** (2017) 012092 doi:10.1088/1757-899X/252/1/012092

early injection, the flames propagated first to the exhaust side, while for SOI 260 the opposite direction was observed. This trend is directly correlated to the fuel jets impingement mechanism; early injection resulted in a rich mixture towards the exhaust valves, while with late injection practically only the jet near the intake valves impinged on the piston (the other five jets impacted the liner). The optimized case of SOI 300 was most likely influenced by tumble motion rather than charge stratification. Again, the effect was more prominent for the alcohols, confirming poor evaporation, especially for butanol.

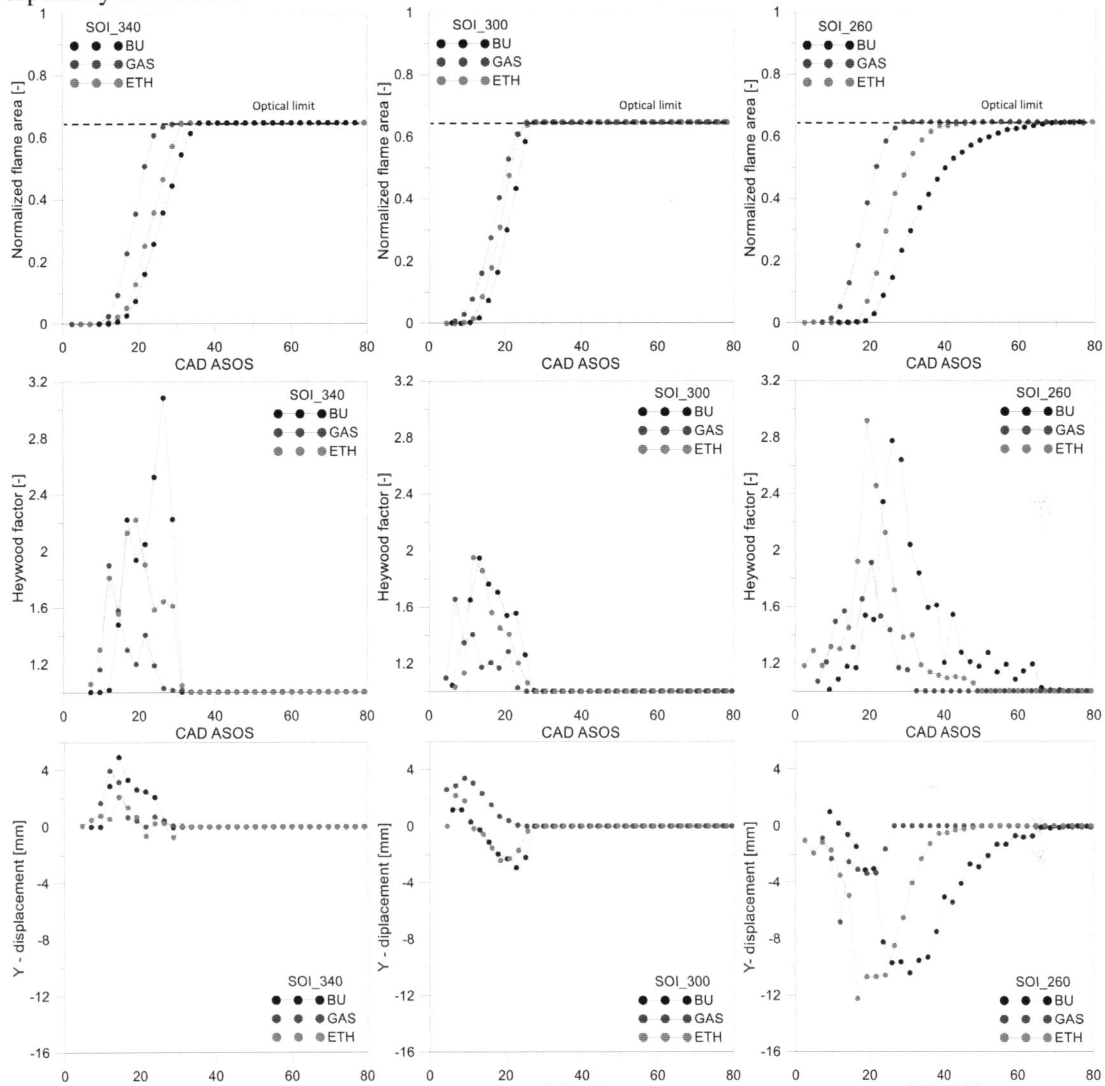

Figure 4. Evolution of the normalized flame area (top), Heywood factor (middle) and displacement along the Y axis (bottom), corresponding to average-like engine cycles.

The analysis of diffusive flames distribution maps (Fig. 5) revealed that both alcohols featured a common impingement pattern towards the intake valves with SOI 260; gasoline resulted in reduced luminous areas, suggesting improved fuel evaporation. For the optimized case of SOI 300, fuel deposits oxidation sites were more centrally located, with butanol featuring the highest spread and frequency. With early injection, all three fuels were characterized by high frequency of diffusive flames in the region of the exhaust valves. These results are in line with the injector-piston-liner geometry and jet impingement mechanism. The fact that butanol was the most sensitive fuel to changes in injection strategy is most evident when looking at the number of fuel deposit oxidation sites (Fig. 6); the actual size of the diffusive flames was also the largest for this alcohol. An interesting observation is that the data correlates relatively well with measured opacity for the butanol and ethanol; with gasoline, on the other hand, a different soot pathway is suggested by these findings, in line with other studies [12][13].

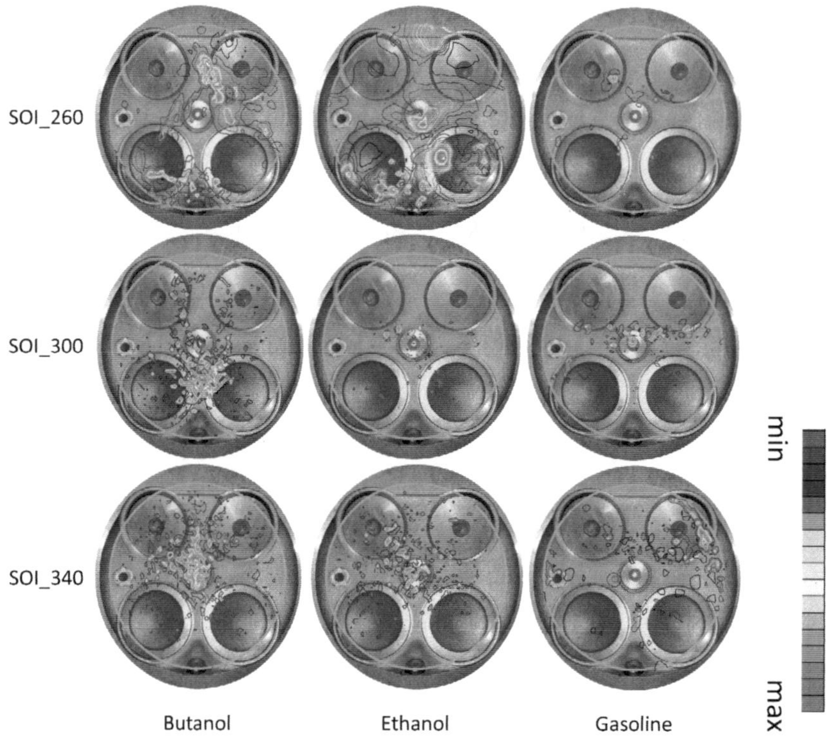

Figure 5. Occurrence frequency map of centroid of the flames induced by fuel deposits.

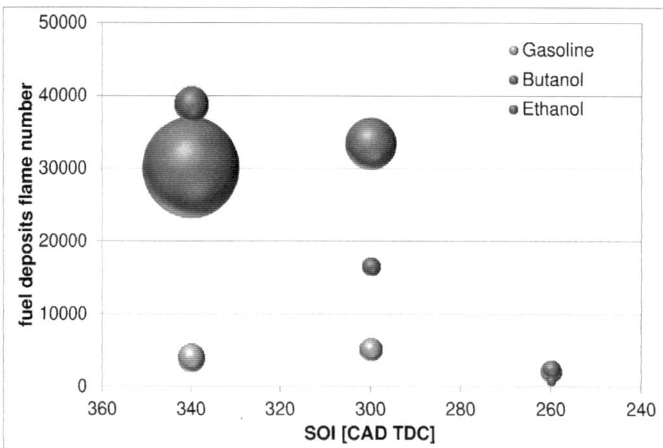

Figure 6. Number of detected flames induced by fuel deposits, weighted with respect to average area.

4. Conclusion

A combined study of thermodynamic and optical data revealed that butanol was most sensitive to changes in injection phasing in a DISI engine running at WOT and medium rotational speed. At this load level, gasoline, butanol and ethanol featured comparable engine performance with optimized injection timing of 300 CAD BTDC, while retarding or advancing fuel delivery reduced engine output. Exhaust gas measurements revealed that opacity was the most sensitive parameter, with early injection featuring the highest opacity values. This was correlated to impingement on the piston crown, with butanol as the 'worst' case, due to evaporative issues. The analysis of flame imaging data confirmed the trends of the thermodynamic values; flame displacement and distribution of fuel deposits suggested significant charge stratification, especially for butanol. All three fuels featured a common trend when changing injection timing, in line with the impingement pattern on the piston and liner.

5. References

[1] Shankar S, 2017, Renewable and Nonrenewable Energy Resources: Bioenergy and Biofuels. In *Principles and Applications of Environmental Biotechnology for a Sustainable Future* (pp. 293-314). Springer Singapore.

[2] Xue C, Zhao XQ, Liu CG, Chen LJ, Bai FW, 2013, Prospective and development of butanol as an advanced biofuel. *Biotechnology advances*, 31(8), 1575-1584

[3] Elfasakhany A, 2014, Experimental study on emissions and performance of an internal combustion engine fueled with gasoline and gasoline/n-butanol blends. *Energy Convers Management* 88, 277–283.

[4] Zhang ZJ, Wanga TY, Jia M, Wei Q, Meng XZ, Shu GQ. 2014, Combustion and particle number emissions of a direct injection spark ignition engine operating on ethanol/gasoline and n-butanol/gasoline blends with exhaust gas recirculation. *Fuel* 130, 177–188.

[5] Di Iorio S, Mancaruso E, Sementa P, Vaglieco BM, 2015, A comprehensive analysis of the impact of biofuels on the performance and emissions from compression and spark-ignition engines. *Int. J. Engine Res.* 16, 680-690, doi:10.1177/1468087415591924.

[6] He X, Ratcliff MA, Zigler BT, 2012, Effects of gasoline direct injection engine operating parameters on particle number emissions, *Energy Fuels* 26, 2014-2027.

[7] Wang C, Xu H, Herreros JM, Wang J, Cracknell R, 2014, Impact of fuel and injection system on particle emissions from a GDI engine, *Appl. Energy* 132, 178-191.

[8] Chen L, Stone R, Richardson D, 2012, A study of mixture preparation and PM emissions using a direct injection engine fuelled with stoichiometric gasoline/ethanol blends. *Fuel* 96, 120–130.

[9] Merola, SS, Irimescu, A, Marchitto L, Tornatore C, Valentino G, 2017, Effect of injection timing on combustion and soot formation in a direct injection spark ignition engine fueled with butanol. *Int. J. Engine Res.* 18, 5-6, 490-504.

[10] Merola SS, Irimescu A, Di Iorio S, Vaglieco BM, 2017, Effect of Fuel Injection Strategy on the Carbonaceous Structure Formation and Nanoparticle Emission in a DISI Engine Fuelled with Butanol. *Energies* 10, 832; doi:10.3390/en10070832.

[11] Breda S, D'Adamo A, Fontanesi S, D'Orrico F, Irimescu A, Merola, SS, Giovannoni N, 2017, Numerical Simulation of Gasoline and n-Butanol Combustion in an Optically Accessible Research Engine. *SAE Int. J. Fuels Lubr.* 2017, 10, 32-55.

[12] Lee K, Seong H, Sakai S, Hageman M, Rothamer D, 2013, Detailed Morphological Properties of Nanoparticles from Gasoline Direct Injection Engine Combustion of Ethanol Blends. SAE Tech Paper n. 2013-24-0185.

[13] Sgro LA, Sementa P, Vaglieco BM, Rusciano G, D'Anna A, Minutolo P, 2012, Investigating the origin of nuclei particles in GDI engine exhausts. *Combust. Flame*, 159, 1687-1692.

Spark discharge and flame inception analysis through spectroscopy in a DISI engine fuelled with gasoline and butanol

A Irimescu, S S Merola*

Istituto Motori – CNR, via G. Marconi – 80125 Napoli (Italy)
*Corresponding author: s.merola@im.cnr.it

Abstract. Extensive application of downsizing, as well as the application of alternative combustion control with respect to well established stoichiometric operation, have determined a continuous increase in the energy that is delivered to the working fluid in order to achieve stable and repeatable ignition. Apart from the complexity of fluid-arc interactions, the extreme thermodynamic conditions of this initial combustion stage make its characterization difficult, both through experimental and numerical techniques. Within this context, the present investigation looks at the analysis of spark discharge and flame kernel formation, through the application of UV-visible spectroscopy. Characterization of the energy transfer from the spark plug's electrodes to the air-fuel mixture was achieved by the evaluation of vibrational and rotational temperatures during ignition, for stoichiometric and lean fuelling of a direct injection spark ignition engine. Optical accessibility was ensured from below the combustion chamber through an elongated piston design, that allowed the central region of the cylinder to be investigated. Fuel effects were evaluated for gasoline and n-butanol; roughly the same load was investigated in throttled and wide-open throttle conditions for both fuels. A brief thermodynamic analysis confirmed that significant gains in efficiency can be obtained with lean fuelling, mainly due to the reduction of pumping losses. Minimal effect of fuel type was observed, while mixture strength was found to have a stronger influence on calculated temperature values, especially during the initial stage of ignition. In-cylinder pressure was found to directly determine emission intensity during ignition, but the vibrational and rotational temperatures featured reduced dependence on this parameter. As expected, at the end of kernel formation, temperature values converged towards those typically found for adiabatic flames. The results show that indeed only a relatively small part of the electrical energy is actually used for promoting chemical reactions and that temperature during the arc and kernel phases are influenced to a reduced extent by fuel concentrations.

Keywords: Direct Injection, Spark Ignition, Butanol, Flame Kernel, Plasma.

1. Introduction

Recent studies demonstrated that alcohol fuels, ethanol and butanol in particular, can be used in spark-ignition (SI) engines with direct injection (DI) systems, as pure fuels or blended with gasoline [1], [2], [3]. High enthalpy of vaporization of alcohols can affect air-fuel mixture formation prior to ignition and may form thicker liquid films on the cylinder wall and piston crown [3]. These liquid films can result in mixture non-homogeneities inside the combustion chamber and hence strongly influence the cyclic variability of early combustion stages. The effect can be amplified when the engine works in lean homogenous air-fuel mixtures. It is well known that the initial phase of combustion has a determining role on how the process evolves and therefore directly influences engine performance. The kernel phase and flame front propagation feature significant cycle-to-cycle variations that are more pronounced for dilute mixtures [4]. The use of butanol introduces further influencing factors due to the evaporative properties that can result in different charge stratification even for low load conditions [3], [4]. Optimization of the ignition process can be realized only through a deeper knowledge of the complex mechanisms that characterise this specific phase. Starting from these considerations, the paper reports an experimental study of the early stages of the combustion process in a single cylinder DISI engine fuelled with commercial gasoline and n-butanol. The engine was optically accessible and equipped with the cylinder head of a commercial power unit for passenger cars, with the same geometrical specifications (bore, stroke, compression ratio). Ultra-violet (UV) natural emission spectroscopy measurements ranging from 230 nm to 470 nm wavelength and simultaneous thermodynamic analysis were used to better understand the effect of butanol content on flame kernel inception and development. Optical investigations allowed to follow the evolution of chemical species that marked the spark discharge (CN and OH radicals), as well as flame front initial growth (OH radicals). Vibrational and rotational temperatures were calculated during the arc and glow phase by the ratio between the emission intensity of CN and OH radicals. Apart from difficulties associated with the study of plasmas due to their high temperature, the application in SI engines also features pressure and temperature variations during ignition, as well as changes in fluid composition and local turbulence intensity. Therefore, by combining in-cylinder pressure measurements with optical techniques, improved insight can be gained into the interactions between the working fluid and the spark that initiates combustion in SI engines. The proposed methodology can be applied for better understanding of phenomena typical of new ignition systems based on microwave discharge, multi-spark discharges, radio frequency discharge, as well as corona and nanosecond pulsed discharge [5].

2. Experimental apparatus and measurement procedures

Experimental investigations were carried out on an optically accessible spark ignition wall guided DISI engine with 79 mm bore and 81.3 mm stroke sketched in Fig.1. More details on the experimental apparatus can be found in [4], [6]. The engine was operated at 2000 rpm and the injection pressure was fixed at 100 bar. Compared to gasoline, injection duration was increased by a factor of 1.21 for butanol, so as to compensate for the alcohols' lower stoichiometric air-fuel ratio.

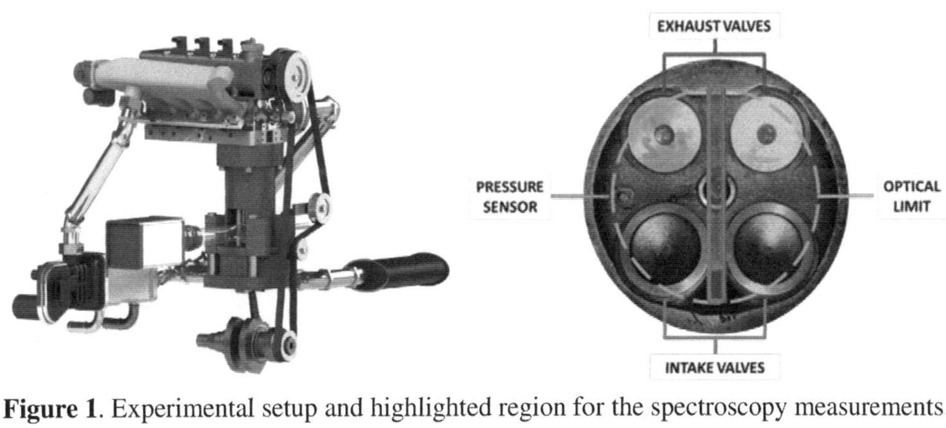

Figure 1. Experimental setup and highlighted region for the spectroscopy measurements.

Mixture strength was monitored with an accuracy of ±1% through measurements with a wide band oxygen gas sensor fitted at the exhaust. Intake pressure was set at 0.7 bar for throttled trials, while during lean operation ($\lambda \approx 1.4$) the pressure was close to 1 bar; the main idea behind this control strategy was to compare engine performance at the same injected fuel quantity, with two quite different combustion conditions. In-cylinder pressure was measured with a crank angle resolution of 0.2 CAD using a flush-installed piezo-electric transducer that provided an accuracy of ±1%. Spark timing was fixed at 30 crank angle degrees (CAD) before top dead centre (BTDC) so that fluid dynamics were roughly the same at the instant of ignition. Start of injection (SOI) was set at 300 CAD BTDC. An automotive ignition system mounting a commercial NGK spark plug (0.8 mm gap) was used. The ignition coil was charged for 3.3 ms in order to achieve the optimum current level at the time of start of spark (SOS); this induced an energy output of 90 mJ for the secondary circuit.

Spark ignition was investigated for both fuels through UV-visible natural emission spectroscopy. Synchronization of control triggers for ignition, injection and optical detector was achieved using an encoder mounted on the crankshaft as an external clock connected to an AVL Engine Timing Unit.

Spectroscopic measurements were performed to follow the evolution of radiative chemical species involved in the combustion process. During the optical investigations, the radiative emission from the combustion chamber was focused by a 78 mm focal length, f/3.8 UV Nikon objective onto the 250 µm micrometer controlled entrance slit of a spectrometer with 150 mm focal length and 300 groove/mm grating (central wavelength 350 nm). From the grating, the radiations were detected through an intensified CCD camera (array size of 1024x1024 pixels, 16-bit pixel digitization and 1MHz sustained repetition rate). Spectroscopic investigations were carried out in the central region of the combustion chamber (highlighted in Fig. 1). The ICCD camera was operated in sequential acquisition mode, thus spectra were detected with fixed gate width (41.6 µs ≈ 0.5 CAD) but variable delay (83.3 µs/cycle ≈ 1.0 CAD/cycle) from the trigger. A routine developed in LabView National Instruments allowed estimating the intensity of OH and CN band systems from the spectra after background subtraction, following the procedure described in [7].

3. Results and discussion

An overall analysis of engine performance was performed based on recorded in-cylinder pressure data averaged over 200 consecutive cycles. The brief overview of indicated mean effective pressure (IMEP) data and its coefficient of variation (COV_{imep}) revealed slightly improved performance for the alcohol when employing throttled operation, but a reversal in this finding when the lean fuelling strategy was used (Fig. 2, with BU for butanol and GAS for gasoline). This was linked to evaporative issues for this alcohol, mainly due to its relatively low saturation pressure within the range usually found for air-fuel mixture formation during the intake stroke [8].

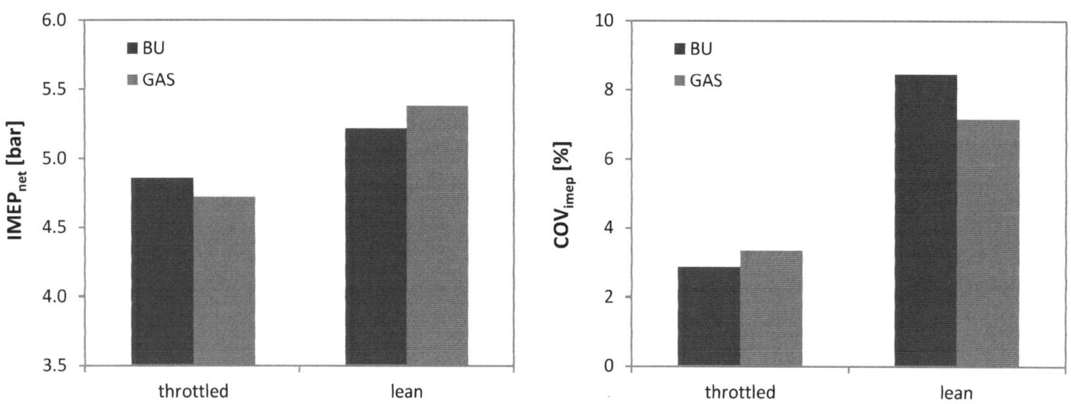

Figure 2. Indicated mean effective pressure (left) and its coefficient of variation (right).

Higher laminar flame speed and resulting faster combustion explains the result of improved engine output with butanol throttled operation. The fact that the opposite was found during lean fuelling suggests that observed variations in IMEP and its COV are less related to fuel impingement issues (given that the injected quantity was the same as with throttling for each fuel) [9], and more to temperature effects. One likely reason is that the higher level of residual gas present in the cylinder in throttled mode, resulted in an increase of fluid temperature during injection, thus improving evaporation. Another explanation is that stoichiometric operation features higher heat flux rates, and therefore determined an increase in walls temperature, enhancing the evaporation of liquid fuel film. Both effects resulted in 'leaner' fuelling when using butanol; this is suggested by the reduction of engine stability compared to gasoline, an effect which is specific for low equivalence ratios. Given the overall results and the observed effects of replacing gasoline with butanol, a more detailed investigation was performed in order to identify whether the ignition part of combustion initiation played a role. An initial evaluation of the secondary current revealed no significant differences between the traces recorded for the two fuels; therefore, optical techniques were applied, in the form of UV-visible spectroscopy.

SI engines combustion is initiated from the electrical discharge produced between the spark plug's electrodes close to the end of the compression phase [10], [11], [12]. The spark ignition features three phases: breakdown, arc discharge, and glow discharge. During breakdown, the stored energy is discharged in less than 100 ns. Breakdown begins with the formation of a thin conductive channel that expands very fast. All molecules in this channel are dissociated and ionized. Plasma temperature is higher than 10^4 K and can reach 10^5 K. Successively, the plasma electrical temperature decreases, due to cooling and recombination phenomena. During the arc phase, the plasma expands, due to heat conduction and diffusion, with a lower ionization level. It should be noted that in the range of 1-10 bar typical of pre-ignition in SI engines, arc phase temperatures are around 5000-10000 K. The final stage of spark plug discharge, called glow, has a duration of around some milliseconds if conventional spark-ignition systems are considered. In the proximity of the channel induced by the glow discharge, the combustible mixture is preheated, up to the adiabatic flame temperature and most likely beyond this value.

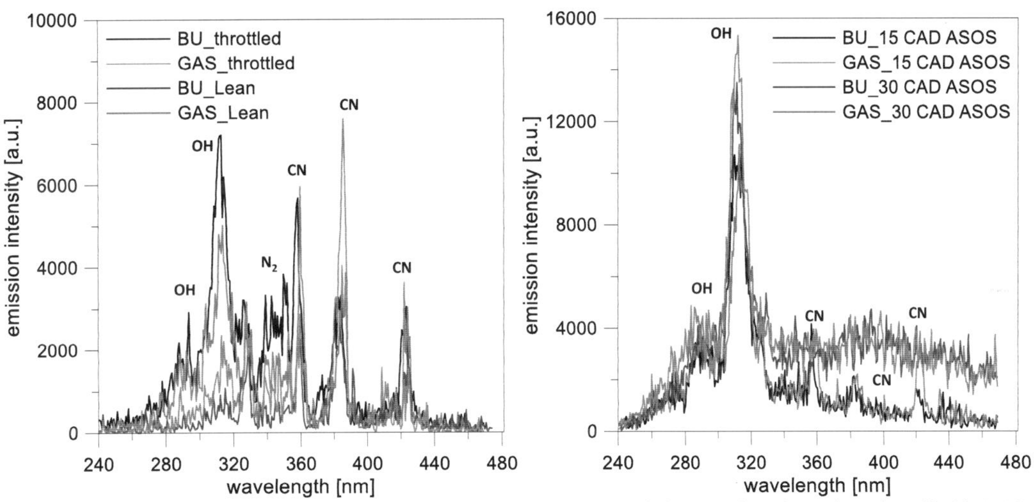

Figure 3. UV-visible emission spectra detected in the center of the combustion chamber (left) 0.42 ms (5 CAD) after spark timing for all selected conditions and (right) at 1.25ms (15 CAD) and 2.5ms (30 CAD) after spark timing, in throttled conditions, for both fuels.

The first stages of the combustion process were investigated through UV natural emission spectroscopy. Fig. 3 (left) shows the spectra detected for both fuels and operative conditions near the spark plug at 5 CAD after spark timing (ASOS), corresponding to an overlapping phase between arc and glow. In all the spectra, emission bands due to CN and OH radicals were well resolved. A weak band system associated to N_2 was identified as well [13], [14], [15]. CN was a recognized marker of spark discharge when ignition occurred in high-pressure conditions (5-10 bar); it could be detected right after breakdown (0.5-1 CAD in this work). OH emissions is instead characteristic for the glow phase and hence can be detected 100-200 μs after spark timing. CN emission bands were due to the violet system determined by the transition $B^2\Sigma^+ \rightarrow X^2\Sigma^+$ with Δv=0 (388 nm), Δv=–1 (421 nm) and Δv=+1 (358 nm) [13], [14]. OH radical transition $A^2\Sigma^+ \rightarrow X^2\Pi$ is typically identified by strong emissions at 309 nm (0,0) and weak band at 283 nm (1,0) [13]. In the arc-glow phase, the bends (0,0) at 337 nm and (1,0) at 315 nm of the second positive system related to N_2 transition $C^3\Pi_u \rightarrow B^3\Pi_g$ resulted sufficiently intense to overcome the weak emission due to OH (0,1) at 347 nm [13], [15]. These effects were generally associated to a plasma temperature higher than 4000 K. A qualitative analysis showed that CN emission intensity was higher for gasoline due to the higher carbon content in the fuel.

At 15 CAD (1.25 ms) after spark timing the glow phase was not ended. In the spectra reported in Fig. 3 (right) and detected at partial load, CN emission bands were still well distinguished and more intense for gasoline compared to butanol. The decrease in the CN emissions was related to the current discharge behaviour [7]. At the same time, the optical evidence of flame kernel development was observed through a sharp increase in the OH emission intensity due to the chemical exothermal reactions that characterize the early stages of combustion processes. The balance of this OH emission with respect to that determined by the glow phase is difficult to evaluate, and thus the radical cannot be considered as an effective marker of spark discharge. Finally, the HCO related broadband signal was observed, typical of the Vaidya band system that extends from 280 nm to 410 nm [16]. The H_2CO Emeléus' system between 300 nm and 500 nm was detected after the end of the spark discharge (Fig. 3, right). The reactions that induced the radiative evidence of formyl radicals and formaldehyde are not completely known. The balance between the two species is complex to evaluate and it is due to different effects related to flame temperature and local air fuel concentration. The assumption that formaldehyde may be involved in HCO formation was based on the observations of Agnew and Agnew [17]; the results suggest cool second-stage rich diethyl ether flames show formaldehyde emission, which is gradually replaced by predominantly HCO emission as the mixture was leaner. After the glow phase, OH emission was detected during the entire combustion process. Moreover, as expected, higher OH emission was measured for gasoline compared to butanol due to higher flame temperature of the fossil fuel.

Spectroscopic investigations allowed the evaluation of the plasma vibrational temperatures of the spark discharge by the emission of CN (*B-X*) and OH (*A-X*) assuming local thermodynamic equilibrium (LTE) [7], [18]. CN emissions permitted to calculate the temperature evolution in the arc phase and in the first stages of glow. The gas rotational temperature was estimated using the emission spectra of OH radicals during the glow phase and flame kernel development. Specifically, the ratio between the emission of the lines corresponding to Δv=0 (388 nm), Δv=+1 (357 nm) for CN was evaluated. Similarly, rotational temperatures were calculated by considering the ratio between the *P* (309 nm) and *R* (306 nm) branches of the OH (*A-X*) Δv=0 transition. Experimental results were compared with those obtained by using LIFBASE [13].

It is known that at fixed gas pressure, the vibrational and rotational temperatures smoothly decrease with decreasing secondary ignition current. On the other hand, both temperatures decrease at increasing pressure, at a given current setting [18], [19]. All these results were evaluated without considering the effect of pressure [13], [20]; the quantum energy levels of an ion/atom can be perturbed by a collision with an atom or by interaction with the electric field of an ion. The resulting width of a spectral line is proportional to the number density of the atoms and hence to pressure, according to the relation $\Delta\lambda \cong a \cdot p \cdot (300/T)^b$, where a and b are constant, p in bar and T in K are the pressure and translational temperature of the system. In this work, the temperature dependence was considered by fixing constant "b" equal to 0.5 [20].

Fig. 4 (left) shows the effect of pressure on CN emissions at fixed collisional broadening parameters (a=1; b=0.5) for different temperatures. It can be observed that only at high pressure (10 bar) the effect of temperature resulted significant. Moreover, at high temperature (>7000 K) the collisional parameters can strongly influence temperature evaluations (Fig. 4, right). The optimal "a" value was obtained by minimizing the difference between the width of the simulated lines and the experimental bands of OH A-X_(0,0) and CN B-X_(0,0) at 0.42 ms after spark timing for each fuel [7], [21].

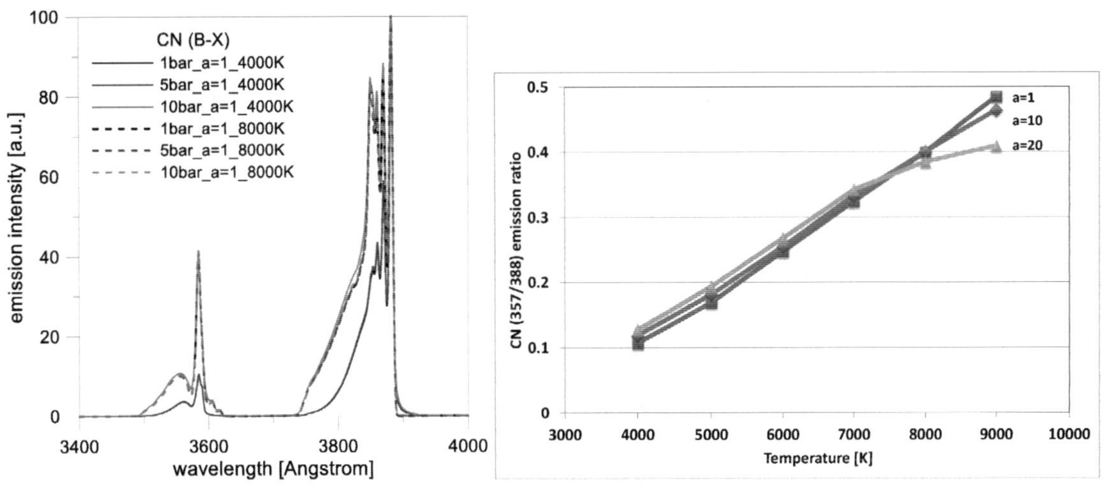

Figure 4. Simulated evolution of CN UV emission for different temperatures and pressure at fixed collisional broadening parameters a=1 and b=0.5 (left); evaluation of the CN emission bands ratio versus gas temperature for different broadening parameters (right) at fixed pressure of 10 bar.

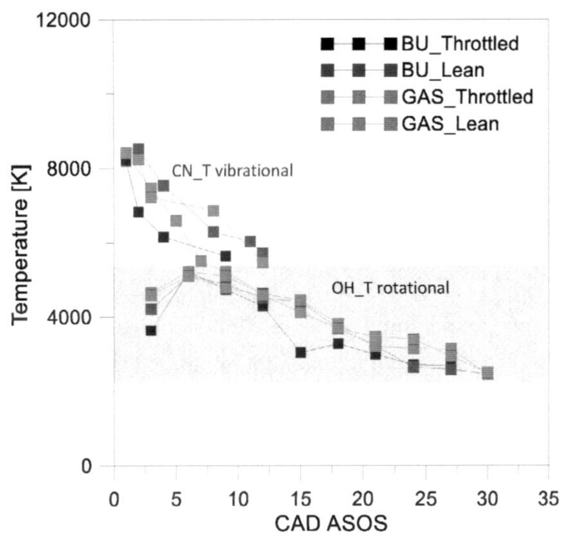

Figure 5. Vibrational temperatures determined based on the evaluation of CN Δv=+1/Δv=0 emission ratio and rotational temperatures from OH Δv=0 *P-R* branches.

Vibrational temperatures were determined based on the evaluation of CN Δv=+1/Δv=0 emission ratios and rotational temperatures from OH Δv=0 *P-R* branches; these two categories of data were calculated for each fuel, considering the measured in-cylinder pressure. Results are reported in Fig. 5. The highest uncertainty resulted approximately ~10% for the vibrational and ~12 % for the rotational temperatures. The vibrational temperatures decreased from the arc phase values (~8000 K) to the glow phase values (~5000 K). According to the literature, the highest glow temperature was about three times the adiabatic flame temperature [22], [23]. Rotational temperatures obtained from OH emission values decreased less abruptly than the vibrational ones. The asymptotic temperatures should converge to the adiabatic flame temperature values, around the end of the glow phase, in correspondence to the flame kernel development.

It should be noted that the temperature behaviours resulted quite similar for both fuels and engine conditions. A slight difference was observed for vibrational temperatures, perhaps due to the different air-fuel mixture near the spark plug at ignition timing, that determined higher values for throttled conditions with respect to lean mixture cases. This occurred because the electrical energy transferred to the working fluid was the same as the quantum energy necessary to induce the ions and radicals excitation. Fuel composition only influenced the relative concentration of excited chemical species and successively the formation of compounds (at ground quantum level) in the flame kernel volume.

4. Conclusion

Energy transfer from the spark plug's electrodes to the working fluid was characterized in an optically accessible wall guided DISI engine. Vibrational and rotational temperatures were determined in throttled stoichiometric conditions, as well as lean wide open throttle, with gasoline and butanol fuelling. Evaluations were based on UV-visible emissions related to CN and OH radicals; emission intensity was found to be directly influenced by local gas density, identified through an increase when in-cylinder pressure was higher. Fuel concentration also had a significant effect, with higher emission intensity for the stoichiometric cases. The two categories of temperature data showed that mixture strength has a determining effect on temperature mostly during the initial stages of ignition. Fuel effects on temperatures were less evident, suggesting that only a small part of the electrical energy is used for promoting chemical reactions and formation of intermediate species. One important conclusion was that only the CN radical can be used as a marker for ignition, given that OH concentrations are also high due to flame chemistry. Overall, the results can be considered as a solid database that can be readily used for validating ignition models and related chemical kinetic simulations. They can also be used for the optimization of ignition systems, in the context of ever wider application of boost levels (and therefore overall pressure during the cycle) and alternative combustion control strategies.

5. References

[1] Chen, Z., Yang, F., Xue, S., Wu, Z., & Liu, J. (2015). Impact of higher n-butanol addition on combustion and performance of GDI engine in stoichiometric combustion. *Energy Conversion and Management*, 106, 385-392.

[2] Papagiannakis, R. G., Rakopoulos, D. C., & Rakopoulos, C. D. (2017). Theoretical Study of the Effects of Spark Timing on the Performance and Emissions of a Light-Duty Spark Ignited Engine Running under Either Gasoline or Ethanol or Butanol Fuel Operating Modes. *Energies*, 10(8), 1198.

[3] Merola, S. S., Irimescu, A., Di Iorio, S., & Vaglieco, B. M. (2017). Effect of Fuel Injection Strategy on the Carbonaceous Structure Formation and Nanoparticle Emission in a DISI Engine Fuelled with Butanol. *Energies* 10(7), 832; doi:10.3390/en10070832

[4] Merola, S. S., Tornatore, C., Irimescu, A., Marchitto, L., & Valentino, G. (2016). Optical diagnostics of early flame development in a DISI (direct injection spark ignition) engine fueled with n-butanol and gasoline. *Energy*, 108, 50-62.

[5] Ju, Y., and Sun, W. (2015) Plasma assisted combustion: Dynamics and chemistry. *Prog. Energ. Combust. Sci.,* 48, 21-83

[6] Breda, S., D'Adamo, A., Fontanesi, S., D'Orrico, F., Irimescu, A., Merola, S., & Giovannoni, N. (2017). Numerical Simulation of Gasoline and n-Butanol Combustion in an Optically Accessible Research Engine. *SAE International Journal of Fuels and Lubricants*, 10, 32-55.

[7] Merola, S.S., Di Iorio, S., Irimescu, A., Sementa, P., Vaglieco, B.M. (2017) Spectroscopic characterization of energy transfer and thermal conditions of the flame kernel in a spark ignition

engine fueled with methane and hydrogen. *International Journal of Hydrogen Energy*, 42 (18), 13276-13288.

[8] Irimescu, A. (2010). Air-fuel h–x diagram for gasoline-isobutanol blends. *Proceedings of the Romanian Academy Series A*, 11, 322-329.

[9] Merola, S.S., Irimescu, A., Marchitto, L., Tornatore, C., Valentino, G. (2017). Effect of injection timing on combustion and soot formation in a direct injection spark ignition engine fueled with butanol. *International Journal of Engine Research*, 18(5-6), 490-504, doi:10.1177/1468087416671017.

[10] Maly, R. (1984). *Spark ignition: its physics and effect on the internal combustion engine* (pp. 91-148). Plenum, New York.

[11] Maly R. (1981). Ignition model for spark discharges and the early phase of flame front growth. *In Symposium (International) on Combustion*, 18(1), 1747-54.

[12] Thiele M, Selle S, Riedel U, Warnatz J, Maas U. (2000). Numerical simulation of spark ignition including ionization. *Proc Combust Inst*, 28(1), 1177-1185.

[13] Luque, J., Crosley, D. R. (1999). *LIFBASE: Database and Spectral Simulation Program* (Version 1.5), SRI International, Rept. MP 99- 009.

[14] Larsson, M., Siegbahn, P. E. M., Agren, H. (1983). A theoretical investigation of the radiative properties of the CN red and violet systems. *Astrophysical J,* 272 ,369-376.

[15] Namihira, T., Sakai, S., Matsuda, M., Wang, D., Kiyan, T., Akiyama, H., Okamoto, K., Toda, K. (2007). Temperature and nitric oxide generation in a pulsed arc discharge plasma. *Plasma Science and Technology*, 9(6), 747-751.

[16] Gaydon, A. G., Wolfhard, H. G. (1953). Mechanism of formation of CH, C2, OH and HCO radicals in flames. In *Symposium (International) on combustion* 4(1), 211-218.

[17] Agnew W. G, Agnew J. T. (1956). Visible Emission Spectra of Two-Stage Flames of Diethyl Ether Produced in Flat-Flame Burner. *Industrial & Engineering Chemistry*, 48(12), 2224-2231.

[18] Kawahara, N., Hashimoto, S., Tomita, E. (2017). Spark discharge ignition process in a spark-ignition engine using a time series of spectra measurements. *Proc. Combust. Inst.* 36(3), 3451-3458.

[19] Ono, R., Nifuku, M., Fujiwara, S., Horiguchi, S., Oda, T. (2005). Gas temperature of capacitance spark discharge in air. *Journal of Applied Physics*, 97(12), 123307.

[20] Griem, H. R. (1974). *Spectral line broadening by plasmas*, Academic, New York.

[21] Merola, S.S., Di Iorio, S., Irimescu, A., Sementa, P., & Vaglieco, B.M. (2017). Spectroscopic investigation of initial combustion stages in a SI engine fuelled with ethanol and gasoline. *SAE Technical Paper* 2017-32-0092.

[22] Shen, H., Hinze, P., Heywood, J. (1994). A Model for Flame Initiation and Early Development in SI Engine and its Application to Cycle-to-Cycle Variations. *SAE Technical Paper 942049.*

[23] Willems, H., Sierens, R. (2003). Modeling the Initial Growth of the Plasma and Flame Kernel in SI Engines. *ASME J Eng Gas Turbines Power* 125(2), 479-484.

AD HOC NETWORKS FOR THE AUTONOMOUS CAR

Davidescu Ron, Eugen Negrus
Universitatea Politehnica din Bucureşti

rondavidescu@gmail.com

Abstract. The future of the vehicle is made of cars, roads and infrastructures connected in a two way automated communication in a holistic system. It is a mandatory to use Encryption to maintain Confidentiality , Integrity and Availability in an ad hoc vehicle network . Vehicle to Vehicle communication, requires multichannel interaction between mobile ,moving and changing parties to insure the full benefit from data sharing and real time decision making , a network of such users referred as mobile ad hoc network (MANET) , however as ad hoc networks were not implemented in such a scale , it is not clear what is the best method and protocol to apply . Furthermore the visibility of secure preferred asymmetric encrypted ad hoc networks in a real time environment of dense moving autonomous vehicles has to be demonstrated, In order to evaluate the performance of Ad Hoc networks in changing conditions a simulation of multiple protocols was performed on large number of mobile nodes. The following common routing protocols were tested , DSDV is a proactive protocol ,every mobile station maintains a routing table with all available destinations , DSR is a reactive routing protocol which allows nodes in the MANET to dynamically discover a source route across multiple network hops , AODV is a reactive routing protocol Instead of being proactive. It minimizes the number of broadcasts by creating routes based on demand , SAODV is a secure version of AODV ,requires heavyweight asymmetric cryptographic , ARIANDE is a routing protocol that relies on highly efficient symmetric cryptography the concept is primarily based on DSR.

A methodical evolution was performed in a various density of transportation, based on known communication bench mark parameters including , Throughput Vs. time , Routing Load per packets and bytes. Out of the none encrypted protocols , It is clear that in terms of performance of throughput and routing load DSR protocol has a clear advantage the high node number mode . The encrypted protocols show lower performance with ARIANDE being superior to SAODV and SRP. Nevertheless all protocol simulation proved it to match required real time performance.

1. Introduction

Still today, security of automotive usually means accident or theft prevention. However as electronics and software in the modern vehicle is growing rapidly to enable the employment of autonomous vehicle and the connected car, safety has become equal with security.
It safe to say that the autonomous vehicle is distinctive in the need of procedure with no tolerance to failure in security, continuity and availability. On top of it, recent demonstrations by research groups

have proven that vehicles can be penetrated remotely through their communication units and ordered to run malicious code that permits the intruder to control remotely the vehicle. Thus, was established that breaches in vehicle security already have dangerous safety effects. As security is always the main concern of all automobiles firms, vehicle manufacturers must make safety the same priority as security.

As vehicles open to external communication networks, they become probable targets of hacker's attacks. More computers and communication interfaces produce larger treats and bring new penetration surfaces .New communication interfaces suffer from classical IT weaknesses and from the fact that cars by nature have to rely on wireless communication with no wired back up.

One of the obvious difficulties in large implementation of the connected vehicle are the opposite demands of strong ,reliable , encryption and description while keeping real time operation in a moving vehicle with low computer resource environment and many times low communication infrastructure.

Vehicle to vehicle communication (V2V) requires multi-channel interaction between mobile, moving and changing parties to insure the full benefit from data sharing and real time decision making, a network of such users referred as mobile ad hoc network (MANET). A Mobile Ad-hoc Wireless Network (MANET) is a collection of autonomous nodes that communicate with each other by forming a multi-hop network, maintaining connectivity in a decentralized manner. It consists of a set of mobile hosts communicating amongst themselves using wireless links, without the use of any other communication support facilities, such as base-stations. The nodes in a MANET can be any device that is capable of transmitting and receiving information. Each node in such a network acts as a host or end system (transmitting and receiving data) and simultaneously as a router. The nodes in a MANET are generally mobile and may go out of range of other nodes in the network [2].

2. None secure Ad Hoc Network Performance Simulation

In order to examine the compatibility of different Ad Hoc protocols in changing conditions, a simulation was performed on large number of mobile nodes. We have examined three common none secure routing protocols for MANET.

DSR is a reactive routing protocol which allows nodes in the MANET to dynamically discover a source route across multiple network hops to any destination. In this proto-col, the mobile nodes are required to maintain route caches or the known routes. The route cache is updated when any new route is known for a particular entry in the route cache.
AODV is a reactive routing protocol instead of being proactive. It minimizes the
Number of broadcasts by creating routes based on demand, which is not the case for DSDV. When any source node wants to send a packet to a destination, it broadcasts a route request (RREQ) packet. The neighboring nodes in turn broadcast the packet to their neighbors and the process continues until the packet reaches the destination [2].
DSDV is a proactive protocol; every mobile station maintains a routing table with all available destinations along with information like next hop, the number of hops to reach to the destination, sequence number of the destination originated by the destination node, etc. DSDV uses both periodic and triggered routing updates to maintain table consistency. Triggered routing updates are used when network topology changes are detected, so that routing information is propagated as quickly as possible [3]

For the simulation of the developed system ViSim 1.0 has been used , ViSim calls ns-2 simulations in a Windows environment , to allow rapid configuration for any MANET routing scenario [2].
Table 1 describes the simulation parameters for none secure as hoc protocols.

Table 1.Simulation Parameters.

```
# Define options
set val(chan)       Channel/WirelessChannel      ;# channel type
set val(prop)       Propagation/TwoRayGround     ;# radio-propagation model
set val(netif)      Phy/WirelessPhy              ;# network interface type
set val(mac)        Mac/802_11                   ;# MAC type
set val(ifq)        Queue/DropTail/PriQueue      ;# interface queue type
set val(ll)         LL                  ;         # link layer type
set val(ant)        Antenna/OmniAntenna          ;# antenna model
set val(ifqlen)     65                           ;# max packet in ifq
set val(nn)         100                          ;# number of mobilenodes
set val(rp)         DSR/AODV/DSDV                ;# routing protocol
set val(x)          1500                         ;# X dimension of topography
set val(y)          750                          ;# Y dimension of topography
set val(stop)       2500                         ;# time of simulation end
```

All three protocols were compared in a 100 mobile nodes in random traffic lanes , the following performance metrics were evaluated to understand the behavior of DSDV,DSR and AODV , Throughput Vs, Time , Routing Load (In terms of packets) & Routing Load (In terms of Bytes).

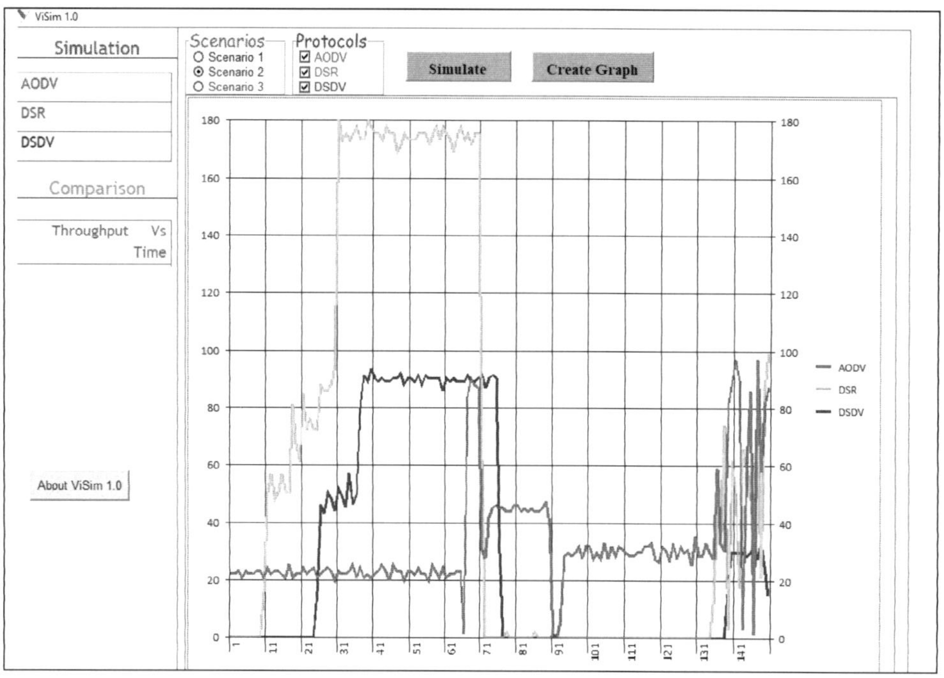

Fig. 1. Throughput Vs, Time results (Simulation results)

Throughput is the number of bytes received by the destination node per second (Data packets and Overhead).

IOP Conf. Series: Materials Science and Engineering **252** (2017) 012094 doi:10.1088/1757-899X/252/1/012094

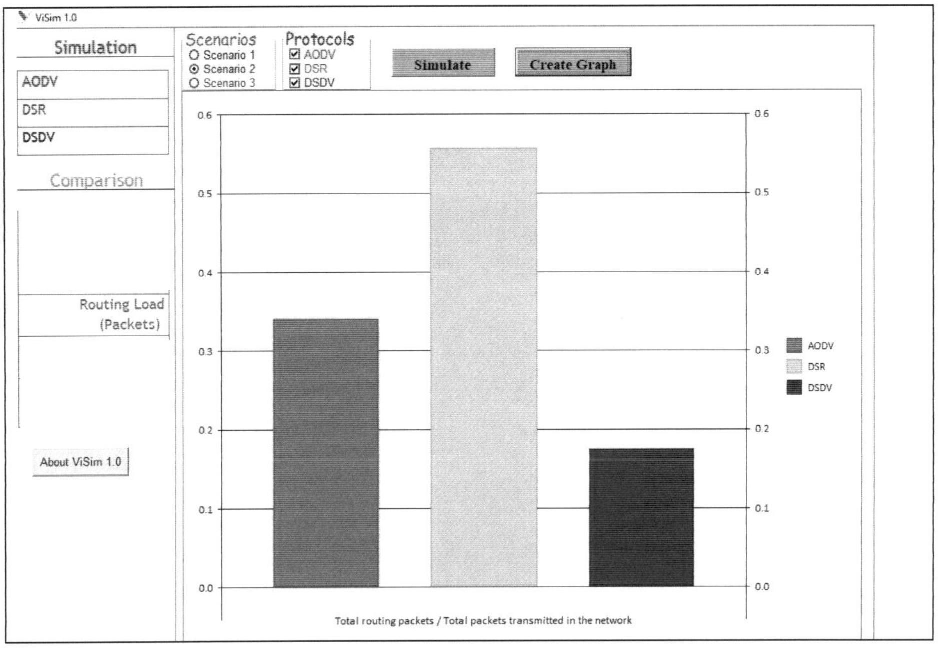

Fig. 2. Routing Load - In terms of packets (Simulation results)

Routing Load (In terms of Packets) is the ratio of the total routing packets that are sent within the network to the total number of packets that are transmitted within the network to reach the destination.

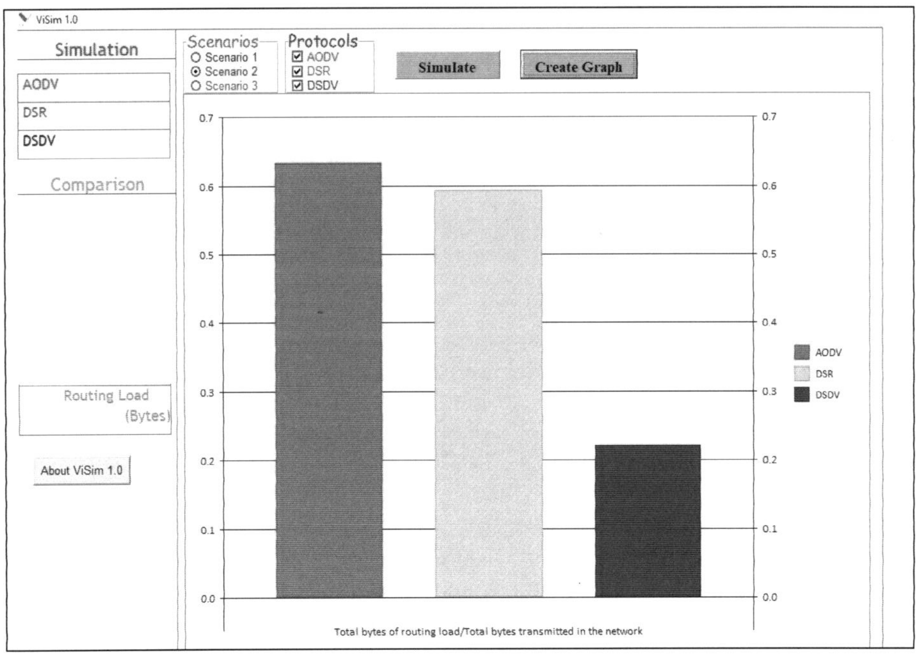

Fig.3. Routing Load - In terms of Bytes (Simulation results)

IOP Conf. Series: Materials Science and Engineering **252** (2017) 012094 doi:10.1088/1757-899X/252/1/012094

Routing Load (In terms of Bytes) is the ratio of the total routing bytes that are sent within the network to the total number of bytes that are transmitted within the network to reach the destination.

It is clear that the DSR protocol has an advantage in Throughput and Routing Load - In terms of packets.
In the third parameter Routing Load - In terms of Bytes it has similar performance to the AODV protocol and superior performance to the DSDV protocol.

3. Secure Ad Hoc Network Performance Simulation

In order to examine the compatibility of different Ad Hoc protocols in a changing conditions , a simulation was performed on multiple number of mobile nodes. We have examined three common secure routing protocols for MANET.

Ariadne : is a routing protocol that relies on highly efficient symmetric cryptography the concept is primarily based on DSR

SAODV is a secure version of AODV, requires heavyweight asymmetric cryptographic.

SRP - The basic idea of SRP is to set up a security association (SA) between a source and a destination node without the need of cryptographic validation . [3]

Table 1.Simulation Parameters.

```
# Define options
set val(chan)       Channel/WirelessChannel     ;# channel type
set val(prop)       Propagation/TwoRayGround    ;# radio-propagation model
set val(netif)      Phy/WirelessPhy             ;# network interface type
set val(mac)        Mac/802_11                  ;# MAC type
set val(ifq)        Queue/DropTail/PriQueue     ;# interface queue type
set val(ll)         LL              ;           # link layer type
set val(ant)        Antenna/OmniAntenna         ;# antenna model
set val(ifqlen)     65                          ;# max packet in ifq
set val(nn)         20/40/60/80/100             ;# number of mobilenodes
set val(rp)         Ariadne / SAODV / SRP       ;# routing protocol
set val(x)          1500                        ;# X dimension of topography
set val(y)          750                         ;# Y dimension of topography
set val(stop)       2500                        ;# time of simulation end
```

All three protocols were compared in a 20,40,60,80 AND 100 mobile nodes in random traffic lanes , the following performance metrics were evaluated to understand the behavior of Ariadne, SAODV and SRP , Max Throughput , Routing Load (In terms of packets) & Routing Load (In terms of Bytes).

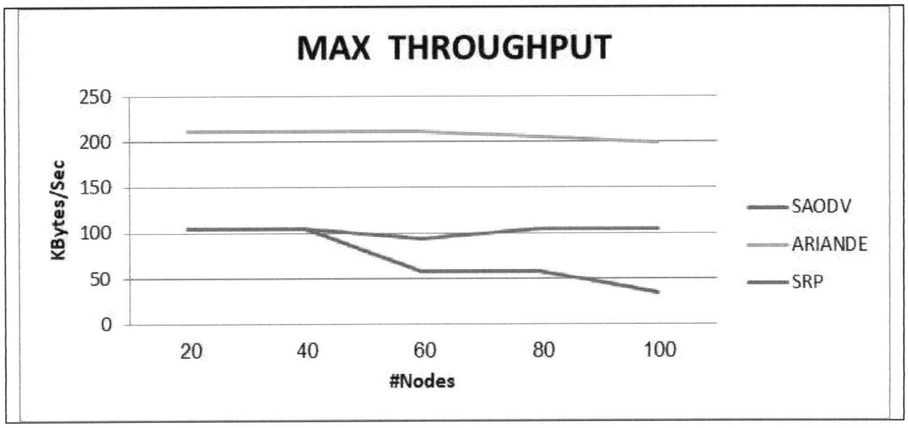

Fig. 4. Max throughput results (Simulation results)

Max Throughput is the max bytes received by the destination node per second (Data packets and Overhead).

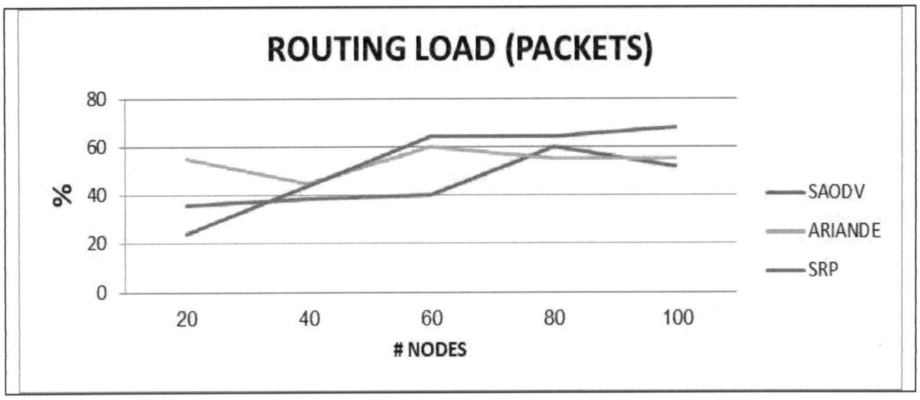

Fig. 5. Routing Load - In terms of packets (Simulation results)

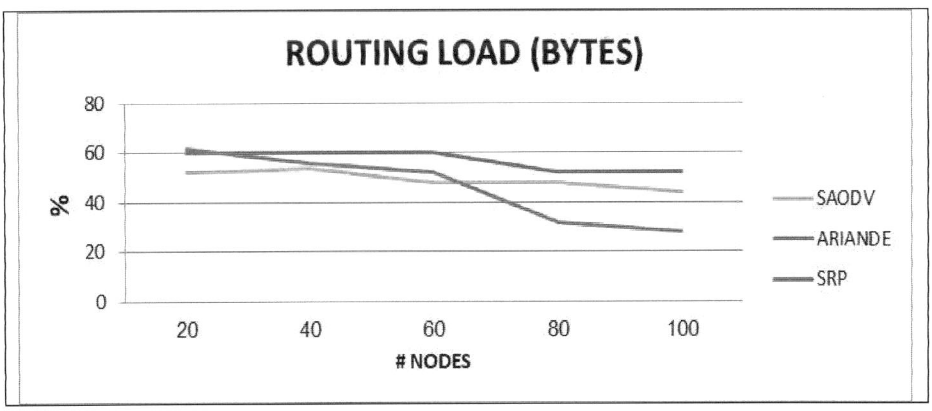

Fig.6. Routing Load - In terms of Bytes (Simulation results)

The ARIANDE protocol shows superior throughput and bytes routing load and similar performance to SRP on the packet routing. Very similar to the none secure DSR protocol behaviour it was developed from.

4. Summary

A full solution of autonomous cars will require Vehicle to Vehicle communication to enhance the local car sensors performance and enable the security and safety required for large connected car implementation and benefits. On top of that it will need integration of older none autonomous cars to share the road.

Never the less any none direct mean of communication will reduce the time to reaction in case of emergency and jeopardize the wellbeing of the passenger. Ad hoc networks allow such communication and create a network of communication that is independent from the vehicle to infrastructure network and that can act as a backup, by its nature the data goes first to the close vehicles. In this paper we evaluated common none secure ad hoc protocols, DSDV, DSR and AODV in terms of throughput and routing Load through computer simulation and concluded that the DSR protocol is superior. Furthermore three secure ad hoc protocols were evaluated in multi node conditions, Ariadne, SAODV and SRP. In that case Ariadne that was developed from the none secure DSR proved better. It is a common believe that autonomous cars will change the world we know for better, however without a standards for V2I and V2V communication that will probably never work. Secure ad hoc networks that will follow an international standard to be implemented on all autonomous and connected cars will resolve the issue.

References

[1] Nivedita Bisht, Sapna Singh : ANALYTICAL STUDY OF DIFFERENT NETWORK TOPOLOGIES : International Research Journal of Engineering and Technology : Volume: 02 Issue: 01 :Mar 2015.

[2] Nazmus Saquib , Md. Sabbir Rahman Sakib : ViSim: A user-friendly graphical simulation tool for performance analysis of MANET routing protocols : Mathematical and Computer Modelling 53 (2011) 2204-2218 .

[3] Sachin Kumar Gupta, R. K. Saket : PERFORMANCE METRIC COMPARISON OF AODV AND DSDV ROUTING PROTOCOLS IN MANETs USING NS-2: International Journal of Research and Reviews in Applied Sciences : June 2011 : Volume 7

[4] Spinder Kaur, Harpreet Kaur : Implementing RSA Algorithm in MANET and Comparison with RSA Digital Signature : INTERNATIONAL JOURNAL FOR ADVANCE RESEARCH IN ENGINEERING AND TECHNOLOGY : Volume 3, Issue V, May 2015

[5] C. Sreedhar, Dr. S. Madhusudhana Verma, Prof. N. Kasiviswanath : A Survey on Security Issues in Wireless Ad hoc Network Routing Protocols : International Journal on Computer Science and Engineering: Vol. 02, No. 02, 2010,

[6] Davide Benetti ,Massimo Merro, Luca Vigan : Model Checking Ad Hoc Network Routing Protocols: ARAN vs. endair A: Software Engineering and Formal Methods (SEFM), 2010 8th IEEE International Conference .

[7] Intelligent Transportation Systems, Joint Program Office, INTELLIGENT TRANSPORTATION SYSTEMS (ITS) Information Security Analysis, U.S. Highway Administration, Department of Transportation, Federal Highway Administration, 1997.

[8] Bryan Parno, Adrian Perrig,: Challenges in Securing Vehicular Networks: http://www.sparrow.ece.cmu.edu.

[9] Maxim Raya, Panos Papadimitratos, : Securing Vehicular Communications: http://www.ece.cmu.edu.

IOP Conf. Series: Materials Science and Engineering **252** (2017) 012095 doi:10.1088/1757-899X/252/1/012095

Real time monitoring system used in route planning for the electric vehicle

LM Ionescu, A Mazare, G Serban, S Ionita
University of Pitesti, Faculty of Electronics, Communication and Computers, Romania

Abstract. The electric vehicle is a new consumer of electricity that is becoming more and more widespread. Under these circumstances, new strategies for optimizing power consumption and increasing vehicle autonomy must be designed. These must include route planning along with consumption, fuelling points and points of interest. The hardware and software solution proposed by us allows: non-invasive monitoring of power consumption, energy autonomy - it does not add any extra consumption, data transmission to a server and data fusion with the route, the points of interest of the route and the power supply points. As a result: an optimal route planning service will be provided to the driver, considering the route, the requirements of the electric vehicle and the consumer profile. The solution can be easily installed on any type of electric car - it does not involve any intervention on the equipment.

1. Introduction

There are several hardware technologies that will be involved in the solution presented in this paper. First, we will use clamp-meter current sensors. These allow for the determination of the current consumption by measuring the electric field around the isolated conductor. The measure range of sensors is between 2 – 600 A (dc), thus they cover a wide capacity area of motor sizes. The clamps can be placed on the main branch to monitor overall consumption (engine and equipment in the car) or on the engine branch. As shown, the sensors are placed on isolated conductors without any intervention on the power supply infrastructure. Secondly, the data acquired by the sensors is transmitted wirelessly to a local collector. This does not require the installation of any cable between the collector and the sensors. The communication is in the free band 2.4GHz and is resistant to disturbances. Finally, the sensors, wireless transmitters as well as the local collector are energetically autonomous. The sensors and wireless transmitters use batteries which can be charged. The local collector also uses a battery pack that loads from a small photovoltaic panel - L 20 cm, l 10 cm. In full darkness, the autonomy is 40 hours for 12 operations per hours which consist in: take data from sensors, pack with location information (from GPS receiver) and send to server via GSM network using secured SSL HTTP protocol. There are also several software technologies and planning methods implemented. To reduce consumption at the local collector, a resource planning method is used: first the gateway for data collection from the sensors is activated and in the next step the gateway is disabled and the GSM modem / GPS receiver for location and data transmission to the server is activated. At the server level, vehicle data are acquired, correlated with GIS (Geographic Information System) information about route and power points and then navigation services are provided to the driver. The application is developed using WSO2 DAS server (for complex event processing module), NodeJs and Angular2 technologies with GoogleMaterials framework for web interface. GIS services will use GoogleMaps – free for 10000 request / day (cover our request of update on 5 min).Paper will present the results with a full functional system solution. Measurements will be made for different consumers. In this sense, a different dynamic load generator will be used. Different routes will be made and will be transmitted to the server. Supply points are also entered the server application. As a result, the server will return the recalculated route considering consumption and power points. We'll have snapshots from the routes proposed by the server.The solution proposed in this paper introduces the concept of on-line monitoring for cars. The novelty is given by: real-time monitoring, using sensors that can be easily attached to power points, autonomous energy solutions for sensors, transmitters and the local collector, as well as special navigation services provided to electric car drivers.

2. State of the art

Research on route planning for electric vehicles was carried out in several papers. For example, the solution presented in the paper [1] or in the patent [2] shown are such solutions. Starting from data on navigation, traffic, etc. they can estimate an optimal route for the electric vehicle. Besides, this is not difficult: given the current navigation services - such as Google Maps - as well as APIs (application programming interface) made available to developers, it is easy to introduce supply stations into an application as intermediate points and then determine the route by estimating the consumption and considering the passage at the supply points. There are several projects and research papers (some have also materialized in commercial applications) dealing with this topic.
However, to increase accuracy in estimation, we need a parameter to be acquired and transmitted from the vehicle: instantaneous power consumption. This is because even for the same route there are many reasons for it to have different values: congestion in traffic, driving style, the presence of additional consumers in the vehicle, etc. Achieving a system that acquire power consumption from the vehicle and to be easy to install on the vehicle raises several challenges:

- Finding a power acquisition solution that does not involve any intervention on the structure of the vehicle (for example: stopping the power supply, removing terminals, etc);
- Finding a solution allowing the transmission of data acquired without intervention on the vehicle (use of data transmission wires, drilling of surfaces for wires etc.);
- The solution must transmit the data remotely for later analysis and transfer to a navigation application;
- The solution must be as energetically autonomous - it does not have to introduce additional significant consumption in the vehicle!

Each of the four points mentioned is a distinct subject of research. First, along with the classical method, whereby an amperemeter must be serial connected with the consumer and the current supply source to measure the power consumption, there are several studies that have resulted in solutions for determining the consumption of current without introducing into the circuit of a measurement device (with all that it means: disconnecting the terminals, soldering yarns). A study presented in paper [3], among others, shows that it is possible to determine the consumption by analysing the electric field loads - whether we are talking about AC or DC. The electric field can be captured by using a capacitive effect as well as an inductive effect by using a Hall sensor as shown in work paper [4]. Here it shows that it is possible to capture this field and accurately measure the current consumption even around the isolated conductor for a continuous current of values ranging from 2 to 600A. At present, there are current measuring equipment around the conductor: clamp meters.
In our case, it is necessary to measure current consumption and transmit it to a local collector (concentrator). At present, there are standards in communications from smart meters: wireless communications. They are used to transmit data in electromagnetic "noisy" environments, as shown in reference [5]. A method of communication that covers these standards is also XBee communication, which is based on the Ad-Hoc ZigBee radio protocol. This radio communication is used to deploy local radio communications networks (area of up to 2 km) using the 2.4 GHz free band. The emission power is low on some of the reasons for the use of the free band and on the other hand also because of the reduction of energy consumption in communication - one of the important elements in Smart Greed communication as shown in paper [6]. Another foreseen element is the protection of data transmitted on radio channels, especially those in the free band. We are talking about ensuring the integrity of data in disturbances, as well as protecting data from unauthorized readings. From this point of view, we apply a series of algorithms that increase the security of communication - algorithms that fit especially in the "lightweight" solutions such as the one presented in reference [7].
Finally, another aspect, involving many studies at present, is about the degree of autonomy of the solution. Any hardware solution proposed for an electric vehicle must involve very low consumption -

IOP Conf. Series: Materials Science and Engineering **252** (2017) 012095 doi:10.1088/1757-899X/252/1/012095

from this point of view, applications for the electric vehicle can be regarded as mobile applications for environments where there are no electricity sources. So, the orientation is towards other alternative energy sources that can be converted into electricity. Of these, the most feasible way in our case is to use light energy to convert to electricity, a solution that applies to wireless grids, as shown in reference study [8]. The system that acquiring electricity from unconventional sources must also be a portable one, which requires an acceptable implementation cost. Technological evolution has reduced the size of the photovoltaic panels and the required controller. Such a "energy harvesting" solution that lends itself to portable smart meter solutions is outlined in paper [9].
All the elements presented in this section have been used to implement our system as described below.

3. System architecture

Figure 1 shows the architecture of the proposed system. The system consists of a mobile component installed on the vehicle and from a server running an application that provides customer navigation services. The mobile component to be installed on the vehicle consists of: current acquisition units and local radio data transmission that are installed on the batteries or other consumers in the vehicle and a "concentrator" unit that takes the packets from the local radio network and transmits them to the server in the form of HTTP packages (HTTP POST requests). The server runs an application that contains several modules - as can be seen in the figure 1: http receiver - responsible for downloading http packages from the concentrators (vehicles), event stream that extracts the data needed for further processing, analytics core responsible for data processing - here is an assessment of current consumption to determine when it will fall below a certain threshold, UI publish that deals with the transmission of prediction results, Geo Dashboard which receives the results provided by UI publish and together with the GIS module translates them into the map, and finally the client responsible UI that displays the results map in a format that the client (the driver) can read.

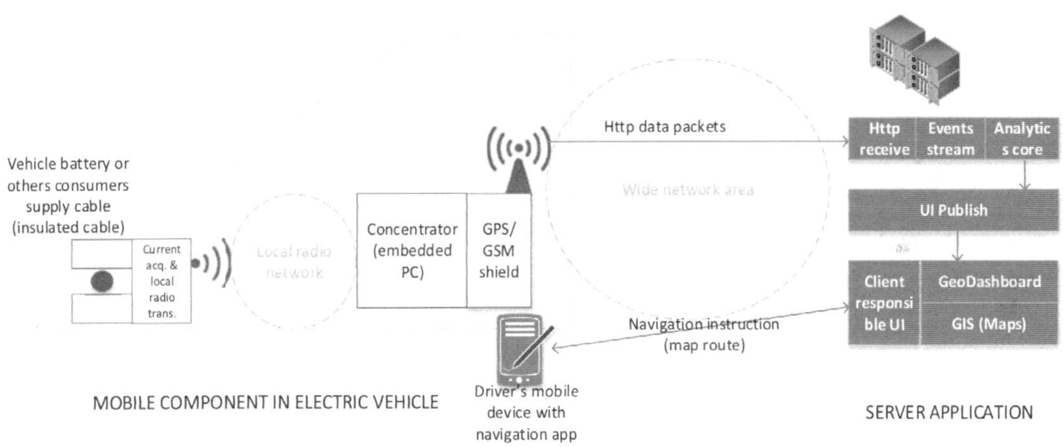

Figure 1. System architecture – block diagram.

The client application running on a mobile device (smartphone, tablet) allows you to enter data related to the desired route and provides, based on the data provided by the current sensor, the planned route. In the following sections, each of the modules involved will be presented.

3.1. Mobile equipment on the vehicle

a) The acquisition unit and local radio transmission unit

It consists of a current sensor (clamp meter) and the wireless data transmission circuit. The circuit allows:

- the acquisition of the current using a clamp sensor, which allows to determine the current consumption by capturing the electric field around the insulated conductor;
- data transmission on a local radio network in the 2.4GHz band (free band);

- the circuit is powered using batteries. The degree of autonomy of the system is one month in operation with the acquisition and transmission of information at 3 minutes. The degree of autonomy can be expanded by increasing the data transmission period.

The format in which the data is transmitted renders communications resistant to electromagnetic disturbances that may also occur in the automotive environment. The circuit can operate in automotive environment between -20 ° C - 55 ° C and 85% humidity.

Figure 2. A capture with experimental mobile system: current consumption sensor (clamp meter) and local radio transceiver – left, concentrator: embedded PC with local radio transceiver gateway, GPS GSM shield and batteries pack with photovoltaic panel – left.

b) The concentrator unit

It consists of a gateway that manages the network and takes over the data from the local radio network, an embedded PC and a GPS / GSM shield that allows data to be transmitted over the GSM network. The embedded PC acts as a data translator - takes the data from the gateway, encapsulates the GPS data of the coordinates and transmits them via the GPS / GSM shield to the server. The unit has the following features:

- Gateway that allows up to 15 units (monitoring points) to acquire and transmit power consumption from the vehicle. Modifying the network configuration is very simple: only a power-on acquisition unit is introduced and the synchronization of the communication with the other units is automatically done. The communication protocol is XBee is based on the use of a unique address of the transmission device. The concentrator has set the interrogation rate of each unit. The transmission envisages using a CSMA / CA protocol to avoid collusion. The experiment and solution presented here uses only one acquisition unit.
- The GSM data transmission rate (and, implicitly, the acquisition rate of the data from the units) is set by the user - the data communication protocol is EDGE GPRS;
- Receiving coordinates, speed and time by using a GPS module. In fact, we used an integrated GPS / GSM module;
- Power supply using batteries charged from a photovoltaic panel (mounted on the vehicle). For the transmission rate set to 3 minutes, the battery's autonomy when it is not charged (the car is in the dark) is 48 hours. The charging time for the batteries is about 5 hours at the average illumination of the panel. Basically, if the panel is cleaned, the concentrator unit has total energy autonomy, including in the cold season for a region at the latitude of Romania (at least).

The concentrator sends the data (current values, coordinates, speed, time, and concentrator ID) as HTTP POST packages, with data attached to the body of the packets. These packages are sent to the server.

3.2. Server

Receives HTTP packets, extracts data from them (creates an event stream), analyses them and estimates the battery discharge rate using instantaneous consumption, speed, current coordinates, time - all taken from the mobile unit - and the information about the route you want and the type of electric vehicle battery - information provided by the driver. The discharge rate of the vehicle battery is associated with the desired route, with the power supply points and map data provided by the GIS module. By using the Geo-DashBoard module and the client interface, is given to the driver important data about the route to be followed. The central element of the server is the Complex Event Processing module of the DAS (Data Analytics Server) WSO2 analytics server. A concept scheme is shown in the figure below:

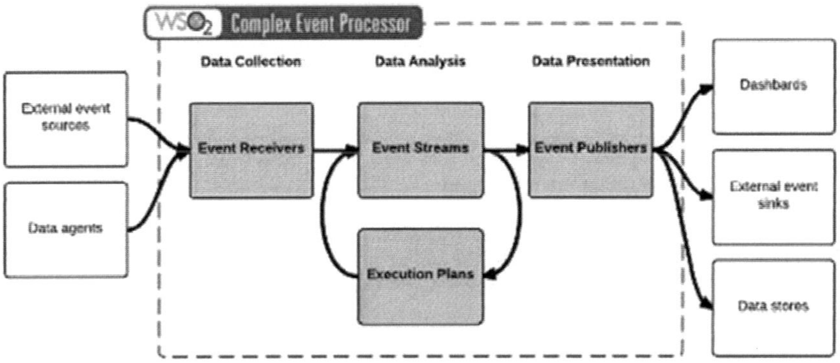

Figure 3. WSO2 CEP module block diagram.

The main components are: event receiver, event flow, execution plan, and publisher. In very short terms:

- the receiver is the one that handles data retrieval - here the HTTP packages sent from mobile units (concentrators) via GSM;
- event stream is the one that allow the data extraction and structuring. Also at this level is data analysis. This is done by using flow plans of execution;
- event publisher is responsible for publishing event results. In our case, the results are sent in the form of coordinates of the route to the Geo Dashboard where they are also displayed. A capture with Geo-DashBoard interface is presented in figure 4.

The result will be a route shown on the driver's device with the intermediate power points. The route will include information about power points, battery status at any of the points on the route (estimated status) and information on the directions that the driver should to follow.

The technologies used at the server application level are follows:

- WSO2 DAS is used to implement the responsible event handler, from the HTTP packets delivered by the concentrator to the GeoDashBoard data delivery. WSO2 DAS contains CEP (Complex Event Processing) component in our system. The data is made available through a data interface called Geo-DashBoard integrated into the DAS. WSO2 DAS is provided free of charge and can be used free of charge in development, testing and production with all the features;
- Google Maps services for all GIS and navigation. Entering intermediate points representing power stations, entering the starting point and destination point, route information, and navigation itself (including voice guidance) are services provided by Google Maps. Access to these services is at the client application level - we are talking about integrating these services into the client application that we have achieved. Google provides the necessary APIs to integrate GMaps into its own applications. For a maximum of 10000 requests per day, services are free of charge;
- Google Materials is a framework of CSS classes that provides formats for all the controls required for the client application: forms, frames where the map, buttons, the application page. Controls are "responsive": portable on both desktop and mobile systems (smaller screens) - without any additional change. The framework is free;

Figure 4. WSO2 Geo-Dashboard interface.

- Angular 2 and Typescript are the technologies by which the custom client application is built. G Materials, Google Maps services, GeoDashboard results are implemented by these two technologies: the first is the concept and framework through which the controls are placed on the client application, and two are the proper language for writing.

Figure 5. List of technologies used for our server application (from left to right): WSO2 DAS, Google Maps, Google Materials, Angular 2 and TypeScript.

4. Implementation

The system presented in this article has been implemented at experimental level with the following equipment:

- A Hall continuous current (CC) sensor connected to a Libelium Waspmote board with the XBee communications shield - this is the local radio acquisition and transmission unit. In our project there have been developed: the shield for data acquisition from the current sensor and the program (firmware) at the Waspmote system for reading data from the sensor and transmitting it to the concentrator. The system is powered by batteries;
- A system consisting of an embedded PC Raspberry PI to which it is connected to the XBee USB gateway and has a Waspmote (serial asynchronous) card connected to a GPS / GSM shield. Developed in our project: the Raspberry PI software for retrieving data from gateway and transmitting it to Waspmote board in NodeJs and the Waspmote firmware for retrieving data from Raspberry and transmitting them to GSM in HTTP Post packages. Here, more optimizations can be made to reduce the number of circuits. This system is the concentrator. The concentrator is powered by batteries that are charged from a photovoltaic panel (batteries 8000 mA / h, and the Waspmote photovoltaic kit – photovoltaic panel and controller).
- A server with the following features running on a VBox virtual machine: i7 2 core processor allocated, 16GB RAM, OS LUbuntu. The WSO2 DAS is installed and started on the server. Here we developed for our solution: implementing the full event stream using the WSO2 DAS interface, implementing the client application in the typescript / angular 2 using the Google Materials framework and the Google Maps API.

In the figures below are presented results displayed for several experiments.

IOP Conf. Series: Materials Science and Engineering **252** (2017) 012095 doi:10.1088/1757-899X/252/1/012095

12AB	19b3b4b	0	1c	1c	2017-5-22 12:37:17
12AB	19abda8	2	2	2	2017-5-22 12:34:6
12AB	19b3b4b	0	1c	1d	2017-5-22 12:34:5
12AB	19abda8	2	2	2	2017-5-22 12:34:36
12AB	19b3b4b	0	1c	1d	2017-5-22 12:34:37
12AB	19abda8	2	2	2	2017-5-22 12:33:34
12AB	19b3b4b	0	1c	1c	2017-5-22 12:33:33
12AB	19abda8	2	2	2	2017-5-22 12:33:2
12AB	19b3b4b	0	1d	1d	2017-5-22 12:33:1
12AB	19abda8	2	2	2	2017-5-22 12:31:27
12AB	19b3b4b	0	1c	1c	2017-5-22 12:31:26
12AB	19abda8	2	2	2	2017-5-22 12:30:56

← Older Newer →

Figure 6. Capture from application: list with current values and a map point with values displayed in real time.

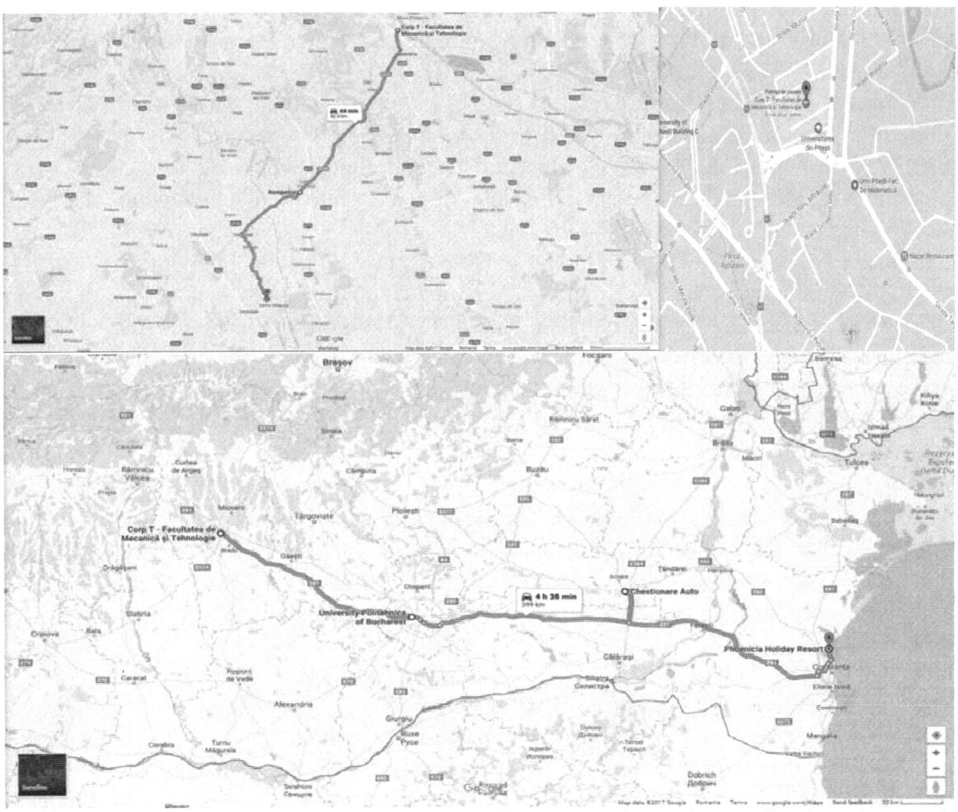

Figure 7. Capture from application: up – setting start point as is usually done in google maps (left), a route between "Corp T Facultatea de Mecanica" start point and "Sarbi Magura" destination point with intermediate point for recharge "Rompetrol" (right); down – another route between "Corp T" and Phoenicia Holiday Resort using 2 intermediate recharge stations: "University Politechnica" and "Chestionare Auto".

5. Conclusions

The system implements a dedicated navigator for electric car owners. Unlike normal navigators, here is the need to ensure the vehicle's power supply - given the lower degree of autonomy that electric vehicles do (yet) have. A system for monitoring the current consumption and transmitting it as an input for the navigation application has been designed for this purpose.
The driver will be provided complete navigation services, which also considers the degree of autonomy of the vehicle. The services are provided depending on the type of battery and its charging status - data that is first entered with the navigation data.
Two key components are identified in this project: the side of monitoring and transmission of instantaneous power consumption - it is a specialized hardware equipment developed here - and the part of the processing and fusion of current consumer related data with those related to navigation and of route planning. The system is currently functioning as an experimental model. As future development directions, we are considering creating a prototype in which to optimize the equipment on the vehicle to obtain a long-lasting energy autonomy.

6. References

[1] Q Guo, S Xin, H Sun, Z Li, B Zhang 2014, Rapid-Charging Navigation of Electric Vehicles Based on Real-Time Power Systems and Traffic Data, *IEEE Transactions on Smart Grid* **5**, 4, Pp 1969 – 1979
[2] LR Klose, W Wittliff 2014, Smart phone control and notification for an electric vehicle charging station, *US Patent* **8**,624,719, 2014 - *Google Patents*
[3] F Allegrini, RW Ebert, G Nicolaou, G Grubbs 2015, Semi-empirical relationships for the energy loss and straggling of 1–50 keV hydrogen ions passing through thin carbon foils, *Nuclear Instruments and Methods in Ph. Res.* pp115–119
[4] C Sun, Y Wen, P Li, W Ye, J Yang, J Qiu,,J Wen 2016, Self-Contained Wireless Hall Current Sensor Applied for Two-Wire Zip-Cords, *IEEE Transactions on Magnetics*, **52**, 7
[5] S Bera, S Misra, MS Obaidat 2014, Energy-Efficient Smart Metering for Green Smart Grid Communication , *Globecom 2014 - Symposium on Selected Areas in Communications: GC14 SAC Green Communication Systems and Networks*,
[6] F Ye, Y Qian, RQ Hu 2015, Energy Efficient Self-Sustaining Wireless Neighborhood Area Network Design for Smart Grid, *Ieee Transactions On Smart Grid*, **6**, 1
[7] K Yu, M Arifuzzaman, Z Wen, D Zhang, T Sato 2015, A Key Management Scheme for Secure Communications of Information Centric Advanced Metering Infrastructure in Smart Grid, *Ieee Transactions On Instrumentation And Measurement*, **64**,8
[8] S Yuan, Y Huang, J Zhou, Q Xu, C Song, P Thompson 2015, Magnetic Field Energy Harvesting Under Overhead Power Lines, *Ieee Transactions on Power Electronics*, **30**, 11
[9] FK Shaikh, S Zeadally 2016, Energy harvesting in wireless sensor networks: A comprehensive review, *Renewable and Sustainable Energy Reviews* **55**, pp. 1041–1054
[10] R Moghe, AR Iyer, FC Lambert, D Divan 2015, A Low-Cost Electric Field Energy Harvester for an MV/HV Asset-Monitoring Smart Sensor, *Ieee Transactions On Industry Applications*, **51**, 2

Acknowledgements

The research that led to the results shown here has received funding from the project “Cost-Efficient Data Collection for Smart Grid and Revenue Assurance (CERA-SG)”, ID: 77594, 2016-19, ERA-Net Smart Grids Plus.

IOP Conf. Series: Materials Science and Engineering **252** (2017) 012096 doi:10.1088/1757-899X/252/1/012096

Current challenges in autonomous driving

I Barabás1*, A Todoruţ1, N Cordoş1, A Molea1

1Technical University of Cluj-Napoca, Faculty of Mechanics, Department of Automotive Engineering and Transports, Cluj-Napoca, Romania
Istvan.barabas@auto.utcluj.ro

Abstract. Nowadays the automotive industry makes a quantum shift to a future, where the driver will have smaller and smaller role in driving his or her vehicle ending up being totally excluded. In this paper, we have investigated the different levels of driving automatization, the prospective effects of these new technologies on the environment and traffic safety, the importance of regulations and their current state, the moral aspects of introducing these technologies and the possible scenarios of deploying the autonomous vehicles. We have found that the self-driving technologies are facing many challenges: a) They must make decisions faster in very diverse conditions which can include many moral dilemmas as well; b) They have an important potential in reducing the environmental pollution by optimizing their routes, driving styles by communicating with other vehicles, infrastructures and their environment; c) There is a considerable gap between the self-drive technology level and the current regulations; fortunately, this gap shows a continuously decreasing trend; d) In case of many types of imminent accidents management there are many concerns about the ability of making the right decision. Considering that this field has an extraordinary speed of development, our study is up to date at the submission deadline. Self-driving technologies become increasingly sophisticated and technically accessible, and in some cases, they can be deployed for commercial vehicles as well. According to the current stage of research and development, it is still unclear how the self-driving technologies will be able to handle extreme and unexpected events including their moral aspects. Since most of the traffic accidents are caused by human error or omission, it is expected that the emergence of the autonomous technologies will reduce these accidents in their number and gravity, but the very few currently available test results have not been able to scientifically underpin this issue yet. The increasing trend in automation of vehicles will radically change the composition of car industry players, as mechatronics will not only be a complementary part of the automobile industry but an indispensable part of it. There is a reasonable expectation that automated cars will perform the same or better in all respects than their conventional counterparts. However, it seems that the current regulations do not keep up with the development of technology and sometimes hinder the development and testing of autonomous technologies.

1. Introduction The first section in your paper

Autonomous vehicles are "such vehicles that are able to perceive their environment and to move on without any intervention of a human driver" [3], [10]. These vehicles are also known as driverless, self-driving, unmanned or robotic vehicles [3], [9].

In 2016, the Society of Automotive Engineers (SAE) International revised its terminology for autonomous vehicles originally released in 2014 and defined their levels as a function of their automation grade (figure 1). As the SAE conceived "these levels are descriptive rather than normative and technical rather than legal". They do not denote any particular order of their market introduction.

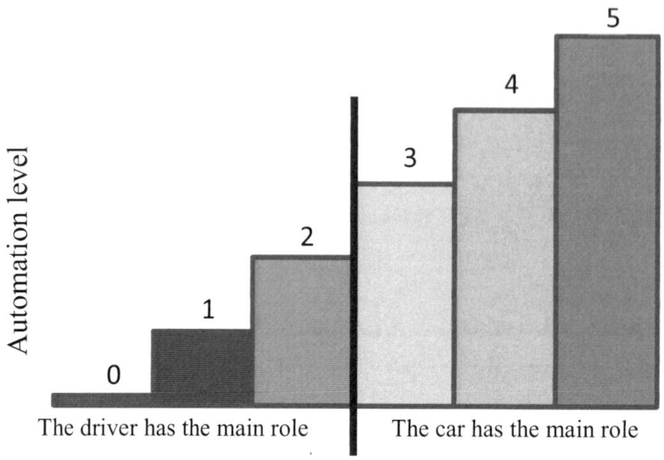

Figure 1. The six levels of automated cars

Each level is described by a set of minimal capabilities of the vehicle and a certain vehicle can operate at different levels, depending on their activated automation features (Table 1).
Autonomous vehicles have to be evaluated taking into account many aspects. Some of them studied in this article are presented in figure 2.

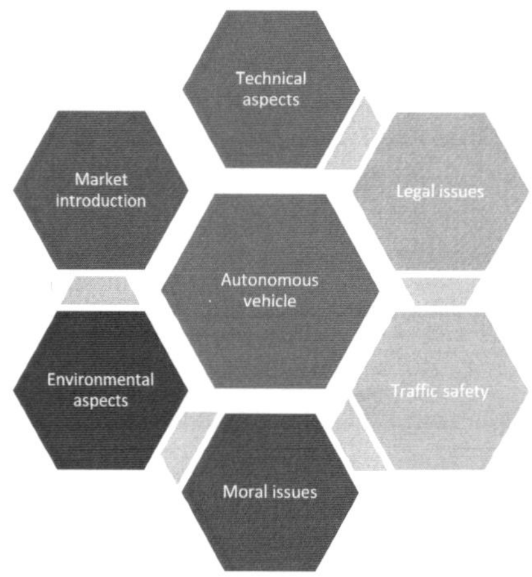

Figure 2. Main challenges of the automated vehicles

The main objectives of this work are all aimed at pointing out the current stage of autonomous vehicle's development and identifying the most important challenges in this field.

2. Environmental and traffic safety benefits

New technologies can significantly reduce vehicle emissions by reducing fuel consumption [2]:

- according to the opinion of several experts, by optimizing the acceleration and braking actions of the driverless vehicle the fuel consumption can be drastically reduced, by up to 60%.;
- as autonomous vehicles are able to communicate with each other and with their environment, they can be organized into platoons with controlled speed; this may result in a reduction of fuel consumption by 5-20%;

– because of the advanced navigation system which uses car to car communication, further reductions in fuel consumption can be obtained by avoiding congestion zones, particularly in crowded urban traffic.

Table 1. Levels of driving automation [1, 3, 4, 9]

SAE Level	Name	Description	Control (steering, acceleration, deceleration)	Monitoring of driving environment, evaluation, and decision	Fall-back performance of dynamic driving task	System capability to manage different driving modes
			The main system of dynamic driving task			
The human driver performs partly or entirely the dynamic driving tasks						
0	There is not any automation	The human driver performs all driving tasks.			–	–
1	Driver assistance	The driver assistance system executes steering OR acceleration and braking operations, by using information about the driving environment. The human driver performs all remaining aspects of the dynamic driving task. The vehicle is fully under the driver's control.				Limited driving modes
2	Partial automation	The driver assistance system undertakes both of steering and acceleration/braking operations, with the expectation that the human driver performs all remaining aspects of the dynamic driving task.	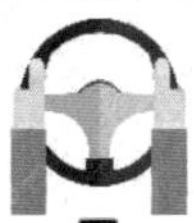 			Limited driving modes
The dynamic driving tasks are managed by the automated driving system						
3	Conditional automation	An automated driving system performs all aspects of the dynamic driving task in certain driving modes. The human driver will respond appropriately to a request to intervene.				Limited driving modes
4	High automation	An automated driving system performs all aspects of the dynamic driving task in certain driving modes even if the human driver will do not respond appropriately to a request to intervene.				Limited driving modes
5	Full automation	An automated driving system performs all aspects of the dynamic driving task in all				Unlimited driving modes

		driving modes even if the human driver will do not respond appropriately to a request to intervene.				

Human errors cause a significant part of traffic accidents. By excluding the driver as the weakest link, it is expected that the automated vehicles will travel more safely, reducing the number of accidents. However, it is also likely that new technologies will cause new types of accidents, especially in cases when the responsibility of driving is transferred from the human driver to the vehicle or vice versa. Further risks may appear because of the presence of conventional vehicles and driverless cars on the roads or because of the incompatibilities between the different autonomous technologies [8].

3. Legal Aspects of Autonomous Driving

In the pursuit to accomplish the fully automated driving, several technical issues have to be solved and the automotive industry seems to be up to this task [11].

As all revolutionary innovations, the technical developments in driverless vehicles are more advanced than the regulatory processes. Worldwide, the regulations regarding all aspects of the road traffic have as the main objective to provide the best road safety, so the autonomous cars must prove that they are safer or at least as safe as their predecessors. Consequently, the legal challenges are among the most critical issues, including the public policies, traffic code, technical standards, and tort law [11].

For example, steer-by-wire technologies have been available for a long time, but could not be integrated into the vehicles because the Convention on Road Traffic signed in Vienna in 1968 [7] stated that, in order to determine the path of the vehicle the steering system should contain a mechanical constraint between the steering wheel and the wheels.

This provision was only revised in 2005 when "Advancing technology, coupled with the wish to improve occupant safety by elimination of the mechanical steering column, and the production advantages associated with easier transfer of the steering control between left- and right-hand drive vehicles, has led to a review of the traditional approach and the Regulation is now amended to take account of the new technologies. Accordingly, it will now be possible to have steering systems in which there is not any positive mechanical connection between the steering control and the road wheels."[6]. A further step was made 2017, when the beforehand mentioned decree was further amended, introducing the concept of autonomous steering system and opening the way for autonomous vehicles.

Also, the initial form of the Convention on Road Traffic contained a requirement that made the autonomous vehicles illegal in the signatory countries of the Convention. In the Article 8 of the Convention states that: (1) “Every moving vehicle or combination of vehicles shall have a driver”, (5) “Every driver shall at all times be able to control his vehicle”. In March 2014, this article was supplemented by (5bis) “Vehicle systems which influence the way vehicles are driven and are not in conformity with the aforementioned conditions of construction, fitting and utilization, shall be deemed to be in conformity with paragraph 5 of this Article and with paragraph 1 of Article 13, when such systems can be overridden or switched off by the driver”. With this modification, the presence of the driver is still mandatory, but the steering of the car can be carried out by an automatic system if the driver can take over it at any time. The amendment thus provides a legal framework for the (semi) autonomous vehicles.

4. Moral and ethical aspects

The self-driven vehicles must make good decisions even in extreme emergency situations. Is this always possible?

Imagine that a vehicle is approaching a pedestrian crossing, the traffic light is green for the car, but suddenly a pedestrian passes to the road. Although it is the rule that the designated pedestrian crossing place should be approached with extreme caution and the speed of the vehicle must be sufficiently low to be able to stop before the pedestrian crossing, if necessary, but the pedestrians can only cross the road if they are convinced that there is no risk. What is the right decision in this case?

Does the car swerve into the traffic from the opposite lane or in the roadside electric pylon, exposing the car owner to unforeseeable consequences? Is it a duty of the car to protect its owner at any cost? Would the dilemma change if not one but two people stepped on the road? Who is responsible for the consequences: the owner, the user or the computer programmer, who stays in his ergonomic chair some thousands of miles away and has no idea what happened? There is, however, a strong need to develop moral algorithms that can solve such situations according to acceptable moral norms.

5. Market

There is no doubt that the interest for driverless cars is increasing. According to a survey conducted by the Boston Consulting Group in 2015, 55% of potential car buyers said they would consider buying a semi-autonomous vehicle, while 44% said that they would consider buying a fully autonomous car [8]. Almost all the major automotive manufacturers are working on meeting this demand. According to the HIS, an international market research company, in 2020 the market share of the 4th and 5th level autonomous vehicles will be 0.004% (4,200 cars), which will increase in 2025 up to 0.5% (578,000 cars) and in 2030 it will reach 3.8% (4,503,000 cars) [5].

The trend of the market penetration of the semi- and fully automated passenger cars, estimated by IHS is shown in figure 3.

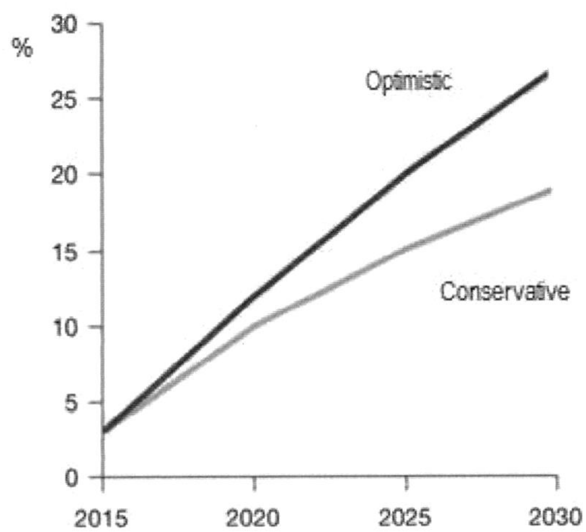

Figure 3. Market penetration of automated vehicles [3], [5]

Autonomous vehicles are marketed following three scenarios: 1) traditional carmakers are integrating more and more automatic components into their products until the vehicle becomes fully autonomous, 2) new market players brake into the market with new concepts, 3) co-operation of traditional car manufacturers with new market players delivering a technology that allows the production of 3rd or 4th cars [3].

The optimistic scenario takes account of less restrictive safety regulation, more pork-barrels, appearance of new market players, the conservative one considers that the present condition will not change.

6. Conclusions

Autonomous technologies are becoming increasingly sophisticated and technically accessible, and in some cases, these can already be installed in commercial vehicles.

As several carmakers have announced that they will start the production of highly automated cars in 2017, it seems realistic that autonomous cars will make their appearance in developed countries in the near future. We can estimate that in 2030 a significant number of driverless vehicles will travel on the roads.

According to today's research and development, it is unclear how these technologies will be able to handle extreme and unexpected events.

There is a concern over the development of several different technologies that will result very different products, which in some situations will not be able to work together and communicate with each other. There are trends that rely solely on the signs of sensors in the car, others need infrastructure improvements that help decision making.

As the cause of most traffic accidents is human error or omission, it is anticipated that the emergence of autonomous technologies will reduce the number of car accidents. There is not enough statistical data to sustain this statement yet.

The increasing trend of automatization level of cars will radically change the composition of car industry players, as with the rise in automation levels, mechatronics will be not only an additional part of the automobile industry but also an indispensable and integral part of it.

There is a reasonable expectation that automated cars will perform the same or better in all aspects than their conventional counterparts. However, it seems that the current regulations do not keep up with the development of technologies and sometimes hinder the development and testing of autonomous technologies.

References

[1] Bartels A, Eberle U and Knapp A 2015 Automated Driving Applications and Technologies for Intelligent vehicles - System Classification and Glossary, Deliverable D2.1. https://www.adaptive-ip.eu/

[2] Learner H 2016 Autonomous Vehicles: Ways to Improve Safety and Accelerate Environmental Progress Together. Entelligent http://elpc.org/tag/autonomous-vehicles/

[3] Varga I and Tettamanti T 2015 A jövô intelligens jármûvei és az infokommunikáció hatása. Magyar-Jövô Internet Konferencia, *Híradástechnika*, **LXXI** 59 63

[4] *** 2015 Automated and Autonomous Driving, Regulation under Uncertainty, Corporate Partnership Board Report, OECD, International Transport Forum, www.internationaltransportforum.org

[5] *** Autonomous vehicle sales forecast to reach 21 mil. globally in 2035, according to IHS Automotive, IHS, 2016.

[6] *** Regulation No 79 of the Economic Commission for Europe of the United Nations (UN/ECE) — Uniform provisions concerning the approval of vehicles with regard to steering equipment. Addendum 78: Regulation No 79, Revision 2.

[7] *** Economic Commission for Europe, Inland Transport Committee, Convention on Road Traffic, Vienna, 1968.

[8] *** Revolution in the Driver's Seat: The Road to Autonomous Vehicles, Boston Consulting Group (BCG), 2015.

[9] *** Taxonomy and Definitions for Terms Related to Driving Automation Systems for On-Road Motor Vehicles. SAE International, Standard J3016_201609, 2016-09-30.

[10] Gehrig S K and Stein F J 1999 *Dead reckoning and cartography using stereo vision for an autonomous car*. IEEE/RSJ International Conference on Intelligent Robots and Systems. **3**. Kyongju. pp. 1507–1512. ISBN 0-7803-5184-3. doi:10.1109/IROS.1999.811692

[11] Bertoncello M and Wee D 2015 Ten ways autonomous driving could redefine the automotive world. McKinsey&Company http://www.mckinsey.com/industries/automotive-and-assembly/our-insights/ten-ways-autonomous-driving-could-redefine-the-automotive-world

Driver drowsiness detection using ANN image processing

T. Vesselenyi[1], S. Moca[1], A. Rus[1], T. Mitran[1], B. Tătaru[1]

[1]University of Oradea

tudor_mitran@yahoo.com

Abstract. The paper presents a study regarding the possibility to develop a drowsiness detection system for car drivers based on three types of methods: EEG and EOG signal processing and driver image analysis. In previous works the authors have described the researches on the first two methods. In this paper the authors have studied the possibility to detect the drowsy or alert state of the driver based on the images taken during driving and by analyzing the state of the driver's eyes: opened, half-opened and closed. For this purpose two kinds of artificial neural networks were employed: a 1 hidden layer network and an autoencoder network.

1. Introduction

Car accidents are one of the major causes for injury or death. Statistics show that car accidents are globally the 9th cause of death: 1.3 million people die in car accidents annually, or 3287 per day. Fatigue at the steering wheel has the following symptoms: frequent yawning, hard to keep the eyes opened and to focus on the road, not remembering what happened in the last few minutes of driving, not keeping the correct distance from the car in front, missing traffic signs and getting too close to the side or center of the road.

Statistics have shown that over 10% of accidents are due to fatigue, most of which occur on highways or after driving on a large number of kilometres. The influence of fatigue on accidents has been proven throughout several studies. According to National Highway Traffic Safety Administration (NHTSA), an annually average from 2009 to 2013, there were over 72,000 police-reported crashes involving drowsy drivers, injuring more than 41,000 people, and killing more than 800.

In recent years, car manufactures have developed systems which aim to reduce all the factors that may lead to accidents. This is how sensors, over-ride warning sensors, the adaptive auto-pilot which keeps a constant distance from the front vehicle and fatigue detection sensors were born. Car manufacturers have studied the possibilities of detecting driver's fatigue for a long time and, of course, the best solutions for the warning on time. Assistance systems now include different types of equipments that can prevent accidents caused by fatigue:

1. In order to prevent serious accidents due to driving fatigue, the Bosch Group provides drivers with a sleepiness detection system. This system identifies the signs of fatigue, through monitoring the movements of the steering wheel, suggesting drivers to take a break for a time. The necessary information is provided either by the vehicle's steering system or by the steering wheel angle sensor, which is part of the ESP system. This way, the sleepiness detection system helps to increase safety while driving.
2. Volvo currently equips some of it's models with a system called Driver Alerts, which is based on recognizing a tired driver driving style, as opposed to that of a fully awake

driver. In order to detect this slightly winding driving pattern, the Swedish system uses a video camera mounted in front of the interior rearview mirror, which constantly transmits the distance between the vehicle and the roadside markings to a computer. When the predetermined limits of the measured oscillations, the driver is warned to take a break, through audio signals, and a coffee break symbol appears on the central display, in order to emphasize the message.

3. "Fatigue assistance" for Mercedes models has the capacity to store data acquired by sensors about the normal reactions of the driver, which can be later compared to eventual unsafe steering actions of the same driver in a state of increased fatigue. The audio and visual warning of the driver can prevent accidents caused by "one second sleep" due to increased fatigue.

2. A concept for a drowsiness detection system

Alternative methods, which have not been tried out on larger scale by the leading automotive companies, that can detect the state of awareness of the driver can be taken into account.

Three of these methods are based on EEG (Electroencephalography) and EOG (Electrooculography) signals measurement and on the eye state (closed or opened) image classification. The EEG method monitors the brain activity through a sensor placed on a specific part of the scalp, the EOG method tracks the eye movements by measuring the signals from the muscles which are acting on the eye and the eye image analysis can monitor the opened or the closed state of the eye. The authors discussed about the development of such a system in [1]. A scheme of the system presented in [1] is shown in figure 1. Each of the methods used in the system has its advantages and disadvantages. For example, the EEG and EOG sensors, electrodes which have to be fixed with a conductive gel and in most devices must transmit the signal by wire, present a major discomfort. Important research in the field of advanced materials and MEMS technology may solve these problems, as for example the use of dry electrodes for EEG, presented in [2].

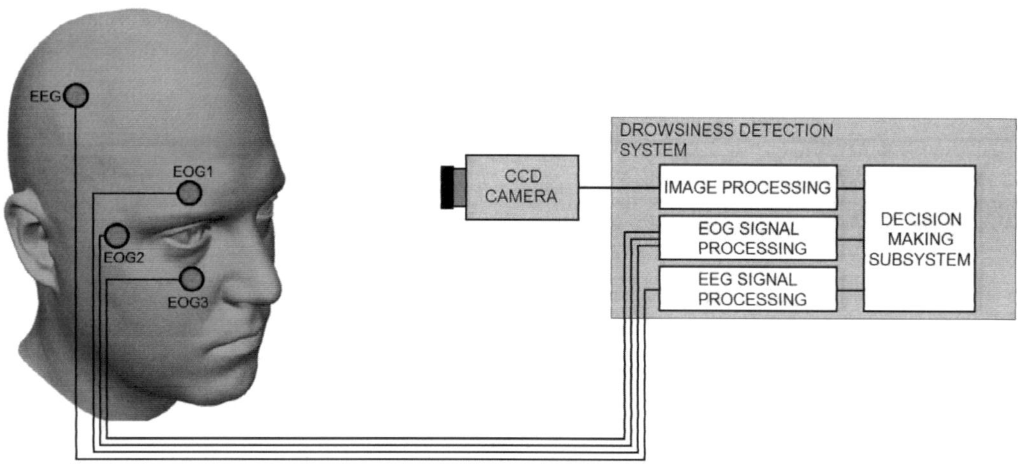

Fig. 1. Drowsiness detection system overview.

Developments in this field are supported by efforts to create brain – computer interfaces [3, 4, 5 and 6] for different applications, including devices that help disabled people.

An application of EEG signal acquisition and processing has been presented in [7, 8 and 9]. In this research the central point is to distinguish between low and high alpha rhythm peaks (as shown in figure 2), which can make the difference between alert and drowsy states.

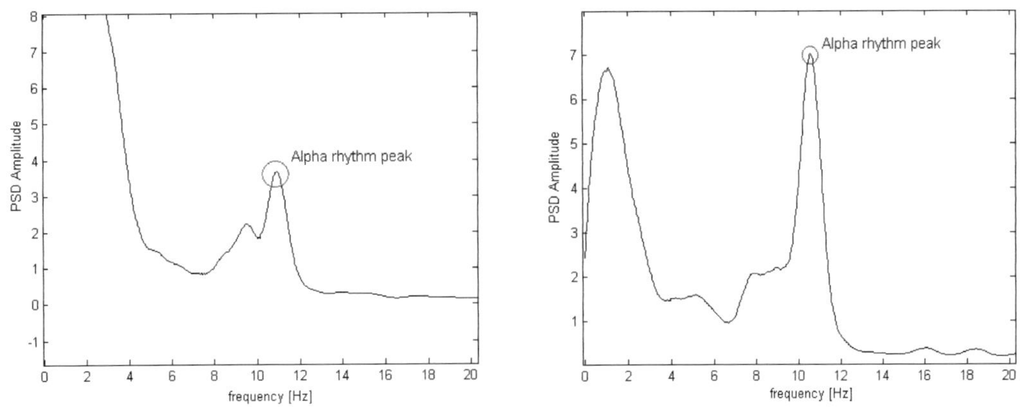

a. **b.**

Fig. 2. Power Spectral Density amplitude versus frequency of an EEG signal, with the alpha rhythm peak in the 10-12 Hz frequency domain: a. alert state; b. drowsy state [7].

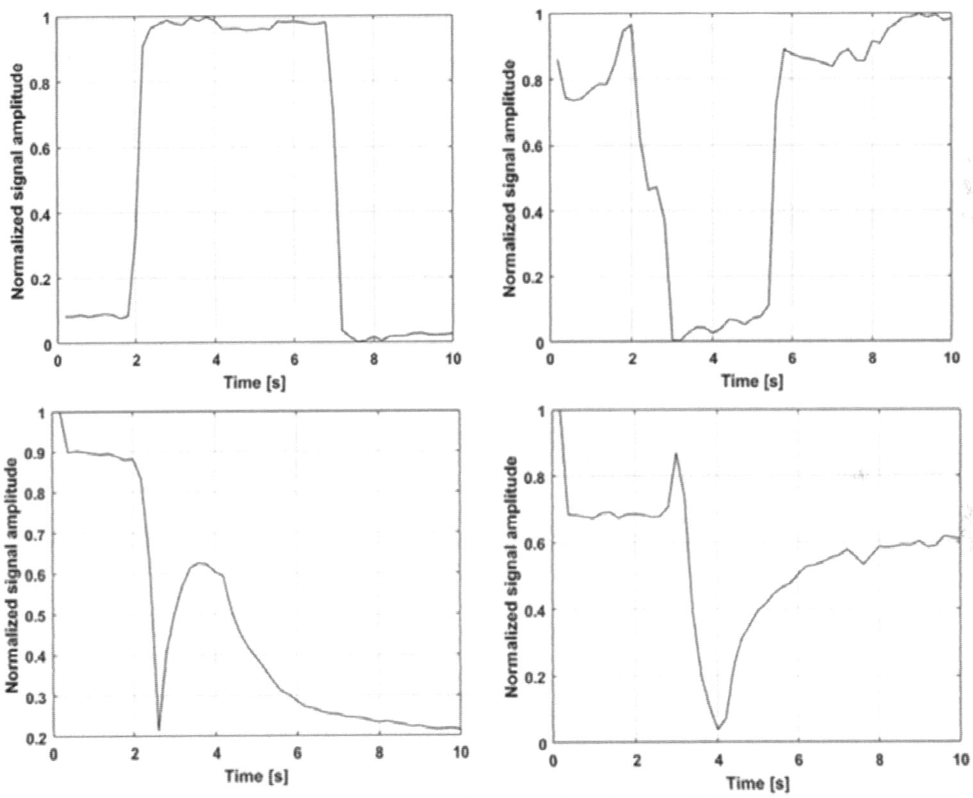

Fig. 3. Four types of signals recorded from EOG sensors (EOG1, EOG2 and EOG3 in figure 1) for different eye movements (up, down, left and right).

In other researches [10, 11] the authors have analyzed the possibility to use EOG signals acquisitioned from 3 sensors (EOG1, EOG2, EOG3 in figure 1). After pre-processing, four types of different signals were identified, which are shown in figure 3. The combination of these four types of signals gives the information to distinguish between left, right, up and down movements of the eye.

In [10] and [11] the possibilities to apply EEG and EOG signals were analysed to detect driver drowsiness. In the present paper the study was extended to analyze driver drowsiness by image processing.

3. Drowsiness detection using the processing of the driver's eye images.

For the classification of the driver's drowsy or alert state, artificial neural networks were used. Artificial neural networks are extensively used for the image classification in the last decades. A new paradigm named Deep Learning has been developed in [12] and [13] and in many other researches. Deep Belief Networks, Restricted Boltzmann Machines and Deep Autoencoders are all methods belonging to the Deep Learning paradigm. These methods are used in a wide range of applications, the image classification being one of the fields in which these are employed with success. For the study presented in this paper, Matlab Neural Network Toolbox with 1 layer ANN and the autoencoder module of the Deep Learning Toolbox were used in order to study if these methods can be applied for image classification of the driver's drowsiness. As a premise, it was assumed that the drowsiness is linked to the images in which the driver has closed eyes and the alert state is linked to the images in which the driver has opened eyes. This module had been used for the analysis [14].

In order to analyze the drowsiness state of the driver 200 images of a driver during a regular driving process were acquisitioned. One hundred of these images contain opened eyes or half opened eyes images and another hundred of the images contain closed eyes images. Examples of these images are presented in figure 4.

In order to avoid memory overload and large processing times, the images had been cropped and down-sampled in order to have a smaller amount of input data for the neural networks (figure 5).

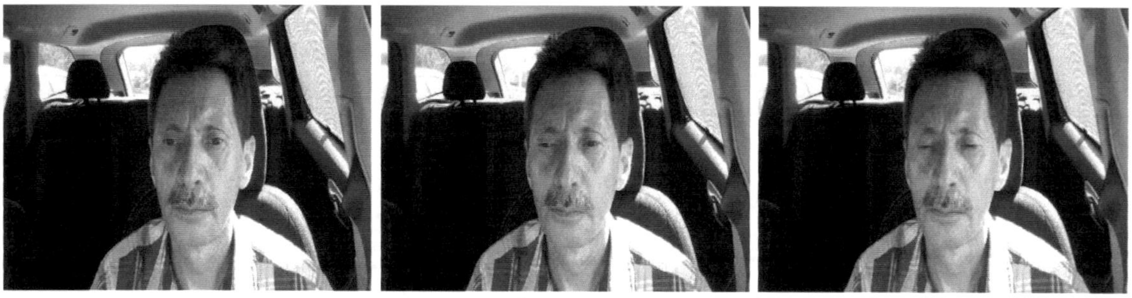

a. **b.** **c.**

Fig. 4. Images of a driver with: a. opened eyes; b. half opened eyes; c. closed eyes.

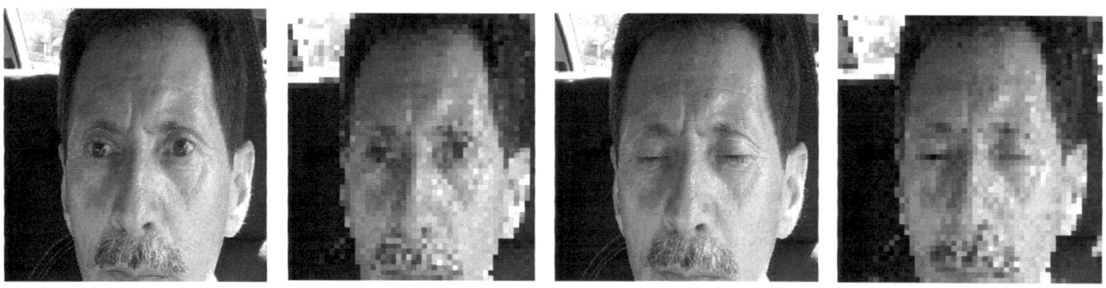

a. **b.** **c.** **d.**

Fig. 5. Cropped images of the driver with opened (a) and closed (c) eyes. Cropped and down-sampled images of the driver with opened (b) and closed (d) eyes.

The down-sampling of the images has produced more unclear images (figure 5, b and d), which ensures a more robust classification. If the network works properly for the down-sampled images, it will surely work well for the initial images.

3.1. One Hidden Layer Artificial Neural Network

Following the acquisition of 200 images, 140 of these had been used to train, validate and test the neural network: 70 with opened eyes or half-opened eyes and 70 with closed eyes. The rest of the images (30 for opened eyes or half-opened eyes and 30 for closed eyes) were kept for testing the network after the completion of the training process. The network was trained with the structure presented in figure 6 (2601 neurons in the input layer, 10 neurons in the hidden layer and 2 neurons in the output layer). The number of neurons in the input layer corresponds to the number of elements of the input vector which is the one column reshaped version of the down-sampled image of the driver (represented by a 51x51 elements gray level matrix). The number of neurons in the output layer corresponds to the number of possible categories in which an image can be classified (2 categories: drowsy or alert).

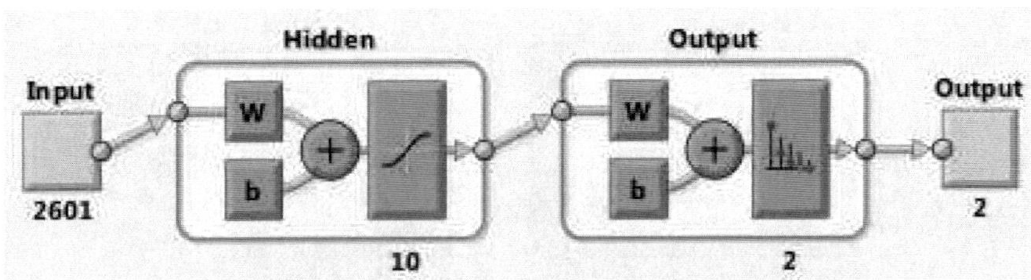

Fig. 6. Network structure for the 1 hidden layer network.

During the training process the training performance diagram (figure 7.a.) and the error histogram (figure 7.b.) of the trained network were obtained. It can be observe that the training performance reaches values of under 10^{-6} at 25 epochs and the histogram boundaries are in a range of $-1.5x10^{-6}$ and $1.46x10^{-6}$, which are very good results.

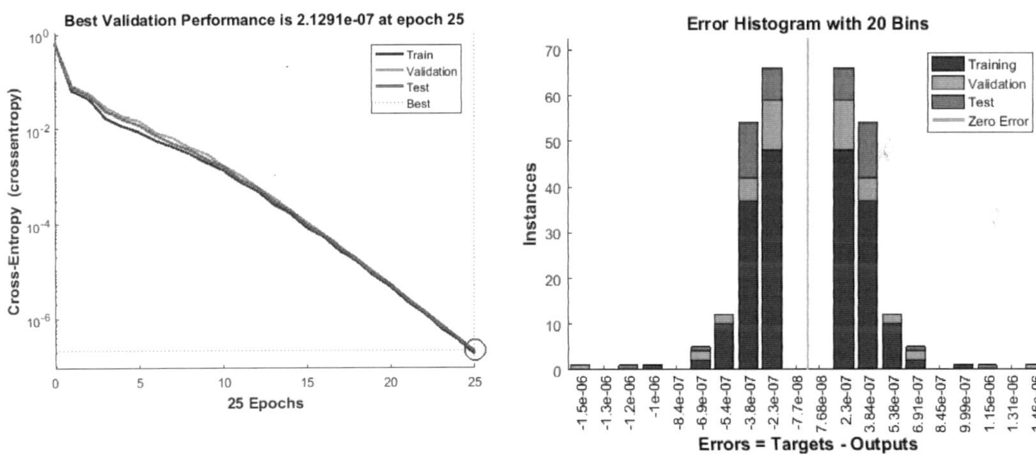

a. b.

Fig. 7. Performance diagram (a) and error histogram (b) of the trained network.

More significant, one can observe the results of training, validation and testing presented in figure 8 in the form of a confusion matrix. In a confusion matrix, each column represents a predicted class while each row represents an actual class. The green squares represent the correctly classified and the red squares represent the incorrectly classified samples. There can be seen that each sample has been correctly classified.

IOP Conf. Series: Materials Science and Engineering **252** (2017) 012097 doi:10.1088/1757-899X/252/1/012097

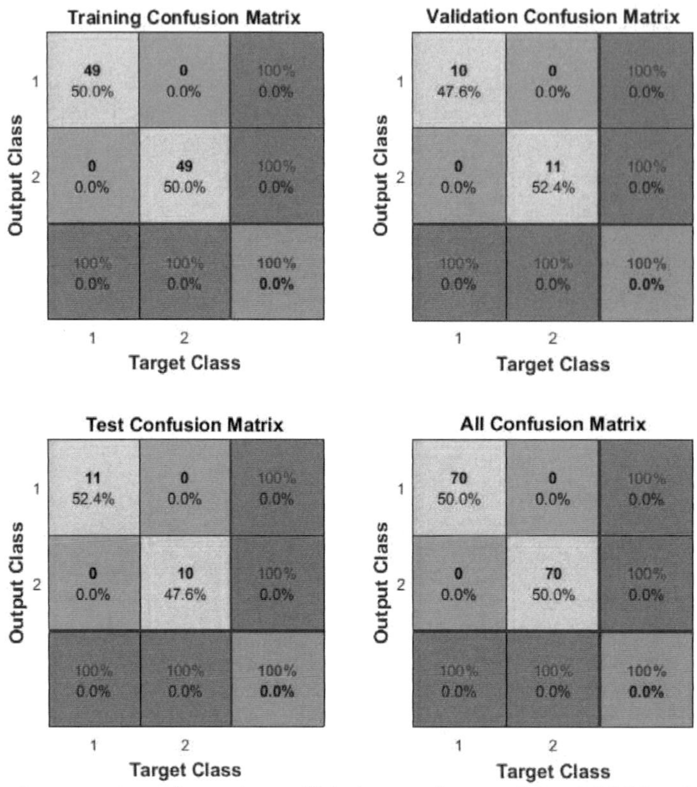

Fig. 8. Confusion matrices for train, validation and test of the 1 hidden layer network.

3.2. Deep learning Autoencoder neural networks

For the autoencoder network the same input and target vectors as in the case of 1 hidden layer network were used. Autoencoders use methods to separately train each layer, then stack them together in a single network with multiple layers and train the final network as a whole. The network structure is presented in figure 9.

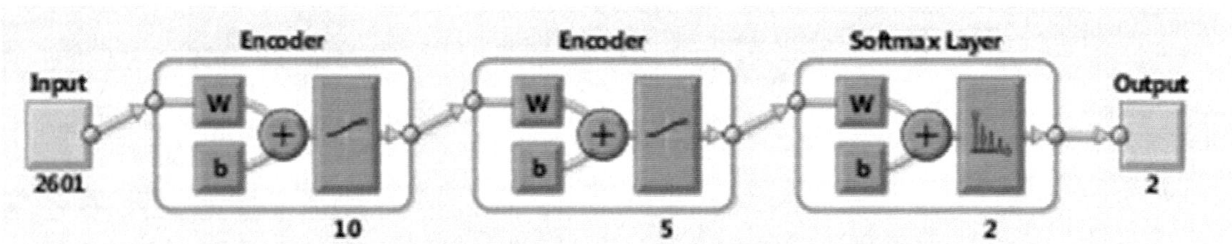

Fig. 9. Network structure for the two hidden layer autoencoder.

The results of the autoencoder's training are shown in figure 10. In figure 10.a. the training performance is presented when it reached the value of less than 0.034 (the smaller the better) after 438 epochs. In figure 10.b. the test results show that no false positives or false negatives had been obtained. This means that every test image (60 images – 30 for each class) has been correctly classified.

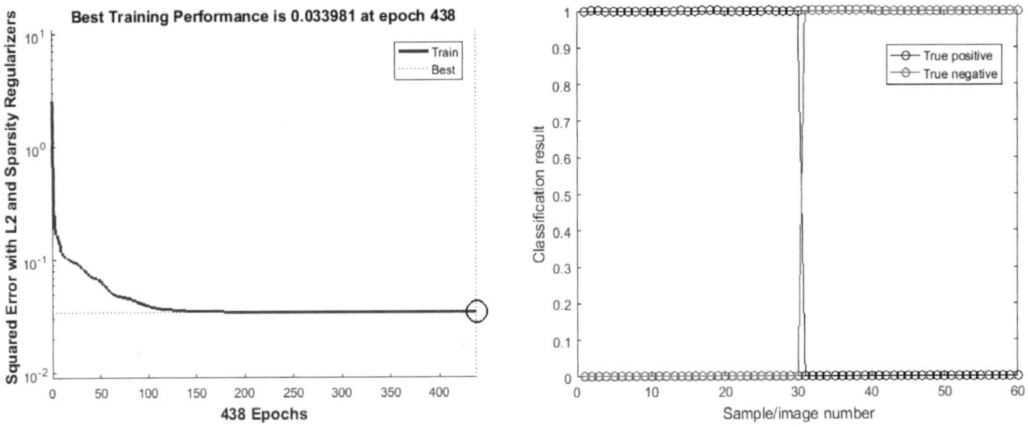

Fig. 10. Training performance and testing results for the autoencoder network.

4. Conclusions

Analyzing the results of applying these networks on the acquisitioned images it can be concluded that both networks had very good results, with 100% positive classification results, which were presented in chapter 3. The small number of neurons used in the hidden layers to successfully classify the images (10 for the 1 hidden layer network and 15 for the autoencoder network) allows the implementation of these networks on compact computing devices, using a very small portion of their memory. Also the processing time is in the order of milliseconds on a Windows based computer, which on a compact device can be more reduced. The training of the network can be done specifically for each driver, thus enhancing the classification success rate.

In future works the authors intend to study the possibility to apply neural network classifiers on more different images, with different drivers, in different positions and lighting conditions. Also the elimination of the cropping and down sampling stages of the image processing will be a goal for future research.

References

[1] Tiberiu Vesselenyi, Alexandru Rus, Tudor Mitran, Bogdan Tataru, Ovidiu Moldovan, VEHICLE DRIVER DROWSINESS MONITORING AND WARNING SYSTEM; 12th International Congress of Automotive and Transport Engineering (CONAT), Brasov, 2016; pag. 873-880.

[2] Sullivan, T.J., Deiss, S.R., Jung, T.P., Cauwenberghs, B., A Brain-Machine Interface using Dry-Contact, Low-Noise EEG Sensors, 2008 IEEE International Symposium on Circuits and Systems Washington, US, 2008

[3] Janis J Daly, Jonathan R., Wolpaw, Brain–Computer Interfaces in Neurological Rehabilitation, Vol. 7, Issue 11, pp. 1032-1043, 2008.

[4] Ting, Jo-Anne, D'Souza, A., Yamamoto, K., Yoshioka, T., Hoffman, Donna, Kakeif, S., Sergio, L., Kalaska, J., Kawato, M., Strick, P., Schaal, S., Variational Bayesian least squares: An application to brain –machine interface data, Neural Networks, Vol. 21, Issue 8, pp. 1112-1131, 2008.

[5] Ting, Jo-Anne, D'Souza, A., Yamamoto, K., Yoshioka, T., Hoffman, Donna, Kakeif, S., Sergio, L., Kalaska, J., Kawato, M., Strick, P., Schaal, S., Variational Bayesian least squares: An application to brain –machine interface data, Neural Networks, Vol. 21, Issue 8, pp. 1112-1131, 2008

[6] Cvetkovic, D., Übeyli, E.D., Cosic, I., Wavelet transform feature extraction from human PPG, ECG, and EEG signal responses to ELF PEMF exposures: A pilot study. Digital Signal

Processing, Vol. 18, Issue 5, pp. 861-874, 2008.
[7] Ting, W., Guo-zheng, Y., Bang-hua, Y., Hong, S., EEG feature extraction based on wavelet packet decomposition for brain computer interface, Measurement, Vol. 41, Issue 6, pp. 618-622, 2008.
[8] Dzitac, I. Vesselenyi, T., Tarca, R. C., Identification of ERD using Fuzzy Inference Systems for Brain Computer Interface, International Journal of Computers Communications & Control, Vol. 6 Issue 3, ISSN 1841-9836, pp. 403-417, 2011.
[9] Dzitac, S., Vesselenyi, T., Popper, L., Moga, I., Secui, C., D. Fuzzy Algorithm for Human Drowsiness Detection Devices, Studies in Informatics and Control, Vol. 19, Issue 4, ISSN 1220-1766, pp. 419-426, 2010.
[10] Vesselenyi, T., Dzitac, I., Dzitac, S., Hora, C., Porumb, C. Preliminary Issues on Brain-Machine Contextual Communication Structure Development, IEEE Conference, 3rd International Workshop on Soft Computing Applications, Szeged, Hungary, pp. 35-40, 2009.
[11] R. B. Nagy, T. Vesselenyi, F. Popentiu-Vladicescu, “Research on recording and filtering electromyogram (EMG) signals,” in Nonconventional Technologies Review, 2015, pp. 21-25.
[12] Robert-Bela Nagy, Popentiu -Vladicescu Florin, Vesselenyi Tiberiu, AN ANALYSIS OF ELECTRO-OCULOGRAM SIGNALS PROCESSING USING AN ARTIFICIAL NEURAL NETWORK, International Scientific Conference - eLearning and Software for Education, 2017, Volume 3, DOI: 10.12753/2066-026X-17-257, Pages: 560-567, 2017.
[13] G. E. Hinton, S. Osindero, and Y.-W. Teh. “A fast learning algorithm for deep belief nets,” in Neural computation 18, 2006, pp. 1527-1554.
[14] Y. Bengio, P. Lamblin, D. Popovici, and H. Larochelle. “Greedy layer-wise training of deep networks,” in Advances in Neural Information Processing Systems, 2007.
[15] ***. MATLAB. – Mathworks, Users Manual, Neural Network Toolbox, Deep Learning, Autoencoders, 2016.
[16] ***. National Highway Traffic Safety Administration - Drowsy Driving Research and Program Plan, March 2016

IOP Conf. Series: Materials Science and Engineering **252** (2017) 012098 doi:10.1088/1757-899X/252/1/012098

Autonomous vehicles: from paradigms to technology

Silviu Ionita
University of Pitesti, Romania
Department of Electronics, Computers and Electrical Engineering
e-mail: silviu.ionita@upit.ro

Abstract. Mobility is a basic necessity of contemporary society and it is a key factor in global economic development. The basic requirements for the transport of people and goods are: safety and duration of travel, but also a number of additional criteria are very important: energy saving, pollution, passenger comfort.

Due to advances in hardware and software, automation has penetrated massively in transport systems both on infrastructure and on vehicles, but man is still the key element in vehicle driving. However, the classic concept of 'human-in-the-loop' in terms of 'hands on' in driving the cars is competing aside from the self-driving startups working towards so-called 'Level 4 autonomy', which is defined as "a self-driving system that does not requires human intervention in most scenarios".

In this paper, a conceptual synthesis of the autonomous vehicle issue is made in connection with the artificial intelligence paradigm. It presents a classification of the tasks that take place during the driving of the vehicle and its modeling from the perspective of traditional control engineering and artificial intelligence. The issue of autonomous vehicle management is addressed on three levels: navigation, movement in traffic, respectively effective maneuver and vehicle dynamics control. Each level is then described in terms of specific tasks, such as: route selection, planning and reconfiguration, recognition of traffic signs and reaction to signaling and traffic events, as well as control of effective speed, distance and direction. The approach will lead to a better understanding of the way technology is moving when talking about autonomous cars, smart/intelligent cars or intelligent transport systems. Keywords: self-driving vehicle, artificial intelligence, deep learning, intelligent transport systems.

1. Introduction

Fundamentally, mobility is ensured by a three-component system: vehicle, driver and transport infrastructure (named in a broader sense environment). Over time, vehicles and infrastructure have evolved spectacularly through technological upgrades and improvements, but the only constant of the system has still remained the human driver. The term of *autonomous* car is equivalent with driverless auto, self-driving car or robotic car. So the autonomy of a vehicle is seen as its capability to run in traffic without a driver and generally without any human inputs. In technical terms, that means replacing the human driver with artificial subsystems which should be able to perform the specific tasks in a similar manner. Under this circumstance, the artificial system should have the right knowledge, have the ability to reason correctly and behave in line with the previous ones, and that means more than traditional automation and adaptive control. However, the recent literature keeps the generic term *automatic* as a basis of the concept in self-driving car.

This paper has three objectives: (i) to present the Advanced Driver-Assistance Systems concept in accordance with the standardization introduced by Society of Automotive Engineers for vehicle automation levels; (ii) to explain the intelligent system concept and the directions for implementing artificial intelligence in autonomous vehicles; (iii) to highlight some recent achievements in the field of driverless vehicles.

The SAE International's new standard J3016: 'Taxonomy and Definitions for Terms Related to On-Road Motor Vehicle Automated Driving Systems' [http://standards.sae.org/j3016_201609/] uses the 'automation' as the key word to classify the levels of driving automation. There are identified six levels of driving automation from 'no automation' to 'full automation' that are defined based on functional aspects of technology. The first three levels, conventionally noted 0, 1 and 2 fall in to the category called 'human driver monitors the environment' and are defined as follow:

- Level 0 named 'no automation' provides only automated on board warnings but has no vehicle control.
- Level 1 named 'driver assistance' is a 'hands on' stage, where the control of some specific tasks can be shared between the driver and automated system. Typical examples for this level are: The Adaptive Cruise Control (ACC) where the driver controls steering and the automated system controls speed of the car, and Parking Assistance, where steering is automated while speed is controlled by the driver.
- Level 2 is named 'partial automation' and provides the so called 'hands off' capabilities when the automated system takes full control of the vehicle dynamics in terms of accelerating, braking, and steering. The driver must monitor the driving and be ready to promptly intervene in due time in case of automated system fails to respond properly.

The second category includes the next three levels representing systems able to monitor the driving environments, from Level 3 to Level 5 as follow:

- Level 3 named 'conditional automation' or literally named 'eyes off' provides functions that make the vehicle able to handle situations when the driver in generally turn their attention away from the current driving tasks. These functions are useful for critical situations that require an immediate response, like emergency braking.
- Level 4 is classified as 'high automation' and literally equated with 'minds off'. This includes the extensions and improvements of functions from Lever 3 providing self-driving capabilities of the car in case of no driver attention exist, for instance the driver may safely go to sleep or leave the driver's seat.
- Level 5 named 'full automation' and so called 'steering-wheel optional'. This level is reached when the system has the total control over the vehicle and no human intervention is required. A generic example might be the robotic taxi.

The above mentioned SAE's document defines also some key terms that make up a matrix of compliance for the proposed automation levels. We reproduce this matrix in Table 1. The terms considered criteria of performance for driving automation systems are presented as they are mentioned by SAE:

- 'Dynamic driving task' includes the operational (steering, braking, accelerating, monitoring the vehicle and roadway) and tactical (responding to events, determining when to change lanes, turn, use signals, etc.) aspects of the driving task, but not the strategic (determining destinations and waypoints) aspect of the driving task.
- 'Request to intervene' is notification by the automated driving system to a human driver that s/he should promptly begin or resume performance of the dynamic driving task.
- 'Driving mode' is a type of driving scenario with characteristic dynamic driving task requirements (e.g., expressway merging, high speed cruising, low speed traffic jam, closed-campus operations, etc.).

Besides standard classification the paradigm of autonomous vehicles is currently generating perspectives and new concepts. For example, at the recent 2017 IEEE Intelligent Vehicles Symposium (www.iv2017.org) the topics explore challenging fields of research like machine vision and interfaces in data fusion platforms for automated driving, multi-sensor fusion and extended object tracking, deep learning for vehicle perception, human factors in intelligent vehicles and cognitively inspired intelligent vehicles. The special topics approached is the hot issue of cyber-security and generally the models of short range vehicular communications and data exchanging between vehicles and the environment.

Table 1. The automation levels: Summary of compliance

Level (SAE)	Type	Dynamic driving task	Monitoring of Driving Environment	Request to intervene (When fallback performance of Dynamic driving task)	Driving mode
0	No automation	*Human driver*	*Human driver*	*Human driver*	*n/a*
1	Driver assistance	*Human driver and system*	*Human driver*	*Human driver*	*Some driving modes*
2	Partial automation	*System*	*Human driver*	*Human driver*	*Some driving modes*
3	Conditional automation	*System*	*System*	*Human driver*	*Some driving modes*
4	High automation	*System*	*System*	*System*	*Some driving modes*
5	Full automation	*System*	*System*	*System*	*All driving modes*

2. The paradigms of intelligent vehicle

The popular approach of the concept 'intelligent system' consists in reporting to the human intellect, so the artificial intelligence (AI) is the key field for modeling and developing intelligent vehicle in general. A system that receive the attribute 'intelligent' should meet simultaneously the following requirements:

i. To learn in 'teaching mode' but also from own experience. This is a necessary condition but not sufficient.
ii. To perform approximate reasoning. This suppose more than true/false logic in formal reasoning, it requires the use of multivalent logic that deals better with the uncertainty.
iii. To behave autonomously. This is the aggregate performance that includes many operational functions based on the first two conditions in order to put in practice the intelligence, which is equivalent with the intelligent behavior.

Now let's nuanced the entire picture of the tasks that are currently performed for driving of the on-road vehicles. In Figure 1 are synthetically presented the main tasks and subordinated functions which are subject of automation and implementation in to the appropriate manner towards smart/intelligent car achieving.

Advanced Driver-Assistance Systems – ADAS briefly, defines the field of technology and concepts that currently guides the development of autonomous vehicle. ADAS functions added to the vehicle are thought to take on more and more tasks of the driver – his intervention became optional or exceptional. The level of functions implemented on vehicles has an impact on the six levels of the autonomous vehicle defined by the SAE. As we observe the Table 1 the latest two levels of automation require really advanced functions that make possible vehicle self-driving without human intervention in most scenario (Level 4) or even in all scenario (Level 5). Definitely, these levels require certain AI models to process the information and to have the human-like behavior.

Next, we review the functions and tasks in Figure 1 to highlight the level of intelligence needed to implement them in autonomous vehicles. At a glance it can be recognized that certain functions can be implemented with typical AI models like machine learning and expert systems, while others are subject to classical automation like error-based control or state-control. We discuss the issue of vehicle driving in terms of two information processing stages: *decision* and *action* (i.e. execution), and three hierarchic levels for problem solving: *strategic*, *tactical* and *direct*. In Figure 2 is depicted a classification of ADAS functions in connection with the levels of decision and execution. The candidate models and time requirement for execution are also specified.

IOP Conf. Series: Materials Science and Engineering **252** (2017) 012098 doi:10.1088/1757-899X/252/1/012098

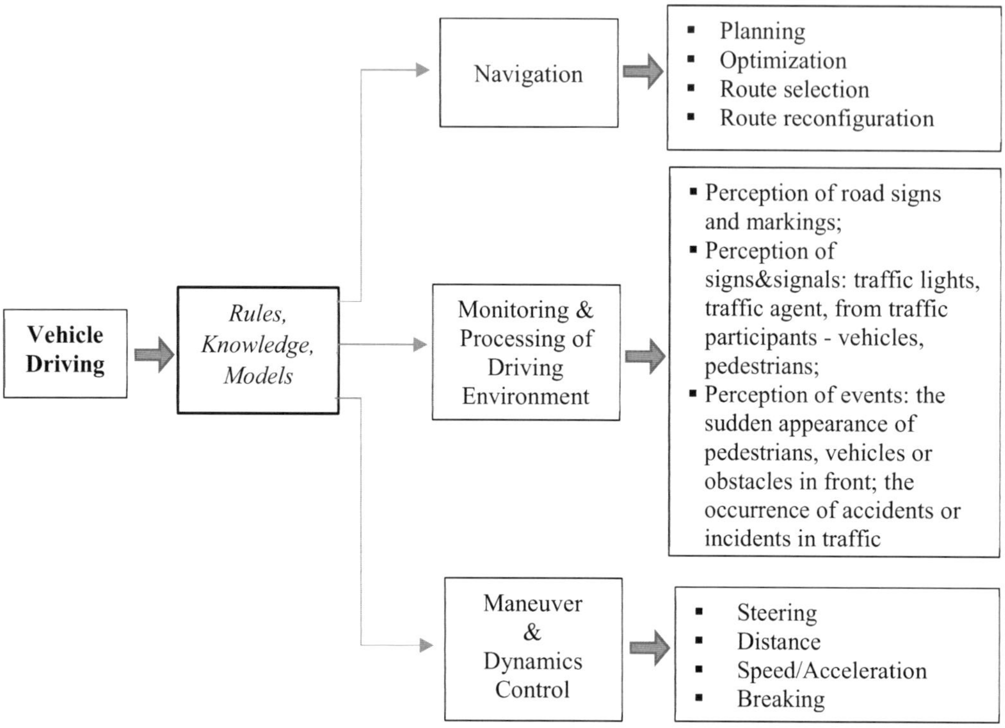

Figure 1. Basic tasks and functions of vehicle driving

Level \ Process	Decision	Action		Function
Strategic	*AI Models*	*Near real time*	→	▪ *Trip/route planning* ▪ *Route optimization and selection* ▪ *Route reconfiguration*
Tactical	*AI Models & Adaptive models*	*Real time*	→	▪ *Maneuver planning, including control of: steering, speed, distance, acceleration/ deceleration* ▪ *All types of current perceptions and events anticipated in the environment*
Direct	*Error-based*	*Reflex (as short as possible)*	→	▪ *Breaking* ▪ *All types of current events sensed in the environment*

Figure 2. Classes of ADAS functions and their requirements for decision and execution

We note the vehicular automation involves the use of AI models as much the decision level is higher – at the strategic and tactical level where the *cognitive decision making* is strongly required. This is typical for the tasks considered difficult like navigation, and also for specific functions that involves perceptions of the environment and maneuver planning when the time requirements are not critical. On the contrary, the functions where time is critical like breaking and maneuver coordination for collision avoidance, involve basic models of automation for a certain and rapid discrimination instead of a nuanced and elaborated decision. This is required in order to implement into the autonomous cars the reflex mode based action specific to the human driver.

The concept of intelligent vehicle is only a part of the problem. In the real scenario, the vehicle itself is like a smart agent that closely interacts with the infrastructure including roads and intersections of transport routes, car parks, and many other facilities for moving vehicles. Under these circumstances the intelligent vehicles are able to exchange information with the infrastructure but also with the other vehicles in traffic. This is a kind of intelligence sharing and social intelligence that pave the road towards intelligent transport systems development.

2.1. Cognitive technologies

The cognitive systems are models for human brain function that try to reproduce in machines the human-like decision and action. Nowadays, the cognitive technologies represent the practical part of artificial intelligence that merge techniques like traditional rule-based algorithms, neural networks, deep learning and natural language processing. These techniques are currently tested and evaluated into the relevant environments on the prototypes of the autonomous vehicle. The key issue of the autonomous vehicle is monitoring and processing of driving environment in real time using an appropriate sensorial system. The massive quantities of data should be provided in real time by artificial vision systems and should be also rapidly processed such as the required actions to be executed in due time. The vision sensorial systems are currently based on video cameras or radar and lidar sensors but the practice shows the combining them is better approach [7]. The vision sensors are deployed on the autonomous vehicle in order to 'see' omnidirectional around it performing environmental mapping. The basic processing tasks are: continuous image acquisition, image processing, features extractions and classification, detection and recognition. At this moment the technology is not validate in relevant environment for any driving mode. The challenge is to perform these tasks in the real traffic context with the lowest error rate. One of the recent reported test on vehicle autonomy comes from Drive.ai, a start-up company whose slogan is 'Building the Brain of Self-Driving Vehicles'. They involve the deep learning technique trying to touch the SAE's Level 4 of autonomy and possibly beyond [1]. Knowing by definition that self-driving with Level 4 is supported only in limited areas ('geofenced') or under special circumstances, like traffic jams, it is remarkable that Drive.ai drivers have forced the use of their system into urban traffic at night, in the rain. (demo available at https://www.drive.ai/). However, the experiment shows that in certain ambiguous situations, the temporary intervention of human driver was necessary.

2.1.1. Deep learning. This is the advanced technique in machine-learning that involves more complex artificial neural networks (ANN) in terms of number of neurons and number of hidden layers. The main distinction of this kind of neural network, named convolutional neural network (CNN) consists in their capacity to make simultaneously both the recognition and detection of the objects into the image, based on a great number of features extracted automatically thanks to a huge number of training images that were preliminary processed. For instance, deep learning is able to detect and classify different moving entities into the scene: cars, pedestrians, pets or a ball rolled down the road. This is a major improvement over ordinary neural networks that are able to recognize only simple object features (shape, orientation) based on supervised learning with manually-entered templates. This makes deep learning perfect for classifying objects within arbitrary scenes like traffic scenes are. Deep learning based on CNN is a major step towards artificial perceptions and so it is suitable to implement the human-like ability of pattern recognition and detection. Basically, their performance depends only on the quantity of data for training, but in practice should be solved few essential problems like dealing with ambiguity and sensors failing.

2.1.2. Knowledge representation. Knowledge are raw materials for AI models just like equations and numbers are for the mathematical models. The knowledge for vehicle driving are usually represented as rules, but in general the human driver is able to processes unstructured knowledge coming from environment as various stimulus, sensorial and perceptual images composed of objects with certain features, facts and events, seemingly arbitrary. The success of the actions is depending by the mode of the rapid interpretation of such unstructured complex information – the capacity to turn them into structured knowledge, as subject of formal reasoning, or to turn them into right actions via the subtle intuitive reasoning. Thus, another challenge in the development of fully autonomous vehicles is the use of deep learning to extract knowledge from data and the formation of specific concepts.

There are few tasks of vehicle driving that can be implemented by classical rule-based systems that are representative for Level 1 to Level 3 of vehicle automation like follows: Adaptive Cruise Control (ACC), Parking Assistance with automated steering, Lane Keeping Assistance (LKA) Type II and a combination of ACC with LKA Type II systems, Traffic Jam Assistance and Key Parking, Traffic Jam Chauffeur. All these tasks include functions that reproduce the human expertise as a limited rule-based inference engine, and eventually need a reduced set of data for training. On the contrary, once the training data becomes huge, as in deep learning technology, the knowledge representation can be scaled and nuanced using a larger set of features and objects' attributes. That is way we consider deep learning technology very promising for intelligent systems engineering and their applications. However, since learning (even deep) is just a necessary but not sufficient condition for the system to be intelligent, then it is imperative to implement a reasoning module with an appropriate basis of inference rules. In Figure 3 are depicted three basic ways in knowledge processing for vehicles driving. We note the bottom path is only that can be completely automatize into a full intelligent driving system on the autonomous vehicles. Formal reasoning is used here in the extended sense, including also analog reasoning and approximate reasoning including fuzzy logic, [3], [4], [6].

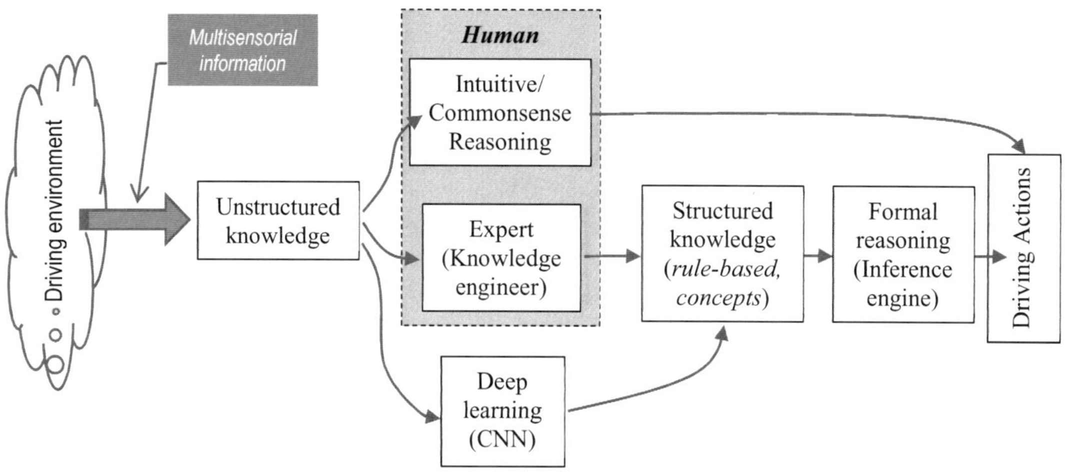

Figure 3. Three ways of knowledge usage in driving actions

2.2. Intelligent vehicle technology

Intelligent vehicle is a largest paradigm that advances the artificial intelligence, the electronics, software and communications in the field of vehicular technologies with impacts on energy saving, reducing pollution, increasing safety and improving comfort of the vehicle and transport. The autonomous vehicle is a part of this paradigm being similar to robotics. In the field of automotive, the pillars of the intelligent technologies can be classified as is following:

a) Intelligent driving systems, including two categories:
   i. Intelligent driver assisting systems (with subcategories: warning systems, infotainment and telematics systems, 'smart' devices like advanced front-lighting system – AFS, etc.)
   ii. Systems capable of taking over the driver's tasks.
b) Communications systems, supporting intelligent technologies (known as V2V, V2I, and V2X).

Category (a) includes the most technologies which is currently implementing ADAS but also the technologies that will come to reproduce human intelligence in driving vehicles. Communication technologies (b) are an essential part on the intelligent vehicle so that the fully autonomous vehicle won't be implemented without a strong data connectivity. Thanks to standardization, V2V and V2I technologies are ready to move from projects and isolated experiments to commercial. SAE has introduced the name *Dedicated Short-Range Communications* (DSRC) for many kinds of communications technologies to serve as the basis for connected vehicle safety and mobility application integration. Vehicle-to-everything (V2X) is a vehicular communication system introduced by the American Society for Testing and Materials under the initial name of Wireless Access in Vehicular Environments (WAVE). This technology is dedicated for vehicular ad-hoc networking integrating more specific types of communications V2V, V2I, V2D (vehicle-to-device) and V2G (vehicle-to-grid). Currently, V2X technology exploits the family of standards IEEE802.11 for WLANs, working in 5.9GHz range and it is tested with promising results in relevant pre-deployment projects in US [10] and EU [8]. Worldwide stakeholders promote new hardware and software solutions and reshape communication technologies to prepare the market for a massive using of vehicular communications. Under these circumstances EU elaborate '5G Action Plan' [2] which see the 5G the next generation system to support communication services of the intelligent transport systems. Electronics and software evolves towards powerful architectures to support internal and external connectivity on the cars. For instance, Harman's propose a scalable approach to meet consumer demands for greater functionality of infotainment and connectivity systems developing a V2X card to be use with the same hardware and software pre-integrated platform [9].

3. Conclusions

Driverless car or autonomous vehicle is a paradigm of robotics and represents an ambitious project which requires the fusion of many technologies like electronics and communications, mechatronics, software engineering and artificial intelligence. However, more disciplines and fields of activity contribute to the development of this aspiring project, which is essentially a multidisciplinary and interdisciplinary one. Not only does the development of the stand-alone vehicle pose problems, but also the development of infrastructure for the movement of self-driving vehicles raises many challenges. The major issue is safety and a huge effort is done to make sure the technologies involved are robust. This effort is mainly concentrated to test the safety of the ADAS but there is also a sustained work on standardization, the development of models and algorithms, the appropriate constructing solutions for implementation, as well as for ethical issues. ADAS systems are currently tested using hardware-in-the-loop methodology, but from safety reasons the human factors should be considered in tests, so the simulations of specific driving scenarios have to use driver-in-the-loop. Introducing the driver in the test loop make possible to see if the ADAS systems respond differently to a human driver and what is comparatively the effect on vehicle's dynamics. As the matter of fact, a strong argument in favor of autonomous vehicle development is the quantity of testing data obtained virtually that will encourage as more introducing of ADAS systems to an increasing degree of vehicle autonomy.

Another issue is related on sensors and their integration on the vehicle. Here are two aspects that aim, on the one hand, at the reliability of the sensors in difficult traffic and weather conditions and, on the other hand, the integration of sensors on the serial car, (e.g. day and night video cameras, microwave radar and laser scanners). Constructive engineering solutions for mounting sensors on the vehicle must take into account sensor safety, maintenance issues and ergonomic aspects. As we can see in prototypes so far, sensory equipment is usually concentrated in a distinct nacelle attached to the roof. This solution is expected to be redesigned for commercially autonomous vehicles to integrate sensors in optimal positions.

The following issue addresses models and algorithms in terms of their robustness, accuracy and completeness. Deep learning has the main advantage of recognizing and detecting in complex scenes and learning is continuous, while the system works. Starting from here, the development of data analysis tools for specific use cases in traffic is very necessary. Practice has also proven the need for models to deal with uncertainty and ambiguity. Therefore, a considerable research effort must be done to develop advanced inference engines for the conversion of sensory information and perceptions into concrete and correct driving actions.

Another issue for intelligent vehicles is the communication. V2X technologies will enable vehicle hyper-connectivity to improve cognitive level for correct decisions and safe traffic actions. Two major challenges are: information management in ad-hoc networks and cybersecurity. The first challenge relates to the use of data for the development of cooperative behavior between vehicles. This will require a lot of research in the area of multi-agent systems and a lot of testing and validation efforts in this context on communication quality of service (QoS), message latency, data perishability, etc. The cybersecurity problem is critical anywhere, but for the domain of autonomous systems is crucial. Cybersecurity strategy is expected to evolve in conjunction with internet of things (IoT) technology, and with the related technologies big-data and cloud computing.

Last but not least, the ethical aspect is a sensitive issue for driverless vehicles. Furthermore, there is currently no legislation on the use of autonomous vehicles. Legal and ethical issues in this area are not yet seriously addressed. The autonomous vehicle can be made to act by the rules, but the rules (laws) are not always moral. For example, a car can park on a free space that is not intended for it, or parked at an entrance. An autonomous car will suddenly stop, for instance when there is a tree branch on the road, but for the drivers coming from behind this is an immoral act. These examples show possible conflicts between legal and ethical correctness. In addition, the technology of autonomous vehicles could conflict with privacy rights of the users. In this case, legislation needs to be reconciled with ethical issues. A notorious aspect is related to civil liability for accidents caused by autonomous vehicles. This issue also addresses the criminal law area as well as the insurance issue. All this requires the creation of new regulations and the alignment of the three components: policies, laws and ethical norms. Some insights on two faces of the artificial intelligence – technology and ethics can be found in [5]. However, things will go on the right way because of in the near future amazing developments are expected in the field of intelligent transport systems: "cars will talk, then collaborate", as guru designers say. Definitely, the autonomous vehicles will just share data at first, but later they will cooperate more actively.

References

[1] Ackerman E March 2017 How Drive.ai Is Mastering Autonomous Driving with Deep Learning, *IEEE Spectrum,* March. (On line: http://spectrum.ieee.org/cars-that-think/transportation/self-driving/how-driveai-is-mastering-autonomous-driving-with-deep-learning).

[2] European Commission 2016 5G for Europe: An Action Plan – COM (2016) 588, (http://ec.europa.eu/newsroom/dae/document.cfm?doc_id=17131).

[3] Ionita S 2005 A Fuzzy Approach On Guiding Model For Interception Flight, in *Fuzzy Systems Engineering Theory and Practice, Series: Studies in Fuzziness and Soft Computing*, Vol. 181, Chapter 4, pp.87-111, Springer.

[4] Ioniţă S 2004 *Elements of knowledge engineering with applications in expert systems* (in Romanian) Ed. MatrixRom.

[5] Ionita S 2015 Artificial Intelligece – Between Technology and Ethics (in Romanian), *Studii şi comunicări – Divizia de Istoria Ştiinţei*, Academia Romana, Vol.VIII/, pp.105-117.

[6] Ioniţă S 2004, *Fuzzy Systems* (in Romanian), University of Pitesti.

[7] Lambert F 2016 Google's self-driving car vs Tesla Autopilot: 1.5M miles in 6 years vs 47M miles in 6 months, (https://electrek.co/2016/04/11/google-self-driving-car-tesla-autopilot/).

[8] Ross E P 2015 Thus Spoke the Autobahn, *IEEE Spectrum*, Jan., pp.50-53.

[9] Scoltock J 2016 Electronics: Infotainment, in *Automotive Engineer*, May, pp.22-23.

[10] US Department of Transportation: Team Report, June 2015, Safety Pilot Model Deployment - Test Conductor, (https://www.nhtsa.gov/sites/nhtsa.dot.gov/files/812171-safetypilotmodeldeploydeltestcondrtmrep.pdf)

Virtual engine management simulator for educational purposes

R Drosescu

Technical University "Gh. Asachi" of Iasi, Romania

Radu.drosescu@tuiasi.ro

Abstract. *This simulator was conceived as a software program capable of generating complex control signals, identical to those in the electronic management systems of modern spark ignition or diesel engines. Speed in rpm and engine load percentage defined by throttle opening angle represent the input variables in the simulation program and are graphically entered by two-meter instruments from the simulator central block diagram. The output signals are divided into four categories: synchronization and position of each cylinder, spark pulses for spark ignition engines, injection pulses and, signals for generating the knock window for each cylinder in the case of a spark ignition engine. The simulation program runs in real-time so each signal evolution reflects the real behavior on a physically thermal engine. In this way, the generated signals (ignition or injection pulses) can be used with additionally drivers to control an engine on the test bench.*

1. Introduction

During the last decade, because of the increasingly restrictive normative requirements which establish gas emissions and fuel consumption, the electronic management systems of internal combustion engines have become increasingly complex, the number and diversity of sensors has increased considerably, and the control strategies and algorithms more and more sophisticated and difficult to test and to calibrate. First goal of the simulator described in this paper is for teaching purposes, and it is therefore compact and easy to program and use, and it allows simulating, controlling and monitoring the main input and control parameters of modern internal combustion engines with spark ignition (with port and direct injection) or diesel engines with direct injection. Virtual engine management simulator (VEMS) is not a Hardware -in-the-Loop (HIL) equipment and makes the presence of the engine, sensors or electronic control units (ECU) unnecessary. The simulation process is going on until the control signals are generated for an engine operating mode, without considering the engine responses (torque, fuel consumption, emission, etc.) to these commands.

The main input parameters of the simulator are the rotational speed pulses and synchronization pulses with the crankshaft, which in real engines are generated by inductive sensors for crank wheels and by hall sensors for camshaft pulse wheels. Another input parameter simulates the throttle valve angle in accordance with the expected output torque.

The control outputs are generated in a synchronous manner with the rotational speed and position pulses of the crankshaft, just like in a real engine in static or transient operation (rotational speed, load), in which both the rotational speed and the load change according to circular motion kinematic relationships.

For a Spark Ignition Engines (SI) outputs are represented by ignition, injection and knock tracking windows pulses, while in the case of Compression Ignition Engines (CI) only Start Injection Angle and injection impulses are considered. If the simulator works properly, the injection and ignition pulses generated by the equipment are identical from the viewpoint of their form and duration with the ones from the electronic control unit of the engine in stable operation, and they may be in their turn used directly to actuate the engine. The basic condition for a good functioning of the simulator is given by a perfect synchronization between input values defining the operating cycle of the engine and output (controls) pulses generated throughout the entire engine range of rotational speeds, just as in case of real engine operation. Therefore, the simulator cannot operate on a PC running the Windows operating system, whereas SDK Android or IOS systems may can prove to be slow to simulate situations in which the rotational speed reaches 6000-9000 rpm.

The system described here is also useful for studying various control strategies applied to internal combustion engines, for instance the generation of multiple injection or ignition pulses per work cycle, of knock analysis algorithms or miss fires. Considering that any other equipment involved in engine management will work synchronously with each engine cycle, the simulator may be equipped with hardware devices and software programs capable to generating other output signals, such as electronic throttle control, fuel-supply control valve for a high-pressure pump in direct injection systems, control for wastegate or bypass valves in the turbocharging systems and controls for variable valve distribution actuating devices. Such complex hardware/software system are present in [2] and [4].

2. Simulator Input Signals

The main function of any internal combustion engine management system consists of the analysis of the signals from the rotational speed and crankshaft position sensors, to find synchronization between every engine cylinder and the crankshaft angular positions during each revolution, and finally to command the generation of injection or ignition pulses, to ensure a proper dosing fuel quantity on each injection pulse. Modern direct injection engines, in which multiple injection events per cycle are generated, require the extreme precision synchronization, below 0.1° crankshaft rotation accuracy. Considering that the duration of the engine cycle (720°) may reach 12 ms, and sometimes even less the simulator's program should loop at 600 kHz in order to attain a 0.1° resolution for a 10000 rpm rotational speed.

The signals generated by these two sensors are of extreme importance, since any malfunction of either of them makes it impossible for the engine management system to work properly. Therefore, modern electronic injection systems also generate an internal synchronization signal, which allows the engine to operate in the limp-home mode and allows the driver to drive his car to the nearest car service shop.

The synchronization signals (from the crankshaft and the camshaft) can come directly from the transducers placed on the test bench, but they may also be generated by the simulator. There is an additional option, in which the simulator generates analogical rotational speed signals like those emitted by a variable reluctance transducer used in almost all IC engines to measure rotational speed. Thus, the simulator is equipped with digital/analogical convertors (DAC) and the signal source for them are represented by Excel files containing samples acquired directly from the two sensors mounted on the engine or from the test bench. In this case, we used a NI-cRIO 9068 controller and a NI 9263 module as analogical output signal generator (4 channels 16-byte resolution, 100 ks/sec, 4 channels with simultaneous update). When only two DAC channels are used the user manual provides a maximum update rate of 200 kHz. As shown in Fig.1, about 50 samples are used for two alternations to generated signal having the same characteristics as the real one**.** Thus, signals with a minimum duration of 250 μs may be achieved. Supposing that the maximum rotational speed is 9000 rpm (150 Hz), it is possible to generate analog signals (AC) like those coming from wheels or ring with a maximum of 25 teeth using DAC hardware.

A series of acquisition equipment based on reconfigurable input/output FPGA technologies which offers high speed acquisition or update time, compact modules, and easy programming are described in [5] and [6].

Fig.2 shows the most commonly used pulse wheels, the former having 35 teeth with 10°RAM distance between them and a missing tooth in position 36, and the latter having 58 teeth with 6°RAM distance between them and 2 missing teeth in positions 59 and 60. Other configuration for crank teeth wheal and cam wheel as well as about synchronization process with crankshaft revolution are extensively described in [1] and [3].

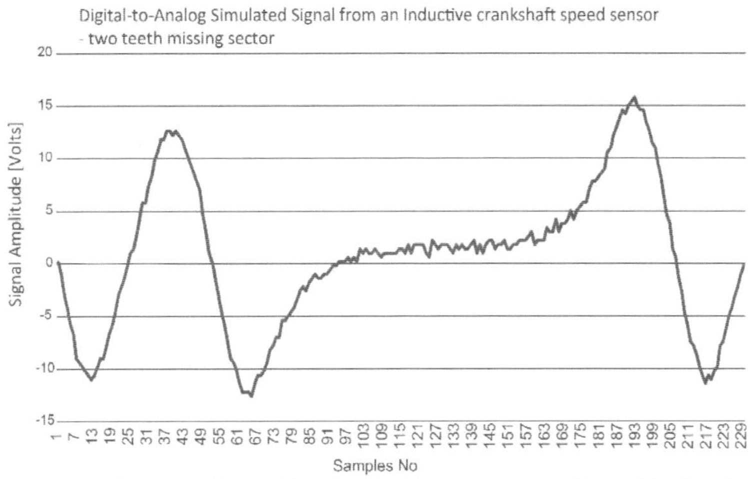

Fig.1 Analog pulse signal generating with a NI 9263 compact configurable DAC module from an Excel file

The missing teeth allow the microcontroller to extract information about the position of crankshaft in relation to engine Top Dead Center (TDC) and help to achieve the synchronization process. To simulate a real transient engine operating cycles that are more difficult to model, it is useful to acquire signals directly from the engine sensors, logging them to excel files and use as source files for DAQ card with high speed (500 000 updates/sec) or with a NI-9236 DAC module but at a resolution of only 25 points for a signal period.

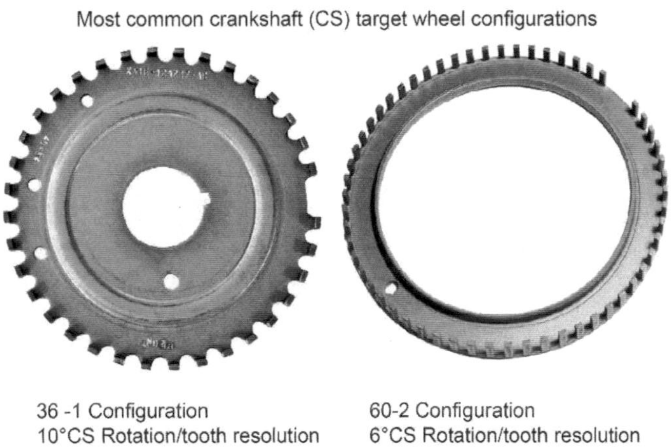

Fig.2 Most common crank wheel configurations

In any engine control ECU, the analogical signal from the crankshaft inductive transducer is converted into a digital signal by means of a comparator, whereas the signal from the camshaft position transducer is sent directly as digital signal due to the use of a Hall-type sensor. Thus, the microcontroller will receive both Crank and Cam signals in digital format. VEMS can receive synchronization signals in digital format in input mode, but it is able to generate its own testing synchronization signals from a multitude of patterns, thus covering all the configurations existing in real engines.

When it is generated by the simulator, the crankshaft synchronization signal allows up to 256 pulses for a rotation, whereas 1, 2 or 3 teeth may be missing. The rarer case in which the missing tooth is replaced by an additional one is also accepted [2]. The position pulse generating gear placed on the camshaft has a multitude of configurations, differing from manufacturer to manufacturer, sometimes even from one model to another of the same manufacturer.

The simulator was tested using several (crankshaft) synchronization and (camshaft) position configurations. In the example below shown in Fig.3 two classical configurations are considered: the firs have 35 synchronization pulses and one missing pulse (36 -1) and the second with 58 synchronization pulses and two missing pulses (60-2). The came wheel have 3, respectively 4-lobes. In the first case, the simulator generates three pulses as Cam positioning signal which correspond to 24, 6, and 6 crankshaft width teeth and three gapes sectors with width of 9, 12, respectively 15 teeth (one tooth corresponds to 10° crankshaft rotation angle). These signals generated by the simulator were recorded by a logical analyser and can be seen in Fig.3 by also with an oscilloscope (see Fig.4).

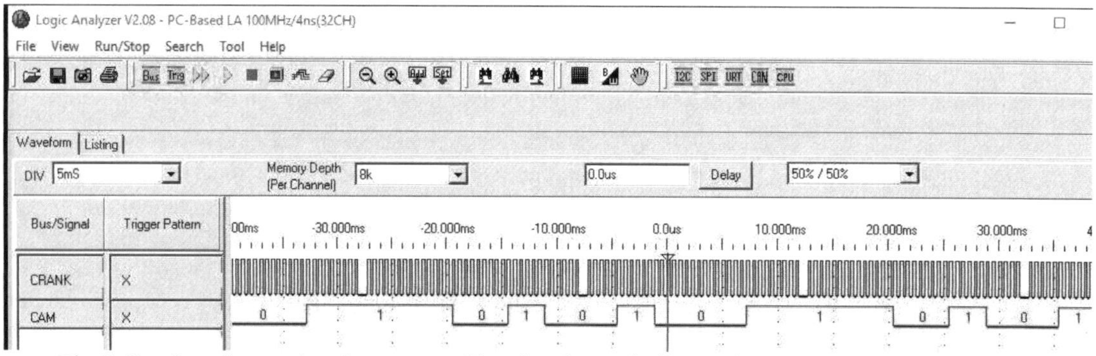

Fig.3 Crank and cam signals generated by simulator similar to those comming from a 36-1 crankshaft wheel (10° resolution) and a 3 lobs cam whel-view from Logic Analizer PLA 2532

In the case of a crankshaft wheel with 58 pulses (60 - 2 configuration) a four-lobe cam has been selected with three equal pulses having a duration corresponding to 6 teeth (36°RAM) and a pulse with a duration corresponding to 24 teeth (144°RAM). The distances between the three equal pulses are 39, 15, 9 and 15 periods corresponding to 6°RAM. The signals generated by the simulator and viewed on the logical analyser are shown in Fig.5.

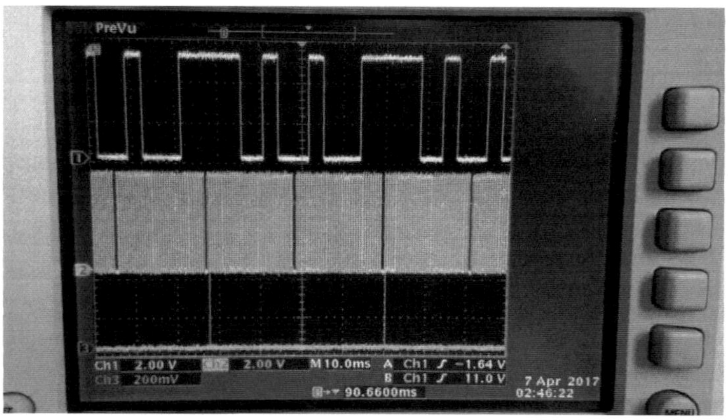

Fig.4 The synchronization signals from the above figure viewed on an oscilloscope

The programming of the pattern of the pulses generating from camshaft wheel especially lob and gape number and width configuration are done in a window of the simulator program starting from a base clock base with programmable frequency. The missing teeth are programmed in a secondary window (removes base pulses), whereas the third window is used for adding teeth programming (adds base pulses). This technique allows to generate any pulses patterns including the case when missing tooth is replaced with an additional tooth, or when a single crank wheel is provided with two sectors with missing tooth/teeth. The graphical programming windows is created using LabView and Veristand development tools.

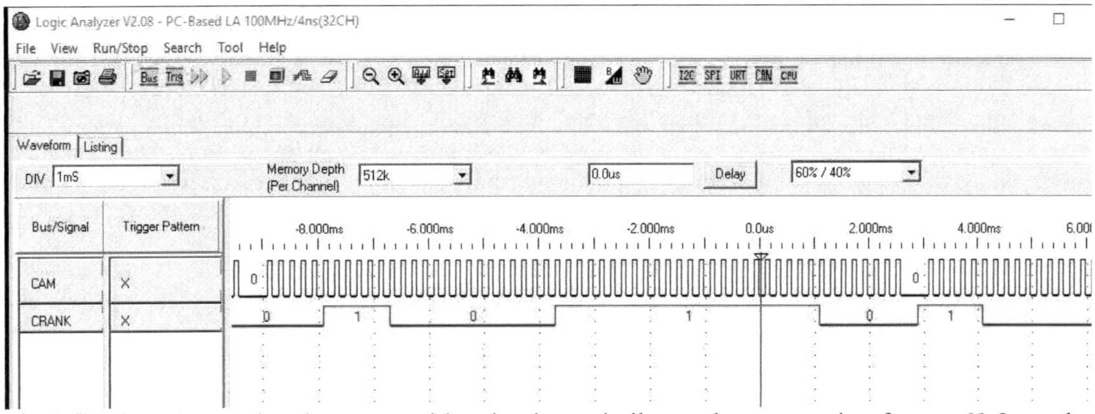

Fig.5 Crank and cam signals generated by simulator similar to those comming from a 60-2 crank wheel (6° crankshaft rotation resolution) and a 3 lobs camshaft whel

Thus, an additional tooth between two teeth of basic pulses is equivalent to a tooth three times thicker, whereas a missing tooth is the equivalent to a gap three times thicker. Fig.6 shows the programming window for the generation of a 36-1 crankshaft wheel pulses, while the same windows is used as in Fig.7 to generate the three lobe camshaft position pulses. A similar configuration program using LabView and Veristand development tools is given in [4] and [5].

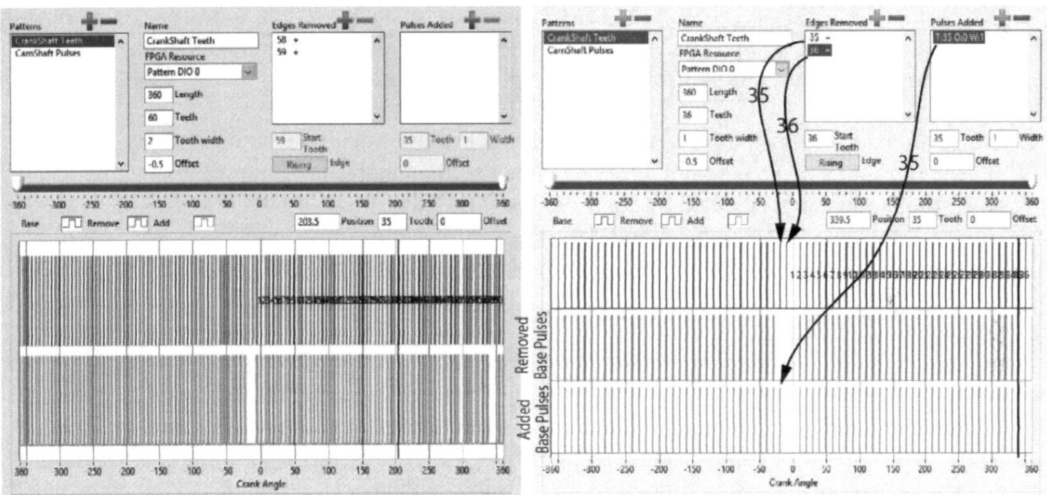

Fig.6 Programming windows for cranckshaft pulse pattern generation

Digital patterns for the two signals can be obtained using a NI-9401 cRIO module or similar PXI DAQ cards. The analog input signals coming from inductive or Hall crankshaft and camshaft sensors can be acquired with simultaneous sample NI-9222 or similar DAQ cards. Any NI-cRIO multiplexed analog input module having a sample rate greater than 250 ks/sec can also be used for crankshaft synchronizing and position signals acquisition. As a real-time controller, a NI 9068 chassis with 8 slots was used.

3. VEMS Synchronization process

The most important and complex function of the simulator consists of the internal synchronization of its central kernel both in time and angularly, with the rotation movement of the crankshaft quantified by the input signals from the two sensors: rotational speed (crankshaft) and position (camshaft). All the simulated output (pulses for manifold port or direct cylinder injection, ignition pulses, knock validation windows) must be perfectly synchronized with the angular position of the crankshaft during the whole 720° cycle, i.e. when the signals are sent by simulator to a real engine, it should "consider" them as coming from its own ECU. In addition to the crankshaft speed and position signals, there are other engine input signals the variation of which is synchronous with the rotation of the engine, namely: the mass air flow, the manifold air pressure, the knock vibrations, the cylinder pressure, and the engine torque. All of these signals simulation should be synchronized with the crankshaft rotation angle (event based) and not in time (time-based).

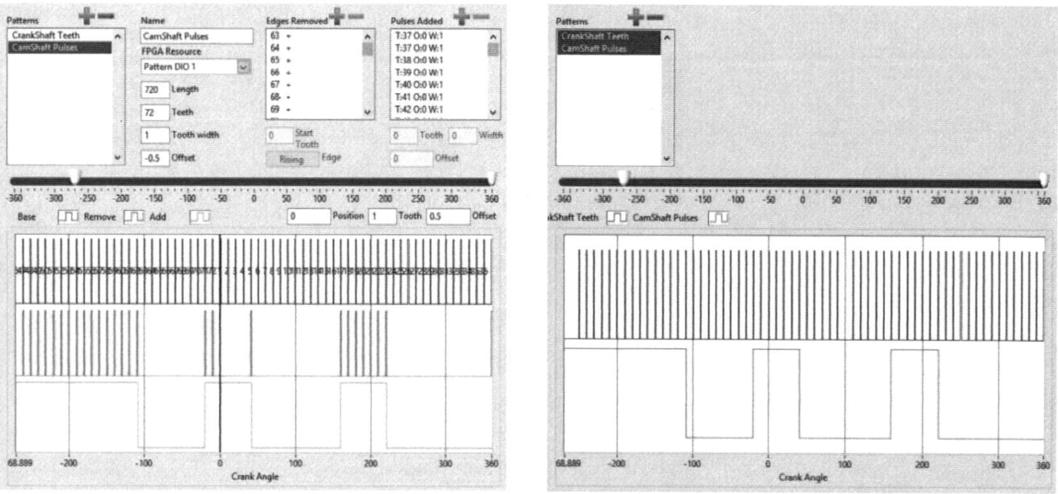

Fig.7 Programming windows for camshaft pulse pattern generation

There are also values which having an asynchronous evolution with the crankshaft revolutions, such as the throttle position, the ambient pressure, the intake air temperature, the engine coolant temperature, the signals from the switches actuated by the car driver, etc. The same characteristic may be extrapolated to the output signals generated by the simulator for all these signals. In certain cases, the control signal should be emitted synchronously with the angular position of crankshaft, but it should last a very clearly defined time, regardless of rotational speed or angle (injection pulses, diesel start of injection duration, dwell time for ignition pulses).

The synchronization pulses generated by the rotational speed sensor are spaced at 10° or 6°, or in other words the simulator receives one pulse for every 10 or 6 degree of crankshaft rotation. The control signals of modern direct injection engines require much finer angular intervals, usually 0.1° RAM, or even smaller. Thus, between 2 consecutive teeth, the engine ECU and also the simulator requires 60 or 100 perfectly spaced pulses, which follow the speed and angular acceleration of the flywheel as a PLL multiplier.

In order to be able to meet these requirements, the simulator is equipped with two counter-timer, an angle marks generator and a state machine which are designed to generate appropriate pulses for a T/C1 reference time base, and a T/C2 programmable generator/counter angular marks. The angular marks generator can generate a configurable 1 to 1024 equally spaced pulses" inserted" between edge transitions coming from two successive teeth of the crankshaft speed transducer. The period of the angle marks counted by T/C2 is rated in time marks generated by T/C1.

For instance, if the toothed wheel on the crankshaft has 36 teeth (35-1 configuration) and a resolution of 0.1°crankshaft rotation is required, on each rotation T/C2 generates 3600 mark pulses for a whole revolution, respectively 7200 pulses for a complete engine cycle. Supposing a 9000 rpm rotational speed, the minimum frequency of the counter clock should be 540 kHz. For the 60 teeth configuration (58-2) and the same 0.1°cranckshaft resolution, we will need 60 angle marks for each tooth, which means also a 540-kHz frequency (see Fig.8). As one may notice, although the configuration and distance between the teeth are different in the two cases, the angle marks with a 0.1°RAM resolution have the same duration, namely 1.85 μs.

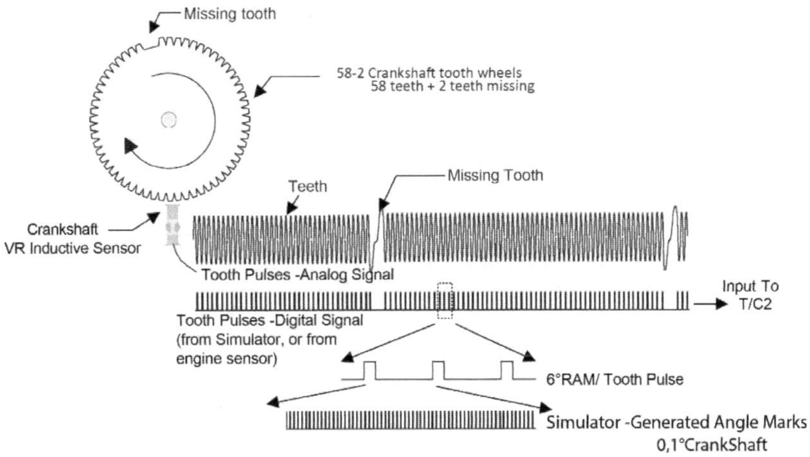

Fig.8 Angular marks pulse generator

Although two successive teeth always have a constant angular deviation, the time elapsed differs with the angular acceleration and speed of the crankshaft. Thus, the frequency of the angle marks will have to change accordingly between two successive teeth in order to achieve an automatic compensation in case of crankshaft acceleration or deceleration. At the same time, when one or two teeth are missing, VEMS will generate its own angular marks signals like those coming from real teeth, so that the injection or ignition pulses should not have interruptions generating lack of ignition or combustion.

The angular marks generator thus becomes the main part of the simulator, its operation being implemented with a state machine which was tested in Simulink Stateflow and then embedded into a configurable Xilinx Zynq 7000 FPGA. For the generation of pulses representing angle marks, the generator has an internal clock, the most commonly used frequency being 100 MHz (10 ns period).

The NI-9361 compact module is equipped with 8 timer/counter with 32 bite resolution, a 100 MHz internal base which allows down prescaller, and a maximum sampling rate (capture of transitions coming on the counting line) of 102.4 kHz. For example, from a rotational speed of the crankshaft of 9000 rpm, the 35-1 pulses have a frequency of 5.4 kHz, whereas the maximum frequency of the 58-2 configuration of the generating gear is 9 kHz; both options are easily covered by the above-mentioned module.

The synchronization process starts by searching the area where one, two or even three teeth are missing. For simplicity reasons and for explanations, the 36-1 configuration is considered. The basic function of any timer/counter consists of the edge detecting (rising, falling or both fronts) of a train of rectangular pulses and the measuring of the period (frequency) by using a large internal frequency base. Thus, it is possible to measure the period between two fronts of the teeth signal and to set a validation window where the next tooth is expected to appear. If the next tooth fails to appear in the validation window, the missing tooth sector may be next, or the crankshaft may suffer a deceleration. Conversely, if the next tooth appears before the expected time calculated in the estimation window, the crankshaft is supposed to have accelerated.

The simulator should detect whether the lack of a pulse in the validation window is due to a missing tooth or to a very strong deceleration, and whether the lack of two successive pulses represents the two missing teeth in a pulse wheel with 60-2 configuration or is the consequence of a missing tooth in the 36-1 configuration and of a sharp deceleration.

As a rule, maximum accelerations are obtained at low crankshaft speed. For example, from the cold starting speed of about 50 rpm, the crankshaft can achieve in a very brief time 5000-7000 rpm. Such a case is presented in Fig. 9 which illustrates the time evolution of pulses coming from a 60-2 crank wheel where a ratio between the teeth period t_A immediately preceding the area with two missing teeth (B sector) and the period t_C between the first two teeth after missing teeth area can reach a 75 value. High crankshaft speed variations result from changing high gear to low gear in a single maneuver. Smaller ratios are obtained in very strong braking maneuvers. At higher crankshaft speeds, much lower variation of the ratios between successive teeth periods are obtained.

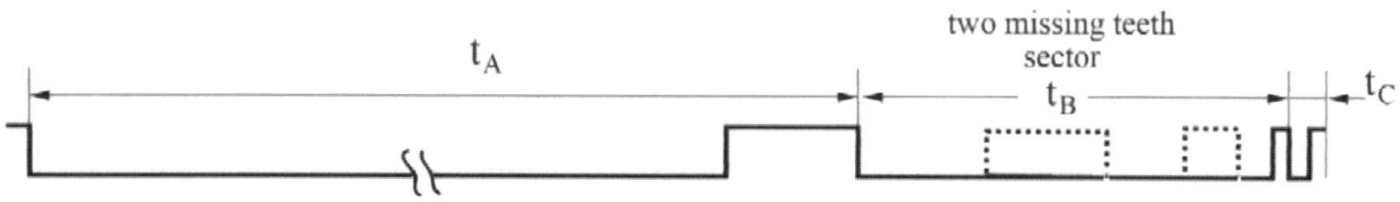

Fig.9 Time evolution for pulses comming from a crank wheel with 60-2 teeth configuration wich is accelerated from 50 rpm to 5000 rpm in a very brief time

The VEMS was designed to accommodate constant crankshaft angular accelerations or decelerations, this assumption being accepted in all calculations regarding crankshaft kinematic and dynamics.

We assume a crankshaft acceleration from initial speed of n_0=900 rpm to the final value of n_f=6000 rpm in one second. Internal timing base of the simulator is driven by a clock frequency of 100 MHz. Crank wheel has 35 teeth and one missing tooth and generates one pulse for every 10° crankshaft rotation. VEMS can generate a maximum of 1024 angle marks for each tooth, but for simplicity of this example a value of 100 is considered resulting a 0,1° crankshaft rotation accuracy.

As in the real case of engine acceleration, where the transition from 900 rpm to 6000 rpm is made up gradually in a specified number of crankshaft successive revolutions, the simulator must have a similar behavior and decrease gradually teeth period from t_0 =1,85 ms to t_f=0,277 ms (277 μs) as resulting from relations (1) and (2).

$$t_0 = \frac{60}{n_0 \cdot z_{teeth}} = \frac{60}{900 \cdot 36} = 1{,}85\, ms \tag{1}$$

$$t_f = \frac{60}{n_f \cdot z_{teeth}} = \frac{60}{6000 \cdot 36} = 0{,}27\, ms \tag{2}$$

Angular marks will gradually change their frequency from initial value of 54 kHz (i.e. 900/60x36x100) to the final value of 360 kHz. Crankshaft and camshaft pulses generated by VEMS for this acceleration are partially shown in Fig.10.

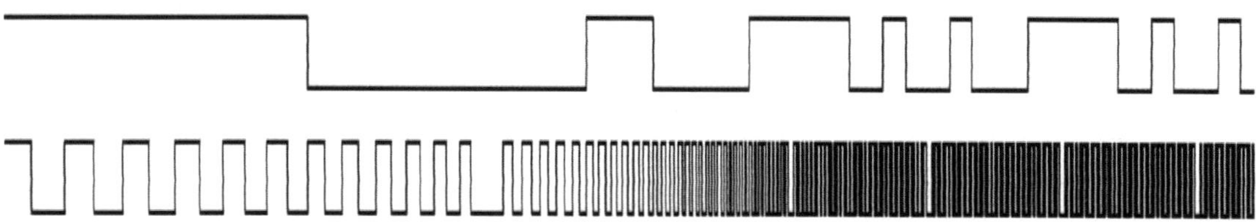

Fig.10 Partial view of a crank and cam pulses evolution for a simulated acceleration of the crankshaft between 900 and 6000 rpm in 1 sec.

From the kinematic equations of circular movement between angular velocity $\omega = \frac{\pi \cdot n[rpm]}{30}$, rotation angle $\theta[rad]$ and angular acceleration $\varepsilon[\frac{rad}{sec^2}] = ct$ it results the change in angular velocity between two succesive teeth separated by a 10° angle:

$$\omega_1^2 - \omega_0^2 = 2 \cdot \varepsilon \cdot \Delta\theta = 2 \cdot \left(\frac{\pi \cdot n_f}{30} - \frac{\pi \cdot n_0}{30}\right) \cdot \frac{1}{\Delta t = 1\, sec} \cdot \frac{10° \cdot \pi}{180°} = 186{,}425\, \frac{rad}{sec^2} \tag{3}$$

Before acceleration, all teeth spin with a constant angular velocity $\omega_0 = 94{,}241\frac{rad}{sec}$, but after acceleration the new angular speed reflected on the first tooth is $\omega_1 = 95{,}231\frac{rad}{sec}$, according to relationship (3). The frequency of the angular marks is changing according to the new velocity but it is maintained to the new value for the rest of whole 10° tooth period.

Thus, the speed difference counted at the first tooth that accelerate is about $\frac{30}{\pi}\cdot(\omega_1 - \omega_0) = 9{,}45$ rpm. Relation (3) can be generalized for any successive teeth in the form given by equation (4).

$$\omega_{i+1}^2 - \omega_i^2 = 186{,}425\,\frac{rad}{sec^2} \quad i=0\div35 \tag{4}$$

Again, using relation (4) to the second tooth gives an angular velocity $\omega_2 = 96{,}205\frac{rad}{sec}$, which means an increasing speed with 9,3 rpm compared with the first tooth, and with 18.75 rpm compared to static angular speed at 900 rpm. As can be seen from this example, the ratio between tooth period at 900 rpm and 6000 rpm is only 6,8, also the faster the crankshaft speed, the less is the tooth period modification.

The basic function of the synchronization process is to generate angle ticks with well-defined frequency (usually for achieving 0,1° or 0,01° crankshaft rotation accuracy) used to clock a timer/counter T/C2 which defines at any time the position of crankshaft and the evolution of each cylinder during on a whole engine cycle (720°) with reference to the TDC of first cylinder. The synchronization process is controlled by a state-machine that can be implemented either in Simulink Stateflow or using LabView StateChart and Control Design Toolboxes.

Synchronization process start with teeth pulses edges (rising edge, falling edge, or both). First edge captured with a Timer/Counter from NI-9361 cRIO timing module denotes that the crankshaft rotates without providing any further information. The detection of the second tooth allows the calculation of the first period between two teeth and the estimation of the validation window for the next coming tooth.

As can be seen from Fig.11 this period can be either between two normally spaced successive teeth or between two teeth separated by a missing tooth (missing teeth) sector.

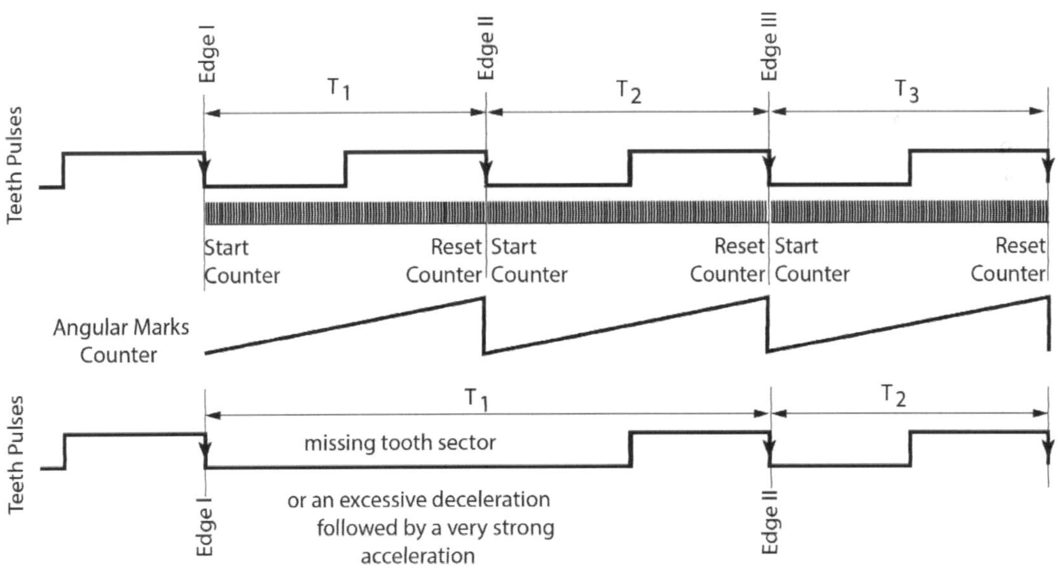

Fig.11 Teeth pulses fronts and missing tooth sector are the key elements in synchronization process

Synchronization can be done only when a missing tooth (teeth) sector is recognized. A simple comparison test between the current and previous tooth periods can identify a possible missing tooth area, but also a sharp deceleration. So, another test is necessary to compare the current period (possible

IOP Conf. Series: Materials Science and Engineering **252** (2017) 012099 doi:10.1088/1757-899X/252/1/012099

missing tooth) with the next tooth period. In the worst case, 35 fronts must be captured and 34 periods must be measured before the transducer can reach the missing tooth sector. On each front, the angular marks counter start counting until the next front is captured, when it is reset and restarted again over and over till the missing area is detected.

A valid test to detect the missing teeth sector is based on the observation that that an acceleration followed by a sharp deceleration and immediately succeeded by another crankshaft acceleration is physically impossible to meet in practice because of the crankshaft inertia. The above example can also prove this supposition. A good test for missing tooth sector is based on two comparisons between three periods corresponding to three successive teeth: $T_{i-1} < \kappa \cdot T_i > T_{i+1}$, where k is an engine related factor which can be set on test bed, common values being in range 0,3-0,8.

Once the sector with a missing tooth (teeth) has been detected, the VEMS control core start the angular marks timer counter T/C2 without resetting it at the next tooth pulse. In this situation there are three options, but only the first two have been implemented on this simulator.

The first variant is useful for simulating simultaneous and/or grouped fuel injection. The angular marks counter starts after the missing tooth (teeth) sector is detected and stop counting after one revolution of the crankshaft. Position of the TDC for each cylinder is well defined but not the phases from an engine cycle (720°).

In the second mode of operation, the T/C2 will count for two successive rotations before being reset and restarted on each engine cycle (720° crankshaft rotation). The both modes of operating are depicted in Fig.12.

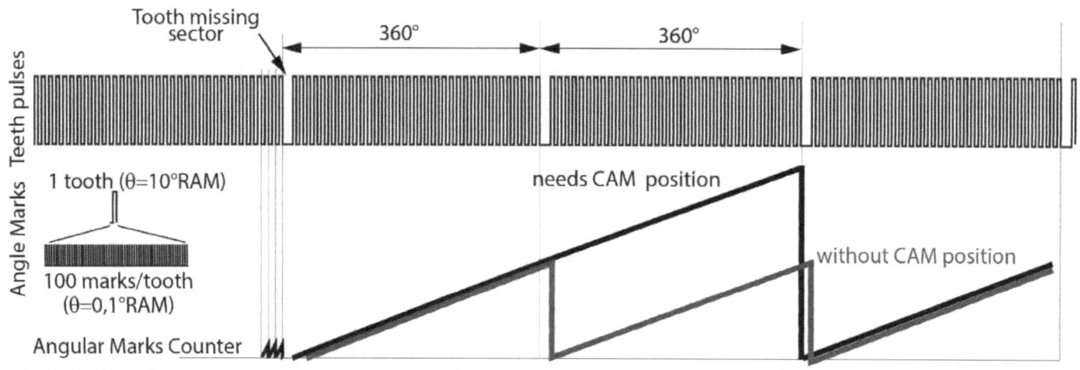

Fig.12 Timer/Counter counts angular marks over a full crankshaft revolution or over a full engine cycle

The third operating mode not yet implemented make possible that angular marks counter will increments on each tick generated by T/C1 with precision of up to 0.01 ° for several consecutive motor cycles thus allowing advance programming of synchronous events related to the crankshaft over multiple cycles. Considering that NI-9361 Compact Module features a 32-bit 8-bit timer/counter, at 9000 rpm, with a crank wheel in the 60-2 tooth configuration and 1024 angular marks for each tooth, it would result that a single counter can count 233 successive engine cycles until overflow and will be automatically reset.

4. VEMS Output Signals

VEMS can generate two types of signals: speed, synchronization and position of the crankshaft, or engine control pulses. In the first category, we find the signals that usually come from engine crank and cam sensors and allow the quantification of the engine crankshaft position to an accuracy of up to 0,01° crankshaft rotations. These signals can be also generated by the simulator and routed to the internal counter inputs, as if coming from external sensors. The simulator also allows the generation of analog signals like those collected from inductive sensors on a real engine. All these signals have been extensively analyzed in Chapter 2 dedicated to simulator input signals and will not be resumed any further. The second category of VEMS output signals comprises injection, ignition and pulses train that generates windows for knock signals detection and analysis.

The simulator can generate pulses for manifold injection or to generate and control current pulses for direct injection.

Pulses for simultaneous and/or grouped fuel injection can be generates on every crankshaft revolution because the fuel quantity required for combustion is injected into two stages: half in one revolution of the crankshaft (360 ° crankshaft rotation) and the rest in the next revolution. For group fuel injection, one injector group (half oh cylinders) injects the total fuel quantity in the first revolution of the crankshaft, and for the next revolution the second group injects. In both cases, there is no need to distinguish between the first rotation and the next one in a motor cycle, and the angular quanta counter is restarted at each detection of the first tooth after the missing teeth sector.

Fig.13 shows an oscilloscope capture with port injection pulses (on the third channel) in synchronization with crankshaft pulses also generated by simulator.

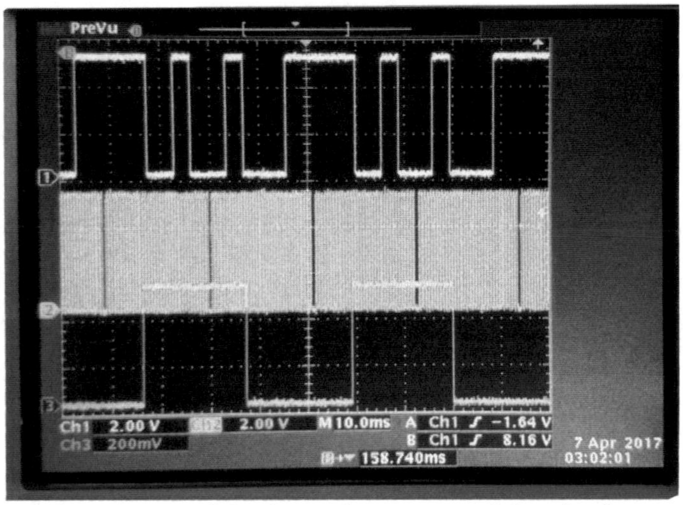

Fig.13 Cam, crankshaft and group injection pulses generated by simulator and viewed on the oscilloscope

Injection pulse width can be programmed in milliseconds with T/C1 but the start of injection pulse and the end of injection pulse positions in relation to crankshaft teeth pulses can be determined as an angular marks values of T/C2.

In case of sequential or individual port injection, but especially for direct injection, it is mandatory to know the position of each cylinder in a complete engine cycle and to make the difference between the first rotation and the next rotation of the crankshaft over a full engine cycle. One NI-cRIO 9758 Port Fuel Injector Driven Module was used to generate pulses for up to eight low pressure injectors.

One of the most complicated task of the VEMS is to generate control signals for high pressure injectors used in direct injection. There are many new features compared to manifold injection:

Several injection pulses (up to five) can be generated each engine cycle for each injection valve;

Each injection must start at a predetermined angle and consists of many phases (up to 5 A-E) as can be seen from the Fig.14;

Each injection event may have different time for the same cylinder and/or from cylinder to cylinder;

Each phase uses a single pulse or a train pulse width modulated (PWM) to control a power transistor such that the current passing through the injector coil complies a certain shape and specific values (see. Fig.14);

During opening phase of the injector (phase A), the voltage has a higher value than battery (40-65V -VBoost). Then, VEMS switches to phase D and reduce the current through injector only to keep it open. At the end of injection, the current decay back to zero.

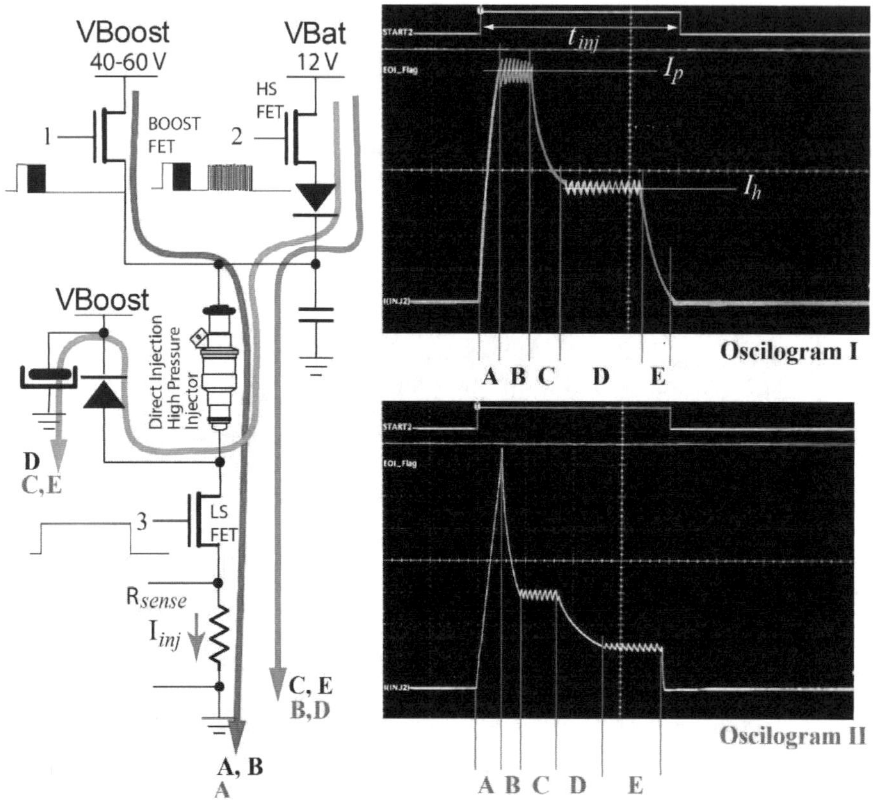

Fig.14 Current shapes used to drive solenoid valves for high pressure direct injection and the power stage driven by VEMS implemented with NI-9751 cRio module

VEMS use a compact RIO NI-9751 able to drive 3 solenoid type gasoline or diesel direct injectors, or piezo actuated injector. As can be seen in Fig.14 two current direct injection shapes (profiles) was generated and recorded on the digital oscilloscope.

Fig.15 is a screen captured from of a logic analyser showing a sequence of four ignition pulses and four injection pulses used to control a direct injection spark ignition engine. All these control signals are set with a high accuracy (0,01°) in relation to the teeth and cam lobes also generated by simulator.

For each cylinder, the simulator delivers one main dwell pulse for ignition coil with a programmed advance angle before TDC (measured in angular marks) followed by a programmable number of secondary ignition pulses. Multi spark ignition technique is used to enhance the combustion of very lean stratified mixture. Such a sequence of four ignition pulses is captured on the oscilloscope and can be visualised in Fig.16.

Spark pulses can be programmed to be active for at least a minimum amount of time to provide a sufficient sparking energy, or to not to exceed a maximum amount of time, which would cause the ignition coil to overheat. Thus, it is possible to provide an appropriate dwell angle, respectively, an optimum spark advance regardless of the accelerations or decelerations to which the crankshaft is subjected. It is also possible to generate more pulses for both ignition and injection events (pilot, main injection, post injection, etc.).

VEMS can generates multiple injection event per engine cycle such as pilot, main, secondary or delayed post- injections. For each injection pulse must provide the following parameters before running the simulation: injection pulse width t_{inj}, time for peak current t_p, time for hold current t_h, peak current I_p and hold current I_h.

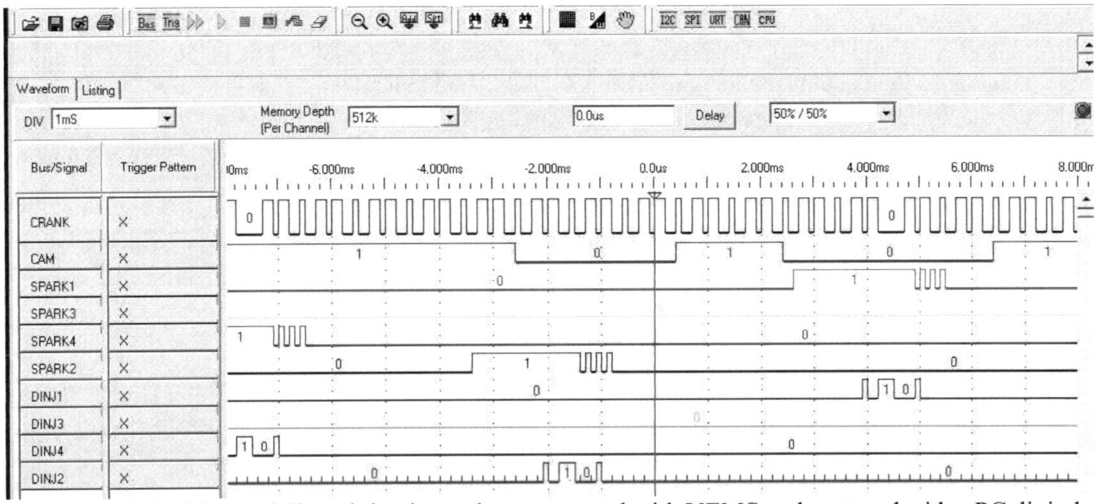

Fig.15 Spark ignition and direct injection pulses generated with VEMS and captured with a PC digital analyzer

Peak current is used for the injector fast opening and it must grow up from o to 20A in a very short time (few microseconds). This is achieved by means of a booster voltage of up to 65V. During pick-up phase t_p battery voltage is applied to the solenoid valve and assist in opening quickly. In this time current I_p is to approx. 20 A. During holding phase the current is dropped to approx. 13A and is entirely supplied from battery.

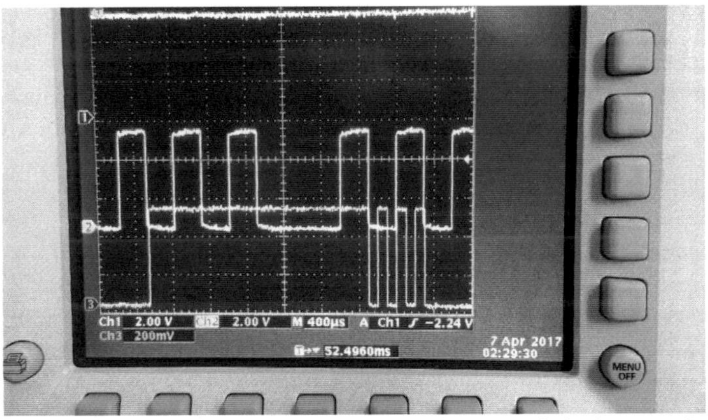

Fig.16 Spark ignition pulses simulated 2500 rpm speed of the crankshaft. VEMS generate one main ignition pulse with 30° crankshaft rotation and 2,4 ms width followed by three secondary pulses

The total time of the injection pulse is a critical parameter and is very well establish for each engine on the test bed and stored in ECU memory as three-dimension map. The simulator can use only a single set of values as a demonstrative tool. The angular distance between the start of injection, angular distance between the pilot and maim injection, or between the main and post-injection pulses can also to be programmed before the start of the simulator.

Fig 17 is another capture of the logic analyser in which are show all the simulator control signals for a gasoline direct injection engine including the knock windows and crankshaft and camshaft synchronization signals.

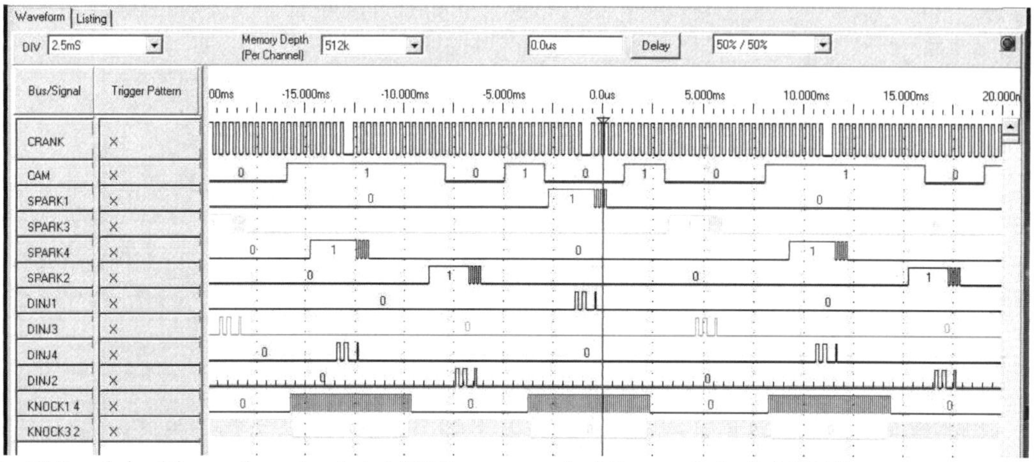

Fig.17 Spark ignition pulses simulated 2500 rpm speed of the crankshaft. VEMS generate one main ignition pulse with 30° crankshaft rotation and 2,4 ms width followed by three secondary pulses

5. Conclusions

Thus, the simulator described in this paper is a hardware/software system comprehensive enough to allow the development of complex projects on the control of any internal combustion engine, without the actual physical presence of the latter. This has several noteworthy advantages such as the simplicity and cost effectiveness of the project, which does not require the actual presence of the engine and of the test bench, at least not during the designing, modelling and simulation stages. The operating conditions that may be modelled here are far more diverse than the ones that may be implemented on a real engine, the control of which is achieved by means of an unchecked electronic control unit prototype, which may actuate the running of the device in dangerous conditions or, on the contrary, cannot cope with extreme working conditions. Moreover, the simulator has the advantage of being able to repeat the same simulated states and conditions without cyclic variations or deviations.

References

[1] Boulat A, Genninasca Y, Charlet A and Higelin P 2001 *ECUTEST-A Real-time Engine Simulator for ECU Development and Testing*, SAE Technical Paper 2001-01-1911

[2] Viele M, Stein L, Gillespie M and Hoekstra G 2004 *A PC and FPGA Hybrid Approach to Hardware-in-the Loop Simulation*, SAE Technical Paper 2004-01-0904

[3] Wang J and Sarlashkar J 2007 Engine Crankshaft Position Tracking Algorithms Applicable for Given Arbitrary Cam- and Crank-Shaft Position Signal Patterns SAE Technical Paper 2007-01-1597

[4] Barrett S and Bouchez M 2015 *Addressing Engine ECU Testing Challenges with FPGA-Based Engine Simulation*, SAE Technical Paper 2015-01-0173

[5] Barrett S 2015 *NI Engine Simulation Toolkit for NI Veristand*, NI Press https://decibel.ni.com/content/docs/DOC-37304

[6] Anand P and Saravanan C G 2012 *Development of Research Engine Control Unit Using FPGA-Bassed Embedded Control System*, Journal of Kones Powertrain and Transport, Vol.19, No.3 2012

[7] Strapko M and Tichanek R *2012 Engine control unit based on NI compactrio platform*, Journal of Middle European Construction and Design of Cars